Be sure to keep up-to-date in 2010!

Register now and **h quarter**
when the newe **online.**

Please provide your name, address, and email below and return this card by mail or
Fax: 1-800-632-6732.

To register online: www.rsmeans411.com/register

Name ____

email ____

Company ____

Street ____

City/Town ____________________ State/Prov. _____ Zip/Postal Code _______

RS**Means**

PLACE
STAMP
HERE

RSMeans
Quarterly Updates
63 Smiths Lane
Kingston, MA 02364-3008
USA

RSMeans Mechanical Cost Data

33RD ANNUAL EDITION

2010

RSMeans
A division of Reed Construction Data
Construction Publishers & Consultants
63 Smiths Lane
Kingston, MA 02364-3008
(781) 422-5000

Printed in the United States of America
ISSN 0748-2698
ISBN 978-0-87629-819-0

Senior Editor
Melville J. Mossman, PE

Contributing Editors
Christopher Babbitt
Ted Baker
Barbara Balboni
Robert A. Bastoni
John H. Chiang, PE
Gary W. Christensen
David G. Drain, PE
Cheryl Elsmore
Robert J. Kuchta
Robert C. McNichols
Jeannene D. Murphy
Stephen C. Plotner
Eugene R. Spencer
Marshall J. Stetson
Phillip R. Waier, PE

Book Design
Norman R. Forgit

President & CEO
Iain Melville

Vice President, Product & Development
Richard Remington

Vice President of Operations
David C. Walsh

Vice President of Sales
Thomas Kesler

Director, Cost Products
Daimon Bridge

Marketing Director
John M. Shea

Engineering Manager
Bob Mewis, CCC

Cover Design
Paul Robertson
Wayne Engebretson

Production Manager
Michael Kokernak

Production Coordinator
Jill Goodman

Production
Paula Reale-Camelio
Sheryl A. Rose
Jonathan Forgit
Debbie Panarelli
Mary Lou Geary

Technical Support
A.K. Dash
Tom Dion
Roger Hancock
Gary L. Hoitt
Genevieve Medeiros
Kathryn S. Rodriguez
Sharon Proulx

This book is printed on recycled paper, using soy-based printing ink. This book is recyclable.

Foreword

Our Mission

Since 1942, RSMeans has been actively engaged in construction cost publishing and consulting throughout North America.

Today, more than 65 years after RSMeans began, our primary objective remains the same: to provide you, the construction and facilities professional, with the most current and comprehensive construction cost data possible.

Whether you are a contractor, owner, architect, engineer, facilities manager, or anyone else who needs a reliable construction cost estimate, you'll find this publication to be a highly useful and necessary tool.

With the constant flow of new construction methods and materials today, it's difficult to find the time to look at and evaluate all the different construction cost possibilities. In addition, because labor and material costs keep changing, last year's cost information is not a reliable basis for today's estimate or budget.

That's why so many construction professionals turn to RSMeans. We keep track of the costs for you, along with a wide range of other key information, from city cost indexes . . . to productivity rates . . . to crew composition . . . to contractor's overhead and profit rates.

RSMeans performs these functions by collecting data from all facets of the industry and organizing it in a format that is instantly accessible to you. From the preliminary budget to the detailed unit price estimate, you'll find the data in this book useful for all phases of construction cost determination.

The Staff, the Organization, and Our Services

When you purchase one of RSMeans' publications, you are, in effect, hiring the services of a full-time staff of construction and engineering professionals.

Our thoroughly experienced and highly qualified staff works daily at collecting, analyzing, and disseminating comprehensive cost information for your needs. These staff members have years of practical construction experience and engineering training prior to joining the firm. As a result, you can count on them not only for accurate cost figures, but also for additional background reference information that will help you create a realistic estimate.

The RSMeans organization is always prepared to help you solve construction problems through its variety of data solutions, including online, CD, and print book formats, as well as cost estimating expertise available via our business solutions, training, and seminars.

Besides a full array of construction cost estimating books, RSMeans also publishes a number of other reference works for the construction industry. Subjects include construction estimating and project and business management, special topics such as green building and job order contracting, and a library of facility management references.

In addition, you can access all of our construction cost data electronically using *Means CostWorks®* CD or on the Web using **MeansCostWorks.com.**

What's more, you can increase your knowledge and improve your construction estimating and management performance with an RSMeans construction seminar or in-house training program. These two-day seminar programs offer unparalleled opportunities for everyone in your organization to become updated on a wide variety of construction-related issues.

RSMeans is also a worldwide provider of construction cost management and analysis services for commercial and government owners.

In short, RSMeans can provide you with the tools and expertise for constructing accurate and dependable construction estimates and budgets in a variety of ways.

Robert Snow Means Established a Tradition of Quality That Continues Today

Robert Snow Means spent years building RSMeans, making certain he always delivered a quality product.

Today, at RSMeans, we do more than talk about the quality of our data and the usefulness of our books. We stand behind all of our data, from historical cost indexes to construction materials and techniques to current costs.

If you have any questions about our products or services, please call us toll-free at 1-800-334-3509. Our customer service representatives will be happy to assist you. You can also visit our Web site at **www.rsmeans.com.**

Table of Contents

Related RSMeans Products and Services

The engineers at RSMeans suggest the following products and services as companion information resources to *RSMeans Mechanical Cost Data*:

Construction Cost Data Books

Electrical Cost Data 2010
Plumbing Cost Data 2010

Reference Books

ADA Compliance Pricing Guide
Building Security: Strategies & Costs
Designing & Building with the IBC
Estimating Building Costs
Estimating Handbook
Green Building: Project Planning & Estimating
How to Estimate with Means Data and CostWorks
Plan Reading & Material Takeoff
Project Scheduling and Management for Construction
Mechanical Estimating Methods

Seminars and In-House Training

Mechanical & Electrical Estimating
Means CostWorks® Training
Means Data for Job Order Contracting (JOC)
Plan Reading & Material Takeoff
Scheduling & Project Management

RSMeans on the Internet

Visit RSMeans at www.rsmeans.com for a one-stop portal for the most reliable and current resources available on the market. Here you can request or download FREE estimating software demos, visit our bookstore for convenient ordering, and learn more about new publications and companion products.

Also visit www.reedconstructiondata.com for RSMeans business solutions, seminars, training, and more.

RSMeans Electronic Data

Get the information found in RSMeans cost books electronically on *Means CostWorks®* CD or on the Web at **MeansCostworks.com**

RSMeans Business Solutions

Cost engineers and analysts conduct facility life cycle and benchmark research studies and predictive cost modeling, as well as offer consultation services for conceptual estimating, real property management, and job order contracting. Research studies are designed with the application of proprietary cost and project data from Reed Construction Data and RSMeans' extensive North American databases. Analysts offer building product manufacturers qualitative and quantitative market research, as well as time/motion studies for new products, and Web-based dashboards for market opportunity analysis. Clients are from both the public and private sectors, including federal, state, and municipal agencies, corporations, institutions, construction management firms, hospitals, and associations.

RSMeans for Job Order Contracting (JOC)

Best practice job order contracting (JOC) services support the owner in the development of their program. RSMeans develops all required contracting documents, works with stakeholders, and develops qualified and experienced contractor lists. Means JOCWorks™ software is the best of class in the functional tools that meet JOC specific requirements. RSMeans data is organized to meet JOC cost estimating requirements.

Construction Costs for Software Applications

Over 25 unit price and assemblies cost databases are available through a number of leading estimating and facilities management software providers (listed below). For more information see the "Other RSMeans Products" pages at the back of this publication.

MeansData™ is also available to federal, state, and local government agencies as multi-year, multi-seat licenses.

- 4Clicks-Solutions, LLC
- ArenaSoft Estimating
- Beck Technology
- BSD – Building Systems Design, Inc.
- CMS – Construction Management Software
- Corecon Technologies, Inc.
- CorVet Systems
- Estimating Systems, Inc.
- Hard Dollar Corporation
- MC²
- Parsons Corporation
- Sage Timberline Office
- UDA Technologies, Inc.
- US Cost, Inc.
- VFA – Vanderweil Facility Advisers
- WinEstimator, Inc.

How the Book Is Built: An Overview

The Construction Specifications Institute (CSI) and Construction Specifications Canada (CSC) have produced the 2004 edition of MasterFormat, a system of titles and numbers used extensively to organize construction information.

All unit price data in the RSMeans cost data books is now arranged in the 50-division MasterFormat 2004 system.

A Powerful Construction Tool

You have in your hands one of the most powerful construction tools available today. A successful project is built on the foundation of an accurate and dependable estimate. This book will enable you to construct just such an estimate.

For the casual user the book is designed to be:

- quickly and easily understood so you can get right to your estimate.
- filled with valuable information so you can understand the necessary factors that go into the cost estimate.

For the regular user, the book is designed to be:

- a handy desk reference that can be quickly referred to for key costs.
- a comprehensive, fully reliable source of current construction costs and productivity rates so you'll be prepared to estimate any project.
- a source book for preliminary project cost, product selections, and alternate materials and methods.

To meet all of these requirements, we have organized the book into the following clearly defined sections.

New: Quick Start

See our new "Quick Start" instructions on the following page to get started right away.

Estimating with RSMeans Unit Price Cost Data

Please refer to these steps for guidance on completing an estimate using RSMeans unit price cost data.

How To Use the Book: The Details

This section contains an in-depth explanation of how the book is arranged and how you can use it to determine a reliable construction cost estimate. It includes information about how we develop our cost figures and how to completely prepare your estimate.

Unit Price Section

All cost data has been divided into the 50 divisions according to the MasterFormat system of classification and numbering. For a listing of these divisions and an outline of their subdivisions, see the Unit Price Section Table of Contents.

Estimating tips are included at the beginning of each division.

Assemblies Section

The cost data in this section has been organized in an "Assemblies" format. These assemblies are the functional elements of a building and are arranged according to the 7 divisions of the UNIFORMAT II classification system. For a complete explanation of a typical "Assemblies" page, see "How To Use the Assemblies Cost Tables."

Reference Section

This section includes information on Equipment Rental Costs, Crew Listings, Historical Cost Indexes, City Cost Indexes, Location Factors, Reference Tables, Change Orders, Square Foot Costs, and a listing of Abbreviations.

Equipment Rental Costs: This section contains the average costs to rent and operate hundreds of pieces of construction equipment.

Crew Listings: This section lists all the crews referenced in the book. For the purposes of this book, a crew is composed of more than one trade classification and/or the addition of power equipment to any trade classification. Power equipment is included in the cost of the crew. Costs are shown both with the bare labor rates and with the installing contractor's overhead and profit added. For each, the total crew cost per eight-hour day and the composite cost per labor-hour are listed.

Historical Cost Indexes: These indexes provide you with data to adjust construction costs over time.

City Cost Indexes: All costs in this book are U.S. national averages. Costs vary because of the regional economy. You can adjust costs by CSI Division to over 316 locations throughout the U.S. and Canada by using the data in this section.

Location Factors: You can adjust total project costs to over 900 locations throughout the U.S. and Canada by using the data in this section.

Reference Tables: At the beginning of selected major classifications in the Unit Price and Assemblies sections are "reference numbers" shown in a shaded box. These numbers refer you to related information in the Reference Section. In this section, you'll find reference tables, explanations, estimating information that support how we develop the unit price data, technical data, and estimating procedures.

Change Orders: This section includes information on the factors that influence the pricing of change orders.

Square Foot Costs: This section contains costs for 59 different building types that allow you to make a rough estimate for the overall cost of a project or its major components.

Abbreviations: A listing of abbreviations used throughout this book, along with the terms they represent, is included in this section.

Index

A comprehensive listing of all terms and subjects in this book will help you quickly find what you need when you are not sure where it falls in MasterFormat.

The Scope of This Book

This book is designed to be as comprehensive and as easy to use as possible. To that end we have made certain assumptions and limited its scope in two key ways:

1. We have established material prices based on a national average.
2. We have computed labor costs based on a 30-city national average of union wage rates.

For a more detailed explanation of how the cost data is developed, see "How To Use the Book: The Details."

Project Size/Type

The material prices in Means data cost books are "contractor's prices." They are the prices that contractors can expect to pay at the lumberyards, suppliers/distributers warehouses, etc. Small orders of speciality items would be higher than the costs shown, while very large orders, such as truckload lots, would be less. The variation would depend on the size, timing, and negotiating power of the contractor. The labor costs are primarily for new construction or major renovations rather than repairs or minor alterations. **With reasonable exercise of judgment, the figures can be used for any building work.**

Absolute Essentials for a Quick Start

If you feel you are ready to use this book and don't think you will need the detailed instructions that begin on the following page, this Absolute Essentials for a Quick Start page is for you. These steps will allow you to get started estimating in a matter of minutes.

1 Scope

Think through the project that you will be estimating, and identify the many individual work tasks that will need to be covered in your estimate.

2 Quantify

Determine the number of units that will be required for each work task that you identified.

3 Pricing

Locate individual Unit Price line items that match the work tasks you identified. The Unit Price Section Table of Contents that begins on page 1 and the Index in the back of the book will help you find these line items.

4 Multiply

Multiply the Total Incl O&P cost for a Unit Price line item in the book by your quantity for that item. The price you calculate will be an estimate for a completed item of work performed by a subcontractor. Keep adding line items in this manner to build your estimate.

5 Project Overhead

Include project overhead items in your estimate. These items are needed to make the job run and are typically, but not always, provided by the General Contractor. They can be found in Division 1. An alternate method of estimating project overhead costs is to apply a percentage of the total project cost.

Include rented tools not included in crews, waste, rubbish handling, and cleanup.

6 Estimate Summary

Include General Contractor's markup on subcontractors, General Contractor's office overhead and profit, and sales tax on materials and equipment.

Adjust your estimate to the project's location by using the City Cost Indexes or Location Factors found in the Reference Section.

Editors' Note: We urge you to spend time reading and understanding the supporting material in the front of this book. An accurate estimate requires experience, knowledge, and careful calculation. The more you know about how we at RSMeans developed the data, the more accurate your estimate will be. In addition, it is important to take into consideration the reference material in the back of the book such as Equipment Listings, Crew Listings, City Cost Indexes, Location Factors, and Reference Numbers.

Estimating with RSMeans Unit Price Cost Data

Following these steps will allow you to complete an accurate estimate using RSMeans Unit Price cost data.

1 Scope out the project

- Identify the individual work tasks that will need to be covered in your estimate.
- The Unit Price data inside this book has been divided into 44 Divisions according to the CSI MasterFormat 2004—their titles are listed on the back cover of your book.
- Think through the project that you will be estimating, and identify those CSI Divisions that you will need to use in your estimate.
- The Unit Price Section Table of Contents that begins on page 1 may also be helpful when scoping out your project.
- Experienced estimators find it helpful to begin an estimate with Division 2, estimating Division 1 after the full scope of the project is known.

2 Quantify

- Determine the number of units that will be required for each work task that you previously identified.
- Experienced estimators will include an allowance for waste in their quantities. (Waste is not included in RSMeans Unit Price line items unless so stated.)

3 Price the quantities

- Use the Unit Price Table of Contents, and the Index, to locate the first individual Unit Price line item for your estimate.
- Reference Numbers indicated within a Unit Price section refer to additional information that you may find useful.
- The crew indicates who is performing the work for that task. Crew codes are expanded in the Crew Listings in the Reference Section to include all trades and equipment that comprise the crew.
- The Daily Output is the amount of work the crew is expected to do in one day.
- The Labor-Hours value is the amount of time it will take for the average crew member to install one unit of measure.
- The abbreviated Unit designation indicates the unit of measure upon which the crew, productivity, and prices are based.
- Bare Costs are shown for materials, labor, and equipment needed to complete the Unit Price line item. Bare costs do not include waste, project overhead, payroll insurance, payroll taxes, main office overhead, or profit.
- The Total Incl O&P cost is the billing rate or invoice amount of the installing contractor or subcontractor who performs the work for the Unit Price line item.

4 Multiply

- Multiply the total number of units needed for your project by the Total Incl O&P cost for the Unit Price line item.
- Be careful that the final unit of measure for your quantity of units matches the unit of measure in the Unit column in the book.
- The price you calculate will be an estimate for a completed item of work.
- Keep scoping individual tasks, determining the number of units required for those tasks, matching them up with individual Unit Price line items in the book, and multiplying quantities by Total Incl O&P costs. In this manner, keep building your estimate.
- An estimate completed to this point, in this manner, will be priced as if a subcontractor, or set of subcontractors, will perform the work. The estimate does not yet include Project Overhead or Estimate Summary components such as General Contractor markups on subcontracted and self-performed work, General Contractor office overhead and profit, contingency, and location factor.

5 Project Overhead

- Include project overhead items from Division 1 – General Requirements.
- These are items that will be needed to make the job run. These are typically, but not always, provided by the General Contractor. They include, but are not limited to, such items as field personnel, insurance, performance bond, permits, testing, temporary utilities, field office and storage facilities, temporary scaffolding and platforms, equipment mobilization and demobilization, temporary roads and sidewalks, winter protection, temporary barricades and fencing, temporary security, temporary signs, field engineering and layout, final cleaning and commissioning.
- These items should be scoped, quantified, matched to individual Unit Price line items in Division 1, priced and added to your estimate.
- An alternate method of estimating project overhead costs is to apply a percentage of the total project cost, usually 5% to 15% with an average of 10% (see General Conditions, page ix).
- Include other project related expenses in your estimate such as:
  - Rented equipment not itemized in the Crew Listings
  - Rubbish handling throughout the project (see 02 41 19.23)

6 Estimate Summary

- Includes sales tax on materials and equipment.
  Note: Sales tax must be added for materials in subcontracted work.
- Include the General Contractor's markup on self-performed work, usually 5% to 15% with an average of 10%.
- Include the General Contractor's markup on subcontracted work, usually 5% to 15% with an average of 10%.
- Include General Contractor's main office overhead and profit.
- RSMeans gives general guidelines on the General Contractor's main office overhead (see section 01 31 13.50 and Reference Number R013113-50).
- RSMeans gives no guidance on the General Contractor's profit.
- Markups will depend on the size of the General Contractor's operations, his projected annual revenue, the level of risk he is taking on, and on the level of competitiveness in the local area and for this project in particular.
- Include a contingency, usually 3% to 5%.
- Adjust your estimate to the project's location by using the City Cost Indexes or the Location Factors in the Reference Section.
- Look at the rules on the pages for How to Use the City Cost Indexes to see how to apply the Indexes for your location.
- When the proper Index or Factor has been identified for the project's location, convert it to a multiplier by dividing it by 100, then multiply that multiplier by your estimate total cost. Your original estimate total cost will now be adjusted up or down from the national average to a total that is appropriate for your location.

Editors' Notes:

1) *We urge you to spend time reading and understanding the supporting material in the front of this book. An accurate estimate requires experience, knowledge, and careful calculation. The more you know about how we at RSMeans developed the data, the more accurate your estimate will be. In addition, it is important to take into consideration the reference material in the back of the book such as Equipment Listings, Crew Listings, City Cost Indexes, Location Factors, and Reference Numbers.*
2) *Contractors who are bidding or are involved in JOC, DOC, SABER, or IDIQ type contracts are cautioned that workers' compensation insurance, federal and state payroll taxes, waste, project supervision, project overhead, main office overhead, and profit are not included in bare costs. Your coefficient or multiplier must cover these costs.*

How to Use the Book: The Details

What's Behind the Numbers? The Development of Cost Data

The staff at RSMeans continuously monitors developments in the construction industry in order to ensure reliable, thorough, and up-to-date cost information.

While *overall* construction costs may vary relative to general economic conditions, price fluctuations within the industry are dependent upon many factors. Individual price variations may, in fact, be opposite to overall economic trends. Therefore, costs are continually monitored and complete updates are published yearly. Also, new items are frequently added in response to changes in materials and methods.

Costs—$ (U.S.)

All costs represent U.S. national averages and are given in U.S. dollars. The RSMeans City Cost Indexes can be used to adjust costs to a particular location. The City Cost Indexes for Canada can be used to adjust U.S. national averages to local costs in Canadian dollars. No exchange rate conversion is necessary.

G The processes or products identified by the green symbol in this publication have been determined to be environmentally responsible and/or resource-efficient solely by the RSMeans engineering staff. The inclusion of the green symbol does not represent compliance with any specific industry association or standard.

Material Costs

The RSMeans staff contacts manufacturers, dealers, distributors, and contractors all across the U.S. and Canada to determine national average material costs. If you have access to current material costs for your specific location, you may wish to make adjustments to reflect differences from the national average. Included within material costs are fasteners for a normal installation. RSMeans engineers use manufacturers' recommendations, written specifications, and/or standard construction practice for size and spacing of fasteners. Adjustments to material costs may be required for your specific application or location. Material costs do not include sales tax.

Labor Costs

Labor costs are based on the average of wage rates from 30 major U.S. cities. Rates are determined from labor union agreements or prevailing wages for construction trades for the current year. Rates, along with overhead and profit markups, are listed on the inside back cover of this book.

- If wage rates in your area vary from those used in this book, or if rate increases are expected within a given year, labor costs should be adjusted accordingly.

Labor costs reflect productivity based on actual working conditions. In addition to actual installation, these figures include time spent during a normal workday on tasks such as, material receiving and handling, mobilization at site, site movement, breaks, and cleanup.

Productivity data is developed over an extended period so as not to be influenced by abnormal variations and reflects a typical average.

Equipment Costs

Equipment costs include not only rental, but also operating costs for equipment under normal use. The operating costs include parts and labor for routine servicing such as repair and replacement of pumps, filters, and worn lines. Normal operating expendables, such as fuel, lubricants, tires, and electricity (where applicable), are also included. Extraordinary operating expendables with highly variable wear patterns, such as diamond bits and blades, are excluded. These costs are included under materials. Equipment rental rates are obtained from industry sources throughout North America—contractors, suppliers, dealers, manufacturers, and distributors.

Equipment costs do not include operators' wages; nor do they include the cost to move equipment to a job site (mobilization) or from a job site (demobilization).

Equipment Cost/Day—The cost of power equipment required for each crew is included in the Crew Listings in the Reference Section (small tools that are considered as essential everyday tools are not listed out separately). The Crew Listings itemize specialized tools and heavy equipment along with labor trades. The daily cost of itemized equipment included in a crew is based on dividing the weekly bare rental rate by 5 (number of working days per week) and then adding the hourly operating cost times 8 (the number of hours per day). This Equipment Cost/Day is shown in the last column of the Equipment Rental Cost pages in the Reference Section.

Mobilization/Demobilization—The cost to move construction equipment from an equipment yard or rental company to the job site and back again is not included in equipment costs. Mobilization (to the site) and demobilization (from the site) costs can be found in Division 01 54 36.50. If a piece of equipment is already at the job site, it is not appropriate to utilize mobil./demob costs again in an estimate.

Overhead and Profit

Total Cost, including O&P for the *Installing Contractor* is shown in the last column on both the Unit Price and the Assemblies pages of this book. This figure is the sum of the bare material cost plus 10% for profit, the bare labor cost plus total overhead and profit, and the bare equipment cost plus 10% for profit. Details for the calculation of Overhead and Profit on labor are shown on the inside back cover and in the Reference Section of this book. (See the "How to Use the Unit Price Pages" for an example of this calculation.)

General Conditions

Cost data in this book is presented in two ways: Bare Costs and Total Cost including O&P (Overhead and Profit). General Conditions, when applicable, should also be added to the Total Cost including O&P. The costs for General Conditions are listed in Division 1 of the Unit Price Section and the Reference Section of this book. General Conditions for the *Installing Contractor* may range from 0% to 10% of the Total Cost including O&P. For the *General* or *Prime Contractor*, costs for General Conditions may range from 5% to 15% of the Total Cost, including O&P, with a figure of 10% as the most typical allowance.

Factors Affecting Costs

Costs can vary depending upon a number of variables. Here's how we have handled the main factors affecting costs.

Quality—The prices for materials and the workmanship upon which productivity is based represent sound construction work. They are also in line with U.S. government specifications.

Overtime—We have made no allowance for overtime. If you anticipate premium time or work beyond normal working hours, be sure to make an appropriate adjustment to your labor costs.

Productivity—The productivity, daily output, and labor-hour figures for each line item are based on working an eight-hour day in daylight hours in moderate temperatures. For work that extends beyond normal work hours or is performed under adverse conditions, productivity may decrease. (See the section in "How To Use the Unit Price Pages" for more on productivity.)

Size of Project—The size, scope of work, and type of construction project will have a significant impact on cost. Economies of scale can reduce costs for large projects. Unit costs can often run higher for small projects. Costs in this book are intended for the size and type of project as previously described in "How the Book Is Built: An Overview." Costs for projects of a significantly different size or type should be adjusted accordingly.

Location—Material prices in this book are for metropolitan areas. However, in dense urban areas, traffic and site storage limitations may increase costs. Beyond a 20-mile radius of large cities, extra trucking or transportation charges may also increase the material costs slightly. On the other hand, lower wage rates may be in effect. Be sure to consider both of these factors when preparing an estimate, particularly if the job site is located in a central city or remote rural location.

In addition, highly specialized subcontract items may require travel and per-diem expenses for mechanics.

Other Factors —

- season of year
- contractor management
- weather conditions
- local union restrictions
- building code requirements
- availability of:
    - adequate energy
    - skilled labor
    - building materials
- owner's special requirements/restrictions
- safety requirements
- environmental considerations

Unpredictable Factors—General business conditions influence "in-place" costs of all items. Substitute materials and construction methods may have to be employed. These may affect the installed cost and/or life cycle costs. Such factors may be difficult to evaluate and cannot necessarily be predicted on the basis of the job's location in a particular section of the country. Thus, where these factors apply, you may find significant but unavoidable cost variations for which you will have to apply a measure of judgment to your estimate.

Rounding of Costs

In general, all unit prices in excess of $5.00 have been rounded to make them easier to use and still maintain adequate precision of the results. The rounding rules we have chosen are in the following table.

Prices from . . .	Rounded to the nearest . . .
$.01 to $5.00	$.01
$5.01 to $20.00	$.05
$20.01 to $100.00	$.50
$100.01 to $300.00	$1.00
$300.01 to $1,000.00	$5.00
$1,000.01 to $10,000.00	$25.00
$10,000.01 to $50,000.00	$100.00
$50,000.01 and above	$500.00

Contingencies

Contingencies: The allowance for contingencies generally is to provide for unforeseen construction difficulties. On alterations or repair jobs, 20% is not too much. If drawings are final and only field contingencies are being considered, 2% or 3% is probably sufficient, and often nothing need be added. As far as the contract is concerned, future changes in plans will be covered by extras. The contractor should consider inflationary price trends and possible material shortages during the course of the job. If drawings are not complete or approved, or a budget cost is wanted, it is wise to add 5% to 10%. Contingencies, then, are a matter of judgment. Additional allowances are shown in Division 01 21 16.50 for contingencies and job conditions and Reference Table R012153-10 for factors to convert prices for repair and remodeling jobs.

Note: For an explanation of "Selective Demolition" versus "Removal for Replacement" see Reference Number R220105-10.

Important Estimating Considerations

One reason for listing a job size "minimum" is to ensure that the construction craftsmen are productive for 8 hours a day on a continuous basis. Otherwise, tasks that require only 6 to 7 hours could be billed as an 8-hour day, thus providing an erroneous productivity estimate. The "productivity," or daily output, of each craftsman includes mobilization and cleanup time, break time, and plan layout time, as well as an allowance to carry stock from the storage trailer or location on the job site up to 200′ into the building and onto the first or second floor. If material has to be transported over greater distances or to higher floors, an additional allowance should be considered by the estimator. An allowance has also been included in the piping and fittings installation time for a leak check and minor tightening.

Equipment installation time includes the following applicable items: positioning, leveling and securing the unit in place, connecting all associated piping, ducts, vents, etc., which shall have been estimated separately, connecting to an adjacent power source, filling/bleeding, startup, adjusting the controls up and down to assure proper response, setting the integral controls/valves/regulators/thermostats for proper operation (does not include external building type control systems, DDC systems, etc.), explaining/training owner's operator, and warranty. A reasonable breakdown of the labor costs is as follows:

1.	Movement into building, installation/ setting of equipment	35%
2.	Connecting to piping/duct/power, etc.	25%
3.	Filling/flushing/cleaning/touchup, etc.	15%
4.	Startup/running adjustments	5%
5.	Training owner's rep.	5%
6.	Warranty/call back/service	15%

Note that cranes or other lifting equipment are not included on any lines in the Mechanical divisions. For example, if a crane is required to lift a heavy piece of pipe into place high above a gym floor, or to put a rooftop unit on the roof of a four-story building, etc., it must be added. Due to the potential for extreme variation—from nothing additional required to a major crane or helicopter—we feel that including a nominal amount for "lifting contingency" would be useless and detract from the accuracy of the estimate. When using equipment rental from RSMeans, remember to include the cost of the operator(s).

Estimating Labor-Hours

The labor-hours expressed in this publication are based on Average Installation time, using an efficiency level of approximately 60–65% (see item 7), which has been found reasonable and acceptable by many contractors.

The book uses this national efficiency average to establish a consistent benchmark.

For bid situations, adjustments to this efficiency level should be the responsibility of the contractor bidding the project.

The unit labor-hour is divided in the following manner. A typical day for a journeyman might be:

1.	Study Plans	3%	14.4 min.
2.	Material Procurement	3%	14.4 min.
3.	Receiving and Storing	3%	14.4 min
4.	Mobilization	5%	24.0 min.
5.	Site Movement	5%	24.0 min.
6.	Layout and Marking	8%	38.4 min.
7.	Actual Installation	64%	307.2 min.
8.	Cleanup	3%	14.4 min.
9.	Breaks, Nonproductive	6%	28.8 min.
		100%	480.0 min.

If any of the percentages expressed in this breakdown do not apply to the particular work or project situation, then that percentage or a portion of it may be deducted from or added to labor-hours.

Final Checklist

Estimating can be a straightforward process provided you remember the basics. Here's a checklist of some of the steps you should remember to complete before finalizing your estimate.

Did you remember to . . .

- factor in the City Cost Index for your locale?
- take into consideration which items have been marked up and by how much?
- mark up the entire estimate sufficiently for your purposes?
- read the background information on techniques and technical matters that could impact your project time span and cost?
- include all components of your project in the final estimate?
- double check your figures for accuracy?
- call RSMeans if you have any questions about your estimate or the data you've found in our publications?

For more information, please see "Tips for Accurate Estimating," R011105-05 in the Reference Section.

Remember, RSMeans stands behind its publications. If you have any questions about your estimate, about the costs you've used from our books, or even about the technical aspects of the job that may affect your estimate, feel free to call the RSMeans editors at 1-800-334-3509.

Unit Price Section

Table of Contents

Table of Contents (cont.)

How to Use the Unit Price Pages

The following is a detailed explanation of a sample entry in the Unit Price Section. Next to each bold number below is the described item with the appropriate component of the sample entry following in parentheses. Some prices are listed as bare costs; others as costs that include overhead and profit of the installing contractor. In most cases, if the work is to be subcontracted, the general contractor will need to add an additional markup (RSMeans suggests using 10%) to the figures in the column "Total Incl. O&P."

Division Number/Title (03 30/Cast-In-Place Concrete)

Use the Unit Price Section Table of Contents to locate specific items. The sections are classified according to the CSI MasterFormat 2004 system.

Line Numbers (03 30 53.40 3920)

Each unit price line item has been assigned a unique 12-digit code based on the CSI MasterFormat classification.

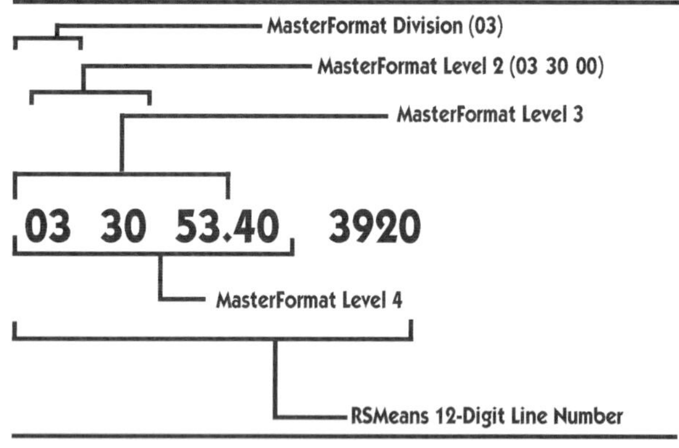

Description (Concrete In Place, etc.)

Each line item is described in detail. Sub-items and additional sizes are indented beneath the appropriate line items. The first line or two after the main item (in boldface) may contain descriptive information that pertains to all line items beneath this boldface listing.

Reference Number Information

R033053 -50 You'll see reference numbers shown in shaded boxes at the beginning of some sections. These refer to related items in the Reference Section, visually identified by a vertical gray bar on the page edges.

The relation may be an estimating procedure that should be read before estimating, or technical information.

The "R" designates the Reference Section. The numbers refer to the MasterFormat 2004 classification system.

It is strongly recommended that you review all reference numbers that appear within the section in which you are working.

03 30 Cast-In-Place Concrete

03 30 53 – Miscellaneous Cast-In-Place Concrete

03 30 53.40 Concrete In Place			Crew	Daily Output	Labor-Hours	Unit	2010 Bare Costs Material	Labor	Equipment	Total	Total Incl O&P
0010	**CONCRETE IN PLACE**										
0020	Including forms (4 uses), reinforcing steel, concrete, placement										
0050	and finishing unless otherwise indicated	R033053-50									
0300	Beams (3500 psi), 5 kip per L.F., 10′ span		C-14A	15.62	12.804	C.Y.	282	535	49.50	866.50	1,200
3850	Over 5 C.Y.		↓	75	1.493		158	59.50	.40	217.90	266
3900	Footings, strip (3000 psi), 18″ x 9″, unreinforced		C-14L	40	2.400		112	93	.77	205.77	267
3920	18″ x 9″, reinforced		C-14C	35	3.200		131	128	.86	259.86	345
3925	20″ x 10″, unreinforced		C-14L	45	2.133		110	82.50	.68	193.18	249
3930	20″ x 10″, reinforced		C-14C	40	2.800		126	112	.76	238.76	310
3935	24″ x 12″, unreinforced		C-14L	55	1.745		110	67.50	.56	178.06	226
3940	24″ x 12″, reinforced		C-14C	48	2.333		126	93	.63	219.63	283

5 Crew (C-14C)

The "Crew" column designates the typical trade or crew used to install the item. If an installation can be accomplished by one trade and requires no power equipment, that trade and the number of workers are listed (for example, "2 Carpenters"). If an installation requires a composite crew, a crew code designation is listed (for example, "C-14C"). You'll find full details on all composite crews in the Crew Listings.

- For a complete list of all trades utilized in this book and their abbreviations, see the inside back cover.

Crews

Crew No.	Bare Costs		Incl. Subs O&P		Cost Per Labor-Hour	
Crew C-14C	Hr.	Daily	Hr.	Daily	Bare Costs	Incl. O&P
1 Carpenter Foreman (out)	$43.55	$348.40	$67.15	$537.20	$39.90	$61.74
6 Carpenters	41.55	1994.40	64.05	3074.40		
2 Rodmen (reinf.)	46.80	748.80	75.40	1206.40		
4 Laborers	33.10	1059.20	51.05	1633.60		
1 Cement Finisher	39.70	317.60	57.95	463.60		
1 Gas Engine Vibrator		30.40		33.44	.27	.30
112 L.H., Daily Totals		$4498.80		$6948.64	$40.17	$62.04

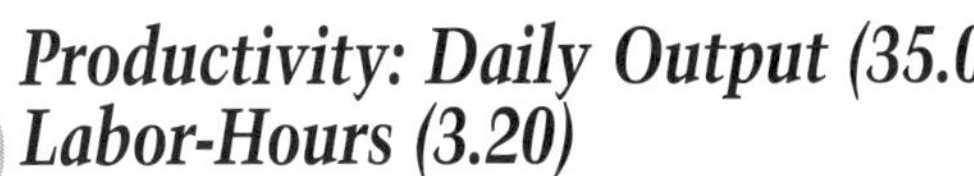

6 Productivity: Daily Output (35.0)/ Labor-Hours (3.20)

The "Daily Output" represents the typical number of units the designated crew will install in a normal 8-hour day. To find out the number of days the given crew would require to complete the installation, divide your quantity by the daily output. For example:

Quantity	÷	Daily Output	=	Duration
100 C.Y.	÷	35.0/ Crew Day	=	2.86 Crew Days

The "Labor-Hours" figure represents the number of labor-hours required to install one unit of work. To find out the number of labor-hours required for your particular task, multiply the quantity of the item times the number of labor-hours shown. For example:

Quantity	x	Productivity Rate	=	Duration
100 C.Y.	x	3.20 Labor-Hours/ C.Y.	=	320 Labor-Hours

7 Unit (C.Y.)

The abbreviated designation indicates the unit of measure upon which the price, production, and crew are based (C.Y. = Cubic Yard). For a complete listing of abbreviations, refer to the Abbreviations Listing in the Reference Section of this book.

8 Bare Costs:

Mat. (Bare Material Cost) (131)

The unit material cost is the "bare" material cost with no overhead or profit included. Costs shown reflect national average material prices for January of the current year and include delivery to the job site. No sales taxes are included.

Labor (128)

The unit labor cost is derived by multiplying bare labor-hour costs for Crew C-14C by labor-hour units. The bare labor-hour cost is found in the Crew Section under C-14C. (If a trade is listed, the hourly labor cost—the wage rate—is found on the inside back cover.)

Labor-Hour Cost Crew C-14C	x	Labor-Hour Units	=	Labor
$39.90	x	3.20		$128.00

Equip. (Equipment) (.86)

Costs for equipment included in each crew are listed in the Crew Listings in the Reference Section. Costs for small tools that are considered as essential everyday tools are covered in the "Overhead" column on the "Installing Contractor's Overhead and Profit" table that lists labor trades, base rates and markups. The unit equipment cost displayed here is derived by multiplying the bare equipment hourly cost in the Crew Listing by the unit Labor-Hours value.

Equipment Cost Crew C-14C	x	Labor-Hour Units	=	Equip.
$.27	x	3.20	=	$.86

Total (259.86)

The total of the bare costs is the arithmetic total of the three previous columns: mat., labor, and equip.

Material	+	Labor	+	Equip.	=	Total
$131	+	$128	+	$.86	=	$259.86

9 Total Costs Including O&P

This figure is the sum of the bare material cost plus 10% for profit; the bare labor cost plus total overhead and profit (per the inside back cover or, if a crew is listed, from the crew listings); and the bare equipment cost plus 10% for profit.

Material is Bare Material Cost + 10% = 131 + 13.10	=	$144.10
Labor for Crew C-14C = Labor-Hour Cost (61.74) x Labor-Hour Units (3.20)	=	$197.57
Equip. is Bare Equip. Cost + 10% = .86 + .09	=	$ .95
Total (Rounded)	=	$ 345

Estimating Tips

01 20 00 Price and Payment Procedures

- Allowances that should be added to estimates to cover contingencies and job conditions that are not included in the national average material and labor costs are shown in section 01 21.
- When estimating historic preservation projects (depending on the condition of the existing structure and the owner's requirements), a 15%–20% contingency or allowance is recommended, regardless of the stage of the drawings.

01 30 00 Administrative Requirements

- Before determining a final cost estimate, it is a good practice to review all the items listed in Subdivisions 01 31 and 01 32 to make final adjustments for items that may need customizing to specific job conditions.
- Requirements for initial and periodic submittals can represent a significant cost to the General Requirements of a job. Thoroughly check the submittal specifications when estimating a project to determine any costs that should be included.

01 40 00 Quality Requirements

- All projects will require some degree of Quality Control. This cost is not included in the unit cost of construction listed in each division. Depending upon the terms of the contract, the various costs of inspection and testing can be the responsibility of either the owner or the contractor. Be sure to include the required costs in your estimate.

01 50 00 Temporary Facilities and Controls

- Barricades, access roads, safety nets, scaffolding, security, and many more requirements for the execution of a safe project are elements of direct cost. These costs can easily be overlooked when preparing an estimate. When looking through the major classifications of this subdivision, determine which items apply to each division in your estimate.
- Construction Equipment Rental Costs can be found in the Reference Section in section 01 54 33. Operators' wages are not included in equipment rental costs.
- Equipment mobilization and demobilization costs are not included in equipment rental costs and must be considered separately in section 01 54 36.50.
- The cost of small tools provided by the installing contractor for his workers is covered in the "Overhead" column on the "Installing Contractor's Overhead and Profit" table that lists labor trades, base rates and markups and, therefore, is included in the "Total Incl. O&P" cost of any Unit Price line item. For those users who are constrained by contract terms to use only bare costs, there are two line items in section 01 54 39.70 for small tools as a percentage of bare labor cost. If some of those users are further constrained by contract terms to refrain from using line items from Division 1, you are advised to cover the cost of small tools within your coefficient or multiplier.

01 70 00 Execution and Closeout Requirements

- When preparing an estimate, thoroughly read the specifications to determine the requirements for Contract Closeout. Final cleaning, record documentation, operation and maintenance data, warranties and bonds, and spare parts and maintenance materials can all be elements of cost for the completion of a contract. Do not overlook these in your estimate.

Reference Numbers

Reference numbers are shown in shaded boxes at the beginning of some major classifications. These numbers refer to related items in the Reference Section. The reference information may be an estimating procedure, an alternate pricing method, or technical information.

Note: Not all subdivisions listed here necessarily appear in this publication.

01 11 Summary of Work

01 11 31 – Professional Consultants

	01 11 31.10 Architectural Fees	Crew	Daily Output	Labor-Hours	Unit	2010 Bare Costs				Total Incl O&P
						Material	Labor	Equipment	Total	
0010	**ARCHITECTURAL FEES** R011110-10									
2000	For "Greening" of building G				Project					3%

01 11 31.30 Engineering Fees

	01 11 31.30 Engineering Fees	Crew	Daily Output	Labor-Hours	Unit	Material	Labor	Equipment	Total	Total Incl O&P
0010	**ENGINEERING FEES** R011110-30									
0020	Educational planning consultant, minimum				Project					.50%
0100	Maximum				″					2.50%
0200	Electrical, minimum				Contrct					4.10%
0300	Maximum									10.10%
0400	Elevator & conveying systems, minimum									2.50%
0500	Maximum									5%
0600	Food service & kitchen equipment, minimum									8%
0700	Maximum									12%
1000	Mechanical (plumbing & HVAC), minimum									4.10%
1100	Maximum									10.10%
4000	Consultant, using DOE software energy analysis, small bldg, min G				SF Flr.					.25
4010	Maximum G									.45
4020	Medium building, minimum G									.15
4030	Maximum G									.35
4040	Large building, minimum G									.05
4050	Maximum G									.25

01 21 Allowances

01 21 16 – Contingency Allowances

01 21 16.50 Contingencies

	01 21 16.50 Contingencies	Crew	Daily Output	Labor-Hours	Unit	Material	Labor	Equipment	Total	Total Incl O&P
0010	**CONTINGENCIES**, Add to estimate									
0020	Conceptual stage				Project					20%
0050	Schematic stage									15%
0100	Preliminary working drawing stage (Design Dev.)									10%
0150	Final working drawing stage									3%

01 21 53 – Factors Allowance

01 21 53.50 Factors

	01 21 53.50 Factors	Crew	Daily Output	Labor-Hours	Unit	Material	Labor	Equipment	Total	Total Incl O&P
0010	**FACTORS** Cost adjustments R012153-10									
0100	Add to construction costs for particular job requirements									
0500	Cut & patch to match existing construction, add, minimum				Costs	2%	3%			
0550	Maximum					5%	9%			
0800	Dust protection, add, minimum					1%	2%			
0850	Maximum					4%	11%			
1100	Equipment usage curtailment, add, minimum					1%	1%			
1150	Maximum					3%	10%			
1400	Material handling & storage limitation, add, minimum					1%	1%			
1450	Maximum					6%	7%			
1700	Protection of existing work, add, minimum					2%	2%			
1750	Maximum					5%	7%			
2000	Shift work requirements, add, minimum						5%			
2050	Maximum						30%			
2300	Temporary shoring and bracing, add, minimum					2%	5%			
2350	Maximum					5%	12%			
2400	Work inside prisons and high security areas, add, minimum						30%			
2450	Maximum						50%			

01 21 Allowances

01 21 55 – Job Conditions Allowance

	01 21 55.50 Job Conditions	Crew	Daily Output	Labor-Hours	Unit	2010 Bare Costs Material	Labor	Equipment	Total	Total Incl O&P
0010	**JOB CONDITIONS** Modifications to applicable									
0020	cost summaries									
0100	Economic conditions, favorable, deduct				Project				2%	2%
0200	Unfavorable, add								5%	5%
0300	Hoisting conditions, favorable, deduct								2%	2%
0400	Unfavorable, add								5%	5%
0700	Labor availability, surplus, deduct								1%	1%
0800	Shortage, add								10%	10%
0900	Material storage area, available, deduct								1%	1%
1000	Not available, add								2%	2%
1100	Subcontractor availability, surplus, deduct								5%	5%
1200	Shortage, add								12%	12%
1300	Work space, available, deduct								2%	2%
1400	Not available, add								5%	5%

01 21 57 – Overtime Allowance

01 21 57.50 Overtime

		Crew	Daily Output	Labor-Hours	Unit	Material	Labor	Equipment	Total	Total Incl O&P
0010	**OVERTIME** for early completion of projects or where R012909-90									
0020	labor shortages exist, add to usual labor, up to				Costs		100%			

01 21 61 – Cost Indexes

01 21 61.10 Construction Cost Index

		Crew	Daily Output	Labor-Hours	Unit	Material	Labor	Equipment	Total	Total Incl O&P
0010	**CONSTRUCTION COST INDEX** (Reference) over 930 zip code locations in									
0020	the U.S. and Canada, total bldg. cost, min. (Guymon, OK)				%					67.40%
0050	Average									100%
0100	Maximum (New York, NY)									133.20%

01 21 61.20 Historical Cost Indexes

		Crew	Daily Output	Labor-Hours	Unit	Material	Labor	Equipment	Total	Total Incl O&P
0010	**HISTORICAL COST INDEXES** (See Reference Section)									

01 21 61.30 Labor Index

		Crew	Daily Output	Labor-Hours	Unit	Material	Labor	Equipment	Total	Total Incl O&P
0010	**LABOR INDEX** (Reference) For over 930 zip code locations in									
0020	the U.S. and Canada, minimum (Del Rio, TX)				%		33.70%			
0050	Average						100%			
0100	Maximum (New York, NY)						166.30%			

01 21 61.50 Material Index

		Crew	Daily Output	Labor-Hours	Unit	Material	Labor	Equipment	Total	Total Incl O&P
0011	**MATERIAL INDEX** For over 930 zip code locations in									
0020	the U.S. and Canada, minimum (Elizabethtown, KY)				%	90%				
0040	Average					100%				
0060	Maximum (Kotchikan, AK)					139.30%				

01 21 63 – Taxes

01 21 63.10 Taxes

		Crew	Daily Output	Labor-Hours	Unit	Material	Labor	Equipment	Total	Total Incl O&P
0010	**TAXES** R012909-80									
0020	Sales tax, State, average				%	5.01%				
0050	Maximum R012909-85					8.25%				
0200	Social Security, on first $106,800 of wages						7.65%			
0300	Unemployment, combined Federal and State, minimum						.80%			
0350	Average						6.20%			
0400	Maximum						10.96%			

01 31 Project Management and Coordination

01 31 13 – Project Coordination

01 31 13.20 Field Personnel

	01 31 13.20 Field Personnel	Crew	Daily Output	Labor-Hours	Unit	Material	2010 Bare Costs Labor	Equipment	Total	Total Incl O&P
0010	**FIELD PERSONNEL**									
0020	Clerk, average				Week		395		395	605
0100	Field engineer, minimum						935		935	1,425
0120	Average						1,215		1,215	1,875
0140	Maximum						1,400		1,400	2,150
0160	General purpose laborer, average						1,300		1,300	2,000
0180	Project manager, minimum						1,725		1,725	2,650
0200	Average						2,000		2,000	3,075
0220	Maximum						2,275		2,275	3,500
0240	Superintendent, minimum						1,675		1,675	2,575
0260	Average						1,850		1,850	2,850
0280	Maximum						2,100		2,100	3,225
0290	Timekeeper, average						1,085		1,085	1,675

01 31 13.30 Insurance

	01 31 13.30 Insurance	Crew	Daily Output	Labor-Hours	Unit	Material	Labor	Equipment	Total	Total Incl O&P
0010	**INSURANCE** R013113-40									
0020	Builders risk, standard, minimum				Job					.24%
0050	Maximum R013113-50									.64%
0200	All-risk type, minimum									.25%
0250	Maximum R013113-60									.62%
0400	Contractor's equipment floater, minimum				Value					.50%
0450	Maximum				"					1.50%
0600	Public liability, average				Job					2.02%
0800	Workers' compensation & employer's liability, average									
0850	by trade, carpentry, general				Payroll		16.88%			
1000	Electrical						6.38%			
1150	Insulation						13.11%			
1450	Plumbing						7.62%			
1550	Sheet metal work (HVAC)						9.14%			

01 31 13.50 General Contractor's Mark-Up

	01 31 13.50 General Contractor's Mark-Up	Crew	Daily Output	Labor-Hours	Unit	Material	Labor	Equipment	Total	Total Incl O&P
0010	**GENERAL CONTRACTOR'S MARK-UP** on Change Orders									
0200	Extra work, by subcontractors, add				%					10%
0250	By General Contractor, add									15%
0400	Omitted work, by subcontractors, deduct all but									5%
0450	By General Contractor, deduct all but									7.50%
0600	Overtime work, by subcontractors, add									15%
0650	By General Contractor, add									10%

01 31 13.60 Installing Contractor's Main Office Overhead

	01 31 13.60 Installing Contractor's Main Office Overhead	Crew	Daily Output	Labor-Hours	Unit	Material	Labor	Equipment	Total	Total Incl O&P
0010	**INSTALLING CONTRACTOR'S MAIN OFFICE OVERHEAD**									
0020	As percent of direct costs, minimum				%				5%	
0050	Average								13%	
0100	Maximum								30%	

01 31 13.80 Overhead and Profit

	01 31 13.80 Overhead and Profit	Crew	Daily Output	Labor-Hours	Unit	Material	Labor	Equipment	Total	Total Incl O&P
0010	**OVERHEAD & PROFIT** Allowance to add to items in this									
0020	book that do not include Subs O&P, average				%				25%	
0100	Allowance to add to items in this book that									
0110	do include Subs O&P, minimum				%					5%
0150	Average									10%
0200	Maximum									15%
0300	Typical, by size of project, under $100,000								30%	
0350	$500,000 project								25%	
0400	$2,000,000 project								20%	
0450	Over $10,000,000 project								15%	

01 31 Project Management and Coordination

01 31 13 – Project Coordination

	01 31 13.90 Performance Bond	Crew	Daily Output	Labor-Hours	Unit	Material	2010 Bare Costs Labor	Equipment	Total	Total Incl O&P
0010	**PERFORMANCE BOND** R013113-80									
0020	For buildings, minimum				Job					.60%
0100	Maximum				"					2.50%

01 32 Construction Progress Documentation

01 32 33 – Photographic Documentation

01 32 33.50 Photographs

		Crew	Daily Output	Labor-Hours	Unit	Material	Labor	Equipment	Total	Total Incl O&P
0010	**PHOTOGRAPHS**									
0020	8" x 10", 4 shots, 2 prints ea., std. mounting				Set	475			475	520
0100	Hinged linen mounts					530			530	580
0200	8" x 10", 4 shots, 2 prints each, in color					415			415	460
0300	For I.D. slugs, add to all above					5.30			5.30	5.85
1500	Time lapse equipment, camera and projector, buy					3,775			3,775	4,175
1550	Rent per month					565			565	620
1700	Cameraman and film, including processing, B.&W.				Day	1,375			1,375	1,525
1720	Color				"	1,375			1,375	1,525

01 41 Regulatory Requirements

01 41 26 – Permits

01 41 26.50 Permits

		Crew	Daily Output	Labor-Hours	Unit	Material	Labor	Equipment	Total	Total Incl O&P
0010	**PERMITS**									
0020	Rule of thumb, most cities, minimum				Job					.50%
0100	Maximum				"					2%

01 45 Quality Control

01 45 23 – Testing and Inspecting Services

01 45 23.50 Testing

		Crew	Daily Output	Labor-Hours	Unit	Material	Labor	Equipment	Total	Total Incl O&P
0010	**TESTING** and Inspecting Services									
9000	Thermographic testing, for bldg envelope heat loss, average 2,000 S.F. G				Ea.					500

01 51 Temporary Utilities

01 51 13 – Temporary Electricity

01 51 13.80 Temporary Utilities

		Crew	Daily Output	Labor-Hours	Unit	Material	Labor	Equipment	Total	Total Incl O&P
0010	**TEMPORARY UTILITIES**									
0100	Heat, incl. fuel and operation, per week, 12 hrs. per day	1 Skwk	100	.080	CSF Flr	27	3.41		30.41	35
0200	24 hrs. per day	"	60	.133		52	5.70		57.70	66.50
0350	Lighting, incl. service lamps, wiring & outlets, minimum	1 Elec	34	.235		2.63	11.55		14.18	20
0360	Maximum	"	17	.471		5.70	23		28.70	41
0400	Power for temp lighting only, per month, min/month 6.6 KWH								.92	1.01
0430	Average/month 11.8 KWH								1.65	1.82
0450	Maximum/month 23.6 KWH								3.30	3.63
0600	Power for job duration incl. elevator, etc., minimum								47	51.70
0650	Maximum								110	121
1000	Toilet, portable, see Equip. Rental 01 54 33 in Reference Section									

01 52 Construction Facilities

01 52 13 – Field Offices and Sheds

01 52 13.20 Office and Storage Space		Crew	Daily Output	Labor-Hours	Unit	Material	2010 Bare Costs Labor	Equipment	Total	Total Incl O&P
0010	**OFFICE AND STORAGE SPACE**									
0020	Trailer, furnished, no hookups, 20′ x 8′, buy	2 Skwk	1	16	Ea.	8,550	680		9,230	10,500
0250	Rent per month					146			146	160
0300	32′ x 8′, buy	2 Skwk	.70	22.857		12,300	975		13,275	15,000
0350	Rent per month					193			193	213
0400	50′ x 10′, buy	2 Skwk	.60	26.667		23,300	1,125		24,425	27,500
0450	Rent per month					286			286	315
0500	50′ x 12′, buy	2 Skwk	.50	32		26,000	1,375		27,375	30,700
0550	Rent per month					360			360	395
0700	For air conditioning, rent per month, add				↓	41			41	45
0800	For delivery, add per mile				Mile	4.50			4.50	4.95
1000	Portable buildings, prefab, on skids, economy, 8′ x 8′	2 Carp	265	.060	S.F.	85	2.51		87.51	97.50
1100	Deluxe, 8′ x 12′	"	150	.107	"	95	4.43		99.43	112
1200	Storage boxes, 20′ x 8′, buy	2 Skwk	1.80	8.889	Ea.	3,225	380		3,605	4,125
1250	Rent per month					69.50			69.50	76.50
1300	40′ x 8′, buy	2 Skwk	1.40	11.429		4,225	485		4,710	5,375
1350	Rent per month				↓	92.50			92.50	102

01 52 13.40 Field Office Expense		Crew	Daily Output	Labor-Hours	Unit	Material	Labor	Equipment	Total	Total Incl O&P
0010	**FIELD OFFICE EXPENSE**									
0100	Office equipment rental average				Month	155			155	171
0120	Office supplies, average				"	85			85	93.50
0125	Office trailer rental, see Div. 01 52 13.20									
0140	Telephone bill; avg. bill/month incl. long dist.				Month	80			80	88
0160	Lights & HVAC				"	150			150	165

01 54 Construction Aids

01 54 16 – Temporary Hoists

01 54 16.50 Weekly Forklift Crew		Crew	Daily Output	Labor-Hours	Unit	Material	Labor	Equipment	Total	Total Incl O&P
0010	**WEEKLY FORKLIFT CREW**									
0100	All-terrain forklift, 45′ lift, 35′ reach, 9000 lb. capacity	A-3P	.20	40	Week		1,650	2,400	4,050	5,100

01 54 19 – Temporary Cranes

01 54 19.50 Daily Crane Crews		Crew	Daily Output	Labor-Hours	Unit	Material	Labor	Equipment	Total	Total Incl O&P
0010	**DAILY CRANE CREWS** for small jobs, portal to portal									
0100	12-ton truck-mounted hydraulic crane	A-3H	1	8	Day		355	870	1,225	1,500
0200	25-ton	A-3I	1	8			355	1,025	1,380	1,650
0300	40-ton	A-3J	1	8			355	1,250	1,605	1,900
0400	55-ton	A-3K	1	16			660	1,775	2,435	2,950
0500	80-ton	A-3L	1	16			660	2,225	2,885	3,425
0600	100-ton	A-3M	1	16	↓		660	2,425	3,085	3,675

01 54 19.60 Monthly Tower Crane Crew		Crew	Daily Output	Labor-Hours	Unit	Material	Labor	Equipment	Total	Total Incl O&P
0010	**MONTHLY TOWER CRANE CREW**, excludes concrete footing									
0100	Static tower crane, 130′ high, 106′ jib, 6200 lb. capacity	A-3N	.05	176	Month		7,825	23,100	30,925	37,100

01 54 23 – Temporary Scaffolding and Platforms

01 54 23.70 Scaffolding		Crew	Daily Output	Labor-Hours	Unit	Material	Labor	Equipment	Total	Total Incl O&P
0010	**SCAFFOLDING** R015423-10									
0015	Steel tube, regular, no plank, labor only to erect & dismantle									
0090	Building exterior, wall face, 1 to 5 stories, 6′-4″ x 5′ frames	3 Carp	8	3	C.S.F.		125		125	192
0200	6 to 12 stories	4 Carp	8	4			166		166	256
0301	13 to 20 stories	5 Clab	8	5	↓		166		166	255

01 54 23 – Temporary Scaffolding and Platforms

01 54 23.70 Scaffolding		Crew	Daily Output	Labor-Hours	Unit	Material	2010 Bare Costs Labor	Equipment	Total	Total Incl O&P
0460	Building interior, wall face area, up to 16′ high	3 Carp	12	2	C.S.F.		83		83	128
0560	16′ to 40′ high		10	2.400			99.50		99.50	154
0800	Building interior floor area, up to 30′ high		150	.160	C.C.F.		6.65		6.65	10.25
0900	Over 30′ high	4 Carp	160	.200	"		8.30		8.30	12.80
0906	Complete system for face of walls, no plank, material only rent/mo				C.S.F.	36			36	39.50
0908	Interior spaces, no plank, material only rent/mo				C.C.F.	3.42			3.42	3.76
0910	Steel tubular, heavy duty shoring, buy									
0920	Frames 5′ high 2′ wide				Ea.	93			93	102
0925	5′ high 4′ wide					105			105	116
0930	6′ high 2′ wide					106			106	117
0935	6′ high 4′ wide					124			124	136
0940	Accessories									
0945	Cross braces				Ea.	18			18	19.80
0950	U-head, 8″ x 8″					21.50			21.50	24
0955	J-head, 4″ x 8″					15.80			15.80	17.40
0960	Base plate, 8″ x 8″					17.60			17.60	19.35
0965	Leveling jack					38			38	41.50
1000	Steel tubular, regular, buy									
1100	Frames 3′ high 5′ wide				Ea.	71			71	78
1150	5′ high 5′ wide					83			83	91.50
1200	6′-4″ high 5′ wide					104			104	114
1350	7′-6″ high 6′ wide					179			179	197
1500	Accessories cross braces					20.50			20.50	22.50
1550	Guardrail post					18.60			18.60	20.50
1600	Guardrail 7′ section					9.90			9.90	10.90
1650	Screw jacks & plates					27			27	29.50
1700	Sidearm brackets					43			43	47.50
1750	8″ casters					35.50			35.50	39
1800	Plank 2″ x 10″ x 16′-0″					51			51	56
1900	Stairway section					325			325	360
1910	Stairway starter bar					36.50			36.50	40
1920	Stairway inside handrail					66			66	72.50
1930	Stairway outside handrail					98			98	108
1940	Walk-thru frame guardrail					47			47	51.50
2000	Steel tubular, regular, rent/mo.									
2100	Frames 3′ high 5′ wide				Ea.	5			5	5.50
2150	5′ high 5′ wide					5			5	5.50
2200	6′-4″ high 5′ wide					5.15			5.15	5.65
2250	7′-6″ high 6′ wide					7			7	7.70
2500	Accessories, cross braces					1			1	1.10
2550	Guardrail post					1			1	1.10
2600	Guardrail 7′ section					1			1	1.10
2650	Screw jacks & plates					2			2	2.20
2700	Sidearm brackets					2			2	2.20
2750	8″ casters					8			8	8.80
2800	Outrigger for rolling tower					3			3	3.30
2850	Plank 2″ x 10″ x 16′-0″					6			6	6.60
2900	Stairway section					40			40	44
2940	Walk-thru frame guardrail					2.50			2.50	2.75
3000	Steel tubular, heavy duty shoring, rent/mo.									
3250	5′ high 2′ & 4′ wide				Ea.	5			5	5.50
3300	6′ high 2′ & 4′ wide					5			5	5.50
3500	Accessories, cross braces					1			1	1.10

01 54 23 – Temporary Scaffolding and Platforms

01 54 23.70 Scaffolding		Crew	Daily Output	Labor-Hours	Unit	Material	2010 Bare Costs Labor	Equipment	Total	Total Incl O&P
3600	U-head, 8" x 8"				Ea.	1			1	1.10
3650	J-head, 4" x 8"					1			1	1.10
3700	Base plate, 8" x 8"					1			1	1.10
3750	Leveling jack					2			2	2.20
5700	Planks, 2" x 10" x 16'-0", labor only to erect & remove to 50' H	3 Carp	72	.333			13.85		13.85	21.50
5800	Over 50' high	4 Carp	80	.400	↓		16.60		16.60	25.50
6000	Heavy duty shoring for elevated slab forms to 8'-2" high, floor area									
6100	Labor only to erect & dismantle	4 Carp	16	2	C.S.F.		83		83	128
6110	Materials only, rent.mo				"	29.50			29.50	32.50
6500	To 14'-8" high									
6600	Labor only to erect & dismantle	4 Carp	10	3.200	C.S.F.		133		133	205
6610	Materials only, rent/mo				"	43			43	47.50

01 54 23.75 Scaffolding Specialties		Crew	Daily Output	Labor-Hours	Unit	Material	Labor	Equipment	Total	Total Incl O&P
0010	**SCAFFOLDING SPECIALTIES**									
1200	Sidewalk bridge, heavy duty steel posts & beams, including									
1210	parapet protection & waterproofing									
1220	8' to 10' wide, 2 posts	3 Carp	15	1.600	L.F.	31.50	66.50		98	137
1230	3 posts	"	10	2.400	"	48.50	99.50		148	208
1500	Sidewalk bridge using tubular steel									
1510	scaffold frames, including planking	3 Carp	45	.533	L.F.	5.60	22		27.60	40
1600	For 2 uses per month, deduct from all above					50%				
1700	For 1 use every 2 months, add to all above					100%				
1900	Catwalks, 20" wide, no guardrails, 7' span, buy				Ea.	145			145	160
2000	10' span, buy					203			203	223
3720	Putlog, standard, 8' span, with hangers, buy					75			75	82.50
3730	Rent per month					10			10	11
3750	12' span, buy					113			113	124
3755	Rent per month					15			15	16.50
3760	Trussed type, 16' span, buy					260			260	286
3770	Rent per month					20			20	22
3790	22' span, buy					310			310	345
3795	Rent per month					30			30	33
3800	Rolling ladders with handrails, 30" wide, buy, 2 step					237			237	261
4000	7 step					725			725	800
4050	10 step					1,000			1,000	1,100
4100	Rolling towers, buy, 5' wide, 7' long, 10' high					1,350			1,350	1,500
4200	For 5' high added sections, to buy, add				↓	207			207	228
4300	Complete incl. wheels, railings, outriggers,									
4350	21' high, to buy				Ea.	2,250			2,250	2,475
4400	Rent/month = 5% of purchase cost				"	113			113	124

01 54 36 – Equipment Mobilization

01 54 36.50 Mobilization		Crew	Daily Output	Labor-Hours	Unit	Material	Labor	Equipment	Total	Total Incl O&P
0010	**MOBILIZATION** (Use line item again for demobilization) R015433-10									
0015	Up to 25 mi haul dist (50 mi RT for mob/demob crew)									
0020	Dozer, loader, backhoe, excav., grader, paver, roller, 70 to 150 H.P.	B-34N	4	2	Ea.		66.50	136	202.50	251
0100	Above 150 HP	B-34K	3	2.667			88.50	305	393.50	475
0900	Shovel or dragline, 3/4 C.Y.	"	3.60	2.222			73.50	256	329.50	395
1100	Small equipment, placed in rear of, or towed by pickup truck	A-3A	8	1			32.50	18.90	51.40	70
1150	Equip up to 70 HP, on flatbed trailer behind pickup truck	A-3D	4	2			64.50	64	128.50	169
2000	Crane, truck-mounted, up to 75 ton, (driver only, one-way)	1 Eqhv	7.20	1.111			49.50		49.50	74
2100	Crane, truck-mounted, over 75 ton	A-3E	2.50	6.400			248	60.50	308.50	440
2200	Crawler-mounted, up to 75 ton	A-3F	2	8	↓		310	475	785	990

01 54 Construction Aids

01 54 36 – Equipment Mobilization

01 54 36.50 Mobilization		Crew	Daily Output	Labor-Hours	Unit	Material	2010 Bare Costs Labor	Equipment	Total	Total Incl O&P
2300	Over 75 ton	A-3G	1.50	10.667	Ea.		415	715	1,130	1,400
2500	For each additional 5 miles haul distance, add						10%	10%		
3000	For large pieces of equipment, allow for assembly/knockdown									
3100	For mob/demob of micro-tunneling equip, see Div. 33 05 23.19									

01 54 39 – Construction Equipment

01 54 39.70 Small Tools

		Crew	Daily Output	Labor-Hours	Unit	Material	Labor	Equipment	Total	Total Incl O&P
0010	**SMALL TOOLS** R013113-50									
0020	As % of contractor's bare labor cost for project, minimum				Total		.50%			
0100	Maximum				"		2%			

01 55 Vehicular Access and Parking

01 55 23 – Temporary Roads

01 55 23.50 Roads and Sidewalks

		Crew	Daily Output	Labor-Hours	Unit	Material	Labor	Equipment	Total	Total Incl O&P
0010	**ROADS AND SIDEWALKS** Temporary									
0050	Roads, gravel fill, no surfacing, 4" gravel depth	B-14	715	.067	S.Y.	4	2.34	.47	6.81	8.50
0100	8" gravel depth	"	615	.078	"	8	2.72	.55	11.27	13.55
1000	Ramp, 3/4" plywood on 2" x 6" joists, 16" O.C.	2 Carp	300	.053	S.F.	.96	2.22		3.18	4.48
1100	On 2" x 10" joists, 16" O.C.	"	275	.058	"	1.30	2.42		3.72	5.15

01 56 Temporary Barriers and Enclosures

01 56 13 – Temporary Air Barriers

01 56 13.60 Tarpaulins

		Crew	Daily Output	Labor-Hours	Unit	Material	Labor	Equipment	Total	Total Incl O&P
0010	**TARPAULINS**									
0020	Cotton duck, 10 oz. to 13.13 oz. per S.Y., minimum				S.F.	.59			.59	.65
0050	Maximum					.53			.53	.58
0100	Polyvinyl coated nylon, 14 oz. to 18 oz., minimum					.48			.48	.53
0150	Maximum					.68			.68	.75
0200	Reinforced polyethylene 3 mils thick, white					.15			.15	.17
0300	4 mils thick, white, clear or black					.20			.20	.22
0400	5.5 mils thick, clear					.14			.14	.15
0500	White, fire retardant					.35			.35	.39
0600	7.5 mils, oil resistant, fire retardant					.40			.40	.44
0700	8.5 mils, black					.53			.53	.58
0710	Woven polyethylene, 6 mils thick					.35			.35	.39
0730	Polyester reinforced w/ integral fastening system 11 mils thick					1.07			1.07	1.18
0740	Mylar polyester, non-reinforced, 7 mils thick				↓	1.17			1.17	1.29

01 56 13.90 Winter Protection

		Crew	Daily Output	Labor-Hours	Unit	Material	Labor	Equipment	Total	Total Incl O&P
0010	**WINTER PROTECTION**									
0100	Framing to close openings	2 Clab	750	.021	S.F.	.39	.71		1.10	1.52
0200	Tarpaulins hung over scaffolding, 8 uses, not incl. scaffolding		1500	.011		.25	.35		.60	.82
0250	Tarpaulin polyester reinf. w/ integral fastening system 11 mils thick		1600	.010		.80	.33		1.13	1.39
0300	Prefab fiberglass panels, steel frame, 8 uses	↓	1200	.013	↓	.85	.44		1.29	1.62

01 56 23 – Temporary Barricades

01 56 23.10 Barricades

		Crew	Daily Output	Labor-Hours	Unit	Material	Labor	Equipment	Total	Total Incl O&P
0010	**BARRICADES**									
0020	5' high, 3 rail @ 2" x 8", fixed	2 Carp	20	.800	L.F.	4.41	33		37.41	56
0150	Movable		30	.533		3.54	22		25.54	38
1000	Guardrail, wooden, 3' high, 1" x 6", on 2" x 4" posts	↓	200	.080	↓	.97	3.32		4.29	6.15

01 56 Temporary Barriers and Enclosures

01 56 23 – Temporary Barricades

01 56 23.10 Barricades		Crew	Daily Output	Labor-Hours	Unit	Material	Labor	Equipment	Total	Total Incl O&P
						2010 Bare Costs				
1100	2" x 6", on 4" x 4" posts	2 Carp	165	.097	L.F.	2	4.03		6.03	8.40
1200	Portable metal with base pads, buy					19.30			19.30	21
1250	Typical installation, assume 10 reuses	2 Carp	600	.027		2.50	1.11		3.61	4.46
1300	Barricade tape, polyethylene, 7 mil, 3" wide x 500' long roll				Ea.	25			25	27.50

01 56 26 – Temporary Fencing

01 56 26.50 Temporary Fencing

Line	Description	Crew	Daily Output	Labor-Hours	Unit	Material	Labor	Equipment	Total	Total Incl O&P
0010	**TEMPORARY FENCING**									
0020	Chain link, 11 ga, 5' high	2 Clab	400	.040	L.F.	2.75	1.32		4.07	5.05
0100	6' high		300	.053		3.25	1.77		5.02	6.30
0200	Rented chain link, 6' high, to 1000' (up to 12 mo.)		400	.040		4.98	1.32		6.30	7.55
0250	Over 1000' (up to 12 mo.)		300	.053		3.98	1.77		5.75	7.10
0350	Plywood, painted, 2" x 4" frame, 4' high	A-4	135	.178		4.89	7.10		11.99	16.20
0400	4" x 4" frame, 8' high	"	110	.218		9.45	8.70		18.15	23.50
0500	Wire mesh on 4" x 4" posts, 4' high	2 Carp	100	.160		9.65	6.65		16.30	21
0550	8' high	"	80	.200		14.75	8.30		23.05	29

01 56 29 – Temporary Protective Walkways

01 56 29.50 Protection

Line	Description	Crew	Daily Output	Labor-Hours	Unit	Material	Labor	Equipment	Total	Total Incl O&P
0010	**PROTECTION**									
0020	Stair tread, 2" x 12" planks, 1 use	1 Carp	75	.107	Tread	3.18	4.43		7.61	10.35
0100	Exterior plywood, 1/2" thick, 1 use		65	.123		1.26	5.10		6.36	9.30
0200	3/4" thick, 1 use		60	.133		1.75	5.55		7.30	10.45
2200	Sidewalks, 2" x 12" planks, 2 uses		350	.023	S.F.	.53	.95		1.48	2.04
2300	Exterior plywood, 2 uses, 1/2" thick		750	.011		.21	.44		.65	.91
2400	5/8" thick		650	.012		.25	.51		.76	1.07
2500	3/4" thick		600	.013		.29	.55		.84	1.17

01 58 Project Identification

01 58 13 – Temporary Project Signage

01 58 13.50 Signs

Line	Description	Crew	Daily Output	Labor-Hours	Unit	Material	Labor	Equipment	Total	Total Incl O&P
0010	**SIGNS**									
0020	High intensity reflectorized, no posts, buy				S.F.	26.50			26.50	29.50

01 74 Cleaning and Waste Management

01 74 13 – Progress Cleaning

01 74 13.20 Cleaning Up

Line	Description	Crew	Daily Output	Labor-Hours	Unit	Material	Labor	Equipment	Total	Total Incl O&P
0010	**CLEANING UP**									
0020	After job completion, allow, minimum				Job					.30%
0040	Maximum				"					1%
0050	Cleanup of floor area, continuous, per day, during const.	A-5	24	.750	M.S.F.	1.70	25	2.50	29.20	42.50
0100	Final by GC at end of job	"	11.50	1.565	"	2.71	51.50	5.20	59.41	88
1000	Mechanical demolition, see Div. 02 41 19									

01 91 Commissioning

01 91 13 – General Commissioning Requirements

01 91 13.50 Commissioning		Crew	Daily Output	Labor-Hours	Unit	Material	2010 Bare Costs Labor	Equipment	Total	Total Incl O&P
0010	**COMMISSIONING** Including documentation of design intent									
0100	performance verification, O&M, training, minimum				Project					1%
0150	Maximum				"					1.25%

01 93 Facility Maintenance

01 93 13 – Facility Maintenance Procedures

01 93 13.15 Mechanical Facilities Maintenance

		Crew	Daily Output	Labor-Hours	Unit	Material	Labor	Equipment	Total	Total Incl O&P
0010	**MECHANICAL FACILITIES MAINTENANCE**									
0100	Air conditioning system maintenance									
0130	Belt, replace	1 Stpi	15	.533	Ea.		27.50		27.50	41.50
0170	Fan, clean		16	.500			26		26	39
0180	Filter, remove, clean, replace		12	.667			34.50		34.50	52
0190	Flexible coupling alignment, inspect		40	.200			10.40		10.40	15.55
0200	Gas leak locate and repair		4	2			104		104	156
0250	Pump packing gland, remove and replace		11	.727			38		38	56.50
0270	Tighten		32	.250			13		13	19.45
0290	Pump disassemble and assemble		4	2			104		104	156
0300	Air pressure regulator, disassemble, clean, assemble	1 Skwk	4	2			85		85	131
0310	Repair or replace part	"	6	1.333			57		57	87.50
0320	Purging system	1 Stpi	16	.500			26		26	39
0400	Compressor, air, remove or install fan wheel	1 Skwk	20	.400			17.05		17.05	26
0410	Disassemble or assemble 2 cylinder, 2 stage		4	2			85		85	131
0420	4 cylinder, 4 stage		1	8			340		340	525
0430	Repair or replace part		2	4			170		170	262
0700	Demolition, for mech. demolition see Div. 23 05 05.10 or 22 05 05.10									
0800	Ductwork, clean									
0810	Rectangular									
0820	6" G	1 Shee	187.50	.043	L.F.		2.10		2.10	3.17
0830	8" G		140.63	.057			2.79		2.79	4.23
0840	10" G		112.50	.071			3.49		3.49	5.30
0850	12" G		93.75	.085			4.19		4.19	6.35
0860	14" G		80.36	.100			4.89		4.89	7.40
0870	16" G		70.31	.114			5.60		5.60	8.45
0900	Round									
0910	4" G	1 Shee	358.10	.022	L.F.		1.10		1.10	1.66
0920	6" G		238.73	.034			1.65		1.65	2.49
0930	8" G		179.05	.045			2.19		2.19	3.32
0940	10" G		143.24	.056			2.74		2.74	4.15
0950	12" G		119.37	.067			3.29		3.29	4.98
0960	16" G		89.52	.089			4.39		4.39	6.65
1000	Expansion joint, not screwed, install or remove	1 Stpi	3	2.667	Ea.		138		138	207
1010	Repack	"	6	1.333	"		69		69	104
1200	Fire protection equipment									
1220	Fire hydrant, replace	Q-1	3	5.333	Ea.		250		250	375
1230	Service, lubricate, inspect, flush, clean	1 Plum	7	1.143			59.50		59.50	89
1240	Test	"	11	.727			38		38	56.50
1310	Inspect valves, pressure, nozzle,	1 Spri	4	2	System		101		101	151
1500	Instrumentation preventive maintenance, rule of thumb									
1510	Annual cost per instrument for calibration & trouble shooting									
1520	Includes flow meters, analytical transmitters, flow control									
1530	chemical feed systems, recorders and distrib. control sys.				Ea.					325

01 93 Facility Maintenance

01 93 13 – Facility Maintenance Procedures

01 93 13.15 Mechanical Facilities Maintenance		Crew	Daily Output	Labor-Hours	Unit	Material	2010 Bare Costs Labor	Equipment	Total	Total Incl O&P
1800	Plumbing fixtures, for installation see Div. 22 41 00									
1801	Plumbing fixtures									
1840	Open drain with snake	1 Plum	13	.615	Ea.		32		32	48
1850	Open drain with toilet auger		16	.500			26		26	39
1860	Grease trap, clean		2	4			208		208	310
1870	Plaster trap, clean		6	1.333			69.50		69.50	104
1881	Clean commode	1 Clab	20	.400		.32	13.25		13.57	21
1882	Clean commode seat		48	.167		.16	5.50		5.66	8.70
1883	Clean single sink		28	.286		.21	9.45		9.66	14.85
1884	Clean double sink		20	.400		.32	13.25		13.57	21
1885	Clean lavatory		28	.286		.21	9.45		9.66	14.85
1886	Clean bathtub		12	.667		.64	22		22.64	34.50
1887	Clean fiberglass shower		12	.667		.64	22		22.64	34.50
1888	Clean fiberglass tub/shower		8	1		.80	33		33.80	52
1889	Clean faucet set		32	.250		.13	8.30		8.43	12.90
1891	Clean shower head		80	.100		.02	3.31		3.33	5.10
1893	Clean water heater		16	.500		1.12	16.55		17.67	26.50
1900	Relief valve, test and adjust	1 Stpi	20	.400			21		21	31
1910	Clean pump, heater, or motor for whirlpool	1 Clab	16	.500		.16	16.55		16.71	25.50
1920	Clean thermal cover for whirlpool	"	16	.500		1.60	16.55		18.15	27.50
2000	Repair or replace-steam trap	1 Stpi	8	1			52		52	78
2020	Y-type or bell strainer		6	1.333			69		69	104
2040	Water trap or vacuum breaker, screwed joints		13	.615			32		32	48
2100	Steam specialties, clean									
2120	Air separator with automatic trap, 1" fittings	1 Stpi	12	.667	Ea.		34.50		34.50	52
2130	Bucket trap, 2" pipe		7	1.143			59.50		59.50	89
2150	Drip leg, 2" fitting		45	.178			9.25		9.25	13.85
2200	Thermodynamic trap, 1" fittings		50	.160			8.30		8.30	12.45
2210	Thermostatic		65	.123			6.40		6.40	9.60
2240	Screen and seat in y-type strainer, plug type.		25	.320			16.60		16.60	25
2242	Screen and seat in y-type strainer, flange type.		12	.667			34.50		34.50	52
2500	Valve, replace broken handwheel		24	.333			17.30		17.30	26
3000	Valve, overhaul, regulator, relief, flushometer, mixing									
3040	Cold water, gas	1 Stpi	5	1.600	Ea.		83		83	124
3050	Hot water, steam		3	2.667			138		138	207
3080	Globe, gate, check up to 4" cold water, gas		10	.800			41.50		41.50	62
3090	Hot water, steam		5	1.600			83		83	124
3100	Over 4" ID hot or cold line		1.40	5.714			297		297	445
3120	Remove and replace, gate, globe or check up to 4"	Q-5	6	2.667			125		125	187
3130	Over 4"	"	2	8			375		375	560
3150	Repack up to 4"	1 Stpi	13	.615			32		32	48
3160	Over 4"	"	4	2			104		104	156

Estimating Tips

02 30 00 Subsurface Investigation

In preparing estimates on structures involving earthwork or foundations, all information concerning soil characteristics should be obtained. Look particularly for hazardous waste, evidence of prior dumping of debris, and previous stream beds.

02 41 00 Demolition and Structure Moving

The costs shown for selective demolition do not include rubbish handling or disposal. These items should be estimated separately using RSMeans data or other sources.

- Historic preservation often requires that the contractor remove materials from the existing structure, rehab them, and replace them. The estimator must be aware of any related measures and precautions that must be taken when doing selective demolition and cutting and patching. Requirements may include special handling and storage, as well as security.
- In addition to Subdivision 02 41 00, you can find selective demolition items in each division. Example: Roofing demolition is in Division 7.

02 42 10 Building Deconstruction

This is a new section on the careful demantling and recycling of most of low-rise building materials.

02 56 13 Containment of Hazardous Waste

This is a new section for disposal of hazardous waste on site.

02 80 00 Hazardous Material Disposal/ Remediation

This subdivision includes information on hazardous waste handling, asbestos remediation, lead remediation, and mold remediation. See reference R028213-20 and R028319-60 for further guidance in using these unit price lines.

02 91 10 Monitoring Chemical Sampling, Testing Analysis

This is a new section for on-site sampling and testing hazardous waste.

Reference Numbers

Reference numbers are shown in shaded boxes at the beginning of some major classifications. These numbers refer to related items in the Reference Section. The reference information may be an estimating procedure, an alternate pricing method, or technical information.

Note: Not all subdivisions listed here necessarily appear in this publication.

02 41 Demolition

02 41 13 – Selective Site Demolition

02 41 13.17 Demolish, Remove Pavement and Curb			Crew	Daily Output	Labor-Hours	Unit	Material	2010 Bare Costs Labor	Equipment	Total	Total Incl O&P
0010	**DEMOLISH, REMOVE PAVEMENT AND CURB**	R024119-10									
5010	Pavement removal, bituminous roads, 3" thick		B-38	690	.058	S.Y.		2.15	1.71	3.86	5.15
5050	4" to 6" thick			420	.095			3.53	2.80	6.33	8.50
5100	Bituminous driveways			640	.063			2.32	1.84	4.16	5.55
5200	Concrete to 6" thick, hydraulic hammer, mesh reinforced			255	.157			5.80	4.62	10.42	13.95
5300	Rod reinforced			200	.200			7.40	5.90	13.30	17.80
5400	Concrete, 7" to 24" thick, plain			33	1.212	C.Y.		45	35.50	80.50	108
5500	Reinforced			24	1.667	"		62	49	111	148
5600	With hand held air equipment, bituminous, to 6" thick		B-39	1900	.025	S.F.		.88	.12	1	1.48
5700	Concrete to 6" thick, no reinforcing			1600	.030			1.04	.14	1.18	1.75
5800	Mesh reinforced			1400	.034			1.19	.16	1.35	2.01
5900	Rod reinforced			765	.063			2.18	.29	2.47	3.67

02 41 13.33 Minor Site Demolition

Line	Description	Ref.	Crew	Daily Output	Labor-Hours	Unit	Material	Labor	Equipment	Total	Total Incl O&P
0010	**MINOR SITE DEMOLITION**	R024119-10									
0015	No hauling, abandon catch basin or manhole		B-6	7	3.429	Ea.		123	48.50	171.50	240
0020	Remove existing catch basin or manhole, masonry			4	6			215	84.50	299.50	425
0030	Catch basin or manhole frames and covers, stored			13	1.846			66	26	92	130
0040	Remove and reset			7	3.429			123	48.50	171.50	240
0900	Hydrants, fire, remove only		B-21A	5	8			330	131	461	645
0950	Remove and reset		"	2	20			825	330	1,155	1,600
4000	Sidewalk removal, bituminous, 2-1/2" thick		B-6	325	.074	S.Y.		2.65	1.04	3.69	5.20
4050	Brick, set in mortar			185	.130			4.65	1.83	6.48	9.10
4100	Concrete, plain, 4"			160	.150			5.35	2.11	7.46	10.50
4200	Mesh reinforced			150	.160			5.75	2.25	8	11.25

02 41 19 – Selective Structure Demolition

02 41 19.13 Selective Building Demolition

Line	Description	Ref.	Crew	Daily Output	Labor-Hours	Unit	Material	Labor	Equipment	Total	Total Incl O&P
0010	**SELECTIVE BUILDING DEMOLITION**										
0020	Costs related to selective demolition of specific building components										
0025	are included under Common Work Results (XX 05 00)										
0030	in the component's appropriate division.										

02 41 19.19 Selective Demolition, Dump Charges

Line	Description	Ref.	Crew	Daily Output	Labor-Hours	Unit	Material	Labor	Equipment	Total	Total Incl O&P
0010	**SELECTIVE DEMOLITION, DUMP CHARGES**	R024119-10									
0020	Dump charges, typical urban city, tipping fees only										
0100	Building construction materials					Ton	90			90	99
0200	Trees, brush, lumber						75			75	83
0300	Rubbish only						80			80	88
0500	Reclamation station, usual charge						90			90	99

02 41 19.23 Selective Demolition, Rubbish Handling

Line	Description	Ref.	Crew	Daily Output	Labor-Hours	Unit	Material	Labor	Equipment	Total	Total Incl O&P
0010	**SELECTIVE DEMOLITION, RUBBISH HANDLING**	R024119-10									
0020	The following are to be added to the demolition prices										
0400	Chute, circular, prefabricated steel, 18" diameter		B-1	40	.600	L.F.	48.50	20.50		69	84.50
0440	30" diameter		"	30	.800	"	50	27		77	96.50
0600	Dumpster, weekly rental, 1 dump/week, 6 C.Y. capacity (2 Tons)					Week	450			450	495
0700	10 C.Y. capacity (4 Tons)						525			525	578
0725	20 C.Y. capacity (8 Tons)	R024119-20					700			700	770
0800	30 C.Y. capacity (10 Tons)						850			850	935
0840	40 C.Y. capacity (13 Tons)						1,000			1,000	1,100
1000	Dust partition, 6 mil polyethylene, 1" x 3" frame		2 Carp	2000	.008	S.F.	.45	.33		.78	1
1080	2" x 4" frame		"	2000	.008	"	.24	.33		.57	.77
2000	Load, haul, dump and return, 50' haul, hand carried		2 Clab	24	.667	C.Y.		22		22	34
2005	Wheeled			37	.432			14.30		14.30	22

02 41 Demolition

02 41 19 – Selective Structure Demolition

02 41 19.23 Selective Demolition, Rubbish Handling		Crew	Daily Output	Labor-Hours	Unit	Material	2010 Bare Costs Labor	Equipment	Total	Total Incl O&P
2040	100′ haul, hand carried	2 Clab	16.50	.970	C.Y.		32		32	49.50
2045	Wheeled		25	.640			21		21	32.50
2080	Over 100′ haul, add per 100 L.F., hand carried		35.50	.451			14.90		14.90	23
2085	Wheeled		54	.296			9.80		9.80	15.15
2120	In elevators, per 10 floors, add		140	.114			3.78		3.78	5.85
2130	Load, haul, dump and return, 50′ haul, incl. 5 riser stair, hand carried		23	.696			23		23	35.50
2135	Wheeled		35	.457			15.15		15.15	23.50
2140	6 – 10 riser stairs, hand carried		22	.727			24		24	37
2145	Wheeled		34	.471			15.60		15.60	24
2150	11 – 20 riser stairs, hand carried		20	.800			26.50		26.50	41
2155	Wheeled		31	.516			17.10		17.10	26.50
2160	21 – 40 riser stairs, hand carried		16	1			33		33	51
2165	Wheeled		24	.667			22		22	34
2170	100′ haul, incl. 5 riser stair, hand carried		15	1.067			35.50		35.50	54.50
2175	Wheeled		23	.696			23		23	35.50
2180	6 – 10 riser stair, hand carried		14	1.143			38		38	58.50
2185	Wheeled		21	.762			25		25	39
2190	11 – 20 riser stair, hand carried		12	1.333			44		44	68
2195	Wheeled		18	.889			29.50		29.50	45.50
2200	21 – 40 riser stair, hand carried		8	2			66		66	102
2205	Wheeled		12	1.333			44		44	68
2210	Over 100′ haul, add per 100 L.F., hand carried		35.50	.451			14.90		14.90	23
2215	Wheeled		54	.296	▼		9.80		9.80	15.15
2220	For each additional flight of stairs, 5 risers		550	.029	Flight		.96		.96	1.49
2225	6 – 10 risers		275	.058			1.93		1.93	2.97
2230	11-20 risers		138	.116			3.84		3.84	5.90
2235	21 – 40 risers	▼	69	.232	▼		7.70		7.70	11.85
3000	Loading & trucking, including 2 mile haul, chute loaded	B-16	45	.711	C.Y.		24	14.85	38.85	53.50
3040	Hand loading truck, 50′ haul	"	48	.667			22.50	13.90	36.40	50
3080	Machine loading truck	B-17	120	.267			9.40	6.15	15.55	21
5000	Haul, per mile, up to 8 C.Y. truck	B-34B	1165	.007			.23	.57	.80	.98
5100	Over 8 C.Y. truck	"	1550	.005	▼		.17	.43	.60	.73

02 41 19.27 Selective Demolition, Torch Cutting

		Crew	Daily Output	Labor-Hours	Unit	Material	Labor	Equipment	Total	Total Incl O&P
0010	**SELECTIVE DEMOLITION, TORCH CUTTING** R024119-10									
0020	Steel, 1″ thick plate	1 Clab	360	.022	L.F.	.23	.74		.97	1.38
0040	1″ diameter bar	"	210	.038	Ea.		1.26		1.26	1.95
1000	Oxygen lance cutting, reinforced concrete walls									
1040	12″ to 16″ thick walls	1 Clab	10	.800	L.F.		26.50		26.50	41
1080	24″ thick walls	"	6	1.333	"		44		44	68

02 65 Underground Storage Tank Removal

02 65 10 – Underground Tank and Contaminated Soil Removal

02 65 10.30 Removal of Underground Storage Tanks

		Crew	Daily Output	Labor-Hours	Unit	Material	Labor	Equipment	Total	Total Incl O&P
0010	**REMOVAL OF UNDERGROUND STORAGE TANKS** R026510-20									
0011	Petroleum storage tanks, non-leaking									
0100	Excavate & load onto trailer									
0110	3000 gal. to 5000 gal. tank G	B-14	4	12	Ea.		420	84.50	504.50	735
0120	6000 gal. to 8000 gal. tank G	B-3A	3	13.333			470	340	810	1,100
0130	9000 gal. to 12000 gal. tank G	"	2	20	▼		700	510	1,210	1,625
0190	Known leaking tank, add				%				100%	100%
0200	Remove sludge, water and remaining product from tank bottom									

02 65 Underground Storage Tank Removal

02 65 10 – Underground Tank and Contaminated Soil Removal

02 65 10.30 Removal of Underground Storage Tanks		Crew	Daily Output	Labor-Hours	Unit	Material	2010 Bare Costs Labor	Equipment	Total	Total Incl O&P
0201	of tank with vacuum truck									
0300	3000 gal. to 5000 gal. tank G	A-13	5	1.600	Ea.		66	155	221	270
0310	6000 gal. to 8000 gal. tank G		4	2			82.50	194	276.50	335
0320	9000 gal. to 12000 gal. tank G		3	2.667			110	259	369	450
0390	Dispose of sludge off-site, average				Gal.				6	6.60
0400	Insert inert solid CO_2 "dry ice" into tank									
0401	For cleaning/transporting tanks (1.5 lbs./100 gal. cap) G	1 Clab	500	.016	Lb.	1.72	.53		2.25	2.71
1020	Haul tank to certified salvage dump, 100 miles round trip									
1023	3000 gal. to 5000 gal. tank				Ea.				700	770
1026	6000 gal. to 8000 gal. tank								800	880
1029	9,000 gal. to 12,000 gal. tank								1,000	1,100
1100	Disposal of contaminated soil to landfill									
1110	Minimum				C.Y.				120	132
1111	Maximum				"				380	418
1120	Disposal of contaminated soil to									
1121	bituminous concrete batch plant									
1130	Minimum				C.Y.				75	83
1131	Maximum				"				120	132
2010	Decontamination of soil on site incl poly tarp on top/bottom									
2011	Soil containment berm, and chemical treatment									
2020	Minimum G	B-11C	100	.160	C.Y.	7.05	6.10	3.38	16.53	20.50
2021	Maximum G	"	100	.160		9.15	6.10	3.38	18.63	23
2050	Disposal of decontaminated soil, minimum								75	83
2055	Maximum								150	165

02 82 Asbestos Remediation

02 82 13 – Asbestos Abatement

02 82 13.41 Asbestos Abatement Equip.

		Crew	Daily Output	Labor-Hours	Unit	Material	Labor	Equipment	Total	Total Incl O&P
0010	**ASBESTOS ABATEMENT EQUIP.** R028213-20									
0011	Equipment and supplies, buy									
0200	Air filtration device, 2000 C.F.M.				Ea.	795			795	875
0250	Large volume air sampling pump, minimum					355			355	390
0260	Maximum					490			490	540
0300	Airless sprayer unit, 2 gun					4,800			4,800	5,300
0350	Light stand, 500 watt					52			52	57.50
0400	Personal respirators									
0410	Negative pressure, 1/2 face, dual operation, min.				Ea.	25.50			25.50	28
0420	Maximum					27.50			27.50	30.50
0450	P.A.P.R., full face, minimum					147			147	162
0460	Maximum					147			147	162
0470	Supplied air, full face, incl. air line, minimum					163			163	179
0480	Maximum					288			288	315
0500	Personnel sampling pump, minimum					221			221	244
1500	Power panel, 20 unit, incl. G.F.I.					570			570	625
1600	Shower unit, including pump and filters					1,175			1,175	1,275
1700	Supplied air system (type C)					9,800			9,800	10,800
1750	Vacuum cleaner, HEPA, 16 gal., stainless steel, wet/dry					1,050			1,050	1,150
1760	55 gallon					1,700			1,700	1,875
1800	Vacuum loader, 9 – 18 ton/hr					90,000			90,000	99,000
1900	Water atomizer unit, including 55 gal. drum					230			230	253
2000	Worker protection, whole body, foot, head cover & gloves, plastic					8.70			8.70	9.60

02 82 Asbestos Remediation

02 82 13 – Asbestos Abatement

02 82 13.41 Asbestos Abatement Equip.		Crew	Daily Output	Labor-Hours	Unit	2010 Bare Costs Material	Labor	Equipment	Total	Total Incl O&P
2500	Respirator, single use				Ea.	28.50			28.50	31.50
2550	Cartridge for respirator					6.90			6.90	7.60
2570	Glove bag, 7 mil, 50" x 64"					17.25			17.25	19
2580	10 mil, 44" x 60"					74.50			74.50	82
2590	6 mil, 44" x 60"					74.50			74.50	82
3000	HEPA vacuum for work area, minimum					1,325			1,325	1,475
3050	Maximum					1,400			1,400	1,525
6000	Disposable polyethylene bags, 6 mil, 3 C.F.					2.26			2.26	2.49
6300	Disposable fiber drums, 3 C.F.					16.30			16.30	17.95
6400	Pressure sensitive caution labels, 3" x 5"					3.11			3.11	3.42
6450	11" x 17"					7.05			7.05	7.75
6500	Negative air machine, 1800 C.F.M.					810			810	890

02 82 13.42 Preparation of Asbestos Containment Area

Line	Description	Crew	Daily Output	Labor-Hours	Unit	Material	Labor	Equipment	Total	Total Incl O&P
0010	**PREPARATION OF ASBESTOS CONTAINMENT AREA** R028213-20									
0100	Pre-cleaning, HEPA vacuum and wet wipe, flat surfaces	A-9	12000	.005	S.F.	.02	.24		.26	.40
0200	Protect carpeted area, 2 layers 6 mil poly on 3/4" plywood	"	1000	.064		2	2.92		4.92	6.75
0300	Separation barrier, 2" x 4" @ 16", 1/2" plywood ea. side, 8' high	2 Carp	400	.040		2	1.66		3.66	4.76
0310	12' high		320	.050		2	2.08		4.08	5.40
0320	16' high		200	.080		2	3.32		5.32	7.30
0400	Personnel decontam. chamber, 2" x 4" @ 16", 3/4" ply ea. side		280	.057		3	2.37		5.37	6.95
0450	Waste decontam. chamber, 2" x 4" studs @ 16", 3/4" ply ea. side		360	.044		3	1.85		4.85	6.15
0500	Cover surfaces with polyethylene sheeting									
0501	Including glue and tape									
0550	Floors, each layer, 6 mil	A-9	8000	.008	S.F.	.05	.36		.41	.63
0551	4 mil		9000	.007		.04	.32		.36	.54
0560	Walls, each layer, 6 mil		6000	.011		.05	.49		.54	.82
0561	4 mil		7000	.009		.04	.42		.46	.69
0570	For heights above 12', add						20%			
0575	For heights above 20', add						30%			
0580	For fire retardant poly, add					100%				
0590	For large open areas, deduct					10%	20%			
0600	Seal floor penetrations with foam firestop to 36 Sq. In.	2 Carp	200	.080	Ea.	7.60	3.32		10.92	13.50
0610	36 Sq. In. to 72 Sq. In.		125	.128		15.25	5.30		20.55	25
0615	72 Sq. In. to 144 Sq. In.		80	.200		30.50	8.30		38.80	46.50
0620	Wall penetrations, to 36 square inches		180	.089		7.60	3.69		11.29	14.10
0630	36 Sq. In. to 72 Sq. In.		100	.160		15.25	6.65		21.90	27
0640	72 Sq. In. to 144 Sq. In.		60	.267		30.50	11.10		41.60	50.50
0800	Caulk seams with latex	1 Carp	230	.035	L.F.	.15	1.45		1.60	2.40
0900	Set up neg. air machine, 1-2k C.F.M. /25 M.C.F. volume	1 Asbe	4.30	1.860	Ea.		84.50		84.50	132

02 82 13.43 Bulk Asbestos Removal

Line	Description	Crew	Daily Output	Labor-Hours	Unit	Material	Labor	Equipment	Total	Total Incl O&P
0010	**BULK ASBESTOS REMOVAL**									
0020	Includes disposable tools and 2 suits and 1respirator filter/day/worker									
0200	Boiler insulation	A-9	480	.133	S.F.	.48	6.10		6.58	9.95
0210	With metal lath, add				%				50%	
0300	Boiler breeching or flue insulation	A-9	520	.123	S.F.	.37	5.60		5.97	9.15
0310	For active boiler, add				%				100%	
0400	Duct or AHU insulation	A-10B	440	.073	S.F.	.22	3.32		3.54	5.40
0500	Duct vibration isolation joints, up to 24 Sq. In. duct	A-9	56	1.143	Ea.	3.48	52		55.48	85
0520	25 Sq. In. to 48 Sq. In. duct		48	1.333		4.06	61		65.06	99
0530	49 Sq. In. to 76 Sq. In. duct		40	1.600		4.87	73		77.87	118
0600	Pipe insulation, air cell type, up to 4" diameter pipe		900	.071	L.F.	.22	3.24		3.46	5.30
0610	4" to 8" diameter pipe		800	.080		.24	3.65		3.89	5.90

02 82 13.43 Bulk Asbestos Removal

	02 82 13.43 Bulk Asbestos Removal	Crew	Daily Output	Labor-Hours	Unit	Material	2010 Bare Costs Labor	Equipment	Total	Total Incl O&P
0620	10" to 12" diameter pipe	A-9	700	.091	L.F.	.28	4.17		4.45	6.80
0630	14" to 16" diameter pipe		550	.116		.35	5.30		5.65	8.65
0650	Over 16" diameter pipe		650	.098	S.F.	.30	4.49		4.79	7.35
0700	With glove bag up to 3" diameter pipe		200	.320	L.F.	6.05	14.60		20.65	29
1000	Pipe fitting insulation up to 4" diameter pipe		320	.200	Ea.	.61	9.10		9.71	14.85
1100	6" to 8" diameter pipe		304	.211		.64	9.60		10.24	15.60
1110	10" to 12" diameter pipe		192	.333		1.01	15.20		16.21	24.50
1120	14" to 16" diameter pipe		128	.500		1.52	23		24.52	37
1130	Over 16" diameter pipe		176	.364	S.F.	1.11	16.60		17.71	27
1200	With glove bag, up to 8" diameter pipe		75	.853	L.F.	6.55	39		45.55	67.50
2000	Scrape foam fireproofing from flat surface		2400	.027	S.F.	.08	1.22		1.30	1.98
2100	Irregular surfaces		1200	.053		.16	2.43		2.59	3.96
3000	Remove cementitious material from flat surface		1800	.036		.11	1.62		1.73	2.64
3100	Irregular surface		1000	.064		.14	2.92		3.06	4.69
6000	Remove contaminated soil from crawl space by hand		400	.160	C.F.	.49	7.30		7.79	11.90
6100	With large production vacuum loader	A-12	700	.091	"	.28	4.17	1.11	5.56	8.05
7000	Radiator backing, not including radiator removal	A-9	1200	.053	S.F.	.16	2.43		2.59	3.96
9000	For type B (supplied air) respirator equipment, add				%					10%

02 82 13.44 Demolition In Asbestos Contaminated Area

	02 82 13.44 Demolition In Asbestos Contaminated Area	Crew	Daily Output	Labor-Hours	Unit	Material	Labor	Equipment	Total	Total Incl O&P
0010	**DEMOLITION IN ASBESTOS CONTAMINATED AREA**									
0200	Ceiling, including suspension system, plaster and lath	A-9	2100	.030	S.F.	.09	1.39		1.48	2.26
0210	Finished plaster, leaving wire lath		585	.109		.33	4.99		5.32	8.10
0220	Suspended acoustical tile		3500	.018		.06	.83		.89	1.36
0230	Concealed tile grid system		3000	.021		.06	.97		1.03	1.58
0240	Metal pan grid system		1500	.043		.13	1.95		2.08	3.16
0250	Gypsum board		2500	.026		.08	1.17		1.25	1.90
0260	Lighting fixtures up to 2' x 4'		72	.889	Ea.	2.70	40.50		43.20	66
0400	Partitions, non load bearing									
0410	Plaster, lath, and studs	A-9	690	.093	S.F.	.85	4.23		5.08	7.55
0450	Gypsum board and studs	"	1390	.046	"	.14	2.10		2.24	3.41
9000	For type B (supplied air) respirator equipment, add				%					10%

02 82 13.45 OSHA Testing

	02 82 13.45 OSHA Testing	Crew	Daily Output	Labor-Hours	Unit	Material	Labor	Equipment	Total	Total Incl O&P
0010	**OSHA TESTING**									
0100	Certified technician, minimum				Day				240	264
0110	Maximum				"				320	352
0200	Personal sampling, PCM analysis, NIOSH 7400, minimum	1 Asbe	8	1	Ea.	2.75	45.50		48.25	74
0210	Maximum	"	4	2	"	3	91		94	145
0300	Industrial hygienist, minimum				Day				320	352
0310	Maximum				"				480	528
1000	Cleaned area samples	1 Asbe	8	1	Ea.	2.63	45.50		48.13	74
1100	PCM air sample analysis, NIOSH 7400, minimum		8	1		30.50	45.50		76	105
1110	Maximum		4	2		3.20	91		94.20	146
1200	TEM air sample analysis, NIOSH 7402, minimum								100	125
1210	Maximum								400	500

02 82 13.46 Decontamination of Asbestos Containment Area

	02 82 13.46 Decontamination of Asbestos Containment Area	Crew	Daily Output	Labor-Hours	Unit	Material	Labor	Equipment	Total	Total Incl O&P
0010	**DECONTAMINATION OF ASBESTOS CONTAINMENT AREA**									
0100	Spray exposed substrate with surfactant (bridging)									
0200	Flat surfaces	A-9	6000	.011	S.F.	.36	.49		.85	1.16
0250	Irregular surfaces		4000	.016	"	.31	.73		1.04	1.47
0300	Pipes, beams, and columns		2000	.032	L.F.	.56	1.46		2.02	2.89
1000	Spray encapsulate polyethylene sheeting		8000	.008	S.F.	.31	.36		.67	.91
1100	Roll down polyethylene sheeting		8000	.008	"		.36		.36	.57

02 82 Asbestos Remediation

02 82 13 – Asbestos Abatement

02 82 13.46 Decontamination of Asbestos Containment Area

	02 82 13.46 Decontamination of Asbestos Containment Area	Crew	Daily Output	Labor-Hours	Unit	2010 Bare Costs Material	Labor	Equipment	Total	Total Incl O&P
1500	Bag polyethylene sheeting	A-9	400	.160	Ea.	.77	7.30		8.07	12.20
2000	Fine clean exposed substrate, with nylon brush		2400	.027	S.F.		1.22		1.22	1.89
2500	Wet wipe substrate		4800	.013			.61		.61	.95
2600	Vacuum surfaces, fine brush		6400	.010			.46		.46	.71
3000	Structural demolition									
3100	Wood stud walls	A-9	2800	.023	S.F.		1.04		1.04	1.62
3500	Window manifolds, not incl. window replacement		4200	.015			.70		.70	1.08
3600	Plywood carpet protection		2000	.032			1.46		1.46	2.27
4000	Remove custom decontamination facility	A-10A	8	3	Ea.	15	137		152	230
4100	Remove portable decontamination facility	3 Asbe	12	2	"	12.75	91		103.75	156
5000	HEPA vacuum, shampoo carpeting	A-9	4800	.013	S.F.	.05	.61		.66	1.01
9000	Final cleaning of protected surfaces	A-10A	8000	.003	"		.14		.14	.21

02 82 13.47 Asbestos Waste Pkg., Handling, and Disp.

	02 82 13.47 Asbestos Waste Pkg., Handling, and Disp.	Crew	Daily Output	Labor-Hours	Unit	Material	Labor	Equipment	Total	Total Incl O&P
0010	**ASBESTOS WASTE PACKAGING, HANDLING, AND DISPOSAL**									
0100	Collect and bag bulk material, 3 C.F. bags, by hand	A-9	400	.160	Ea.	2.26	7.30		9.56	13.85
0200	Large production vacuum loader	A-12	880	.073		.80	3.32	.88	5	7
1000	Double bag and decontaminate	A-9	960	.067		2.26	3.04		5.30	7.20
2000	Containerize bagged material in drums, per 3 C.F. drum	"	800	.080		16.30	3.65		19.95	23.50
3000	Cart bags 50′ to dumpster	2 Asbe	400	.040			1.82		1.82	2.83
5000	Disposal charges, not including haul, minimum				C.Y.				50	55
5020	Maximum				"				170	187
5100	Remove refrigerant from system	1 Stpi	40	.200	Lb.		10.40		10.40	15.55
9000	For type B (supplied air) respirator equipment, add				%					10%

02 82 13.48 Asbestos Encapsulation With Sealants

	02 82 13.48 Asbestos Encapsulation With Sealants	Crew	Daily Output	Labor-Hours	Unit	Material	Labor	Equipment	Total	Total Incl O&P
0010	**ASBESTOS ENCAPSULATION WITH SEALANTS**									
0100	Ceilings and walls, minimum	A-9	21000	.003	S.F.	.27	.14		.41	.52
0110	Maximum		10600	.006	"	.42	.28		.70	.89
0300	Pipes to 12″ diameter including minor repairs, minimum		800	.080	L.F.	.37	3.65		4.02	6.05
0310	Maximum		400	.160	"	1.04	7.30		8.34	12.50

02 85 Mold Remediation

02 85 16 – Mold Remediation Preparation and Containment

02 85 16.50 Preparation of Mold Containment Area

	02 85 16.50 Preparation of Mold Containment Area	Crew	Daily Output	Labor-Hours	Unit	Material	Labor	Equipment	Total	Total Incl O&P
0010	**PREPARATION OF MOLD CONTAINMENT AREA** R028213-20									
0100	Pre-cleaning, HEPA vacuum and wet wipe, flat surfaces	A-9	12000	.005	S.F.	.02	.74		76	40
0300	Separation barrier, 2″ x 4″ @ 16″, 1/2″ plywood ea. side, 8′ high	2 Carp	400	.040		2	1.66		3.66	4.76
0310	12′ high		320	.050		2	2.08		4.08	5.40
0320	16′ high		200	.080		2	3.32		5.32	7.30
0400	Personnel decontam. chamber, 2″ x 4″ @ 16″, 3/4″ ply ea. side		280	.057		3	2.37		5.37	6.95
0450	Waste decontam. chamber, 2″ x 4″ studs @ 16″, 3/4″ ply each side		360	.044		3	1.85		4.85	6.15
0500	Cover surfaces with polyethylene sheeting									
0501	Including glue and tape									
0550	Floors, each layer, 6 mil	A-9	8000	.008	S.F.	.06	.36		.42	.64
0551	4 mil		9000	.007		.04	.32		.36	.55
0560	Walls, each layer, 6 mil		6000	.011		.09	.49		.58	.86
0561	4 mil		7000	.009		.07	.42		.49	.73
0570	For heights above 12′, add						20%			
0575	For heights above 20′, add						30%			
0580	For fire retardant poly, add					100%				
0590	For large open areas, deduct					10%	20%			

02 85 Mold Remediation

02 85 16 – Mold Remediation Preparation and Containment

02 85 16.50 Preparation of Mold Containment Area		Crew	Daily Output	Labor-Hours	Unit	Material	2010 Bare Costs Labor	Equipment	Total	Total Incl O&P
0600	Seal floor penetrations with foam firestop to 36 sq. in.	2 Carp	200	.080	Ea.	7.60	3.32		10.92	13.50
0610	36 sq. in. to 72 sq. in.		125	.128		15.25	5.30		20.55	25
0615	72 sq. in. to 144 sq. in.		80	.200		30.50	8.30		38.80	46.50
0620	Wall penetrations, to 36 square inches		180	.089		7.60	3.69		11.29	14.10
0630	36 sq. in. to 72 sq. in.		100	.160		15.25	6.65		21.90	27
0640	72 Sq. in. to 144 sq. in.		60	.267		30.50	11.10		41.60	50.50
0800	Caulk seams with latex caulk	1 Carp	230	.035	L.F.	.15	1.45		1.60	2.40
0900	Set up neg. air machine, 1-2k C.F.M. /25 M.C.F. volume	1 Asbe	4.30	1.860	Ea.		84.50		84.50	132

02 85 33 – Removal and Disposal of Materials with Mold

02 85 33.50 Demolition in Mold Contaminated Area		Crew	Daily Output	Labor-Hours	Unit	Material	2010 Bare Costs Labor	Equipment	Total	Total Incl O&P
0010	**DEMOLITION IN MOLD CONTAMINATED AREA**									
0200	Ceiling, including suspension system, plaster and lath	A-9	2100	.030	S.F.	.09	1.39		1.48	2.26
0210	Finished plaster, leaving wire lath		585	.109		.33	4.99		5.32	8.10
0220	Suspended acoustical tile		3500	.018		.06	.83		.89	1.36
0230	Concealed tile grid system		3000	.021		.06	.97		1.03	1.58
0240	Metal pan grid system		1500	.043		.13	1.95		2.08	3.16
0250	Gypsum board		2500	.026		.08	1.17		1.25	1.90
0255	Plywood		2500	.026		.08	1.17		1.25	1.90
0260	Lighting fixtures up to 2' x 4'		72	.889	Ea.	2.70	40.50		43.20	66
0400	Partitions, non load bearing									
0410	Plaster, lath, and studs	A-9	690	.093	S.F.	.85	4.23		5.08	7.55
0450	Gypsum board and studs		1390	.046		.14	2.10		2.24	3.41
0465	Carpet & pad		1390	.046		.14	2.10		2.24	3.41
0600	Pipe insulation, air cell type, up to 4" diameter pipe		900	.071	L.F.	.22	3.24		3.46	5.30
0610	4" to 8" diameter pipe		800	.080		.24	3.65		3.89	5.90
0620	10" to 12" diameter pipe		700	.091		.28	4.17		4.45	6.80
0630	14" to 16" diameter pipe		550	.116		.35	5.30		5.65	8.65
0650	Over 16" diameter pipe		650	.098	S.F.	.30	4.49		4.79	7.35
9000	For type B (supplied air) respirator equipment, add				%					10%

Estimating Tips

General

- Carefully check all the plans and specifications. Concrete often appears on drawings other than structural drawings, including mechanical and electrical drawings for equipment pads. The cost of cutting and patching is often difficult to estimate. See Subdivision 03 81 for Concrete Cutting, Subdivision 02 41 19.16 for Cutout Demolition, Subdivision 03 05 05.10 for Concrete Demolition, and Subdivision 02 41 19.23 for Rubbish Handling (handling, loading and hauling of debris).
- Always obtain concrete prices from suppliers near the job site. A volume discount can often be negotiated, depending upon competition in the area. Remember to add for waste, particularly for slabs and footings on grade.

03 10 00 Concrete Forming and Accessories

- A primary cost for concrete construction is forming. Most jobs today are constructed with prefabricated forms. The selection of the forms best suited for the job and the total square feet of forms required for efficient concrete forming and placing are key elements in estimating concrete construction. Enough forms must be available for erection to make efficient use of the concrete placing equipment and crew.
- Concrete accessories for forming and placing depend upon the systems used. Study the plans and specifications to ensure that all special accessory requirements have been included in the cost estimate, such as anchor bolts, inserts, and hangers.
- Included within costs for forms-in-place are all necessary bracing and shoring.

03 20 00 Concrete Reinforcing

- Ascertain that the reinforcing steel supplier has included all accessories, cutting, bending, and an allowance for lapping, splicing, and waste. A good rule of thumb is 10% for lapping, splicing, and waste. Also, 10% waste should be allowed for welded wire fabric.
- The unit price items in the subdivision for Reinforcing In Place include the labor to install accessories such as beam and slab bolsters, high chairs, and bar ties and tie wire. The material cost for these accessories is not included; they may be obtained from the Accessories Division.

03 30 00 Cast-In-Place Concrete

- When estimating structural concrete, pay particular attention to requirements for concrete additives, curing methods, and surface treatments. Special consideration for climate, hot or cold, must be included in your estimate. Be sure to include requirements for concrete placing equipment, and concrete finishing.
- For accurate concrete estimating, the estimator must consider each of the following major components individually: forms, reinforcing steel, ready-mix concrete, placement of the concrete, and finishing of the top surface. For faster estimating, Subdivision 03 30 53.40 for Concrete-In-Place can be used; here, various items of concrete work are presented that include the costs of all five major components (unless specifically stated otherwise).

03 40 00 Precast Concrete

03 50 00 Cast Decks and Underlayment

- The cost of hauling precast concrete structural members is often an important factor. For this reason, it is important to get a quote from the nearest supplier. It may become economically feasible to set up precasting beds on the site if the hauling costs are prohibitive.

Reference Numbers

Reference numbers are shown in shaded boxes at the beginning of some major classifications. These numbers refer to related items in the Reference Section. The reference information may be an estimating procedure, an alternate pricing method, or technical information.

Note: Not all subdivisions listed here necessarily appear in this publication.

03 01 Maintenance of Concrete

03 01 30 – Maintenance of Cast-In-Place Concrete

03 01 30.62 Concrete Patching		Crew	Daily Output	Labor-Hours	Unit	Material	Labor	Equipment	Total	Total Incl O&P
						2010 Bare Costs				
0010	CONCRETE PATCHING									
0100	Floors, 1/4" thick, small areas, regular grout	1 Cefi	170	.047	S.F.	1.18	1.87		3.05	4.03
0150	Epoxy grout	"	100	.080	"	7.20	3.18		10.38	12.55
2000	Walls, including chipping, cleaning and epoxy grout									
2100	1/4" deep	1 Cefi	65	.123	S.F.	7.50	4.89		12.39	15.40
2150	1/2" deep		50	.160		15.05	6.35		21.40	26
2200	3/4" deep	↓	40	.200	↓	22.50	7.95		30.45	36.50

03 11 Concrete Forming

03 11 13 – Structural Cast-In-Place Concrete Forming

03 11 13.40 Forms In Place, Equipment Foundations

		Crew	Daily Output	Labor-Hours	Unit	Material	Labor	Equipment	Total	Total Incl O&P
0010	FORMS IN PLACE, EQUIPMENT FOUNDATIONS									
0020	1 use	C-2	160	.300	SFCA	2.58	12.15		14.73	21.50
0050	2 use		190	.253		1.42	10.25		11.67	17.30
0100	3 use		200	.240		1.03	9.70		10.73	16.15
0150	4 use	↓	205	.234	↓	.84	9.50		10.34	15.55

03 11 13.45 Forms In Place, Footings

		Crew	Daily Output	Labor-Hours	Unit	Material	Labor	Equipment	Total	Total Incl O&P
0010	FORMS IN PLACE, FOOTINGS									
0020	Continuous wall, plywood, 1 use	C-1	375	.085	SFCA	5.25	3.37		8.62	10.95
0050	2 use		440	.073		2.88	2.87		5.75	7.60
0100	3 use		470	.068		2.09	2.69		4.78	6.45
0150	4 use		485	.066		1.70	2.60		4.30	5.90
5000	Spread footings, job-built lumber, 1 use		305	.105		1.54	4.14		5.68	8.10
5050	2 use		371	.086		.86	3.40		4.26	6.20
5100	3 use		401	.080		.62	3.15		3.77	5.55
5150	4 use	↓	414	.077	↓	.50	3.05		3.55	5.25

03 11 13.65 Forms In Place, Slab On Grade

		Crew	Daily Output	Labor-Hours	Unit	Material	Labor	Equipment	Total	Total Incl O&P
0010	FORMS IN PLACE, SLAB ON GRADE									
3000	Edge forms, wood, 4 use, on grade, to 6" high	C-1	600	.053	L.F.	.27	2.10		2.37	3.54
6000	Trench forms in floor, wood, 1 use		160	.200	SFCA	1.13	7.90		9.03	13.40
6050	2 use		175	.183		.62	7.20		7.82	11.80
6100	3 use		180	.178		.45	7		7.45	11.30
6150	4 use		185	.173	↓	.37	6.80		7.17	10.90
8760	Void form, corrugated fiberboard, 6" x 12", 10' long [G]	↓	240	.133	S.F.	.95	5.25		6.20	9.15

03 15 Concrete Accessories

03 15 05 – Concrete Forming Accessories

03 15 05.75 Sleeves and Chases

		Crew	Daily Output	Labor-Hours	Unit	Material	Labor	Equipment	Total	Total Incl O&P
0010	SLEEVES AND CHASES									
0100	Plastic, 1 use, 9" long, 2" diameter	1 Carp	100	.080	Ea.	1.36	3.32		4.68	6.60
0150	4" diameter		90	.089		4.01	3.69		7.70	10.10
0200	6" diameter		75	.107		7.05	4.43		11.48	14.65
0250	12" diameter		60	.133		19.85	5.55		25.40	30.50
5000	Sheet metal, 2" diameter [G]		100	.080		1.29	3.32		4.61	6.50
5100	4" diameter [G]		90	.089		1.61	3.69		5.30	7.45
5150	6" diameter [G]		75	.107		1.61	4.43		6.04	8.60
5200	12" diameter [G]		60	.133		3.21	5.55		8.76	12.10
6000	Steel pipe, 2" diameter [G]		100	.080		4.13	3.32		7.45	9.65
6100	4" diameter [G]	↓	90	.089	↓	12.25	3.69		15.94	19.15

03 15 Concrete Accessories

03 15 05 – Concrete Forming Accessories

03 15 05.75 Sleeves and Chases			Crew	Daily Output	Labor-Hours	Unit	Material	Labor	Equipment	Total	Total Incl O&P
							2010 Bare Costs				
6150	6" diameter	G	1 Carp	75	.107	Ea.	21.50	4.43		25.93	30.50
6200	12" diameter	G	↓	60	.133	↓	60.50	5.55		66.05	75.50

03 21 Reinforcing Steel

03 21 10 – Uncoated Reinforcing Steel

03 21 10.60 Reinforcing In Place

			Crew	Daily Output	Labor-Hours	Unit	Material	Labor	Equipment	Total	Total Incl O&P
0015	**REINFORCING IN PLACE**, 50-60 ton lots, A615 Grade 60										
0020	Includes labor, but not material cost, to install accessories										
0030	Made from recycled materials	G									
0502	Footings, #4 to #7	G	4 Rodm	4200	.008	Lb.	.42	.36		.78	1.03
0552	#8 to #18	G		7200	.004		.42	.21		.63	.79
0602	Slab on grade, #3 to #7	G	↓	4200	.008	↓	.40	.36		.76	1.01
0900	For other than 50-60 ton lots										
1000	Under 10 ton job, #3 to #7, add					Ton	25%	10%			
1010	#8 to #18, add						20%	10%			
1050	10 – 50 ton job, #3 to #7, add						10%				
1060	#8 to #18, add						5%				
1100	60 – 100 ton job, #3 - #7, deduct						5%				
1110	#8 to #18, deduct						10%				
1150	Over 100 ton job, #3 - #7, deduct						10%				
1160	#8 - #18, deduct					↓	15%				

03 22 Welded Wire Fabric Reinforcing

03 22 05 – Uncoated Welded Wire Fabric

03 22 05.50 Welded Wire Fabric

			Crew	Daily Output	Labor-Hours	Unit	Material	Labor	Equipment	Total	Total Incl O&P
0010	**WELDED WIRE FABRIC** ASTM A185										
0030	Made from recycled materials	G									
0050	Sheets										
0100	6 x 6 - W1.4 x W1.4 (10 x 10) 21 lb. per C.S.F.	G	2 Rodm	35	.457	C.S.F.	12	21.50		33.50	47.50

03 30 Cast-In-Place Concrete

03 30 53 – Miscellaneous Cast-In-Place Concrete

03 30 53.40 Concrete In Place

		Crew	Daily Output	Labor-Hours	Unit	Material	Labor	Equipment	Total	Total Incl O&P
0010	**CONCRETE IN PLACE**									
0020	Including forms (4 uses), reinforcing steel, concrete, placement,									
0050	and finishing unless otherwise indicated									
0500	Chimney foundations (5000 psi), industrial, minimum	C-14C	32.22	3.476	C.Y.	140	139	.94	279.94	370
0510	Maximum	"	23.71	4.724	"	162	188	1.28	351.28	470
3540	Equipment pad (3000 psi), 3' x 3' x 6" thick	C-14H	45	1.067	Ea.	42.50	44	.67	87.17	115
3550	4' x 4' x 6" thick		30	1.600		62.50	65.50	1.01	129.01	171
3560	5' x 5' x 8" thick		18	2.667		108	109	1.68	218.68	290
3570	6' x 6' x 8" thick		14	3.429		146	141	2.16	289.16	380
3580	8' x 8' x 10" thick		8	6		310	246	3.78	559.78	725
3590	10' x 10' x 12" thick	↓	5	9.600	↓	530	395	6.05	931.05	1,200
3800	Footings (3000 psi), spread under 1 C.Y.	C-14C	28	4	C.Y.	150	160	1.08	311.08	415
3825	1 C.Y to 5 C.Y.		43	2.605		173	104	.70	277.70	350
3850	Over 5 C.Y.	↓	75	1.493	↓	158	59.50	.40	217.90	266

03 30 Cast-In-Place Concrete

03 30 53 – Miscellaneous Cast-In-Place Concrete

03 30 53.40 Concrete In Place		Crew	Daily Output	Labor-Hours	Unit	2010 Bare Costs Material	Labor	Equipment	Total	Total Incl O&P
3900	Footings, strip (3000 psi), 18" x 9", unreinforced	C-14L	40	2.400	C.Y.	112	93	.77	205.77	267
3920	18" x 9", reinforced	C-14C	35	3.200		131	128	.86	259.86	345
3925	20" x 10", unreinforced	C-14L	45	2.133		110	82.50	.68	193.18	249
3930	20" x 10", reinforced	C-14C	40	2.800		126	112	.76	238.76	310
3935	24" x 12", unreinforced	C-14L	55	1.745		110	67.50	.56	178.06	226
3940	24" x 12", reinforced	C-14C	48	2.333		126	93	.63	219.63	283
3945	36" x 12", unreinforced	C-14L	70	1.371		107	53	.44	160.44	200
3950	36" x 12", reinforced	C-14C	60	1.867		122	74.50	.50	197	250
4000	Foundation mat (3000 psi), under 10 C.Y.		38.67	2.896		180	116	.78	296.78	380
4050	Over 20 C.Y.	↓	56.40	1.986		159	79	.54	238.54	299
4650	Slab on grade (3500 psi), not including finish, 4" thick	C-14E	60.75	1.449		117	59.50	.51	177.01	222
4700	6" thick	"	92	.957	↓	113	39.50	.33	152.83	186
4701	Thickened slab edge (3500 psi), for slab on grade poured									
4702	monolithically with slab; depth is in addition to slab thickness;									
4703	formed vertical outside edge, earthen bottom and inside slope									
4705	8" deep x 8" wide bottom, unreinforced	C-14L	2190	.044	L.F.	3.07	1.70	.01	4.78	6
4710	8" x 8", reinforced	C-14C	1670	.067		4.95	2.68	.02	7.65	9.60
4715	12" deep x 12" wide bottom, unreinforced	C-14L	1800	.053		6.40	2.07	.02	8.49	10.25
4720	12" x 12", reinforced	C-14C	1310	.086		9.80	3.41	.02	13.23	16.15
4725	16" deep x 16" wide bottom, unreinforced	C-14L	1440	.067		10.90	2.58	.02	13.50	16
4730	16" x 16", reinforced	C-14C	1120	.100		15	3.99	.03	19.02	22.50
4735	20" deep x 20" wide bottom, unreinforced	C-14L	1150	.083		16.65	3.23	.03	19.91	23.50
4740	20" x 20", reinforced	C-14C	920	.122		22	4.86	.03	26.89	31.50
4745	24" deep x 24" wide bottom, unreinforced	C-14L	930	.103		23.50	4	.03	27.53	32
4750	24" x 24", reinforced	C-14C	740	.151	↓	30.50	6.05	.04	36.59	43
4751	Slab on grade (3500 psi), incl. troweled finish, not incl. forms									
4760	or reinforcing, over 10,000 S.F., 4" thick	C-14F	3425	.021	S.F.	1.29	.79	.01	2.09	2.61
4820	6" thick		3350	.021		1.89	.81	.01	2.71	3.29
4840	8" thick		3184	.023		2.59	.85	.01	3.45	4.13
4900	12" thick		2734	.026		3.88	.99	.01	4.88	5.75
4950	15" thick	↓	2505	.029	↓	4.88	1.08	.01	5.97	6.95

03 31 Structural Concrete

03 31 05 – Normal Weight Structural Concrete

03 31 05.35 Normal Weight Concrete, Ready Mix		Crew	Daily Output	Labor-Hours	Unit	Material	Labor	Equipment	Total	Total Incl O&P
0010	**NORMAL WEIGHT CONCRETE, READY MIX**, delivered									
0012	Includes local aggregate, sand, Portland cement, and water									
0015	Excludes all additives and treatments									
0020	2000 psi				C.Y.	91.50			91.50	101
0100	2500 psi					94			94	103
0150	3000 psi					97			97	107
0200	3500 psi					99.50			99.50	110
0300	4000 psi					103			103	113
0350	4500 psi					106			106	116
0400	5000 psi					109			109	120
0411	6000 psi					124			124	137
0412	8000 psi					203			203	223
0413	10,000 psi					288			288	315
0414	12,000 psi					350			350	380
1000	For high early strength cement, add					10%				
1300	For winter concrete (hot water), add				↓	4.25			4.25	4.68

03 31 Structural Concrete

03 31 05 – Normal Weight Structural Concrete

03 31 05.35 Normal Weight Concrete, Ready Mix		Crew	Daily Output	Labor-Hours	Unit	2010 Bare Costs Material	Labor	Equipment	Total	Total Incl O&P
1400	For hot weather concrete (ice), add				C.Y.	9.10			9.10	10
1410	For mid-range water reducer, add					4.13			4.13	4.54
1420	For high-range water reducer/superplasticizer, add					6.35			6.35	6.95
1430	For retarder, add					2.71			2.71	2.98
1440	For non-Chloride accelerator, add					4.83			4.83	5.30
1450	For Chloride accelerator, per 1%, add					3.28			3.28	3.61
1460	For fiber reinforcing, synthetic (1 Lb./C.Y.), add					6.65			6.65	7.30
1500	For Saturday delivery, add				↓	8.85			8.85	9.70
1510	For truck holding/waiting time past 1st hour per load, add				Hr.	87.50			87.50	96.50
1520	For short load (less than 4 C.Y.), add per load				Ea.	112			112	124
2000	For all lightweight aggregate, add				C.Y.	45%				

03 31 05.70 Placing Concrete		Crew	Daily Output	Labor-Hours	Unit	Material	Labor	Equipment	Total	Total Incl O&P
0010	**PLACING CONCRETE**									
0020	Includes labor and equipment to place, strike off and consolidate									
1900	Footings, continuous, shallow, direct chute	C-6	120	.400	C.Y.		13.80	.51	14.31	21.50
1950	Pumped	C-20	150	.427			15.10	5.35	20.45	29
2000	With crane and bucket	C-7	90	.800			28.50	14.35	42.85	59.50
2100	Footings, continuous, deep, direct chute	C-6	140	.343			11.85	.44	12.29	18.55
2150	Pumped	C-20	160	.400			14.15	5	19.15	27
2200	With crane and bucket	C-7	110	.655			23.50	11.75	35.25	48.50
2400	Footings, spread, under 1 C.Y., direct chute	C-6	55	.873			30	1.11	31.11	47
2450	Pumped	C-20	65	.985			35	12.30	47.30	66.50
2500	With crane and bucket	C-7	45	1.600			57	28.50	85.50	119
2600	Over 5 C.Y., direct chute	C-6	120	.400			13.80	.51	14.31	21.50
2650	Pumped	C-20	150	.427			15.10	5.35	20.45	29
2700	With crane and bucket	C-7	100	.720			25.50	12.90	38.40	53
2900	Foundation mats, over 20 C.Y., direct chute	C-6	350	.137			4.74	.17	4.91	7.45
2950	Pumped	C-20	400	.160			5.65	2	7.65	10.85
3000	With crane and bucket	C-7	300	.240	↓		8.60	4.30	12.90	17.80

03 35 Concrete Finishing

03 35 29 – Tooled Concrete Finishing

03 35 29.30 Finishing Floors		Crew	Daily Output	Labor-Hours	Unit	Material	Labor	Equipment	Total	Total Incl O&P
0010	**FINISHING FLOORS**									
0012	Finishing requires that concrete first be placed, struck off & consolidated									
0015	Basic finishing for various unspecified flatwork									
0100	Bull float only	C-10	4000	.006	S.F.		.23		.23	.33
0125	Bull float & manual float		2000	.012			.45		.45	.67
0150	Bull float, manual float, & broom finish, w/ edging & joints		1850	.013			.49		.49	.72
0200	Bull float, manual float & manual steel trowel	↓	1265	.019	↓		.71		.71	1.06
0210	For specified Random Access Floors in ACI Classes 1, 2, 3 and 4 to achieve									
0215	Composite Overall Floor Flatness and Levelness values up to F35/F25									
0250	Bull float, machine float & machine trowel (walk-behind)	C-10C	1715	.014	S.F.		.52	.03	.55	.81
0300	Power screed, bull float, machine float & trowel (walk-behind)	C-10D	2400	.010			.38	.05	.43	.61
0350	Power screed, bull float, machine float & trowel (ride-on)	C-10E	4000	.006	↓		.23	.06	.29	.40
0352	For specified Random Access Floors in ACI Classes 5, 6, 7 and 8 to achieve									
0354	Composite Overall Floor Flatness and Levelness values up to F50/F50									
0356	Add for two-dimensional restraightening after power float	C-10	6000	.004	S.F.		.15		.15	.22
0358	For specified Random or Defined Access Floors in ACI Class 9 to achieve									
0360	Composite Overall Floor Flatness and Levelness values up to F100/F100									
0362	Add for two-dimensional restraightening after bull float & power float	C-10	3000	.008	S.F.		.30		.30	.45

03 35 Concrete Finishing

03 35 29 – Tooled Concrete Finishing

03 35 29.30 Finishing Floors		Crew	Daily Output	Labor-Hours	Unit	Material	2010 Bare Costs Labor	Equipment	Total	Total Incl O&P
0364	For specified Superflat Defined Access Floors in ACI Class 9 to achieve									
0366	Minimum Floor Flatness and Levelness values of F100/F100									
0368	Add for 2-dim'l restraightening after bull float, power float, power trowel	C-10	2000	.012	S.F.		.45		.45	.67

03 35 29.35 Control Joints, Saw Cut		Crew	Daily Output	Labor-Hours	Unit	Material	Labor	Equipment	Total	Total Incl O&P
0010	**CONTROL JOINTS, SAW CUT**									
0100	Sawcut in green concrete									
0120	1″ depth	C-27	2000	.008	L.F.	.06	.32	.08	.46	.62
0140	1-1/2″ depth		1800	.009		.09	.35	.09	.53	.72
0160	2″ depth		1600	.010		.12	.40	.10	.62	.82
0200	Blow construction debris out of control joint	C-28	6000	.001			.05		.05	.08
0300	Joint sealant									
0320	Backer rod, polyethylene, 1/4″ diameter	1 Cefi	460	.017	L.F.	.03	.69		.72	1.04
0340	Sealant, polyurethane									
0360	1/4″ x 1/4″ (308 LF/Gal)	1 Cefi	270	.030	L.F.	.16	1.18		1.34	1.90
0380	1/4″ x 1/2″ (154 LF/Gal)	″	255	.031	″	.32	1.25		1.57	2.18

03 54 Cast Underlayment

03 54 16 – Hydraulic Cement Underlayment

03 54 16.50 Cement Underlayment		Crew	Daily Output	Labor-Hours	Unit	Material	Labor	Equipment	Total	Total Incl O&P
0010	**CEMENT UNDERLAYMENT**									
2510	Underlayment, P.C based self-leveling, 4100 psi, pumped, 1/4″	C-8	20000	.003	S.F.	1.52	.10	.04	1.66	1.87
2520	1/2″		19000	.003		3.04	.11	.04	3.19	3.55
2530	3/4″		18000	.003		4.56	.11	.04	4.71	5.20
2540	1″		17000	.003		6.10	.12	.04	6.26	6.95
2550	1-1/2″		15000	.004		9.10	.14	.05	9.29	10.30
2560	Hand mix, 1/2″	C-18	4000	.002		3.04	.08	.01	3.13	3.49
2610	Topping, P.C. based self-level/dry 6100 psi, pumped, 1/4″	C-8	20000	.003		2.34	.10	.04	2.48	2.77
2620	1/2″		19000	.003		4.68	.11	.04	4.83	5.35
2630	3/4″		18000	.003		7	.11	.04	7.15	7.90
2660	1″		17000	.003		9.35	.12	.04	9.51	10.55
2670	1-1/2″		15000	.004		14.05	.14	.05	14.24	15.70
2680	Hand mix, 1/2″	C-18	4000	.002		4.68	.08	.01	4.77	5.30

03 63 Epoxy Grouting

03 63 05 – Grouting of Dowels and Fasteners

03 63 05.10 Epoxy Only		Crew	Daily Output	Labor-Hours	Unit	Material	Labor	Equipment	Total	Total Incl O&P
0010	**EPOXY ONLY**									
1500	Chemical anchoring, epoxy cartridge, excludes layout, drilling, fastener									
1530	For fastener 3/4″ diam. x 6″ embedment	2 Skwk	72	.222	Ea.	4.57	9.45		14.02	19.60
1535	1″ diam. x 8″ embedment		66	.242		6.85	10.35		17.20	23.50
1540	1-1/4″ diam. x 10″ embedment		60	.267		13.70	11.35		25.05	32.50
1545	1-3/4″ diam. x 12″ embedment		54	.296		23	12.60		35.60	44.50
1550	14″ embedment		48	.333		27.50	14.20		41.70	52
1555	2″ diam. x 12″ embedment		42	.381		36.50	16.25		52.75	65
1560	18″ embedment		32	.500		45.50	21.50		67	83.50

03 82 Concrete Boring

03 82 13 – Concrete Core Drilling

03 82 13.10 Core Drilling		Crew	Daily Output	Labor-Hours	Unit	Material	2010 Bare Costs Labor	Equipment	Total	Total Incl O&P
0010	**CORE DRILLING**									
0020	Reinf. conc slab, up to 6" thick, incl. bit, layout & set up									
0100	1" diameter core	B-89A	17	.941	Ea.	3.65	35.50	6.70	45.85	66.50
0150	Each added inch thick in same hole, add		1440	.011		.68	.42	.08	1.18	1.49
0300	3" diameter core		16	1		12.20	38	7.10	57.30	79.50
0350	Each added inch thick in same hole, add		720	.022		1.40	.84	.16	2.40	3.01
0500	4" diameter core		15	1.067		15.45	40.50	7.60	63.55	87.50
0550	Each added inch thick in same hole, add		480	.033		1.72	1.26	.24	3.22	4.09
0700	6" diameter core		14	1.143		20	43.50	8.15	71.65	97.50
0750	Each added inch thick in same hole, add		360	.044		2.73	1.68	.32	4.73	5.95
0900	8" diameter core		13	1.231		33	46.50	8.75	88.25	118
0950	Each added inch thick in same hole, add		288	.056		3.75	2.10	.40	6.25	7.80
1100	10" diameter core		12	1.333		26.50	50.50	9.50	86.50	117
1150	Each added inch thick in same hole, add		240	.067		3.75	2.52	.47	6.74	8.55
1300	12" diameter core		11	1.455		31	55	10.35	96.35	130
1350	Each added inch thick in same hole, add		206	.078		4.55	2.94	.55	8.04	10.15
1500	14" diameter core		10	1.600		37	60.50	11.40	108.90	147
1550	Each added inch thick in same hole, add		180	.089		6.20	3.36	.63	10.19	12.70
1700	18" diameter core		9	1.778		45.50	67.50	12.65	125.65	168
1750	Each added inch thick in same hole, add		144	.111		8.80	4.21	.79	13.80	17.05
1760	For horizontal holes, add to above						20%	20%		
1770	Prestressed hollow core plank, 8" thick									
1780	1" diameter core	B-89A	17.50	.914	Ea.	2.37	34.50	6.50	43.37	63.50
1790	Each added inch thick in same hole, add		3840	.004		.47	.16	.03	.66	.79
1800	3" diameter core		17	.941		4.95	35.50	6.70	47.15	68
1810	Each added inch thick in same hole, add		1920	.008		.85	.32	.06	1.23	1.50
1820	4" diameter core		16.50	.970		6.80	36.50	6.90	50.20	71.50
1830	Each added inch thick in same hole, add		1280	.013		1.25	.47	.09	1.81	2.21
1840	6" diameter core		15.50	1.032		8.35	39	7.35	54.70	77.50
1850	Each added inch thick in same hole, add		960	.017		1.42	.63	.12	2.17	2.66
1860	8" diameter core		15	1.067		11.85	40.50	7.60	59.95	83.50
1870	Each added inch thick in same hole, add		768	.021		1.99	.79	.15	2.93	3.56
1880	10" diameter core		14	1.143		15.70	43.50	8.15	67.35	92.50
1890	Each added inch thick in same hole, add		640	.025		2.07	.95	.18	3.20	3.94
1900	12" diameter core		13.50	1.185		18.35	45	8.45	71.80	98.50
1910	Each added inch thick in same hole, add		548	.029		3.32	1.11	.21	4.64	5.60
1999	Drilling, core, minimum labor/equipment charge		4	4	Job		151	28.50	179.50	265

03 82 16 – Concrete Drilling

03 82 16.10 Concrete Impact Drilling		Crew	Daily Output	Labor-Hours	Unit	Material	Labor	Equipment	Total	Total Incl O&P
0010	**CONCRETE IMPACT DRILLING**									
0050	Up to 4" deep in conc/brick floor/wall, incl. bit & layout, no anchor									
0100	Holes, 1/4" diameter	1 Carp	75	.107	Ea.	.07	4.43		4.50	6.90
0150	For each additional inch of depth, add		430	.019		.02	.77		.79	1.21
0200	3/8" diameter		63	.127		.06	5.30		5.36	8.20
0250	For each additional inch of depth, add		340	.024		.02	.98		1	1.53
0300	1/2" diameter		50	.160		.06	6.65		6.71	10.30
0350	For each additional inch of depth, add		250	.032		.02	1.33		1.35	2.07
0400	5/8" diameter		48	.167		.09	6.95		7.04	10.80
0450	For each additional inch of depth, add		240	.033		.02	1.38		1.40	2.16
0500	3/4" diameter		45	.178		.11	7.40		7.51	11.50
0550	For each additional inch of depth, add		220	.036		.03	1.51		1.54	2.36
0600	7/8" diameter		43	.186		.13	7.75		7.88	12.05

03 82 Concrete Boring

03 82 16 – Concrete Drilling

03 82 16.10 Concrete Impact Drilling		Crew	Daily Output	Labor-Hours	Unit	2010 Bare Costs Material	Labor	Equipment	Total	Total Incl O&P
0650	For each additional inch of depth, add	1 Carp	210	.038	Ea.	.03	1.58		1.61	2.48
0700	1" diameter		40	.200		.15	8.30		8.45	12.95
0750	For each additional inch of depth, add		190	.042		.04	1.75		1.79	2.74
0800	1-1/4" diameter		38	.211		.22	8.75		8.97	13.75
0850	For each additional inch of depth, add		180	.044		.06	1.85		1.91	2.91
0900	1-1/2" diameter		35	.229		.32	9.50		9.82	15
0950	For each additional inch of depth, add	▼	165	.048	▼	.08	2.01		2.09	3.20
1000	For ceiling installations, add						40%			

Estimating Tips

04 05 00 Common Work Results for Masonry

- The terms *mortar* and *grout* are often used interchangeably, and incorrectly. Mortar is used to bed masonry units, seal the entry of air and moisture, provide architectural appearance, and allow for size variations in the units. Grout is used primarily in reinforced masonry construction and is used to bond the masonry to the reinforcing steel. Common mortar types are M(2500 psi), S(1800 psi), N(750 psi), and O(350 psi), and conform to ASTM C270. Grout is either fine or coarse and conforms to ASTM C476, and in-place strengths generally exceed 2500 psi. Mortar and grout are different components of masonry construction and are placed by entirely different methods. An estimator should be aware of their unique uses and costs.
- Waste, specifically the loss/droppings of mortar and the breakage of brick and block, is included in all masonry assemblies in this division. A factor of 25% is added for mortar and 3% for brick and concrete masonry units.
- Scaffolding or staging is not included in any of the Division 4 costs. Refer to Subdivision 01 54 23 for scaffolding and staging costs.

04 20 00 Unit Masonry

- The most common types of unit masonry are brick and concrete masonry. The major classifications of brick are building brick (ASTM C62), facing brick (ASTM C216), glazed brick, fire brick, and pavers. Many varieties of texture and appearance can exist within these classifications, and the estimator would be wise to check local custom and availability within the project area. For repair and remodeling jobs, matching the existing brick may be the most important criteria.
- Brick and concrete block are priced by the piece and then converted into a price per square foot of wall. Openings less than two square feet are generally ignored by the estimator because any savings in units used is offset by the cutting and trimming required.
- It is often difficult and expensive to find and purchase small lots of historic brick. Costs can vary widely. Many design issues affect costs, selection of mortar mix, and repairs or replacement of masonry materials. Cleaning techniques must be reflected in the estimate.
- All masonry walls, whether interior or exterior, require bracing. The cost of bracing walls during construction should be included by the estimator, and this bracing must remain in place until permanent bracing is complete. Permanent bracing of masonry walls is accomplished by masonry itself, in the form of pilasters or abutting wall corners, or by anchoring the walls to the structural frame. Accessories in the form of anchors, anchor slots, and ties are used, but their supply and installation can be by different trades. For instance, anchor slots on spandrel beams and columns are supplied and welded in place by the steel fabricator, but the ties from the slots into the masonry are installed by the bricklayer. Regardless of the installation method, the estimator must be certain that these accessories are accounted for in pricing.

Reference Numbers

Reference numbers are shown in shaded boxes at the beginning of some major classifications. These numbers refer to related items in the Reference Section. The reference information may be an estimating procedure, an alternate pricing method, or technical information.

Note: Not all subdivisions listed here necessarily appear in this publication.

04 05 Common Work Results for Masonry

04 05 23 – Masonry Accessories

04 05 23.19 Vent Box		Crew	Daily Output	Labor-Hours	Unit	Material	Labor	Equipment	Total	Total Incl O&P
						2010 Bare Costs				
0010	**VENT BOX**									
0020	Extruded aluminum, 4" deep, 2-3/8" x 8-1/8"	1 Bric	30	.267	Ea.	30.50	11.15		41.65	50.50
0050	5" x 8-1/8"		25	.320		40	13.35		53.35	64
0100	2-1/4" x 25"		25	.320		69.50	13.35		82.85	96.50
0150	5" x 16-1/2"		22	.364		58.50	15.20		73.70	87
0200	6" x 16-1/2"		22	.364		81	15.20		96.20	112
0250	7-3/4" x 16-1/2"		20	.400		69	16.70		85.70	101
0400	For baked enamel finish, add					35%				
0500	For cast aluminum, painted, add					60%				
1000	Stainless steel ventilators, 6" x 6"	1 Bric	25	.320		126	13.35		139.35	158
1050	8" x 8"		24	.333		132	13.90		145.90	167
1100	12" x 12"		23	.348		153	14.50		167.50	190
1150	12" x 6"		24	.333		134	13.90		147.90	168
1200	Foundation block vent, galv., 1-1/4" thk, 8" high, 16" long, no damper		30	.267		24.50	11.15		35.65	44
1250	For damper, add					8.20			8.20	9

04 21 Clay Unit Masonry

04 21 13 – Brick Masonry

04 21 13.15 Chimney

		Crew	Daily Output	Labor-Hours	Unit	Material	Labor	Equipment	Total	Total Incl O&P
0010	**CHIMNEY**, excludes foundation, scaffolding, grout and reinforcing									
0100	Brick, 16" x 16", 8" flue	D-1	18.20	.879	V.L.F.	21	33		54	73.50
0150	16" x 20" with one 8" x 12" flue		16	1		33.50	37.50		71	94
0200	16" x 24" with two 8" x 8" flues		14	1.143		49	43		92	119
0250	20" x 20" with one 12" x 12" flue		13.70	1.168		38.50	44		82.50	109
0300	20" x 24" with two 8" x 12" flues		12	1.333		55.50	50.50		106	137
0350	20" x 32" with two 12" x 12" flues		10	1.600		68	60.50		128.50	166
1800	Metal, high temp. steel jacket, factory lining, 24" diam.	E-2	65	.862		210	39.50	24.50	274	325
1900	60" diameter	"	30	1.867		765	85	53.50	903.50	1,050
2100	Poured concrete, brick lining, 200' high x 10' diam.					7,275			7,275	8,000
2800	500' x 20' diameter					12,700			12,700	14,000

Estimating Tips

05 05 00 Common Work Results for Metals

- Nuts, bolts, washers, connection angles, and plates can add a significant amount to both the tonnage of a structural steel job and the estimated cost. As a rule of thumb, add 10% to the total weight to account for these accessories.
- Type 2 steel construction, commonly referred to as "simple construction," consists generally of field-bolted connections with lateral bracing supplied by other elements of the building, such as masonry walls or x-bracing. The estimator should be aware, however, that shop connections may be accomplished by welding or bolting. The method may be particular to the fabrication shop and may have an impact on the estimated cost.

05 12 23 Structural Steel

- Steel items can be obtained from two sources: a fabrication shop or a metals service center. Fabrication shops can fabricate items under more controlled conditions than can crews in the field. They are also more efficient and can produce items more economically. Metal service centers serve as a source of long mill shapes to both fabrication shops and contractors.
- Most line items in this structural steel subdivision, and most items in 05 50 00 Metal Fabrications, are indicated as being shop fabricated. The bare material cost for these shop fabricated items is the "Invoice Cost" from the shop and includes the mill base price of steel plus mill extras, transportation to the shop, shop drawings and detailing where warranted, shop fabrication and handling, sandblasting and a shop coat of primer paint, all necessary structural bolts, and delivery to the job site. The bare labor cost and bare equipment cost for these shop fabricated items is for field installation or erection.
- Line items in Subdivision 05 12 23.40 Lightweight Framing, and other items scattered in Division 5, are indicated as being field fabricated. The bare material cost for these field fabricated items is the "Invoice Cost" from the metals service center and includes the mill base price of steel plus mill extras, transportation to the metals service center, material handling, and delivery of long lengths of mill shapes to the job site. Material costs for structural bolts and welding rods should be added to the estimate. The bare labor cost and bare equipment cost for these items is for both field fabrication and field installation or erection, and include time for cutting, welding and drilling in the fabricated metal items. Drilling into concrete and fasteners to fasten field fabricated items to other work are not included and should be added to the estimate.

05 20 00 Steel Joist Framing

- In any given project the total weight of open web steel joists is determined by the loads to be supported and the design. However, economies can be realized in minimizing the amount of labor used to place the joists. This is done by maximizing the joist spacing, and therefore minimizing the number of joists required to be installed on the job. Certain spacings and locations may be required by the design, but in other cases maximizing the spacing and keeping it as uniform as possible will keep the costs down.

05 30 00 Steel Decking

- The takeoff and estimating of metal deck involves more than simply the area of the floor or roof and the type of deck specified or shown on the drawings. Many different sizes and types of openings may exist. Small openings for individual pipes or conduits may be drilled after the floor/roof is installed, but larger openings may require special deck lengths as well as reinforcing or structural support. The estimator should determine who will be supplying this reinforcing. Additionally, some deck terminations are part of the deck package, such as screed angles and pour stops, and others will be part of the steel contract, such as angles attached to structural members and cast-in-place angles and plates. The estimator must ensure that all pieces are accounted for in the complete estimate.

05 50 00 Metal Fabrications

- The most economical steel stairs are those that use common materials, standard details, and most importantly, a uniform and relatively simple method of field assembly. Commonly available A36 channels and plates are very good choices for the main stringers of the stairs, as are angles and tees for the carrier members. Risers and treads are usually made by specialty shops, and it is most economical to use a typical detail in as many places as possible. The stairs should be pre-assembled and shipped directly to the site. The field connections should be simple and straightforward to be accomplished efficiently, and with minimum equipment and labor.

Reference Numbers

Reference numbers are shown in shaded boxes at the beginning of some major classifications. These numbers refer to related items in the Reference Section. The reference information may be an estimating procedure, an alternate pricing method, or technical information. *Note:* Not all subdivisions listed here necessarily appear in this publication.

05 05 Common Work Results for Metals

05 05 21 – Fastening Methods for Metal

05 05 21.15 Drilling Steel			Crew	Daily Output	Labor-Hours	Unit	Material	2010 Bare Costs Labor	Equipment	Total	Total Incl O&P
0010	**DRILLING STEEL**										
1910	Drilling & layout for steel, up to 1/4" deep, no anchor										
1920	Holes, 1/4" diameter		1 Sswk	112	.071	Ea.	.09	3.35		3.44	6
1925	For each additional 1/4" depth, add			336	.024		.09	1.12		1.21	2.06
1930	3/8" diameter			104	.077		.10	3.61		3.71	6.45
1935	For each additional 1/4" depth, add			312	.026		.10	1.20		1.30	2.22
1940	1/2" diameter			96	.083		.11	3.91		4.02	6.95
1945	For each additional 1/4" depth, add			288	.028		.11	1.30		1.41	2.41
1950	5/8" diameter			88	.091		.15	4.26		4.41	7.65
1955	For each additional 1/4" depth, add			264	.030		.15	1.42		1.57	2.66
1960	3/4" diameter			80	.100		.17	4.69		4.86	8.45
1965	For each additional 1/4" depth, add			240	.033		.17	1.56		1.73	2.93
1970	7/8" diameter			72	.111		.20	5.20		5.40	9.35
1975	For each additional 1/4" depth, add			216	.037		.20	1.74		1.94	3.26
1980	1" diameter			64	.125		.22	5.85		6.07	10.55
1985	For each additional 1/4" depth, add			192	.042		.22	1.95		2.17	3.69
1990	For drilling up, add							40%			

05 05 23 – Metal Fastenings

05 05 23.05 Anchor Bolts			Crew	Daily Output	Labor-Hours	Unit	Material	Labor	Equipment	Total	Total Incl O&P
0010	**ANCHOR BOLTS**										
0015	Made from recycled materials	G									
0100	J-type, incl. hex nut & washer, 1/2" diameter x 6" long	G	2 Carp	70	.229	Ea.	1.22	9.50		10.72	16
0110	12" long	G		65	.246		1.53	10.25		11.78	17.45
0120	18" long	G		60	.267		1.99	11.10		13.09	19.30
0130	3/4" diameter x 8" long	G		50	.320		3.20	13.30		16.50	24
0140	12" long	G		45	.356		4	14.75		18.75	27.50
0150	18" long	G		40	.400		5.20	16.60		21.80	31
0160	1" diameter x 12" long	G		35	.457		8.40	19		27.40	39
0170	18" long	G		30	.533		10.15	22		32.15	45
0180	24" long	G		25	.640		12.35	26.50		38.85	54.50
0190	36" long	G		20	.800		16.95	33		49.95	69.50
0200	1-1/2" diameter x 18" long	G		22	.727		28	30		58	77
0210	24" long	G		16	1		33	41.50		74.50	101
0300	L-type, incl. hex nut & washer, 3/4" diameter x 12" long	G		45	.356		3.72	14.75		18.47	27
0310	18" long	G		40	.400		4.77	16.60		21.37	31
0320	24" long	G		35	.457		5.80	19		24.80	36
0330	30" long	G		30	.533		7.40	22		29.40	42
0340	36" long	G		25	.640		8.45	26.50		34.95	50.50
0350	1" diameter x 12" long	G		35	.457		6.75	19		25.75	37
0360	18" long	G		30	.533		8.40	22		30.40	43
0370	24" long	G		25	.640		10.35	26.50		36.85	52.50
0380	30" long	G		23	.696		12.20	29		41.20	58
0390	36" long	G		20	.800		13.95	33		46.95	66.50
0400	42" long	G		18	.889		16.95	37		53.95	75.50
0410	48" long	G		15	1.067		19.05	44.50		63.55	89.50
0420	1-1/4" diameter x 18" long	G		25	.640		13.30	26.50		39.80	55.50
0430	24" long	G		22	.727		15.80	30		45.80	64
0440	30" long	G		20	.800		18.30	33		51.30	71
0450	36" long	G		18	.889		21	37		58	80
0460	42" long	G		16	1		23.50	41.50		65	90
0470	48" long	G		14	1.143		27	47.50		74.50	103
0480	54" long	G		12	1.333		31.50	55.50		87	121

05 05 23 – Metal Fastenings

05 05 23.05 Anchor Bolts			Crew	Daily Output	Labor-Hours	Unit	Material	Labor	Equipment	Total	Total Incl O&P
							2010 Bare Costs				
0490	60" long	G	2 Carp	10	1.600	Ea.	34.50	66.50		101	140
0500	1-1/2" diameter x 18" long	G		22	.727		19.80	30		49.80	68.50
0510	24" long	G		19	.842		23	35		58	79.50
0520	30" long	G		17	.941		26	39		65	89
0540	42" long	G		15	1.067		34	44.50		78.50	106
0550	48" long	G		13	1.231		38.50	51		89.50	121
0560	54" long	G		11	1.455		47	60.50		107.50	145
0570	60" long	G		9	1.778		51.50	74		125.50	171
0580	1-3/4" diameter x 18" long	G		20	.800		28.50	33		61.50	82
0590	24" long	G		18	.889		33.50	37		70.50	93.50
0600	30" long	G		17	.941		38.50	39		77.50	103
0610	36" long	G		16	1		44	41.50		85.50	113
0620	42" long	G		14	1.143		49.50	47.50		97	128
0630	48" long	G		12	1.333		54.50	55.50		110	146
0640	54" long	G		10	1.600		67.50	66.50		134	177
0650	60" long	G		8	2		73	83		156	209
0660	2" diameter x 24" long	G		17	.941		45	39		84	110
0670	30" long	G		15	1.067		50.50	44.50		95	125
0680	36" long	G		13	1.231		55.50	51		106.50	141
0690	42" long	G		11	1.455		62	60.50		122.50	162
0700	48" long	G		10	1.600		71.50	66.50		138	181
0710	54" long	G		9	1.778		85	74		159	208
0720	60" long	G		8	2		91.50	83		174.50	229
0730	66" long	G		7	2.286		98	95		193	254
0740	72" long	G		6	2.667		107	111		218	289
0990	For galvanized, add						75%				

05 05 23.15 Chemical Anchors		Crew	Daily Output	Labor-Hours	Unit	Material	Labor	Equipment	Total	Total Incl O&P
0010	**CHEMICAL ANCHORS**									
0020	Includes layout & drilling									
1430	Chemical anchor, w/rod & epoxy cartridge, 3/4" diam. x 9-1/2" long	B-89A	27	.593	Ea.	9.45	22.50	4.21	36.16	49.50
1435	1" diameter x 11-3/4" long		24	.667		16.55	25	4.74	46.29	62.50
1440	1-1/4" diameter x 14" long		21	.762		29	29	5.40	63.40	82
1445	1-3/4" diameter x 15" long		20	.800		50.50	30.50	5.70	86.70	108
1450	18" long		17	.941		60.50	35.50	6.70	102.70	129
1455	2" diameter x 18" long		16	1		82.50	38	7.10	127.60	157
1460	24" long		15	1.067		107	40.50	7.60	155.10	187

05 05 23.20 Expansion Anchors			Crew	Daily Output	Labor-Hours	Unit	Material	Labor	Equipment	Total	Total Incl O&P
0010	**EXPANSION ANCHORS**										
0100	Anchors for concrete, brick or stone, no layout and drilling										
0200	Expansion shields, zinc, 1/4" diameter, 1-5/16" long, single	G	1 Carp	90	.089	Ea.	.34	3.69		4.03	6.05
0300	1-3/8" long, double	G		85	.094		.47	3.91		4.38	6.55
0400	3/8" diameter, 1-1/2" long, single	G		85	.094		.63	3.91		4.54	6.75
0500	2" long, double	G		80	.100		1.17	4.16		5.33	7.70
0600	1/2" diameter, 2-1/16" long, single	G		80	.100		1.16	4.16		5.32	7.70
0700	2-1/2" long, double	G		75	.107		1.63	4.43		6.06	8.65
0800	5/8" diameter, 2-5/8" long, single	G		75	.107		2.26	4.43		6.69	9.35
0900	2-3/4" long, double	G		70	.114		2.33	4.75		7.08	9.85
1000	3/4" diameter, 2-3/4" long, single	G		70	.114		2.07	4.75		6.82	9.60
1100	3-15/16" long, double	G		65	.123		4.51	5.10		9.61	12.85
2100	Hollow wall anchors for gypsum wall board, plaster or tile										
2300	1/8" diameter, short	G	1 Carp	160	.050	Ea.	.18	2.08		2.26	3.40
2400	Long	G		150	.053		.21	2.22		2.43	3.65

05 05 23 – Metal Fastenings

05 05 23.20 Expansion Anchors			Crew	Daily Output	Labor-Hours	Unit	Material	2010 Bare Costs Labor	Equipment	Total	Total Incl O&P
2500	3/16" diameter, short	G	1 Carp	150	.053	Ea.	.32	2.22		2.54	3.77
2600	Long	G		140	.057		.46	2.37		2.83	4.17
2700	1/4" diameter, short	G		140	.057		.54	2.37		2.91	4.25
2800	Long	G		130	.062		.48	2.56		3.04	4.47
3000	Toggle bolts, bright steel, 1/8" diameter, 2" long	G		85	.094		.18	3.91		4.09	6.25
3100	4" long	G		80	.100		.23	4.16		4.39	6.65
3200	3/16" diameter, 3" long	G		80	.100		.22	4.16		4.38	6.65
3300	6" long	G		75	.107		.31	4.43		4.74	7.20
3400	1/4" diameter, 3" long	G		75	.107		.29	4.43		4.72	7.15
3500	6" long	G		70	.114		.46	4.75		5.21	7.80
3600	3/8" diameter, 3" long	G		70	.114		.70	4.75		5.45	8.05
3700	6" long	G		60	.133		1.20	5.55		6.75	9.85
3800	1/2" diameter, 4" long	G		60	.133		1.69	5.55		7.24	10.40
3900	6" long	G	↓	50	.160	↓	2.02	6.65		8.67	12.45
5000	Screw anchors for concrete, masonry,										
5100	stone & tile, no layout or drilling included										
5700	Lag screw shields, 1/4" diameter, short	G	1 Carp	90	.089	Ea.	.29	3.69		3.98	6
5800	Long	G		85	.094		.35	3.91		4.26	6.45
5900	3/8" diameter, short	G		85	.094		.59	3.91		4.50	6.70
6000	Long	G		80	.100		.72	4.16		4.88	7.20
6100	1/2" diameter, short	G		80	.100		.92	4.16		5.08	7.40
6200	Long	G		75	.107		1.15	4.43		5.58	8.10
6300	5/8" diameter, short	G		70	.114		1.45	4.75		6.20	8.90
6400	Long	G		65	.123		1.94	5.10		7.04	10.05
6600	Lead, #6 & #8, 3/4" long	G		260	.031		.12	1.28		1.40	2.10
6700	#10 - #14, 1-1/2" long	G		200	.040		.18	1.66		1.84	2.76
6800	#16 & #18, 1-1/2" long	G		160	.050		.29	2.08		2.37	3.52
6900	Plastic, #6 & #8, 3/4" long			260	.031		.03	1.28		1.31	2
7000	#8 & #10, 7/8" long			240	.033		.02	1.38		1.40	2.15
7100	#10 & #12, 1" long			220	.036		.03	1.51		1.54	2.36
7200	#14 & #16, 1-1/2" long		↓	160	.050	↓	.05	2.08		2.13	3.26
8000	Wedge anchors, not including layout or drilling										
8050	Carbon steel, 1/4" diameter, 1-3/4" long	G	1 Carp	150	.053	Ea.	.29	2.22		2.51	3.74
8100	3-1/4" long	G		140	.057		.38	2.37		2.75	4.08
8150	3/8" diameter, 2-1/4" long	G		145	.055		.34	2.29		2.63	3.91
8200	5" long	G		140	.057		.60	2.37		2.97	4.32
8250	1/2" diameter, 2-3/4" long	G		140	.057		.73	2.37		3.10	4.46
8300	7" long	G		125	.064		1.24	2.66		3.90	5.45
8350	5/8" diameter, 3-1/2" long	G		130	.062		1.15	2.56		3.71	5.20
8400	8-1/2" long	G		115	.070		2.45	2.89		5.34	7.15
8450	3/4" diameter, 4-1/4" long	G		115	.070		2.02	2.89		4.91	6.70
8500	10" long	G		95	.084		4.60	3.50		8.10	10.45
8550	1" diameter, 6" long	G		100	.080		7.70	3.32		11.02	13.60
8575	9" long	G		85	.094		10	3.91		13.91	17.05
8600	12" long	G		75	.107		10.80	4.43		15.23	18.75
8650	1-1/4" diameter, 9" long	G		70	.114		23	4.75		27.75	33
8700	12" long	G	↓	60	.133	↓	29.50	5.55		35.05	41
8750	For type 303 stainless steel, add						350%				
8800	For type 316 stainless steel, add						450%				
8950	Self-drilling concrete screw, hex washer head, 3/16" diam. x 1-3/4" long	G	1 Carp	300	.027	Ea.	.19	1.11		1.30	1.92
8960	2-1/4" long	G		250	.032		.20	1.33		1.53	2.27
8970	Phillips flat head, 3/16" diam. x 1-3/4" long	G		300	.027		.17	1.11		1.28	1.90
8980	2-1/4" long	G	↓	250	.032	↓	.20	1.33		1.53	2.27

05 05 Common Work Results for Metals

05 05 23 – Metal Fastenings

05 05 23.30 Lag Screws		Crew	Daily Output	Labor-Hours	Unit	2010 Bare Costs Material	Labor	Equipment	Total	Total Incl O&P
0010	**LAG SCREWS**									
0020	Steel, 1/4" diameter, 2" long	G	1 Carp	200	.040	Ea.	.09	1.66		1.75	2.66
0100	3/8" diameter, 3" long	G		150	.053		.26	2.22		2.48	3.71
0200	1/2" diameter, 3" long	G		130	.062		.48	2.56		3.04	4.47
0300	5/8" diameter, 3" long	G		120	.067		1.11	2.77		3.88	5.50

05 05 23.35 Machine Screws

Line	Description		Crew	Daily Output	Labor-Hours	Unit	Material	Labor	Equipment	Total	Total Incl O&P
0010	**MACHINE SCREWS**										
0020	Steel, round head, #8 x 1" long	G	1 Carp	4.80	1.667	C	3.45	69.50		72.95	111
0110	#8 x 2" long	G		2.40	3.333		4.34	139		143.34	219
0200	#10 x 1" long	G		4	2		5.90	83		88.90	135
0300	#10 x 2" long	G		2	4		7.25	166		173.25	264

05 05 23.40 Machinery Anchors

Line	Description		Crew	Daily Output	Labor-Hours	Unit	Material	Labor	Equipment	Total	Total Incl O&P
0010	**MACHINERY ANCHORS**, heavy duty, incl. sleeve, floating base nut,										
0020	lower stud & coupling nut, fiber plug, connecting stud, washer & nut.										
0030	For flush mounted embedment in poured concrete heavy equip. pads.										
0200	Stud & bolt, 1/2" diameter	G	E-16	40	.400	Ea.	68.50	19.15	3.64	91.29	113
0300	5/8" diameter	G		35	.457		76	22	4.16	102.16	127
0500	3/4" diameter	G		30	.533		87.50	25.50	4.86	117.86	147
0600	7/8" diameter	G		25	.640		95.50	30.50	5.85	131.85	165
0800	1" diameter	G		20	.800		100	38.50	7.30	145.80	186
0900	1-1/4" diameter	G		15	1.067		133	51	9.70	193.70	248

05 05 23.50 Powder Actuated Tools and Fasteners

Line	Description		Crew	Daily Output	Labor-Hours	Unit	Material	Labor	Equipment	Total	Total Incl O&P
0010	**POWDER ACTUATED TOOLS & FASTENERS**										
0020	Stud driver, .22 caliber, buy, minimum					Ea.	219			219	241
0100	Maximum					"	370			370	405
0300	Powder charges for above, low velocity					C	4.99			4.99	5.50
0400	Standard velocity						5.95			5.95	6.55
0600	Drive pins & studs, 1/4" & 3/8" diam., to 3" long, minimum	G	1 Carp	4.80	1.667		4.38	69.50		73.88	112
0700	Maximum	G	"	4	2		9.40	83		92.40	138

05 05 23.55 Rivets

Line	Description		Crew	Daily Output	Labor-Hours	Unit	Material	Labor	Equipment	Total	Total Incl O&P
0010	**RIVETS**										
0100	Aluminum rivet & mandrel, 1/2" grip length x 1/8" diameter	G	1 Carp	4.80	1.667	C	6.35	69.50		75.85	114
0200	3/16" diameter	G		4	2		11.15	83		94.15	140
0300	Aluminum rivet, steel mandrel, 1/8" diameter	G		4.80	1.667		6.65	69.50		76.15	114
0400	3/16" diameter	G		4	2		10.55	83		93.55	140
0500	Copper rivet, steel mandrel, 1/8" diameter	G		4.80	1.667		7.75	69.50		77.25	116
1200	Steel rivet and mandrel, 1/8" diameter	G		4.80	1.667		6.50	69.50		76	114
1300	3/16" diameter	G		4	2		10.65	83		93.65	140
1400	Hand riveting tool, minimum					Ea.	40			40	44
1500	Maximum						227			227	250
1600	Power riveting tool, minimum						450			450	495
1700	Maximum						2,275			2,275	2,525

05 05 23.70 Structural Blind Bolts

Line	Description		Crew	Daily Output	Labor-Hours	Unit	Material	Labor	Equipment	Total	Total Incl O&P
0010	**STRUCTURAL BLIND BOLTS**										
0100	1/4" diameter x 1/4" grip	G	1 Sswk	240	.033	Ea.	1.24	1.56		2.80	4.11
0150	1/2" grip	G		216	.037		.95	1.74		2.69	4.10
0200	3/8" diameter x 1/2" grip	G		232	.034		1.75	1.62		3.37	4.77
0250	3/4" grip	G		208	.038		1.83	1.80		3.63	5.20
0300	1/2" diameter x 1/2" grip	G		224	.036		5.15	1.67		6.82	8.60
0350	3/4" grip	G		200	.040		5.60	1.88		7.48	9.45
0400	5/8" diameter x 3/4" grip	G		216	.037		8.25	1.74		9.99	12.15

05 05 Common Work Results for Metals

05 05 23 – Metal Fastenings

05 05 23.70 Structural Blind Bolts		Crew	Daily Output	Labor-Hours	Unit	Material	2010 Bare Costs Labor	Equipment	Total	Total Incl O&P
0450	1" grip G	1 Sswk	192	.042	Ea.	9.05	1.95		11	13.40

05 12 Structural Steel Framing

05 12 23 – Structural Steel for Buildings

05 12 23.40 Lightweight Framing

		Crew	Daily Output	Labor-Hours	Unit	Material	Labor	Equipment	Total	Total Incl O&P
0010	**LIGHTWEIGHT FRAMING**									
0015	Made from recycled materials G									
0400	Angle framing, field fabricated, 4" and larger G	E-3	440	.055	Lb.	.64	2.59	.33	3.56	5.60
0450	Less than 4" angles G		265	.091		.66	4.31	.55	5.52	8.90
0600	Channel framing, field fabricated, 8" and larger G		500	.048		.66	2.28	.29	3.23	5.05
0650	Less than 8" channels G	↓	335	.072		.66	3.41	.44	4.51	7.20
1000	Continuous slotted channel framing system, shop fab, minimum G	2 Sswk	2400	.007		3.41	.31		3.72	4.30
1200	Maximum G	"	1600	.010		3.85	.47		4.32	5.05
1250	Plate & bar stock for reinforcing beams and trusses G					1.21			1.21	1.33
1300	Cross bracing, rods, shop fabricated, 3/4" diameter G	E-3	700	.034		1.32	1.63	.21	3.16	4.55
1310	7/8" diameter G		850	.028		1.32	1.34	.17	2.83	4
1320	1" diameter G		1000	.024		1.32	1.14	.15	2.61	3.62
1330	Angle, 5" x 5" x 3/8" G		2800	.009		1.32	.41	.05	1.78	2.23
1350	Hanging lintels, shop fabricated, average G	↓	850	.028		1.32	1.34	.17	2.83	4
1380	Roof frames, shop fabricated, 3'-0" square, 5' span G	E-2	4200	.013		1.32	.61	.38	2.31	2.90
1400	Tie rod, not upset, 1-1/2" to 4" diameter, with turnbuckle G	2 Sswk	800	.020		1.43	.94		2.37	3.22
1420	No turnbuckle G		700	.023		1.38	1.07		2.45	3.39
1500	Upset, 1-3/4" to 4" diameter, with turnbuckle G		800	.020		1.43	.94		2.37	3.22
1520	No turnbuckle G	↓	700	.023	↓	1.38	1.07		2.45	3.39

05 12 23.60 Pipe Support Framing

		Crew	Daily Output	Labor-Hours	Unit	Material	Labor	Equipment	Total	Total Incl O&P
0010	**PIPE SUPPORT FRAMING**									
0020	Under 10#/L.F., shop fabricated G	E-4	3900	.008	Lb.	1.47	.39	.04	1.90	2.34
0200	10.1 to 15#/L.F. G		4300	.007		1.45	.35	.03	1.83	2.26
0400	15.1 to 20#/L.F. G		4800	.007		1.43	.32	.03	1.78	2.16
0600	Over 20#/L.F. G	↓	5400	.006	↓	1.41	.28	.03	1.72	2.07

Estimating Tips

06 05 00 Common Work Results for Wood, Plastics, and Composites

- Common to any wood-framed structure are the accessory connector items such as screws, nails, adhesives, hangers, connector plates, straps, angles, and hold-downs. For typical wood-framed buildings, such as residential projects, the aggregate total for these items can be significant, especially in areas where seismic loading is a concern. For floor and wall framing, the material cost is based on 10 to 25 lbs. per MBF. Hold-downs, hangers, and other connectors should be taken off by the piece.

06 10 00 Carpentry

- Lumber is a traded commodity and therefore sensitive to supply and demand in the marketplace. Even in "budgetary" estimating of wood-framed projects, it is advisable to call local suppliers for the latest market pricing.
- Common quantity units for wood-framed projects are "thousand board feet" (MBF). A board foot is a volume of wood, 1" x 1' x 1', or 144 cubic inches. Board foot quantities are generally calculated using nominal material dimensions—dressed sizes are ignored. Board foot per lineal foot of any stick of lumber can be calculated by dividing the nominal cross-sectional area by 12. As an example, 2,000 lineal feet of 2 x 12 equates to 4 MBF by dividing the nominal area, 2 x 12, by 12, which equals 2, and multiplying by 2,000 to give 4,000 board feet. This simple rule applies to all nominal dimensioned lumber.
- Waste is an issue of concern at the quantity takeoff for any area of construction. Framing lumber is sold in even foot lengths, i.e., 10', 12', 14', 16', and depending on spans, wall heights, and the grade of lumber, waste is inevitable. A rule of thumb for lumber waste is 5%–10% depending on material quality and the complexity of the framing.
- Wood in various forms and shapes is used in many projects, even where the main structural framing is steel, concrete, or masonry. Plywood as a back-up partition material and 2x boards used as blocking and cant strips around roof edges are two common examples. The estimator should ensure that the costs of all wood materials are included in the final estimate.

06 20 00 Finish Carpentry

- It is necessary to consider the grade of workmanship when estimating labor costs for erecting millwork and interior finish. In practice, there are three grades: premium, custom, and economy. The RSMeans daily output for base and case moldings is in the range of 200 to 250 L.F. per carpenter per day. This is appropriate for most average custom-grade projects. For premium projects, an adjustment to productivity of 25%–50% should be made, depending on the complexity of the job.

Reference Numbers

Reference numbers are shown in shaded boxes at the beginning of some major classifications. These numbers refer to related items in the Reference Section. The reference information may be an estimating procedure, an alternate pricing method, or technical information.

Note: Not all subdivisions listed here necessarily appear in this publication.

06 05 23 – Wood, Plastic, and Composite Fastenings

06 05 23.10 Nails		Crew	Daily Output	Labor-Hours	Unit	Material	Labor	Equipment	Total	Total Incl O&P
0010	**NAILS**, material only, based upon 50# box purchase									
0020	Copper nails, plain				Lb.	10			10	11
0400	Stainless steel, plain					7.55			7.55	8.30
0500	Box, 3d to 20d, bright					1.26			1.26	1.39
0520	Galvanized					1.77			1.77	1.95
0600	Common, 3d to 60d, plain					1.30			1.30	1.43
0700	Galvanized					1.56			1.56	1.72
0800	Aluminum					4.55			4.55	5
1000	Annular or spiral thread, 4d to 60d, plain					1.80			1.80	1.98
1200	Galvanized					3			3	3.30
1400	Drywall nails, plain					.85			.85	.94
1600	Galvanized					1.75			1.75	1.93
1800	Finish nails, 4d to 10d, plain					1.40			1.40	1.54
2000	Galvanized					1.79			1.79	1.97
2100	Aluminum					4.10			4.10	4.51
2300	Flooring nails, hardened steel, 2d to 10d, plain					2.55			2.55	2.81
2400	Galvanized					3.88			3.88	4.27
2500	Gypsum lath nails, 1-1/8", 13 ga. flathead, blued					2.68			2.68	2.95
2600	Masonry nails, hardened steel, 3/4" to 3" long, plain					2.45			2.45	2.70
2700	Galvanized					3.72			3.72	4.09
5000	Add to prices above for cement coating					.11			.11	.12
5200	Zinc or tin plating					.14			.14	.15
5500	Vinyl coated sinkers, 8d to 16d				↓	.62			.62	.68

06 05 23.40 Sheet Metal Screws		Crew	Daily Output	Labor-Hours	Unit	Material	Labor	Equipment	Total	Total Incl O&P
0010	**SHEET METAL SCREWS**									
0020	Steel, standard, #8 x 3/4", plain				C	3.42			3.42	3.76
0100	Galvanized					3.42			3.42	3.76
0300	#10 x 1", plain					4.70			4.70	5.15
0400	Galvanized					4.70			4.70	5.15
0600	With washers, #14 x 1", plain					16.05			16.05	17.65
0700	Galvanized					16.05			16.05	17.65
0900	#14 x 2", plain					21			21	23
1000	Galvanized					21			21	23
1500	Self-drilling, with washers, (pinch point) #8 x 3/4", plain					12.25			12.25	13.45
1600	Galvanized					12.25			12.25	13.45
1800	#10 x 3/4", plain					14.50			14.50	15.95
1900	Galvanized					14.50			14.50	15.95
3000	Stainless steel w/aluminum or neoprene washers, #14 x 1", plain					35			35	38.50
3100	#14 x 2", plain				↓	44			44	48.50

06 05 23.50 Wood Screws		Crew	Daily Output	Labor-Hours	Unit	Material	Labor	Equipment	Total	Total Incl O&P
0010	**WOOD SCREWS**									
0020	#8 x 1" long, steel				C	5.10			5.10	5.60
0100	Brass					10.30			10.30	11.35
0200	#8, 2" long, steel					5.10			5.10	5.60
0300	Brass					17.40			17.40	19.10
0400	#10, 1" long, steel					3.63			3.63	3.99
0500	Brass					15.80			15.80	17.40
0600	#10, 2" long, steel					6.15			6.15	6.80
0700	Brass					24			24	26.50
0800	#10, 3" long, steel					10.30			10.30	11.30
1000	#12, 2" long, steel					9.05			9.05	9.95
1100	Brass				↓	31.50			31.50	34.50

Labor, Equipment, Material and Total columns are under "2010 Bare Costs".

06 05 Common Work Results for Wood, Plastics, and Composites

06 05 23 – Wood, Plastic, and Composite Fastenings

06 05 23.50 Wood Screws		Crew	Daily Output	Labor-Hours	Unit	2010 Bare Costs Material	Labor	Equipment	Total	Total Incl O&P
1500	#12, 3" long, steel				C	12.10			12.10	13.30
2000	#12, 4" long, steel				↓	18.15			18.15	19.95

06 16 Sheathing

06 16 36 – Wood Panel Product Sheathing

06 16 36.10 Sheathing

		Crew	Daily Output	Labor-Hours	Unit	Material	Labor	Equipment	Total	Total Incl O&P
0010	**SHEATHING**									
0012	Plywood on roofs, CDX									
0030	5/16" thick	2 Carp	1600	.010	S.F.	.53	.42		.95	1.22
0035	Pneumatic nailed		1952	.008		.53	.34		.87	1.11
0050	3/8" thick		1525	.010		.36	.44		.80	1.07
0055	Pneumatic nailed		1860	.009		.36	.36		.72	.95
0100	1/2" thick		1400	.011		.42	.47		.89	1.19
0105	Pneumatic nailed		1708	.009		.42	.39		.81	1.06
0200	5/8" thick		1300	.012		.51	.51		1.02	1.35
0205	Pneumatic nailed		1586	.010		.51	.42		.93	1.21
0300	3/4" thick		1200	.013		.58	.55		1.13	1.49
0305	Pneumatic nailed		1464	.011		.58	.45		1.03	1.34
0500	Plywood on walls with exterior CDX, 3/8" thick		1200	.013		.36	.55		.91	1.25
0505	Pneumatic nailed		1488	.011		.36	.45		.81	1.09
0600	1/2" thick		1125	.014		.42	.59		1.01	1.37
0605	Pneumatic nailed		1395	.011		.42	.48		.90	1.19
0700	5/8" thick		1050	.015		.51	.63		1.14	1.54
0705	Pneumatic nailed		1302	.012		.51	.51		1.02	1.35
0800	3/4" thick		975	.016		.58	.68		1.26	1.69
0805	Pneumatic nailed	↓	1209	.013	↓	.58	.55		1.13	1.49

Division Notes

		CREW	DAILY OUTPUT	LABOR-HOURS	UNIT	2010 BARE COSTS MAT.	LABOR	EQUIP.	TOTAL	TOTAL INCL O&P

Estimating Tips

07 10 00 Dampproofing and Waterproofing

- Be sure of the job specifications before pricing this subdivision. The difference in cost between waterproofing and dampproofing can be great. Waterproofing will hold back standing water. Dampproofing prevents the transmission of water vapor. Also included in this section are vapor retarding membranes.

07 20 00 Thermal Protection

- Insulation and fireproofing products are measured by area, thickness, volume or R-value. Specifications may give only what the specific R-value should be in a certain situation. The estimator may need to choose the type of insulation to meet that R-value.

07 30 00 Steep Slope Roofing
07 40 00 Roofing and Siding Panels

- Many roofing and siding products are bought and sold by the square. One square is equal to an area that measures 100 square feet.

This simple change in unit of measure could create a large error if the estimator is not observant. Accessories necessary for a complete installation must be figured into any calculations for both material and labor.

07 50 00 Membrane Roofing
07 60 00 Flashing and Sheet Metal
07 70 00 Roofing and Wall Specialties and Accessories

- The items in these subdivisions compose a roofing system. No one component completes the installation, and all must be estimated. Built-up or single-ply membrane roofing systems are made up of many products and installation trades. Wood blocking at roof perimeters or penetrations, parapet coverings, reglets, roof drains, gutters, downspouts, sheet metal flashing, skylights, smoke vents, and roof hatches all need to be considered along with the roofing material. Several different installation trades will need to work together on the roofing system. Inherent difficulties in the scheduling and coordination of various trades must be accounted for when estimating labor costs.

07 90 00 Joint Protection

- To complete the weather-tight shell, the sealants and caulkings must be estimated. Where different materials meet—at expansion joints, at flashing penetrations, and at hundreds of other locations throughout a construction project—they provide another line of defense against water penetration. Often, an entire system is based on the proper location and placement of caulking or sealants. The detailed drawings that are included as part of a set of architectural plans show typical locations for these materials. When caulking or sealants are shown at typical locations, this means the estimator must include them for all the locations where this detail is applicable. Be careful to keep different types of sealants separate, and remember to consider backer rods and primers if necessary.

Reference Numbers

Reference numbers are shown in shaded boxes at the beginning of some major classifications. These numbers refer to related items in the Reference Section. The reference information may be an estimating procedure, an alternate pricing method, or technical information.

Note: Not all subdivisions listed here necessarily appear in this publication.

07 65 Flexible Flashing

07 65 10 – Sheet Metal Flashing

07 65 10.10 Sheet Metal Flashing and Counter Flashing

	07 65 10.10 Sheet Metal Flashing and Counter Flashing	Crew	Daily Output	Labor-Hours	Unit	2010 Bare Costs Material	Labor	Equipment	Total	Total Incl O&P
0010	**SHEET METAL FLASHING AND COUNTER FLASHING**									
0011	Including up to 4 bends									
0020	Aluminum, mill finish, .013" thick	1 Rofc	145	.055	S.F.	.67	1.95		2.62	3.99
0030	.016" thick		145	.055		.79	1.95		2.74	4.12
0060	.019" thick		145	.055		1.02	1.95		2.97	4.37
0100	.032" thick		145	.055		1.48	1.95		3.43	4.88
0200	.040" thick		145	.055		2.21	1.95		4.16	5.70
0300	.050" thick		145	.055		2.52	1.95		4.47	6
0400	Painted finish, add					.37			.37	.41
1600	Copper, 16 oz, sheets, under 1000 lbs.	1 Rofc	115	.070		7.55	2.46		10.01	12.40
1700	Over 4000 lbs.		155	.052		6.75	1.83		8.58	10.50
1900	20 oz sheets, under 1000 lbs.		110	.073		9.05	2.57		11.62	14.25
2000	Over 4000 lbs.		145	.055		8.15	1.95		10.10	12.20
2200	24 oz sheets, under 1000 lbs.		105	.076		11.30	2.70		14	16.95
2300	Over 4000 lbs.		135	.059		10.15	2.10		12.25	14.70
2500	32 oz sheets, under 1000 lbs.		100	.080		14.10	2.83		16.93	20
2600	Over 4000 lbs.		130	.062		12.70	2.18		14.88	17.55
5800	Lead, 2.5 lb. per SF, up to 12" wide		135	.059		4.55	2.10		6.65	8.50
5900	Over 12" wide		135	.059		4.55	2.10		6.65	8.50
8650	Copper, 16 oz		100	.080		6	2.83		8.83	11.30
8900	Stainless steel sheets, 32 ga, .010" thick		155	.052		3.79	1.83		5.62	7.20
9000	28 ga, .015" thick		155	.052		4.70	1.83		6.53	8.20
9100	26 ga, .018" thick		155	.052		5.70	1.83		7.53	9.30
9200	24 ga, .025" thick		155	.052		7.40	1.83		9.23	11.20
9400	Terne coated stainless steel, .015" thick, 28 ga		155	.052		7.35	1.83		9.18	11.15
9500	.018" thick, 26 ga		155	.052		8.30	1.83		10.13	12.15
9600	Zinc and copper alloy (brass), .020" thick		155	.052		5.15	1.83		6.98	8.70
9700	.027" thick		155	.052		6.90	1.83		8.73	10.65
9800	.032" thick		155	.052		8.05	1.83		9.88	11.90
9900	.040" thick		155	.052		9.85	1.83		11.68	13.85

07 65 13 – Laminated Sheet Flashing

07 65 13.10 Laminated Sheet Flashing

	07 65 13.10 Laminated Sheet Flashing	Crew	Daily Output	Labor-Hours	Unit	Material	Labor	Equipment	Total	Total Incl O&P
0010	**LAMINATED SHEET FLASHING**, Including up to 4 bends									
4300	Copper-clad stainless steel, .015" thick, under 500 lbs.	1 Rofc	115	.070	S.F.	5.25	2.46		7.71	9.85
4400	Over 2000 lbs.		155	.052		5.05	1.83		6.88	8.60
4600	.018" thick, under 500 lbs.		100	.080		6.90	2.83		9.73	12.30
4700	Over 2000 lbs.		145	.055		5.05	1.95		7	8.80
8550	3 ply copper and fabric, 3 oz		155	.052		2.31	1.83		4.14	5.60
8600	7 oz		155	.052		3.25	1.83		5.08	6.60
8700	Lead on copper and fabric, 5 oz		155	.052		2.78	1.83		4.61	6.10
8800	7 oz		155	.052		5.25	1.83		7.08	8.85

07 65 19 – Plastic Sheet Flashing

07 65 19.10 Plastic Sheet Flashing and Counter Flashing

	07 65 19.10 Plastic Sheet Flashing and Counter Flashing	Crew	Daily Output	Labor-Hours	Unit	Material	Labor	Equipment	Total	Total Incl O&P
0010	**PLASTIC SHEET FLASHING AND COUNTER FLASHING**									
7300	Polyvinyl chloride, black, .010" thick	1 Rofc	285	.028	S.F.	.23	.99		1.22	1.90
7400	.020" thick		285	.028		.33	.99		1.32	2.01
7600	.030" thick		285	.028		.42	.99		1.41	2.11
7700	.056" thick		285	.028		1.03	.99		2.02	2.78
7900	Black or white for exposed roofs, .060" thick		285	.028		2.25	.99		3.24	4.13
8060	PVC tape, 5" x 45 mils, for joint covers, 100 L.F./roll				Ea.	120			120	132
8850	Polyvinyl chloride, .030" thick	1 Rofc	160	.050	S.F.	.46	1.77		2.23	3.45

07 71 Roof Specialties

07 71 23 – Manufactured Gutters and Downspouts

07 71 23.10 Downspouts		Crew	Daily Output	Labor-Hours	Unit	2010 Bare Costs Material	Labor	Equipment	Total	Total Incl O&P
0010	**DOWNSPOUTS**									
0020	Aluminum 2" x 3", .020" thick, embossed	1 Shee	190	.042	L.F.	1.04	2.07		3.11	4.27
0100	Enameled		190	.042		1.59	2.07		3.66	4.88
4800	Steel, galvanized, round, corrugated, 2" or 3" diameter, 28 gauge		190	.042		2.24	2.07		4.31	5.60
4900	4" diameter, 28 gauge		145	.055		2.73	2.71		5.44	7.10
5100	5" diameter, 28 gauge		130	.062		3.53	3.02		6.55	8.45
5200	26 gauge		130	.062		3.75	3.02		6.77	8.70
5400	6" diameter, 28 gauge		105	.076		4	3.74		7.74	10.05
5500	26 gauge	↓	105	.076	↓	5.45	3.74		9.19	11.65

07 72 Roof Accessories

07 72 26 – Ridge Vents

07 72 26.10 Ridge Vents and Accessories

		Crew	Daily Output	Labor-Hours	Unit	Material	Labor	Equipment	Total	Total Incl O&P
0010	**RIDGE VENTS AND ACCESSORIES**									
2300	Ridge vent strip, mill finish	1 Shee	155	.052	L.F.	2.73	2.53		5.26	6.85

07 72 53 – Snow Guards

07 72 53.10 Snow Guard Options

		Crew	Daily Output	Labor-Hours	Unit	Material	Labor	Equipment	Total	Total Incl O&P
0010	**SNOW GUARD OPTIONS**									
0100	Slate & asphalt shingle roofs, fastened with nails	1 Rofc	160	.050	Ea.	9.40	1.77		11.17	13.30
0200	Standing seam metal roofs, fastened with set screws		48	.167		14.60	5.90		20.50	26
0300	Surface mount for metal roofs, fastened with solder		48	.167	↓	7.70	5.90		13.60	18.25
0400	Double rail pipe type, including pipe	↓	130	.062	L.F.	22.50	2.18		24.68	28.50

07 72 73 – Pitch Pockets

07 72 73.10 Pitch Pockets, Variable Sizes

		Crew	Daily Output	Labor-Hours	Unit	Material	Labor	Equipment	Total	Total Incl O&P
0010	**PITCH POCKETS, VARIABLE SIZES**									
0100	Adjustable, 4" to 7", welded corners, 4" deep	1 Rofc	48	.167	Ea.	12.90	5.90		18.80	24
0200	Side extenders, 6"	"	240	.033	"	1.97	1.18		3.15	4.13

07 84 Firestopping

07 84 13 – Penetration Firestopping

07 84 13.10 Firestopping

		Crew	Daily Output	Labor-Hours	Unit	Material	Labor	Equipment	Total	Total Incl O&P
0010	**FIRESTOPPING** R078413-30									
0100	Metallic piping, non insulated									
0110	Through walls, 2" diameter	1 Carp	16	.500	Ea.	12.40	21		33.40	45.50
0120	4" diameter		14	.571		18.90	23.50		42.40	57.50
0130	6" diameter		12	.667		25.50	27.50		53	70.50
0140	12" diameter		10	.800		45	33		78	101
0150	Through floors, 2" diameter		32	.250		7.50	10.40		17.90	24.50
0160	4" diameter		28	.286		10.80	11.85		22.65	30
0170	6" diameter		24	.333		14.20	13.85		28.05	37
0180	12" diameter	↓	20	.400	↓	24	16.60		40.60	52
0190	Metallic piping, insulated									
0200	Through walls, 2" diameter	1 Carp	16	.500	Ea.	17.55	21		38.55	51.50
0210	4" diameter		14	.571		24	23.50		47.50	63
0220	6" diameter		12	.667		30.50	27.50		58	76
0230	12" diameter		10	.800		50	33		83	106
0240	Through floors, 2" diameter		32	.250		12.70	10.40		23.10	30
0250	4" diameter	↓	28	.286	↓	15.95	11.85		27.80	36

07 84 13.10 Firestopping		Crew	Daily Output	Labor-Hours	Unit	Material	2010 Bare Costs Labor	Equipment	Total	Total Incl O&P
0260	6" diameter	1 Carp	24	.333	Ea.	19.35	13.85		33.20	43
0270	12" diameter	↓	20	.400	↓	24	16.60		40.60	52
0280	Non metallic piping, non insulated									
0290	Through walls, 2" diameter	1 Carp	12	.667	Ea.	51	27.50		78.50	98.50
0300	4" diameter		10	.800		64	33		97	122
0310	6" diameter		8	1		89	41.50		130.50	162
0330	Through floors, 2" diameter		16	.500		40	21		61	76
0340	4" diameter		6	1.333		49.50	55.50		105	140
0350	6" diameter	↓	6	1.333	↓	59.50	55.50		115	151
0370	Ductwork, insulated & non insulated, round									
0380	Through walls, 6" diameter	1 Carp	12	.667	Ea.	26	27.50		53.50	71
0390	12" diameter		10	.800		51.50	33		84.50	108
0400	18" diameter		8	1		84	41.50		125.50	157
0410	Through floors, 6" diameter		16	.500		14.20	21		35.20	47.50
0420	12" diameter		14	.571		26	23.50		49.50	65
0430	18" diameter	↓	12	.667	↓	45	27.50		72.50	92
0440	Ductwork, insulated & non insulated, rectangular									
0450	With stiffener/closure angle, through walls, 6" x 12"	1 Carp	8	1	Ea.	21.50	41.50		63	87.50
0460	12" x 24"		6	1.333		28.50	55.50		84	117
0470	24" x 48"		4	2		81.50	83		164.50	218
0480	With stiffener/closure angle, through floors, 6" x 12"		10	.800		11.65	33		44.65	64
0490	12" x 24"		8	1		21	41.50		62.50	87
0500	24" x 48"	↓	6	1.333	↓	41	55.50		96.50	131
0510	Multi trade openings									
0520	Through walls, 6" x 12"	1 Carp	2	4	Ea.	45	166		211	305
0530	12" x 24"	"	1	8		182	330		512	710
0540	24" x 48"	2 Carp	1	16		730	665		1,395	1,825
0550	48" x 96"	"	.75	21.333		2,925	885		3,810	4,600
0560	Through floors, 6" x 12"	1 Carp	2	4		45	166		211	305
0570	12" x 24"	"	1	8		182	330		512	710
0580	24" x 48"	2 Carp	.75	21.333		730	885		1,615	2,175
0590	48" x 96"	"	.50	32	↓	2,925	1,325		4,250	5,275
0600	Structural penetrations, through walls									
0610	Steel beams, W8 x 10	1 Carp	8	1	Ea.	28.50	41.50		70	95.50
0620	W12 x 14		6	1.333		45	55.50		100.50	135
0630	W21 x 44		5	1.600		90.50	66.50		157	202
0640	W36 x 135		3	2.667		220	111		331	415
0650	Bar joists, 18" deep		6	1.333		41.50	55.50		97	131
0660	24" deep		6	1.333		51.50	55.50		107	143
0670	36" deep		5	1.600		77.50	66.50		144	188
0680	48" deep	↓	4	2	↓	90.50	83		173.50	228
0690	Construction joints, floor slab at exterior wall									
0700	Precast, brick, block or drywall exterior									
0710	2" wide joint	1 Carp	125	.064	L.F.	6.45	2.66		9.11	11.20
0720	4" wide joint	"	75	.107	"	12.90	4.43		17.33	21
0730	Metal panel, glass or curtain wall exterior									
0740	2" wide joint	1 Carp	40	.200	L.F.	15.30	8.30		23.60	29.50
0750	4" wide joint	"	25	.320	"	21	13.30		34.30	43.50
0760	Floor slab to drywall partition									
0770	Flat joint	1 Carp	100	.080	L.F.	6.35	3.32		9.67	12.05
0780	Fluted joint		50	.160		12.90	6.65		19.55	24.50
0790	Etched fluted joint	↓	75	.107	↓	8.40	4.43		12.83	16.10
0800	Floor slab to concrete/masonry partition									

07 84 Firestopping

07 84 13 – Penetration Firestopping

07 84 13.10 Firestopping		Crew	Daily Output	Labor-Hours	Unit	Material	Labor	Equipment	Total	Total Incl O&P
						2010 Bare Costs				
0810	Flat joint	1 Carp	75	.107	L.F.	14.20	4.43		18.63	22.50
0820	Fluted joint	"	50	.160	"	16.80	6.65		23.45	29
0830	Concrete/CMU wall joints									
0840	1" wide	1 Carp	100	.080	L.F.	7.75	3.32		11.07	13.65
0850	2" wide		75	.107		14.20	4.43		18.63	22.50
0860	4" wide	↓	50	.160	↓	27	6.65		33.65	40.50
0870	Concrete/CMU floor joints									
0880	1" wide	1 Carp	200	.040	L.F.	3.88	1.66		5.54	6.85
0890	2" wide		150	.053		7.10	2.22		9.32	11.20
0900	4" wide	↓	100	.080	↓	13.55	3.32		16.87	20

07 91 Pre-formed Joint Seals

07 91 13 – Compression Seals

07 91 13.10 Compression Seals		Crew	Daily Output	Labor-Hours	Unit	Material	Labor	Equipment	Total	Total Incl O&P
0010	**COMPRESSION SEALS**									
4900	Compression seals, O-ring type cord, 1/4"	1 Bric	472	.017	L.F.	.47	.71		1.18	1.59
4910	1/2"		440	.018		1.78	.76		2.54	3.10
4920	3/4"		424	.019		3.61	.79		4.40	5.15
4930	1"		408	.020		6.95	.82		7.77	8.90
4940	1-1/4"		384	.021		13.60	.87		14.47	16.25
4950	1-1/2"		368	.022		17.20	.91		18.11	20.50
4960	1-3/4"		352	.023		36	.95		36.95	41
4970	2"	↓	344	.023	↓	63.50	.97		64.47	71.50

07 91 16 – Joint Gaskets

07 91 16.10 Joint Gaskets		Crew	Daily Output	Labor-Hours	Unit	Material	Labor	Equipment	Total	Total Incl O&P
0010	**JOINT GASKETS**									
4400	Joint gaskets, neoprene, closed cell w/adh, 1/8" x 3/8"	1 Bric	240	.033	L.F.	.26	1.39		1.65	2.39
4500	1/4" x 3/4"		215	.037		.58	1.55		2.13	2.98
4700	1/2" x 1"		200	.040		1.30	1.67		2.97	3.95
4800	3/4" x 1-1/2"	↓	165	.048	↓	1.44	2.02		3.46	4.63

07 91 23 – Backer Rods

07 91 23.10 Backer Rods		Crew	Daily Output	Labor-Hours	Unit	Material	Labor	Equipment	Total	Total Incl O&P
0010	**BACKER RODS**									
0030	Backer rod, polyethylene, 1/4" diameter	1 Bric	4.60	1.739	C.L.F.	2.82	72.50		75.32	112
0050	1/2" diameter		4.60	1.739		6.60	72.50		79.10	116
0070	3/4" diameter		4.60	1.739		10.10	72.50		82.60	120
0090	1" diameter	↓	4.60	1.739	↓	18.70	72.50		91.20	130

07 91 26 – Joint Fillers

07 91 26.10 Joint Fillers		Crew	Daily Output	Labor-Hours	Unit	Material	Labor	Equipment	Total	Total Incl O&P
0010	**JOINT FILLERS**									
4360	Butyl rubber filler, 1/4" x 1/4"	1 Bric	290	.028	L.F.	.14	1.15		1.29	1.90
4365	1/2" x 1/2"		250	.032		.57	1.34		1.91	2.63
4370	1/2" x 3/4"		210	.038		.85	1.59		2.44	3.34
4375	3/4" x 3/4"		230	.035		1.28	1.45		2.73	3.60
4380	1" x 1"	↓	180	.044	↓	1.71	1.86		3.57	4.68
4390	For coloring, add					12%				
4980	Polyethylene joint backing, 1/4" x 2"	1 Bric	2.08	3.846	C.L.F.	12	161		173	255
4990	1/4" x 6"		1.28	6.250	"	36	261		297	435
5600	Silicone, room temp vulcanizing foam seal, 1/4" x 1/2"	↓	1312	.006	L.F.	.29	.25		.54	.70

07 91 Pre-formed Joint Seals

07 91 26 – Joint Fillers

07 91 26.10 Joint Fillers		Crew	Daily Output	Labor-Hours	Unit	Material	Labor	Equipment	Total	Total Incl O&P
5610	1/2" x 1/2"	1 Bric	656	.012	L.F.	.57	.51		1.08	1.40
5620	1/2" x 3/4"		442	.018		.86	.76		1.62	2.09
5630	3/4" x 3/4"		328	.024		1.29	1.02		2.31	2.96
5640	1/8" x 1"		1312	.006		.29	.25		.54	.70
5650	1/8" x 3"		442	.018		.86	.76		1.62	2.09
5670	1/4" x 3"		295	.027		1.72	1.13		2.85	3.60
5680	1/4" x 6"		148	.054		3.44	2.26		5.70	7.20
5690	1/2" x 6"		82	.098		6.90	4.07		10.97	13.70
5700	1/2" x 9"		52.50	.152		10.30	6.35		16.65	21
5710	1/2" x 12"	↓	33	.242	↓	13.75	10.10		23.85	30.50

07 92 Joint Sealants

07 92 13 – Elastomeric Joint Sealants

07 92 13.20 Caulking and Sealant Options

Line	Description	Crew	Daily Output	Labor-Hours	Unit	Material	Labor	Equipment	Total	Total Incl O&P
0010	**CAULKING AND SEALANT OPTIONS**									
0050	Latex acrylic based, bulk				Gal.	27			27	29.50
0055	Bulk in place 1/4" x 1/4" bead	1 Bric	300	.027	L.F.	.09	1.11		1.20	1.77
0060	1/4" x 3/8"		294	.027		.14	1.14		1.28	1.87
0065	1/4" x 1/2"		288	.028		.19	1.16		1.35	1.96
0075	3/8" x 3/8"		284	.028		.21	1.18		1.39	2
0080	3/8" x 1/2"		280	.029		.28	1.19		1.47	2.11
0085	3/8" x 5/8"		276	.029		.35	1.21		1.56	2.21
0095	3/8" x 3/4"		272	.029		.42	1.23		1.65	2.32
0100	1/2" x 1/2"		275	.029		.38	1.21		1.59	2.25
0105	1/2" x 5/8"		269	.030		.47	1.24		1.71	2.39
0110	1/2" x 3/4"		263	.030		.57	1.27		1.84	2.54
0115	1/2" x 7/8"		256	.031		.66	1.30		1.96	2.70
0120	1/2" x 1"		250	.032		.76	1.34		2.10	2.84
0125	3/4" x 3/4"		244	.033		.85	1.37		2.22	3
0130	3/4" x 1"		225	.036		1.14	1.48		2.62	3.49
0135	1" x 1"	↓	200	.040	↓	1.52	1.67		3.19	4.19
0190	Cartridges				Gal.	22.50			22.50	24.50
0200	11 fl. oz cartridge				Ea.	1.93			1.93	2.12
0500	1/4" x 1/2"	1 Bric	288	.028	L.F.	.16	1.16		1.32	1.92
0600	1/2" x 1/2"		275	.029		.32	1.21		1.53	2.18
0800	3/4" x 3/4"		244	.033		.71	1.37		2.08	2.84
0900	3/4" x 1"		225	.036		.95	1.48		2.43	3.28
1000	1" x 1"	↓	200	.040	↓	1.18	1.67		2.85	3.82
1400	Butyl based, bulk				Gal.	26			26	28.50
1500	Cartridges				"	32			32	35
1700	1/4" x 1/2", 154 L.F./gal.	1 Bric	288	.028	L.F.	.17	1.16		1.33	1.94
1800	1/2" x 1/2", 77 L.F./gal.	"	275	.029	"	.34	1.21		1.55	2.20
2300	Polysulfide compounds, 1 component, bulk				Gal.	51			51	56
2600	1 or 2 component, in place, 1/4" x 1/4", 308 L.F./gal.	1 Bric	300	.027	L.F.	.17	1.11		1.28	1.86
2700	1/2" x 1/4", 154 L.F./gal.		288	.028		.33	1.16		1.49	2.12
2900	3/4" x 3/8", 68 L.F./gal.		272	.029		.75	1.23		1.98	2.68
3000	1" x 1/2", 38 L.F./gal.	↓	250	.032	↓	1.35	1.34		2.69	3.49
3200	Polyurethane, 1 or 2 component				Gal.	50			50	55
3300	Cartridges				"	64			64	70.50
3500	Bulk, in place, 1/4" x 1/4"	1 Bric	300	.027	L.F.	.16	1.11		1.27	1.86
3655	1/2" x 1/4"	↓	288	.028	↓	.32	1.16		1.48	2.11

07 92 13 – Elastomeric Joint Sealants

07 92 13.20 Caulking and Sealant Options		Crew	Daily Output	Labor-Hours	Unit	Material	Labor	Equipment	Total	Total Incl O&P
						2010 Bare Costs				
3800	3/4" x 3/8", 68 L.F./gal.	1 Bric	272	.029	L.F.	.74	1.23		1.97	2.66
3900	1" x 1/2"	↓	250	.032	↓	1.30	1.34		2.64	3.44
4100	Silicone rubber, bulk				Gal.	41			41	45
4200	Cartridges				"	41			41	45

07 92 16 – Rigid Joint Sealants

07 92 16.10 Rigid Joint Sealants		Crew	Daily Output	Labor-Hours	Unit	Material	Labor	Equipment	Total	Total Incl O&P
0010	**RIGID JOINT SEALANTS**									
5800	Tapes, sealant, P.V.C. foam adhesive, 1/16" x 1/4"				C.L.F.	5.65			5.65	6.20
5900	1/16" x 1/2"					8.35			8.35	9.20
5950	1/16" x 1"					13.85			13.85	15.25
6000	1/8" x 1/2"				↓	9.35			9.35	10.30

07 92 19 – Acoustical Joint Sealants

07 92 19.10 Acoustical Sealant		Crew	Daily Output	Labor-Hours	Unit	Material	Labor	Equipment	Total	Total Incl O&P
0010	**ACOUSTICAL SEALANT**									
0020	Acoustical sealant, elastomeric, cartridges				Ea.	2.25			2.25	2.48
0025	In place, 1/4" x 1/4"	1 Bric	300	.027	L.F.	.09	1.11		1.20	1.78
0030	1/4" x 1/2"		288	.028		.18	1.16		1.34	1.95
0035	1/2" x 1/2"		275	.029		.37	1.21		1.58	2.23
0040	1/2" x 3/4"		263	.030		.55	1.27		1.82	2.52
0045	3/4" x 3/4"		244	.033		.83	1.37		2.20	2.97
0050	1" x 1"	↓	200	.040	↓	1.47	1.67		3.14	4.14

Division Notes

	CREW	DAILY OUTPUT	LABOR-HOURS	UNIT	2010 BARE COSTS MAT.	LABOR	EQUIP.	TOTAL	TOTAL INCL O&P

Estimating Tips

08 10 00 Doors and Frames

All exterior doors should be addressed for their energy conservation.

- Most metal doors and frames look alike, but there may be significant differences among them. When estimating these items, be sure to choose the line item that most closely compares to the specification or door schedule requirements regarding:
  - — type of metal
  - — metal gauge
  - — door core material
  - — fire rating
  - — finish
- Wood and plastic doors vary considerably in price. The primary determinant is the veneer material. Lauan, birch, and oak are the most common veneers. Other variables include the following:
  - — hollow or solid core
  - — fire rating
  - — flush or raised panel
  - — finish

08 30 00 Specialty Doors and Frames

- There are many varieties of special doors, and they are usually priced per each. Add frames, hardware, or operators required for a complete installation.

08 40 00 Entrances, Storefronts, and Curtain Walls

- Glazed curtain walls consist of the metal tube framing and the glazing material. The cost data in this subdivision is presented for the metal tube framing alone or the composite wall. If your estimate requires a detailed takeoff of the framing, be sure to add the glazing cost.

08 50 00 Windows

- Most metal windows are delivered preglazed. However, some metal windows are priced without glass. Refer to 08 80 00 Glazing for glass pricing. The grade C indicates commercial grade windows, usually ASTM C-35.
- All wood windows and vinyl are priced preglazed. The glazing is insulating glass. Some wood windows may have single pane float glass. Add the cost of screens and grills if required, and not already included.

08 70 00 Hardware

- Hardware costs add considerably to the cost of a door. The most efficient method to determine the hardware requirements for a project is to review the door schedule.
- Door hinges are priced by the pair, with most doors requiring 1-1/2 pairs per door. The hinge prices do not include installation labor because it is included in door installation. Hinges are classified according to the frequency of use.

08 80 00 Glazing

- Different openings require different types of glass. The three most common types are:
  - — float
  - — tempered
  - — insulating
- Most exterior windows are glazed with insulating glass. Entrance doors and window walls, where the glass is less than 18″ from the floor, are generally glazed with tempered glass. Interior windows and some residential windows are glazed with float glass.
- Coastal communities are starting to require the use of impact-resistant glass.
- The insulation or 'u' value is a strong consideration, along with solar heat gain.

Reference Numbers

Reference numbers are shown in shaded boxes at the beginning of some major classifications. These numbers refer to related items in the Reference Section. The reference information may be an estimating procedure, an alternate pricing method, or technical information.

Note: Not all subdivisions listed here necessarily appear in this publication.

08 34 Special Function Doors

08 34 13 – Cold Storage Doors

08 34 13.10 Doors for Cold Area Storage		Crew	Daily Output	Labor-Hours	Unit	Material	2010 Bare Costs Labor	Equipment	Total	Total Incl O&P
0010	**DOORS FOR COLD AREA STORAGE**									
0020	Single, 20 ga. galvanized steel									
0300	Horizontal sliding, 5′ x 7′, manual operation, 3.5″ thick	2 Carp	2	8	Ea.	3,150	330		3,480	3,950
0400	4″ thick		2	8		3,825	330		4,155	4,725
0500	6″ thick		2	8		3,200	330		3,530	4,025
0800	5′ x 7′, power operation, 2″ thick		1.90	8.421		5,225	350		5,575	6,300
0900	4″ thick		1.90	8.421		5,350	350		5,700	6,425
1000	6″ thick		1.90	8.421		6,100	350		6,450	7,250
1300	9′ x 10′, manual operation, 2″ insulation		1.70	9.412		4,250	390		4,640	5,275
1400	4″ insulation		1.70	9.412		4,350	390		4,740	5,375
1500	6″ insulation		1.70	9.412		5,225	390		5,615	6,350
1800	Power operation, 2″ insulation		1.60	10		7,200	415		7,615	8,575
1900	4″ insulation		1.60	10		7,425	415		7,840	8,800
2000	6″ insulation	↓	1.70	9.412	↓	8,400	390		8,790	9,850
2300	For stainless steel face, add					22%				
3000	Hinged, lightweight, 3′ x 7′-0″, galvanized 1 face, 2″ thick	2 Carp	2	8	Ea.	1,350	330		1,680	1,975
3050	4″ thick		1.90	8.421		1,700	350		2,050	2,400
3300	Aluminum doors, 3′ x 7′-0″, 4″ thick		1.90	8.421		1,250	350		1,600	1,925
3350	6″ thick		1.40	11.429		2,200	475		2,675	3,150
3600	Stainless steel, 3′ x 7′-0″, 4″ thick		1.90	8.421		1,625	350		1,975	2,325
3650	6″ thick		1.40	11.429		2,775	475		3,250	3,775
3900	Painted, 3′ x 7′-0″, 4″ thick		1.90	8.421		1,200	350		1,550	1,875
3950	6″ thick	↓	1.40	11.429	↓	2,200	475		2,675	3,150
5000	Bi-parting, electric operated									
5010	6′ x 8′ opening, galv. faces, 4″ thick for cooler	2 Carp	.80	20	Opng.	6,800	830		7,630	8,750
5050	For freezer, 4″ thick		.80	20		7,500	830		8,330	9,525
5300	For door buck framing and door protection, add		2.50	6.400		600	266		866	1,075
6000	Galvanized batten door, galvanized hinges, 4′ x 7′		2	8		1,700	330		2,030	2,375
6050	6′ x 8′		1.80	8.889		2,325	370		2,695	3,125
6500	Fire door, 3 hr., 6′ x 8′, single slide		.80	20		7,800	830		8,630	9,850
6550	Double, bi-parting	↓	.70	22.857	↓	12,200	950		13,150	14,900

08 91 Louvers

08 91 19 – Fixed Louvers

08 91 19.10 Aluminum Louvers

		Crew	Daily Output	Labor-Hours	Unit	Material	Labor	Equipment	Total	Total Incl O&P
0010	**ALUMINUM LOUVERS**									
0020	Aluminum with screen, residential, 8″ x 8″	1 Carp	38	.211	Ea.	10.95	8.75		19.70	25.50
0100	12″ x 12″		38	.211		14	8.75		22.75	29
0200	12″ x 18″		35	.229		13.95	9.50		23.45	30
0250	14″ x 24″		30	.267		25	11.10		36.10	44.50
0300	18″ x 24″		27	.296		27	12.30		39.30	48.50
0500	24″ x 30″		24	.333		32	13.85		45.85	56.50
0700	Triangle, adjustable, small		20	.400		32.50	16.60		49.10	61.50
0800	Large	↓	15	.533	↓	40.50	22		62.50	78.50
1200	Extruded aluminum, see Div. 23 37 15.40									
2100	Midget, aluminum, 3/4″ deep, 1″ diameter	1 Carp	85	.094	Ea.	.73	3.91		4.64	6.85
2150	3″ diameter		60	.133		1.78	5.55		7.33	10.50
2200	4″ diameter		50	.160		3.64	6.65		10.29	14.25
2250	6″ diameter	↓	30	.267	↓	3.74	11.10		14.84	21

08 95 Vents

08 95 13 – Soffit Vents

08 95 13.10 Wall Louvers		Crew	Daily Output	Labor-Hours	Unit	Material	Labor	Equipment	Total	Total Incl O&P
						2010 Bare Costs				
0010	**WALL LOUVERS**									
2330	Soffit vent, continuous, 3" wide, aluminum, mill finish	1 Carp	200	.040	L.F.	.48	1.66		2.14	3.09
2340	Baked enamel finish		200	.040	"	.52	1.66		2.18	3.13
2400	Under eaves vent, aluminum, mill finish, 16" x 4"		48	.167	Ea.	1.33	6.95		8.28	12.15
2500	16" x 8"	↓	48	.167	"	1.54	6.95		8.49	12.40

08 95 16 – Wall Vents

08 95 16.10 Louvers		Crew	Daily Output	Labor-Hours	Unit	Material	Labor	Equipment	Total	Total Incl O&P
0010	**LOUVERS**									
0020	Redwood, 2'-0" diameter, full circle	1 Carp	16	.500	Ea.	169	21		190	218
0100	Half circle		16	.500		189	21		210	240
0200	Octagonal		16	.500		129	21		150	174
0300	Triangular, 5/12 pitch, 5'-0" at base		16	.500		189	21		210	240
1100	Rectangular, 1'-4" x 1'-8"		16	.500		25.50	21		46.50	60
1200	1'-4" x 2'-2"		15	.533		29.50	22		51.50	66
1300	1'-9" x 2'-2"		15	.533		35	22		57	72.50
1400	2'-3" x 2'-2"		14	.571		47	23.50		70.50	88
1700	2'-4" x 2'-11"		13	.615		47	25.50		72.50	91
2000	Aluminum, 12" x 16"		25	.320		13.45	13.30		26.75	35.50
2010	16" x 20"		25	.320		22.50	13.30		35.80	45
2020	24" x 30"		25	.320		50.50	13.30		63.80	76
2100	6' triangle		12	.667		162	27.50		189.50	221
7000	Vinyl gable vent, 8" x 8"		38	.211		11.55	8.75		20.30	26
7020	12" x 12"		38	.211		24	8.75		32.75	40
7080	12" x 18"		35	.229		31	9.50		40.50	48.50
7200	18" x 24"	↓	30	.267	↓	37.50	11.10		48.60	58.50

Division Notes

	CREW	DAILY OUTPUT	LABOR-HOURS	UNIT	2010 BARE COSTS MAT.	LABOR	EQUIP.	TOTAL	TOTAL INCL O&P

Estimating Tips

General

- Room Finish Schedule: A complete set of plans should contain a room finish schedule. If one is not available, it would be well worth the time and effort to obtain one.

09 20 00 Plaster and Gypsum Board

- Lath is estimated by the square yard plus a 5% allowance for waste. Furring, channels, and accessories are measured by the linear foot. An extra foot should be allowed for each accessory miter or stop.
- Plaster is also estimated by the square yard. Deductions for openings vary by preference, from zero deduction to 50% of all openings over 2 feet in width. The estimator should allow one extra square foot for each linear foot of horizontal interior or exterior angle located below the ceiling level. Also, double the areas of small radius work.
- Drywall accessories, studs, track, and acoustical caulking are all measured by the linear foot. Drywall taping is figured by the square foot. Gypsum wallboard is estimated by the square foot. No material deductions should be made for door or window openings under 32 S.F.

09 60 00 Flooring

- Tile and terrazzo areas are taken off on a square foot basis. Trim and base materials are measured by the linear foot. Accent tiles are listed per each. Two basic methods of installation are used. Mud set is approximately 30% more expensive than thin set. In terrazzo work, be sure to include the linear footage of embedded decorative strips, grounds, machine rubbing, and power cleanup.
- Wood flooring is available in strip, parquet, or block configuration. The latter two types are set in adhesives with quantities estimated by the square foot. The laying pattern will influence labor costs and material waste. In addition to the material and labor for laying wood floors, the estimator must make allowances for sanding and finishing these areas unless the flooring is prefinished.
- Sheet flooring is measured by the square yard. Roll widths vary, so consideration should be given to use the most economical width, as waste must be figured into the total quantity. Consider also the installation methods available, direct glue down or stretched.

09 70 00 Wall Finishes

- Wall coverings are estimated by the square foot. The area to be covered is measured, length by height of wall above baseboards, to calculate the square footage of each wall. This figure is divided by the number of square feet in the single roll which is being used. Deduct, in full, the areas of openings such as doors and windows. Where a pattern match is required allow 25%–30% waste.

09 80 00 Acoustic Treatment

- Acoustical systems fall into several categories. The takeoff of these materials should be by the square foot of area with a 5% allowance for waste. Do not forget about scaffolding, if applicable, when estimating these systems.

09 90 00 Painting and Coating

- A major portion of the work in painting involves surface preparation. Be sure to include cleaning, sanding, filling, and masking costs in the estimate.
- Protection of adjacent surfaces is not included in painting costs. When considering the method of paint application, an important factor is the amount of protection and masking required. These must be estimated separately and may be the determining factor in choosing the method of application.

Reference Numbers

Reference numbers are shown in shaded boxes at the beginning of some major classifications. These numbers refer to related items in the Reference Section. The reference information may be an estimating procedure, an alternate pricing method, or technical information.

Note: Not all subdivisions listed here necessarily appear in this publication.

09 22 Supports for Plaster and Gypsum Board

09 22 03 – Fastening Methods for Finishes

09 22 03.20 Drilling Plaster/Drywall		Crew	Daily Output	Labor-Hours	Unit	Material	2010 Bare Costs Labor	Equipment	Total	Total Incl O&P
0010	**DRILLING PLASTER/DRYWALL**									
1100	Drilling & layout for drywall/plaster walls, up to 1" deep, no anchor									
1200	Holes, 1/4" diameter	1 Carp	150	.053	Ea.	.01	2.22		2.23	3.43
1300	3/8" diameter		140	.057		.01	2.37		2.38	3.67
1400	1/2" diameter		130	.062		.01	2.56		2.57	3.95
1500	3/4" diameter		120	.067		.01	2.77		2.78	4.29
1600	1" diameter		110	.073		.02	3.02		3.04	4.68
1700	1-1/4" diameter		100	.080		.03	3.32		3.35	5.15
1800	1-1/2" diameter		90	.089		.04	3.69		3.73	5.75
1900	For ceiling installations, add						40%			

09 69 Access Flooring

09 69 13 – Rigid-Grid Access Flooring

09 69 13.10 Access Floors

		Crew	Daily Output	Labor-Hours	Unit	Material	Labor	Equipment	Total	Total Incl O&P
0010	**ACCESS FLOORS**									
0015	Access floor package including panel, pedestal, stringers & laminate cover									
0100	Computer room, greater than 6,000 S.F.	4 Carp	750	.043	S.F.	8.65	1.77		10.42	12.25
0110	Less than 6,000 S.F.	2 Carp	375	.043		9.95	1.77		11.72	13.70
0120	Office, greater than 6,000 S.F.	4 Carp	1050	.030		4.56	1.27		5.83	6.95
0250	Panels, particle board or steel, 1250# load, no covering, under 6,000 S.F.	2 Carp	600	.027		4.24	1.11		5.35	6.35
0300	Over 6,000 S.F.		640	.025		3.72	1.04		4.76	5.70
0400	Aluminum, 24" panels		500	.032		31.50	1.33		32.83	37
0600	For carpet covering, add					8.55			8.55	9.40
0700	For vinyl floor covering, add					6.70			6.70	7.35
0900	For high pressure laminate covering, add					5.45			5.45	6
0910	For snap on stringer system, add	2 Carp	1000	.016		1.44	.66		2.10	2.60
0950	Office applications, steel or concrete panels,									
0960	no covering, over 6,000 S.F.	2 Carp	960	.017	S.F.	10.10	.69		10.79	12.15
1000	Machine cutouts after initial installation	1 Carp	50	.160	Ea.	5.05	6.65		11.70	15.80
1050	Pedestals, 6" to 12"	2 Carp	85	.188		7.95	7.80		15.75	21
1100	Air conditioning grilles, 4" x 12"	1 Carp	17	.471		66	19.55		85.55	103
1150	4" x 18"	"	14	.571		90.50	23.50		114	136
1200	Approach ramps, minimum	2 Carp	60	.267	S.F.	24.50	11.10		35.60	44
1300	Maximum	"	40	.400	"	33.50	16.60		50.10	62
1500	Handrail, 2 rail, aluminum	1 Carp	15	.533	L.F.	94.50	22		116.50	138

09 91 Painting

09 91 23 – Interior Painting

09 91 23.52 Miscellaneous, Interior

		Crew	Daily Output	Labor-Hours	Unit	Material	Labor	Equipment	Total	Total Incl O&P
0010	**MISCELLANEOUS, INTERIOR** R099100-10									
3800	Grilles, per side, oil base, primer coat, brushwork	1 Pord	520	.015	S.F.	.14	.56		.70	.99
3850	Spray		1140	.007		.15	.26		.41	.54
3880	Paint 1 coat, brushwork		520	.015		.21	.56		.77	1.06
3900	Spray		1140	.007		.24	.26		.50	.64
3920	Paint 2 coats, brushwork		325	.025		.41	.89		1.30	1.78
3940	Spray		650	.012		.47	.45		.92	1.19
3950	Prime & paint 1 coat		325	.025		.35	.89		1.24	1.72
3960	Prime & paint 2 coats		270	.030		.35	1.08		1.43	1.98
4500	Louvers, one side, primer, brushwork		524	.015		.06	.56		.62	.90

09 91 23.52 Miscellaneous, Interior		Crew	Daily Output	Labor-Hours	Unit	Material	Labor	Equipment	Total	Total Incl O&P
						2010 Bare Costs				
4520	Paint one coat, brushwork	1 Pord	520	.015	S.F.	.07	.56		.63	.91
4530	Spray		1140	.007		.08	.26		.34	.47
4540	Paint two coats, brushwork		325	.025		.14	.89		1.03	1.48
4550	Spray		650	.012		.15	.45		.60	.84
4560	Paint three coats, brushwork		270	.030		.21	1.08		1.29	1.83
4570	Spray	↓	500	.016	↓	.23	.58		.81	1.12
5000	Pipe, 1" - 4" diameter, primer or sealer coat, oil base, brushwork	2 Pord	1250	.013	L.F.	.06	.47		.53	.76
5100	Spray		2165	.007		.06	.27		.33	.47
5200	Paint 1 coat, brushwork		1250	.013		.08	.47		.55	.78
5300	Spray		2165	.007		.07	.27		.34	.48
5350	Paint 2 coats, brushwork		775	.021		.15	.75		.90	1.28
5400	Spray		1240	.013		.16	.47		.63	.88
5420	Paint 3 coats, brushwork		775	.021		.22	.75		.97	1.36
5450	5" - 8" diameter, primer or sealer coat, brushwork		620	.026		.13	.94		1.07	1.54
5500	Spray		1085	.015		.21	.54		.75	1.03
5550	Paint 1 coat, brushwork		620	.026		.22	.94		1.16	1.64
5600	Spray		1085	.015		.25	.54		.79	1.07
5650	Paint 2 coats, brushwork		385	.042		.29	1.51		1.80	2.57
5700	Spray		620	.026		.32	.94		1.26	1.76
5720	Paint 3 coats, brushwork		385	.042		.43	1.51		1.94	2.73
5750	9" - 12" diameter, primer or sealer coat, brushwork		415	.039		.19	1.40		1.59	2.30
5800	Spray		725	.022		.26	.80		1.06	1.48
5850	Paint 1 coat, brushwork		415	.039		.22	1.40		1.62	2.34
6000	Spray		725	.022		.25	.80		1.05	1.46
6200	Paint 2 coats, brushwork		260	.062		.44	2.24		2.68	3.81
6250	Spray		415	.039		.48	1.40		1.88	2.62
6270	Paint 3 coats, brushwork		260	.062		.65	2.24		2.89	4.04
6300	13" - 16" diameter, primer or sealer coat, brushwork		310	.052		.25	1.88		2.13	3.07
6350	Spray		540	.030		.28	1.08		1.36	1.91
6400	Paint 1 coat, brushwork		310	.052		.30	1.88		2.18	3.12
6450	Spray		540	.030		.33	1.08		1.41	1.97
6500	Paint 2 coats, brushwork		195	.082		.58	2.98		3.56	5.10
6550	Spray	↓	310	.052	↓	.65	1.88		2.53	3.50
6600	Radiators, per side, primer, brushwork	1 Pord	520	.015	S.F.	.06	.56		.62	.90
6620	Paint, one coat		520	.015		.06	.56		.62	.89
6640	Two coats		340	.024		.14	.86		1	1.42
6660	Three coats	↓	283	.028	↓	.21	1.03		1.24	1.76

Division Notes

		CREW	DAILY OUTPUT	LABOR-HOURS	UNIT	2010 BARE COSTS MAT.	LABOR	EQUIP.	TOTAL	TOTAL INCL O&P

Estimating Tips

General

- The items in this division are usually priced per square foot or each.
- Many items in Division 10 require some type of support system or special anchors that are not usually furnished with the item. The required anchors must be added to the estimate in the appropriate division.
- Some items in Division 10, such as lockers, may require assembly before installation. Verify the amount of assembly required. Assembly can often exceed installation time.

10 20 00 Interior Specialties

- Support angles and blocking are not included in the installation of toilet compartments, shower/dressing compartments, or cubicles. Appropriate line items from Divisions 5 or 6 may need to be added to support the installations.
- Toilet partitions are priced by the stall. A stall consists of a side wall, pilaster, and door with hardware. Toilet tissue holders and grab bars are extra.
- The required acoustical rating of a folding partition can have a significant impact on costs. Verify the sound transmission coefficient rating of the panel priced to the specification requirements.
- Grab bar installation does not include supplemental blocking or backing to support the required load. When grab bars are installed at an existing facility, provisions must be made to attach the grab bars to solid structure.

Reference Numbers

Reference numbers are shown in shaded boxes at the beginning of some major classifications. These numbers refer to related items in the Reference Section. The reference information may be an estimating procedure, an alternate pricing method, or technical information.

Note: Not all subdivisions listed here necessarily appear in this publication.

10 31 Manufactured Fireplaces

10 31 13 – Manufactured Fireplace Chimneys

10 31 13.10 Fireplace Chimneys		Crew	Daily Output	Labor-Hours	Unit	Material	2010 Bare Costs Labor	Equipment	Total	Total Incl O&P
0010	**FIREPLACE CHIMNEYS**									
0500	Chimney dbl. wall, all stainless, over 8'-6", 7" diam., add to fireplace	1 Carp	33	.242	V.L.F.	71.50	10.05		81.55	94
0600	10" diameter, add to fireplace		32	.250		109	10.40		119.40	136
0700	12" diameter, add to fireplace		31	.258		136	10.70		146.70	166
0800	14" diameter, add to fireplace		30	.267		183	11.10		194.10	218
1000	Simulated brick chimney top, 4' high, 16" x 16"		10	.800	Ea.	258	33		291	335
1100	24" x 24"		7	1.143	"	480	47.50		527.50	605

10 31 23 – Prefabricated Fireplaces

10 31 23.10 Fireplace, Prefabricated		Crew	Daily Output	Labor-Hours	Unit	Material	Labor	Equipment	Total	Total Incl O&P
0010	**FIREPLACE, PREFABRICATED**, free standing or wall hung									
0100	With hood & screen, minimum	1 Carp	1.30	6.154	Ea.	1,200	256		1,456	1,725
0150	Average		1	8		1,700	330		2,030	2,350
0200	Maximum		.90	8.889		3,375	370		3,745	4,275
1500	Simulated logs, gas fired, 40,000 BTU, 2' long, minimum		7	1.143	Set	450	47.50		497.50	570
1600	Maximum		6	1.333		980	55.50		1,035.50	1,150
1700	Electric, 1,500 BTU, 1'-6" long, minimum		7	1.143		175	47.50		222.50	266
1800	11,500 BTU, maximum		6	1.333		395	55.50		450.50	515

10 35 Stoves

10 35 13 – Heating Stoves

10 35 13.10 Woodburning Stoves		Crew	Daily Output	Labor-Hours	Unit	Material	Labor	Equipment	Total	Total Incl O&P
0010	**WOODBURNING STOVES**									
0015	Cast iron, minimum	2 Carp	1.30	12.308	Ea.	1,175	510		1,685	2,100
0020	Average		1	16		1,600	665		2,265	2,775
0030	Maximum		.80	20		2,850	830		3,680	4,425
0050	For gas log lighter, add					42			42	46.50

Estimating Tips

General

- The items in this division are usually priced per square foot or each. Many of these items are purchased by the owner for installation by the contractor. Check the specifications for responsibilities and include time for receiving, storage, installation, and mechanical and electrical hookups in the appropriate divisions.
- Many items in Division 11 require some type of support system that is not usually furnished with the item. Examples of these systems include blocking for the attachment of casework and support angles for ceiling-hung projection screens. The required blocking or supports must be added to the estimate in the appropriate division.
- Some items in Division 11 may require assembly or electrical hookups. Verify the amount of assembly required or the need for a hard electrical connection and add the appropriate costs.

Reference Numbers

Reference numbers are shown in shaded boxes at the beginning of some major classifications. These numbers refer to related items in the Reference Section. The reference information may be an estimating procedure, an alternate pricing method, or technical information.

Note: Not all subdivisions listed here necessarily appear in this publication.

11 11 Vehicle Service Equipment

11 11 13 – Compressed-Air Vehicle Service Equipment

11 11 13.10 Compressed Air Equipment		Crew	Daily Output	Labor-Hours	Unit	Material	2010 Bare Costs Labor	Equipment	Total	Total Incl O&P
0010	**COMPRESSED AIR EQUIPMENT**									
0030	Compressors, electric, 1-1/2 H.P., standard controls	L-4	1.50	16	Ea.	400	625		1,025	1,400
0550	Dual controls		1.50	16		715	625		1,340	1,750
0600	5 H.P., 115/230 volt, standard controls		1	24		2,050	935		2,985	3,675
0650	Dual controls	↓	1	24	↓	2,175	935		3,110	3,825

11 11 19 – Vehicle Lubrication Equipment

11 11 19.10 Lubrication Equipment		Crew	Daily Output	Labor-Hours	Unit	Material	Labor	Equipment	Total	Total Incl O&P
0010	**LUBRICATION EQUIPMENT**									
3000	Lube equipment, 3 reel type, with pumps, not including piping	L-4	.50	48	Set	7,900	1,875		9,775	11,600
3700	Pump lubrication, pneumatic, not incl. air compressor									
3710	Oil/gear lube	Q-1	9.60	1.667	Ea.	910	78		988	1,125
3720	Grease	"	9.60	1.667	"	945	78		1,023	1,175

11 11 33 – Vehicle Spray Painting Equipment

11 11 33.10 Spray Painting Equipment		Crew	Daily Output	Labor-Hours	Unit	Material	Labor	Equipment	Total	Total Incl O&P
0010	**SPRAY PAINTING EQUIPMENT**									
4000	Spray painting booth, 26' long, complete	L-4	.40	60	Ea.	15,900	2,325		18,225	21,000

11 23 Commercial Laundry and Dry Cleaning Equipment

11 23 16 – Drying and Conditioning Equipment

11 23 16.13 Dryers		Crew	Daily Output	Labor-Hours	Unit	Material	Labor	Equipment	Total	Total Incl O&P
0010	**DRYERS**, Not including rough-in									
1500	Industrial, 30 lb. capacity	1 Plum	2	4	Ea.	2,975	208		3,183	3,575
1600	50 lb. capacity	"	1.70	4.706	"	3,200	245		3,445	3,900

11 23 26 – Commercial Washers and Extractors

11 23 26.13 Washers and Extractors		Crew	Daily Output	Labor-Hours	Unit	Material	Labor	Equipment	Total	Total Incl O&P
0010	**WASHERS AND EXTRACTORS**, not including rough-in									
6000	Combination washer/extractor, 20 lb. capacity	L-6	1.50	8	Ea.	5,350	410		5,760	6,500
6100	30 lb. capacity		.80	15		8,500	765		9,265	10,500
6200	50 lb. capacity		.68	17.647		9,925	900		10,825	12,300
6300	75 lb. capacity		.30	40		18,600	2,050		20,650	23,600
6350	125 lb. capacity	↓	.16	75	↓	24,900	3,825		28,725	33,100

11 23 33 – Coin-Operated Laundry Equipment

11 23 33.13 Coin Operated Washers and Dryers		Crew	Daily Output	Labor-Hours	Unit	Material	Labor	Equipment	Total	Total Incl O&P
0010	**COIN OPERATED WASHERS AND DRYERS**									
0990	Dryer, gas fired									
1000	Commercial, 30 lb. capacity, coin operated, single	1 Plum	3	2.667	Ea.	3,100	139		3,239	3,600
1100	Double stacked	"	2	4	"	6,275	208		6,483	7,200
5290	Clothes washer									
5300	Commercial, coin operated, average	1 Plum	3	2.667	Ea.	1,125	139		1,264	1,450

11 24 Maintenance Equipment

11 24 19 – Vacuum Cleaning Systems

11 24 19.10 Vacuum Cleaning		Crew	Daily Output	Labor-Hours	Unit	Material	2010 Bare Costs Labor	Equipment	Total	Total Incl O&P
0010	**VACUUM CLEANING**									
0020	Central, 3 inlet, residential	1 Skwk	.90	8.889	Total	1,025	380		1,405	1,725
0200	Commercial		.70	11.429		1,225	485		1,710	2,075
0400	5 inlet system, residential		.50	16		1,450	680		2,130	2,650
0600	7 inlet system, commercial		.40	20		1,625	850		2,475	3,075
0800	9 inlet system, residential	↓	.30	26.667		3,600	1,125		4,725	5,700
4010	Rule of thumb: First 1200 S.F., installed				↓				1,325	1,450
4020	For each additional S.F., add				S.F.					.24

11 26 Unit Kitchens

11 26 13 – Metal Unit Kitchens

11 26 13.10 Commercial Unit Kitchens		Crew	Daily Output	Labor-Hours	Unit	Material	Labor	Equipment	Total	Total Incl O&P
0010	**COMMERCIAL UNIT KITCHENS**									
1500	Combination range, refrigerator and sink, 30″ wide, minimum	L-1	2	8	Ea.	925	405		1,330	1,625
1550	Maximum		1	16		3,675	810		4,485	5,250
1570	60″ wide, average		1.40	11.429		3,225	575		3,800	4,375
1590	72″ wide, average		1.20	13.333		4,200	675		4,875	5,625
1600	Office model, 48″ wide		2	8		2,650	405		3,055	3,500
1620	Refrigerator and sink only	↓	2.40	6.667	↓	2,725	335		3,060	3,475
1640	Combination range, refrigerator, sink, microwave									
1660	Oven and ice maker	L-1	.80	20	Ea.	4,450	1,000		5,450	6,400

11 27 Photographic Processing Equipment

11 27 13 – Darkroom Processing Equipment

11 27 13.10 Darkroom Equipment		Crew	Daily Output	Labor-Hours	Unit	Material	Labor	Equipment	Total	Total Incl O&P
0010	**DARKROOM EQUIPMENT**									
0020	Developing sink, 5″ deep, 24″ x 48″	Q-1	2	8	Ea.	5,625	375		6,000	6,725
0050	48″ x 52″		1.70	9.412		5,675	440		6,115	6,875
0200	10″ deep, 24″ x 48″		1.70	9.412		7,775	440		8,215	9,200
0250	24″ x 108″		1.50	10.667		10,200	500		10,700	12,100
3500	Washers, round, minimum sheet 11″ x 14″		2	8		3,400	375		3,775	4,300
3550	Maximum sheet 20″ x 24″		1	16		3,475	750		4,225	4,950
3800	Square, minimum sheet 20″ x 24″		1	16		3,125	750		3,875	4,550
3900	Maximum sheet 50″ x 56″	↓	.80	20	↓	4,850	935		5,785	6,725
4500	Combination tank sink, tray sink, washers, with									
4510	Dry side tables, average	Q-1	.45	35.556	Ea.	10,600	1,675		12,275	14,200

11 31 Residential Appliances

11 31 13 – Residential Kitchen Appliances

11 31 13.13 Cooking Equipment		Crew	Daily Output	Labor-Hours	Unit	Material	Labor	Equipment	Total	Total Incl O&P
0010	**COOKING EQUIPMENT**									
0020	Cooking range, 30″ free standing, 1 oven, minimum	2 Clab	10	1.600	Ea.	345	53		398	460
0050	Maximum		4	4		1,900	132		2,032	2,300
0150	2 oven, minimum		10	1.600		1,400	53		1,453	1,600
0200	Maximum	↓	10	1.600	↓	1,600	53		1,653	1,850

11 31 Residential Appliances

11 31 13 – Residential Kitchen Appliances

11 31 13.23 Refrigeration Equipment		Crew	Daily Output	Labor-Hours	Unit	Material	2010 Bare Costs Labor	Equipment	Total	Total Incl O&P
0010	**REFRIGERATION EQUIPMENT**									
5200	Icemaker, automatic, 20 lb. per day	1 Plum	7	1.143	Ea.	880	59.50		939.50	1,050
5350	51 lb. per day	"	2	4	"	1,250	208		1,458	1,675

11 31 13.33 Kitchen Cleaning Equipment

		Crew	Daily Output	Labor-Hours	Unit	Material	Labor	Equipment	Total	Total Incl O&P
0010	**KITCHEN CLEANING EQUIPMENT**									
2750	Dishwasher, built-in, 2 cycles, minimum	L-1	4	4	Ea.	212	202		414	535
2800	Maximum		2	8		320	405		725	955
2950	4 or more cycles, minimum		4	4		279	202		481	605
2960	Average		4	4		370	202		572	710
3000	Maximum		2	8		1,050	405		1,455	1,750

11 31 13.43 Waste Disposal Equipment

		Crew	Daily Output	Labor-Hours	Unit	Material	Labor	Equipment	Total	Total Incl O&P
0010	**WASTE DISPOSAL EQUIPMENT**									
3300	Garbage disposal, sink type, minimum	L-1	10	1.600	Ea.	72.50	81		153.50	201
3350	Maximum	"	10	1.600	"	191	81		272	330

11 31 13.53 Kitchen Ventilation Equipment

		Crew	Daily Output	Labor-Hours	Unit	Material	Labor	Equipment	Total	Total Incl O&P
0010	**KITCHEN VENTILATION EQUIPMENT**									
4150	Hood for range, 2 speed, vented, 30" wide, minimum	L-3	5	3.200	Ea.	53.50	145		198.50	279
4200	Maximum		3	5.333		820	242		1,062	1,275
4300	42" wide, minimum		5	3.200		234	145		379	475
4330	Custom		5	3.200		1,500	145		1,645	1,900
4350	Maximum		3	5.333		1,850	242		2,092	2,400
4500	For ventless hood, 2 speed, add					17.05			17.05	18.75
4650	For vented 1 speed, deduct from maximum					45.50			45.50	50

11 31 23 – Residential Laundry Appliances

11 31 23.13 Washers

		Crew	Daily Output	Labor-Hours	Unit	Material	Labor	Equipment	Total	Total Incl O&P
0010	**WASHERS**									
5000	Residential, 4 cycle, average	1 Plum	3	2.667	Ea.	800	139		939	1,100
6650	Washing machine, automatic, minimum		3	2.667		355	139		494	600
6700	Maximum		1	8		1,000	415		1,415	1,750

11 31 23.23 Dryers

		Crew	Daily Output	Labor-Hours	Unit	Material	Labor	Equipment	Total	Total Incl O&P
0010	**DRYERS**									
0500	Gas fired residential, 16 lb. capacity, average	1 Plum	3	2.667	Ea.	625	139		764	900
7450	Vent kits for dryers	1 Carp	10	.800	"	25.50	33		58.50	79

11 31 33 – Miscellaneous Residential Appliances

11 31 33.13 Sump Pumps

		Crew	Daily Output	Labor-Hours	Unit	Material	Labor	Equipment	Total	Total Incl O&P
0010	**SUMP PUMPS**									
6400	Cellar drainer, pedestal, 1/3 H.P., molded PVC base	1 Plum	3	2.667	Ea.	119	139		258	340
6450	Solid brass	"	2	4	"	263	208		471	600
6460	Sump pump, see also Div. 22 14 29.16									

11 31 33.23 Water Heaters

		Crew	Daily Output	Labor-Hours	Unit	Material	Labor	Equipment	Total	Total Incl O&P
0010	**WATER HEATERS**									
6900	Electric, glass lined, 30 gallon, minimum	L-1	5	3.200	Ea.	520	162		682	810
6950	Maximum		3	5.333		720	269		989	1,200
7100	80 gallon, minimum		2	8		935	405		1,340	1,625
7150	Maximum		1	16		1,300	810		2,110	2,625
7180	Gas, glass lined, 30 gallon, minimum	2 Plum	5	3.200		805	167		972	1,125
7220	Maximum		3	5.333		1,125	278		1,403	1,650
7260	50 gallon, minimum		2.50	6.400		845	335		1,180	1,425
7300	Maximum		1.50	10.667		1,175	555		1,730	2,100

11 31 Residential Appliances

11 31 33 – Miscellaneous Residential Appliances

11 31 33.43 Air Quality		Crew	Daily Output	Labor-Hours	Unit	Material	2010 Bare Costs Labor	Equipment	Total	Total Incl O&P
0010	AIR QUALITY									
2450	Dehumidifier, portable, automatic, 15 pint				Ea.	154			154	170
2550	40 pint					171			171	188
3550	Heater, electric, built-in, 1250 watt, ceiling type, minimum	1 Elec	4	2		96	98		194	251
3600	Maximum		3	2.667		156	131		287	365
3700	Wall type, minimum		4	2		154	98		252	315
3750	Maximum		3	2.667		166	131		297	375
3900	1500 watt wall type, with blower		4	2		154	98		252	315
3950	3000 watt		3	2.667		315	131		446	540
4850	Humidifier, portable, 8 gallons per day					165			165	182
5000	15 gallons per day					198			198	218

11 41 Food Storage Equipment

11 41 13 – Refrigerated Food Storage Cases

11 41 13.10 Refrigerated Food Cases

		Crew	Daily Output	Labor-Hours	Unit	Material	Labor	Equipment	Total	Total Incl O&P
0010	REFRIGERATED FOOD CASES									
0030	Dairy, multi-deck, 12' long	Q-5	3	5.333	Ea.	10,400	249		10,649	11,900
0100	For rear sliding doors, add					1,500			1,500	1,650
0200	Delicatessen case, service deli, 12' long, single deck	Q-5	3.90	4.103		7,025	192		7,217	8,000
0300	Multi-deck, 18 S.F. shelf display		3	5.333		6,450	249		6,699	7,450
0400	Freezer, self-contained, chest-type, 30 C.F.		3.90	4.103		7,550	192		7,742	8,575
0500	Glass door, upright, 78 C.F.		3.30	4.848		9,750	226		9,976	11,000
0600	Frozen food, chest type, 12' long		3.30	4.848		7,050	226		7,276	8,100
0700	Glass door, reach-in, 5 door		3	5.333		13,500	249		13,749	15,200
0800	Island case, 12' long, single deck		3.30	4.848		7,975	226		8,201	9,125
0900	Multi-deck		3	5.333		16,900	249		17,149	19,000
1000	Meat case, 12' long, single deck		3.30	4.848		5,800	226		6,026	6,725
1050	Multi-deck		3.10	5.161		9,975	241		10,216	11,400
1100	Produce, 12' long, single deck		3.30	4.848		7,675	226		7,901	8,775
1200	Multi-deck		3.10	5.161		8,375	241		8,616	9,550

11 41 13.20 Refrigerated Food Storage Equipment

		Crew	Daily Output	Labor-Hours	Unit	Material	Labor	Equipment	Total	Total Incl O&P
0010	REFRIGERATED FOOD STORAGE EQUIPMENT									
2350	Cooler, reach-in, beverage, 6' long	Q-1	6	2.667	Ea.	4,525	125		4,650	5,150
4300	Freezers, reach-in, 44 C.F.		4	4		2,700	187		2,887	3,250
4500	68 C.F.		3	5.333		3,650	250		3,900	4,375

11 44 Food Cooking Equipment

11 44 13 – Commercial Ranges

11 44 13.10 Cooking Equipment

		Crew	Daily Output	Labor-Hours	Unit	Material	Labor	Equipment	Total	Total Incl O&P
0010	COOKING EQUIPMENT									
0020	Bake oven, gas, one section	Q-1	8	2	Ea.	4,650	93.50		4,743.50	5,275
0300	Two sections		7	2.286		10,500	107		10,607	11,700
0600	Three sections		6	2.667		13,500	125		13,625	15,000
0900	Electric convection, single deck	L-7	4	7		6,425	281		6,706	7,475
6350	Kettle, w/steam jacket, tilting, w/positive lock, SS, 20 gallons		7	4		7,125	161		7,286	8,075
6600	60 gallons		6	4.667		9,225	188		9,413	10,500

11 47 Ice Machines

11 47 10 – Commercial Ice Machines

	11 47 10.10 Commercial Ice Equipment	Crew	Daily Output	Labor-Hours	Unit	Material	2010 Bare Costs Labor	Equipment	Total	Total Incl O&P
0010	**COMMERCIAL ICE EQUIPMENT**									
5800	Ice cube maker, 50 pounds per day	Q-1	6	2.667	Ea.	1,550	125		1,675	1,875
6050	500 pounds per day	"	4	4	"	3,125	187		3,312	3,700

11 48 Cleaning and Disposal Equipment

11 48 13 – Commercial Dishwashers

11 48 13.10 Dishwashers

		Crew	Daily Output	Labor-Hours	Unit	Material	Labor	Equipment	Total	Total Incl O&P
0010	**DISHWASHERS**									
2700	Dishwasher, commercial, rack type									
2720	10 to 12 racks per hour	Q-1	3.20	5	Ea.	4,300	234		4,534	5,075
2730	Energy star rated, 35-40 racks/hour G		1.30	12.308		4,650	575		5,225	5,975
2740	50-60 racks/hour G	↓	1.30	12.308		8,525	575		9,100	10,300
2800	Automatic, 190 to 230 racks per hour	L-6	.35	34.286		13,100	1,750		14,850	17,000
2820	235 to 275 racks per hour		.25	48		30,300	2,450		32,750	37,000
2840	8,750 to 12,500 dishes per hour	↓	.10	120	↓	49,500	6,125		55,625	63,500

11 53 Laboratory Equipment

11 53 13 – Laboratory Fume Hoods

11 53 13.13 Recirculating Laboratory Fume Hoods

		Crew	Daily Output	Labor-Hours	Unit	Material	Labor	Equipment	Total	Total Incl O&P
0010	**RECIRCULATING LABORATORY FUME HOODS**									
0600	Fume hood, with countertop & base, not including HVAC									
0610	Simple, minimum	2 Carp	5.40	2.963	L.F.	605	123		728	855
0620	Complex, including fixtures		2.40	6.667		1,800	277		2,077	2,400
0630	Special, maximum	↓	1.70	9.412	↓	1,825	390		2,215	2,600
0670	Service fixtures, average				Ea.	273			273	300
0680	For sink assembly with hot and cold water, add	1 Plum	1.40	5.714	"	705	297		1,002	1,225

11 53 13.23 Exhaust Hoods

		Crew	Daily Output	Labor-Hours	Unit	Material	Labor	Equipment	Total	Total Incl O&P
0010	**EXHAUST HOODS**									
0650	Ductwork, minimum	2 Shee	1	16	Hood	3,575	785		4,360	5,125
0660	Maximum	"	.50	32	"	5,500	1,575		7,075	8,425

11 53 19 – Laboratory Sterilizers

11 53 19.13 Sterilizers

		Crew	Daily Output	Labor-Hours	Unit	Material	Labor	Equipment	Total	Total Incl O&P
0010	**STERILIZERS**									
0700	Glassware washer, undercounter, minimum	L-1	1.80	8.889	Ea.	5,650	450		6,100	6,900
0710	Maximum	"	1	16		11,800	810		12,610	14,200
1850	Utensil washer-sanitizer	1 Plum	2	4	↓	10,900	208		11,108	12,300

11 53 33 – Emergency Safety Appliances

11 53 33.13 Emergency Equipment

		Crew	Daily Output	Labor-Hours	Unit	Material	Labor	Equipment	Total	Total Incl O&P
0010	**EMERGENCY EQUIPMENT**									
1400	Safety equipment, eye wash, hand held				Ea.	405			405	445
1450	Deluge shower				"	740			740	815

11 53 43 – Service Fittings and Accessories

11 53 43.13 Fittings

		Crew	Daily Output	Labor-Hours	Unit	Material	Labor	Equipment	Total	Total Incl O&P
0010	**FITTINGS**									
1600	Sink, one piece plastic, flask wash, hose, free standing	1 Plum	1.60	5	Ea.	1,800	260		2,060	2,375
1610	Epoxy resin sink, 25" x 16" x 10"	"	2	4	"	198	208		406	530
8000	Alternate pricing method: as percent of lab furniture									

11 53 Laboratory Equipment

11 53 43 – Service Fittings and Accessories

11 53 43.13 Fittings		Crew	Daily Output	Labor-Hours	Unit	Material	Labor	Equipment	Total	Total Incl O&P
						2010 Bare Costs				
8050	Installation, not incl. plumbing & duct work				% Furn.				20%	22%
8100	Plumbing, final connections, simple system								9.09%	10%
8110	Moderately complex system								13.64%	15%
8120	Complex system								18.18%	20%
8150	Electrical, simple system								9.09%	10%
8160	Moderately complex system								18.18%	20%
8170	Complex system								31.80%	35%

11 71 Medical Sterilizing Equipment

11 71 10 – Medical Sterilizers & Distillers

11 71 10.10 Sterilizers and Distillers

		Crew	Daily Output	Labor-Hours	Unit	Material	Labor	Equipment	Total	Total Incl O&P
0010	**STERILIZERS AND DISTILLERS**									
0700	Distiller, water, steam heated, 50 gal. capacity	1 Plum	1.40	5.714	Ea.	18,500	297		18,797	20,700
5600	Sterilizers, floor loading, 26" x 62" x 42", single door, steam					158,500			158,500	174,500
5650	Double door, steam					204,000			204,000	224,000
5800	General purpose, 20" x 20" x 38", single door					15,300			15,300	16,800
6000	Portable, counter top, steam, minimum					3,800			3,800	4,175
6020	Maximum					5,950			5,950	6,550
6050	Portable, counter top, gas, 17" x 15" x 32-1/2"					39,300			39,300	43,200
6150	Manual washer/sterilizer, 16" x 16" x 26"	1 Plum	2	4		54,000	208		54,208	59,500
6200	Steam generators, electric 10 kW to 180 kW, freestanding									
6250	Minimum	1 Elec	3	2.667	Ea.	8,000	131		8,131	9,000
6300	Maximum	"	.70	11.429		28,800	560		29,360	32,500
8200	Bed pan washer-sanitizer	1 Plum	2	4		7,125	208		7,333	8,150

11 73 Patient Care Equipment

11 73 10 – Patient Treatment Equipment

11 73 10.10 Treatment Equipment

		Crew	Daily Output	Labor-Hours	Unit	Material	Labor	Equipment	Total	Total Incl O&P
0010	**TREATMENT EQUIPMENT**									
1800	Heat therapy unit, humidified, 26" x 78" x 28"				Ea.	3,400			3,400	3,725
8400	Whirlpool bath, mobile, sst, 18" x 24" x 60"					4,500			4,500	4,950
8450	Fixed, incl. mixing valves	1 Plum	2	4		9,200	208		9,408	10,400

11 74 Dental Equipment

11 74 10 – Dental Office Equipment

11 74 10.10 Diagnostic and Treatment Equipment

		Crew	Daily Output	Labor-Hours	Unit	Material	Labor	Equipment	Total	Total Incl O&P
0010	**DIAGNOSTIC AND TREATMENT EQUIPMENT**									
0020	Central suction system, minimum	1 Plum	1.20	6.667	Ea.	1,550	345		1,895	2,225
0100	Maximum	"	.90	8.889		4,250	465		4,715	5,375
0600	Chair, electric or hydraulic, minimum	1 Skwk	.50	16		2,550	680		3,230	3,875
0700	Maximum		.25	32		14,400	1,375		15,775	18,000
2000	Light, ceiling mounted, minimum		8	1		1,225	42.50		1,267.50	1,425
2100	Maximum		8	1		1,625	42.50		1,667.50	1,875
2200	Unit light, minimum	2 Skwk	5.33	3.002		600	128		728	855
2210	Maximum		5.33	3.002		1,625	128		1,753	2,000
2220	Track light, minimum		3.20	5		1,525	213		1,738	2,000
2230	Maximum		3.20	5		3,725	213		3,938	4,425

11 74 Dental Equipment

11 74 10 – Dental Office Equipment

11 74 10.10 Diagnostic and Treatment Equipment		Crew	Daily Output	Labor-Hours	Unit	2010 Bare Costs Material	Labor	Equipment	Total	Total Incl O&P
2300	Sterilizers, steam portable, minimum				Ea.	1,850			1,850	2,050
2350	Maximum					10,400			10,400	11,400
2600	Steam, institutional					4,250			4,250	4,675
2650	Dry heat, electric, portable, 3 trays				↓	1,225			1,225	1,325

11 78 Mortuary Equipment

11 78 13 – Mortuary Refrigerators

11 78 13.10 Mortuary and Autopsy Equipment

		Crew	Daily Output	Labor-Hours	Unit	Material	Labor	Equipment	Total	Total Incl O&P
0010	**MORTUARY AND AUTOPSY EQUIPMENT**									
0015	Autopsy table, standard	1 Plum	1	8	Ea.	8,850	415		9,265	10,400
0020	Deluxe	"	.60	13.333		17,700	695		18,395	20,500
3200	Mortuary refrigerator, end operated, 2 capacity					12,800			12,800	14,100
3300	6 capacity				↓	23,900			23,900	26,300

11 78 16 – Crematorium Equipment

11 78 16.10 Crematory

		Crew	Daily Output	Labor-Hours	Unit	Material	Labor	Equipment	Total	Total Incl O&P
0010	**CREMATORY**									
1500	Crematory, not including building, 1 place	Q-3	.20	160	Ea.	61,500	7,925		69,425	79,500
1750	2 place	"	.10	320	"	87,500	15,900		103,400	120,500

11 82 Solid Waste Handling Equipment

11 82 19 – Packaged Incinerators

11 82 19.10 Packaged Gas Fired Incinerators

		Crew	Daily Output	Labor-Hours	Unit	Material	Labor	Equipment	Total	Total Incl O&P
0010	**PACKAGED GAS FIRED INCINERATORS**									
4400	Incinerator, gas, not incl. chimney, elec. or pipe, 50#/hr., minimum	Q-3	.80	40	Ea.	11,200	1,975		13,175	15,300
4420	Maximum		.70	45.714		36,100	2,275		38,375	43,100
4440	200 lb. per hr., minimum (batch type)		.60	53.333		62,000	2,650		64,650	72,000
4460	Maximum (with feeder)		.50	64		87,500	3,175		90,675	101,500
4480	400 lb. per hr., minimum (batch type)		.30	106		68,500	5,300		73,800	83,500
4500	Maximum (with feeder)		.25	128		92,500	6,350		98,850	111,500
4520	800 lb. per hr., with feeder, minimum		.20	160		112,000	7,925		119,925	135,000
4540	Maximum		.17	188		154,500	9,325		163,825	184,000
4560	1,200 lb. per hr., with feeder, minimum		.15	213		150,500	10,600		161,100	181,500
4580	Maximum		.11	290		179,500	14,400		193,900	219,000
4600	2,000 lb. per hr., with feeder, minimum		.10	320		368,000	15,900		383,900	429,000
4620	Maximum		.05	640		504,000	31,700		535,700	602,000
4700	For heat recovery system, add, minimum		.25	128		71,000	6,350		77,350	87,500
4710	Add, maximum		.11	290		227,000	14,400		241,400	271,500
4720	For automatic ash conveyer, add		.50	64	↓	29,800	3,175		32,975	37,600
4750	Large municipal incinerators, incl. stack, minimum		.25	128	Ton/day	18,100	6,350		24,450	29,500
4850	Maximum	↓	.10	320	"	48,200	15,900		64,100	77,000

11 82 26 – Waste Compactors and Destructors

11 82 26.10 Compactors

		Crew	Daily Output	Labor-Hours	Unit	Material	Labor	Equipment	Total	Total Incl O&P
0010	**COMPACTORS**									
0020	Compactors, 115 volt, 250#/hr., chute fed	L-4	1	24	Ea.	11,200	935		12,135	13,800
0100	Hand fed		2.40	10		8,225	390		8,615	9,625
0300	Multi-bag, 230 volt, 600#/hr, chute fed		1	24		10,200	935		11,135	12,600
0400	Hand fed		1	24		8,750	935		9,685	11,100
0500	Containerized, hand fed, 2 to 6 C.Y. containers, 250#/hr.	↓	1	24	↓	10,800	935		11,735	13,300

11 82 26 – Waste Compactors and Destructors

11 82 26.10 Compactors		Crew	Daily Output	Labor-Hours	Unit	2010 Bare Costs Material	Labor	Equipment	Total	Total Incl O&P
0550	For chute fed, add per floor	L-4	1	24	Ea.	1,250	935		2,185	2,800
1000	Heavy duty industrial compactor, 0.5 C.Y. capacity		1	24		7,300	935		8,235	9,475
1050	1.0 C.Y. capacity		1	24		11,100	935		12,035	13,700
1100	3.0 C.Y. capacity		.50	48		15,200	1,875		17,075	19,600
1150	5.0 C.Y. capacity		.50	48		18,800	1,875		20,675	23,600
1200	Combination shredder/compactor (5,000 lbs./hr.)		.50	48		37,200	1,875		39,075	43,800
1400	For handling hazardous waste materials, 55 gallon drum packer, std.					17,300			17,300	19,000
1410	55 gallon drum packer w/HEPA filter					21,600			21,600	23,700
1420	55 gallon drum packer w/charcoal & HEPA filter					28,800			28,800	31,700
1430	All of the above made explosion proof, add					13,100			13,100	14,500
5800	Shredder, industrial, minimum					21,200			21,200	23,400
5850	Maximum					114,000			114,000	125,500
5900	Baler, industrial, minimum					8,500			8,500	9,350
5950	Maximum					497,000			497,000	547,000

Division Notes

		CREW	DAILY OUTPUT	LABOR-HOURS	UNIT	2010 BARE COSTS MAT.	LABOR	EQUIP.	TOTAL	TOTAL INCL O&P

Estimating Tips

General

- The items and systems in this division are usually estimated, purchased, supplied, and installed as a unit by one or more subcontractors. The estimator must ensure that all parties are operating from the same set of specifications and assumptions, and that all necessary items are estimated and will be provided. Many times the complex items and systems are covered, but the more common ones, such as excavation or a crane, are overlooked for the very reason that everyone assumes nobody could miss them. The estimator should be the central focus and be able to ensure that all systems are complete.
- Another area where problems can develop in this division is at the interface between systems. The estimator must ensure, for instance, that anchor bolts, nuts, and washers are estimated and included for the air-supported structures and pre-engineered buildings to be bolted to their foundations. Utility supply is a common area where essential items or pieces of equipment can be missed or overlooked due to the fact that each subcontractor may feel it is another's responsibility. The estimator should also be aware of certain items which may be supplied as part of a package but installed by others, and ensure that the installing contractor's estimate includes the cost of installation. Conversely, the estimator must also ensure that items are not costed by two different subcontractors, resulting in an inflated overall estimate.

13 30 00 Special Structures

- The foundations and floor slab, as well as rough mechanical and electrical, should be estimated, as this work is required for the assembly and erection of the structure. Generally, as noted in the book, the pre-engineered building comes as a shell. Pricing is based on the size and structural design parameters stated in the reference section. Additional features, such as windows and doors, must also be included by the estimator. Here again, the estimator must have a clear understanding of the scope of each portion of the work and all the necessary interfaces.

Reference Numbers

Reference numbers are shown in shaded boxes at the beginning of some major classifications. These numbers refer to related items in the Reference Section. The reference information may be an estimating procedure, an alternate pricing method, or technical information.

Note: Not all subdivisions listed here necessarily appear in this publication.

13 11 Swimming Pools

13 11 13 – Below-Grade Swimming Pools

13 11 13.50 Swimming Pools		Crew	Daily Output	Labor-Hours	Unit	Material	2010 Bare Costs Labor	Equipment	Total	Total Incl O&P
0010	**SWIMMING POOLS** Residential in-ground, vinyl lined, concrete									
0020	Sides including equipment, sand bottom	B-52	300	.187	SF Surf	13.90	7.15	1.80	22.85	28
0100	Metal or polystyrene sides	B-14	410	.117		11.60	4.07	.82	16.49	19.90
0200	Add for vermiculite bottom					.89			.89	.98
0500	Gunite bottom and sides, white plaster finish									
0600	12′ x 30′ pool	B-52	145	.386	SF Surf	26	14.85	3.72	44.57	55
0720	16′ x 32′ pool		155	.361		23.50	13.90	3.48	40.88	51
0750	20′ x 40′ pool		250	.224		21	8.60	2.15	31.75	38.50
0810	Concrete bottom and sides, tile finish									
0820	12′ x 30′ pool	B-52	80	.700	SF Surf	26	27	6.75	59.75	77
0830	16′ x 32′ pool		95	.589		21.50	22.50	5.65	49.65	64.50
0840	20′ x 40′ pool		130	.431		17.15	16.55	4.14	37.84	49
1100	Motel, gunite with plaster finish, incl. medium									
1150	capacity filtration & chlorination	B-52	115	.487	SF Surf	32	18.70	4.68	55.38	68.50
1200	Municipal, gunite with plaster finish, incl. high									
1250	capacity filtration & chlorination	B-52	100	.560	SF Surf	41	21.50	5.40	67.90	84.50
1350	Add for formed gutters				L.F.	60.50			60.50	66.50
1360	Add for stainless steel gutters				"	179			179	197
1700	Filtration and deck equipment only, as % of total				Total				20%	20%
1800	Deck equipment, rule of thumb, 20′ x 40′ pool				SF Pool				1.18	1.30
1900	5000 S.F. pool				"				1.73	1.90
3000	Painting pools, preparation + 3 coats, 20′ x 40′ pool, epoxy	2 Pord	.33	48.485	Total	1,625	1,750		3,375	4,400
3100	Rubber base paint, 18 gallons	"	.33	48.485		1,100	1,750		2,850	3,825
3500	42′ x 82′ pool, 75 gallons, epoxy paint	3 Pord	.14	171		6,850	6,225		13,075	16,800
3600	Rubber base paint	"	.14	171		4,500	6,225		10,725	14,200

13 11 46 – Swimming Pool Accessories

13 11 46.50 Swimming Pool Equipment		Crew	Daily Output	Labor-Hours	Unit	Material	Labor	Equipment	Total	Total Incl O&P
0010	**SWIMMING POOL EQUIPMENT**									
0020	Diving stand, stainless steel, 3 meter	2 Carp	.40	40	Ea.	10,500	1,650		12,150	14,100
0300	1 meter	"	2.70	5.926	"	6,175	246		6,421	7,175
1100	Gutter system, stainless steel, with grating, stock,									
1110	contains supply and drainage system	E-1	20	1.200	L.F.	268	55	7.30	330.30	395
1120	Integral gutter and 5′ high wall system, stainless steel	"	10	2.400	"	360	110	14.60	484.60	595
2100	Lights, underwater, 12 volt, with transformer, 300 watt	1 Elec	1	8	Ea.	263	390		653	875
2200	110 volt, 500 watt, standard		1	8		209	390		599	815
2400	Low water cutoff type		1	8		254	390		644	865
2800	Heaters, see Div. 23 52 28.10									

13 18 Ice Rinks

13 18 13 – Ice Rink Floor Systems

13 18 13.50 Ice Skating		Crew	Daily Output	Labor-Hours	Unit	Material	Labor	Equipment	Total	Total Incl O&P
0010	**ICE SKATING** Equipment incl. refrigeration, plumbing & cooling									
0020	coils & concrete slab, 85′ x 200′ rink									
0300	55° system, 5 mos., 100 ton				Total	575,000			575,000	632,500
0700	90° system, 12 mos., 135 ton				"	600,000			600,000	660,000
1200	Subsoil heating system (recycled from compressor), 85′ x 200′	Q-7	.27	118	Ea.	25,000	5,850		30,850	36,300
1300	Subsoil insulation, 2 lb. polystyrene with vapor barrier, 85′ x 200′	2 Carp	.14	114	"	30,000	4,750		34,750	40,300

13 18 Ice Rinks

13 18 16 – Ice Rink Dasher Boards

	13 18 16.50 Ice Rink Dasher Boards	Crew	Daily Output	Labor-Hours	Unit	Material	2010 Bare Costs Labor	Equipment	Total	Total Incl O&P
0010	**ICE RINK DASHER BOARDS**									
1000	Dasher boards, 1/2″ H.D. polyethylene faced steel frame, 3′ acrylic									
1020	screen at sides, 5′ acrylic ends, 85′ x 200′	F-5	.06	533	Ea.	125,000	22,400		147,400	172,000
1100	Fiberglass & aluminum construction, same sides and ends	″	.06	533	″	125,000	22,400		147,400	172,000

13 21 Controlled Environment Rooms

13 21 13 – Clean Rooms

13 21 13.50 Clean Room Components

	Description	Crew	Daily Output	Labor-Hours	Unit	Material	Labor	Equipment	Total	Total Incl O&P
0010	**CLEAN ROOM COMPONENTS**									
1100	Clean room, soft wall, 12′ x 12′, Class 100	1 Carp	.18	44.444	Ea.	15,500	1,850		17,350	20,000
1110	Class 1,000		.18	44.444		12,600	1,850		14,450	16,800
1120	Class 10,000		.21	38.095		11,100	1,575		12,675	14,700
1130	Class 100,000		.21	38.095		10,400	1,575		11,975	14,000
2800	Ceiling grid support, slotted channel struts 4′-0″ O.C., ea. way				S.F.				5.91	6.50
3000	Ceiling panel, vinyl coated foil on mineral substrate									
3020	Sealed, non-perforated				S.F.				1.27	1.40
4000	Ceiling panel seal, silicone sealant, 150 L.F./gal.	1 Carp	150	.053	L.F.	.28	2.22		2.50	3.73
4100	Two sided adhesive tape	″	240	.033	″	.13	1.38		1.51	2.27
4200	Clips, one per panel				Ea.	1.09			1.09	1.20
6000	HEPA filter, 2′ x 4′, 99.97% eff., 3″ dp beveled frame (silicone seal)					365			365	400
6040	6″ deep skirted frame (channel seal)					390			390	430
6100	99.99% efficient, 3″ deep beveled frame (silicone seal)					380			380	420
6140	6″ deep skirted frame (channel seal)					405			405	445
6200	99.999% efficient, 3″ deep beveled frame (silicone seal)					395			395	435
6240	6″ deep skirted frame (channel seal)					420			420	465
7000	Wall panel systems, including channel strut framing									
7020	Polyester coated aluminum, particle board				S.F.				18.18	20
7100	Porcelain coated aluminum, particle board								31.82	35
7400	Wall panel support, slotted channel struts, to 12′ high								16.36	18

13 21 26 – Cold Storage Rooms

13 21 26.50 Refrigeration

	Description	Crew	Daily Output	Labor-Hours	Unit	Material	Labor	Equipment	Total	Total Incl O&P
0010	**REFRIGERATION**									
0020	Curbs, 12″ high, 4″ thick, concrete	2 Carp	58	.276	L.F.	4.37	11.45		15.82	22.50
1000	Doors, see Div. 08 34 13.10									
2400	Finishes, 2 coat portland cement plaster, 1/2″ thick	1 Plas	48	.167	S.F.	1.20	6.20		7.40	10.60
2500	For galvanized reinforcing mesh, add	1 Lath	335	.024		.80	.88		1.68	2.18
2700	3/16″ thick latex cement	1 Plas	88	.091		2.05	3.39		5.44	7.30
2900	For glass cloth reinforced ceilings, add	″	450	.018		.49	.66		1.15	1.53
3100	Fiberglass panels, 1/8″ thick	1 Carp	149.45	.054		2.71	2.22		4.93	6.40
3200	Polystyrene, plastic finish ceiling, 1″ thick		274	.029		2.50	1.21		3.71	4.62
3400	2″ thick		274	.029		2.85	1.21		4.06	5
3500	4″ thick		219	.037		3.17	1.52		4.69	5.85
3800	Floors, concrete, 4″ thick	1 Cefi	93	.086		1.19	3.42		4.61	6.30
3900	6″ thick	″	85	.094		1.77	3.74		5.51	7.40
4000	Insulation, 1″ to 6″ thick, cork				B.F.	1.16			1.16	1.28
4100	Urethane					1.14			1.14	1.25
4300	Polystyrene, regular					.77			.77	.85
4400	Bead board					.58			.58	.64
4600	Installation of above, add per layer	2 Carp	657.60	.024	S.F.	.38	1.01		1.39	1.98
4700	Wall and ceiling juncture		298.90	.054	L.F.	1.87	2.22		4.09	5.50

13 21 Controlled Environment Rooms

13 21 26 – Cold Storage Rooms

13 21 26.50 Refrigeration		Crew	Daily Output	Labor-Hours	Unit	2010 Bare Costs Material	Labor	Equipment	Total	Total Incl O&P
4900	Partitions, galvanized sandwich panels, 4" thick, stock	2 Carp	219.20	.073	S.F.	7.85	3.03		10.88	13.30
5000	Aluminum or fiberglass		219.20	.073	″	8.55	3.03		11.58	14.15
5200	Prefab walk-in, 7'-6" high, aluminum, incl. door & floors,									
5210	not incl. partitions or refrigeration, 6' x 6' O.D. nominal	2 Carp	54.80	.292	SF Flr.	140	12.15		152.15	173
5500	10' x 10' O.D. nominal		82.20	.195		113	8.10		121.10	136
5700	12' x 14' O.D. nominal		109.60	.146		101	6.05		107.05	120
5800	12' x 20' O.D. nominal		109.60	.146		88	6.05		94.05	106
6100	For 8'-6" high, add					5%				
6300	Rule of thumb for complete units, w/o doors & refrigeration, cooler	2 Carp	146	.110		127	4.55		131.55	147
6400	Freezer		109.60	.146		150	6.05		156.05	174
6600	Shelving, plated or galvanized, steel wire type		360	.044	SF Hor.	11.10	1.85		12.95	15.10
6700	Slat shelf type		375	.043		13.70	1.77		15.47	17.85
6900	For stainless steel shelving, add					300%				
7000	Vapor barrier, on wood walls	2 Carp	1644	.010	S.F.	.16	.40		.56	.80
7200	On masonry walls	″	1315	.012	″	.42	.51		.93	1.24
7500	For air curtain doors, see Div. 23 34 33.10									

13 24 Special Activity Rooms

13 24 16 – Saunas

13 24 16.50 Saunas and Heaters		Crew	Daily Output	Labor-Hours	Unit	Material	Labor	Equipment	Total	Total Incl O&P
0010	**SAUNAS AND HEATERS**									
0020	Prefabricated, incl. heater & controls, 7' high, 6' x 4', C/C	L-7	2.20	12.727	Ea.	4,625	510		5,135	5,875
0050	6' x 4', C/P		2	14		4,275	565		4,840	5,550
0400	6' x 5', C/C		2	14		5,175	565		5,740	6,525
0450	6' x 5', C/P		2	14		4,800	565		5,365	6,150
0600	6' x 6', C/C		1.80	15.556		5,475	625		6,100	6,975
0650	6' x 6', C/P		1.80	15.556		5,125	625		5,750	6,575
0800	6' x 9', C/C		1.60	17.500		6,800	705		7,505	8,550
0850	6' x 9', C/P		1.60	17.500		6,300	705		7,005	8,025
1000	8' x 12', C/C		1.10	25.455		10,600	1,025		11,625	13,300
1050	8' x 12', C/P		1.10	25.455		9,725	1,025		10,750	12,300
1200	8' x 8', C/C		1.40	20		8,100	805		8,905	10,100
1250	8' x 8', C/P		1.40	20		7,600	805		8,405	9,600
1400	8' x 10', C/C		1.20	23.333		8,975	940		9,915	11,300
1450	8' x 10', C/P		1.20	23.333		8,350	940		9,290	10,600
1600	10' x 12', C/C		1	28		11,200	1,125		12,325	14,000
1650	10' x 12', C/P		1	28		10,100	1,125		11,225	12,800
2500	Heaters only (incl. above), wall mounted, to 200 C.F.					605			605	665
2750	To 300 C.F.					770			770	845
3000	Floor standing, to 720 C.F., 10,000 watts, w/controls	1 Elec	3	2.667		2,025	131		2,156	2,425
3250	To 1,000 C.F., 16,000 watts	″	3	2.667		2,525	131		2,656	2,975

13 24 26 – Steam Baths

13 24 26.50 Steam Baths and Components		Crew	Daily Output	Labor-Hours	Unit	Material	Labor	Equipment	Total	Total Incl O&P
0010	**STEAM BATHS AND COMPONENTS**									
0020	Heater, timer & head, single, to 140 C.F.	1 Plum	1.20	6.667	Ea.	1,400	345		1,745	2,075
0500	To 300 C.F.		1.10	7.273		1,625	380		2,005	2,350
1000	Commercial size, with blow-down assembly, to 800 C.F.		.90	8.889		5,200	465		5,665	6,425
1500	To 2500 C.F.		.80	10		8,025	520		8,545	9,600
2000	Multiple, motels, apts., 2 baths, w/ blow-down assm., 500 C.F.	Q-1	1.30	12.308		5,900	575		6,475	7,375
2500	4 baths	″	.70	22.857		7,450	1,075		8,525	9,800

13 24 Special Activity Rooms

13 24 26 – Steam Baths

13 24 26.50 Steam Baths and Components		Crew	Daily Output	Labor-Hours	Unit	2010 Bare Costs Material	Labor	Equipment	Total	Total Incl O&P
2700	Conversion unit for residential tub, including door				Ea.	4,100			4,100	4,500

13 34 Fabricated Engineered Structures

13 34 23 – Fabricated Structures

13 34 23.10 Comfort Stations

		Crew	Daily Output	Labor-Hours	Unit	Material	Labor	Equipment	Total	Total Incl O&P
0010	**COMFORT STATIONS** Prefab., stock, w/doors, windows & fixt.									
0100	Not incl. interior finish or electrical									
0300	Mobile, on steel frame, minimum				S.F.	46.50			46.50	51
0350	Maximum					75			75	82.50
0400	Permanent, including concrete slab, minimum	B-12J	50	.320		199	12.40	16.25	227.65	256
0500	Maximum	"	43	.372	▼	288	14.40	18.90	321.30	360
0600	Alternate pricing method, mobile, minimum				Fixture	2,150			2,150	2,375
0650	Maximum					3,175			3,175	3,500
0700	Permanent, minimum	B-12J	.70	22.857		12,200	885	1,150	14,235	16,100
0750	Maximum	"	.50	32	▼	20,400	1,250	1,625	23,275	26,100

13 34 23.16 Fabricated Control Booths

		Crew	Daily Output	Labor-Hours	Unit	Material	Labor	Equipment	Total	Total Incl O&P
0010	**FABRICATED CONTROL BOOTHS**									
0100	Guard House, prefab conc w/bullet resistant doors & windows, roof & wiring									
0110	8′ x 8′, Level III	L-10	1	24	Ea.	22,000	1,125	655	23,780	26,800
0120	8′ x 8′, Level IV	"	1	24	"	27,500	1,125	655	29,280	32,900

Division Notes

		CREW	DAILY OUTPUT	LABOR-HOURS	UNIT	2010 BARE COSTS MAT.	LABOR	EQUIP.	TOTAL	TOTAL INCL O&P

Estimating Tips

General

- Many products in Division 14 will require some type of support or blocking for installation not included with the item itself. Examples are supports for conveyors or tube systems, attachment points for lifts, and footings for hoists or cranes. Add these supports in the appropriate division.

14 10 00 Dumbwaiters
14 20 00 Elevators

- Dumbwaiters and elevators are estimated and purchased in a method similar to buying a car. The manufacturer has a base unit with standard features. Added to this base unit price will be whatever options the owner or specifications require. Increased load capacity, additional vertical travel, additional stops, higher speed, and cab finish options are items to be considered. When developing an estimate for dumbwaiters and elevators, remember that some items needed by the installers may have to be included as part of the general contract.

  Examples are:
  - — shaftway
  - — rail support brackets
  - — machine room
  - — electrical supply
  - — sill angles
  - — electrical connections
  - — pits
  - — roof penthouses
  - — pit ladders

Check the job specifications and drawings before pricing.

- Installation of elevators and handicapped lifts in historic structures can require significant additional costs. The associated structural requirements may involve cutting into and repairing finishes, mouldings, flooring, etc. The estimator must account for these special conditions.

14 30 00 Escalators and Moving Walks

- Escalators and moving walks are specialty items installed by specialty contractors. There are numerous options associated with these items. For specific options, contact a manufacturer or contractor. In a method similar to estimating dumbwaiters and elevators, you should verify the extent of general contract work and add items as necessary.

14 40 00 Lifts
14 90 00 Other Conveying Equipment

- Products such as correspondence lifts, chutes, and pneumatic tube systems, as well as other items specified in this subdivision, may require trained installers. The general contractor might not have any choice as to who will perform the installation or when it will be performed. Long lead times are often required for these products, making early decisions in scheduling necessary.

Reference Numbers

Reference numbers are shown in shaded boxes at the beginning of some major classifications. These numbers refer to related items in the Reference Section. The reference information may be an estimating procedure, an alternate pricing method, or technical information.

Note: Not all subdivisions listed here necessarily appear in this publication.

14 91 Facility Chutes

14 91 33 – Laundry and Linen Chutes

14 91 33.10 Chutes		Crew	Daily Output	Labor-Hours	Unit	Material	2010 Bare Costs Labor	Equipment	Total	Total Incl O&P
0011	**CHUTES**, linen, trash or refuse									
0050	Aluminized steel, 16 ga., 18" diameter	2 Shee	3.50	4.571	Floor	1,275	224		1,499	1,775
0100	24" diameter		3.20	5		1,350	246		1,596	1,875
0200	30" diameter		3	5.333		1,550	262		1,812	2,100
0300	36" diameter		2.80	5.714		1,725	281		2,006	2,325
0400	Galvanized steel, 16 ga., 18" diameter		3.50	4.571		1,000	224		1,224	1,450
0500	24" diameter		3.20	5		1,125	246		1,371	1,625
0600	30" diameter		3	5.333		1,275	262		1,537	1,800
0700	36" diameter		2.80	5.714		1,500	281		1,781	2,075
0800	Stainless steel, 18" diameter		3.50	4.571		2,650	224		2,874	3,250
0900	24" diameter		3.20	5		2,875	246		3,121	3,525
1000	30" diameter		3	5.333		3,550	262		3,812	4,300
1005	36" diameter		2.80	5.714		3,975	281		4,256	4,800
1200	Linen chute bottom collector, aluminized steel		4	4	Ea.	1,400	196		1,596	1,850
1300	Stainless steel		4	4		1,800	196		1,996	2,275
1500	Refuse, bottom hopper, aluminized steel, 18" diameter		3	5.333		985	262		1,247	1,475
1600	24" diameter		3	5.333		1,200	262		1,462	1,700
1800	36" diameter		3	5.333		2,050	262		2,312	2,650

14 91 82 – Trash Chutes

14 91 82.10 Trash Chutes and Accessories		Crew	Daily Output	Labor-Hours	Unit	Material	Labor	Equipment	Total	Total Incl O&P
0010	**TRASH CHUTES AND ACCESSORIES**									
2900	Package chutes, spiral type, minimum	2 Shee	4.50	3.556	Floor	2,425	175		2,600	2,950
3000	Maximum	"	1.50	10.667	"	6,325	525		6,850	7,775

14 92 Pneumatic Tube Systems

14 92 10 – Conventional, Automatic and Computer Controlled Pneumatic Tube Systems

14 92 10.10 Pneumatic Tube Systems		Crew	Daily Output	Labor-Hours	Unit	Material	Labor	Equipment	Total	Total Incl O&P
0010	**PNEUMATIC TUBE SYSTEMS**									
0020	100' long, single tube, 2 stations, stock									
0100	3" diameter	2 Stpi	.12	133	Total	6,600	6,925		13,525	17,700
0300	4" diameter	"	.09	177	"	7,450	9,225		16,675	22,000
0400	Twin tube, two stations or more, conventional system									
0600	2-1/2" round	2 Stpi	62.50	.256	L.F.	13.30	13.30		26.60	34.50
0700	3" round		46	.348		15.25	18.05		33.30	44
0900	4" round		49.60	.323		16.90	16.75		33.65	43.50
1000	4" x 7" oval		37.60	.426		24.50	22		46.50	60
1050	Add for blower		2	8	System	5,200	415		5,615	6,350
1110	Plus for each round station, add		7.50	2.133	Ea.	585	111		696	810
1150	Plus for each oval station, add		7.50	2.133	"	585	111		696	810
1200	Alternate pricing method: base cost, minimum		.75	21.333	Total	6,375	1,100		7,475	8,675
1300	Maximum		.25	64	"	12,700	3,325		16,025	18,900
1500	Plus total system length, add, minimum		93.40	.171	L.F.	8.20	8.90		17.10	22.50
1600	Maximum		37.60	.426	"	24.50	22		46.50	59.50
1800	Completely automatic system, 4" round, 15 to 50 stations		.29	55.172	Station	20,100	2,875		22,975	26,400
2200	51 to 144 stations		.32	50		15,600	2,600		18,200	21,100
2400	6" round or 4" x 7" oval, 15 to 50 stations		.24	66.667		25,200	3,450		28,650	32,900
2800	51 to 144 stations		.23	69.565		21,100	3,600		24,700	28,600

Estimating Tips

22 10 00 Plumbing Piping and Pumps

This subdivision is primarily basic pipe and related materials. The pipe may be used by any of the mechanical disciplines, i.e., plumbing, fire protection, heating, and air conditioning.

- The labor adjustment factors listed in Subdivision 22 01 02.20 apply throughout Divisions 21, 22, and 23. CAUTION: the correct percentage may vary for the same items. For example, the percentage add for the basic pipe installation should be based on the maximum height that the craftsman must install for that particular section. If the pipe is to be located 14′ above the floor but it is suspended on threaded rod from beams, the bottom flange of which is 18′ high (4′ rods), then the height is actually 18′ and the add is 20%. The pipe coverer, however, does not have to go above the 14′, and so his or her add should be 10%.
- Most pipe is priced first as straight pipe with a joint (coupling, weld, etc.) every 10′ and a hanger usually every 10′. There are exceptions with hanger spacing such as for cast iron pipe (5′) and plastic pipe (3 per 10′). Following each type of pipe there are several lines listing sizes and the amount to be subtracted to delete couplings and hangers. This is for pipe that is to be buried or supported together on trapeze hangers. The reason that the couplings are deleted is that these runs are usually long, and frequently longer lengths of pipe are used. By deleting the couplings, the estimator is expected to look up and add back the correct reduced number of couplings.
- When preparing an estimate, it may be necessary to approximate the fittings. Fittings usually run between 25% and 50% of the cost of the pipe. The lower percentage is for simpler runs, and the higher number is for complex areas, such as mechanical rooms.
- For historic restoration projects, the systems must be as invisible as possible, and pathways must be sought for pipes, conduit, and ductwork. While installations in accessible spaces (such as basements and attics) are relatively straightforward to estimate, labor costs may be more difficult to determine when delivery systems must be concealed.

22 40 00 Plumbing Fixtures

- Plumbing fixture costs usually require two lines: the fixture itself and its "rough-in, supply, and waste."
- In the Assemblies Section (Plumbing D2010) for the desired fixture, the System Components Group at the center of the page shows the fixture on the first line. The rest of the list (fittings, pipe, tubing, etc.) will total up to what we refer to in the Unit Price section as "Rough-in, supply, waste, and vent." Note that for most fixtures we allow a nominal 5′ of tubing to reach from the fixture to a main or riser.
- Remember that gas- and oil-fired units need venting.

Reference Numbers

Reference numbers are shown in shaded boxes at the beginning of some major classifications. These numbers refer to related items in the Reference Section. The reference information may be an estimating procedure, an alternate pricing method, or technical information.

Note: Not all subdivisions listed here necessarily appear in this publication.

Note: **Trade Service,** *in part, has been used as a reference source for some of the material prices used in Division 22.*

22 01 02 – Labor Adjustments

22 01 02.10 Boilers, General		Crew	Daily Output	Labor-Hours	Unit	Material	2010 Bare Costs Labor	Equipment	Total	Total Incl O&P
0010	**BOILERS, GENERAL**, Prices do not include flue piping, elec. wiring,									
0020	gas or oil piping, boiler base, pad, or tankless unless noted									
0100	Boiler H.P.: 10 KW = 34 lbs/steam/hr = 33,475 BTU/hr. R235000-50									
0150	To convert SFR to BTU rating: Hot water, 150 x SFR;									
0160	Forced hot water, 180 x SFR; steam, 240 x SFR									

22 01 02.20 Labor Adjustment Factors		Crew	Daily Output	Labor-Hours	Unit	Material	2010 Bare Costs Labor	Equipment	Total	Total Incl O&P
0010	**LABOR ADJUSTMENT FACTORS**, (For Div. 21,22 and 23)									
1000	Add to labor for elevated installation (Above floor level)									
1080	10' to 14.5' high R221113-70						10%			
1100	15' to 19.5' high						20%			
1120	20' to 24.5' high						25%			
1140	25' to 29.5' high						35%			
1160	30' to 34.5' high						40%			
1180	35' to 39.5' high						50%			
1200	40' and higher						55%			
2000	Add to labor for crawl space									
2100	3' high						40%			
2140	4' high						30%			
3000	Add to labor for multi-story building									
3100	Add per floor for floors 3 thru 19						2%			
3140	Add per floor for floors 20 and up						4%			
4000	Add to labor for working in existing occupied buildings									
4100	Hospital						35%			
4140	Office building						25%			
4180	School						20%			
4220	Factory or warehouse						15%			
4260	Multi dwelling						15%			
5000	Add to labor, miscellaneous									
5100	Cramped shaft						35%			
5140	Congested area						15%			
5180	Excessive heat or cold						30%			
9000	Labor factors, The above are reasonable suggestions, however									
9010	each project should be evaluated for its own peculiarities.									
9100	Other factors to be considered are:									
9140	Movement of material and equipment through finished areas									
9180	Equipment room									
9220	Attic space									
9260	No service road									
9300	Poor unloading/storage area									
9340	Congested site area/heavy traffic									

22 05 05 – Selective Plumbing Demolition

22 05 05.10 Plumbing Demolition		Crew	Daily Output	Labor-Hours	Unit	Material	2010 Bare Costs Labor	Equipment	Total	Total Incl O&P
0010	**PLUMBING DEMOLITION** R220105-10									
0400	Air compressor, up thru 2 H.P.	Q-1	10	1.600	Ea.		75		75	112
0410	3 H.P. thru 7-1/2 H.P.		5.60	2.857			134		134	201
0420	10 H.P. thru 15 H.P.		1.40	11.429			535		535	805
0430	20 H.P. thru 30 H.P.	Q-2	1.30	18.462			895		895	1,350
0500	Backflow preventer, up thru 2" diameter	1 Plum	17	.471			24.50		24.50	36.50
0510	2-1/2" thru 3" diameter	Q-1	10	1.600			75		75	112
0520	4" thru 6" diameter	"	5	3.200			150		150	225
0530	8" thru 10" diameter	Q-2	3	8			390		390	585
0700	Carriers and supports									
0710	Fountains, sinks, lavatories and urinals	1 Plum	14	.571	Ea.		29.50		29.50	44.50
0720	Water closets	"	12	.667	"		34.50		34.50	52
0730	Grinder pump or sewage ejector system									
0732	Simplex	Q-1	7	2.286	Ea.		107		107	161
0734	Duplex	"	2.80	5.714			268		268	400
0738	Hot water dispenser	1 Plum	36	.222			11.55		11.55	17.35
0740	Hydrant, wall		26	.308			16		16	24
0744	Ground		12	.667			34.50		34.50	52
0760	Cleanouts and drains, up thru 4" pipe diameter		10	.800			41.50		41.50	62.50
0764	5" thru 8" pipe diameter	Q-1	10	1.600			75		75	112
0780	Industrial safety fixtures	1 Plum	8	1			52		52	78
1020	Fixtures, including 10' piping									
1100	Bathtubs, cast iron	1 Plum	4	2	Ea.		104		104	156
1120	Fiberglass		6	1.333			69.50		69.50	104
1140	Steel		5	1.600			83.50		83.50	125
1150	Bidet	Q-1	7	2.286			107		107	161
1200	Lavatory, wall hung	1 Plum	10	.800			41.50		41.50	62.50
1220	Counter top		8	1			52		52	78
1300	Sink, single compartment		8	1			52		52	78
1320	Double compartment		7	1.143			59.50		59.50	89
1340	Shower, stall and receptor	Q-1	6	2.667			125		125	187
1350	Group	"	7	2.286			107		107	161
1400	Water closet, floor mounted	1 Plum	8	1			52		52	78
1420	Wall mounted	"	7	1.143			59.50		59.50	89
1440	Wash fountain, 36" diameter	Q-2	8	3			146		146	218
1442	54" diameter	"	7	3.429			167		167	250
1500	Urinal, floor mounted	1 Plum	4	2			104		104	156
1520	Wall mounted	"	7	1.143			59.50		59.50	89
1590	Whirl pool or hot tub	Q-1	2.60	6.154			288		288	430
1600	Water fountains, free standing	1 Plum	8	1			52		52	78
1620	Wall or deck mounted		6	1.333			69.50		69.50	104
1800	Medical gas specialties		8	1			52		52	78
1810	Plumbing demo, Floor drain, Remove		12	.667			34.50		34.50	52
1900	Piping fittings, single connection, up thru 1-1/2" diameter		30	.267			13.90		13.90	21
1910	2" thru 4" diameter		14	.571			29.50		29.50	44.50
1980	Pipe hanger/support removal		80	.100			5.20		5.20	7.80
1990	Glass pipe with fittings, 1" thru 3" diameter		200	.040	L.F.		2.08		2.08	3.12
1992	4" thru 6" diameter		150	.053			2.78		2.78	4.16
2000	Piping, metal, up thru 1-1/2" diameter		200	.040			2.08		2.08	3.12
2050	2" thru 3-1/2" diameter		150	.053			2.78		2.78	4.16
2100	4" thru 6" diameter	2 Plum	100	.160			8.35		8.35	12.50
2150	8" thru 14" diameter	"	60	.267			13.90		13.90	21
2153	16" thru 20" diameter	Q-18	70	.343			16.60	.84	17.44	26

22 05 05.10 Plumbing Demolition		Crew	Daily Output	Labor-Hours	Unit	Material	2010 Bare Costs Labor	Equipment	Total	Total Incl O&P
2155	24" thru 26" diameter	Q-18	55	.436	L.F.		21	1.06	22.06	32.50
2156	30" thru 36" diameter		40	.600			29	1.46	30.46	45
2160	Plastic pipe with fittings, up thru 1-1/2" diameter	1 Plum	250	.032			1.67		1.67	2.50
2162	2" thru 3" diameter	"	200	.040			2.08		2.08	3.12
2164	4" thru 6" diameter	Q-1	200	.080			3.75		3.75	5.60
2166	8" thru 14" diameter		150	.107			5		5	7.50
2168	16" diameter		100	.160			7.50		7.50	11.25
2170	Prison fixtures, lavatory or sink		18	.889	Ea.		41.50		41.50	62.50
2172	Shower		5.60	2.857			134		134	201
2174	Urinal or water closet		13	1.231			57.50		57.50	86.50
2180	Pumps, all fractional horse-power		12	1.333			62.50		62.50	93.50
2184	1 H.P. thru 5 H.P.		6	2.667			125		125	187
2186	7-1/2 H.P. thru 15 H.P.		2.50	6.400			300		300	450
2188	20 H.P. thru 25 H.P.	Q-2	4	6			291		291	435
2190	30 H.P. thru 60 H.P.		.80	30			1,450		1,450	2,175
2192	75 H.P. thru 100 H.P.		.60	40			1,950		1,950	2,925
2194	150 H.P.		.50	48			2,325		2,325	3,500
2198	Pump, sump or submersible	1 Plum	12	.667			34.50		34.50	52
2200	Receptors and interceptors, up thru 20 GPM	"	8	1			52		52	78
2204	25 thru 100 GPM	Q-1	6	2.667			125		125	187
2208	125 thru 300 GPM	"	2.40	6.667			310		310	470
2211	325 thru 500 GPM	Q-2	2.60	9.231			450		450	670
2212	Deduct for salvage, aluminum scrap				Ton				728	800
2214	Brass scrap								2,000	2,200
2216	Copper scrap								2,820	3,100
2218	Lead scrap								635	700
2220	Steel scrap								180	200
2230	Temperature maintenance cable	1 Plum	1200	.007	L.F.		.35		.35	.52
2250	Water heater, 40 gal.	"	6	1.333	Ea.		69.50		69.50	104
3100	Tanks, water heaters and liquid containers									
3110	Up thru 45 gallons	Q-1	22	.727	Ea.		34		34	51
3120	50 thru 120 gallons		14	1.143			53.50		53.50	80.50
3130	130 thru 240 gallons		7.60	2.105			98.50		98.50	148
3140	250 thru 500 gallons		5.40	2.963			139		139	208
3150	600 thru 1000 gallons	Q-2	1.60	15			730		730	1,100
3160	1100 thru 2000 gallons		.70	34.286			1,675		1,675	2,500
3170	2100 thru 4000 gallons		.50	48			2,325		2,325	3,500
6000	Remove and reset fixtures, minimum	1 Plum	6	1.333			69.50		69.50	104
6100	Maximum	"	4	2			104		104	156
9100	Valve, metal valves, strainers and similar, up thru 1-1/2" diameter	1 Stpi	28	.286			14.85		14.85	22
9110	2" thru 3" diameter	Q-1	11	1.455			68		68	102
9120	4" thru 6" diameter	"	8	2			93.50		93.50	140
9130	8" thru 14" diameter	Q-2	8	3			146		146	218
9140	16" thru 20" diameter		2	12			585		585	875
9150	24" diameter		1.20	20			970		970	1,450
9200	Valve, plastic, up thru 1-1/2" diameter	1 Plum	42	.190			9.90		9.90	14.85
9210	2" thru 3" diameter		15	.533			28		28	41.50
9220	4" thru 6" diameter		12	.667			34.50		34.50	52
9300	Vent flashing and caps		55	.145			7.55		7.55	11.35
9350	Water filter, commercial, 1" thru 1-1/2"	Q-1	2	8			375		375	560
9360	2" thru 2-1/2"	"	1.60	10			470		470	700
9400	Water heaters									
9410	Up thru 245 GPH	Q-1	2.40	6.667	Ea.		310		310	470

22 05 05 – Selective Plumbing Demolition

22 05 05.10 Plumbing Demolition		Crew	Daily Output	Labor-Hours	Unit	2010 Bare Costs Material	Labor	Equipment	Total	Total Incl O&P
9420	250 thru 756 GPH	Q-1	1.60	10	Ea.		470		470	700
9430	775 thru 1640 GPH	"	.80	20			935		935	1,400
9440	1650 thru 4000 GPH	Q-2	.50	48			2,325		2,325	3,500
9470	Water softener	Q-1	2	8	↓		375		375	560

22 05 23 – General-Duty Valves for Plumbing Piping

22 05 23.10 Valves, Brass

		Crew	Daily Output	Labor-Hours	Unit	Material	Labor	Equipment	Total	Total Incl O&P
0010	**VALVES, BRASS**									
0032	For motorized valves, see Div. 23 09 53.10									
0500	Gas cocks, threaded									
0510	1/4"	1 Plum	26	.308	Ea.	9.25	16		25.25	34
0520	3/8"		24	.333		9.25	17.35		26.60	36
0530	1/2"		24	.333		8.90	17.35		26.25	36
0540	3/4"		22	.364		11.20	18.95		30.15	41
0550	1"		19	.421		17.65	22		39.65	52.50
0560	1-1/4"		15	.533		38	28		66	83.50
0570	1-1/2"		13	.615		51	32		83	104
0580	2"	↓	11	.727	↓	75	38		113	139
0672	For larger sizes use lubricated plug valve, Div. 23 05 23.70									

22 05 23.20 Valves, Bronze

		Crew	Daily Output	Labor-Hours	Unit	Material	Labor	Equipment	Total	Total Incl O&P
0010	**VALVES, BRONZE** R220523-90									
1020	Angle, 150 lb., rising stem, threaded									
1030	1/8"	1 Plum	24	.333	Ea.	89.50	17.35		106.85	125
1040	1/4"		24	.333		92.50	17.35		109.85	128
1050	3/8"		24	.333		92.50	17.35		109.85	128
1060	1/2"		22	.364		92.50	18.95		111.45	131
1070	3/4"		20	.400		127	21		148	170
1080	1"		19	.421		182	22		204	234
1090	1-1/4"		15	.533		234	28		262	299
1100	1-1/2"		13	.615		305	32		337	385
1110	2"	↓	11	.727	↓	490	38		528	595
1380	Ball, 150 psi, threaded									
1400	1/4"	1 Plum	24	.333	Ea.	10.80	17.35		28.15	38
1430	3/8"		24	.333		10.80	17.35		28.15	38
1450	1/2"		22	.364		10.80	18.95		29.75	40.50
1460	3/4"		20	.400		17.80	21		38.80	50.50
1470	1"		19	.421		22.50	22		44.50	57.50
1480	1-1/4"		15	.533		42	28		70	87.50
1490	1-1/2"		13	.615		54	32		86	108
1500	2"		11	.727		65.50	38		103.50	129
1510	2-1/2"		9	.889		220	46.50		266.50	310
1520	3"	↓	8	1	↓	335	52		387	445
1600	Butterfly, 175 psi, full port, solder or threaded ends									
1610	Stainless steel disc and stem									
1620	1/4"	1 Plum	24	.333	Ea.	12.30	17.35		29.65	39.50
1630	3/8"		24	.333		12.30	17.35		29.65	39.50
1640	1/2"		22	.364		12.95	18.95		31.90	43
1650	3/4"		20	.400		21	21		42	54
1660	1"		19	.421		25.50	22		47.50	61
1670	1-1/4"		15	.533		40.50	28		68.50	86
1680	1-1/2"		13	.615		53	32		85	107
1690	2"	↓	11	.727	↓	66.50	38		104.50	130
1750	Check, swing, class 150, regrinding disc, threaded									

22 05 23 – General-Duty Valves for Plumbing Piping

22 05 23.20 Valves, Bronze		Crew	Daily Output	Labor-Hours	Unit	Material	Labor	Equipment	Total	Total Incl O&P
						2010 Bare Costs				
1800	1/8"	1 Plum	24	.333	Ea.	43.50	17.35		60.85	73.50
1830	1/4"		24	.333		43.50	17.35		60.85	73.50
1840	3/8"		24	.333		45.50	17.35		62.85	76
1850	1/2"		24	.333		45.50	17.35		62.85	76
1860	3/4"		20	.400		65.50	21		86.50	103
1870	1"		19	.421		94.50	22		116.50	137
1880	1-1/4"		15	.533		136	28		164	192
1890	1-1/2"		13	.615		158	32		190	222
1900	2"		11	.727		233	38		271	315
1910	2-1/2"	Q-1	15	1.067		520	50		570	650
1920	3"	"	13	1.231		695	57.50		752.50	850
2000	For 200 lb, add					5%	10%			
2040	For 300 lb, add					15%	15%			
2060	Check swing, 300#, sweat, 3/8" size	1 Plum	24	.333	Ea.	32	17.35		49.35	61
2070	1/2"		24	.333		32	17.35		49.35	61
2080	3/4"		20	.400		40.50	21		61.50	75.50
2090	1"		19	.421		56	22		78	95
2100	1-1/4"		15	.533		73	28		101	122
2110	1-1/2"		13	.615		96.50	32		128.50	154
2120	2"		11	.727		146	38		184	217
2130	2-1/2"	Q-1	15	1.067		320	50		370	430
2140	3"	"	13	1.231		460	57.50		517.50	590
2350	Check, lift, class 150, horizontal composition disc, threaded									
2430	1/4"	1 Plum	24	.333	Ea.	130	17.35		147.35	169
2440	3/8"		24	.333		165	17.35		182.35	207
2450	1/2"		24	.333		143	17.35		160.35	183
2460	3/4"		20	.400		175	21		196	223
2470	1"		19	.421		248	22		270	305
2480	1-1/4"		15	.533		335	28		363	410
2490	1-1/2"		13	.615		400	32		432	490
2500	2"		11	.727		665	38		703	785
2850	Gate, N.R.S., soldered, 125 psi									
2900	3/8"	1 Plum	24	.333	Ea.	38	17.35		55.35	67.50
2920	1/2"		24	.333		32	17.35		49.35	61
2940	3/4"		20	.400		36.50	21		57.50	71
2950	1"		19	.421		52	22		74	90
2960	1-1/4"		15	.533		82.50	28		110.50	133
2970	1-1/2"		13	.615		88.50	32		120.50	145
2980	2"		11	.727		125	38		163	195
2990	2-1/2"	Q-1	15	1.067		315	50		365	420
3000	3"	"	13	1.231		410	57.50		467.50	535
3350	Threaded, class 150									
3410	1/4"	1 Plum	24	.333	Ea.	54	17.35		71.35	85.50
3420	3/8"		24	.333		54	17.35		71.35	85.50
3430	1/2"		24	.333		50.50	17.35		67.85	81.50
3440	3/4"		20	.400		58	21		79	94.50
3450	1"		19	.421		73.50	22		95.50	114
3460	1-1/4" size		15	.533		98.50	28		126.50	150
3470	1-1/2"		13	.615		142	32		174	204
3480	2"		11	.727		169	38		207	243
3490	2-1/2"	Q-1	15	1.067		395	50		445	510
3500	3" size	"	13	1.231		560	57.50		617.50	700
3600	Gate, flanged, 150 lb.									

22 05 23 – General-Duty Valves for Plumbing Piping

22 05 23.20 Valves, Bronze		Crew	Daily Output	Labor-Hours	Unit	2010 Bare Costs Material	Labor	Equipment	Total	Total Incl O&P
3610	1″	1 Plum	7	1.143	Ea.	1,075	59.50		1,134.50	1,275
3620	1-1/2″		6	1.333		1,325	69.50		1,394.50	1,550
3630	2″		5	1.600		1,950	83.50		2,033.50	2,275
3634	2-1/2″	Q-1	5	3.200		3,025	150		3,175	3,550
3640	3″	″	4.50	3.556		3,525	167		3,692	4,125
3850	Rising stem, soldered, 300 psi									
3900	3/8″	1 Plum	24	.333	Ea.	86	17.35		103.35	121
3920	1/2″		24	.333		86	17.35		103.35	121
3940	3/4″		20	.400		93.50	21		114.50	134
3950	1″		19	.421		136	22		158	183
3960	1-1/4″		15	.533		189	28		217	249
3970	1-1/2″		13	.615		222	32		254	292
3980	2″		11	.727		360	38		398	450
3990	2-1/2″	Q-1	15	1.067		805	50		855	960
4000	3″	″	13	1.231		1,150	57.50		1,207.50	1,325
4250	Threaded, class 150									
4310	1/4″	1 Plum	24	.333	Ea.	48	17.35		65.35	79
4320	3/8″		24	.333		48	17.35		65.35	79
4330	1/2″		24	.333		45	17.35		62.35	76
4340	3/4″		20	.400		52.50	21		73.50	89
4350	1″		19	.421		70.50	22		92.50	111
4360	1-1/4″		15	.533		95	28		123	146
4370	1-1/2″		13	.615		120	32		152	180
4380	2″		11	.727		162	38		200	236
4390	2-1/2″	Q-1	15	1.067		375	50		425	490
4400	3″	″	13	1.231		525	57.50		582.50	665
4500	For 300 psi, threaded, add					100%	15%			
4540	For chain operated type, add					15%				
4850	Globe, class 150, rising stem, threaded									
4920	1/4″	1 Plum	24	.333	Ea.	69	17.35		86.35	102
4940	3/8″ size		24	.333		69	17.35		86.35	102
4950	1/2″		24	.333		69	17.35		86.35	102
4960	3/4″		20	.400		92.50	21		113.50	133
4970	1″		19	.421		145	22		167	193
4980	1-1/4″		15	.533		231	28		259	296
4990	1-1/2″		13	.615		281	32		313	360
5000	2″		11	.727		420	38		458	520
5010	2-1/2″	Q-1	15	1.067		850	50		900	1,000
5020	3″	″	13	1.231		1,200	57.50		1,257.50	1,400
5120	For 300 lb threaded, add					50%	15%			
5130	Globe, 300 lb., sweat, 3/8″ size	1 Plum	24	.333	Ea.	86	17.35		103.35	121
5140	1/2″		24	.333		88.50	17.35		105.85	123
5150	3/4″		20	.400		120	21		141	163
5160	1″		19	.421		166	22		188	216
5170	1-1/4″		15	.533		263	28		291	330
5180	1-1/2″		13	.615		320	32		352	400
5190	2″		11	.727		470	38		508	570
5200	2-1/2″	Q-1	15	1.067		915	50		965	1,075
5210	3″	″	13	1.231		1,175	57.50		1,232.50	1,375
5600	Relief, pressure & temperature, self-closing, ASME, threaded									
5640	3/4″	1 Plum	28	.286	Ea.	124	14.85		138.85	160
5650	1″		24	.333		190	17.35		207.35	234
5660	1-1/4″		20	.400		360	21		381	425

22 05 23.20 Valves, Bronze		Crew	Daily Output	Labor-Hours	Unit	Material	Labor	Equipment	Total	Total Incl O&P
						2010 Bare Costs				
5670	1-1/2"	1 Plum	18	.444	Ea.	735	23		758	845
5680	2"		16	.500		755	26		781	870
5950	Pressure, poppet type, threaded									
6000	1/2"	1 Plum	30	.267	Ea.	33	13.90		46.90	57
6040	3/4"	"	28	.286	"	35	14.85		49.85	61
6400	Pressure, water, ASME, threaded									
6440	3/4"	1 Plum	28	.286	Ea.	70	14.85		84.85	99.50
6450	1"		24	.333		149	17.35		166.35	190
6460	1-1/4"		20	.400		236	21		257	290
6470	1-1/2"		18	.444		325	23		348	395
6480	2"		16	.500		470	26		496	560
6490	2-1/2"		15	.533		2,375	28		2,403	2,650
6900	Reducing, water pressure									
6920	300 psi to 25-75 psi, threaded or sweat									
6940	1/2"	1 Plum	24	.333	Ea.	225	17.35		242.35	273
6950	3/4"		20	.400		225	21		246	278
6960	1"		19	.421		350	22		372	420
6970	1-1/4"		15	.533		625	28		653	725
6980	1-1/2"		13	.615		940	32		972	1,075
6990	2"		11	.727		1,350	38		1,388	1,550
7100	For built-in by-pass or 10-35 psi, add					11.25			11.25	12.40
7700	High capacity, 250 psi to 25-75 psi, threaded									
7740	1/2"	1 Plum	24	.333	Ea.	225	17.35		242.35	273
7780	3/4"		20	.400		225	21		246	278
7790	1"		19	.421		350	22		372	420
7800	1-1/4"		15	.533		635	28		663	740
7810	1-1/2"		13	.615		940	32		972	1,075
7820	2"		11	.727		1,350	38		1,388	1,550
7830	2-1/2"		9	.889		1,825	46.50		1,871.50	2,075
7840	3"		8	1		2,125	52		2,177	2,400
7850	3" flanged (iron body)	Q-1	10	1.600		2,375	75		2,450	2,700
7860	4" flanged (iron body)	"	8	2		3,625	93.50		3,718.50	4,125
7920	For higher pressure, add					25%				
8000	Silent check, bronze trim									
8010	Compact wafer type, for 125 or 150 lb. flanges									
8020	1-1/2"	1 Plum	11	.727	Ea.	297	38		335	380
8021	2"	"	9	.889		315	46.50		361.50	415
8022	2-1/2"	Q-1	9	1.778		330	83.50		413.50	490
8023	3"		8	2		370	93.50		463.50	545
8024	4"		5	3.200		630	150		780	920
8025	5"	Q-2	6	4		835	194		1,029	1,200
8026	6"		5	4.800		985	233		1,218	1,425
8027	8"		4.50	5.333		2,100	259		2,359	2,725
8028	10"		4	6		3,850	291		4,141	4,650
8029	12"		3	8		8,700	390		9,090	10,200
8050	For 250 or 300 lb flanges, thru 6" no change									
8051	For 8" and 10", add				Ea.	40%	10%			
8060	Full flange wafer type, 150 lb.									
8061	1"	1 Plum	14	.571	Ea.	310	29.50		339.50	390
8062	1-1/4"		12	.667		325	34.50		359.50	405
8063	1-1/2"		11	.727		390	38		428	485
8064	2"		9	.889		525	46.50		571.50	650
8065	2-1/2"	Q-1	9	1.778		595	83.50		678.50	780

22 05 23 – General-Duty Valves for Plumbing Piping

22 05 23.20 Valves, Bronze		Crew	Daily Output	Labor-Hours	Unit	Material	Labor	Equipment	Total	Total Incl O&P
8066	3"	Q-1	8	2	Ea.	660	93.50		753.50	865
8067	4"		5	3.200		970	150		1,120	1,300
8068	5"	Q-2	6	4		1,250	194		1,444	1,675
8069	6"	"	5	4.800		1,650	233		1,883	2,175
8080	For 300 lb., add					40%	10%			
8100	Globe type, 150 lb.									
8110	2"	1 Plum	9	.889	Ea.	615	46.50		661.50	745
8111	2-1/2"	Q-1	9	1.778		760	83.50		843.50	960
8112	3"		8	2		910	93.50		1,003.50	1,150
8113	4"		5	3.200		1,250	150		1,400	1,600
8114	5"	Q-2	6	4		1,550	194		1,744	2,000
8115	6"	"	5	4.800		2,150	233		2,383	2,700
8130	For 300 lb., add					20%	10%			
8140	Screwed end type, 250 lb.									
8141	1/2"	1 Plum	24	.333	Ea.	55	17.35		72.35	86.50
8142	3/4"		20	.400		55	21		76	91.50
8143	1"		19	.421		63	22		85	103
8144	1-1/4"		15	.533		87	28		115	138
8145	1-1/2"		13	.615		97.50	32		129.50	155
8146	2"		11	.727		132	38		170	202
8350	Tempering, water, sweat connections									
8400	1/2"	1 Plum	24	.333	Ea.	74.50	17.35		91.85	108
8440	3/4"	"	20	.400	"	91	21		112	131
8650	Threaded connections									
8700	1/2"	1 Plum	24	.333	Ea.	112	17.35		129.35	149
8740	3/4"		20	.400		390	21		411	460
8750	1"		19	.421		430	22		452	510
8760	1-1/4"		15	.533		685	28		713	790
8770	1-1/2"		13	.615		745	32		777	870
8780	2"		11	.727		1,125	38		1,163	1,275

22 05 23.60 Valves, Plastic		Crew	Daily Output	Labor-Hours	Unit	Material	Labor	Equipment	Total	Total Incl O&P
0010	**VALVES, PLASTIC** R220523-90									
1100	Angle, PVC, threaded									
1110	1/4"	1 Plum	26	.308	Ea.	61	16		77	91
1120	1/2"		26	.308		61	16		77	91
1130	3/4"		25	.320		72.50	16.65		89.15	105
1140	1"		23	.348		88	18.10		106.10	124
1150	Ball, PVC, socket or threaded, single union									
1230	1/2"	1 Plum	26	.308	Ea.	21	16		37	47
1240	3/4"		25	.320		24	16.65		40.65	51.50
1250	1"		23	.348		30.50	18.10		48.60	60.50
1260	1-1/4"		21	.381		41	19.85		60.85	74.50
1270	1-1/2"		20	.400		48.50	21		69.50	84.50
1280	2"		17	.471		70.50	24.50		95	114
1290	2-1/2"	Q-1	26	.615		297	29		326	370
1300	3"		24	.667		360	31		391	440
1310	4"		20	.800		610	37.50		647.50	725
1360	For PVC, flanged, add					100%	15%			
1450	Double union 1/2"	1 Plum	26	.308		32	16		48	59
1460	3/4"		25	.320		38	16.65		54.65	66.50
1470	1"		23	.348		45	18.10		63.10	76.50
1480	1-1/4"		21	.381		62.50	19.85		82.35	98

22 05 23 – General-Duty Valves for Plumbing Piping

22 05 23.60	Valves, Plastic	Crew	Daily Output	Labor-Hours	Unit	Material	2010 Bare Costs Labor	Equipment	Total	Total Incl O&P
1490	1-1/2"	1 Plum	20	.400	Ea.	75	21		96	114
1500	2"		17	.471		99	24.50		123.50	146
1650	CPVC, socket or threaded, single union									
1700	1/2"	1 Plum	26	.308	Ea.	40	16		56	68
1720	3/4"		25	.320		50.50	16.65		67.15	80.50
1730	1"		23	.348		60	18.10		78.10	93
1750	1-1/4"		21	.381		95.50	19.85		115.35	135
1760	1-1/2"		20	.400		95.50	21		116.50	136
1770	2"		17	.471		132	24.50		156.50	182
1780	3"	Q-1	24	.667		380	31		411	465
1840	For CPVC, flanged, add					65%	15%			
1880	For true union, socket or threaded, add					50%	5%			
2050	Polypropylene, threaded									
2100	1/4"	1 Plum	26	.308	Ea.	42.50	16		58.50	70.50
2120	3/8"		26	.308		42.50	16		58.50	70.50
2130	1/2"		26	.308		42.50	16		58.50	70.50
2140	3/4"		25	.320		53	16.65		69.65	83
2150	1"		23	.348		61.50	18.10		79.60	95
2160	1-1/4"		21	.381		90.50	19.85		110.35	129
2170	1-1/2"		20	.400		104	21		125	145
2180	2"		17	.471		141	24.50		165.50	192
2190	3"	Q-1	24	.667		375	31		406	455
2200	4"	"	20	.800		620	37.50		657.50	735
2550	PVC, three way, socket or threaded									
2600	1/2"	1 Plum	26	.308	Ea.	48	16		64	76.50
2640	3/4"		25	.320		54.50	16.65		71.15	85
2650	1"		23	.348		58.50	18.10		76.60	91.50
2660	1-1/2"		20	.400		119	21		140	162
2670	2"		17	.471		159	24.50		183.50	211
2680	3"	Q-1	24	.667		345	31		376	425
2740	For flanged, add					60%	15%			
3150	Ball check, PVC, socket or threaded									
3200	1/4"	1 Plum	26	.308	Ea.	33	16		49	60.50
3220	3/8"		26	.308		33	16		49	60.50
3240	1/2"		26	.308		34.50	16		50.50	62
3250	3/4"		25	.320		37	16.65		53.65	66
3260	1"		23	.348		46.50	18.10		64.60	78
3270	1-1/4"		21	.381		78	19.85		97.85	115
3280	1-1/2"		20	.400		78	21		99	117
3290	2"		17	.471		106	24.50		130.50	154
3310	3"	Q-1	24	.667		294	31		325	370
3320	4"	"	20	.800		415	37.50		452.50	515
3360	For PVC, flanged, add					50%	15%			
3750	CPVC, socket or threaded									
3800	1/2"	1 Plum	26	.308	Ea.	48.50	16		64.50	77
3840	3/4"		25	.320		58	16.65		74.65	88.50
3850	1"		23	.348		68	18.10		86.10	102
3860	1-1/2"		20	.400		117	21		138	160
3870	2"		17	.471		158	24.50		182.50	211
3880	3"	Q-1	24	.667		425	31		456	515
3920	4"	"	20	.800		575	37.50		612.50	685
3930	For CPVC, flanged, add					40%	15%			
4340	Polypropylene, threaded									

22 05 23 – General-Duty Valves for Plumbing Piping

22 05 23.60 Valves, Plastic		Crew	Daily Output	Labor-Hours	Unit	Material	Labor	Equipment	Total	Total Incl O&P
						2010 Bare Costs				
4360	1/2"	1 Plum	26	.308	Ea.	35	16		51	62.50
4400	3/4"		25	.320		41.50	16.65		58.15	71
4440	1"		23	.348		53	18.10		71.10	85
4450	1-1/2"		20	.400		102	21		123	143
4460	2"		17	.471		128	24.50		152.50	178
4500	For polypropylene flanged, add					200%	15%			
4850	Foot valve, PVC, socket or threaded									
4900	1/2"	1 Plum	34	.235	Ea.	61.50	12.25		73.75	86.50
4930	3/4"		32	.250		69.50	13		82.50	96
4940	1"		28	.286		90.50	14.85		105.35	122
4950	1-1/4"		27	.296		174	15.40		189.40	214
4960	1-1/2"		26	.308		174	16		190	215
4970	2"		24	.333		202	17.35		219.35	248
4980	3"		20	.400		480	21		501	560
4990	4"		18	.444		845	23		868	965
5000	For flanged, add					25%	10%			
5050	CPVC, socket or threaded									
5060	1/2"	1 Plum	34	.235	Ea.	59.50	12.25		71.75	84
5070	3/4"		32	.250		69.50	13		82.50	96
5080	1"		28	.286		85	14.85		99.85	116
5090	1-1/4"		27	.296		136	15.40		151.40	172
5100	1-1/2"		26	.308		136	16		152	173
5110	2"		24	.333		174	17.35		191.35	218
5120	3"		20	.400		355	21		376	420
5130	4"		18	.444		645	23		668	745
5140	For flanged, add					25%	10%			
5280	Needle valve, PVC, threaded									
5300	1/4"	1 Plum	26	.308	Ea.	44	16		60	72.50
5340	3/8"		26	.308		51	16		67	80
5360	1/2"		26	.308		51	16		67	80
5380	For polypropylene, add					10%				
5800	Y check, PVC, socket or threaded									
5820	1/2"	1 Plum	26	.308	Ea.	68	16		84	98.50
5840	3/4"		25	.320		78	16.65		94.65	111
5850	1"		23	.348		79.50	18.10		97.60	115
5860	1-1/4"		21	.381		125	19.85		144.85	168
5870	1-1/2"		20	.400		137	21		158	181
5880	2"		17	.471		170	24.50		194.50	224
5890	2-1/2"		15	.533		360	28		388	435
5900	3"	Q-1	24	.667		425	31		456	515
5910	4"	"	20	.800		585	37.50		622.50	700
5960	For PVC flanged, add					45%	15%			
6350	Y sediment strainer, PVC, socket or threaded									
6400	1/2"	1 Plum	26	.308	Ea.	41	16		57	69
6440	3/4"		24	.333		44	17.35		61.35	74.50
6450	1"		23	.348		53	18.10		71.10	85
6460	1-1/4"		21	.381		87	19.85		106.85	126
6470	1-1/2"		20	.400		87	21		108	127
6480	2"		17	.471		107	24.50		131.50	155
6490	2-1/2"		15	.533		262	28		290	330
6500	3"	Q-1	24	.667		262	31		293	335
6510	4"	"	20	.800		435	37.50		472.50	535
6560	For PVC, flanged, add					55%	15%			

22 05 29 – Hangers and Supports for Plumbing Piping and Equipment

22 05 29.10 Hangers & Supp. for Plumb'g/HVAC Pipe/Equip.		Crew	Daily Output	Labor-Hours	Unit	2010 Bare Costs Material	Labor	Equipment	Total	Total Incl O&P
0010	**HANGERS AND SUPPORTS FOR PLUMB'G/HVAC PIPE/EQUIP.**									
0011	TYPE numbers per MSS-SP58									
0050	Brackets									
0060	Beam side or wall, malleable iron, TYPE 34									
0070	3/8" threaded rod size	1 Plum	48	.167	Ea.	3.01	8.70		11.71	16.30
0080	1/2" threaded rod size		48	.167		5.35	8.70		14.05	18.85
0090	5/8" threaded rod size		48	.167		8.20	8.70		16.90	22
0100	3/4" threaded rod size		48	.167		9.85	8.70		18.55	24
0110	7/8" threaded rod size		48	.167		11.50	8.70		20.20	25.50
0120	For concrete installation, add						30%			
0150	Wall, welded steel, medium, TYPE 32									
0160	0 size, 12" wide, 18" deep	1 Plum	34	.235	Ea.	204	12.25		216.25	242
0170	1 size, 18" wide, 24" deep		34	.235		242	12.25		254.25	285
0180	2 size, 24" wide, 30" deep		34	.235		320	12.25		332.25	375
0200	Beam attachment, welded, TYPE 22									
0202	3/8"	Q-15	80	.200	Ea.	6.95	9.35	.73	17.03	22.50
0203	1/2"		76	.211		6.95	9.85	.77	17.57	23.50
0204	5/8"		72	.222		7.30	10.40	.81	18.51	24.50
0205	3/4"		68	.235		7.95	11	.86	19.81	26
0206	7/8"		64	.250		11.55	11.70	.92	24.17	31.50
0207	1"		56	.286		16.50	13.40	1.05	30.95	39.50
0300	Clamps									
0310	C-clamp, for mounting on steel beam flange, w/locknut, TYPE 23									
0320	3/8" threaded rod size	1 Plum	160	.050	Ea.	2.83	2.60		5.43	7
0330	1/2" threaded rod size		160	.050		3.50	2.60		6.10	7.75
0340	5/8" threaded rod size		160	.050		5.35	2.60		7.95	9.80
0350	3/4" threaded rod size		160	.050		7.25	2.60		9.85	11.90
0352	7/8" threaded rod size		140	.057		23	2.97		25.97	30
0400	High temperature to 1050°F, alloy steel									
0410	4" pipe size	Q-1	106	.151	Ea.	22.50	7.05		29.55	35
0420	6" pipe size		106	.151		37.50	7.05		44.55	52
0430	8" pipe size		97	.165		42.50	7.75		50.25	58.50
0440	10" pipe size		84	.190		69.50	8.90		78.40	90
0450	12" pipe size		72	.222		80.50	10.40		90.90	104
0460	14" pipe size		64	.250		218	11.70		229.70	258
0470	16" pipe size		56	.286		228	13.40		241.40	271
0480	Beam clamp, flange type, TYPE 25									
0482	For 3/8" bolt	1 Plum	48	.167	Ea.	4.86	8.70		13.56	18.35
0483	For 1/2" bolt		44	.182		6.70	9.45		16.15	21.50
0484	For 5/8" bolt		40	.200		6.70	10.40		17.10	23
0485	For 3/4" bolt		36	.222		6.70	11.55		18.25	24.50
0486	For 1" bolt		32	.250		6.70	13		19.70	27
0500	I-beam, for mounting on bottom flange, strap iron, TYPE 21									
0510	2" flange size	1 Plum	96	.083	Ea.	13.20	4.34		17.54	21
0520	3" flange size		95	.084		15.70	4.38		20.08	24
0530	4" flange size		93	.086		6.90	4.48		11.38	14.30
0540	5" flange size		92	.087		7.60	4.53		12.13	15.15
0550	6" flange size		90	.089		8.70	4.63		13.33	16.50
0560	7" flange size		88	.091		9.55	4.73		14.28	17.60
0570	8" flange size		86	.093		10.20	4.84		15.04	18.45
0600	One hole, vertical mounting, malleable iron									
0610	1/2" pipe size	1 Plum	160	.050	Ea.	1.02	2.60		3.62	5

22 05 Common Work Results for Plumbing

22 05 29 – Hangers and Supports for Plumbing Piping and Equipment

22 05 29.10 Hangers & Supp. for Plumb'g/HVAC Pipe/Equip.		Crew	Daily Output	Labor-Hours	Unit	Material	2010 Bare Costs Labor	Equipment	Total	Total Incl O&P
0620	3/4" pipe size	1 Plum	145	.055	Ea.	1.26	2.87		4.13	5.70
0630	1" pipe size		136	.059		1.72	3.06		4.78	6.50
0640	1-1/4" pipe size		128	.063		2.07	3.25		5.32	7.15
0650	1-1/2" pipe size		120	.067		2.31	3.47		5.78	7.75
0660	2" pipe size		112	.071		3.36	3.72		7.08	9.25
0670	2-1/2" pipe size		104	.077		6.55	4		10.55	13.20
0680	3" pipe size		96	.083		9.30	4.34		13.64	16.75
0690	3-1/2" pipe size		90	.089		15.15	4.63		19.78	23.50
0700	4" pipe size		84	.095		19.55	4.96		24.51	29
0750	Riser or extension pipe, carbon steel, TYPE 8									
0756	1/2" pipe size	1 Plum	52	.154	Ea.	4.09	8		12.09	16.50
0760	3/4" pipe size		48	.167		4.09	8.70		12.79	17.50
0770	1" pipe size		47	.170		4.20	8.85		13.05	17.90
0780	1-1/4" pipe size		46	.174		5.25	9.05		14.30	19.30
0790	1-1/2" pipe size		45	.178		5.60	9.25		14.85	20
0800	2" pipe size		43	.186		5.80	9.70		15.50	21
0810	2-1/2" pipe size		41	.195		6.20	10.15		16.35	22
0820	3" pipe size		40	.200		6.85	10.40		17.25	23
0830	3-1/2" pipe size		39	.205		12.35	10.70		23.05	29.50
0840	4" pipe size		38	.211		8.75	10.95		19.70	26
0850	5" pipe size		37	.216		11.65	11.25		22.90	29.50
0860	6" pipe size		36	.222		14.65	11.55		26.20	33.50
0870	8" pipe size		34	.235		24	12.25		36.25	45
0880	10" pipe size		32	.250		33.50	13		46.50	56
0890	12" pipe size		28	.286		47.50	14.85		62.35	74.50
0900	For plastic coating 3/4" to 4", add					190%				
0910	For copper plating 3/4" to 4", add					58%				
0950	Two piece, complete, carbon steel, medium weight, TYPE 4									
0960	1/2" pipe size	Q-1	137	.117	Ea.	2.95	5.45		8.40	11.45
0970	3/4" pipe size		134	.119		2.95	5.60		8.55	11.65
0980	1" pipe size		132	.121		2.98	5.70		8.68	11.80
0990	1-1/4" pipe size		130	.123		3.91	5.75		9.66	12.95
1000	1-1/2" pipe size		126	.127		3.91	5.95		9.86	13.20
1010	2" pipe size		124	.129		4.56	6.05		10.61	14.05
1020	2-1/2" pipe size		120	.133		4.85	6.25		11.10	14.70
1030	3" pipe size		117	.137		5.40	6.40		11.80	15.55
1040	3-1/2" pipe size		114	.140		7.65	6.60		14.25	18.25
1050	4" pipe size		110	.145		7.65	6.80		14.45	18.60
1060	5" pipe size		106	.151		14.70	7.05		21.75	27
1070	6" pipe size		104	.154		18.70	7.20		25.90	31.50
1080	8" pipe size		100	.160		22.50	7.50		30	36
1090	10" pipe size		96	.167		40	7.80		47.80	55.50
1100	12" pipe size		89	.180		51.50	8.40		59.90	69
1110	14" pipe size		82	.195		99.50	9.15		108.65	123
1120	16" pipe size		68	.235		107	11		118	135
1130	For galvanized, add					45%				
1150	Insert, concrete									
1160	Wedge type, carbon steel body, malleable iron nut									
1170	1/4" threaded rod size	1 Plum	96	.083	Ea.	1.35	4.34		5.69	8
1180	3/8" threaded rod size		96	.083		1.36	4.34		5.70	8
1190	1/2" threaded rod size		96	.083		1.44	4.34		5.78	8.10
1200	5/8" threaded rod size		96	.083		1.47	4.34		5.81	8.10
1210	3/4" threaded rod size		96	.083		1.73	4.34		6.07	8.40

	22 05 29.10 Hangers & Supp. for Plumb'g/HVAC Pipe/Equip.	Crew	Daily Output	Labor-Hours	Unit	2010 Bare Costs Material	Labor	Equipment	Total	Total Incl O&P
1220	7/8" threaded rod size	1 Plum	96	.083	Ea.	2.17	4.34		6.51	8.90
1230	For galvanized, add					.70			.70	.77
1250	Pipe guide sized for insulation									
1260	No. 1, 1" pipe size, 1" thick insulation	1 Stpi	26	.308	Ea.	137	15.95		152.95	174
1270	No. 2, 1-1/4"-2" pipe size, 1" thick insulation		23	.348		161	18.05		179.05	204
1280	No. 3, 1-1/4"-2" pipe size, 1-1/2" thick insulation		21	.381		161	19.75		180.75	207
1290	No. 4, 2-1/2"-3-1/2" pipe size, 1-1/2" thick insulation		18	.444		161	23		184	212
1300	No. 5, 4"-5" pipe size, 1-1/2" thick insulation		16	.500		179	26		205	236
1310	No. 6, 5"-6" pipe size, 2" thick insulation	Q-5	21	.762		199	35.50		234.50	273
1320	No. 7, 8" pipe size, 2" thick insulation		16	1		265	46.50		311.50	360
1330	No. 8, 10" pipe size, 2" thick insulation		12	1.333		440	62.50		502.50	580
1340	No. 9, 12" pipe size, 2" thick insulation	Q-6	17	1.412		440	68.50		508.50	585
1350	No. 10, 12"-14" pipe size, 2-1/2" thick insulation		16	1.500		490	72.50		562.50	650
1360	No. 11, 16" pipe size, 2-1/2" thick insulation		10.50	2.286		490	111		601	705
1370	No. 12, 16"-18" pipe size, 3" thick insulation		9	2.667		690	129		819	955
1380	No. 13, 20" pipe size, 3" thick insulation		7.50	3.200		690	155		845	990
1390	No. 14, 24" pipe size, 3" thick insulation		7	3.429		940	166		1,106	1,275
1400	Bands									
1410	Adjustable band, carbon steel, for non-insulated pipe, TYPE 7									
1420	1/2" pipe size	Q-1	142	.113	Ea.	.65	5.30		5.95	8.60
1430	3/4" pipe size		140	.114		.67	5.35		6.02	8.80
1440	1" pipe size		137	.117		.65	5.45		6.10	8.90
1450	1-1/4" pipe size		134	.119		.69	5.60		6.29	9.15
1460	1-1/2" pipe size		131	.122		.70	5.70		6.40	9.35
1470	2" pipe size		129	.124		.70	5.80		6.50	9.45
1480	2-1/2" pipe size		125	.128		1.26	6		7.26	10.40
1490	3" pipe size		122	.131		1.38	6.15		7.53	10.70
1500	3-1/2" pipe size		119	.134		2.12	6.30		8.42	11.80
1510	4" pipe size		114	.140		2.12	6.60		8.72	12.20
1520	5" pipe size		110	.145		3.74	6.80		10.54	14.30
1530	6" pipe size		108	.148		4.18	6.95		11.13	15
1540	8" pipe size		104	.154		5.60	7.20		12.80	16.95
1550	For copper plated, add					50%				
1560	For galvanized, add					30%				
1570	For plastic coating, add					30%				
1600	Adjusting nut malleable iron, steel band, TYPE 9									
1610	1/2" pipe size, galvanized band	Q-1	137	.117	Ea.	1.17	5.45		6.62	9.50
1620	3/4" pipe size, galvanized band		135	.119		1.31	5.55		6.86	9.75
1630	1" pipe size, galvanized band		132	.121		1.31	5.70		7.01	9.95
1640	1-1/4" pipe size, galvanized band		129	.124		1.40	5.80		7.20	10.25
1650	1-1/2" pipe size, galvanized band		126	.127		1.40	5.95		7.35	10.45
1660	2" pipe size, galvanized band		124	.129		1.49	6.05		7.54	10.70
1670	2-1/2" pipe size, galvanized band		120	.133		2.48	6.25		8.73	12.10
1680	3" pipe size, galvanized band		117	.137		2.57	6.40		8.97	12.45
1690	3-1/2" pipe size, galvanized band		114	.140		3.42	6.60		10.02	13.60
1700	4" pipe size, cadmium plated band		110	.145		3.42	6.80		10.22	13.95
1740	For plastic coated band, add					35%				
1750	For completely copper coated, add					45%				
1800	Clevis, adjustable, carbon steel, for non-insulated pipe, TYPE 1									
1810	1/2" pipe size	Q-1	137	.117	Ea.	1.45	5.45		6.90	9.80
1820	3/4" pipe size		135	.119		1.45	5.55		7	9.90
1830	1" pipe size		132	.121		1.50	5.70		7.20	10.15
1840	1-1/4" pipe size		129	.124		1.58	5.80		7.38	10.45

22 05 29 – Hangers and Supports for Plumbing Piping and Equipment

22 05 29.10 Hangers & Supp. for Plumb'g/HVAC Pipe/Equip.		Crew	Daily Output	Labor-Hours	Unit	2010 Bare Costs Material	Labor	Equipment	Total	Total Incl O&P
1850	1-1/2" pipe size	Q-1	126	.127	Ea.	1.66	5.95		7.61	10.75
1860	2" pipe size		124	.129		1.99	6.05		8.04	11.25
1870	2-1/2" pipe size		120	.133		3.21	6.25		9.46	12.90
1880	3" pipe size		117	.137		3.92	6.40		10.32	13.90
1890	3-1/2" pipe size		114	.140		4.24	6.60		10.84	14.50
1900	4" pipe size		110	.145		4.78	6.80		11.58	15.45
1910	5" pipe size		106	.151		6.65	7.05		13.70	17.95
1920	6" pipe size		104	.154		8.25	7.20		15.45	19.90
1930	8" pipe size		100	.160		13.60	7.50		21.10	26
1940	10" pipe size		96	.167		23.50	7.80		31.30	37.50
1950	12" pipe size		89	.180		31	8.40		39.40	46.50
1960	14" pipe size		82	.195		41	9.15		50.15	58.50
1970	16" pipe size		68	.235		63.50	11		74.50	86.50
1971	18" pipe size		54	.296		77	13.90		90.90	106
1972	20" pipe size	↓	38	.421	↓	146	19.75		165.75	190
1980	For galvanized, add					66%				
1990	For copper plated 1/2" to 4", add					77%				
2000	For light weight 1/2" to 4", deduct					13%				
2010	Insulated pipe type, 3/4" to 12" pipe, add					180%				
2020	Insulated pipe type, chrome-moly U-strap, add					530%				
2250	Split ring, malleable iron, for non-insulated pipe, TYPE 11									
2260	1/2" pipe size	Q-1	137	.117	Ea.	5.35	5.45		10.80	14.05
2270	3/4" pipe size		135	.119		5.35	5.55		10.90	14.20
2280	1" pipe size		132	.121		5.70	5.70		11.40	14.75
2290	1-1/4" pipe size		129	.124		7.05	5.80		12.85	16.50
2300	1-1/2" pipe size		126	.127		8.60	5.95		14.55	18.35
2310	2" pipe size		124	.129		9.70	6.05		15.75	19.75
2320	2-1/2" pipe size		120	.133		14.05	6.25		20.30	25
2330	3" pipe size		117	.137		17.50	6.40		23.90	29
2340	3-1/2" pipe size		114	.140		18.30	6.60		24.90	30
2350	4" pipe size		110	.145		18.30	6.80		25.10	30
2360	5" pipe size		106	.151		24.50	7.05		31.55	37
2370	6" pipe size		104	.154		54.50	7.20		61.70	71
2380	8" pipe size	↓	100	.160	↓	89.50	7.50		97	110
2390	For copper plated, add					8%				
2500	Washer, flat steel									
2502	3/8"	1 Plum	240	.033	Ea.	.09	1.73		1.82	2.70
2503	1/2"		220	.036		.23	1.89		2.12	3.09
2504	5/8"		200	.040		.41	2.08		2.49	3.57
2505	3/4"		180	.044		.54	2.31		2.85	4.06
2506	7/8"		160	.050		.82	2.60		3.42	4.80
2507	1"		140	.057		1.01	2.97		3.98	5.55
2508	1-1/4"	↓	120	.067	↓	2.27	3.47		5.74	7.70
2520	Nut, steel, hex									
2522	3/8"	1 Plum	200	.040	Ea.	.38	2.08		2.46	3.54
2523	1/2"		180	.044		.54	2.31		2.85	4.06
2524	5/8"		160	.050		1.14	2.60		3.74	5.15
2525	3/4"		140	.057		1.67	2.97		4.64	6.30
2526	7/8"		120	.067		2.11	3.47		5.58	7.50
2527	1"		100	.080		3.43	4.16		7.59	10
2528	1-1/4"	↓	80	.100	↓	4.32	5.20		9.52	12.55
2532	Turnbuckle, TYPE 13									
2534	3/8"	1 Plum	80	.100	Ea.	17.70	5.20		22.90	27.50

22 05 29 – Hangers and Supports for Plumbing Piping and Equipment

22 05 29.10 Hangers & Supp. for Plumb'g/HVAC Pipe/Equip.		Crew	Daily Output	Labor-Hours	Unit	Material	2010 Bare Costs Labor	Equipment	Total	Total Incl O&P
2535	1/2"	1 Plum	72	.111	Ea.	23.50	5.80		29.30	34
2536	5/8"		64	.125		34.50	6.50		41	48
2537	3/4"		56	.143		60	7.45		67.45	77
2538	7/8"		48	.167		74	8.70		82.70	94.50
2539	1"		40	.200		103	10.40		113.40	130
2540	1-1/4"		32	.250		120	13		133	152
2544	Eye rod, welded									
2546	3/8"	1 Plum	56	.143	Ea.	16.90	7.45		24.35	30
2547	1/2"		48	.167		25	8.70		33.70	40.50
2548	5/8"		40	.200		38	10.40		48.40	57
2549	3/4"		32	.250		48	13		61	72.50
2550	7/8"		24	.333		59	17.35		76.35	90.50
2551	1"		16	.500		77	26		103	124
2556	Eye rod, welded, linked									
2558	3/8"	1 Plum	56	.143	Ea.	91	7.45		98.45	111
2559	1/2"		48	.167		122	8.70		130.70	147
2560	5/8"		40	.200		174	10.40		184.40	208
2561	3/4"		32	.250		205	13		218	245
2562	7/8"		24	.333		281	17.35		298.35	335
2563	1"		16	.500		355	26		381	430
2650	Rods, carbon steel									
2660	Continuous thread									
2670	1/4" thread size	1 Plum	144	.056	L.F.	.16	2.89		3.05	4.51
2680	3/8" thread size		144	.056		.25	2.89		3.14	4.61
2690	1/2" thread size		144	.056		.45	2.89		3.34	4.83
2700	5/8" thread size		144	.056		.82	2.89		3.71	5.25
2710	3/4" thread size		144	.056		1.20	2.89		4.09	5.65
2720	7/8" thread size		144	.056		1.73	2.89		4.62	6.25
2721	1" thread size	Q-1	160	.100		2.45	4.69		7.14	9.70
2722	1-1/8" thread size	"	120	.133		12.50	6.25		18.75	23
2725	1/4" thread size, bright finish	1 Plum	144	.056		1.76	2.89		4.65	6.25
2726	1/2" thread size, bright finish	"	144	.056		3.01	2.89		5.90	7.65
2730	For galvanized, add					40%				
2750	Both ends machine threaded 18" length									
2760	3/8" thread size	1 Plum	240	.033	Ea.	5.80	1.73		7.53	9
2770	1/2" thread size		240	.033		10.70	1.73		12.43	14.40
2780	5/8" thread size		240	.033		15.40	1.73		17.13	19.55
2790	3/4" thread size		240	.033		24.50	1.73		26.23	29.50
2800	7/8" thread size		240	.033		33.50	1.73		35.23	39
2810	1" thread size		240	.033		39.50	1.73		41.23	46
2820	Rod couplings									
2821	3/8"	1 Plum	60	.133	Ea.	1.64	6.95		8.59	12.20
2822	1/2"		54	.148		2.19	7.70		9.89	13.95
2823	5/8"		48	.167		2.51	8.70		11.21	15.75
2824	3/4"		44	.182		3.45	9.45		12.90	18
2825	7/8"		40	.200		5.80	10.40		16.20	22
2826	1"		34	.235		7.75	12.25		20	27
2827	1-1/8"		30	.267		38.50	13.90		52.40	63
2860	Pipe hanger assy, adj. clevis, saddle, rod, clamp, insul. allowance									
2864	1/2" pipe size	Q-5	35	.457	Ea.	39.50	21.50		61	75.50
2866	3/4" pipe size		34.80	.460		41	21.50		62.50	77
2868	1" pipe size		34.60	.462		34.50	21.50		56	70.50
2869	1-1/4" pipe size		34.30	.466		34.50	22		56.50	70.50

22 05 29 – Hangers and Supports for Plumbing Piping and Equipment

22 05 29.10 Hangers & Supp. for Plumb'g/HVAC Pipe/Equip.		Crew	Daily Output	Labor-Hours	Unit	2010 Bare Costs Material	Labor	Equipment	Total	Total Incl O&P
2870	1-1/2" pipe size	Q-5	33.90	.472	Ea.	35.50	22		57.50	72
2872	2" pipe size		33.30	.480		40	22.50		62.50	77.50
2874	2-1/2" pipe size		32.30	.495		46	23		69	85
2876	3" pipe size		31.20	.513		55	24		79	96.50
2880	4" pipe size		30.70	.521		59	24.50		83.50	102
2884	6" pipe size		29.80	.537		74	25		99	119
2888	8" pipe size		28	.571		104	26.50		130.50	154
2892	10" pipe size		25.20	.635		117	29.50		146.50	173
2896	12" pipe size		23.20	.690		169	32		201	235
2900	Rolls									
2910	Adjustable yoke, carbon steel with CI roll, TYPE 43									
2918	2" pipe size	Q-1	140	.114	Ea.	14.60	5.35		19.95	24
2920	2-1/2" pipe size		137	.117		14.60	5.45		20.05	24.50
2930	3" pipe size		131	.122		16.25	5.70		21.95	26.50
2940	3-1/2" pipe size		124	.129		21.50	6.05		27.55	33
2950	4" pipe size		117	.137		21.50	6.40		27.90	33.50
2960	5" pipe size		110	.145		25.50	6.80		32.30	38
2970	6" pipe size		104	.154		33.50	7.20		40.70	47.50
2980	8" pipe size		96	.167		47	7.80		54.80	63
2990	10" pipe size		80	.200		59.50	9.35		68.85	79.50
3000	12" pipe size		68	.235		87.50	11		98.50	113
3010	14" pipe size		56	.286		219	13.40		232.40	261
3020	16" pipe size		48	.333		269	15.60		284.60	320
3050	Chair, carbon steel with CI roll									
3060	2" pipe size	1 Plum	68	.118	Ea.	19.20	6.10		25.30	30
3070	2-1/2" pipe size		65	.123		21	6.40		27.40	32.50
3080	3" pipe size		62	.129		21.50	6.70		28.20	33.50
3090	3-1/2" pipe size		60	.133		27	6.95		33.95	40.50
3100	4" pipe size		58	.138		29	7.20		36.20	43
3110	5" pipe size		56	.143		31.50	7.45		38.95	46
3120	6" pipe size		53	.151		44	7.85		51.85	60.50
3130	8" pipe size		50	.160		59.50	8.35		67.85	78
3140	10" pipe size		48	.167		77.50	8.70		86.20	98
3150	12" pipe size		46	.174		109	9.05		118.05	134
3170	Single pipe roll, (see line 2650 for rods), TYPE 41, 1" pipe size	Q-1	137	.117		12.80	5.45		18.25	22.50
3180	1-1/4" pipe size		131	.122		13.25	5.70		18.95	23
3190	1-1/2" pipe size		129	.124		13.70	5.80		19.50	24
3200	2" pipe size		124	.129		14.15	6.05		20.20	24.50
3210	2-1/2" pipe size		118	.136		15.20	6.35		21.55	26.50
3220	3" pipe size		115	.139		16.05	6.50		22.55	27.50
3230	3-1/2" pipe size		113	.142		17.80	6.65		24.45	29.50
3240	4" pipe size		112	.143		17.80	6.70		24.50	29.50
3250	5" pipe size		110	.145		32.50	6.80		39.30	46
3260	6" pipe size		101	.158		24.50	7.40		31.90	38
3270	8" pipe size		90	.178		45.50	8.35		53.85	62.50
3280	10" pipe size		80	.200		54	9.35		63.35	73
3290	12" pipe size		68	.235		83	11		94	108
3291	14" pipe size		56	.286		129	13.40		142.40	162
3292	16" pipe size		48	.333		175	15.60		190.60	217
3293	18" pipe size		40	.400		188	18.75		206.75	235
3294	20" pipe size		35	.457		258	21.50		279.50	315
3296	24" pipe size		30	.533		410	25		435	490
3297	30" pipe size		25	.640		805	30		835	930

22 05 29.10 Hangers & Supp. for Plumb'g/HVAC Pipe/Equip.		Crew	Daily Output	Labor-Hours	Unit	2010 Bare Costs Material	Labor	Equipment	Total	Total Incl O&P
3298	36" pipe size	Q-1	20	.800	Ea.	990	37.50		1,027.50	1,150
3300	Saddles (add vertical pipe riser, usually 3" diameter)									
3310	Pipe support, complete, adjust., CI saddle, TYPE 36									
3320	2-1/2" pipe size	1 Plum	96	.083	Ea.	126	4.34		130.34	146
3330	3" pipe size		88	.091		128	4.73		132.73	148
3340	3-1/2" pipe size		79	.101		125	5.25		130.25	146
3350	4" pipe size		68	.118		167	6.10		173.10	192
3360	5" pipe size		64	.125		169	6.50		175.50	196
3370	6" pipe size		59	.136		175	7.05		182.05	203
3380	8" pipe size		53	.151		182	7.85		189.85	213
3390	10" pipe size		50	.160		181	8.35		189.35	212
3400	12" pipe size		48	.167		209	8.70		217.70	243
3450	For standard pipe support, one piece, CI deduct					34%				
3460	For stanchion support, CI with steel yoke, deduct					60%				
3550	Insulation shield 1" thick, 1/2" pipe size, TYPE 40	1 Asbe	100	.080	Ea.	5.45	3.64		9.09	11.65
3560	3/4" pipe size		100	.080		6.30	3.64		9.94	12.60
3570	1" pipe size		98	.082		6.75	3.72		10.47	13.25
3580	1-1/4" pipe size		98	.082		7.25	3.72		10.97	13.80
3590	1-1/2" pipe size		96	.083		7.25	3.80		11.05	13.90
3600	2" pipe size		96	.083		7.50	3.80		11.30	14.15
3610	2-1/2" pipe size		94	.085		7.75	3.88		11.63	14.60
3620	3" pipe size		94	.085		8.30	3.88		12.18	15.15
3630	2" thick, 3-1/2" pipe size		92	.087		9.30	3.96		13.26	16.35
3640	4" pipe size		92	.087		13.25	3.96		17.21	20.50
3650	5" pipe size		90	.089		17.10	4.05		21.15	25
3660	6" pipe size		90	.089		19.55	4.05		23.60	28
3670	8" pipe size		88	.091		29.50	4.14		33.64	39
3680	10" pipe size		88	.091		40	4.14		44.14	50.50
3690	12" pipe size		86	.093		45.50	4.24		49.74	57
3700	14" pipe size		86	.093		88	4.24		92.24	104
3710	16" pipe size		84	.095		94	4.34		98.34	110
3720	18" pipe size		84	.095		104	4.34		108.34	122
3730	20" pipe size		82	.098		114	4.44		118.44	133
3732	24" pipe size		80	.100		139	4.56		143.56	160
3733	30" pipe size		78	.103		460	4.67		464.67	510
3735	36" pipe size		75	.107		695	4.86		699.86	770
3750	Covering protection saddle, TYPE 39									
3760	1" covering size									
3770	3/4" pipe size	1 Plum	68	.118	Ea.	9.90	6.10		16	20
3780	1" pipe size		68	.118		9.90	6.10		16	20
3790	1-1/4" pipe size		68	.118		9.90	6.10		16	20
3800	1-1/2" pipe size		66	.121		10.75	6.30		17.05	21.50
3810	2" pipe size		66	.121		10.75	6.30		17.05	21.50
3820	2-1/2" pipe size		64	.125		10.75	6.50		17.25	21.50
3830	3" pipe size		64	.125		15.05	6.50		21.55	26.50
3840	3-1/2" pipe size		62	.129		16	6.70		22.70	27.50
3850	4" pipe size		62	.129		16	6.70		22.70	27.50
3860	5" pipe size		60	.133		16	6.95		22.95	28
3870	6" pipe size		60	.133		19.15	6.95		26.10	31.50
3900	1-1/2" covering size									
3910	3/4" pipe size	1 Plum	68	.118	Ea.	12.05	6.10		18.15	22.50
3920	1" pipe size		68	.118		12.05	6.10		18.15	22.50
3930	1-1/4" pipe size		68	.118		12.05	6.10		18.15	22.50

22 05 29 – Hangers and Supports for Plumbing Piping and Equipment

22 05 29.10 Hangers & Supp. for Plumb'g/HVAC Pipe/Equip.		Crew	Daily Output	Labor-Hours	Unit	Material	2010 Bare Costs Labor	Equipment	Total	Total Incl O&P
3940	1-1/2" pipe size	1 Plum	66	.121	Ea.	12.05	6.30		18.35	22.50
3950	2" pipe size		66	.121		13.10	6.30		19.40	24
3960	2-1/2" pipe size		64	.125		15.45	6.50		21.95	27
3970	3" pipe size		64	.125		15.45	6.50		21.95	27
3980	3-1/2" pipe size		62	.129		24	6.70		30.70	36.50
3990	4" pipe size		62	.129		15.65	6.70		22.35	27.50
4000	5" pipe size		60	.133		15.65	6.95		22.60	27.50
4010	6" pipe size		60	.133		21	6.95		27.95	33.50
4020	8" pipe size		58	.138		28.50	7.20		35.70	42
4022	10" pipe size		56	.143		28.50	7.45		35.95	42
4024	12" pipe size		54	.148		50.50	7.70		58.20	67
4028	2" covering size									
4029	2-1/2" pipe size	1 Plum	62	.129	Ea.	14.80	6.70		21.50	26.50
4032	3" pipe size		60	.133		17.40	6.95		24.35	29.50
4033	4" pipe size		58	.138		17.40	7.20		24.60	30
4034	6" pipe size		56	.143		25	7.45		32.45	38.50
4035	8" pipe size		54	.148		29.50	7.70		37.20	44
4080	10" pipe size		58	.138		31.50	7.20		38.70	45.50
4090	12" pipe size		56	.143		56.50	7.45		63.95	73
4100	14" pipe size		56	.143		56.50	7.45		63.95	73
4110	16" pipe size		54	.148		79	7.70		86.70	98.50
4120	18" pipe size		54	.148		79	7.70		86.70	98.50
4130	20" pipe size		52	.154		88	8		96	109
4150	24" pipe size		50	.160		103	8.35		111.35	126
4160	30" pipe size		48	.167		113	8.70		121.70	137
4180	36" pipe size		45	.178		125	9.25		134.25	151
4186	2-1/2" covering size									
4187	3" pipe size	1 Plum	58	.138	Ea.	16.80	7.20		24	29
4188	4" pipe size		56	.143		18.25	7.45		25.70	31
4189	6" pipe size		52	.154		28	8		36	42.50
4190	8" pipe size		48	.167		34	8.70		42.70	50.50
4191	10" pipe size		44	.182		37	9.45		46.45	55
4192	12" pipe size		40	.200		61.50	10.40		71.90	83
4193	14" pipe size		36	.222		61.50	11.55		73.05	85
4194	16" pipe size		32	.250		83	13		96	111
4195	18" pipe size		28	.286		92.50	14.85		107.35	125
4200	Sockets									
4210	Rod end, malleable iron, TYPE 16									
4220	1/4" thread size	1 Plum	240	.033	Ea.	1.66	1.73		3.39	4.43
4230	3/8" thread size		240	.033		1.81	1.73		3.54	4.59
4240	1/2" thread size		230	.035		2.16	1.81		3.97	5.10
4250	5/8" thread size		225	.036		4.26	1.85		6.11	7.45
4260	3/4" thread size		220	.036		6.95	1.89		8.84	10.50
4270	7/8" thread size		210	.038		12.65	1.98		14.63	16.85
4290	Strap, 1/2" pipe size, TYPE 26	Q-1	142	.113		3.80	5.30		9.10	12.10
4300	3/4" pipe size		140	.114		3.88	5.35		9.23	12.30
4310	1" pipe size		137	.117		4.06	5.45		9.51	12.65
4320	1-1/4" pipe size		134	.119		4.47	5.60		10.07	13.30
4330	1-1/2" pipe size		131	.122		5.65	5.70		11.35	14.80
4340	2" pipe size		129	.124		5.80	5.80		11.60	15.10
4350	2-1/2" pipe size		125	.128		10.45	6		16.45	20.50
4360	3" pipe size		122	.131		11	6.15		17.15	21.50
4370	3-1/2" pipe size		119	.134		13.10	6.30		19.40	24

22 05 29 – Hangers and Supports for Plumbing Piping and Equipment

22 05 29.10 Hangers & Supp. for Plumb'g/HVAC Pipe/Equip.		Crew	Daily Output	Labor-Hours	Unit	Material	2010 Bare Costs Labor	Equipment	Total	Total Incl O&P
4380	4" pipe size	Q-1	114	.140	Ea.	13.90	6.60		20.50	25
4400	U-bolt, carbon steel									
4410	Standard, with nuts, TYPE 42									
4420	1/2" pipe size	1 Plum	160	.050	Ea.	1.66	2.60		4.26	5.75
4430	3/4" pipe size		158	.051		1.66	2.64		4.30	5.80
4450	1" pipe size		152	.053		1.69	2.74		4.43	5.95
4460	1-1/4" pipe size		148	.054		2.01	2.81		4.82	6.45
4470	1-1/2" pipe size		143	.056		2.16	2.91		5.07	6.75
4480	2" pipe size		139	.058		2.36	3		5.36	7.10
4490	2-1/2" pipe size		134	.060		3.88	3.11		6.99	8.95
4500	3" pipe size		128	.063		4.09	3.25		7.34	9.40
4510	3-1/2" pipe size		122	.066		4.29	3.41		7.70	9.80
4520	4" pipe size		117	.068		4.38	3.56		7.94	10.15
4530	5" pipe size		114	.070		4.41	3.65		8.06	10.30
4540	6" pipe size		111	.072		9.05	3.75		12.80	15.55
4550	8" pipe size		109	.073		10.90	3.82		14.72	17.70
4560	10" pipe size		107	.075		19.65	3.89		23.54	27.50
4570	12" pipe size	↓	104	.077	↓	27	4		31	36
4580	For plastic coating on 1/2" thru 6" size, add					150%				
4700	U-hook, carbon steel, requires mounting screws or bolts									
4710	3/4" thru 2" pipe size									
4720	6" long	1 Plum	96	.083	Ea.	.63	4.34		4.97	7.20
4730	8" long		96	.083		.81	4.34		5.15	7.40
4740	10" long		96	.083		.91	4.34		5.25	7.50
4750	12" long	↓	96	.083		.99	4.34		5.33	7.60
4760	For copper plated, add					50%				
8000	Pipe clamp, plastic, 1/2" CTS	1 Plum	80	.100		.22	5.20		5.42	8.05
8010	3/4" CTS		73	.110		.23	5.70		5.93	8.80
8020	1" CTS		68	.118		.52	6.10		6.62	9.75
8080	Economy clamp, 1/4" CTS		175	.046		.05	2.38		2.43	3.63
8090	3/8" CTS		168	.048		.07	2.48		2.55	3.79
8100	1/2" CTS		160	.050		.07	2.60		2.67	3.98
8110	3/4" CTS		145	.055		.14	2.87		3.01	4.45
8200	Half clamp, 1/2" CTS		80	.100		.08	5.20		5.28	7.90
8210	3/4" CTS		73	.110		.12	5.70		5.82	8.70
8300	Suspension clamp, 1/2" CTS		80	.100		.22	5.20		5.42	8.05
8310	3/4" CTS		73	.110		.23	5.70		5.93	8.80
8320	1" CTS		68	.118		.52	6.10		6.62	9.75
8400	Insulator, 1/2" CTS		80	.100		.36	5.20		5.56	8.20
8410	3/4" CTS		73	.110		.37	5.70		6.07	8.95
8420	1" CTS		68	.118		.38	6.10		6.48	9.60
8500	J hook clamp with nail 1/2" CTS		240	.033		.13	1.73		1.86	2.74
8501	3/4" CTS	↓	240	.033	↓	.13	1.73		1.86	2.74
8800	Wire cable support system									
8810	Cable with hook terminal and locking device									
8830	2 mm, (.079 ") dia cable, (100 lb. cap.)									
8840	1 m, (3.3') length, with hook	1 Shee	96	.083	Ea.	2.86	4.09		6.95	9.35
8850	2 m, (6.6') length, with hook		84	.095		3.33	4.68		8.01	10.75
8860	3 m, (9.9') length, with hook	↓	72	.111		3.74	5.45		9.19	12.35
8870	5 m, (16.4') length, with hook	Q-9	60	.267		4.69	11.80		16.49	23
8880	10 m, (32.8') length, with hook	"	30	.533	↓	6.80	23.50		30.30	43
8900	3mm, (.118 ") dia cable, (200 lb. cap.)									
8910	1 m, (3.3') length, with hook	1 Shee	96	.083	Ea.	3.66	4.09		7.75	10.25

22 05 29 – Hangers and Supports for Plumbing Piping and Equipment

22 05 29.10 Hangers & Supp. for Plumb'g/HVAC Pipe/Equip.		Crew	Daily Output	Labor-Hours	Unit	Material	2010 Bare Costs Labor	Equipment	Total	Total Incl O&P
8920	2 m, (6.6') length, with hook	1 Shee	84	.095	Ea.	4.11	4.68		8.79	11.60
8930	3 m, (9.9') length, with hook	↓	72	.111		4.54	5.45		9.99	13.25
8940	5 m, (16.4') length, with hook	Q-9	60	.267		5.60	11.80		17.40	24
8950	10 m, (32.8') length, with hook	"	30	.533	↓	8.10	23.50		31.60	44.50
9000	Cable system accessories									
9010	Anchor bolt, 3/8", with nut	1 Shee	140	.057	Ea.	1.10	2.81		3.91	5.45
9020	Air duct corner protector		160	.050		.49	2.46		2.95	4.26
9030	Air duct support attachment	↓	140	.057	↓	.94	2.81		3.75	5.30
9040	Flange clip, hammer-on style									
9044	For flange thickness 3/32" - 9/64", 160 lb. cap.	1 Shee	180	.044	Ea.	.25	2.18		2.43	3.58
9048	For flange thickness 1/8" - 1/4", 200 lb. cap.		160	.050		.25	2.46		2.71	4
9052	For flange thickness 5/16" - 1/2", 200 lb. cap.		150	.053		.52	2.62		3.14	4.54
9056	For flange thickness 9/16" - 3/4", 200 lb. cap.		140	.057	↓	.71	2.81		3.52	5.05
9060	Wire insulation protection tube	↓	180	.044	L.F.	.32	2.18		2.50	3.65
9070	Wire cutter				Ea.	30.50			30.50	33.50

22 05 48 – Vibration and Seismic Controls for Plumbing Piping and Equipment

22 05 48.40 Vibration Absorbers		Crew	Daily Output	Labor-Hours	Unit	Material	Labor	Equipment	Total	Total Incl O&P
0010	**VIBRATION ABSORBERS**									
0100	Hangers, neoprene flex									
0200	10 – 120 lb. capacity				Ea.	22.50			22.50	24.50
0220	75 – 550 lb. capacity					29			29	32
0240	250 – 1100 lb. capacity					59			59	65
0260	1000 – 4000 lb. capacity					93.50			93.50	103
0500	Spring flex, 60 lb. capacity					31.50			31.50	34.50
0520	450 lb. capacity					39			39	43
0540	900 lb. capacity					51.50			51.50	57
0560	1100 – 1300 lb. capacity				↓	51.50			51.50	57
0600	Rubber in shear									
0610	45 – 340 lb., up to 1/2" rod size	1 Stpi	22	.364	Ea.	14.30	18.85		33.15	44.50
0620	130 – 700 lb., up to 3/4" rod size		20	.400		31	21		52	65
0630	50 – 1000 lb., up to 3/4" rod size	↓	18	.444		36.50	23		59.50	74.50
1000	Mounts, neoprene, 45 – 380 lb. capacity					14.85			14.85	16.30
1020	250 – 1100 lb. capacity					39			39	43
1040	1000 – 4000 lb. capacity					87			87	96
1100	Spring flex, 60 lb. capacity					61			61	67
1120	165 lb. capacity					61			61	67
1140	260 lb. capacity					61			61	67
1160	450 lb. capacity					90			90	99
1180	600 lb. capacity					90			90	99
1200	750 lb. capacity					90			90	99
1220	900 lb. capacity					93			93	102
1240	1100 lb. capacity					88.50			88.50	97
1260	1300 lb. capacity					88.50			88.50	97
1280	1500 lb. capacity					129			129	142
1300	1800 lb. capacity					137			137	150
1320	2200 lb. capacity					136			136	149
1340	2600 lb. capacity				↓	139			139	153
1399	Spring type									
1400	2 Piece									
1410	50 – 1000 lb.	1 Stpi	12	.667	Ea.	130	34.50		164.50	195
1420	1100 – 1600 lb.	"	12	.667	"	141	34.50		175.50	207
1500	Double spring open									

22 05 Common Work Results for Plumbing

22 05 48 – Vibration and Seismic Controls for Plumbing Piping and Equipment

22 05 48.40 Vibration Absorbers		Crew	Daily Output	Labor-Hours	Unit	Material	2010 Bare Costs Labor	Equipment	Total	Total Incl O&P
1510	150 – 450 lb.	1 Stpi	24	.333	Ea.	89	17.30		106.30	124
1520	500 – 1000 lb.		24	.333		89	17.30		106.30	124
1530	1100 – 1600 lb.		24	.333		89	17.30		106.30	124
1540	1700 – 2400 lb.		24	.333		97	17.30		114.30	132
1550	2500 – 3400 lb.	↓	24	.333		165	17.30		182.30	208
2000	Pads, cork rib, 18″ x 18″ x 1″, 10-50 psi					141			141	155
2020	18″ x 36″ x 1″, 10-50 psi					280			280	310
2100	Shear flexible pads, 18″ x 18″ x 3/8″, 20-70 psi					66.50			66.50	73
2120	18″ x 36″ x 3/8″, 20-70 psi				↓	140			140	154
2150	Laminated neoprene and cork									
2160	1″ thick	1 Stpi	16	.500	S.F.	54	26		80	98.50
3000	Note overlap in capacities due to deflections									

22 05 53 – Identification for Plumbing Piping and Equipment

22 05 53.10 Piping System Identification Labels		Crew	Daily Output	Labor-Hours	Unit	Material	2010 Bare Costs Labor	Equipment	Total	Total Incl O&P
0010	**PIPING SYSTEM IDENTIFICATION LABELS**									
0100	Indicate contents and flow direction									
0106	Pipe markers									
0110	Plastic snap around									
0114	1/2″ pipe	1 Plum	80	.100	Ea.	3.90	5.20		9.10	12.10
0116	3/4″ pipe		80	.100		3.90	5.20		9.10	12.10
0118	1″ pipe		80	.100		3.90	5.20		9.10	12.10
0120	2″ pipe		75	.107		5.30	5.55		10.85	14.15
0122	3″ pipe		70	.114		9.90	5.95		15.85	19.80
0124	4″ pipe		60	.133		9.90	6.95		16.85	21.50
0126	6″ pipe		60	.133		9.90	6.95		16.85	21.50
0128	8″ pipe		56	.143		15	7.45		22.45	27.50
0130	10″ pipe		56	.143		15	7.45		22.45	27.50
0200	Over 10″ pipe size	↓	50	.160	↓	18.30	8.35		26.65	32.50
1110	Self adhesive									
1114	1″ pipe	1 Plum	80	.100	Ea.	2.90	5.20		8.10	11
1116	2″ pipe		75	.107		2.90	5.55		8.45	11.50
1118	3″ pipe		70	.114		4.60	5.95		10.55	13.95
1120	4″ pipe		60	.133		4.60	6.95		11.55	15.45
1122	6″ pipe		60	.133		4.60	6.95		11.55	15.45
1124	8″ pipe		56	.143		4.70	7.45		12.15	16.30
1126	10″ pipe		56	.143		9.40	7.45		16.85	21.50
1200	Over 10″ pipe size	↓	50	.160	↓	9.40	8.35		17.75	23
2000	Valve tags									
2010	Numbered plus identifying legend									
2100	Brass, 2″ diameter	1 Plum	40	.200	Ea.	2.70	10.40		13.10	18.55
2200	Plastic, 1-1/2″ Dia.	″	40	.200	″	3.30	10.40		13.70	19.25

22 07 16 – Plumbing Equipment Insulation

22 07 16.10 Insulation for Plumbing Equipment			Crew	Daily Output	Labor-Hours	Unit	Material	2010 Bare Costs Labor	Equipment	Total	Total Incl O&P
0010	INSULATION FOR PLUMBING EQUIPMENT										
2900	Domestic water heater wrap kit										
2920	1-1/2" with vinyl jacket, 20-60 gal.	G	1 Plum	8	1	Ea.	16.05	52		68.05	95.50

22 07 19 – Plumbing Piping Insulation

22 07 19.10 Piping Insulation			Crew	Daily Output	Labor-Hours	Unit	Material	2010 Bare Costs Labor	Equipment	Total	Total Incl O&P
0010	PIPING INSULATION										
0100	Rule of thumb, as a percentage of total mechanical costs					Job				10%	
0110	Insulation req'd is based on the surface size/area to be covered										
2930	Insulated protectors, (ADA)										
2935	For exposed piping under sinks or lavatories										
2940	Vinyl coated foam, velcro tabs	♿									
2945	P Trap, 1-1/4" or 1-1/2"		1 Plum	32	.250	Ea.	18.50	13		31.50	40
2960	Valve and supply cover										
2965	1/2", 3/8", and 7/16" pipe size		1 Plum	32	.250	Ea.	17.60	13		30.60	39
2970	Extension drain cover										
2975	1-1/4", or 1-1/2" pipe size		1 Plum	32	.250	Ea.	19.30	13		32.30	41
2980	Tailpiece offset (wheelchair)										
2985	1-1/4" pipe size		1 Plum	32	.250	Ea.	21.50	13		34.50	43
4000	Pipe covering (price copper tube one size less than IPS)										
4280	Cellular glass, closed cell foam, all service jacket, sealant,										
4281	working temp. (-450°F to +900°F), 0 water vapor transmission										
4284	1" wall,										
4286	1/2" iron pipe size	G	Q-14	120	.133	L.F.	1.91	5.45		7.36	10.60
4300	1-1/2" wall,										
4301	1" iron pipe size	G	Q-14	105	.152	L.F.	3.07	6.25		9.32	13.10
4304	2-1/2" iron pipe size	G		90	.178		5.15	7.30		12.45	17.05
4306	3" iron pipe size	G		85	.188		5.25	7.70		12.95	17.75
4308	4" iron pipe size	G		70	.229		6.95	9.35		16.30	22
4310	5" iron pipe size	G		65	.246		8	10.10		18.10	24.50
4320	2" wall,										
4322	1" iron pipe size	G	Q-14	100	.160	L.F.	4.57	6.55		11.12	15.25
4324	2-1/2" iron pipe size	G		85	.188		6.75	7.70		14.45	19.45
4326	3" iron pipe size	G		80	.200		6.95	8.20		15.15	20.50
4328	4" iron pipe size	G		65	.246		7.80	10.10		17.90	24.50
4330	5" iron pipe size	G		60	.267		9.35	10.95		20.30	27.50
4332	6" iron pipe size	G		50	.320		11.05	13.10		24.15	32.50
4336	8" iron pipe size	G		40	.400		14.30	16.40		30.70	41
4338	10" iron pipe size	G		35	.457		14.75	18.75		33.50	45.50
4350	2-1/2" wall,										
4360	12" iron pipe size	G	Q-14	32	.500	L.F.	23.50	20.50		44	57.50
4362	14" iron pipe size	G	"	28	.571	"	25	23.50		48.50	64
4370	3" wall,										
4378	6" iron pipe size	G	Q-14	48	.333	L.F.	14.75	13.65		28.40	37.50
4380	8" iron pipe size	G		38	.421		17.20	17.25		34.45	46
4382	10" iron pipe size	G		33	.485		22.50	19.90		42.40	56
4384	16" iron pipe size	G		25	.640		29.50	26		55.50	73.50
4386	18" iron pipe size	G		22	.727		32	30		62	82
4388	20" iron pipe size	G		20	.800		38	33		71	92.50
4400	3-1/2" wall,										
4412	12" iron pipe size	G	Q-14	27	.593	L.F.	33	24.50		57.50	74.50
4414	14" iron pipe size	G	"	25	.640	"	35	26		61	79.50
4430	4" wall,										

22 07 Plumbing Insulation

22 07 19 – Plumbing Piping Insulation

22 07 19.10 Piping Insulation			Crew	Daily Output	Labor-Hours	Unit	Material	Labor	Equipment	Total	Total Incl O&P
							2010 Bare Costs				
4446	16″ iron pipe size	G	Q-14	22	.727	L.F.	35.50	30		65.50	85.50
4448	18″ iron pipe size	G		20	.800		46	33		79	102
4450	20″ iron pipe size	G		18	.889		48	36.50		84.50	110
4480	Fittings, average with fabric and mastic										
4484	1″ wall,										
4486	1/2″ iron pipe size	G	1 Asbe	40	.200	Ea.	3.39	9.10		12.49	17.90
4500	1-1/2″ wall,										
4502	1″ iron pipe size	G	1 Asbe	38	.211	Ea.	5.85	9.60		15.45	21.50
4504	2-1/2″ iron pipe size	G		32	.250		7.25	11.40		18.65	25.50
4506	3″ iron pipe size	G		30	.267		8	12.15		20.15	27.50
4508	4″ iron pipe size	G		28	.286		12.30	13		25.30	33.50
4510	5″ iron pipe size	G		24	.333		13.55	15.20		28.75	38.50
4520	2″ wall,										
4522	1″ iron pipe size	G	1 Asbe	36	.222	Ea.	6.75	10.10		16.85	23
4524	2-1/2″ iron pipe size	G		30	.267		8	12.15		20.15	27.50
4526	3″ iron pipe size	G		28	.286		9.60	13		22.60	30.50
4528	4″ iron pipe size	G		24	.333		12.90	15.20		28.10	37.50
4530	5″ iron pipe size	G		22	.364		16.30	16.55		32.85	44
4532	6″ iron pipe size	G		20	.400		19.05	18.20		37.25	49.50
4536	8″ iron pipe size	G		12	.667		29	30.50		59.50	79
4538	10″ iron pipe size	G		8	1		35.50	45.50		81	110
4550	2-1/2″ wall,										
4560	12″ iron pipe size	G	1 Asbe	6	1.333	Ea.	61.50	60.50		122	162
4562	14″ iron pipe size	G	″	4	2	″	71.50	91		162.50	221
4570	3″ wall,										
4578	6″ iron pipe size	G	1 Asbe	16	.500	Ea.	22	23		45	60
4580	8″ iron pipe size	G		10	.800		32.50	36.50		69	92.50
4582	10″ iron pipe size	G		6	1.333		46	60.50		106.50	145
4584	16″ iron pipe size	G		2	4		111	182		293	405
4586	18″ iron pipe size	G		2	4		143	182		325	440
4588	20″ iron pipe size	G		2	4		168	182		350	470
4600	3-1/2″ wall,										
4612	12″ iron pipe size	G	1 Asbe	4	2	Ea.	80	91		171	230
4614	14″ iron pipe size	G	″	2	4	″	88.50	182		270.50	380
4630	4″ wall,										
4646	16″ iron pipe size	G	1 Asbe	2	4	Ea.	133	182		315	430
4648	18″ iron pipe size	G		2	4		162	182		344	460
4650	20″ iron pipe size	G		2	4		189	182		371	490
4900	Calcium silicate, with cover										
5100	1″ wall, 1/2″ iron pipe size	G	Q-14	170	.094	L.F.	2.82	3.86		6.68	9.10
5130	3/4″ iron pipe size	G		170	.094		2.84	3.86		6.70	9.10
5140	1″ iron pipe size	G		170	.094		2.75	3.86		6.61	9.05
5150	1-1/4″ iron pipe size	G		165	.097		2.82	3.98		6.80	9.30
5160	1-1/2″ iron pipe size	G		165	.097		2.86	3.98		6.84	9.35
5170	2″ iron pipe size	G		160	.100		3.28	4.10		7.38	9.95
5180	2-1/2″ iron pipe size	G		160	.100		3.49	4.10		7.59	10.20
5190	3″ iron pipe size	G		150	.107		3.77	4.37		8.14	10.95
5200	4″ iron pipe size	G		140	.114		4.68	4.69		9.37	12.45
5210	5″ iron pipe size	G		135	.119		4.93	4.86		9.79	12.95
5220	6″ iron pipe size	G		130	.123		5.25	5.05		10.30	13.60
5280	1-1/2″ wall, 1/2″ iron pipe size	G		150	.107		3.05	4.37		7.42	10.15
5310	3/4″ iron pipe size	G		150	.107		3.10	4.37		7.47	10.20
5320	1″ iron pipe size	G		150	.107		3.38	4.37		7.75	10.50

22 07 19.10 Piping Insulation			Crew	Daily Output	Labor-Hours	Unit	Material	2010 Bare Costs Labor	Equipment	Total	Total Incl O&P
5330	1-1/4" iron pipe size	G	Q-14	145	.110	L.F.	3.63	4.52		8.15	11.05
5340	1-1/2" iron pipe size	G		145	.110		3.89	4.52		8.41	11.35
5350	2" iron pipe size	G		140	.114		4.29	4.69		8.98	12
5360	2-1/2" iron pipe size	G		140	.114		4.68	4.69		9.37	12.45
5370	3" iron pipe size	G		135	.119		4.92	4.86		9.78	12.95
5380	4" iron pipe size	G		125	.128		5.70	5.25		10.95	14.40
5390	5" iron pipe size	G		120	.133		6.40	5.45		11.85	15.55
5400	6" iron pipe size	G		110	.145		6.60	5.95		12.55	16.55
5402	8" iron pipe size	G		95	.168		8.70	6.90		15.60	20.50
5404	10" iron pipe size	G		85	.188		11.35	7.70		19.05	24.50
5406	12" iron pipe size	G		80	.200		13.40	8.20		21.60	27.50
5408	14" iron pipe size	G		75	.213		15.10	8.75		23.85	30
5410	16" iron pipe size	G		70	.229		16.90	9.35		26.25	33
5412	18" iron pipe size	G		65	.246		18.65	10.10		28.75	36
5460	2" wall, 1/2" iron pipe size	G		135	.119		4.70	4.86		9.56	12.70
5490	3/4" iron pipe size	G		135	.119		4.94	4.86		9.80	13
5500	1" iron pipe size	G		135	.119		5.20	4.86		10.06	13.25
5510	1-1/4" iron pipe size	G		130	.123		5.60	5.05		10.65	14
5520	1-1/2" iron pipe size	G		130	.123		5.80	5.05		10.85	14.25
5530	2" iron pipe size	G		125	.128		6.10	5.25		11.35	14.90
5540	2-1/2" iron pipe size	G		125	.128		7.30	5.25		12.55	16.15
5550	3" iron pipe size	G		120	.133		7.35	5.45		12.80	16.60
5560	4" iron pipe size	G		115	.139		8.50	5.70		14.20	18.20
5570	5" iron pipe size	G		110	.145		9.70	5.95		15.65	19.90
5580	6" iron pipe size	G		105	.152		10.60	6.25		16.85	21.50
5581	8" iron pipe size	G		95	.168		12.60	6.90		19.50	24.50
5582	10" iron pipe size	G		85	.188		15.75	7.70		23.45	29.50
5583	12" iron pipe size	G		80	.200		17.50	8.20		25.70	32
5584	14" iron pipe size	G		75	.213		19.25	8.75		28	34.50
5585	16" iron pipe size	G		70	.229		21	9.35		30.35	38
5586	18" iron pipe size	G		65	.246		23	10.10		33.10	41
5587	3" wall, 1-1/4" iron pipe size	G		100	.160		8.60	6.55		15.15	19.70
5588	1-1/2" iron pipe size	G		100	.160		8.70	6.55		15.25	19.75
5589	2" iron pipe size	G		95	.168		9	6.90		15.90	20.50
5590	2-1/2" iron pipe size	G		95	.168		10.50	6.90		17.40	22.50
5591	3" iron pipe size	G		90	.178		10.65	7.30		17.95	23
5592	4" iron pipe size	G		85	.188		13.45	7.70		21.15	27
5593	6" iron pipe size	G		80	.200		16.35	8.20		24.55	31
5594	8" iron pipe size	G		75	.213		19.45	8.75		28.20	35
5595	10" iron pipe size	G		65	.246		23.50	10.10		33.60	41.50
5596	12" iron pipe size	G		60	.267		26	10.95		36.95	45.50
5597	14" iron pipe size	G		55	.291		29	11.95		40.95	50.50
5598	16" iron pipe size	G		50	.320		32	13.10		45.10	55.50
5599	18" iron pipe size	G	↓	45	.356	↓	35	14.60		49.60	61
5600	Calcium silicate, no cover										
5720	1" wall, 1/2" iron pipe size	G	Q-14	180	.089	L.F.	2.56	3.64		6.20	8.45
5740	3/4" iron pipe size	G		180	.089		2.56	3.64		6.20	8.45
5750	1" iron pipe size	G		180	.089		2.45	3.64		6.09	8.35
5760	1-1/4" iron pipe size	G		175	.091		2.50	3.75		6.25	8.60
5770	1-1/2" iron pipe size	G		175	.091		2.52	3.75		6.27	8.60
5780	2" iron pipe size	G		170	.094		2.90	3.86		6.76	9.20
5790	2-1/2" iron pipe size	G		170	.094		3.07	3.86		6.93	9.40
5800	3" iron pipe size	G	↓	160	.100	↓	3.30	4.10		7.40	10

22 07 19.10 Piping Insulation			Crew	Daily Output	Labor-Hours	Unit	2010 Bare Costs Material	Labor	Equipment	Total	Total Incl O&P
5810	4" iron pipe size	G	Q-14	150	.107	L.F.	4.13	4.37		8.50	11.35
5820	5" iron pipe size	G		145	.110		4.29	4.52		8.81	11.75
5830	6" iron pipe size	G		140	.114		4.51	4.69		9.20	12.25
5900	1-1/2" wall, 1/2" iron pipe size	G		160	.100		2.71	4.10		6.81	9.35
5920	3/4" iron pipe size	G		160	.100		2.75	4.10		6.85	9.40
5930	1" iron pipe size	G		160	.100		3	4.10		7.10	9.65
5940	1-1/4" iron pipe size	G		155	.103		3.23	4.23		7.46	10.15
5950	1-1/2" iron pipe size	G		155	.103		3.48	4.23		7.71	10.45
5960	2" iron pipe size	G		150	.107		3.83	4.37		8.20	11
5970	2-1/2" iron pipe size	G		150	.107		4.18	4.37		8.55	11.40
5980	3" iron pipe size	G		145	.110		4.37	4.52		8.89	11.85
5990	4" iron pipe size	G		135	.119		5.05	4.86		9.91	13.10
6000	5" iron pipe size	G		130	.123		5.70	5.05		10.75	14.10
6010	6" iron pipe size	G		120	.133		5.80	5.45		11.25	14.90
6020	7" iron pipe size	G		115	.139		6.85	5.70		12.55	16.35
6030	8" iron pipe size	G		105	.152		7.70	6.25		13.95	18.20
6040	9" iron pipe size	G		100	.160		9.25	6.55		15.80	20.50
6050	10" iron pipe size	G		95	.168		10.20	6.90		17.10	22
6060	12" iron pipe size	G		90	.178		12.10	7.30		19.40	24.50
6070	14" iron pipe size	G		85	.188		13.70	7.70		21.40	27
6080	16" iron pipe size	G		80	.200		15.30	8.20		23.50	29.50
6090	18" iron pipe size	G		75	.213		16.95	8.75		25.70	32.50
6120	2" wall, 1/2" iron pipe size	G		145	.110		4.28	4.52		8.80	11.75
6140	3/4" iron pipe size	G		145	.110		4.51	4.52		9.03	12
6150	1" iron pipe size	G		145	.110		4.75	4.52		9.27	12.30
6160	1-1/4" iron pipe size	G		140	.114		5.10	4.69		9.79	12.90
6170	1-1/2" iron pipe size	G		140	.114		5.30	4.69		9.99	13.15
6180	2" iron pipe size	G		135	.119		5.60	4.86		10.46	13.70
6190	2-1/2" iron pipe size	G		135	.119		6.70	4.86		11.56	14.95
6200	3" iron pipe size	G		130	.123		6.75	5.05		11.80	15.25
6210	4" iron pipe size	G		125	.128		7.80	5.25		13.05	16.70
6220	5" iron pipe size	G		120	.133		8.90	5.45		14.35	18.30
6230	6" iron pipe size	G		115	.139		9.70	5.70		15.40	19.55
6240	7" iron pipe size	G		110	.145		10.55	5.95		16.50	21
6250	8" iron pipe size	G		105	.152		11.55	6.25		17.80	22.50
6260	9" iron pipe size	G		100	.160		13.10	6.55		19.65	24.50
6270	10" iron pipe size	G		95	.168		14.55	6.90		21.45	27
6280	12" iron pipe size	G		90	.178		16.15	7.30		23.45	29
6290	14" iron pipe size	G		85	.188		17.75	7.70		25.45	31.50
6300	16" iron pipe size	G		80	.200		19.55	8.20		27.75	34.50
6310	18" iron pipe size	G		75	.213		21.50	8.75		30.25	37
6320	20" iron pipe size	G		65	.246		26.50	10.10		36.60	44.50
6330	22" iron pipe size	G		60	.267		29	10.95		39.95	49
6340	24" iron pipe size	G		55	.291		30	11.95		41.95	51
6360	3" wall, 1/2" iron pipe size	G		115	.139		7.75	5.70		13.45	17.40
6380	3/4" iron pipe size	G		115	.139		7.80	5.70		13.50	17.45
6390	1" iron pipe size	G		115	.139		7.85	5.70		13.55	17.50
6400	1-1/4" iron pipe size	G		110	.145		8	5.95		13.95	18.05
6410	1-1/2" iron pipe size	G		110	.145		8.05	5.95		14	18.10
6420	2" iron pipe size	G		105	.152		8.30	6.25		14.55	18.80
6430	2-1/2" iron pipe size	G		105	.152		9.75	6.25		16	20.50
6440	3" iron pipe size	G		100	.160		9.85	6.55		16.40	21
6450	4" iron pipe size	G		95	.168		12.55	6.90		19.45	24.50

22 07 19.10 Piping Insulation			Crew	Daily Output	Labor-Hours	Unit	2010 Bare Costs Material	Labor	Equipment	Total	Total Incl O&P
6460	5″ iron pipe size	G	Q-14	90	.178	L.F.	13.60	7.30		20.90	26.50
6470	6″ iron pipe size	G		90	.178		15.30	7.30		22.60	28
6480	7″ iron pipe size	G		85	.188		17.05	7.70		24.75	31
6490	8″ iron pipe size	G		85	.188		18.25	7.70		25.95	32
6500	9″ iron pipe size	G		80	.200		20.50	8.20		28.70	36
6510	10″ iron pipe size	G		75	.213		22	8.75		30.75	37.50
6520	12″ iron pipe size	G		70	.229		24.50	9.35		33.85	41.50
6530	14″ iron pipe size	G		65	.246		27.50	10.10		37.60	45.50
6540	16″ iron pipe size	G		60	.267		30	10.95		40.95	50
6550	18″ iron pipe size	G		55	.291		33	11.95		44.95	55
6560	20″ iron pipe size	G		50	.320		39.50	13.10		52.60	64
6570	22″ iron pipe size	G		45	.356		42.50	14.60		57.10	69.50
6580	24″ iron pipe size	G		40	.400		46	16.40		62.40	76
6600	Fiberglass, with all service jacket										
6640	1/2″ wall, 1/2″ iron pipe size	G	Q-14	250	.064	L.F.	.79	2.62		3.41	4.95
6660	3/4″ iron pipe size	G		240	.067		.89	2.73		3.62	5.25
6670	1″ iron pipe size	G		230	.070		.92	2.85		3.77	5.45
6680	1-1/4″ iron pipe size	G		220	.073		.98	2.98		3.96	5.70
6690	1-1/2″ iron pipe size	G		220	.073		1.13	2.98		4.11	5.90
6700	2″ iron pipe size	G		210	.076		1.21	3.12		4.33	6.20
6710	2-1/2″ iron pipe size	G		200	.080		1.26	3.28		4.54	6.50
6840	1″ wall, 1/2″ iron pipe size	G		240	.067		.95	2.73		3.68	5.30
6860	3/4″ iron pipe size	G		230	.070		1.04	2.85		3.89	5.55
6870	1″ iron pipe size	G		220	.073		1.12	2.98		4.10	5.85
6880	1-1/4″ iron pipe size	G		210	.076		1.21	3.12		4.33	6.20
6890	1-1/2″ iron pipe size	G		210	.076		1.31	3.12		4.43	6.30
6900	2″ iron pipe size	G		200	.080		1.41	3.28		4.69	6.65
6910	2-1/2″ iron pipe size	G		190	.084		1.61	3.45		5.06	7.10
6920	3″ iron pipe size	G		180	.089		1.72	3.64		5.36	7.55
6930	3-1/2″ iron pipe size	G		170	.094		1.86	3.86		5.72	8.05
6940	4″ iron pipe size	G		150	.107		2.27	4.37		6.64	9.30
6950	5″ iron pipe size	G		140	.114		2.57	4.69		7.26	10.15
6960	6″ iron pipe size	G		120	.133		2.72	5.45		8.17	11.50
6970	7″ iron pipe size	G		110	.145		3.24	5.95		9.19	12.80
6980	8″ iron pipe size	G		100	.160		4.45	6.55		11	15.10
6990	9″ iron pipe size	G		90	.178		4.69	7.30		11.99	16.50
7000	10″ iron pipe size	G		90	.178		4.72	7.30		12.02	16.55
7010	12″ iron pipe size	G		00	.200		5.15	8.20		13.35	18.40
7020	14″ iron pipe size	G		80	.200		6.15	8.20		14.35	19.55
7030	16″ iron pipe size	G		70	.229		7.85	9.35		17.20	23
7040	18″ iron pipe size	G		70	.229		8.70	9.35		18.05	24
7050	20″ iron pipe size	G		60	.267		9.70	10.95		20.65	27.50
7060	24″ iron pipe size	G		60	.267		11.80	10.95		22.75	30
7080	1-1/2″ wall, 1/2″ iron pipe size	G		230	.070		1.81	2.85		4.66	6.40
7100	3/4″ iron pipe size	G		220	.073		1.82	2.98		4.80	6.65
7110	1″ iron pipe size	G		210	.076		1.94	3.12		5.06	7
7120	1-1/4″ iron pipe size	G		200	.080		2.12	3.28		5.40	7.45
7130	1-1/2″ iron pipe size	G		200	.080		2.21	3.28		5.49	7.55
7140	2″ iron pipe size	G		190	.084		2.44	3.45		5.89	8.05
7150	2-1/2″ iron pipe size	G		180	.089		2.62	3.64		6.26	8.55
7160	3″ iron pipe size	G		170	.094		2.74	3.86		6.60	9
7170	3-1/2″ iron pipe size	G		160	.100		3	4.10		7.10	9.65
7180	4″ iron pipe size	G		140	.114		3.11	4.69		7.80	10.70

22 07 19 – Plumbing Piping Insulation

22 07 19.10 Piping Insulation			Crew	Daily Output	Labor-Hours	Unit	Material	Labor	Equipment	Total	Total Incl O&P
							2010 Bare Costs				
7190	5" iron pipe size	G	Q-14	130	.123	L.F.	3.49	5.05		8.54	11.70
7200	6" iron pipe size	G		110	.145		3.68	5.95		9.63	13.30
7210	7" iron pipe size	G		100	.160		4.09	6.55		10.64	14.70
7220	8" iron pipe size	G		90	.178		5.15	7.30		12.45	17.05
7230	9" iron pipe size	G		85	.188		5.40	7.70		13.10	17.90
7240	10" iron pipe size	G		80	.200		5.55	8.20		13.75	18.90
7250	12" iron pipe size	G		75	.213		6.30	8.75		15.05	20.50
7260	14" iron pipe size	G		70	.229		7.55	9.35		16.90	23
7270	16" iron pipe size	G		65	.246		9.80	10.10		19.90	26.50
7280	18" iron pipe size	G		60	.267		11.05	10.95		22	29
7290	20" iron pipe size	G		55	.291		11.25	11.95		23.20	31
7300	24" iron pipe size	G		50	.320		13.80	13.10		26.90	35.50
7320	2" wall, 1/2" iron pipe size	G		220	.073		2.81	2.98		5.79	7.75
7340	3/4" iron pipe size	G		210	.076		2.90	3.12		6.02	8.05
7350	1" iron pipe size	G		200	.080		3.08	3.28		6.36	8.50
7360	1-1/4" iron pipe size	G		190	.084		3.26	3.45		6.71	8.95
7370	1-1/2" iron pipe size	G		190	.084		3.40	3.45		6.85	9.10
7380	2" iron pipe size	G		180	.089		3.58	3.64		7.22	9.60
7390	2-1/2" iron pipe size	G		170	.094		3.86	3.86		7.72	10.25
7400	3" iron pipe size	G		160	.100		4.09	4.10		8.19	10.85
7410	3-1/2" iron pipe size	G		150	.107		4.43	4.37		8.80	11.65
7420	4" iron pipe size	G		130	.123		4.77	5.05		9.82	13.10
7430	5" iron pipe size	G		120	.133		5.45	5.45		10.90	14.50
7440	6" iron pipe size	G		100	.160		5.65	6.55		12.20	16.40
7450	7" iron pipe size	G		90	.178		6.40	7.30		13.70	18.40
7460	8" iron pipe size	G		80	.200		6.85	8.20		15.05	20.50
7470	9" iron pipe size	G		75	.213		7.50	8.75		16.25	22
7480	10" iron pipe size	G		70	.229		8.20	9.35		17.55	23.50
7490	12" iron pipe size	G		65	.246		9.20	10.10		19.30	26
7500	14" iron pipe size	G		60	.267		11.80	10.95		22.75	30
7510	16" iron pipe size	G		55	.291		12.95	11.95		24.90	33
7520	18" iron pipe size	G		50	.320		14.50	13.10		27.60	36.50
7530	20" iron pipe size	G		45	.356		16.45	14.60		31.05	40.50
7540	24" iron pipe size	G		40	.400		17.65	16.40		34.05	45
7560	2-1/2" wall, 1/2" iron pipe size	G		210	.076		3.32	3.12		6.44	8.50
7562	3/4" iron pipe size	G		200	.080		3.47	3.28		6.75	8.90
7564	1" iron pipe size	G		190	.084		3.62	3.45		7.07	9.35
7566	1-1/4" iron pipe size	G		185	.086		3.75	3.55		7.30	9.65
7568	1-1/2" iron pipe size	G		180	.089		3.95	3.64		7.59	10
7570	2" iron pipe size	G		170	.094		4.14	3.86		8	10.55
7572	2-1/2" iron pipe size	G		160	.100		4.76	4.10		8.86	11.60
7574	3" iron pipe size	G		150	.107		5	4.37		9.37	12.30
7576	3-1/2" iron pipe size	G		140	.114		5.45	4.69		10.14	13.30
7578	4" iron pipe size	G		120	.133		5.75	5.45		11.20	14.85
7580	5" iron pipe size	G		110	.145		6.85	5.95		12.80	16.80
7582	6" iron pipe size	G		90	.178		8.25	7.30		15.55	20.50
7584	7" iron pipe size	G		80	.200		8.30	8.20		16.50	22
7586	8" iron pipe size	G		70	.229		8.80	9.35		18.15	24.50
7588	9" iron pipe size	G		65	.246		9.65	10.10		19.75	26.50
7590	10" iron pipe size	G		60	.267		10.50	10.95		21.45	28.50
7592	12" iron pipe size	G		55	.291		12.75	11.95		24.70	32.50
7594	14" iron pipe size	G		50	.320		14.90	13.10		28	37
7596	16" iron pipe size	G		45	.356		17.10	14.60		31.70	41.50

22 07 19.10 Piping Insulation			Crew	Daily Output	Labor-Hours	Unit	2010 Bare Costs Material	Labor	Equipment	Total	Total Incl O&P
7598	18" iron pipe size	G	Q-14	40	.400	L.F.	18.55	16.40		34.95	46
7602	24" iron pipe size	G		30	.533		24.50	22		46.50	61
7620	3" wall, 1/2" iron pipe size	G		200	.080		4.25	3.28		7.53	9.80
7622	3/4" iron pipe size	G		190	.084		4.50	3.45		7.95	10.30
7624	1" iron pipe size	G		180	.089		4.76	3.64		8.40	10.90
7626	1-1/4" iron pipe size	G		175	.091		4.86	3.75		8.61	11.20
7628	1-1/2" iron pipe size	G		170	.094		5.10	3.86		8.96	11.60
7630	2" iron pipe size	G		160	.100		5.50	4.10		9.60	12.40
7632	2-1/2" iron pipe size	G		150	.107		5.70	4.37		10.07	13.10
7634	3" iron pipe size	G		140	.114		6.10	4.69		10.79	14
7636	3-1/2" iron pipe size	G		130	.123		6.70	5.05		11.75	15.25
7638	4" iron pipe size	G		110	.145		7.20	5.95		13.15	17.20
7640	5" iron pipe size	G		100	.160		8.15	6.55		14.70	19.20
7642	6" iron pipe size	G		80	.200		8.70	8.20		16.90	22.50
7644	7" iron pipe size	G		70	.229		10	9.35		19.35	25.50
7646	8" iron pipe size	G		60	.267		10.85	10.95		21.80	29
7648	9" iron pipe size	G		55	.291		11.70	11.95		23.65	31.50
7650	10" iron pipe size	G		50	.320		12.55	13.10		25.65	34.50
7652	12" iron pipe size	G		45	.356		15.65	14.60		30.25	40
7654	14" iron pipe size	G		40	.400		18.05	16.40		34.45	45.50
7656	16" iron pipe size	G		35	.457		20.50	18.75		39.25	51.50
7658	18" iron pipe size	G		32	.500		22	20.50		42.50	56
7660	20" iron pipe size	G		30	.533		24	22		46	60
7662	24" iron pipe size	G		28	.571		30	23.50		53.50	69.50
7664	26" iron pipe size	G		26	.615		33.50	25		58.50	75.50
7666	30" iron pipe size	G		24	.667		38	27.50		65.50	84
7800	For fiberglass with standard canvas jacket, deduct						5%				
7802	For fittings, add 3 L.F. for each fitting										
7804	plus 4 L.F. for each flange of the fitting										
7810	Finishes										
7814	For single layer of felt, add						10%	10%			
7816	For roofing paper, 45 lb to 55 lb, add						25%	10%			
7820	Polyethylene tubing flexible closed cell foam, UV resistant										
7828	Standard temperature (-90°F to +212°F)										
7830	3/8" wall, 1/8" iron pipe size	G	1 Asbe	130	.062	L.F.	.19	2.80		2.99	4.57
7831	1/4" iron pipe size	G		130	.062		.20	2.80		3	4.58
7832	3/8" iron pipe size	G		130	.062		.22	2.80		3.02	4.60
7833	1/2" iron pipe size	G		126	.063		.25	2.89		3.14	4.78
7834	3/4" iron pipe size	G		122	.066		.29	2.99		3.28	4.96
7835	1" iron pipe size	G		120	.067		.33	3.04		3.37	5.10
7836	1-1/4" iron pipe size	G		118	.068		.41	3.09		3.50	5.25
7837	1-1/2" iron pipe size	G		118	.068		.50	3.09		3.59	5.35
7838	2" iron pipe size	G		116	.069		.60	3.14		3.74	5.55
7839	2-1/2" iron pipe size	G		114	.070		.82	3.20		4.02	5.85
7840	3" iron pipe size	G		112	.071		1.35	3.25		4.60	6.55
7842	1/2" wall, 1/8" iron pipe size	G		120	.067		.28	3.04		3.32	5.05
7843	1/4" iron pipe size	G		120	.067		.30	3.04		3.34	5.05
7844	3/8" iron pipe size	G		120	.067		.33	3.04		3.37	5.10
7845	1/2" iron pipe size	G		118	.068		.37	3.09		3.46	5.20
7846	3/4" iron pipe size	G		116	.069		.42	3.14		3.56	5.35
7847	1" iron pipe size	G		114	.070		.47	3.20		3.67	5.50
7848	1-1/4" iron pipe size	G		112	.071		.55	3.25		3.80	5.65
7849	1-1/2" iron pipe size	G		110	.073		.79	3.31		4.10	6

22 07 19 – Plumbing Piping Insulation

22 07 19.10 Piping Insulation			Crew	Daily Output	Labor-Hours	Unit	Material	Labor	Equipment	Total	Total Incl O&P
							2010 Bare Costs				
7850	2" iron pipe size	G	1 Asbe	108	.074	L.F.	.95	3.37		4.32	6.30
7851	2-1/2" iron pipe size	G		106	.075		1.07	3.44		4.51	6.55
7852	3" iron pipe size	G		104	.077		1.42	3.50		4.92	7
7853	3-1/2" iron pipe size	G		102	.078		1.63	3.57		5.20	7.35
7854	4" iron pipe size	G		100	.080		1.87	3.64		5.51	7.70
7855	3/4" wall, 1/8" iron pipe size	G		110	.073		.42	3.31		3.73	5.60
7856	1/4" iron pipe size	G		110	.073		.45	3.31		3.76	5.65
7857	3/8" iron pipe size	G		108	.074		.53	3.37		3.90	5.85
7858	1/2" iron pipe size	G		106	.075		.63	3.44		4.07	6.05
7859	3/4" iron pipe size	G		104	.077		.76	3.50		4.26	6.30
7860	1" iron pipe size	G		102	.078		.87	3.57		4.44	6.50
7861	1-1/4" iron pipe size	G		100	.080		1.18	3.64		4.82	6.95
7862	1-1/2" iron pipe size	G		100	.080		1.55	3.64		5.19	7.35
7863	2" iron pipe size	G		98	.082		1.72	3.72		5.44	7.70
7864	2-1/2" iron pipe size	G		96	.083		1.83	3.80		5.63	7.90
7865	3" iron pipe size	G		94	.085		2.54	3.88		6.42	8.85
7866	3-1/2" iron pipe size	G		92	.087		2.78	3.96		6.74	9.20
7867	4" iron pipe size	G		90	.089		3.12	4.05		7.17	9.75
7868	1" wall, 1/4" iron pipe size	G		100	.080		.93	3.64		4.57	6.65
7869	3/8" iron pipe size	G		98	.082		1.04	3.72		4.76	6.95
7870	1/2" iron pipe size	G		96	.083		1.16	3.80		4.96	7.20
7871	3/4" iron pipe size	G		94	.085		1.36	3.88		5.24	7.55
7872	1" iron pipe size	G		92	.087		1.66	3.96		5.62	8
7873	1-1/4" iron pipe size	G		90	.089		1.90	4.05		5.95	8.40
7874	1-1/2" iron pipe size	G		90	.089		2.38	4.05		6.43	8.90
7875	2" iron pipe size	G		88	.091		2.85	4.14		6.99	9.60
7876	2-1/2" iron pipe size	G		86	.093		3.22	4.24		7.46	10.15
7877	3" iron pipe size	G	↓	84	.095	↓	4.06	4.34		8.40	11.20
7878	Contact cement, quart can	G				Ea.	9.85			9.85	10.85
7879	Rubber tubing, flexible closed cell foam										
7880	3/8" wall, 1/4" iron pipe size	G	1 Asbe	120	.067	L.F.	.55	3.04		3.59	5.35
7900	3/8" iron pipe size	G		120	.067		.68	3.04		3.72	5.45
7910	1/2" iron pipe size	G		115	.070		.70	3.17		3.87	5.70
7920	3/4" iron pipe size	G		115	.070		.85	3.17		4.02	5.85
7930	1" iron pipe size	G		110	.073		.98	3.31		4.29	6.25
7940	1-1/4" iron pipe size	G		110	.073		1.20	3.31		4.51	6.45
7950	1-1/2" iron pipe size	G		110	.073		1.30	3.31		4.61	6.60
8100	1/2" wall, 1/4" iron pipe size	G		90	.089		.50	4.05		4.55	6.85
8120	3/8" iron pipe size	G		90	.089		.55	4.05		4.60	6.90
8130	1/2" iron pipe size	G		89	.090		.61	4.09		4.70	7
8140	3/4" iron pipe size	G		89	.090		.68	4.09		4.77	7.10
8150	1" iron pipe size	G		88	.091		.75	4.14		4.89	7.30
8160	1-1/4" iron pipe size	G		87	.092		.87	4.19		5.06	7.45
8170	1-1/2" iron pipe size	G		87	.092		1.06	4.19		5.25	7.65
8180	2" iron pipe size	G		86	.093		1.38	4.24		5.62	8.10
8190	2-1/2" iron pipe size	G		86	.093		1.74	4.24		5.98	8.50
8200	3" iron pipe size	G		85	.094		1.95	4.29		6.24	8.80
8210	3-1/2" iron pipe size	G		85	.094		2.70	4.29		6.99	9.60
8220	4" iron pipe size	G		80	.100		2.89	4.56		7.45	10.30
8230	5" iron pipe size	G		80	.100		3.97	4.56		8.53	11.45
8240	6" iron pipe size	G		75	.107		3.97	4.86		8.83	11.90
8300	3/4" wall, 1/4" iron pipe size	G		90	.089		.77	4.05		4.82	7.15
8320	3/8" iron pipe size	G	↓	90	.089	↓	.84	4.05		4.89	7.20

22 07 19.10 Piping Insulation			Crew	Daily Output	Labor-Hours	Unit	Material	Labor	Equipment	Total	Total Incl O&P
							2010 Bare Costs				
8330	1/2" iron pipe size	G	1 Asbe	89	.090	L.F.	1.01	4.09		5.10	7.45
8340	3/4" iron pipe size	G		89	.090		1.23	4.09		5.32	7.70
8350	1" iron pipe size	G		88	.091		1.41	4.14		5.55	8
8360	1-1/4" iron pipe size	G		87	.092		1.90	4.19		6.09	8.60
8370	1-1/2" iron pipe size	G		87	.092		2.15	4.19		6.34	8.85
8380	2" iron pipe size	G		86	.093		2.52	4.24		6.76	9.35
8390	2-1/2" iron pipe size	G		86	.093		3.38	4.24		7.62	10.30
8400	3" iron pipe size	G		85	.094		3.85	4.29		8.14	10.90
8410	3-1/2" iron pipe size	G		85	.094		4.58	4.29		8.87	11.70
8420	4" iron pipe size	G		80	.100		4.88	4.56		9.44	12.45
8430	5" iron pipe size	G		80	.100		5.90	4.56		10.46	13.55
8440	6" iron pipe size	G		80	.100		7.10	4.56		11.66	14.90
8444	1" wall, 1/2" iron pipe size	G		86	.093		1.93	4.24		6.17	8.70
8445	3/4" iron pipe size	G		84	.095		2.34	4.34		6.68	9.30
8446	1" iron pipe size	G		84	.095		2.72	4.34		7.06	9.75
8447	1-1/4" iron pipe size	G		82	.098		3.08	4.44		7.52	10.30
8448	1-1/2" iron pipe size	G		82	.098		3.58	4.44		8.02	10.85
8449	2" iron pipe size	G		80	.100		4.78	4.56		9.34	12.35
8450	2-1/2" iron pipe size	G	↓	80	.100	↓	6.25	4.56		10.81	13.95
8456	Rubber insulation tape, 1/8" x 2" x 30'	G				Ea.	11.40			11.40	12.50
8460	Polyolefin tubing, flexible closed cell foam, UV stabilized, work										
8462	temp.-165°F to +210°F, 0 water vapor transmission										
8464	3/8" wall, 1/8" iron pipe size	G	1 Asbe	140	.057	L.F.	.37	2.60		2.97	4.46
8466	1/4" iron pipe size	G		140	.057		.39	2.60		2.99	4.48
8468	3/8" iron pipe size	G		140	.057		.42	2.60		3.02	4.51
8470	1/2" iron pipe size	G		136	.059		.48	2.68		3.16	4.69
8472	3/4" iron pipe size	G		132	.061		.55	2.76		3.31	4.90
8474	1" iron pipe size	G		130	.062		.60	2.80		3.40	5
8476	1-1/4" iron pipe size	G		128	.063		.78	2.85		3.63	5.30
8478	1-1/2" iron pipe size	G		128	.063		.96	2.85		3.81	5.50
8480	2" iron pipe size	G		126	.063		1.07	2.89		3.96	5.70
8482	2-1/2" iron pipe size	G		123	.065		1.56	2.96		4.52	6.30
8484	3" iron pipe size	G		121	.066		2.03	3.01		5.04	6.90
8486	4" iron pipe size	G		118	.068		3.34	3.09		6.43	8.45
8500	1/2" wall, 1/8" iron pipe size	G		130	.062		.54	2.80		3.34	4.95
8502	1/4" iron pipe size	G		130	.062		.58	2.80		3.38	5
8504	3/8" iron pipe size	G		130	.062		.63	2.80		3.43	5.05
8506	1/2" iron pipe size	G		128	.063		.68	2.85		3.53	5.20
8508	3/4" iron pipe size	G		126	.063		.79	2.89		3.68	5.35
8510	1" iron pipe size	G		123	.065		.86	2.96		3.82	5.55
8512	1-1/4" iron pipe size	G		121	.066		1.07	3.01		4.08	5.85
8514	1-1/2" iron pipe size	G		119	.067		1.29	3.06		4.35	6.20
8516	2" iron pipe size	G		117	.068		1.50	3.11		4.61	6.50
8518	2-1/2" iron pipe size	G		114	.070		2.04	3.20		5.24	7.20
8520	3" iron pipe size	G		112	.071		2.90	3.25		6.15	8.25
8522	4" iron pipe size	G		110	.073		3.60	3.31		6.91	9.10
8534	3/4" wall, 1/8" iron pipe size	G		120	.067		.83	3.04		3.87	5.65
8536	1/4" iron pipe size	G		120	.067		.88	3.04		3.92	5.70
8538	3/8" iron pipe size	G		117	.068		1.01	3.11		4.12	5.95
8540	1/2" iron pipe size	G		114	.070		1.12	3.20		4.32	6.20
8542	3/4" iron pipe size	G		112	.071		1.45	3.25		4.70	6.65
8544	1" iron pipe size	G		110	.073		1.67	3.31		4.98	7
8546	1-1/4" iron pipe size	G	↓	108	.074	↓	2.26	3.37		5.63	7.75

22 07 19.10 Piping Insulation			Crew	Daily Output	Labor-Hours	Unit	2010 Bare Costs Material	Labor	Equipment	Total	Total Incl O&P
8548	1-1/2" iron pipe size	G	1 Asbe	108	.074	L.F.	2.62	3.37		5.99	8.15
8550	2" iron pipe size	G		106	.075		3.10	3.44		6.54	8.75
8552	2-1/2" iron pipe size	G		104	.077		3.90	3.50		7.40	9.75
8554	3" iron pipe size	G		102	.078		4.44	3.57		8.01	10.45
8556	4" iron pipe size	G		100	.080		6.30	3.64		9.94	12.60
8570	1" wall, 1/8" iron pipe size	G		110	.073		1.53	3.31		4.84	6.85
8572	1/4" iron pipe size	G		108	.074		1.60	3.37		4.97	7
8574	3/8" iron pipe size	G		106	.075		1.69	3.44		5.13	7.20
8576	1/2" iron pipe size	G		104	.077		1.96	3.50		5.46	7.60
8578	3/4" iron pipe size	G		102	.078		2.18	3.57		5.75	7.95
8580	1" iron pipe size	G		100	.080		2.34	3.64		5.98	8.20
8582	1-1/4" iron pipe size	G		97	.082		2.52	3.76		6.28	8.60
8584	1-1/2" iron pipe size	G		97	.082		3.27	3.76		7.03	9.45
8586	2" iron pipe size	G		95	.084		4.29	3.84		8.13	10.65
8588	2-1/2" iron pipe size	G		93	.086		5.60	3.92		9.52	12.25
8590	3" iron pipe size	G		91	.088		6.70	4		10.70	13.55
8606	Contact adhesive (R-320)	G				Qt.	16.70			16.70	18.40
8608	Contact adhesive (R-320)	G				Gal.	53.50			53.50	58.50
8610	NOTE: Preslit/preglued vs unslit, same price										

22 07 19.30 Piping Insulation Protective Jacketing, PVC		Crew	Daily Output	Labor-Hours	Unit	Material	Labor	Equipment	Total	Total Incl O&P
0010	**PIPING INSULATION PROTECTIVE JACKETING, PVC**									
0100	PVC, white, 48" lengths cut from roll goods									
0120	20 mil thick									
0140	Size based on OD of insulation									
0150	1-1/2" ID	Q-14	270	.059	L.F.	.26	2.43		2.69	4.07
0152	2" ID		260	.062		.32	2.52		2.84	4.27
0154	2-1/2" ID		250	.064		.40	2.62		3.02	4.52
0156	3" ID		240	.067		.47	2.73		3.20	4.77
0158	3-1/2" ID		230	.070		.54	2.85		3.39	5
0160	4" ID		220	.073		.60	2.98		3.58	5.30
0162	4-1/2" ID		210	.076		.67	3.12		3.79	5.60
0164	5" ID		200	.080		.75	3.28		4.03	5.95
0166	5-1/2" ID		190	.084		.82	3.45		4.27	6.25
0168	6" ID		180	.089		.88	3.64		4.52	6.60
0170	6-1/2" ID		175	.091		.95	3.75		4.70	6.90
0172	7" ID		170	.094		1.03	3.86		4.89	7.15
0174	7-1/2" ID		164	.098		1.10	4		5.10	7.40
0176	8" ID		161	.099		1.18	4.07		5.25	7.65
0178	8-1/2" ID		158	.101		1.24	4.15		5.39	7.80
0180	9" ID		155	.103		1.31	4.23		5.54	8.05
0182	9-1/2" ID		152	.105		1.38	4.32		5.70	8.20
0184	10" ID		149	.107		1.46	4.40		5.86	8.45
0186	10-1/2" ID		146	.110		1.52	4.49		6.01	8.65
0188	11" ID		143	.112		1.59	4.59		6.18	8.90
0190	11-1/2" ID		140	.114		1.66	4.69		6.35	9.15
0192	12" ID		137	.117		1.74	4.79		6.53	9.35
0194	12-1/2" ID		134	.119		1.80	4.90		6.70	9.60
0195	13" ID		132	.121		1.88	4.97		6.85	9.75
0196	13-1/2" ID		132	.121		1.95	4.97		6.92	9.85
0198	14" ID		130	.123		2.02	5.05		7.07	10.05
0200	15" ID		128	.125		2.15	5.15		7.30	10.30
0202	16" ID		126	.127		2.30	5.20		7.50	10.65

22 07 19.30 Piping Insulation Protective Jacketing, PVC		Crew	Daily Output	Labor-Hours	Unit	Material	Labor	Equipment	Total	Total Incl O&P
						2010 Bare Costs				
0204	17" ID	Q-14	124	.129	L.F.	2.43	5.30		7.73	10.85
0206	18" ID		122	.131		2.58	5.40		7.98	11.20
0208	19" ID		120	.133		2.72	5.45		8.17	11.50
0210	20" ID		118	.136		2.86	5.55		8.41	11.80
0212	21" ID		116	.138		3	5.65		8.65	12.10
0214	22" ID		114	.140		3.14	5.75		8.89	12.40
0216	23" ID		112	.143		3.28	5.85		9.13	12.70
0218	24" ID		110	.145		3.43	5.95		9.38	13
0220	25" ID		108	.148		3.56	6.05		9.61	13.35
0222	26" ID		106	.151		3.71	6.20		9.91	13.70
0224	27" ID		104	.154		3.84	6.30		10.14	14
0226	28" ID		102	.157		3.99	6.45		10.44	14.40
0228	29" ID		100	.160		4.13	6.55		10.68	14.75
0230	30" ID	↓	98	.163	↓	4.27	6.70		10.97	15.10
0300	For colors, add				Ea.	10%				
1000	30 mil thick									
1010	Size based on OD of insulation									
1020	2" ID	Q-14	260	.062	L.F.	.50	2.52		3.02	4.47
1022	2-1/2" ID		250	.064		.60	2.62		3.22	4.74
1024	3" ID		240	.067		.72	2.73		3.45	5.05
1026	3-1/2" ID		230	.070		.82	2.85		3.67	5.35
1028	4" ID		220	.073		.92	2.98		3.90	5.65
1030	4-1/2" ID		210	.076		1.03	3.12		4.15	6
1032	5" ID		200	.080		1.13	3.28		4.41	6.35
1034	5-1/2" ID		190	.084		1.24	3.45		4.69	6.70
1036	6" ID		180	.089		1.34	3.64		4.98	7.10
1038	6-1/2" ID		175	.091		1.46	3.75		5.21	7.45
1040	7" ID		170	.094		1.56	3.86		5.42	7.70
1042	7-1/2" ID		164	.098		1.68	4		5.68	8.05
1044	8" ID		161	.099		1.78	4.07		5.85	8.30
1046	8-1/2" ID		158	.101		1.89	4.15		6.04	8.55
1048	9" ID		155	.103		1.99	4.23		6.22	8.80
1050	9-1/2" ID		152	.105		2.11	4.32		6.43	9
1052	10" ID		149	.107		2.21	4.40		6.61	9.30
1054	10-1/2" ID		146	.110		2.32	4.49		6.81	9.55
1056	11" ID		143	.112		2.42	4.59		7.01	9.80
1058	11-1/2" ID		140	.114		2.53	4.69		7.22	10.10
1060	12" ID		137	.117		2.63	4.79		7.42	10.35
1062	12-1/2" ID		134	.119		2.74	4.90		7.64	10.60
1063	13" ID		132	.121		2.86	4.97		7.83	10.85
1064	13-1/2" ID		132	.121		2.96	4.97		7.93	10.95
1066	14" ID		130	.123		3.07	5.05		8.12	11.25
1068	15" ID		128	.125		3.28	5.15		8.43	11.55
1070	16" ID		126	.127		3.49	5.20		8.69	11.95
1072	17" ID		124	.129		3.71	5.30		9.01	12.30
1074	18" ID		122	.131		3.92	5.40		9.32	12.65
1076	19" ID		120	.133		4.13	5.45		9.58	13.05
1078	20" ID		118	.136		4.35	5.55		9.90	13.45
1080	21" ID		116	.138		4.56	5.65		10.21	13.80
1082	22" ID		114	.140		4.77	5.75		10.52	14.20
1084	23" ID		112	.143		4.98	5.85		10.83	14.60
1086	24" ID		110	.145		5.20	5.95		11.15	15
1088	25" ID	↓	108	.148	↓	5.40	6.05		11.45	15.40

22 07 19 – Plumbing Piping Insulation

22 07 19.30 Piping Insulation Protective Jacketing, PVC		Crew	Daily Output	Labor-Hours	Unit	Material	2010 Bare Costs Labor	Equipment	Total	Total Incl O&P
1090	26" ID	Q-14	106	.151	L.F.	5.65	6.20		11.85	15.80
1092	27" ID		104	.154		5.85	6.30		12.15	16.25
1094	28" ID		102	.157		6.05	6.45		12.50	16.65
1096	29" ID		100	.160		6.30	6.55		12.85	17.10
1098	30" ID		98	.163		6.50	6.70		13.20	17.55
1300	For colors, add				Ea.	10%				
2000	PVC, white, fitting covers									
2020	Fiberglass insulation inserts included with sizes 1-3/4" thru 9-3/4"									
2030	Size is based on OD of insulation									
2040	90° Elbow fitting									
2060	1-3/4"	Q-14	135	.119	Ea.	.49	4.86		5.35	8.10
2062	2"		130	.123		.60	5.05		5.65	8.50
2064	2-1/4"		128	.125		.66	5.15		5.81	8.70
2068	2-1/2"		126	.127		.74	5.20		5.94	8.90
2070	2-3/4"		123	.130		.82	5.35		6.17	9.20
2072	3"		120	.133		.83	5.45		6.28	9.40
2074	3-3/8"		116	.138		.95	5.65		6.60	9.85
2076	3-3/4"		113	.142		1.08	5.80		6.88	10.20
2078	4-1/8"		110	.145		1.34	5.95		7.29	10.70
2080	4-3/4"		105	.152		1.61	6.25		7.86	11.45
2082	5-1/4"		100	.160		1.88	6.55		8.43	12.25
2084	5-3/4"		95	.168		2.52	6.90		9.42	13.50
2086	6-1/4"		90	.178		4.24	7.30		11.54	16
2088	6-3/4"		87	.184		4.49	7.55		12.04	16.65
2090	7-1/4"		85	.188		5.60	7.70		13.30	18.20
2092	7-3/4"		83	.193		5.80	7.90		13.70	18.70
2094	8-3/4"		80	.200		7.60	8.20		15.80	21
2096	9-3/4"		77	.208		10.15	8.50		18.65	24.50
2098	10-7/8"		74	.216		11.30	8.85		20.15	26.50
2100	11-7/8"		71	.225		12.95	9.25		22.20	28.50
2102	12-7/8"		68	.235		16.70	9.65		26.35	33.50
2104	14-1/8"		66	.242		18.55	9.95		28.50	36
2106	15-1/8"		64	.250		20	10.25		30.25	38
2108	16-1/8"		63	.254		22	10.40		32.40	40
2110	17-1/8"		62	.258		24.50	10.60		35.10	43.50
2112	18-1/8"		61	.262		30.50	10.75		41.25	50
2114	19-1/8"		60	.267		42.50	10.95		53.45	64
2116	20-1/8"		59	.271		55	11.10		66.10	78
2200	45° Elbow fitting									
2220	1-3/4" thru 9-3/4" same price as 90° Elbow fitting									
2320	10-7/8"	Q-14	74	.216	Ea.	11.30	8.85		20.15	26.50
2322	11-7/8"		71	.225		12.50	9.25		21.75	28
2324	12-7/8"		68	.235		13.90	9.65		23.55	30.50
2326	14-1/8"		66	.242		15.85	9.95		25.80	33
2328	15-1/8"		64	.250		17.05	10.25		27.30	34.50
2330	16-1/8"		63	.254		19.50	10.40		29.90	37.50
2332	17-1/8"		62	.258		22	10.60		32.60	40.50
2334	18-1/8"		61	.262		27	10.75		37.75	46
2336	19-1/8"		60	.267		36.50	10.95		47.45	57
2338	20-1/8"		59	.271		41.50	11.10		52.60	63
2400	Tee fitting									
2410	1-3/4"	Q-14	96	.167	Ea.	.93	6.85		7.78	11.60
2412	2"		94	.170		1.03	7		8.03	12

22 07 19.30 Piping Insulation Protective Jacketing, PVC		Crew	Daily Output	Labor-Hours	Unit	Material	Labor	Equipment	Total	Total Incl O&P
						2010 Bare Costs				
2414	2-1/4"	Q-14	91	.176	Ea.	1.15	7.20		8.35	12.45
2416	2-1/2"		88	.182		1.25	7.45		8.70	13
2418	2-3/4"		85	.188		1.37	7.70		9.07	13.50
2420	3"		82	.195		1.49	8		9.49	14.10
2422	3-3/8"		79	.203		1.71	8.30		10.01	14.80
2424	3-3/4"		76	.211		1.96	8.65		10.61	15.55
2426	4-1/8"		73	.219		2.30	9		11.30	16.50
2428	4-3/4"		70	.229		2.88	9.35		12.23	17.70
2430	5-1/4"		67	.239		3.45	9.80		13.25	19
2432	5-3/4"		63	.254		4.57	10.40		14.97	21.50
2434	6-1/4"		60	.267		6	10.95		16.95	23.50
2436	6-3/4"		59	.271		7.45	11.10		18.55	25.50
2438	7-1/4"		57	.281		12.05	11.50		23.55	31
2440	7-3/4"		54	.296		13.20	12.15		25.35	33.50
2442	8-3/4"		52	.308		16.05	12.60		28.65	37.50
2444	9-3/4"		50	.320		18.95	13.10		32.05	41.50
2446	10-7/8"		48	.333		19.15	13.65		32.80	42
2448	11-7/8"		47	.340		21.50	13.95		35.45	45
2450	12-7/8"		46	.348		23.50	14.25		37.75	48
2452	14-1/8"		45	.356		25.50	14.60		40.10	50.50
2454	15-1/8"		44	.364		27.50	14.90		42.40	53.50
2456	16-1/8"		43	.372		30	15.25		45.25	56.50
2458	17-1/8"		42	.381		32	15.60		47.60	59.50
2460	18-1/8"		41	.390		35	16		51	63.50
2462	19-1/8"		40	.400		38.50	16.40		54.90	67.50
2464	20-1/8"		39	.410		41.50	16.80		58.30	71.50
4000	Mechanical grooved fitting cover, including insert									
4020	90° Elbow fitting									
4030	3/4" & 1"	Q-14	140	.114	Ea.	6.75	4.69		11.44	14.75
4040	1-1/4" & 1-1/2"		135	.119		7.05	4.86		11.91	15.30
4042	2"		130	.123		7.75	5.05		12.80	16.35
4044	2-1/2"		125	.128		8.60	5.25		13.85	17.60
4046	3"		120	.133		9.65	5.45		15.10	19.15
4048	3-1/2"		115	.139		11.20	5.70		16.90	21
4050	4"		110	.145		12.50	5.95		18.45	23
4052	5"		100	.160		15.45	6.55		22	27
4054	6"		90	.178		22	7.30		29.30	35.50
4056	8"		80	.200		25	8.20		33.20	40.50
4058	10"		75	.213		32	8.75		40.75	48.50
4060	12"		68	.235		46	9.65		55.65	65.50
4062	14"		65	.246		57	10.10		67.10	78
4064	16"		63	.254		77.50	10.40		87.90	101
4066	18"		61	.262		106	10.75		116.75	134
4100	45° Elbow fitting									
4120	3/4" & 1"	Q-14	140	.114	Ea.	6.45	4.69		11.14	14.40
4130	1-1/4" & 1-1/2"		135	.119		6.75	4.86		11.61	15
4140	2"		130	.123		7.10	5.05		12.15	15.65
4142	2-1/2"		125	.128		7.85	5.25		13.10	16.80
4144	3"		120	.133		8.65	5.45		14.10	18.05
4146	3-1/2"		115	.139		10.10	5.70		15.80	20
4148	4"		110	.145		10.45	5.95		16.40	21
4150	5"		100	.160		13.95	6.55		20.50	25.50
4152	6"		90	.178		19.40	7.30		26.70	33

22 07 19 – Plumbing Piping Insulation

22 07 19.30 Piping Insulation Protective Jacketing, PVC		Crew	Daily Output	Labor-Hours	Unit	Material	2010 Bare Costs Labor	Equipment	Total	Total Incl O&P
4154	8"	Q-14	80	.200	Ea.	22	8.20		30.20	37
4156	10"		75	.213		28	8.75		36.75	44
4158	12"		68	.235		39.50	9.65		49.15	58.50
4160	14"		65	.246		44.50	10.10		54.60	64.50
4162	16"		63	.254		65.50	10.40		75.90	88.50
4164	18"		61	.262		94	10.75		104.75	120
4200	Tee fitting									
4220	3/4" & 1"	Q-14	93	.172	Ea.	9.30	7.05		16.35	21
4230	1-1/4" & 1-1/2"		90	.178		9.90	7.30		17.20	22
4240	2"		87	.184		10.95	7.55		18.50	24
4242	2-1/2"		84	.190		12.20	7.80		20	25.50
4244	3"		80	.200		13.10	8.20		21.30	27
4246	3-1/2"		77	.208		15.25	8.50		23.75	30
4248	4"		73	.219		17.95	9		26.95	33.50
4250	5"		67	.239		21	9.80		30.80	38
4252	6"		60	.267		25	10.95		35.95	44.50
4254	8"		54	.296		31	12.15		43.15	53
4256	10"		50	.320		39.50	13.10		52.60	64
4258	12"		46	.348		52.50	14.25		66.75	80
4260	14"		43	.372		67	15.25		82.25	97
4262	16"		42	.381		89	15.60		104.60	123
4264	18"		41	.390		118	16		134	154

22 07 19.40 Pipe Insulation Protective Jacketing, Aluminum

		Crew	Daily Output	Labor-Hours	Unit	Material	Labor	Equipment	Total	Total Incl O&P
0010	**PIPE INSULATION PROTECTIVE JACKETING, ALUMINUM**									
0100	Metal roll jacketing									
0120	Aluminum with polykraft moisture barrier									
0140	Smooth, based on OD of insulation, .010" thick									
0180	1/2" ID	Q-14	220	.073	L.F.	.24	2.98		3.22	4.90
0190	3/4" ID		215	.074		.32	3.05		3.37	5.10
0200	1" ID		210	.076		.38	3.12		3.50	5.30
0210	1-1/4" ID		205	.078		.46	3.20		3.66	5.50
0220	1-1/2" ID		202	.079		.53	3.25		3.78	5.65
0230	1-3/4" ID		199	.080		.61	3.30		3.91	5.75
0240	2" ID		195	.082		.68	3.36		4.04	6
0250	2-1/4" ID		191	.084		.75	3.43		4.18	6.20
0260	2-1/2" ID		187	.086		.83	3.51		4.34	6.35
0270	2-3/4" ID		184	.087		.90	3.57		4.47	6.55
0280	3" ID		180	.089		.97	3.64		4.61	6.70
0290	3-1/4" ID		176	.091		1.04	3.73		4.77	6.95
0300	3-1/2" ID		172	.093		1.12	3.81		4.93	7.20
0310	3-3/4" ID		169	.095		1.19	3.88		5.07	7.35
0320	4" ID		165	.097		1.26	3.98		5.24	7.60
0330	4-1/4" ID		161	.099		1.34	4.07		5.41	7.80
0340	4-1/2" ID		157	.102		1.41	4.18		5.59	8.05
0350	4-3/4" ID		154	.104		1.49	4.26		5.75	8.25
0360	5" ID		150	.107		1.55	4.37		5.92	8.50
0370	5-1/4" ID		146	.110		1.63	4.49		6.12	8.80
0380	5-1/2" ID		143	.112		1.70	4.59		6.29	9
0390	5-3/4" ID		139	.115		1.78	4.72		6.50	9.30
0400	6" ID		135	.119		1.85	4.86		6.71	9.60
0410	6-1/4" ID		133	.120		1.92	4.93		6.85	9.75
0420	6-1/2" ID		131	.122		2	5		7	10

22 07 19.40 Pipe Insulation Protective Jacketing, Aluminum		Crew	Daily Output	Labor-Hours	Unit	Material	2010 Bare Costs Labor	Equipment	Total	Total Incl O&P
0430	7" ID	Q-14	128	.125	L.F.	2.14	5.15		7.29	10.30
0440	7-1/4" ID		125	.128		2.21	5.25		7.46	10.60
0450	7-1/2" ID		123	.130		2.29	5.35		7.64	10.80
0460	8" ID		121	.132		2.43	5.40		7.83	11.10
0470	8-1/2" ID		119	.134		2.58	5.50		8.08	11.40
0480	9" ID		116	.138		2.72	5.65		8.37	11.80
0490	9-1/2" ID		114	.140		2.87	5.75		8.62	12.10
0500	10" ID		112	.143		3.02	5.85		8.87	12.40
0510	10-1/2" ID		110	.145		3.17	5.95		9.12	12.75
0520	11" ID		107	.150		3.32	6.15		9.47	13.20
0530	11-1/2" ID		105	.152		3.46	6.25		9.71	13.50
0540	12" ID		103	.155		3.61	6.35		9.96	13.85
0550	12-1/2" ID		100	.160		3.75	6.55		10.30	14.35
0560	13" ID		99	.162		3.89	6.65		10.54	14.60
0570	14" ID		98	.163		4.19	6.70		10.89	15
0580	15" ID		96	.167		4.48	6.85		11.33	15.55
0590	16" ID		95	.168		4.78	6.90		11.68	16
0600	17" ID		93	.172		5.05	7.05		12.10	16.55
0610	18" ID		92	.174		5.25	7.15		12.40	16.90
0620	19" ID		90	.178		5.65	7.30		12.95	17.60
0630	20" ID		89	.180		5.95	7.35		13.30	18
0640	21" ID		87	.184		6.25	7.55		13.80	18.55
0650	22" ID		86	.186		6.55	7.65		14.20	19.05
0660	23" ID		84	.190		6.85	7.80		14.65	19.65
0670	24" ID		83	.193		7.10	7.90		15	20
0710	For smooth .020" thick, add					27%	10%			
0720	For smooth .024" thick, add					52%	20%			
0730	For smooth .032" thick, add					104%	33%			
0800	For stucco embossed, add					1%				
0820	For corrugated, add					2.50%				
0900	White aluminum with polysurlyn moisture barrier									
0910	Smooth, % is an add to polykraft lines of same thickness									
0940	For smooth .016" thick, add				L.F.	35%				
0960	For smooth .024" thick, add				"	22%				
1000	Aluminum fitting covers									
1010	Size is based on OD of insulation									
1020	90 LR elbow, 2 piece									
1100	1-1/2"	Q-14	140	.114	Ea.	3.05	4.69		7.74	10.65
1110	1-3/4"		135	.119		3.05	4.86		7.91	10.90
1120	2"		130	.123		3.83	5.05		8.88	12.05
1130	2-1/4"		128	.125		3.83	5.15		8.98	12.15
1140	2-1/2"		126	.127		3.83	5.20		9.03	12.30
1150	2-3/4"		123	.130		3.83	5.35		9.18	12.50
1160	3"		120	.133		4.16	5.45		9.61	13.10
1170	3-1/4"		117	.137		4.16	5.60		9.76	13.30
1180	3-1/2"		115	.139		4.91	5.70		10.61	14.25
1190	3-3/4"		113	.142		5.05	5.80		10.85	14.55
1200	4"		110	.145		5.25	5.95		11.20	15.05
1210	4-1/4"		108	.148		5.40	6.05		11.45	15.40
1220	4-1/2"		106	.151		5.40	6.20		11.60	15.55
1230	4-3/4"		104	.154		5.40	6.30		11.70	15.75
1240	5"		102	.157		6.10	6.45		12.55	16.70
1250	5-1/4"		100	.160		7.05	6.55		13.60	17.95

22 07 19.40 Pipe Insulation Protective Jacketing, Aluminum		Crew	Daily Output	Labor-Hours	Unit	Material	2010 Bare Costs Labor	Equipment	Total	Total Incl O&P
1260	5-1/2"	Q-14	97	.165	Ea.	8.80	6.75		15.55	20
1270	5-3/4"		95	.168		8.80	6.90		15.70	20.50
1280	6"		92	.174		7.50	7.15		14.65	19.35
1290	6-1/4"		90	.178		7.50	7.30		14.80	19.60
1300	6-1/2"		87	.184		9	7.55		16.55	21.50
1310	7"		85	.188		10.95	7.70		18.65	24
1320	7-1/4"		84	.190		10.95	7.80		18.75	24
1330	7-1/2"		83	.193		16.10	7.90		24	30
1340	8"		82	.195		12	8		20	25.50
1350	8-1/2"		80	.200		22	8.20		30.20	37.50
1360	9"		78	.205		22	8.40		30.40	37.50
1370	9-1/2"		77	.208		16.30	8.50		24.80	31
1380	10"		76	.211		16.30	8.65		24.95	31.50
1390	10-1/2"		75	.213		16.65	8.75		25.40	32
1400	11"		74	.216		16.65	8.85		25.50	32
1410	11-1/2"		72	.222		18.95	9.10		28.05	35
1420	12"		71	.225		18.95	9.25		28.20	35.50
1430	12-1/2"		69	.232		33.50	9.50		43	52
1440	13"		68	.235		33.50	9.65		43.15	52
1450	14"		66	.242		45	9.95		54.95	65.50
1460	15"		64	.250		47.50	10.25		57.75	68
1470	16"	↓	63	.254	↓	51.50	10.40		61.90	73
2000	45 Elbow, 2 piece									
2010	2-1/2"	Q-14	126	.127	Ea.	3.17	5.20		8.37	11.60
2020	2-3/4"		123	.130		3.17	5.35		8.52	11.80
2030	3"		120	.133		3.58	5.45		9.03	12.45
2040	3-1/4"		117	.137		3.58	5.60		9.18	12.65
2050	3-1/2"		115	.139		4.08	5.70		9.78	13.35
2060	3-3/4"		113	.142		4.08	5.80		9.88	13.50
2070	4"		110	.145		4.15	5.95		10.10	13.80
2080	4-1/4"		108	.148		4.73	6.05		10.78	14.65
2090	4-1/2"		106	.151		4.73	6.20		10.93	14.80
2100	4-3/4"		104	.154		4.73	6.30		11.03	15
2110	5"		102	.157		5.55	6.45		12	16.10
2120	5-1/4"		100	.160		5.55	6.55		12.10	16.30
2130	5-1/2"		97	.165		5.85	6.75		12.60	16.90
2140	6"		92	.174		5.85	7.15		13	17.50
2150	6-1/2"		87	.184		8.05	7.55		15.60	20.50
2160	7"		85	.188		8.05	7.70		15.75	21
2170	7-1/2"		83	.193		8.10	7.90		16	21.50
2180	8"		82	.195		8.10	8		16.10	21.50
2190	8-1/2"		80	.200		10.10	8.20		18.30	24
2200	9"		78	.205		10.10	8.40		18.50	24
2210	9-1/2"		77	.208		14.05	8.50		22.55	28.50
2220	10"		76	.211		14.05	8.65		22.70	29
2230	10-1/2"		75	.213		13.10	8.75		21.85	28
2240	11"		74	.216		13.10	8.85		21.95	28.50
2250	11-1/2"		72	.222		15.75	9.10		24.85	31.50
2260	12"		71	.225		15.75	9.25		25	31.50
2270	13"		68	.235		18.50	9.65		28.15	35.50
2280	14"		66	.242		22.50	9.95		32.45	40.50
2290	15"		64	.250		35	10.25		45.25	54.50
2300	16"	↓	63	.254	↓	38	10.40		48.40	57.50

22 07 19.40 Pipe Insulation Protective Jacketing, Aluminum		Crew	Daily Output	Labor-Hours	Unit	Material	2010 Bare Costs Labor	Equipment	Total	Total Incl O&P
2310	17″	Q-14	62	.258	Ea.	37	10.60		47.60	57
2320	18″		61	.262		42	10.75		52.75	63
2330	19″		60	.267		51.50	10.95		62.45	73.50
2340	20″		59	.271		50	11.10		61.10	72.50
2350	21″		58	.276		54	11.30		65.30	76.50
3000	Tee, 4 piece									
3010	2-1/2″	Q-14	88	.182	Ea.	24.50	7.45		31.95	38.50
3020	2-3/4″		86	.186		24.50	7.65		32.15	39
3030	3″		84	.190		28	7.80		35.80	42.50
3040	3-1/4″		82	.195		28	8		36	43
3050	3-1/2″		80	.200		29	8.20		37.20	45
3060	4″		78	.205		29.50	8.40		37.90	45.50
3070	4-1/4″		76	.211		31	8.65		39.65	47.50
3080	4-1/2″		74	.216		31	8.85		39.85	48
3090	4-3/4″		72	.222		31	9.10		40.10	48
3100	5″		70	.229		32	9.35		41.35	50
3110	5-1/4″		68	.235		32	9.65		41.65	50.50
3120	5-1/2″		66	.242		33.50	9.95		43.45	52.50
3130	6″		64	.250		33.50	10.25		43.75	53
3140	6-1/2″		60	.267		37	10.95		47.95	57.50
3150	7″		58	.276		37	11.30		48.30	58
3160	7-1/2″		56	.286		41	11.70		52.70	63
3170	8″		54	.296		41	12.15		53.15	64
3180	8-1/2″		52	.308		42	12.60		54.60	65.50
3190	9″		50	.320		42	13.10		55.10	66.50
3200	9-1/2″		49	.327		31	13.40		44.40	55.50
3210	10″		48	.333		31	13.65		44.65	55.50
3220	10-1/2″		47	.340		33	13.95		46.95	58
3230	11″		46	.348		33	14.25		47.25	58.50
3240	11-1/2″		45	.356		35	14.60		49.60	61
3250	12″		44	.364		35	14.90		49.90	61.50
3260	13″		43	.372		37	15.25		52.25	64
3270	14″		42	.381		38.50	15.60		54.10	67
3280	15″		41	.390		41.50	16		57.50	71
3290	16″		40	.400		43	16.40		59.40	72.50
3300	17″		39	.410		49	16.80		65.80	80
3310	18″		38	.421		51	17.25		68.25	83
3320	19″		37	.432		56.50	17.75		74.25	89.50
3330	20″		36	.444		57.50	18.20		75.70	92
3340	22″		35	.457		73.50	18.75		92.25	110
3350	23″		34	.471		75.50	19.30		94.80	113
3360	24″		31	.516		77.50	21		98.50	118

22 07 19.50 Pipe Insulation Protective Jacketing, St. Stl.

		Crew	Daily Output	Labor-Hours	Unit	Material	Labor	Equipment	Total	Total Incl O&P
0010	**PIPE INSULATION PROTECTIVE JACKETING, STAINLESS STEEL**									
0100	Metal roll jacketing									
0120	Type 304 with moisture barrier									
0140	Smooth, based on OD of insulation, .010″ thick									
0260	2-1/2″ ID	Q-14	250	.064	L.F.	2.30	2.62		4.92	6.60
0270	2-3/4″ ID		245	.065		2.50	2.68		5.18	6.90
0280	3″ ID		240	.067		2.70	2.73		5.43	7.20
0290	3-1/4″ ID		235	.068		2.91	2.79		5.70	7.55
0300	3-1/2″ ID		230	.070		3.11	2.85		5.96	7.85

22 07 19.50 Pipe Insulation Protective Jacketing, St. Stl.		Crew	Daily Output	Labor-Hours	Unit	Material	2010 Bare Costs Labor	Equipment	Total	Total Incl O&P
0310	3-3/4" ID	Q-14	225	.071	L.F.	3.32	2.92		6.24	8.20
0320	4" ID		220	.073		3.52	2.98		6.50	8.50
0330	4-1/4" ID		215	.074		3.72	3.05		6.77	8.85
0340	4-1/2" ID		210	.076		3.93	3.12		7.05	9.20
0350	5" ID		200	.080		4.34	3.28		7.62	9.85
0360	5-1/2" ID		190	.084		4.74	3.45		8.19	10.55
0370	6" ID		180	.089		5.15	3.64		8.79	11.30
0380	6-1/2" ID		175	.091		5.55	3.75		9.30	11.95
0390	7" ID		170	.094		5.95	3.86		9.81	12.55
0400	7-1/2" ID		164	.098		6.35	4		10.35	13.20
0410	8" ID		161	.099		6.80	4.07		10.87	13.80
0420	8-1/2" ID		158	.101		7.20	4.15		11.35	14.35
0430	9" ID		155	.103		7.60	4.23		11.83	14.95
0440	9-1/2" ID		152	.105		8	4.32		12.32	15.50
0450	10" ID		149	.107		8.40	4.40		12.80	16.10
0460	10-1/2" ID		146	.110		8.80	4.49		13.29	16.70
0470	11" ID		143	.112		9.25	4.59		13.84	17.30
0480	12" ID		137	.117		10.05	4.79		14.84	18.50
0490	13" ID		132	.121		10.85	4.97		15.82	19.65
0500	14" ID		130	.123		11.65	5.05		16.70	20.50
0700	For smooth .016" thick, add					45%	33%			
1000	Stainless steel, Type 316, fitting covers									
1010	Size is based on OD of insulation									
1020	90 LR Elbow, 2 piece									
1100	1-1/2"	Q-14	126	.127	Ea.	19.55	5.20		24.75	29.50
1110	2-3/4"		123	.130		12.50	5.35		17.85	22
1120	3"		120	.133		13.05	5.45		18.50	23
1130	3-1/4"		117	.137		13.05	5.60		18.65	23
1140	3-1/2"		115	.139		13.75	5.70		19.45	24
1150	3-3/4"		113	.142		14.50	5.80		20.30	25
1160	4"		110	.145		15.95	5.95		21.90	27
1170	4-1/4"		108	.148		21	6.05		27.05	33
1180	4-1/2"		106	.151		21	6.20		27.20	33
1190	5"		102	.157		21.50	6.45		27.95	34
1200	5-1/2"		97	.165		32.50	6.75		39.25	46
1210	6"		92	.174		35.50	7.15		42.65	50
1220	6-1/2"		87	.184		49	7.55		56.55	65
1230	7"		85	.188		49	7.70		56.70	65.50
1240	7-1/2"		83	.193		57	7.90		64.90	75
1250	8"		80	.200		57	8.20		65.20	75.50
1260	8-1/2"		80	.200		59	8.20		67.20	78
1270	9"		78	.205		89.50	8.40		97.90	112
1280	9-1/2"		77	.208		88	8.50		96.50	110
1290	10"		76	.211		88	8.65		96.65	110
1300	10-1/2"		75	.213		105	8.75		113.75	130
1310	11"		74	.216		101	8.85		109.85	125
1320	12"		71	.225		114	9.25		123.25	140
1330	13"		68	.235		158	9.65		167.65	189
1340	14"		66	.242		159	9.95		168.95	190
2000	45 Elbow, 2 piece									
2010	2-1/2"	Q-14	126	.127	Ea.	10.40	5.20		15.60	19.55
2020	2-3/4"		123	.130		10.40	5.35		15.75	19.75
2030	3"		120	.133		11.20	5.45		16.65	21

22 07 19 – Plumbing Piping Insulation

22 07 19.50 Pipe Insulation Protective Jacketing, St. Stl.		Crew	Daily Output	Labor-Hours	Unit	Material	2010 Bare Costs Labor	Equipment	Total	Total Incl O&P
2040	3-1/4"	Q-14	117	.137	Ea.	11.20	5.60		16.80	21
2050	3-1/2"		115	.139		11.35	5.70		17.05	21.50
2060	3-3/4"		113	.142		11.35	5.80		17.15	21.50
2070	4"		110	.145		14.60	5.95		20.55	25.50
2080	4-1/4"		108	.148		20.50	6.05		26.55	32
2090	4-1/2"		106	.151		20.50	6.20		26.70	32
2100	4-3/4"		104	.154		20.50	6.30		26.80	32.50
2110	5"		102	.157		21	6.45		27.45	33
2120	5-1/2"		97	.165		21	6.75		27.75	33.50
2130	6"		92	.174		24.50	7.15		31.65	38
2140	6-1/2"		87	.184		24.50	7.55		32.05	38.50
2150	7"		85	.188		42	7.70		49.70	58
2160	7-1/2"		83	.193		42.50	7.90		50.40	59.50
2170	8"		82	.195		42.50	8		50.50	59.50
2180	8-1/2"		80	.200		50	8.20		58.20	68
2190	9"		78	.205		50	8.40		58.40	68
2200	9-1/2"		77	.208		58.50	8.50		67	78
2210	10"		76	.211		58.50	8.65		67.15	78
2220	10-1/2"		75	.213		70.50	8.75		79.25	91
2230	11"		74	.216		70.50	8.85		79.35	91.50
2240	12"		71	.225		76.50	9.25		85.75	98.50
2250	13"	↓	68	.235	↓	90	9.65		99.65	114

22 11 Facility Water Distribution

22 11 13 – Facility Water Distribution Piping

22 11 13.14 Pipe, Brass

		Crew	Daily Output	Labor-Hours	Unit	Material	Labor	Equipment	Total	Total Incl O&P
0010	**PIPE, BRASS**, Plain end									
0900	Field threaded, coupling & clevis hanger assembly 10' O.C.									
0920	Regular weight									
1120	1/2" diameter	1 Plum	48	.167	L.F.	5.80	8.70		14.50	19.40
1140	3/4" diameter		46	.174		7.85	9.05		16.90	22
1160	1" diameter	↓	43	.186		11.45	9.70		21.15	27
1180	1-1/4" diameter	Q-1	72	.222		17.50	10.40		27.90	35
1200	1-1/2" diameter		65	.246		21	11.55		32.55	40.50
1220	2" diameter		53	.302		29	14.15		43.15	53
1240	2-1/2" diameter		41	.390		49	18.30		67.30	81.50
1260	3" diameter	↓	31	.516		66.50	24		90.50	110
1300	4" diameter	Q-2	37	.649	↓	124	31.50		155.50	183
1930	To delete coupling & hanger, subtract									
1940	1/8" diam. to 1/2" diam.					14%	46%			
1950	3/4" diam. to 1-1/2" diam.					8%	47%			
1960	2" diam. to 4" diam.					10%	37%			

22 11 13.16 Pipe Fittings, Brass

		Crew	Daily Output	Labor-Hours	Unit	Material	Labor	Equipment	Total	Total Incl O&P
0010	**PIPE FITTINGS, BRASS**, Rough bronze, threaded.									
1000	Standard wt., 90° Elbow									
1040	1/8"	1 Plum	13	.615	Ea.	14.05	32		46.05	63.50
1060	1/4"		13	.615		14.05	32		46.05	63.50
1080	3/8"		13	.615		14.05	32		46.05	63.50
1100	1/2"		12	.667		14.05	34.50		48.55	67.50
1120	3/4"		11	.727		18.75	38		56.75	77
1140	1"	↓	10	.800	↓	30.50	41.50		72	96

22 11 13.16 Pipe Fittings, Brass		Crew	Daily Output	Labor-Hours	Unit	2010 Bare Costs Material	Labor	Equipment	Total	Total Incl O&P
1160	1-1/4"	Q-1	17	.941	Ea.	49	44		93	120
1180	1-1/2"		16	1		61	47		108	137
1200	2"		14	1.143		98.50	53.50		152	189
1220	2-1/2"		11	1.455		238	68		306	365
1240	3"		8	2		365	93.50		458.50	540
1260	4"	Q-2	11	2.182		735	106		841	970
1280	5"		8	3		2,025	146		2,171	2,475
1300	6"		7	3.429		3,000	167		3,167	3,550
1500	45° Elbow, 1/8"	1 Plum	13	.615		17.25	32		49.25	67
1540	1/4"		13	.615		17.25	32		49.25	67
1560	3/8"		13	.615		17.25	32		49.25	67
1580	1/2"		12	.667		17.25	34.50		51.75	71
1600	3/4"		11	.727		25	38		63	84
1620	1"		10	.800		42	41.50		83.50	109
1640	1-1/4"	Q-1	17	.941		66.50	44		110.50	139
1660	1-1/2"		16	1		83.50	47		130.50	162
1680	2"		14	1.143		135	53.50		188.50	230
1700	2-1/2"		11	1.455		258	68		326	385
1720	3"		8	2		395	93.50		488.50	575
1740	4"	Q-2	11	2.182		900	106		1,006	1,150
1760	5"		8	3		1,650	146		1,796	2,050
1780	6"		7	3.429		2,300	167		2,467	2,775
2000	Tee, 1/8"	1 Plum	9	.889		16.45	46.50		62.95	87.50
2040	1/4"		9	.889		16.45	46.50		62.95	87.50
2060	3/8"		9	.889		16.45	46.50		62.95	87.50
2080	1/2"		8	1		16.45	52		68.45	96
2100	3/4"		7	1.143		23.50	59.50		83	115
2120	1"		6	1.333		42	69.50		111.50	151
2140	1-1/4"	Q-1	10	1.600		72.50	75		147.50	192
2160	1-1/2"		9	1.778		82	83.50		165.50	216
2180	2"		8	2		136	93.50		229.50	290
2200	2-1/2"		7	2.286		325	107		432	515
2220	3"		5	3.200		495	150		645	770
2240	4"	Q-2	7	3.429		1,225	167		1,392	1,600
2260	5"		5	4.800		2,575	233		2,808	3,175
2280	6"		4	6		4,000	291		4,291	4,825
2500	Coupling, 1/8"	1 Plum	26	.308		11.75	16		27.75	37
2540	1/4"		22	.364		11.75	18.95		30.70	41.50
2560	3/8"		18	.444		11.75	23		34.75	47.50
2580	1/2"		15	.533		11.75	28		39.75	54.50
2600	3/4"		14	.571		16.45	29.50		45.95	62.50
2620	1"		13	.615		28	32		60	79
2640	1-1/4"	Q-1	22	.727		47	34		81	103
2660	1-1/2"		20	.800		61	37.50		98.50	123
2680	2"		18	.889		101	41.50		142.50	174
2700	2-1/2"		14	1.143		211	53.50		264.50	315
2720	3"		10	1.600		268	75		343	405
2740	4"	Q-2	12	2		625	97		722	835
2760	5"		10	2.400		1,175	117		1,292	1,475
2780	6"		9	2.667		1,700	130		1,830	2,075
3000	Union, 125 lb									
3020	1/8"	1 Plum	12	.667	Ea.	39	34.50		73.50	95
3040	1/4"		12	.667		39	34.50		73.50	95

22 11 13 – Facility Water Distribution Piping

22 11 13.16 Pipe Fittings, Brass		Crew	Daily Output	Labor-Hours	Unit	Material	Labor	Equipment	Total	Total Incl O&P
						2010 Bare Costs				
3060	3/8"	1 Plum	12	.667	Ea.	39	34.50		73.50	95
3080	1/2"		11	.727		39	38		77	99.50
3100	3/4"		10	.800		55	41.50		96.50	123
3120	1"		9	.889		82.50	46.50		129	161
3140	1-1/4"	Q-1	16	1		120	47		167	202
3160	1-1/2"		15	1.067		143	50		193	232
3180	2"		13	1.231		192	57.50		249.50	298
3200	2-1/2"		10	1.600		525	75		600	685
3220	3"		7	2.286		815	107		922	1,050
3240	4"	Q-2	10	2.400		2,550	117		2,667	2,975
3320	For 250 lb. (navy pattern), add					100%				

22 11 13.23 Pipe/Tube, Copper		Crew	Daily Output	Labor-Hours	Unit	Material	Labor	Equipment	Total	Total Incl O&P
0010	**PIPE/TUBE, COPPER**, Solder joints R221113-50									
0100	Solder									
0120	Solder, lead free, roll				Lb.	15.35			15.35	16.85
1000	Type K tubing, couplings & clevis hanger assemblies 10' O.C.									
1100	1/4" diameter	1 Plum	84	.095	L.F.	2.80	4.96		7.76	10.55
1120	3/8" diameter		82	.098		2.92	5.10		8.02	10.80
1140	1/2" diameter		78	.103		3.36	5.35		8.71	11.70
1160	5/8" diameter		77	.104		4.32	5.40		9.72	12.85
1180	3/4" diameter		74	.108		6	5.65		11.65	15.05
1200	1" diameter		66	.121		7.95	6.30		14.25	18.20
1220	1-1/4" diameter		56	.143		10	7.45		17.45	22
1240	1-1/2" diameter		50	.160		13	8.35		21.35	27
1260	2" diameter		40	.200		20	10.40		30.40	37.50
1280	2-1/2" diameter	Q-1	60	.267		30.50	12.50		43	52.50
1300	3" diameter		54	.296		43	13.90		56.90	68
1320	3-1/2" diameter		42	.381		58.50	17.85		76.35	91.50
1330	4" diameter		38	.421		73	19.75		92.75	110
1340	5" diameter		32	.500		163	23.50		186.50	215
1360	6" diameter	Q-2	38	.632		239	30.50		269.50	310
1380	8" diameter	"	34	.706		400	34.50		434.50	490
1390	For other than full hard temper, add					13%				
1440	For silver solder, add						15%			
1800	For medical clean, (oxygen class), add					12%				
1950	To delete cplgs. & hngrs., 1/4"-1" pipe, subtract					27%	60%			
1960	1-1/4" -3" pipe, subtract					14%	52%			
1970	3 1/2"-5" pipe, subtract					10%	60%			
1980	6"-8" pipe, subtract					19%	53%			
2000	Type L tubing, couplings & clevis hanger assemblies 10' O.C.									
2100	1/4" diameter	1 Plum	88	.091	L.F.	1.60	4.73		6.33	8.85
2120	3/8" diameter		84	.095		2.27	4.96		7.23	9.95
2140	1/2" diameter		81	.099		2.49	5.15		7.64	10.45
2160	5/8" diameter		79	.101		3.87	5.25		9.12	12.15
2180	3/4" diameter		76	.105		3.83	5.50		9.33	12.40
2200	1" diameter		68	.118		5.95	6.10		12.05	15.75
2220	1-1/4" diameter		58	.138		8.35	7.20		15.55	19.95
2240	1-1/2" diameter		52	.154		10.75	8		18.75	24
2260	2" diameter		42	.190		16.70	9.90		26.60	33.50
2280	2-1/2" diameter	Q-1	62	.258		26	12.10		38.10	46.50
2300	3" diameter		56	.286		35.50	13.40		48.90	59
2320	3-1/2" diameter		43	.372		49	17.45		66.45	80

22 11 13 – Facility Water Distribution Piping

22 11 13.23 Pipe/Tube, Copper		Crew	Daily Output	Labor-Hours	Unit	Material	2010 Bare Costs Labor	Equipment	Total	Total Incl O&P
2340	4″ diameter	Q-1	39	.410	L.F.	60.50	19.20		79.70	95.50
2360	5″ diameter	↓	34	.471		130	22		152	176
2380	6″ diameter	Q-2	40	.600		182	29		211	244
2400	8″ diameter	″	36	.667		305	32.50		337.50	385
2410	For other than full hard temper, add				↓	21%				
2590	For silver solder, add						15%			
2900	For medical clean, (oxygen class), add					12%				
2940	To delete cplgs. & hngrs., 1/4″-1″ pipe, subtract					37%	63%			
2960	1-1/4″-3″ pipe, subtract					12%	53%			
2970	3-1/2″-5″ pipe, subtract					12%	63%			
2980	6″-8″ pipe, subtract					24%	55%			
3000	Type M tubing, couplings & clevis hanger assemblies 10′ O.C.									
3100	1/4″ diameter	1 Plum	90	.089	L.F.	1.85	4.63		6.48	9
3120	3/8″ diameter		87	.092		1.89	4.79		6.68	9.20
3140	1/2″ diameter		84	.095		1.86	4.96		6.82	9.50
3160	5/8″ diameter		81	.099		3.03	5.15		8.18	11.05
3180	3/4″ diameter		78	.103		2.93	5.35		8.28	11.20
3200	1″ diameter		70	.114		4.72	5.95		10.67	14.10
3220	1-1/4″ diameter		60	.133		6.95	6.95		13.90	18.05
3240	1-1/2″ diameter		54	.148		9.45	7.70		17.15	22
3260	2″ diameter	↓	44	.182		14.80	9.45		24.25	30.50
3280	2-1/2″ diameter	Q-1	64	.250		22.50	11.70		34.20	42.50
3300	3″ diameter		58	.276		30.50	12.90		43.40	53
3320	3-1/2″ diameter		45	.356		44.50	16.65		61.15	74
3340	4″ diameter		40	.400		55.50	18.75		74.25	89
3360	5″ diameter	↓	36	.444		130	21		151	174
3370	6″ diameter	Q-2	42	.571		227	28		255	292
3380	8″ diameter	″	38	.632	↓	300	30.50		330.50	380
3440	For silver solder, add						15%			
3960	To delete cplgs. & hngrs., 1/4″-1″ pipe, subtract					35%	65%			
3970	1-1/4″-3″ pipe, subtract					19%	56%			
3980	3-1/2″-5″ pipe, subtract					13%	65%			
3990	6″-8″ pipe, subtract					28%	58%			
4000	Type DWV tubing, couplings & clevis hanger assemblies 10′ O.C.									
4100	1-1/4″ diameter	1 Plum	60	.133	L.F.	9.80	6.95		16.75	21
4120	1-1/2″ diameter		54	.148		9.25	7.70		16.95	22
4140	2″ diameter	↓	44	.182		12.55	9.45		22	28
4160	3″ diameter	Q-1	58	.276		23.50	12.90		36.40	45.50
4180	4″ diameter		40	.400		42	18.75		60.75	74
4200	5″ diameter	↓	36	.444		117	21		138	160
4220	6″ diameter	Q-2	42	.571		169	28		197	228
4240	8″ diameter	″	38	.632	↓	410	30.50		440.50	495
4730	To delete cplgs. & hngrs., 1-1/4″-2″ pipe, subtract					16%	53%			
4740	3″-4″ pipe, subtract					13%	60%			
4750	5″-8″ pipe, subtract					23%	58%			
5200	ACR tubing, type L, hard temper, cleaned and									
5220	capped, no couplings or hangers									
5240	3/8″ OD				L.F.	1.28			1.28	1.41
5250	1/2″ OD					2.05			2.05	2.26
5260	5/8″ OD					2.48			2.48	2.73
5270	3/4″ OD					3.52			3.52	3.87
5280	7/8″ OD					3.90			3.90	4.29
5290	1-1/8″ OD				↓	5.65			5.65	6.20

22 11 13 – Facility Water Distribution Piping

22 11 13.23 Pipe/Tube, Copper		Crew	Daily Output	Labor-Hours	Unit	Material	Labor	Equipment	Total	Total Incl O&P
						2010 Bare Costs				
5300	1-3/8" OD				L.F.	7.55			7.55	8.35
5310	1-5/8" OD					9.70			9.70	10.65
5320	2-1/8" OD					15.25			15.25	16.80
5330	2-5/8" OD					22.50			22.50	25
5340	3-1/8" OD					30.50			30.50	33.50
5350	3-5/8" OD					40			40	44
5360	4-1/8" OD				↓	51			51	56
5380	ACR tubing, type L, hard, cleaned and capped									
5381	No couplings or hangers									
5384	3/8"	1 Stpi	160	.050	L.F.	1.28	2.60		3.88	5.30
5385	1/2"		160	.050		2.05	2.60		4.65	6.15
5386	5/8"		160	.050		2.48	2.60		5.08	6.60
5387	3/4"		130	.062		3.52	3.19		6.71	8.65
5388	7/8"		130	.062		3.90	3.19		7.09	9.10
5389	1-1/8"		115	.070		5.65	3.61		9.26	11.60
5390	1-3/8"		100	.080		7.55	4.15		11.70	14.55
5391	1-5/8"		90	.089		9.70	4.61		14.31	17.55
5392	2-1/8"	↓	80	.100		15.25	5.20		20.45	24.50
5393	2-5/8"	Q-5	125	.128		22.50	6		28.50	34
5394	3-1/8"		105	.152		30.50	7.10		37.60	44
5395	4-1/8"	↓	95	.168	↓	51	7.85		58.85	68
5800	Refrigeration tubing, dryseal, 60' coils									
5840	1/8" OD				Coil	26			26	28.50
5850	3/16" OD					30.50			30.50	33.50
5860	1/4" OD					35			35	38.50
5870	5/16" OD					46.50			46.50	51.50
5880	3/8" OD					49.50			49.50	54.50
5890	1/2" OD					70			70	77
5900	5/8" OD					95			95	104
5910	3/4" OD					114			114	126
5920	7/8" OD					170			170	187
5930	1-1/8" OD					245			245	270
5940	1-3/8" OD					385			385	420
5950	1-5/8" OD				↓	485			485	535
9400	Sub assemblies used in assembly systems									
9410	Chilled water unit, coil connections per unit under 10 ton	Q-5	.80	20	System	1,175	935		2,110	2,700
9420	Chilled water unit, coil connections per unit 10 ton and up		1	16		1,750	745		2,495	3,050
9430	Chilled water dist. piping per ton, less than 61 ton systems	↓	26	.615		16.25	28.50		44.75	61
9440	Chilled water dist. piping per ton, 61 through 120 ton systems	Q-6	31	.774		32	37.50		69.50	91
9450	Chilled water dist. piping/ton, 135 ton systems and up	Q-8	25.40	1.260		43.50	62.50	2.31	108.31	144
9510	Refrigerant piping/ton of cooling for remote condensers	Q-5	2	8		335	375		710	930
9520	Refrigerant piping per ton up to 10 ton w/remote condensing unit		2.40	6.667		137	310		447	615
9530	Refrigerant piping per ton, 20 ton w/remote condensing unit		2	8		208	375		583	790
9540	Refrigerant piping per ton, 40 ton w/remote condensing unit	↓	1.90	8.421		280	395		675	900
9550	Refrigerant piping per ton, 75-80 ton w/remote condensing unit	Q-6	2.40	10		420	485		905	1,175
9560	Refrigerant piping per ton, 100 ton w/remote condensing unit	"	2.20	10.909	↓	560	530		1,090	1,400

22 11 13.25 Pipe/Tube Fittings, Copper		Crew	Daily Output	Labor-Hours	Unit	Material	Labor	Equipment	Total	Total Incl O&P
0010	**PIPE/TUBE FITTINGS, COPPER**, Wrought unless otherwise noted									
0040	Solder joints, copper x copper									
0070	90° elbow, 1/4"	1 Plum	22	.364	Ea.	4.59	18.95		23.54	33.50
0090	3/8"		22	.364		4.36	18.95		23.31	33.50
0100	1/2"	↓	20	.400	↓	1.45	21		22.45	32.50

22 11 13 – Facility Water Distribution Piping

22 11 13.25 Pipe/Tube Fittings, Copper		Crew	Daily Output	Labor-Hours	Unit	Material	2010 Bare Costs Labor	Equipment	Total	Total Incl O&P
0110	5/8"	1 Plum	19	.421	Ea.	4.88	22		26.88	38.50
0120	3/4"		19	.421		3.27	22		25.27	36.50
0130	1"		16	.500		8	26		34	48
0140	1-1/4"		15	.533		12.10	28		40.10	55
0150	1-1/2"		13	.615		18.90	32		50.90	69
0160	2"		11	.727		34.50	38		72.50	94.50
0170	2-1/2"	Q-1	13	1.231		69	57.50		126.50	163
0180	3"		11	1.455		92	68		160	203
0190	3-1/2"		10	1.600		320	75		395	465
0200	4"		9	1.778		236	83.50		319.50	385
0210	5"		6	2.667		985	125		1,110	1,250
0220	6"	Q-2	9	2.667		1,325	130		1,455	1,650
0230	8"	"	8	3		4,875	146		5,021	5,575
0250	45° elbow, 1/4"	1 Plum	22	.364		8.60	18.95		27.55	38
0270	3/8"		22	.364		7	18.95		25.95	36
0280	1/2"		20	.400		2.67	21		23.67	34
0290	5/8"		19	.421		13.35	22		35.35	47.50
0300	3/4"		19	.421		4.68	22		26.68	38
0310	1"		16	.500		11.75	26		37.75	52
0320	1-1/4"		15	.533		15.90	28		43.90	59
0330	1-1/2"		13	.615		19.15	32		51.15	69
0340	2"		11	.727		32	38		70	91.50
0350	2-1/2"	Q-1	13	1.231		68	57.50		125.50	162
0360	3"		13	1.231		101	57.50		158.50	198
0370	3-1/2"		10	1.600		170	75		245	299
0380	4"		9	1.778		215	83.50		298.50	360
0390	5"		6	2.667		835	125		960	1,100
0400	6"	Q-2	9	2.667		1,325	130		1,455	1,650
0410	8"	"	8	3		4,475	146		4,621	5,150
0450	Tee, 1/4"	1 Plum	14	.571		9.65	29.50		39.15	55
0470	3/8"		14	.571		7.35	29.50		36.85	52.50
0480	1/2"		13	.615		2.49	32		34.49	50.50
0490	5/8"		12	.667		16	34.50		50.50	69.50
0500	3/4"		12	.667		6	34.50		40.50	58.50
0510	1"		10	.800		18.55	41.50		60.05	83
0520	1-1/4"		9	.889		25	46.50		71.50	97
0530	1-1/2"		8	1		39	52		91	121
0540	2"		7	1.143		61	59.50		120.50	156
0550	2-1/2"	Q-1	8	2		123	93.50		216.50	275
0560	3"		7	2.286		187	107		294	365
0570	3-1/2"		6	2.667		545	125		670	780
0580	4"		5	3.200		455	150		605	725
0590	5"		4	4		1,475	187		1,662	1,900
0600	6"	Q-2	6	4		2,025	194		2,219	2,525
0610	8"	"	5	4.800		7,825	233		8,058	8,975
0612	Tee, reducing on the outlet, 1/4"	1 Plum	15	.533		17.65	28		45.65	61
0613	3/8"		15	.533		14.55	28		42.55	57.50
0614	1/2"		14	.571		15.15	29.50		44.65	61
0615	5/8"		13	.615		26	32		58	76.50
0616	3/4"		12	.667		8.50	34.50		43	61.50
0617	1"		11	.727		23	38		61	81.50
0618	1-1/4"		10	.800		32	41.50		73.50	97.50
0619	1-1/2"		9	.889		29	46.50		75.50	102

22 11 13.25 Pipe/Tube Fittings, Copper		Crew	Daily Output	Labor-Hours	Unit	2010 Bare Costs Material	Labor	Equipment	Total	Total Incl O&P
0620	2"	1 Plum	8	1	Ea.	52.50	52		104.50	136
0621	2-1/2"	Q-1	9	1.778		143	83.50		226.50	283
0622	3"		8	2		158	93.50		251.50	315
0623	4"		6	2.667		305	125		430	520
0624	5"		5	3.200		1,475	150		1,625	1,850
0625	6"	Q-2	7	3.429		1,925	167		2,092	2,375
0626	8"	"	6	4		7,825	194		8,019	8,925
0630	Tee, reducing on the run, 1/4"	1 Plum	15	.533		19.45	28		47.45	63
0631	3/8"		15	.533		26	28		54	70
0632	1/2"		14	.571		17.90	29.50		47.40	64
0633	5/8"		13	.615		26	32		58	76.50
0634	3/4"		12	.667		7.05	34.50		41.55	60
0635	1"		11	.727		22	38		60	81
0636	1-1/4"		10	.800		36	41.50		77.50	102
0637	1-1/2"		9	.889		62	46.50		108.50	138
0638	2"		8	1		86	52		138	173
0639	2-1/2"	Q-1	9	1.778		186	83.50		269.50	330
0640	3"		8	2		287	93.50		380.50	455
0641	4"		6	2.667		605	125		730	850
0642	5"		5	3.200		1,400	150		1,550	1,775
0643	6"	Q-2	7	3.429		2,150	167		2,317	2,600
0644	8"	"	6	4		7,825	194		8,019	8,925
0650	Coupling, 1/4"	1 Plum	24	.333		1.06	17.35		18.41	27
0670	3/8"		24	.333		1.41	17.35		18.76	27.50
0680	1/2"		22	.364		1.10	18.95		20.05	29.50
0690	5/8"		21	.381		3.32	19.85		23.17	33
0700	3/4"		21	.381		2.36	19.85		22.21	32
0710	1"		18	.444		4.41	23		27.41	39.50
0715	1-1/4"		17	.471		7.85	24.50		32.35	45
0716	1-1/2"		15	.533		10.35	28		38.35	53
0718	2"		13	.615		17.30	32		49.30	67
0721	2-1/2"	Q-1	15	1.067		37	50		87	116
0722	3"		13	1.231		55.50	57.50		113	148
0724	3-1/2"		8	2		106	93.50		199.50	257
0726	4"		7	2.286		116	107		223	288
0728	5"		6	2.667		286	125		411	500
0731	6"	Q-2	8	3		460	146		606	725
0732	8"	"	7	3.429		1,525	167		1,692	1,925
0741	Coupling, reducing, concentric									
0743	1/2"	1 Plum	23	.348	Ea.	2.80	18.10		20.90	30
0745	3/4"		21.50	.372		4.96	19.35		24.31	34.50
0747	1"		19.50	.410		8.60	21.50		30.10	41.50
0748	1-1/4"		18	.444		10.20	23		33.20	46
0749	1-1/2"		16	.500		16.85	26		42.85	57.50
0751	2"		14	.571		26	29.50		55.50	73.50
0752	2-1/2"		13	.615		67	32		99	122
0753	3"	Q-1	14	1.143		76	53.50		129.50	164
0755	4"	"	8	2		178	93.50		271.50	335
0757	5"	Q-2	7.50	3.200		615	155		770	910
0759	6"		7	3.429		1,000	167		1,167	1,350
0761	8"		6.50	3.692		1,650	179		1,829	2,100
0771	Cap, sweat									
0773	1/2"	1 Plum	40	.200	Ea.	1.06	10.40		11.46	16.75

22 11 13 – Facility Water Distribution Piping

22 11 13.25 Pipe/Tube Fittings, Copper		Crew	Daily Output	Labor-Hours	Unit	Material	2010 Bare Costs Labor	Equipment	Total	Total Incl O&P
0775	3/4"	1 Plum	38	.211	Ea.	1.97	10.95		12.92	18.55
0777	1"		32	.250		4.68	13		17.68	24.50
0778	1-1/4"		29	.276		6.25	14.35		20.60	28.50
0779	1-1/2"		26	.308		9.15	16		25.15	34
0781	2"		22	.364		16.90	18.95		35.85	47
0791	Flange, sweat									
0793	3"	Q-1	22	.727	Ea.	187	34		221	257
0795	4"		18	.889		261	41.50		302.50	350
0797	5"		12	1.333		490	62.50		552.50	630
0799	6"	Q-2	18	1.333		510	65		575	660
0801	8"	"	16	1.500		1,525	73		1,598	1,775
0850	Unions, 1/4"	1 Plum	21	.381		31	19.85		50.85	64
0870	3/8"		21	.381		31.50	19.85		51.35	64
0880	1/2"		19	.421		18.20	22		40.20	53
0890	5/8"		18	.444		71.50	23		94.50	113
0900	3/4"		18	.444		21	23		44	57.50
0910	1"		15	.533		36	28		64	81
0920	1-1/4"		14	.571		62.50	29.50		92	113
0930	1-1/2"		12	.667		82.50	34.50		117	143
0940	2"		10	.800		140	41.50		181.50	217
0950	2-1/2"	Q-1	12	1.333		305	62.50		367.50	435
0960	3"	"	10	1.600		795	75		870	985
0980	Adapter, copper x male IPS, 1/4"	1 Plum	20	.400		15.15	21		36.15	47.50
0990	3/8"		20	.400		7.55	21		28.55	39.50
1000	1/2"		18	.444		3.04	23		26.04	38
1010	3/4"		17	.471		5.10	24.50		29.60	42
1020	1"		15	.533		13.20	28		41.20	56
1030	1-1/4"		13	.615		19.55	32		51.55	69.50
1040	1-1/2"		12	.667		22.50	34.50		57	76.50
1050	2"		11	.727		38	38		76	98.50
1060	2-1/2"	Q-1	10.50	1.524		146	71.50		217.50	267
1070	3"		10	1.600		193	75		268	325
1080	3-1/2"		9	1.778		251	83.50		334.50	400
1090	4"		8	2		276	93.50		369.50	445
1200	5"		6	2.667		1,300	125		1,425	1,600
1210	6"	Q-2	8.50	2.824		1,575	137		1,712	1,950
1214	Adapter, copper x female IPS									
1216	1/2"	1 Plum	18	.444	Ea.	4.83	23		27.83	40
1218	3/4"		17	.471		6.65	24.50		31.15	44
1220	1"		15	.533		15.35	28		43.35	58.50
1221	1-1/4"		13	.615		22.50	32		54.50	72.50
1222	1-1/2"		12	.667		35	34.50		69.50	90.50
1224	2"		11	.727		47.50	38		85.50	109
1250	Cross, 1/2"		10	.800		22.50	41.50		64	87.50
1260	3/4"		9.50	.842		44	44		88	114
1270	1"		8	1		75	52		127	160
1280	1-1/4"		7.50	1.067		107	55.50		162.50	201
1290	1-1/2"		6.50	1.231		153	64		217	264
1300	2"		5.50	1.455		289	75.50		364.50	435
1310	2-1/2"	Q-1	6.50	2.462		665	115		780	910
1320	3"	"	5.50	2.909		840	136		976	1,125
1500	Tee fitting, mechanically formed, (Type 1, 'branch sizes up to 2 in.')									
1520	1/2" run size, 3/8" to 1/2" branch size	1 Plum	80	.100	Ea.		5.20		5.20	7.80

22 11 13 – Facility Water Distribution Piping

22 11 13.25 Pipe/Tube Fittings, Copper		Crew	Daily Output	Labor-Hours	Unit	2010 Bare Costs Material	Labor	Equipment	Total	Total Incl O&P
1530	3/4" run size, 3/8" to 3/4" branch size	1 Plum	60	.133	Ea.		6.95		6.95	10.40
1540	1" run size, 3/8" to 1" branch size		54	.148			7.70		7.70	11.55
1550	1-1/4" run size, 3/8" to 1-1/4" branch size		48	.167			8.70		8.70	13
1560	1-1/2" run size, 3/8" to 1-1/2" branch size		40	.200			10.40		10.40	15.60
1570	2" run size, 3/8" to 2" branch size		35	.229			11.90		11.90	17.85
1580	2-1/2" run size, 1/2" to 2" branch size		32	.250			13		13	19.50
1590	3" run size, 1" to 2" branch size		26	.308			16		16	24
1600	4" run size, 1" to 2" branch size	↓	24	.333	↓		17.35		17.35	26
1640	Tee fitting, mechanically formed, (Type 2, branches 2-1/2" thru 4")									
1650	2-1/2" run size, 2-1/2" branch size	1 Plum	12.50	.640	Ea.		33.50		33.50	50
1660	3" run size, 2-1/2" to 3" branch size		12	.667			34.50		34.50	52
1670	3-1/2" run size, 2-1/2" to 3-1/2" branch size		11	.727			38		38	56.50
1680	4" run size, 2-1/2" to 4" branch size		10.50	.762			39.50		39.50	59.50
1698	5" run size, 2" to 4" branch size		9.50	.842			44		44	65.50
1700	6" run size, 2" to 4" branch size		8.50	.941			49		49	73.50
1710	8" run size, 2" to 4" branch size	↓	7	1.143	↓		59.50		59.50	89
1800	ACR fittings, OD size									
1802	Tee, straight									
1808	5/8"	1 Stpi	12	.667	Ea.	2.49	34.50		36.99	54.50
1810	3/4"		12	.667		16	34.50		50.50	69.50
1812	7/8"		10	.800		6	41.50		47.50	68.50
1813	1"		10	.800		43.50	41.50		85	110
1814	1-1/8"		10	.800		18.55	41.50		60.05	82.50
1816	1-3/8"		9	.889		25	46		71	96.50
1818	1-5/8"		8	1		39	52		91	121
1820	2-1/8"	↓	7	1.143		61	59.50		120.50	156
1822	2-5/8"	Q-5	8	2		123	93.50		216.50	275
1824	3-1/8"		7	2.286		187	107		294	365
1826	4-1/8"	↓	5	3.200	↓	445	149		594	715
1830	90° elbow									
1836	5/8"	1 Stpi	19	.421	Ea.	4.88	22		26.88	38.50
1838	3/4"		19	.421		8.85	22		30.85	42.50
1840	7/8"		16	.500		8.80	26		34.80	48.50
1842	1-1/8"		16	.500		11.75	26		37.75	52
1844	1-3/8"		15	.533		12.10	27.50		39.60	55
1846	1-5/8"		13	.615		18.90	32		50.90	69
1848	2-1/8"	↓	11	.727		34.50	38		72.50	94.50
1850	2-5/8"	Q-5	13	1.231		69	57.50		126.50	162
1852	3-1/8"		11	1.455		92	68		160	203
1854	4-1/8"	↓	9	1.778	↓	236	83		319	385
1860	Coupling									
1866	5/8"	1 Stpi	21	.381	Ea.	1.10	19.75		20.85	30.50
1868	3/4"		21	.381		3.32	19.75		23.07	33
1870	7/8"		18	.444		2.21	23		25.21	37
1871	1"		18	.444		7.55	23		30.55	43
1872	1-1/8"		18	.444		4.41	23		27.41	39.50
1874	1-3/8"		17	.471		7.85	24.50		32.35	45
1876	1-5/8"		15	.533		10.35	27.50		37.85	53
1878	2-1/8"	↓	13	.615		17.30	32		49.30	67
1880	2-5/8"	Q-5	15	1.067		37	50		87	115
1882	3-1/8"		13	1.231		55.50	57.50		113	147
1884	4-1/8"	↓	7	2.286	↓	119	107		226	291
2000	DWV, solder joints, copper x copper									

22 11 13.25 Pipe/Tube Fittings, Copper		Crew	Daily Output	Labor-Hours	Unit	2010 Bare Costs Material	Labor	Equipment	Total	Total Incl O&P
2030	90° Elbow, 1-1/4"	1 Plum	13	.615	Ea.	14.05	32		46.05	63.50
2050	1-1/2"		12	.667		21	34.50		55.50	75
2070	2"		10	.800		27.50	41.50		69	93
2090	3"	Q-1	10	1.600		68	75		143	187
2100	4"	"	9	1.778		390	83.50		473.50	555
2150	45° Elbow, 1-1/4"	1 Plum	13	.615		12.95	32		44.95	62.50
2170	1-1/2"		12	.667		10.70	34.50		45.20	64
2180	2"		10	.800		24.50	41.50		66	89.50
2190	3"	Q-1	10	1.600		52.50	75		127.50	170
2200	4"	"	9	1.778		251	83.50		334.50	400
2250	Tee, Sanitary, 1-1/4"	1 Plum	9	.889		28	46.50		74.50	100
2270	1-1/2"		8	1		34.50	52		86.50	116
2290	2"		7	1.143		40.50	59.50		100	134
2310	3"	Q-1	7	2.286		164	107		271	340
2330	4"	"	6	2.667		415	125		540	645
2400	Coupling, 1-1/4"	1 Plum	14	.571		6.60	29.50		36.10	52
2420	1-1/2"		13	.615		8.20	32		40.20	57
2440	2"		11	.727		11.35	38		49.35	69
2460	3"	Q-1	11	1.455		22	68		90	126
2480	4"	"	10	1.600		70	75		145	189
2602	Traps, see Div. 22 13 16.60									
3500	Compression joint fittings									
3510	As used for plumbing and oil burner work									
3520	Fitting price includes nuts and sleeves									
3540	Sleeve, 1/8"				Ea.	.15			.15	.17
3550	3/16"					.15			.15	.17
3560	1/4"					.06			.06	.07
3570	5/16"					.16			.16	.18
3580	3/8"					.31			.31	.34
3600	1/2"					.44			.44	.48
3620	Nut, 1/8"					.23			.23	.25
3630	3/16"					.24			.24	.26
3640	1/4"					.23			.23	.25
3650	5/16"					.35			.35	.39
3660	3/8"					.38			.38	.42
3670	1/2"					.68			.68	.75
3710	Union, 1/8"	1 Plum	26	.308		1.42	16		17.42	25.50
3720	3/16"		24	.333		1.36	17.35		18.71	27.50
3730	1/4"		24	.333		1.43	17.35		18.78	27.50
3740	5/16"		23	.348		1.82	18.10		19.92	29
3750	3/8"		22	.364		1.84	18.95		20.79	30.50
3760	1/2"		22	.364		2.83	18.95		21.78	31.50
3780	5/8"		21	.381		3.14	19.85		22.99	33
3820	Union tee, 1/8"		17	.471		5.40	24.50		29.90	42.50
3830	3/16"		16	.500		3.13	26		29.13	42.50
3840	1/4"		15	.533		2.83	28		30.83	44.50
3850	5/16"		15	.533		3.98	28		31.98	46
3860	3/8"		15	.533		4.05	28		32.05	46
3870	1/2"		15	.533		6.80	28		34.80	49
3910	Union elbow, 1/4"		24	.333		1.74	17.35		19.09	28
3920	5/16"		23	.348		3.02	18.10		21.12	30.50
3930	3/8"		22	.364		2.99	18.95		21.94	32
3940	1/2"		22	.364		4.84	18.95		23.79	34

22 11 13.25 Pipe/Tube Fittings, Copper		Crew	Daily Output	Labor-Hours	Unit	Material	Labor	Equipment	Total	Total Incl O&P
						2010 Bare Costs				
3980	Female connector, 1/8"	1 Plum	26	.308	Ea.	1.10	16		17.10	25
4000	3/16" x 1/8"		24	.333		1.40	17.35		18.75	27.50
4010	1/4" x 1/8"		24	.333		1.24	17.35		18.59	27.50
4020	1/4"		24	.333		1.67	17.35		19.02	28
4030	3/8" x 1/4"		22	.364		1.67	18.95		20.62	30.50
4040	1/2" x 3/8"		22	.364		2.66	18.95		21.61	31.50
4050	5/8" x 1/2"		21	.381		3.74	19.85		23.59	33.50
4090	Male connector, 1/8"		26	.308		.89	16		16.89	25
4100	3/16" x 1/8"		24	.333		.92	17.35		18.27	27
4110	1/4" x 1/8"		24	.333		.95	17.35		18.30	27
4120	1/4"		24	.333		.95	17.35		18.30	27
4130	5/16" x 1/8"		23	.348		1.12	18.10		19.22	28
4140	5/16" x 1/4"		23	.348		1.26	18.10		19.36	28.50
4150	3/8" x 1/8"		22	.364		1.32	18.95		20.27	30
4160	3/8" x 1/4"		22	.364		1.52	18.95		20.47	30
4170	3/8"		22	.364		1.52	18.95		20.47	30
4180	3/8" x 1/2"		22	.364		2.03	18.95		20.98	30.50
4190	1/2" x 3/8"		22	.364		2.03	18.95		20.98	30.50
4200	1/2"		22	.364		2.27	18.95		21.22	31
4210	5/8" x 1/2"		21	.381		2.62	19.85		22.47	32.50
4240	Male elbow, 1/8"		26	.308		1.82	16		17.82	26
4250	3/16" x 1/8"		24	.333		2.01	17.35		19.36	28
4260	1/4" x 1/8"		24	.333		1.52	17.35		18.87	27.50
4270	1/4"		24	.333		1.73	17.35		19.08	28
4280	3/8" x 1/4"		22	.364		2.20	18.95		21.15	31
4290	3/8"		22	.364		3.77	18.95		22.72	32.50
4300	1/2" x 1/4"		22	.364		3.64	18.95		22.59	32.50
4310	1/2" x 3/8"		22	.364		5.75	18.95		24.70	35
4340	Female elbow, 1/8"		26	.308		3.92	16		19.92	28.50
4350	1/4" x 1/8"		24	.333		2.22	17.35		19.57	28.50
4360	1/4"		24	.333		2.93	17.35		20.28	29
4370	3/8" x 1/4"		22	.364		2.88	18.95		21.83	31.50
4380	1/2" x 3/8"		22	.364		4.92	18.95		23.87	34
4390	1/2"		22	.364		6.60	18.95		25.55	36
4420	Male run tee, 1/4" x 1/8"		15	.533		2.73	28		30.73	44.50
4430	5/16" x 1/8"		15	.533		5.80	28		33.80	48
4440	3/8" x 1/4"		15	.533		4.22	28		32.22	46
4480	Male branch tee, 1/4" x 1/8"		15	.533		3.17	28		31.17	45
4490	1/4"		15	.533		3.41	28		31.41	45.50
4500	3/8" x 1/4"		15	.533		3.96	28		31.96	46
4510	1/2" x 3/8"		15	.533		6.15	28		34.15	48.50
4520	1/2"		15	.533		9.85	28		37.85	52.50
4800	Flare joint fittings									
4810	Refrigeration fittings									
4820	Flare joint nuts and labor not incl. in price. Add 1 nut per jnt.									
4830	90° Elbow, 1/4"				Ea.	1.94			1.94	2.13
4840	3/8"					3.03			3.03	3.33
4850	1/2"					4.41			4.41	4.85
4860	5/8"					6.45			6.45	7.10
4870	3/4"					8.05			8.05	8.85
5030	Tee, 1/4"					2.46			2.46	2.71
5040	5/16"					3.15			3.15	3.47
5050	3/8"					3.66			3.66	4.03

22 11 13.25 Pipe/Tube Fittings, Copper		Crew	Daily Output	Labor-Hours	Unit	Material	Labor	Equipment	Total	Total Incl O&P
						2010 Bare Costs				
5060	1/2"				Ea.	5.55			5.55	6.15
5070	5/8"					8.65			8.65	9.50
5080	3/4"					13.10			13.10	14.40
5140	Union, 3/16"					.95			.95	1.05
5150	1/4"					.87			.87	.96
5160	5/16"					1.08			1.08	1.19
5170	3/8"					1.12			1.12	1.23
5180	1/2"					1.67			1.67	1.84
5190	5/8"					2.68			2.68	2.95
5200	3/4"					6.70			6.70	7.35
5260	Long flare nut, 3/16"	1 Stpi	42	.190		1.57	9.90		11.47	16.55
5270	1/4"		41	.195		.73	10.15		10.88	16
5280	5/16"		40	.200		1.31	10.40		11.71	17
5290	3/8"		39	.205		1.52	10.65		12.17	17.60
5300	1/2"		38	.211		2.36	10.95		13.31	19
5310	5/8"		37	.216		4.13	11.20		15.33	21.50
5320	3/4"		34	.235		10.60	12.20		22.80	30
5380	Short flare nut, 3/16"		42	.190		.48	9.90		10.38	15.35
5390	1/4"		41	.195		.51	10.15		10.66	15.75
5400	5/16"		40	.200		.61	10.40		11.01	16.20
5410	3/8"		39	.205		.80	10.65		11.45	16.85
5420	1/2"		38	.211		1.19	10.95		12.14	17.70
5430	5/8"		36	.222		1.66	11.55		13.21	19.15
5440	3/4"		34	.235		5.55	12.20		17.75	24.50
5500	90° Elbow flare by MIPS, 1/4"					1.96			1.96	2.16
5510	3/8"					1.98			1.98	2.18
5520	1/2"					3.17			3.17	3.49
5530	5/8"					7.75			7.75	8.50
5540	3/4"					8.75			8.75	9.65
5600	Flare by FIPS, 1/4"					3			3	3.30
5610	3/8"					3.29			3.29	3.62
5620	1/2"					4.68			4.68	5.15
5670	Flare by sweat, 1/4"					1.86			1.86	2.05
5680	3/8"					2.15			2.15	2.37
5690	1/2"					3.83			3.83	4.21
5700	5/8"					14.80			14.80	16.30
5760	Tee flare by IPS, 1/4"					3.78			3.78	4.16
5770	3/8"					5.30			5.30	5.85
5780	1/2"					8.90			8.90	9.80
5790	5/8"					6.95			6.95	7.60
5850	Connector, 1/4"					.65			.65	.72
5860	3/8"					.97			.97	1.07
5870	1/2"					1.57			1.57	1.73
5880	5/8"					1.55			1.55	1.71
5890	3/4"					5.10			5.10	5.65
5950	Seal cap, 1/4"					.34			.34	.37
5960	3/8"					.66			.66	.73
5970	1/2"					.74			.74	.81
5980	5/8"					1.11			1.11	1.22
5990	3/4"					3.26			3.26	3.59
6000	Water service fittings									
6010	Flare joints nut and labor are included in the fitting price.									
6020	90° Elbow, C x C, 3/8"	1 Plum	19	.421	Ea.	56	22		78	94.50

22 11 13 – Facility Water Distribution Piping

22 11 13.25 Pipe/Tube Fittings, Copper		Crew	Daily Output	Labor-Hours	Unit	2010 Bare Costs Material	Labor	Equipment	Total	Total Incl O&P
6030	1/2"	1 Plum	18	.444	Ea.	56	23		79	96
6040	3/4"		16	.500		69	26		95	115
6050	1"		15	.533		135	28		163	191
6080	2"		10	.800		465	41.50		506.50	580
6090	90° Elbow, C x MPT, 3/8"		19	.421		40.50	22		62.50	77.50
6100	1/2"		18	.444		40.50	23		63.50	79
6110	3/4"		16	.500		47	26		73	90.50
6120	1"		15	.533		128	28		156	182
6130	1-1/4"		13	.615		263	32		295	335
6140	1-1/2"		12	.667		263	34.50		297.50	340
6150	2"		10	.800		345	41.50		386.50	440
6160	90° Elbow, C x FPT, 3/8"		19	.421		43.50	22		65.50	80.50
6170	1/2"		18	.444		43.50	23		66.50	82
6180	3/4"		16	.500		53.50	26		79.50	98
6190	1"		15	.533		124	28		152	178
6200	1-1/4"		13	.615		138	32		170	200
6210	1-1/2"		12	.667		530	34.50		564.50	635
6220	2"		10	.800		765	41.50		806.50	905
6230	Tee, C x C x C, 3/8"		13	.615		82	32		114	138
6240	1/2"		12	.667		82	34.50		116.50	142
6250	3/4"		11	.727		96	38		134	163
6260	1"		10	.800		170	41.50		211.50	250
6330	Tube nut, C x nut seat, 3/8"		40	.200		16.60	10.40		27	34
6340	1/2"		38	.211		16.60	10.95		27.55	34.50
6350	3/4"		34	.235		16.60	12.25		28.85	36.50
6360	1"		32	.250		30	13		43	52.50
6380	Coupling, C x C, 3/8"		19	.421		50.50	22		72.50	88.50
6390	1/2"		18	.444		50.50	23		73.50	90
6400	3/4"		16	.500		63.50	26		89.50	109
6410	1"		15	.533		118	28		146	171
6420	1-1/4"		12	.667		190	34.50		224.50	261
6430	1-1/2"		12	.667		288	34.50		322.50	365
6440	2"		10	.800		435	41.50		476.50	540
6450	Adapter, C x FPT, 3/8"		19	.421		38	22		60	75
6460	1/2"		18	.444		38	23		61	76.50
6470	3/4"		16	.500		46	26		72	89.50
6480	1"		15	.533		98	28		126	150
6490	1-1/4"		13	.615		208	32		240	277
6500	1-1/2"		12	.667		208	34.50		242.50	281
6510	2"		10	.800		271	41.50		312.50	360
6520	Adapter, C x MPT, 3/8"		19	.421		34	22		56	70.50
6530	1/2"		18	.444		34	23		57	72
6540	3/4"		16	.500		47	26		73	90.50
6550	1"		15	.533		82.50	28		110.50	133
6560	1-1/4"		13	.615		207	32		239	275
6570	1-1/2"		12	.667		207	34.50		241.50	279
6580	2"	↓	10	.800	↓	280	41.50		321.50	375
6992	Polybutylene/polyethylene pipe, See Div. 22 11 13.76 for plastic ftng.									
7000	Insert type Brass/copper, 100 psi @ 180°F, CTS									
7010	Adapter MPT 3/8" x 3/8" CTS	1 Plum	29	.276	Ea.	1.74	14.35		16.09	23.50
7020	1/2" x 1/2"		26	.308		1.74	16		17.74	26
7030	3/4" x 1/2"		26	.308		2.96	16		18.96	27.50
7040	3/4" x 3/4"	↓	25	.320	↓	2.56	16.65		19.21	28

22 11 13.25 Pipe/Tube Fittings, Copper		Crew	Daily Output	Labor-Hours	Unit	Material	2010 Bare Costs Labor	Equipment	Total	Total Incl O&P
7050	Adapter CTS 1/2" x 1/2" sweat	1 Plum	24	.333	Ea.	.66	17.35		18.01	26.50
7060	3/4" x 3/4" sweat		22	.364		.82	18.95		19.77	29.50
7070	Coupler center set 3/8" CTS		25	.320		.84	16.65		17.49	26
7080	1/2" CTS		23	.348		.66	18.10		18.76	27.50
7090	3/4" CTS		22	.364		.82	18.95		19.77	29.50
7100	Elbow 90°, copper 3/8"		25	.320		1.86	16.65		18.51	27
7110	1/2" CTS		23	.348		1.19	18.10		19.29	28.50
7120	3/4" CTS		22	.364		1.44	18.95		20.39	30
7130	Tee copper 3/8" CTS		17	.471		2.21	24.50		26.71	39
7140	1/2" CTS		15	.533		1.51	28		29.51	43
7150	3/4" CTS		14	.571		2.33	29.50		31.83	47
7160	3/8" x 3/8" x 1/2"		16	.500		1.51	26		27.51	40.50
7170	1/2" x 3/8" x 1/2"		15	.533		1.51	28		29.51	43
7180	3/4" x 1/2" x 3/4"		14	.571		2.33	29.50		31.83	47

22 11 13.27 Pipe/Tube, Grooved Joint for Copper		Crew	Daily Output	Labor-Hours	Unit	Material	Labor	Equipment	Total	Total Incl O&P
0010	**PIPE/TUBE, GROOVED JOINT FOR COPPER**									
4000	Fittings: coupling material required at joints not incl. in fitting price.									
4001	Add 1 selected coupling, material only, per joint for installed price.									
4010	Coupling, rigid style									
4018	2" diameter	1 Plum	50	.160	Ea.	13	8.35		21.35	27
4020	2-1/2" diameter	Q-1	80	.200		14.90	9.35		24.25	30.50
4022	3" diameter		67	.239		16.40	11.20		27.60	35
4024	4" diameter		50	.320		25	15		40	50
4026	5" diameter		40	.400		42.50	18.75		61.25	75
4028	6" diameter	Q-2	50	.480		56.50	23.50		80	97
4100	Elbow, 90° or 45°									
4108	2" diameter	1 Plum	25	.320	Ea.	23.50	16.65		40.15	50.50
4110	2-1/2" diameter	Q-1	40	.400		25.50	18.75		44.25	56
4112	3" diameter		33	.485		35.50	22.50		58	73.50
4114	4" diameter		25	.640		82.50	30		112.50	136
4116	5" diameter		20	.800		235	37.50		272.50	315
4118	6" diameter	Q-2	25	.960		375	46.50		421.50	485
4200	Tee									
4208	2" diameter	1 Plum	17	.471	Ea.	38.50	24.50		63	79
4210	2-1/2" diameter	Q-1	27	.593		41	28		69	87
4212	3" diameter		22	.727		61.50	34		95.50	119
4214	4" diameter		17	.941		135	44		179	215
4216	5" diameter		13	1.231		380	57.50		437.50	500
4218	6" diameter	Q-2	17	1.412		470	68.50		538.50	620
4300	Reducer, concentric									
4310	3" x 2-1/2" diameter	Q-1	35	.457	Ea.	34	21.50		55.50	69
4312	4" x 2-1/2" diameter		32	.500		69.50	23.50		93	112
4314	4" x 3" diameter		29	.552		69.50	26		95.50	116
4316	5" x 3" diameter		25	.640		197	30		227	261
4318	5" x 4" diameter		22	.727		197	34		231	267
4320	6" x 3" diameter	Q-2	28	.857		213	41.50		254.50	297
4322	6" x 4" diameter		26	.923		213	45		258	300
4324	6" x 5" diameter		24	1		213	48.50		261.50	305
4350	Flange, w/groove gasket									
4351	ANSI class 125 and 150									
4355	2" diameter	1 Plum	23	.348	Ea.	103	18.10		121.10	140
4356	2-1/2" diameter	Q-1	37	.432		107	20.50		127.50	149

22 11 13.27 Pipe/Tube, Grooved Joint for Copper		Crew	Daily Output	Labor-Hours	Unit	Material	Labor	Equipment	Total	Total Incl O&P
						2010 Bare Costs				
4358	3″ diameter	Q-1	31	.516	Ea.	112	24		136	160
4360	4″ diameter		23	.696		123	32.50		155.50	184
4362	5″ diameter		19	.842		165	39.50		204.50	241
4364	6″ diameter	Q-2	23	1.043		180	50.50		230.50	274

22 11 13.44 Pipe, Steel

		Crew	Daily Output	Labor-Hours	Unit	Material	Labor	Equipment	Total	Total Incl O&P
0010	**PIPE, STEEL** R221113-50									
0020	All pipe sizes are to Spec. A-53 unless noted otherwise R221113-70									
0032	Schedule 10, see Div. 22 11 13.48 0500									
0050	Schedule 40, threaded, with couplings, and clevis hanger									
0060	assemblies sized for covering, 10′ O.C.									
0540	Black, 1/4″ diameter	1 Plum	66	.121	L.F.	2.78	6.30		9.08	12.50
0550	3/8″ diameter		65	.123		3.16	6.40		9.56	13.10
0560	1/2″ diameter		63	.127		3.61	6.60		10.21	13.85
0570	3/4″ diameter		61	.131		4.28	6.85		11.13	14.95
0580	1″ diameter		53	.151		6.25	7.85		14.10	18.65
0590	1-1/4″ diameter	Q-1	89	.180		7.95	8.40		16.35	21.50
0600	1-1/2″ diameter		80	.200		9.45	9.35		18.80	24.50
0610	2″ diameter		64	.250		12.60	11.70		24.30	31.50
0620	2-1/2″ diameter		50	.320		19.45	15		34.45	44
0630	3″ diameter		43	.372		25	17.45		42.45	53.50
0640	3-1/2″ diameter		40	.400		34	18.75		52.75	65.50
0650	4″ diameter		36	.444		36.50	21		57.50	71.50
0809	A-106, gr. A/B, seamless w/cplgs. & clevis hanger assemblies									
0811	1/4″ diameter	1 Plum	66	.121	L.F.	7.60	6.30		13.90	17.85
0812	3/8″ diameter		65	.123		7.15	6.40		13.55	17.45
0813	1/2″ diameter		63	.127		7.20	6.60		13.80	17.80
0814	3/4″ diameter		61	.131		8.85	6.85		15.70	19.95
0815	1″ diameter		53	.151		9.80	7.85		17.65	22.50
0816	1-1/4″ diameter	Q-1	89	.180		11.85	8.40		20.25	25.50
0817	1-1/2″ diameter		80	.200		21	9.35		30.35	37
0819	2″ diameter		64	.250		24	11.70		35.70	44
0821	2-1/2″ diameter		50	.320		29	15		44	54
0822	3″ diameter		43	.372		37.50	17.45		54.95	67.50
0823	4″ diameter		36	.444		56.50	21		77.50	93.50
1220	To delete coupling & hanger, subtract									
1230	1/4″ diam. to 3/4″ diam.					31%	56%			
1240	1″ diam. to 1-1/2″ diam.					23%	51%			
1250	2″ diam. to 4″ diam.					23%	41%			
1280	All pipe sizes are to Spec. A-53 unless noted otherwise									
1281	Schedule 40, threaded, with couplings and clevis hanger									
1282	assemblies sized for covering, 10′ O. C.									
1290	Galvanized, 1/4″ diameter	1 Plum	66	.121	L.F.	3.80	6.30		10.10	13.65
1300	3/8″ diameter		65	.123		4.22	6.40		10.62	14.25
1310	1/2″ diameter		63	.127		5.15	6.60		11.75	15.55
1320	3/4″ diameter		61	.131		5	6.85		11.85	15.75
1330	1″ diameter		53	.151		8.45	7.85		16.30	21
1340	1-1/4″ diameter	Q-1	89	.180		10.90	8.40		19.30	24.50
1350	1-1/2″ diameter		80	.200		12.90	9.35		22.25	28.50
1360	2″ diameter		64	.250		17.25	11.70		28.95	36.50
1370	2-1/2″ diameter		50	.320		29	15		44	54
1380	3″ diameter		43	.372		37	17.45		54.45	66.50
1390	3-1/2″ diameter		40	.400		45.50	18.75		64.25	78

22 11 13.44 Pipe, Steel		Crew	Daily Output	Labor-Hours	Unit	Material	2010 Bare Costs Labor	Equipment	Total	Total Incl O&P
1400	4" diameter	Q-1	36	.444	L.F.	52	21		73	88.50
1750	To delete coupling & hanger, subtract									
1760	1/4" diam. to 3/4" diam.					31%	56%			
1770	1" diam. to 1-1/2" diam.					23%	51%			
1780	2" diam. to 4" diam.					23%	41%			
2000	Welded, sch. 40, on yoke & roll hanger assy's, sized for covering, 10' O.C.									
2040	Black, 1" diameter	Q-15	93	.172	L.F.	3.75	8.05	.63	12.43	16.90
2050	1-1/4" diameter		84	.190		4.95	8.90	.70	14.55	19.60
2060	1-1/2" diameter		76	.211		5.45	9.85	.77	16.07	21.50
2070	2" diameter		61	.262		6.55	12.30	.96	19.81	26.50
2080	2-1/2" diameter		47	.340		9.35	15.95	1.25	26.55	35.50
2090	3" diameter		43	.372		11.40	17.45	1.36	30.21	40
2100	3-1/2" diameter		39	.410		13.95	19.20	1.50	34.65	46
2110	4" diameter		37	.432		15.95	20.50	1.58	38.03	50
2120	5" diameter	↓	32	.500		21.50	23.50	1.83	46.83	61
2130	6" diameter	Q-16	36	.667		26.50	32.50	1.63	60.63	79.50
2140	8" diameter		29	.828		39	40	2.02	81.02	105
2150	10" diameter		24	1		55	48.50	2.44	105.94	136
2160	12" diameter		19	1.263		83	61.50	3.08	147.58	187
2170	14" diameter, (two rod roll type hanger for 14" dia. and up)		15	1.600		86	77.50	3.90	167.40	216
2180	16" diameter, (two rod roll type hanger)		13	1.846		99	89.50	4.50	193	248
2190	18" diameter, (two rod roll type hanger)		11	2.182		115	106	5.30	226.30	291
2200	20" diameter, (two rod roll type hanger)		9	2.667		140	130	6.50	276.50	355
2220	24" diameter, (two rod roll type hanger)	↓	8	3	↓	199	146	7.30	352.30	445
2345	Sch. 40, A-53, gr. A/B, ERW, welded w/hngrs.									
2346	2" diameter	Q-15	61	.262	L.F.	5.70	12.30	.96	18.96	26
2347	2-1/2" diameter		47	.340		8.60	15.95	1.25	25.80	35
2348	3" diameter		43	.372		10.95	17.45	1.36	29.76	39.50
2349	4" diameter	↓	38	.421		15.15	19.75	1.54	36.44	48
2350	6" diameter	Q-16	37	.649		62	31.50	1.58	95.08	117
2351	8" diameter		29	.828		81	40	2.02	123.02	152
2352	10" diameter	↓	24	1	↓	115	48.50	2.44	165.94	202
2363	.375" wall, A-53, gr. A/B, ERW, welded w/hngrs.									
2364	12" diameter	Q-16	20	1.200	L.F.	138	58.50	2.93	199.43	242
2365	14" diameter		17	1.412		151	68.50	3.44	222.94	273
2366	16" diameter	↓	14	1.714	↓	202	83.50	4.18	289.68	355
2560	To delete hanger, subtract									
2570	1" diam. to 1-1/2" diam.					15%	34%			
2580	2" diam. to 3-1/2" diam.					9%	21%			
2590	4" diam. to 12" diam.					5%	12%			
2596	14" diam. to 36" diam.					3%	10%			
2640	Galvanized, 1" diameter	Q-15	93	.172	L.F.	8.75	8.05	.63	17.43	22.50
2650	1-1/4" diameter		84	.190		11.35	8.90	.70	20.95	26.50
2660	1-1/2" diameter		76	.211		13.05	9.85	.77	23.67	30
2670	2" diameter		61	.262		16.75	12.30	.96	30.01	38
2680	2-1/2" diameter		47	.340		25	15.95	1.25	42.20	53
2690	3" diameter		43	.372		32	17.45	1.36	50.81	62.50
2700	3-1/2" diameter		39	.410		47.50	19.20	1.50	68.20	82.50
2710	4" diameter		37	.432		44.50	20.50	1.58	66.58	81
2720	5" diameter	↓	32	.500		50.50	23.50	1.83	75.83	92.50
2730	6" diameter	Q-16	36	.667		59.50	32.50	1.63	93.63	115
2740	8" diameter		29	.828		68.50	40	2.02	110.52	138
2750	10" diameter	↓	24	1	↓	107	48.50	2.44	157.94	194

22 11 13 – Facility Water Distribution Piping

22 11 13.44 Pipe, Steel		Crew	Daily Output	Labor-Hours	Unit	Material	Labor	Equipment	Total	Total Incl O&P
						2010 Bare Costs				
2760	12″ diameter	Q-16	19	1.263	L.F.	137	61.50	3.08	201.58	246
3160	To delete hanger, subtract									
3170	1″ diam. to 1-1/2″ diam.					30%	34%			
3180	2″ diam. to 3-1/2″ diam.					19%	21%			
3190	4″ diam. to 12″ diam.					10%	12%			
3250	Flanged, 150 lb. weld neck, on yoke & roll hangers									
3260	sized for covering, 10′ O.C.									
3290	Black, 1″ diameter	Q-15	70	.229	L.F.	9.05	10.70	.84	20.59	27
3300	1-1/4″ diameter		64	.250		10.25	11.70	.92	22.87	30
3310	1-1/2″ diameter		58	.276		10.75	12.90	1.01	24.66	32.50
3320	2″ diameter		45	.356		12.80	16.65	1.30	30.75	40.50
3330	2-1/2″ diameter		36	.444		16.15	21	1.63	38.78	50.50
3340	3″ diameter		32	.500		18.80	23.50	1.83	44.13	57.50
3350	3-1/2″ diameter		29	.552		23	26	2.02	51.02	66.50
3360	4″ diameter		26	.615		25	29	2.25	56.25	73
3370	5″ diameter		21	.762		36.50	35.50	2.79	74.79	96.50
3380	6″ diameter	Q-16	25	.960		42	46.50	2.34	90.84	119
3390	8″ diameter		19	1.263		65.50	61.50	3.08	130.08	168
3400	10″ diameter		16	1.500		107	73	3.66	183.66	231
3410	12″ diameter		14	1.714		144	83.50	4.18	231.68	288
3470	For 300 lb. flanges, add					63%				
3480	For 600 lb. flanges, add					310%				
3960	To delete flanges & hanger, subtract									
3970	1″ diam. to 2″ diam.					76%	65%			
3980	2-1/2″ diam. to 4″ diam.					62%	59%			
3990	5″ diam. to 12″ diam.					60%	46%			
4040	Galvanized, 1″ diameter	Q-15	70	.229	L.F.	14.05	10.70	.84	25.59	32.50
4050	1-1/4″ diameter		64	.250		16.60	11.70	.92	29.22	37
4060	1-1/2″ diameter		58	.276		18.35	12.90	1.01	32.26	40.50
4070	2″ diameter		45	.356		22.50	16.65	1.30	40.45	51
4080	2-1/2″ diameter		36	.444		31.50	21	1.63	54.13	67.50
4090	3″ diameter		32	.500		39	23.50	1.83	64.33	80
4100	3-1/2″ diameter		29	.552		55	26	2.02	83.02	102
4110	4″ diameter		26	.615		54	29	2.25	85.25	105
4120	5″ diameter		21	.762		65.50	35.50	2.79	103.79	129
4130	6″ diameter	Q-16	25	.960		74	46.50	2.34	122.84	154
4140	8″ diameter		19	1.263		83	61.50	3.08	147.58	186
4150	10″ diameter		16	1.500		133	73	3.66	209.66	259
4160	12″ diameter		14	1.714		168	83.50	4.18	255.68	315
4220	For 300 lb. flanges, add					65%				
4240	For 600 lb. flanges, add					350%				
4660	To delete flanges & hanger, subtract									
4670	1″ diam. to 2″ diam.					73%	65%			
4680	2-1/2″ diam. to 4″ diam.					58%	59%			
4690	5″ diam. to 12″ diam.					43%	46%			
4750	Schedule 80, threaded, with couplings, and clevis hanger assemblies									
4760	sized for covering, 10′ O.C.									
4790	Black, 1/4″ diameter	1 Plum	54	.148	L.F.	3.64	7.70		11.34	15.55
4800	3/8″ diameter		53	.151		4.63	7.85		12.48	16.85
4810	1/2″ diameter		52	.154		5.45	8		13.45	18
4820	3/4″ diameter		50	.160		6.70	8.35		15.05	19.90
4830	1″ diameter		45	.178		9.25	9.25		18.50	24
4840	1-1/4″ diameter	Q-1	75	.213		12.30	10		22.30	28.50

22 11 13.44 Pipe, Steel

22 11 13.44 Pipe, Steel		Crew	Daily Output	Labor-Hours	Unit	2010 Bare Costs Material	Labor	Equipment	Total	Total Incl O&P
4850	1-1/2" diameter	Q-1	69	.232	L.F.	14.50	10.85		25.35	32.50
4860	2" diameter		56	.286		19.70	13.40		33.10	41.50
4870	2-1/2" diameter		44	.364		30.50	17.05		47.55	59
4880	3" diameter		38	.421		40.50	19.75		60.25	74
4890	3-1/2" diameter		35	.457		49.50	21.50		71	86.50
4900	4" diameter		32	.500		46	23.50		69.50	85.50
5061	A-106, gr. A/B seamless with cplgs. & clevis hanger assemblies, 1/4" dia.	1 Plum	63	.127		7.95	6.60		14.55	18.65
5062	3/8" diameter		62	.129		9.35	6.70		16.05	20.50
5063	1/2" diameter		61	.131		8.10	6.85		14.95	19.15
5064	3/4" diameter		57	.140		9.45	7.30		16.75	21.50
5065	1" diameter		51	.157		11.30	8.15		19.45	24.50
5066	1-1/4" diameter	Q-1	85	.188		12.95	8.80		21.75	27.50
5067	1-1/2" diameter		77	.208		25.50	9.75		35.25	42.50
5071	2" diameter		61	.262		23	12.30		35.30	44
5072	2-1/2" diameter		48	.333		34	15.60		49.60	61
5073	3" diameter		41	.390		45	18.30		63.30	77
5074	4" diameter		35	.457		66	21.50		87.50	105
5430	To delete coupling & hanger, subtract									
5440	1/4" diam. to 1/2" diam.					31%	54%			
5450	3/4" diam. to 1-1/2" diam.					28%	49%			
5460	2" diam. to 4" diam.					21%	40%			
5510	Galvanized, 1/4" diameter	1 Plum	54	.148	L.F.	7	7.70		14.70	19.25
5520	3/8" diameter		53	.151		7	7.85		14.85	19.45
5530	1/2" diameter		52	.154		7.15	8		15.15	19.85
5540	3/4" diameter		50	.160		8.75	8.35		17.10	22
5550	1" diameter		45	.178		12.25	9.25		21.50	27.50
5560	1-1/4" diameter	Q-1	75	.213		16.40	10		26.40	33
5570	1-1/2" diameter		69	.232		19.45	10.85		30.30	38
5580	2" diameter		56	.286		27	13.40		40.40	49.50
5590	2-1/2" diameter		44	.364		42	17.05		59.05	71.50
5600	3" diameter		38	.421		55.50	19.75		75.25	90.50
5610	3-1/2" diameter		35	.457		66	21.50		87.50	105
5620	4" diameter		32	.500		78.50	23.50		102	121
5930	To delete coupling & hanger, subtract									
5940	1/4" diam. to 1/2" diam.					31%	54%			
5950	3/4" diam. to 1-1/2" diam.					28%	49%			
5960	2" diam. to 4" diam.					21%	40%			
6000	Welded, on yoke & roller hangers									
6010	sized for covering, 10' O.C.									
6040	Black, 1" diameter	Q-15	85	.188	L.F.	9.95	8.80	.69	19.44	25
6050	1-1/4" diameter		79	.203		13.20	9.50	.74	23.44	29.50
6060	1-1/2" diameter		72	.222		15.35	10.40	.81	26.56	33.50
6070	2" diameter		57	.281		20	13.15	1.03	34.18	43
6080	2-1/2" diameter		44	.364		30.50	17.05	1.33	48.88	60.50
6090	3" diameter		40	.400		40	18.75	1.46	60.21	73.50
6100	3-1/2" diameter		34	.471		49	22	1.72	72.72	89
6110	4" diameter		33	.485		45	22.50	1.77	69.27	85.50
6120	5" diameter		26	.615		55.50	29	2.25	86.75	106
6130	6" diameter	Q-16	30	.800		67	39	1.95	107.95	135
6140	8" diameter		25	.960		117	46.50	2.34	165.84	202
6150	10" diameter		20	1.200		202	58.50	2.93	263.43	315
6160	12" diameter		15	1.600		355	77.50	3.90	436.40	510
6540	To delete hanger, subtract									

22 11 13 – Facility Water Distribution Piping

22 11 13.44 Pipe, Steel		Crew	Daily Output	Labor-Hours	Unit	Material	2010 Bare Costs Labor	Equipment	Total	Total Incl O&P
6550	1" diam. to 1-1/2" diam.					30%	14%			
6560	2" diam. to 3" diam.					23%	9%			
6570	3-1/2" diam. to 5" diam.					12%	6%			
6580	6" diam. to 12" diam.					10%	4%			
6610	Galvanized, 1" diameter	Q-15	85	.188	L.F.	12.95	8.80	.69	22.44	28
6650	1-1/4" diameter		79	.203		17.30	9.50	.74	27.54	34
6660	1-1/2" diameter		72	.222		20.50	10.40	.81	31.71	39
6670	2" diameter		57	.281		27.50	13.15	1.03	41.68	51
6680	2-1/2" diameter		44	.364		41.50	17.05	1.33	59.88	73
6690	3" diameter		40	.400		55	18.75	1.46	75.21	90
6700	3-1/2" diameter		34	.471		65	22	1.72	88.72	106
6710	4" diameter		33	.485		77	22.50	1.77	101.27	121
6720	5" diameter		26	.615		107	29	2.25	138.25	162
6730	6" diameter	Q-16	30	.800		118	39	1.95	158.95	191
6740	8" diameter		25	.960		177	46.50	2.34	225.84	268
6750	10" diameter		20	1.200		325	58.50	2.93	386.43	450
6760	12" diameter		15	1.600		470	77.50	3.90	551.40	635
7150	To delete hanger, subtract									
7160	1" diam. to 1-1/2" diam.					26%	14%			
7170	2" diam. to 3" diam.					16%	9%			
7180	3-1/2" diam. to 5" diam.					10%	6%			
7190	6" diam. to 12" diam.					7%	4%			
7250	Flanged, 300 lb. weld neck, on yoke & roll hangers									
7260	sized for covering, 10' O.C.									
7290	Black, 1" diameter	Q-15	66	.242	L.F.	16.35	11.35	.89	28.59	36
7300	1-1/4" diameter		61	.262		19.55	12.30	.96	32.81	41
7310	1-1/2" diameter		54	.296		21.50	13.90	1.08	36.48	46
7320	2" diameter		42	.381		28	17.85	1.39	47.24	59.50
7330	2-1/2" diameter		33	.485		39.50	22.50	1.77	63.77	79.50
7340	3" diameter		29	.552		49.50	26	2.02	77.52	95.50
7350	3-1/2" diameter		24	.667		64.50	31	2.44	97.94	121
7360	4" diameter		23	.696		60.50	32.50	2.55	95.55	118
7370	5" diameter		19	.842		79	39.50	3.08	121.58	149
7380	6" diameter	Q-16	23	1.043		87.50	50.50	2.55	140.55	175
7390	8" diameter		17	1.412		154	68.50	3.44	225.94	277
7400	10" diameter		14	1.714		276	83.50	4.18	363.68	435
7410	12" diameter		12	2		450	97	4.88	551.88	645
7470	For 600 lb. flanges, add					100%				
7940	To delete flanges & hanger, subtract									
7950	1" diam. to 1-1/2" diam.					75%	66%			
7960	2" diam. to 3" diam.					62%	60%			
7970	3-1/2" diam. to 5" diam.					54%	66%			
7980	6" diam. to 12" diam.					55%	62%			
8040	Galvanized, 1" diameter	Q-15	66	.242	L.F.	19.35	11.35	.89	31.59	39.50
8050	1-1/4" diameter		61	.262		23.50	12.30	.96	36.76	45.50
8060	1-1/2" diameter		54	.296		26.50	13.90	1.08	41.48	51.50
8070	2" diameter		42	.381		35.50	17.85	1.39	54.74	67.50
8080	2-1/2" diameter		33	.485		51	22.50	1.77	75.27	92
8090	3" diameter		29	.552		64	26	2.02	92.02	112
8100	3-1/2" diameter		24	.667		80.50	31	2.44	113.94	138
8110	4" diameter		23	.696		92.50	32.50	2.55	127.55	154
8120	5" diameter		19	.842		130	39.50	3.08	172.58	205
8130	6" diameter	Q-16	23	1.043		139	50.50	2.55	192.05	232

22 11 13.44 Pipe, Steel		Crew	Daily Output	Labor-Hours	Unit	Material	2010 Bare Costs Labor	Equipment	Total	Total Incl O&P
8140	8" diameter	Q-16	17	1.412	L.F.	214	68.50	3.44	285.94	345
8150	10" diameter		14	1.714		400	83.50	4.18	487.68	570
8160	12" diameter		12	2		565	97	4.88	666.88	770
8240	For 600 lb. flanges, add					100%				
8900	To delete flanges & hangers, subtract									
8910	1" diam. to 1-1/2" diam.					72%	66%			
8920	2" diam. to 3" diam.					59%	60%			
8930	3-1/2" diam. to 5" diam.					51%	66%			
8940	6" diam. to 12" diam.					49%	62%			
9000	Threading pipe labor, one end, all schedules through 80									
9010	1/4" through 3/4" pipe size	1 Plum	80	.100	Ea.		5.20		5.20	7.80
9020	1" through 2" pipe size		73	.110			5.70		5.70	8.55
9030	2-1/2" pipe size		53	.151			7.85		7.85	11.75
9040	3" pipe size		50	.160			8.35		8.35	12.50
9050	3-1/2" pipe size	Q-1	89	.180			8.40		8.40	12.65
9060	4" pipe size		73	.219			10.25		10.25	15.40
9070	5" pipe size		53	.302			14.15		14.15	21
9080	6" pipe size		46	.348			16.30		16.30	24.50
9090	8" pipe size		29	.552			26		26	39
9100	10" pipe size		21	.762			35.50		35.50	53.50
9110	12" pipe size		13	1.231			57.50		57.50	86.50
9200	Welding labor per joint									
9210	Schedule 40,									
9230	1/2" pipe size	Q-15	32	.500	Ea.		23.50	1.83	25.33	37
9240	3/4" pipe size		27	.593			28	2.17	30.17	44
9250	1" pipe size		23	.696			32.50	2.55	35.05	52
9260	1-1/4" pipe size		20	.800			37.50	2.93	40.43	59
9270	1-1/2" pipe size		19	.842			39.50	3.08	42.58	62.50
9280	2" pipe size		16	1			47	3.66	50.66	74
9290	2-1/2" pipe size		13	1.231			57.50	4.50	62	91.50
9300	3" pipe size		12	1.333			62.50	4.88	67.38	99
9310	4" pipe size		10	1.600			75	5.85	80.85	118
9320	5" pipe size		9	1.778			83.50	6.50	90	132
9330	6" pipe size		8	2			93.50	7.30	100.80	148
9340	8" pipe size		5	3.200			150	11.70	161.70	238
9350	10" pipe size		4	4			187	14.65	201.65	297
9360	12" pipe size		3	5.333			250	19.50	269.50	395
9370	14" pipe size		2.60	6.154			288	22.50	310.50	455
9380	16" pipe size		2.20	7.273			340	26.50	366.50	540
9390	18" pipe size		2	8			375	29.50	404.50	590
9400	20" pipe size		1.80	8.889			415	32.50	447.50	660
9410	22" pipe size		1.70	9.412			440	34.50	474.50	700
9420	24" pipe size		1.50	10.667			500	39	539	795
9450	Schedule 80,									
9460	1/2" pipe size	Q-15	27	.593	Ea.		28	2.17	30.17	44
9470	3/4" pipe size		23	.696			32.50	2.55	35.05	52
9480	1" pipe size		20	.800			37.50	2.93	40.43	59
9490	1-1/4" pipe size		19	.842			39.50	3.08	42.58	62.50
9500	1-1/2" pipe size		18	.889			41.50	3.25	44.75	66
9510	2" pipe size		15	1.067			50	3.90	53.90	79.50
9520	2-1/2" pipe size		12	1.333			62.50	4.88	67.38	99
9530	3" pipe size		11	1.455			68	5.30	73.30	108
9540	4" pipe size		8	2			93.50	7.30	100.80	148

22 11 13 – Facility Water Distribution Piping

22 11 13.44 Pipe, Steel		Crew	Daily Output	Labor-Hours	Unit	Material	2010 Bare Costs Labor	Equipment	Total	Total Incl O&P
9550	5" pipe size	Q-15	6	2.667	Ea.		125	9.75	134.75	198
9560	6" pipe size		5	3.200			150	11.70	161.70	238
9570	8" pipe size		4	4			187	14.65	201.65	297
9580	10" pipe size		3	5.333			250	19.50	269.50	395
9590	12" pipe size		2	8			375	29.50	404.50	590
9600	14" pipe size	Q-16	2.60	9.231			450	22.50	472.50	695
9610	16" pipe size		2.30	10.435			505	25.50	530.50	790
9620	18" pipe size		2	12			585	29.50	614.50	905
9630	20" pipe size		1.80	13.333			650	32.50	682.50	1,000
9640	22" pipe size		1.60	15			730	36.50	766.50	1,150
9650	24" pipe size		1.50	16			775	39	814	1,225

22 11 13.45 Pipe Fittings, Steel, Threaded		Crew	Daily Output	Labor-Hours	Unit	Material	Labor	Equipment	Total	Total Incl O&P
0010	**PIPE FITTINGS, STEEL, THREADED**									
0020	Cast Iron									
0040	Standard weight, black									
0060	90° Elbow, straight									
0070	1/4"	1 Plum	16	.500	Ea.	6.95	26		32.95	46.50
0080	3/8"		16	.500		10.05	26		36.05	50
0090	1/2"		15	.533		4.41	28		32.41	46.50
0100	3/4"		14	.571		4.60	29.50		34.10	49.50
0110	1"		13	.615		5.45	32		37.45	54
0120	1-1/4"	Q-1	22	.727		7.75	34		41.75	59.50
0130	1-1/2"		20	.800		10.70	37.50		48.20	68
0140	2"		18	.889		16.70	41.50		58.20	81
0150	2-1/2"		14	1.143		40	53.50		93.50	125
0160	3"		10	1.600		65.50	75		140.50	184
0170	3-1/2"		8	2		178	93.50		271.50	335
0180	4"		6	2.667		122	125		247	320
0250	45° Elbow, straight									
0260	1/4"	1 Plum	16	.500	Ea.	8.35	26		34.35	48
0270	3/8"		16	.500		8.95	26		34.95	49
0280	1/2"		15	.533		6.75	28		34.75	49
0300	3/4"		14	.571		6.80	29.50		36.30	52
0320	1"		13	.615		7.95	32		39.95	57
0330	1-1/4"	Q-1	22	.727		10.70	34		44.70	63
0340	1-1/2"		20	.800		17.75	37.50		55.25	75.50
0350	2"		18	.889		20.50	41.50		62	85
0360	2-1/2"		14	1.143		53	53.50		106.50	139
0370	3"		10	1.600		84.50	75		159.50	205
0380	3-1/2"		8	2		187	93.50		280.50	345
0400	4"		6	2.667		175	125		300	380
0500	Tee, straight									
0510	1/4"	1 Plum	10	.800	Ea.	10.90	41.50		52.40	74.50
0520	3/8"		10	.800		10.60	41.50		52.10	74
0530	1/2"		9	.889		6.90	46.50		53.40	77
0540	3/4"		9	.889		8	46.50		54.50	78.50
0550	1"		8	1		7.15	52		59.15	86
0560	1-1/4"	Q-1	14	1.143		13	53.50		66.50	95
0570	1-1/2"		13	1.231		16.90	57.50		74.40	105
0580	2"		11	1.455		23.50	68		91.50	128
0590	2-1/2"		9	1.778		61	83.50		144.50	193
0600	3"		6	2.667		94	125		219	290

22 11 13 – Facility Water Distribution Piping

22 11 13.45 Pipe Fittings, Steel, Threaded		Crew	Daily Output	Labor-Hours	Unit	2010 Bare Costs Material	Labor	Equipment	Total	Total Incl O&P
0610	3-1/2″	Q-1	5	3.200	Ea.	190	150		340	435
0620	4″		4	4		183	187		370	485
0660	Tee, reducing, run or outlet									
0661	1/2″	1 Plum	9	.889	Ea.	16.70	46.50		63.20	88
0662	3/4″		9	.889		18.45	46.50		64.95	90
0663	1″		8	1		22	52		74	102
0664	1-1/4″	Q-1	14	1.143		23.50	53.50		77	106
0665	1-1/2″		13	1.231		27.50	57.50		85	117
0666	2″		11	1.455		36.50	68		104.50	143
0667	2-1/2″		9	1.778		80.50	83.50		164	214
0668	3″		6	2.667		159	125		284	360
0669	3-1/2″		5	3.200		205	150		355	450
0670	4″		4	4		296	187		483	605
0674	Reducer, concentric									
0675	3/4″	1 Plum	18	.444	Ea.	12.25	23		35.25	48
0676	1″	″	15	.533		8.35	28		36.35	50.50
0677	1-1/4″	Q-1	26	.615		24	29		53	69.50
0678	1-1/2″		24	.667		38	31		69	89
0679	2″		21	.762		44.50	35.50		80	103
0680	2-1/2″		18	.889		63.50	41.50		105	132
0681	3″		14	1.143		88	53.50		141.50	177
0682	3-1/2″		12	1.333		166	62.50		228.50	277
0683	4″		10	1.600		180	75		255	310
0687	Reducer, eccentric									
0688	3/4″	1 Plum	16	.500	Ea.	26	26		52	67.50
0689	1″	″	14	.571		28	29.50		57.50	75.50
0690	1-1/4″	Q-1	25	.640		41.50	30		71.50	91
0691	1-1/2″		22	.727		59	34		93	116
0692	2″		20	.800		83	37.50		120.50	147
0693	2-1/2″		16	1		114	47		161	195
0694	3″		12	1.333		180	62.50		242.50	292
0695	3-1/2″		10	1.600		250	75		325	385
0696	4″		9	1.778		325	83.50		408.50	485
0700	Standard weight, galvanized cast iron									
0720	90° Elbow, straight									
0730	1/4″	1 Plum	16	.500	Ea.	11.65	26		37.65	52
0740	3/8″		16	.500		11.65	26		37.65	52
0750	1/2″		15	.533		8.85	28		36.85	51
0760	3/4″		14	.571		13.05	29.50		42.55	59
0770	1″		13	.615		15.05	32		47.05	64.50
0780	1-1/4″	Q-1	22	.727		23.50	34		57.50	77
0790	1-1/2″		20	.800		32	37.50		69.50	91.50
0800	2″		18	.889		47.50	41.50		89	115
0810	2-1/2″		14	1.143		97.50	53.50		151	188
0820	3″		10	1.600		149	75		224	275
0830	3-1/2″		8	2		237	93.50		330.50	400
0840	4″		6	2.667		272	125		397	485
0900	45° Elbow, straight									
0910	1/4″	1 Plum	16	.500	Ea.	16.70	26		42.70	57.50
0920	3/8″		16	.500		14.90	26		40.90	55.50
0930	1/2″		15	.533		13.30	28		41.30	56
0940	3/4″		14	.571		15.25	29.50		44.75	61.50
0950	1″		13	.615		15.90	32		47.90	65.50

22 11 13.45 Pipe Fittings, Steel, Threaded		Crew	Daily Output	Labor-Hours	Unit	Material	Labor	Equipment	Total	Total Incl O&P
						2010 Bare Costs				
0960	1-1/4"	Q-1	22	.727	Ea.	21.50	34		55.50	74.50
0970	1-1/2"		20	.800		39.50	37.50		77	99.50
0980	2"		18	.889		55.50	41.50		97	124
0990	2-1/2"		14	1.143		109	53.50		162.50	201
1000	3"		10	1.600		174	75		249	305
1010	3-1/2"		8	2		278	93.50		371.50	445
1020	4"		6	2.667		320	125		445	535
1100	Tee, straight									
1110	1/4"	1 Plum	10	.800	Ea.	22	41.50		63.50	86.50
1120	3/8"		10	.800		14.90	41.50		56.40	79
1130	1/2"		9	.889		13.75	46.50		60.25	84.50
1140	3/4"		9	.889		18.75	46.50		65.25	90
1150	1"		8	1		20.50	52		72.50	101
1160	1-1/4"	Q-1	14	1.143		35.50	53.50		89	120
1170	1-1/2"		13	1.231		47	57.50		104.50	139
1180	2"		11	1.455		59	68		127	167
1190	2-1/2"		9	1.778		127	83.50		210.50	264
1200	3"		6	2.667		330	125		455	545
1210	3-1/2"		5	3.200		335	150		485	595
1220	4"		4	4		320	187		507	635
1300	Extra heavy weight, black									
1310	Couplings, steel straight									
1320	1/4"	1 Plum	19	.421	Ea.	2.71	22		24.71	36
1330	3/8"		19	.421		2.95	22		24.95	36.50
1340	1/2"		19	.421		3.99	22		25.99	37.50
1350	3/4"		18	.444		4.26	23		27.26	39
1360	1"		15	.533		5.45	28		33.45	47.50
1370	1-1/4"	Q-1	26	.615		8.70	29		37.70	52.50
1380	1-1/2"		24	.667		8.70	31		39.70	56.50
1390	2"		21	.762		13.30	35.50		48.80	68
1400	2-1/2"		18	.889		19.70	41.50		61.20	84
1410	3"		14	1.143		23.50	53.50		77	107
1420	3-1/2"		12	1.333		31.50	62.50		94	128
1430	4"		10	1.600		35.50	75		110.50	151
1510	90° Elbow, straight									
1520	1/2"	1 Plum	15	.533	Ea.	22	28		50	65.50
1530	3/4"		14	.571		23	29.50		52.50	70
1540	1"		13	.615		28	32		60	79
1550	1-1/4"	Q-1	22	.727		41.50	34		75.50	97
1560	1-1/2"		20	.800		51.50	37.50		89	113
1580	2"		18	.889		63.50	41.50		105	133
1590	2-1/2"		14	1.143		149	53.50		202.50	245
1600	3"		10	1.600		197	75		272	330
1610	4"		6	2.667		410	125		535	640
1650	45° Elbow, straight									
1660	1/2"	1 Plum	15	.533	Ea.	31	28		59	76
1670	3/4"		14	.571		30	29.50		59.50	77.50
1680	1"		13	.615		36	32		68	88
1690	1-1/4"	Q-1	22	.727		59.50	34		93.50	117
1700	1-1/2"		20	.800		65.50	37.50		103	128
1710	2"		18	.889		93	41.50		134.50	166
1720	2-1/2"		14	1.143		161	53.50		214.50	258
1800	Tee, straight									

22 11 13.45 Pipe Fittings, Steel, Threaded		Crew	Daily Output	Labor-Hours	Unit	2010 Bare Costs Material	Labor	Equipment	Total	Total Incl O&P
1810	1/2"	1 Plum	9	.889	Ea.	34.50	46.50		81	108
1820	3/4"		9	.889		34.50	46.50		81	108
1830	1"		8	1		41.50	52		93.50	124
1840	1-1/4"	Q-1	14	1.143		62	53.50		115.50	149
1850	1-1/2"		13	1.231		80	57.50		137.50	175
1860	2"		11	1.455		99	68		167	211
1870	2-1/2"		9	1.778		211	83.50		294.50	360
1880	3"		6	2.667		288	125		413	500
1890	4"		4	4		560	187		747	895
4000	Standard weight, black									
4010	Couplings, steel straight, merchants									
4030	1/4"	1 Plum	19	.421	Ea.	.69	22		22.69	34
4040	3/8"		19	.421		.83	22		22.83	34
4050	1/2"		19	.421		.83	22		22.83	34
4060	3/4"		18	.444		.95	23		23.95	35.50
4070	1"		15	.533		1.33	28		29.33	43
4080	1-1/4"	Q-1	26	.615		1.69	29		30.69	45
4090	1-1/2"		24	.667		2.15	31		33.15	49.50
4100	2"		21	.762		3.08	35.50		38.58	57
4110	2-1/2"		18	.889		8.05	41.50		49.55	71.50
4120	3"		14	1.143		11.35	53.50		64.85	93
4130	3-1/2"		12	1.333		20	62.50		82.50	116
4140	4"		10	1.600		20	75		95	134
4166	Plug, 1/4"	1 Plum	38	.211		1.96	10.95		12.91	18.55
4167	3/8"		38	.211		1.96	10.95		12.91	18.55
4168	1/2"		38	.211		1.96	10.95		12.91	18.55
4169	3/4"		32	.250		5.90	13		18.90	26
4170	1"		30	.267		6.30	13.90		20.20	28
4171	1-1/4"	Q-1	52	.308		7.25	14.40		21.65	29.50
4172	1-1/2"		48	.333		10.30	15.60		25.90	35
4173	2"		42	.381		13.30	17.85		31.15	41.50
4176	2-1/2"		36	.444		19.65	21		40.65	52.50
4180	4"		20	.800		34	37.50		71.50	93.50
4200	Standard weight, galvanized									
4210	Couplings, steel straight, merchants									
4230	1/4"	1 Plum	19	.421	Ea.	.80	22		22.80	34
4240	3/8"		19	.421		1	22		23	34
4250	1/2"		19	.421		1.03	22		23.03	34
4260	3/4"		18	.444		1.14	23		24.14	36
4270	1"		15	.533		1.59	28		29.59	43.50
4280	1-1/4"	Q-1	26	.615		2.05	29		31.05	45.50
4290	1-1/2"		24	.667		2.53	31		33.53	50
4300	2"		21	.762		3.79	35.50		39.29	57.50
4310	2-1/2"		18	.889		14.80	41.50		56.30	79
4320	3"		14	1.143		20.50	53.50		74	103
4330	3-1/2"		12	1.333		23	62.50		85.50	119
4340	4"		10	1.600		23	75		98	138
4370	Plug, galvanized, square head									
4374	1/2"	1 Plum	38	.211	Ea.	4.73	10.95		15.68	21.50
4375	3/4"		32	.250		4.73	13		17.73	24.50
4376	1"		30	.267		4.73	13.90		18.63	26
4377	1-1/4"	Q-1	52	.308		7.90	14.40		22.30	30
4378	1-1/2"		48	.333		10.70	15.60		26.30	35.50

22 11 13.45 Pipe Fittings, Steel, Threaded		Crew	Daily Output	Labor-Hours	Unit	Material	2010 Bare Costs Labor	Equipment	Total	Total Incl O&P
4379	2"	Q-1	42	.381	Ea.	13.25	17.85		31.10	41.50
4380	2-1/2"		36	.444		28	21		49	62
4381	3"		28	.571		36	27		63	79.50
4382	4"		20	.800		99.50	37.50		137	165
4700	Nipple, black									
4710	1/2" x 4" long	1 Plum	19	.421	Ea.	1.54	22		23.54	34.50
4712	3/4" x 4" long		18	.444		1.87	23		24.87	36.50
4714	1" x 4" long		15	.533		2.59	28		30.59	44.50
4716	1-1/4" x 4" long	Q-1	26	.615		3.25	29		32.25	46.50
4718	1-1/2" x 4" long		24	.667		3.84	31		34.84	51
4720	2" x 4" long		21	.762		5.35	35.50		40.85	59.50
4722	2-1/2" x 4" long		18	.889		14.60	41.50		56.10	78.50
4724	3" x 4" long		14	1.143		18.55	53.50		72.05	101
4726	4" x 4" long		10	1.600		25	75		100	140
4800	Nipple, galvanized									
4810	1/2" x 4" long	1 Plum	19	.421	Ea.	1.90	22		23.90	35
4812	3/4" x 4" long		18	.444		2.34	23		25.34	37
4814	1" x 4" long		15	.533		3.17	28		31.17	45
4816	1-1/4" x 4" long	Q-1	26	.615		3.89	29		32.89	47.50
4818	1-1/2" x 4" long		24	.667		4.93	31		35.93	52.50
4820	2" x 4" long		21	.762		6.25	35.50		41.75	60.50
4822	2-1/2" x 4" long		18	.889		16.45	41.50		57.95	80.50
4824	3" x 4" long		14	1.143		22	53.50		75.50	105
4826	4" x 4" long		10	1.600		29.50	75		104.50	144
5000	Malleable iron, 150 lb.									
5020	Black									
5040	90° elbow, straight									
5060	1/4"	1 Plum	16	.500	Ea.	3.42	26		29.42	43
5070	3/8"		16	.500		3.42	26		29.42	43
5080	1/2"		15	.533		2.37	28		30.37	44
5090	3/4"		14	.571		2.87	29.50		32.37	47.50
5100	1"		13	.615		4.98	32		36.98	53.50
5110	1-1/4"	Q-1	22	.727		8.20	34		42.20	60
5120	1-1/2"		20	.800		10.80	37.50		48.30	68
5130	2"		18	.889		18.60	41.50		60.10	83
5140	2-1/2"		14	1.143		41.50	53.50		95	126
5150	3"		10	1.600		60.50	75		135.50	179
5160	3-1/2"		8	2		162	93.50		255.50	320
5170	4"		6	2.667		130	125		255	330
5250	45° elbow, straight									
5270	1/4"	1 Plum	16	.500	Ea.	5.15	26		31.15	44.50
5280	3/8"		16	.500		5.15	26		31.15	44.50
5290	1/2"		15	.533		3.92	28		31.92	46
5300	3/4"		14	.571		4.85	29.50		34.35	50
5310	1"		13	.615		6.10	32		38.10	54.50
5320	1-1/4"	Q-1	22	.727		10.80	34		44.80	63
5330	1-1/2"		20	.800		13.35	37.50		50.85	70.50
5340	2"		18	.889		20	41.50		61.50	84.50
5350	2-1/2"		14	1.143		58.50	53.50		112	145
5360	3"		10	1.600		76	75		151	196
5370	3-1/2"		8	2		145	93.50		238.50	299
5380	4"		6	2.667		149	125		274	350
5450	Tee, straight									

22 11 13 – Facility Water Distribution Piping

22 11 13.45 Pipe Fittings, Steel, Threaded		Crew	Daily Output	Labor-Hours	Unit	Material	2010 Bare Costs Labor	Equipment	Total	Total Incl O&P
5470	1/4"	1 Plum	10	.800	Ea.	4.98	41.50		46.48	68
5480	3/8"		10	.800		4.98	41.50		46.48	68
5490	1/2"		9	.889		3.23	46.50		49.73	73
5500	3/4"		9	.889		4.55	46.50		51.05	74.50
5510	1"		8	1		7.80	52		59.80	86.50
5520	1-1/4"	Q-1	14	1.143		12.65	53.50		66.15	94.50
5530	1-1/2"		13	1.231		15.70	57.50		73.20	104
5540	2"		11	1.455		27	68		95	132
5550	2-1/2"		9	1.778		57.50	83.50		141	189
5560	3"		6	2.667		85	125		210	281
5570	3-1/2"		5	3.200		191	150		341	435
5580	4"		4	4		205	187		392	505
5601	Tee, reducing, on outlet									
5602	1/2"	1 Plum	9	.889	Ea.	7.85	46.50		54.35	78
5603	3/4"		9	.889		6.75	46.50		53.25	77
5604	1"		8	1		12.50	52		64.50	92
5605	1-1/4"	Q-1	14	1.143		21.50	53.50		75	104
5606	1-1/2"		13	1.231		22	57.50		79.50	111
5607	2"		11	1.455		30.50	68		98.50	136
5608	2-1/2"		9	1.778		87.50	83.50		171	222
5609	3"		6	2.667		119	125		244	320
5610	3-1/2"		5	3.200		265	150		415	515
5611	4"		4	4		265	187		452	575
5650	Coupling									
5670	1/4"	1 Plum	19	.421	Ea.	4.26	22		26.26	37.50
5680	3/8"		19	.421		4.26	22		26.26	37.50
5690	1/2"		19	.421		3.28	22		25.28	36.50
5700	3/4"		18	.444		3.84	23		26.84	38.50
5710	1"		15	.533		5.75	28		33.75	48
5720	1-1/4"	Q-1	26	.615		6.40	29		35.40	50
5730	1-1/2"		24	.667		10.05	31		41.05	58
5740	2"		21	.762		14.90	35.50		50.40	70
5750	2-1/2"		18	.889		41	41.50		82.50	108
5760	3"		14	1.143		56	53.50		109.50	142
5770	3-1/2"		12	1.333		113	62.50		175.50	218
5780	4"		10	1.600		112	75		187	235
5840	Reducer, concentric, 1/4"	1 Plum	19	.421		4.48	22		26.48	38
5850	3/8"		19	.421		5.75	22		27.75	39.50
5860	1/2"		19	.421		4.54	22		26.54	38
5870	3/4"		16	.500		5.15	26		31.15	44.50
5880	1"		15	.533		9.05	28		37.05	51.50
5890	1-1/4"	Q-1	26	.615		10.15	29		39.15	54
5900	1-1/2"		24	.667		14.50	31		45.50	63
5910	2"		21	.762		21	35.50		56.50	76.50
5911	2-1/2"		18	.889		46.50	41.50		88	114
5912	3"		14	1.143		59	53.50		112.50	146
5913	3-1/2"		12	1.333		175	62.50		237.50	286
5914	4"		10	1.600		133	75		208	258
5981	Bushing, 1/4"	1 Plum	19	.421		3.52	22		25.52	37
5982	3/8"		19	.421		4.17	22		26.17	37.50
5983	1/2"		19	.421		6.50	22		28.50	40
5984	3/4"		16	.500		3.85	26		29.85	43
5985	1"		15	.533		6.25	28		34.25	48.50

22 11 13 – Facility Water Distribution Piping

22 11 13.45 Pipe Fittings, Steel, Threaded		Crew	Daily Output	Labor-Hours	Unit	Material	Labor	Equipment	Total	Total Incl O&P
						2010 Bare Costs				
5986	1-1/4"	Q-1	26	.615	Ea.	7.75	29		36.75	51.50
5987	1-1/2"		24	.667		8.15	31		39.15	56
5988	2"		21	.762		8.45	35.50		43.95	63
5989	Cap, 1/4"	1 Plum	38	.211		4.07	10.95		15.02	21
5991	3/8"		38	.211		2.71	10.95		13.66	19.40
5992	1/2"		38	.211		2.40	10.95		13.35	19.05
5993	3/4"		32	.250		3.25	13		16.25	23
5994	1"		30	.267		3.93	13.90		17.83	25.50
5995	1-1/4"	Q-1	52	.308		5.20	14.40		19.60	27
5996	1-1/2"		48	.333		7.10	15.60		22.70	31.50
5997	2"		42	.381		10.40	17.85		28.25	38.50
6000	For galvanized elbows, tees, and couplings add					20%				
6058	For galvanized reducers, caps and bushings add					20%				
6100	90° Elbow, galvanized, 150 lb., reducing									
6110	3/4" x 1/2"	1 Plum	15.40	.519	Ea.	6.80	27		33.80	48
6112	1" x 3/4"		14	.571		8.75	29.50		38.25	54
6114	1" x 1/2"		14.50	.552		9.30	28.50		37.80	53
6116	1-1/4" x 1"	Q-1	24.20	.661		14.65	31		45.65	62.50
6118	1-1/4" x 3/4"		25.40	.630		17.65	29.50		47.15	63.50
6120	1-1/4" x 1/2"		26.20	.611		18.70	28.50		47.20	63.50
6122	1-1/2" x 1-1/4"		21.60	.741		23.50	34.50		58	78
6124	1-1/2" x 1"		23.50	.681		23.50	32		55.50	74
6126	1-1/2" x 3/4"		24.60	.650		23.50	30.50		54	71.50
6128	2" x 1-1/2"		20.50	.780		27.50	36.50		64	85
6130	2" x 1-1/4"		21	.762		31.50	35.50		67	88
6132	2" x 1"		22.80	.702		32.50	33		65.50	85.50
6134	2" x 3/4"		23.90	.669		33.50	31.50		65	83.50
6136	2-1/2" x 2"		12.30	1.301		93	61		154	194
6138	2-1/2" x 1-1/2"		12.50	1.280		104	60		164	204
6140	3" x 2-1/2"		8.60	1.860		164	87		251	310
6142	3" x 2"		11.80	1.356		147	63.50		210.50	256
6144	4" x 3"		8.20	1.951		370	91.50		461.50	545
6160	90° Elbow, black, 150 lb., reducing									
6170	1" x 3/4"	1 Plum	14	.571	Ea.	6	29.50		35.50	51
6174	1-1/2" x 1"	Q-1	23.50	.681		14.15	32		46.15	63.50
6178	1-1/2" x 3/4"		24.60	.650		16.20	30.50		46.70	63.50
6182	2" x 1-1/2"		20.50	.780		20.50	36.50		57	77.50
6186	2" x 1"		22.80	.702		23.50	33		56.50	75
6190	2" x 3/4"		23.90	.669		24.50	31.50		56	74
6194	2-1/2" x 2"		12.30	1.301		56.50	61		117.50	154
7000	Union, with brass seat									
7010	1/4"	1 Plum	15	.533	Ea.	16.80	28		44.80	60
7020	3/8"		15	.533		11.60	28		39.60	54.50
7030	1/2"		14	.571		10.50	29.50		40	56
7040	3/4"		13	.615		12.05	32		44.05	61.50
7050	1"		12	.667		15.70	34.50		50.20	69.50
7060	1-1/4"	Q-1	21	.762		22.50	35.50		58	78.50
7070	1-1/2"		19	.842		28	39.50		67.50	90
7080	2"		17	.941		32.50	44		76.50	102
7090	2-1/2"		13	1.231		97.50	57.50		155	194
7100	3"		9	1.778		117	83.50		200.50	254
7120	Union, galvanized									
7124	1/2"	1 Plum	14	.571	Ea.	13.95	29.50		43.45	60

22 11 13 – Facility Water Distribution Piping

22 11 13.45 Pipe Fittings, Steel, Threaded		Crew	Daily Output	Labor-Hours	Unit	Material	2010 Bare Costs Labor	Equipment	Total	Total Incl O&P
7125	3/4"	1 Plum	13	.615	Ea.	16	32		48	65.50
7126	1"		12	.667		21	34.50		55.50	75
7127	1-1/4"	Q-1	21	.762		30.50	35.50		66	87
7128	1-1/2"		19	.842		36.50	39.50		76	99.50
7129	2"		17	.941		42	44		86	113
7130	2-1/2"		13	1.231		147	57.50		204.50	248
7131	3"		9	1.778		205	83.50		288.50	350
7500	Malleable iron, 300 lb									
7520	Black									
7540	90° Elbow, straight, 1/4"	1 Plum	16	.500	Ea.	12.05	26		38.05	52.50
7560	3/8"		16	.500		10.65	26		36.65	51
7570	1/2"		15	.533		13.85	28		41.85	56.50
7580	3/4"		14	.571		15.65	29.50		45.15	61.50
7590	1"		13	.615		20	32		52	70
7600	1-1/4"	Q-1	22	.727		29	34		63	83
7610	1-1/2"		20	.800		34.50	37.50		72	94
7620	2"		18	.889		49.50	41.50		91	117
7630	2-1/2"		14	1.143		128	53.50		181.50	222
7640	3"		10	1.600		147	75		222	274
7650	4"		6	2.667		335	125		460	555
7700	45° Elbow, straight, 1/4"	1 Plum	16	.500		17.90	26		43.90	58.50
7720	3/8"		16	.500		17.90	26		43.90	58.50
7730	1/2"		15	.533		19.90	28		47.90	63.50
7740	3/4"		14	.571		22	29.50		51.50	69
7750	1"		13	.615		24.50	32		56.50	75
7760	1-1/4"	Q-1	22	.727		39.50	34		73.50	94
7770	1-1/2"		20	.800		51	37.50		88.50	113
7780	2"		18	.889		77.50	41.50		119	148
7790	2-1/2"		14	1.143		168	53.50		221.50	266
7800	3"		10	1.600		221	75		296	355
7810	4"		6	2.667		500	125		625	735
7850	Tee, straight, 1/4"	1 Plum	10	.800		15.15	41.50		56.65	79
7870	3/8"		10	.800		16.05	41.50		57.55	80
7880	1/2"		9	.889		20.50	46.50		67	92
7890	3/4"		9	.889		22	46.50		68.50	93.50
7900	1"		8	1		26.50	52		78.50	107
7910	1-1/4"	Q-1	14	1.143		38	53.50		91.50	123
7920	1-1/2"		13	1.231		44.50	57.50		102	135
7930	2"		11	1.455		65.50	68		133.50	174
7940	2-1/2"		9	1.778		155	83.50		238.50	295
7950	3"		6	2.667		227	125		352	435
7960	4"		4	4		630	187		817	970
8050	Couplings, straight, 1/4"	1 Plum	19	.421		10.85	22		32.85	45
8070	3/8"		19	.421		10.85	22		32.85	45
8080	1/2"		19	.421		12.10	22		34.10	46.50
8090	3/4"		18	.444		13.90	23		36.90	50
8100	1"		15	.533		15.90	28		43.90	59
8110	1-1/4"	Q-1	26	.615		18.90	29		47.90	64
8120	1-1/2"		24	.667		28.50	31		59.50	78
8130	2"		21	.762		41	35.50		76.50	98.50
8140	2-1/2"		18	.889		73	41.50		114.50	143
8150	3"		14	1.143		105	53.50		158.50	197
8160	4"		10	1.600		132	75		207	257

22 11 13.45 Pipe Fittings, Steel, Threaded		Crew	Daily Output	Labor-Hours	Unit	2010 Bare Costs Material	Labor	Equipment	Total	Total Incl O&P
8200	Galvanized									
8220	90° Elbow, straight, 1/4"	1 Plum	16	.500	Ea.	21.50	26		47.50	62.50
8222	3/8"		16	.500		22	26		48	63
8224	1/2"		15	.533		26	28		54	70.50
8226	3/4"		14	.571		29.50	29.50		59	77
8228	1"		13	.615		37.50	32		69.50	89.50
8230	1-1/4"	Q-1	22	.727		59	34		93	116
8232	1-1/2"		20	.800		64	37.50		101.50	127
8234	2"		18	.889		108	41.50		149.50	181
8236	2-1/2"		14	1.143		231	53.50		284.50	335
8238	3"		10	1.600		283	75		358	420
8240	4"		6	2.667		825	125		950	1,100
8280	45° Elbow, straight									
8282	1/2"	1 Plum	15	.533	Ea.	39	28		67	84
8284	3/4"		14	.571		45	29.50		74.50	94
8286	1"		13	.615		49	32		81	102
8288	1-1/4"	Q-1	22	.727		76	34		110	135
8290	1-1/2"		20	.800		98.50	37.50		136	164
8292	2"		18	.889		136	41.50		177.50	213
8310	Tee, straight, 1/4"	1 Plum	10	.800		30.50	41.50		72	96
8312	3/8"		10	.800		31	41.50		72.50	97
8314	1/2"		9	.889		39	46.50		85.50	113
8316	3/4"		9	.889		43	46.50		89.50	117
8318	1"		8	1		53	52		105	137
8320	1-1/4"	Q-1	14	1.143		75.50	53.50		129	164
8322	1-1/2"		13	1.231		78.50	57.50		136	173
8324	2"		11	1.455		126	68		194	240
8326	2-1/2"		9	1.778		385	83.50		468.50	550
8328	3"		6	2.667		450	125		575	680
8330	4"		4	4		1,125	187		1,312	1,525
8380	Couplings, straight, 1/4"	1 Plum	19	.421		14.10	22		36.10	48.50
8382	3/8"		19	.421		20.50	22		42.50	55.50
8384	1/2"		19	.421		21	22		43	56
8386	3/4"		18	.444		23.50	23		46.50	60.50
8388	1"		15	.533		31	28		59	76
8390	1-1/4"	Q-1	26	.615		40.50	29		69.50	87.50
8392	1-1/2"		24	.667		57.50	31		88.50	110
8394	2"		21	.762		69.50	35.50		105	130
8396	2-1/2"		18	.889		177	41.50		218.50	258
8398	3"		14	1.143		211	53.50		264.50	315
8399	4"		10	1.600		220	75		295	355
8529	Black									
8530	Reducer, concentric, 1/4"	1 Plum	19	.421	Ea.	14.30	22		36.30	49
8531	3/8"		19	.421		14.35	22		36.35	49
8532	1/2"		17	.471		17.25	24.50		41.75	55.50
8533	3/4"		16	.500		24	26		50	65.50
8534	1"		15	.533		28.50	28		56.50	73
8535	1-1/4"	Q-1	26	.615		43	29		72	90
8536	1-1/2"		24	.667		44.50	31		75.50	96
8537	2"		21	.762		64	35.50		99.50	124
8550	Cap, 1/4"	1 Plum	38	.211		11.10	10.95		22.05	28.50
8551	3/8"		38	.211		11.10	10.95		22.05	28.50
8552	1/2"		34	.235		11	12.25		23.25	30.50

22 11 13 – Facility Water Distribution Piping

22 11 13.45 Pipe Fittings, Steel, Threaded		Crew	Daily Output	Labor-Hours	Unit	Material	Labor	Equipment	Total	Total Incl O&P
						2010 Bare Costs				
8553	3/4"	1 Plum	32	.250	Ea.	13.90	13		26.90	35
8554	1"		30	.267		18.25	13.90		32.15	41
8555	1-1/4"	Q-1	52	.308		20	14.40		34.40	44
8556	1-1/2"		48	.333		29.50	15.60		45.10	56
8557	2"		42	.381		41	17.85		58.85	72
8570	Plug, 1/4"	1 Plum	38	.211		2.51	10.95		13.46	19.15
8571	3/8"		38	.211		2.51	10.95		13.46	19.15
8572	1/2"		34	.235		3.22	12.25		15.47	22
8573	3/4"		32	.250		4.01	13		17.01	24
8574	1"		30	.267		6.15	13.90		20.05	28
8575	1-1/4"	Q-1	52	.308		12.65	14.40		27.05	35.50
8576	1-1/2"		48	.333		14.45	15.60		30.05	39.50
8577	2"		42	.381		23	17.85		40.85	52
9500	Union with brass seat, 1/4"	1 Plum	15	.533		27	28		55	71
9530	3/8"		15	.533		24.50	28		52.50	68.50
9540	1/2"		14	.571		25	29.50		54.50	72
9550	3/4"		13	.615		22.50	32		54.50	72.50
9560	1"		12	.667		29	34.50		63.50	84
9570	1-1/4"	Q-1	21	.762		47.50	35.50		83	106
9580	1-1/2"		19	.842		49.50	39.50		89	114
9590	2"		17	.941		61.50	44		105.50	134
9600	2-1/2"		13	1.231		198	57.50		255.50	305
9610	3"		9	1.778		256	83.50		339.50	405
9620	4"		5	3.200		815	150		965	1,125
9630	Union, all iron, 1/4"	1 Plum	15	.533		38	28		66	83.50
9650	3/8"		15	.533		38	28		66	83.50
9660	1/2"		14	.571		40	29.50		69.50	88.50
9670	3/4"		13	.615		52.50	32		84.50	106
9680	1"		12	.667		53.50	34.50		88	111
9690	1-1/4"	Q-1	21	.762		81	35.50		116.50	143
9700	1-1/2"		19	.842		99	39.50		138.50	168
9710	2"		17	.941		125	44		169	204
9720	2-1/2"		13	1.231		222	57.50		279.50	330
9730	3"		9	1.778		420	83.50		503.50	590
9750	For galvanized unions, add					15%				
9757	Forged steel, 3000 lb.									
9758	Black									
9760	90° Elbow, 1/4"	1 Plum	16	.500	Ea.	17.65	26		43.65	58.50
9761	3/8"		16	.500		17.65	26		43.65	58.50
9762	1/2"		15	.533		13.55	28		41.55	56.50
9763	3/4"		14	.571		17	29.50		46.50	63
9764	1"		13	.615		25	32		57	76
9765	1-1/4"	Q-1	22	.727		48	34		82	104
9766	1-1/2"		20	.800		62	37.50		99.50	124
9767	2"		18	.889		75	41.50		116.50	145
9780	45° Elbow 1/4"	1 Plum	16	.500		22.50	26		48.50	63.50
9781	3/8"		16	.500		22.50	26		48.50	63.50
9782	1/2"		15	.533		22	28		50	66
9783	3/4"		14	.571		25.50	29.50		55	72.50
9784	1"		13	.615		35	32		67	86.50
9785	1-1/4"	Q-1	22	.727		48	34		82	104
9786	1-1/2"		20	.800		69	37.50		106.50	132
9787	2"		18	.889		95	41.50		136.50	168

22 11 13.45 Pipe Fittings, Steel, Threaded		Crew	Daily Output	Labor-Hours	Unit	2010 Bare Costs Material	Labor	Equipment	Total	Total Incl O&P
9800	Tee, 1/4"	1 Plum	10	.800	Ea.	21.50	41.50		63	86
9801	3/8"		10	.800		21.50	41.50		63	86
9802	1/2"		9	.889		19.30	46.50		65.80	90.50
9803	3/4"		9	.889		26	46.50		72.50	98
9804	1"		8	1		34.50	52		86.50	116
9805	1-1/4"	Q-1	14	1.143		66.50	53.50		120	154
9806	1-1/2"		13	1.231		78	57.50		135.50	172
9807	2"		11	1.455		101	68		169	213
9820	Reducer, concentric, 1/4"	1 Plum	19	.421		10.55	22		32.55	44.50
9821	3/8"		19	.421		11	22		33	45
9822	1/2"		17	.471		11	24.50		35.50	48.50
9823	3/4"		16	.500		13.05	26		39.05	53.50
9824	1"		15	.533		17	28		45	60
9825	1-1/4"	Q-1	26	.615		30	29		59	76
9826	1-1/2"		24	.667		32.50	31		63.50	82.50
9827	2"		21	.762		47	35.50		82.50	105
9840	Cap, 1/4"	1 Plum	38	.211		6.50	10.95		17.45	23.50
9841	3/8"		38	.211		6.50	10.95		17.45	23.50
9842	1/2"		34	.235		6.30	12.25		18.55	25.50
9843	3/4"		32	.250		8.80	13		21.80	29
9844	1"		30	.267		13.50	13.90		27.40	36
9845	1-1/4"	Q-1	52	.308		21.50	14.40		35.90	45.50
9846	1-1/2"		48	.333		26	15.60		41.60	52
9847	2"		42	.381		37.50	17.85		55.35	68
9860	Plug, 1/4"	1 Plum	38	.211		2.93	10.95		13.88	19.60
9861	3/8"		38	.211		3.07	10.95		14.02	19.80
9862	1/2"		34	.235		3.22	12.25		15.47	22
9863	3/4"		32	.250		4.01	13		17.01	24
9864	1"		30	.267		6.15	13.90		20.05	28
9865	1-1/4"	Q-1	52	.308		12.65	14.40		27.05	35.50
9866	1-1/2"		48	.333		14.45	15.60		30.05	39.50
9867	2"		42	.381		22.50	17.85		40.35	52
9880	Union, bronze seat, 1/4"	1 Plum	15	.533		48	28		76	94
9881	3/8"		15	.533		48	28		76	94
9882	1/2"		14	.571		46	29.50		75.50	95
9883	3/4"		13	.615		61.50	32		93.50	116
9884	1"		12	.667		69.50	34.50		104	129
9885	1-1/4"	Q-1	21	.762		148	35.50		183.50	217
9886	1-1/2"		19	.842		139	39.50		178.50	212
9887	2"		17	.941		164	44		208	247
9900	Coupling, 1/4"	1 Plum	19	.421		6.40	22		28.40	40
9901	3/8"		19	.421		6.40	22		28.40	40
9902	1/2"		17	.471		5.55	24.50		30.05	42.50
9903	3/4"		16	.500		7.20	26		33.20	47
9904	1"		15	.533		12.75	28		40.75	55.50
9905	1-1/4"	Q-1	26	.615		21	29		50	66
9906	1-1/2"		24	.667		27	31		58	76.50
9907	2"		21	.762		33.50	35.50		69	90.50

22 11 13.47 Pipe Fittings, Steel	
0010	**PIPE FITTINGS, STEEL**, Flanged, Welded & Special
0020	Flanged joints, C.I., standard weight, black. One gasket & bolt
0040	set, mat'l only, required at each joint, not included (see line 0620)

22 11 13 – Facility Water Distribution Piping

22 11 13.47 Pipe Fittings, Steel		Crew	Daily Output	Labor-Hours	Unit	Material	2010 Bare Costs Labor	Equipment	Total	Total Incl O&P
0060	90° Elbow, straight, 1-1/2" pipe size	Q-1	14	1.143	Ea.	380	53.50		433.50	500
0080	2" pipe size		13	1.231		232	57.50		289.50	340
0090	2-1/2" pipe size		12	1.333		251	62.50		313.50	370
0100	3" pipe size		11	1.455		209	68		277	330
0110	4" pipe size		8	2		259	93.50		352.50	425
0120	5" pipe size		7	2.286		615	107		722	835
0130	6" pipe size	Q-2	9	2.667		405	130		535	640
0140	8" pipe size		8	3		700	146		846	990
0150	10" pipe size		7	3.429		1,525	167		1,692	1,950
0160	12" pipe size		6	4		3,100	194		3,294	3,700
0171	90° Elbow, reducing									
0172	2-1/2" by 2" pipe size	Q-1	12	1.333	Ea.	740	62.50		802.50	905
0173	3" by 2-1/2" pipe size		11	1.455		775	68		843	955
0174	4" by 3" pipe size		8	2		575	93.50		668.50	770
0175	5" by 3" pipe size		7	2.286		1,200	107		1,307	1,450
0176	6" by 4" pipe size	Q-2	9	2.667		715	130		845	980
0177	8" by 6" pipe size		8	3		1,050	146		1,196	1,375
0178	10" by 8" pipe size		7	3.429		1,975	167		2,142	2,425
0179	12" by 10" pipe size		6	4		3,800	194		3,994	4,475
0200	45° Elbow, straight, 1-1/2" pipe size	Q-1	14	1.143		465	53.50		518.50	590
0220	2" pipe size		13	1.231		335	57.50		392.50	455
0230	2-1/2" pipe size		12	1.333		360	62.50		422.50	490
0240	3" pipe size		11	1.455		350	68		418	485
0250	4" pipe size		8	2		390	93.50		483.50	570
0260	5" pipe size		7	2.286		940	107		1,047	1,175
0270	6" pipe size	Q-2	9	2.667		640	130		770	900
0280	8" pipe size		8	3		940	146		1,086	1,250
0290	10" pipe size		7	3.429		1,975	167		2,142	2,425
0300	12" pipe size		6	4		3,025	194		3,219	3,625
0310	Cross, straight									
0311	2-1/2" pipe size	Q-1	6	2.667	Ea.	735	125		860	995
0312	3" pipe size		5	3.200		775	150		925	1,075
0313	4" pipe size		4	4		1,000	187		1,187	1,375
0314	5" pipe size		3	5.333		2,125	250		2,375	2,700
0315	6" pipe size	Q-2	5	4.800		2,125	233		2,358	2,675
0316	8" pipe size		4	6		3,400	291		3,691	4,175
0317	10" pipe size		3	8		4,050	390		4,440	5,025
0318	12" pipe size		2	12		6,175	585		6,760	7,675
0350	Tee, straight, 1-1/2" pipe size	Q-1	10	1.600		435	75		510	590
0370	2" pipe size		9	1.778		254	83.50		337.50	405
0380	2-1/2" pipe size		8	2		370	93.50		463.50	545
0390	3" pipe size		7	2.286		259	107		366	445
0400	4" pipe size		5	3.200		395	150		545	660
0410	5" pipe size		4	4		1,050	187		1,237	1,425
0420	6" pipe size	Q-2	6	4		570	194		764	920
0430	8" pipe size		5	4.800		980	233		1,213	1,425
0440	10" pipe size		4	6		2,625	291		2,916	3,325
0450	12" pipe size		3	8		4,125	390		4,515	5,125
0459	Tee, reducing on outlet									
0460	2-1/2" by 2" pipe size	Q-1	8	2	Ea.	705	93.50		798.50	915
0461	3" by 2-1/2" pipe size		7	2.286		770	107		877	1,000
0462	4" by 3" pipe size		5	3.200		810	150		960	1,125
0463	5" by 4" pipe size		4	4		1,700	187		1,887	2,150

22 11 13.47 Pipe Fittings, Steel		Crew	Daily Output	Labor-Hours	Unit	Material	2010 Bare Costs Labor	Equipment	Total	Total Incl O&P
0464	6" by 4" pipe size	Q-2	6	4	Ea.	780	194		974	1,150
0465	8" by 6" pipe size		5	4.800		1,225	233		1,458	1,700
0466	10" by 8" pipe size		4	6		2,800	291		3,091	3,500
0467	12" by 10" pipe size		3	8		4,825	390		5,215	5,900
0476	Reducer, concentric									
0477	3" by 2-1/2"	Q-1	12	1.333	Ea.	465	62.50		527.50	605
0478	4" by 3"		9	1.778		520	83.50		603.50	700
0479	5" by 4"		8	2		810	93.50		903.50	1,025
0480	6" by 4"	Q-2	10	2.400		700	117		817	945
0481	8" by 6"		9	2.667		880	130		1,010	1,175
0482	10" by 8"		8	3		1,825	146		1,971	2,225
0483	12" by 10"		7	3.429		3,125	167		3,292	3,700
0492	Reducer, eccentric									
0493	4" by 3"	Q-1	8	2	Ea.	850	93.50		943.50	1,075
0494	5" by 4"	"	7	2.286		1,250	107		1,357	1,525
0495	6" by 4"	Q-2	9	2.667		800	130		930	1,075
0496	8" by 6"		8	3		1,025	146		1,171	1,350
0497	10" by 8"		7	3.429		2,575	167		2,742	3,100
0498	12" by 10"		6	4		3,425	194		3,619	4,075
0500	For galvanized elbows and tees, add					100%				
0520	For extra heavy weight elbows and tees, add					140%				
0620	Gasket and bolt set, 150#, 1/2" pipe size	1 Plum	20	.400		2.16	21		23.16	33.50
0622	3/4" pipe size		19	.421		2.82	22		24.82	36
0624	1" pipe size		18	.444		2.31	23		25.31	37
0626	1-1/4" pipe size		17	.471		2.43	24.50		26.93	39
0628	1-1/2" pipe size		15	.533		2.58	28		30.58	44.50
0630	2" pipe size		13	.615		4.69	32		36.69	53
0640	2-1/2" pipe size		12	.667		4.84	34.50		39.34	57.50
0650	3" pipe size		11	.727		5	38		43	62
0660	3-1/2" pipe size		9	.889		8.15	46.50		54.65	78.50
0670	4" pipe size		8	1		9	52		61	88
0680	5" pipe size		7	1.143		14.80	59.50		74.30	105
0690	6" pipe size		6	1.333		14.80	69.50		84.30	120
0700	8" pipe size		5	1.600		16.95	83.50		100.45	144
0710	10" pipe size		4.50	1.778		30.50	92.50		123	173
0720	12" pipe size		4.20	1.905		32.50	99		131.50	185
0730	14" pipe size		4	2		34.50	104		138.50	194
0740	16" pipe size		3	2.667		36	139		175	248
0750	18" pipe size		2.70	2.963		70	154		224	310
0760	20" pipe size		2.30	3.478		111	181		292	395
0780	24" pipe size		1.90	4.211		142	219		361	485
0790	26" pipe size		1.60	5		192	260		452	600
0810	30" pipe size		1.40	5.714		375	297		672	855
0830	36" pipe size		1.10	7.273		700	380		1,080	1,325
0850	For 300 lb gasket set, add					40%				
2000	Flanged unions, 125 lb., black, 1/2" pipe size	1 Plum	17	.471	Ea.	56	24.50		80.50	98
2040	3/4" pipe size		17	.471		74	24.50		98.50	118
2050	1" pipe size		16	.500		71.50	26		97.50	118
2060	1-1/4" pipe size	Q-1	28	.571		85	27		112	134
2070	1-1/2" pipe size		27	.593		78.50	28		106.50	128
2080	2" pipe size		26	.615		92	29		121	144
2090	2-1/2" pipe size		24	.667		125	31		156	184
2100	3" pipe size		22	.727		141	34		175	206

22 11 13.47 Pipe Fittings, Steel		Crew	Daily Output	Labor-Hours	Unit	Material	2010 Bare Costs Labor	Equipment	Total	Total Incl O&P
2110	3-1/2" pipe size	Q-1	18	.889	Ea.	239	41.50		280.50	325
2120	4" pipe size		16	1		192	47		239	281
2130	5" pipe size		14	1.143		445	53.50		498.50	570
2140	6" pipe size	Q-2	19	1.263		420	61.50		481.50	550
2150	8" pipe size	"	16	1.500		935	73		1,008	1,125
2200	For galvanized unions, add					150%				
2290	Threaded flange									
2300	Cast iron									
2310	Black, 125 lb., per flange									
2320	1" pipe size	1 Plum	27	.296	Ea.	32	15.40		47.40	58.50
2330	1-1/4" pipe size	Q-1	44	.364		38.50	17.05		55.55	67.50
2340	1-1/2" pipe size		40	.400		35.50	18.75		54.25	67
2350	2" pipe size		36	.444		35.50	21		56.50	70.50
2360	2-1/2" pipe size		28	.571		41.50	27		68.50	85.50
2370	3" pipe size		20	.800		53.50	37.50		91	115
2380	3-1/2" pipe size		16	1		75	47		122	153
2390	4" pipe size		12	1.333		72	62.50		134.50	173
2400	5" pipe size		10	1.600		102	75		177	224
2410	6" pipe size	Q-2	14	1.714		115	83.50		198.50	251
2420	8" pipe size		12	2		181	97		278	345
2430	10" pipe size		10	2.400		320	117		437	530
2440	12" pipe size		8	3		690	146		836	975
2460	For galvanized flanges, add					95%				
2490	Blind flange									
2492	Cast iron									
2494	Black, 125 lb., per flange									
2496	1" pipe size	1 Plum	27	.296	Ea.	51	15.40		66.40	79
2500	1-1/2" pipe size	Q-1	40	.400		57.50	18.75		76.25	91
2502	2" pipe size		36	.444		64.50	21		85.50	102
2504	2-1/2" pipe size		28	.571		71	27		98	119
2506	3" pipe size		20	.800		85.50	37.50		123	150
2508	4" pipe size		12	1.333		109	62.50		171.50	214
2510	5" pipe size		10	1.600		177	75		252	305
2512	6" pipe size	Q-2	14	1.714		193	83.50		276.50	335
2514	8" pipe size		12	2		305	97		402	480
2516	10" pipe size		10	2.400		450	117		567	670
2518	12" pipe size		8	3		815	146		961	1,125
2520	For galvanized flanges, add					80%				
2570	Threaded flange									
2580	Forged steel,									
2590	Black 150 lb., per flange									
2600	1/2" pipe size	1 Plum	30	.267	Ea.	26.50	13.90		40.40	50
2610	3/4" pipe size		28	.286		26.50	14.85		41.35	51.50
2620	1" pipe size		27	.296		26.50	15.40		41.90	52
2630	1-1/4" pipe size	Q-1	44	.364		26.50	17.05		43.55	54.50
2640	1-1/2" pipe size		40	.400		26.50	18.75		45.25	57
2650	2" pipe size		36	.444		29.50	21		50.50	63.50
2660	2-1/2" pipe size		28	.571		36	27		63	80
2670	3" pipe size		20	.800		36.50	37.50		74	96
2690	4" pipe size		12	1.333		42.50	62.50		105	140
2700	5" pipe size		10	1.600		66.50	75		141.50	186
2710	6" pipe size	Q-2	14	1.714		73	83.50		156.50	206
2720	8" pipe size		12	2		126	97		223	284

22 11 13 – Facility Water Distribution Piping

22 11 13.47 Pipe Fittings, Steel		Crew	Daily Output	Labor-Hours	Unit	Material	2010 Bare Costs Labor	Equipment	Total	Total Incl O&P
2730	10" pipe size	Q-2	10	2.400	Ea.	228	117		345	425
2860	Black 300 lb., per flange									
2870	1/2" pipe size	1 Plum	30	.267	Ea.	29	13.90		42.90	53
2880	3/4" pipe size		28	.286		29	14.85		43.85	54.50
2890	1" pipe size		27	.296		29	15.40		44.40	55
2900	1-1/4" pipe size	Q-1	44	.364		29	17.05		46.05	57.50
2910	1-1/2" pipe size		40	.400		29	18.75		47.75	60
2920	2" pipe size		36	.444		34	21		55	68.50
2930	2-1/2" pipe size		28	.571		47.50	27		74.50	92
2940	3" pipe size		20	.800		50.50	37.50		88	112
2960	4" pipe size		12	1.333		73	62.50		135.50	174
2970	6" pipe size	Q-2	14	1.714		136	83.50		219.50	275
3000	Weld joint, butt, carbon steel, standard weight									
3040	90° elbow, long radius									
3050	1/2" pipe size	Q-15	16	1	Ea.	45	47	3.66	95.66	124
3060	3/4" pipe size		16	1		45	47	3.66	95.66	124
3070	1" pipe size		16	1		21	47	3.66	71.66	97
3080	1-1/4" pipe size		14	1.143		21	53.50	4.18	78.68	108
3090	1-1/2" pipe size		13	1.231		21	57.50	4.50	83	114
3100	2" pipe size		10	1.600		22.50	75	5.85	103.35	143
3110	2-1/2" pipe size		8	2		27.50	93.50	7.30	128.30	179
3120	3" pipe size		7	2.286		33.50	107	8.35	148.85	207
3130	4" pipe size		5	3.200		48	150	11.70	209.70	291
3136	5" pipe size		4	4		75	187	14.65	276.65	380
3140	6" pipe size	Q-16	5	4.800		106	233	11.70	350.70	480
3150	8" pipe size		3.75	6.400		200	310	15.60	525.60	700
3160	10" pipe size		3	8		400	390	19.50	809.50	1,050
3170	12" pipe size		2.50	9.600		680	465	23.50	1,168.50	1,475
3180	14" pipe size		2	12		970	585	29.50	1,584.50	1,975
3190	16" pipe size		1.50	16		1,325	775	39	2,139	2,700
3191	18" pipe size		1.25	19.200		1,775	935	47	2,757	3,400
3192	20" pipe size		1.15	20.870		2,450	1,025	51	3,526	4,275
3194	24" pipe size		1.02	23.529		3,450	1,150	57.50	4,657.50	5,600
3200	45° Elbow, long									
3210	1/2" pipe size	Q-15	16	1	Ea.	62.50	47	3.66	113.16	143
3220	3/4" pipe size		16	1		62.50	47	3.66	113.16	143
3230	1" pipe size		16	1		22	47	3.66	72.66	98.50
3240	1-1/4" pipe size		14	1.143		22	53.50	4.18	79.68	110
3250	1-1/2" pipe size		13	1.231		22	57.50	4.50	84	116
3260	2" pipe size		10	1.600		22	75	5.85	102.85	143
3270	2-1/2" pipe size		8	2		26.50	93.50	7.30	127.30	177
3280	3" pipe size		7	2.286		27.50	107	8.35	142.85	201
3290	4" pipe size		5	3.200		49.50	150	11.70	211.20	292
3296	5" pipe size		4	4		75	187	14.65	276.65	380
3300	6" pipe size	Q-16	5	4.800		97.50	233	11.70	342.20	470
3310	8" pipe size		3.75	6.400		163	310	15.60	488.60	660
3320	10" pipe size		3	8		325	390	19.50	734.50	960
3330	12" pipe size		2.50	9.600		460	465	23.50	948.50	1,225
3340	14" pipe size		2	12		615	585	29.50	1,229.50	1,575
3341	16" pipe size		1.50	16		1,100	775	39	1,914	2,450
3342	18" pipe size		1.25	19.200		1,550	935	47	2,532	3,175
3343	20" pipe size		1.15	20.870		1,625	1,025	51	2,701	3,350
3345	24" pipe size		1.05	22.857		2,300	1,100	56	3,456	4,250

22 11 13 – Facility Water Distribution Piping

22 11 13.47 Pipe Fittings, Steel		Crew	Daily Output	Labor-Hours	Unit	Material	Labor	Equipment	Total	Total Incl O&P
						2010 Bare Costs				
3346	26" pipe size	Q-16	.85	28.235	Ea.	2,700	1,375	69	4,144	5,100
3347	30" pipe size		.45	53.333		2,975	2,600	130	5,705	7,300
3349	36" pipe size		.38	63.158		3,275	3,075	154	6,504	8,400
3350	Tee, straight									
3352	For reducing tees and concentrics see starting line 4600									
3360	1/2" pipe size	Q-15	10	1.600	Ea.	110	75	5.85	190.85	239
3370	3/4" pipe size		10	1.600		110	75	5.85	190.85	239
3380	1" pipe size		10	1.600		54.50	75	5.85	135.35	178
3390	1-1/4" pipe size		9	1.778		68.50	83.50	6.50	158.50	208
3400	1-1/2" pipe size		8	2		68.50	93.50	7.30	169.30	224
3410	2" pipe size		6	2.667		54.50	125	9.75	189.25	258
3420	2-1/2" pipe size		5	3.200		75.50	150	11.70	237.20	320
3430	3" pipe size		4	4		84	187	14.65	285.65	390
3440	4" pipe size		3	5.333		118	250	19.50	387.50	525
3446	5" pipe size		2.50	6.400		195	300	23.50	518.50	690
3450	6" pipe size	Q-16	3	8		203	390	19.50	612.50	830
3460	8" pipe size		2.50	9.600		355	465	23.50	843.50	1,125
3470	10" pipe size		2	12		695	585	29.50	1,309.50	1,675
3480	12" pipe size		1.60	15		975	730	36.50	1,741.50	2,225
3481	14" pipe size		1.30	18.462		1,700	895	45	2,640	3,275
3482	16" pipe size		1	24		1,900	1,175	58.50	3,133.50	3,925
3483	18" pipe size		.80	30		3,025	1,450	73	4,548	5,575
3484	20" pipe size		.75	32		4,750	1,550	78	6,378	7,625
3486	24" pipe size		.70	34.286		6,150	1,675	83.50	7,908.50	9,350
3487	26" pipe size		.55	43.636		6,775	2,125	106	9,006	10,700
3488	30" pipe size		.30	80		7,425	3,875	195	11,495	14,200
3490	36" pipe size		.25	96		8,200	4,675	234	13,109	16,300
3491	Eccentric reducer, 1-1/2" pipe size	Q-15	14	1.143		30.50	53.50	4.18	88.18	119
3492	2" pipe size		11	1.455		43	68	5.30	116.30	155
3493	2-1/2" pipe size		9	1.778		50	83.50	6.50	140	187
3494	3" pipe size		8	2		59	93.50	7.30	159.80	213
3495	4" pipe size		6	2.667		75.50	125	9.75	210.25	281
3496	6" pipe size	Q-16	5	4.800		232	233	11.70	476.70	620
3497	8" pipe size		4	6		345	291	14.65	650.65	830
3498	10" pipe size		3	8		435	390	19.50	844.50	1,075
3499	12" pipe size		2.50	9.600		385	465	23.50	873.50	1,150
3501	Cap, 1-1/2" pipe size	Q-15	28	.571		25.50	27	2.09	54.59	70.50
3502	2" pipe size		22	.727		25.50	34	2.66	62.16	82
3503	2-1/2" pipe size		18	.889		26.50	41.50	3.25	71.25	95
3504	3" pipe size		16	1		26.50	47	3.66	77.16	103
3505	4" pipe size		12	1.333		38.50	62.50	4.88	105.88	141
3506	6" pipe size	Q-16	10	2.400		65.50	117	5.85	188.35	253
3507	8" pipe size		8	3		99	146	7.30	252.30	335
3508	10" pipe size		6	4		180	194	9.75	383.75	500
3509	12" pipe size		5	4.800		272	233	11.70	516.70	660
3511	14" pipe size		4	6		360	291	14.65	665.65	850
3512	16" pipe size		4	6		360	291	14.65	665.65	850
3513	18" pipe size		3	8		575	390	19.50	984.50	1,250
3517	Weld joint, butt, carbon steel, extra strong									
3519	90° elbow, long									
3520	1/2" pipe size	Q-15	13	1.231	Ea.	55.50	57.50	4.50	117.50	153
3530	3/4" pipe size		12	1.333		55.50	62.50	4.88	122.88	160
3540	1" pipe size		11	1.455		27	68	5.30	100.30	138

22 11 13.47 Pipe Fittings, Steel		Crew	Daily Output	Labor-Hours	Unit	Material	2010 Bare Costs Labor	Equipment	Total	Total Incl O&P
3550	1-1/4" pipe size	Q-15	10	1.600	Ea.	27	75	5.85	107.85	148
3560	1-1/2" pipe size		9	1.778		27	83.50	6.50	117	162
3570	2" pipe size		8	2		27.50	93.50	7.30	128.30	179
3580	2-1/2" pipe size		7	2.286		38.50	107	8.35	153.85	213
3590	3" pipe size		6	2.667		49.50	125	9.75	184.25	252
3600	4" pipe size		4	4		81.50	187	14.65	283.15	385
3606	5" pipe size		3.50	4.571		195	214	16.75	425.75	550
3610	6" pipe size	Q-16	4.50	5.333		206	259	13	478	630
3620	8" pipe size		3.50	6.857		395	335	16.75	746.75	950
3630	10" pipe size		2.50	9.600		830	465	23.50	1,318.50	1,625
3640	12" pipe size		2.25	10.667		1,025	520	26	1,571	1,925
3650	45° Elbow, long									
3660	1/2" pipe size	Q-15	13	1.231	Ea.	62	57.50	4.50	124	159
3670	3/4" pipe size		12	1.333		62	62.50	4.88	129.38	167
3680	1" pipe size		11	1.455		29	68	5.30	102.30	139
3690	1-1/4" pipe size		10	1.600		29	75	5.85	109.85	150
3700	1-1/2" pipe size		9	1.778		29	83.50	6.50	119	164
3710	2" pipe size		8	2		29	93.50	7.30	129.80	180
3720	2-1/2" pipe size		7	2.286		64	107	8.35	179.35	241
3730	3" pipe size		6	2.667		37	125	9.75	171.75	239
3740	4" pipe size		4	4		58.50	187	14.65	260.15	360
3746	5" pipe size		3.50	4.571		139	214	16.75	369.75	490
3750	6" pipe size	Q-16	4.50	5.333		160	259	13	432	580
3760	8" pipe size		3.50	6.857		276	335	16.75	627.75	825
3770	10" pipe size		2.50	9.600		540	465	23.50	1,028.50	1,325
3780	12" pipe size		2.25	10.667		790	520	26	1,336	1,675
3800	Tee, straight									
3810	1/2" pipe size	Q-15	9	1.778	Ea.	136	83.50	6.50	226	281
3820	3/4" pipe size		8.50	1.882		131	88	6.90	225.90	284
3830	1" pipe size		8	2		72.50	93.50	7.30	173.30	228
3840	1-1/4" pipe size		7	2.286		67	107	8.35	182.35	244
3850	1-1/2" pipe size		6	2.667		81.50	125	9.75	216.25	288
3860	2" pipe size		5	3.200		68.50	150	11.70	230.20	315
3870	2-1/2" pipe size		4	4		114	187	14.65	315.65	420
3880	3" pipe size		3.50	4.571		106	214	16.75	336.75	455
3890	4" pipe size		2.50	6.400		171	300	23.50	494.50	665
3896	5" pipe size		2.25	7.111		375	335	26	736	940
3900	6" pipe size	Q-16	2.25	10.667		288	520	26	834	1,125
3910	8" pipe size		2	12		545	585	29.50	1,159.50	1,500
3920	10" pipe size		1.75	13.714		965	665	33.50	1,663.50	2,100
3930	12" pipe size		1.50	16		1,400	775	39	2,214	2,775
4000	Eccentric reducer, 1-1/2" pipe size	Q-15	10	1.600		17.30	75	5.85	98.15	137
4010	2" pipe size		9	1.778		47.50	83.50	6.50	137.50	185
4020	2-1/2" pipe size		8	2		71.50	93.50	7.30	172.30	227
4030	3" pipe size		7	2.286		59	107	8.35	174.35	235
4040	4" pipe size		5	3.200		95.50	150	11.70	257.20	345
4046	5" pipe size		4.70	3.404		245	159	12.45	416.45	520
4050	6" pipe size	Q-16	4.50	5.333		259	259	13	531	690
4060	8" pipe size		3.50	6.857		390	335	16.75	741.75	950
4070	10" pipe size		2.50	9.600		665	465	23.50	1,153.50	1,450
4080	12" pipe size		2.25	10.667		855	520	26	1,401	1,750
4090	14" pipe size		2.10	11.429		1,575	555	28	2,158	2,575
4100	16" pipe size		1.90	12.632		2,025	615	31	2,671	3,175

22 11 13.47 Pipe Fittings, Steel		Crew	Daily Output	Labor-Hours	Unit	Material	Labor	Equipment	Total	Total Incl O&P
						2010 Bare Costs				
4151	Cap, 1-1/2" pipe size	Q-15	24	.667	Ea.	31.50	31	2.44	64.94	84.50
4152	2" pipe size		18	.889		25.50	41.50	3.25	70.25	94
4153	2-1/2" pipe size		16	1		34.50	47	3.66	85.16	112
4154	3" pipe size		14	1.143		38.50	53.50	4.18	96.18	127
4155	4" pipe size		10	1.600		89	75	5.85	169.85	216
4156	6" pipe size	Q-16	9	2.667		107	130	6.50	243.50	320
4157	8" pipe size		7	3.429		160	167	8.35	335.35	435
4158	10" pipe size		5	4.800		245	233	11.70	489.70	630
4159	12" pipe size		4	6		330	291	14.65	635.65	810
4190	Weld fittings, reducing, standard weight									
4200	Welding ring w/spacer pins, 2" pipe size				Ea.	3.11			3.11	3.42
4210	2-1/2" pipe size					3.76			3.76	4.14
4220	3" pipe size					3.89			3.89	4.28
4230	4" pipe size					4.04			4.04	4.44
4236	5" pipe size					4.76			4.76	5.25
4240	6" pipe size					4.76			4.76	5.25
4250	8" pipe size					5.60			5.60	6.20
4260	10" pipe size					6.90			6.90	7.55
4270	12" pipe size					7.90			7.90	8.70
4280	14" pipe size					9.20			9.20	10.10
4290	16" pipe size					10.35			10.35	11.40
4300	18" pipe size					11.95			11.95	13.15
4310	20" pipe size					12.65			12.65	13.95
4330	24" pipe size					14.75			14.75	16.25
4340	26" pipe size					19.70			19.70	21.50
4350	30" pipe size					19.70			19.70	21.50
4370	36" pipe size					22			22	24
4600	Tee, reducing on outlet									
4601	2-1/2" x 2" pipe size	Q-15	5	3.200	Ea.	98	150	11.70	259.70	345
4602	3" x 2-1/2" pipe size		4	4		118	187	14.65	319.65	425
4604	4" x 3" pipe size		3	5.333		124	250	19.50	393.50	535
4605	5" x 4" pipe size		2.50	6.400		268	300	23.50	591.50	770
4606	6" x 5" pipe size	Q-16	3	8		370	390	19.50	779.50	1,025
4607	8" x 6" pipe size		2.50	9.600		460	465	23.50	948.50	1,225
4608	10" x 8" pipe size		2	12		850	585	29.50	1,464.50	1,850
4609	12" x 10" pipe size		1.60	15		1,250	730	36.50	2,016.50	2,525
4610	16" x 12" pipe size		1.50	16		2,000	775	39	2,814	3,425
4611	14" x 12" pipe size		1.52	15.789		1,900	765	38.50	2,703.50	3,275
4618	Reducer, concentric									
4619	2-1/2" by 2" pipe size	Q-15	10	1.600	Ea.	42.50	75	5.85	123.35	165
4620	3" by 2-1/2" pipe size		9	1.778		35	83.50	6.50	125	171
4621	3-1/2" by 3" pipe size		8	2		89.50	93.50	7.30	190.30	246
4622	4" by 2-1/2" pipe size		7	2.286		54.50	107	8.35	169.85	230
4623	5" by 3" pipe size		7	2.286		114	107	8.35	229.35	295
4624	6" by 4" pipe size	Q-16	6	4		94	194	9.75	297.75	405
4625	8" by 6" pipe size		5	4.800		124	233	11.70	368.70	500
4626	10" by 8" pipe size		4	6		256	291	14.65	561.65	730
4627	12" by 10" pipe size		3	8		296	390	19.50	705.50	930
4660	Reducer, eccentric									
4662	3" x 2" pipe size	Q-15	8	2	Ea.	59	93.50	7.30	159.80	213
4664	4" x 3" pipe size		6	2.667		71.50	125	9.75	206.25	277
4666	4" x 2" pipe size		6	2.667		87.50	125	9.75	222.25	294
4670	6" x 4" pipe size	Q-16	5	4.800		148	233	11.70	392.70	525

22 11 13.47	Pipe Fittings, Steel	Crew	Daily Output	Labor-Hours	Unit	Material	2010 Bare Costs Labor	Equipment	Total	Total Incl O&P
4672	6" x 3" pipe size	Q-16	5	4.800	Ea.	232	233	11.70	476.70	620
4676	8" x 6" pipe size		4	6		188	291	14.65	493.65	660
4678	8" x 4" pipe size		4	6		345	291	14.65	650.65	830
4682	10" x 8" pipe size		3	8		247	390	19.50	656.50	880
4684	10" x 6" pipe size		3	8		435	390	19.50	844.50	1,075
4688	12" x 10" pipe size		2.50	9.600		385	465	23.50	873.50	1,150
4690	12" x 8" pipe size		2.50	9.600		605	465	23.50	1,093.50	1,400
4691	14" x 12" pipe size		2.20	10.909		675	530	26.50	1,231.50	1,575
4693	16" x 14" pipe size		1.80	13.333		1,025	650	32.50	1,707.50	2,125
4694	16" x 12" pipe size		2	12		1,375	585	29.50	1,989.50	2,400
4696	18" x 16" pipe size		1.60	15		1,875	730	36.50	2,641.50	3,225
5000	Weld joint, socket, forged steel, 3000 lb., schedule 40 pipe									
5010	90° elbow, straight									
5020	1/4" pipe size	Q-15	22	.727	Ea.	21.50	34	2.66	58.16	78
5030	3/8" pipe size		22	.727		21.50	34	2.66	58.16	78
5040	1/2" pipe size		20	.800		12.05	37.50	2.93	52.48	72.50
5050	3/4" pipe size		20	.800		12.55	37.50	2.93	52.98	73
5060	1" pipe size		20	.800		16.10	37.50	2.93	56.53	77
5070	1-1/4" pipe size		18	.889		31	41.50	3.25	75.75	100
5080	1-1/2" pipe size		16	1		36	47	3.66	86.66	114
5090	2" pipe size		12	1.333		53	62.50	4.88	120.38	157
5100	2-1/2" pipe size		10	1.600		150	75	5.85	230.85	283
5110	3" pipe size		8	2		259	93.50	7.30	359.80	435
5120	4" pipe size		6	2.667		685	125	9.75	819.75	955
5130	45° Elbow, straight									
5134	1/4" pipe size	Q-15	22	.727	Ea.	21.50	34	2.66	58.16	78
5135	3/8" pipe size		22	.727		21.50	34	2.66	58.16	78
5136	1/2" pipe size		20	.800		16.05	37.50	2.93	56.48	77
5137	3/4" pipe size		20	.800		18.65	37.50	2.93	59.08	79.50
5140	1" pipe size		20	.800		24.50	37.50	2.93	64.93	86
5150	1-1/4" pipe size		18	.889		33.50	41.50	3.25	78.25	103
5160	1-1/2" pipe size		16	1		41	47	3.66	91.66	119
5170	2" pipe size		12	1.333		66	62.50	4.88	133.38	171
5180	2-1/2" pipe size		10	1.600		173	75	5.85	253.85	310
5190	3" pipe size		8	2		288	93.50	7.30	388.80	465
5200	4" pipe size		6	2.667		565	125	9.75	699.75	825
5250	Tee, straight									
5254	1/4" pipe size	Q-15	15	1.067	Ea.	24	50	3.90	77.90	105
5255	3/8" pipe size		15	1.067		24	50	3.90	77.90	105
5256	1/2" pipe size		13	1.231		14.90	57.50	4.50	76.90	108
5257	3/4" pipe size		13	1.231		18.25	57.50	4.50	80.25	111
5260	1" pipe size		13	1.231		25	57.50	4.50	87	119
5270	1-1/4" pipe size		12	1.333		38	62.50	4.88	105.38	141
5280	1-1/2" pipe size		11	1.455		50.50	68	5.30	123.80	163
5290	2" pipe size		8	2		73	93.50	7.30	173.80	228
5300	2-1/2" pipe size		6	2.667		209	125	9.75	343.75	430
5310	3" pipe size		5	3.200		500	150	11.70	661.70	790
5320	4" pipe size		4	4		805	187	14.65	1,006.65	1,175
5350	For reducing sizes, add					60%				
5450	Couplings									
5451	1/4" pipe size	Q-15	23	.696	Ea.	13.40	32.50	2.55	48.45	66.50
5452	3/8" pipe size		23	.696		13.40	32.50	2.55	48.45	66.50
5453	1/2" pipe size		21	.762		6.20	35.50	2.79	44.49	63.50

22 11 13.47 Pipe Fittings, Steel		Crew	Daily Output	Labor-Hours	Unit	Material	2010 Bare Costs Labor	Equipment	Total	Total Incl O&P
5454	3/4" pipe size	Q-15	21	.762	Ea.	8	35.50	2.79	46.29	65.50
5460	1" pipe size		20	.800		8.85	37.50	2.93	49.28	69
5470	1-1/4" pipe size		20	.800		15.65	37.50	2.93	56.08	76.50
5480	1-1/2" pipe size		18	.889		17.35	41.50	3.25	62.10	85
5490	2" pipe size		14	1.143		27.50	53.50	4.18	85.18	116
5500	2-1/2" pipe size		12	1.333		62	62.50	4.88	129.38	167
5510	3" pipe size		9	1.778		139	83.50	6.50	229	285
5520	4" pipe size		7	2.286		209	107	8.35	324.35	400
5570	Union, 1/4" pipe size		21	.762		29	35.50	2.79	67.29	88
5571	3/8" pipe size		21	.762		29	35.50	2.79	67.29	88
5572	1/2" pipe size		19	.842		24	39.50	3.08	66.58	88.50
5573	3/4" pipe size		19	.842		28	39.50	3.08	70.58	93.50
5574	1" pipe size		19	.842		36	39.50	3.08	78.58	102
5575	1-1/4" pipe size		17	.941		58.50	44	3.44	105.94	134
5576	1-1/2" pipe size		15	1.067		63.50	50	3.90	117.40	149
5577	2" pipe size		11	1.455		91	68	5.30	164.30	208
5600	Reducer, 1/4" pipe size		23	.696		31.50	32.50	2.55	66.55	86.50
5601	3/8" pipe size		23	.696		32.50	32.50	2.55	67.55	87.50
5602	1/2" pipe size		21	.762		22	35.50	2.79	60.29	80.50
5603	3/4" pipe size		21	.762		22	35.50	2.79	60.29	80.50
5604	1" pipe size		21	.762		25.50	35.50	2.79	63.79	84.50
5605	1-1/4" pipe size		19	.842		39	39.50	3.08	81.58	105
5607	1-1/2" pipe size		17	.941		42	44	3.44	89.44	116
5608	2" pipe size		13	1.231		46.50	57.50	4.50	108.50	142
5612	Cap, 1/4" pipe size		46	.348		12.85	16.30	1.27	30.42	40
5613	3/8" pipe size		46	.348		12.85	16.30	1.27	30.42	40
5614	1/2" pipe size		42	.381		7.80	17.85	1.39	27.04	37
5615	3/4" pipe size		42	.381		9.10	17.85	1.39	28.34	38.50
5616	1" pipe size		42	.381		13.80	17.85	1.39	33.04	43.50
5617	1-1/4" pipe size		38	.421		16.25	19.75	1.54	37.54	49
5618	1-1/2" pipe size		34	.471		22.50	22	1.72	46.22	60
5619	2" pipe size		26	.615		35	29	2.25	66.25	84
5630	T-O-L, 1/4" pipe size, nozzle		23	.696		6.70	32.50	2.55	41.75	59
5631	3/8" pipe size, nozzle		23	.696		6.80	32.50	2.55	41.85	59.50
5632	1/2" pipe size, nozzle		22	.727		6.70	34	2.66	43.36	61.50
5633	3/4" pipe size, nozzle		21	.762		7.70	35.50	2.79	45.99	65
5634	1" pipe size, nozzle		20	.800		9	37.50	2.93	49.43	69
5635	1-1/4" pipe size, nozzle		18	.889		13.85	41.50	3.25	58.60	81.50
5636	1-1/2" pipe size, nozzle		16	1		14.25	47	3.66	64.91	89.50
5637	2" pipe size, nozzle		12	1.333		14.25	62.50	4.88	81.63	115
5638	2-1/2" pipe size, nozzle		10	1.600		54.50	75	5.85	135.35	178
5639	4" pipe size, nozzle		6	2.667		123	125	9.75	257.75	335
5640	W-O-L, 1/4" pipe size, nozzle		23	.696		15.85	32.50	2.55	50.90	69
5641	3/8" pipe size, nozzle		23	.696		14.95	32.50	2.55	50	68.50
5642	1/2" pipe size, nozzle		22	.727		13.90	34	2.66	50.56	69
5643	3/4" pipe size, nozzle		21	.762		14.65	35.50	2.79	52.94	72.50
5644	1" pipe size, nozzle		20	.800		15.30	37.50	2.93	55.73	76
5645	1-1/4" pipe size, nozzle		18	.889		18.15	41.50	3.25	62.90	86
5646	1-1/2" pipe size, nozzle		16	1		18.15	47	3.66	68.81	94
5647	2" pipe size, nozzle		12	1.333		18.35	62.50	4.88	85.73	119
5648	2-1/2" pipe size, nozzle		10	1.600		42	75	5.85	122.85	164
5649	3" pipe size, nozzle		8	2		46	93.50	7.30	146.80	199
5650	4" pipe size, nozzle		6	2.667		58	125	9.75	192.75	262

22 11 13.47 Pipe Fittings, Steel		Crew	Daily Output	Labor-Hours	Unit	Material	Labor	Equipment	Total	Total Incl O&P
						2010 Bare Costs				
5651	5" pipe size, nozzle	Q-15	5	3.200	Ea.	143	150	11.70	304.70	395
5652	6" pipe size, nozzle		4	4		164	187	14.65	365.65	480
5653	8" pipe size, nozzle		3	5.333		315	250	19.50	584.50	740
5654	10" pipe size, nozzle		2.60	6.154		455	288	22.50	765.50	955
5655	12" pipe size, nozzle		2.20	7.273		865	340	26.50	1,231.50	1,500
5656	14" pipe size, nozzle		2	8		2,800	375	29.50	3,204.50	3,675
5674	S-O-L, 1/4" pipe size, outlet		23	.696		8.90	32.50	2.55	43.95	61.50
5675	3/8" pipe size, outlet		23	.696		8.90	32.50	2.55	43.95	61.50
5676	1/2" pipe size, outlet		22	.727		8.30	34	2.66	44.96	63
5677	3/4" pipe size, outlet		21	.762		8.45	35.50	2.79	46.74	66
5678	1" pipe size, outlet		20	.800		9.35	37.50	2.93	49.78	69.50
5679	1-1/4" pipe size, outlet		18	.889		15.60	41.50	3.25	60.35	83
5680	1-1/2" pipe size, outlet		16	1		15.60	47	3.66	66.26	91
5681	2" pipe size, outlet	↓	12	1.333	↓	17.75	62.50	4.88	85.13	118
6000	Weld-on flange, forged steel									
6020	Slip-on, 150 lb. flange (welded front and back)									
6050	1/2" pipe size	Q-15	18	.889	Ea.	17.50	41.50	3.25	62.25	85.50
6060	3/4" pipe size		18	.889		17.50	41.50	3.25	62.25	85.50
6070	1" pipe size		17	.941		17.50	44	3.44	64.94	89
6080	1-1/4" pipe size		16	1		17.50	47	3.66	68.16	93.50
6090	1-1/2" pipe size		15	1.067		17.50	50	3.90	71.40	98.50
6100	2" pipe size		12	1.333		19.45	62.50	4.88	86.83	120
6110	2-1/2" pipe size		10	1.600		33.50	75	5.85	114.35	155
6120	3" pipe size		9	1.778		26.50	83.50	6.50	116.50	161
6130	3-1/2" pipe size		7	2.286		33	107	8.35	148.35	207
6140	4" pipe size		6	2.667		33	125	9.75	167.75	234
6150	5" pipe size	↓	5	3.200		57.50	150	11.70	219.20	300
6160	6" pipe size	Q-16	6	4		54.50	194	9.75	258.25	360
6170	8" pipe size		5	4.800		82.50	233	11.70	327.20	455
6180	10" pipe size		4	6		143	291	14.65	448.65	610
6190	12" pipe size		3	8		212	390	19.50	621.50	840
6191	14" pipe size		2.50	9.600		281	465	23.50	769.50	1,025
6192	16" pipe size	↓	1.80	13.333	↓	440	650	32.50	1,122.50	1,500
6200	300 lb. flange									
6210	1/2" pipe size	Q-15	17	.941	Ea.	23.50	44	3.44	70.94	95.50
6220	3/4" pipe size		17	.941		23.50	44	3.44	70.94	95.50
6230	1" pipe size		16	1		23.50	47	3.66	74.16	99.50
6240	1-1/4" pipe size		13	1.231		23.50	57.50	4.50	85.50	117
6250	1-1/2" pipe size		12	1.333		23.50	62.50	4.88	90.88	124
6260	2" pipe size		11	1.455		31	68	5.30	104.30	142
6270	2-1/2" pipe size		9	1.778		35	83.50	6.50	125	171
6280	3" pipe size		7	2.286		38	107	8.35	153.35	212
6290	4" pipe size		6	2.667		55.50	125	9.75	190.25	259
6300	5" pipe size	↓	4	4		95	187	14.65	296.65	400
6310	6" pipe size	Q-16	5	4.800		95.50	233	11.70	340.20	470
6320	8" pipe size		4	6		163	291	14.65	468.65	630
6330	10" pipe size		3.40	7.059		278	345	17.20	640.20	840
6340	12" pipe size	↓	2.80	8.571	↓	340	415	21	776	1,025
6400	Welding neck, 150 lb. flange									
6410	1/2" pipe size	Q-15	40	.400	Ea.	25.50	18.75	1.46	45.71	57.50
6420	3/4" pipe size		36	.444		25.50	21	1.63	48.13	61
6430	1" pipe size		32	.500		25.50	23.50	1.83	50.83	65
6440	1-1/4" pipe size	↓	29	.552	↓	25.50	26	2.02	53.52	69

22 11 13.47 Pipe Fittings, Steel		Crew	Daily Output	Labor-Hours	Unit	Material	Labor	Equipment	Total	Total Incl O&P
6450	1-1/2" pipe size	Q-15	26	.615	Ea.	25.50	29	2.25	56.75	73.50
6460	2" pipe size		20	.800		29.50	37.50	2.93	69.93	91
6470	2-1/2" pipe size		16	1		32	47	3.66	82.66	109
6480	3" pipe size		14	1.143		35	53.50	4.18	92.68	124
6500	4" pipe size		10	1.600		42.50	75	5.85	123.35	165
6510	5" pipe size		8	2		66.50	93.50	7.30	167.30	222
6520	6" pipe size	Q-16	10	2.400		64	117	5.85	186.85	251
6530	8" pipe size		7	3.429		113	167	8.35	288.35	385
6540	10" pipe size		6	4		180	194	9.75	383.75	500
6550	12" pipe size		5	4.800		262	233	11.70	506.70	650
6551	14" pipe size		4.50	5.333		380	259	13	652	820
6552	16" pipe size		3	8		570	390	19.50	979.50	1,225
6553	18" pipe size		2.50	9.600		800	465	23.50	1,288.50	1,600
6554	20" pipe size		2.30	10.435		975	505	25.50	1,505.50	1,875
6556	24" pipe size		2	12		1,300	585	29.50	1,914.50	2,350
6557	26" pipe size		1.70	14.118		1,425	685	34.50	2,144.50	2,650
6558	30" pipe size		.90	26.667		1,650	1,300	65	3,015	3,825
6559	36" pipe size		.75	32		1,900	1,550	78	3,528	4,475
6560	300 lb. flange									
6570	1/2" pipe size	Q-15	36	.444	Ea.	31	21	1.63	53.63	67
6580	3/4" pipe size		34	.471		31	22	1.72	54.72	69
6590	1" pipe size		30	.533		31	25	1.95	57.95	73.50
6600	1-1/4" pipe size		28	.571		31	27	2.09	60.09	76.50
6610	1-1/2" pipe size		24	.667		31	31	2.44	64.44	83.50
6620	2" pipe size		18	.889		38.50	41.50	3.25	83.25	108
6630	2-1/2" pipe size		14	1.143		44	53.50	4.18	101.68	134
6640	3" pipe size		12	1.333		44.50	62.50	4.88	111.88	148
6650	4" pipe size		8	2		73	93.50	7.30	173.80	228
6660	5" pipe size		7	2.286		111	107	8.35	226.35	292
6670	6" pipe size	Q-16	9	2.667		113	130	6.50	249.50	325
6680	8" pipe size		6	4		195	194	9.75	398.75	515
6690	10" pipe size		5	4.800		355	233	11.70	599.70	755
6700	12" pipe size		4	6		455	291	14.65	760.65	950
6710	14" pipe size		3.50	6.857		980	335	16.75	1,331.75	1,600
6720	16" pipe size		2	12		1,150	585	29.50	1,764.50	2,175
7740	Plain ends for plain end pipe, mechanically coupled									
7750	Cplg & labor required at joints not included, add 1 per									
7760	joint for installed price, see line 9180									
7770	Malleable iron, painted, unless noted otherwise									
7800	90° Elbow 1"				Ea.	75.50			75.50	83.50
7810	1-1/2"					90			90	99
7820	2"					134			134	147
7830	2-1/2"					157			157	172
7840	3"					163			163	179
7860	4"					185			185	203
7870	5" welded steel					212			212	233
7880	6"					268			268	295
7890	8" welded steel					505			505	555
7900	10" welded steel					665			665	735
7910	12" welded steel					740			740	815
7970	45° Elbow 1"					45			45	49.50
7980	1-1/2"					67.50			67.50	74.50
7990	2"					142			142	156

Material, Labor, Equipment and Total columns: 2010 Bare Costs

22 11 13.47 Pipe Fittings, Steel		Crew	Daily Output	Labor-Hours	Unit	Material	2010 Bare Costs Labor	Equipment	Total	Total Incl O&P
8000	2-1/2"				Ea.	142			142	156
8010	3"					163			163	179
8030	4"					170			170	187
8040	5" welded steel					212			212	233
8050	6"					243			243	267
8060	8"					285			285	315
8070	10" welded steel					305			305	335
8080	12" welded steel					555			555	610
8140	Tee, straight 1"					85			85	93.50
8150	1-1/2"					109			109	120
8160	2"					109			109	120
8170	2-1/2"					142			142	157
8180	3"					219			219	241
8200	4"					315			315	345
8210	5" welded steel					430			430	475
8220	6"					375			375	410
8230	8" welded steel					540			540	595
8240	10" welded steel					845			845	930
8250	12" welded steel				▼	1,025			1,025	1,125
8340	Segmentally welded steel, painted									
8390	Wye 2"				Ea.	136			136	150
8400	2-1/2"					136			136	150
8410	3"					153			153	168
8430	4"					228			228	251
8440	5"					277			277	305
8450	6"					380			380	420
8460	8"					495			495	545
8470	10"					730			730	800
8480	12"					1,125			1,125	1,250
8540	Wye, lateral 2"					148			148	162
8550	2-1/2"					170			170	187
8560	3"					202			202	222
8580	4"					278			278	305
8590	5"					470			470	520
8600	6"					485			485	535
8610	8"					820			820	900
8620	10"					850			850	935
8630	12"					1,550			1,550	1,725
8690	Cross, 2"					144			144	159
8700	2-1/2"					144			144	159
8710	3"					172			172	189
8730	4"					235			235	259
8740	5"					335			335	365
8750	6"					440			440	485
8760	8"					565			565	620
8770	10"					825			825	905
8780	12"					1,200			1,200	1,300
8800	Tees, reducing 2" x 1"					93.50			93.50	103
8810	2" x 1-1/2"					93.50			93.50	103
8820	3" x 1"					93.50			93.50	103
8830	3" x 1-1/2"					93.50			93.50	103
8840	3" x 2"					93.50			93.50	103
8850	4" x 1"				▼	136			136	149

22 11 13 – Facility Water Distribution Piping

22 11 13.47 Pipe Fittings, Steel		Crew	Daily Output	Labor-Hours	Unit	Material	Labor	Equipment	Total	Total Incl O&P
						2010 Bare Costs				
8860	4" x 1-1/2"				Ea.	136			136	149
8870	4" x 2"					136			136	149
8880	4" x 2-1/2"					139			139	153
8890	4" x 3"					139			139	153
8900	6" x 2"					201			201	221
8910	6" x 3"					216			216	238
8920	6" x 4"					263			263	289
8930	8" x 2"					274			274	300
8940	8" x 3"					280			280	310
8950	8" x 4"					280			280	310
8960	8" x 5"					289			289	320
8970	8" x 6"					315			315	350
8980	10" x 4"					415			415	455
8990	10" x 6"					415			415	455
9000	10" x 8"					440			440	485
9010	12" x 6"					640			640	700
9020	12" x 8"					640			640	705
9030	12" x 10"				↓	665			665	730
9080	Adapter nipples 3" long									
9090	1"				Ea.	12.60			12.60	13.85
9100	1-1/2"					12.60			12.60	13.85
9110	2"					12.60			12.60	13.85
9120	2-1/2"					14.70			14.70	16.15
9130	3"					18			18	19.80
9140	4"					30			30	33
9150	6"				↓	79.50			79.50	87.50
9180	Coupling, mechanical, plain end pipe to plain end pipe or fitting									
9190	1"	Q-1	29	.552	Ea.	45.50	26		71.50	89
9200	1-1/2"		28	.571		45.50	27		72.50	90
9210	2"		27	.593		45.50	28		73.50	91.50
9220	2-1/2"		26	.615		45.50	29		74.50	93
9230	3"		25	.640		66.50	30		96.50	119
9240	3-1/2"		24	.667		77.50	31		108.50	132
9250	4"	↓	22	.727		77.50	34		111.50	136
9260	5"	Q-2	28	.857		110	41.50		151.50	184
9270	6"		24	1		135	48.50		183.50	221
9280	8"		19	1.263		238	61.50		299.50	355
9290	10"		16	1.500		310	73		383	450
9300	12"	↓	12	2	↓	390	97		487	570
9310	Outlets for precut holes through pipe wall									
9331	Strapless type, with gasket									
9332	4" to 8" pipe x 1/2"	1 Plum	13	.615	Ea.	49.50	32		81.50	103
9333	4" to 8" pipe x 3/4"		13	.615		53.50	32		85.50	107
9334	10" pipe and larger x 1/2"		11	.727		49.50	38		87.50	111
9335	10" pipe and larger x 3/4"	↓	11	.727	↓	53.50	38		91.50	115
9341	Thermometer wells with gasket									
9342	4" to 8" pipe, 6" stem	1 Plum	14	.571	Ea.	75.50	29.50		105	128
9343	8" pipe and larger, 6" stem	"	13	.615	"	75.50	32		107.50	131
9940	For galvanized fittings for plain end pipe, add					20%				

22 11 13.48 Pipe, Fittings and Valves, Steel, Grooved-Joint	
0010	**PIPE, FITTINGS AND VALVES, STEEL, GROOVED-JOINT**
0012	Fittings are ductile iron. Steel fittings noted.

22 11 13 – Facility Water Distribution Piping

22 11 13.48 Pipe, Fittings and Valves, Steel, Grooved-Joint		Crew	Daily Output	Labor-Hours	Unit	Material	2010 Bare Costs Labor	Equipment	Total	Total Incl O&P
0020	Pipe includes coupling & clevis type hanger assemblies, 10′ O.C.									
0500	Schedule 10, black									
0550	2″ diameter	1 Plum	43	.186	L.F.	5.10	9.70		14.80	20
0560	2-1/2″ diameter	Q-1	61	.262		6.55	12.30		18.85	25.50
0570	3″ diameter		55	.291		7.80	13.65		21.45	29
0580	3-1/2″ diameter		53	.302		9.05	14.15		23.20	31
0590	4″ diameter		49	.327		10.40	15.30		25.70	34.50
0600	5″ diameter		40	.400		12.35	18.75		31.10	41.50
0610	6″ diameter	Q-2	46	.522		15.95	25.50		41.45	55.50
0620	8″ diameter	″	41	.585		22	28.50		50.50	66.50
0700	To delete couplings & hangers, subtract									
0710	2″ diam. to 5″ diam.					25%	20%			
0720	6″ diam. to 8″ diam.					27%	15%			
1000	Schedule 40, black									
1040	3/4″ diameter	1 Plum	71	.113	L.F.	5.30	5.85		11.15	14.60
1050	1″ diameter		63	.127		3.38	6.60		9.98	13.60
1060	1-1/4″ diameter		58	.138		4.43	7.20		11.63	15.60
1070	1-1/2″ diameter		51	.157		5.10	8.15		13.25	17.85
1080	2″ diameter		40	.200		6.30	10.40		16.70	22.50
1090	2-1/2″ diameter	Q-1	57	.281		9.15	13.15		22.30	30
1100	3″ diameter		50	.320		11.50	15		26.50	35
1110	4″ diameter		45	.356		16.35	16.65		33	43
1120	5″ diameter		37	.432		22.50	20.50		43	55.50
1130	6″ diameter	Q-2	42	.571		28.50	28		56.50	73
1140	8″ diameter		37	.649		44	31.50		75.50	95.50
1150	10″ diameter		31	.774		65	37.50		102.50	128
1160	12″ diameter		27	.889		82.50	43		125.50	156
1170	14″ diameter		20	1.200		86	58.50		144.50	183
1180	16″ diameter		17	1.412		104	68.50		172.50	217
1190	18″ diameter		14	1.714		119	83.50		202.50	255
1200	20″ diameter		12	2		145	97		242	305
1210	24″ diameter		10	2.400		175	117		292	365
1740	To delete coupling & hanger, subtract									
1750	3/4″ diam. to 2″ diam.					65%	27%			
1760	2-1/2″ diam. to 5″ diam.					41%	18%			
1770	6″ diam. to 12″ diam.					31%	13%			
1780	14″ diam. to 24″ diam.					35%	10%			
1800	Galvanized									
1840	3/4″ diameter	1 Plum	71	.113	L.F.	6	5.85		11.85	15.40
1850	1″ diameter		63	.127		8.35	6.60		14.95	19.10
1860	1-1/4″ diameter		58	.138		10.80	7.20		18	22.50
1870	1-1/2″ diameter		51	.157		12.70	8.15		20.85	26
1880	2″ diameter		40	.200		16.50	10.40		26.90	34
1890	2-1/2″ diameter	Q-1	57	.281		25	13.15		38.15	47
1900	3″ diameter		50	.320		32.50	15		47.50	58
1910	4″ diameter		45	.356		45	16.65		61.65	74.50
1920	5″ diameter		37	.432		51.50	20.50		72	87
1930	6″ diameter	Q-2	42	.571		60	28		88	108
1940	8″ diameter		37	.649		71	31.50		102.50	125
1950	10″ diameter		31	.774		104	37.50		141.50	172
1960	12″ diameter		27	.889		137	43		180	215
2540	To delete coupling & hanger, subtract									
2550	3/4″ diam. to 2″ diam.					36%	27%			

22 11 13.48 Pipe, Fittings and Valves, Steel, Grooved-Joint		Crew	Daily Output	Labor-Hours	Unit	2010 Bare Costs Material	Labor	Equipment	Total	Total Incl O&P
2560	2-1/2" diam. to 5" diam.					19%	18%			
2570	6" diam. to 12" diam.					14%	13%			
2600	Schedule 80, black									
2610	3/4" diameter	1 Plum	65	.123	L.F.	7.40	6.40		13.80	17.70
2650	1" diameter		61	.131		9.60	6.85		16.45	21
2660	1-1/4" diameter		55	.145		12.70	7.55		20.25	25.50
2670	1-1/2" diameter		49	.163		15.05	8.50		23.55	29.50
2680	2" diameter		38	.211		19.90	10.95		30.85	38.50
2690	2-1/2" diameter	Q-1	54	.296		30.50	13.90		44.40	54.50
2700	3" diameter		48	.333		40.50	15.60		56.10	68
2710	4" diameter		44	.364		45.50	17.05		62.55	75.50
2720	5" diameter		35	.457		56.50	21.50		78	94
2730	6" diameter	Q-2	40	.600		64.50	29		93.50	115
2740	8" diameter		35	.686		116	33.50		149.50	177
2750	10" diameter		29	.828		199	40		239	280
2760	12" diameter		24	1		350	48.50		398.50	460
3240	To delete coupling & hanger, subtract									
3250	3/4" diam. to 2" diam.					30%	25%			
3260	2-1/2" diam. to 5" diam.					14%	17%			
3270	6" diam. to 12" diam.					12%	12%			
3300	Galvanized									
3310	3/4" diameter	1 Plum	65	.123	L.F.	9.40	6.40		15.80	19.95
3350	1" diameter		61	.131		12.65	6.85		19.50	24
3360	1-1/4" diameter		55	.145		16.80	7.55		24.35	30
3370	1-1/2" diameter		46	.174		20	9.05		29.05	35.50
3380	2" diameter		38	.211		27	10.95		37.95	46
3390	2-1/2" diameter	Q-1	54	.296		41.50	13.90		55.40	66.50
3400	3" diameter		48	.333		55	15.60		70.60	84
3410	4" diameter		44	.364		77.50	17.05		94.55	111
3420	5" diameter		35	.457		107	21.50		128.50	149
3430	6" diameter	Q-2	40	.600		115	29		144	170
3440	8" diameter		35	.686		174	33.50		207.50	242
3450	10" diameter		29	.828		325	40		365	415
3460	12" diameter		24	1		460	48.50		508.50	585
3920	To delete coupling & hanger, subtract									
3930	3/4" diam. to 2" diam.					30%	25%			
3940	2-1/2" diam. to 5" diam.					15%	17%			
3950	6" diam. to 12" diam.					11%	12%			
3990	Fittings: coupling material required at joints not incl. in fitting price.									
3994	Add 1 selected coupling, material only, per joint for installed price.									
4000	Elbow, 90° or 45°, painted									
4030	3/4" diameter	1 Plum	50	.160	Ea.	40	8.35		48.35	56.50
4040	1" diameter		50	.160		21.50	8.35		29.85	36
4050	1-1/4" diameter		40	.200		21.50	10.40		31.90	39
4060	1-1/2" diameter		33	.242		21.50	12.60		34.10	42.50
4070	2" diameter		25	.320		21.50	16.65		38.15	48.50
4080	2-1/2" diameter	Q-1	40	.400		21.50	18.75		40.25	51.50
4090	3" diameter		33	.485		38	22.50		60.50	75.50
4100	4" diameter		25	.640		41.50	30		71.50	90.50
4110	5" diameter		20	.800		99	37.50		136.50	165
4120	6" diameter	Q-2	25	.960		116	46.50		162.50	198
4130	8" diameter		21	1.143		243	55.50		298.50	350
4140	10" diameter		18	1.333		440	65		505	580

22 11 13.48 Pipe, Fittings and Valves, Steel, Grooved-Joint		Crew	Daily Output	Labor-Hours	Unit	2010 Bare Costs Material	Labor	Equipment	Total	Total Incl O&P
4150	12" diameter	Q-2	15	1.600	Ea.	705	77.50		782.50	890
4170	14" diameter		12	2		835	97		932	1,075
4180	16" diameter		11	2.182		1,075	106		1,181	1,350
4190	18" diameter	Q-3	15	2.133		1,375	106		1,481	1,675
4200	20" diameter		13	2.462		1,825	122		1,947	2,175
4210	24" diameter		11	2.909		2,625	144		2,769	3,125
4250	For galvanized elbows, add					26%				
4690	Tee, painted									
4700	3/4" diameter	1 Plum	38	.211	Ea.	43.50	10.95		54.45	64
4740	1" diameter		33	.242		33	12.60		45.60	55.50
4750	1-1/4" diameter		27	.296		33	15.40		48.40	59.50
4760	1-1/2" diameter		22	.364		33	18.95		51.95	65
4770	2" diameter		17	.471		33	24.50		57.50	73
4780	2-1/2" diameter	Q-1	27	.593		33	28		61	78
4790	3" diameter		22	.727		46	34		80	102
4800	4" diameter		17	.941		70	44		114	143
4810	5" diameter		13	1.231		163	57.50		220.50	267
4820	6" diameter	Q-2	17	1.412		189	68.50		257.50	310
4830	8" diameter		14	1.714		415	83.50		498.50	580
4840	10" diameter		12	2		860	97		957	1,100
4850	12" diameter		10	2.400		1,200	117		1,317	1,500
4851	14" diameter		9	2.667		1,200	130		1,330	1,525
4852	16" diameter		8	3		1,275	146		1,421	1,625
4853	18" diameter	Q-3	11	2.909		1,575	144		1,719	1,950
4854	20" diameter		10	3.200		2,250	159		2,409	2,750
4855	24" diameter		8	4		3,450	198		3,648	4,100
4900	For galvanized tees, add					24%				
4906	Couplings, rigid style, painted									
4908	1" diameter	1 Plum	100	.080	Ea.	16	4.16		20.16	24
4909	1-1/4" diameter		100	.080		16	4.16		20.16	24
4910	1-1/2" diameter		67	.119		16	6.20		22.20	27
4912	2" diameter		50	.160		16.40	8.35		24.75	30.50
4914	2-1/2" diameter	Q-1	80	.200		18.90	9.35		28.25	35
4916	3" diameter		67	.239		22	11.20		33.20	41
4918	4" diameter		50	.320		31	15		46	56.50
4920	5" diameter		40	.400		40	18.75		58.75	72
4922	6" diameter	Q-2	50	.480		53	23.50		76.50	93.50
4924	8" diameter		42	.571		83.50	28		111.50	134
4926	10" diameter		35	.686		149	33.50		182.50	214
4928	12" diameter		32	.750		167	36.50		203.50	238
4930	14" diameter		24	1		192	48.50		240.50	284
4931	16" diameter		20	1.200		250	58.50		308.50	365
4932	18" diameter		18	1.333		289	65		354	415
4933	20" diameter		16	1.500		395	73		468	545
4934	24" diameter		13	1.846		505	89.50		594.50	690
4940	Flexible, standard, painted									
4950	3/4" diameter	1 Plum	100	.080	Ea.	11.50	4.16		15.66	18.90
4960	1" diameter		100	.080		11.50	4.16		15.66	18.90
4970	1-1/4" diameter		80	.100		15.30	5.20		20.50	24.50
4980	1-1/2" diameter		67	.119		16.80	6.20		23	28
4990	2" diameter		50	.160		17.80	8.35		26.15	32
5000	2-1/2" diameter	Q-1	80	.200		21	9.35		30.35	37
5010	3" diameter		67	.239		23	11.20		34.20	42.50

22 11 13.48 Pipe, Fittings and Valves, Steel, Grooved-Joint		Crew	Daily Output	Labor-Hours	Unit	Material	2010 Bare Costs Labor	Equipment	Total	Total Incl O&P
5020	3-1/2" diameter	Q-1	57	.281	Ea.	33.50	13.15		46.65	56.50
5030	4" diameter		50	.320		33.50	15		48.50	59.50
5040	5" diameter		40	.400		51	18.75		69.75	84
5050	6" diameter	Q-2	50	.480		60	23.50		83.50	101
5070	8" diameter		42	.571		98	28		126	150
5090	10" diameter		35	.686		162	33.50		195.50	228
5110	12" diameter		32	.750		185	36.50		221.50	258
5120	14" diameter		24	1		198	48.50		246.50	290
5130	16" diameter		20	1.200		259	58.50		317.50	375
5140	18" diameter		18	1.333		305	65		370	430
5150	20" diameter		16	1.500		405	73		478	555
5160	24" diameter		13	1.846		525	89.50		614.50	710
5176	Lightweight style, painted									
5178	1-1/2" diameter	1 Plum	67	.119	Ea.	14.70	6.20		20.90	25.50
5180	2" diameter	"	50	.160		15	8.35		23.35	29
5182	2-1/2" diameter	Q-1	80	.200		17.30	9.35		26.65	33
5184	3" diameter		67	.239		20	11.20		31.20	39
5186	3-1/2" diameter		57	.281		28	13.15		41.15	50.50
5188	4" diameter		50	.320		28	15		43	53.50
5190	5" diameter		40	.400		40.50	18.75		59.25	72.50
5192	6" diameter	Q-2	50	.480		48.50	23.50		72	88
5194	8" diameter		42	.571		76	28		104	125
5196	10" diameter		35	.686		203	33.50		236.50	273
5198	12" diameter		32	.750		226	36.50		262.50	305
5200	For galvanized couplings, add					33%				
5220	Tee, reducing, painted									
5225	2" x 1-1/2" diameter	Q-1	38	.421	Ea.	70.50	19.75		90.25	107
5226	2-1/2" x 2" diameter		28	.571		70.50	27		97.50	118
5227	3" x 2-1/2" diameter		23	.696		62.50	32.50		95	118
5228	4" x 3" diameter		18	.889		84	41.50		125.50	155
5229	5" x 4" diameter		15	1.067		180	50		230	273
5230	6" x 4" diameter	Q-2	18	1.333		198	65		263	315
5231	8" x 6" diameter		15	1.600		415	77.50		492.50	570
5232	10" x 8" diameter		13	1.846		540	89.50		629.50	730
5233	12" x 10" diameter		11	2.182		825	106		931	1,075
5234	14" x 12" diameter		10	2.400		815	117		932	1,075
5235	16" x 12" diameter		9	2.667		1,000	130		1,130	1,300
5236	18" x 12" diameter	Q-3	12	2.667		1,200	132		1,332	1,525
5237	18" x 16" diameter		11	2.909		1,525	144		1,669	1,900
5238	20" x 16" diameter		10	3.200		2,000	159		2,159	2,450
5239	24" x 20" diameter		9	3.556		3,150	176		3,326	3,750
5240	Reducer, concentric, painted									
5241	2-1/2" x 2" diameter	Q-1	43	.372	Ea.	25	17.45		42.45	53.50
5242	3" x 2-1/2" diameter		35	.457		30	21.50		51.50	65
5243	4" x 3" diameter		29	.552		36.50	26		62.50	79
5244	5" x 4" diameter		22	.727		50.50	34		84.50	107
5245	6" x 4" diameter	Q-2	26	.923		58.50	45		103.50	132
5246	8" x 6" diameter		23	1.043		152	50.50		202.50	243
5247	10" x 8" diameter		20	1.200		310	58.50		368.50	430
5248	12" x 10" diameter		16	1.500		555	73		628	720
5255	Eccentric, painted									
5256	2-1/2" x 2" diameter	Q-1	42	.381	Ea.	52.50	17.85		70.35	84.50
5257	3" x 2-1/2" diameter		34	.471		60	22		82	99

22 11 13.48 Pipe, Fittings and Valves, Steel, Grooved-Joint		Crew	Daily Output	Labor-Hours	Unit	Material	2010 Bare Costs Labor	Equipment	Total	Total Incl O&P
5258	4" x 3" diameter	Q-1	28	.571	Ea.	73	27		100	121
5259	5" x 4" diameter	↓	21	.762		99.50	35.50		135	164
5260	6" x 4" diameter	Q-2	25	.960		116	46.50		162.50	198
5261	8" x 6" diameter		22	1.091		235	53		288	340
5262	10" x 8" diameter		19	1.263		645	61.50		706.50	795
5263	12" x 10" diameter	↓	15	1.600	↓	885	77.50		962.50	1,100
5270	Coupling, reducing, painted									
5272	2" x 1-1/2" diameter	1 Plum	52	.154	Ea.	25	8		33	39.50
5274	2-1/2" x 2" diameter	Q-1	82	.195		32.50	9.15		41.65	49.50
5276	3" x 2" diameter		69	.232		37	10.85		47.85	57.50
5278	4" x 2" diameter		52	.308		59	14.40		73.40	86
5280	5" x 4" diameter	↓	42	.381		66.50	17.85		84.35	100
5282	6" x 4" diameter	Q-2	52	.462		100	22.50		122.50	144
5284	8" x 6" diameter	"	44	.545	↓	150	26.50		176.50	205
5290	Outlet coupling, painted									
5294	1-1/2" x 1" pipe size	1 Plum	65	.123	Ea.	32	6.40		38.40	44.50
5296	2" x 1" pipe size	"	48	.167		32.50	8.70		41.20	49
5298	2-1/2" x 1" pipe size	Q-1	78	.205		50.50	9.60		60.10	70
5300	2-1/2" x 1-1/4" pipe size	1 Plum	70	.114		57	5.95		62.95	71.50
5302	3" x 1" pipe size	Q-1	65	.246		64	11.55		75.55	88
5304	4" x 3/4" pipe size		48	.333		71.50	15.60		87.10	102
5306	4" x 1-1/2" pipe size	↓	46	.348		101	16.30		117.30	136
5308	6" x 1-1/2" pipe size	Q-2	44	.545	↓	143	26.50		169.50	198
5750	Flange, w/groove gasket, black steel									
5754	See Div. 22 11 13.44 0620 for gasket & bolt set									
5760	ANSI class 125 and 150, painted									
5780	2" pipe size	1 Plum	23	.348	Ea.	71.50	18.10		89.60	106
5790	2-1/2" pipe size	Q-1	37	.432		89	20.50		109.50	128
5800	3" pipe size		31	.516		96	24		120	142
5820	4" pipe size		23	.696		128	32.50		160.50	190
5830	5" pipe size	↓	19	.842		149	39.50		188.50	223
5840	6" pipe size	Q-2	23	1.043		162	50.50		212.50	255
5850	8" pipe size		17	1.412		183	68.50		251.50	305
5860	10" pipe size		14	1.714		290	83.50		373.50	445
5870	12" pipe size		12	2		380	97		477	565
5880	14" pipe size		10	2.400		770	117		887	1,025
5890	16" pipe size		9	2.667		890	130		1,020	1,175
5900	18" pipe size		6	4		1,100	194		1,294	1,500
5910	20" pipe size		5	4.800		1,325	233		1,558	1,800
5920	24" pipe size	↓	4.50	5.333	↓	1,700	259		1,959	2,250
5940	ANSI class 350, painted									
5946	2" pipe size	1 Plum	23	.348	Ea.	89.50	18.10		107.60	126
5948	2-1/2" pipe size	Q-1	37	.432		104	20.50		124.50	145
5950	3" pipe size		31	.516		142	24		166	193
5952	4" pipe size		23	.696		189	32.50		221.50	256
5954	5" pipe size	↓	19	.842		215	39.50		254.50	295
5956	6" pipe size	Q-2	23	1.043		250	50.50		300.50	350
5958	8" pipe size		17	1.412		286	68.50		354.50	420
5960	10" pipe size	↓	14	1.714		455	83.50		538.50	625
5962	12" pipe size	1 Plum	12	.667	↓	485	34.50		519.50	585
6100	Cross, painted									
6110	2" diameter	1 Plum	12.50	.640	Ea.	59.50	33.50		93	116
6112	2-1/2" diameter	Q-1	20	.800	↓	59.50	37.50		97	122

22 11 13 – Facility Water Distribution Piping

22 11 13.48 Pipe, Fittings and Valves, Steel, Grooved-Joint		Crew	Daily Output	Labor-Hours	Unit	Material	2010 Bare Costs Labor	Equipment	Total	Total Incl O&P
6114	3" diameter	Q-1	15.50	1.032	Ea.	106	48.50		154.50	190
6116	4" diameter	↓	12.50	1.280		177	60		237	284
6118	6" diameter	Q-2	12.50	1.920		465	93.50		558.50	650
6120	8" diameter		10.50	2.286		610	111		721	835
6122	10" diameter		9	2.667		1,050	130		1,180	1,350
6124	12" diameter	↓	7.50	3.200	↓	1,500	155		1,655	1,875
7400	Suction diffuser									
7402	Grooved end inlet x flanged outlet									
7410	3" x 3"	Q-1	50	.320	Ea.	620	15		635	705
7412	4" x 4"		38	.421		840	19.75		859.75	955
7414	5" x 5"	↓	30	.533		985	25		1,010	1,125
7416	6" x 6"	Q-2	38	.632		1,250	30.50		1,280.50	1,425
7418	8" x 8"		27	.889		2,300	43		2,343	2,625
7420	10" x 10"		20	1.200		3,150	58.50		3,208.50	3,550
7422	12" x 12"		16	1.500		5,175	73		5,248	5,775
7424	14" x 14"		15	1.600		6,025	77.50		6,102.50	6,775
7426	16" x 14"	↓	14	1.714	↓	6,225	83.50		6,308.50	6,950
7500	Strainer, tee type, painted									
7506	2" pipe size	1 Plum	38	.211	Ea.	420	10.95		430.95	475
7508	2-1/2" pipe size	Q-1	62	.258		440	12.10		452.10	505
7510	3" pipe size		50	.320		495	15		510	565
7512	4" pipe size		38	.421		560	19.75		579.75	645
7514	5" pipe size	↓	30	.533		810	25		835	930
7516	6" pipe size	Q-2	38	.632		875	30.50		905.50	1,000
7518	8" pipe size		27	.889		1,350	43		1,393	1,550
7520	10" pipe size		20	1.200		1,975	58.50		2,033.50	2,275
7522	12" pipe size		16	1.500		2,550	73		2,623	2,900
7524	14" pipe size		15	1.600		8,150	77.50		8,227.50	9,075
7526	16" pipe size	↓	14	1.714	↓	10,100	83.50		10,183.50	11,200
7570	Expansion joint, max. 3" travel									
7572	2" diameter	1 Plum	38	.211	Ea.	490	10.95		500.95	555
7574	3" diameter	Q-1	50	.320		565	15		580	645
7576	4" diameter	"	38	.421		740	19.75		759.75	845
7578	6" diameter	Q-2	38	.632	↓	1,150	30.50		1,180.50	1,325
7800	Ball valve w/handle, carbon steel trim									
7810	1-1/2" pipe size	1 Plum	50	.160	Ea.	119	8.35		127.35	144
7812	2" pipe size	"	38	.211		132	10.95		142.95	161
7814	2-1/2" pipe size	Q-1	62	.258		288	12.10		300.10	335
7816	3" pipe size		50	.320		465	15		480	540
7818	4" pipe size	↓	38	.421		720	19.75		739.75	825
7820	6" pipe size	Q-2	30	.800	↓	2,225	39		2,264	2,475
7830	With gear operator									
7834	2-1/2" pipe size	Q-1	62	.258	Ea.	585	12.10		597.10	660
7836	3" pipe size		50	.320		825	15		840	935
7838	4" pipe size	↓	38	.421		1,050	19.75		1,069.75	1,175
7840	6" pipe size	Q-2	30	.800	↓	2,475	39		2,514	2,775
7870	Check valve									
7874	2-1/2" pipe size	Q-1	62	.258	Ea.	199	12.10		211.10	237
7876	3" pipe size		50	.320		236	15		251	282
7878	4" pipe size		38	.421		249	19.75		268.75	305
7880	5" pipe size	↓	30	.533		415	25		440	495
7882	6" pipe size	Q-2	38	.632		490	30.50		520.50	585
7884	8" pipe size	↓	27	.889	↓	670	43		713	805

22 11 13 – Facility Water Distribution Piping

22 11 13.48 Pipe, Fittings and Valves, Steel, Grooved-Joint		Crew	Daily Output	Labor-Hours	Unit	2010 Bare Costs Material	Labor	Equipment	Total	Total Incl O&P
7886	10" pipe size	Q-2	20	1.200	Ea.	1,950	58.50		2,008.50	2,225
7888	12" pipe size	↓	16	1.500	↓	2,300	73		2,373	2,625
7900	Plug valve, balancing, w/lever operator									
7906	3" pipe size	Q-1	50	.320	Ea.	465	15		480	535
7908	4" pipe size	"	38	.421		570	19.75		589.75	660
7909	6" pipe size	Q-2	30	.800	↓	875	39		914	1,025
7916	With gear operator									
7920	3" pipe size	Q-1	50	.320	Ea.	880	15		895	995
7922	4" pipe size	"	38	.421		915	19.75		934.75	1,025
7924	6" pipe size	Q-2	38	.632		1,225	30.50		1,255.50	1,400
7926	8" pipe size		27	.889		1,900	43		1,943	2,150
7928	10" pipe size		20	1.200		2,575	58.50		2,633.50	2,950
7930	12" pipe size	↓	16	1.500	↓	3,925	73		3,998	4,425
8000	Butterfly valve, 2 position handle, with standard trim									
8010	1-1/2" pipe size	1 Plum	50	.160	Ea.	176	8.35		184.35	207
8020	2" pipe size	"	38	.211		176	10.95		186.95	210
8030	3" pipe size	Q-1	50	.320		253	15		268	300
8050	4" pipe size	"	38	.421		278	19.75		297.75	335
8070	6" pipe size	Q-2	38	.632		560	30.50		590.50	660
8080	8" pipe size		27	.889		805	43		848	950
8090	10" pipe size	↓	20	1.200	↓	1,425	58.50		1,483.50	1,675
8200	With stainless steel trim									
8240	1-1/2" pipe size	1 Plum	50	.160	Ea.	223	8.35		231.35	259
8250	2" pipe size	"	38	.211		223	10.95		233.95	262
8270	3" pipe size	Q-1	50	.320		300	15		315	355
8280	4" pipe size	"	38	.421		325	19.75		344.75	390
8300	6" pipe size	Q-2	38	.632		610	30.50		640.50	715
8310	8" pipe size		27	.889		1,600	43		1,643	1,850
8320	10" pipe size		20	1.200		2,900	58.50		2,958.50	3,275
8322	12" pipe size		16	1.500		3,325	73		3,398	3,750
8324	14" pipe size		15	1.600		4,025	77.50		4,102.50	4,550
8326	16" pipe size	↓	14	1.714		5,825	83.50		5,908.50	6,550
8328	18" pipe size	Q-3	12	2.667		7,175	132		7,307	8,100
8330	20" pipe size		11	2.909		8,825	144		8,969	9,950
8332	24" pipe size	↓	10	3.200	↓	11,800	159		11,959	13,200
8336	Note: sizes 8" up w/manual gear operator									
9000	Cut one groove, labor									
9010	3/4" pipe size	Q-1	152	.105	Ea.		4.93		4.93	7.40
9020	1" pipe size		140	.114			5.35		5.35	8.05
9030	1-1/4" pipe size		124	.129			6.05		6.05	9.05
9040	1-1/2" pipe size		114	.140			6.60		6.60	9.85
9050	2" pipe size		104	.154			7.20		7.20	10.80
9060	2-1/2" pipe size		96	.167			7.80		7.80	11.70
9070	3" pipe size		88	.182			8.50		8.50	12.75
9080	3-1/2" pipe size		83	.193			9.05		9.05	13.55
9090	4" pipe size		78	.205			9.60		9.60	14.40
9100	5" pipe size		72	.222			10.40		10.40	15.60
9110	6" pipe size		70	.229			10.70		10.70	16.05
9120	8" pipe size		54	.296			13.90		13.90	21
9130	10" pipe size		38	.421			19.75		19.75	29.50
9140	12" pipe size		30	.533			25		25	37.50
9150	14" pipe size		20	.800			37.50		37.50	56
9160	16" pipe size	↓	19	.842	↓		39.50		39.50	59

22 11 13.48 Pipe, Fittings and Valves, Steel, Grooved-Joint		Crew	Daily Output	Labor-Hours	Unit	Material	2010 Bare Costs Labor	Equipment	Total	Total Incl O&P
9170	18" pipe size	Q-1	18	.889	Ea.		41.50		41.50	62.50
9180	20" pipe size		17	.941			44		44	66
9190	24" pipe size		15	1.067			50		50	75
9210	Roll one groove									
9220	3/4" pipe size	Q-1	266	.060	Ea.		2.82		2.82	4.22
9230	1" pipe size		228	.070			3.29		3.29	4.93
9240	1-1/4" pipe size		200	.080			3.75		3.75	5.60
9250	1-1/2" pipe size		178	.090			4.21		4.21	6.30
9260	2" pipe size		116	.138			6.45		6.45	9.70
9270	2-1/2" pipe size		110	.145			6.80		6.80	10.20
9280	3" pipe size		100	.160			7.50		7.50	11.25
9290	3-1/2" pipe size		94	.170			7.95		7.95	11.95
9300	4" pipe size		86	.186			8.70		8.70	13.05
9310	5" pipe size		84	.190			8.90		8.90	13.40
9320	6" pipe size		80	.200			9.35		9.35	14.05
9330	8" pipe size		66	.242			11.35		11.35	17.05
9340	10" pipe size		58	.276			12.90		12.90	19.35
9350	12" pipe size		46	.348			16.30		16.30	24.50
9360	14" pipe size		30	.533			25		25	37.50
9370	16" pipe size		28	.571			27		27	40
9380	18" pipe size		27	.593			28		28	41.50
9390	20" pipe size		25	.640			30		30	45
9400	24" pipe size		23	.696			32.50		32.50	49

22 11 13.60 Tubing, Stainless Steel

		Crew	Daily Output	Labor-Hours	Unit	Material	Labor	Equipment	Total	Total Incl O&P
0010	**TUBING, STAINLESS STEEL**									
5010	Type 304, no joints, no hangers									
5020	.035 wall									
5021	1/4"	1 Plum	160	.050	L.F.	3.21	2.60		5.81	7.45
5022	3/8"		160	.050		4.05	2.60		6.65	8.35
5023	1/2"		160	.050		5.20	2.60		7.80	9.65
5024	5/8"		160	.050		6.40	2.60		9	10.95
5025	3/4"		133	.060		7.70	3.13		10.83	13.15
5026	7/8"		133	.060		9.15	3.13		12.28	14.75
5027	1"		114	.070		10.95	3.65		14.60	17.50
5040	.049 wall									
5041	1/4"	1 Plum	160	.050	L.F.	3.77	2.60		6.37	8.05
5042	3/8"		160	.050		5.20	2.60		7.80	9.65
5043	1/2"		160	.050		5.80	2.60		8.40	10.25
5044	5/8"		160	.050		7.80	2.60		10.40	12.50
5045	3/4"		133	.060		8.30	3.13		11.43	13.85
5046	7/8"		133	.060		9.75	3.13		12.88	15.45
5047	1"		114	.070		10.35	3.65		14	16.85
5060	.065 wall									
5061	1/4"	1 Plum	160	.050	L.F.	4.23	2.60		6.83	8.55
5062	3/8"		160	.050		5.85	2.60		8.45	10.30
5063	1/2"		160	.050		7.05	2.60		9.65	11.65
5064	5/8"		160	.050		8.60	2.60		11.20	13.35
5065	3/4"		133	.060		9.10	3.13		12.23	14.70
5066	7/8"		133	.060		11.75	3.13		14.88	17.60
5067	1"		114	.070		11.50	3.65		15.15	18.10
5210	Type 316									
5220	.035 wall									

22 11 13.60 Tubing, Stainless Steel

22 11 13.60 Tubing, Stainless Steel		Crew	Daily Output	Labor-Hours	Unit	Material	Labor	Equipment	Total	Total Incl O&P
						2010 Bare Costs				
5221	1/4"	1 Plum	160	.050	L.F.	4.38	2.60		6.98	8.70
5222	3/8"		160	.050		4.98	2.60		7.58	9.40
5223	1/2"		160	.050		7.40	2.60		10	12.05
5224	5/8"		160	.050		10.20	2.60		12.80	15.10
5225	3/4"		133	.060		9.05	3.13		12.18	14.65
5226	7/8"		133	.060		17.70	3.13		20.83	24
5227	1"	↓	114	.070	↓	11.50	3.65		15.15	18.10
5240	.049 wall									
5241	1/4"	1 Plum	160	.050	L.F.	5.30	2.60		7.90	9.75
5242	3/8"		160	.050		6.10	2.60		8.70	10.60
5243	1/2"		160	.050		7.50	2.60		10.10	12.15
5244	5/8"		160	.050		10.55	2.60		13.15	15.55
5245	3/4"		133	.060		10.30	3.13		13.43	16.05
5246	7/8"		133	.060		14.80	3.13		17.93	21
5247	1"	↓	114	.070	↓	14.90	3.65		18.55	22
5260	.065 wall									
5261	1/4"	1 Plum	160	.050	L.F.	5.05	2.60		7.65	9.45
5262	3/8"		160	.050		10.65	2.60		13.25	15.60
5263	1/2"		160	.050		10.05	2.60		12.65	14.95
5264	5/8"		160	.050		12.75	2.60		15.35	17.95
5265	3/4"		133	.060		12.70	3.13		15.83	18.65
5266	7/8"		133	.060		17.35	3.13		20.48	23.50
5267	1"	↓	114	.070	↓	16.45	3.65		20.10	23.50

22 11 13.61 Tubing Fittings, Stainless Steel

22 11 13.61 Tubing Fittings, Stainless Steel		Crew	Daily Output	Labor-Hours	Unit	Material	Labor	Equipment	Total	Total Incl O&P
0010	**TUBING FITTINGS, STAINLESS STEEL**									
8200	Tube fittings, compression type									
8202	Type 316									
8204	90° elbow									
8206	1/4"	1 Plum	24	.333	Ea.	14.40	17.35		31.75	42
8207	3/8"		22	.364		17.60	18.95		36.55	48
8208	1/2"		22	.364		29	18.95		47.95	60.50
8209	5/8"		21	.381		32.50	19.85		52.35	65.50
8210	3/4"		21	.381		51.50	19.85		71.35	86
8211	7/8"		20	.400		78.50	21		99.50	118
8212	1"	↓	20	.400	↓	98	21		119	139
8220	Union tee									
8222	1/4"	1 Plum	15	.533	Ea.	20.50	28		40.50	64
8224	3/8"		15	.533		26	28		54	70
8225	1/2"		15	.533		40.50	28		68.50	86
8226	5/8"		14	.571		45	29.50		74.50	94
8227	3/4"		14	.571		60.50	29.50		90	111
8228	7/8"		13	.615		108	32		140	167
8229	1"	↓	13	.615	↓	129	32		161	190
8234	Union									
8236	1/4"	1 Plum	24	.333	Ea.	9.95	17.35		27.30	37
8237	3/8"		22	.364		14.15	18.95		33.10	44
8238	1/2"		22	.364		21	18.95		39.95	51.50
8239	5/8"		21	.381		27.50	19.85		47.35	59.50
8240	3/4"		21	.381		34	19.85		53.85	67
8241	7/8"		20	.400		56	21		77	92.50
8242	1"	↓	20	.400	↓	59	21		80	96
8250	Male connector									

22 11 13.61 Tubing Fittings, Stainless Steel		Crew	Daily Output	Labor-Hours	Unit	Material	2010 Bare Costs Labor	Equipment	Total	Total Incl O&P
8252	1/4" x 1/4"	1 Plum	24	.333	Ea.	6.40	17.35		23.75	33
8253	3/8" x 3/8"		22	.364		9.95	18.95		28.90	39.50
8254	1/2" x 1/2"		22	.364		14.70	18.95		33.65	44.50
8256	3/4" x 3/4"		21	.381		22.50	19.85		42.35	54.50
8258	1" x 1"		20	.400		39.50	21		60.50	74.50

22 11 13.64 Pipe, Stainless Steel		Crew	Daily Output	Labor-Hours	Unit	Material	Labor	Equipment	Total	Total Incl O&P
0010	**PIPE, STAINLESS STEEL**									
0020	Welded, with clevis type hanger assemblies, 10' O.C.									
0500	Schedule 5, type 304									
0540	1/2" diameter	Q-15	128	.125	L.F.	7.35	5.85	.46	13.66	17.40
0550	3/4" diameter		116	.138		9.40	6.45	.50	16.35	20.50
0560	1" diameter		103	.155		11.80	7.30	.57	19.67	24.50
0570	1-1/4" diameter		93	.172		16.55	8.05	.63	25.23	31
0580	1-1/2" diameter		85	.188		22.50	8.80	.69	31.99	38.50
0590	2" diameter		69	.232		37.50	10.85	.85	49.20	58
0600	2-1/2" diameter		53	.302		41	14.15	1.10	56.25	67.50
0610	3" diameter		48	.333		50	15.60	1.22	66.82	80
0620	4" diameter		44	.364		63.50	17.05	1.33	81.88	97
0630	5" diameter		36	.444		130	21	1.63	152.63	176
0640	6" diameter	Q-16	42	.571		120	28	1.39	149.39	175
0650	8" diameter		34	.706		185	34.50	1.72	221.22	257
0660	10" diameter		26	.923		256	45	2.25	303.25	350
0670	12" diameter		21	1.143		340	55.50	2.79	398.29	455
0700	To delete hangers, subtract									
0710	1/2" diam. to 1-1/2" diam.					8%	19%			
0720	2" diam. to 5" diam.					4%	9%			
0730	6" diam. to 12" diam.					3%	4%			
0750	For small quantities, add				L.F.	10%				
1250	Schedule 5, type 316									
1290	1/2" diameter	Q-15	128	.125	L.F.	10.35	5.85	.46	16.66	20.50
1300	3/4" diameter		116	.138		13	6.45	.50	19.95	24.50
1310	1" diameter		103	.155		16.80	7.30	.57	24.67	30
1320	1-1/4" diameter		93	.172		20.50	8.05	.63	29.18	35.50
1330	1-1/2" diameter		85	.188		29	8.80	.69	38.49	46
1340	2" diameter		69	.232		49.50	10.85	.85	61.20	71.50
1350	2-1/2" diameter		53	.302		58	14.15	1.10	73.25	86
1360	3" diameter		48	.333		74	15.60	1.22	90.82	106
1370	4" diameter		44	.364		92	17.05	1.33	110.38	128
1380	5" diameter		36	.444		176	21	1.63	198.63	227
1390	6" diameter	Q-16	42	.571		171	28	1.39	200.39	231
1400	8" diameter		34	.706		258	34.50	1.72	294.22	335
1410	10" diameter		26	.923		370	45	2.25	417.25	475
1420	12" diameter		21	1.143		475	55.50	2.79	533.29	610
1490	For small quantities, add					10%				
1940	To delete hanger, subtract									
1950	1/2" diam. to 1-1/2" diam.					5%	19%			
1960	2" diam. to 5" diam.					3%	9%			
1970	6" diam. to 12" diam.					2%	4%			
2000	Schedule 10, type 304									
2040	1/4" diameter	Q-15	131	.122	L.F.	4.52	5.70	.45	10.67	14.05
2050	3/8" diameter		128	.125		5.15	5.85	.46	11.46	14.95
2060	1/2" diameter		125	.128		5.25	6	.47	11.72	15.25

22 11 13.64 Pipe, Stainless Steel		Crew	Daily Output	Labor-Hours	Unit	Material	Labor	Equipment	Total	Total Incl O&P
						2010 Bare Costs				
2070	3/4" diameter	Q-15	113	.142	L.F.	6.60	6.65	.52	13.77	17.75
2080	1" diameter		100	.160		10.45	7.50	.59	18.54	23.50
2090	1-1/4" diameter		91	.176		10.65	8.25	.64	19.54	25
2100	1-1/2" diameter		83	.193		11.70	9.05	.71	21.46	27
2110	2" diameter		67	.239		14.25	11.20	.87	26.32	33.50
2120	2-1/2" diameter		51	.314		18.75	14.70	1.15	34.60	44
2130	3" diameter		46	.348		23	16.30	1.27	40.57	51.50
2140	4" diameter		42	.381		34	17.85	1.39	53.24	66
2150	5" diameter		35	.457		46	21.50	1.67	69.17	84.50
2160	6" diameter	Q-16	40	.600		50	29	1.46	80.46	100
2170	8" diameter		33	.727		85.50	35.50	1.77	122.77	149
2180	10" diameter		25	.960		119	46.50	2.34	167.84	204
2190	12" diameter		21	1.143		149	55.50	2.79	207.29	250
2250	For small quantities, add					10%				
2650	To delete hanger, subtract									
2660	1/4" diam. to 3/4" diam.					9%	22%			
2670	1" diam. to 2" diam.					4%	15%			
2680	2-1/2" diam. to 5" diam.					3%	8%			
2690	6" diam. to 12" diam.					3%	4%			
2750	Schedule 10, type 316									
2790	1/4" diameter	Q-15	131	.122	L.F.	5.35	5.70	.45	11.50	15
2800	3/8" diameter		128	.125		6.55	5.85	.46	12.86	16.50
2810	1/2" diameter		125	.128		6.95	6	.47	13.42	17.15
2820	3/4" diameter		113	.142		9.60	6.65	.52	16.77	21
2830	1" diameter		100	.160		12.80	7.50	.59	20.89	26
2840	1-1/4" diameter		91	.176		15.80	8.25	.64	24.69	30.50
2850	1-1/2" diameter		83	.193		21.50	9.05	.71	31.26	38
2860	2" diameter		67	.239		22.50	11.20	.87	34.57	42.50
2870	2-1/2" diameter		51	.314		29	14.70	1.15	44.85	55.50
2880	3" diameter		46	.348		36	16.30	1.27	53.57	65.50
2890	4" diameter		42	.381		44.50	17.85	1.39	63.74	77.50
2900	5" diameter		35	.457		63	21.50	1.67	86.17	103
2910	6" diameter	Q-16	40	.600		73	29	1.46	103.46	126
2920	8" diameter		33	.727		123	35.50	1.77	160.27	190
2930	10" diameter		25	.960		162	46.50	2.34	210.84	251
2940	12" diameter		21	1.143		203	55.50	2.79	261.29	310
2990	For small quantities, add					10%				
3430	To delete hangor, subtract									
3440	1/4" diam. to 3/4" diam.					6%	22%			
3450	1" diam. to 2" diam.					3%	15%			
3460	2-1/2" diam. to 5" diam.					2%	8%			
3470	6" diam. to 12" diam.					2%	4%			
3500	Threaded, couplings and clevis hanger assemblies, 10' O.C.									
3520	Schedule 40, type 304									
3540	1/4" diameter	1 Plum	54	.148	L.F.	7.35	7.70		15.05	19.65
3550	3/8" diameter		53	.151		7.45	7.85		15.30	19.95
3560	1/2" diameter		52	.154		8.25	8		16.25	21
3570	3/4" diameter		51	.157		10.45	8.15		18.60	24
3580	1" diameter		45	.178		14.70	9.25		23.95	30
3590	1-1/4" diameter	Q-1	76	.211		20	9.85		29.85	37
3600	1-1/2" diameter		69	.232		26	10.85		36.85	45
3610	2" diameter		57	.281		32.50	13.15		45.65	55
3620	2-1/2" diameter		44	.364		66	17.05		83.05	98.50

22 11 13.64 Pipe, Stainless Steel		Crew	Daily Output	Labor-Hours	Unit	Material	2010 Bare Costs Labor	Equipment	Total	Total Incl O&P
3630	3" diameter	Q-1	38	.421	L.F.	78.50	19.75		98.25	116
3640	4" diameter	Q-2	51	.471		102	23		125	147
3740	For small quantities, add					10%				
4200	To delete couplings & hangers, subtract									
4210	1/4" diam. to 3/4" diam.					15%	56%			
4220	1" diam. to 2" diam.					18%	49%			
4230	2-1/2" diam. to 4" diam.					34%	40%			
4250	Schedule 40, type 316									
4290	1/4" diameter	1 Plum	54	.148	L.F.	8.90	7.70		16.60	21.50
4300	3/8" diameter		53	.151		9.60	7.85		17.45	22.50
4310	1/2" diameter		52	.154		13.20	8		21.20	26.50
4320	3/4" diameter		51	.157		17.20	8.15		25.35	31
4330	1" diameter		45	.178		25	9.25		34.25	41.50
4340	1-1/4" diameter	Q-1	76	.211		35.50	9.85		45.35	54
4350	1-1/2" diameter		69	.232		39	10.85		49.85	59.50
4360	2" diameter		57	.281		55.50	13.15		68.65	81
4370	2-1/2" diameter		44	.364		93	17.05		110.05	128
4380	3" diameter		38	.421		122	19.75		141.75	165
4390	4" diameter	Q-2	51	.471		167	23		190	218
4490	For small quantities, add					10%				
4900	To delete couplings & hangers, subtract									
4910	1/4" diam. to 3/4" diam.					12%	56%			
4920	1" diam. to 2" diam.					14%	49%			
4930	2-1/2" diam. to 4" diam.					27%	40%			
5000	Schedule 80, type 304									
5040	1/4" diameter	1 Plum	53	.151	L.F.	16.50	7.85		24.35	30
5050	3/8" diameter		52	.154		17.85	8		25.85	31.50
5060	1/2" diameter		51	.157		23	8.15		31.15	37.50
5070	3/4" diameter		48	.167		27	8.70		35.70	42.50
5080	1" diameter		43	.186		32.50	9.70		42.20	50
5090	1-1/4" diameter	Q-1	73	.219		42.50	10.25		52.75	62.50
5100	1-1/2" diameter		67	.239		53.50	11.20		64.70	75.50
5110	2" diameter		54	.296		70.50	13.90		84.40	98.50
5190	For small quantities, add					10%				
5700	To delete couplings & hangers, subtract									
5710	1/4" diam. to 3/4" diam.					10%	53%			
5720	1" diam. to 2" diam.					14%	47%			
5750	Schedule 80, type 316									
5790	1/4" diameter	1 Plum	53	.151	L.F.	19.15	7.85		27	33
5800	3/8" diameter		52	.154		20.50	8		28.50	34.50
5810	1/2" diameter		51	.157		27	8.15		35.15	42
5820	3/4" diameter		48	.167		36.50	8.70		45.20	53
5830	1" diameter		43	.186		43	9.70		52.70	62
5840	1-1/4" diameter	Q-1	73	.219		62	10.25		72.25	84
5850	1-1/2" diameter		67	.239		69.50	11.20		80.70	93
5860	2" diameter		54	.296		98.50	13.90		112.40	129
5950	For small quantities, add					10%				
7000	To delete couplings & hangers, subtract									
7010	1/4" diam. to 3/4" diam.					9%	53%			
7020	1" diam. to 2" diam.					14%	47%			
8000	Weld joints with clevis type hanger assemblies, 10' O.C.									
8010	Schedule 40, type 304									
8050	1/8" pipe size	Q-15	126	.127	L.F.	5.85	5.95	.46	12.26	15.85

22 11 13 – Facility Water Distribution Piping

22 11 13.64 Pipe, Stainless Steel		Crew	Daily Output	Labor-Hours	Unit	2010 Bare Costs Material	Labor	Equipment	Total	Total Incl O&P
8060	1/4" pipe size	Q-15	125	.128	L.F.	6.05	6	.47	12.52	16.20
8070	3/8" pipe size		122	.131		5.95	6.15	.48	12.58	16.30
8080	1/2" pipe size		118	.136		6.25	6.35	.50	13.10	16.90
8090	3/4" pipe size		109	.147		7.80	6.90	.54	15.24	19.45
8100	1" pipe size		95	.168		10.35	7.90	.62	18.87	24
8110	1-1/4" pipe size		86	.186		13.25	8.70	.68	22.63	28.50
8120	1-1/2" pipe size		78	.205		18.05	9.60	.75	28.40	35
8130	2" pipe size		62	.258		20	12.10	.94	33.04	41
8140	2-1/2" pipe size		49	.327		37.50	15.30	1.20	54	66
8150	3" pipe size		44	.364		39.50	17.05	1.33	57.88	70.50
8160	3-1/2" pipe size		44	.364		42	17.05	1.33	60.38	73
8170	4" pipe size		39	.410		44.50	19.20	1.50	65.20	79.50
8180	5" pipe size		32	.500		89.50	23.50	1.83	114.83	136
8190	6" pipe size	Q-16	37	.649		109	31.50	1.58	142.08	169
8200	8" pipe size		29	.828		153	40	2.02	195.02	231
8210	10" pipe size		24	1		259	48.50	2.44	309.94	360
8220	12" pipe size		20	1.200		293	58.50	2.93	354.43	410
8300	Schedule 40, type 316									
8310	1/8" pipe size	Q-15	126	.127	L.F.	8.05	5.95	.46	14.46	18.25
8320	1/4" pipe size		125	.128		7.40	6	.47	13.87	17.60
8330	3/8" pipe size		122	.131		7.80	6.15	.48	14.43	18.35
8340	1/2" pipe size		118	.136		10.75	6.35	.50	17.60	22
8350	3/4" pipe size		109	.147		13.95	6.90	.54	21.39	26
8360	1" pipe size		95	.168		19.75	7.90	.62	28.27	34
8370	1-1/4" pipe size		86	.186		27	8.70	.68	36.38	44
8380	1-1/2" pipe size		78	.205		29.50	9.60	.75	39.85	47.50
8390	2" pipe size		62	.258		41	12.10	.94	54.04	64
8400	2-1/2" pipe size		49	.327		59	15.30	1.20	75.50	89
8410	3" pipe size		44	.364		75.50	17.05	1.33	93.88	110
8420	3-1/2" pipe size		44	.364		89	17.05	1.33	107.38	125
8430	4" pipe size		39	.410		98	19.20	1.50	118.70	139
8440	5" pipe size		32	.500		132	23.50	1.83	157.33	182
8450	6" pipe size	Q-16	37	.649		154	31.50	1.58	187.08	219
8460	8" pipe size		29	.828		290	40	2.02	332.02	385
8470	10" pipe size		24	1		385	48.50	2.44	435.94	500
8480	12" pipe size		20	1.200		525	58.50	2.93	586.43	670
8500	Schedule 80, type 304									
8510	1/4" pipe size	Q-15	110	.145	L.F.	15.05	6.80	.53	22.38	27.50
8520	3/8" pipe size		109	.147		16.15	6.90	.54	23.59	28.50
8530	1/2" pipe size		106	.151		21	7.05	.55	28.60	34
8540	3/4" pipe size		96	.167		24	7.80	.61	32.41	39
8550	1" pipe size		87	.184		28	8.60	.67	37.27	44
8560	1-1/4" pipe size		81	.198		31.50	9.25	.72	41.47	49
8570	1-1/2" pipe size		74	.216		40.50	10.15	.79	51.44	60.50
8580	2" pipe size		58	.276		53.50	12.90	1.01	67.41	79.50
8590	2-1/2" pipe size		46	.348		109	16.30	1.27	126.57	146
8600	3" pipe size		41	.390		147	18.30	1.43	166.73	191
8610	4" pipe size		33	.485		230	22.50	1.77	254.27	289
8630	6" pipe size	Q-16	30	.800		505	39	1.95	545.95	620
8640	Schedule 80, type 316									
8650	1/4" pipe size	Q-15	110	.145	L.F.	17.45	6.80	.53	24.78	30
8660	3/8" pipe size		109	.147		18.65	6.90	.54	26.09	31.50
8670	1/2" pipe size		106	.151		24.50	7.05	.55	32.10	38

22 11 13.64 Pipe, Stainless Steel		Crew	Daily Output	Labor-Hours	Unit	Material	2010 Bare Costs Labor	Equipment	Total	Total Incl O&P
8680	3/4" pipe size	Q-15	96	.167	L.F.	32.50	7.80	.61	40.91	48.50
8690	1" pipe size		87	.184		37	8.60	.67	46.27	54
8700	1-1/4" pipe size		81	.198		48	9.25	.72	57.97	67
8710	1-1/2" pipe size		74	.216		52.50	10.15	.79	63.44	74
8720	2" pipe size		58	.276		75.50	12.90	1.01	89.41	103
8730	2-1/2" pipe size		46	.348		203	16.30	1.27	220.57	250
8740	3" pipe size		41	.390		237	18.30	1.43	256.73	290
8760	4" pipe size		33	.485		330	22.50	1.77	354.27	400
8770	6" pipe size	Q-16	30	.800		890	39	1.95	930.95	1,050
9100	Threading pipe labor, sst, one end, schedules 40 & 80									
9110	1/4" through 3/4" pipe size	1 Plum	61.50	.130	Ea.		6.75		6.75	10.15
9120	1" through 2" pipe size		55.90	.143			7.45		7.45	11.15
9130	2-1/2" pipe size		41.50	.193			10.05		10.05	15.05
9140	3" pipe size		38.50	.208			10.80		10.80	16.20
9150	3-1/2" pipe size	Q-1	68.40	.234			10.95		10.95	16.45
9160	4" pipe size		73	.219			10.25		10.25	15.40
9170	5" pipe size		40.70	.393			18.40		18.40	27.50
9180	6" pipe size		35.40	.452			21		21	31.50
9190	8" pipe size		22.30	.717			33.50		33.50	50.50
9200	10" pipe size		16.10	.994			46.50		46.50	70
9210	12" pipe size		12.30	1.301			61		61	91.50
9250	Welding labor per joint for stainless steel									
9260	Schedule 5 and 10									
9270	1/4" pipe size	Q-15	36	.444	Ea.		21	1.63	22.63	33
9280	3/8" pipe size		35	.457			21.50	1.67	23.17	34
9290	1/2" pipe size		35	.457			21.50	1.67	23.17	34
9300	3/4" pipe size		28	.571			27	2.09	29.09	42.50
9310	1" pipe size		25	.640			30	2.34	32.34	47.50
9320	1-1/4" pipe size		22	.727			34	2.66	36.66	54
9330	1-1/2" pipe size		21	.762			35.50	2.79	38.29	56.50
9340	2" pipe size		18	.889			41.50	3.25	44.75	66
9350	2-1/2" pipe size		12	1.333			62.50	4.88	67.38	99
9360	3" pipe size		9.73	1.644			77	6	83	122
9370	4" pipe size		7.37	2.171			102	7.95	109.95	161
9380	5" pipe size		6.15	2.602			122	9.50	131.50	194
9390	6" pipe size		5.71	2.802			131	10.25	141.25	208
9400	8" pipe size		3.69	4.336			203	15.85	218.85	320
9410	10" pipe size		2.91	5.498			258	20	278	405
9420	12" pipe size		2.31	6.926			325	25.50	350.50	515
9500	Schedule 40									
9510	1/4" pipe size	Q-15	28	.571	Ea.		27	2.09	29.09	42.50
9520	3/8" pipe size		27	.593			28	2.17	30.17	44
9530	1/2" pipe size		25.40	.630			29.50	2.31	31.81	46.50
9540	3/4" pipe size		22.22	.720			33.50	2.64	36.14	53.50
9550	1" pipe size		20.25	.790			37	2.89	39.89	58.50
9560	1-1/4" pipe size		18.82	.850			40	3.11	43.11	63
9570	1-1/2" pipe size		17.78	.900			42	3.29	45.29	66.50
9580	2" pipe size		15.09	1.060			49.50	3.88	53.38	79
9590	2-1/2" pipe size		7.96	2.010			94	7.35	101.35	149
9600	3" pipe size		6.43	2.488			117	9.10	126.10	185
9610	4" pipe size		4.88	3.279			154	12	166	243
9620	5" pipe size		4.26	3.756			176	13.75	189.75	279
9630	6" pipe size		3.77	4.244			199	15.55	214.55	315

22 11 13.64 Pipe, Stainless Steel		Crew	Daily Output	Labor-Hours	Unit	2010 Bare Costs Material	Labor	Equipment	Total	Total Incl O&P
9640	8" pipe size	Q-15	2.44	6.557	Ea.		305	24	329	485
9650	10" pipe size		1.92	8.333			390	30.50	420.50	620
9660	12" pipe size		1.52	10.526			495	38.50	533.50	785
9750	Schedule 80									
9760	1/4" pipe size	Q-15	21.55	.742	Ea.		35	2.72	37.72	55
9770	3/8" pipe size		20.75	.771			36	2.82	38.82	57
9780	1/2" pipe size		19.54	.819			38.50	3	41.50	61
9790	3/4" pipe size		17.09	.936			44	3.43	47.43	70
9800	1" pipe size		15.58	1.027			48	3.76	51.76	76
9810	1-1/4" pipe size		14.48	1.105			52	4.04	56.04	82
9820	1-1/2" pipe size		13.68	1.170			55	4.28	59.28	86.50
9830	2" pipe size		11.61	1.378			64.50	5.05	69.55	103
9840	2-1/2" pipe size		6.12	2.614			122	9.55	131.55	195
9850	3" pipe size		4.94	3.239			152	11.85	163.85	240
9860	4" pipe size		3.75	4.267			200	15.60	215.60	315
9870	5" pipe size		3.27	4.893			229	17.90	246.90	365
9880	6" pipe size		2.90	5.517			258	20	278	405
9890	8" pipe size		1.87	8.556			400	31.50	431.50	635
9900	10" pipe size		1.48	10.811			505	39.50	544.50	805
9910	12" pipe size		1.17	13.675			640	50	690	1,025
9920	Schedule 160, 1/2" pipe size		17	.941			44	3.44	47.44	70
9930	3/4" pipe size		14.81	1.080			50.50	3.95	54.45	80.50
9940	1" pipe size		13.50	1.185			55.50	4.34	59.84	88
9950	1-1/4" pipe size		12.55	1.275			59.50	4.67	64.17	94.50
9960	1-1/2" pipe size		11.85	1.350			63.50	4.94	68.44	100
9970	2" pipe size		10	1.600			75	5.85	80.85	118
9980	3" pipe size		4.28	3.738			175	13.70	188.70	278
9990	4" pipe size		3.25	4.923			231	18	249	365

22 11 13.66 Pipe Fittings, Stainless Steel		Crew	Daily Output	Labor-Hours	Unit	Material	Labor	Equipment	Total	Total Incl O&P
0010	**PIPE FITTINGS, STAINLESS STEEL**									
0100	Butt weld joint, schedule 5, type 304									
0120	90° Elbow, long									
0140	1/2"	Q-15	17.50	.914	Ea.	19.80	43	3.35	66.15	89.50
0150	3/4"		14	1.143		19.80	53.50	4.18	77.48	107
0160	1"		12.50	1.280		21	60	4.68	85.68	118
0170	1-1/4"		11	1.455		28.50	68	5.30	101.80	139
0180	1-1/2"		10.50	1.524		24	71.50	5.60	101.10	140
0190	2"		9	1.778		28.50	83.50	6.50	118.50	164
0200	2-1/2"		6	2.667		66	125	9.75	200.75	270
0210	3"		4.86	3.292		72.50	154	12.05	238.55	325
0220	3-1/2"		4.27	3.747		182	176	13.70	371.70	480
0230	4"		3.69	4.336		99	203	15.85	317.85	430
0240	5"		3.08	5.195		385	243	19	647	810
0250	6"	Q-16	4.29	5.594		297	272	13.65	582.65	745
0260	8"		2.76	8.696		625	420	21	1,066	1,350
0270	10"		2.18	11.009		970	535	27	1,532	1,900
0280	12"		1.73	13.873		1,375	675	34	2,084	2,550
0320	For schedule 5, type 316, add					30%				
0600	45° Elbow, long									
0620	1/2"	Q-15	17.50	.914	Ea.	19.80	43	3.35	66.15	89.50
0630	3/4"		14	1.143		19.80	53.50	4.18	77.48	107
0640	1"		12.50	1.280		21	60	4.68	85.68	118

22 11 13 – Facility Water Distribution Piping

22 11 13.66 Pipe Fittings, Stainless Steel		Crew	Daily Output	Labor-Hours	Unit	Material	Labor	Equipment	Total	Total Incl O&P
						2010 Bare Costs				
0650	1-1/4"	Q-15	11	1.455	Ea.	28.50	68	5.30	101.80	139
0660	1-1/2"		10.50	1.524		24	71.50	5.60	101.10	140
0670	2"		9	1.778		28.50	83.50	6.50	118.50	164
0680	2-1/2"		6	2.667		66	125	9.75	200.75	270
0690	3"		4.86	3.292		72.50	154	12.05	238.55	325
0700	3-1/2"		4.27	3.747		182	176	13.70	371.70	480
0710	4"		3.69	4.336		83.50	203	15.85	302.35	415
0720	5"		3.08	5.195		310	243	19	572	725
0730	6"	Q-16	4.29	5.594		209	272	13.65	494.65	650
0740	8"		2.76	8.696		440	420	21	881	1,150
0750	10"		2.18	11.009		770	535	27	1,332	1,675
0760	12"		1.73	13.873		965	675	34	1,674	2,075
0800	For schedule 5, type 316, add					25%				
1100	Tee, straight									
1130	1/2"	Q-15	11.66	1.372	Ea.	59.50	64.50	5	129	168
1140	3/4"		9.33	1.715		59.50	80.50	6.30	146.30	192
1150	1"		8.33	1.921		62.50	90	7.05	159.55	212
1160	1-1/4"		7.33	2.183		50.50	102	8	160.50	217
1170	1-1/2"		7	2.286		49.50	107	8.35	164.85	225
1180	2"		6	2.667		51.50	125	9.75	186.25	255
1190	2-1/2"		4	4		124	187	14.65	325.65	435
1200	3"		3.24	4.938		198	231	18.05	447.05	585
1210	3-1/2"		2.85	5.614		260	263	20.50	543.50	705
1220	4"		2.46	6.504		141	305	24	470	635
1230	5"		2	8		445	375	29.50	849.50	1,075
1240	6"	Q-16	2.85	8.421		350	410	20.50	780.50	1,025
1250	8"		1.84	13.043		750	635	32	1,417	1,800
1260	10"		1.45	16.552		1,200	805	40.50	2,045.50	2,575
1270	12"		1.15	20.870		1,675	1,025	51	2,751	3,425
1320	For schedule 5, type 316, add					25%				
2000	Butt weld joint, schedule 10, type 304									
2020	90° elbow, long									
2040	1/2"	Q-15	17	.941	Ea.	14.85	44	3.44	62.29	86
2050	3/4"		14	1.143		16.35	53.50	4.18	74.03	103
2060	1"		12.50	1.280		12.70	60	4.68	77.38	109
2070	1-1/4"		11	1.455		21.50	68	5.30	94.80	131
2080	1-1/2"		10.50	1.524		18.15	71.50	5.60	95.25	133
2090	2"		9	1.778		21.50	83.50	6.50	111.50	156
2100	2-1/2"		6	2.667		49.50	125	9.75	184.25	252
2110	3"		4.86	3.292		46	154	12.05	212.05	295
2120	3-1/2"		4.27	3.747		136	176	13.70	325.70	430
2130	4"		3.69	4.336		77.50	203	15.85	296.35	410
2140	5"		3.08	5.195		289	243	19	551	705
2150	6"	Q-16	4.29	5.594		223	272	13.65	508.65	665
2160	8"		2.76	8.696		470	420	21	911	1,175
2170	10"		2.18	11.009		725	535	27	1,287	1,625
2180	12"		1.73	13.873		1,025	675	34	1,734	2,150
2500	45° elbow, long									
2520	1/2"	Q-15	17.50	.914	Ea.	14.85	43	3.35	61.20	84
2530	3/4"		14	1.143		14.85	53.50	4.18	72.53	101
2540	1"		12.50	1.280		15.70	60	4.68	80.38	112
2550	1-1/4"		11	1.455		21.50	68	5.30	94.80	131
2560	1-1/2"		10.50	1.524		18.15	71.50	5.60	95.25	133

22 11 13.66 Pipe Fittings, Stainless Steel		Crew	Daily Output	Labor-Hours	Unit	Material	Labor	Equipment	Total	Total Incl O&P
						2010 Bare Costs				
2570	2"	Q-15	9	1.778	Ea.	21.50	83.50	6.50	111.50	156
2580	2-1/2"		6	2.667		49.50	125	9.75	184.25	252
2590	3"		4.86	3.292		37	154	12.05	203.05	285
2600	3-1/2"		4.27	3.747		136	176	13.70	325.70	430
2610	4"		3.69	4.336		62.50	203	15.85	281.35	390
2620	5"		3.08	5.195		231	243	19	493	640
2630	6"	Q-16	4.29	5.594		157	272	13.65	442.65	590
2640	8"		2.76	8.696		330	420	21	771	1,025
2650	10"		2.18	11.009		580	535	27	1,142	1,475
2660	12"		1.73	13.873		720	675	34	1,429	1,825
2670	Reducer, concentric									
2674	1" x 3/4"	Q-15	13.25	1.208	Ea.	32	56.50	4.42	92.92	125
2676	2" x 1-1/2"	"	9.75	1.641		26.50	77	6	109.50	151
2678	6" x 4"	Q-16	4.91	4.888		79	237	11.95	327.95	455
2680	Caps									
2682	1"	Q-15	25	.640	Ea.	33	30	2.34	65.34	84
2684	1-1/2"		21	.762		39.50	35.50	2.79	77.79	100
2685	2"		18	.889		37	41.50	3.25	81.75	107
2686	4"		7.38	2.168		59.50	102	7.95	169.45	226
2687	6"	Q-16	8.58	2.797		95	136	6.85	237.85	315
3000	Tee, straight									
3030	1/2"	Q-15	11.66	1.372	Ea.	44.50	64.50	5	114	151
3040	3/4"		9.33	1.715		44.50	80.50	6.30	131.30	176
3050	1"		8.33	1.921		47	90	7.05	144.05	194
3060	1-1/4"		7.33	2.183		57.50	102	8	167.50	225
3070	1-1/2"		7	2.286		37	107	8.35	152.35	211
3080	2"		6	2.667		39	125	9.75	173.75	240
3090	2-1/2"		4	4		93	187	14.65	294.65	400
3100	3"		3.24	4.938		73.50	231	18.05	322.55	445
3110	3-1/2"		2.85	5.614		108	263	20.50	391.50	535
3120	4"		2.46	6.504		106	305	24	435	595
3130	5"		2	8		335	375	29.50	739.50	960
3140	6"	Q-16	2.85	8.421		264	410	20.50	694.50	930
3150	8"		1.84	13.043		560	635	32	1,227	1,600
3151	10"		1.45	16.552		910	805	40.50	1,755.50	2,250
3152	12"		1.15	20.870		1,250	1,025	51	2,326	2,975
3154	For schedule 10, type 316, add					25%				
3281	Butt weld joint, schedule 40, type 304									
3284	90° Elbow, long, 1/2"	Q-15	12.70	1.260	Ea.	17.35	59	4.61	80.96	113
3288	3/4"		11.10	1.441		17.35	67.50	5.30	90.15	126
3289	1"		10.13	1.579		18.15	74	5.80	97.95	137
3290	1-1/4"		9.40	1.702		24	79.50	6.25	109.75	153
3300	1-1/2"		8.89	1.800		19	84.50	6.60	110.10	154
3310	2"		7.55	2.119		27	99.50	7.75	134.25	188
3320	2-1/2"		3.98	4.020		52	188	14.70	254.70	355
3330	3"		3.21	4.984		67	234	18.25	319.25	445
3340	3-1/2"		2.83	5.654		248	265	20.50	533.50	690
3350	4"		2.44	6.557		116	305	24	445	615
3360	5"		2.13	7.512		390	350	27.50	767.50	985
3370	6"	Q-16	2.83	8.481		340	410	20.50	770.50	1,025
3380	8"		1.83	13.115		675	635	32	1,342	1,725
3390	10"		1.44	16.667		1,425	810	40.50	2,275.50	2,850
3400	12"		1.14	21.053		1,825	1,025	51.50	2,901.50	3,575

22 11 13 – Facility Water Distribution Piping

22 11 13.66 Pipe Fittings, Stainless Steel	Crew	Daily Output	Labor-Hours	Unit	Material	2010 Bare Costs Labor	Equipment	Total	Total Incl O&P
3410 For schedule 40, type 316, add					25%				
3460 45° Elbow, long, 1/2"	Q-15	12.70	1.260	Ea.	17.35	59	4.61	80.96	113
3470 3/4"		11.10	1.441		17.35	67.50	5.30	90.15	126
3480 1"		10.13	1.579		18.15	74	5.80	97.95	137
3490 1-1/4"		9.40	1.702		24	79.50	6.25	109.75	153
3500 1-1/2"		8.89	1.800		19	84.50	6.60	110.10	154
3510 2"		7.55	2.119		27	99.50	7.75	134.25	188
3520 2-1/2"		3.98	4.020		52	188	14.70	254.70	355
3530 3"		3.21	4.984		53	234	18.25	305.25	430
3540 3-1/2"		2.83	5.654		248	265	20.50	533.50	690
3550 4"		2.44	6.557		82.50	305	24	411.50	580
3560 5"		2.13	7.512		272	350	27.50	649.50	860
3570 6"	Q-16	2.83	8.481		239	410	20.50	669.50	905
3580 8"		1.83	13.115		475	635	32	1,142	1,500
3590 10"		1.44	16.667		1,000	810	40.50	1,850.50	2,375
3600 12"		1.14	21.053		1,275	1,025	51.50	2,351.50	2,975
3610 For schedule 40, type 316, add					25%				
3660 Tee, straight 1/2"	Q-15	8.46	1.891	Ea.	44.50	88.50	6.90	139.90	190
3670 3/4"		7.40	2.162		44.50	101	7.90	153.40	210
3680 1"		6.74	2.374		47	111	8.70	166.70	228
3690 1-1/4"		6.27	2.552		110	120	9.35	239.35	310
3700 1-1/2"		5.92	2.703		50.50	127	9.90	187.40	256
3710 2"		5.03	3.181		75	149	11.65	235.65	320
3720 2-1/2"		2.65	6.038		95	283	22	400	555
3730 3"		2.14	7.477		95.50	350	27.50	473	660
3740 3-1/2"		1.88	8.511		181	400	31	612	835
3750 4"		1.62	9.877		181	465	36	682	935
3760 5"		1.42	11.268		415	530	41	986	1,300
3770 6"	Q-16	1.88	12.766		370	620	31	1,021	1,375
3780 8"		1.22	19.672		810	955	48	1,813	2,375
3790 10"		.96	25		1,575	1,225	61	2,861	3,625
3800 12"		.76	31.579		2,100	1,525	77	3,702	4,675
3810 For schedule 40, type 316, add					25%				
3820 Tee, reducing on outlet, 3/4" x 1/2"	Q-15	7.73	2.070	Ea.	59.50	97	7.60	164.10	219
3822 1" x 1/2"		7.24	2.210		64.50	104	8.10	176.60	235
3824 1" x 3/4"		6.96	2.299		59.50	108	8.40	175.90	236
3826 1-1/4" x 1"		6.43	2.488		145	117	9.10	271.10	345
3828 1-1/2" x 1/2"		6.58	2.432		87.50	114	8.90	210.40	277
3830 1-1/2" x 3/4"		6.35	2.520		81.50	118	9.20	208.70	277
3832 1-1/2" x 1"		6.18	2.589		62.50	121	9.50	193	261
3834 2" x 1"		5.50	2.909		112	136	10.65	258.65	340
3836 2" x 1-1/2"		5.30	3.019		94	141	11.05	246.05	325
3838 2-1/2" x 2"		3.15	5.079		142	238	18.60	398.60	530
3840 3" x 1-1/2"		2.72	5.882		144	276	21.50	441.50	595
3842 3" x 2"		2.65	6.038		120	283	22	425	580
3844 4" x 2"		2.10	7.619		299	355	28	682	895
3846 4" x 3"		1.77	9.040		217	425	33	675	910
3848 5" x 4"		1.48	10.811		500	505	39.50	1,044.50	1,350
3850 6" x 3"	Q-16	2.19	10.959		555	530	26.50	1,111.50	1,450
3852 6" x 4"		2.04	11.765		480	570	28.50	1,078.50	1,400
3854 8" x 4"		1.46	16.438		1,125	800	40	1,965	2,475
3856 10" x 8"		.69	34.783		1,875	1,700	85	3,660	4,700
3858 12" x 10"		.55	43.636		2,525	2,125	106	4,756	6,075

22 11 13 – Facility Water Distribution Piping

22 11 13.66 Pipe Fittings, Stainless Steel		Crew	Daily Output	Labor-Hours	Unit	Material	Labor	Equipment	Total	Total Incl O&P
						2010 Bare Costs				
3950	Reducer, concentric, 3/4" x 1/2"	Q-15	11.85	1.350	Ea.	33	63.50	4.94	101.44	137
3952	1" x 3/4"		10.60	1.509		34.50	70.50	5.50	110.50	150
3954	1-1/4" x 3/4"		10.19	1.570		85	73.50	5.75	164.25	210
3956	1-1/4" x 1"		9.76	1.639		43	77	6	126	169
3958	1-1/2" x 3/4"		9.88	1.619		72	76	5.95	153.95	200
3960	1-1/2" x 1"		9.47	1.690		58	79	6.20	143.20	189
3962	2" x 1"		8.65	1.850		25	86.50	6.75	118.25	164
3964	2" x 1-1/2"		8.16	1.961		29	92	7.20	128.20	178
3966	2-1/2" x 1"		5.71	2.802		125	131	10.25	266.25	345
3968	2-1/2" x 2"		5.21	3.071		63.50	144	11.25	218.75	298
3970	3" x 1"		4.88	3.279		86	154	12	252	340
3972	3" x 1-1/2"		4.72	3.390		43.50	159	12.40	214.90	300
3974	3" x 2"		4.51	3.548		41.50	166	13	220.50	310
3976	4" x 2"		3.69	4.336		58.50	203	15.85	277.35	385
3978	4" x 3"		2.77	5.776		43.50	271	21	335.50	475
3980	5" x 3"		2.56	6.250		261	293	23	577	750
3982	5" x 4"		2.27	7.048		210	330	26	566	755
3984	6" x 3"	Q-16	3.57	6.723		134	325	16.40	475.40	655
3986	6" x 4"		3.19	7.524		111	365	18.35	494.35	690
3988	8" x 4"		2.44	9.836		345	480	24	849	1,125
3990	8" x 6"		2.22	10.811		259	525	26.50	810.50	1,100
3992	10" x 6"		1.91	12.565		485	610	30.50	1,125.50	1,475
3994	10" x 8"		1.61	14.907		400	725	36.50	1,161.50	1,550
3995	12" x 6"		1.63	14.724		745	715	36	1,496	1,925
3996	12" x 8"		1.41	17.021		680	825	41.50	1,546.50	2,050
3997	12" x 10"		1.27	18.898		480	920	46	1,446	1,950
4000	Socket weld joint, 3000 lb., type 304									
4100	90° Elbow									
4140	1/4"	Q-15	13.47	1.188	Ea.	46	55.50	4.35	105.85	139
4150	3/8"		12.97	1.234		60	58	4.52	122.52	157
4160	1/2"		12.21	1.310		62.50	61.50	4.80	128.80	166
4170	3/4"		10.68	1.498		75	70	5.50	150.50	194
4180	1"		9.74	1.643		113	77	6	196	247
4190	1-1/4"		9.05	1.768		197	83	6.45	286.45	350
4200	1-1/2"		8.55	1.871		240	87.50	6.85	334.35	405
4210	2"		7.26	2.204		385	103	8.05	496.05	590
4300	45° Elbow									
4340	1/4"	Q-15	13.47	1.188	Ea.	86.50	55.50	4.35	146.35	183
4350	3/8"		12.97	1.234		86.50	58	4.52	149.02	186
4360	1/2"		12.21	1.310		86.50	61.50	4.80	152.80	192
4370	3/4"		10.68	1.498		98	70	5.50	173.50	219
4380	1"		9.74	1.643		142	77	6	225	278
4390	1-1/4"		9.05	1.768		234	83	6.45	323.45	390
4400	1-1/2"		8.55	1.871		234	87.50	6.85	328.35	395
4410	2"		7.26	2.204		430	103	8.05	541.05	635
4500	Tee									
4540	1/4"	Q-15	8.97	1.784	Ea.	72.50	83.50	6.55	162.55	212
4550	3/8"		8.64	1.852		72.50	87	6.80	166.30	217
4560	1/2"		8.13	1.968		89	92	7.20	188.20	244
4570	3/4"		7.12	2.247		103	105	8.20	216.20	280
4580	1"		6.48	2.469		138	116	9.05	263.05	335
4590	1-1/4"		6.03	2.653		245	124	9.70	378.70	465
4600	1-1/2"		5.69	2.812		350	132	10.30	492.30	595

22 11 13.66 Pipe Fittings, Stainless Steel		Crew	Daily Output	Labor-Hours	Unit	Material	Labor	Equipment	Total	Total Incl O&P
						2010 Bare Costs				
4610	2"	Q-15	4.83	3.313	Ea.	530	155	12.10	697.10	830
5000	Socket weld joint, 3000 lb., type 316									
5100	90° Elbow									
5140	1/4"	Q-15	13.47	1.188	Ea.	56.50	55.50	4.35	116.35	150
5150	3/8"		12.97	1.234		66	58	4.52	128.52	164
5160	1/2"		12.21	1.310		79	61.50	4.80	145.30	184
5170	3/4"		10.68	1.498		104	70	5.50	179.50	226
5180	1"		9.74	1.643		148	77	6	231	285
5190	1-1/4"		9.05	1.768		264	83	6.45	353.45	420
5200	1-1/2"		8.55	1.871		299	87.50	6.85	393.35	470
5210	2"	↓	7.26	2.204	↓	510	103	8.05	621.05	725
5300	45° Elbow									
5340	1/4"	Q-15	13.47	1.188	Ea.	114	55.50	4.35	173.85	213
5350	3/8"		12.97	1.234		114	58	4.52	176.52	216
5360	1/2"		12.21	1.310		114	61.50	4.80	180.30	222
5370	3/4"		10.68	1.498		126	70	5.50	201.50	250
5380	1"		9.74	1.643		190	77	6	273	330
5390	1-1/4"		9.05	1.768		272	83	6.45	361.45	430
5400	1-1/2"		8.55	1.871		305	87.50	6.85	399.35	475
5410	2"	↓	7.26	2.204	↓	455	103	8.05	566.05	665
5500	Tee									
5540	1/4"	Q-15	8.97	1.784	Ea.	76.50	83.50	6.55	166.55	216
5550	3/8"		8.64	1.852		93.50	87	6.80	187.30	240
5560	1/2"		8.13	1.968		104	92	7.20	203.20	260
5570	3/4"		7.12	2.247		129	105	8.20	242.20	310
5580	1"		6.48	2.469		196	116	9.05	321.05	400
5590	1-1/4"		6.03	2.653		310	124	9.70	443.70	540
5600	1-1/2"		5.69	2.812		440	132	10.30	582.30	695
5610	2"	↓	4.83	3.313		705	155	12.10	872.10	1,025
5700	For socket weld joint, 6000 lb., type 304 and 316, add				↓	100%				
6000	Threaded companion flange									
6010	Stainless steel, 150 lb., type 304									
6020	1/2" diam.	1 Plum	30	.267	Ea.	38.50	13.90		52.40	63.50
6030	3/4" diam.		28	.286		43	14.85		57.85	69.50
6040	1" diam.	↓	27	.296		47.50	15.40		62.90	75
6050	1-1/4" diam.	Q-1	44	.364		60.50	17.05		77.55	92
6060	1-1/2" diam.		40	.400		60.50	18.75		79.25	94.50
6070	2" diam.		36	.444		79	21		100	118
6080	2-1/2" diam.		28	.571		110	27		137	161
6090	3" diam.		20	.800		116	37.50		153.50	183
6110	4" diam.	↓	12	1.333		158	62.50		220.50	268
6130	6" diam.	Q-2	14	1.714		285	83.50		368.50	440
6140	8" diam.	"	12	2	↓	550	97		647	750
6150	For type 316 add					40%				
6260	Weld flanges, stainless steel, type 304									
6270	Slip on, 150 lb. (welded, front and back)									
6280	1/2" diam.	Q-15	12.70	1.260	Ea.	34	59	4.61	97.61	131
6290	3/4" diam.		11.11	1.440		35	67.50	5.25	107.75	145
6300	1" diam.		10.13	1.579		38.50	74	5.80	118.30	160
6310	1-1/4" diam.		9.41	1.700		53	79.50	6.20	138.70	184
6320	1-1/2" diam.		8.89	1.800		53	84.50	6.60	144.10	191
6330	2" diam.		7.55	2.119		67	99.50	7.75	174.25	232
6340	2-1/2" diam.	↓	3.98	4.020	↓	94.50	188	14.70	297.20	400

22 11 13.66 Pipe Fittings, Stainless Steel		Crew	Daily Output	Labor-Hours	Unit	2010 Bare Costs Material	Labor	Equipment	Total	Total Incl O&P
6350	3" diam.	Q-15	3.21	4.984	Ea.	101	234	18.25	353.25	480
6370	4" diam.		2.44	6.557		138	305	24	467	640
6390	6" diam.	Q-16	1.89	12.698		209	615	31	855	1,200
6400	8" diam.	"	1.22	19.672		395	955	48	1,398	1,925
6410	For type 316, add					40%				
6530	Weld neck 150 lb.									
6540	1/2" diam.	Q-15	25.40	.630	Ea.	16.85	29.50	2.31	48.66	65
6550	3/4" diam.		22.22	.720		21.50	33.50	2.64	57.64	77
6560	1" diam.		20.25	.790		26	37	2.89	65.89	87.50
6570	1-1/4" diam.		18.82	.850		43	40	3.11	86.11	110
6580	1-1/2" diam.		17.78	.900		38.50	42	3.29	83.79	109
6590	2" diam.		15.09	1.060		44.50	49.50	3.88	97.88	128
6600	2-1/2" diam.		7.96	2.010		64	94	7.35	165.35	220
6610	3" diam.		6.43	2.488		72	117	9.10	198.10	265
6630	4" diam.		4.88	3.279		107	154	12	273	360
6640	5" diam.		4.26	3.756		148	176	13.75	337.75	440
6650	6" diam.	Q-16	5.66	4.240		189	206	10.35	405.35	530
6652	8" diam.		3.65	6.575		330	320	16.05	666.05	860
6654	10" diam.		2.88	8.333		465	405	20.50	890.50	1,150
6656	12" diam.		2.28	10.526		720	510	25.50	1,255.50	1,600
6670	For type 316 add					23%				
7000	Threaded joint, 150 lb., type 304									
7030	90° elbow									
7040	1/8"	1 Plum	13	.615	Ea.	26.50	32		58.50	77
7050	1/4"		13	.615		26.50	32		58.50	77
7070	3/8"		13	.615		29	32		61	80
7080	1/2"		12	.667		26	34.50		60.50	80.50
7090	3/4"		11	.727		29.50	38		67.50	89
7100	1"		10	.800		39.50	41.50		81	106
7110	1-1/4"	Q-1	17	.941		61	44		105	133
7120	1-1/2"		16	1		67	47		114	144
7130	2"		14	1.143		96.50	53.50		150	187
7140	2-1/2"		11	1.455		267	68		335	395
7150	3"		8	2		370	93.50		463.50	545
7160	4"	Q-2	11	2.182		495	106		601	705
7180	45° elbow									
7190	1/8"	1 Plum	13	.615	Ea.	39	32		71	90.50
7200	1/4"		13	.615		39	32		71	90.50
7210	3/8"		13	.615		37	32		69	88.50
7220	1/2"		12	.667		37.50	34.50		72	93
7230	3/4"		11	.727		29.50	38		67.50	89
7240	1"		10	.800		44.50	41.50		86	112
7250	1-1/4"	Q-1	17	.941		57.50	44		101.50	129
7260	1-1/2"		16	1		84	47		131	163
7270	2"		14	1.143		105	53.50		158.50	197
7280	2-1/2"		11	1.455		375	68		443	515
7290	3"		8	2		550	93.50		643.50	745
7300	4"	Q-2	11	2.182		995	106		1,101	1,250
7320	Tee, straight									
7330	1/8"	1 Plum	9	.889	Ea.	40.50	46.50		87	115
7340	1/4"		9	.889		40.50	46.50		87	115
7350	3/8"		9	.889		41	46.50		87.50	115
7360	1/2"		8	1		39	52		91	121

22 11 13.66 Pipe Fittings, Stainless Steel		Crew	Daily Output	Labor-Hours	Unit	Material	Labor	Equipment	Total	Total Incl O&P
						2010 Bare Costs				
7370	3/4"	1 Plum	7	1.143	Ea.	41.50	59.50		101	135
7380	1"		6.50	1.231		49	64		113	150
7390	1-1/4"	Q-1	11	1.455		82.50	68		150.50	193
7400	1-1/2"		10	1.600		101	75		176	223
7410	2"		9	1.778		126	83.50		209.50	263
7420	2-1/2"		7	2.286		385	107		492	580
7430	3"		5	3.200		585	150		735	865
7440	4"	Q-2	7	3.429		1,450	167		1,617	1,850
7460	Coupling, straight									
7470	1/8"	1 Plum	19	.421	Ea.	10.70	22		32.70	45
7480	1/4"		19	.421		12.65	22		34.65	47
7490	3/8"		19	.421		15.15	22		37.15	49.50
7500	1/2"		19	.421		20	22		42	55
7510	3/4"		18	.444		27	23		50	64
7520	1"		15	.533		43.50	28		71.50	89
7530	1-1/4"	Q-1	26	.615		70	29		99	120
7540	1-1/2"		24	.667		78	31		109	133
7550	2"		21	.762		123	35.50		158.50	189
7560	2-1/2"		18	.889		286	41.50		327.50	380
7570	3"		14	1.143		390	53.50		443.50	510
7580	4"	Q-2	16	1.500		575	73		648	740
7600	Reducer, concentric, 1/2"	1 Plum	12	.667		20	34.50		54.50	74
7610	3/4"		11	.727		27	38		65	86
7612	1"		10	.800		46.50	41.50		88	114
7614	1-1/4"	Q-1	17	.941		98.50	44		142.50	174
7616	1-1/2"		16	1		108	47		155	189
7618	2"		14	1.143		126	53.50		179.50	220
7620	2-1/2"		11	1.455		415	68		483	560
7622	3"		8	2		465	93.50		558.50	650
7624	4"	Q-2	11	2.182		770	106		876	1,000
7710	Union									
7720	1/8"	1 Plum	12	.667	Ea.	49.50	34.50		84	107
7730	1/4"		12	.667		49.50	34.50		84	107
7740	3/8"		12	.667		58	34.50		92.50	116
7750	1/2"		11	.727		72.50	38		110.50	137
7760	3/4"		10	.800		99.50	41.50		141	172
7770	1"		9	.889		144	46.50		190.50	229
7780	1-1/4"	Q-1	16	1		335	47		382	440
7790	1-1/2"		15	1.067		365	50		415	475
7800	2"		13	1.231		460	57.50		517.50	590
7810	2-1/2"		10	1.600		870	75		945	1,075
7820	3"		7	2.286		1,150	107		1,257	1,425
7830	4"	Q-2	10	2.400		1,600	117		1,717	1,925
7838	Caps									
7840	1/2"	1 Plum	24	.333	Ea.	13.40	17.35		30.75	41
7841	3/4"		22	.364		19.65	18.95		38.60	50
7842	1"		20	.400		31.50	21		52.50	65.50
7843	1-1/2"	Q-1	32	.500		75	23.50		98.50	117
7844	2"	"	28	.571		94.50	27		121.50	144
7845	4"	Q-2	22	1.091		370	53		423	490
7850	For 150 lb., type 316, add					25%				
8750	Threaded joint, 2000 lb., type 304									
8770	90° Elbow									

22 11 13 – Facility Water Distribution Piping

22 11 13.66 Pipe Fittings, Stainless Steel		Crew	Daily Output	Labor-Hours	Unit	Material	Labor	Equipment	Total	Total Incl O&P
						2010 Bare Costs				
8780	1/8"	1 Plum	13	.615	Ea.	34	32		66	85
8790	1/4"		13	.615		34	32		66	85
8800	3/8"		13	.615		42	32		74	94
8810	1/2"		12	.667		57	34.50		91.50	115
8820	3/4"		11	.727		66	38		104	129
8830	1"		10	.800		87.50	41.50		129	159
8840	1-1/4"	Q-1	17	.941		142	44		186	222
8850	1-1/2"		16	1		228	47		275	320
8860	2"		14	1.143		320	53.50		373.50	435
8880	45° Elbow									
8890	1/8"	1 Plum	13	.615	Ea.	70.50	32		102.50	126
8900	1/4"		13	.615		70.50	32		102.50	126
8910	3/8"		13	.615		88.50	32		120.50	145
8920	1/2"		12	.667		88.50	34.50		123	149
8930	3/4"		11	.727		97.50	38		135.50	164
8940	1"		10	.800		116	41.50		157.50	191
8950	1-1/4"	Q-1	17	.941		203	44		247	289
8960	1-1/2"		16	1		272	47		319	370
8970	2"		14	1.143		315	53.50		368.50	425
8990	Tee, straight									
9000	1/8"	1 Plum	9	.889	Ea.	44	46.50		90.50	118
9010	1/4"		9	.889		44	46.50		90.50	118
9020	3/8"		9	.889		56.50	46.50		103	132
9030	1/2"		8	1		72	52		124	158
9040	3/4"		7	1.143		82	59.50		141.50	179
9050	1"		6.50	1.231		113	64		177	220
9060	1-1/4"	Q-1	11	1.455		191	68		259	310
9070	1-1/2"		10	1.600		315	75		390	460
9080	2"		9	1.778		440	83.50		523.50	610
9100	For couplings and unions use 3000 lb., type 304									
9120	2000 lb., type 316									
9130	90° Elbow									
9140	1/8"	1 Plum	13	.615	Ea.	42	32		74	94.50
9150	1/4"		13	.615		42	32		74	94.50
9160	3/8"		13	.615		49.50	32		81.50	103
9170	1/2"		12	.667		64	34.50		98.50	123
9180	3/4"		11	.727		78.50	38		116.50	143
9190	1"		10	.800		94.50	41.50		136	167
9200	1-1/4"	Q-1	17	.941		219	44		263	305
9210	1-1/2"		16	1		243	47		290	335
9220	2"		14	1.143		415	53.50		468.50	535
9240	45° Elbow									
9250	1/8"	1 Plum	13	.615	Ea.	89.50	32		121.50	147
9260	1/4"		13	.615		89.50	32		121.50	147
9270	3/8"		13	.615		89.50	32		121.50	147
9280	1/2"		12	.667		89.50	34.50		124	151
9300	3/4"		11	.727		101	38		139	168
9310	1"		10	.800		152	41.50		193.50	230
9320	1-1/4"	Q-1	17	.941		217	44		261	305
9330	1-1/2"		16	1		297	47		344	395
9340	2"		14	1.143		365	53.50		418.50	480
9360	Tee, straight									
9370	1/8"	1 Plum	9	.889	Ea.	53.50	46.50		100	129

22 11 13 – Facility Water Distribution Piping

22 11 13.66 Pipe Fittings, Stainless Steel		Crew	Daily Output	Labor-Hours	Unit	Material	Labor	Equipment	Total	Total Incl O&P
						2010 Bare Costs				
9380	1/4"	1 Plum	9	.889	Ea.	53.50	46.50		100	129
9390	3/8"		9	.889		65.50	46.50		112	142
9400	1/2"		8	1		81.50	52		133.50	168
9410	3/4"		7	1.143		101	59.50		160.50	200
9420	1"		6.50	1.231		154	64		218	266
9430	1-1/4"	Q-1	11	1.455		250	68		318	375
9440	1-1/2"		10	1.600		360	75		435	505
9450	2"		9	1.778		570	83.50		653.50	755
9470	For couplings and unions use 3000 lb., type 316									
9490	3000 lb., type 304									
9510	Coupling									
9520	1/8"	1 Plum	19	.421	Ea.	13.35	22		35.35	47.50
9530	1/4"		19	.421		14.55	22		36.55	49
9540	3/8"		19	.421		16.90	22		38.90	51.50
9550	1/2"		19	.421		20.50	22		42.50	55.50
9560	3/4"		18	.444		27	23		50	64.50
9570	1"		15	.533		46	28		74	92
9580	1-1/4"	Q-1	26	.615		112	29		141	166
9590	1-1/2"		24	.667		131	31		162	191
9600	2"		21	.762		172	35.50		207.50	244
9620	Union									
9630	1/8"	1 Plum	12	.667	Ea.	102	34.50		136.50	165
9640	1/4"		12	.667		102	34.50		136.50	165
9650	3/8"		12	.667		109	34.50		143.50	172
9660	1/2"		11	.727		109	38		147	177
9670	3/4"		10	.800		134	41.50		175.50	211
9680	1"		9	.889		207	46.50		253.50	297
9690	1-1/4"	Q-1	16	1		380	47		427	490
9700	1-1/2"		15	1.067		430	50		480	550
9710	2"		13	1.231		575	57.50		632.50	720
9730	3000 lb., type 316									
9750	Coupling									
9770	1/8"	1 Plum	19	.421	Ea.	15.10	22		37.10	49.50
9780	1/4"		19	.421		16.95	22		38.95	51.50
9790	3/8"		19	.421		18.20	22		40.20	53
9800	1/2"		19	.421		26	22		48	61.50
9810	3/4"		18	.444		38	23		61	76
9820	1"		15	.533		63.50	28		91.50	112
9830	1-1/4"	Q-1	26	.615		142	29		171	200
9840	1-1/2"		24	.667		166	31		197	229
9850	2"		21	.762		229	35.50		264.50	305
9870	Union									
9880	1/8"	1 Plum	12	.667	Ea.	112	34.50		146.50	176
9890	1/4"		12	.667		112	34.50		146.50	176
9900	3/8"		12	.667		130	34.50		164.50	195
9910	1/2"		11	.727		130	38		168	200
9920	3/4"		10	.800		170	41.50		211.50	250
9930	1"		9	.889		268	46.50		314.50	365
9940	1-1/4"	Q-1	16	1		470	47		517	585
9950	1-1/2"		15	1.067		580	50		630	710
9960	2"		13	1.231		735	57.50		792.50	895

22 11 13 – Facility Water Distribution Piping

22 11 13.74 Pipe, Plastic		Crew	Daily Output	Labor-Hours	Unit	Material	Labor	Equipment	Total	Total Incl O&P
						2010 Bare Costs				
0010	**PIPE, PLASTIC**									
0020	Fiberglass reinforced, couplings 10′ O.C., clevis hanger assy's, 3 per 10′									
0080	General service									
0120	2″ diameter	Q-1	59	.271	L.F.	6.35	12.70		19.05	26
0140	3″ diameter		52	.308		8.75	14.40		23.15	31
0150	4″ diameter		48	.333		11.80	15.60		27.40	36.50
0160	6″ diameter		39	.410		21.50	19.20		40.70	52.50
0170	8″ diameter	Q-2	49	.490		35	24		59	74
0180	10″ diameter		41	.585		53.50	28.50		82	101
0190	12″ diameter		36	.667		69.50	32.50		102	125
0600	PVC, high impact/pressure, cplgs. 10′ O.C., clevis hanger assy's, 3 per 10′									
1020	Schedule 80									
1070	1/2″ diameter	1 Plum	50	.160	L.F.	3.34	8.35		11.69	16.15
1080	3/4″ diameter		47	.170		4.18	8.85		13.03	17.90
1090	1″ diameter		43	.186		5.55	9.70		15.25	20.50
1100	1-1/4″ diameter		39	.205		7.40	10.70		18.10	24
1110	1-1/2″ diameter		34	.235		8.70	12.25		20.95	28
1120	2″ diameter	Q-1	55	.291		11.45	13.65		25.10	33
1140	3″ diameter		50	.320		23.50	15		38.50	48
1150	4″ diameter		46	.348		33.50	16.30		49.80	61.50
1170	6″ diameter		38	.421		75.50	19.75		95.25	113
1730	To delete coupling & hangers, subtract									
1740	1/4″ diam. to 1/2″ diam.					62%	80%			
1750	3/4″ diam. to 1-1/4″ diam.					58%	73%			
1760	1-1/2″ diam. to 6″ diam.					40%	57%			
1770	8″ diam. to 12″ diam.					34%	50%			
1800	PVC, couplings 10′ O.C., clevis hanger assemblies, 3 per 10′									
1820	Schedule 40									
1860	1/2″ diameter	1 Plum	54	.148	L.F.	1.65	7.70		9.35	13.35
1870	3/4″ diameter		51	.157		1.89	8.15		10.04	14.35
1880	1″ diameter		46	.174		2.18	9.05		11.23	15.95
1890	1-1/4″ diameter		42	.190		2.72	9.90		12.62	17.85
1900	1-1/2″ diameter		36	.222		2.92	11.55		14.47	20.50
1910	2″ diameter	Q-1	59	.271		3.51	12.70		16.21	23
1920	2-1/2″ diameter		56	.286		5.40	13.40		18.80	26
1930	3″ diameter		53	.302		6.95	14.15		21.10	28.50
1940	4″ diameter		48	.333		9.60	15.60		25.20	34
1950	5″ diameter		43	.372		13.95	17.45		31.40	41.50
1960	6″ diameter		39	.410		17.80	19.20		37	48.50
1970	8″ diameter	Q-2	48	.500		26	24.50		50.50	65
1980	10″ diameter		43	.558		57.50	27		84.50	104
1990	12″ diameter		42	.571		69	28		97	118
2000	14″ diameter		31	.774		116	37.50		153.50	184
2010	16″ diameter		23	1.043		169	50.50		219.50	261
2340	To delete coupling & hangers, subtract									
2360	1/2″ diam. to 1-1/4″ diam.					65%	74%			
2370	1-1/2″ diam. to 6″ diam.					44%	57%			
2380	8″ diam. to 12″ diam.					41%	53%			
2390	14″ diam. to 16″ diam.					48%	45%			
2420	Schedule 80									
2440	1/4″ diameter	1 Plum	58	.138	L.F.	1.30	7.20		8.50	12.20
2450	3/8″ diameter		55	.145		1.30	7.55		8.85	12.80

22 11 13.74 Pipe, Plastic		Crew	Daily Output	Labor-Hours	Unit	Material	2010 Bare Costs Labor	Equipment	Total	Total Incl O&P
2460	1/2" diameter	1 Plum	50	.160	L.F.	1.73	8.35		10.08	14.40
2470	3/4" diameter		47	.170		2.02	8.85		10.87	15.55
2480	1" diameter		43	.186		2.36	9.70		12.06	17.10
2490	1-1/4" diameter		39	.205		3	10.70		13.70	19.30
2500	1-1/2" diameter		34	.235		3.25	12.25		15.50	22
2510	2" diameter	Q-1	55	.291		3.93	13.65		17.58	25
2520	2-1/2" diameter		52	.308		5.05	14.40		19.45	27
2530	3" diameter		50	.320		7.95	15		22.95	31.50
2540	4" diameter		46	.348		11.15	16.30		27.45	37
2550	5" diameter		42	.381		16.55	17.85		34.40	45
2560	6" diameter		38	.421		24.50	19.75		44.25	56.50
2570	8" diameter	Q-2	47	.511		31.50	25		56.50	71.50
2580	10" diameter		42	.571		66	28		94	114
2590	12" diameter		38	.632		82	30.50		112.50	137
2830	To delete coupling & hangers, subtract									
2840	1/4" diam. to 1/2" diam.					66%	80%			
2850	3/4" diam. to 1-1/4" diam.					61%	73%			
2860	1-1/2" diam. to 6" diam.					41%	57%			
2870	8" diam. to 12" diam.					31%	50%			
2900	Schedule 120									
2910	1/2" diameter	1 Plum	50	.160	L.F.	1.84	8.35		10.19	14.50
2950	3/4" diameter		47	.170		2.14	8.85		10.99	15.65
2960	1" diameter		43	.186		2.54	9.70		12.24	17.30
2970	1-1/4" diameter		39	.205		3.25	10.70		13.95	19.55
2980	1-1/2" diameter		33	.242		3.64	12.60		16.24	23
2990	2" diameter	Q-1	54	.296		4.52	13.90		18.42	26
3000	2-1/2" diameter		52	.308		7.45	14.40		21.85	29.50
3010	3" diameter		49	.327		9.15	15.30		24.45	33
3020	4" diameter		45	.356		14.15	16.65		30.80	40.50
3030	6" diameter		37	.432		26	20.50		46.50	59
3240	To delete coupling & hangers, subtract									
3250	1/2" diam. to 1-1/4" diam.					52%	74%			
3260	1-1/2" diam. to 4" diam.					30%	57%			
3270	6" diam.					17%	50%			
3300	PVC, pressure, couplings 10' O.C., clevis hanger assy's, 3 per 10'									
3310	SDR 26, 160 psi									
3350	1-1/4" diameter	1 Plum	42	.190	L.F.	2.29	9.90		12.19	17.35
3360	1-1/2" diameter	"	36	.222		2.47	11.55		14.02	20
3370	2" diameter	Q-1	59	.271		2.93	12.70		15.63	22.50
3380	2-1/2" diameter		56	.286		4.39	13.40		17.79	25
3390	3" diameter		53	.302		5.95	14.15		20.10	27.50
3400	4" diameter		48	.333		8.55	15.60		24.15	33
3420	6" diameter		39	.410		17.15	19.20		36.35	48
3430	8" diameter	Q-2	48	.500		26	24.50		50.50	65.50
3660	To delete coupling & clevis hanger assy's, subtract									
3670	1-1/4" diam.					63%	68%			
3680	1-1/2" diam. to 4" diam.					48%	57%			
3690	6" diam. to 8" diam.					60%	54%			
3720	SDR 21, 200 psi, 1/2" diameter	1 Plum	54	.148	L.F.	1.55	7.70		9.25	13.25
3740	3/4" diameter		51	.157		1.71	8.15		9.86	14.15
3750	1" diameter		46	.174		1.87	9.05		10.92	15.60
3760	1-1/4" diameter		42	.190		2.38	9.90		12.28	17.45
3770	1-1/2" diameter		36	.222		2.58	11.55		14.13	20

22 11 13.74 Pipe, Plastic		Crew	Daily Output	Labor-Hours	Unit	Material	2010 Bare Costs Labor	Equipment	Total	Total Incl O&P
3780	2" diameter	Q-1	59	.271	L.F.	3.10	12.70		15.80	22.50
3790	2-1/2" diameter		56	.286		5.60	13.40		19	26
3800	3" diameter		53	.302		6.35	14.15		20.50	28
3810	4" diameter		48	.333		10.35	15.60		25.95	35
3830	6" diameter		39	.410		18.65	19.20		37.85	49.50
3840	8" diameter	Q-2	48	.500		28.50	24.50		53	68
4000	To delete coupling & hangers, subtract									
4010	1/2" diam. to 3/4" diam.					71%	77%			
4020	1" diam. to 1-1/4" diam.					63%	70%			
4030	1-1/2" diam. to 6" diam.					44%	57%			
4040	8" diam.					46%	54%			
4100	DWV type, schedule 40, couplings 10' O.C., clevis hanger assy's, 3 per 10'									
4210	ABS, schedule 40, foam core type									
4212	Plain end black									
4214	1-1/2" diameter	1 Plum	39	.205	L.F.	1.93	10.70		12.63	18.15
4216	2" diameter	Q-1	62	.258		2.35	12.10		14.45	20.50
4218	3" diameter		56	.286		4.42	13.40		17.82	25
4220	4" diameter		51	.314		6.10	14.70		20.80	28.50
4222	6" diameter		42	.381		12.35	17.85		30.20	40.50
4240	To delete coupling & hangers, subtract									
4244	1-1/2" diam. to 6" diam.					43%	48%			
4400	PVC									
4410	1-1/4" diameter	1 Plum	42	.190	L.F.	1.99	9.90		11.89	17.05
4420	1-1/2" diameter	"	36	.222		2	11.55		13.55	19.55
4460	2" diameter	Q-1	59	.271		2.41	12.70		15.11	21.50
4470	3" diameter		53	.302		4.51	14.15		18.66	26
4480	4" diameter		48	.333		6.10	15.60		21.70	30.50
4490	6" diameter		39	.410		11.10	19.20		30.30	41
4500	8" diameter	Q-2	48	.500		17.95	24.50		42.45	56.50
4510	To delete coupling & hangers, subtract									
4520	1-1/4" diam. to 1-1/2" diam.					48%	60%			
4530	2" diam. to 8" diam.					42%	54%			
4550	PVC, schedule 40, foam core type									
4552	Plain end, white									
4554	1-1/2" diameter	1 Plum	39	.205	L.F.	1.74	10.70		12.44	17.90
4556	2" diameter	Q-1	62	.258		2.09	12.10		14.19	20.50
4558	3" diameter		56	.286		3.90	13.40		17.30	24.50
4560	4" diameter		51	.314		5.20	14.70		19.90	27.50
4562	6" diameter		42	.381		9.25	17.85		27.10	37
4564	8" diameter	Q-2	51	.471		15.05	23		38.05	51
4568	10" diameter		48	.500		19.70	24.50		44.20	58
4570	12" diameter		46	.522		24.50	25.50		50	65
4580	To delete coupling & hangers, subtract									
4582	1-1/2" dia to 2" dia					58%	54%			
4584	3" dia to 12" dia					46%	42%			
4800	PVC, clear pipe, cplgs. 10' O.C., clevis hanger assy's 3 per 10', Sched. 40									
4840	1/4" diameter	1 Plum	59	.136	L.F.	1.76	7.05		8.81	12.55
4850	3/8" diameter		56	.143		2.04	7.45		9.49	13.40
4860	1/2" diameter		54	.148		2.59	7.70		10.29	14.40
4870	3/4" diameter		51	.157		3.16	8.15		11.31	15.75
4880	1" diameter		46	.174		4.40	9.05		13.45	18.40
4890	1-1/4" diameter		42	.190		5.50	9.90		15.40	21
4900	1-1/2" diameter		36	.222		6.40	11.55		17.95	24.50

22 11 13.74 Pipe, Plastic		Crew	Daily Output	Labor-Hours	Unit	Material	Labor	Equipment	Total	Total Incl O&P
						2010 Bare Costs				
4910	2" diameter	Q-1	59	.271	L.F.	8.40	12.70		21.10	28.50
4920	2-1/2" diameter		56	.286		13.20	13.40		26.60	34.50
4930	3" diameter		53	.302		17.05	14.15		31.20	40
4940	3-1/2" diameter		50	.320		22.50	15		37.50	47.50
4950	4" diameter		48	.333		21.50	15.60		37.10	47.50
5250	To delete coupling & hangers, subtract									
5260	1/4" diam. to 3/8" diam.					60%	81%			
5270	1/2" diam. to 3/4" diam.					41%	77%			
5280	1" diam. to 1-1/2" diam.					26%	67%			
5290	2" diam. to 4" diam.					16%	58%			
5300	CPVC, socket joint, couplings 10' O.C., clevis hanger assemblies, 3 per 10'									
5302	Schedule 40									
5304	1/2" diameter	1 Plum	54	.148	L.F.	2.35	7.70		10.05	14.15
5305	3/4" diameter		51	.157		2.85	8.15		11	15.40
5306	1" diameter		46	.174		3.76	9.05		12.81	17.70
5307	1-1/4" diameter		42	.190		4.88	9.90		14.78	20
5308	1-1/2" diameter		36	.222		5.75	11.55		17.30	23.50
5309	2" diameter	Q-1	59	.271		7.15	12.70		19.85	27
5310	2-1/2" diameter		56	.286		12.25	13.40		25.65	33.50
5311	3" diameter		53	.302		14.90	14.15		29.05	37.50
5312	4" diameter		48	.333		20.50	15.60		36.10	46
5314	6" diameter		43	.372		40	17.45		57.45	70
5318	To delete coupling & hangers, subtract									
5319	1/2" diam. to 3/4" diam.					37%	77%			
5320	1" diam. to 1-1/4" diam.					27%	70%			
5321	1-1/2" diam. to 3" diam.					21%	57%			
5322	4" diam. to 6" diam.					16%	57%			
5324	Schedule 80									
5325	1/2" diameter	1 Plum	50	.160	L.F.	2.53	8.35		10.88	15.30
5326	3/4" diameter		47	.170		3.13	8.85		11.98	16.75
5327	1" diameter		43	.186		4.14	9.70		13.84	19.05
5328	1-1/4" diameter		39	.205		5.50	10.70		16.20	22
5329	1-1/2" diameter		34	.235		6.55	12.25		18.80	25.50
5330	2" diameter	Q-1	55	.291		8.45	13.65		22.10	30
5331	2-1/2" diameter		52	.308		13.90	14.40		28.30	37
5332	3" diameter		50	.320		17.35	15		32.35	41.50
5333	4" diameter		46	.348		24	16.30		40.30	51
5334	6" diameter		38	.421		49	19.75		68.75	83
5335	8" diameter	Q-2	47	.511		93.50	25		118.50	140
5339	To delete couplings & hangers, subtract									
5340	1/2" diam. to 3/4" diam.					44%	77%			
5341	1" diam. to 1-1/4" diam.					32%	71%			
5342	1-1/2" diam. to 4" diam.					25%	58%			
5343	6" diam. to 8" diam.					20%	53%			
5360	CPVC, threaded, couplings 10' O.C., clevis hanger assemblies, 3 per 10'									
5380	Schedule 40									
5460	1/2" diameter	1 Plum	54	.148	L.F.	3.28	7.70		10.98	15.15
5470	3/4" diameter		51	.157		3.88	8.15		12.03	16.50
5480	1" diameter		46	.174		4.76	9.05		13.81	18.80
5490	1-1/4" diameter		42	.190		5.85	9.90		15.75	21.50
5500	1-1/2" diameter		36	.222		6.60	11.55		18.15	24.50
5510	2" diameter	Q-1	59	.271		8.20	12.70		20.90	28
5520	2-1/2" diameter		56	.286		13.35	13.40		26.75	34.50

22 11 13.74 Pipe, Plastic		Crew	Daily Output	Labor-Hours	Unit	2010 Bare Costs Material	Labor	Equipment	Total	Total Incl O&P
5530	3" diameter	Q-1	53	.302	L.F.	16.50	14.15		30.65	39
5540	4" diameter		48	.333		26.50	15.60		42.10	53
5550	6" diameter		43	.372		43	17.45		60.45	73.50
5730	To delete coupling & hangers, subtract									
5740	1/2" diam. to 3/4" diam.					37%	77%			
5750	1" diam. to 1-1/4" diam.					27%	70%			
5760	1-1/2" diam. to 3" diam.					21%	57%			
5770	4" diam. to 6" diam.					16%	57%			
5800	Schedule 80									
5860	1/2" diameter	1 Plum	50	.160	L.F.	3.46	8.35		11.81	16.30
5870	3/4" diameter		47	.170		4.16	8.85		13.01	17.90
5880	1" diameter		43	.186		5.15	9.70		14.85	20
5890	1-1/4" diameter		39	.205		6.45	10.70		17.15	23
5900	1-1/2" diameter		34	.235		7.40	12.25		19.65	26.50
5910	2" diameter	Q-1	55	.291		9.45	13.65		23.10	31
5920	2-1/2" diameter		52	.308		15.05	14.40		29.45	38
5930	3" diameter		50	.320		18.95	15		33.95	43.50
5940	4" diameter		46	.348		30.50	16.30		46.80	58
5950	6" diameter		38	.421		52	19.75		71.75	86.50
5960	8" diameter	Q-2	47	.511		93.50	25		118.50	140
6060	To delete couplings & hangers, subtract									
6070	1/2" diam. to 3/4" diam.					44%	77%			
6080	1" diam. to 1-1/4" diam.					32%	71%			
6090	1-1/2" diam. to 4" diam.					25%	58%			
6100	6" diam. to 8" diam.					20%	53%			
6240	CTS, 1/2" diameter	1 Plum	54	.148	L.F.	1.56	7.70		9.26	13.25
6250	3/4" diameter		51	.157		2.14	8.15		10.29	14.60
6260	1" diameter		46	.174		3.48	9.05		12.53	17.40
6270	1 1/4"		42	.190		4.70	9.90		14.60	20
6280	1 1/2" diameter		36	.222		5.95	11.55		17.50	24
6290	2" diameter	Q-1	59	.271		9.60	12.70		22.30	29.50
6370	To delete coupling & hangers, subtract									
6380	1/2" diam.					51%	79%			
6390	3/4" diam.					40%	76%			
6392	1" thru 2" diam.					72%	68%			
7280	PEX, flexible, no couplings or hangers									
7282	Note: For labor costs add 25% to the couplings and fittings labor total.									
7300	Non-barrier type, hot/cold tubing rolls									
7310	1/4" diameter x 100'				L.F.	.44			.44	.48
7350	3/8" diameter x 100'					.49			.49	.54
7360	1/2" diameter x 100'					.55			.55	.61
7370	1/2" diameter x 500'					.55			.55	.61
7380	1/2" diameter x 1000'					.55			.55	.61
7400	3/4" diameter x 100'					1			1	1.10
7410	3/4" diameter x 500'					1			1	1.10
7420	3/4" diameter x 1000'					1			1	1.10
7460	1" diameter x 100'					1.71			1.71	1.88
7470	1" diameter x 300'					1.71			1.71	1.88
7480	1" diameter x 500'					1.71			1.71	1.88
7500	1-1/4" diameter x 100'					2.90			2.90	3.19
7510	1-1/4" diameter x 300'					2.90			2.90	3.19
7540	1-1/2" diameter x 100'					3.64			3.64	4
7550	1-1/2" diameter x 300'					3.40			3.40	3.74

22 11 13 – Facility Water Distribution Piping

22 11 13.74 Pipe, Plastic		Crew	Daily Output	Labor-Hours	Unit	2010 Bare Costs Material	Labor	Equipment	Total	Total Incl O&P
7596	Most sizes available in red or blue									
7700	Non-barrier type, hot/cold tubing straight lengths									
7710	1/2" diameter x 20'				L.F.	.55			.55	.61
7750	3/4" diameter x 20'					.99			.99	1.09
7760	1" diameter x 20'					1.71			1.71	1.88
7770	1-1/4" diameter x 20'					2.90			2.90	3.19
7780	1-1/2" diameter x 20'					3.64			3.64	4
7796	Most sizes available in red or blue									

22 11 13.76 Pipe Fittings, Plastic		Crew	Daily Output	Labor-Hours	Unit	Material	Labor	Equipment	Total	Total Incl O&P
0010	**PIPE FITTINGS, PLASTIC**									
0030	Epoxy resin, fiberglass reinforced, general service									
0090	Elbow, 90°, 2"	Q-1	33.10	.483	Ea.	82	22.50		104.50	124
0100	3"		20.80	.769		94.50	36		130.50	158
0110	4"		16.50	.970		130	45.50		175.50	211
0120	6"		10.10	1.584		188	74		262	320
0130	8"	Q-2	9.30	2.581		345	125		470	570
0140	10"		8.50	2.824		435	137		572	685
0150	12"		7.60	3.158		625	153		778	920
0160	45° Elbow, same as 90°									
0170	Elbow, 90°, flanged									
0172	2"	Q-1	23	.696	Ea.	127	32.50		159.50	188
0173	3"		16	1		145	47		192	230
0174	4"		13	1.231		189	57.50		246.50	295
0176	6"		8	2		345	93.50		438.50	515
0177	8"	Q-2	9	2.667		620	130		750	880
0178	10"		7	3.429		845	167		1,012	1,175
0179	12"		5	4.800		1,150	233		1,383	1,600
0186	Elbow, 45°, flanged									
0188	2"	Q-1	23	.696	Ea.	127	32.50		159.50	188
0189	3"		16	1		145	47		192	230
0190	4"		13	1.231		189	57.50		246.50	295
0192	6"		8	2		345	93.50		438.50	515
0193	8"	Q-2	9	2.667		590	130		720	840
0194	10"		7	3.429		745	167		912	1,075
0195	12"		5	4.800		955	233		1,188	1,400
0290	Tee, 2"	Q-1	20	.800		45	37.50		82.50	106
0300	3"		13.90	1.151		54	54		108	141
0310	4"		11	1.455		74.50	68		142.50	184
0320	6"		6.70	2.388		213	112		325	405
0330	8"	Q-2	6.20	3.871		1,325	188		1,513	1,750
0340	10"		5.70	4.211		1,500	205		1,705	1,975
0350	12"		5.10	4.706		1,850	229		2,079	2,375
0352	Tee, flanged									
0354	2"	Q-1	17	.941	Ea.	171	44		215	255
0355	3"		10	1.600		227	75		302	360
0356	4"		8	2		254	93.50		347.50	420
0358	6"		5	3.200		435	150		585	705
0359	8"	Q-2	6	4		830	194		1,024	1,200
0360	10"		5	4.800		1,200	233		1,433	1,675
0361	12"		4	6		1,650	291		1,941	2,250
0365	Wye, flanged									
0367	2"	Q-1	17	.941	Ea.	340	44		384	440

22 11 13 – Facility Water Distribution Piping

22 11 13.76 Pipe Fittings, Plastic		Crew	Daily Output	Labor-Hours	Unit	Material	2010 Bare Costs Labor	Equipment	Total	Total Incl O&P
0368	3"	Q-1	10	1.600	Ea.	465	75		540	625
0369	4"		8	2		625	93.50		718.50	825
0371	6"		5	3.200		875	150		1,025	1,175
0372	8"	Q-2	6	4		1,450	194		1,644	1,875
0373	10"		5	4.800		2,225	233		2,458	2,800
0374	12"		4	6		3,250	291		3,541	4,000
0380	Couplings									
0410	2"	Q-1	33.10	.483	Ea.	14.70	22.50		37.20	50
0420	3"		20.80	.769		15.75	36		51.75	71.50
0430	4"		16.50	.970		21	45.50		66.50	91
0440	6"		10.10	1.584		50.50	74		124.50	167
0450	8"	Q-2	9.30	2.581		85.50	125		210.50	282
0460	10"		8.50	2.824		126	137		263	345
0470	12"		7.60	3.158		168	153		321	415
0473	High corrosion resistant couplings, add					30%				
0474	Reducer, concentric, flanged									
0475	2" x 1-1/2"	Q-1	30	.533	Ea.	330	25		355	400
0476	3" x 2"		24	.667		360	31		391	440
0477	4" x 3"		19	.842		405	39.50		444.50	505
0479	6" x 4"		15	1.067		405	50		455	520
0480	8" x 6"	Q-2	16	1.500		660	73		733	840
0481	10" x 8"		13	1.846		615	89.50		704.50	810
0482	12" x 10"		11	2.182		950	106		1,056	1,200
0486	Adapter, bell x male or female									
0488	2"	Q-1	28	.571	Ea.	16.40	27		43.40	58
0489	3"		20	.800		24	37.50		61.50	82.50
0491	4"		17	.941		32.50	44		76.50	102
0492	6"		12	1.333		68.50	62.50		131	169
0493	8"	Q-2	15	1.600		99	77.50		176.50	226
0494	10"	"	11	2.182		143	106		249	315
0528	Flange									
0532	2"	Q-1	46	.348	Ea.	24	16.30		40.30	51
0533	3"		32	.500		29.50	23.50		53	67.50
0534	4"		26	.615		39	29		68	85.50
0536	6"		16	1		73.50	47		120.50	151
0537	8"	Q-2	18	1.333		112	65		177	221
0538	10"		14	1.714		160	83.50		243.50	300
0539	12"		10	2.400		201	117		318	395
2100	PVC schedule 80, socket joint									
2110	90° elbow, 1/2"	1 Plum	30.30	.264	Ea.	2.12	13.75		15.87	23
2130	3/4"		26	.308		2.71	16		18.71	27
2140	1"		22.70	.352		4.07	18.35		22.42	32
2150	1-1/4"		20.20	.396		5.85	20.50		26.35	37.50
2160	1-1/2"		18.20	.440		6.25	23		29.25	41.50
2170	2"	Q-1	33.10	.483		7.60	22.50		30.10	42.50
2180	3"		20.80	.769		19.90	36		55.90	76
2190	4"		16.50	.970		30.50	45.50		76	102
2200	6"		10.10	1.584		86	74		160	206
2210	8"	Q-2	9.30	2.581		238	125		363	450
2250	45° elbow, 1/2"	1 Plum	30.30	.264		4	13.75		17.75	25
2270	3/4"		26	.308		6.05	16		22.05	30.50
2280	1"		22.70	.352		9.15	18.35		27.50	37.50
2290	1-1/4"		20.20	.396		11.65	20.50		32.15	44

22 11 13 – Facility Water Distribution Piping

22 11 13.76 Pipe Fittings, Plastic		Crew	Daily Output	Labor-Hours	Unit	2010 Bare Costs Material	Labor	Equipment	Total	Total Incl O&P
2300	1-1/2"	1 Plum	18.20	.440	Ea.	13.80	23		36.80	49.50
2310	2"	Q-1	33.10	.483		17.85	22.50		40.35	53.50
2320	3"		20.80	.769		22.50	36		58.50	79
2330	4"		16.50	.970		82	45.50		127.50	159
2340	6"		10.10	1.584		104	74		178	225
2350	8"	Q-2	9.30	2.581		225	125		350	435
2400	Tee, 1/2"	1 Plum	20.20	.396		6	20.50		26.50	37.50
2420	3/4"		17.30	.462		6.25	24		30.25	43
2430	1"		15.20	.526		7.85	27.50		35.35	49.50
2440	1-1/4"		13.50	.593		21.50	31		52.50	69.50
2450	1-1/2"		12.10	.661		21.50	34.50		56	75
2460	2"	Q-1	20	.800		27	37.50		64.50	85.50
2470	3"		13.90	1.151		36.50	54		90.50	122
2480	4"		11	1.455		42.50	68		110.50	149
2490	6"		6.70	2.388		145	112		257	325
2500	8"	Q-2	6.20	3.871		300	188		488	610
2510	Flange, socket, 150 lb., 1/2"	1 Plum	55.60	.144		11.65	7.50		19.15	24
2514	3/4"		47.60	.168		12.40	8.75		21.15	27
2518	1"		41.70	.192		13.85	10		23.85	30
2522	1-1/2"		33.30	.240		14.55	12.50		27.05	35
2526	2"	Q-1	60.60	.264		19.40	12.35		31.75	40
2530	4"		30.30	.528		42	24.50		66.50	83
2534	6"		18.50	.865		66	40.50		106.50	133
2538	8"	Q-2	17.10	1.404		118	68		186	232
2550	Coupling, 1/2"	1 Plum	30.30	.264		3.84	13.75		17.59	24.50
2570	3/4"		26	.308		5.20	16		21.20	30
2580	1"		22.70	.352		5.35	18.35		23.70	33.50
2590	1-1/4"		20.20	.396		8.15	20.50		28.65	40
2600	1-1/2"		18.20	.440		8.80	23		31.80	44
2610	2"	Q-1	33.10	.483		9.40	22.50		31.90	44.50
2620	3"		20.80	.769		23	36		59	79.50
2630	4"		16.50	.970		33.50	45.50		79	105
2640	6"		10.10	1.584		72	74		146	190
2650	8"	Q-2	9.30	2.581		97.50	125		222.50	295
2660	10"		8.50	2.824		335	137		472	575
2670	12"		7.60	3.158		385	153		538	655
2700	PVC (white), schedule 40, socket joints									
2760	90° elbow, 1/2"	1 Plum	33.30	.240	Ea.	.41	12.50		12.91	19.20
2770	3/4"		28.60	.280		.46	14.55		15.01	22.50
2780	1"		25	.320		.83	16.65		17.48	26
2790	1-1/4"		22.20	.360		1.47	18.75		20.22	29.50
2800	1-1/2"		20	.400		1.58	21		22.58	32.50
2810	2"	Q-1	36.40	.440		2.47	20.50		22.97	33.50
2820	2-1/2"		26.70	.599		7.55	28		35.55	50.50
2830	3"		22.90	.699		9	32.50		41.50	59
2840	4"		18.20	.879		16.10	41		57.10	79.50
2850	5"		12.10	1.322		41.50	62		103.50	139
2860	6"		11.10	1.441		51	67.50		118.50	158
2870	8"	Q-2	10.30	2.330		132	113		245	315
2980	45° elbow, 1/2"	1 Plum	33.30	.240		.68	12.50		13.18	19.50
2990	3/4"		28.60	.280		1.06	14.55		15.61	23
3000	1"		25	.320		1.27	16.65		17.92	26.50
3010	1-1/4"		22.20	.360		1.77	18.75		20.52	30

22 11 13.76 Pipe Fittings, Plastic		Crew	Daily Output	Labor-Hours	Unit	Material	Labor	Equipment	Total	Total Incl O&P
						2010 Bare Costs				
3020	1-1/2"	1 Plum	20	.400	Ea.	2.22	21		23.22	33.50
3030	2"	Q-1	36.40	.440		2.90	20.50		23.40	34
3040	2-1/2"		26.70	.599		7.55	28		35.55	50.50
3050	3"		22.90	.699		11.70	32.50		44.20	62
3060	4"		18.20	.879		21	41		62	84.50
3070	5"		12.10	1.322		41.50	62		103.50	139
3080	6"		11.10	1.441		52	67.50		119.50	158
3090	8"	Q-2	10.30	2.330		125	113		238	305
3180	Tee, 1/2"	1 Plum	22.20	.360		.51	18.75		19.26	28.50
3190	3/4"		19	.421		.59	22		22.59	33.50
3200	1"		16.70	.479		1.10	25		26.10	38.50
3210	1-1/4"		14.80	.541		1.71	28		29.71	44
3220	1-1/2"		13.30	.602		2.09	31.50		33.59	49.50
3230	2"	Q-1	24.20	.661		3.05	31		34.05	50
3240	2-1/2"		17.80	.899		10.05	42		52.05	74
3250	3"		15.20	1.053		13.25	49.50		62.75	88.50
3260	4"		12.10	1.322		24	62		86	120
3270	5"		8.10	1.975		58	92.50		150.50	203
3280	6"		7.40	2.162		80.50	101		181.50	241
3290	8"	Q-2	6.80	3.529		187	171		358	460
3380	Coupling, 1/2"	1 Plum	33.30	.240		.28	12.50		12.78	19.05
3390	3/4"		28.60	.280		.38	14.55		14.93	22.50
3400	1"		25	.320		.68	16.65		17.33	26
3410	1-1/4"		22.20	.360		.90	18.75		19.65	29
3420	1-1/2"		20	.400		.97	21		21.97	32
3430	2"	Q-1	36.40	.440		1.48	20.50		21.98	32.50
3440	2-1/2"		26.70	.599		3.27	28		31.27	45.50
3450	3"		22.90	.699		5.10	32.50		37.60	54.50
3460	4"		18.20	.879		7.40	41		48.40	69.50
3470	5"		12.10	1.322		13.55	62		75.55	108
3480	6"		11.10	1.441		23.50	67.50		91	127
3490	8"	Q-2	10.30	2.330		43.50	113		156.50	218
3710	Reducing insert, schedule 40, socket weld									
3712	3/4"	1 Plum	31.50	.254	Ea.	.43	13.20		13.63	20.50
3713	1"		27.50	.291		.78	15.15		15.93	23.50
3715	1-1/2"		22	.364		1.12	18.95		20.07	29.50
3716	2"	Q-1	40	.400		1.85	18.75		20.60	30
3717	4"		20	.800		9.80	37.50		47.30	67
3718	6"		12.20	1.311		24	61.50		85.50	119
3719	8"	Q-2	11.30	2.124		95.50	103		198.50	260
3730	Reducing insert, socket weld x female/male thread									
3732	1/2"	1 Plum	38.30	.209	Ea.	1.96	10.85		12.81	18.45
3733	3/4"		32.90	.243		1.21	12.65		13.86	20.50
3734	1"		28.80	.278		1.70	14.45		16.15	23.50
3736	1-1/2"		23	.348		3.05	18.10		21.15	30.50
3737	2"	Q-1	41.90	.382		3.27	17.90		21.17	30.50
3738	4"	"	20.90	.766		12.40	36		48.40	67.50
3742	Male adapter, socket weld x male thread									
3744	1/2"	1 Plum	38.30	.209	Ea.	.38	10.85		11.23	16.70
3745	3/4"		32.90	.243		.41	12.65		13.06	19.40
3746	1"		28.80	.278		.74	14.45		15.19	22.50
3748	1-1/2"		23	.348		1.21	18.10		19.31	28.50
3749	2"	Q-1	41.90	.382		1.53	17.90		19.43	28.50

22 11 13 – Facility Water Distribution Piping

22 11 13.76 Pipe Fittings, Plastic		Crew	Daily Output	Labor-Hours	Unit	Material	Labor	Equipment	Total	Total Incl O&P
						2010 Bare Costs				
3750	4"	Q-1	20.90	.766	Ea.	9.10	36		45.10	64
3754	Female adapter, socket weld x female thread									
3756	1/2"	1 Plum	38.30	.209	Ea.	.46	10.85		11.31	16.80
3757	3/4"		32.90	.243		.59	12.65		13.24	19.60
3758	1"		28.80	.278		.68	14.45		15.13	22.50
3760	1-1/2"		23	.348		1.21	18.10		19.31	28.50
3761	2"	Q-1	41.90	.382		1.62	17.90		19.52	29
3762	4"	"	20.90	.766		9.15	36		45.15	64
3800	PVC, schedule 80, socket joints									
3810	Reducing insert									
3812	3/4"	1 Plum	28.60	.280	Ea.	1.24	14.55		15.79	23.50
3813	1"		25	.320		3.54	16.65		20.19	29
3815	1-1/2"		20	.400		7.60	21		28.60	39.50
3816	2"	Q-1	36.40	.440		10.80	20.50		31.30	43
3817	4"		18.20	.879		41	41		82	107
3818	6"		11.10	1.441		57.50	67.50		125	164
3819	8"	Q-2	10.20	2.353		330	114		444	535
3830	Reducing insert, socket weld x female/male thread									
3832	1/2"	1 Plum	34.80	.230	Ea.	7.75	11.95		19.70	26.50
3833	3/4"		29.90	.268		4.75	13.95		18.70	26.50
3834	1"		26.10	.307		7.45	15.95		23.40	32
3836	1-1/2"		20.90	.383		9.35	19.90		29.25	40.50
3837	2"	Q-1	38	.421		13.70	19.75		33.45	44.50
3838	4"	"	19	.842		66.50	39.50		106	132
3844	Adapter, male socket x male thread									
3846	1/2"	1 Plum	34.80	.230	Ea.	3	11.95		14.95	21.50
3847	3/4"		29.90	.268		3.30	13.95		17.25	24.50
3848	1"		26.10	.307		5.70	15.95		21.65	30.50
3850	1-1/2"		20.90	.383		9.60	19.90		29.50	40.50
3851	2"	Q-1	38	.421		13.85	19.75		33.60	45
3852	4"	"	19	.842		31	39.50		70.50	93.50
3860	Adapter, female socket x female thread									
3862	1/2"	1 Plum	34.80	.230	Ea.	3.59	11.95		15.54	22
3863	3/4"		29.90	.268		5.35	13.95		19.30	27
3864	1"		26.10	.307		7.90	15.95		23.85	32.50
3866	1-1/2"		20.90	.383		15.65	19.90		35.55	47
3867	2"	Q-1	38	.421		27.50	19.75		47.25	59.50
3868	4"	"	19	.842		83.50	39.50		123	151
3872	Union, socket joints									
3874	1/2"	1 Plum	25.80	.310	Ea.	7.90	16.15		24.05	32.50
3875	3/4"		22.10	.362		10	18.85		28.85	39
3876	1"		19.30	.415		11.40	21.50		32.90	45
3878	1-1/2"		15.50	.516		25.50	27		52.50	69
3879	2"	Q-1	28.10	.569		35	26.50		61.50	78.50
3888	Cap									
3890	1/2"	1 Plum	54.50	.147	Ea.	3.77	7.65		11.42	15.60
3891	3/4"		46.70	.171		3.96	8.90		12.86	17.70
3892	1"		41	.195		7.05	10.15		17.20	23
3894	1-1/2"		32.80	.244		8.50	12.70		21.20	28.50
3895	2"	Q-1	59.50	.269		22.50	12.60		35.10	43.50
3896	4"		30	.533		67.50	25		92.50	112
3897	6"		18.20	.879		168	41		209	246
3898	8"	Q-2	16.70	1.437		216	70		286	340

22 11 13.76 Pipe Fittings, Plastic		Crew	Daily Output	Labor-Hours	Unit	Material	2010 Bare Costs Labor	Equipment	Total	Total Incl O&P
4500	DWV, ABS, non pressure, socket joints									
4540	1/4 Bend, 1-1/4"	1 Plum	20.20	.396	Ea.	2.16	20.50		22.66	33.50
4560	1-1/2"	"	18.20	.440		1.56	23		24.56	36
4570	2"	Q-1	33.10	.483		2.30	22.50		24.80	36.50
4580	3"		20.80	.769		5.80	36		41.80	60.50
4590	4"		16.50	.970		11	45.50		56.50	80
4600	6"		10.10	1.584		52	74		126	168
4650	1/8 Bend, same as 1/4 Bend									
4800	Tee, sanitary									
4820	1-1/4"	1 Plum	13.50	.593	Ea.	2.58	31		33.58	49
4830	1-1/2"	"	12.10	.661		2.08	34.50		36.58	54
4840	2"	Q-1	20	.800		3.21	37.50		40.71	59.50
4850	3"		13.90	1.151		8.80	54		62.80	90.50
4860	4"		11	1.455		16.15	68		84.15	120
4862	Tee, sanitary, reducing, 2" x 1-1/2"		22	.727		3.06	34		37.06	54.50
4864	3" x 2"		15.30	1.046		6.40	49		55.40	80.50
4868	4" x 3"		12.10	1.322		16	62		78	111
4870	Combination Y and 1/8 bend									
4872	1-1/2"	1 Plum	12.10	.661	Ea.	4.85	34.50		39.35	57
4874	2"	Q-1	20	.800		5.90	37.50		43.40	62.50
4876	3"		13.90	1.151		12.70	54		66.70	95
4878	4"		11	1.455		26	68		94	131
4880	3" x 1-1/2"		15.50	1.032		12.10	48.50		60.60	86
4882	4" x 3"		12.10	1.322		20.50	62		82.50	116
4900	Wye, 1-1/4"	1 Plum	13.50	.593		2.97	31		33.97	49.50
4902	1-1/2"	"	12.10	.661		3.12	34.50		37.62	55
4904	2"	Q-1	20	.800		4.29	37.50		41.79	60.50
4906	3"		13.90	1.151		9.80	54		63.80	92
4908	4"		11	1.455		21	68		89	125
4910	6"		6.70	2.388		64.50	112		176.50	239
4918	3" x 1-1/2"		15.50	1.032		8.10	48.50		56.60	81.50
4920	4" x 3"		12.10	1.322		16.25	62		78.25	111
4922	6" x 4"		6.90	2.319		53.50	109		162.50	222
4930	Double Wye, 1-1/2"	1 Plum	9.10	.879		6.95	46		52.95	76
4932	2"	Q-1	16.60	.964		8.30	45		53.30	76.50
4934	3"		10.40	1.538		21.50	72		93.50	132
4936	4"		8.25	1.939		43.50	91		134.50	184
4940	2" x 1-1/2"		16.80	.952		8.30	44.50		52.80	76
4942	3" x 2"		10.60	1.509		15.95	70.50		86.45	124
4944	4" x 3"		8.45	1.893		34.50	88.50		123	171
4946	6" x 4"		7.25	2.207		68	103		171	230
4950	Reducer bushing, 2" x 1-1/2"		36.40	.440		1.15	20.50		21.65	32.50
4952	3" x 1-1/2"		27.30	.586		4.33	27.50		31.83	46
4954	4" x 2"		18.20	.879		8.85	41		49.85	71
4956	6" x 4"		11.10	1.441		23.50	67.50		91	127
4960	Couplings, 1-1/2"	1 Plum	18.20	.440		.65	23		23.65	35
4962	2"	Q-1	33.10	.483		.96	22.50		23.46	35
4963	3"		20.80	.769		2.53	36		38.53	57
4964	4"		16.50	.970		4.92	45.50		50.42	73.50
4966	6"		10.10	1.584		19.35	74		93.35	133
4970	2" x 1-1/2"		33.30	.480		1.91	22.50		24.41	35.50
4972	3" x 1-1/2"		21	.762		5.30	35.50		40.80	59.50
4974	4" x 3"		16.70	.958		9.40	45		54.40	78

22 11 13 – Facility Water Distribution Piping

22 11 13.76 Pipe Fittings, Plastic		Crew	Daily Output	Labor-Hours	Unit	2010 Bare Costs Material	Labor	Equipment	Total	Total Incl O&P
4978	Closet flange, 4″	1 Plum	32	.250	Ea.	4.46	13		17.46	24.50
4980	4″ x 3″	″	34	.235		5.05	12.25		17.30	24
5000	DWV, PVC, schedule 40, socket joints									
5040	1/4 bend, 1-1/4″	1 Plum	20.20	.396	Ea.	3.69	20.50		24.19	35
5060	1-1/2″	″	18.20	.440		1.10	23		24.10	35.50
5070	2″	Q-1	33.10	.483		1.74	22.50		24.24	36
5080	3″		20.80	.769		5.15	36		41.15	59.50
5090	4″		16.50	.970		10.10	45.50		55.60	79
5100	6″		10.10	1.584		35.50	74		109.50	150
5105	8″	Q-2	9.30	2.581		72	125		197	267
5106	10″	″	8.50	2.824		161	137		298	385
5110	1/4 bend, long sweep, 1-1/2″	1 Plum	18.20	.440		2.65	23		25.65	37.50
5112	2″	Q-1	33.10	.483		2.86	22.50		25.36	37
5114	3″		20.80	.769		6.75	36		42.75	61.50
5116	4″		16.50	.970		12.90	45.50		58.40	82
5150	1/8 bend, 1-1/4″	1 Plum	20.20	.396		2.09	20.50		22.59	33.50
5170	1-1/2″	″	18.20	.440		1.09	23		24.09	35.50
5180	2″	Q-1	33.10	.483		1.55	22.50		24.05	35.50
5190	3″		20.80	.769		4.69	36		40.69	59
5200	4″		16.50	.970		8.05	45.50		53.55	77
5210	6″		10.10	1.584		32.50	74		106.50	147
5215	8″	Q-2	9.30	2.581		54	125		179	248
5216	10″		8.50	2.824		135	137		272	355
5217	12″		7.60	3.158		238	153		391	490
5250	Tee, sanitary 1-1/4″	1 Plum	13.50	.593		3.19	31		34.19	49.50
5254	1-1/2″	″	12.10	.661		2	34.50		36.50	53.50
5255	2″	Q-1	20	.800		2.94	37.50		40.44	59
5256	3″		13.90	1.151		7.55	54		61.55	89.50
5257	4″		11	1.455		13.10	68		81.10	116
5259	6″		6.70	2.388		57.50	112		169.50	231
5261	8″	Q-2	6.20	3.871		190	188		378	490
5276	Tee, sanitary, reducing									
5281	2″ x 1-1/2″ x 1-1/2″	Q-1	23	.696	Ea.	2.59	32.50		35.09	52
5282	2″ x 1-1/2″ x 2″		22	.727		3.13	34		37.13	54.50
5283	2″ x 2″ x 1-1/2″		22	.727		2.60	34		36.60	54
5284	3″ x 3″ x 1-1/2″		15.50	1.032		5.45	48.50		53.95	78.50
5285	3″ x 3″ x 2″		15.30	1.046		5.75	49		54.75	80
5286	4″ x 4″ x 1-1/2″		12.30	1.301		13.60	61		74.60	106
5287	4″ x 4″ x 2″		12.20	1.311		12.25	61.50		73.75	106
5288	4″ x 4″ x 3″		12.10	1.322		16.65	62		78.65	111
5291	6″ x 6″ x 4″		6.90	2.319		55.50	109		164.50	224
5294	Tee, double sanitary									
5295	1-1/2″	1 Plum	9.10	.879	Ea.	4.08	46		50.08	73
5296	2″	Q-1	16.60	.964		6	45		51	74
5297	3″		10.40	1.538		16.75	72		88.75	126
5298	4″		8.25	1.939		27	91		118	166
5303	Wye, reducing									
5304	2″ x 1-1/2″ x 1-1/2″	Q-1	23	.696	Ea.	5.05	32.50		37.55	54.50
5305	2″ x 2″ x 1-1/2″		22	.727		4.40	34		38.40	56
5306	3″ x 3″ x 2″		15.30	1.046		14.30	49		63.30	89.50
5307	4″ x 4″ x 2″		12.20	1.311		10.55	61.50		72.05	104
5309	4″ x 4″ x 3″		12.10	1.322		14.25	62		76.25	109
5314	Combination Y & 1/8 bend, 1-1/2″	1 Plum	12.10	.661		4.34	34.50		38.84	56.50

22 11 13.76 Pipe Fittings, Plastic		Crew	Daily Output	Labor-Hours	Unit	Material	Labor	Equipment	Total	Total Incl O&P
						2010 Bare Costs				
5315	2"	Q-1	20	.800	Ea.	5.75	37.50		43.25	62.50
5317	3"		13.90	1.151		14.30	54		68.30	96.50
5318	4"		11	1.455		28.50	68		96.50	133
5319	6"		6.70	2.388		145	112		257	330
5320	8"	Q-2	6.20	3.871		270	188		458	580
5324	Combination Y & 1/8 bend, reducing									
5325	2" x 2" x 1-1/2"	Q-1	22	.727	Ea.	6.25	34		40.25	58
5327	3" x 3" x 1-1/2"		15.50	1.032		10.80	48.50		59.30	84.50
5328	3" x 3" x 2"		15.30	1.046		8.50	49		57.50	83
5329	4" x 4" x 2"		12.20	1.311		13.80	61.50		75.30	107
5331	Wye, 1-1/4"	1 Plum	13.50	.593		4.06	31		35.06	50.50
5332	1-1/2"	"	12.10	.661		3.06	34.50		37.56	55
5333	2"	Q-1	20	.800		3.58	37.50		41.08	60
5334	3"		13.90	1.151		9.65	54		63.65	91.50
5335	4"		11	1.455		17.55	68		85.55	121
5336	6"		6.70	2.388		53	112		165	226
5337	8"	Q-2	6.20	3.871		155	188		343	455
5338	10"		5.70	4.211		281	205		486	615
5339	12"		5.10	4.706		455	229		684	845
5341	2" x 1-1/2"	Q-1	22	.727		4.39	34		38.39	56
5342	3" x 1-1/2"		15.50	1.032		6.30	48.50		54.80	79.50
5343	4" x 3"		12.10	1.322		14.25	62		76.25	109
5344	6" x 4"		6.90	2.319		39	109		148	206
5345	8" x 6"	Q-2	6.40	3.750		117	182		299	400
5347	Double wye, 1-1/2"	1 Plum	9.10	.879		6.90	46		52.90	76
5348	2"	Q-1	16.60	.964		7.45	45		52.45	75.50
5349	3"		10.40	1.538		19.20	72		91.20	129
5350	4"		8.25	1.939		39	91		130	179
5353	Double wye, reducing									
5354	2" x 2" x 1-1/2" x 1-1/2"	Q-1	16.80	.952	Ea.	6.80	44.50		51.30	74.50
5355	3" x 3" x 2" x 2"		10.60	1.509		14.35	70.50		84.85	122
5356	4" x 4" x 3" x 3"		8.45	1.893		31	88.50		119.50	167
5357	6" x 6" x 4" x 4"		7.25	2.207		81.50	103		184.50	245
5374	Coupling, 1-1/4"	1 Plum	20.20	.396		1.93	20.50		22.43	33
5376	1-1/2"	"	18.20	.440		.55	23		23.55	35
5378	2"	Q-1	33.10	.483		.71	22.50		23.21	35
5380	3"		20.80	.769		2.46	36		38.46	56.50
5390	4"		16.50	.970		4.37	45.50		49.87	73
5400	6"		10.10	1.584		14.20	74		88.20	127
5402	8"	Q-2	9.30	2.581		35	125		160	227
5404	2" x 1-1/2"	Q-1	33.30	.480		1.59	22.50		24.09	35.50
5406	3" x 1-1/2"		21	.762		4.86	35.50		40.36	59
5408	4" x 3"		16.70	.958		7.85	45		52.85	76
5410	Reducer bushing, 2" x 1-1/4"		36.50	.438		1.95	20.50		22.45	33
5411	2" x 1-1/2"		36.40	.440		.94	20.50		21.44	32
5412	3" x 1-1/2"		27.30	.586		4.53	27.50		32.03	46
5413	3" x 2"		27.10	.590		2.63	27.50		30.13	44.50
5414	4" x 2"		18.20	.879		8.75	41		49.75	71
5415	4" x 3"		16.70	.958		4.47	45		49.47	72.50
5416	6" x 4"		11.10	1.441		23.50	67.50		91	127
5418	8" x 6"	Q-2	10.20	2.353		50.50	114		164.50	227
5425	Closet flange 4"	Q-1	32	.500		5.25	23.50		28.75	41
5426	4" x 3"	"	34	.471		4.50	22		26.50	38

22 11 13.76 Pipe Fittings, Plastic		Crew	Daily Output	Labor-Hours	Unit	Material	2010 Bare Costs Labor	Equipment	Total	Total Incl O&P
5450	Solvent cement for PVC, industrial grade, per quart				Qt.	19.20			19.20	21
5500	CPVC, Schedule 80, threaded joints									
5540	90° Elbow, 1/4"	1 Plum	32	.250	Ea.	10.40	13		23.40	31
5560	1/2"		30.30	.264		6.05	13.75		19.80	27
5570	3/4"		26	.308		9.05	16		25.05	34
5580	1"		22.70	.352		12.65	18.35		31	41.50
5590	1-1/4"		20.20	.396		24.50	20.50		45	58
5600	1-1/2"		18.20	.440		26.50	23		49.50	63.50
5610	2"	Q-1	33.10	.483		35	22.50		57.50	72.50
5620	2-1/2"		24.20	.661		109	31		140	167
5630	3"		20.80	.769		117	36		153	183
5640	4"		16.50	.970		183	45.50		228.50	270
5650	6"		10.10	1.584		216	74		290	350
5660	45° Elbow same as 90° Elbow									
5700	Tee, 1/4"	1 Plum	22	.364	Ea.	20	18.95		38.95	50.50
5702	1/2"		20.20	.396		20	20.50		40.50	53
5704	3/4"		17.30	.462		29	24		53	68
5706	1"		15.20	.526		31.50	27.50		59	75.50
5708	1-1/4"		13.50	.593		31.50	31		62.50	81
5710	1-1/2"		12.10	.661		33	34.50		67.50	88
5712	2"	Q-1	20	.800		37	37.50		74.50	96.50
5714	2-1/2"		16.20	.988		181	46.50		227.50	269
5716	3"		13.90	1.151		210	54		264	310
5718	4"		11	1.455		495	68		563	645
5720	6"		6.70	2.388		590	112		702	820
5730	Coupling, 1/4"	1 Plum	32	.250		13.25	13		26.25	34
5732	1/2"		30.30	.264		13.25	13.75		27	35
5734	3/4"		26	.308		15.90	16		31.90	41.50
5736	1"		22.70	.352		17.60	18.35		35.95	47
5738	1-1/4"		20.20	.396		21	20.50		41.50	54.50
5740	1-1/2"		18.20	.440		23	23		46	59.50
5742	2"	Q-1	33.10	.483		27	22.50		49.50	63.50
5744	2-1/2"		24.20	.661		48	31		79	99.50
5746	3"		20.80	.769		56	36		92	116
5748	4"		16.50	.970		114	45.50		159.50	194
5750	6"		10.10	1.584		156	74		230	282
5752	8"	Q-2	9.30	2.581		335	125		460	555
5900	CPVC, Schedule 80, socket joints									
5904	90° Elbow, 1/4"	1 Plum	32	.250	Ea.	9.70	13		22.70	30
5906	1/2"		30.30	.264		3.81	13.75		17.56	24.50
5908	3/4"		26	.308		4.85	16		20.85	29.50
5910	1"		22.70	.352		7.70	18.35		26.05	36
5912	1-1/4"		20.20	.396		16.65	20.50		37.15	49.50
5914	1-1/2"		18.20	.440		18.60	23		41.60	55
5916	2"	Q-1	33.10	.483		22.50	22.50		45	58.50
5918	2-1/2"		24.20	.661		51.50	31		82.50	104
5920	3"		20.80	.769		58.50	36		94.50	119
5922	4"		16.50	.970		105	45.50		150.50	184
5924	6"		10.10	1.584		212	74		286	345
5926	8"		9.30	1.720		520	80.50		600.50	690
5930	45° Elbow, 1/4"	1 Plum	32	.250		14.40	13		27.40	35.50
5932	1/2"		30.30	.264		4.65	13.75		18.40	25.50
5934	3/4"		26	.308		6.70	16		22.70	31.50

22 11 13.76 Pipe Fittings, Plastic		Crew	Daily Output	Labor-Hours	Unit	Material	2010 Bare Costs Labor	Equipment	Total	Total Incl O&P
5936	1"	1 Plum	22.70	.352	Ea.	10.70	18.35		29.05	39.50
5938	1-1/4"		20.20	.396		21	20.50		41.50	54
5940	1-1/2"		18.20	.440		21.50	23		44.50	58
5942	2"	Q-1	33.10	.483		24	22.50		46.50	60.50
5944	2-1/2"		24.20	.661		49.50	31		80.50	101
5946	3"		20.80	.769		63.50	36		99.50	124
5948	4"		16.50	.970		87	45.50		132.50	164
5950	6"		10.10	1.584		268	74		342	405
5952	8"		9.30	1.720		560	80.50		640.50	735
5960	Tee, 1/4"	1 Plum	22	.364		8.90	18.95		27.85	38.50
5962	1/2"		20.20	.396		8.90	20.50		29.40	41
5964	3/4"		17.30	.462		9.10	24		33.10	46
5966	1"		15.20	.526		11.10	27.50		38.60	53.50
5968	1-1/4"		13.50	.593		23.50	31		54.50	72
5970	1-1/2"		12.10	.661		27	34.50		61.50	81
5972	2"	Q-1	20	.800		30	37.50		67.50	89
5974	2-1/2"		16.20	.988		76	46.50		122.50	153
5976	3"		13.90	1.151		76	54		130	165
5978	4"		11	1.455		101	68		169	213
5980	6"		6.70	2.388		264	112		376	460
5982	8"	Q-2	6.20	3.871		755	188		943	1,100
5990	Coupling, 1/4"	1 Plum	32	.250		10.35	13		23.35	31
5992	1/2"		30.30	.264		4.02	13.75		17.77	25
5994	3/4"		26	.308		5.60	16		21.60	30
5996	1"		22.70	.352		7.55	18.35		25.90	36
5998	1-1/4"		20.20	.396		11.35	20.50		31.85	43.50
6000	1-1/2"		18.20	.440		14.25	23		37.25	50
6002	2"	Q-1	33.10	.483		16.55	22.50		39.05	52.50
6004	2-1/2"		24.20	.661		37	31		68	87
6006	3"		20.80	.769		40	36		76	98
6008	4"		16.50	.970		52.50	45.50		98	126
6010	6"		10.10	1.584		124	74		198	247
6012	8"	Q-2	9.30	2.581		335	125		460	560
6200	CTS, 100 psi at 180°F, hot and cold water									
6230	90° Elbow, 1/2"	1 Plum	20	.400	Ea.	.32	21		21.32	31.50
6250	3/4"		19	.421		.53	22		22.53	33.50
6251	1"		16	.500		1.47	26		27.47	40.50
6252	1-1/4"		15	.533		2.52	28		30.52	44.50
6253	1-1/2"		14	.571		4.56	29.50		34.06	49.50
6254	2"	Q-1	23	.696		8.75	32.50		41.25	58.50
6260	45° Elbow, 1/2"	1 Plum	20	.400		.41	21		21.41	31.50
6280	3/4"		19	.421		.70	22		22.70	34
6281	1"		16	.500		1.47	26		27.47	40.50
6282	1-1/4"		15	.533		2.92	28		30.92	44.50
6283	1-1/2"		14	.571		4.23	29.50		33.73	49
6284	2"	Q-1	23	.696		10.85	32.50		43.35	61
6290	Tee, 1/2"	1 Plum	13	.615		.41	32		32.41	48.50
6310	3/4"		12	.667		.77	34.50		35.27	53
6311	1"		11	.727		3.32	38		41.32	60
6312	1-1/4"		10	.800		5.10	41.50		46.60	68
6313	1-1/2"		10	.800		6.65	41.50		48.15	70
6314	2"	Q-1	17	.941		10.75	44		54.75	78
6320	Coupling, 1/2"	1 Plum	22	.364		.26	18.95		19.21	29

22 11 13 – Facility Water Distribution Piping

22 11 13.76 Pipe Fittings, Plastic		Crew	Daily Output	Labor-Hours	Unit	Material	2010 Bare Costs Labor	Equipment	Total	Total Incl O&P
6340	3/4"	1 Plum	21	.381	Ea.	.35	19.85		20.20	30
6341	1"		18	.444		1.29	23		24.29	36
6342	1-1/4"		17	.471		1.56	24.50		26.06	38
6343	1-1/2"		16	.500		2.20	26		28.20	41.50
6344	2"	Q-1	28	.571		4.48	27		31.48	45
6360	Solvent cement for CPVC, commercial grade, per quart				Qt.	25.50			25.50	28
7550	Union, schedule 40, socket joints, 1/2"	1 Plum	19	.421	Ea.	4.05	22		26.05	37.50
7560	3/4"		18	.444		4.61	23		27.61	39.50
7570	1"		15	.533		4.74	28		32.74	46.50
7580	1-1/4"		14	.571		14.25	29.50		43.75	60
7590	1-1/2"		13	.615		15.30	32		47.30	65
7600	2"	Q-1	20	.800		22	37.50		59.50	80
7992	Polybutyl/polyethyl pipe, for copper fittings see Div. 22 11 13.25 7000									
8000	Compression type, PVC, 160 psi cold water									
8010	Coupling, 3/4" CTS	1 Plum	21	.381	Ea.	3.20	19.85		23.05	33
8020	1" CTS		18	.444		3.98	23		26.98	39
8030	1-1/4" CTS		17	.471		5.60	24.50		30.10	42.50
8040	1-1/2" CTS		16	.500		7.65	26		33.65	47.50
8050	2" CTS		15	.533		10.65	28		38.65	53
8060	Female adapter, 3/4" FPT x 3/4" CTS		23	.348		3.54	18.10		21.64	31
8070	3/4" FPT x 1" CTS		21	.381		4.32	19.85		24.17	34.50
8080	1" FPT x 1" CTS		20	.400		5.65	21		26.65	37.50
8090	1-1/4" FPT x 1-1/4" CTS		18	.444		7.40	23		30.40	42.50
8100	1-1/2" FPT x 1-1/2" CTS		16	.500		8.45	26		34.45	48.50
8110	2" FPT x 2" CTS		13	.615		10.25	32		42.25	59.50
8130	Male adapter, 3/4" MPT x 3/4" CTS		23	.348		3.42	18.10		21.52	31
8140	3/4" MPT x 1" CTS		21	.381		3.69	19.85		23.54	33.50
8150	1" MPT x 1" CTS		20	.400		4.78	21		25.78	36.50
8160	1-1/4" MPT x 1-1/4" CTS		18	.444		6.50	23		29.50	41.50
8170	1-1/2" MPT x 1-1/2" CTS		16	.500		7.85	26		33.85	47.50
8180	2" MPT x 2" CTS		13	.615		10.15	32		42.15	59
8200	Spigot adapter, 3/4" IPS x 3/4" CTS		23	.348		2.48	18.10		20.58	29.50
8210	3/4" IPS x 1" CTS		21	.381		3.05	19.85		22.90	33
8220	1" IPS x 1" CTS		20	.400		3.05	21		24.05	34.50
8230	1-1/4" IPS x 1-1/4" CTS		18	.444		4.59	23		27.59	39.50
8240	1-1/2" IPS x 1-1/2" CTS		16	.500		4.82	26		30.82	44.50
8250	2" IPS x 2" CTS		13	.615		5.95	32		37.95	54.50
8270	Price includes insert stiffeners									
8280	250 psi is same price as 160 psi									
8300	Insert type, nylon, 160 & 250 psi, cold water									
8310	Clamp ring stainless steel, 3/4" IPS	1 Plum	115	.070	Ea.	2.25	3.62		5.87	7.95
8320	1" IPS		107	.075		2.30	3.89		6.19	8.40
8330	1-1/4" IPS		101	.079		2.32	4.12		6.44	8.75
8340	1-1/2" IPS		95	.084		3.11	4.38		7.49	9.95
8350	2" IPS		85	.094		3.57	4.90		8.47	11.30
8370	Coupling, 3/4" IPS		22	.364		.76	18.95		19.71	29.50
8390	1-1/4" IPS		18	.444		1.17	23		24.17	36
8400	1-1/2" IPS		17	.471		1.39	24.50		25.89	38
8410	2" IPS		16	.500		2.63	26		28.63	42
8430	Elbow, 90°, 3/4" IPS		22	.364		1.52	18.95		20.47	30
8440	1" IPS		19	.421		1.67	22		23.67	35
8450	1-1/4" IPS		18	.444		1.87	23		24.87	36.50
8460	1-1/2" IPS		17	.471		2.21	24.50		26.71	39

22 11 13.76 Pipe Fittings, Plastic		Crew	Daily Output	Labor-Hours	Unit	Material	2010 Bare Costs Labor	Equipment	Total	Total Incl O&P
8470	2" IPS	1 Plum	16	.500	Ea.	3.08	26		29.08	42.50
8490	Male adapter, 3/4" IPS x 3/4" MPT		25	.320		.76	16.65		17.41	26
8500	1" IPS x 1" MPT		21	.381		.78	19.85		20.63	30.50
8510	1-1/4" IPS x 1-1/4" MPT		20	.400		1.24	21		22.24	32.50
8520	1-1/2" IPS x 1-1/2" MPT		18	.444		1.39	23		24.39	36
8530	2" IPS x 2" MPT		15	.533		2.66	28		30.66	44.50
8550	Tee, 3/4" IPS		14	.571		1.47	29.50		30.97	46
8560	1" IPS		13	.615		1.91	32		33.91	50
8570	1-1/4" IPS		12	.667		2.99	34.50		37.49	55.50
8580	1-1/2" IPS		11	.727		3.38	38		41.38	60
8590	2" IPS	↓	10	.800	↓	6.65	41.50		48.15	70
8610	Insert type, PVC, 100 psi @ 180°F, hot & cold water									
8620	Coupler, male, 3/8" CTS x 3/8" MPT	1 Plum	29	.276	Ea.	.62	14.35		14.97	22
8630	3/8" CTS x 1/2" MPT		28	.286		.62	14.85		15.47	23
8640	1/2" CTS x 1/2" MPT		27	.296		.62	15.40		16.02	23.50
8650	1/2" CTS x 3/4" MPT		26	.308		1.97	16		17.97	26
8660	3/4" CTS x 1/2" MPT		25	.320		1.77	16.65		18.42	27
8670	3/4" CTS x 3/4" MPT		25	.320		.76	16.65		17.41	26
8700	Coupling, 3/8" CTS x 1/2" CTS		25	.320		3.21	16.65		19.86	28.50
8710	1/2" CTS		23	.348		3.75	18.10		21.85	31
8730	3/4" CTS		22	.364		5.90	18.95		24.85	35
8750	Elbow 90°, 3/8" CTS		25	.320		3.81	16.65		20.46	29
8760	1/2" CTS		23	.348		4.51	18.10		22.61	32
8770	3/4" CTS		22	.364		6.90	18.95		25.85	36
8800	Rings, crimp, copper, 3/8" CTS		120	.067		.16	3.47		3.63	5.40
8810	1/2" CTS		117	.068		.16	3.56		3.72	5.55
8820	3/4" CTS		115	.070		.21	3.62		3.83	5.70
8850	Reducer tee, 3/8" x 3/8" x 1/2" CTS		17	.471		1.29	24.50		25.79	38
8860	1/2" x 3/8" x 1/2" CTS		15	.533		1.29	28		29.29	43
8870	3/4" x 1/2" x 1/2" CTS		14	.571		1.98	29.50		31.48	46.50
8890	3/4" x 3/4" x 1/2" CTS		14	.571		1.98	29.50		31.48	46.50
8900	3/4" x 1/2" x 3/8" CTS		14	.571		1.98	29.50		31.48	46.50
8930	Tee, 3/8" CTS		17	.471		1.50	24.50		26	38
8940	1/2" CTS		15	.533		1.69	28		29.69	43.50
8950	3/4" CTS	↓	14	.571	↓	1.84	29.50		31.34	46.50
8960	Copper rings included in fitting price									
9000	Flare type, assembled, acetal, hot & cold water									
9010	Coupling, 1/4" & 3/8" CTS	1 Plum	24	.333	Ea.	3.20	17.35		20.55	29.50
9020	1/2" CTS		22	.364		3.65	18.95		22.60	32.50
9030	3/4" CTS		21	.381		5.40	19.85		25.25	35.50
9040	1" CTS		18	.444		6.90	23		29.90	42
9050	Elbow 90°, 1/4" CTS		26	.308		3.56	16		19.56	28
9060	3/8" CTS		24	.333		3.81	17.35		21.16	30
9070	1/2" CTS		22	.364		4.51	18.95		23.46	33.50
9080	3/4" CTS		21	.381		6.90	19.85		26.75	37
9090	1" CTS		18	.444		8.35	23		31.35	43.50
9110	Tee ,1/4" & 3/8" CTS		15	.533		3.93	28		31.93	46
9120	1/2" CTS		14	.571		5.15	29.50		34.65	50
9130	3/4" CTS		13	.615		7.85	32		39.85	56.50
9140	1" CTS	↓	12	.667	↓	10.50	34.50		45	63.50
9552	For plastic hangers see Div. 22 05 29.10 8000									
9562	For copper/brass fittings see Div. 22 11 13.25 7000									

22 11 13 – Facility Water Distribution Piping

22 11 13.78 Pipe, High Density Polyethylene Plastic (HDPE)		Crew	Daily Output	Labor-Hours	Unit	2010 Bare Costs Material	Labor	Equipment	Total	Total Incl O&P
0010	**PIPE, HIGH DENSITY POLYETHYLENE PLASTIC (HDPE)**									
0020	Not incl. hangers, trenching, backfill, hoisting or digging equipment.									
0030	Standard length is 40′, add a weld for each joint									
0040	Single wall									
0050	Straight									
0054	1″ diameter DR 11				L.F.	.48			.48	.53
0058	1-1/2″ diameter DR 11					.60			.60	.66
0062	2″ diameter DR 11					1			1	1.10
0066	3″ diameter DR 11					1.20			1.20	1.32
0070	3″ diameter DR 17					.96			.96	1.06
0074	4″ diameter DR 11					2			2	2.20
0078	4″ diameter DR 17					2			2	2.20
0082	6″ diameter DR 11					5			5	5.50
0086	6″ diameter DR 17					3.30			3.30	3.63
0090	8″ diameter DR 11					8.30			8.30	9.15
0094	8″ diameter DR 26					3.90			3.90	4.29
0098	10″ diameter DR 11					13			13	14.30
0102	10″ diameter DR 26					6			6	6.60
0106	12″ diameter DR 11					19			19	21
0110	12″ diameter DR 26					9			9	9.90
0114	16″ diameter DR 11					29			29	32
0118	16″ diameter DR 26					13			13	14.30
0122	18″ diameter DR 11					37			37	40.50
0126	18″ diameter DR 26					17			17	18.70
0130	20″ diameter DR 11					45			45	49.50
0134	20″ diameter DR 26					20			20	22
0138	22″ diameter DR 11					55			55	60.50
0142	22″ diameter DR 26					25			25	27.50
0146	24″ diameter DR 11					65			65	71.50
0150	24″ diameter DR 26					29			29	32
0154	28″ diameter DR 17					60			60	66
0158	28″ diameter DR 26					40			40	44
0162	30″ diameter DR 21					56			56	61.50
0166	30″ diameter DR 26					46			46	50.50
0170	36″ diameter DR 26					46			46	50.50
0174	42″ diameter DR 26					89			89	98
0178	48″ diameter DR 26					117			117	129
0182	54″ diameter DR 26				▼	147			147	162
0300	90° Elbow									
0304	1″ diameter DR 11				Ea.	4.40			4.40	4.84
0308	1-1/2″ diameter DR 11					5.50			5.50	6.05
0312	2″ diameter DR 11					5.50			5.50	6.05
0316	3″ diameter DR 11					11			11	12.10
0320	3″ diameter DR 17					11			11	12.10
0324	4″ diameter DR 11					15.40			15.40	16.95
0328	4″ diameter DR 17					15.40			15.40	16.95
0332	6″ diameter DR 11					35			35	38.50
0336	6″ diameter DR 17					35			35	38.50
0340	8″ diameter DR 11					87			87	95.50
0344	8″ diameter DR 26					77			77	84.50
0348	10″ diameter DR 11					325			325	360
0352	10″ diameter DR 26				▼	300			300	330

22 11 13.78 Pipe, High Density Polyethylene Plastic (HDPE)		Crew	Daily Output	Labor-Hours	Unit	2010 Bare Costs Material	Labor	Equipment	Total	Total Incl O&P
0356	12" diameter DR 11				Ea.	345			345	375
0360	12" diameter DR 26					310			310	340
0364	16" diameter DR 11					405			405	445
0368	16" diameter DR 26					400			400	440
0372	18" diameter DR 11					510			510	560
0376	18" diameter DR 26					480			480	530
0380	20" diameter DR 11					600			600	660
0384	20" diameter DR 26					580			580	640
0388	22" diameter DR 11					620			620	680
0392	22" diameter DR 26					600			600	660
0396	24" diameter DR 11					700			700	770
0400	24" diameter DR 26					680			680	750
0404	28" diameter DR 17					850			850	935
0408	28" diameter DR 26					800			800	880
0412	30" diameter DR 17					1,200			1,200	1,325
0416	30" diameter DR 26					1,100			1,100	1,200
0420	36" diameter DR 26					1,400			1,400	1,550
0424	42" diameter DR 26					1,800			1,800	1,975
0428	48" diameter DR 26					2,100			2,100	2,300
0432	54" diameter DR 26				▼	5,000			5,000	5,500
0500	45° Elbow									
0512	2" diameter DR 11				Ea.	4.40			4.40	4.84
0516	3" diameter DR 11					11			11	12.10
0520	3" diameter DR 17					11			11	12.10
0524	4" diameter DR 11					15.40			15.40	16.95
0528	4" diameter DR 17					15.40			15.40	16.95
0532	6" diameter DR 11					35			35	38.50
0536	6" diameter DR 17					35			35	38.50
0540	8" diameter DR 11					87			87	95.50
0544	8" diameter DR 26					50.50			50.50	55.50
0548	10" diameter DR 11					325			325	360
0552	10" diameter DR 26					300			300	330
0556	12" diameter DR 11					345			345	375
0560	12" diameter DR 26					310			310	340
0564	16" diameter DR 11					180			180	198
0568	16" diameter DR 26					170			170	187
0572	18" diameter DR 11					191			191	210
0576	18" diameter DR 26					175			175	193
0580	20" diameter DR 11					290			290	320
0584	20" diameter DR 26					275			275	305
0588	22" diameter DR 11					400			400	440
0592	22" diameter DR 26					380			380	420
0596	24" diameter DR 11					490			490	540
0600	24" diameter DR 26					470			470	515
0604	28" diameter DR 17					555			555	610
0608	28" diameter DR 26					540			540	595
0612	30" diameter DR 17					690			690	760
0616	30" diameter DR 26					670			670	735
0620	36" diameter DR 26					850			850	935
0624	42" diameter DR 26					1,100			1,100	1,200
0628	48" diameter DR 26					1,200			1,200	1,300
0632	54" diameter DR 26				▼	1,600			1,600	1,750
0700	Tee									

22 11 13 – Facility Water Distribution Piping

22 11 13.78 Pipe, High Density Polyethylene Plastic (HDPE)		Crew	Daily Output	Labor-Hours	Unit	2010 Bare Costs Material	Labor	Equipment	Total	Total Incl O&P
0704	1" diameter DR 11				Ea.	5.75			5.75	6.35
0708	1-1/2" diameter DR 11					8.05			8.05	8.85
0712	2" diameter DR 11					6.90			6.90	7.60
0716	3" diameter DR 11					12.65			12.65	13.90
0720	3" diameter DR 17					12.65			12.65	13.90
0724	4" diameter DR 11					18.40			18.40	20
0728	4" diameter DR 17					18.40			18.40	20
0732	6" diameter DR 11					46			46	50.50
0736	6" diameter DR 17					46			46	50.50
0740	8" diameter DR 11					114			114	125
0744	8" diameter DR 17					114			114	125
0748	10" diameter DR 11					340			340	375
0752	10" diameter DR 17					340			340	375
0756	12" diameter DR 11					455			455	500
0760	12" diameter DR 17					455			455	500
0764	16" diameter DR 11					239			239	263
0768	16" diameter DR 17					197			197	217
0772	18" diameter DR 11					335			335	370
0776	18" diameter DR 17					277			277	305
0780	20" diameter DR 11					410			410	455
0784	20" diameter DR 17					335			335	370
0788	22" diameter DR 11					530			530	580
0792	22" diameter DR 17					415			415	460
0796	24" diameter DR 11					655			655	720
0800	24" diameter DR 17					550			550	605
0804	28" diameter DR 17					1,075			1,075	1,175
0812	30" diameter DR 17					1,225			1,225	1,350
0820	36" diameter DR 17					2,025			2,025	2,225
0824	42" diameter DR 26					2,250			2,250	2,475
0828	48" diameter DR 26				▼	2,425			2,425	2,650
1000	Flange adptr, w/back-up ring and 1/2 cost of plated bolt set									
1004	1" diameter DR 11				Ea.	22			22	24
1008	1-1/2" diameter DR 11					22			22	24
1012	2" diameter DR 11					13.80			13.80	15.20
1016	3" diameter DR 11					16.10			16.10	17.70
1020	3" diameter DR 17					16.10			16.10	17.70
1024	4" diameter DR 11					22			22	24
1028	4" diameter DR 17					22			22	24
1032	6" diameter DR 11					31			31	34
1036	6" diameter DR 17					31			31	34
1040	8" diameter DR 11					45			45	49.50
1044	8" diameter DR 26					45			45	49.50
1048	10" diameter DR 11					71.50			71.50	78.50
1052	10" diameter DR 26					71.50			71.50	78.50
1056	12" diameter DR 11					105			105	115
1060	12" diameter DR 26					105			105	115
1064	16" diameter DR 11					224			224	247
1068	16" diameter DR 26					224			224	247
1072	18" diameter DR 11					291			291	320
1076	18" diameter DR 26					291			291	320
1080	20" diameter DR 11					405			405	445
1084	20" diameter DR 26					405			405	445
1088	22" diameter DR 11				▼	440			440	480

22 11 13 – Facility Water Distribution Piping

22 11 13.78 Pipe, High Density Polyethylene Plastic (HDPE)		Crew	Daily Output	Labor-Hours	Unit	Material	Labor	Equipment	Total	Total Incl O&P
						2010 Bare Costs				
1092	22" diameter DR 26				Ea.	435			435	480
1096	24" diameter DR 17					475			475	525
1100	24" diameter DR 32.5					475			475	525
1104	28" diameter DR 15.5					645			645	710
1108	28" diameter DR 32.5					645			645	710
1112	30" diameter DR 11					750			750	820
1116	30" diameter DR 21					750			750	820
1120	36" diameter DR 26					815			815	900
1124	42" diameter DR 26					930			930	1,025
1128	48" diameter DR 26					1,150			1,150	1,250
1132	54" diameter DR 26				▼	1,375			1,375	1,525
1200	Reducer									
1208	2" x 1-1/2" diameter DR 11				Ea.	6.60			6.60	7.25
1212	3" x 2" diameter DR 11					6.60			6.60	7.25
1216	4" x 2" diameter DR 11					7.70			7.70	8.45
1220	4" x 3" diameter DR 11					9.90			9.90	10.90
1224	6" x 4" diameter DR 11					23			23	25.50
1228	8" x 6" diameter DR 11					35			35	38.50
1232	10" x 8" diameter DR 11					60.50			60.50	66.50
1236	12" x 8" diameter DR 11					99			99	109
1240	12" x 10" diameter DR 11					79			79	87
1244	14" x 12" diameter DR 11					88			88	97
1248	16" x 14" diameter DR 11					112			112	123
1252	18" x 16" diameter DR 11					136			136	150
1256	20" x 18" diameter DR 11					271			271	298
1260	22" x 20" diameter DR 11					330			330	365
1264	24" x 22" diameter DR 11					375			375	410
1268	26" x 24" diameter DR 11					440			440	485
1272	28" x 24" diameter DR 11					560			560	615
1276	32" x 28" diameter DR 17					725			725	800
1280	36" x 32" diameter DR 17				▼	990			990	1,100
4000	Welding labor per joint, not including welding machine									
4010	Pipe joint size (cost based on thickest wall for each dia.)									
4030	1" pipe size	4 Skwk	273	.117	Ea.		4.99		4.99	7.70
4040	1-1/2" pipe size		175	.183			7.80		7.80	12
4050	2" pipe size		128	.250			10.65		10.65	16.40
4060	3" pipe size		100	.320			13.65		13.65	21
4070	4" pipe size	▼	77	.416			17.70		17.70	27
4080	6" pipe size	5 Skwk	63	.635			27		27	41.50
4090	8" pipe size		48	.833			35.50		35.50	54.50
4100	10" pipe size	▼	40	1			42.50		42.50	65.50
4110	12" pipe size	6 Skwk	41	1.171			50		50	76.50
4120	16" pipe size		34	1.412			60		60	92.50
4130	18" pipe size	▼	32	1.500			64		64	98.50
4140	20" pipe size	8 Skwk	37	1.730			73.50		73.50	113
4150	22" pipe size		35	1.829			78		78	120
4160	24" pipe size		34	1.882			80		80	123
4170	28" pipe size		33	1.939			82.50		82.50	127
4180	30" pipe size		32	2			85		85	131
4190	36" pipe size		31	2.065			88		88	135
4200	42" pipe size	▼	30	2.133			91		91	140
4210	48" pipe size	9 Skwk	33	2.182			93		93	143
4220	54" pipe size	"	31	2.323	▼		99		99	152

22 11 13 – Facility Water Distribution Piping

22 11 13.78 Pipe, High Density Polyethylene Plastic (HDPE)		Crew	Daily Output	Labor-Hours	Unit	Material	2010 Bare Costs Labor	Equipment	Total	Total Incl O&P
4300	Note: Cost for set up each time welder is moved.									
4301	Add 50% of a weld cost									
4310	Welder usually remains stationary with pipe moved through it.									
4340	Weld machine, rental per day based on dia. capacity									
4350	1" thru 2" diameter				Ea.			40.25	40.25	44.25
4360	3" thru 4" diameter							46	46	50.50
4370	6" thru 8" diameter							103	103	113
4380	10" thru 12" diameter							178	178	196
4390	16" thru 18" diameter							259	259	285
4400	20" thru 24" diameter							500	500	550
4410	28" thru 32" diameter							545	545	600
4420	36" diameter							570	570	627
4430	42" thru 54" diameter				▼			890	890	980
5000	Dual wall contained pipe									
5040	Straight									
5054	1" DR 11 x 3" DR 11				L.F.	6.90			6.90	7.60
5058	1" DR 11 x 4" DR 11					7.35			7.35	8.05
5062	1-1/2" DR 11 x 4" DR 17					7.55			7.55	8.35
5066	2" DR 11 x 4" DR 17					8.15			8.15	8.95
5070	2" DR 11 x 6" DR 17					12.20			12.20	13.40
5074	3" DR 11 x 6" DR 17					13.60			13.60	14.95
5078	3" DR 11 x 6" DR 26					11.70			11.70	12.85
5086	3" DR 17 x 8" DR 17					17.25			17.25	19
5090	4" DR 11 x 8" DR 17					20			20	22
5094	4" DR 17 x 8" DR 26					15.95			15.95	17.55
5098	6" DR 11 x 10" DR 17					31.50			31.50	34.50
5102	6" DR 17 x 10" DR 26					23.50			23.50	26
5106	6" DR 26 x 10" DR 26					22			22	24.50
5110	8" DR 17 x 12" DR 26					33.50			33.50	36.50
5114	8" DR 26 x 12" DR 32.5					28			28	31
5118	10" DR 17 x 14" DR 26					46.50			46.50	51
5122	10" DR 17 x 16" DR 26					51			51	56
5126	10" DR 26 x 16" DR 26					46			46	50.50
5130	12" DR 26 x 16" DR 26					52.50			52.50	57.50
5134	12" DR 17 x 18" DR 26					67			67	73.50
5138	12" DR 26 x 18" DR 26					60			60	66
5142	14" DR 26 x 20" DR 32.5					65.50			65.50	72
5146	16" DR 26 x 22" DR 32.5					99			99	109
5150	18" DR 26 x 24" DR 32.5					91			91	100
5154	20" DR 32.5 x 28" DR 32.5					118			118	130
5158	22" DR 32.5 x 30" DR 32.5					118			118	130
5162	24" DR 32.5 x 32" DR 32.5					151			151	166
5166	36" DR 32.5 x 42" DR 32.5				▼	254			254	280
5300	Force transfer coupling									
5354	1" DR 11 x 3" DR 11				Ea.	211			211	232
5358	1" DR 11 x 4" DR 17					221			221	244
5362	1-1/2" DR 11 x 4" DR 17					232			232	255
5366	2" DR 11 x 4" DR 17					243			243	267
5370	2" DR 11 x 6" DR 17					350			350	385
5374	3" DR 11 x 6" DR 17					350			350	385
5378	3" DR 11 x 6" DR 26					350			350	385
5382	3" DR 11 x 8" DR 11					370			370	410
5386	3" DR 11 x 8" DR 17				▼	370			370	410

22 11 13.78 Pipe, High Density Polyethylene Plastic (HDPE)		Crew	Daily Output	Labor-Hours	Unit	2010 Bare Costs Material	Labor	Equipment	Total	Total Incl O&P
5390	4" DR 11 x 8" DR 17				Ea.	370			370	410
5394	4" DR 17 x 8" DR 26					325			325	360
5398	6" DR 11 x 10" DR 17					470			470	520
5402	6" DR 17 x 10" DR 26					390			390	430
5406	6" DR 26 x 10" DR 26					390			390	430
5410	8" DR 17 x 12" DR 26					515			515	570
5414	8" DR 26 x 12" DR 32.5					415			415	460
5418	10" DR 17 x 14" DR 26					675			675	745
5422	10" DR 17 x 16" DR 26					675			675	745
5426	10" DR 26 x 16" DR 26					675			675	745
5430	12" DR 26 x 16" DR 26					850			850	935
5434	12" DR 17 x 18" DR 26					870			870	955
5438	12" DR 26 x 18" DR 26					870			870	955
5442	14" DR 26 x 20" DR 32.5					920			920	1,000
5446	16" DR 26 x 22" DR 32.5					945			945	1,050
5450	18" DR 26 x 24" DR 32.5					1,025			1,025	1,125
5454	20" DR 32.5 x 28" DR 32.5					1,250			1,250	1,375
5458	22" DR 32.5 x 30" DR 32.5					1,375			1,375	1,525
5462	24" DR 32.5 x 32" DR 32.5					1,500			1,500	1,650
5466	36" DR 32.5 x 42" DR 32.5					2,475			2,475	2,725
5600	90° Elbow									
5654	1" DR 11 x 3" DR 11				Ea.	174			174	192
5658	1" DR 11 x 4" DR 17					167			167	184
5662	1-1/2" DR 11 x 4" DR 17					182			182	201
5666	2" DR 11 x 4" DR 17					193			193	213
5670	2" DR 11 x 6" DR 17					246			246	270
5674	3" DR 11 x 6" DR 17					282			282	310
5678	3" DR 17 x 6" DR 26					224			224	247
5682	3" DR 17 x 8" DR 11					455			455	500
5686	3" DR 17 x 8" DR 17					355			355	390
5690	4" DR 11 x 8" DR 17					415			415	455
5694	4" DR 17 x 8" DR 26					330			330	365
5698	6" DR 11 x 10" DR 17					555			555	610
5702	6" DR 17 x 10" DR 26					425			425	465
5706	6" DR 26 x 10" DR 26					395			395	435
5710	8" DR 17 x 12" DR 26					630			630	695
5714	8" DR 26 x 12" DR 32.5					585			585	640
5718	10" DR 17 x 14" DR 26					990			990	1,100
5722	10" DR 17 x 16" DR 26					950			950	1,050
5726	10" DR 26 x 16" DR 26					880			880	965
5730	12" DR 26 x 16" DR 26					975			975	1,075
5734	12" DR 17 x 18" DR 26					1,150			1,150	1,275
5738	12" DR 26 x 18" DR 26					1,075			1,075	1,175
5742	14" DR 26 x 20" DR 32.5					1,400			1,400	1,550
5746	16" DR 26 x 22" DR 32.5					1,425			1,425	1,550
5750	18" DR 26 x 24" DR 32.5					1,725			1,725	1,900
5754	20" DR 32.5 x 28" DR 32.5					2,000			2,000	2,200
5758	22" DR 32.5 x 30" DR 32.5					2,725			2,725	3,000
5762	24" DR 32.5 x 32" DR 32.5					2,650			2,650	2,925
5766	36" DR 32.5 x 42" DR 32.5					4,325			4,325	4,750
5800	45° Elbow									
5804	1" DR 11 x 3" DR 11				Ea.	104			104	115
5808	1" DR 11 x 4" DR 17					104			104	114

22 11 13 – Facility Water Distribution Piping

22 11 13.78 Pipe, High Density Polyethylene Plastic (HDPE)		Crew	Daily Output	Labor-Hours	Unit	Material	2010 Bare Costs Labor	Equipment	Total	Total Incl O&P
5812	1-1/2" DR 11 x 4" DR 17				Ea.	111			111	122
5816	2" DR 11 x 4" DR 17					121			121	134
5820	2" DR 11 x 6" DR 17					153			153	168
5824	3" DR 11 x 6" DR 17					169			169	186
5828	3" DR 17 x 6" DR 26					140			140	154
5832	3" DR 17 x 8" DR 11					259			259	285
5836	3" DR 17 x 8" DR 17					207			207	228
5840	4" DR 11 x 8" DR 17					242			242	266
5844	4" DR 17 x 8" DR 26					197			197	217
5848	6" DR 11 x 10" DR 17					320			320	350
5852	6" DR 17 x 10" DR 26					250			250	275
5856	6" DR 26 x 10" DR 26					236			236	260
5860	8" DR 17 x 12" DR 26					380			380	420
5864	8" DR 26 x 12" DR 32.5					355			355	390
5868	10" DR 17 x 14" DR 26					595			595	655
5872	10" DR 17 x 16" DR 26					585			585	640
5876	10" DR 26 x 16" DR 26					545			545	600
5880	12" DR 26 x 16" DR 26					615			615	680
5884	12" DR 17 x 18" DR 26					725			725	800
5888	12" DR 26 x 18" DR 26					675			675	745
5892	14" DR 26 x 20" DR 32.5					875			875	960
5896	16" DR 26 x 22" DR 32.5					915			915	1,000
5900	18" DR 26 x 24" DR 32.5					1,075			1,075	1,200
5904	20" DR 32.5 x 28" DR 32.5					1,325			1,325	1,450
5908	22" DR 32.5 x 30" DR 32.5					1,675			1,675	1,850
5912	24" DR 32.5 x 32" DR 32.5					1,675			1,675	1,825
5916	36" DR 32.5 x 42" DR 32.5				↓	2,600			2,600	2,850
6000	Access port with 4" riser									
6050	1" DR 11 x 4" DR 17				Ea.	228			228	251
6054	1-1/2" DR 11 x 4" DR 17					232			232	256
6058	2" DR 11 x 6" DR 17					284			284	310
6062	3" DR 11 x 6" DR 17					285			285	315
6066	3" DR 17 x 6" DR 26					280			280	310
6070	3" DR 17 x 8" DR 11					320			320	350
6074	3" DR 17 x 8" DR 17					305			305	335
6078	4" DR 11 x 8" DR 17					315			315	345
6082	4" DR 17 x 8" DR 26					305			305	335
6086	6" DR 11 x 10" DR 17					395			395	435
6090	6" DR 17 x 10" DR 26					365			365	400
6094	6" DR 26 x 10" DR 26					355			355	390
6098	8" DR 17 x 12" DR 26					380			380	420
6102	8" DR 26 x 12" DR 32.5				↓	365			365	400
6200	End termination with vent plug									
6204	1" DR 11 x 3" DR 11				Ea.	189			189	207
6208	1" DR 11 x 4" DR 17					202			202	222
6212	1-1/2" DR 11 x 4" DR 17					202			202	222
6216	2" DR 11 x 4" DR 17					219			219	241
6220	2" DR 11 x 6" DR 17					250			250	275
6224	3" DR 11 x 6" DR 17					250			250	275
6228	3" DR 17 x 6" DR 26					250			250	275
6232	3" DR 17 x 8" DR 11					345			345	380
6236	3" DR 17 x 8" DR 17					345			345	380
6240	4" DR 11 x 8" DR 17				↓	345			345	380

22 11 13.78 Pipe, High Density Polyethylene Plastic (HDPE)		Crew	Daily Output	Labor-Hours	Unit	Material	2010 Bare Costs Labor	Equipment	Total	Total Incl O&P
6244	4" DR 17 x 8" DR 26				Ea.	300			300	330
6248	6" DR 11 x 10" DR 17					410			410	450
6252	6" DR 17 x 10" DR 26					335			335	370
6256	6" DR 26 x 10" DR 26					335			335	370
6260	8" DR 17 x 12" DR 26					500			500	550
6264	8" DR 26 x 12" DR 32.5					400			400	440
6268	10" DR 17 x 14" DR 26					520			520	570
6272	10" DR 17 x 16" DR 26					520			520	570
6276	10" DR 26 x 16" DR 26					520			520	570
6280	12" DR 26 x 16" DR 26					520			520	570
6284	12" DR 17 x 18" DR 26					685			685	750
6288	12" DR 26 x 18" DR 26					685			685	750
6292	14" DR 26 x 20" DR 32.5					695			695	765
6296	16" DR 26 x 22" DR 32.5					810			810	890
6300	18" DR 26 x 24" DR 32.5					835			835	915
6304	20" DR 32.5 x 28" DR 32.5					1,075			1,075	1,175
6308	22" DR 32.5 x 30" DR 32.5					1,175			1,175	1,300
6312	24" DR 32.5 x 32" DR 32.5					1,300			1,300	1,425
6316	36" DR 32.5 x 42" DR 32.5				▼	1,900			1,900	2,075
6600	Tee									
6604	1" DR 11 x 3" DR 11				Ea.	192			192	211
6608	1" DR 11 x 4" DR 17					238			238	262
6612	1-1/2" DR 11 x 4" DR 17					265			265	292
6616	2" DR 11 x 4" DR 17					290			290	320
6620	2" DR 11 x 6" DR 17					350			350	385
6624	3" DR 11 x 6" DR 17					405			405	445
6628	3" DR 17 x 6" DR 26					395			395	435
6632	3" DR 17 x 8" DR 11					475			475	525
6636	3" DR 17 x 8" DR 17					455			455	500
6640	4" DR 11 x 8" DR 17					495			495	545
6644	4" DR 17 x 8" DR 26					475			475	520
6648	6" DR 11 x 10" DR 17					620			620	680
6652	6" DR 17 x 10" DR 26					585			585	645
6656	6" DR 26 x 10" DR 26					580			580	640
6660	8" DR 17 x 12" DR 26					580			580	640
6664	8" DR 26 x 12" DR 32.5					755			755	830
6668	10" DR 17 x 14" DR 26					875			875	965
6672	10" DR 17 x 16" DR 26					920			920	1,000
6676	10" DR 26 x 16" DR 26					920			920	1,000
6680	12" DR 26 x 16" DR 26					990			990	1,100
6684	12" DR 17 x 18" DR 26					990			990	1,100
6688	12" DR 26 x 18" DR 26					1,100			1,100	1,200
6692	14" DR 26 x 20" DR 32.5					1,450			1,450	1,600
6696	16" DR 26 x 22" DR 32.5					1,775			1,775	1,950
6700	18" DR 26 x 24" DR 32.5					1,925			1,925	2,125
6704	20" DR 32.5 x 28" DR 32.5					2,450			2,450	2,675
6708	22" DR 32.5 x 30" DR 32.5					2,900			2,900	3,200
6712	24" DR 32.5 x 32" DR 32.5					3,475			3,475	3,825
6716	36" DR 32.5 x 42" DR 32.5				▼	7,075			7,075	7,775
6800	Wye									
6816	2" DR 11 x 4" DR 17				Ea.	365			365	400
6820	2" DR 11 x 6" DR 17					415			415	455
6824	3" DR 11 x 6" DR 17				▼	420			420	460

22 11 13 – Facility Water Distribution Piping

22 11 13.78 Pipe, High Density Polyethylene Plastic (HDPE)		Crew	Daily Output	Labor-Hours	Unit	Material	Labor	Equipment	Total	Total Incl O&P
						2010 Bare Costs				
6828	3″ DR 17 x 6″ DR 26				Ea.	405			405	445
6832	3″ DR 17 x 8″ DR 11					510			510	560
6836	3″ DR 17 x 8″ DR 17					475			475	525
6840	4″ DR 11 x 8″ DR 17					520			520	570
6844	4″ DR 17 x 8″ DR 26					495			495	540
6848	6″ DR 11 x 10″ DR 17					645			645	705
6852	6″ DR 17 x 10″ DR 26					595			595	655
6856	6″ DR 26 x 10″ DR 26					595			595	655
6860	8″ DR 17 x 12″ DR 26					850			850	935
6864	8″ DR 26 x 12″ DR 32.5					820			820	905
6868	10″ DR 17 x 14″ DR 26					1,025			1,025	1,125
6872	10″ DR 17 x 16″ DR 26					1,100			1,100	1,200
6876	10″ DR 26 x 16″ DR 26					1,075			1,075	1,175
6880	12″ DR 26 x 16″ DR 26					1,200			1,200	1,325
6884	12″ DR 17 x 18″ DR 26					1,350			1,350	1,500
6888	12″ DR 26 x 18″ DR 26					1,375			1,375	1,525
6892	14″ DR 26 x 20″ DR 32.5					1,675			1,675	1,825
6896	16″ DR 26 x 22″ DR 32.5					2,400			2,400	2,650
6900	18″ DR 26 x 24″ DR 32.5					2,825			2,825	3,125
6904	20″ DR 32.5 x 28″ DR 32.5					3,550			3,550	3,900
6908	22″ DR 32.5 x 30″ DR 32.5					4,050			4,050	4,450
6912	24″ DR 32.5 x 32″ DR 32.5				↓	4,600			4,600	5,075
9000	Welding labor per joint, not including welding machine									
9010	Pipe joint size, outer pipe (cost based on the thickest walls)									
9020	Straight pipe									
9050	3″ pipe size	4 Skwk	96	.333	Ea.		14.20		14.20	22
9060	4″ pipe size	″	77	.416			17.70		17.70	27
9070	6″ pipe size	5 Skwk	60	.667			28.50		28.50	43.50
9080	8″ pipe size	″	40	1			42.50		42.50	65.50
9090	10″ pipe size	6 Skwk	41	1.171			50		50	76.50
9100	12″ pipe size		39	1.231			52.50		52.50	80.50
9110	14″ pipe size		38	1.263			54		54	82.50
9120	16″ pipe size	↓	35	1.371			58.50		58.50	90
9130	18″ pipe size	8 Skwk	45	1.422			60.50		60.50	93
9140	20″ pipe size		42	1.524			65		65	100
9150	22″ pipe size		40	1.600			68		68	105
9160	24″ pipe size		38	1.684			72		72	110
9170	28″ pipe size		37	1.730			73.50		73.50	113
9180	30″ pipe size		36	1.778			75.50		75.50	116
9190	32″ pipe size		35	1.829			78		78	120
9200	42″ pipe size	↓	32	2	↓		85		85	131
9300	Note: Cost for set up each time welder is moved.									
9301	Add 100% of weld labor cost									
9310	For handling between fitting welds add 50% of weld labor cost									
9320	Welder usually remains stationary with pipe moved through it									
9360	Weld machine, rental per day based on dia. capacity									
9380	3″ thru 4″ diameter				Ea.			63.50	63.50	70
9390	6″ thru 8″ diameter							207	207	228
9400	10″ thru 12″ diameter							310	310	341
9410	14″ thru 18″ diameter							420	420	462
9420	20″ thru 24″ diameter							748	748	823
9430	28″ thru 32″ diameter							880	880	968
9440	42″ thru 58″ diameter				↓			910	910	1,000

22 11 19 – Domestic Water Piping Specialties

22 11 19.10 Flexible Connectors		Crew	Daily Output	Labor-Hours	Unit	2010 Bare Costs Material	Labor	Equipment	Total	Total Incl O&P
0010	**FLEXIBLE CONNECTORS**, Corrugated, 7/8" O.D., 1/2" I.D.									
0050	Gas, seamless brass, steel fittings									
0200	12" long	1 Plum	36	.222	Ea.	15.10	11.55		26.65	34
0220	18" long		36	.222		18.70	11.55		30.25	38
0240	24" long		34	.235		22	12.25		34.25	43
0260	30" long		34	.235		24	12.25		36.25	45
0280	36" long		32	.250		26.50	13		39.50	48.50
0320	48" long		30	.267		33.50	13.90		47.40	58
0340	60" long		30	.267		40	13.90		53.90	65
0360	72" long	↓	30	.267	↓	46	13.90		59.90	71.50
2000	Water, copper tubing, dielectric separators									
2100	12" long	1 Plum	36	.222	Ea.	14	11.55		25.55	33
2220	15" long		36	.222		15.45	11.55		27	34.50
2240	18" long		36	.222		20.50	11.55		32.05	40
2260	24" long	↓	34	.235	↓	20.50	12.25		32.75	41

22 11 19.14 Flexible Metal Hose		Crew	Daily Output	Labor-Hours	Unit	Material	Labor	Equipment	Total	Total Incl O&P
0010	**FLEXIBLE METAL HOSE**, Connectors, standard lengths									
0100	Bronze braided, bronze ends									
0120	3/8" diameter x 12"	1 Stpi	26	.308	Ea.	18.30	15.95		34.25	44
0140	1/2" diameter x 12"		24	.333		17.40	17.30		34.70	45
0160	3/4" diameter x 12"		20	.400		25.50	21		46.50	59
0180	1" diameter x 18"		19	.421		33	22		55	69
0200	1-1/2" diameter x 18"		13	.615		52	32		84	106
0220	2" diameter x 18"	↓	11	.727	↓	63	38		101	126
1000	Carbon steel ends									
1020	1/4" diameter x 12"	1 Stpi	28	.286	Ea.	13.60	14.85		28.45	37
1040	3/8" diameter x 12"		26	.308		13.95	15.95		29.90	39.50
1060	1/2" diameter x 12"		24	.333		15.40	17.30		32.70	43
1080	1/2" diameter x 24"		24	.333		37	17.30		54.30	66.50
1100	1/2" diameter x 36"		24	.333		60.50	17.30		77.80	92.50
1120	3/4" diameter x 12"		20	.400		25	21		46	58.50
1140	3/4" diameter x 24"		20	.400		46.50	21		67.50	82
1160	3/4" diameter x 36"		20	.400		55	21		76	91.50
1180	1" diameter x 18"		19	.421		31.50	22		53.50	67.50
1200	1" diameter x 30"		19	.421		64.50	22		86.50	104
1220	1" diameter x 36"		19	.421		73	22		95	114
1240	1-1/4" diameter x 18"		15	.533		43	27.50		70.50	89
1260	1-1/4" diameter x 36"		15	.533		84.50	27.50		112	135
1280	1-1/2" diameter x 18"		13	.615		61	32		93	115
1300	1-1/2" diameter x 36"		13	.615		93	32		125	150
1320	2" diameter x 24"		11	.727		88.50	38		126.50	154
1340	2" diameter x 36"		11	.727		117	38		155	185
1360	2-1/2" diameter x 24"		9	.889		202	46		248	292
1380	2-1/2" diameter x 36"		9	.889		263	46		309	360
1400	3" diameter x 24"		7	1.143		279	59.50		338.50	395
1420	3" diameter x 36"	↓	7	1.143	↓	375	59.50		434.50	500
2000	Carbon steel braid, carbon steel solid ends									
2100	1/2" diameter x 12"	1 Stpi	24	.333	Ea.	34.50	17.30		51.80	64
2120	3/4" diameter x 12"		20	.400		52.50	21		73.50	89
2140	1" diameter x 12"		19	.421		74.50	22		96.50	115
2160	1-1/4" diameter x 12"		15	.533		56.50	27.50		84	104
2180	1-1/2" diameter x 12"	↓	13	.615	↓	58.50	32		90.50	113

22 11 19 – Domestic Water Piping Specialties

22 11 19.14 Flexible Metal Hose		Crew	Daily Output	Labor-Hours	Unit	2010 Bare Costs Material	Labor	Equipment	Total	Total Incl O&P
3000	Stainless steel braid, welded on carbon steel ends									
3100	1/2" diameter x 12"	1 Stpi	24	.333	Ea.	45.50	17.30		62.80	76
3120	3/4" diameter x 12"		20	.400		57.50	21		78.50	94
3140	3/4" diameter x 24"		20	.400		66	21		87	104
3160	3/4" diameter x 36"		20	.400		74.50	21		95.50	113
3180	1" diameter x 12"		19	.421		68.50	22		90.50	108
3200	1" diameter x 24"		19	.421		80.50	22		102.50	122
3220	1" diameter x 36"		19	.421		92.50	22		114.50	135
3240	1-1/4" diameter x 12"		15	.533		101	27.50		128.50	153
3260	1-1/4" diameter x 24"		15	.533		108	27.50		135.50	161
3280	1-1/4" diameter x 36"		15	.533		123	27.50		150.50	177
3300	1-1/2" diameter x 12"		13	.615		108	32		140	167
3320	1-1/2" diameter x 24"		13	.615		120	32		152	180
3340	1-1/2" diameter x 36"		13	.615		141	32		173	203
3400	Metal stainless steel braid, over corrugated stainless steel, flanged ends									
3410	150 PSI									
3420	1/2" diameter x 12"	1 Stpi	24	.333	Ea.	37	17.30		54.30	67
3430	1" diameter x 12"		20	.400		114	21		135	157
3440	1-1/2" diameter x 12"		15	.533		204	27.50		231.50	266
3450	2-1/2" diameter x 9"		12	.667		75	34.50		109.50	135
3460	3" diameter x 9"		9	.889		90	46		136	168
3470	4" diameter x 9"		7	1.143		110	59.50		169.50	210
3480	4" diameter x 30"		5	1.600		560	83		643	740
3490	4" diameter x 36"		4.80	1.667		580	86.50		666.50	770
3500	6" diameter x 11"		5	1.600		183	83		266	325
3510	6" diameter x 36"		3.80	2.105		505	109		614	725
3520	8" diameter x 12"		4	2		360	104		464	550
3530	10" diameter x 13"		3	2.667		490	138		628	745
3540	12" diameter x 14"	Q-5	4	4		670	187		857	1,025
6000	Molded rubber with helical wire reinforcement									
6010	150 PSI									
6020	1-1/2" diameter x 12"	1 Stpi	15	.533	Ea.	82	27.50		109.50	132
6030	2" diameter x 12"		12	.667		234	34.50		268.50	310
6040	3" diameter x 12"		8	1		244	52		296	345
6050	4" diameter x 12"		6	1.333		310	69		379	445
6060	6" diameter x 18"		4	2		465	104		569	665
6070	8" diameter x 24"		3	2.667		650	138		788	920
6080	10" diameter x 24"		2	4		790	208		998	1,175
6090	12" diameter x 24"	Q-5	3	5.333		900	249		1,149	1,375
7000	Molded teflon with stainless steel flanges									
7010	150 PSI									
7020	2-1/2" diameter x 3-3/16"	Q-1	7.80	2.051	Ea.	2,275	96		2,371	2,650
7030	3" diameter x 3-5/8"		6.50	2.462		1,575	115		1,690	1,925
7040	4" diameter x 3-5/8"		5	3.200		2,050	150		2,200	2,475
7050	6" diameter x 4"		4.30	3.721		2,875	174		3,049	3,425
7060	8" diameter x 6"		3.80	4.211		4,575	197		4,772	5,350

22 11 19.18 Mixing Valve		Crew	Daily Output	Labor-Hours	Unit	Material	Labor	Equipment	Total	Total Incl O&P
0010	**MIXING VALVE**, Automatic, water tempering.									
0040	1/2" size	1 Stpi	19	.421	Ea.	470	22		492	555
0050	3/4" size		18	.444		470	23		493	555
0100	1" size		16	.500		705	26		731	815
0120	1-1/4" size		13	.615		975	32		1,007	1,125

22 11 19 – Domestic Water Piping Specialties

22 11 19.18 Mixing Valve		Crew	Daily Output	Labor-Hours	Unit	Material	2010 Bare Costs Labor	Equipment	Total	Total Incl O&P
0140	1-1/2" size	1 Stpi	10	.800	Ea.	1,150	41.50		1,191.50	1,325
0160	2" size		8	1		1,450	52		1,502	1,675
0170	2-1/2" size		6	1.333		1,450	69		1,519	1,700
0180	3" size		4	2		3,450	104		3,554	3,950
0190	4" size	↓	3	2.667	↓	3,450	138		3,588	4,000
22 11 19.22 Pressure Reducing Valve										
0010	**PRESSURE REDUCING VALVE**, Steam, pilot operated.									
0100	Threaded, iron body									
0200	1-1/2" size	1 Stpi	8	1	Ea.	1,700	52		1,752	1,950
0220	2" size	"	5	1.600	"	1,950	83		2,033	2,275
1000	Flanged, iron body, 125 lb. flanges									
1020	2" size	1 Stpi	8	1	Ea.	1,975	52		2,027	2,250
1040	2-1/2" size	"	4	2		2,375	104		2,479	2,775
1060	3" size	Q-5	4.50	3.556		2,875	166		3,041	3,425
1080	4" size	"	3	5.333		4,150	249		4,399	4,950
1500	For 250 lb. flanges, add				↓	5%				
22 11 19.26 Pressure Regulators										
0010	**PRESSURE REGULATORS**									
0100	Gas appliance regulators									
0106	Main burner and pilot applications									
0108	Rubber seat poppet type									
0109	1/8" pipe size	1 Stpi	24	.333	Ea.	13.70	17.30		31	41
0110	1/4" pipe size		24	.333		15.75	17.30		33.05	43.50
0112	3/8" pipe size		24	.333		16.75	17.30		34.05	44.50
0113	1/2" pipe size		24	.333		18.50	17.30		35.80	46.50
0114	3/4" pipe size	↓	20	.400	↓	22	21		43	55
0122	Lever action type									
0123	3/8" pipe size	1 Stpi	24	.333	Ea.	21	17.30		38.30	49
0124	1/2" pipe size		24	.333		21	17.30		38.30	49
0125	3/4" pipe size		20	.400		41.50	21		62.50	77
0126	1" pipe size	↓	19	.421	↓	41.50	22		63.50	79
0132	Double diaphragm type									
0133	3/8" pipe size	1 Stpi	24	.333	Ea.	30	17.30		47.30	59
0134	1/2" pipe size		24	.333		42.50	17.30		59.80	72.50
0135	3/4" pipe size		20	.400		71	21		92	109
0136	1" pipe size		19	.421		86.50	22		108.50	128
0137	1-1/4" pipe size		15	.533		300	27.50		327.50	370
0138	1-1/2" pipe size		13	.615		555	32		587	660
0139	2" pipe size	↓	11	.727		555	38		593	665
0140	2-1/2" pipe size	Q-5	15	1.067		1,075	50		1,125	1,250
0141	3" pipe size		13	1.231		1,075	57.50		1,132.50	1,250
0142	4" pipe size (flanged)	↓	8	2	↓	1,900	93.50		1,993.50	2,250
0160	Main burner only									
0162	Straight-thru-flow design									
0163	1/2" pipe size	1 Stpi	24	.333	Ea.	34.50	17.30		51.80	64
0164	3/4" pipe size		20	.400		47	21		68	83
0165	1" pipe size		19	.421		67	22		89	107
0166	1-1/4" pipe size	↓	15	.533	↓	67	27.50		94.50	115
0200	Oil, light, hot water, ordinary steam, threaded									
0220	Bronze body, 1/4" size	1 Stpi	24	.333	Ea.	169	17.30		186.30	212
0230	3/8" size		24	.333		183	17.30		200.30	227
0240	1/2" size	↓	24	.333	↓	227	17.30		244.30	276

22 11 19 – Domestic Water Piping Specialties

22 11 19.26 Pressure Regulators		Crew	Daily Output	Labor-Hours	Unit	Material	2010 Bare Costs Labor	Equipment	Total	Total Incl O&P
0250	3/4" size	1 Stpi	20	.400	Ea.	271	21		292	330
0260	1" size		19	.421		410	22		432	490
0270	1-1/4" size		15	.533		555	27.50		582.50	650
0280	1-1/2" size		13	.615		595	32		627	705
0290	2" size		11	.727		1,075	38		1,113	1,250
0320	Iron body, 1/4" size		24	.333		132	17.30		149.30	171
0330	3/8" size		24	.333		156	17.30		173.30	197
0340	1/2" size		24	.333		165	17.30		182.30	207
0350	3/4" size		20	.400		200	21		221	251
0360	1" size		19	.421		261	22		283	320
0370	1-1/4" size		15	.533		360	27.50		387.50	440
0380	1-1/2" size		13	.615		405	32		437	495
0390	2" size		11	.727		585	38		623	700
0500	Oil, heavy, viscous fluids, threaded									
0520	Bronze body, 3/8" size	1 Stpi	24	.333	Ea.	291	17.30		308.30	345
0530	1/2" size		24	.333		375	17.30		392.30	435
0540	3/4" size		20	.400		415	21		436	485
0550	1" size		19	.421		520	22		542	605
0560	1-1/4" size		15	.533		735	27.50		762.50	845
0570	1-1/2" size		13	.615		840	32		872	975
0600	Iron body, 3/8" size		24	.333		231	17.30		248.30	280
0620	1/2" size		24	.333		280	17.30		297.30	335
0630	3/4" size		20	.400		320	21		341	380
0640	1" size		19	.421		375	22		397	450
0650	1-1/4" size		15	.533		545	27.50		572.50	640
0660	1-1/2" size		13	.615		580	32		612	690
0800	Process steam, wet or super heated, monel trim, threaded									
0820	Bronze body, 1/4" size	1 Stpi	24	.333	Ea.	610	17.30		627.30	695
0830	3/8" size		24	.333		610	17.30		627.30	695
0840	1/2" size		24	.333		700	17.30		717.30	795
0850	3/4" size		20	.400		825	21		846	935
0860	1" size		19	.421		1,050	22		1,072	1,175
0870	1-1/4" size		15	.533		1,300	27.50		1,327.50	1,475
0880	1-1/2" size		13	.615		1,550	32		1,582	1,750
0920	Iron body, max 125 PSIG press out, 1/4" size		24	.333		132	17.30		149.30	171
0930	3/8" size		24	.333		156	17.30		173.30	197
0940	1/2" size		24	.333		165	17.30		182.30	207
0950	3/4" size		20	.400		200	21		221	251
0960	1" size		19	.421		261	22		283	320
0970	1-1/4" size		15	.533		360	27.50		387.50	440
0980	1-1/2" size		13	.615		405	32		437	495
0990	2" size		11	.727		585	38		623	700
1000	Flanged, Class 125									
1006	2-1/2" size	Q-5	10	1.600	Ea.	2,375	74.50		2,449.50	2,725
1008	3" size		9	1.778		2,875	83		2,958	3,300
1010	4" size		8	2		4,150	93.50		4,243.50	4,725
1020	6" size		7	2.286		10,200	107		10,307	11,400
3000	Steam, high capacity, bronze body, stainless steel trim									
3020	Threaded, 1/2" diameter	1 Stpi	24	.333	Ea.	1,375	17.30		1,392.30	1,525
3030	3/4" diameter		24	.333		1,375	17.30		1,392.30	1,525
3040	1" diameter		19	.421		1,525	22		1,547	1,700
3060	1-1/4" diameter		15	.533		1,675	27.50		1,702.50	1,900
3080	1-1/2" diameter		13	.615		1,925	32		1,957	2,175

22 11 19 – Domestic Water Piping Specialties

22 11 19.26 Pressure Regulators		Crew	Daily Output	Labor-Hours	Unit	2010 Bare Costs Material	Labor	Equipment	Total	Total Incl O&P
3100	2" diameter	1 Stpi	11	.727	Ea.	2,350	38		2,388	2,650
3120	2-1/2" diameter	Q-5	12	1.333		2,950	62.50		3,012.50	3,350
3140	3" diameter	"	11	1.455		3,375	68		3,443	3,800
3500	Flanged connection, iron body, 125 lb. W.S.P.									
3520	3" diameter	Q-5	11	1.455	Ea.	3,700	68		3,768	4,150
3540	4" diameter	"	5	3.200	"	4,650	149		4,799	5,350
9002	For water pressure regulators, see Div. 22 05 23.20									

22 11 19.30 Pressure and Temperature Safety Plug		Crew	Daily Output	Labor-Hours	Unit	Material	Labor	Equipment	Total	Total Incl O&P
0010	**PRESSURE & TEMPERATURE SAFETY PLUG**									
1000	3/4" external thread, 3/8" diam. element									
1020	Carbon steel									
1050	7-1/2" insertion	1 Stpi	32	.250	Ea.	87	13		100	115
1120	304 stainless steel									
1150	7-1/2" insertion	1 Stpi	32	.250	Ea.	81.50	13		94.50	109
1220	316 Stainless steel									
1250	7-1/2" insertion	1 Stpi	32	.250	Ea.	85.50	13		98.50	113

22 11 19.34 Sleeves and Escutcheons		Crew	Daily Output	Labor-Hours	Unit	Material	Labor	Equipment	Total	Total Incl O&P
0010	**SLEEVES & ESCUTCHEONS**									
0100	Pipe sleeve									
0110	Steel, w/water stop, 12" long, with link seal									
0120	2" diam. for 1/2" carrier pipe	1 Plum	8.40	.952	Ea.	45.50	49.50		95	125
0130	2-1/2" diam. for 3/4" carrier pipe		8	1		51.50	52		103.50	135
0140	2-1/2" diam. for 1" carrier pipe		8	1		48.50	52		100.50	132
0150	3" diam. for 1-1/4" carrier pipe		7.20	1.111		63	58		121	156
0160	3-1/2" diam. for 1-1/2" carrier pipe		6.80	1.176		63	61		124	161
0170	4" diam. for 2" carrier pipe		6	1.333		68	69.50		137.50	179
0180	4" diam. for 2-1/2" carrier pipe		6	1.333		69	69.50		138.50	180
0190	5" diam. for 3" carrier pipe		5.40	1.481		81.50	77		158.50	206
0200	6" diam. for 4" carrier pipe		4.80	1.667		92	87		179	231
0210	10" diam. for 6" carrier pipe	Q-1	8	2		159	93.50		252.50	315
0220	12" diam. for 8" carrier pipe		7.20	2.222		222	104		326	400
0230	14" diam. for 10" carrier pipe		6.40	2.500		256	117		373	455
0240	16" diam. for 12" carrier pipe		5.80	2.759		292	129		421	515
0250	18" diam. for 14" carrier pipe		5.20	3.077		430	144		574	685
0260	24" diam. for 18" carrier pipe		4	4		670	187		857	1,025
0270	24" diam. for 20" carrier pipe		4	4		585	187		772	920
0280	30" diam. for 24" carrier pipe		3.20	5		965	234		1,199	1,425
0500	Wall sleeve									
0510	Ductile iron with rubber gasket seal									
0520	3"	1 Plum	8.40	.952	Ea.	700	49.50		749.50	845
0530	4"		7.20	1.111		750	58		808	910
0540	6"		6	1.333		920	69.50		989.50	1,125
0550	8"		4	2		1,125	104		1,229	1,400
0560	10"		3	2.667		1,375	139		1,514	1,725
0570	12"		2.40	3.333		1,625	174		1,799	2,050
5000	Escutcheon									
5100	Split ring, pipe									
5110	Chrome plated									
5120	1/2"	1 Plum	160	.050	Ea.	1.05	2.60		3.65	5.05
5130	3/4"		160	.050		1.16	2.60		3.76	5.20
5140	1"		135	.059		1.17	3.08		4.25	5.90
5150	1-1/2"		115	.070		1.76	3.62		5.38	7.40

22 11 19.34 Sleeves and Escutcheons		Crew	Daily Output	Labor-Hours	Unit	2010 Bare Costs Material	Labor	Equipment	Total	Total Incl O&P
5160	2″	1 Plum	100	.080	Ea.	1.96	4.16		6.12	8.40
5170	4″		80	.100		4.35	5.20		9.55	12.60
5180	6″		68	.118		7.25	6.10		13.35	17.20
5400	Shallow flange type									
5410	Chrome plated steel									
5420	1/2″ CTS	1 Plum	180	.044	Ea.	.32	2.31		2.63	3.82
5430	3/4″ CTS		180	.044		.29	2.31		2.60	3.79
5440	1/2″ IPS		180	.044		.33	2.31		2.64	3.83
5450	3/4″ IPS		180	.044		.28	2.31		2.59	3.78
5460	1″ IPS		175	.046		.46	2.38		2.84	4.08
5470	1-1/2″ IPS		170	.047		.72	2.45		3.17	4.46
5480	2″ IPS		160	.050		.86	2.60		3.46	4.85

22 11 19.38 Water Supply Meters

		Crew	Daily Output	Labor-Hours	Unit	Material	Labor	Equipment	Total	Total Incl O&P
0010	**WATER SUPPLY METERS**									
1000	Detector, serves dual systems such as fire and domestic or									
1020	process water, wide range cap., UL and FM approved									
1100	3″ mainline x 2″ by-pass, 400 GPM	Q-1	3.60	4.444	Ea.	6,250	208		6,458	7,175
1140	4″ mainline x 2″ by-pass, 700 GPM	″	2.50	6.400		6,250	300		6,550	7,325
1180	6″ mainline x 3″ by-pass, 1600 GPM	Q-2	2.60	9.231		9,550	450		10,000	11,200
1220	8″ mainline x 4″ by-pass, 2800 GPM		2.10	11.429		14,200	555		14,755	16,400
1260	10″ mainline x 6″ by-pass, 4400 GPM		2	12		20,300	585		20,885	23,200
1300	10″x12″ mainlines x 6″ by-pass,5400 GPM		1.70	14.118		27,500	685		28,185	31,200
2000	Domestic/commercial, bronze									
2020	Threaded									
2060	5/8″ diameter, to 20 GPM	1 Plum	16	.500	Ea.	42	26		68	85
2080	3/4″ diameter, to 30 GPM		14	.571		76.50	29.50		106	129
2100	1″ diameter, to 50 GPM		12	.667		116	34.50		150.50	180
2300	Threaded/flanged									
2340	1-1/2″ diameter, to 100 GPM	1 Plum	8	1	Ea.	284	52		336	390
2360	2″ diameter, to 160 GPM	″	6	1.333	″	385	69.50		454.50	530
2600	Flanged, compound									
2640	3″ diameter, 320 GPM	Q-1	3	5.333	Ea.	2,625	250		2,875	3,250
2660	4″ diameter, to 500 GPM		1.50	10.667		4,200	500		4,700	5,350
2680	6″ diameter, to 1,000 GPM		1	16		6,675	750		7,425	8,475
2700	8″ diameter, to 1,800 GPM		.80	20		10,500	935		11,435	12,900
7000	Turbine									
7260	Flanged									
7300	2″ diameter, to 160 GPM	1 Plum	7	1.143	Ea.	510	59.50		569.50	655
7320	3″ diameter, to 450 GPM	Q-1	3.60	4.444		1,075	208		1,283	1,475
7340	4″ diameter, to 650 GPM	″	2.50	6.400		1,775	300		2,075	2,400
7360	6″ diameter, to 1800 GPM	Q-2	2.60	9.231		3,150	450		3,600	4,150
7380	8″ diameter, to 2500 GPM		2.10	11.429		5,000	555		5,555	6,350
7400	10″ diameter, to 5500 GPM		1.70	14.118		6,775	685		7,460	8,475

22 11 19.42 Backflow Preventers

		Crew	Daily Output	Labor-Hours	Unit	Material	Labor	Equipment	Total	Total Incl O&P
0010	**BACKFLOW PREVENTERS**, Includes valves									
0020	and four test cocks, corrosion resistant, automatic operation									
1000	Double check principle									
1010	Threaded, with ball valves									
1020	3/4″ pipe size	1 Plum	16	.500	Ea.	245	26		271	310
1030	1″ pipe size		14	.571		282	29.50		311.50	355
1040	1-1/2″ pipe size		10	.800		740	41.50		781.50	875
1050	2″ pipe size		7	1.143		880	59.50		939.50	1,050

22 11 19.42 Backflow Preventers		Crew	Daily Output	Labor-Hours	Unit	Material	Labor	Equipment	Total	Total Incl O&P
						2010 Bare Costs				
1080	Threaded, with gate valves									
1100	3/4" pipe size	1 Plum	16	.500	Ea.	945	26		971	1,100
1120	1" pipe size		14	.571		955	29.50		984.50	1,100
1140	1-1/2" pipe size		10	.800		1,225	41.50		1,266.50	1,425
1160	2" pipe size		7	1.143		1,500	59.50		1,559.50	1,750
1200	Flanged, valves are gate									
1210	3" pipe size	Q-1	4.50	3.556	Ea.	2,200	167		2,367	2,675
1220	4" pipe size	"	3	5.333		2,700	250		2,950	3,350
1230	6" pipe size	Q-2	3	8		3,975	390		4,365	4,950
1240	8" pipe size		2	12		7,250	585		7,835	8,850
1250	10" pipe size		1	24		10,700	1,175		11,875	13,600
1300	Flanged, valves are OS&Y									
1370	1" pipe size	1 Plum	5	1.600	Ea.	1,025	83.50		1,108.50	1,250
1374	1-1/2" pipe size		5	1.600		1,300	83.50		1,383.50	1,575
1378	2" pipe size		4.80	1.667		1,600	87		1,687	1,875
1380	3" pipe size	Q-1	4.50	3.556		3,875	167		4,042	4,525
1400	4" pipe size	"	3	5.333		4,325	250		4,575	5,125
1420	6" pipe size	Q-2	3	8		6,750	390		7,140	8,000
1430	8" pipe size	"	2	12		13,300	585		13,885	15,600
4000	Reduced pressure principle									
4100	Threaded, bronze, valves are ball									
4120	3/4" pipe size	1 Plum	16	.500	Ea.	365	26		391	440
4140	1" pipe size		14	.571		390	29.50		419.50	475
4150	1-1/4" pipe size		12	.667		670	34.50		704.50	790
4160	1-1/2" pipe size		10	.800		735	41.50		776.50	875
4180	2" pipe size		7	1.143		825	59.50		884.50	1,000
5000	Flanged, bronze, valves are OS&Y									
5060	2-1/2" pipe size	Q-1	5	3.200	Ea.	3,525	150		3,675	4,100
5080	3" pipe size		4.50	3.556		3,700	167		3,867	4,325
5100	4" pipe size		3	5.333		4,650	250		4,900	5,475
5120	6" pipe size	Q-2	3	8		6,725	390		7,115	7,975
5200	Flanged, iron, valves are gate									
5210	2-1/2" pipe size	Q-1	5	3.200	Ea.	2,475	150		2,625	2,950
5220	3" pipe size		4.50	3.556		2,575	167		2,742	3,075
5230	4" pipe size		3	5.333		3,500	250		3,750	4,225
5240	6" pipe size	Q-2	3	8		4,925	390		5,315	6,000
5250	8" pipe size		2	12		8,825	585		9,410	10,600
5260	10" pipe size		1	24		12,500	1,175		13,675	15,500
5600	Flanged, iron, valves are OS&Y									
5660	2-1/2" pipe size	Q-1	5	3.200	Ea.	1,925	150		2,075	2,350
5680	3" pipe size		4.50	3.556		2,950	167		3,117	3,500
5700	4" pipe size		3	5.333		3,800	250		4,050	4,550
5720	6" pipe size	Q-2	3	8		5,375	390		5,765	6,475
5740	8" pipe size		2	12		9,450	585		10,035	11,300
5760	10" pipe size		1	24		12,700	1,175		13,875	15,700

22 11 19.50 Vacuum Breakers		Crew	Daily Output	Labor-Hours	Unit	Material	Labor	Equipment	Total	Total Incl O&P
0010	**VACUUM BREAKERS**									
0013	See also backflow preventers Div. 22 11 19.42									
1000	Anti-siphon continuous pressure type									
1010	Max. 150 PSI - 210°F									
1020	Bronze body									
1030	1/2" size	1 Stpi	24	.333	Ea.	172	17.30		189.30	215

22 11 19.50 Vacuum Breakers		Crew	Daily Output	Labor-Hours	Unit	Material	2010 Bare Costs Labor	Equipment	Total	Total Incl O&P
1040	3/4" size	1 Stpi	20	.400	Ea.	172	21		193	220
1050	1" size		19	.421		178	22		200	228
1060	1-1/4" size		15	.533		355	27.50		382.50	430
1070	1-1/2" size		13	.615		425	32		457	520
1080	2" size		11	.727		440	38		478	540
1200	Max. 125 PSI with atmospheric vent									
1210	Brass, in-line construction									
1220	1/4" size	1 Stpi	24	.333	Ea.	59.50	17.30		76.80	91.50
1230	3/8" size	"	24	.333		59.50	17.30		76.80	91.50
1260	For polished chrome finish, add					13%				
2000	Anti-siphon, non-continuous pressure type									
2010	Hot or cold water 125 PSI - 210°F									
2020	Bronze body									
2030	1/4" size	1 Stpi	24	.333	Ea.	38	17.30		55.30	67.50
2040	3/8" size		24	.333		38	17.30		55.30	67.50
2050	1/2" size		24	.333		43	17.30		60.30	73.50
2060	3/4" size		20	.400		52.50	21		73.50	88.50
2070	1" size		19	.421		79.50	22		101.50	121
2080	1-1/4" size		15	.533		139	27.50		166.50	195
2090	1-1/2" size		13	.615		164	32		196	228
2100	2" size		11	.727		255	38		293	335
2110	2-1/2" size		8	1		730	52		782	885
2120	3" size		6	1.333		975	69		1,044	1,175
2150	For polished chrome finish, add					50%				

22 11 19.54 Water Hammer Arresters/Shock Absorbers		Crew	Daily Output	Labor-Hours	Unit	Material	Labor	Equipment	Total	Total Incl O&P
0010	**WATER HAMMER ARRESTERS/SHOCK ABSORBERS**									
0490	Copper									
0500	3/4" male I.P.S. For 1 to 11 fixtures	1 Plum	12	.667	Ea.	20.50	34.50		55	74.50
0600	1" male I.P.S., For 12 to 32 fixtures		8	1		41.50	52		93.50	124
0700	1-1/4" male I.P.S. For 33 to 60 fixtures		8	1		44	52		96	127
0800	1-1/2" male I.P.S. For 61 to 113 fixtures		8	1		63	52		115	148
0900	2" male I.P.S.For 114 to 154 fixtures		8	1		92	52		144	179
1000	2-1/2" male I.P.S. For 155 to 330 fixtures		4	2		285	104		389	470
4000	Bellows type									
4010	3/4" FNPT, to 11 fixture units	1 Plum	10	.800	Ea.	205	41.50		246.50	289
4020	1" FNPT, to 32 fixture units		8.80	.909		415	47.50		462.50	525
4030	1" FNPT, to 60 fixture units		8.80	.909		620	47.50		667.50	755
4040	1" FNPT, to 113 fixture units		8.80	.909		1,550	47.50		1,597.50	1,800
4050	1" FNPT, to 154 fixture units		6.60	1.212		1,750	63		1,813	2,025
4060	1-1/2" FNPT, to 300 fixture units		5	1.600		2,150	83.50		2,233.50	2,500

22 11 19.64 Hydrants		Crew	Daily Output	Labor-Hours	Unit	Material	Labor	Equipment	Total	Total Incl O&P
0010	**HYDRANTS**									
0050	Wall type, moderate climate, bronze, encased									
0200	3/4" IPS connection	1 Plum	16	.500	Ea.	600	26		626	700
0300	1" IPS connection	"	14	.571		690	29.50		719.50	805
0500	Anti-siphon type, 3/4" connection					520			520	570
1000	Non-freeze, bronze, exposed									
1100	3/4" IPS connection, 4" to 9" thick wall	1 Plum	14	.571	Ea.	405	29.50		434.50	490
1120	10" to 14" thick wall		12	.667		520	34.50		554.50	620
1140	15" to 19" thick wall		12	.667		520	34.50		554.50	620
1160	20" to 24" thick wall		10	.800		530	41.50		571.50	650
1200	For 1" IPS connection, add					15%	10%			

22 11 19 – Domestic Water Piping Specialties

22 11 19.64 Hydrants		Crew	Daily Output	Labor-Hours	Unit	2010 Bare Costs Material	Labor	Equipment	Total	Total Incl O&P
1240	For 3/4" adapter type vacuum breaker, add				Ea.	51			51	56
1280	For anti-siphon type, add				"	106			106	117
2000	Non-freeze bronze, encased, anti-siphon type									
2100	3/4" IPS connection, 5" to 9" thick wall	1 Plum	14	.571	Ea.	1,025	29.50		1,054.50	1,175
2120	10" to 14" thick wall		12	.667		1,050	34.50		1,084.50	1,225
2140	15" to 19" thick wall		12	.667		1,100	34.50		1,134.50	1,275
2160	20" to 24" thick wall		10	.800		1,150	41.50		1,191.50	1,350
2200	For 1" IPS connection, add					10%	10%			
3000	Ground box type, bronze frame, 3/4" IPS connection									
3080	Non-freeze, all bronze, polished face, set flush									
3100	2 feet depth of bury	1 Plum	8	1	Ea.	765	52		817	925
3120	3 feet depth of bury		8	1		825	52		877	985
3140	4 feet depth of bury		8	1		885	52		937	1,050
3160	5 feet depth of bury		7	1.143		945	59.50		1,004.50	1,150
3180	6 feet depth of bury		7	1.143		1,000	59.50		1,059.50	1,200
3200	7 feet depth of bury		6	1.333		1,050	69.50		1,119.50	1,275
3220	8 feet depth of bury		5	1.600		1,125	83.50		1,208.50	1,350
3240	9 feet depth of bury		4	2		1,175	104		1,279	1,450
3260	10 feet depth of bury		4	2		1,225	104		1,329	1,500
3400	For 1" IPS connection, add					15%	10%			
3450	For 1-1/4" IPS connection, add					325%	14%			
3500	For 1-1/2" connection, add					370%	18%			
3550	For 2" connection, add					445%	24%			
3600	For tapped drain port in box, add					69.50			69.50	76
4000	Non-freeze, CI body, bronze frame & scoriated cover									
4010	with hose storage									
4100	2 feet depth of bury	1 Plum	7	1.143	Ea.	1,425	59.50		1,484.50	1,650
4120	3 feet depth of bury		7	1.143		1,475	59.50		1,534.50	1,725
4140	4 feet depth of bury		7	1.143		1,525	59.50		1,584.50	1,775
4160	5 feet depth of bury		6.50	1.231		1,550	64		1,614	1,800
4180	6 feet depth of bury		6	1.333		1,575	69.50		1,644.50	1,825
4200	7 feet depth of bury		5.50	1.455		1,625	75.50		1,700.50	1,925
4220	8 feet depth of bury		5	1.600		1,700	83.50		1,783.50	1,975
4240	9 feet depth of bury		4.50	1.778		1,750	92.50		1,842.50	2,075
4260	10 feet depth of bury		4	2		1,800	104		1,904	2,125
4280	For 1" IPS connection, add					310			310	345
4300	For tapped drain port in box, add					69.50			69.50	76
5000	Moderate climate, all bronze, polished face									
5020	and scoriated cover, set flush									
5100	3/4" IPS connection	1 Plum	16	.500	Ea.	530	26		556	625
5120	1" IPS connection	"	14	.571		655	29.50		684.50	765
5200	For tapped drain port in box, add					69.50			69.50	76
6000	Ground post type, all non-freeze, all bronze, aluminum casing									
6010	guard, exposed head, 3/4" IPS connection									
6100	2 feet depth of bury	1 Plum	8	1	Ea.	750	52		802	905
6120	3 feet depth of bury		8	1		810	52		862	970
6140	4 feet depth of bury		8	1		875	52		927	1,050
6160	5 feet depth of bury		7	1.143		935	59.50		994.50	1,125
6180	6 feet depth of bury		7	1.143		1,000	59.50		1,059.50	1,200
6200	7 feet depth of bury		6	1.333		1,075	69.50		1,144.50	1,275
6220	8 feet depth of bury		5	1.600		1,125	83.50		1,208.50	1,375
6240	9 feet depth of bury		4	2		1,200	104		1,304	1,475
6260	10 feet depth of bury		4	2		1,275	104		1,379	1,550

22 11 19 – Domestic Water Piping Specialties

22 11 19.64 Hydrants		Crew	Daily Output	Labor-Hours	Unit	Material	Labor	Equipment	Total	Total Incl O&P
						2010 Bare Costs				
6300	For 1" IPS connection, add					40%	10%			
6350	For 1-1/4" IPS connection, add					140%	14%			
6400	For 1-1/2" IPS connection, add					225%	18%			
6450	For 2" IPS connection, add					315%	24%			

22 11 23 – Domestic Water Pumps

22 11 23.10 General Utility Pumps		Crew	Daily Output	Labor-Hours	Unit	Material	Labor	Equipment	Total	Total Incl O&P
0010	**GENERAL UTILITY PUMPS**									
2000	Single stage									
3000	Double suction,									
3190	75 HP, to 2500 GPM	Q-3	.28	114	Ea.	16,400	5,675		22,075	26,500
3220	100 HP, to 3000 GPM		.26	123		20,600	6,100		26,700	31,800
3240	150 HP, to 4000 GPM		.24	133		31,700	6,600		38,300	44,800
4000	Centrifugal, end suction, mounted on base									
4010	Horizontal mounted, with drip proof motor, rated @ 100' head									
4020	Vertical split case, single stage									
4040	100 GPM, 5 HP, 1-1/2" discharge	Q-1	1.70	9.412	Ea.	3,250	440		3,690	4,225
4050	200 GPM, 10 HP, 2" discharge		1.30	12.308		3,550	575		4,125	4,775
4060	250 GPM, 10 HP, 3" discharge		1.28	12.500		3,975	585		4,560	5,250
4070	300 GPM, 15 HP, 2" discharge	Q-2	1.56	15.385		4,725	745		5,470	6,325
4080	500 GPM, 20 HP, 4" discharge		1.44	16.667		5,025	810		5,835	6,750
4090	750 GPM, 30 HP, 4" discharge		1.20	20		5,075	970		6,045	7,025
4100	1050 GPM, 40 HP, 5" discharge		1	24		7,025	1,175		8,200	9,475
4110	1500 GPM, 60 HP, 6" discharge		.60	40		10,900	1,950		12,850	14,900
4120	2000 GPM, 75 HP, 6" discharge		.50	48		10,900	2,325		13,225	15,500
4130	3000 GPM, 100 HP, 8" discharge		.40	60		15,500	2,925		18,425	21,400
4200	Horizontal split case, single stage									
4210	100 GPM, 7.5 HP, 1-1/2" discharge	Q-1	1.70	9.412	Ea.	3,850	440		4,290	4,875
4220	250 GPM, 15 HP, 2-1/2" discharge	"	1.30	12.308		6,825	575		7,400	8,400
4230	500 GPM, 20 HP, 4" discharge	Q-2	1.60	15		8,250	730		8,980	10,200
4240	750 GPM, 25 HP, 5" discharge		1.54	15.584		8,500	755		9,255	10,500
4250	1000 GPM, 40 HP, 5" discharge		1.20	20		11,300	970		12,270	13,900
4260	1500 GPM, 50 HP, 6" discharge	Q-3	1.42	22.535		14,000	1,125		15,125	17,100
4270	2000 GPM, 75 HP, 8" discharge		1.14	28.070		18,900	1,400		20,300	22,800
4280	3000 GPM, 100 HP, 10" discharge		.96	33.333		25,200	1,650		26,850	30,200
4290	3500 GPM, 150 HP, 10" discharge		.86	37.209		25,400	1,850		27,250	30,800
4300	4000 GPM, 200 HP, 10" discharge		.66	48.485		34,600	2,400		37,000	41,700
4330	Horizontal split case, two stage, 500' head									
4340	100 GPM, 40 HP, 1-1/2" discharge	Q-2	1.70	14.118	Ea.	14,200	685		14,885	16,600
4350	200 GPM, 50 HP, 1-1/2" discharge	"	1.44	16.667		14,200	810		15,010	16,800
4360	300 GPM, 75 HP, 2" discharge	Q-3	1.57	20.382		20,400	1,000		21,400	24,000
4370	400 GPM, 100 HP, 3" discharge		1.14	28.070		20,400	1,400		21,800	24,600
4380	800 GPM, 200 HP, 4" discharge		.86	37.209		32,400	1,850		34,250	38,400
5000	Centrifugal, in-line									
5006	Vertical mount, iron body, 125 lb. flgd, 3550 RPM TEFC mtr									
5010	Single stage									
5020	50 GPM, 3 HP, 1-1/2" discharge	Q-1	2.30	6.957	Ea.	1,025	325		1,350	1,625
5030	75 GPM, 5 HP, 1-1/2" discharge		1.60	10		1,100	470		1,570	1,900
5040	100 GPM, 7.5 HP, 1-1/2" discharge		1.30	12.308		1,350	575		1,925	2,375
5050	125 GPM, 10 HP, 1-1/2" discharge	Q-2	1.70	14.118		1,575	685		2,260	2,750
5060	150 GPM, 15 HP, 1-1/2" discharge		1.60	15		2,650	730		3,380	4,000
5080	250 GPM, 40 HP, 2" discharge		1.40	17.143		3,950	835		4,785	5,575
5090	300 GPM, 50 HP, 3" discharge		1.30	18.462		4,225	895		5,120	6,000

22 11 23 – Domestic Water Pumps

22 11 23.10 General Utility Pumps		Crew	Daily Output	Labor-Hours	Unit	Material	2010 Bare Costs Labor	Equipment	Total	Total Incl O&P
5100	400 GPM, 75 HP, 3" discharge	Q-3	.60	53.333	Ea.	5,350	2,650		8,000	9,850
5110	600 GPM, 100 HP, 3" discharge	"	.50	64		5,625	3,175		8,800	11,000

22 11 23.11 Miscellaneous Pumps

Code	Description	Crew	Daily Output	Labor-Hours	Unit	Material	Labor	Equipment	Total	Total Incl O&P
0010	**MISCELLANEOUS PUMPS**									
0020	Water pump, portable, gasoline powered									
0100	6000 GPH, 2" discharge	Q-1	11	1.455	Ea.	730	68		798	905
0110	8000 GPH, 3" discharge		10.50	1.524		695	71.50		766.50	870
0120	10,000 GPH, 4" discharge		10	1.600		1,350	75		1,425	1,575
0500	Pump, propylene body, housing and impeller									
0510	22 GPM, 1/3 HP, 40' HD	1 Plum	5	1.600	Ea.	480	83.50		563.50	655
0520	33 GPM, 1/2 HP, 40' HD	Q-1	5	3.200		535	150		685	815
0530	53 GPM, 3/4 HP, 40' HD	"	4	4		900	187		1,087	1,275
0600	Rotary pump, CI									
0614	202 GPH, 3/4 HP, 1" discharge	Q-1	4.50	3.556	Ea.	930	167		1,097	1,275
0618	277 GPH, 1 HP, 1" discharge		4	4		930	187		1,117	1,300
0624	1100 GPH, 1.5 HP, 1-1/4" discharge		3.60	4.444		2,325	208		2,533	2,850
0628	1900 GPH, 2 HP, 1-1/4" discharge		3.20	5		3,200	234		3,434	3,850
1000	Turbine pump, CI									
1010	50 GPM, 2 HP, 3" discharge	Q-1	.80	20	Ea.	3,325	935		4,260	5,050
1020	100 GPM, 3 HP, 4" discharge	Q-2	.96	25		5,225	1,225		6,450	7,575
1030	250 GPM, 15 HP, 6" discharge	"	.94	25.532		6,475	1,250		7,725	8,975
1040	500 GPM, 25 HP, 6" discharge	Q-3	1.22	26.230		6,525	1,300		7,825	9,125
1050	1000 GPM, 50 HP, 8" discharge		1.14	28.070		9,000	1,400		10,400	12,000
1060	2000 GPM, 100 HP, 10" discharge		1	32		10,700	1,575		12,275	14,100
1070	3000 GPM, 150 HP, 10" discharge		.80	40		17,200	1,975		19,175	21,900
1080	4000 GPM, 200 HP, 12" discharge		.70	45.714		18,800	2,275		21,075	24,100
1090	6000 GPM, 300 HP, 14" discharge		.60	53.333		28,600	2,650		31,250	35,500
1100	10,000 GPM, 300 HP, 18" discharge		.58	55.172		33,100	2,725		35,825	40,500
2000	Centrifugal stainless steel pumps									
2100	100 GPM, 100' TDH, 3 HP	Q-1	1.80	8.889	Ea.	9,450	415		9,865	11,000
2130	250 GPM, 100' TDH, 10 HP	"	1.28	12.500		13,200	585		13,785	15,400
2160	500 GPM, 100' TDH, 20 HP	Q-2	1.44	16.667		15,100	810		15,910	17,800
2200	Vertical turbine stainless steel pumps									
2220	100 GPM, 100' TDH, 7.5 HP	Q-2	.95	25.263	Ea.	53,000	1,225		54,225	60,000
2240	250 GPM, 100' TDH, 15 HP		.94	25.532		58,500	1,250		59,750	66,500
2260	500 GPM, 100' TDH, 20 HP		.93	25.806		64,000	1,250		65,250	72,500
2280	750 GPM, 100' TDH, 30 HP	Q-3	1.19	26.891		70,000	1,325		71,325	79,000
2300	1000 GPM, 100' TDH, 40 HP	"	1.16	27.586		75,500	1,375		76,875	85,000
2320	100 GPM, 200' TDH, 15 HP	Q-2	.94	25.532		81,000	1,250		82,250	91,500
2340	250 GPM, 200' TDH, 25 HP	Q-3	1.22	26.230		87,000	1,300		88,300	97,500
2360	500 GPM, 200' TDH, 40 HP		1.16	27.586		92,500	1,375		93,875	104,000
2380	750 GPM, 200' TDH, 60 HP		1.13	28.319		93,000	1,400		94,400	104,500
2400	1000 GPM, 200' TDH, 75 HP		1.12	28.571		96,000	1,425		97,425	108,000

22 11 23.13 Domestic-Water Packaged Booster Pumps

Code	Description	Crew	Daily Output	Labor-Hours	Unit	Material	Labor	Equipment	Total	Total Incl O&P
0010	**DOMESTIC-WATER PACKAGED BOOSTER PUMPS**									
0200	Pump system, with diaphragm tank, control, press. switch									
0300	1 HP pump	Q-1	1.30	12.308	Ea.	5,225	575		5,800	6,600
0400	1-1/2 HP pump		1.25	12.800		5,275	600		5,875	6,700
0420	2 HP pump		1.20	13.333		5,400	625		6,025	6,850
0440	3 HP pump		1.10	14.545		5,475	680		6,155	7,050
0460	5 HP pump	Q-2	1.50	16		6,050	775		6,825	7,850
0480	7-1/2 HP pump		1.42	16.901		6,750	820		7,570	8,650

22 11 Facility Water Distribution

22 11 23 – Domestic Water Pumps

22 11 23.13 Domestic-Water Packaged Booster Pumps		Crew	Daily Output	Labor-Hours	Unit	2010 Bare Costs Material	Labor	Equipment	Total	Total Incl O&P
0500	10 HP pump	Q-2	1.34	17.910	Ea.	7,050	870		7,920	9,075
2000	Pump system, variable speed, base, controls, starter									
2010	Duplex, 100′ head									
2020	400 GPM, 7-1/2HP, 4″ discharge	Q-2	.70	34.286	Ea.	48,400	1,675		50,075	55,500
2025	Triplex, 100′ head									
2030	1000 GPM, 15HP, 6″ discharge	Q-2	.50	48	Ea.	69,000	2,325		71,325	79,500
2040	1700 GPM, 30HP, 6″ discharge	″	.30	80	″	78,000	3,875		81,875	91,500

22 12 Facility Potable-Water Storage Tanks

22 12 23 – Facility Indoor Potable-Water Storage Tanks

22 12 23.13 Facility Steel, Indoor Pot.-Water Storage Tanks

		Crew	Daily Output	Labor-Hours	Unit	Material	Labor	Equipment	Total	Total Incl O&P
0010	**FACILITY STEEL, INDOOR POT.-WATER STORAGE TANKS**									
2000	Galvanized steel, 15 gal., 14″ diam., 26″ LOA	1 Plum	12	.667	Ea.	1,175	34.50		1,209.50	1,325
2060	30 gal., 14″ diam. x 49″ LOA		11	.727		1,375	38		1,413	1,575
2080	80 gal., 20″ diam. x 64″ LOA		9	.889		2,150	46.50		2,196.50	2,425
2100	135 gal., 24″ diam. x 75″ LOA		6	1.333		3,125	69.50		3,194.50	3,550
2120	240 gal., 30″ diam. x 86″ LOA		4	2		6,100	104		6,204	6,875
2140	300 gal., 36″ diam. x 76″ LOA	↓	3	2.667		8,675	139		8,814	9,750
2160	400 gal., 36″ diam. x 100″ LOA	Q-1	4	4		10,700	187		10,887	12,100
2180	500 gal., 36″ diam., x 126″ LOA	″	3	5.333		13,200	250		13,450	14,900
3000	Glass lined, P.E., 80 gal., 20″ diam. x 60″ LOA	1 Plum	9	.889		2,600	46.50		2,646.50	2,925
3060	140 gal., 24″ diam. x 80″ LOA		6	1.333		3,600	69.50		3,669.50	4,050
3080	225 gal., 30″ diam. x 78″ LOA		4	2		4,450	104		4,554	5,050
3100	325 gal., 36″ diam. x 81″ LOA	↓	3	2.667		5,700	139		5,839	6,450
3120	460 gal., 42″ diam. x 84″ LOA	Q-1	4	4		6,000	187		6,187	6,875
3140	605 gal., 48″ diam. x 87″ LOA		3	5.333		11,400	250		11,650	13,000
3160	740 gal., 54″ diam. x 91″ LOA		3	5.333		12,900	250		13,150	14,600
3180	940 gal., 60″ diam. x 93″ LOA		2.50	6.400		14,900	300		15,200	16,900
3200	1330 gal., 66″ diam. x 107″ LOA		2	8		20,400	375		20,775	23,000
3220	1615 gal., 72″ diam. x 110″ LOA		1.50	10.667		23,200	500		23,700	26,300
3240	2285 gal., 84″ diam. x 128″ LOA	↓	1	16		28,700	750		29,450	32,600
3260	3440 gal., 96″ diam. x 157″ LOA	Q-2	1.50	16	↓	42,400	775		43,175	47,900

22 13 Facility Sanitary Sewerage

22 13 16 – Sanitary Waste and Vent Piping

22 13 16.40 Pipe Fittings, Cast Iron for Drainage

		Crew	Daily Output	Labor-Hours	Unit	Material	Labor	Equipment	Total	Total Incl O&P
0010	**PIPE FITTINGS, CAST IRON FOR DRAINAGE**, Special									
1000	Drip pan elbow, (safety valve discharge elbow)									
1010	Cast iron, threaded inlet									
1013	2″	Q-1	9	1.778	Ea.	148	83.50		231.50	287
1014	2-1/2″		8	2		900	93.50		993.50	1,125
1015	3″		6.40	2.500		970	117		1,087	1,250
1017	4″	↓	4.80	3.333		1,300	156		1,456	1,650
1019	6″	Q-2	3.60	6.667		1,375	325		1,700	1,975
1020	8″	″	2.60	9.231	↓	2,000	450		2,450	2,875

22 13 16.60 Traps

		Crew	Daily Output	Labor-Hours	Unit	Material	Labor	Equipment	Total	Total Incl O&P
0010	**TRAPS**									
4700	Copper, drainage, drum trap									

22 13 16 – Sanitary Waste and Vent Piping

22 13 16.60 Traps		Crew	Daily Output	Labor-Hours	Unit	2010 Bare Costs Material	Labor	Equipment	Total	Total Incl O&P
4800	3" x 5" solid, 1-1/2" pipe size	1 Plum	16	.500	Ea.	127	26		153	178
4840	3" x 6" swivel, 1-1/2" pipe size	"	16	.500	"	201	26		227	260
5100	P trap, standard pattern									
5200	1-1/4" pipe size	1 Plum	18	.444	Ea.	93.50	23		116.50	138
5240	1-1/2" pipe size		17	.471		90	24.50		114.50	136
5260	2" pipe size		15	.533		194	28		222	255
5280	3" pipe size		11	.727		335	38		373	425
5340	With cleanout and slip joint									
5360	1-1/4" pipe size	1 Plum	18	.444	Ea.	68.50	23		91.50	110
5400	1-1/2" pipe size		17	.471		138	24.50		162.50	188
5420	2" pipe size		15	.533		222	28		250	286
5460	For swivel, 1-1/2"					201			201	221
6710	ABS DWV P trap, solvent weld joint									
6720	1-1/2" pipe size	1 Plum	18	.444	Ea.	4.74	23		27.74	39.50
6722	2" pipe size		17	.471		6.40	24.50		30.90	43.50
6724	3" pipe size		15	.533		24.50	28		52.50	68.50
6726	4" pipe size		14	.571		50.50	29.50		80	100
6732	PVC DWV P trap, solvent weld joint									
6733	1-1/2" pipe size	1 Plum	18	.444	Ea.	3.83	23		26.83	38.50
6734	2" pipe size		17	.471		5.15	24.50		29.65	42
6735	3" pipe size		15	.533		17.50	28		45.50	61
6736	4" pipe size		14	.571		40	29.50		69.50	88.50
6760	PP DWV, dilution trap, 1-1/2" pipe size		16	.500		186	26		212	244
6770	P trap, 1-1/2" pipe size		17	.471		46	24.50		70.50	87
6780	2" pipe size		16	.500		68	26		94	114
6790	3" pipe size		14	.571		125	29.50		154.50	183
6800	4" pipe size		13	.615		210	32		242	279
6830	S trap, 1-1/2" pipe size		16	.500		38	26		64	81
6840	2" pipe size		15	.533		57	28		85	104
6850	Universal trap, 1-1/2" pipe size		14	.571		75	29.50		104.50	127
6860	PVC DWV hub x hub, basin trap, 1-1/4" pipe size		18	.444		7.50	23		30.50	43
6870	Sink P trap, 1-1/2" pipe size		18	.444		7.50	23		30.50	43
6880	Tubular S trap, 1-1/2" pipe size		17	.471		13.90	24.50		38.40	52
6890	PVC sch. 40 DWV, drum trap									
6900	1-1/2" pipe size	1 Plum	16	.500	Ea.	17.25	26		43.25	58
6910	P trap, 1-1/2" pipe size		18	.444		4.26	23		27.26	39
6920	2" pipe size		17	.471		5.75	24.50		30.25	43
6930	3" pipe size		15	.533		19.45	28		47.45	63
6940	4" pipe size		14	.571		44.50	29.50		74	93.50
6950	P trap w/clean out, 1-1/2" pipe size		18	.444		7.05	23		30.05	42.50
6960	2" pipe size		17	.471		11.90	24.50		36.40	49.50
6970	P trap adjustable, 1-1/2" pipe size		17	.471		5.45	24.50		29.95	42.50
6980	P trap adj. w/union & cleanout, 1-1/2" pipe size		16	.500		15.20	26		41.20	56

22 13 16.80 Vent Flashing and Caps		Crew	Daily Output	Labor-Hours	Unit	Material	Labor	Equipment	Total	Total Incl O&P
0010	**VENT FLASHING AND CAPS**									
0120	Vent caps									
0140	Cast iron									
0180	2-1/2" - 3-5/8" pipe	1 Plum	21	.381	Ea.	40	19.85		59.85	73.50
0190	4" - 4-1/8" pipe	"	19	.421	"	48	22		70	86
0900	Vent flashing									
1000	Aluminum with lead ring									
1020	1-1/4" pipe	1 Plum	20	.400	Ea.	10.35	21		31.35	42.50

22 13 Facility Sanitary Sewerage

22 13 16 – Sanitary Waste and Vent Piping

22 13 16.80 Vent Flashing and Caps		Crew	Daily Output	Labor-Hours	Unit	Material	2010 Bare Costs Labor	Equipment	Total	Total Incl O&P
1030	1-1/2" pipe	1 Plum	20	.400	Ea.	10.60	21		31.60	42.50
1040	2" pipe		18	.444		11.10	23		34.10	47
1050	3" pipe		17	.471		12.30	24.50		36.80	50
1060	4" pipe		16	.500		14.85	26		40.85	55.50
1350	Copper with neoprene ring									
1400	1-1/4" pipe	1 Plum	20	.400	Ea.	19.95	21		40.95	53
1430	1-1/2" pipe		20	.400		19.95	21		40.95	53
1440	2" pipe		18	.444		21	23		44	57.50
1450	3" pipe		17	.471		24.50	24.50		49	63.50
1460	4" pipe		16	.500		27	26		53	69
2000	Galvanized with neoprene ring									
2020	1-1/4" pipe	1 Plum	20	.400	Ea.	14.20	21		35.20	46.50
2030	1-1/2" pipe		20	.400		14.20	21		35.20	46.50
2040	2" pipe		18	.444		14.70	23		37.70	50.50
2050	3" pipe		17	.471		15.40	24.50		39.90	53.50
2060	4" pipe		16	.500		17.90	26		43.90	58.50
2980	Neoprene, one piece									
3000	1-1/4" pipe	1 Plum	24	.333	Ea.	7.10	17.35		24.45	34
3030	1-1/2" pipe		24	.333		7.75	17.35		25.10	34.50
3040	2" pipe		23	.348		7.10	18.10		25.20	35
3050	3" pipe		21	.381		8.20	19.85		28.05	38.50
3060	4" pipe		20	.400		12.15	21		33.15	44.50
4000	Lead, 4#, 8" skirt, VTR									
4100	2" pipe	1 Plum	18	.444	Ea.	38	23		61	76
4110	3" pipe		17	.471		45	24.50		69.50	86
4120	4" pipe		16	.500		51.50	26		77.50	95.50
4130	6" pipe		14	.571		73.50	29.50		103	126

22 13 26 – Sanitary Waste Separators

22 13 26.10 Separators

		Crew	Daily Output	Labor-Hours	Unit	Material	Labor	Equipment	Total	Total Incl O&P
0010	**SEPARATORS**, Entrainment eliminator, steel body, 150 PSIG.									
0100	1/4" size	1 Stpi	24	.333	Ea.	246	17.30		263.30	297
0120	1/2" size		24	.333		254	17.30		271.30	305
0140	3/4" size		20	.400		266	21		287	325
0160	1" size		19	.421		274	22		296	335
0180	1-1/4" size		15	.533		291	27.50		318.50	360
0200	1-1/2" size		13	.615		320	32		352	405
0220	2" size		11	.727		355	38		393	450
0240	2-1/2" size	Q-5	15	1.067		1,775	50		1,825	2,025
0260	3" size		13	1.231		2,050	57.50		2,107.50	2,325
0280	4" size		10	1.600		2,275	74.50		2,349.50	2,600
0300	5" size		6	2.667		2,600	125		2,725	3,025
0320	6" size		3	5.333		2,850	249		3,099	3,525
0340	8" size	Q-6	4.40	5.455		3,500	264		3,764	4,250
0360	10" size	"	4	6		5,650	291		5,941	6,650
1000	For 300 PSIG, add					15%				

22 13 29 – Sanitary Sewerage Pumps

22 13 29.13 Wet-Pit-Mounted, Vertical Sewerage Pumps

		Crew	Daily Output	Labor-Hours	Unit	Material	Labor	Equipment	Total	Total Incl O&P
0010	**WET-PIT-MOUNTED, VERTICAL SEWERAGE PUMPS**									
0020	Controls incl. alarm/disconnect panel w/wire. Excavation not included									
0260	Simplex, 9 GPM at 60 PSIG, 91 gal. tank				Ea.	2,975			2,975	3,275
0300	Unit with manway, 26" I.D., 18" high					3,375			3,375	3,700
0340	26" I.D., 36" high					3,500			3,500	3,850

22 13 29 – Sanitary Sewerage Pumps

22 13 29.13 Wet-Pit-Mounted, Vertical Sewerage Pumps		Crew	Daily Output	Labor-Hours	Unit	2010 Bare Costs Material	Labor	Equipment	Total	Total Incl O&P
0380	43" I.D., 4' high				Ea.	3,575			3,575	3,925
0600	Simplex, 9 GPM at 60 PSIG, 150 gal. tank, indoor					3,175			3,175	3,475
0700	Unit with manway, 26" I.D., 36" high					3,875			3,875	4,275
0740	26" I.D., 4' high					4,000			4,000	4,400
2000	Duplex, 18 GPM at 60 PSIG, 150 gal. tank, indoor					6,125			6,125	6,725
2060	Unit with manway, 43" I.D., 4' high					7,125			7,125	7,825
2400	For core only					1,600			1,600	1,750
3000	Indoor residential type installation									
3020	Simplex, 9 GPM at 60 PSIG, 91 gal. HDPE tank				Ea.	2,950			2,950	3,225

22 13 29.14 Sewage Ejector Pumps		Crew	Daily Output	Labor-Hours	Unit	Material	Labor	Equipment	Total	Total Incl O&P
0010	**SEWAGE EJECTOR PUMPS**, With operating and level controls									
0100	Simplex system incl. tank, cover, pump 15' head									
0500	37 gal PE tank, 12 GPM, 1/2 HP, 2" discharge	Q-1	3.20	5	Ea.	440	234		674	835
0510	3" discharge		3.10	5.161		480	242		722	885
0530	87 GPM, .7 HP, 2" discharge		3.20	5		670	234		904	1,075
0540	3" discharge		3.10	5.161		725	242		967	1,150
0600	45 gal. coated stl tank, 12 GPM, 1/2 HP, 2" discharge		3	5.333		780	250		1,030	1,225
0610	3" discharge		2.90	5.517		810	258		1,068	1,275
0630	87 GPM, .7 HP, 2" discharge		3	5.333		1,000	250		1,250	1,475
0640	3" discharge		2.90	5.517		1,050	258		1,308	1,525
0660	134 GPM, 1 HP, 2" discharge		2.80	5.714		1,075	268		1,343	1,600
0680	3" discharge		2.70	5.926		1,150	278		1,428	1,675
0700	70 gal. PE tank, 12 GPM, 1/2 HP, 2" discharge		2.60	6.154		855	288		1,143	1,375
0710	3" discharge		2.40	6.667		910	310		1,220	1,475
0730	87 GPM, 0.7 HP, 2" discharge		2.50	6.400		1,100	300		1,400	1,650
0740	3" discharge		2.30	6.957		1,150	325		1,475	1,775
0760	134 GPM, 1 HP, 2" discharge		2.20	7.273		1,200	340		1,540	1,800
0770	3" discharge		2	8		1,275	375		1,650	1,950
0800	75 gal. coated stl. tank, 12 GPM, 1/2 HP, 2 " discharge		2.40	6.667		935	310		1,245	1,500
0810	3" discharge		2.20	7.273		970	340		1,310	1,575
0830	87 GPM, .7 HP, 2" discharge		2.30	6.957		1,175	325		1,500	1,800
0840	3" discharge		2.10	7.619		1,225	355		1,580	1,875
0860	134 GPM, 1 HP, 2" discharge		2	8		1,250	375		1,625	1,925
0880	3" discharge		1.80	8.889		1,325	415		1,740	2,075
1040	Duplex system incl. tank, covers, pumps									
1060	110 gal. fiberglass tank, 24 GPM, 1/2 HP, 2" discharge	Q-1	1.60	10	Ea.	1,700	470		2,170	2,575
1080	3" discharge		1.40	11.429		1,800	535		2,335	2,775
1100	174 GPM, .7 HP, 2" discharge		1.50	10.667		2,200	500		2,700	3,150
1120	3" discharge		1.30	12.308		2,275	575		2,850	3,375
1140	268 GPM, 1 HP, 2" discharge		1.20	13.333		2,375	625		3,000	3,550
1160	3" discharge		1	16		2,475	750		3,225	3,825
1260	135 gal. coated stl. tank, 24 GPM, 1/2 HP, 2" discharge	Q-2	1.70	14.118		1,750	685		2,435	2,950
2000	3" discharge		1.60	15		1,850	730		2,580	3,125
2640	174 GPM, .7 HP, 2" discharge		1.60	15		2,275	730		3,005	3,600
2660	3" discharge		1.50	16		2,400	775		3,175	3,800
2700	268 GPM, 1 HP, 2" discharge		1.30	18.462		2,475	895		3,370	4,075
3040	3" discharge		1.10	21.818		2,625	1,050		3,675	4,500
3060	275 gal. coated stl. tank, 24 GPM, 1/2 HP, 2" discharge		1.50	16		2,175	775		2,950	3,575
3080	3" discharge		1.40	17.143		2,200	835		3,035	3,675
3100	174 GPM, .7 HP, 2" discharge		1.40	17.143		2,800	835		3,635	4,325
3120	3" discharge		1.30	18.462		2,975	895		3,870	4,625
3140	268 GPM, 1 HP, 2" discharge		1.10	21.818		3,075	1,050		4,125	4,975

22 13 Facility Sanitary Sewerage

22 13 29 – Sanitary Sewerage Pumps

22 13 29.14 Sewage Ejector Pumps		Crew	Daily Output	Labor-Hours	Unit	2010 Bare Costs Material	Labor	Equipment	Total	Total Incl O&P
3160	3″ discharge	Q-2	.90	26.667	Ea.	3,250	1,300		4,550	5,525
3260	Pump system accessories, add									
3300	Alarm horn and lights, 115V mercury switch	Q-1	8	2	Ea.	93	93.50		186.50	242
3340	Switch, mag. contactor, alarm bell, light, 3 level control		5	3.200		455	150		605	730
3380	Alternator, mercury switch activated		4	4		825	187		1,012	1,175

22 14 Facility Storm Drainage

22 14 29 – Sump Pumps

22 14 29.13 Wet-Pit-Mounted, Vertical Sump Pumps

		Crew	Daily Output	Labor-Hours	Unit	Material	Labor	Equipment	Total	Total Incl O&P
0010	**WET-PIT-MOUNTED, VERTICAL SUMP PUMPS**									
0400	Molded PVC base, 21 GPM at 15′ head, 1/3 HP	1 Plum	5	1.600	Ea.	119	83.50		202.50	256
0800	Iron base, 21 GPM at 15′ head, 1/3 HP		5	1.600		150	83.50		233.50	290
1200	Solid brass, 21 GPM at 15′ head, 1/3 HP		5	1.600		263	83.50		346.50	415

22 14 29.16 Submersible Sump Pumps

		Crew	Daily Output	Labor-Hours	Unit	Material	Labor	Equipment	Total	Total Incl O&P
0010	**SUBMERSIBLE SUMP PUMPS**									
7000	Sump pump, automatic									
7100	Plastic, 1-1/4″ discharge, 1/4 HP	1 Plum	6	1.333	Ea.	215	69.50		284.50	340
7140	1/3 HP		5	1.600		175	83.50		258.50	315
7160	1/2 HP		5	1.600		215	83.50		298.50	360
7180	1-1/2″ discharge, 1/2 HP		4	2		238	104		342	420
7500	Cast iron, 1-1/4″ discharge, 1/4 HP		6	1.333		166	69.50		235.50	286
7540	1/3 HP		6	1.333		195	69.50		264.50	320
7560	1/2 HP		5	1.600		236	83.50		319.50	385

22 15 General Service Compressed-Air Systems

22 15 13 – General Service Compressed-Air Piping

22 15 13.10 Compressor Accessories

		Crew	Daily Output	Labor-Hours	Unit	Material	Labor	Equipment	Total	Total Incl O&P
0010	**COMPRESSOR ACCESSORIES**									
1700	Refrigerated air dryers with ambient air filters									
1710	10 CFM	Q-5	8	2	Ea.	675	93.50		768.50	885
1720	25 CFM		6.60	2.424		920	113		1,033	1,200
1730	50 CFM		6.20	2.581		1,300	121		1,421	1,600
1740	75 CFM		5.80	2.759		2,375	129		2,504	2,825
1750	100 CFM		5.60	2.857		2,700	133		2,833	3,175
4000	Couplers, air line, sleeve type									
4010	Female, connection size NPT									
4020	1/4″	1 Stpi	38	.211	Ea.	8.15	10.95		19.10	25.50
4030	3/8″		36	.222		13.65	11.55		25.20	32.50
4040	1/2″		35	.229		19.85	11.85		31.70	40
4050	3/4″		34	.235		19.85	12.20		32.05	40.50
4100	Male									
4110	1/4″	1 Stpi	38	.211	Ea.	8.50	10.95		19.45	26
4120	3/8″		36	.222		9.35	11.55		20.90	27.50
4130	1/2″		35	.229		19.85	11.85		31.70	40
4140	3/4″		34	.235		19.85	12.20		32.05	40.50
4150	Coupler, combined male and female halves									
4160	1/2″	1 Stpi	17	.471	Ea.	39.50	24.50		64	80
4170	3/4″	″	15	.533	″	39.50	27.50		67	85

22 15 19 – General Service Packaged Air Compressors and Receivers

22 15 19.10 Air Compressors		Crew	Daily Output	Labor-Hours	Unit	Material	2010 Bare Costs Labor	Equipment	Total	Total Incl O&P
0010	**AIR COMPRESSORS**									
5250	Air, reciprocating air cooled, splash lubricated, tank mounted									
5300	Single stage, 1 phase, 140 psi									
5303	1/2 HP, 30 gal tank	1 Stpi	3	2.667	Ea.	1,600	138		1,738	1,975
5305	3/4 HP, 30 gal tank		2.60	3.077		1,650	160		1,810	2,075
5307	1 HP, 30 gal tank		2.20	3.636		2,000	189		2,189	2,475
5309	2 HP, 30 gal tank	Q-5	4	4		2,475	187		2,662	3,000
5310	3 HP, 30 gal tank		3.60	4.444		2,925	208		3,133	3,525
5314	3 HP, 60 gal tank		3.50	4.571		3,300	213		3,513	3,950
5320	5 HP, 60 gal tank		3.20	5		3,475	234		3,709	4,175
5330	5 HP, 80 gal tank		3	5.333		3,725	249		3,974	4,475
5340	7.5 HP, 80 gal tank		2.60	6.154		4,875	287		5,162	5,800
5600	2 stage pkg., 3 phase									
5650	6 CFM at 125 psi 1-1/2 HP, 60 gal tank	Q-5	3	5.333	Ea.	2,950	249		3,199	3,625
5670	10.9 CFM at 125 psi, 3 HP, 80 gal tank		1.50	10.667		3,500	500		4,000	4,600
5680	38.7 CFM at 125 psi, 10 HP, 120 gal tank		.60	26.667		5,925	1,250		7,175	8,400
5690	105 CFM at 125 psi, 25 HP, 250 gal tank	Q-6	.60	40		12,300	1,925		14,225	16,400
5800	With single stage pump									
5850	8.3 CFM at 125 psi, 2 HP, 80 gal tank	Q-6	3.50	6.857	Ea.	3,425	330		3,755	4,275
5860	38.7 CFM at 125 psi, 10 HP, 120 gal tank	″	.90	26.667	″	5,925	1,300		7,225	8,450
6000	Reciprocating, 2 stage, tank mtd, 3 Ph., Cap. rated @175 PSIG									
6050	Pressure lubricated, hvy duty, 9.7 CFM, 3 HP, 120 Gal. tank	Q-5	1.30	12.308	Ea.	5,225	575		5,800	6,600
6054	5 CFM, 1-1/2 HP, 80 gal tank		2.80	5.714		3,150	267		3,417	3,875
6056	6.4 CFM, 2 HP, 80 gal tank		2	8		3,425	375		3,800	4,325
6058	8.1 CFM, 3 HP, 80 gal tank		1.70	9.412		3,500	440		3,940	4,500
6059	14.8 CFM, 5 HP, 80 gal tank		1	16		3,725	745		4,470	5,225
6060	16.5 CFM, 5 HP, 120 gal. tank		1	16		5,325	745		6,070	7,000
6063	13 CFM, 6 HP, 80 gal tank		.90	17.778		5,000	830		5,830	6,725
6066	19.8 CFM, 7.5 HP, 80 gal tank		.80	20		5,425	935		6,360	7,350
6070	25.8 CFM, 7-1/2 HP, 120 gal. tank		.80	20		7,800	935		8,735	9,975
6078	34.8 CFM, 10 HP, 80 gal. tank		.70	22.857		6,975	1,075		8,050	9,275
6080	34.8 CFM, 10 HP, 120 gal. tank		.60	26.667		7,375	1,250		8,625	10,000
6090	53.7 CFM, 15 HP, 120 gal. tank	Q-6	.80	30		8,900	1,450		10,350	12,000
6100	76.7 CFM, 20 HP, 120 gal. tank		.70	34.286		11,700	1,650		13,350	15,400
6104	76.7 CFM, 20 HP, 240 gal tank		.68	35.294		12,800	1,700		14,500	16,700
6110	90.1 CFM, 25 HP, 120 gal. tank		.63	38.095		12,200	1,850		14,050	16,300
6120	101 CFM, 30 HP, 120 gal. tank		.57	42.105		13,100	2,050		15,150	17,600
6130	101 CFM, 30 HP, 250 gal. tank		.52	46.154		14,300	2,225		16,525	19,100
6200	Oil-less, 13.6 CFM, 5 HP, 120 gal. tank	Q-5	.88	18.182		14,600	850		15,450	17,400
6210	13.6 CFM, 5 HP, 250 gal. tank		.80	20		16,000	935		16,935	19,000
6220	18.2 CFM, 7.5 HP, 120 gal. tank		.73	21.918		15,100	1,025		16,125	18,100
6230	18.2 CFM, 7.5 HP, 250 gal. tank		.67	23.881		16,200	1,125		17,325	19,500
6250	30.5 CFM, 10 HP, 120 gal. tank		.57	28.070		17,600	1,300		18,900	21,300
6260	30.5 CFM, 10 HP, 250 gal. tank		.53	30.189		18,700	1,400		20,100	22,700
6270	41.3 CFM, 15 HP, 120 gal. tank	Q-6	.70	34.286		19,100	1,650		20,750	23,500
6280	41.3 CFM, 15 HP, 250 gal. tank	″	.67	35.821		20,200	1,725		21,925	24,800

22 31 Domestic Water Softeners

22 31 13 – Residential Domestic Water Softeners

22 31 13.10 Residential Water Softeners		Crew	Daily Output	Labor-Hours	Unit	Material	2010 Bare Costs Labor	Equipment	Total	Total Incl O&P
0010	**RESIDENTIAL WATER SOFTENERS**									
7350	Water softener, automatic, to 30 grains per gallon	2 Plum	5	3.200	Ea.	380	167		547	670
7400	To 100 grains per gallon	"	4	4	"	750	208		958	1,125

22 35 Domestic Water Heat Exchangers

22 35 30 – Water Heating by Steam

22 35 30.10 Water Heating Transfer Package

		Crew	Daily Output	Labor-Hours	Unit	Material	Labor	Equipment	Total	Total Incl O&P
0010	**WATER HEATING TRANSFER PACKAGE**, Complete controls,									
0020	expansion tank, converter, air separator									
1000	Hot water, 180°F enter, 200°F leaving, 15# steam									
1010	One pump system, 28 GPM	Q-6	.75	32	Ea.	15,800	1,550		17,350	19,700
1020	35 GPM		.70	34.286		17,200	1,650		18,850	21,400
1040	55 GPM		.65	36.923		20,300	1,800		22,100	25,000
1060	130 GPM		.55	43.636		25,000	2,125		27,125	30,700
1080	255 GPM		.40	60		33,800	2,900		36,700	41,600
1100	550 GPM		.30	80		42,700	3,875		46,575	53,000
1120	800 GPM		.25	96		50,500	4,650		55,150	63,000
1220	Two pump system, 28 GPM		.70	34.286		21,400	1,650		23,050	26,100
1240	35 GPM		.65	36.923		25,000	1,800		26,800	30,200
1260	55 GPM		.60	40		27,000	1,925		28,925	32,600
1280	130 GPM		.50	48		34,100	2,325		36,425	41,000
1300	255 GPM		.35	68.571		45,500	3,325		48,825	55,000
1320	550 GPM		.25	96		57,000	4,650		61,650	69,500
1340	800 GPM	↓	.20	120	↓	66,000	5,800		71,800	81,000

22 41 Residential Plumbing Fixtures

22 41 39 – Residential Faucets, Supplies and Trim

22 41 39.10 Faucets and Fittings

		Crew	Daily Output	Labor-Hours	Unit	Material	Labor	Equipment	Total	Total Incl O&P
0010	**FAUCETS AND FITTINGS**									
5000	Sillcock, compact, brass, IPS or copper to hose	1 Plum	24	.333	Ea.	8.05	17.35		25.40	35
6000	Stop and waste valves, bronze									
6100	Angle, solder end 1/2"	1 Plum	24	.333	Ea.	8.85	17.35		26.20	35.50
6110	3/4"		20	.400		10.35	21		31.35	42.50
6300	Straightway, solder end 3/8"		24	.333		6.15	17.35		23.50	33
6310	1/2"		24	.333		2.81	17.35		20.16	29
6320	3/4"		20	.400		3.48	21		24.48	35
6410	Straightway, threaded 1/2"		24	.333		3.82	17.35		21.17	30
6420	3/4"		20	.400		6.35	21		27.35	38
6430	1"	↓	19	.421	↓	12.60	22		34.60	47

22 51 Swimming Pool Plumbing Systems

22 51 19 – Swimming Pool Water Treatment Equipment

22 51 19.50 Swimming Pool Filtration Equipment		Crew	Daily Output	Labor-Hours	Unit	Material	2010 Bare Costs Labor	Equipment	Total	Total Incl O&P
0010	SWIMMING POOL FILTRATION EQUIPMENT									
0900	Filter system, sand or diatomite type, incl. pump, 6,000 gal./hr.	2 Plum	1.80	8.889	Total	1,700	465		2,165	2,575
1020	Add for chlorination system, 800 S.F. pool		3	5.333	Ea.	103	278		381	530
1040	5,000 S.F. pool	↓	3	5.333	"	2,050	278		2,328	2,675

22 52 Fountain Plumbing Systems

22 52 16 – Fountain Pumps

22 52 16.10 Fountain Water Pumps

		Crew	Daily Output	Labor-Hours	Unit	Material	Labor	Equipment	Total	Total Incl O&P
0010	FOUNTAIN WATER PUMPS									
0100	Pump w/controls									
0200	Single phase, 100' cord, 1/2 H.P. pump	2 Skwk	4.40	3.636	Ea.	1,825	155		1,980	2,250
0300	3/4 H.P. pump		4.30	3.721		2,075	159		2,234	2,525
0400	1 H.P. pump		4.20	3.810		3,850	162		4,012	4,475
0500	1-1/2 H.P. pump		4.10	3.902		3,525	166		3,691	4,150
0600	2 H.P. pump		4	4		4,775	170		4,945	5,500
0700	Three phase, 200' cord, 5 H.P. pump		3.90	4.103		5,050	175		5,225	5,825
0800	7-1/2 H.P. pump		3.80	4.211		8,575	179		8,754	9,725
0900	10 H.P. pump		3.70	4.324		9,375	184		9,559	10,600
1000	15 H.P. pump	↓	3.60	4.444	↓	11,400	189		11,589	12,900
2000	DESIGN NOTE: Use two horsepower per surface acre.									

22 52 33 – Fountain Ancillary

22 52 33.10 Fountain Miscellaneous

		Crew	Daily Output	Labor-Hours	Unit	Material	Labor	Equipment	Total	Total Incl O&P
0010	FOUNTAIN MISCELLANEOUS									
1100	Nozzles, minimum	2 Skwk	8	2	Ea.	151	85		236	298
1200	Maximum		8	2		305	85		390	465
1300	Lights w/mounting kits, 200 watt		18	.889		1,000	38		1,038	1,150
1400	300 watt		18	.889		1,200	38		1,238	1,375
1500	500 watt		18	.889		1,325	38		1,363	1,500
1600	Color blender	↓	12	1.333	↓	535	57		592	680

22 62 Vacuum Systems for Laboratory and Healthcare Facilities

22 62 19 – Vacuum Equipment for Laboratory and Healthcare Facilities

22 62 19.70 Healthcare Vacuum Equipment

		Crew	Daily Output	Labor-Hours	Unit	Material	Labor	Equipment	Total	Total Incl O&P
0010	HEALTHCARE VACUUM EQUIPMENT									
0100	Medical, with receiver									
0110	Duplex									
0120	20 SCFM	Q-1	1.14	14.035	Ea.	16,000	660		16,660	18,600
0130	60 SCFM, 10 HP	Q-2	1.20	20	"	29,000	970		29,970	33,400
0200	Triplex									
0220	180 SCFM	Q-2	.86	27.907	Ea.	45,000	1,350		46,350	51,500
0300	Dental oral									
0310	Duplex									
0330	165 SCFM with 77 Gal separator	Q-2	1.30	18.462	Ea.	47,000	895		47,895	53,000
1100	Vacuum system									
1110	Vacuum outlet alarm panel	1 Plum	3.20	2.500	Ea.	1,000	130		1,130	1,300

22 63 Gas Systems for Laboratory and Healthcare Facilities

22 63 13 – Gas Piping for Laboratory and Healthcare Facilities

22 63 13.70 Healthcare Gas Piping		Crew	Daily Output	Labor-Hours	Unit	2010 Bare Costs Material	Labor	Equipment	Total	Total Incl O&P
0010	**HEALTHCARE GAS PIPING**									
1000	Nitrogen or oxygen system									
1010	Cylinder manifold									
1020	5 cylinder	1 Plum	.80	10	Ea.	5,400	520		5,920	6,725
1026	10 cylinder	"	.40	20	"	6,500	1,050		7,550	8,700
3000	Outlets and valves									
3010	Recessed, wall mounted									
3012	Single outlet	1 Plum	3.20	2.500	Ea.	65	130		195	267
3100	Ceiling outlet									
3200	Zone valve with box									
3210	2"	1 Plum	3.20	2.500	Ea.	600	130		730	855
4000	Alarm panel, medical gases and vacuum									
4010	Alarm panel	1 Plum	3.20	2.500	Ea.	1,000	130		1,130	1,300

Estimating Tips

The labor adjustment factors listed in Subdivision 22 01 02.20 also apply to Division 23.

23 10 00 Facility Fuel Systems

- The prices in this subdivision for above- and below-ground storage tanks do not include foundations or hold-down slabs, unless noted. The estimator should refer to Divisions 3 and 31 for foundation system pricing. In addition to the foundations, required tank accessories, such as tank gauges, leak detection devices, and additional manholes and piping, must be added to the tank prices.

23 50 00 Central Heating Equipment

- When estimating the cost of an HVAC system, check to see who is responsible for providing and installing the temperature control system. It is possible to overlook controls, assuming that they would be included in the electrical estimate.
- When looking up a boiler, be careful on specified capacity. Some manufacturers rate their products on output while others use input.
- Include HVAC insulation for pipe, boiler, and duct (wrap and liner).
- Be careful when looking up mechanical items to get the correct pressure rating and connection type (thread, weld, flange).

23 70 00 Central HVAC Equipment

- Combination heating and cooling units are sized by the air conditioning requirements. (See Reference No. R236000-20 for preliminary sizing guide.)
- A ton of air conditioning is nominally 400 CFM.
- Rectangular duct is taken off by the linear foot for each size, but its cost is usually estimated by the pound. Remember that SMACNA standards now base duct on internal pressure.
- Prefabricated duct is estimated and purchased like pipe: straight sections and fittings.
- Note that cranes or other lifting equipment are not included on any lines in Division 23. For example, if a crane is required to lift a heavy piece of pipe into place high above a gym floor, or to put a rooftop unit on the roof of a four-story building, etc., it must be added. Due to the potential for extreme variation—from nothing additional required to a major crane or helicopter—we feel that including a nominal amount for "lifting contingency" would be useless and detract from the accuracy of the estimate. When using equipment rental cost data from RSMeans, do not forget to include the cost of the operator(s).

Reference Numbers

Reference numbers are shown in shaded boxes at the beginning of some major classifications. These numbers refer to related items in the Reference Section. The reference information may be an estimating procedure, an alternate pricing method, or technical information.

Note: Not all subdivisions listed here necessarily appear in this publication.

Note: **Trade Service,** *in part, has been used as a reference source for some of the material prices used in Division 23.*

23 05 05 – Selective HVAC Demolition

23 05 05.10 HVAC Demolition		Crew	Daily Output	Labor-Hours	Unit	Material	2010 Bare Costs Labor	Equipment	Total	Total Incl O&P
0010	**HVAC DEMOLITION** R220105-10									
0100	Air conditioner, split unit, 3 ton	Q-5	2	8	Ea.		375		375	560
0150	Package unit, 3 ton	Q-6	3	8			385		385	580
0190	Rooftop, self contained, up to 5 ton	1 Plum	1.20	6.667			345		345	520
0250	Air curtain	Q-9	20	.800	L.F.		35.50		35.50	53.50
0254	Air filters, up thru 16,000 CFM		20	.800	Ea.		35.50		35.50	53.50
0256	20,000 thru 60,000 CFM		16	1			44		44	67
0297	Boiler blowdown	Q-5	8	2			93.50		93.50	140
0298	Boilers									
0300	Electric, up thru 148 kW	Q-19	2	12	Ea.		570		570	850
0310	150 thru 518 kW	"	1	24			1,150		1,150	1,700
0320	550 thru 2000 kW	Q-21	.40	80			3,875		3,875	5,825
0330	2070 kW and up	"	.30	106			5,175		5,175	7,750
0340	Gas and/or oil, up thru 150 MBH	Q-7	2.20	14.545			720		720	1,075
0350	160 thru 2000 MBH		.80	40			1,975		1,975	2,975
0360	2100 thru 4500 MBH		.50	64			3,175		3,175	4,750
0370	4600 thru 7000 MBH		.30	106			5,275		5,275	7,900
0380	7100 thru 12,000 MBH		.16	200			9,875		9,875	14,800
0390	12,200 thru 25,000 MBH		.12	266			13,200		13,200	19,800
0400	Central station air handler unit, up thru 15 ton	Q-5	1.60	10			465		465	700
0410	17.5 thru 30 ton	"	.80	20			935		935	1,400
0430	Computer room unit									
0434	Air cooled split, up thru 10 ton	Q-5	.67	23.881	Ea.		1,125		1,125	1,675
0436	12 thru 23 ton		.53	30.189			1,400		1,400	2,125
0440	Chilled water, up thru 10 ton		1.30	12.308			575		575	860
0444	12 thru 23 ton		1	16			745		745	1,125
0450	Glycol system, up thru 10 ton		.53	30.189			1,400		1,400	2,125
0454	12 thru 23 ton		.40	40			1,875		1,875	2,800
0460	Water cooled, not including condenser, up thru 10 ton		.80	20			935		935	1,400
0464	12 thru 23 ton		.60	26.667			1,250		1,250	1,875
0600	Condenser, up thru 50 ton		1	16			745		745	1,125
0610	51 thru 100 ton	Q-6	.80	30			1,450		1,450	2,175
0620	101 thru 1000 ton	"	.16	150			7,275		7,275	10,900
0660	Condensing unit, up thru 10 ton	Q-5	1.25	12.800			600		600	895
0670	11 thru 50 ton	"	.40	40			1,875		1,875	2,800
0680	60 thru 100 ton	Q-6	.30	80			3,875		3,875	5,800
0700	Cooling tower, up thru 400 ton		.80	30			1,450		1,450	2,175
0710	450 thru 600 ton		.53	45.283			2,200		2,200	3,300
0720	700 thru 1300 ton		.40	60			2,900		2,900	4,350
0780	Dehumidifier, up thru 155 lb./hr.	Q-1	8	2			93.50		93.50	140
0790	240 lb./hr. and up	"	2	8			375		375	560
1560	Ductwork									
1570	Metal, steel, sst, fabricated	Q-9	1000	.016	Lb.		.71		.71	1.07
1580	Aluminum, fabricated		485	.033	"		1.46		1.46	2.21
1590	Spiral, prefabricated		400	.040	L.F.		1.77		1.77	2.68
1600	Fiberglass, prefabricated		400	.040			1.77		1.77	2.68
1610	Flex, prefabricated		500	.032			1.41		1.41	2.14
1620	Glass fiber reinforced plastic, prefabricated		280	.057			2.53		2.53	3.82
1630	Diffusers, registers or grills, up thru 20" max dimension	1 Shee	50	.160	Ea.		7.85		7.85	11.90
1640	21 thru 36" max dimension		36	.222			10.90		10.90	16.50
1650	Above 36" max dimension		30	.267			13.10		13.10	19.85
1700	Evaporator, up thru 12,000 BTUH	Q-5	5.30	3.019			141		141	211
1710	12,500 thru 30,000 BTUH	"	2.70	5.926			277		277	415

23 05 05 – Selective HVAC Demolition

23 05 05.10 HVAC Demolition		Crew	Daily Output	Labor-Hours	Unit	Material	2010 Bare Costs Labor	Equipment	Total	Total Incl O&P
1720	31,000 BTUH and up	Q-6	1.50	16	Ea.		775		775	1,150
1730	Evaporative cooler, up thru 5 H.P.	Q-9	2.70	5.926			262		262	395
1740	10 thru 30 H.P.	"	.67	23.881			1,050		1,050	1,600
1750	Exhaust systems									
1760	Exhaust components	1 Shee	8	1	System		49		49	74.50
1770	Weld fume hoods	"	20	.400	Ea.		19.65		19.65	29.50
2120	Fans, up thru 1 H.P. or 2000 CFM	Q-9	8	2			88.50		88.50	134
2124	1-1/2 thru 10 H.P. or 20,000 CFM		5.30	3.019			133		133	202
2128	15 thru 30 H.P. or above 20,000 CFM		4	4			177		177	268
2150	Fan coil air conditioner, chilled water, up thru 7.5 ton	Q-5	14	1.143			53.50		53.50	80
2154	Direct expansion, up thru 10 ton		8	2			93.50		93.50	140
2158	11 thru 30 ton		2	8			375		375	560
2170	Flue shutter damper	Q-9	8	2			88.50		88.50	134
2200	Furnace, electric	Q-20	2	10			450		450	680
2300	Gas or oil, under 120 MBH	Q-9	4	4			177		177	268
2340	Over 120 MBH	"	3	5.333			236		236	355
2730	Heating and ventilating unit	Q-5	2.70	5.926			277		277	415
2740	Heater, electric, wall, baseboard and quartz	1 Elec	10	.800			39		39	58.50
2750	Heater, electric, unit, cabinet, fan and convector	"	8	1			49		49	73
2760	Heat exchanger, shell and tube type	Q-5	1.60	10			465		465	700
2770	Plate type	Q-6	.60	40			1,925		1,925	2,900
2810	Heat pump									
2820	Air source, split, up thru 10 ton	Q-5	.90	17.778	Ea.		830		830	1,250
2830	15 thru 25 ton	Q-6	.80	30			1,450		1,450	2,175
2850	Single package, up thru 12 ton	Q-5	1	16			745		745	1,125
2860	Water source, up thru 15 ton	"	.90	17.778			830		830	1,250
2870	20 thru 50 ton	Q-6	.80	30			1,450		1,450	2,175
2910	Heat recovery package, up thru 20,000 CFM	Q-5	2	8			375		375	560
2920	25,000 CFM and up		1.20	13.333			625		625	935
2930	Heat transfer package, up thru 130 GPM		.80	20			935		935	1,400
2934	255 thru 800 GPM		.42	38.095			1,775		1,775	2,675
2940	Humidifier		10.60	1.509			70.50		70.50	106
2961	Hydronic unit heaters, up thru 200 MBH		14	1.143			53.50		53.50	80
2962	Above 200 MBH		8	2			93.50		93.50	140
2964	Valance units		32	.500			23.50		23.50	35
2966	Radiant floor heating									
2967	System valves, controls, manifolds	Q-5	16	1	Ea.		46.50		46.50	70
2968	Per room distribution		8	2			93.50		93.50	140
2970	Hydronic heating, baseboard radiation		16	1			46.50		46.50	70
2976	Convectors and free standing radiators		18	.889			41.50		41.50	62
2980	Induced draft fan, up thru 1 H.P.	Q-9	4.60	3.478			154		154	233
2984	1-1/2 H.P. thru 7-1/2 H.P.	"	2.20	7.273			320		320	485
2988	Infra-red unit	Q-5	16	1			46.50		46.50	70
2992	Louvers	1 Shee	46	.174	S.F.		8.55		8.55	12.95
3000	Mechanical equipment, light items. Unit is weight, not cooling.	Q-5	.90	17.778	Ton		830		830	1,250
3600	Heavy items	"	1.10	14.545	"		680		680	1,025
3700	Deduct for salvage (when applicable), minimum				Job				73	80
3710	Maximum				"				455	500
3720	Make up air unit, up thru 6000 CFM	Q-5	3	5.333	Ea.		249		249	375
3730	6500 thru 30,000 CFM	"	1.60	10			465		465	700
3740	35,000 thru 75,000 CFM	Q-6	1	24			1,150		1,150	1,750
3800	Mixing boxes, constant and VAV	Q-9	18	.889			39.50		39.50	59.50
4000	Packaged terminal air conditioner, up thru 18,000 BTUH	Q-5	8	2			93.50		93.50	140

23 05 05 – Selective HVAC Demolition

23 05 05.10 HVAC Demolition		Crew	Daily Output	Labor-Hours	Unit	Material	Labor	Equipment	Total	Total Incl O&P
						2010 Bare Costs				
4010	24,000 thru 48,000 BTUH	Q-5	2.80	5.714	Ea.		267		267	400
5000	Refrigerant compressor, reciprocating or scroll									
5010	Up thru 5 ton	1 Stpi	6	1.333	Ea.		69		69	104
5020	5.08 thru 10 ton	Q-5	6	2.667			125		125	187
5030	15 thru 50 ton	"	.40	40			1,875		1,875	2,800
5040	60 thru 130 ton	Q-6	.48	50			2,425		2,425	3,625
5100	Roof top air conditioner, up thru 10 ton	Q-5	1.40	11.429			535		535	800
5110	12 thru 40 ton	Q-6	1	24			1,150		1,150	1,750
5120	50 thru 140 ton		.50	48			2,325		2,325	3,475
5130	150 thru 300 ton		.30	80			3,875		3,875	5,800
6000	Self contained single package air conditioner, up thru 10 ton	Q-5	1.60	10			465		465	700
6010	15 thru 60 ton	Q-6	1.20	20			970		970	1,450
6100	Space heaters, up thru 200 MBH	Q-5	10	1.600			74.50		74.50	112
6110	Over 200 MBH		5	3.200			149		149	224
6200	Split ductless, both sections		8	2			93.50		93.50	140
6300	Steam condensate meter	1 Stpi	11	.727			38		38	56.50
6600	Thru-the-wall air conditioner	L-2	8	2			73		73	113
7000	Vent chimney, prefabricated, up thru 12" diameter	Q-9	94	.170	V.L.F.		7.50		7.50	11.40
7010	14" thru 36" diameter		40	.400			17.70		17.70	27
7020	38" thru 48" diameter		32	.500			22		22	33.50
7030	54" thru 60" diameter	Q-10	14	1.714			78.50		78.50	119
7400	Ventilators, up thru 14" neck diameter	Q-9	58	.276	Ea.		12.20		12.20	18.45
7410	16" thru 50" neck diameter		40	.400			17.70		17.70	27
7450	Relief vent, up thru 24" x 96"		22	.727			32		32	48.50
7460	48" x 60" thru 96" x 144"		10	1.600			70.50		70.50	107
8000	Water chiller up thru 10 ton	Q-5	2.50	6.400			299		299	450
8010	15 thru 100 ton	Q-6	.48	50			2,425		2,425	3,625
8020	110 thru 500 ton	Q-7	.29	110			5,450		5,450	8,175
8030	600 thru 1000 ton		.23	139			6,875		6,875	10,300
8040	1100 ton and up		.20	160			7,900		7,900	11,900
8400	Window air conditioner	1 Carp	16	.500			21		21	32

23 05 23 – General-Duty Valves for HVAC Piping

23 05 23.30 Valves, Iron Body		Crew	Daily Output	Labor-Hours	Unit	Material	Labor	Equipment	Total	Total Incl O&P
0010	**VALVES, IRON BODY** R220523-90									
0022	For grooved joint, see Div. 22 11 13.48									
0100	Angle, 125 lb.									
0110	Flanged									
0116	2"	1 Plum	5	1.600	Ea.	930	83.50		1,013.50	1,150
0118	4"	Q-1	3	5.333		1,550	250		1,800	2,075
0120	6"	Q-2	3	8		3,000	390		3,390	3,875
0122	8"	"	2.50	9.600		5,375	465		5,840	6,625
0560	Butterfly, lug type, pneumatic operator, 2" size	1 Stpi	14	.571		201	29.50		230.50	266
0570	3"	Q-1	8	2		213	93.50		306.50	375
0580	4"	"	5	3.200		233	150		383	480
0590	6"	Q-2	5	4.800		320	233		553	705
0600	8"		4.50	5.333		415	259		674	845
0610	10"		4	6		520	291		811	1,000
0620	12"		3	8		675	390		1,065	1,325
0630	14"		2.30	10.435		1,275	505		1,780	2,150
0640	18"		1.50	16		2,000	775		2,775	3,400
0650	20"		1	24		2,450	1,175		3,625	4,450
0790	Butterfly, lug type, gear operated, 2" size, 200 lb. except noted	1 Plum	14	.571		261	29.50		290.50	330

23 05 23.30 Valves, Iron Body		Crew	Daily Output	Labor-Hours	Unit	Material	Labor	Equipment	Total	Total Incl O&P
						2010 Bare Costs				
0800	2-1/2"	Q-1	9	1.778	Ea.	263	83.50		346.50	415
0810	3"		8	2		276	93.50		369.50	445
0820	4"		5	3.200		310	150		460	570
0830	5"	Q-2	5	4.800		390	233		623	775
0840	6"		5	4.800		430	233		663	825
0850	8"		4.50	5.333		555	259		814	1,000
0860	10"		4	6		735	291		1,026	1,250
0870	12"		3	8		1,075	390		1,465	1,750
0880	14", 150 lb.		2.30	10.435		2,050	505		2,555	3,000
0890	16", 150 lb.		1.75	13.714		2,625	665		3,290	3,875
0900	18", 150 lb.		1.50	16		3,725	775		4,500	5,275
0910	20", 150 lb.		1	24		4,700	1,175		5,875	6,925
0930	24", 150 lb.		.75	32		8,450	1,550		10,000	11,600
1020	Butterfly, wafer type, gear actuator, 200 lb.									
1030	2"	1 Plum	14	.571	Ea.	167	29.50		196.50	228
1040	2-1/2"	Q-1	9	1.778		172	83.50		255.50	315
1050	3"		8	2		178	93.50		271.50	335
1060	4"		5	3.200		202	150		352	450
1070	5"	Q-2	5	4.800		269	233		502	645
1080	6"		5	4.800		305	233		538	685
1090	8"		4.50	5.333		405	259		664	835
1100	10"		4	6		505	291		796	995
1110	12"		3	8		705	390		1,095	1,350
1200	Wafer type, lever actuator, 200 lb.									
1220	2"	1 Plum	14	.571	Ea.	116	29.50		145.50	172
1230	2-1/2"	Q-1	9	1.778		119	83.50		202.50	256
1240	3"		8	2		126	93.50		219.50	279
1250	4"		5	3.200		154	150		304	395
1260	5"	Q-2	5	4.800		213	233		446	585
1270	6"		5	4.800		257	233		490	635
1280	8"		4.50	5.333		380	259		639	805
1290	10"		4	6		530	291		821	1,025
1300	12"		3	8		955	390		1,345	1,625
1600	Gate, threaded, 125 lb.									
1603	1-1/2"	1 Plum	13	.615	Ea.	320	32		352	405
1604	2"		11	.727		770	38		808	900
1605	2-1/2"		10	.800		835	41.50		876.50	985
1606	3"		8	1		995	52		1,047	1,175
1607	4"		5	1.600		1,400	83.50		1,483.50	1,675
1650	Gate, 125 lb., N.R.S.									
2150	Flanged									
2200	2"	1 Plum	5	1.600	Ea.	525	83.50		608.50	705
2240	2-1/2"	Q-1	5	3.200		540	150		690	820
2260	3"		4.50	3.556		605	167		772	915
2280	4"		3	5.333		865	250		1,115	1,325
2290	5"	Q-2	3.40	7.059		1,475	345		1,820	2,150
2300	6"		3	8		1,475	390		1,865	2,200
2320	8"		2.50	9.600		2,525	465		2,990	3,475
2340	10"		2.20	10.909		4,450	530		4,980	5,700
2360	12"		1.70	14.118		6,100	685		6,785	7,750
2370	14"		1.30	18.462		8,125	895		9,020	10,300
2380	16"		1	24		11,300	1,175		12,475	14,300
2420	For 250 lb flanged, add					200%	10%			

23 05 23.30 Valves, Iron Body	Crew	Daily Output	Labor-Hours	Unit	Material	2010 Bare Costs Labor	Equipment	Total	Total Incl O&P
3500 OS&Y, 125 lb., (225# noted), threaded									
3504 3/4" (225 lb.)	1 Plum	18	.444	Ea.	440	23		463	520
3505 1" (225 lb.)		16	.500		500	26		526	590
3506 1-1/2" (225 lb.)		13	.615		730	32		762	850
3507 2"		11	.727		900	38		938	1,050
3550 OS&Y, 125 lb., flanged									
3600 2"	1 Plum	5	1.600	Ea.	320	83.50		403.50	475
3640 2-1/2"	Q-1	5	3.200		330	150		480	585
3660 3"		4.50	3.556		370	167		537	655
3670 3-1/2"		3	5.333		415	250		665	835
3680 4"		3	5.333		525	250		775	950
3690 5"	Q-2	3.40	7.059		865	345		1,210	1,475
3700 6"		3	8		865	390		1,255	1,550
3720 8"		2.50	9.600		1,550	465		2,015	2,400
3740 10"		2.20	10.909		2,825	530		3,355	3,900
3760 12"		1.70	14.118		3,750	685		4,435	5,150
3770 14"		1.30	18.462		7,175	895		8,070	9,250
3780 16"		1	24		10,800	1,175		11,975	13,600
3790 18"		.80	30		13,900	1,450		15,350	17,500
3800 20"		.60	40		20,100	1,950		22,050	25,000
3830 24"		.50	48		28,700	2,325		31,025	35,000
3900 For 175 lb, flanged, add					200%	10%			
4350 Globe, OS&Y									
4540 Class 125, flanged									
4550 2"	1 Plum	5	1.600	Ea.	650	83.50		733.50	840
4560 2-1/2"	Q-1	5	3.200		655	150		805	945
4570 3"		4.50	3.556		795	167		962	1,125
4580 4"		3	5.333		1,150	250		1,400	1,625
4590 5"	Q-2	3.40	7.059		2,075	345		2,420	2,800
4600 6"		3	8		2,075	390		2,465	2,850
4610 8"		2.50	9.600		4,075	465		4,540	5,175
4612 10"		2.20	10.909		6,100	530		6,630	7,525
4614 12"		1.70	14.118		6,450	685		7,135	8,100
5040 Class 250, flanged									
5050 2"	1 Plum	4.50	1.778	Ea.	1,050	92.50		1,142.50	1,300
5060 2-1/2"	Q-1	4.50	3.556		1,350	167		1,517	1,750
5070 3"		4	4		1,400	187		1,587	1,825
5080 4"		2.70	5.926		2,050	278		2,328	2,675
5090 5"	Q-2	3	8		3,675	390		4,065	4,625
5100 6"		2.70	8.889		3,675	430		4,105	4,700
5110 8"		2.20	10.909		6,250	530		6,780	7,675
5120 10"		2	12		9,475	585		10,060	11,300
5130 12"		1.60	15		14,200	730		14,930	16,700
5240 Valve sprocket rim w/chain, for 2" valve	1 Stpi	30	.267		118	13.85		131.85	151
5250 2-1/2" valve		27	.296		118	15.40		133.40	153
5260 3-1/2" valve		25	.320		118	16.60		134.60	155
5270 6" valve		20	.400		118	21		139	161
5280 8" valve		18	.444		158	23		181	209
5290 12" valve		16	.500		158	26		184	213
5300 16" valve		12	.667		218	34.50		252.50	292
5310 20" valve		10	.800		218	41.50		259.50	300
5320 36" valve		8	1		415	52		467	535
5450 Swing check, 125 lb., threaded									

23 05 23.30 Valves, Iron Body		Crew	Daily Output	Labor-Hours	Unit	Material	2010 Bare Costs Labor	Equipment	Total	Total Incl O&P
5470	1"	1 Plum	13	.615	Ea.	575	32		607	685
5500	2"	"	11	.727		675	38		713	795
5540	2-1/2"	Q-1	15	1.067		660	50		710	805
5550	3"		13	1.231		735	57.50		792.50	895
5560	4"		10	1.600		1,250	75		1,325	1,475
5950	Flanged									
5994	1"	1 Plum	7	1.143	Ea.	153	59.50		212.50	257
5998	1-1/2"		6	1.333		214	69.50		283.50	340
6000	2"		5	1.600		237	83.50		320.50	385
6040	2-1/2"	Q-1	5	3.200		265	150		415	515
6050	3"		4.50	3.556		285	167		452	565
6060	4"		3	5.333		445	250		695	865
6070	6"	Q-2	3	8		765	390		1,155	1,425
6080	8"		2.50	9.600		1,425	465		1,890	2,275
6090	10"		2.20	10.909		2,450	530		2,980	3,500
6100	12"		1.70	14.118		3,975	685		4,660	5,400
6110	18"		1.30	18.462		12,100	895		12,995	14,700
6114	24"		.75	32		22,900	1,550		24,450	27,500
6160	For 250 lb flanged, add					200%	20%			
6600	Silent check, bronze trim									
6610	Compact wafer type, for 125 or 150 lb. flanges									
6630	1-1/2"	1 Plum	11	.727	Ea.	117	38		155	186
6640	2"	"	9	.889		140	46.50		186.50	224
6650	2-1/2"	Q-1	9	1.778		153	83.50		236.50	294
6660	3"		8	2		165	93.50		258.50	320
6670	4"		5	3.200		215	150		365	460
6680	5"	Q-2	6	4		274	194		468	590
6690	6"		6	4		370	194		564	695
6700	8"		4.50	5.333		655	259		914	1,100
6710	10"		4	6		1,125	291		1,416	1,675
6720	12"		3	8		2,175	390		2,565	2,950
6740	For 250 or 300 lb. flanges, thru 6" no change									
6741	For 8" and 10", add				Ea.	11%	10%			
6750	Twin disc									
6752	2"	1 Plum	9	.889	Ea.	226	46.50		272.50	320
6754	4"	Q-1	5	3.200		370	150		520	630
6756	6"	Q-2	5	4.800		550	233		783	955
6758	8"		4.50	5.333		840	259		1,099	1,325
6760	10"		4	6		1,350	291		1,641	1,900
6762	12"		3	8		1,750	390		2,140	2,500
6764	18"		1.50	16		7,525	775		8,300	9,450
6766	24"		.75	32		10,600	1,550		12,150	14,000
6800	Full flange type, 150 lb.									
6810	1"	1 Plum	14	.571	Ea.	137	29.50		166.50	196
6811	1-1/4"		12	.667		145	34.50		179.50	211
6812	1-1/2"		11	.727		168	38		206	241
6813	2"		9	.889		192	46.50		238.50	282
6814	2-1/2"	Q-1	9	1.778		233	83.50		316.50	380
6815	3"		8	2		241	93.50		334.50	405
6816	4"		5	3.200		299	150		449	555
6817	5"	Q-2	6	4		470	194		664	810
6818	6"		5	4.800		620	233		853	1,025
6819	8"		4.50	5.333		1,050	259		1,309	1,550

23 05 23 – General-Duty Valves for HVAC Piping

23 05 23.30	Valves, Iron Body	Crew	Daily Output	Labor-Hours	Unit	2010 Bare Costs Material	Labor	Equipment	Total	Total Incl O&P
6820	10"	Q-2	4	6	Ea.	1,325	291		1,616	1,875
6840	For 250 lb., add					30%	10%			
6900	Globe type, 125 lb.									
6910	2"	1 Plum	9	.889	Ea.	315	46.50		361.50	415
6911	2-1/2"	Q-1	9	1.778		345	83.50		428.50	505
6912	3"		8	2		375	93.50		468.50	550
6913	4"		5	3.200		505	150		655	780
6914	5"	Q-2	6	4		645	194		839	1,000
6915	6"		5	4.800		810	233		1,043	1,250
6916	8"		4.50	5.333		1,475	259		1,734	2,025
6917	10"		4	6		1,875	291		2,166	2,475
6918	12"		3	8		3,100	390		3,490	4,000
6919	14"		2.30	10.435		4,225	505		4,730	5,400
6920	16"		1.75	13.714		6,100	665		6,765	7,700
6921	18"		1.50	16		10,700	775		11,475	13,000
6922	20"		1	24		11,200	1,175		12,375	14,100
6923	24"		.75	32		13,800	1,550		15,350	17,500
6940	For 250 lb., add					40%	10%			
6980	Screwed end type, 125 lb.									
6981	1"	1 Plum	19	.421	Ea.	48.50	22		70.50	86
6982	1-1/4"		15	.533		63	28		91	111
6983	1-1/2"		13	.615		77.50	32		109.50	133
6984	2"		11	.727		106	38		144	174

23 05 23.50 Valves, Multipurpose

Line	Description	Crew	Daily Output	Labor-Hours	Unit	Material	Labor	Equipment	Total	Total Incl O&P
0010	**VALVES, MULTIPURPOSE**									
0100	Functions as a shut off, balancing, check & metering valve									
1000	Cast iron body									
1010	Threaded									
1020	1-1/2" size	1 Stpi	11	.727	Ea.	335	38		373	425
1030	2" size	"	8	1		405	52		457	525
1040	2-1/2" size	Q-5	5	3.200		440	149		589	710
1050	3" size	"	4.50	3.556		505	166		671	805
1200	Flanged									
1210	3" size	Q-5	4.20	3.810	Ea.	505	178		683	820
1220	4" size	"	3	5.333		1,025	249		1,274	1,500
1230	5" size	Q-6	3.80	6.316		1,200	305		1,505	1,775
1240	6" size		3	8		1,500	385		1,885	2,225
1250	8" size		2.50	9.600		2,625	465		3,090	3,575
1260	10" size		2.20	10.909		3,750	530		4,280	4,925
1270	12" size		2.10	11.429		10,300	555		10,855	12,100
1280	14" size		2	12		14,200	580		14,780	16,600

23 05 23.70 Valves, Semi-Steel

Line	Description	Crew	Daily Output	Labor-Hours	Unit	Material	Labor	Equipment	Total	Total Incl O&P
0010	**VALVES, SEMI-STEEL** R220523-90									
1020	Lubricated plug valve, threaded, 200 psi									
1030	1/2"	1 Plum	18	.444	Ea.	73	23		96	115
1040	3/4"		16	.500		73	26		99	119
1050	1"		14	.571		93	29.50		122.50	147
1060	1-1/4"		12	.667		110	34.50		144.50	173
1070	1-1/2"		11	.727		119	38		157	188
1080	2"		8	1		142	52		194	234
1090	2-1/2"	Q-1	5	3.200		218	150		368	465
1100	3"	"	4.50	3.556		268	167		435	545

23 05 23 – General-Duty Valves for HVAC Piping

23 05 23.70 Valves, Semi-Steel		Crew	Daily Output	Labor-Hours	Unit	Material (2010 Bare Costs)	Labor (2010 Bare Costs)	Equipment (2010 Bare Costs)	Total (2010 Bare Costs)	Total Incl O&P
6990	Flanged, 200 psi									
7000	2″	1 Plum	8	1	Ea.	176	52		228	272
7010	2-1/2″	Q-1	5	3.200		253	150		403	505
7020	3″		4.50	3.556		305	167		472	585
7030	4″		3	5.333		385	250		635	795
7036	5″		2.50	6.400		840	300		1,140	1,375
7040	6″	Q-2	3	8		630	390		1,020	1,275
7050	8″		2.50	9.600		1,225	465		1,690	2,050
7060	10″		2.20	10.909		1,950	530		2,480	2,925
7070	12″		1.70	14.118		4,425	685		5,110	5,875

23 05 23.80 Valves, Steel		Crew	Daily Output	Labor-Hours	Unit	Material	Labor	Equipment	Total	Total Incl O&P
0010	**VALVES, STEEL** R220523-90									
0800	Cast									
1350	Check valve, swing type, 150 lb., flanged									
1370	1″	1 Plum	10	.800	Ea.	525	41.50		566.50	640
1400	2″	″	8	1		660	52		712	805
1440	2-1/2″	Q-1	5	3.200		765	150		915	1,075
1450	3″		4.50	3.556		780	167		947	1,100
1460	4″		3	5.333		1,125	250		1,375	1,600
1470	6″	Q-2	3	8		1,750	390		2,140	2,500
1480	8″		2.50	9.600		2,850	465		3,315	3,850
1490	10″		2.20	10.909		4,225	530		4,755	5,450
1500	12″		1.70	14.118		6,025	685		6,710	7,675
1510	14″		1.30	18.462		8,725	895		9,620	10,900
1520	16″		1	24		9,875	1,175		11,050	12,700
1548	For 600 lb., flanged, add					110%	20%			
1571	300 lb., 2″	1 Plum	7.40	1.081		700	56.50		756.50	855
1572	2-1/2″	Q-1	4.20	3.810		1,050	178		1,228	1,425
1573	3″		4	4		1,050	187		1,237	1,425
1574	4″		2.80	5.714		1,375	268		1,643	1,900
1575	6″	Q-2	2.90	8.276		2,650	400		3,050	3,525
1576	8″		2.40	10		3,825	485		4,310	4,950
1577	10″		2.10	11.429		5,675	555		6,230	7,075
1578	12″		1.60	15		7,950	730		8,680	9,825
1579	14″		1.20	20		12,000	970		12,970	14,700
1581	16″		.90	26.667		15,500	1,300		16,800	19,100
1950	Gate valve, 150 lb., flanged									
2000	2″	1 Plum	8	1	Ea.	710	52		762	860
2040	2-1/2″	Q-1	5	3.200		1,000	150		1,150	1,325
2050	3″		4.50	3.556		1,000	167		1,167	1,350
2060	4″		3	5.333		1,250	250		1,500	1,750
2070	6″	Q-2	3	8		1,850	390		2,240	2,600
2080	8″		2.50	9.600		1,925	465		2,390	2,825
2090	10″		2.20	10.909		4,400	530		4,930	5,625
2100	12″		1.70	14.118		5,975	685		6,660	7,600
2110	14″		1.30	18.462		9,500	895		10,395	11,900
2120	16″		1	24		12,300	1,175		13,475	15,300
2130	18″		.80	30		14,900	1,450		16,350	18,600
2140	20″		.60	40		17,900	1,950		19,850	22,600
2650	300 lb., flanged									
2700	2″	1 Plum	7.40	1.081	Ea.	965	56.50		1,021.50	1,125
2740	2-1/2″	Q-1	4.20	3.810		1,300	178		1,478	1,700

23 05 23 – General-Duty Valves for HVAC Piping

23 05 23.80 Valves, Steel		Crew	Daily Output	Labor-Hours	Unit	2010 Bare Costs Material	Labor	Equipment	Total	Total Incl O&P
2750	3″	Q-1	4	4	Ea.	1,300	187		1,487	1,700
2760	4″		2.80	5.714		1,825	268		2,093	2,400
2770	6″	Q-2	2.90	8.276		3,075	400		3,475	3,975
2780	8″		2.40	10		4,800	485		5,285	6,000
2790	10″		2.10	11.429		6,450	555		7,005	7,925
2800	12″		1.60	15		9,300	730		10,030	11,300
2810	14″		1.20	20		17,600	970		18,570	20,900
2820	16″		.90	26.667		22,100	1,300		23,400	26,300
2830	18″		.70	34.286		31,500	1,675		33,175	37,200
2840	20″		.50	48		35,100	2,325		37,425	42,100
3650	Globe valve, 150 lb., flanged									
3700	2″	1 Plum	8	1	Ea.	890	52		942	1,050
3740	2-1/2″	Q-1	5	3.200		1,125	150		1,275	1,475
3750	3″		4.50	3.556		1,125	167		1,292	1,500
3760	4″		3	5.333		1,650	250		1,900	2,200
3770	6″	Q-2	3	8		2,600	390		2,990	3,425
3780	8″		2.50	9.600		4,750	465		5,215	5,950
3790	10″		2.20	10.909		8,425	530		8,955	10,100
3800	12″		1.70	14.118		11,400	685		12,085	13,500
4080	300 lb., flanged									
4100	2″	1 Plum	7.40	1.081	Ea.	1,200	56.50		1,256.50	1,375
4140	2-1/2″	Q-1	4.20	3.810		1,625	178		1,803	2,050
4150	3″		4	4		1,625	187		1,812	2,050
4160	4″		2.80	5.714		2,250	268		2,518	2,875
4170	6″	Q-2	2.90	8.276		4,050	400		4,450	5,050
4180	8″		2.40	10		6,625	485		7,110	8,000
4190	10″		2.10	11.429		12,500	555		13,055	14,500
4200	12″		1.60	15		14,500	730		15,230	17,000
4680	600 lb., flanged									
4700	2″	1 Plum	7	1.143	Ea.	1,650	59.50		1,709.50	1,925
4740	2-1/2″	Q-1	4	4		2,475	187		2,662	3,000
4750	3″		3.60	4.444		2,475	208		2,683	3,025
4760	4″		2.50	6.400		4,000	300		4,300	4,850
4770	6″	Q-2	2.60	9.231		8,325	450		8,775	9,825
4780	8″	″	2.10	11.429		13,500	555		14,055	15,700
4800	Silent check, 316 S.S. trim									
4810	Full flange type, 150 lb.									
4811	1″	1 Plum	14	.571	Ea.	485	29.50		514.50	580
4812	1-1/4″		12	.667		485	34.50		519.50	585
4813	1-1/2″		11	.727		530	38		568	640
4814	2″		9	.889		640	46.50		686.50	775
4815	2-1/2″	Q-1	9	1.778		775	83.50		858.50	975
4816	3″		8	2		900	93.50		993.50	1,125
4817	4″		5	3.200		1,225	150		1,375	1,575
4818	5″	Q-2	6	4		1,550	194		1,744	2,000
4819	6″		5	4.800		2,000	233		2,233	2,550
4820	8″		4.50	5.333		2,875	259		3,134	3,575
4821	10″		4	6		4,075	291		4,366	4,900
4840	For 300 lb., add					20%	10%			
4860	For 600 lb., add					50%	15%			
4900	Globe type, 150 lb.									
4910	2″	1 Plum	9	.889	Ea.	765	46.50		811.50	910
4911	2-1/2″	Q-1	9	1.778		920	83.50		1,003.50	1,125

23 05 23 – General-Duty Valves for HVAC Piping

23 05 23.80 Valves, Steel		Crew	Daily Output	Labor-Hours	Unit	Material	2010 Bare Costs Labor	Equipment	Total	Total Incl O&P
4912	3"	Q-1	8	2	Ea.	1,025	93.50		1,118.50	1,275
4913	4"		5	3.200		1,350	150		1,500	1,725
4914	5"	Q-2	6	4		1,700	194		1,894	2,175
4915	6"		5	4.800		2,100	233		2,333	2,675
4916	8"		4.50	5.333		2,950	259		3,209	3,650
4917	10"		4	6		5,100	291		5,391	6,050
4918	12"		3	8		6,275	390		6,665	7,475
4919	14"		2.30	10.435		9,475	505		9,980	11,200
4920	16"		1.75	13.714		13,200	665		13,865	15,600
4921	18"		1.50	16		13,900	775		14,675	16,500
4940	For 300 lb., add					30%	10%			
4960	For 600 lb., add					60%	15%			
5150	Forged									
5340	Ball valve, 1500 psi, threaded, 1/4" size	1 Plum	24	.333	Ea.	56	17.35		73.35	88
5350	3/8"		24	.333		56	17.35		73.35	88
5360	1/2"		24	.333		69.50	17.35		86.85	103
5370	3/4"		20	.400		92.50	21		113.50	133
5380	1"		19	.421		115	22		137	160
5390	1-1/4"		15	.533		154	28		182	211
5400	1-1/2"		13	.615		201	32		233	269
5410	2"		11	.727		266	38		304	350
5460	Ball valve, 800 lb., socket weld, 1/4" size	Q-15	19	.842		56	39.50	3.08	98.58	124
5470	3/8"		19	.842		56	39.50	3.08	98.58	124
5480	1/2"		19	.842		69.50	39.50	3.08	112.08	139
5490	3/4"		19	.842		92.50	39.50	3.08	135.08	164
5500	1"		15	1.067		115	50	3.90	168.90	206
5510	1-1/4"		13	1.231		154	57.50	4.50	216	260
5520	1-1/2"		11	1.455		201	68	5.30	274.30	330
5530	2"		8.50	1.882		266	88	6.90	360.90	430
5550	Ball valve, 150 lb., flanged									
5560	4"	Q-1	3	5.333	Ea.	485	250		735	905
5570	6"	Q-2	3	8		1,150	390		1,540	1,850
5580	8"	"	2.50	9.600		3,050	465		3,515	4,050
5650	Check valve, class 800, horizontal, socket									
5652	1/4"	Q-15	19	.842	Ea.	88	39.50	3.08	130.58	159
5654	3/8"		19	.842		88	39.50	3.08	130.58	159
5656	1/2"		19	.842		88	39.50	3.08	130.58	159
5658	3/4"		19	.842		94.50	39.50	3.08	137.08	166
5660	1"		15	1.067		111	50	3.90	164.90	201
5662	1-1/4"		13	1.231		217	57.50	4.50	279	330
5664	1-1/2"		11	1.455		217	68	5.30	290.30	345
5666	2"		8.50	1.882		305	88	6.90	399.90	475
5698	Threaded									
5700	1/4"	1 Plum	24	.333	Ea.	88	17.35		105.35	123
5720	3/8"		24	.333		88	17.35		105.35	123
5730	1/2"		24	.333		88	17.35		105.35	123
5740	3/4"		20	.400		94.50	21		115.50	135
5750	1"		19	.421		111	22		133	155
5760	1-1/4"		15	.533		217	28		245	281
5770	1-1/2"		13	.615		217	32		249	287
5780	2"		11	.727		305	38		343	390
5840	For class 150, flanged, add					100%	15%			
5860	For class 300, flanged, add					120%	20%			

23 05 23.80 Valves, Steel		Crew	Daily Output	Labor-Hours	Unit	Material	2010 Bare Costs Labor	Equipment	Total	Total Incl O&P
6100	Gate, class 800, OS&Y, socket									
6102	3/8"	Q-15	19	.842	Ea.	61	39.50	3.08	103.58	130
6103	1/2"		19	.842		61	39.50	3.08	103.58	130
6104	3/4"		19	.842		67	39.50	3.08	109.58	136
6105	1"		15	1.067		81.50	50	3.90	135.40	169
6106	1-1/4"		13	1.231		154	57.50	4.50	216	260
6107	1-1/2"		11	1.455		154	68	5.30	227.30	277
6108	2"		8.50	1.882		200	88	6.90	294.90	360
6118	Threaded									
6120	3/8"	1 Plum	24	.333	Ea.	61	17.35		78.35	93.50
6130	1/2"		24	.333		61	17.35		78.35	93.50
6140	3/4"		20	.400		67	21		88	105
6150	1"		19	.421		81.50	22		103.50	123
6160	1-1/4"		15	.533		154	28		182	211
6170	1-1/2"		13	.615		154	32		186	217
6180	2"		11	.727		200	38		238	277
6260	For OS&Y, flanged, add					100%	20%			
6700	Globe, OS&Y, class 800, socket									
6710	1/4"	Q-15	19	.842	Ea.	94.50	39.50	3.08	137.08	166
6720	3/8"		19	.842		94.50	39.50	3.08	137.08	166
6730	1/2"		19	.842		94.50	39.50	3.08	137.08	166
6740	3/4"		19	.842		108	39.50	3.08	150.58	181
6750	1"		15	1.067		141	50	3.90	194.90	234
6760	1-1/4"		13	1.231		276	57.50	4.50	338	395
6770	1-1/2"		11	1.455		276	68	5.30	349.30	415
6780	2"		8.50	1.882		355	88	6.90	449.90	530
6860	For OS&Y, flanged, add					300%	20%			
6880	Threaded									
6882	1/4"	1 Plum	24	.333	Ea.	94.50	17.35		111.85	130
6884	3/8"		24	.333		94.50	17.35		111.85	130
6886	1/2"		24	.333		94.50	17.35		111.85	130
6888	3/4"		20	.400		108	21		129	150
6890	1"		19	.421		141	22		163	188
6892	1-1/4"		15	.533		276	28		304	345
6894	1-1/2"		13	.615		276	32		308	355
6896	2"		11	.727		355	38		393	445

23 05 23.90 Valves, Stainless Steel		Crew	Daily Output	Labor-Hours	Unit	Material	Labor	Equipment	Total	Total Incl O&P
0010	**VALVES, STAINLESS STEEL** R220523-90									
1610	Ball, threaded 1/4"	1 Stpi	24	.333	Ea.	42.50	17.30		59.80	73
1620	3/8"		24	.333		42.50	17.30		59.80	73
1630	1/2"		22	.364		42.50	18.85		61.35	75.50
1640	3/4"		20	.400		70.50	21		91.50	109
1650	1"		19	.421		86	22		108	128
1660	1-1/4		15	.533		165	27.50		192.50	223
1670	1-1/2"		13	.615		171	32		203	237
1680	2"		11	.727		223	38		261	300
1700	Check, 200 lb., threaded									
1710	1/4"	1 Plum	24	.333	Ea.	130	17.35		147.35	169
1720	1/2"		22	.364		130	18.95		148.95	172
1730	3/4"		20	.400		142	21		163	187
1750	1"		19	.421		182	22		204	233
1760	1-1/2"		13	.615		350	32		382	435

23 05 23 – General-Duty Valves for HVAC Piping

23 05 23.90 Valves, Stainless Steel		Crew	Daily Output	Labor-Hours	Unit	Material	2010 Bare Costs Labor	Equipment	Total	Total Incl O&P
1770	2"	1 Plum	11	.727	Ea.	585	38		623	700
1800	150 lb., flanged									
1810	2-1/2"	Q-1	5	3.200	Ea.	1,400	150		1,550	1,775
1820	3"		4.50	3.556		1,400	167		1,567	1,800
1830	4"		3	5.333		2,100	250		2,350	2,675
1840	6"	Q-2	3	8		3,725	390		4,115	4,675
1850	8"	"	2.50	9.600		6,250	465		6,715	7,575
2100	Gate, OS&Y, 150 lb., flanged									
2120	1/2"	1 Plum	18	.444	Ea.	415	23		438	490
2140	3/4"		16	.500		440	26		466	525
2150	1"		14	.571		540	29.50		569.50	640
2160	1-1/2"		11	.727		750	38		788	885
2170	2"		8	1		845	52		897	1,000
2180	2-1/2"	Q-1	5	3.200		1,400	150		1,550	1,750
2190	3"		4.50	3.556		2,050	167		2,217	2,500
2200	4"		3	5.333		2,050	250		2,300	2,625
2205	5"		2.80	5.714		3,550	268		3,818	4,325
2210	6"	Q-2	3	8		3,550	390		3,940	4,500
2220	8"		2.50	9.600		5,725	465		6,190	7,000
2230	10"		2.30	10.435		9,375	505		9,880	11,100
2240	12"		1.90	12.632		12,000	615		12,615	14,100
2260	For 300 lb., flanged, add					120%	15%			
2600	600 lb., flanged									
2620	1/2"	1 Plum	16	.500	Ea.	199	26		225	258
2640	3/4"		14	.571		215	29.50		244.50	282
2650	1"		12	.667		259	34.50		293.50	335
2660	1-1/2"		10	.800		415	41.50		456.50	520
2670	2"		7	1.143		570	59.50		629.50	715
2680	2-1/2"	Q-1	4	4		4,950	187		5,137	5,725
2690	3"	"	3.60	4.444		4,950	208		5,158	5,750
3100	Globe, OS&Y, 150 lb., flanged									
3120	1/2"	1 Plum	18	.444	Ea.	415	23		438	490
3140	3/4"		16	.500		460	26		486	545
3150	1"		14	.571		570	29.50		599.50	670
3160	1-1/2"		11	.727		845	38		883	985
3170	2"		8	1		1,025	52		1,077	1,200
3180	2-1/2"	Q-1	5	3.200		2,225	150		2,375	2,675
3190	3"		4.50	3.556		2,225	167		2,392	2,700
3200	4"		3	5.333		2,600	250		2,850	3,250
3210	6"	Q-2	3	8		4,575	390		4,965	5,625
5000	Silent check, 316 S.S. body and trim									
5010	Compact wafer type, 300 lb.									
5020	1"	1 Plum	14	.571	Ea.	805	29.50		834.50	930
5021	1-1/4"		12	.667		870	34.50		904.50	1,000
5022	1-1/2"		11	.727		900	38		938	1,050
5023	2"		9	.889		1,025	46.50		1,071.50	1,200
5024	2-1/2"	Q-1	9	1.778		1,400	83.50		1,483.50	1,675
5025	3"	"	8	2		1,625	93.50		1,718.50	1,950
5100	Full flange wafer type, 150 lb.									
5110	1"	1 Plum	14	.571	Ea.	565	29.50		594.50	665
5111	1-1/4"		12	.667		610	34.50		644.50	720
5112	1-1/2"		11	.727		695	38		733	820
5113	2"		9	.889		910	46.50		956.50	1,075

23 05 23 – General-Duty Valves for HVAC Piping

23 05 23.90 Valves, Stainless Steel		Crew	Daily Output	Labor-Hours	Unit	Material	2010 Bare Costs Labor	Equipment	Total	Total Incl O&P
5114	2-1/2"	Q-1	9	1.778	Ea.	1,075	83.50		1,158.50	1,300
5115	3"		8	2		1,225	93.50		1,318.50	1,500
5116	4"		5	3.200		1,800	150		1,950	2,225
5117	5"	Q-2	6	4		2,100	194		2,294	2,625
5118	6"		5	4.800		2,650	233		2,883	3,250
5119	8"		4.50	5.333		4,400	259		4,659	5,250
5120	10"		4	6		6,100	291		6,391	7,125
5200	Globe type, 300 lb.									
5210	2"	1 Plum	9	.889	Ea.	1,025	46.50		1,071.50	1,225
5211	2-1/2"	Q-1	9	1.778		1,275	83.50		1,358.50	1,525
5212	3"		8	2		1,575	93.50		1,668.50	1,875
5213	4"		5	3.200		2,175	150		2,325	2,625
5214	5"	Q-2	6	4		3,050	194		3,244	3,650
5215	6"		5	4.800		3,450	233		3,683	4,150
5216	8"		4.50	5.333		4,875	259		5,134	5,775
5217	10"		4	6		6,925	291		7,216	8,025
5300	Screwed end type, 300 lb.									
5310	1/2"	1 Plum	24	.333	Ea.	82	17.35		99.35	116
5311	3/4"		20	.400		97.50	21		118.50	138
5312	1"		19	.421		114	22		136	158
5313	1-1/4"		15	.533		150	28		178	207
5314	1-1/2"		13	.615		161	32		193	225
5315	2"		11	.727		201	38		239	278

23 05 93 – Testing, Adjusting, and Balancing for HVAC

23 05 93.10 Balancing, Air		Crew	Daily Output	Labor-Hours	Unit	Material	Labor	Equipment	Total	Total Incl O&P
0010	**BALANCING, AIR** (Subcontractor's quote incl. material and labor) R230500-10									
0900	Heating and ventilating equipment									
1000	Centrifugal fans, utility sets				Ea.				325	355
1100	Heating and ventilating unit								485	535
1200	In-line fan								485	535
1300	Propeller and wall fan								92	101
1400	Roof exhaust fan								217	238
2000	Air conditioning equipment, central station								705	775
2100	Built-up low pressure unit								650	715
2200	Built-up high pressure unit								760	835
2300	Built-up high pressure dual duct								1,200	1,300
2400	Built-up variable volume								1,400	1,550
2500	Multi-zone A.C. and heating unit								485	535
2600	For each zone over one, add								108	119
2700	Package A.C. unit								270	297
2800	Rooftop heating and cooling unit								379	415
3000	Supply, return, exhaust, registers & diffusers, avg. height ceiling								65	71
3100	High ceiling								97	107
3200	Floor height								54	59
3300	Off mixing box								43	48
3500	Induction unit								70	77
3600	Lab fume hood								325	355
3700	Linear supply								162	178
3800	Linear supply high								189	208
4000	Linear return								54	59
4100	Light troffers								65	71
4200	Moduline - master								65	71

23 05 93 – Testing, Adjusting, and Balancing for HVAC

23 05 93.10 Balancing, Air		Crew	Daily Output	Labor-Hours	Unit	Material	2010 Bare Costs Labor	Equipment	Total	Total Incl O&P
4300	Moduline - slaves				Ea.				31.50	35.50
4400	Regenerators								435	475
4500	Taps into ceiling plenums								81	89
4600	Variable volume boxes								65	71.50
23 05 93.20 Balancing, Water										
0010	**BALANCING, WATER** (Subcontractor's quote incl. material and labor)									
0050	Air cooled condenser				Ea.				198	218
0080	Boiler								400	440
0100	Cabinet unit heater R230500-10								68	74.50
0200	Chiller								480	530
0300	Convector								56.50	62
0400	Converter								283	310
0500	Cooling tower								370	405
0600	Fan coil unit, unit ventilator								101	112
0700	Fin tube and radiant panels								113	124
0800	Main and duct re-heat coils								105	115
0810	Heat exchanger								105	115
0900	Main balancing cocks								85	93.50
1000	Pumps								249	274
1100	Unit heater								79.50	87
23 05 93.50 Piping, Testing										
0010	**PIPING, TESTING**									
0100	Nondestructive testing									
0110	Nondestructive hydraulic pressure test, isolate & 1 hr. hold									
0120	1" - 4" pipe									
0140	0 – 250 L.F.	1 Stpi	1.33	6.015	Ea.		310		310	470
0160	250 – 500 L.F.	"	.80	10			520		520	780
0180	500 – 1000 L.F.	Q-5	1.14	14.035			655		655	980
0200	1000 – 2000 L.F.	"	.80	20			935		935	1,400
0300	6" - 10" pipe									
0320	0 – 250 L.F.	Q-5	1	16	Ea.		745		745	1,125
0340	250 – 500 L.F.		.73	21.918			1,025		1,025	1,525
0360	500 – 1000 L.F.		.53	30.189			1,400		1,400	2,125
0380	1000 – 2000 L.F.		.38	42.105			1,975		1,975	2,950
1000	Pneumatic pressure test, includes soaping joints									
1120	1" - 4" pipe									
1140	0 – 250 L.F.	Q-5	2.67	5.993	Ea.	7	280		287	430
1160	250 – 500 L.F.		1.33	12.030		14.05	560		574.05	855
1180	500 – 1000 L.F.		.80	20		21	935		956	1,425
1200	1000 – 2000 L.F.		.50	32		28	1,500		1,528	2,275
1300	6" - 10" pipe									
1320	0 – 250 L.F.	Q-5	1.33	12.030	Ea.	7	560		567	850
1340	250 – 500 L.F.		.67	23.881		14.05	1,125		1,139.05	1,700
1360	500 – 1000 L.F.		.40	40		28	1,875		1,903	2,825
1380	1000 – 2000 L.F.		.25	64		35	3,000		3,035	4,525
2000	X-Ray of welds									
2110	2" diam.	1 Stpi	8	1	Ea.	11	52		63	90
2120	3" diam.		8	1		11	52		63	90
2130	4" diam.		8	1		16.50	52		68.50	96
2140	6" diam.		8	1		16.50	52		68.50	96
2150	8" diam.		6.60	1.212		16.50	63		79.50	113
2160	10" diam.		6	1.333		22	69		91	128

23 05 93 – Testing, Adjusting, and Balancing for HVAC

23 05 93.50 Piping, Testing		Crew	Daily Output	Labor-Hours	Unit	2010 Bare Costs Material	Labor	Equipment	Total	Total Incl O&P
3000	Liquid penetration of welds									
3110	2" diam.	1 Stpi	14	.571	Ea.	3.05	29.50		32.55	48
3120	3" diam.		13.60	.588		3.05	30.50		33.55	49.50
3130	4" diam.		13.40	.597		3.05	31		34.05	50
3140	6" diam.		13.20	.606		3.05	31.50		34.55	50.50
3150	8" diam.		13	.615		4.58	32		36.58	53
3160	10" diam.	↓	12.80	.625	↓	4.58	32.50		37.08	53.50

23 07 HVAC Insulation

23 07 13 – Duct Insulation

23 07 13.10 Duct Thermal Insulation		Crew	Daily Output	Labor-Hours	Unit	Material	Labor	Equipment	Total	Total Incl O&P
0010	**DUCT THERMAL INSULATION**									
0100	Rule of thumb, as a percentage of total mechanical costs				Job				10%	
0110	Insulation req'd is based on the surface size/area to be covered									
3000	Ductwork									
3020	Blanket type, fiberglass, flexible									
3140	FSK vapor barrier wrap, .75 lb. density									
3160	1" thick G	Q-14	350	.046	S.F.	.20	1.87		2.07	3.13
3170	1-1/2" thick G		320	.050		.23	2.05		2.28	3.44
3180	2" thick G		300	.053		.28	2.19		2.47	3.71
3190	3" thick G		260	.062		.40	2.52		2.92	4.36
3200	4" thick G		242	.066		.60	2.71		3.31	4.87
3212	Vinyl Jacket, .75 lb. density, 1-1/2" thick G	↓	320	.050	↓	.23	2.05		2.28	3.44
3280	Unfaced, 1 lb. density									
3310	1" thick G	Q-14	360	.044	S.F.	.19	1.82		2.01	3.04
3320	1-1/2" thick G		330	.048		.23	1.99		2.22	3.34
3330	2" thick G	↓	310	.052	↓	.31	2.12		2.43	3.63
3400	FSK facing, 1 lb. density									
3420	1-1/2" thick G	Q-14	310	.052	S.F.	.18	2.12		2.30	3.49
3430	2" thick G	"	300	.053	"	.20	2.19		2.39	3.62
3450	FSK facing, 1.5 lb. density									
3470	1-1/2" thick G	Q-14	300	.053	S.F.	.27	2.19		2.46	3.70
3480	2" thick G	"	290	.055	"	.33	2.26		2.59	3.88
3730	Sheet insulation									
3760	Polyethylene foam, closed cell, UV resistant									
3770	Standard temperature (-90°F to +212°F)									
3771	1/4" thick G	Q-14	450	.036	S.F.	1.38	1.46		2.84	3.79
3772	3/8" thick G		440	.036		1.97	1.49		3.46	4.49
3773	1/2" thick G		420	.038		2.42	1.56		3.98	5.10
3774	3/4" thick G		400	.040		3.46	1.64		5.10	6.35
3775	1" thick G		380	.042		4.67	1.73		6.40	7.85
3776	1-1/2" thick G		360	.044		7.40	1.82		9.22	11
3777	2" thick G		340	.047		9.90	1.93		11.83	13.90
3778	2-1/2" thick G	↓	320	.050	↓	12.55	2.05		14.60	17
3779	Adhesive (see line 7878)									
3780	Foam, rubber									
3782	1" thick G	1 Stpi	50	.160	S.F.	4.15	8.30		12.45	17
3795	Finishes									
3800	Stainless steel woven mesh	Q-14	100	.160	S.F.	.84	6.55		7.39	11.10
3820	18 oz. fiberglass cloth, pasted on		170	.094		.77	3.86		4.63	6.85
3900	8 oz. canvas, pasted on	↓	180	.089	↓	.49	3.64		4.13	6.20

23 07 HVAC Insulation

23 07 13 – Duct Insulation

23 07 13.10 Duct Thermal Insulation			Crew	Daily Output	Labor-Hours	Unit	Material	2010 Bare Costs Labor	Equipment	Total	Total Incl O&P
7878	Contact cement, quart can					Ea.	9.85			9.85	10.85
9600	Minimum labor/equipment charge		1 Stpi	4	2	Job		104		104	156

23 07 16 – HVAC Equipment Insulation

23 07 16.10 HVAC Equipment Thermal Insulation			Crew	Daily Output	Labor-Hours	Unit	Material	Labor	Equipment	Total	Total Incl O&P
0010	**HVAC EQUIPMENT THERMAL INSULATION**										
0100	Rule of thumb, as a percentage of total mechanical costs					Job				10%	
0110	Insulation req'd is based on the surface size/area to be covered										
1000	Boiler, 1-1/2" calcium silicate only	G	Q-14	110	.145	S.F.	2.92	5.95		8.87	12.45
1020	Plus 2" fiberglass	G	"	80	.200	"	3.82	8.20		12.02	16.95
2000	Breeching, 2" calcium silicate										
2020	Rectangular	G	Q-14	42	.381	S.F.	5.70	15.60		21.30	31
2040	Round	G	"	38.70	.413	"	5.95	16.95		22.90	33
2300	Calcium silicate block, + 200°F to + 1200°F										
2310	On irregular surfaces, valves and fittings										
2340	1" thick	G	Q-14	30	.533	S.F.	2.59	22		24.59	37
2360	1-1/2" thick	G		25	.640		2.92	26		28.92	44
2380	2" thick	G		22	.727		3.83	30		33.83	50.50
2400	3" thick	G	↓	18	.889	↓	5.90	36.50		42.40	63
2410	On plane surfaces										
2420	1" thick	G	Q-14	126	.127	S.F.	2.59	5.20		7.79	10.95
2430	1-1/2" thick	G		120	.133		2.92	5.45		8.37	11.70
2440	2" thick	G		100	.160		3.83	6.55		10.38	14.40
2450	3" thick	G	↓	70	.229	↓	5.90	9.35		15.25	21

23 09 Instrumentation and Control for HVAC

23 09 13 – Instrumentation and Control Devices for HVAC

23 09 13.60 Water Level Controls		Crew	Daily Output	Labor-Hours	Unit	Material	Labor	Equipment	Total	Total Incl O&P
0010	**WATER LEVEL CONTROLS**									
1000	Electric water feeder	1 Stpi	12	.667	Ea.	242	34.50		276.50	320
2000	Feeder cut-off combination									
2100	Steam system up to 5000 sq. ft.	1 Stpi	12	.667	Ea.	440	34.50		474.50	535
2200	Steam system above 5000 sq. ft.		12	.667		640	34.50		674.50	755
2300	Steam and hot water, high pressure	↓	10	.800	↓	755	41.50		796.50	890
3000	Low water cut-off for hot water boiler, 50 psi maximum									
3100	1" top & bottom equalizing pipes, manual reset	1 Stpi	14	.571	Ea.	291	29.50		320.50	365
3200	1" top & bottom equalizing pipes		14	.571		291	29.50		320.50	365
3300	2-1/2" side connection for nipple-to-boiler	↓	14	.571	↓	263	29.50		292.50	335
4000	Low water cut-off for low pressure steam with quick hook-up ftgs.									
4100	For installation in gauge glass tappings	1 Stpi	16	.500	Ea.	210	26		236	270
4200	Built-in type, 2-1/2" tap - 3-1/8" insertion		16	.500		176	26		202	232
4300	Built-in type, 2-1/2" tap - 1-3/4" insertion		16	.500		187	26		213	244
4400	Side connection to 2-1/2" tapping		16	.500		200	26		226	259
5000	Pump control, low water cut-off and alarm switch	↓	14	.571	↓	455	29.50		484.50	545
9000	Water gauges, complete									
9010	Rough brass, wheel type									
9020	125 PSI at 350°F									
9030	3/8" pipe size	1 Stpi	11	.727	Ea.	52.50	38		90.50	114
9040	1/2" pipe size	"	10	.800	"	57	41.50		98.50	125
9060	200 PSI at 400°F									
9070	3/8" pipe size	1 Stpi	11	.727	Ea.	65	38		103	128

23 09 13 – Instrumentation and Control Devices for HVAC

23 09 13.60 Water Level Controls		Crew	Daily Output	Labor-Hours	Unit	Material	2010 Bare Costs Labor	Equipment	Total	Total Incl O&P
9080	1/2″ pipe size	1 Stpi	10	.800	Ea.	65	41.50		106.50	134
9090	3/4″ pipe size	↓	9	.889	↓	83	46		129	161
9130	Rough brass, chain lever type									
9140	250 PSI at 400°F									
9200	Polished brass, wheel type									
9210	200 PSI at 400°F									
9220	3/8″ pipe size	1 Stpi	11	.727	Ea.	123	38		161	192
9230	1/2″ pipe size		10	.800		124	41.50		165.50	199
9240	3/4″ pipe size	↓	9	.889	↓	132	46		178	214
9260	Polished brass, chain lever type									
9270	250 PSI at 400°F									
9280	1/2″ pipe size	1 Stpi	10	.800	Ea.	170	41.50		211.50	249
9290	3/4″ pipe size	″	9	.889	″	185	46		231	272
9400	Bronze, high pressure, ASME									
9410	1/2″ pipe size	1 Stpi	10	.800	Ea.	214	41.50		255.50	297
9420	3/4″ pipe size	″	9	.889	″	237	46		283	330
9460	Chain lever type									
9470	1/2″ pipe size	1 Stpi	10	.800	Ea.	287	41.50		328.50	375
9480	3/4″ pipe size	″	9	.889	″	300	46		346	400
9500	316 stainless steel, high pressure, ASME									
9510	500 PSI at 450°F									
9520	1/2″ pipe size	1 Stpi	10	.800	Ea.	685	41.50		726.50	815
9530	3/4″ pipe size	″	9	.889	″	670	46		716	805

23 09 23 – Direct-Digital Control System for HVAC

23 09 23.10 Control Components/DDC Systems		Crew	Daily Output	Labor-Hours	Unit	Material	Labor	Equipment	Total	Total Incl O&P
0010	**CONTROL COMPONENTS/DDC SYSTEMS** (Sub's quote incl. M & L)									
0100	Analog inputs									
0110	Sensors (avg. 50′ run in 1/2″ EMT)									
0120	Duct temperature				Ea.				360	396
0130	Space temperature								577	634.64
0140	Duct humidity, +/- 3%								605.32	665.85
0150	Space humidity, +/- 2%								926.90	1,019.59
0160	Duct static pressure								491.82	541
0170	C.F.M./Transducer								662.07	728.28
0172	Water temp. (see Div. 23 21 20 for well tap add)								567.64	624.24
0174	Water flow (see Div. 23 21 20 for circuit sensor add)								2,080.80	2,288.88
0176	Water pressure differential (see Div. 23 21 20 for tap add)								851.24	936.36
0177	Steam flow (see Div. 23 21 20 for circuit sensor add)								2,080.80	2,288.88
0178	Steam pressure (see Div. 23 21 20 for tap add)								889.06	977.97
0180	K.W./Transducer								1,182.27	1,300.50
0182	K.W.H. totalization (not incl. elec. meter pulse xmtr.)								543.38	597.72
0190	Space static pressure				↓				927.27	1,020
1000	Analog outputs (avg. 50′ run in 1/2″ EMT)									
1010	P/I Transducer				Ea.				548.57	603.43
1020	Analog output, matl. in MUX								264.83	291.31
1030	Pneumatic (not incl. control device)								558	613.83
1040	Electric (not incl control device)				↓				331	364.14
2000	Status (Alarms)									
2100	Digital inputs (avg. 50′ run in 1/2″ EMT)									
2110	Freeze				Ea.				378.33	416.16
2120	Fire								340.49	374.54
2130	Differential pressure, (air)				↓				520.20	572.22

23 09 23 – Direct-Digital Control System for HVAC

23 09 23.10 Control Components/DDC Systems		Crew	Daily Output	Labor-Hours	Unit	2010 Bare Costs Material	Labor	Equipment	Total	Total Incl O&P
2140	Differential pressure, (water)				Ea.				850	935
2150	Current sensor								378.33	416.16
2160	Duct high temperature thermostat								496.55	546.21
2170	Duct smoke detector								614.78	676.26
2200	Digital output (avg. 50′ run in 1/2″ EMT)									
2210	Start/stop				Ea.				296.73	326.40
2220	On/off (maintained contact)				″				510	561
3000	Controller M.U.X. panel, incl. function boards									
3100	48 point				Ea.				4,587.22	5,045.94
3110	128 point				″				6,290	6,918.66
3200	D.D.C. controller (avg. 50′ run in conduit)									
3210	Mechanical room									
3214	16 point controller (incl. 120v/1ph power supply)				Ea.				1,937.45	3,121.20
3229	32 point controller (incl. 120v/1ph power supply)				″				4,729	5,202
3230	Includes software programming and checkout									
3260	Space									
3266	V.A.V. terminal box (incl. space temp. sensor)				Ea.				733	806.31
3280	Host computer (avg. 50′ run in conduit)									
3281	Package complete with PC, keyboard,									
3282	printer, color CRT, modem, basic software				Ea.				8,512.36	9,363.60
4000	Front end costs									
4100	Computer (P.C.)/software program				Ea.				5,675	6,242.40
4200	Color graphics software								3,405	3,745.44
4300	Color graphics slides								426	468.18
4350	Additional dot matrix printer								851	936.36
4400	Communications trunk cable				L.F.				3.31	3.64
4500	Engineering labor, (not incl. dftg.)				Point				71.88	79.07
4600	Calibration labor								71.88	79.07
4700	Start-up, checkout labor								108.76	119.64
4800	Drafting labor, as req'd									
5000	Communications bus (data transmission cable)									
5010	#18 twisted shielded pair in 1/2″ EMT conduit				C.L.F.				331.04	364.14
8000	Applications software									
8050	Basic maintenance manager software (not incl. data base entry)				Ea.				1,702	1,872.72
8100	Time program				Point				5.95	6.55
8120	Duty cycle								11.86	13.05
8140	Optimum start/stop								35.94	39.53
8160	Demand limiting								17.78	19.56
8100	Enthalpy program								35.94	39.53
8200	Boiler optimization				Ea.				1,064	1,170.45
8220	Chiller optimization				″				1,419	1,560.60
8240	Custom applications									
8260	Cost varies with complexity									

23 09 33 – Electric and Electronic Control System for HVAC

23 09 33.10 Electronic Control Systems		Crew	Daily Output	Labor-Hours	Unit	Material	Labor	Equipment	Total	Total Incl O&P
0010	**ELECTRONIC CONTROL SYSTEMS** R230500-10									
0020	For electronic costs, add to Div. 23 09 43.10				Ea.					15%

23 09 43 – Pneumatic Control System for HVAC

23 09 43.10 Pneumatic Control Systems		Crew	Daily Output	Labor-Hours	Unit	Material	Labor	Equipment	Total	Total Incl O&P
0010	**PNEUMATIC CONTROL SYSTEMS**									
0011	Including a nominal 50 Ft. of tubing. Add control panelboard if req'd.									
0100	Heating and ventilating, split system R230500-10									

23 09 43 – Pneumatic Control System for HVAC

23 09 43.10 Pneumatic Control Systems		Crew	Daily Output	Labor-Hours	Unit	Material	Labor	Equipment	Total	Total Incl O&P
						2010 Bare Costs				
0200	Mixed air control, economizer cycle, panel readout, tubing									
0220	Up to 10 tons G	Q-19	.68	35.294	Ea.	3,575	1,675		5,250	6,425
0240	For 10 to 20 tons G		.63	37.915		3,825	1,800		5,625	6,900
0260	For over 20 tons G		.58	41.096		4,150	1,950		6,100	7,475
0270	Enthalpy cycle, up to 10 tons		.50	48.387		3,950	2,300		6,250	7,775
0280	For 10 to 20 tons		.46	52.174		4,250	2,475		6,725	8,375
0290	For over 20 tons		.42	56.604		4,625	2,675		7,300	9,100
0300	Heating coil, hot water, 3 way valve,									
0320	Freezestat, limit control on discharge, readout	Q-5	.69	23.088	Ea.	2,650	1,075		3,725	4,550
0500	Cooling coil, chilled water, room									
0520	Thermostat, 3 way valve	Q-5	2	8	Ea.	1,175	375		1,550	1,850
0600	Cooling tower, fan cycle, damper control,									
0620	Control system including water readout in/out at panel	Q-19	.67	35.821	Ea.	4,700	1,700		6,400	7,725
1000	Unit ventilator, day/night operation,									
1100	freezestat, ASHRAE, cycle 2	Q-19	.91	26.374	Ea.	2,600	1,250		3,850	4,725
2000	Compensated hot water from boiler, valve control,									
2100	readout and reset at panel, up to 60 GPM	Q-19	.55	43.956	Ea.	4,875	2,075		6,950	8,475
2120	For 120 GPM		.51	47.059		5,200	2,225		7,425	9,075
2140	For 240 GPM		.49	49.180		5,450	2,325		7,775	9,500
3000	Boiler room combustion air, damper to 5 SF, controls		1.37	17.582		2,350	835		3,185	3,825
3500	Fan coil, heating and cooling valves, 4 pipe control system		3	8		1,050	380		1,430	1,750
3600	Heat exchanger system controls		.86	27.907		2,275	1,325		3,600	4,475
3900	Multizone control (one per zone), includes thermostat, damper									
3910	motor and reset of discharge temperature	Q-5	.51	31.373	Ea.	2,350	1,475		3,825	4,775
4000	Pneumatic thermostat, including controlling room radiator valve	"	2.43	6.593		705	310		1,015	1,225
4040	Program energy saving optimizer G	Q-19	1.21	19.786		5,875	940		6,815	7,850
4060	Pump control system	"	3	8		1,075	380		1,455	1,775
4080	Reheat coil control system, not incl coil	Q-5	2.43	6.593		920	310		1,230	1,450
4500	Air supply for pneumatic control system									
4600	Tank mounted duplex compressor, starter, alternator,									
4620	piping, dryer, PRV station and filter									
4630	1/2 HP	Q-19	.68	35.139	Ea.	8,725	1,675		10,400	12,100
4640	3/4 HP		.64	37.383		9,150	1,775		10,925	12,800
4650	1 HP		.61	39.539		10,000	1,875		11,875	13,800
4660	1-1/2 HP		.58	41.739		10,700	1,975		12,675	14,700
4680	3 HP		.55	43.956		14,500	2,075		16,575	19,100
4690	5 HP		.42	57.143		25,300	2,725		28,025	32,000
4800	Main air supply, includes 3/8" copper main and labor	Q-5	1.82	8.791	C.L.F.	293	410		703	940
4810	If poly tubing used, deduct									30%
7000	Static pressure control for air handling unit, includes pressure									
7010	sensor, receiver controller, readout and damper motors	Q-19	.64	37.383	Ea.	6,950	1,775		8,725	10,300
7020	If return air fan requires control, add									70%
8600	VAV boxes, incl. thermostat, damper motor, reheat coil & tubing	Q-5	1.46	10.989		1,100	515		1,615	1,975
8610	If no reheat coil, deduct									204
9400	Sub assemblies for assembly systems									

23 09 53 – Pneumatic and Electric Control System for HVAC

23 09 53.10 Control Components		Crew	Daily Output	Labor-Hours	Unit	Material	Labor	Equipment	Total	Total Incl O&P
0010	**CONTROL COMPONENTS**									
0500	Aquastats									
0508	Immersion type									
0510	High/low limit, breaks contact w/temp rise	1 Stpi	4	2	Ea.	330	104		434	520
0514	Sequencing, break 2 switches w/temp rise		4	2		251	104		355	430

23 09 53.10 Control Components		Crew	Daily Output	Labor-Hours	Unit	Material	2010 Bare Costs Labor	Equipment	Total	Total Incl O&P
0518	Circulating, makes contact w/temp rise	1 Stpi	4	2	Ea.	225	104		329	405
0600	Carbon monoxide detector system									
0606	Panel	1 Stpi	4	2	Ea.	1,350	104		1,454	1,650
0610	Sensor	"	7.30	1.096	"	785	57		842	950
0700	Controller, receiver									
0730	Pneumatic, panel mount, single input	1 Plum	8	1	Ea.	410	52		462	530
0740	With conversion mounting bracket		8	1		410	52		462	530
0750	Dual input, with control point adjustment		7	1.143		565	59.50		624.50	710
0850	Electric, single snap switch	1 Elec	4	2		450	98		548	640
0860	Dual snap switches		3	2.667		605	131		736	860
0870	Humidity controller		8	1		210	49		259	305
0880	Load limiting controller		8	1		775	49		824	930
0890	Temperature controller		8	1		435	49		484	555
0900	Control panel readout									
0910	Panel up to 12 indicators	1 Stpi	1.20	6.667	Ea.	264	345		609	810
0914	Panel up to 24 indicators		.86	9.302		355	485		840	1,125
0918	Panel up to 48 indicators		.50	16		860	830		1,690	2,200
1000	Enthalpy control, boiler water temperature control									
1010	governed by outdoor temperature, with timer	1 Elec	3	2.667	Ea.	330	131		461	560
1300	Energy control/monitor									
1320	BTU computer meter and controller	1 Stpi	1	8	Ea.	1,425	415		1,840	2,175
1600	Flow meters									
1610	Gas	1 Stpi	4	2	Ea.	112	104		216	279
1620	Liquid	"	4	2	"	112	104		216	279
1640	Freezestat									
1644	20′ sensing element, adjustable	1 Stpi	3.50	2.286	Ea.	62	119		181	246
2000	Gauges, pressure or vacuum									
2100	2″ diameter dial	1 Stpi	32	.250	Ea.	24.50	13		37.50	46
2200	2-1/2″ diameter dial		32	.250		26.50	13		39.50	49
2300	3-1/2″ diameter dial		32	.250		35.50	13		48.50	59
2400	4-1/2″ diameter dial		32	.250		52.50	13		65.50	77.50
2700	Flanged iron case, black ring									
2800	3-1/2″ diameter dial	1 Stpi	32	.250	Ea.	102	13		115	131
2900	4-1/2″ diameter dial		32	.250		105	13		118	135
3000	6″ diameter dial		32	.250		167	13		180	202
3010	Steel case, 0 – 300 psi									
3012	2″ diam. dial	1 Stpi	16	.500	Ea.	5.60	26		31.60	45
3014	4″ diam. dial	"	16	.500	"	57	26		83	102
3020	Aluminum case, 0 – 300 psi									
3022	3-1/2″ diam. dial	1 Stpi	16	.500	Ea.	57	26		83	102
3024	4-1/2″ diam. dial		16	.500		134	26		160	187
3026	6″ diam. dial		16	.500		211	26		237	271
3028	8-1/2″ diam. dial		16	.500		305	26		331	375
3030	Brass case, 0 – 300 psi									
3032	2″ diam. dial	1 Stpi	16	.500	Ea.	34.50	26		60.50	77
3034	4-1/2″ diam. dial	"	16	.500	"	88	26		114	136
3040	Steel case, high pressure, 0 -10,000 psi									
3042	4-1/2″ diam. dial	1 Stpi	16	.500	Ea.	205	26		231	265
3044	6-1/2″ diam. dial		16	.500		211	26		237	271
3046	8-1/2″ diam. dial		16	.500		305	26		331	375
3080	Pressure gauge, differential, magnehelic									
3084	0 – 2″ W.C., with air filter kit	1 Stpi	6	1.333	Ea.	149	69		218	268
3300	For compound pressure-vacuum, add					18%				

23 09 53 – Pneumatic and Electric Control System for HVAC

23 09 53.10 Control Components		Crew	Daily Output	Labor-Hours	Unit	2010 Bare Costs Material	Labor	Equipment	Total	Total Incl O&P
3350	Humidistat									
3360	Pneumatic operation									
3361	Room humidistat, direct acting	1 Stpi	12	.667	Ea.	289	34.50		323.50	370
3362	Room humidistat, reverse acting		12	.667		289	34.50		323.50	370
3363	Room humidity transmitter		17	.471		310	24.50		334.50	375
3364	Duct mounted controller		12	.667		340	34.50		374.50	420
3365	Duct mounted transmitter		12	.667		310	34.50		344.50	390
3366	Humidity indicator, 3-1/2"		28	.286		123	14.85		137.85	157
3390	Electric operated	1 Shee	8	1		76.50	49		125.50	159
3400	Relays									
3430	Pneumatic/electric	1 Plum	16	.500	Ea.	310	26		336	380
3440	Pneumatic proportioning		8	1		209	52		261	310
3450	Pneumatic switching		12	.667		142	34.50		176.50	208
3460	Selector, 3 point		6	1.333		101	69.50		170.50	215
3470	Pneumatic time delay		8	1		264	52		316	370
3500	Sensor, air operated									
3520	Humidity	1 Plum	16	.500	Ea.	340	26		366	410
3540	Pressure		16	.500		58	26		84	103
3560	Temperature		12	.667		152	34.50		186.50	219
3600	Electric operated									
3620	Humidity	1 Elec	8	1	Ea.	85	49		134	167
3650	Pressure		8	1		1,500	49		1,549	1,725
3680	Temperature		10	.800		93.50	39		132.50	162
3700	Switches									
3710	Minimum position									
3720	Electrical	1 Stpi	4	2	Ea.	19.45	104		123.45	178
3730	Pneumatic	"	4	2	"	78.50	104		182.50	243
4000	Thermometers									
4100	Dial type, 3-1/2" diameter, vapor type, union connection	1 Stpi	32	.250	Ea.	227	13		240	269
4120	Liquid type, union connection		32	.250		440	13		453	505
4130	Remote reading, 15' capillary		32	.250		172	13		185	208
4500	Stem type, 6-1/2" case, 2" stem, 1/2" NPT		32	.250		60.50	13		73.50	86
4520	4" stem, 1/2" NPT		32	.250		74	13		87	101
4600	9" case, 3-1/2" stem, 3/4" NPT		28	.286		83	14.85		97.85	114
4620	6" stem, 3/4" NPT		28	.286		98	14.85		112.85	130
4640	8" stem, 3/4" NPT		28	.286		169	14.85		183.85	208
4660	12" stem, 1" NPT		26	.308		185	15.95		200.95	228
4670	Bi-metal, dial type, steel case brass stem									
4672	2" dial, 4" - 9" stem	1 Stpi	16	.500	Ea.	29.50	26		55.50	71.50
4673	2-1/2" dial, 4" - 9" stem		16	.500		29.50	26		55.50	71.50
4674	3-1/2" dial, 4" - 9" stem		16	.500		33	26		59	75
4680	Mercury filled, industrial, union connection type									
4682	Angle stem, 7" scale	1 Stpi	16	.500	Ea.	63	26		89	109
4683	9" scale		16	.500		169	26		195	225
4684	12" scale		16	.500		218	26		244	279
4686	Straight stem, 7" scale		16	.500		149	26		175	203
4687	9" scale		16	.500		151	26		177	205
4688	12" scale		16	.500		183	26		209	240
4690	Mercury filled, industrial, separable socket type									
4692	Angle stem, with socket, 7" scale	1 Stpi	16	.500	Ea.	59	26		85	104
4693	9" scale		16	.500		68	26		94	114
4694	12" scale		16	.500		139	26		165	192
4696	Straight stem, with socket, 7" scale		16	.500		29	26		55	71

23 09 53 – Pneumatic and Electric Control System for HVAC

23 09 53.10 Control Components		Crew	Daily Output	Labor-Hours	Unit	2010 Bare Costs Material	Labor	Equipment	Total	Total Incl O&P
4697	9″ scale	1 Stpi	16	.500	Ea.	46	26		72	89.50
4698	12″ scale	↓	16	.500	↓	98	26		124	147
5000	Thermostats									
5030	Manual	1 Shee	8	1	Ea.	44	49		93	123
5040	1 set back, electric, timed G		8	1		33.50	49		82.50	111
5050	2 set back, electric, timed G	↓	8	1		143	49		192	232
5100	Locking cover					17.10			17.10	18.80
5200	24 hour, automatic, clock G	1 Shee	8	1		133	49		182	221
5220	Electric, low voltage, 2 wire	1 Elec	13	.615		18	30		48	65
5230	3 wire		10	.800		22	39		61	83
5236	Heating/cooling, low voltage, with clock	↓	8	1	↓	181	49		230	273
5240	Pneumatic									
5250	Single temp., single pressure	1 Stpi	8	1	Ea.	200	52		252	297
5251	Dual pressure		8	1		280	52		332	390
5252	Dual temp., dual pressure		8	1		257	52		309	360
5253	Reverse acting w/averaging element		8	1		198	52		250	296
5254	Heating-cooling w/deadband		8	1		445	52		497	570
5255	Integral w/piston top valve actuator		8	1		177	52		229	273
5256	Dual temp., dual pressure		8	1		181	52		233	277
5257	Low limit, 8′ averaging element		8	1		171	52		223	266
5258	Room single temp. proportional		8	1		71.50	52		123.50	157
5260	Dual temp, direct acting for VAV fan		8	1		271	52		323	375
5262	Capillary tube type, 20′, 30°F to 100°F	↓	8	1	↓	209	52		261	310
5300	Transmitter, pneumatic									
5320	Temperature averaging element	Q-1	8	2	Ea.	116	93.50		209.50	267
5350	Pressure differential	1 Plum	7	1.143		975	59.50		1,034.50	1,175
5370	Humidity, duct		8	1		310	52		362	420
5380	Room		12	.667		310	34.50		344.50	390
5390	Temperature, with averaging element	↓	6	1.333	↓	164	69.50		233.50	284
5500	Timer/time clocks									
5510	7 day, 12 hr. battery	1 Stpi	2.60	3.077	Ea.	180	160		340	435
5600	Recorders									
5610	Hydrograph humidity, wall mtd., aluminum case, grad. chart									
5612	8″	1 Stpi	1.60	5	Ea.	635	260		895	1,075
5614	10″	″	1.60	5	″	700	260		960	1,150
5620	Thermo-hydrograph, wall mtd., grad. chart, 5′ SS probe									
5622	8″	1 Stpi	1.60	5	Ea.	560	260		820	1,000
5624	10″	″	1.60	5	″	1,125	260		1,385	1,650
5630	Time of operation, wall mtd., 1 pen cap, 24 hr. clock									
5632	6″	1 Stpi	1.60	5	Ea.	420	260		680	850
5634	8″	″	1.60	5	″	630	260		890	1,075
5640	Pressure & vacuum, bourdon tube type									
5642	Flush mount, 1 pen									
5644	4″	1 Stpi	1.60	5	Ea.	435	260		695	865
5645	6″		1.60	5		420	260		680	850
5646	8″		1.60	5		700	260		960	1,150
5648	10″	↓	1.60	5	↓	970	260		1,230	1,475
6000	Valves, motorized zone									
6100	Sweat connections, 1/2″ C x C	1 Stpi	20	.400	Ea.	156	21		177	203
6110	3/4″ C x C		20	.400		157	21		178	204
6120	1″ C x C		19	.421		204	22		226	257
6140	1/2″ C x C, with end switch, 2 wire		20	.400		154	21		175	201
6150	3/4″ C x C, with end switch, 2 wire	↓	20	.400	↓	165	21		186	213

23 09 53 – Pneumatic and Electric Control System for HVAC

23 09 53.10 Control Components		Crew	Daily Output	Labor-Hours	Unit	Material	2010 Bare Costs Labor	Equipment	Total	Total Incl O&P
6160	1" C x C, with end switch, 2 wire	1 Stpi	19	.421	Ea.	188	22		210	240
7090	Valves, motor controlled, including actuator									
7100	Electric motor actuated									
7200	Brass, two way, screwed									
7210	1/2" pipe size	L-6	36	.333	Ea.	208	17		225	254
7220	3/4" pipe size		30	.400		267	20.50		287.50	325
7230	1" pipe size		28	.429		310	22		332	380
7240	1-1/2" pipe size		19	.632		405	32		437	500
7250	2" pipe size		16	.750		675	38.50		713.50	800
7350	Brass, three way, screwed									
7360	1/2" pipe size	L-6	33	.364	Ea.	254	18.55		272.55	310
7370	3/4" pipe size		27	.444		296	22.50		318.50	360
7380	1" pipe size		25.50	.471		385	24		409	460
7384	1-1/4" pipe size		21	.571		490	29		519	585
7390	1-1/2" pipe size		17	.706		560	36		596	670
7400	2" pipe size		14	.857		745	43.50		788.50	885
7550	Iron body, two way, flanged									
7560	2-1/2" pipe size	L-6	4	3	Ea.	1,075	153		1,228	1,400
7570	3" pipe size		3	4		1,150	204		1,354	1,575
7580	4" pipe size		2	6		2,025	305		2,330	2,675
7850	Iron body, three way, flanged									
7860	2-1/2" pipe size	L-6	3	4	Ea.	1,125	204		1,329	1,525
7870	3" pipe size		2.50	4.800		1,250	245		1,495	1,775
7880	4" pipe size		2	6		1,625	305		1,930	2,225
8000	Pneumatic, air operated									
8050	Brass, two way, screwed									
8060	1/2" pipe size, class 250	1 Plum	24	.333	Ea.	164	17.35		181.35	206
8070	3/4" pipe size, class 250		20	.400		196	21		217	247
8080	1" pipe size, class 250		19	.421		229	22		251	285
8090	1-1/4" pipe size, class 125		15	.533		285	28		313	355
8100	1-1/2" pipe size, class 125		13	.615		365	32		397	450
8110	2" pipe size, class 125		11	.727		420	38		458	520
8180	Brass, three way, screwed									
8190	1/2" pipe size, class 250	1 Plum	22	.364	Ea.	180	18.95		198.95	227
8200	3/4" pipe size, class 250		18	.444		223	23		246	280
8210	1" pipe size, class 250		17	.471		262	24.50		286.50	325
8214	1-1/4" pipe size, class 250		14	.571		365	29.50		394.50	450
8220	1-1/2" pipe size, class 125		11	.727		435	38		473	535
8230	2" pipe size, class 125		9	.889		525	46.50		571.50	645
8450	Iron body, two way, flanged									
8460	2-1/2" pipe size, 250 lb. flanges	Q-1	5	3.200	Ea.	1,750	150		1,900	2,150
8470	3" pipe size, 250 lb. flanges		4.50	3.556		1,850	167		2,017	2,300
8480	4" pipe size, 250 lb. flanges		3	5.333		2,175	250		2,425	2,775
8510	2-1/2" pipe size, class 125		5	3.200		890	150		1,040	1,200
8520	3" pipe size, class 125		4.50	3.556		975	167		1,142	1,325
8530	4" pipe size, class 125		3	5.333		1,450	250		1,700	1,975
8540	5" pipe size, class 125	Q-2	3.40	7.059		2,950	345		3,295	3,775
8550	6" pipe size, class 125	"	3	8		3,400	390		3,790	4,300
8560	Iron body, three way, flanged									
8570	2-1/2" pipe size, class 125	Q-1	4.50	3.556	Ea.	960	167		1,127	1,300
8580	3" pipe size, class 125		4	4		1,100	187		1,287	1,500
8590	4" pipe size, class 125		2.50	6.400		2,225	300		2,525	2,900
8600	6" pipe size, class 125	Q-2	3	8		3,400	390		3,790	4,300

23 09 53.10 Control Components		Crew	Daily Output	Labor-Hours	Unit	Material	2010 Bare Costs Labor	Equipment	Total	Total Incl O&P
9005	Pneumatic system misc. components									
9010	Adding/subtracting repeater	1 Stpi	20	.400	Ea.	156	21		177	203
9020	Adjustable ratio network		16	.500		147	26		173	201
9030	Comparator		20	.400		300	21		321	360
9040	Cumulator									
9041	Air switching	1 Stpi	20	.400	Ea.	142	21		163	187
9042	Averaging		20	.400		128	21		149	171
9043	2:1 ratio		20	.400		288	21		309	345
9044	Sequencing		20	.400		90.50	21		111.50	131
9045	Two-position		16	.500		177	26		203	233
9046	Two-position pilot		16	.500		340	26		366	410
9050	Damper actuator									
9051	For smoke control	1 Stpi	8	1	Ea.	134	52		186	225
9052	For ventilation	"	8	1	"	193	52		245	290
9053	Series duplex pedestal mounted									
9054	For inlet vanes on fans, compressors	1 Stpi	6	1.333	Ea.	1,475	69		1,544	1,725
9060	Enthalpy logic center		16	.500		287	26		313	355
9070	High/low pressure selector		22	.364		57	18.85		75.85	91.50
9080	Signal limiter		20	.400		90.50	21		111.50	131
9081	Transmitter		28	.286		34.50	14.85		49.35	60
9090	Optimal start									
9092	Mass temperature transmitter	1 Stpi	18	.444	Ea.	135	23		158	184
9100	Step controller with positioner, time delay restrictor									
9110	recycler air valve, switches and cam settings									
9112	With cabinet									
9113	6 points	1 Stpi	4	2	Ea.	885	104		989	1,125
9114	6 points, 6 manual sequences		2	4		775	208		983	1,150
9115	8 points		3	2.667		965	138		1,103	1,275
9200	Pressure controller and switches									
9210	High static pressure limit	1 Stpi	8	1	Ea.	202	52		254	300
9220	Pressure transmitter		8	1		213	52		265	310
9230	Differential pressure transmitter		8	1		835	52		887	1,000
9240	Static pressure transmitter		8	1		835	52		887	1,000
9250	Proportional-only control, single input		6	1.333		410	69		479	555
9260	Proportional plus integral, single input		6	1.333		610	69		679	775
9270	Proportional-only, dual input		6	1.333		565	69		634	725
9280	Proportional plus integral, dual input		6	1.333		745	69		814	925
9281	Time delay for above proportional units		16	.500		143	26		169	197
9290	Proportional/2 position controller units		6	1.333		430	69		499	580
9300	Differential pressure control direct/reverse		6	1.333		365	69		434	505
9310	Air pressure reducing valve									
9311	1/8" size	1 Stpi	18	.444	Ea.	31.50	23		54.50	69
9315	Precision valve, 1/4" size		17	.471		176	24.50		200.50	230
9320	Air flow controller		12	.667		164	34.50		198.50	232
9330	Booster relay volume amplifier		18	.444		97	23		120	142
9340	Series restrictor, straight		60	.133		4.99	6.90		11.89	15.85
9341	T-fitting		40	.200		7.25	10.40		17.65	23.50
9342	In-line adjustable		48	.167		32	8.65		40.65	48.50
9350	Diode tee		40	.200		16.70	10.40		27.10	34
9351	Restrictor tee		40	.200		16.70	10.40		27.10	34
9360	Pneumatic gradual switch		8	1		279	52		331	385
9361	Selector switch		8	1		101	52		153	189
9370	Electro-pneumatic motor driven servo		10	.800		880	41.50		921.50	1,025

23 09 53 – Pneumatic and Electric Control System for HVAC

23 09 53.10 Control Components		Crew	Daily Output	Labor-Hours	Unit	2010 Bare Costs Material	Labor	Equipment	Total	Total Incl O&P
9390	Fan control switch and mounting base	1 Stpi	16	.500	Ea.	97.50	26		123.50	146
9400	Circulating pump sequencer	↓	14	.571	↓	1,150	29.50		1,179.50	1,325
9410	Pneumatic tubing, fittings and accessories									
9414	Tubing, urethane									
9415	1/8″ OD x 1/16″ ID	1 Stpi	120	.067	L.F.	.66	3.46		4.12	5.95
9416	1/4″ OD x 1/8″ ID		115	.070		1.30	3.61		4.91	6.85
9417	5/32″ OD x 3/32″ ID	↓	110	.073	↓	.74	3.77		4.51	6.45
9420	Coupling, straight									
9422	Barb x barb									
9423	1/4″ x 5/32″	1 Stpi	160	.050	Ea.	.74	2.60		3.34	4.70
9424	1/4″ x 1/4″		158	.051		.38	2.63		3.01	4.36
9425	3/8″ x 3/8″		154	.052		.51	2.70		3.21	4.60
9426	3/8″ x 1/4″		150	.053		.51	2.77		3.28	4.71
9427	1/2″ x 1/4″		148	.054		4.73	2.81		7.54	9.40
9428	1/2″ x 3/8″		144	.056		.99	2.88		3.87	5.40
9429	1/2″ x 1/2″	↓	140	.057	↓	.65	2.97		3.62	5.15
9440	Tube x tube									
9441	1/4″ x 1/4″	1 Stpi	100	.080	Ea.	2.03	4.15		6.18	8.45
9442	3/8″ x 1/4″		96	.083		3.77	4.32		8.09	10.65
9443	3/8″ x 3/8″		92	.087		2.52	4.51		7.03	9.50
9444	1/2″ x 3/8″		88	.091		4.05	4.72		8.77	11.50
9445	1/2″ x 1/2″	↓	84	.095	↓	4.62	4.94		9.56	12.50
9450	Elbow coupling									
9452	Barb x barb									
9454	1/4″ x 1/4″	1 Stpi	158	.051	Ea.	.87	2.63		3.50	4.90
9455	3/8″ x 3/8″		154	.052		2.37	2.70		5.07	6.65
9456	1/2″ x 1/2″	↓	140	.057	↓	1.32	2.97		4.29	5.90
9460	Tube x tube									
9462	1/2″ x 1/2″	1 Stpi	84	.095	Ea.	6.40	4.94		11.34	14.45
9470	Tee coupling									
9472	Barb x barb x barb									
9474	1/4″ x 1/4″ x 5/32″	1 Stpi	108	.074	Ea.	1.31	3.84		5.15	7.20
9475	1/4″ x 1/4″ x 1/4″		104	.077		.89	3.99		4.88	7
9476	3/8″ x 3/8″ x 5/32″		102	.078		2.84	4.07		6.91	9.20
9477	3/8″ x 3/8″ x 1/4″		99	.081		1.31	4.19		5.50	7.75
9478	3/8″ x 3/8″ x 3/8″		98	.082		1.40	4.24		5.64	7.90
9479	1/2″ x 1/2″ x 1/4″		96	.083		3.57	4.32		7.89	10.45
9480	1/2″ x 1/2″ x 3/8″		95	.084		2.97	4.37		7.34	9.80
9481	1/2″ x 1/2″ x 1/2″	↓	92	.087	↓	1.59	4.51		6.10	8.50
9484	Tube x tube x tube									
9485	1/4″ x 1/4″ x 1/4″	1 Stpi	66	.121	Ea.	4.03	6.30		10.33	13.90
9486	3/8″ x 1/4″ x 1/4″		64	.125		12.30	6.50		18.80	23.50
9487	3/8″ x 3/8″ x 1/4″		61	.131		5.85	6.80		12.65	16.60
9488	3/8″ x 3/8″ x 3/8″		60	.133		6.10	6.90		13	17.10
9489	1/2″ x 1/2″ x 3/8″		58	.138		12.75	7.15		19.90	25
9490	1/2″ x 1/2″ x 1/2″	↓	56	.143	↓	8.60	7.40		16	20.50
9492	Needle valve									
9494	Tube x tube									
9495	1/4″ x 1/4″	1 Stpi	86	.093	Ea.	7.15	4.83		11.98	15.15
9496	3/8″ x 3/8″	″	82	.098	″	10.60	5.05		15.65	19.30
9600	Electronic system misc. components									
9610	Electric motor damper actuator	1 Elec	8	1	Ea.	605	49		654	740
9620	Damper position indicator	↓	10	.800	↓	114	39		153	185

23 09 Instrumentation and Control for HVAC

23 09 53 – Pneumatic and Electric Control System for HVAC

23 09 53.10 Control Components		Crew	Daily Output	Labor-Hours	Unit	Material	2010 Bare Costs Labor	Equipment	Total	Total Incl O&P
9630	Pneumatic-electronic transducer	1 Elec	12	.667	Ea.	59.50	32.50		92	114
9700	Modulating step controller									
9701	2-5 steps	1 Elec	5.30	1.509	Ea.	2,000	74		2,074	2,300
9702	6-10 steps		3.20	2.500		2,575	123		2,698	3,025
9703	11-15 steps		2.70	2.963		3,125	145		3,270	3,675
9704	16-20 steps	↓	1.60	5	↓	3,625	245		3,870	4,375
9710	Staging network									
9740	Accessory unit regulator									
9750	Accessory power supply	1 Elec	18	.444	Ea.	175	22		197	225
9760	Step down transformer	"	16	.500	"	181	24.50		205.50	236

23 11 Facility Fuel Piping

23 11 13 – Facility Fuel-Oil Piping

23 11 13.10 Fuel Oil Specialties		Crew	Daily Output	Labor-Hours	Unit	Material	Labor	Equipment	Total	Total Incl O&P
0010	**FUEL OIL SPECIALTIES**									
0020	Foot valve, single poppet, metal to metal construction									
0040	Bevel seat, 1/2" diameter	1 Stpi	20	.400	Ea.	57	21		78	94
0060	3/4" diameter		18	.444		57	23		80	97
0080	1" diameter		16	.500		68	26		94	114
0100	1-1/4" diameter		15	.533		88	27.50		115.50	139
0120	1-1/2" diameter		13	.615		122	32		154	182
0140	2" diameter	↓	11	.727	↓	126	38		164	195
0160	Foot valve, double poppet, metal to metal construction									
0164	1" diameter	1 Stpi	15	.533	Ea.	152	27.50		179.50	209
0166	1-1/2" diameter	"	12	.667	"	171	34.50		205.50	240
0400	Fuel fill box, flush type									
0408	Nonlocking, watertight									
0410	1-1/2" diameter	1 Stpi	12	.667	Ea.	19.40	34.50		53.90	73.50
0440	2" diameter		10	.800		16.75	41.50		58.25	80.50
0450	2-1/2" diameter		9	.889		56	46		102	131
0460	3" diameter		7	1.143		54	59.50		113.50	149
0470	4" diameter	↓	5	1.600	↓	75	83		158	207
0500	Locking inner cover									
0510	2" diameter	1 Stpi	8	1	Ea.	82.50	52		134.50	169
0520	2-1/2" diameter		7	1.143		110	59.50		169.50	210
0530	3" diameter		5	1.600		118	83		201	254
0540	4" diameter	↓	4	2	↓	152	104		256	325
0600	Fuel system components									
0620	Spill container	1 Stpi	4	2	Ea.	620	104		724	835
0640	Fill adapter, 4", straight drop		8	1		72	52		124	157
0680	Fill cap, 4"		30	.267		35.50	13.85		49.35	60.50
0700	Extractor fitting, 4" x 1-1/2"		8	1		330	52		382	440
0740	Vapor hose adapter, 4"	↓	8	1		102	52		154	190
0760	Wood gage stick, 10'					14.55			14.55	16
1000	Oil filters, 3/8" IPT., 20 gal. per hour	1 Stpi	20	.400		19.25	21		40.25	52
1020	32 gal. per hour		18	.444		34.50	23		57.50	72.50
1040	40 gal. per hour		16	.500		37	26		63	80
1060	50 gal. per hour	↓	14	.571	↓	38.50	29.50		68	86.50
2000	Remote tank gauging system, self contained									
2100	Single tank kit/8 sensors inputs w/printer	1 Stpi	2.50	3.200	Ea.	3,400	166		3,566	4,000
2120	Two tank kit/8 sensors inputs w/printer	↓	2	4	↓	4,600	208		4,808	5,350

23 11 13 – Facility Fuel-Oil Piping

23 11 13.10 Fuel Oil Specialties		Crew	Daily Output	Labor-Hours	Unit	2010 Bare Costs Material	Labor	Equipment	Total	Total Incl O&P
3000	Valve, ball check, globe type, 3/8" diameter	1 Stpi	24	.333	Ea.	12.20	17.30		29.50	39.50
3500	Fusible, 3/8" diameter		24	.333		12.50	17.30		29.80	40
3600	1/2" diameter		24	.333		30.50	17.30		47.80	59.50
3610	3/4" diameter		20	.400		67	21		88	105
3620	1" diameter		19	.421		195	22		217	248
4000	Nonfusible, 3/8" diameter		24	.333		21	17.30		38.30	49.50
4500	Shutoff, gate type, lever handle, spring-fusible kit									
4520	1/4" diameter	1 Stpi	14	.571	Ea.	33	29.50		62.50	81
4540	3/8" diameter		12	.667		32	34.50		66.50	87
4560	1/2" diameter		10	.800		47	41.50		88.50	114
4570	3/4" diameter		8	1		82	52		134	168
4580	Lever handle, requires weight and fusible kit									
4600	1" diameter	1 Stpi	9	.889	Ea.	164	46		210	249
4620	1-1/4" diameter		8	1		189	52		241	286
4640	1-1/2" diameter		7	1.143		244	59.50		303.50	355
4660	2" diameter		6	1.333		305	69		374	440
4680	For fusible link, weight and braided wire, add					5%				
5000	Vent alarm, whistling signal					26			26	28.50
5500	Vent protector/breather, 1-1/4" diameter	1 Stpi	32	.250		10.40	13		23.40	31
5520	1-1/2" diameter		32	.250		12	13		25	32.50
5540	2" diameter		32	.250		21	13		34	42.50
5560	3" diameter		28	.286		33.50	14.85		48.35	59
5580	4" diameter		24	.333		37.50	17.30		54.80	67.50
5600	Dust cap, breather, 2"		40	.200		11.70	10.40		22.10	28.50
8000	Fuel oil and tank heaters									
8020	Electric, capacity rated at 230 volts									
8040	Immersion element in steel manifold									
8060	96 GPH at 50°F rise	Q-5	6.40	2.500	Ea.	870	117		987	1,125
8070	128 GPH at 50°F rise		6.20	2.581		900	121		1,021	1,175
8080	160 GPH at 50°F rise		5.90	2.712		1,000	127		1,127	1,300
8090	192 GPH at 50°F rise		5.50	2.909		1,100	136		1,236	1,400
8100	240 GPH at 50°F rise		5.10	3.137		1,250	147		1,397	1,600
8110	288 GPH at 50°F rise		4.60	3.478		1,425	162		1,587	1,800
8120	384 GPH at 50°F rise		3.10	5.161		1,650	241		1,891	2,175
8130	480 GPH at 50°F rise		2.30	6.957		2,075	325		2,400	2,750
8140	576 GPH at 50°F rise		2.10	7.619		2,075	355		2,430	2,800
8300	Suction stub, immersion type									
8320	75" long, 750 watts	1 Stpi	14	.571	Ea.	590	29.50		619.50	695
8330	99" long, 2000 watts		12	.667		830	34.50		864.50	965
8340	123" long, 3000 watts		10	.800		895	41.50		936.50	1,050
8660	Steam, cross flow, rated at 5 PSIG									
8680	42 GPH	Q-5	7	2.286	Ea.	1,375	107		1,482	1,650
8690	73 GPH		6.70	2.388		1,600	112		1,712	1,925
8700	112 GPH		6.20	2.581		1,650	121		1,771	2,000
8710	158 GPH		5.80	2.759		1,800	129		1,929	2,175
8720	187 GPH		4	4		2,475	187		2,662	3,000
8730	270 GPH		3.60	4.444		2,675	208		2,883	3,225
8740	365 GPH	Q-6	4.90	4.898		3,125	237		3,362	3,800
8750	635 GPH		3.70	6.486		3,950	315		4,265	4,825
8760	845 GPH		2.50	9.600		5,800	465		6,265	7,075
8770	1420 GPH		1.60	15		9,425	725		10,150	11,500
8780	2100 GPH		1.10	21.818		12,900	1,050		13,950	15,800

23 11 Facility Fuel Piping

23 11 23 – Facility Natural-Gas Piping

23 11 23.10 Gas Meters		Crew	Daily Output	Labor-Hours	Unit	2010 Bare Costs Material	Labor	Equipment	Total	Total Incl O&P
0010	**GAS METERS**									
4000	Residential									
4010	Gas meter, residential, 3/4" pipe size	1 Plum	14	.571	Ea.	233	29.50		262.50	300
4020	Gas meter, residential, 1" pipe size		12	.667		233	34.50		267.50	310
4030	Gas meter, residential, 1-1/4" pipe size	↓	10	.800	↓	242	41.50		283.50	330

23 12 Facility Fuel Pumps

23 12 13 – Facility Fuel-Oil Pumps

23 12 13.10 Pump and Motor Sets

		Crew	Daily Output	Labor-Hours	Unit	Material	Labor	Equipment	Total	Total Incl O&P
0010	**PUMP AND MOTOR SETS**									
1810	Light fuel and diesel oils									
1820	20 GPH 1/3 HP	Q-5	6	2.667	Ea.	790	125		915	1,050
1830	27 GPH, 1/3 HP		6	2.667		790	125		915	1,050
1840	80 GPH, 1/3 HP		5	3.200		790	149		939	1,100
1850	145 GPH, 1/2 HP		4	4		835	187		1,022	1,200
1860	277 GPH, 1 HP		4	4		930	187		1,117	1,300
1870	700 GPH, 1-1/2 HP		3	5.333		1,925	249		2,174	2,500
1880	1000 GPH, 2 HP		3	5.333		2,325	249		2,574	2,925
1890	1800 GPH, 5 HP	↓	1.80	8.889	↓	3,200	415		3,615	4,125

23 12 16 – Facility Gasoline Dispensing Pumps

23 12 16.20 Fuel Dispensing Equipment

		Crew	Daily Output	Labor-Hours	Unit	Material	Labor	Equipment	Total	Total Incl O&P
0010	**FUEL DISPENSING EQUIPMENT**									
1100	Product dispenser with vapor recovery for 6 nozzles, installed, not									
1110	including piping to storage tanks				Ea.	22,500			22,500	24,800

23 13 Facility Fuel-Storage Tanks

23 13 13 – Facility Underground Fuel-Oil, Storage Tanks

23 13 13.09 Single-Wall Steel Fuel-Oil Tanks

		Crew	Daily Output	Labor-Hours	Unit	Material	Labor	Equipment	Total	Total Incl O&P
0010	**SINGLE-WALL STEEL FUEL-OIL TANKS**									
5000	Steel underground, sti-P3, set in place, not incl. hold-down bars.									
5500	Excavation, pad, pumps and piping not included									
5510	Single wall, 500 gallon capacity, 7 gauge shell	Q-5	2.70	5.926	Ea.	1,975	277		2,252	2,600
5520	1,000 gallon capacity, 7 gauge shell	"	2.50	6.400		3,150	299		3,449	3,925
5530	2,000 gallon capacity, 1/4" thick shell	Q-7	4.60	6.957		5,125	345		5,470	6,150
5535	2,500 gallon capacity, 7 gauge shell	Q-5	3	5.333		5,650	249		5,899	6,600
5540	5,000 gallon capacity, 1/4" thick shell	Q-7	3.20	10		8,250	495		8,745	9,825
5560	10,000 gallon capacity, 1/4" thick shell		2	16		10,200	790		10,990	12,400
5580	15,000 gallon capacity, 5/16" thick shell		1.70	18.824		12,700	930		13,630	15,300
5600	20,000 gallon capacity, 5/16" thick shell		1.50	21.333		21,100	1,050		22,150	24,800
5610	25,000 gallon capacity, 3/8" thick shell		1.30	24.615		25,700	1,225		26,925	30,100
5620	30,000 gallon capacity, 3/8" thick shell		1.10	29.091		31,700	1,450		33,150	37,100
5630	40,000 gallon capacity, 3/8" thick shell		.90	35.556		34,600	1,750		36,350	40,600
5640	50,000 gallon capacity, 3/8" thick shell	↓	.80	40	↓	38,400	1,975		40,375	45,200

23 13 13.13 Dbl-Wall Steel, Undrgrnd Fuel-Oil, Stor. Tanks

		Crew	Daily Output	Labor-Hours	Unit	Material	Labor	Equipment	Total	Total Incl O&P
0010	**DOUBLE-WALL STEEL, UNDERGROUND FUEL-OIL, STORAGE TANKS**									
6200	Steel, underground, 360°, double wall, U.L. listed,									
6210	with sti-P3 corrosion protection,									
6220	(dielectric coating, cathodic protection, electrical									

23 13 13 – Facility Underground Fuel-Oil, Storage Tanks

23 13 13.13 Dbl-Wall Steel, Undrgrnd Fuel-Oil, Stor. Tanks		Crew	Daily Output	Labor-Hours	Unit	2010 Bare Costs Material	Labor	Equipment	Total	Total Incl O&P
6230	isolation) 30 year warranty,									
6240	not incl. manholes or hold-downs.									
6250	500 gallon capacity	Q-5	2.40	6.667	Ea.	3,900	310		4,210	4,775
6260	1,000 gallon capactiy	″	2.25	7.111		5,700	330		6,030	6,775
6270	2,000 gallon capacity	Q-7	4.16	7.692		7,475	380		7,855	8,800
6280	3,000 gallon capacity		3.90	8.205		10,600	405		11,005	12,200
6290	4,000 gallon capacity		3.64	8.791		12,500	435		12,935	14,500
6300	5,000 gallon capacity		2.91	10.997		14,100	545		14,645	16,300
6310	6,000 gallon capacity		2.42	13.223		16,000	655		16,655	18,600
6320	8,000 gallon capacity		2.08	15.385		17,800	760		18,560	20,800
6330	10,000 gallon capacity		1.82	17.582		21,900	870		22,770	25,400
6340	12,000 gallon capacity		1.70	18.824		23,700	930		24,630	27,500
6350	15,000 gallon capacity		1.33	24.060		29,000	1,200		30,200	33,700
6360	20,000 gallon capacity		1.33	24.060		38,400	1,200		39,600	44,000
6370	25,000 gallon capacity		1.16	27.586		60,000	1,375		61,375	68,000
6380	30,000 gallon capacity		1.03	31.068		72,500	1,525		74,025	82,000
6390	40,000 gallon capacity		.80	40		93,000	1,975		94,975	105,000
6395	50,000 gallon capacity		.73	43.836		113,000	2,175		115,175	127,500
6400	For hold-downs 500-2000 gal, add		16	2	Set	151	99		250	315
6410	For hold-downs 3000-6000 gal, add		12	2.667		305	132		437	535
6420	For hold-downs 8000-12,000 gal, add		11	2.909		365	144		509	620
6430	For hold-downs 15,000 gal, add		9	3.556		530	176		706	845
6440	For hold-downs 20,000 gal, add		8	4		610	198		808	965
6450	For hold-downs 20,000 gal plus, add		6	5.333		790	264		1,054	1,275
6500	For manways, add				Ea.	1,500			1,500	1,650
6600	In place with hold-downs									
6652	550 gallon capacity	Q-5	1.84	8.696	Ea.	4,050	405		4,455	5,075
23 13 13.23 Glass-Fiber-Reinfcd-Plastic, Fuel-Oil, Storage										
0010	**GLASS-FIBER-REINFCD-PLASTIC, UNDERGRND FUEL-OIL, STORAGE**									
0210	Fiberglass, underground, single wall, U.L. listed, not including									
0220	manway or hold-down strap									
0225	550 gallon capacity	Q-5	2.67	5.993	Ea.	3,000	280		3,280	3,725
0230	1,000 gallon capacity	″	2.46	6.504		3,650	305		3,955	4,475
0240	2,000 gallon capacity	Q-7	4.57	7.002		5,750	345		6,095	6,850
0250	4,000 gallon capacity		3.55	9.014		8,150	445		8,595	9,650
0260	6,000 gallon capacity		2.67	11.985		9,050	590		9,640	10,900
0270	8,000 gallon capacity		2.29	13.974		10,800	690		11,490	12,800
0280	10,000 gallon capacity		2	16		12,200	790		12,990	14,600
0282	12,000 gallon capacity		1.88	17.021		20,500	840		21,340	23,900
0284	15,000 gallon capacity		1.68	19.048		22,600	940		23,540	26,200
0290	20,000 gallon capacity		1.45	22.069		24,400	1,100		25,500	28,400
0300	25,000 gallon capacity		1.28	25		36,100	1,225		37,325	41,600
0320	30,000 gallon capacity		1.14	28.070		43,500	1,400		44,900	49,900
0340	40,000 gallon capacity		.89	35.955		62,000	1,775		63,775	70,500
0360	48,000 gallon capacity		.81	39.506		73,000	1,950		74,950	83,000
0500	For manway, fittings and hold-downs, add					20%	15%			
0600	For manways, add					1,800			1,800	1,975
1000	For helical heating coil, add	Q-5	2.50	6.400		3,700	299		3,999	4,500
1020	Fiberglass, underground, double wall, U.L. listed									
1030	includes manways, not incl. hold-down straps									
1040	600 gallon capacity	Q-5	2.42	6.612	Ea.	6,700	310		7,010	7,850
1050	1,000 gallon capacity	″	2.25	7.111		8,475	330		8,805	9,825

23 13 Facility Fuel-Storage Tanks

23 13 13 – Facility Underground Fuel-Oil, Storage Tanks

23 13 13.23 Glass-Fiber-Reinfcd-Plastic, Fuel-Oil, Storage		Crew	Daily Output	Labor-Hours	Unit	2010 Bare Costs Material	Labor	Equipment	Total	Total Incl O&P
1060	2,500 gallon capacity	Q-7	4.16	7.692	Ea.	12,900	380		13,280	14,800
1070	3,000 gallon capacity		3.90	8.205		13,900	405		14,305	15,900
1080	4,000 gallon capacity		3.64	8.791		15,700	435		16,135	18,000
1090	6,000 gallon capacity		2.42	13.223		15,700	655		16,355	18,300
1100	8,000 gallon capacity		2.08	15.385		20,100	760		20,860	23,300
1110	10,000 gallon capacity		1.82	17.582		23,300	870		24,170	26,900
1120	12,000 gallon capacity		1.70	18.824		28,700	930		29,630	32,900
1122	15,000 gallon capacity		1.52	21.053		37,200	1,050		38,250	42,500
1124	20,000 gallon capacity		1.33	24.060		44,600	1,200		45,800	51,000
1126	25,000 gallon capacity		1.16	27.586		59,500	1,375		60,875	67,500
1128	30,000 gallon capacity	↓	1.03	31.068		71,500	1,525		73,025	81,000
1140	For hold-down straps, add				↓	2%	10%			
1150	For hold-downs 500-4000 gal., add	Q-7	16	2	Set	425	99		524	620
1160	For hold-downs 5000-15000 gal., add		8	4		850	198		1,048	1,225
1170	For hold-downs 20,000 gal., add		5.33	6.004		1,275	297		1,572	1,850
1180	For hold-downs 25,000 gal., add		4	8		1,700	395		2,095	2,475
1190	For hold-downs 30,000 gal., add	↓	2.60	12.308	↓	2,550	610		3,160	3,700
2210	Fiberglass, underground, single wall, U.L. listed, including									
2220	hold-down straps, no manways									
2225	550 gallon capacity	Q-5	2	8	Ea.	3,425	375		3,800	4,300
2230	1,000 gallon capacity	"	1.88	8.511		4,075	395		4,470	5,075
2240	2,000 gallon capacity	Q-7	3.55	9.014		6,175	445		6,620	7,450
2250	4,000 gallon capacity		2.90	11.034		8,575	545		9,120	10,300
2260	6,000 gallon capacity		2	16		9,900	790		10,690	12,100
2270	8,000 gallon capacity		1.78	17.978		11,600	890		12,490	14,100
2280	10,000 gallon capacity		1.60	20		13,000	990		13,990	15,800
2282	12,000 gallon capacity		1.52	21.053		21,400	1,050		22,450	25,100
2284	15,000 gallon capacity		1.39	23.022		23,400	1,150		24,550	27,500
2290	20,000 gallon capacity		1.14	28.070		25,600	1,400		27,000	30,300
2300	25,000 gallon capacity		.96	33.333		37,800	1,650		39,450	44,000
2320	30,000 gallon capacity	↓	.80	40	↓	46,000	1,975		47,975	53,500
3020	Fiberglass, underground, double wall, U.L. listed									
3030	includes manways and hold-down straps									
3040	600 gallon capacity	Q-5	1.86	8.602	Ea.	7,125	400		7,525	8,450
3050	1,000 gallon capacity	"	1.70	9.412		8,900	440		9,340	10,500
3060	2,500 gallon capacity	Q-7	3.29	9.726		13,300	480		13,780	15,300
3070	3,000 gallon capacity		3.13	10.224		14,300	505		14,805	16,600
3080	4,000 gallon capacity		2.93	10.922		16,100	540		16,640	18,600
3090	6,000 gallon capacity		1.86	17.204		16,500	850		17,350	19,500
3100	8,000 gallon capacity		1.65	19.394		21,000	960		21,960	24,500
3110	10,000 gallon capacity		1.48	21.622		24,100	1,075		25,175	28,100
3120	12,000 gallon capacity		1.40	22.857		29,500	1,125		30,625	34,200
3122	15,000 gallon capacity		1.28	25		38,000	1,225		39,225	43,700
3124	20,000 gallon capacity		1.06	30.189		45,900	1,500		47,400	52,500
3126	25,000 gallon capacity		.90	35.556		61,000	1,750		62,750	70,000
3128	30,000 gallon capacity	↓	.74	43.243	↓	74,000	2,150		76,150	84,500

23 13 23 – Facility Aboveground Fuel-Oil, Storage Tanks

23 13 23.13 Vertical, Steel, Abvground Fuel-Oil, Stor. Tanks		Crew	Daily Output	Labor-Hours	Unit	Material	Labor	Equipment	Total	Total Incl O&P
0010	**VERTICAL, STEEL, ABOVEGROUND FUEL-OIL, STORAGE TANKS**									
4000	Fixed roof oil storage tanks, steel, (1 BBL=42 GAL w/foundation 3'D x 1'W)									
4200	5,000 barrels				Ea.				194,000	213,500
4300	24,000 barrels				↓				333,500	367,000

23 13 Facility Fuel-Storage Tanks

23 13 23 – Facility Aboveground Fuel-Oil, Storage Tanks

23 13 23.13 Vertical, Steel, Abvground Fuel-Oil, Stor. Tanks		Crew	Daily Output	Labor-Hours	Unit	2010 Bare Costs Material	Labor	Equipment	Total	Total Incl O&P
4500	56,000 barrels				Ea.				729,000	802,000
4600	110,000 barrels								1,060,000	1,166,000
4800	143,000 barrels								1,250,000	1,375,000
4900	224,000 barrels								1,360,000	1,496,000
5100	Floating roof gasoline tanks, steel, 5,000 barrels (w/foundation 3'D x 1'W)								204,000	225,000
5200	25,000 barrels								381,000	419,000
5400	55,000 barrels								839,000	923,000
5500	100,000 barrels								1,253,000	1,379,000
5700	150,000 barrels								1,532,000	1,685,000
5800	225,000 barrels								2,300,000	2,783,000

23 13 23.16 Horizontal, Stl, Abvgrd Fuel-Oil, Storage Tanks

		Crew	Daily Output	Labor-Hours	Unit	Material	Labor	Equipment	Total	Total Incl O&P
0010	**HORIZONTAL, STEEL, ABOVEGROUND FUEL-OIL, STORAGE TANKS**									
3000	Steel, storage, above ground, including cradles, coating,									
3020	fittings, not including foundation, pumps or piping									
3040	Single wall, 275 gallon	Q-5	5	3.200	Ea.	375	149		524	635
3060	550 gallon	"	2.70	5.926		2,250	277		2,527	2,900
3080	1,000 gallon	Q-7	5	6.400		2,825	315		3,140	3,600
3100	1,500 gallon		4.75	6.737		4,275	335		4,610	5,200
3120	2,000 gallon		4.60	6.957		4,875	345		5,220	5,875
3140	5,000 gallon		3.20	10		6,075	495		6,570	7,450
3150	10,000 gallon		2	16		18,200	790		18,990	21,200
3160	15,000 gallon		1.70	18.824		20,700	930		21,630	24,200
3170	20,000 gallon		1.45	22.069		26,000	1,100		27,100	30,200
3180	25,000 gallon		1.30	24.615		31,900	1,225		33,125	36,900
3190	30,000 gallon		1.10	29.091		35,700	1,450		37,150	41,500
3320	Double wall, 500 gallon capacity	Q-5	2.40	6.667		2,375	310		2,685	3,075
3330	2000 gallon capacity	Q-7	4.15	7.711		8,250	380		8,630	9,650
3340	4000 gallon capacity		3.60	8.889		14,700	440		15,140	16,900
3350	6000 gallon capacity		2.40	13.333		17,400	660		18,060	20,100
3360	8000 gallon capacity		2	16		22,300	790		23,090	25,700
3370	10000 gallon capacity		1.80	17.778		25,000	880		25,880	28,800
3380	15000 gallon capacity		1.50	21.333		37,900	1,050		38,950	43,300
3390	20000 gallon capacity		1.30	24.615		43,200	1,225		44,425	49,300
3400	25000 gallon capacity		1.15	27.826		52,500	1,375		53,875	59,500
3410	30000 gallon capacity		1	32		57,500	1,575		59,075	66,000

23 21 Hydronic Piping and Pumps

23 21 20 – Hydronic HVAC Piping Specialties

23 21 20.10 Air Control

		Crew	Daily Output	Labor-Hours	Unit	Material	Labor	Equipment	Total	Total Incl O&P
0010	**AIR CONTROL**									
0030	Air separator, with strainer									
0040	2" diameter	Q-5	6	2.667	Ea.	835	125		960	1,100
0080	2-1/2" diameter		5	3.200		945	149		1,094	1,275
0100	3" diameter		4	4		1,450	187		1,637	1,850
0120	4" diameter		3	5.333		2,075	249		2,324	2,675
0130	5" diameter	Q-6	3.60	6.667		2,650	325		2,975	3,400
0140	6" diameter		3.40	7.059		3,175	340		3,515	4,000
0160	8" diameter		3	8		4,750	385		5,135	5,800
0180	10" diameter		2.20	10.909		7,375	530		7,905	8,900
0200	12" diameter		1.70	14.118		12,400	685		13,085	14,600
0210	14" diameter		1.30	18.462		14,700	895		15,595	17,600

23 21 20 – Hydronic HVAC Piping Specialties

23 21 20.10 Air Control		Crew	Daily Output	Labor-Hours	Unit	Material	Labor	Equipment	Total	Total Incl O&P
						2010 Bare Costs				
0220	16" diameter	Q-6	1	24	Ea.	22,100	1,150		23,250	26,200
0230	18" diameter		.80	30		29,100	1,450		30,550	34,200
0240	20" diameter	↓	.60	40	↓	32,600	1,925		34,525	38,800
0300	Without strainer									
0310	2" diameter	Q-5	6	2.667	Ea.	645	125		770	895
0320	2-1/2" diameter		5	3.200		775	149		924	1,075
0330	3" diameter		4	4		1,075	187		1,262	1,475
0340	4" diameter		3	5.333		1,700	249		1,949	2,250
0350	5" diameter	↓	2.40	6.667		2,300	310		2,610	3,025
0360	6" diameter	Q-6	3.40	7.059		2,650	340		2,990	3,425
0370	8" diameter		3	8		3,650	385		4,035	4,575
0380	10" diameter		2.20	10.909		5,350	530		5,880	6,700
0390	12" diameter		1.70	14.118		8,250	685		8,935	10,100
0400	14" diameter		1.30	18.462		12,300	895		13,195	15,000
0410	16" diameter		1	24		17,300	1,150		18,450	20,800
0420	18" diameter		.80	30		22,500	1,450		23,950	27,000
0430	20" diameter	↓	.60	40	↓	27,000	1,925		28,925	32,600
1000	Micro-bubble separator for total air removal, closed loop system									
1010	Requires bladder type tank in system.									
1020	Water (hot or chilled) or glycol system									
1030	Threaded									
1040	3/4" diameter	1 Stpi	20	.400	Ea.	83.50	21		104.50	123
1050	1" diameter		19	.421		94	22		116	136
1060	1-1/4" diameter		16	.500		130	26		156	182
1070	1-1/2" diameter		13	.615		169	32		201	234
1080	2" diameter	↓	11	.727		820	38		858	955
1090	2-1/2" diameter	Q-5	15	1.067		910	50		960	1,075
1100	3" diameter		13	1.231		1,300	57.50		1,357.50	1,500
1110	4" diameter	↓	10	1.600	↓	1,400	74.50		1,474.50	1,650
1230	Flanged									
1250	2" diameter	1 Stpi	8	1	Ea.	1,050	52		1,102	1,225
1260	2-1/2" diameter	Q-5	5	3.200		1,125	149		1,274	1,475
1270	3" diameter		4.50	3.556		1,525	166		1,691	1,925
1280	4" diameter	↓	3	5.333		1,675	249		1,924	2,225
1290	5" diameter	Q-6	3.40	7.059		2,700	340		3,040	3,475
1300	6" diameter	"	3	8	↓	3,275	385		3,660	4,175
1400	Larger sizes available									
1590	With extended tank dirt catcher									
1600	Flanged									
1620	2" diameter	1 Stpi	7.40	1.081	Ea.	1,375	56		1,431	1,575
1630	2-1/2" diameter	Q-5	4.20	3.810		1,450	178		1,628	1,875
1640	3" diameter		4	4		1,900	187		2,087	2,350
1650	4" diameter	↓	2.80	5.714		2,050	267		2,317	2,650
1660	5" diameter	Q-6	3.10	7.742		3,025	375		3,400	3,875
1670	6" diameter	"	2.90	8.276	↓	3,725	400		4,125	4,675
1800	With drain/dismantle/cleaning access flange									
1810	Flanged									
1820	2" diameter	1 Stpi	7	1.143	Ea.	2,150	59.50		2,209.50	2,475
1830	2-1/2" diameter	Q-5	4	4		2,300	187		2,487	2,825
1840	3" diameter		3.60	4.444		3,125	208		3,333	3,750
1850	4" diameter	↓	2.50	6.400		3,450	299		3,749	4,250
1860	5" diameter	Q-6	2.80	8.571		5,150	415		5,565	6,300
1870	6" diameter	"	2.60	9.231	↓	6,725	445		7,170	8,075

23 21 20 – Hydronic HVAC Piping Specialties

23 21 20.10 Air Control		Crew	Daily Output	Labor-Hours	Unit	Material	2010 Bare Costs Labor	Equipment	Total	Total Incl O&P
2000	Boiler air fitting, separator									
2010	1-1/4"	Q-5	22	.727	Ea.	186	34		220	255
2020	1-1/2"		20	.800		355	37.50		392.50	450
2021	2"		18	.889		615	41.50		656.50	740
2400	Compression tank air fitting									
2410	For tanks 9" to 24" diameter	1 Stpi	15	.533	Ea.	58	27.50		85.50	106
2420	For tanks 100 gallon or larger	"	12	.667	"	231	34.50		265.50	305

23 21 20.14 Air Purging Scoop		Crew	Daily Output	Labor-Hours	Unit	Material	Labor	Equipment	Total	Total Incl O&P
0010	**AIR PURGING SCOOP**, With tappings.									
0020	For air vent and expansion tank connection									
0100	1" pipe size, threaded	1 Stpi	19	.421	Ea.	21.50	22		43.50	56.50
0110	1-1/4" pipe size, threaded		15	.533		21.50	27.50		49	65
0120	1-1/2" pipe size, threaded		13	.615		47	32		79	99.50
0130	2" pipe size, threaded		11	.727		54	38		92	116
0140	2-1/2" pipe size, threaded	Q-5	15	1.067		115	50		165	201
0150	3" pipe size, threaded		13	1.231		150	57.50		207.50	251
0160	4" 150 lb. flanges		3	5.333		335	249		584	745

23 21 20.18 Automatic Air Vent		Crew	Daily Output	Labor-Hours	Unit	Material	Labor	Equipment	Total	Total Incl O&P
0010	**AUTOMATIC AIR VENT**									
0020	Cast iron body, stainless steel internals, float type									
0060	1/2" NPT inlet, 300 psi	1 Stpi	12	.667	Ea.	88	34.50		122.50	149
0140	3/4" NPT inlet, 300 psi		12	.667		88	34.50		122.50	149
0180	1/2" NPT inlet, 250 psi		10	.800		275	41.50		316.50	360
0220	3/4" NPT inlet, 250 psi		10	.800		275	41.50		316.50	360
0260	1" NPT inlet, 250 psi		10	.800		410	41.50		451.50	515
0340	1-1/2" NPT inlet, 250 psi	Q-5	12	1.333		870	62.50		932.50	1,050
0380	2" NPT inlet, 250 psi	"	12	1.333		870	62.50		932.50	1,050
0600	Forged steel body, stainless steel internals, float type									
0640	1/2" NPT inlet, 750 psi	1 Stpi	12	.667	Ea.	920	34.50		954.50	1,050
0680	3/4" NPT inlet, 750 psi		12	.667		920	34.50		954.50	1,050
0760	3/4" NPT inlet, 1000 psi		10	.800		1,375	41.50		1,416.50	1,575
0800	1" NPT inlet, 1000 psi	Q-5	12	1.333		1,375	62.50		1,437.50	1,625
0880	1-1/2" NPT inlet, 1000 psi		10	1.600		3,875	74.50		3,949.50	4,375
0920	2" NPT inlet, 1000 psi		10	1.600		3,875	74.50		3,949.50	4,375
1100	Formed steel body, noncorrosive									
1110	1/8" NPT inlet 150 psi	1 Stpi	32	.250	Ea.	10.95	13		23.95	31.50
1120	1/4" NPT inlet 150 psi		32	.250		37	13		50	60
1130	3/4" NPT inlet 150 psi		32	.250		37	13		50	60
1300	Chrome plated brass, automatic/manual, for radiators									
1310	1/8" NPT inlet, nickel plated brass	1 Stpi	32	.250	Ea.	6.30	13		19.30	26.50

23 21 20.22 Circuit Sensor		Crew	Daily Output	Labor-Hours	Unit	Material	Labor	Equipment	Total	Total Incl O&P
0010	**CIRCUIT SENSOR**, Flow meter									
0020	Metering stations									
0040	Wafer orifice insert type									
0060	2-1/2" pipe size	Q-5	12	1.333	Ea.	175	62.50		237.50	287
0100	3" pipe size		11	1.455		196	68		264	320
0140	4" pipe size		8	2		224	93.50		317.50	385
0180	5" pipe size		7.30	2.192		287	102		389	470
0220	6" pipe size		6.40	2.500		345	117		462	550
0260	8" pipe size	Q-6	5.30	4.528		475	219		694	855
0280	10" pipe size		4.60	5.217		545	253		798	975
0360	12" pipe size		4.20	5.714		840	277		1,117	1,350

23 21 20.22 Circuit Sensor		Crew	Daily Output	Labor-Hours	Unit	Material	2010 Bare Costs Labor	Equipment	Total	Total Incl O&P
2000	In-line probe type									
2200	Copper, soldered									
2210	1/2" size	1 Stpi	24	.333	Ea.	164	17.30		181.30	206
2220	3/4" size		20	.400		170	21		191	218
2230	1" size		19	.421		193	22		215	246
2240	1-1/4" size		15	.533		203	27.50		230.50	265
2250	1-1/2" size		13	.615		242	32		274	315
2260	2" size		11	.727		269	38		307	355
2270	2-1/2" size	Q-5	15	1.067		284	50		334	385
2280	3" size	"	13	1.231		305	57.50		362.50	425
2500	Brass, threaded									
2510	1" size	1 Stpi	19	.421	Ea.	203	22		225	256
2520	1-1/4" size		15	.533		224	27.50		251.50	288
2530	1-1/2" size		13	.615		269	32		301	345
2540	2" size		11	.727		284	38		322	365
2550	2-1/2" size	Q-5	15	1.067		305	50		355	415
2560	3" size	"	13	1.231		315	57.50		372.50	430
2700	Steel, threaded									
2710	1" size	1 Stpi	19	.421	Ea.	195	22		217	247
2720	1-1/4" size		15	.533		211	27.50		238.50	274
2730	1-1/2" size		13	.615		240	32		272	310
2740	2" size		11	.727		267	38		305	350
2750	2-1/2" size	Q-5	15	1.067		284	50		334	385
2760	3" size	"	13	1.231		310	57.50		367.50	425
3000	Pitot tube probe type									
3100	Weld-on mounting with probe									
3110	2" size	Q-17	18	.889	Ea.	220	41.50	3.25	264.75	310
3120	2-1/2" size		18	.889		224	41.50	3.25	268.75	310
3130	3" size		18	.889		237	41.50	3.25	281.75	325
3140	4" size		18	.889		315	41.50	3.25	359.75	410
3150	5" size		17	.941		355	44	3.44	402.44	460
3160	6" size		17	.941		460	44	3.44	507.44	575
3170	8" size		16	1		650	46.50	3.66	700.16	795
3180	10" size		15	1.067		705	50	3.90	758.90	855
3190	12" size		15	1.067		755	50	3.90	808.90	910
3200	14" size		14	1.143		865	53.50	4.18	922.68	1,025
3210	16" size		14	1.143		1,025	53.50	4.18	1,082.68	1,200
3220	18" size		13	1.231		1,100	57.50	4.50	1,162	1,300
3230	20" size		13	1.231		1,175	57.50	4.50	1,237	1,400
3240	24" size		12	1.333		1,250	62.50	4.88	1,317.38	1,475
3400	Weld-on wet tap with probe									
3410	2" size, 5/16" probe dia.	Q-17	14	1.143	Ea.	535	53.50	4.18	592.68	670
3420	2-1/2" size, 5/16" probe dia.		14	1.143		540	53.50	4.18	597.68	680
3430	3" size, 5/16" probe dia.		14	1.143		575	53.50	4.18	632.68	720
3440	4" size, 5/16" probe dia.		14	1.143		670	53.50	4.18	727.68	820
3450	5" size, 3/8" probe dia.		13	1.231		685	57.50	4.50	747	845
3460	6" size, 3/8" probe dia.		13	1.231		655	57.50	4.50	717	810
3470	8" size, 1/2" probe dia.		12	1.333		680	62.50	4.88	747.38	850
3480	10" size, 3/4" probe dia.		11	1.455		750	68	5.30	823.30	935
3490	12" size, 3/4" probe dia.		11	1.455		795	68	5.30	868.30	980
3500	14" size, 1" probe dia.		10	1.600		895	74.50	5.85	975.35	1,100
3510	16" size, 1" probe dia.		10	1.600		1,075	74.50	5.85	1,155.35	1,300
3520	18" size, 1" probe dia.		9	1.778		1,150	83	6.50	1,239.50	1,400

23 21 20 – Hydronic HVAC Piping Specialties

23 21 20.22 Circuit Sensor		Crew	Daily Output	Labor-Hours	Unit	Material	2010 Bare Costs Labor	Equipment	Total	Total Incl O&P
3530	20" size, 1" probe dia.	Q-17	8	2	Ea.	1,225	93.50	7.30	1,325.80	1,500
3540	24" size, 1" probe dia.	↓	8	2	↓	1,375	93.50	7.30	1,475.80	1,675
3700	Clamp-on wet tap with probe									
3710	2" size, 5/16" probe dia.	Q-5	16	1	Ea.	765	46.50		811.50	910
3720	2-1/2" size, 5/16" probe dia.		16	1		665	46.50		711.50	805
3730	3" size, 5/16" probe dia.		15	1.067		665	50		715	810
3740	4" size, 5/16" probe dia.		15	1.067		770	50		820	925
3750	5" size, 3/8" probe dia.		14	1.143		815	53.50		868.50	975
3760	6" size, 3/8" probe dia.		14	1.143		865	53.50		918.50	1,025
3770	8" size, 1/2" probe dia.		13	1.231		950	57.50		1,007.50	1,125
3780	10" size, 3/4" probe dia.		12	1.333		995	62.50		1,057.50	1,200
3790	12" size, 3/4" probe dia.		12	1.333		1,150	62.50		1,212.50	1,375
3800	14" size, 1" probe dia.		11	1.455		1,500	68		1,568	1,750
3810	16" size, 1" probe dia.		11	1.455		1,725	68		1,793	2,000
3820	18" size, 1" probe dia.		10	1.600		1,875	74.50		1,949.50	2,150
3830	20" size, 1" probe dia		9	1.778		2,125	83		2,208	2,475
3840	24" size, 1" probe dia.	↓	9	1.778	↓	2,200	83		2,283	2,550
4000	Wet tap drills									
4010	3/8" dia. for 5/16" probe				Ea.	128			128	141
4020	1/2" dia. for 3/8" probe					132			132	146
4030	5/8" dia for 1/2" probe					158			158	173
4040	7/8" dia for 3/4" probe					205			205	226
4050	1-3/16" dia for 1" probe				↓	219			219	241
4100	Wet tap punch									
4110	5/16" dia for 5/16" probe				Ea.	132			132	145
4120	3/8" dia for 3/8" probe					142			142	156
4130	1/2" dia for 1/2" probe				↓	170			170	187
4150	Note: labor for wet tap drill or punch									
4160	is included with the mounting and probe assembly									
9000	Readout instruments									
9200	Gauges									
9220	GPM gauge				Ea.	2,150			2,150	2,375
9230	Dual gauge kit				"	2,250			2,250	2,475

23 21 20.26 Circuit Setter		Crew	Daily Output	Labor-Hours	Unit	Material	Labor	Equipment	Total	Total Incl O&P
0010	**CIRCUIT SETTER**, Balance valve									
0012	Bronze body, soldered									
0013	1/2" size	1 Stpi	24	.333	Ea.	64.50	17.30		81.80	97
0014	3/4" size	"	20	.400	"	68	21		89	106
0018	Threaded									
0019	1/2" pipe size	1 Stpi	22	.364	Ea.	66.50	18.85		85.35	102
0020	3/4" pipe size		20	.400		71.50	21		92.50	110
0040	1" pipe size		18	.444		92.50	23		115.50	137
0050	1-1/4" pipe size		15	.533		136	27.50		163.50	191
0060	1-1/2" pipe size		12	.667		160	34.50		194.50	228
0080	2" pipe size	↓	10	.800		229	41.50		270.50	315
0100	2-1/2" pipe size	Q-5	15	1.067		520	50		570	645
0120	3" pipe size	"	10	1.600	↓	730	74.50		804.50	910
0130	Cast iron body, flanged									
0136	3" pipe size, flanged	Q-5	4	4	Ea.	730	187		917	1,075
0140	4" pipe size	"	3	5.333		1,100	249		1,349	1,575
0200	For differential meter, accurate to 1%, add					840			840	925
0300	For differential meter 400 psi/200°F continuous, add				↓	2,075			2,075	2,275

	23 21 20.30 Cocks, Drains and Specialties	Crew	Daily Output	Labor-Hours	Unit	Material	2010 Bare Costs Labor	Equipment	Total	Total Incl O&P
0010	**COCKS, DRAINS AND SPECIALTIES**									
1000	Boiler drain									
1010	Pipe thread to hose									
1020	Bronze									
1030	1/2" size	1 Stpi	36	.222	Ea.	8.70	11.55		20.25	27
1040	3/4" size	"	34	.235	"	9.45	12.20		21.65	28.50
1100	Solder to hose									
1110	Bronze									
1120	1/2" size	1 Stpi	46	.174	Ea.	7.60	9.05		16.65	22
1130	3/4" size	"	44	.182	"	8.30	9.45		17.75	23.50
1600	With built-in vacuum breaker									
1610	1/2" IP or solder	1 Stpi	36	.222	Ea.	28.50	11.55		40.05	48.50
1630	With tamper proof vacuum breaker									
1640	1/2" IP or solder	1 Stpi	36	.222	Ea.	32	11.55		43.55	53
1650	3/4" IP or solder	"	34	.235	"	32.50	12.20		44.70	54.50
3000	Cocks									
3010	Air, lever or tee handle									
3020	Bronze, single thread									
3030	1/8" size	1 Stpi	52	.154	Ea.	8.25	8		16.25	21
3040	1/4" size		46	.174		8.70	9.05		17.75	23
3050	3/8" size		40	.200		8.80	10.40		19.20	25.50
3060	1/2" size		36	.222		10.40	11.55		21.95	29
3100	Bronze, double thread									
3110	1/8" size	1 Stpi	26	.308	Ea.	10.35	15.95		26.30	35.50
3120	1/4" size		22	.364		10.85	18.85		29.70	40.50
3130	3/8" size		18	.444		11.25	23		34.25	47
3140	1/2" size		15	.533		13.80	27.50		41.30	56.50
4000	Steam									
4010	Bronze									
4020	1/8" size	1 Stpi	26	.308	Ea.	31.50	15.95		47.45	58.50
4030	1/4" size		22	.364		31.50	18.85		50.35	63
4040	3/8" size		18	.444		39.50	23		62.50	78
4050	1/2" size		15	.533		42	27.50		69.50	87.50
4060	3/4" size		14	.571		50	29.50		79.50	99.50
4070	1" size		13	.615		64.50	32		96.50	119
4080	1-1/4" size		11	.727		99	38		137	166
4090	1-1/2" size		10	.800		128	41.50		169.50	202
4100	2" size		9	.889		212	46		258	300
4300	Bronze, 3 way									
4320	1/4" size	1 Stpi	15	.533	Ea.	47	27.50		74.50	93
4330	3/8" size		12	.667		47	34.50		81.50	104
4340	1/2" size		10	.800		47	41.50		88.50	114
4350	3/4" size		9.50	.842		80.50	43.50		124	154
4360	1" size		8.50	.941		97	49		146	180
4370	1-1/4" size		7.50	1.067		145	55.50		200.50	243
4380	1-1/2" size		6.50	1.231		169	64		233	282
4390	2" size		6	1.333		250	69		319	380
4500	Gauge cock, brass									
4510	1/4" FPT	1 Stpi	24	.333	Ea.	8.70	17.30		26	35.50
4512	1/4" MPT	"	24	.333	"	11	17.30		28.30	38
4600	Pigtail, steam syphon									
4604	1/4"	1 Stpi	24	.333	Ea.	15.95	17.30		33.25	43.50

23 21 20.30 Cocks, Drains and Specialties		Crew	Daily Output	Labor-Hours	Unit	Material	2010 Bare Costs Labor	Equipment	Total	Total Incl O&P
4650	Snubber valve									
4654	1/4"	1 Stpi	22	.364	Ea.	12.15	18.85		31	42
4660	Nipple, black steel									
4664	1/4" x 3"	1 Stpi	37	.216	Ea.	.99	11.20		12.19	17.90

23 21 20.34 Dielectric Unions		Crew	Daily Output	Labor-Hours	Unit	Material	Labor	Equipment	Total	Total Incl O&P
0010	**DIELECTRIC UNIONS**, Standard gaskets for water and air.									
0020	250 psi maximum pressure									
0280	Female IPT to sweat, straight									
0300	1/2" pipe size	1 Plum	24	.333	Ea.	4.16	17.35		21.51	30.50
0340	3/4" pipe size		20	.400		5.05	21		26.05	36.50
0360	1" pipe size		19	.421		6.95	22		28.95	40.50
0380	1-1/4" pipe size		15	.533		10.80	28		38.80	53.50
0400	1-1/2" pipe size		13	.615		16.20	32		48.20	66
0420	2" pipe size	↓	11	.727	↓	22	38		60	80.50
0580	Female IPT to brass pipe thread, straight									
0600	1/2" pipe size	1 Plum	24	.333	Ea.	9.85	17.35		27.20	37
0640	3/4" pipe size		20	.400		10.85	21		31.85	43
0660	1" pipe size		19	.421		20.50	22		42.50	55.50
0680	1-1/4" pipe size		15	.533		25	28		53	69
0700	1-1/2" pipe size		13	.615		36	32		68	87.50
0720	2" pipe size	↓	11	.727	↓	70.50	38		108.50	134
0780	Female IPT to female IPT, straight									
0800	1/2" pipe size	1 Plum	24	.333	Ea.	9.20	17.35		26.55	36
0840	3/4" pipe size		20	.400		10.40	21		31.40	42.50
0860	1" pipe size		19	.421		13.90	22		35.90	48.50
0880	1-1/4" pipe size		15	.533		18.80	28		46.80	62
0900	1-1/2" pipe size		13	.615		29	32		61	79.50
0920	2" pipe size	↓	11	.727	↓	42	38		80	103
2000	175 psi maximum pressure									
2180	Female IPT to sweat									
2240	2" pipe size	1 Plum	9	.889	Ea.	122	46.50		168.50	204
2260	2-1/2" pipe size	Q-1	15	1.067		133	50		183	221
2280	3" pipe size		14	1.143		182	53.50		235.50	281
2300	4" pipe size	↓	11	1.455	↓	480	68		548	630
2480	Female IPT to brass pipe									
2500	1-1/2" pipe size	1 Plum	11	.727	Ea.	152	38		190	224
2540	2" pipe size	"	9	.889		179	46.50		225.50	267
2560	2-1/2" pipe size	Q-1	15	1.067		262	50		312	365
2580	3" pipe size		14	1.143		305	53.50		358.50	415
2600	4" pipe size	↓	11	1.455	↓	550	68		618	705

23 21 20.38 Expansion Couplings		Crew	Daily Output	Labor-Hours	Unit	Material	Labor	Equipment	Total	Total Incl O&P
0010	**EXPANSION COUPLINGS**, Hydronic									
0100	Copper to copper, sweat									
1000	Baseboard riser fitting, 5" stub by coupling 12" long									
1020	1/2" diameter	1 Stpi	24	.333	Ea.	16.35	17.30		33.65	44
1040	3/4" diameter		20	.400		23.50	21		44.50	56.50
1060	1" diameter		19	.421		32.50	22		54.50	68.50
1080	1-1/4" diameter	↓	15	.533	↓	42.50	27.50		70	88
1180	9" Stub by tubing 8" long									
1200	1/2" diameter	1 Stpi	24	.333	Ea.	14.70	17.30		32	42
1220	3/4" diameter		20	.400		19.45	21		40.45	52.50
1240	1" diameter	↓	19	.421	↓	27	22		49	62.50

23 21 20.38 Expansion Couplings		Crew	Daily Output	Labor-Hours	Unit	Material	2010 Bare Costs Labor	Equipment	Total	Total Incl O&P
1260	1-1/4" diameter	1 Stpi	15	.533	Ea.	38	27.50		65.50	83

23 21 20.42 Expansion Joints		Crew	Daily Output	Labor-Hours	Unit	Material	Labor	Equipment	Total	Total Incl O&P
0010	**EXPANSION JOINTS**									
0100	Bellows type, neoprene cover, flanged spool									
0140	6" face to face, 1-1/4" diameter	1 Stpi	11	.727	Ea.	248	38		286	330
0160	1-1/2" diameter	"	10.60	.755		248	39		287	330
0180	2" diameter	Q-5	13.30	1.203		251	56		307	360
0190	2-1/2" diameter		12.40	1.290		260	60.50		320.50	375
0200	3" diameter		11.40	1.404		291	65.50		356.50	420
0210	4" diameter		8.40	1.905		315	89		404	480
0220	5" diameter		7.60	2.105		385	98.50		483.50	570
0230	6" diameter		6.80	2.353		395	110		505	600
0240	8" diameter		5.40	2.963		460	138		598	710
0250	10" diameter		5	3.200		635	149		784	925
0260	12" diameter		4.60	3.478		730	162		892	1,050
0480	10" face to face, 2" diameter		13	1.231		360	57.50		417.50	480
0500	2-1/2" diameter		12	1.333		380	62.50		442.50	515
0520	3" diameter		11	1.455		390	68		458	525
0540	4" diameter		8	2		440	93.50		533.50	625
0560	5" diameter		7	2.286		525	107		632	740
0580	6" diameter		6	2.667		545	125		670	785
0600	8" diameter		5	3.200		650	149		799	940
0620	10" diameter		4.60	3.478		720	162		882	1,025
0640	12" diameter		4	4		890	187		1,077	1,250
0660	14" diameter		3.80	4.211		1,100	197		1,297	1,500
0680	16" diameter		2.90	5.517		1,275	258		1,533	1,775
0700	18" diameter		2.50	6.400		1,425	299		1,724	2,025
0720	20" diameter		2.10	7.619		1,500	355		1,855	2,175
0740	24" diameter		1.80	8.889		1,750	415		2,165	2,550
0760	26" diameter		1.40	11.429		1,975	535		2,510	2,950
0780	30" diameter		1.20	13.333		2,200	625		2,825	3,350
0800	36" diameter		1	16		2,725	745		3,470	4,125
1000	Bellows with internal sleeves and external covers									
1010	Stainless steel, 150 lb.									
1020	With male threads									
1040	3/4" diameter	1 Stpi	20	.400	Ea.	132	21		153	177
1050	1" diameter		19	.421		140	22		162	187
1060	1-1/4" diameter		16	.500		146	26		172	200
1070	1-1/2" diameter		13	.615		178	32		210	244
1080	2" diameter		11	.727		228	38		266	305
1110	With flanged ends									
1120	1-1/4" diameter	1 Stpi	12	.667	Ea.	182	34.50		216.50	252
1130	1-1/2" diameter		11	.727		223	38		261	300
1140	2" diameter		9	.889		283	46		329	380
1150	3" diameter	Q-5	8	2		495	93.50		588.50	685
1160	4" diameter	"	5	3.200		840	149		989	1,150
1170	5" diameter	Q-6	6	4		995	194		1,189	1,400
1180	6" diameter		5	4.800		1,175	232		1,407	1,650
1190	8" diameter		4.50	5.333		1,675	258		1,933	2,225
1200	10" diameter		4	6		2,025	291		2,316	2,650
1210	12" diameter		3.60	6.667		3,150	325		3,475	3,950

23 21 20 – Hydronic HVAC Piping Specialties

23 21 20.46 Expansion Tanks		Crew	Daily Output	Labor-Hours	Unit	Material	Labor	Equipment	Total	Total Incl O&P
						2010 Bare Costs				
0010	**EXPANSION TANKS**									
1400	Plastic, corrosion resistant, see Plumbing Cost Data									
1507	Fiberglass and steel single/double wall storage, see Div. 23 13 13									
1512	Tank leak detection systems, see Div. 28 33 33.50									
2000	Steel, liquid expansion, ASME, painted, 15 gallon capacity	Q-5	17	.941	Ea.	520	44		564	640
2020	24 gallon capacity		14	1.143		555	53.50		608.50	690
2040	30 gallon capacity		12	1.333		580	62.50		642.50	735
2060	40 gallon capacity		10	1.600		680	74.50		754.50	860
2080	60 gallon capacity		8	2		815	93.50		908.50	1,050
2100	80 gallon capacity		7	2.286		875	107		982	1,125
2120	100 gallon capacity		6	2.667		1,175	125		1,300	1,475
2130	120 gallon capacity		5	3.200		1,275	149		1,424	1,625
2140	135 gallon capacity		4.50	3.556		1,325	166		1,491	1,700
2150	175 gallon capacity		4	4		2,075	187		2,262	2,550
2160	220 gallon capacity		3.60	4.444		2,350	208		2,558	2,875
2170	240 gallon capacity		3.30	4.848		2,450	226		2,676	3,050
2180	305 gallon capacity		3	5.333		3,450	249		3,699	4,175
2190	400 gallon capacity	↓	2.80	5.714	↓	4,250	267		4,517	5,075
2360	Galvanized									
2370	15 gallon capacity	Q-5	17	.941	Ea.	915	44		959	1,075
2380	24 gallon capacity		14	1.143		1,025	53.50		1,078.50	1,200
2390	30 gallon capacity		12	1.333		1,100	62.50		1,162.50	1,300
2400	40 gallon capacity		10	1.600		1,275	74.50		1,349.50	1,525
2410	60 gallon capacity		8	2		1,475	93.50		1,568.50	1,775
2420	80 gallon capacity		7	2.286		1,700	107		1,807	2,025
2430	100 gallon capacity		6	2.667		2,200	125		2,325	2,600
2440	120 gallon capacity		5	3.200		2,375	149		2,524	2,825
2450	135 gallon capacity		4.50	3.556		2,500	166		2,666	3,000
2460	175 gallon capacity		4	4		4,050	187		4,237	4,725
2470	220 gallon capacity		3.60	4.444		4,650	208		4,858	5,425
2480	240 gallon capacity		3.30	4.848		4,850	226		5,076	5,675
2490	305 gallon capacity		3	5.333		7,025	249		7,274	8,100
2500	400 gallon capacity		2.80	5.714		8,650	267		8,917	9,925
3000	Steel ASME expansion, rubber diaphragm, 19 gal. cap. accept.		12	1.333		2,150	62.50		2,212.50	2,475
3020	31 gallon capacity		8	2		2,375	93.50		2,468.50	2,775
3040	61 gallon capacity		6	2.667		3,350	125		3,475	3,875
3060	79 gallon capacity		5	3.200		3,425	149		3,574	4,000
3080	119 gallon capacity		4	4		3,625	187		3,812	4,250
3100	158 gallon capacity		3.80	4.211		5,025	197		5,222	5,825
3120	211 gallon capacity		3.30	4.848		5,800	226		6,026	6,750
3140	317 gallon capacity		2.80	5.714		7,600	267		7,867	8,750
3160	422 gallon capacity		2.60	6.154		11,200	287		11,487	12,800
3180	528 gallon capacity	↓	2.40	6.667	↓	12,300	310		12,610	14,100

23 21 20.50 Float Valves

		Crew	Daily Output	Labor-Hours	Unit	Material	Labor	Equipment	Total	Total Incl O&P
0010	**FLOAT VALVES**									
0020	With ball and bracket									
0030	Single seat, threaded									
0040	Brass body									
0050	1/2"	1 Stpi	11	.727	Ea.	74.50	38		112.50	139
0060	3/4"		9	.889		83	46		129	161
0070	1"		7	1.143		107	59.50		166.50	207
0080	1-1/2"	↓	4.50	1.778	↓	157	92.50		249.50	310

23 21 20 – Hydronic HVAC Piping Specialties

23 21 20.50 Float Valves		Crew	Daily Output	Labor-Hours	Unit	Material	Labor	Equipment	Total	Total Incl O&P
						2010 Bare Costs				
0090	2"	1 Stpi	3.60	2.222	Ea.	160	115		275	350
0300	For condensate receivers, CI, in-line mount									
0320	1" inlet	1 Stpi	7	1.143	Ea.	123	59.50		182.50	224
0360	For condensate receiver, CI, external float, flanged tank mount									
0370	3/4" inlet	1 Stpi	5	1.600	Ea.	99	83		182	233
23 21 20.54 Flow Check Control										
0010	**FLOW CHECK CONTROL**									
0100	Bronze body, soldered									
0110	3/4" size	1 Stpi	20	.400	Ea.	56.50	21		77.50	93
0120	1" size	"	19	.421	"	67.50	22		89.50	107
0200	Cast iron body, threaded									
0210	3/4" size	1 Stpi	20	.400	Ea.	43.50	21		64.50	78.50
0220	1" size		19	.421		48.50	22		70.50	86
0230	1-1/4" size		15	.533		59.50	27.50		87	107
0240	1-1/2" size		13	.615		90.50	32		122.50	148
0250	2" size	↓	11	.727	↓	130	38		168	200
0300	Flanged inlet, threaded outlet									
0310	2-1/2" size	1 Stpi	8	1	Ea.	291	52		343	400
0320	3" size	Q-5	9	1.778		345	83		428	505
0330	4" size	"	7	2.286	↓	675	107		782	905
23 21 20.58 Hydronic Heating Control Valves										
0010	**HYDRONIC HEATING CONTROL VALVES**									
0050	Hot water, nonelectric, thermostatic									
0100	Radiator supply, 1/2" diameter	1 Stpi	24	.333	Ea.	53.50	17.30		70.80	85
0120	3/4" diameter		20	.400		57	21		78	94
0140	1" diameter		19	.421		72	22		94	112
0160	1-1/4" diameter	↓	15	.533	↓	103	27.50		130.50	155
0500	For low pressure steam, add					25%				
1000	Manual, radiator supply									
1010	1/2" pipe size, angle union	1 Stpi	24	.333	Ea.	35.50	17.30		52.80	65
1020	3/4" pipe size, angle union		20	.400		45	21		66	80.50
1030	1" pipe size, angle union	↓	19	.421	↓	58.50	22		80.50	97
1100	Radiator, balancing, straight, sweat connections									
1110	1/2" pipe size	1 Stpi	24	.333	Ea.	13.65	17.30		30.95	41
1120	3/4" pipe size		20	.400		19.05	21		40.05	52
1130	1" pipe size	↓	19	.421	↓	31	22		53	67
1200	Steam, radiator, supply									
1210	1/2" pipe size, angle union	1 Stpi	24	.333	Ea.	35	17.30		52.30	64.50
1220	3/4" pipe size, angle union		20	.400		38	21		59	73
1230	1" pipe size, angle union		19	.421		44	22		66	81.50
1240	1-1/4" pipe size, angle union	↓	15	.533	↓	56.50	27.50		84	104
8000	System balancing and shut-off									
8020	Butterfly, quarter turn, calibrated, threaded or solder									
8040	Bronze, -30°F to +350°F, pressure to 175 psi									
8060	1/2" size	1 Stpi	22	.364	Ea.	12.95	18.85		31.80	43
8070	3/4" size		20	.400		21	21		42	54
8080	1" size		19	.421		25.50	22		47.50	61
8090	1-1/4" size		15	.533		40.50	27.50		68	86
8100	1-1/2" size		13	.615		53	32		85	107
8110	2" size	↓	11	.727	↓	66.50	38		104.50	130

23 21 20 – Hydronic HVAC Piping Specialties

	23 21 20.62 Liquid Drainers	Crew	Daily Output	Labor-Hours	Unit	2010 Bare Costs Material	Labor	Equipment	Total	Total Incl O&P
0010	**LIQUID DRAINERS**									
0100	Guided lever type									
0110	Cast iron body, threaded									
0120	1/2" pipe size	1 Stpi	22	.364	Ea.	90	18.85		108.85	128
0130	3/4" pipe size		20	.400		240	21		261	295
0140	1" pipe size		19	.421		395	22		417	470
0150	1-1/2" pipe size		11	.727		835	38		873	970
0160	2" pipe size		8	1		835	52		887	995
0200	Forged steel body, threaded									
0210	1/2" pipe size	1 Stpi	18	.444	Ea.	890	23		913	1,025
0220	3/4" pipe size		16	.500		890	26		916	1,025
0230	1" pipe size		14	.571		1,325	29.50		1,354.50	1,525
0240	1-1/2" pipe size	Q-5	10	1.600		3,875	74.50		3,949.50	4,375
0250	2" pipe size	"	8	2		3,875	93.50		3,968.50	4,425
0300	Socket weld									
0310	1/2" pipe size	Q-17	16	1	Ea.	1,025	46.50	3.66	1,075.16	1,200
0320	3/4" pipe size		14	1.143		1,025	53.50	4.18	1,082.68	1,200
0330	1" pipe size		12	1.333		1,500	62.50	4.88	1,567.38	1,750
0340	1-1/2" pipe size		9	1.778		1,500	83	6.50	1,589.50	1,775
0350	2" pipe size		7	2.286		4,025	107	8.35	4,140.35	4,625
0400	Flanged									
0410	1/2" pipe size	1 Stpi	12	.667	Ea.	1,775	34.50		1,809.50	2,000
0420	3/4" pipe size		10	.800		1,775	41.50		1,816.50	2,000
0430	1" pipe size		7	1.143		2,275	59.50		2,334.50	2,600
0440	1-1/2" pipe size	Q-5	9	1.778		5,025	83		5,108	5,650
0450	2" pipe size	"	7	2.286		5,025	107		5,132	5,675
1000	Fixed lever and snap action type									
1100	Cast iron body, threaded									
1110	1/2" pipe size	1 Stpi	22	.364	Ea.	141	18.85		159.85	184
1120	3/4" pipe size		20	.400		370	21		391	440
1130	1" pipe size		19	.421		370	22		392	445
1200	Forged steel body, threaded									
1210	1/2" pipe size	1 Stpi	18	.444	Ea.	965	23		988	1,075
1220	3/4" pipe size		16	.500		965	26		991	1,100
1230	1" pipe size		14	.571		2,375	29.50		2,404.50	2,675
1240	1-1/4" pipe size		12	.667		2,375	34.50		2,409.50	2,675
2000	High pressure high leverage type									
2100	Forged steel body, threaded									
2110	1/2" pipe size	1 Stpi	18	.444	Ea.	1,900	23		1,923	2,100
2120	3/4" pipe size		16	.500		1,900	26		1,926	2,125
2130	1" pipe size		14	.571		3,175	29.50		3,204.50	3,525
2140	1-1/4" pipe size		12	.667		3,175	34.50		3,209.50	3,525
2150	1-1/2" pipe size	Q-5	10	1.600		4,775	74.50		4,849.50	5,350
2160	2" pipe size	"	8	2		4,775	93.50		4,868.50	5,400
2200	Socket weld									
2210	1/2" pipe size	Q-17	16	1	Ea.	2,050	46.50	3.66	2,100.16	2,325
2220	3/4" pipe size		14	1.143		2,050	53.50	4.18	2,107.68	2,325
2230	1" pipe size		12	1.333		3,325	62.50	4.88	3,392.38	3,750
2240	1-1/4" pipe size		11	1.455		3,325	68	5.30	3,398.30	3,750
2250	1-1/2" pipe size		9	1.778		4,925	83	6.50	5,014.50	5,550
2260	2" pipe size		7	2.286		4,925	107	8.35	5,040.35	5,600
2300	Flanged									

23 21 20 – Hydronic HVAC Piping Specialties

23 21 20.62 Liquid Drainers		Crew	Daily Output	Labor-Hours	Unit	Material	2010 Bare Costs Labor	Equipment	Total	Total Incl O&P
2310	1/2" pipe size	1 Stpi	12	.667	Ea.	2,750	34.50		2,784.50	3,075
2320	3/4" pipe size		10	.800		2,750	41.50		2,791.50	3,075
2330	1" pipe size		7	1.143		4,325	59.50		4,384.50	4,850
2340	1-1/4" pipe size		6	1.333		4,325	69		4,394	4,850
2350	1-1/2" pipe size	Q-5	9	1.778		5,925	83		6,008	6,625
2360	2" pipe size	"	7	2.286		5,925	107		6,032	6,650
3000	Guided lever dual gravity type									
3100	Cast iron body, threaded									
3110	1/2" pipe size	1 Stpi	22	.364	Ea.	345	18.85		363.85	410
3120	3/4" pipe size		20	.400		345	21		366	410
3130	1" pipe size		19	.421		505	22		527	590
3140	1-1/2" pipe size		11	.727		970	38		1,008	1,125
3150	2" pipe size		8	1		970	52		1,022	1,150
3200	Forged steel body, threaded									
3210	1/2" pipe size	1 Stpi	18	.444	Ea.	1,025	23		1,048	1,150
3220	3/4" pipe size		16	.500		1,025	26		1,051	1,175
3230	1" pipe size		14	.571		1,525	29.50		1,554.50	1,725
3240	1-1/2" pipe size	Q-5	10	1.600		4,075	74.50		4,149.50	4,600
3250	2" pipe size	"	8	2		4,075	93.50		4,168.50	4,650
3300	Socket weld									
3310	1/2" pipe size	Q-17	16	1	Ea.	1,175	46.50	3.66	1,225.16	1,350
3320	3/4" pipe size		14	1.143		1,175	53.50	4.18	1,232.68	1,350
3330	1" pipe size		12	1.333		1,675	62.50	4.88	1,742.38	1,950
3340	1-1/2" pipe size		9	1.778		4,225	83	6.50	4,314.50	4,775
3350	2" pipe size		7	2.286		4,225	107	8.35	4,340.35	4,825
3400	Flanged									
3410	1/2" pipe size	1 Stpi	12	.667	Ea.	1,875	34.50		1,909.50	2,125
3420	3/4" pipe size		10	.800		1,875	41.50		1,916.50	2,125
3430	1" pipe size		7	1.143		2,450	59.50		2,509.50	2,800
3440	1-1/2" pipe size	Q-5	9	1.778		5,225	83		5,308	5,875
3450	2" pipe size	"	7	2.286		5,225	107		5,332	5,900

23 21 20.66 Monoflow Tee Fitting		Crew	Daily Output	Labor-Hours	Unit	Material	Labor	Equipment	Total	Total Incl O&P
0010	**MONOFLOW TEE FITTING**									
1100	For one pipe hydronic, supply and return									
1110	Copper, soldered									
1120	3/4" x 1/2" size	1 Stpi	13	.615	Ea.	17.25	32		49.25	67
1130	1" x 1/2" size		12	.667		17.25	34.50		51.75	71
1140	1" x 3/4" size		11	.727		17.25	38		55.25	75.50
1150	1-1/4" x 1/2" size		11	.727		18	38		56	76.50
1160	1-1/4" x 3/4" size		10	.800		18	41.50		59.50	82
1170	1-1/2" x 3/4" size		10	.800		25	41.50		66.50	89.50
1180	1-1/2" x 1" size		9	.889		25	46		71	96.50
1190	2" x 3/4" size		9	.889		39	46		85	112
1200	2" x 1" size		8	1		39	52		91	121

23 21 20.70 Steam Traps		Crew	Daily Output	Labor-Hours	Unit	Material	Labor	Equipment	Total	Total Incl O&P
0010	**STEAM TRAPS**									
0030	Cast iron body, threaded									
0040	Inverted bucket									
0050	1/2" pipe size	1 Stpi	12	.667	Ea.	131	34.50		165.50	196
0070	3/4" pipe size		10	.800		228	41.50		269.50	310
0100	1" pipe size		9	.889		350	46		396	455
0120	1-1/4" pipe size		8	1		530	52		582	660

23 21 20.70 Steam Traps	Crew	Daily Output	Labor-Hours	Unit	Material	2010 Bare Costs Labor	Equipment	Total	Total Incl O&P
0130 1-1/2" pipe size	1 Stpi	7	1.143	Ea.	665	59.50		724.50	825
0140 2" pipe size	↓	6	1.333	↓	1,050	69		1,119	1,250
0200 With thermic vent & check valve									
0202 1/2" pipe size	1 Stpi	12	.667	Ea.	154	34.50		188.50	221
0204 3/4" pipe size		10	.800		261	41.50		302.50	350
0206 1" pipe size		9	.889		380	46		426	485
0208 1-1/4" pipe size		8	1		575	52		627	710
0210 1-1/2" pipe size		7	1.143		635	59.50		694.50	790
0212 2" pipe size	↓	6	1.333	↓	975	69		1,044	1,175
1000 Float & thermostatic, 15 psi									
1010 3/4" pipe size	1 Stpi	16	.500	Ea.	105	26		131	155
1020 1" pipe size		15	.533		127	27.50		154.50	181
1030 1-1/4" pipe size		13	.615		154	32		186	217
1040 1-1/2" pipe size		9	.889		223	46		269	315
1060 2" pipe size	↓	6	1.333		410	69		479	555
1100 For vacuum breaker, add				↓	56			56	61.50
1110 Differential condensate controller									
1120 With union and manual metering valve									
1130 1/2" pipe size	1 Stpi	24	.333	Ea.	490	17.30		507.30	565
1140 3/4" pipe size		20	.400		575	21		596	660
1150 1" pipe size		19	.421		770	22		792	880
1160 1-1/4" pipe size		15	.533		1,025	27.50		1,052.50	1,200
1170 1-1/2" pipe size		13	.615		1,250	32		1,282	1,425
1180 2" pipe size	↓	11	.727	↓	1,625	38		1,663	1,850
1290 Brass body, threaded									
1300 Thermostatic, angle union, 25 psi									
1310 1/2" pipe size	1 Stpi	24	.333	Ea.	49.50	17.30		66.80	80.50
1320 3/4" pipe size		20	.400		75	21		96	114
1330 1" pipe size	↓	19	.421	↓	111	22		133	155
1990 Carbon steel body, threaded									
2000 Controlled disc, integral strainer & blow down									
2010 Threaded									
2020 3/8" pipe size	1 Stpi	24	.333	Ea.	221	17.30		238.30	269
2030 1/2" pipe size		24	.333		221	17.30		238.30	269
2040 3/4" pipe size		20	.400		296	21		317	355
2050 1" pipe size	↓	19	.421	↓	380	22		402	455
2100 Socket weld									
2110 3/8" pipe size	Q-17	17	.941	Ea.	263	44	3.44	310.44	360
2120 1/2" pipe size		16	1		263	46.50	3.66	313.16	365
2130 3/4" pipe size		14	1.143		345	53.50	4.18	402.68	460
2140 1" pipe size	↓	12	1.333	↓	430	62.50	4.88	497.38	575
2170 Flanged, 600 lb. ASA									
2180 3/8" pipe size	1 Stpi	12	.667	Ea.	595	34.50		629.50	705
2190 1/2" pipe size		12	.667		595	34.50		629.50	705
2200 3/4" pipe size		10	.800		710	41.50		751.50	840
2210 1" pipe size	↓	7	1.143	↓	840	59.50		899.50	1,025
5000 Forged steel body									
5010 Inverted bucket									
5020 Threaded									
5030 1/2" pipe size	1 Stpi	12	.667	Ea.	395	34.50		429.50	485
5040 3/4" pipe size		10	.800		680	41.50		721.50	810
5050 1" pipe size		9	.889		1,025	46		1,071	1,200
5060 1-1/4" pipe size	↓	8	1	↓	1,475	52		1,527	1,700

23 21 20 – Hydronic HVAC Piping Specialties

23 21 20.70 Steam Traps		Crew	Daily Output	Labor-Hours	Unit	Material	2010 Bare Costs Labor	Equipment	Total	Total Incl O&P
5070	1-1/2" pipe size	1 Stpi	7	1.143	Ea.	2,100	59.50		2,159.50	2,400
5080	2" pipe size	↓	6	1.333	↓	3,700	69		3,769	4,175
5100	Socket weld									
5110	1/2" pipe size	Q-17	16	1	Ea.	475	46.50	3.66	525.16	595
5120	3/4" pipe size		14	1.143		760	53.50	4.18	817.68	925
5130	1" pipe size		12	1.333		1,100	62.50	4.88	1,167.38	1,325
5140	1-1/4" pipe size		10	1.600		1,575	74.50	5.85	1,655.35	1,850
5150	1-1/2" pipe size		9	1.778		2,175	83	6.50	2,264.50	2,525
5160	2" pipe size	↓	7	2.286	↓	3,800	107	8.35	3,915.35	4,350
5200	Flanged									
5210	1/2" pipe size	1 Stpi	12	.667	Ea.	730	34.50		764.50	855
5220	3/4" pipe size		10	.800		1,200	41.50		1,241.50	1,375
5230	1" pipe size		7	1.143		1,625	59.50		1,684.50	1,900
5240	1-1/4" pipe size	↓	6	1.333		2,225	69		2,294	2,550
5250	1-1/2" pipe size	Q-5	9	1.778		2,825	83		2,908	3,225
5260	2" pipe size	"	7	2.286	↓	4,425	107		4,532	5,025
6000	Stainless steel body									
6010	Inverted bucket									
6020	Threaded									
6030	1/2" pipe size	1 Stpi	12	.667	Ea.	130	34.50		164.50	195
6040	3/4" pipe size		10	.800		171	41.50		212.50	250
6050	1" pipe size	↓	9	.889	↓	605	46		651	735
6100	Threaded with thermic vent									
6110	1/2" pipe size	1 Stpi	12	.667	Ea.	148	34.50		182.50	215
6120	3/4" pipe size		10	.800		191	41.50		232.50	272
6130	1" pipe size	↓	9	.889	↓	630	46		676	765
6200	Threaded with check valve									
6210	1/2" pipe size	1 Stpi	12	.667	Ea.	186	34.50		220.50	256
6220	3/4" pipe size	"	10	.800	"	191	41.50		232.50	272
6300	Socket weld									
6310	1/2" pipe size	Q-17	16	1	Ea.	150	46.50	3.66	200.16	239
6320	3/4" pipe size	"	14	1.143	"	195	53.50	4.18	252.68	299

23 21 20.74 Strainers, Basket Type

		Crew	Daily Output	Labor-Hours	Unit	Material	Labor	Equipment	Total	Total Incl O&P
0010	**STRAINERS, BASKET TYPE**, Perforated stainless steel basket									
0100	Brass or monel available									
2000	Simplex style									
2300	Bronze body									
2320	Screwed, 3/8" pipe size	1 Stpi	22	.364	Ea.	152	18.85		170.85	196
2340	1/2" pipe size		20	.400		155	21		176	202
2360	3/4" pipe size		17	.471		159	24.50		183.50	212
2380	1" pipe size		15	.533		208	27.50		235.50	270
2400	1-1/4" pipe size		13	.615		310	32		342	395
2420	1-1/2" pipe size		12	.667		320	34.50		354.50	405
2440	2" pipe size	↓	10	.800		475	41.50		516.50	585
2460	2-1/2" pipe size	Q-5	15	1.067		685	50		735	825
2480	3" pipe size	"	14	1.143		985	53.50		1,038.50	1,150
2600	Flanged, 2" pipe size	1 Stpi	6	1.333		725	69		794	905
2620	2-1/2" pipe size	Q-5	4.50	3.556		1,125	166		1,291	1,500
2640	3" pipe size		3.50	4.571		1,275	213		1,488	1,725
2660	4" pipe size	↓	3	5.333		2,175	249		2,424	2,750
2680	5" pipe size	Q-6	3.40	7.059		3,325	340		3,665	4,150
2700	6" pipe size	↓	3	8	↓	4,225	385		4,610	5,200

23 21 20.74 Strainers, Basket Type		Crew	Daily Output	Labor-Hours	Unit	Material	2010 Bare Costs Labor	Equipment	Total	Total Incl O&P
2710	8″ pipe size	Q-6	2.50	9.600	Ea.	6,875	465		7,340	8,250
3600	Iron body									
3700	Screwed, 3/8″ pipe size	1 Stpi	22	.364	Ea.	86	18.85		104.85	123
3720	1/2″ pipe size		20	.400		88.50	21		109.50	129
3740	3/4″ pipe size		17	.471		113	24.50		137.50	161
3760	1″ pipe size		15	.533		115	27.50		142.50	168
3780	1-1/4″ pipe size		13	.615		150	32		182	213
3800	1-1/2″ pipe size		12	.667		166	34.50		200.50	234
3820	2″ pipe size		10	.800		198	41.50		239.50	280
3840	2-1/2″ pipe size	Q-5	15	1.067		265	50		315	365
3860	3″ pipe size	″	14	1.143		320	53.50		373.50	435
4000	Flanged, 2″ pipe size	1 Stpi	6	1.333		305	69		374	440
4020	2-1/2″ pipe size	Q-5	4.50	3.556		415	166		581	705
4040	3″ pipe size		3.50	4.571		435	213		648	795
4060	4″ pipe size		3	5.333		665	249		914	1,100
4080	5″ pipe size	Q-6	3.40	7.059		1,000	340		1,340	1,600
4100	6″ pipe size		3	8		1,275	385		1,660	2,000
4120	8″ pipe size		2.50	9.600		2,325	465		2,790	3,250
4140	10″ pipe size		2.20	10.909		5,100	530		5,630	6,425
6000	Cast steel body									
6400	Screwed, 1″ pipe size	1 Stpi	15	.533	Ea.	184	27.50		211.50	244
6410	1-1/4″ pipe size		13	.615		278	32		310	355
6420	1-1/2″ pipe size		12	.667		278	34.50		312.50	355
6440	2″ pipe size		10	.800		380	41.50		421.50	480
6460	2-1/2″ pipe size	Q-5	15	1.067		540	50		590	670
6480	3″ pipe size	″	14	1.143		705	53.50		758.50	855
6560	Flanged, 2″ pipe size	1 Stpi	6	1.333		615	69		684	785
6580	2-1/2″ pipe size	Q-5	4.50	3.556		925	166		1,091	1,275
6600	3″ pipe size		3.50	4.571		990	213		1,203	1,425
6620	4″ pipe size		3	5.333		1,400	249		1,649	1,925
6640	6″ pipe size	Q-6	3	8		2,575	385		2,960	3,400
6660	8″ pipe size	″	2.50	9.600		4,150	465		4,615	5,275
7000	Stainless steel body									
7200	Screwed, 1″ pipe size	1 Stpi	15	.533	Ea.	278	27.50		305.50	345
7210	1-1/4″ pipe size		13	.615		430	32		462	525
7220	1-1/2″ pipe size		12	.667		430	34.50		464.50	525
7240	2″ pipe size		10	.800		645	41.50		686.50	765
7260	2-1/2″ pipe size	Q-5	15	1.067		910	50		960	1,075
7280	3″ pipe size	″	14	1.143		1,250	53.50		1,303.50	1,450
7400	Flanged, 2″ pipe size	1 Stpi	6	1.333		1,025	69		1,094	1,225
7420	2-1/2″ pipe size	Q-5	4.50	3.556		1,850	166		2,016	2,275
7440	3″ pipe size		3.50	4.571		1,900	213		2,113	2,400
7460	4″ pipe size		3	5.333		2,975	249		3,224	3,650
7480	6″ pipe size	Q-6	3	8		5,175	385		5,560	6,250
7500	8″ pipe size	″	2.50	9.600		7,525	465		7,990	8,975
8100	Duplex style									
8200	Bronze body									
8240	Screwed, 3/4″ pipe size	1 Stpi	16	.500	Ea.	1,075	26		1,101	1,225
8260	1″ pipe size		14	.571		1,075	29.50		1,104.50	1,225
8280	1-1/4″ pipe size		12	.667		2,100	34.50		2,134.50	2,350
8300	1-1/2″ pipe size		11	.727		2,100	38		2,138	2,350
8320	2″ pipe size		9	.889		3,400	46		3,446	3,800
8340	2-1/2″ pipe size	Q-5	14	1.143		4,400	53.50		4,453.50	4,900

23 21 20.74 Strainers, Basket Type		Crew	Daily Output	Labor-Hours	Unit	Material	2010 Bare Costs Labor	Equipment	Total	Total Incl O&P
8420	Flanged, 2" pipe size	1 Stpi	6	1.333	Ea.	3,650	69		3,719	4,125
8440	2-1/2" pipe size	Q-5	4.50	3.556		5,050	166		5,216	5,800
8460	3" pipe size		3.50	4.571		5,500	213		5,713	6,375
8480	4" pipe size		3	5.333		7,900	249		8,149	9,050
8500	5" pipe size	Q-6	3.40	7.059		17,600	340		17,940	19,800
8520	6" pipe size	"	3	8		17,600	385		17,985	19,900
8700	Iron body									
8740	Screwed, 3/4" pipe size	1 Stpi	16	.500	Ea.	575	26		601	675
8760	1" pipe size		14	.571		575	29.50		604.50	680
8780	1-1/4" pipe size		12	.667		1,025	34.50		1,059.50	1,175
8800	1-1/2" pipe size		11	.727		1,025	38		1,063	1,175
8820	2" pipe size		9	.889		1,725	46		1,771	1,950
8840	2-1/2" pipe size	Q-5	14	1.143		1,900	53.50		1,953.50	2,150
9000	Flanged, 2" pipe size	1 Stpi	6	1.333		2,050	69		2,119	2,375
9020	2-1/2" pipe size	Q-5	4.50	3.556		1,950	166		2,116	2,400
9040	3" pipe size		3.50	4.571		2,125	213		2,338	2,675
9060	4" pipe size		3	5.333		3,600	249		3,849	4,325
9080	5" pipe size	Q-6	3.40	7.059		7,150	340		7,490	8,350
9100	6" pipe size		3	8		7,150	385		7,535	8,425
9120	8" pipe size		2.50	9.600		12,500	465		12,965	14,400
9140	10" pipe size		2.20	10.909		17,300	530		17,830	19,900
9160	12" pipe size		1.70	14.118		19,200	685		19,885	22,200
9170	14" pipe size		1.40	17.143		23,300	830		24,130	26,900
9180	16" pipe size		1	24		29,200	1,150		30,350	34,000
9300	Cast steel body									
9340	Screwed, 1" pipe size	1 Stpi	14	.571	Ea.	1,225	29.50		1,254.50	1,400
9360	1-1/2" pipe size		11	.727		1,975	38		2,013	2,225
9380	2" pipe size		9	.889		2,600	46		2,646	2,950
9460	Flanged, 2" pipe size		6	1.333		2,800	69		2,869	3,175
9480	2-1/2" pipe size	Q-5	4.50	3.556		4,500	166		4,666	5,225
9500	3" pipe size		3.50	4.571		4,925	213		5,138	5,725
9520	4" pipe size		3	5.333		6,075	249		6,324	7,050
9540	6" pipe size	Q-6	3	8		11,400	385		11,785	13,100
9560	8" pipe size	"	2.50	9.600		27,400	465		27,865	30,800
9700	Stainless steel body									
9740	Screwed, 1" pipe size	1 Stpi	14	.571	Ea.	1,575	29.50		1,604.50	1,800
9760	1-1/2" pipe size		11	.727		2,375	38		2,413	2,675
9780	2" pipe size		9	.889		3,325	46		3,371	3,725
9860	Flanged, 2" pipe size		6	1.333		3,625	69		3,694	4,100
9880	2-1/2" pipe size	Q-5	4.50	3.556		6,075	166		6,241	6,950
9900	3" pipe size		3.50	4.571		6,625	213		6,838	7,625
9920	4" pipe size		3	5.333		8,625	249		8,874	9,875
9940	6" pipe size	Q-6	3	8		13,200	385		13,585	15,100
9960	8" pipe size	"	2.50	9.600		36,800	465		37,265	41,200

23 21 20.76 Strainers, Y Type, Bronze Body		Crew	Daily Output	Labor-Hours	Unit	Material	Labor	Equipment	Total	Total Incl O&P
0010	**STRAINERS, Y TYPE, BRONZE BODY**									
0050	Screwed, 150 lb., 1/4" pipe size	1 Stpi	24	.333	Ea.	17.30	17.30		34.60	45
0070	3/8" pipe size		24	.333		23	17.30		40.30	51
0100	1/2" pipe size		20	.400		23	21		44	56
0120	3/4" pipe size		19	.421		26	22		48	61.50
0140	1" pipe size		17	.471		30.50	24.50		55	70.50
0150	1-1/4" pipe size		15	.533		62	27.50		89.50	110

23 21 20 – Hydronic HVAC Piping Specialties

23 21 20.76 Strainers, Y Type, Bronze Body		Crew	Daily Output	Labor-Hours	Unit	Material	2010 Bare Costs Labor	Equipment	Total	Total Incl O&P
0160	1-1/2" pipe size	1 Stpi	14	.571	Ea.	66	29.50		95.50	118
0180	2" pipe size		13	.615		88.50	32		120.50	145
0182	3" pipe size		12	.667		670	34.50		704.50	785
0200	300 lb., 2-1/2" pipe size	Q-5	17	.941		420	44		464	530
0220	3" pipe size		16	1		835	46.50		881.50	990
0240	4" pipe size		15	1.067		1,900	50		1,950	2,175
0500	For 300 lb rating 1/4" thru 2", add					15%				
1000	Flanged, 150 lb., 1-1/2" pipe size	1 Stpi	11	.727	Ea.	435	38		473	535
1020	2" pipe size	"	8	1		590	52		642	730
1030	2-1/2" pipe size	Q-5	5	3.200		875	149		1,024	1,175
1040	3" pipe size		4.50	3.556		1,075	166		1,241	1,425
1060	4" pipe size		3	5.333		1,625	249		1,874	2,175
1080	5" pipe size	Q-6	3.40	7.059		2,050	340		2,390	2,750
1100	6" pipe size		3	8		3,125	385		3,510	4,000
1106	8" pipe size		2.60	9.231		3,425	445		3,870	4,450
1500	For 300 lb rating, add					40%				

23 21 20.78 Strainers, Y Type, Iron Body		Crew	Daily Output	Labor-Hours	Unit	Material	Labor	Equipment	Total	Total Incl O&P
0010	**STRAINERS, Y TYPE, IRON BODY**									
0050	Screwed, 250 lb., 1/4" pipe size	1 Stpi	20	.400	Ea.	9.60	21		30.60	41.50
0070	3/8" pipe size		20	.400		9.60	21		30.60	41.50
0100	1/2" pipe size		20	.400		9.60	21		30.60	41.50
0120	3/4" pipe size		18	.444		11.50	23		34.50	47
0140	1" pipe size		16	.500		15.85	26		41.85	56.50
0150	1-1/4" pipe size		15	.533		21	27.50		48.50	64.50
0160	1-1/2" pipe size		12	.667		26	34.50		60.50	80.50
0180	2" pipe size		8	1		39	52		91	121
0200	2-1/2" pipe size	Q-5	12	1.333		232	62.50		294.50	350
0220	3" pipe size		11	1.455		251	68		319	380
0240	4" pipe size		5	3.200		425	149		574	690
0500	For galvanized body, add					50%				
1000	Flanged, 125 lb., 1-1/2" pipe size	1 Stpi	11	.727	Ea.	135	38		173	205
1020	2" pipe size	"	8	1		142	52		194	234
1030	2-1/2" pipe size	Q-5	5	3.200		102	149		251	335
1040	3" pipe size		4.50	3.556		133	166		299	395
1060	4" pipe size		3	5.333		242	249		491	640
1080	5" pipe size	Q-6	3.40	7.059		380	340		720	925
1100	6" pipe size		3	8		460	385		845	1,100
1120	8" pipe size		2.50	9.600		775	465		1,240	1,550
1140	10" pipe size		2	12		1,475	580		2,055	2,500
1160	12" pipe size		1.70	14.118		2,225	685		2,910	3,475
1170	14" pipe size		1.30	18.462		4,125	895		5,020	5,875
1180	16" pipe size		1	24		5,825	1,150		6,975	8,150
1500	For 250 lb rating, add					20%				
2000	For galvanized body, add					50%				
2500	For steel body, add					40%				

23 21 20.80 Suction Diffusers		Crew	Daily Output	Labor-Hours	Unit	Material	Labor	Equipment	Total	Total Incl O&P
0010	**SUCTION DIFFUSERS**									
0100	Cast iron body with integral straightening vanes, strainer									
1000	Flanged									
1010	2" inlet, 1-1/2" pump side	1 Stpi	6	1.333	Ea.	283	69		352	415
1020	2" pump side	"	5	1.600		293	83		376	450
1030	3" inlet, 2" pump side	Q-5	6.50	2.462		315	115		430	520

23 21 20 – Hydronic HVAC Piping Specialties

	23 21 20.80 Suction Diffusers	Crew	Daily Output	Labor-Hours	Unit	Material	Labor (2010 Bare Costs)	Equipment	Total	Total Incl O&P
1040	2-1/2" pump side	Q-5	5.30	3.019	Ea.	525	141		666	790
1050	3" pump side		4.50	3.556		530	166		696	835
1060	4" inlet, 3" pump side		3.50	4.571		620	213		833	1,000
1070	4" pump side		3	5.333		720	249		969	1,175
1080	5" inlet, 4" pump side	Q-6	3.80	6.316		820	305		1,125	1,375
1090	5" pump side		3.40	7.059		980	340		1,320	1,575
1100	6" inlet, 4" pump side		3.30	7.273		855	350		1,205	1,475
1110	5" pump side		3.10	7.742		1,025	375		1,400	1,675
1120	6" pump side		3	8		1,075	385		1,460	1,775
1130	8" inlet, 6" pump side		2.70	8.889		1,175	430		1,605	1,925
1140	8" pump side		2.50	9.600		2,025	465		2,490	2,925
1150	10" inlet, 6" pump side		2.40	10		2,800	485		3,285	3,800
1160	8" pump side		2.30	10.435		3,125	505		3,630	4,200
1170	10" pump side		2.20	10.909		3,375	530		3,905	4,500
1180	12" inlet, 8" pump side		2	12		3,575	580		4,155	4,800
1190	10" pump side		1.80	13.333		3,950	645		4,595	5,325
1200	12" pump side		1.70	14.118		4,450	685		5,135	5,925
1210	14" inlet, 12" pump side		1.50	16		5,750	775		6,525	7,500
1220	14" pump side		1.30	18.462		5,875	895		6,770	7,800

23 21 20.84 Thermoflo Indicator

		Crew	Daily Output	Labor-Hours	Unit	Material	Labor	Equipment	Total	Total Incl O&P
0010	**THERMOFLO INDICATOR**, For balancing									
1000	Sweat connections, 1-1/4" pipe size	1 Stpi	12	.667	Ea.	535	34.50		569.50	640
1020	1-1/2" pipe size		10	.800		545	41.50		586.50	660
1040	2" pipe size		8	1		570	52		622	710
1060	2-1/2" pipe size		7	1.143		875	59.50		934.50	1,050
2000	Flange connections, 3" pipe size	Q-5	5	3.200		1,050	149		1,199	1,375
2020	4" pipe size		4	4		1,250	187		1,437	1,650
2030	5" pipe size		3.50	4.571		1,575	213		1,788	2,050
2040	6" pipe size		3	5.333		1,675	249		1,924	2,225
2060	8" pipe size		2	8		2,050	375		2,425	2,800

23 21 20.88 Venturi Flow

		Crew	Daily Output	Labor-Hours	Unit	Material	Labor	Equipment	Total	Total Incl O&P
0010	**VENTURI FLOW**, Measuring device									
0050	1/2" diameter	1 Stpi	24	.333	Ea.	281	17.30		298.30	335
0100	3/4" diameter		20	.400		257	21		278	315
0120	1" diameter		19	.421		276	22		298	340
0140	1-1/4" diameter		15	.533		340	27.50		367.50	415
0160	1-1/2" diameter		13	.615		355	32		387	440
0180	2" diameter		11	.727		365	38		403	460
0200	2-1/2" diameter	Q-5	16	1		500	46.50		546.50	620
0220	3" diameter		14	1.143		515	53.50		568.50	650
0240	4" diameter		11	1.455		775	68		843	950
0260	5" diameter	Q-6	4	6		1,025	291		1,316	1,550
0280	6" diameter		3.50	6.857		1,125	330		1,455	1,750
0300	8" diameter		3	8		1,450	385		1,835	2,175
0320	10" diameter		2	12		3,425	580		4,005	4,650
0330	12" diameter		1.80	13.333		4,950	645		5,595	6,425
0340	14" diameter		1.60	15		5,600	725		6,325	7,275
0350	16" diameter		1.40	17.143		6,250	830		7,080	8,125
0500	For meter, add					2,125			2,125	2,350

23 21 20.94 Weld End Ball Joints

		Crew	Daily Output	Labor-Hours	Unit	Material	Labor	Equipment	Total	Total Incl O&P
0010	**WELD END BALL JOINTS**, Steel									
0050	2-1/2" diameter	Q-17	13	1.231	Ea.	615	57.50	4.50	677	770

23 21 20 – Hydronic HVAC Piping Specialties

23 21 20.94 Weld End Ball Joints		Crew	Daily Output	Labor-Hours	Unit	Material	2010 Bare Costs Labor	Equipment	Total	Total Incl O&P
0100	3″ diameter	Q-17	12	1.333	Ea.	720	62.50	4.88	787.38	890
0120	4″ diameter		11	1.455		1,175	68	5.30	1,248.30	1,375
0140	5″ diameter	Q-18	14	1.714		1,350	83	4.18	1,437.18	1,600
0160	6″ diameter		12	2		2,000	97	4.88	2,101.88	2,350
0180	8″ diameter		9	2.667		3,025	129	6.50	3,160.50	3,525
0200	10″ diameter		8	3		4,075	145	7.30	4,227.30	4,700
0220	12″ diameter		6	4		6,500	194	9.75	6,703.75	7,450

23 21 20.98 Zone Valves		Crew	Daily Output	Labor-Hours	Unit	Material	Labor	Equipment	Total	Total Incl O&P
0010	**ZONE VALVES**									
1000	Bronze body									
1010	2 way, 125 psi									
1020	Standard, 65′ pump head									
1030	1/2″ soldered	1 Stpi	24	.333	Ea.	112	17.30		129.30	149
1040	3/4″ soldered		20	.400		112	21		133	154
1050	1″ soldered		19	.421		120	22		142	165
1060	1-1/4″ soldered		15	.533		141	27.50		168.50	197
1200	High head, 125′ pump head									
1210	1/2″ soldered	1 Stpi	24	.333	Ea.	153	17.30		170.30	194
1220	3/4″ soldered		20	.400		158	21		179	205
1230	1″ soldered		19	.421		172	22		194	222
1300	Geothermal, 150′ pump head									
1310	3/4″ soldered	1 Stpi	20	.400	Ea.	181	21		202	230
1320	1″ soldered		19	.421		193	22		215	245
1330	3/4″ threaded		20	.400		181	21		202	230
3000	3 way, 125 psi									
3200	Bypass, 65′ pump head									
3210	1/2″ soldered	1 Stpi	24	.333	Ea.	160	17.30		177.30	202
3220	3/4″ soldered		20	.400		165	21		186	212
3230	1″ soldered		19	.421		188	22		210	239
9000	Transformer, for up to 3 zone valves		20	.400		23	21		44	56.50

23 21 23 – Hydronic Pumps

23 21 23.13 In-Line Centrifugal Hydronic Pumps		Crew	Daily Output	Labor-Hours	Unit	Material	Labor	Equipment	Total	Total Incl O&P
0010	**IN-LINE CENTRIFUGAL HYDRONIC PUMPS**									
0600	Bronze, sweat connections, 1/40 HP, in line									
0640	3/4″ size	Q-1	16	1	Ea.	192	47		239	281
1000	Flange connection, 3/4″ to 1-1/2″ size									
1040	1/12 HP	Q-1	6	2.667	Ea.	495	125		620	730
1060	1/8 HP		6	2.667		855	125		980	1,125
1100	1/3 HP		6	2.667		955	125		1,080	1,225
1140	2″ size, 1/6 HP		5	3.200		1,025	150		1,175	1,350
1180	2-1/2″ size, 1/4 HP		5	3.200		1,525	150		1,675	1,925
1220	3″ size, 1/4 HP		4	4		1,600	187		1,787	2,050
1260	1/3 HP		4	4		1,975	187		2,162	2,450
1300	1/2 HP		4	4		2,000	187		2,187	2,475
1340	3/4 HP		4	4		2,200	187		2,387	2,675
1380	1 HP		4	4		3,550	187		3,737	4,175
2000	Cast iron, flange connection									
2040	3/4″ to 1-1/2″ size, in line, 1/12 HP	Q-1	6	2.667	Ea.	320	125		445	535
2060	1/8 HP		6	2.667		535	125		660	775
2100	1/3 HP		6	2.667		595	125		720	840
2140	2″ size, 1/6 HP		5	3.200		655	150		805	945
2180	2-1/2″ size, 1/4 HP		5	3.200		830	150		980	1,125

23 21 23 – Hydronic Pumps

23 21 23.13 In-Line Centrifugal Hydronic Pumps		Crew	Daily Output	Labor-Hours	Unit	2010 Bare Costs Material	Labor	Equipment	Total	Total Incl O&P
2220	3" size, 1/4 HP	Q-1	4	4	Ea.	840	187		1,027	1,200
2260	1/3 HP		4	4		1,150	187		1,337	1,525
2300	1/2 HP		4	4		1,175	187		1,362	1,575
2340	3/4 HP		4	4		1,375	187		1,562	1,775
2380	1 HP	↓	4	4	↓	1,975	187		2,162	2,450
2600	For non-ferrous impeller, add					3%				
3000	High head, bronze impeller									
3030	1-1/2" size 1/2 HP	Q-1	5	3.200	Ea.	1,025	150		1,175	1,375
3040	1-1/2" size 3/4 HP		5	3.200		1,100	150		1,250	1,450
3050	2" size 1 HP		4	4		1,325	187		1,512	1,750
3090	2" size 1-1/2 HP	↓	4	4	↓	1,625	187		1,812	2,075
4000	Close coupled, end suction, bronze impeller									
4040	1-1/2" size, 1-1/2 HP, to 40 GPM	Q-1	3	5.333	Ea.	1,775	250		2,025	2,325
4090	2" size, 2 HP, to 50 GPM		3	5.333		2,150	250		2,400	2,725
4100	2" size, 3 HP, to 90 GPM		2.30	6.957		2,225	325		2,550	2,950
4190	2-1/2" size, 3 HP, to 150 GPM		2	8		2,400	375		2,775	3,200
4300	3" size, 5 HP, to 225 GPM		1.80	8.889		2,775	415		3,190	3,675
4410	3" size, 10 HP, to 350 GPM		1.60	10		3,700	470		4,170	4,775
4420	4" size, 7-1/2 HP, to 350 GPM	↓	1.60	10		3,600	470		4,070	4,650
4520	4" size, 10 HP, to 600 GPM	Q-2	1.70	14.118		4,100	685		4,785	5,550
4530	5" size, 15 HP, to 1000 GPM		1.70	14.118		4,125	685		4,810	5,575
4610	5" size, 20 HP, to 1350 GPM		1.50	16		4,375	775		5,150	6,000
4620	5" size, 25 HP, to 1550 GPM	↓	1.50	16	↓	6,000	775		6,775	7,775
5000	Base mounted, bronze impeller, coupling guard									
5040	1-1/2" size, 1-1/2 HP, to 40 GPM	Q-1	2.30	6.957	Ea.	5,100	325		5,425	6,100
5090	2" size, 2 HP, to 50 GPM		2.30	6.957		5,625	325		5,950	6,700
5100	2" size, 3 HP, to 90 GPM		2	8		5,650	375		6,025	6,750
5190	2-1/2" size, 3 HP, to 150 GPM		1.80	8.889		6,150	415		6,565	7,400
5300	3" size, 5 HP, to 225 GPM		1.60	10		6,175	470		6,645	7,500
5410	4" size, 5 HP, to 350 GPM		1.50	10.667		6,925	500		7,425	8,375
5420	4" size, 7-1/2 HP, to 350 GPM	↓	1.50	10.667		6,975	500		7,475	8,425
5520	5" size, 10 HP, to 600 GPM	Q-2	1.60	15		8,400	730		9,130	10,400
5530	5" size, 15 HP, to 1000 GPM		1.60	15		9,200	730		9,930	11,200
5610	6" size, 20 HP, to 1350 GPM		1.40	17.143		10,900	835		11,735	13,300
5620	6" size, 25 HP, to 1550 GPM	↓	1.40	17.143	↓	12,000	835		12,835	14,500
5800	The above pump capacities are based on 1800 RPM,									
5810	at a 60 foot head. Increasing the RPM									
5820	or decreasing the head will increase the GPM.									

23 21 29 – Automatic Condensate Pump Units

23 21 29.10 Condensate Removal Pump System		Crew	Daily Output	Labor-Hours	Unit	Material	Labor	Equipment	Total	Total Incl O&P
0010	**CONDENSATE REMOVAL PUMP SYSTEM**									
0020	Pump with 1 Gal. ABS tank G									
0100	115 V G									
0120	1/50 HP, 200 GPH G	1 Stpi	12	.667	Ea.	148	34.50		182.50	215
0140	1/18 HP, 270 GPH G		10	.800		164	41.50		205.50	242
0160	1/5 HP, 450 GPH G	↓	8	1	↓	355	52		407	470
0200	230 V G									
0240	1/18 HP, 270 GPH G	1 Stpi	10	.800	Ea.	164	41.50		205.50	242
0260	1/5 HP, 450 GPH G	"	8	1	"	390	52		442	510

23 22 Steam and Condensate Piping and Pumps

23 22 13 – Steam and Condensate Heating Piping

23 22 13.23 Aboveground Steam and Condensate Piping		Crew	Daily Output	Labor-Hours	Unit	2010 Bare Costs Material	Labor	Equipment	Total	Total Incl O&P
0010	**ABOVEGROUND STEAM AND CONDENSATE HEATING PIPING**									
0020	Condensate meter									
0100	500 lb. per hour	1 Stpi	14	.571	Ea.	2,350	29.50		2,379.50	2,650
0140	1500 lb. per hour		7	1.143		2,550	59.50		2,609.50	2,900
0160	3000 lb. per hour		5	1.600		3,425	83		3,508	3,875
0200	12,000 lb. per hour	Q-5	3.50	4.571		4,575	213		4,788	5,350

23 22 23 – Steam Condensate Pumps

23 22 23.10 Condensate Return System

Line	Description	Crew	Daily Output	Labor-Hours	Unit	Material	Labor	Equipment	Total	Total Incl O&P
0010	**CONDENSATE RETURN SYSTEM**									
2000	Simplex									
2010	With pump, motor, CI receiver, float switch									
2020	3/4 HP, 15 GPM	Q-1	1.80	8.889	Ea.	5,550	415		5,965	6,725
2100	Duplex									
2110	With 2 pumps and motors, CI receiver, float switch, alternator									
2120	3/4 HP, 15 GPM, 15 Gal CI rcvr	Q-1	1.40	11.429	Ea.	6,050	535		6,585	7,450
2130	1 HP, 25 GPM		1.20	13.333		7,250	625		7,875	8,900
2140	1-1/2 HP, 45 GPM		1	16		8,425	750		9,175	10,400
2150	1-1/2 HP, 60 GPM		1	16		9,450	750		10,200	11,500

23 23 Refrigerant Piping

23 23 13 – Refrigerant Piping Valves

23 23 13.10 Valves

Line	Description	Crew	Daily Output	Labor-Hours	Unit	Material	Labor	Equipment	Total	Total Incl O&P
0010	**VALVES**									
8100	Check valve, soldered									
8110	5/8"	1 Stpi	36	.222	Ea.	115	11.55		126.55	143
8114	7/8"		26	.308		132	15.95		147.95	170
8118	1-1/8"		18	.444		158	23		181	208
8122	1-3/8"		14	.571		217	29.50		246.50	284
8126	1-5/8"		13	.615		279	32		311	355
8130	2-1/8"		12	.667		310	34.50		344.50	390
8134	2-5/8"	Q-5	22	.727		465	34		499	560
8138	3-1/8"	"	20	.800		565	37.50		602.50	680
8500	Refrigeration valve, packless, soldered									
8510	1/2"	1 Stpi	38	.211	Ea.	33	10.95		43.95	53
8514	5/8"		36	.222		33	11.55		44.55	54
8518	7/8"		26	.308		132	15.95		147.95	169
8520	Packed, soldered									
8522	1-1/8"	1 Stpi	18	.444	Ea.	161	23		184	212
8526	1-3/8"		14	.571		284	29.50		313.50	355
8530	1-5/8"		13	.615		315	32		347	395
8534	2-1/8"		12	.667		500	34.50		534.50	600
8538	2-5/8"	Q-5	22	.727		680	34		714	795
8542	3-1/8"		20	.800		785	37.50		822.50	920
8546	4-1/8"		18	.889		1,375	41.50		1,416.50	1,575
8600	Solenoid valve, flange/solder									
8610	1/2"	1 Stpi	38	.211	Ea.	163	10.95		173.95	196
8614	5/8"		36	.222		205	11.55		216.55	242
8618	3/4"		30	.267		256	13.85		269.85	300
8622	7/8"		26	.308		320	15.95		335.95	380
8626	1-1/8"		18	.444		420	23		443	495

23 23 Refrigerant Piping

23 23 13 – Refrigerant Piping Valves

23 23 13.10 Valves		Crew	Daily Output	Labor-Hours	Unit	Material	2010 Bare Costs Labor	Equipment	Total	Total Incl O&P
8630	1-3/8"	1 Stpi	14	.571	Ea.	510	29.50		539.50	605
8634	1-5/8"		13	.615		715	32		747	835
8638	2-1/8"		12	.667		785	34.50		819.50	915
8800	Thermostatic valve, flange/solder									
8810	1/2 – 3 ton, 3/8" x 5/8"	1 Stpi	9	.889	Ea.	131	46		177	213
8814	4 – 5 ton, 1/2" x 7/8"		7	1.143		151	59.50		210.50	255
8818	6 – 8 ton, 5/8" x 7/8"		5	1.600		151	83		234	290
8822	7 – 12 ton, 7/8" x 1-1/8"		4	2		176	104		280	350
8826	15 – 20 ton, 7/8" x 1-3/8"		3.20	2.500		176	130		306	390

23 23 16 – Refrigerant Piping Specialties

23 23 16.10 Refrigerant Piping Component Specialties

Line	Description	Crew	Daily Output	Labor-Hours	Unit	Material	Labor	Equipment	Total	Total Incl O&P
0010	**REFRIGERANT PIPING COMPONENT SPECIALTIES**									
0600	Accumulator									
0610	3/4"	1 Stpi	8.80	.909	Ea.	70.50	47		117.50	148
0614	7/8"		6.40	1.250		80.50	65		145.50	186
0618	1-1/8"		4.80	1.667		111	86.50		197.50	252
0622	1-3/8"		4	2		145	104		249	315
0626	1-5/8"		3.20	2.500		157	130		287	365
0630	2-1/8"		2.40	3.333		375	173		548	675
0700	Condensate drip/drain pan		24	.333		132	17.30		149.30	171
1000	Filter dryer									
1010	Replaceable core type, solder									
1020	1/2"	1 Stpi	20	.400	Ea.	130	21		151	174
1030	5/8"		19	.421		142	22		164	189
1040	7/8"		18	.444		142	23		165	191
1050	1-1/8"		15	.533		148	27.50		175.50	205
1060	1-3/8"		14	.571		148	29.50		177.50	208
1070	1-5/8"		12	.667		152	34.50		186.50	220
1080	2-1/8"		10	.800		155	41.50		196.50	232
1090	2-5/8"		9	.889		650	46		696	785
1100	3-1/8"		8	1		695	52		747	845
1200	Sealed in-line, solder									
1210	1/4", 3 cubic inches	1 Stpi	22	.364	Ea.	8.10	18.85		26.95	37.50
1220	3/8", 5 cubic inches		21	.381		7.40	19.75		27.15	37.50
1230	1/2", 9 cubic inches		20	.400		12	21		33	44
1240	1/2", 16 cubic inches		20	.400		15.60	21		36.60	48
1250	5/8", 16 cubic inches		19	.421		16.35	22		38.35	51
1260	5/8", 30 cubic inches		19	.421		26.50	22		48.50	62
1270	7/8", 30 cubic inches		18	.444		28	23		51	65.50
1280	7/8", 41 cubic inches		17	.471		37.50	24.50		62	78
1290	1-1/8", 60 cubic inches		15	.533		52	27.50		79.50	99
4000	P-Trap, suction line, solder									
4010	5/8"	1 Stpi	19	.421	Ea.	56	22		78	94.50
4020	3/4"		19	.421		56	22		78	94.50
4030	7/8"		18	.444		60.50	23		83.50	101
4040	1-1/8"		15	.533		63.50	27.50		91	112
4050	1-3/8"		14	.571		115	29.50		144.50	172
4060	1-5/8"		12	.667		175	34.50		209.50	245
4070	2-1/8"		10	.800		360	41.50		401.50	455
5000	Sightglass									
5010	Moisture and liquid indicator, solder									
5020	1/4"	1 Stpi	22	.364	Ea.	17.80	18.85		36.65	48

23 23 Refrigerant Piping

23 23 16 – Refrigerant Piping Specialties

23 23 16.10 Refrigerant Piping Component Specialties		Crew	Daily Output	Labor-Hours	Unit	Material	Labor	Equipment	Total	Total Incl O&P
						2010 Bare Costs				
5030	3/8"	1 Stpi	21	.381	Ea.	19.05	19.75		38.80	50.50
5040	1/2"		20	.400		23	21		44	56
5050	5/8"		19	.421		23.50	22		45.50	59
5060	7/8"		18	.444		35	23		58	73
5070	1-1/8"		15	.533		37.50	27.50		65	83
5080	1-3/8"		14	.571		56	29.50		85.50	107
5090	1-5/8"		12	.667		63.50	34.50		98	122
5100	2-1/8"		10	.800		79.50	41.50		121	150
7400	Vacuum pump set									
7410	Two stage, high vacuum continuous duty	1 Stpi	8	1	Ea.	3,225	52		3,277	3,625

23 23 23 – Refrigerants

23 23 23.10 Anti-Freeze

		Crew	Daily Output	Labor-Hours	Unit	Material	Labor	Equipment	Total	Total Incl O&P
0010	**ANTI-FREEZE**, Inhibited									
0900	Ethylene glycol concentrated									
1000	55 gallon drums, small quantities				Gal.	7.40			7.40	8.15
1200	Large quantities					6.90			6.90	7.60
2000	Propylene glycol, for solar heat, small quantities					12.85			12.85	14.15
2100	Large quantities					11.50			11.50	12.65

23 23 23.20 Refrigerant

		Crew	Daily Output	Labor-Hours	Unit	Material	Labor	Equipment	Total	Total Incl O&P
0010	**REFRIGERANT**									
4420	Refrigerant, R-22, 30 lb. disposable cylinder				Lb.	12.50			12.50	13.75
4428	Refrigerant, R-134A, 30 lb. disposable cylinder					17.75			17.75	19.50
4434	Refrigerant, R-407C, 30 lb. disposable cylinder					17.35			17.35	19.05
4440	Refrigerant, R-408A, 25 lb. disposable cylinder					24			24	26.50
4450	Refrigerant, R-410A, 25 lb. disposable cylinder					19.65			19.65	21.50
4470	Refrigerant, R-507, 25 lb. disposable cylinder					19.95			19.95	22

23 31 HVAC Ducts and Casings

23 31 13 – Metal Ducts

23 31 13.13 Rectangular Metal Ducts

		Crew	Daily Output	Labor-Hours	Unit	Material	Labor	Equipment	Total	Total Incl O&P
0010	**RECTANGULAR METAL DUCTS** R233100-20									
0020	Fabricated rectangular, includes fittings, joints, supports,									
0030	allowance for flexible connections, no insulation R233100-30									
0031	NOTE: Fabrication and installation are combined									
0040	as LABOR cost. Approx. 25% fittings assumed. R233100-40									
0042	Fabrication/Inst. is to commercial quality standards									
0043	(SMACNA or equiv.) for structure, sealing, leak testing, etc. R233100-50									
0050	Add to labor for elevated installation									
0051	of fabricated ductwork									
0052	10′ to 15′ high						6%			
0053	15′ to 20′ high						12%			
0054	20′ to 25′ high						15%			
0055	25′ to 30′ high						21%			
0056	30′ to 35′ high						24%			
0057	35′ to 40′ high						30%			
0058	Over 40′ high						33%			
0072	For duct insulation see Div. 23 07 13.10 3000									
0100	Aluminum, alloy 3003-H14, under 100 lb.	Q-10	75	.320	Lb.	3.72	14.65		18.37	26
0110	100 to 500 lb.		80	.300		2.48	13.75		16.23	23.50
0120	500 to 1,000 lb.		95	.253		2.31	11.60		13.91	20

23 31 13.13 Rectangular Metal Ducts		Crew	Daily Output	Labor-Hours	Unit	Material	2010 Bare Costs Labor	Equipment	Total	Total Incl O&P
0140	1,000 to 2,000 lb.	Q-10	120	.200	Lb.	2.15	9.15		11.30	16.25
0150	2,000 to 5,000 lb.		130	.185		2.15	8.45		10.60	15.15
0160	Over 5,000 lb.		145	.166		2.15	7.60		9.75	13.85
0500	Galvanized steel, under 200 lb.		235	.102		.47	4.68		5.15	7.60
0520	200 to 500 lb.		245	.098		.46	4.49		4.95	7.30
0540	500 to 1,000 lb.		255	.094		.45	4.31		4.76	7.05
0560	1,000 to 2,000 lb.		265	.091		.45	4.15		4.60	6.80
0570	2,000 to 5,000 lb.		275	.087		.44	4		4.44	6.55
0580	Over 5,000 lb.		285	.084		.43	3.86		4.29	6.30
0600	For large quantities special prices available from supplier									
1000	Stainless steel, type 304, under 100 lb.	Q-10	165	.145	Lb.	4.35	6.65		11	14.90
1020	100 to 500 lb.		175	.137		4.02	6.30		10.32	13.90
1030	500 to 1,000 lb.		190	.126		3.68	5.80		9.48	12.80
1040	1,000 to 2,000 lb.		200	.120		3.65	5.50		9.15	12.35
1050	2,000 to 5,000 lb.		225	.107		3.31	4.89		8.20	11.05
1060	Over 5,000 lb.		235	.102		3.24	4.68		7.92	10.65
1080	Note: Minimum order cost exceeds per lb. cost for min. wt.									
1100	For medium pressure ductwork, add				Lb.		15%			
1200	For high pressure ductwork, add						40%			
1210	For welded ductwork, add						85%			
1220	For 30% fittings, add						11%			
1224	For 40% fittings, add						34%			
1228	For 50% fittings, add						56%			
1232	For 60% fittings, add						79%			
1236	For 70% fittings, add						101%			
1240	For 80% fittings, add						124%			
1244	For 90% fittings, add						147%			
1248	For 100% fittings, add						169%			
1252	Note: Fittings add includes time for detailing and installation.									
1276	Duct tape, 2" x 60 yd roll				Ea.	5.20			5.20	5.70

23 31 13.16 Round and Flat-Oval Spiral Ducts		Crew	Daily Output	Labor-Hours	Unit	Material	Labor	Equipment	Total	Total Incl O&P
0010	**ROUND AND FLAT-OVAL SPIRAL DUCTS**									
1280	Add to labor for elevated installation									
1282	of prefabricated (purchased) ductwork									
1283	10' to 15' high						10%			
1284	15' to 20' high						20%			
1285	20' to 25' high						25%			
1286	25' to 30' high						35%			
1287	30' to 35' high						40%			
1288	35' to 40' high						50%			
1289	Over 40' high						55%			
5400	Spiral preformed, steel, galv., straight lengths, Max 10" spwg.									
5410	4" diameter, 26 ga.	Q-9	360	.044	L.F.	1.61	1.96		3.57	4.74
5416	5" diameter, 26 ga.		320	.050		1.61	2.21		3.82	5.10
5420	6" diameter, 26 ga.		280	.057		1.61	2.53		4.14	5.60
5425	7" diameter, 26 ga.		240	.067		2.05	2.95		5	6.70
5430	8" diameter, 26 ga.		200	.080		2.14	3.54		5.68	7.70
5440	10" diameter, 26 ga.		160	.100		2.67	4.42		7.09	9.65
5450	12" diameter, 26 ga.		120	.133		3.21	5.90		9.11	12.45
5460	14" diameter, 24 ga.		80	.200		4.20	8.85		13.05	18
5480	16" diameter, 24 ga.		60	.267		4.81	11.80		16.61	23
5490	18" diameter, 24 ga.		50	.320		5.40	14.15		19.55	27.50

23 31 13.16	Round and Flat-Oval Spiral Ducts	Crew	Daily Output	Labor-Hours	Unit	2010 Bare Costs Material	Labor	Equipment	Total	Total Incl O&P
5500	20″ diameter, 24 ga.	Q-10	65	.369	L.F.	6	16.90		22.90	32
5510	22″ diameter, 24 ga.		60	.400		6.60	18.35		24.95	35.50
5520	24″ diameter, 24 ga.		55	.436		7.20	20		27.20	38.50
5540	30″ diameter, 22 ga.		45	.533		10.80	24.50		35.30	49
5600	36″ diameter, 22 ga.		40	.600		12.95	27.50		40.45	56
5800	Connector, 4″ diameter	Q-9	100	.160	Ea.	1.95	7.05		9	12.85
5810	5″ diameter		94	.170		2.10	7.50		9.60	13.70
5820	6″ diameter		88	.182		2.20	8.05		10.25	14.55
5840	8″ diameter		78	.205		2.45	9.05		11.50	16.45
5860	10″ diameter		70	.229		2.95	10.10		13.05	18.55
5880	12″ diameter		50	.320		3.45	14.15		17.60	25.50
5900	14″ diameter		44	.364		3.65	16.05		19.70	28.50
5920	16″ diameter		40	.400		4	17.70		21.70	31.50
5930	18″ diameter		37	.432		4.25	19.10		23.35	33.50
5940	20″ diameter		34	.471		5	21		26	37
5950	22″ diameter		31	.516		6.05	23		29.05	41
5960	24″ diameter		28	.571		6.45	25.50		31.95	45.50
5980	30″ diameter		22	.727		7.60	32		39.60	57
6000	36″ diameter		18	.889		9.10	39.50		48.60	69.50
6300	Elbow, 45°, 4″ diameter		60	.267		6.05	11.80		17.85	24.50
6310	5″ diameter		52	.308		6.30	13.60		19.90	27.50
6320	6″ diameter		44	.364		6.70	16.05		22.75	32
6340	8″ diameter		28	.571		8.10	25.50		33.60	47.50
6360	10″ diameter		18	.889		8.50	39.50		48	69
6380	12″ diameter		13	1.231		9.20	54.50		63.70	92.50
6400	14″ diameter		11	1.455		12.75	64.50		77.25	112
6420	16″ diameter		10	1.600		18.30	70.50		88.80	127
6430	18″ diameter		9.60	1.667		24	73.50		97.50	139
6440	20″ diameter	Q-10	14	1.714		30.50	78.50		109	153
6450	22″ diameter		13	1.846		51.50	84.50		136	185
6460	24″ diameter		12	2		63.50	91.50		155	209
6480	30″ diameter		9	2.667		91.50	122		213.50	286
6500	36″ diameter		7	3.429		111	157		268	360
6600	Elbow, 90°, 4″ diameter	Q-9	60	.267		4.60	11.80		16.40	23
6610	5″ diameter		52	.308		4.60	13.60		18.20	25.50
6620	6″ diameter		44	.364		4.60	16.05		20.65	29.50
6625	7″ diameter		36	.444		5	19.65		24.65	35.50
6630	8″ diameter		28	.571		5.65	25.50		31.15	44.50
6640	10″ diameter		18	.889		7.40	39.50		46.90	67.50
6650	12″ diameter		13	1.231		9.25	54.50		63.75	92.50
6660	14″ diameter		11	1.455		15.85	64.50		80.35	115
6670	16″ diameter		10	1.600		23	70.50		93.50	132
6676	18″ diameter		9.60	1.667		30	73.50		103.50	145
6680	20″ diameter	Q-10	14	1.714		38	78.50		116.50	161
6684	22″ diameter		13	1.846		62	84.50		146.50	196
6690	24″ diameter		12	2		76	91.50		167.50	223
6700	30″ diameter		9	2.667		192	122		314	395
6710	36″ diameter		7	3.429		230	157		387	490
6720	End cap									
6722	4″	Q-9	110	.145	Ea.	4.38	6.45		10.83	14.55
6724	6″		94	.170		4.38	7.50		11.88	16.20
6725	8″		88	.182		5.50	8.05		13.55	18.20
6726	10″		78	.205		6.85	9.05		15.90	21.50

23 31 13.16 Round and Flat-Oval Spiral Ducts		Crew	Daily Output	Labor-Hours	Unit	Material	Labor	Equipment	Total	Total Incl O&P
						2010 Bare Costs				
6727	12"	Q-9	70	.229	Ea.	7.45	10.10		17.55	23.50
6728	14"		50	.320		10.45	14.15		24.60	33
6729	16"		44	.364		11.65	16.05		27.70	37.50
6730	18"		40	.400		16	17.70		33.70	44.50
6731	20"		37	.432		32.50	19.10		51.60	65
6732	22"		34	.471		36.50	21		57.50	72
6733	24"		30	.533		41	23.50		64.50	80.50
6800	Reducing coupling, 6" x 4"		46	.348		13	15.35		28.35	38
6820	8" x 6"		40	.400		12.15	17.70		29.85	40.50
6840	10" x 8"		32	.500		13	22		35	48
6860	12" x 10"		24	.667		14.75	29.50		44.25	61
6880	14" x 12"		20	.800		15.50	35.50		51	70.50
6900	16" x 14"		18	.889		19.85	39.50		59.35	81.50
6920	18" x 16"		16	1		22	44		66	91
6940	20" x 18"	Q-10	24	1		35	46		81	108
6950	22" x 20"		23	1.043		40	48		88	117
6960	24" x 22"		22	1.091		45	50		95	125
6980	30" x 28"		18	1.333		113	61		174	217
7000	36" x 34"		16	1.500		140	69		209	258
7100	Tee, 90°, 4" diameter	Q-9	40	.400		21	17.70		38.70	50
7110	5" diameter		34.60	.462		31.50	20.50		52	66
7120	6" diameter		30	.533		31.50	23.50		55	70.50
7130	8" diameter		19	.842		41	37		78	102
7140	10" diameter		12	1.333		56.50	59		115.50	151
7150	12" diameter		9	1.778		68.50	78.50		147	195
7160	14" diameter		7	2.286		104	101		205	267
7170	16" diameter		6.70	2.388		120	106		226	292
7176	18" diameter		6	2.667		175	118		293	370
7180	20" diameter	Q-10	8.70	2.759		208	126		334	420
7190	24" diameter		7	3.429		325	157		482	600
7200	30" diameter		6	4		400	183		583	720
7210	36" diameter		5	4.800		510	220		730	895
7400	Steel, PVC coated both sides, straight lengths									
7410	4" diameter, 26 ga.	Q-9	360	.044	L.F.	3.64	1.96		5.60	6.95
7416	5" diameter, 26 ga.		240	.067		3.64	2.95		6.59	8.45
7420	6" diameter, 26 ga.		280	.057		3.64	2.53		6.17	7.80
7440	8" diameter, 26 ga.		200	.080		4.85	3.54		8.39	10.70
7460	10" diameter, 26 ga.		160	.100		6.05	4.42		10.47	13.35
7480	12" diameter, 26 ga.		120	.133		7.25	5.90		13.15	16.90
7500	14" diameter, 26 ga.		80	.200		8.50	8.85		17.35	23
7520	16" diameter, 24 ga.		60	.267		9.70	11.80		21.50	28.50
7540	18" diameter, 24 ga.		45	.356		10.90	15.70		26.60	36
7560	20" diameter, 24 ga.	Q-10	65	.369		12.10	16.90		29	39
7580	24" diameter, 24 ga.		55	.436		14.55	20		34.55	46.50
7600	30" diameter, 22 ga.		45	.533		22	24.50		46.50	61
7620	36" diameter, 22 ga.		40	.600		26.50	27.50		54	70.50
7890	Connector, 4" diameter	Q-9	100	.160	Ea.	2.60	7.05		9.65	13.55
7896	5" diameter		83	.193		3	8.50		11.50	16.20
7900	6" diameter		88	.182		3.30	8.05		11.35	15.80
7920	8" diameter		78	.205		3.90	9.05		12.95	18.05
7940	10" diameter		70	.229		4.50	10.10		14.60	20.50
7960	12" diameter		50	.320		5.30	14.15		19.45	27.50
7980	14" diameter		44	.364		5.60	16.05		21.65	30.50

23 31 13.16 Round and Flat-Oval Spiral Ducts		Crew	Daily Output	Labor-Hours	Unit	Material	2010 Bare Costs Labor	Equipment	Total	Total Incl O&P
8000	16″ diameter	Q-9	40	.400	Ea.	6.10	17.70		23.80	33.50
8020	18″ diameter		37	.432		6.80	19.10		25.90	36.50
8040	20″ diameter		34	.471		8.90	21		29.90	41.50
8060	24″ diameter		28	.571		11	25.50		36.50	50.50
8080	30″ diameter		22	.727		14.70	32		46.70	64.50
8100	36″ diameter		18	.889		17.90	39.50		57.40	79
8390	Elbow, 45°, 4″ diameter		60	.267		9.45	11.80		21.25	28.50
8396	5″ diameter		36	.444		9.75	19.65		29.40	41
8400	6″ diameter		44	.364		10.05	16.05		26.10	35.50
8420	8″ diameter		28	.571		10.90	25.50		36.40	50.50
8440	10″ diameter		18	.889		11.50	39.50		51	72
8460	12″ diameter		13	1.231		12.75	54.50		67.25	96.50
8480	14″ diameter		11	1.455		13.70	64.50		78.20	113
8500	16″ diameter		10	1.600		19.75	70.50		90.25	129
8520	18″ diameter	↓	9	1.778		25	78.50		103.50	147
8540	20″ diameter	Q-10	13	1.846		57	84.50		141.50	191
8560	24″ diameter		11	2.182		81.50	100		181.50	241
8580	30″ diameter		9	2.667		129	122		251	325
8600	36″ diameter	↓	7	3.429		157	157		314	410
8800	Elbow, 90°, 4″ diameter	Q-9	60	.267		13.35	11.80		25.15	32.50
8804	5″ diameter		52	.308		13.90	13.60		27.50	36
8806	6″ diameter		44	.364		13.95	16.05		30	40
8808	8″ diameter		28	.571		14.55	25.50		40.05	54.50
8810	10″ diameter		18	.889		15.15	39.50		54.65	76
8812	12″ diameter		13	1.231		15.75	54.50		70.25	100
8814	14″ diameter		11	1.455		22	64.50		86.50	122
8816	16″ diameter		10	1.600		29	70.50		99.50	139
8818	18″ diameter	↓	9.60	1.667		35.50	73.50		109	151
8820	20″ diameter	Q-10	14	1.714		45	78.50		123.50	169
8822	24″ diameter		12	2		98	91.50		189.50	247
8824	30″ diameter		9	2.667		258	122		380	470
8826	36″ diameter	↓	7	3.429	↓	315	157		472	585
9000	Reducing coupling									
9040	6″ x 4″	Q-9	46	.348	Ea.	11.20	15.35		26.55	36
9060	8″ x 6″		40	.400		14.20	17.70		31.90	42.50
9080	10″ x 8″		32	.500		16.60	22		38.60	52
9100	12″ x 10″		24	.667		18.80	29.50		48.30	65
9120	14″ x 12″		20	.800		21.50	35.50		57	77.50
9140	16″ x 14″		18	.889		25	39.50		64.50	87
9160	18″ x 16″	↓	16	1		30	44		74	100
9180	20″ x 18″	Q-10	24	1		36.50	46		82.50	110
9200	24″ x 22″		22	1.091		45	50		95	125
9220	30″ x 28″		18	1.333		99	61		160	202
9240	36″ x 34″	↓	16	1.500	↓	115	69		184	231
9800	Steel, stainless, straight lengths									
9810	3″ diameter, 26 Ga.	Q-9	400	.040	L.F.	14.20	1.77		15.97	18.30
9820	4″ diameter, 26 Ga.		360	.044		16.45	1.96		18.41	21
9830	5″ diameter, 26 Ga.		320	.050		17.75	2.21		19.96	23
9840	6″ diameter, 26 Ga.		280	.057		17.75	2.53		20.28	23.50
9850	7″ diameter, 26 Ga.		240	.067		19.55	2.95		22.50	26
9860	8″ diameter, 26 Ga.		200	.080		23	3.54		26.54	31
9870	9″ diameter, 26 Ga.		180	.089		33.50	3.93		37.43	43
9880	10″ diameter, 26 Ga.	↓	160	.100	↓	43.50	4.42		47.92	54.50

23 31 13 – Metal Ducts

23 31 13.16 Round and Flat-Oval Spiral Ducts		Crew	Daily Output	Labor-Hours	Unit	Material	2010 Bare Costs Labor	Equipment	Total	Total Incl O&P
9890	12" diameter, 26 Ga.	Q-9	120	.133	L.F.	61.50	5.90		67.40	76.50
9900	14" diameter, 26 Ga..	↓	80	.200	↓	65	8.85		73.85	85

23 31 13.17 Round Grease Duct		Crew	Daily Output	Labor-Hours	Unit	Material	Labor	Equipment	Total	Total Incl O&P
0010	**ROUND GREASE DUCT**									
1280	Add to labor for elevated installation									
1282	of prefabricated (purchased) ductwork									
1283	10' to 15' high						10%			
1284	15' to 20' high						20%			
1285	20' to 25' high						25%			
1286	25' to 30' high						35%			
1287	30' to 35' high						40%			
1288	35' to 40' high						50%			
1289	Over 40' high						55%			
4020	Zero clearance, 2 hr. fire rated, UL listed, 3" double wall									
4030	304 interior, aluminized exterior, all sizes are ID									
4032	For 316 interior and aluminized exterior, add					6%				
4034	For 304 interior and 304 exterior, add					21%				
4036	For 316 interior and 316 exterior, add					40%				
4040	Straight									
4050	5"	Q-9	36.48	.439	L.F.	62.50	19.40		81.90	98
4052	6"		34.20	.468		79	20.50		99.50	118
4054	8"		29.64	.540		90.50	24		114.50	136
4056	10"		27.36	.585		102	26		128	151
4058	12"		25.10	.637		116	28		144	171
4060	14"		23.94	.668		134	29.50		163.50	192
4062	16"		22.80	.702		140	31		171	201
4064	18"	↓	21.66	.739		172	32.50		204.50	240
4066	20"	Q-10	20.52	1.170		198	53.50		251.50	299
4068	22"		19.38	1.238		225	57		282	335
4070	24"		18.24	1.316		256	60.50		316.50	375
4072	26"		17.67	1.358		271	62.50		333.50	395
4074	28"		17.10	1.404		285	64.50		349.50	415
4076	30"		15.96	1.504		310	69		379	445
4078	32"		15.39	1.559		335	71.50		406.50	480
4080	36"	↓	14.25	1.684	↓	375	77		452	525
4100	Adjustable section, 30" long									
4102	5"	Q-9	18.24	.877	Ea.	196	39		235	274
4104	6"		17.10	.936		245	41.50		286.50	330
4106	8"		14.82	1.080		250	47.50		297.50	350
4108	10"		13.68	1.170		320	51.50		371.50	430
4110	12"		12.54	1.276		365	56.50		421.50	485
4112	14"		11.97	1.337		420	59		479	550
4114	16"		11.40	1.404		435	62		497	575
4116	18"	↓	10.83	1.477		535	65.50		600.50	690
4118	20"	Q-10	10.26	2.339		620	107		727	840
4120	22"		9.69	2.477		705	114		819	945
4122	24"		9.12	2.632		800	121		921	1,075
4124	26"		8.83	2.718		845	125		970	1,125
4126	28"		8.55	2.807		890	129		1,019	1,175
4128	30"		8.26	2.906		965	133		1,098	1,275
4130	32"		7.98	3.008		1,050	138		1,188	1,350
4132	36"	↓	6.84	3.509	↓	1,150	161		1,311	1,500

23 31 13.17 Round Grease Duct		Crew	Daily Output	Labor-Hours	Unit	2010 Bare Costs Material	Labor	Equipment	Total	Total Incl O&P
4140	Tee, 90°, grease									
4142	5″	Q-9	14.25	1.123	Ea.	520	49.50		569.50	645
4144	6″		13.68	1.170		610	51.50		661.50	750
4146	8″		12.54	1.276		700	56.50		756.50	855
4148	10″		11.97	1.337		820	59		879	990
4150	12″		11.40	1.404		955	62		1,017	1,150
4152	14″		10.26	1.559		1,125	69		1,194	1,350
4154	16″		9.12	1.754		1,275	77.50		1,352.50	1,525
4156	18″		7.98	2.005		1,500	88.50		1,588.50	1,775
4158	20″		9.69	1.651		1,675	73		1,748	1,950
4160	22″		8.26	1.937		1,875	85.50		1,960.50	2,200
4162	24″		6.84	2.339		2,075	103		2,178	2,425
4164	26″		6.55	2.443		2,350	108		2,458	2,750
4166	28″	Q-10	6.27	3.828		2,600	175		2,775	3,125
4168	30″		5.98	4.013		3,075	184		3,259	3,675
4170	32″		5.70	4.211		3,550	193		3,743	4,225
4172	36″		5.13	4.678		4,150	214		4,364	4,875
4180	Cleanout Tee Cap									
4182	5″	Q-9	22.80	.702	Ea.	152	31		183	215
4184	6″		21.09	.759		180	33.50		213.50	249
4186	8″		19.38	.826		183	36.50		219.50	257
4188	10″		18.24	.877		201	39		240	281
4190	12″		17.10	.936		210	41.50		251.50	294
4192	14″		15.96	1.003		242	44.50		286.50	335
4194	16″		14.25	1.123		265	49.50		314.50	365
4196	18″		13.68	1.170		299	51.50		350.50	410
4198	20″	Q-10	15.39	1.559		325	71.50		396.50	465
4200	22″		12.54	1.914		365	87.50		452.50	535
4202	24″		11.97	2.005		410	92		502	590
4204	26″		11.40	2.105		455	96.50		551.50	650
4206	28″		10.83	2.216		505	102		607	710
4208	30″		10.26	2.339		560	107		667	780
4210	32″		9.69	2.477		620	114		734	850
4212	36″		8.55	2.807		770	129		899	1,050
4220	Elbow, 45°									
4222	5″	Q-9	18.24	.877	Ea.	284	39		323	370
4224	6″		17.10	.936		335	41.50		376.50	430
4226	8″		14.82	1.080		380	47.50		427.50	490
4228	10″		13.68	1.170		425	51.50		476.50	545
4230	12″		12.54	1.276		485	56.50		541.50	615
4232	14″		11.97	1.337		555	59		614	700
4234	16″		11.40	1.404		625	62		687	785
4236	18″		10.83	1.477		710	65.50		775.50	880
4238	20″	Q-10	10.26	2.339		810	107		917	1,050
4240	22″		9.69	2.477		915	114		1,029	1,175
4242	24″		9.12	2.632		1,050	121		1,171	1,325
4244	26″		8.83	2.718		1,100	125		1,225	1,425
4246	28″		8.55	2.807		1,175	129		1,304	1,500
4248	30″		8.26	2.906		1,375	133		1,508	1,725
4250	32″		7.98	3.008		1,575	138		1,713	1,950
4252	36″		6.84	3.509		2,225	161		2,386	2,700
4260	Elbow, 90°									
4262	5″	Q-9	18.24	.877	Ea.	450	39		489	555

23 31 13.17 Round Grease Duct		Crew	Daily Output	Labor-Hours	Unit	Material	2010 Bare Costs Labor	Equipment	Total	Total Incl O&P
4264	6"	Q-9	17.10	.936	Ea.	530	41.50		571.50	645
4266	8"		14.82	1.080		580	47.50		627.50	715
4268	10"		13.68	1.170		670	51.50		721.50	815
4270	12"		12.54	1.276		760	56.50		816.50	920
4272	14"		11.97	1.337		890	59		949	1,075
4274	16"		11.40	1.404		990	62		1,052	1,200
4276	18"		10.83	1.477		1,150	65.50		1,215.50	1,375
4278	20"	Q-10	10.26	2.339		1,275	107		1,382	1,575
4280	22"		9.69	2.477		1,450	114		1,564	1,750
4282	24"		9.12	2.632		1,600	121		1,721	1,925
4284	26"		8.83	2.718		1,800	125		1,925	2,175
4286	28"		8.55	2.807		2,000	129		2,129	2,400
4288	30"		8.26	2.906		2,325	133		2,458	2,750
4290	32"		7.98	3.008		2,650	138		2,788	3,125
4292	36"		6.84	3.509		3,250	161		3,411	3,825
4300	Support strap									
4302	5"	Q-9	27	.593	Ea.	22.50	26		48.50	64
4304	6"		25.50	.627		22.50	27.50		50	66.50
4306	8"		22	.727		23.50	32		55.50	74
4308	10"		20.50	.780		23.50	34.50		58	77.50
4310	12"		18.80	.851		23.50	37.50		61	82.50
4312	14"		17.90	.894		23.50	39.50		63	85.50
4314	16"		17	.941		23.50	41.50		65	88.50
4316	18"		16.20	.988		24.50	43.50		68	93
4318	20"	Q-10	15.40	1.558		25.50	71.50		97	136
4320	22"	"	14.50	1.655		25.50	76		101.50	143
4340	Plate Support Assembly									
4342	5"	Q-9	17.10	.936	Ea.	106	41.50		147.50	179
4344	6"		14.82	1.080		124	47.50		171.50	209
4346	8"		12.54	1.276		139	56.50		195.50	239
4348	10"		11.40	1.404		152	62		214	262
4350	12"		10.26	1.559		158	69		227	278
4352	14"		9.69	1.651		193	73		266	325
4354	16"		9.12	1.754		201	77.50		278.50	340
4356	18"		8.55	1.871		214	82.50		296.50	360
4358	20"	Q-10	9.12	2.632		225	121		346	430
4360	22"		8.55	2.807		232	129		361	450
4362	24"		7.98	3.008		238	138		376	470
4364	26"		7.69	3.121		292	143		435	535
4366	28"		7.41	3.239		330	148		478	590
4368	30"		7.12	3.371		335	154		489	605
4370	32"		6.84	3.509		380	161		541	660
4372	36"		5.70	4.211		475	193		668	810
4380	Wall Guide Assembly									
4382	5"	Q-9	20.20	.792	Ea.	124	35		159	189
4384	6"		19.10	.838		135	37		172	204
4386	8"		16.50	.970		141	43		184	220
4388	10"		15.30	1.046		148	46		194	233
4390	12"		14.10	1.135		155	50		205	246
4392	14"		13.40	1.194		167	53		220	264
4394	16"		12.75	1.255		169	55.50		224.50	270
4396	18"		12.10	1.322		177	58.50		235.50	284
4398	20"	Q-10	11.50	2.087		186	95.50		281.50	350

23 31 13.17 Round Grease Duct		Crew	Daily Output	Labor-Hours	Unit	2010 Bare Costs Material	Labor	Equipment	Total	Total Incl O&P
4400	22″	Q-10	10.90	2.202	Ea.	191	101		292	365
4402	24″		10.20	2.353		193	108		301	375
4404	26″		9.90	2.424		201	111		312	390
4406	28″		9.60	2.500		216	115		331	410
4408	30″		9.30	2.581		225	118		343	425
4410	32″		9	2.667		235	122		357	445
4412	36″		7.60	3.158		256	145		401	500
4420	Hood Transition, Flanged or Unflanged									
4422	5″	Q-9	16.60	.964	Ea.	42.50	42.50		85	111
4424	6″		15.60	1.026		50	45.50		95.50	124
4426	8″		13.50	1.185		57	52.50		109.50	142
4428	10″		12.50	1.280		63.50	56.50		120	156
4430	12″		11.40	1.404		73.50	62		135.50	175
4432	14″		10.90	1.468		84.50	65		149.50	192
4434	16″		10.40	1.538		94.50	68		162.50	207
4436	18″		9.80	1.633		109	72		181	229
4438	20″	Q-10	9.40	2.553		127	117		244	315
4440	22″		8.80	2.727		149	125		274	355
4442	24″		8.30	2.892		168	133		301	385
4444	26″		8.10	2.963		179	136		315	405
4446	28″		7.80	3.077		193	141		334	425
4448	30″		7.50	3.200		208	147		355	450
4450	32″		7.30	3.288		218	151		369	470
4452	36″		6.20	3.871		264	177		441	560
4460	Fan Adapter									
4462	5″	Q-9	16	1	Ea.	240	44		284	330
4464	6″		15	1.067		290	47		337	390
4466	8″		13	1.231		325	54.50		379.50	440
4468	10″		12	1.333		360	59		419	490
4470	12″		11	1.455		415	64.50		479.50	560
4472	14″		10.50	1.524		480	67.50		547.50	630
4474	16″		10	1.600		550	70.50		620.50	705
4476	18″		9.50	1.684		625	74.50		699.50	800
4478	20″	Q-10	9	2.667		730	122		852	985
4480	22″		8.50	2.824		850	129		979	1,125
4482	24″		8	3		875	137		1,012	1,175
4484	26″		7.70	3.117		905	143		1,048	1,200
4486	28″		7.50	3.200		925	147		1,072	1,250
4488	30″		7.20	3.333		955	153		1,108	1,275
4490	32″		7	3.429		985	157		1,142	1,325
4492	36″		6	4		1,025	183		1,208	1,400
4500	Tapered Increaser/Reducer									
4510	5″ x 10″ dia	Q-9	13.68	1.170	Ea.	465	51.50		516.50	590
4520	5″ x 20″ dia		10.30	1.553		725	68.50		793.50	905
4530	6″ x 10″ dia		13.40	1.194		465	53		518	590
4540	6″ x 20″ dia		10.10	1.584		765	70		835	945
4550	8″ x 10″ dia		13.20	1.212		440	53.50		493.50	560
4560	8″ x 20″ dia		9.90	1.616		765	71.50		836.50	955
4570	10″ x 20″ dia		10.60	1.509		825	66.50		891.50	1,000
4580	10″ x 28″ dia	Q-10	8.60	2.791		1,125	128		1,253	1,450
4590	12″ x 20″ dia		8.80	2.727		870	125		995	1,150
4600	14″ x 28″ dia		8.40	2.857		1,300	131		1,431	1,625
4610	16″ x 20″ dia		8.60	2.791		870	128		998	1,150

23 31 13 – Metal Ducts

23 31 13.17 Round Grease Duct		Crew	Daily Output	Labor-Hours	Unit	Material	2010 Bare Costs Labor	Equipment	Total	Total Incl O&P
4620	16" x 30" dia	Q-10	8.30	2.892	Ea.	1,425	133		1,558	1,775
4630	18" x 24" dia		8.80	2.727		1,050	125		1,175	1,350
4640	18" x 30" dia		8.10	2.963		1,425	136		1,561	1,775
4650	20" x 24" dia		8.30	2.892		995	133		1,128	1,300
4660	20" x 32" dia		8	3		1,525	137		1,662	1,875
4670	24" mm x 26" dia		8	3		995	137		1,132	1,300
4680	24" mm x 36" dia		7.30	3.288		1,750	151		1,901	2,150
4690	28" x 30" dia		7.90	3.038		1,200	139		1,339	1,525
4700	28" x 36" dia		7.10	3.380		1,750	155		1,905	2,150
4710	30" x 36" dia	↓	6.90	3.478	↓	1,800	159		1,959	2,225
4730	Many intermediate standard sizes of tapered fittings are available									

23 31 13.19 Metal Duct Fittings		Crew	Daily Output	Labor-Hours	Unit	Material	Labor	Equipment	Total	Total Incl O&P
0010	**METAL DUCT FITTINGS**									
0050	Air extractors, 12" x 4"	1 Shee	24	.333	Ea.	16.10	16.35		32.45	43
0100	8" x 6"		22	.364		16.10	17.85		33.95	45
0120	12" x 6"		21	.381		22.50	18.70		41.20	53
0140	16" x 6"		20	.400		28	19.65		47.65	60.50
0160	24" x 6"		18	.444		46.50	22		68.50	84.50
0180	12" x 8"		20	.400		25	19.65		44.65	57
0200	20" x 8"		16	.500		36.50	24.50		61	77
0240	18" x 10"		14	.571		35.50	28		63.50	81.50
0260	24" x 10"		12	.667		46	32.50		78.50	100
0280	24" x 12"		10	.800		49.50	39.50		89	114
0300	30" x 12"	↓	8	1	↓	62	49		111	143
0350	Duct collar, spin-in type									
0352	Without damper									
0354	Round									
0356	4"	1 Shee	32	.250	Ea.	2.07	12.30		14.37	21
0357	5"		30	.267		2.07	13.10		15.17	22
0358	6"		28	.286		2.07	14.05		16.12	23.50
0359	7"		26	.308		2.38	15.10		17.48	25.50
0360	8"		24	.333		2.38	16.35		18.73	27.50
0361	9"		22	.364		3.60	17.85		21.45	31
0362	10"		20	.400		3.60	19.65		23.25	33.50
0363	12"		17	.471	↓	4.61	23		27.61	40
2000	Fabrics for flexible connections, with metal edge		100	.080	L.F.	2.13	3.93		6.06	8.30
2100	Without metal edge	↓	160	.050	"	1.53	2.46		3.99	5.40

23 31 16 – Nonmetal Ducts

23 31 16.13 Fibrous-Glass Ducts		Crew	Daily Output	Labor-Hours	Unit	Material	Labor	Equipment	Total	Total Incl O&P
0010	**FIBROUS-GLASS DUCTS** R233100-20									
1280	Add to labor for elevated installation									
1282	of prefabricated (purchased) ductwork									
1283	10' to 15' high						10%			
1284	15' to 20' high						20%			
1285	20' to 25' high						25%			
1286	25' to 30' high						35%			
1287	30' to 35' high						40%			
1288	35' to 40' high						50%			
1289	Over 40' high						55%			
3490	Rigid fiberglass duct board, foil reinf. kraft facing									
3500	Rectangular, 1" thick, alum. faced, (FRK), std. weight	Q-10	350	.069	SF Surf	.77	3.14		3.91	5.60

23 31 16 – Nonmetal Ducts

23 31 16.16 Thermoset Fiberglass-Reinforced Plastic Ducts		Crew	Daily Output	Labor-Hours	Unit	2010 Bare Costs Material	Labor	Equipment	Total	Total Incl O&P
0010	**THERMOSET FIBERGLASS-REINFORCED PLASTIC DUCTS** R233100-20									
1280	Add to labor for elevated installation									
1282	of prefabricated (purchased) ductwork									
1283	10′ to 15′ high						10%			
1284	15′ to 20′ high						20%			
1285	20′ to 25′ high						25%			
1286	25′ to 30′ high						35%			
1287	30′ to 35′ high						40%			
1288	35′ to 40′ high						50%			
1289	Over 40′ high						55%			
3550	Rigid fiberglass reinforced plastic, FM approved									
3552	for acid fume and smoke exhaust system, nonflammable									
3554	Straight, 4″ diameter	Q-9	200	.080	L.F.	10.60	3.54		14.14	17
3555	6″ diameter		146	.110		13.90	4.84		18.74	22.50
3556	8″ diameter		106	.151		16.85	6.65		23.50	28.50
3557	10″ diameter		85	.188		20	8.30		28.30	34.50
3558	12″ diameter		68	.235		23.50	10.40		33.90	42
3559	14″ diameter		50	.320		35	14.15		49.15	60
3560	16″ diameter		40	.400		39	17.70		56.70	70
3561	18″ diameter		31	.516		43.50	23		66.50	82
3562	20″ diameter	Q-10	44	.545		48	25		73	90.50
3563	22″ diameter		40	.600		51.50	27.50		79	98.50
3564	24″ diameter		37	.649		55.50	29.50		85	106
3565	26″ diameter		34	.706		60	32.50		92.50	115
3566	28″ diameter		28	.857		64.50	39.50		104	131
3567	30″ diameter		24	1		68.50	46		114.50	145
3568	32″ diameter		22.60	1.062		73	48.50		121.50	154
3569	34″ diameter		21.60	1.111		77	51		128	162
3570	36″ diameter		20.60	1.165		81.50	53.50		135	171
3571	38″ diameter		19.80	1.212		125	55.50		180.50	222
3572	42″ diameter		18.30	1.311		140	60		200	245
3573	46″ diameter		17	1.412		155	64.50		219.50	268
3574	50″ diameter		15.90	1.509		168	69		237	290
3575	54″ diameter		14.90	1.611		185	74		259	315
3576	60″ diameter		13.60	1.765		212	81		293	355
3577	64″ diameter		12.80	1.875		310	86		396	470
3578	68″ diameter		12.20	1.967		330	90		420	500
3579	72″ diameter		11.90	2.017		355	92.50		447.50	530
3584	Note: joints are cemented with									
3586	fiberglass resin, included in material cost.									
3590	Elbow, 90°, 4″ diameter	Q-9	20.30	.788	Ea.	51	35		86	110
3591	6″ diameter		13.80	1.159		71.50	51.50		123	156
3592	8″ diameter		10	1.600		79	70.50		149.50	194
3593	10″ diameter		7.80	2.051		103	90.50		193.50	251
3594	12″ diameter		6.50	2.462		131	109		240	310
3595	14″ diameter		5.50	2.909		161	129		290	370
3596	16″ diameter		4.80	3.333		191	147		338	435
3597	18″ diameter		4.40	3.636		274	161		435	545
3598	20″ diameter	Q-10	5.90	4.068		289	186		475	595
3599	22″ diameter		5.40	4.444		355	204		559	700
3600	24″ diameter		4.90	4.898		410	224		634	790
3601	26″ diameter		4.60	5.217		415	239		654	815

23 31 16.16 Thermoset Fiberglass-Reinforced Plastic Ducts		Crew	Daily Output	Labor-Hours	Unit	Material	2010 Bare Costs Labor	Equipment	Total	Total Incl O&P
3602	28″ diameter	Q-10	4.20	5.714	Ea.	515	262		777	965
3603	30″ diameter		3.90	6.154		575	282		857	1,050
3604	32″ diameter		3.70	6.486		640	297		937	1,150
3605	34″ diameter		3.45	6.957		755	320		1,075	1,325
3606	36″ diameter		3.20	7.500		865	345		1,210	1,475
3607	38″ diameter		3.07	7.818		1,225	360		1,585	1,900
3608	42″ diameter		2.70	8.889		1,475	405		1,880	2,250
3609	46″ diameter		2.54	9.449		1,725	435		2,160	2,550
3610	50″ diameter		2.33	10.300		2,000	470		2,470	2,925
3611	54″ diameter		2.14	11.215		2,300	515		2,815	3,300
3612	60″ diameter		1.94	12.371		3,050	565		3,615	4,200
3613	64″ diameter		1.82	13.187		4,425	605		5,030	5,800
3614	68″ diameter		1.71	14.035		4,925	645		5,570	6,400
3615	72″ diameter	↓	1.62	14.815		5,450	680		6,130	7,000
3626	Elbow, 45°, 4″ diameter	Q-9	22.30	.717		49	31.50		80.50	102
3627	6″ diameter		15.20	1.053		68	46.50		114.50	146
3628	8″ diameter		11	1.455		72.50	64.50		137	177
3629	10″ diameter		8.60	1.860		76.50	82		158.50	209
3630	12″ diameter		7.20	2.222		83.50	98		181.50	241
3631	14″ diameter		6.10	2.623		102	116		218	289
3632	16″ diameter		5.28	3.030		122	134		256	335
3633	18″ diameter	↓	4.84	3.306		143	146		289	380
3634	20″ diameter	Q-10	6.50	3.692		165	169		334	440
3635	22″ diameter		5.94	4.040		226	185		411	530
3636	24″ diameter		5.40	4.444		257	204		461	595
3637	26″ diameter		5.06	4.743		291	217		508	650
3638	28″ diameter		4.62	5.195		315	238		553	705
3639	30″ diameter		4.30	5.581		365	256		621	785
3640	32″ diameter		4.07	5.897		405	270		675	855
3641	34″ diameter		3.80	6.316		480	289		769	965
3642	36″ diameter		3.52	6.818		550	310		860	1,075
3643	38″ diameter		3.38	7.101		775	325		1,100	1,350
3644	42″ diameter		2.97	8.081		925	370		1,295	1,575
3645	46″ diameter		2.80	8.571		1,075	395		1,470	1,800
3646	50″ diameter		2.56	9.375		1,250	430		1,680	2,025
3647	54″ diameter		2.35	10.213		1,425	470		1,895	2,275
3648	60″ diameter		2.13	11.268		1,925	515		2,440	2,875
3649	64″ diameter		2	12		2,775	550		3,325	3,875
3650	68″ diameter		1.88	12.766		3,075	585		3,660	4,250
3651	72″ diameter	↓	1.78	13.483		3,400	620		4,020	4,650
3660	Tee, 90°, 4″ diameter	Q-9	14.06	1.138		28	50.50		78.50	107
3661	6″ diameter		9.52	1.681		43	74.50		117.50	159
3662	8″ diameter		6.98	2.292		58.50	101		159.50	218
3663	10″ diameter		5.45	2.936		76.50	130		206.50	280
3664	12″ diameter		4.60	3.478		95.50	154		249.50	340
3665	14″ diameter		3.86	4.145		139	183		322	430
3666	16″ diameter		3.40	4.706		165	208		373	495
3667	18″ diameter	↓	3.05	5.246		196	232		428	565
3668	20″ diameter	Q-10	4.12	5.825		229	267		496	655
3669	22″ diameter		3.78	6.349		262	291		553	730
3670	24″ diameter		3.46	6.936		295	320		615	805
3671	26″ diameter		3.20	7.500		335	345		680	890
3672	28″ diameter	↓	2.96	8.108	↓	375	370		745	975

23 31 16.16 Thermoset Fiberglass-Reinforced Plastic Ducts		Crew	Daily Output	Labor-Hours	Unit	Material	2010 Bare Costs Labor	Equipment	Total	Total Incl O&P
3673	30″ diameter	Q-10	2.75	8.727	Ea.	420	400		820	1,075
3674	32″ diameter		2.54	9.449		465	435		900	1,175
3675	34″ diameter		2.40	10		505	460		965	1,250
3676	36″ diameter		2.27	10.573		555	485		1,040	1,350
3677	38″ diameter		2.15	11.163		835	510		1,345	1,700
3678	42″ diameter		1.91	12.565		1,000	575		1,575	1,975
3679	46″ diameter		1.78	13.483		1,175	620		1,795	2,225
3680	50″ diameter		1.63	14.724		1,375	675		2,050	2,550
3681	54″ diameter		1.50	16		1,600	735		2,335	2,850
3682	60″ diameter		1.35	17.778		2,000	815		2,815	3,425
3683	64″ diameter		1.27	18.898		3,000	865		3,865	4,600
3684	68″ diameter		1.20	20		3,350	915		4,265	5,075
3685	72″ diameter		1.13	21.239		3,750	975		4,725	5,600
3690	For Y @ 45°, add					22%				
3700	Blast gate, 4″ diameter	Q-9	22.30	.717		136	31.50		167.50	197
3701	6″ diameter		15.20	1.053		154	46.50		200.50	240
3702	8″ diameter		11	1.455		207	64.50		271.50	325
3703	10″ diameter		8.60	1.860		236	82		318	385
3704	12″ diameter		7.20	2.222		289	98		387	465
3705	14″ diameter		6.10	2.623		320	116		436	525
3706	16″ diameter		5.28	3.030		355	134		489	595
3707	18″ diameter		4.84	3.306		390	146		536	650
3708	20″ diameter	Q-10	6.50	3.692		420	169		589	715
3709	22″ diameter		5.94	4.040		455	185		640	780
3710	24″ diameter		5.40	4.444		485	204		689	845
3711	26″ diameter		5.06	4.743		520	217		737	905
3712	28″ diameter		4.62	5.195		560	238		798	975
3713	30″ diameter		4.30	5.581		595	256		851	1,050
3714	32″ diameter		4.07	5.897		650	270		920	1,125
3715	34″ diameter		3.80	6.316		710	289		999	1,225
3716	36″ diameter		3.52	6.818		765	310		1,075	1,325
3717	38″ diameter		3.38	7.101		1,025	325		1,350	1,625
3718	42″ diameter		2.97	8.081		1,175	370		1,545	1,850
3719	46″ diameter		2.80	8.571		1,350	395		1,745	2,100
3720	50″ diameter		2.56	9.375		1,550	430		1,980	2,350
3721	54″ diameter		2.35	10.213		1,750	470		2,220	2,625
3722	60″ diameter		2.13	11.268		2,050	515		2,565	3,025
3723	64″ diameter		2	12		2,275	550		2,825	3,325
3724	68″ diameter		1.88	12.766		2,500	585		3,085	3,625
3725	72″ diameter		1.78	13.483		2,750	620		3,370	3,950
3740	Reducers, 4″ diameter	Q-9	24.50	.653		34	29		63	81
3741	6″ diameter		16.70	.958		50.50	42.50		93	120
3742	8″ diameter		12	1.333		68.50	59		127.50	165
3743	10″ diameter		9.50	1.684		84.50	74.50		159	206
3744	12″ diameter		7.90	2.025		101	89.50		190.50	248
3745	14″ diameter		6.70	2.388		118	106		224	290
3746	16″ diameter		5.80	2.759		135	122		257	335
3747	18″ diameter		5.30	3.019		153	133		286	370
3748	20″ diameter	Q-10	7.20	3.333		169	153		322	415
3749	22″ diameter		6.50	3.692		273	169		442	555
3750	24″ diameter		5.90	4.068		295	186		481	605
3751	26″ diameter		5.60	4.286		320	196		516	645
3752	28″ diameter		5.10	4.706		350	216		566	710

23 31 16 – Nonmetal Ducts

23 31 16.16 Thermoset Fiberglass-Reinforced Plastic Ducts		Crew	Daily Output	Labor-Hours	Unit	2010 Bare Costs Material	Labor	Equipment	Total	Total Incl O&P
3753	30" diameter	Q-10	4.70	5.106	Ea.	375	234		609	765
3754	32" diameter		4.50	5.333		400	244		644	810
3755	34" diameter		4.20	5.714		425	262		687	860
3756	36" diameter		3.90	6.154		450	282		732	920
3757	38" diameter		3.70	6.486		680	297		977	1,200
3758	42" diameter		3.30	7.273		750	335		1,085	1,325
3759	46" diameter		3.10	7.742		820	355		1,175	1,450
3760	50" diameter		2.80	8.571		890	395		1,285	1,575
3761	54" diameter		2.60	9.231		965	425		1,390	1,700
3762	60" diameter		2.34	10.256		1,075	470		1,545	1,900
3763	64" diameter		2.20	10.909		1,675	500		2,175	2,600
3764	68" diameter		2.06	11.650		1,775	535		2,310	2,750
3765	72" diameter		1.96	12.245		1,900	560		2,460	2,925
3780	Flange, per each, 4" diameter	Q-9	40.60	.394		82	17.40		99.40	117
3781	6" diameter		27.70	.578		96	25.50		121.50	145
3782	8" diameter		20.50	.780		120	34.50		154.50	184
3783	10" diameter		16.30	.982		144	43.50		187.50	224
3784	12" diameter		13.90	1.151		165	51		216	259
3785	14" diameter		11.30	1.416		191	62.50		253.50	305
3786	16" diameter		9.70	1.649		215	73		288	345
3787	18" diameter		8.30	1.928		238	85		323	390
3788	20" diameter	Q-10	11.30	2.124		262	97.50		359.50	435
3789	22" diameter		10.40	2.308		286	106		392	475
3790	24" diameter		9.50	2.526		310	116		426	515
3791	26" diameter		8.80	2.727		335	125		460	560
3792	28" diameter		8.20	2.927		360	134		494	600
3793	30" diameter		7.60	3.158		380	145		525	635
3794	32" diameter		7.20	3.333		405	153		558	675
3795	34" diameter		6.80	3.529		430	162		592	715
3796	36" diameter		6.40	3.750		455	172		627	760
3797	38" diameter		6.10	3.934		480	180		660	800
3798	42" diameter		5.50	4.364		520	200		720	880
3799	46" diameter		5	4.800		570	220		790	965
3800	50" diameter		4.60	5.217		620	239		859	1,050
3801	54" diameter		4.20	5.714		665	262		927	1,125
3802	60" diameter		3.80	6.316		740	289		1,029	1,250
3803	64" diameter		3.60	6.667		785	305		1,090	1,325
3804	68" diameter		3.40	7.059		830	325		1,155	1,400
3805	72" diameter		3.20	7.500		880	345		1,225	1,500

23 31 16.19 PVC Ducts		Crew	Daily Output	Labor-Hours	Unit	Material	Labor	Equipment	Total	Total Incl O&P
0010	**PVC DUCTS**									
4000	Rigid plastic, corrosive fume resistant PVC									
4020	Straight, 6" diameter	Q-9	220	.073	L.F.	7.95	3.21		11.16	13.60
4030	7" diameter		160	.100		11.80	4.42		16.22	19.65
4040	8" diameter		160	.100		11.80	4.42		16.22	19.65
4050	9" diameter		140	.114		13.30	5.05		18.35	22.50
4060	10" diameter		120	.133		14.85	5.90		20.75	25.50
4070	12" diameter		100	.160		17.35	7.05		24.40	30
4080	14" diameter		70	.229		22.50	10.10		32.60	40.50
4090	16" diameter		62	.258		26	11.40		37.40	46
4100	18" diameter		58	.276		40.50	12.20		52.70	63
4110	20" diameter	Q-10	75	.320		52	14.65		66.65	79.50

23 31 16 – Nonmetal Ducts

23 31 16.19 PVC Ducts		Crew	Daily Output	Labor-Hours	Unit	Material	2010 Bare Costs Labor	Equipment	Total	Total Incl O&P
4120	22" diameter	Q-10	60	.400	L.F.	59.50	18.35		77.85	93.50
4130	24" diameter		55	.436		67	20		87	104
4140	26" diameter		42	.571		76	26		102	123
4150	28" diameter		35	.686		81.50	31.50		113	137
4160	30" diameter		28	.857		88	39.50		127.50	157
4170	36" diameter		23	1.043		105	48		153	189
4180	42" diameter		20	1.200		123	55		178	219
4200	48" diameter		18	1.333		140	61		201	247
4250	Coupling, 6" diameter	Q-9	88	.182	Ea.	10.45	8.05		18.50	23.50
4260	7" diameter		82	.195		11.20	8.60		19.80	25.50
4270	8" diameter		78	.205		11.70	9.05		20.75	26.50
4280	9" diameter		74	.216		12.75	9.55		22.30	28.50
4290	10" diameter		70	.229		12.75	10.10		22.85	29.50
4300	12" diameter		55	.291		13.65	12.85		26.50	34.50
4310	14" diameter		44	.364		16.60	16.05		32.65	43
4320	16" diameter		40	.400		19.65	17.70		37.35	48.50
4330	18" diameter		38	.421		23	18.60		41.60	53
4340	20" diameter	Q-10	55	.436		39.50	20		59.50	74
4350	22" diameter		50	.480		45.50	22		67.50	83.50
4360	24" diameter		45	.533		51.50	24.50		76	93.50
4370	26" diameter		39	.615		54	28		82	102
4380	28" diameter		33	.727		57	33.50		90.50	113
4390	30" diameter		27	.889		62	40.50		102.50	130
4400	36" diameter		25	.960		98	44		142	175
4410	42" diameter		23	1.043		110	48		158	194
4420	48" diameter		21	1.143		128	52.50		180.50	221
4470	Elbow, 90°, 6" diameter	Q-9	44	.364		83	16.05		99.05	116
4480	7" diameter		36	.444		89	19.65		108.65	128
4490	8" diameter		28	.571		89	25.50		114.50	136
4500	9" diameter		22	.727		122	32		154	183
4510	10" diameter		18	.889		144	39.50		183.50	218
4520	12" diameter		15	1.067		196	47		243	288
4530	14" diameter		11	1.455		269	64.50		333.50	395
4540	16" diameter		10	1.600		325	70.50		395.50	460
4550	18" diameter	Q-10	15	1.600		350	73.50		423.50	495
4560	20" diameter		14	1.714		370	78.50		448.50	530
4570	22" diameter		13	1.846		475	84.50		559.50	650
4580	24" diameter		12	2		535	91.50		626.50	725
4590	26" diameter		11	2.182		610	100		710	820
4600	28" diameter		10	2.400		725	110		835	960
4610	30" diameter		9	2.667		850	122		972	1,125
4620	36" diameter		7	3.429		1,175	157		1,332	1,550
4630	42" diameter		5	4.800		1,600	220		1,820	2,100
4640	48" diameter		4	6		2,125	275		2,400	2,750
4750	Elbow 45°, use 90° and deduct					45%				

23 33 13 – Dampers

23 33 13.13 Volume-Control Dampers		Crew	Daily Output	Labor-Hours	Unit	Material	Labor	Equipment	Total	Total Incl O&P
						2010 Bare Costs				
0010	**VOLUME-CONTROL DAMPERS**									
5990	Multi-blade dampers, opposed blade, 8″ x 6″	1 Shee	24	.333	Ea.	20	16.35		36.35	47
5994	8″ x 8″		22	.364		21	17.85		38.85	50
5996	10″ x 10″		21	.381		24.50	18.70		43.20	55.50
6000	12″ x 12″		21	.381		27.50	18.70		46.20	58.50
6020	12″ x 18″		18	.444		36.50	22		58.50	73.50
6030	14″ x 10″		20	.400		26.50	19.65		46.15	59
6031	14″ x 14″		17	.471		32.50	23		55.50	70.50
6032	16″ x 10″		18	.444		29	22		51	64.50
6033	16″ x 12″		17	.471		32.50	23		55.50	70.50
6035	16″ x 16″		16	.500		40.50	24.50		65	81.50
6036	18″ x 14″		16	.500		40.50	24.50		65	81.50
6037	18″ x 16″		15	.533		44.50	26		70.50	88.50
6038	18″ x 18″		15	.533		48	26		74	92.50
6040	18″ x 24″		12	.667		62	32.50		94.50	118
6060	18″ x 28″		10	.800		72	39.50		111.50	139
6068	20″ x 6″		18	.444		26	22		48	61.50
6069	20″ x 8″		16	.500		30	24.50		54.50	70.50
6070	20″ x 16″		14	.571		48	28		76	95.50
6071	20″ x 18″		13	.615		52	30		82	103
6072	20″ x 20″		13	.615		57.50	30		87.50	110
6073	22″ x 6″		20	.400		26.50	19.65		46.15	59
6074	22″ x 18″		14	.571		57.50	28		85.50	106
6075	22″ x 20″		12	.667		62.50	32.50		95	119
6076	24″ x 16″		11	.727		57	35.50		92.50	117
6077	22″ x 22″		10	.800		67	39.50		106.50	133
6078	24″ x 20″		8	1		67	49		116	148
6080	24″ x 24″		8	1		78.50	49		127.50	161
6100	24″ x 28″		6	1.333		92	65.50		157.50	200
6110	26″ x 26″		6	1.333		86.50	65.50		152	194
6120	28″ x 28″	Q-9	11	1.455		98	64.50		162.50	206
6130	30″ x 18″		10	1.600		83.50	70.50		154	199
6132	30″ x 24″		7	2.286		104	101		205	268
6133	30″ x 30″		6.60	2.424		124	107		231	298
6135	32″ x 32″		6.40	2.500		144	111		255	325
6150	34″ x 34″		6	2.667		161	118		279	355
6151	36″ x 12″		10	1.600		65.50	70.50		136	179
6152	36″ x 16″		8	2		89.50	88.50		178	232
6156	36″ x 32″		6.30	2.540		161	112		273	345
6157	36″ x 34″		6.20	2.581		168	114		282	360
6158	36″ x 36″		6	2.667		176	118		294	370
6160	44″ x 28″		5.80	2.759		181	122		303	385
6180	48″ x 36″		5.60	2.857		237	126		363	450
6200	56″ x 36″		5.40	2.963		287	131		418	515
6220	60″ x 36″		5.20	3.077		305	136		441	540
6240	60″ x 44″		5	3.200		360	141		501	610
7500	Variable volume modulating motorized damper, incl. elect. mtr.									
7504	8″ x 6″	1 Shee	15	.533	Ea.	109	26		135	160
7506	10″ x 6″		14	.571		109	28		137	163
7510	10″ x 10″		13	.615		112	30		142	170
7520	12″ x 12″		12	.667		117	32.50		149.50	179
7522	12″ x 16″		11	.727		118	35.50		153.50	184
7524	16″ x 10″		12	.667		115	32.50		147.50	177

23 33 13 – Dampers

23 33 13.13 Volume-Control Dampers		Crew	Daily Output	Labor-Hours	Unit	Material	Labor	Equipment	Total	Total Incl O&P
						2010 Bare Costs				
7526	16" x 14"	1 Shee	10	.800	Ea.	120	39.50		159.50	191
7528	16" x 18"		9	.889		125	43.50		168.50	203
7540	18" x 12"		10	.800		120	39.50		159.50	192
7542	18" x 18"		8	1		127	49		176	215
7544	20" x 14"		8	1		130	49		179	218
7546	20" x 18"		7	1.143		136	56		192	235
7560	24" x 12"		8	1		132	49		181	220
7562	24" x 18"		7	1.143		143	56		199	242
7568	28" x 10"		7	1.143		132	56		188	230
7580	28" x 16"		6	1.333		149	65.50		214.50	263
7590	30" x 14"		5	1.600		140	78.50		218.50	273
7600	30" x 18"		4	2		180	98		278	345
7610	30" x 24"		3.80	2.105		239	103		342	420
7690	48" x 48"	Q-9	6	2.667		760	118		878	1,025
7694	6' x 14' w/2 motors	"	3	5.333		3,250	236		3,486	3,925
7700	For thermostat, add	1 Shee	8	1		40.50	49		89.50	119
7800	For transformer 40 VA capacity, add	"	16	.500		18	24.50		42.50	57
8000	Multi-blade dampers, parallel blade									
8100	8" x 8"	1 Shee	24	.333	Ea.	73	16.35		89.35	106
8120	12" x 8"		22	.364		73	17.85		90.85	108
8140	16" x 10"		20	.400		94	19.65		113.65	133
8160	18" x 12"		18	.444		107	22		129	151
8180	22" x 12"		15	.533		109	26		135	160
8200	24" x 16"		11	.727		120	35.50		155.50	185
8220	28" x 16"		10	.800		134	39.50		173.50	207
8240	30" x 16"		8	1		138	49		187	227
8260	30" x 18"		7	1.143		164	56		220	265
8400	Round damper, butterfly, vol. control w/lever lock reg.									
8410	6" diam.	1 Shee	22	.364	Ea.	23	17.85		40.85	52.50
8412	7" diam.		21	.381		24	18.70		42.70	54.50
8414	8" diam.		20	.400		24.50	19.65		44.15	56.50
8416	9" diam.		19	.421		25.50	20.50		46	59.50
8418	10" diam.		18	.444		28	22		50	64
8420	12" diam.		16	.500		29	24.50		53.50	68.50
8422	14" diam.		14	.571		37.50	28		65.50	83.50
8424	16" diam.		13	.615		40.50	30		70.50	90.50
8426	18" diam.		12	.667		49.50	32.50		82	104
8428	20" diam.		11	.727		60.50	35.50		96	121
8430	24" diam.		10	.800		75.50	39.50		115	143
8432	30" diam.		9	.889		106	43.50		149.50	182
8434	36" diam.		8	1		141	49		190	230
8500	Round motor operated damper									
8510	6" dia.	1 Shee	22	.364	Ea.	57	17.85		74.85	89.50
8512	7" dia.		21	.381		60	18.70		78.70	94
8514	8" dia.		20	.400		62	19.65		81.65	97.50
8516	10" dia.		18	.444		68.50	22		90.50	108
8518	12" dia.		16	.500		73.50	24.50		98	118
8520	14" dia.		14	.571		83.50	28		111.50	135

23 33 13.16 Fire Dampers		Crew	Daily Output	Labor-Hours	Unit	Material	Labor	Equipment	Total	Total Incl O&P
0010	**FIRE DAMPERS**									
3000	Fire damper, curtain type, 1-1/2 hr rated, vertical, 6" x 6"	1 Shee	24	.333	Ea.	21.50	16.35		37.85	49
3020	8" x 6"		22	.364		21.50	17.85		39.35	51

23 33 13.16 Fire Dampers		Crew	Daily Output	Labor-Hours	Unit	Material	2010 Bare Costs Labor	Equipment	Total	Total Incl O&P
3021	8" x 8"	1 Shee	22	.364	Ea.	21.50	17.85		39.35	51
3030	10" x 10"		22	.364		21.50	17.85		39.35	51
3040	12" x 6"		22	.364		21.50	17.85		39.35	51
3044	12" x 12"		20	.400		21.50	19.65		41.15	53.50
3060	20" x 6"		18	.444		25	22		47	60.50
3080	12" x 8"		22	.364		25	17.85		42.85	54.50
3100	24" x 8"		16	.500		30	24.50		54.50	70
3120	12" x 10"		21	.381		21.50	18.70		40.20	52.50
3140	24" x 10"		15	.533		31.50	26		57.50	74.50
3160	36" x 10"		12	.667		41	32.50		73.50	94.50
3180	16" x 12"		20	.400		24	19.65		43.65	55.50
3200	24" x 12"		13	.615		33.50	30		63.50	83
3220	48" x 12"		10	.800		51	39.50		90.50	116
3238	14" x 14"		19	.421		29.50	20.50		50	64
3240	16" x 14"		18	.444		29.50	22		51.50	65.50
3260	24" x 14"		12	.667		36.50	32.50		69	89.50
3280	30" x 14"		11	.727		43	35.50		78.50	102
3298	16" x 16"		18	.444		33.50	22		55.50	70
3300	18" x 16"		17	.471		35.50	23		58.50	74.50
3320	24" x 16"		11	.727		38.50	35.50		74	96.50
3340	36" x 16"		10	.800		51.50	39.50		91	116
3356	18" x 18"		16	.500		37	24.50		61.50	77.50
3360	24" x 18"		10	.800		45	39.50		84.50	109
3380	48" x 18"		8	1		64.50	49		113.50	146
3398	20" x 20"		10	.800		38.50	39.50		78	102
3400	24" x 20"		8	1		42	49		91	121
3420	36" x 20"		7	1.143		55	56		111	146
3440	24" x 22"		7	1.143		43	56		99	133
3460	30" x 22"		6	1.333		46	65.50		111.50	150
3478	24" x 24"		8	1		44	49		93	123
3480	26" x 24"		7	1.143		47	56		103	137
3484	32" x 24"	Q-9	13	1.231		55	54.50		109.50	143
3500	48" x 24"		12	1.333		72	59		131	168
3520	28" x 26"		13	1.231		52.50	54.50		107	141
3540	30" x 28"		12	1.333		58.50	59		117.50	153
3560	48" x 30"		11	1.455		85	64.50		149.50	191
3580	48" x 36"		10	1.600		93	70.50		163.50	209
3590	32" x 40"		10	1.600		73.50	70.50		144	188
3600	44" x 44"		10	1.600		97	70.50		167.50	214
3620	48" x 48"		8	2		127	88.50		215.50	274
3700	U.L. label included in above									
3800	For horizontal operation, add				Ea.	20%				
3900	For cap for blades out of air stream, add					20%				
4000	For 10" 22 ga., U.L. approved sleeve, add					35%				
4100	For 10" 22 ga., U.L. sleeve, 100% free area, add					65%				
4200	For oversize openings group dampers									
4210	Fire damper, round, Type A									
4214	4" diam.	1 Shee	20	.400	Ea.	56.50	19.65		76.15	92
4216	5" diam.		16	.500		51	24.50		75.50	93
4218	6" diam.		14.30	.559		51	27.50		78.50	97.50
4220	7" diam.		13.30	.602		51	29.50		80.50	101
4222	8" diam.		11.40	.702		51	34.50		85.50	108
4224	9" diam.		10.80	.741		55.50	36.50		92	116

23 33 13 – Dampers

23 33 13.16 Fire Dampers		Crew	Daily Output	Labor-Hours	Unit	Material	Labor	Equipment	Total	Total Incl O&P
						2010 Bare Costs				
4226	10" diam.	1 Shee	10	.800	Ea.	58	39.50		97.50	123
4228	12" diam.		9.50	.842		63.50	41.50		105	133
4230	14" diam.		8.20	.976		80	48		128	161
4232	16" diam.		7.60	1.053		80	51.50		131.50	167
4234	18" diam.		7.20	1.111		95.50	54.50		150	188
4236	20" diam.		6.40	1.250		101	61.50		162.50	204
4238	22" diam.		6.10	1.311		109	64.50		173.50	218
4240	24" diam.		5.70	1.404		114	69		183	229
4242	26" diam.		5.50	1.455		128	71.50		199.50	249
4244	28" diam.		5.40	1.481		134	72.50		206.50	258
4246	30" diam.		5.30	1.509		140	74		214	266
4500	Fire/smoke combination damper, louver type, UL									
4506	6" x 6"	1 Shee	24	.333	Ea.	37	16.35		53.35	65.50
4510	8" x 8"		22	.364		130	17.85		147.85	170
4520	16" x 8"		20	.400		142	19.65		161.65	186
4540	18" x 8"		18	.444		142	22		164	189
4560	20" x 8"		16	.500		152	24.50		176.50	204
4580	10" x 10"		21	.381		147	18.70		165.70	191
4600	24" x 10"		15	.533		173	26		199	230
4620	30" x 10"		12	.667		179	32.50		211.50	247
4640	12" x 12"		20	.400		153	19.65		172.65	199
4660	18" x 12"		18	.444		156	22		178	205
4680	24" x 12"		13	.615		165	30		195	227
4700	30" x 12"		11	.727		171	35.50		206.50	242
4720	14" x 14"		17	.471		165	23		188	217
4740	16" x 14"		18	.444		170	22		192	220
4760	20" x 14"		14	.571		173	28		201	233
4780	24" x 14"		12	.667		177	32.50		209.50	245
4800	30" x 14"		10	.800		188	39.50		227.50	266
4820	16" x 16"		16	.500		183	24.50		207.50	238
4840	20" x 16"		14	.571		188	28		216	250
4860	24" x 16"		11	.727		194	35.50		229.50	267
4880	30" x 16"		8	1		202	49		251	298
4900	18" x 18"		15	.533		181	26		207	239
5000	24" x 18"		10	.800		193	39.50		232.50	273
5020	36" x 18"		7	1.143		212	56		268	320
5040	20" x 20"		13	.615		202	30		232	269
5060	24" x 20"		8	1		207	49		256	300
5080	30" x 20"		8	1		218	49		267	315
5100	36" x 20"		7	1.143		229	56		285	335
5120	24" x 24"		8	1		224	49		273	320
5130	30" x 24"		7.60	1.053		212	51.50		263.50	310
5140	36" x 24"	Q-9	12	1.333		244	59		303	360
5141	36" x 30"		10	1.600		258	70.50		328.50	390
5142	40" x 36"		8	2		460	88.50		548.50	640
5143	48" x 48"		4	4		895	177		1,072	1,250
5150	Damper operator motor, 24 or 120 volt	1 Shee	16	.500		245	24.50		269.50	305

23 33 13.25 Exhaust Vent Damper		Crew	Daily Output	Labor-Hours	Unit	Material	Labor	Equipment	Total	Total Incl O&P
0010	**EXHAUST VENT DAMPER**, Auto, OB with elect. actuator.									
1110	6" x 6"	1 Shee	24	.333	Ea.	59	16.35		75.35	90
1114	8" x 6"		23	.348		60.50	17.10		77.60	92.50
1116	8" x 8"		22	.364		62.50	17.85		80.35	96

23 33 13 – Dampers

23 33 13.25 Exhaust Vent Damper		Crew	Daily Output	Labor-Hours	Unit	Material	2010 Bare Costs Labor	Equipment	Total	Total Incl O&P
1118	10" x 6"	1 Shee	23	.348	Ea.	62.50	17.10		79.60	95
1120	10" x 10"		22	.364		67	17.85		84.85	101
1122	12" x 6"		22	.364		63.50	17.85		81.35	96.50
1124	12" x 10"		21	.381		70.50	18.70		89.20	106
1126	12" x 12"		20	.400		74	19.65		93.65	111
1128	14" x 6"		21	.381		65.50	18.70		84.20	101
1130	14" x 10"		19	.421		74	20.50		94.50	113
1132	16" x 6"		21	.381		66	18.70		84.70	102
1134	16" x 8"		20	.400		73.50	19.65		93.15	111
1136	18" x 6"		20	.400		70	19.65		89.65	107
1138	18" x 8"		19	.421		74	20.50		94.50	113

23 33 13.28 Splitter Damper Assembly		Crew	Daily Output	Labor-Hours	Unit	Material	Labor	Equipment	Total	Total Incl O&P
0009	**SPLITTER DAMPER ASSEMBLY**									
0010	Self locking, 1' rod	1 Shee	24	.333	Ea.	21.50	16.35		37.85	49
7020	3' rod		22	.364		28	17.85		45.85	58
7040	4' rod		20	.400		31.50	19.65		51.15	64
7060	6' rod		18	.444		38	22		60	75

23 33 13.32 Relief Damper		Crew	Daily Output	Labor-Hours	Unit	Material	Labor	Equipment	Total	Total Incl O&P
0010	**RELIEF DAMPER**, Electronic bypass with tight seal									
8310	8" x 6"	1 Shee	22	.364	Ea.	163	17.85		180.85	207
8314	10" x 6"		22	.364		163	17.85		180.85	207
8318	10" x 10"		21	.381		170	18.70		188.70	216
8322	12" x 12"		20	.400		173	19.65		192.65	220
8326	12" x 16"		19	.421		180	20.50		200.50	230
8330	16" x 10"		20	.400		189	19.65		208.65	238
8334	16" x 14"		18	.444		194	22		216	246
8338	16" x 18"		17	.471		211	23		234	267
8342	18" x 12"		18	.444		211	22		233	265
8346	18" x 18"		15	.533		199	26		225	259
8350	20" x 14"		14	.571		196	28		224	259
8354	20" x 18"		13	.615		205	30		235	271
8358	24" x 12"		13	.615		200	30		230	266
8362	24" x 18"		10	.800		156	39.50		195.50	231
8363	24" x 24"		10	.800		218	39.50		257.50	300
8364	24" x 36"		9	.889		259	43.50		302.50	350
8365	24" x 48"		6	1.333		290	65.50		355.50	420
8366	28" x 10"		13	.615		144	30		174	204
8370	28" x 16"		10	.800		221	39.50		260.50	305
8374	30" x 14"		11	.727		151	35.50		186.50	220
8378	30" x 18"		8	1		228	49		277	325
8382	30" x 24"		6	1.333		242	65.50		307.50	365
8390	46" x 36"		4	2		320	98		418	505
8394	48" x 48"		3	2.667		350	131		481	585
8396	54" x 36"		2	4		370	196		566	700

23 33 19 – Duct Silencers

23 33 19.10 Duct Silencers		Crew	Daily Output	Labor-Hours	Unit	Material	Labor	Equipment	Total	Total Incl O&P
0009	**DUCT SILENCERS**									
0010	Silencers, noise control for air flow, duct				MCFM	55.50			55.50	61
9004	Duct sound trap, packaged									
9010	12" x 18" x 36", 2000 CFM	1 Shee	12	.667	Ea.	430	32.50		462.50	525
9011	24" x 18" x 36", 9000 CFM	Q-9	80	.200		495	8.85		503.85	560
9012	24" x 24" x 36", 9000 CFM		60	.267		745	11.80		756.80	835

23 33 Air Duct Accessories

23 33 19 – Duct Silencers

23 33 19.10 Duct Silencers		Crew	Daily Output	Labor-Hours	Unit	Material	2010 Bare Costs Labor	Equipment	Total	Total Incl O&P
9013	24" x 30" x 36", 9000 CFM	Q-9	50	.320	Ea.	795	14.15		809.15	890
9014	24" x 36" x 36", 9000 CFM		40	.400		990	17.70		1,007.70	1,125
9015	24" x 48" x 36", 9000 CFM		28	.571		1,300	25.50		1,325.50	1,475
9018	36" x 18" x 36", 6300 CFM		42	.381		990	16.85		1,006.85	1,125
9019	36" x 36" x 36", 6300 CFM		21	.762		1,975	33.50		2,008.50	2,225
9020	36" x 48" x 36", 6300 CFM		14	1.143		2,575	50.50		2,625.50	2,900
9021	36" x 60" x 36", 6300 CFM		13	1.231		3,275	54.50		3,329.50	3,675
9022	48" x 48" x 36", 6300 CFM		12	1.333		3,475	59		3,534	3,925
9023	24" x 24" x 60", 6000 CFM		3.60	4.444		1,150	196		1,346	1,550
9026	24" x 30" x 60", 7000 CFM		3.50	4.571		1,275	202		1,477	1,700
9030	24" x 36" x 60", 8000 CFM		3.40	4.706		1,400	208		1,608	1,875
9034	24" x 48" x 60", 11,000 CFM		3.30	4.848		1,200	214		1,414	1,650
9038	36" x 18" x 60", 6300 CFM		3.50	4.571		950	202		1,152	1,350
9042	36" x 36" x 60", 12,000 CFM		3.30	4.848		1,525	214		1,739	2,000
9046	36" x 48" x 60", 17,000 CFM		3.10	5.161		2,000	228		2,228	2,550
9050	36" x 60" x 60", 20,000 CFM		2.80	5.714		2,450	253		2,703	3,075
9054	48" x 48" x 60", 23,000 CFM		2.60	6.154		2,825	272		3,097	3,500
9056	48" x 60" x 60", 23,000 CFM	↓	2.10	7.619	↓	3,525	335		3,860	4,375

23 33 23 – Turning Vanes

23 33 23.13 Air Turning Vanes		Crew	Daily Output	Labor-Hours	Unit	Material	Labor	Equipment	Total	Total Incl O&P
0010	**AIR TURNING VANES**									
9400	Turning vane components									
9410	Turning vane rail	1 Shee	160	.050	L.F.	.75	2.46		3.21	4.55
9420	Double thick, factory fab. vane		300	.027		1.06	1.31		2.37	3.15
9428	12" high set		170	.047		1.56	2.31		3.87	5.20
9432	14" high set		160	.050		1.82	2.46		4.28	5.70
9434	16" high set		150	.053		2.08	2.62		4.70	6.25
9436	18" high set		144	.056		2.34	2.73		5.07	6.70
9438	20" high set		138	.058		2.60	2.85		5.45	7.15
9440	22" high set		130	.062		2.86	3.02		5.88	7.75
9442	24" high set		124	.065		3.12	3.17		6.29	8.25
9444	26" high set		116	.069		3.38	3.39		6.77	8.85
9446	30" high set	↓	112	.071	↓	3.90	3.51		7.41	9.60

23 33 33 – Duct-Mounting Access Doors

23 33 33.13 Duct Access Doors		Crew	Daily Output	Labor-Hours	Unit	Material	Labor	Equipment	Total	Total Incl O&P
0010	**DUCT ACCESS DOORS**									
1000	Duct access door, insulated, 6" x 6"	1 Shee	14	.571	Ea.	15	28		43	59
1020	10" x 10"		11	.727		18.15	35.50		53.65	74
1040	12" x 12"		10	.800		19.40	39.50		58.90	81
1050	12" x 18"		9	.889		34.50	43.50		78	104
1060	16" x 12"		9	.889		27.50	43.50		71	96
1070	18" x 18"		8	1		34.50	49		83.50	113
1074	24" x 18"		8	1		41	49		90	120
1080	24" x 24"	↓	8	1	↓	43.50	49		92.50	123

23 33 46 – Flexible Ducts

23 33 46.10 Flexible Air Ducts		Crew	Daily Output	Labor-Hours	Unit	Material	Labor	Equipment	Total	Total Incl O&P
0010	**FLEXIBLE AIR DUCTS** R233100-20									
1280	Add to labor for elevated installation									
1282	of prefabricated (purchased) ductwork									
1283	10' to 15' high						10%			
1284	15' to 20' high						20%			

23 33 46 – Flexible Ducts

23 33 46.10 Flexible Air Ducts		Crew	Daily Output	Labor-Hours	Unit	Material	2010 Bare Costs Labor	Equipment	Total	Total Incl O&P
1285	20′ to 25′ high						25%			
1286	25′ to 30′ high						35%			
1287	30′ to 35′ high						40%			
1288	35′ to 40′ high						50%			
1289	Over 40′ high						55%			
1300	Flexible, coated fiberglass fabric on corr. resist. metal helix									
1400	pressure to 12″ (WG) UL-181									
1500	Non-insulated, 3″ diameter	Q-9	400	.040	L.F.	1.02	1.77		2.79	3.80
1520	4″ diameter		360	.044		1.02	1.96		2.98	4.09
1540	5″ diameter		320	.050		1.25	2.21		3.46	4.73
1560	6″ diameter		280	.057		1.53	2.53		4.06	5.50
1580	7″ diameter		240	.067		1.82	2.95		4.77	6.45
1600	8″ diameter		200	.080		2.10	3.54		5.64	7.65
1620	9″ diameter		180	.089		2.49	3.93		6.42	8.70
1640	10″ diameter		160	.100		2.61	4.42		7.03	9.55
1660	12″ diameter		120	.133		3.17	5.90		9.07	12.40
1680	14″ diameter		80	.200		3.85	8.85		12.70	17.65
1700	16″ diameter	↓	60	.267	↓	5.50	11.80		17.30	24
1800	For adjustable clamps, add				Ea.	.72			.72	.79
1900	Insulated, 1″ thick, PE jacket, 3″ diameter G	Q-9	380	.042	L.F.	2.15	1.86		4.01	5.20
1910	4″ diameter G		340	.047		2.15	2.08		4.23	5.50
1920	5″ diameter G		300	.053		2.15	2.36		4.51	5.95
1940	6″ diameter G		260	.062		2.38	2.72		5.10	6.75
1960	7″ diameter G		220	.073		2.78	3.21		5.99	7.95
1980	8″ diameter G		180	.089		3	3.93		6.93	9.25
2000	9″ diameter G		160	.100		3.40	4.42		7.82	10.45
2020	10″ diameter G		140	.114		3.51	5.05		8.56	11.50
2040	12″ diameter G		100	.160		4.36	7.05		11.41	15.50
2060	14″ diameter G		80	.200		5.25	8.85		14.10	19.20
2080	16″ diameter G		60	.267		6.55	11.80		18.35	25
2100	18″ diameter G	↓	45	.356		8.35	15.70		24.05	33
2120	20″ diameter G	Q-10	65	.369	↓	9.10	16.90		26	35.50
2500	Insulated, heavy duty, coated fiberglass fabric									
2520	4″ diameter G	Q-9	340	.047	L.F.	3.45	2.08		5.53	6.95
2540	5″ diameter G		300	.053		3.69	2.36		6.05	7.65
2560	6″ diameter G		260	.062		4.14	2.72		6.86	8.65
2580	7″ diameter G		220	.073		4.59	3.21		7.80	9.90
2600	8″ diameter G		180	.089		5.20	3.93		9.13	11.70
2620	9″ diameter G		160	.100		5.55	4.42		9.97	12.80
2640	10″ diameter G		140	.114		6.75	5.05		11.80	15.05
2660	12″ diameter G		100	.160		7.20	7.05		14.25	18.60
2680	14″ diameter G		80	.200		9.40	8.85		18.25	24
2700	16″ diameter G		60	.267		14.10	11.80		25.90	33.50
2720	18″ diameter G	↓	45	.356	↓	17.30	15.70		33	43
2800	Flexible, aluminum, pressure to 12″ (WG) UL-181									
2820	Non-insulated									
2830	3″ diameter	Q-9	400	.040	L.F.	1.14	1.77		2.91	3.93
2831	4″ diameter		360	.044		1.36	1.96		3.32	4.47
2832	5″ diameter		320	.050		1.75	2.21		3.96	5.30
2833	6″ diameter		280	.057		2.23	2.53		4.76	6.25
2834	7″ diameter		240	.067		2.68	2.95		5.63	7.40
2835	8″ diameter		200	.080		3.07	3.54		6.61	8.75
2836	9″ diameter	↓	180	.089	↓	3.65	3.93		7.58	9.95

23 33 46 – Flexible Ducts

23 33 46.10 Flexible Air Ducts		Crew	Daily Output	Labor-Hours	Unit	Material	Labor	Equipment	Total	Total Incl O&P
						2010 Bare Costs				
2837	10″ diameter	Q-9	160	.100	L.F.	4.30	4.42		8.72	11.45
2838	12″ diameter		120	.133		5.30	5.90		11.20	14.70
2839	14″ diameter		80	.200		6.50	8.85		15.35	20.50
2840	15″ diameter		70	.229		8.25	10.10		18.35	24.50
2841	16″ diameter		60	.267		10.15	11.80		21.95	29
2842	18″ diameter		45	.356		11.30	15.70		27	36.50
2843	20″ diameter	Q-10	65	.369		12.50	16.90		29.40	39.50
2880	Insulated, 1″ thick with 3/4 lb., PE jacket									
2890	3″ diameter G	Q-9	380	.042	L.F.	1.97	1.86		3.83	4.99
2891	4″ diameter G		340	.047		2.27	2.08		4.35	5.65
2892	5″ diameter G		300	.053		3.12	2.36		5.48	7
2893	6″ diameter G		260	.062		3.45	2.72		6.17	7.90
2894	7″ diameter G		220	.073		4.25	3.21		7.46	9.55
2895	8″ diameter G		180	.089		4.55	3.93		8.48	10.95
2896	9″ diameter G		160	.100		5.15	4.42		9.57	12.35
2897	10″ diameter G		140	.114		6.30	5.05		11.35	14.60
2898	12″ diameter G		100	.160		6.95	7.05		14	18.35
2899	14″ diameter G		80	.200		8.45	8.85		17.30	22.50
2900	15″ diameter G		70	.229		10.90	10.10		21	27.50
2901	16″ diameter G		60	.267		12.60	11.80		24.40	31.50
2902	18″ diameter G		45	.356		14.70	15.70		30.40	40
3000	Transitions, 3″ deep, square to round									
3010	6″ x 6″ to round	Q-9	44	.364	Ea.	8.50	16.05		24.55	34
3014	6″ x 12″ to round		42	.381		15.85	16.85		32.70	43
3018	8″ x 8″ to round		40	.400		10.20	17.70		27.90	38
3022	9″ x 9″ to round		38	.421		10.20	18.60		28.80	39
3026	9″ x 12″ to round		37	.432		20.50	19.10		39.60	51.50
3030	9″ x 15″ to round		36	.444		24	19.65		43.65	56
3034	10″ x 10″ to round		34	.471		12.45	21		33.45	45
3038	10″ x 22″ to round		32	.500		29	22		51	65.50
3042	12″ x 12″ to round		30	.533		12.45	23.50		35.95	49
3046	12″ x 18″ to round		28	.571		29	25.50		54.50	70.50
3050	12″ x 24″ to round		26	.615		29	27		56	73
3054	15″ x 15″ to round		25	.640		15.85	28.50		44.35	60.50
3058	15″ x 18″ to round		24	.667		31.50	29.50		61	79.50
3062	18″ x 18″ to round		22	.727		19.85	32		51.85	70.50
3066	21″ x 21″ to round		21	.762		34	33.50		67.50	88.50
3070	22″ x 22″ to round		20	.800		44	35.50		79.50	102
3074	24″ x 24″ to round		18	.889		44	39.50		83.50	108

23 33 53 – Duct Liners

23 33 53.10 Duct Liner Board		Crew	Daily Output	Labor-Hours	Unit	Material	Labor	Equipment	Total	Total Incl O&P
0010	**DUCT LINER BOARD**									
3340	Board type fiberglass liner, FSK, 1-1/2 lb. density									
3344	1″ thick G	Q-14	150	.107	S.F.	.57	4.37		4.94	7.45
3345	1-1/2″ thick G		130	.123		.62	5.05		5.67	8.55
3346	2″ thick G		120	.133		.73	5.45		6.18	9.30
3348	3″ thick G		110	.145		.94	5.95		6.89	10.30
3350	4″ thick G		100	.160		1.15	6.55		7.70	11.45
3356	3 lb. density, 1″ thick G		150	.107		.77	4.37		5.14	7.65
3358	1-1/2″ thick G		130	.123		.97	5.05		6.02	8.90
3360	2″ thick G		120	.133		1.18	5.45		6.63	9.80
3362	2-1/2″ thick G		110	.145		1.38	5.95		7.33	10.75

23 33 Air Duct Accessories

23 33 53 – Duct Liners

23 33 53.10 Duct Liner Board			Crew	Daily Output	Labor-Hours	Unit	Material	2010 Bare Costs Labor	Equipment	Total	Total Incl O&P
3364	3" thick	G	Q-14	100	.160	S.F.	1.59	6.55		8.14	11.95
3366	4" thick	G		90	.178		2	7.30		9.30	13.55
3370	6 lb. density, 1" thick	G		140	.114		1.09	4.69		5.78	8.50
3374	1-1/2" thick	G		120	.133		1.46	5.45		6.91	10.10
3378	2" thick	G		100	.160		1.82	6.55		8.37	12.20
3490	Board type, fiberglass liner, 3 lb. density										
3500	Fire resistant, black pigmented, 1 side										
3520	1" thick	G	Q-14	150	.107	S.F.	.77	4.37		5.14	7.65
3540	1-1/2" thick	G		130	.123		.97	5.05		6.02	8.90
3560	2" thick	G		120	.133		1.18	5.45		6.63	9.80
3600	FSK vapor barrier										
3620	1" thick	G	Q-14	150	.107	S.F.	.77	4.37		5.14	7.65
3630	1-1/2" thick	G		130	.123		.97	5.05		6.02	8.90
3640	2" thick	G		120	.133		1.18	5.45		6.63	9.80
3680	No finish										
3700	1" thick	G	Q-14	170	.094	S.F.	.40	3.86		4.26	6.45
3710	1-1/2" thick	G		140	.114		.60	4.69		5.29	7.95
3720	2" thick	G		130	.123		.81	5.05		5.86	8.75
3940	Board type, non-fibrous foam										
3950	Temperature, bacteria and fungi resistant										
3960	1" thick	G	Q-14	150	.107	S.F.	2.25	4.37		6.62	9.30
3970	1-1/2" thick	G		130	.123		3	5.05		8.05	11.15
3980	2" thick	G		120	.133		3.60	5.45		9.05	12.45

23 34 HVAC Fans

23 34 13 – Axial HVAC Fans

23 34 13.10 Axial Flow HVAC Fans			Crew	Daily Output	Labor-Hours	Unit	Material	Labor	Equipment	Total	Total Incl O&P
0010	**AXIAL FLOW HVAC FANS**	R233100-10									
0020	Air conditioning and process air handling										
0030	Axial flow, compact, low sound, 2.5" S.P.	R233400-10									
0050	3,800 CFM, 5 HP		Q-20	3.40	5.882	Ea.	4,500	266		4,766	5,350
0080	6,400 CFM, 5 HP			2.80	7.143		5,025	325		5,350	6,000
0100	10,500 CFM, 7-1/2 HP			2.40	8.333		6,250	375		6,625	7,450
0120	15,600 CFM, 10 HP			1.60	12.500		7,875	565		8,440	9,525
0140	23,000 CFM, 15 HP			.70	28.571		12,100	1,300		13,400	15,300
0160	28,000 CFM, 20 HP			.40	50		13,500	2,250		15,750	18,200
0500	Axial flow, constant speed										
0505	Direct drive, 1/8" S.P.										
0510	12", 1060 CFM, 1/6 HP		Q-20	3	6.667	Ea.	560	300		860	1,075
0514	12", 2095 CFM, 1/2 HP			3	6.667		660	300		960	1,175
0518	16", 2490 CFM, 1/3 HP			2.80	7.143		660	325		985	1,200
0522	20", 4130 CFM, 3/4 HP			2.60	7.692		785	345		1,130	1,400
0526	22", 4700 CFM, 3/4 HP			2.60	7.692		1,175	345		1,520	1,800
0530	24", 5850 CFM, 1 HP			2.50	8		1,200	360		1,560	1,875
0534	24", 7925 CFM, 1-1/2 HP			2.40	8.333		1,475	375		1,850	2,200
0538	30", 10640 CFM, 2 HP			2.20	9.091		1,900	410		2,310	2,725
0542	30", 14765 CFM, 2-1/2 HP			2.10	9.524		2,225	430		2,655	3,100
0546	36", 16780 CFM, 2 HP			2.10	9.524		2,275	430		2,705	3,175
0550	36", 22920 CFM, 5 HP			1.80	11.111		2,500	500		3,000	3,525
0560	Belt drive, 1/8" S.P.										
0562	15", 2800 CFM, 1/3 HP		Q-20	3.20	6.250	Ea.	870	282		1,152	1,375

23 34 HVAC Fans

23 34 13 – Axial HVAC Fans

23 34 13.10 Axial Flow HVAC Fans		Crew	Daily Output	Labor-Hours	Unit	Material	2010 Bare Costs Labor	Equipment	Total	Total Incl O&P
0564	15", 3400 CFM, 1/2 HP	Q-20	3	6.667	Ea.	895	300		1,195	1,450
0568	18", 3280 CFM, 1/3 HP		2.80	7.143		950	325		1,275	1,525
0572	18", 3900 CFM, 1/2 HP		2.80	7.143		970	325		1,295	1,550
0576	18", 5250 CFM, 1 HP		2.70	7.407		1,075	335		1,410	1,700
0584	24", 6430 CFM, 1 HP		2.50	8		1,275	360		1,635	1,950
0588	24", 8860 CFM, 2 HP		2.40	8.333		1,325	375		1,700	2,025
0592	30", 9250 CFM, 1 HP		2.20	9.091		1,675	410		2,085	2,450
0596	30", 16900 CFM, 5 HP		2	10		1,825	450		2,275	2,675
0604	36", 14475 CFM, 2 HP		2.30	8.696		1,950	395		2,345	2,725
0608	36", 20080 CFM, 5 HP		1.80	11.111		2,050	500		2,550	3,025
0612	36", 14475 CFM, 7-1/2 HP		1.60	12.500		2,250	565		2,815	3,325
0616	42", 29000 CFM, 7-1/2 HP		1.40	14.286		2,775	645		3,420	4,025
1000	Return air fan									
1010	9200 CFM	Q-20	1.40	14.286	Ea.	3,675	645		4,320	5,025
1020	13,200 CFM		1.30	15.385		5,275	695		5,970	6,850
1030	16,500 CFM		1.20	16.667		6,575	755		7,330	8,375
1040	19,500 CFM		1	20		7,775	905		8,680	9,900
1500	Vaneaxial, low pressure, 2000 CFM, 1/2 HP		3.60	5.556		1,950	251		2,201	2,525
1520	4,000 CFM, 1 HP		3.20	6.250		2,250	282		2,532	2,900
1540	8,000 CFM, 2 HP		2.80	7.143		2,900	325		3,225	3,675
1560	16,000 CFM, 5 HP		2.40	8.333		4,175	375		4,550	5,150

23 34 14 – Blower HVAC Fans

23 34 14.10 Blower Type HVAC Fans		Crew	Daily Output	Labor-Hours	Unit	Material	Labor	Equipment	Total	Total Incl O&P
0010	**BLOWER TYPE HVAC FANS**									
2000	Blowers, direct drive with motor, complete									
2020	1045 CFM @ .5" S.P., 1/5 HP	Q-20	18	1.111	Ea.	244	50		294	345
2040	1385 CFM @ .5" S.P., 1/4 HP		18	1.111		248	50		298	350
2060	1640 CFM @ .5" S.P., 1/3 HP		18	1.111		225	50		275	325
2080	1760 CFM @ .5" S.P., 1/2 HP		18	1.111		235	50		285	335
2090	4 speed									
2100	1164 to 1739 CFM @ .5" S.P., 1/3 HP	Q-20	16	1.250	Ea.	241	56.50		297.50	350
2120	1467 to 2218 CFM @ 1.0" S.P., 3/4 HP	"	14	1.429	"	280	64.50		344.50	410
2500	Ceiling fan, right angle, extra quiet, 0.10" S.P.									
2520	95 CFM	Q-20	20	1	Ea.	240	45		285	330
2540	210 CFM		19	1.053		283	47.50		330.50	380
2560	385 CFM		18	1.111		360	50		410	470
2580	885 CFM		16	1.250		710	56.50		766.50	865
2600	1,650 CFM		13	1.538		980	69.50		1,049.50	1,175
2620	2,960 CFM		11	1.818		1,300	82		1,382	1,575
2640	For wall or roof cap, add	1 Shee	16	.500		240	24.50		264.50	300
2660	For straight thru fan, add					10%				
2680	For speed control switch, add	1 Elec	16	.500		131	24.50		155.50	181
7500	Utility set, steel construction, pedestal, 1/4" S.P.									
7520	Direct drive, 150 CFM, 1/8 HP	Q-20	6.40	3.125	Ea.	755	141		896	1,050
7540	485 CFM, 1/6 HP		5.80	3.448		950	156		1,106	1,275
7560	1950 CFM, 1/2 HP		4.80	4.167		1,125	188		1,313	1,500
7580	2410 CFM, 3/4 HP		4.40	4.545		2,050	205		2,255	2,550
7600	3328 CFM, 1-1/2 HP		3	6.667		2,300	300		2,600	2,975
7680	V-belt drive, drive cover, 3 phase									
7700	800 CFM, 1/4 HP	Q-20	6	3.333	Ea.	775	151		926	1,075
7720	1,300 CFM, 1/3 HP		5	4		815	181		996	1,175
7740	2,000 CFM, 1 HP		4.60	4.348		965	196		1,161	1,350

23 34 14 – Blower HVAC Fans

23 34 14.10 Blower Type HVAC Fans		Crew	Daily Output	Labor-Hours	Unit	Material	2010 Bare Costs Labor	Equipment	Total	Total Incl O&P
7760	2,900 CFM, 3/4 HP	Q-20	4.20	4.762	Ea.	1,300	215		1,515	1,750
7780	3,600 CFM, 3/4 HP		4	5		1,600	226		1,826	2,100
7800	4,800 CFM, 1 HP		3.50	5.714		1,900	258		2,158	2,475
7820	6,700 CFM, 1-1/2 HP		3	6.667		2,325	300		2,625	3,000
7830	7,500 CFM, 2 HP		2.50	8		3,175	360		3,535	4,025
7840	11,000 CFM, 3 HP		2	10		4,225	450		4,675	5,325
7860	13,000 CFM, 3 HP		1.60	12.500		4,300	565		4,865	5,575
7880	15,000 CFM, 5 HP		1	20		4,450	905		5,355	6,225
7900	17,000 CFM, 7-1/2 HP		.80	25		4,750	1,125		5,875	6,925
7920	20,000 CFM, 7-1/2 HP		.80	25		5,675	1,125		6,800	7,925

23 34 16 – Centrifugal HVAC Fans

23 34 16.10 Centrifugal Type HVAC Fans		Crew	Daily Output	Labor-Hours	Unit	Material	Labor	Equipment	Total	Total Incl O&P
0010	**CENTRIFUGAL TYPE HVAC FANS**									
0200	In-line centrifugal, supply/exhaust booster									
0220	aluminum wheel/hub, disconnect switch, 1/4" S.P.									
0240	500 CFM, 10" diameter connection	Q-20	3	6.667	Ea.	1,150	300		1,450	1,725
0260	1,380 CFM, 12" diameter connection		2	10		1,225	450		1,675	2,025
0280	1,520 CFM, 16" diameter connection		2	10		1,325	450		1,775	2,125
0300	2,560 CFM, 18" diameter connection		1	20		1,450	905		2,355	2,950
0320	3,480 CFM, 20" diameter connection		.80	25		1,725	1,125		2,850	3,575
0326	5,080 CFM, 20" diameter connection		.75	26.667		1,875	1,200		3,075	3,875
0340	7,500 CFM, 22" diameter connection		.70	28.571		2,225	1,300		3,525	4,400
0350	10,000 CFM, 27" diameter connection		.65	30.769		2,700	1,400		4,100	5,075
3500	Centrifugal, airfoil, motor and drive, complete									
3520	1000 CFM, 1/2 HP	Q-20	2.50	8	Ea.	1,650	360		2,010	2,375
3540	2,000 CFM, 1 HP		2	10		1,850	450		2,300	2,700
3560	4,000 CFM, 3 HP		1.80	11.111		2,300	500		2,800	3,275
3580	8,000 CFM, 7-1/2 HP		1.40	14.286		3,575	645		4,220	4,900
3600	12,000 CFM, 10 HP		1	20		4,300	905		5,205	6,075
4000	Single width, belt drive, not incl. motor, capacities									
4020	at 2000 FPM, 2.5" S.P. for indicated motor									
4040	6900 CFM, 5 HP	Q-9	2.40	6.667	Ea.	3,300	295		3,595	4,075
4060	10,340 CFM, 7-1/2 HP		2.20	7.273		4,575	320		4,895	5,500
4080	15,320 CFM, 10 HP		2	8		5,375	355		5,730	6,450
4100	22,780 CFM, 15 HP		1.80	8.889		8,025	395		8,420	9,425
4120	33,840 CFM, 20 HP		1.60	10		10,600	440		11,040	12,400
4140	41,400 CFM, 25 HP		1.40	11.429		14,100	505		14,605	16,300
4160	50,100 CFM, 30 HP		.80	20		17,900	885		18,785	21,100
4200	Double width wheel, 12,420 CFM, 7.5 HP		2.20	7.273		4,700	320		5,020	5,650
4220	18,620 CFM, 15 HP		2	8		6,725	355		7,080	7,925
4240	27,580 CFM, 20 HP		1.80	8.889		8,225	395		8,620	9,650
4260	40,980 CFM, 25 HP		1.50	10.667		12,900	470		13,370	14,900
4280	60,920 CFM, 40 HP		1	16		17,300	705		18,005	20,200
4300	74,520 CFM, 50 HP		.80	20		20,500	885		21,385	24,000
4320	90,160 CFM, 50 HP		.70	22.857		28,100	1,000		29,100	32,400
4340	110,300 CFM, 60 HP		.50	32		38,100	1,425		39,525	44,100
4360	134,960 CFM, 75 HP		.40	40		50,500	1,775		52,275	58,000
4500	Corrosive fume resistant, plastic									
4600	roof ventilators, centrifugal, V belt drive, motor									
4620	1/4" S.P., 250 CFM, 1/4 HP	Q-20	6	3.333	Ea.	3,875	151		4,026	4,475
4640	895 CFM, 1/3 HP		5	4		4,200	181		4,381	4,900
4660	1630 CFM, 1/2 HP		4	5		4,975	226		5,201	5,825

23 34 16 – Centrifugal HVAC Fans

23 34 16.10 Centrifugal Type HVAC Fans		Crew	Daily Output	Labor-Hours	Unit	2010 Bare Costs Material	Labor	Equipment	Total	Total Incl O&P
4680	2240 CFM, 1 HP	Q-20	3	6.667	Ea.	5,175	300		5,475	6,150
4700	3810 CFM, 2 HP		2	10		5,750	450		6,200	7,000
4710	5000 CFM, 2 HP		1.80	11.111		12,100	500		12,600	14,100
4715	8000 CFM, 5 HP		1.40	14.286		13,500	645		14,145	15,800
4720	11760 CFM, 5 HP		1	20		13,700	905		14,605	16,400
4740	18810 CFM, 10 HP		.70	28.571		14,200	1,300		15,500	17,600
4800	For intermediate capacity, motors may be varied									
4810	For explosion proof motor, add				Ea.	15%				
5000	Utility set, centrifugal, V belt drive, motor									
5020	1/4″ S.P., 1200 CFM, 1/4 HP	Q-20	6	3.333	Ea.	4,275	151		4,426	4,925
5040	1520 CFM, 1/3 HP		5	4		1,650	181		1,831	2,075
5060	1850 CFM, 1/2 HP		4	5		1,700	226		1,926	2,225
5080	2180 CFM, 3/4 HP		3	6.667		1,950	300		2,250	2,600
5100	1/2″ S.P., 3600 CFM, 1 HP		2	10		2,225	450		2,675	3,125
5120	4250 CFM, 1-1/2 HP		1.60	12.500		2,475	565		3,040	3,575
5140	4800 CFM, 2 HP		1.40	14.286		2,725	645		3,370	3,975
5160	6920 CFM, 5 HP		1.30	15.385		3,475	695		4,170	4,875
5180	7700 CFM, 7-1/2 HP		1.20	16.667		4,400	755		5,155	5,975
5200	For explosion proof motor, add					15%				
5300	Fume exhauster without hose and intake nozzle									
5310	630 CFM, 1-1/2 HP	Q-20	2	10	Ea.	980	450		1,430	1,750
5320	Hose extension kit 5′	″	20	1	″	72.50	45		117.50	148
5500	Fans, industrial exhauster, for air which may contain granular matl.									
5520	1000 CFM, 1-1/2 HP	Q-20	2.50	8	Ea.	2,450	360		2,810	3,225
5540	2000 CFM, 3 HP		2	10		3,025	450		3,475	4,000
5560	4000 CFM, 7-1/2 HP		1.80	11.111		4,300	500		4,800	5,475
5580	8000 CFM, 15 HP		1.40	14.286		5,525	645		6,170	7,050
5600	12,000 CFM, 30 HP		1	20		8,950	905		9,855	11,200
7000	Roof exhauster, centrifugal, aluminum housing, 12″ galvanized									
7020	curb, bird screen, back draft damper, 1/4″ S.P.									
7100	Direct drive, 320 CFM, 11″ sq. damper	Q-20	7	2.857	Ea.	600	129		729	855
7120	600 CFM, 11″ sq. damper		6	3.333		750	151		901	1,050
7140	815 CFM, 13″ sq. damper		5	4		750	181		931	1,100
7160	1450 CFM, 13″ sq. damper		4.20	4.762		1,125	215		1,340	1,550
7180	2050 CFM, 16″ sq. damper		4	5		1,450	226		1,676	1,925
7200	V-belt drive, 1650 CFM, 12″ sq. damper		6	3.333		1,075	151		1,226	1,425
7220	2750 CFM, 21″ sq. damper		5	4		1,300	181		1,481	1,700
7230	3500 CFM, 21″ sq. damper		4.50	4.444		1,450	201		1,651	1,900
7240	4910 CFM, 23″ sq. damper		4	5		1,775	226		2,001	2,300
7260	8525 CFM, 28″ sq. damper		3	6.667		2,400	300		2,700	3,075
7280	13,760 CFM, 35″ sq. damper		2	10		3,200	450		3,650	4,200
7300	20,558 CFM, 43″ sq. damper		1	20		6,625	905		7,530	8,650
7320	For 2 speed winding, add					15%				
7340	For explosionproof motor, add					495			495	545
7360	For belt driven, top discharge, add					15%				
7400	Roof mounted kitchen exhaust, aluminum, centrifugal									
7410	Direct drive, 2 speed, temp to 200°F									
7412	1/12 HP, 9-3/4″	Q-20	8	2.500	Ea.	540	113		653	765
7414	1/3 HP, 12-5/8″		7	2.857		665	129		794	930
7416	1/2 HP, 13-1/2″		6	3.333		735	151		886	1,025
7418	3/4 HP, 15″		5	4		1,050	181		1,231	1,425
7424	Belt drive, temp to 250°F									
7426	3/4 HP, 20″	Q-20	5	4	Ea.	2,325	181		2,506	2,825

23 34 HVAC Fans

23 34 16 – Centrifugal HVAC Fans

23 34 16.10 Centrifugal Type HVAC Fans		Crew	Daily Output	Labor-Hours	Unit	Material	2010 Bare Costs Labor	Equipment	Total	Total Incl O&P
7428	1-1/2 HP, 24-1/2"	Q-20	4	5	Ea.	2,550	226		2,776	3,175
7430	1-1/2 HP, 30"	↓	4	5	↓	3,375	226		3,601	4,075
7450	Upblast, propeller, w/BDD, BS, + C									
7454	30,300 CFM @ 3/8" S.P., 5 HP	Q-20	1.60	12.500	Ea.	3,900	565		4,465	5,150
7458	36,000 CFM @ 1/2" S.P., 15 HP	"	1.60	12.500	"	5,550	565		6,115	6,975
8500	Wall exhausters, centrifugal, auto damper, 1/8" S.P.									
8520	Direct drive, 610 CFM, 1/20 HP	Q-20	14	1.429	Ea.	350	64.50		414.50	485
8540	796 CFM, 1/12 HP		13	1.538		735	69.50		804.50	910
8560	822 CFM, 1/6 HP		12	1.667		885	75.50		960.50	1,075
8580	1,320 CFM, 1/4 HP		12	1.667		975	75.50		1,050.50	1,200
8600	1756 CFM, 1/4 HP		11	1.818		1,025	82		1,107	1,275
8620	1983 CFM, 1/4 HP		10	2		1,075	90.50		1,165.50	1,300
8640	2900 CFM, 1/2 HP		9	2.222		1,125	100		1,225	1,400
8660	3307 CFM, 3/4 HP		8	2.500		1,225	113		1,338	1,525
8670	5940 CFM, 1 HP	↓	7	2.857	↓	1,325	129		1,454	1,675
9500	V-belt drive, 3 phase									
9520	2,800 CFM, 1/4 HP	Q-20	9	2.222	Ea.	1,575	100		1,675	1,900
9540	3,740 CFM, 1/2 HP		8	2.500		1,650	113		1,763	1,975
9560	4400 CFM, 3/4 HP		7	2.857		1,650	129		1,779	2,025
9580	5700 CFM, 1-1/2 HP	↓	6	3.333	↓	1,725	151		1,876	2,125

23 34 23 – HVAC Power Ventilators

23 34 23.10 HVAC Power Circulators and Ventilators		Crew	Daily Output	Labor-Hours	Unit	Material	Labor	Equipment	Total	Total Incl O&P
0010	**HVAC POWER CIRCULATORS AND VENTILATORS**									
3000	Paddle blade air circulator, 3 speed switch									
3020	42", 5,000 CFM high, 3000 CFM low G	1 Elec	2.40	3.333	Ea.	130	163		293	385
3040	52", 6,500 CFM high, 4000 CFM low G	"	2.20	3.636	"	142	178		320	420
3100	For antique white motor, same cost									
3200	For brass plated motor, same cost									
3300	For light adaptor kit, add G				Ea.	38.50			38.50	42.50
3310	Industrial grade, reversible, 4 blade									
3312	5500 CFM	Q-20	5	4	Ea.	170	181		351	460
3314	7600 CFM		5	4		185	181		366	475
3316	21,015 CFM	↓	4	5	↓	240	226		466	605
6000	Propeller exhaust, wall shutter, 1/4" S.P.									
6020	Direct drive, two speed									
6100	375 CFM, 1/10 HP	Q-20	10	2	Ea.	430	90.50		520.50	610
6120	730 CFM, 1/7 HP		9	2.222		480	100		580	675
6140	1000 CFM, 1/8 HP		8	2.500		665	113		778	905
6160	1890 CFM, 1/4 HP		7	2.857		680	129		809	940
6180	3275 CFM, 1/2 HP		6	3.333		690	151		841	985
6200	4720 CFM, 1 HP	↓	5	4	↓	1,075	181		1,256	1,450
6300	V-belt drive, 3 phase									
6320	6175 CFM, 3/4 HP	Q-20	5	4	Ea.	870	181		1,051	1,225
6340	7500 CFM, 3/4 HP		5	4		935	181		1,116	1,300
6360	10,100 CFM, 1 HP		4.50	4.444		1,000	201		1,201	1,400
6380	14,300 CFM, 1-1/2 HP		4	5		1,075	226		1,301	1,525
6400	19,800 CFM, 2 HP		3	6.667		1,425	300		1,725	2,000
6420	26,250 CFM, 3 HP		2.60	7.692		1,875	345		2,220	2,575
6440	38,500 CFM, 5 HP		2.20	9.091		2,200	410		2,610	3,050
6460	46,000 CFM, 7-1/2 HP		2	10		2,350	450		2,800	3,275
6480	51,500 CFM, 10 HP	↓	1.80	11.111	↓	2,450	500		2,950	3,450
6490	V-belt drive, 115V., residential, whole house									

23 34 23 – HVAC Power Ventilators

23 34 23.10 HVAC Power Circulators and Ventilators		Crew	Daily Output	Labor-Hours	Unit	Material	2010 Bare Costs Labor	Equipment	Total	Total Incl O&P
6500	Ceiling-wall, 5200 CFM, 1/4 HP, 30" x 30"	1 Shee	6	1.333	Ea.	470	65.50		535.50	615
6510	7500 CFM, 1/3 HP, 36"x36"		5	1.600		490	78.50		568.50	660
6520	10,500 CFM, 1/3 HP, 42"x42"		5	1.600		510	78.50		588.50	685
6530	13,200 CFM, 1/3 HP, 48" x 48"		4	2		565	98		663	775
6540	15,445 CFM, 1/2 HP, 48" x 48"		4	2		590	98		688	795
6550	17,025 CFM, 1/2 HP, 54" x 54"		4	2		1,125	98		1,223	1,400
6560	For two speed motor, add					20%				
6570	Shutter, automatic, ceiling/wall									
6580	30" x 30"	1 Shee	8	1	Ea.	153	49		202	243
6590	36" x 36"		8	1		174	49		223	267
6600	42" x 42"		8	1		218	49		267	315
6610	48" x 48"		7	1.143		236	56		292	345
6620	54" x 54"		6	1.333		310	65.50		375.50	445
6630	Timer, shut off, to 12 Hr.		20	.400		50	19.65		69.65	84.50
6650	Residential, bath exhaust, grille, back draft damper									
6660	50 CFM	Q-20	24	.833	Ea.	44.50	37.50		82	106
6670	110 CFM		22	.909		71.50	41		112.50	141
6672	180 CFM		22	.909		144	41		185	221
6673	210 CFM		22	.909		194	41		235	275
6674	260 CFM		22	.909		217	41		258	300
6675	300 CFM		22	.909		296	41		337	385
6680	Light combination, squirrel cage, 100 watt, 70 CFM		24	.833		80.50	37.50		118	146
6700	Light/heater combination, ceiling mounted									
6710	70 CFM, 1450 watt	Q-20	24	.833	Ea.	114	37.50		151.50	182
6800	Heater combination, recessed, 70 CFM		24	.833		46.50	37.50		84	108
6820	With 2 infrared bulbs		23	.870		70	39.50		109.50	136
6840	Wall mount, 170 CFM		22	.909		223	41		264	305
6846	Ceiling mount, 180 CFM		22	.909		225	41		266	310
6900	Kitchen exhaust, grille, complete, 160 CFM		22	.909		83.50	41		124.50	154
6910	180 CFM		20	1		70.50	45		115.50	146
6920	270 CFM		18	1.111		128	50		178	217
6930	350 CFM		16	1.250		99.50	56.50		156	195
6940	Residential roof jacks and wall caps									
6944	Wall cap with back draft damper									
6946	3" & 4" dia. round duct	1 Shee	11	.727	Ea.	19.05	35.50		54.55	75
6948	6" dia. round duct	"	11	.727	"	46	35.50		81.50	105
6958	Roof jack with bird screen and back draft damper									
6960	3" & 4" dia. round duct	1 Shee	11	.727	Ea.	18.30	35.50		53.80	74
6962	3-1/4" x 10" rectangular duct	"	10	.800	"	34	39.50		73.50	97
6980	Transition									
6982	3-1/4" x 10" to 6" dia. round	1 Shee	20	.400	Ea.	19.95	19.65		39.60	51.50
8020	Attic, roof type									
8030	Aluminum dome, damper & curb									
8040	6" diameter, 300 CFM	1 Elec	16	.500	Ea.	400	24.50		424.50	475
8050	7" diameter, 450 CFM		15	.533		435	26		461	520
8060	9" diameter, 900 CFM		14	.571		480	28		508	565
8080	12" diameter, 1000 CFM (gravity)		10	.800		495	39		534	605
8090	16" diameter, 1500 CFM (gravity)		9	.889		595	43.50		638.50	720
8100	20" diameter, 2500 CFM (gravity)		8	1		730	49		779	880
8110	26" diameter, 4000 CFM (gravity)		7	1.143		885	56		941	1,050
8120	32" diameter, 6500 CFM (gravity)		6	1.333		1,225	65.50		1,290.50	1,450
8130	38" diameter, 8000 CFM (gravity)		5	1.600		1,800	78.50		1,878.50	2,100
8140	50" diameter, 13,000 CFM (gravity)		4	2		2,625	98		2,723	3,025

23 34 23 – HVAC Power Ventilators

23 34 23.10 HVAC Power Circulators and Ventilators		Crew	Daily Output	Labor-Hours	Unit	2010 Bare Costs Material	Labor	Equipment	Total	Total Incl O&P
8160	Plastic, ABS dome									
8180	1050 CFM	1 Elec	14	.571	Ea.	146	28		174	202
8200	1600 CFM	"	12	.667	"	218	32.50		250.50	289
8240	Attic, wall type, with shutter, one speed									
8250	12" diameter, 1000 CFM	1 Elec	14	.571	Ea.	315	28		343	385
8260	14" diameter, 1500 CFM		12	.667		340	32.50		372.50	425
8270	16" diameter, 2000 CFM	↓	9	.889	↓	385	43.50		428.50	490
8290	Whole house, wall type, with shutter, one speed									
8300	30" diameter, 4800 CFM	1 Elec	7	1.143	Ea.	825	56		881	990
8310	36" diameter, 7000 CFM		6	1.333		900	65.50		965.50	1,075
8320	42" diameter, 10,000 CFM		5	1.600		1,000	78.50		1,078.50	1,225
8330	48" diameter, 16,000 CFM	↓	4	2		1,250	98		1,348	1,525
8340	For two speed, add				↓	75.50			75.50	83
8350	Whole house, lay-down type, with shutter, one speed									
8360	30" diameter, 4500 CFM	1 Elec	8	1	Ea.	880	49		929	1,050
8370	36" diameter, 6500 CFM		7	1.143		945	56		1,001	1,125
8380	42" diameter, 9000 CFM		6	1.333		1,050	65.50		1,115.50	1,250
8390	48" diameter, 12,000 CFM	↓	5	1.600		1,175	78.50		1,253.50	1,425
8440	For two speed, add					56.50			56.50	62
8450	For 12 hour timer switch, add	1 Elec	32	.250	↓	56.50	12.25		68.75	80

23 34 33 – Air Curtains

23 34 33.10 Air Barrier Curtains		Crew	Daily Output	Labor-Hours	Unit	Material	Labor	Equipment	Total	Total Incl O&P
0010	**AIR BARRIER CURTAINS**, Incl. motor starters, transformers,									
0050	door switches & temperature controls									
0100	Shipping and receiving doors, unheated, minimal wind stoppage									
0150	8' high, multiples of 3' wide	2 Shee	6	2.667	L.F.	335	131		466	570
0160	5' wide		10	1.600		320	78.50		398.50	475
0210	10' high, multiples of 4' wide		8	2		193	98		291	360
0250	12' high, 3'-6" wide		7	2.286		214	112		326	405
0260	12' wide		6	2.667		223	131		354	445
0350	16' high, 3'-6" wide		7	2.286		213	112		325	405
0360	12' wide	↓	6	2.667	↓	223	131		354	445
1500	Customer entrance doors, unheated, minimal wind stoppage									
1550	10' high, multiples of 3' wide	2 Shee	6	2.667	L.F.	335	131		466	570
1560	5' wide		10	1.600		300	78.50		378.50	450
1650	Maximum wind stoppage, 12' high, multiples of 4' wide	↓	8	2	↓	350	98		448	535
1700	Heated, minimal wind stoppage, electric heat									
1750	8' high, multiples of 3' wide	2 Shee	6	2.667	L.F.	242	131		373	465
1850	10' high, multiples of 3' wide		6	2.667		855	131		986	1,150
1860	Multiples of 5' wide	↓	10	1.600	↓	560	78.50		638.50	735
1950	Maximum wind stoppage, steam heat									
1960	12' high, multiples of 4' wide	2 Shee	8	2	L.F.	630	98		728	840
2000	Walk-in coolers and freezers, ambient air, minimal wind stoppage									
2050	8' high, multiples of 3' wide	2 Shee	6	2.667	L.F.	365	131		496	600
2060	Multiples of 5' wide		10	1.600		400	78.50		478.50	560
2250	Maximum wind stoppage, 12' high, multiples of 3' wide		6	2.667		445	131		576	690
2450	Conveyor openings or service windows, unheated, 5' high		5	3.200		171	157		328	425
2460	Heated, electric, 5' high, 2'-6" wide	↓	5	3.200	↓	146	157		303	400

23 35 16.10 Engine Exhaust Removal Systems		Crew	Daily Output	Labor-Hours	Unit	Material	2010 Bare Costs Labor	Equipment	Total	Total Incl O&P
0010	**ENGINE EXHAUST REMOVAL SYSTEMS** D3090–320									
0500	Engine exhaust, garage, in-floor system									
0510	Single tube outlet assemblies									
0520	For transite pipe ducting, self-storing tube									
0530	3" tubing adapter plate	1 Shee	16	.500	Ea.	262	24.50		286.50	325
0540	4" tubing adapter plate		16	.500		264	24.50		288.50	325
0550	5" tubing adapter plate		16	.500		265	24.50		289.50	330
0600	For vitrified tile ducting									
0610	3" tubing adapter plate, self-storing tube	1 Shee	16	.500	Ea.	262	24.50		286.50	325
0620	4" tubing adapter plate, self-storing tube		16	.500		262	24.50		286.50	325
0660	5" tubing adapter plate, self-storing tube		16	.500		262	24.50		286.50	325
0800	Two tube outlet assemblies									
0810	For transite pipe ducting, self-storing tube									
0820	3" tubing, dual exhaust adapter plate	1 Shee	16	.500	Ea.	276	24.50		300.50	340
0850	For vitrified tile ducting									
0860	3" tubing, dual exhaust, self-storing tube	1 Shee	16	.500	Ea.	276	24.50		300.50	340
0870	3" tubing, double outlet, non-storing tubes	"	16	.500	"	276	24.50		300.50	340
0900	Accessories for metal tubing, (overhead systems also)									
0910	Adapters, for metal tubing end									
0920	3" tail pipe type				Ea.	54			54	59.50
0930	4" tail pipe type					58			58	64
0940	5" tail pipe type					59			59	65
0990	5" diesel stack type					350			350	385
1000	6" diesel stack type					360			360	395
1100	Bullnose (guide) required for in-floor assemblies									
1110	3" tubing size				Ea.	30			30	33
1120	4" tubing size					32			32	35
1130	5" tubing size					34			34	37.50
1150	Plain rings, for tubing end									
1160	3" tubing size				Ea.	26			26	28.50
1170	4" tubing size				"	29			29	32
1200	Tubing, galvanized, flexible, (for overhead systems also)									
1210	3" ID				L.F.	12.70			12.70	13.95
1220	4" ID					15.60			15.60	17.15
1230	5" ID					18.40			18.40	20
1240	6" ID					21.50			21.50	23.50
1250	Stainless steel, flexible, (for overhead system, also)									
1260	3" ID				L.F.	2.83			2.83	3.11
1270	4" ID					3.87			3.87	4.26
1280	5" ID					4.38			4.38	4.82
1290	6" ID					5.10			5.10	5.60
1500	Engine exhaust, garage, overhead components, for neoprene tubing									
1510	Alternate metal tubing & accessories see above									
1550	Adapters, for neoprene tubing end									
1560	3" tail pipe, adjustable, neoprene				Ea.	52			52	57
1570	3" tail pipe, heavy wall neoprene					58			58	64
1580	4" tail pipe, heavy wall neoprene					111			111	122
1590	5" tail pipe, heavy wall neoprene					117			117	129
1650	Connectors, tubing									
1660	3" interior, aluminum				Ea.	26			26	28.50
1670	4" interior, aluminum					29			29	32
1710	5" interior, neoprene					56			56	61.50
1750	3" spiralock, neoprene					26			26	28.50

23 35 16 – Engine Exhaust Systems

23 35 16.10 Engine Exhaust Removal Systems		Crew	Daily Output	Labor-Hours	Unit	2010 Bare Costs Material	Labor	Equipment	Total	Total Incl O&P
1760	4" spiralock, neoprene				Ea.	29			29	32
1780	Y for 3" ID tubing, neoprene, dual exhaust					218			218	240
1790	Y for 4" ID tubing, aluminum, dual exhaust					218			218	240
1850	Elbows, aluminum, splice into tubing for strap									
1860	3" neoprene tubing size				Ea.	54			54	59.50
1870	4" neoprene tubing size					45			45	49.50
1900	Flange assemblies, connect tubing to overhead duct					41.50			41.50	45.50
2000	Hardware and accessories									
2020	Cable, galvanized, 1/8" diameter				L.F.	.44			.44	.48
2040	Cleat, tie down cable or rope				Ea.	5			5	5.50
2060	Pulley					6.90			6.90	7.60
2080	Pulley hook, universal					5			5	5.50
2100	Rope, nylon, 1/4" diameter				L.F.	.37			.37	.41
2120	Winch, 1" diameter				Ea.	107			107	117
2150	Lifting strap, mounts on neoprene									
2160	3" tubing size				Ea.	25			25	27.50
2170	4" tubing size					25			25	27.50
2180	5" tubing size					25			25	27.50
2190	6" tubing size					25			25	27.50
2200	Tubing, neoprene, 11' lengths									
2210	3" ID				L.F.	10.70			10.70	11.80
2220	4" ID					17.45			17.45	19.20
2230	5" ID					24.50			24.50	27
2500	Engine exhaust, thru-door outlet									
2510	3" tube size	1 Carp	16	.500	Ea.	59	21		80	97
2530	4" tube size	"	16	.500	"	59	21		80	97
3000	Tubing, exhaust, flex hose, with									
3010	coupler, damper and tail pipe adapter									
3020	Neoprene									
3040	3" x 20'	1 Shee	6	1.333	Ea.	325	65.50		390.50	455
3050	4" x 15'		5.40	1.481		430	72.50		502.50	580
3060	4" x 20'		5	1.600		515	78.50		593.50	690
3070	5" x 15'		4.40	1.818		575	89.50		664.50	765
3100	Galvanized									
3110	3" x 20'	1 Shee	6	1.333	Ea.	360	65.50		425.50	495
3120	4" x 17'		5.60	1.429		380	70		450	520
3130	4" x 20'		5	1.600		425	78.50		503.50	590
3140	5" x 17'		4.60	1.739		460	85.50		545.50	635
8000	Blower, for tailpipe exhaust system									
8010	Direct drive									
8012	495 CFM, 1/3 HP	Q-9	5	3.200	Ea.	1,750	141		1,891	2,150
8014	1445 CFM, 3/4 HP		4	4		2,175	177		2,352	2,650
8016	1840 CFM, 1-1/2 HP		3	5.333		2,575	236		2,811	3,175
8030	Beltdrive									
8032	1400 CFM, 1 HP	Q-9	4	4	Ea.	2,175	177		2,352	2,650
8034	2023 CFM, 1-1/2 HP		3	5.333		2,925	236		3,161	3,575
8036	2750 CFM, 2 HP		2	8		3,500	355		3,855	4,375
8038	4400 CFM, 3 HP		1.80	8.889		5,150	395		5,545	6,275
8040	7060 CFM, 5 HP		1.40	11.429		7,725	505		8,230	9,275

23 35 Special Exhaust Systems

23 35 43 – Welding Fume Elimination Systems

23 35 43.10 Welding Fume Elimination System Components		Crew	Daily Output	Labor-Hours	Unit	2010 Bare Costs Material	Labor	Equipment	Total	Total Incl O&P
0010	WELDING FUME ELIMINATION SYSTEM COMPONENTS									
7500	Welding fume elimination accessories for garage exhaust systems									
7600	Cut off (blast gate)									
7610	3" tubing size, 3" x 6" opening	1 Shee	24	.333	Ea.	26	16.35		42.35	53.50
7620	4" tubing size, 4" x 8" opening		24	.333		27	16.35		43.35	54.50
7630	5" tubing size, 5" x 10" opening		24	.333		32	16.35		48.35	60
7640	6" tubing size		24	.333		34	16.35		50.35	62.50
7650	8" tubing size	↓	24	.333	↓	47	16.35		63.35	76.50
7700	Hoods, magnetic, with handle & screen									
7710	3" tubing size, 3" x 6" opening	1 Shee	24	.333	Ea.	94	16.35		110.35	129
7720	4" tubing size, 4" x 8" opening		24	.333		94	16.35		110.35	129
7730	5" tubing size, 5" x 10" opening	↓	24	.333	↓	94	16.35		110.35	129

23 36 Air Terminal Units

23 36 13 – Constant-Air-Volume Units

23 36 13.10 Constant Volume Mixing Boxes		Crew	Daily Output	Labor-Hours	Unit	Material	Labor	Equipment	Total	Total Incl O&P
0010	CONSTANT VOLUME MIXING BOXES									
5180	Mixing box, includes electric or pneumatic motor									
5192	Recommend use with silencer, see Div. 23 33 19.10 0010									
5200	Constant volume, 150 to 270 CFM	Q-9	12	1.333	Ea.	635	59		694	785
5210	270 to 600 CFM		11	1.455		650	64.50		714.50	815
5230	550 to 1000 CFM		9	1.778		650	78.50		728.50	835
5240	1000 to 1600 CFM		8	2		665	88.50		753.50	870
5250	1300 to 1900 CFM		6	2.667		680	118		798	930
5260	550 to 2640 CFM		5.60	2.857		750	126		876	1,025
5270	650 to 3120 CFM	↓	5.20	3.077	↓	800	136		936	1,075

23 36 16 – Variable-Air-Volume Units

23 36 16.10 Variable Volume Mixing Boxes		Crew	Daily Output	Labor-Hours	Unit	Material	Labor	Equipment	Total	Total Incl O&P
0010	VARIABLE VOLUME MIXING BOXES									
5180	Mixing box, includes electric or pneumatic motor									
5192	Recommend use with attenuator, see Div. 23 33 19.10 0010									
5500	VAV Cool only, pneumatic, pressure independent 300 to 600 CFM	Q-9	11	1.455	Ea.	410	64.50		474.50	555
5510	500 to 1000 CFM		9	1.778		425	78.50		503.50	585
5520	800 to 1600 CFM		9	1.778		435	78.50		513.50	600
5530	1100 to 2000 CFM		8	2		445	88.50		533.50	625
5540	1500 to 3000 CFM		7	2.286		475	101		576	680
5550	2000 to 4000 CFM	↓	6	2.667		480	118		598	710
5560	For electric, w/thermostat, pressure dependent, add				↓	23			23	25.50
5600	VAV Cool & HW coils, damper, actuator and t'stat									
5610	200 CFM	Q-9	11	1.455	Ea.	715	64.50		779.50	885
5620	400 CFM		10	1.600		720	70.50		790.50	895
5630	600 CFM		10	1.600		720	70.50		790.50	895
5640	800 CFM		8	2		740	88.50		828.50	950
5650	1000 CFM		8	2		740	88.50		828.50	950
5660	1250 CFM		6	2.667		810	118		928	1,075
5670	1500 CFM		6	2.667		810	118		928	1,075
5680	2000 CFM		4	4		880	177		1,057	1,250
5684	3000 CFM	↓	3.80	4.211	↓	960	186		1,146	1,325
5700	VAV Cool only, fan powered, damper, actuator, thermostat									
5710	200 CFM	Q-9	10	1.600	Ea.	975	70.50		1,045.50	1,175

23 36 Air Terminal Units

23 36 16 – Variable-Air-Volume Units

23 36 16.10 Variable Volume Mixing Boxes		Crew	Daily Output	Labor-Hours	Unit	Material	Labor	Equipment	Total	Total Incl O&P
						2010 Bare Costs				
5720	400 CFM	Q-9	9	1.778	Ea.	1,000	78.50		1,078.50	1,250
5730	600 CFM		9	1.778		1,000	78.50		1,078.50	1,250
5740	800 CFM		7	2.286		1,075	101		1,176	1,350
5750	1000 CFM		7	2.286		1,075	101		1,176	1,350
5760	1250 CFM		5	3.200		1,175	141		1,316	1,500
5770	1500 CFM		5	3.200		1,175	141		1,316	1,500
5780	2000 CFM	↓	4	4	↓	1,300	177		1,477	1,700
5800	VAV Fan powr'd, cooling, with HW coils, dampers, actuators, t'stat									
5810	200 CFM	Q-9	10	1.600	Ea.	1,175	70.50		1,245.50	1,400
5820	400 CFM		9	1.778		1,200	78.50		1,278.50	1,450
5830	600 CFM		9	1.778		1,200	78.50		1,278.50	1,450
5840	800 CFM		7	2.286		1,275	101		1,376	1,550
5850	1000 CFM		7	2.286		1,275	101		1,376	1,550
5860	1250 CFM		5	3.200		1,450	141		1,591	1,825
5870	1500 CFM		5	3.200		1,450	141		1,591	1,825
5880	2000 CFM	↓	3.50	4.571	↓	1,575	202		1,777	2,025

23 37 Air Outlets and Inlets

23 37 13 – Diffusers, Registers, and Grilles

23 37 13.10 Diffusers

			Crew	Daily Output	Labor-Hours	Unit	Material	Labor	Equipment	Total	Total Incl O&P
0010	**DIFFUSERS**, Aluminum, opposed blade damper unless noted	R233100-10									
0100	Ceiling, linear, also for sidewall										
0120	2" wide	R233100-30	1 Shee	32	.250	L.F.	43	12.30		55.30	65.50
0140	3" wide			30	.267		49.50	13.10		62.60	74
0160	4" wide	R233700-60		26	.308		57.50	15.10		72.60	86.50
0180	6" wide			24	.333		71.50	16.35		87.85	104
0200	8" wide			22	.364		83.50	17.85		101.35	119
0220	10" wide			20	.400		95	19.65		114.65	135
0240	12" wide		↓	18	.444	↓	112	22		134	156
0260	For floor or sill application, add						15%				
0500	Perforated, 24" x 24" lay-in panel size, 6" x 6"		1 Shee	16	.500	Ea.	57.50	24.50		82	101
0520	8" x 8"			15	.533		59	26		85	105
0530	9" x 9"			14	.571		63	28		91	112
0540	10" x 10"			14	.571		65	28		93	114
0560	12" x 12"			12	.667		66	32.50		98.50	122
0580	15" x 15"			11	.727		131	35.50		166.50	198
0590	16" x 16"			11	.727		109	35.50		144.50	174
0600	18" x 18"			10	.800		115	39.50		154.50	187
0610	20" x 20"			10	.800		133	39.50		172.50	206
0620	24" x 24"			9	.889		165	43.50		208.50	248
1000	Rectangular, 1 to 4 way blow, 6" x 6"			16	.500		46.50	24.50		71	88
1010	8" x 8"			15	.533		55.50	26		81.50	101
1014	9" x 9"			15	.533		61	26		87	107
1016	10" x 10"			15	.533		65	26		91	111
1020	12" x 6"			15	.533		67.50	26		93.50	114
1040	12" x 9"			14	.571		73.50	28		101.50	124
1060	12" x 12"			12	.667		73	32.50		105.50	130
1070	14" x 6"			13	.615		73	30		103	127
1074	14" x 14"			12	.667		103	32.50		135.50	163
1080	18" x 12"			11	.727		130	35.50		165.50	197
1120	15" x 15"		↓	10	.800	↓	106	39.50		145.50	177

23 37 13.10 Diffusers		Crew	Daily Output	Labor-Hours	Unit	Material	2010 Bare Costs Labor	Equipment	Total	Total Incl O&P
1140	18" x 15"	1 Shee	9	.889	Ea.	152	43.50		195.50	234
1150	18" x 18"		9	.889		128	43.50		171.50	207
1160	21" x 21"		8	1		183	49		232	276
1170	24" x 12"		10	.800		136	39.50		175.50	210
1180	24" x 24"		7	1.143		263	56		319	375
1500	Round, butterfly damper, steel, diffuser size, 6" diameter		18	.444		23	22		45	58.50
1520	8" diameter		16	.500		25	24.50		49.50	64.50
1540	10" diameter		14	.571		30.50	28		58.50	76
1560	12" diameter		12	.667		40.50	32.50		73	94
1580	14" diameter		10	.800		50.50	39.50		90	115
1600	18" diameter		9	.889		124	43.50		167.50	202
1610	20" diameter		9	.889		135	43.50		178.50	215
1620	22" diameter		8	1		148	49		197	238
1630	24" diameter		8	1		168	49		217	259
1640	28" diameter		7	1.143		206	56		262	310
1650	32" diameter		6	1.333		231	65.50		296.50	355
1700	Round, steel, adjustable core, annular segmented damper, diffuser size									
1704	6" diameter	1 Shee	18	.444	Ea.	77	22		99	118
1708	8" diameter		16	.500		82.50	24.50		107	128
1712	10" diameter		14	.571		93	28		121	145
1716	12" diameter		12	.667		103	32.50		135.50	164
1720	14" diameter		10	.800		115	39.50		154.50	186
1724	18" diameter		9	.889		175	43.50		218.50	259
1728	20" diameter		9	.889		203	43.50		246.50	289
1732	24" diameter		8	1		315	49		364	425
1736	30" diameter		7	1.143		550	56		606	690
1740	36" diameter		6	1.333		670	65.50		735.50	840
2000	T bar mounting, 24" x 24" lay-in frame, 6" x 6"		16	.500		42.50	24.50		67	84
2020	8" x 8"		14	.571		44.50	28		72.50	91
2040	12" x 12"		12	.667		52	32.50		84.50	107
2060	16" x 16"		11	.727		70	35.50		105.50	131
2080	18" x 18"		10	.800		78	39.50		117.50	146
2500	Combination supply and return									
2520	21" x 21" supply, 15" x 15" return	Q-9	10	1.600	Ea.	310	70.50		380.50	445
2540	24" x 24" supply, 18" x 18" return		9.50	1.684		390	74.50		464.50	545
2560	27" x 27" supply, 18" x 18" return		9	1.778		480	78.50		558.50	645
2580	30" x 30" supply, 21" x 21" return		8.50	1.882		605	83		688	790
2600	33" x 33" supply, 24" x 24" return		8	2		655	88.50		743.50	855
2620	36" x 36" supply, 24" x 24" return		7.50	2.133		830	94.50		924.50	1,050
3000	Baseboard, white enameled steel									
3100	18" long	1 Shee	20	.400	Ea.	12.70	19.65		32.35	43.50
3120	24" long		18	.444		24.50	22		46.50	60
3140	48" long		16	.500		45.50	24.50		70	87
3400	For matching return, deduct					10%				
4000	Floor, steel, adjustable pattern									
4100	2" x 10"	1 Shee	34	.235	Ea.	17	11.55		28.55	36
4120	2" x 12"		32	.250		18.65	12.30		30.95	39
4140	2" x 14"		30	.267		21	13.10		34.10	43.50
4200	4" x 10"		28	.286		19.70	14.05		33.75	42.50
4220	4" x 12"		26	.308		22	15.10		37.10	47
4240	4" x 14"		25	.320		24.50	15.70		40.20	51
4260	6" x 10"		26	.308		29.50	15.10		44.60	55.50
4280	6" x 12"		24	.333		31	16.35		47.35	59

23 37 13 – Diffusers, Registers, and Grilles

23 37 13.10 Diffusers		Crew	Daily Output	Labor-Hours	Unit	Material	2010 Bare Costs Labor	Equipment	Total	Total Incl O&P
4300	6" x 14"	1 Shee	22	.364	Ea.	32.50	17.85		50.35	62.50
4500	Linear bar diffuser in wall, sill or ceiling									
4510	2" wide	1 Shee	32	.250	L.F.	35	12.30		47.30	57
4512	3" wide		30	.267		37.50	13.10		50.60	61.50
4514	4" wide		26	.308		46.50	15.10		61.60	74
4516	6" wide		24	.333		57.50	16.35		73.85	88
4520	10" wide		20	.400		76.50	19.65		96.15	114
4522	12" wide		18	.444		86	22		108	128
5000	Sidewall, aluminum, 3 way dispersion									
5100	8" x 4"	1 Shee	24	.333	Ea.	9.30	16.35		25.65	35.50
5110	8" x 6"		24	.333		9.95	16.35		26.30	36
5120	10" x 4"		22	.364		9.80	17.85		27.65	38
5130	10" x 6"		22	.364		10.50	17.85		28.35	38.50
5140	10" x 8"		20	.400		14.30	19.65		33.95	45.50
5160	12" x 4"		18	.444		10.55	22		32.55	44.50
5170	12" x 6"		17	.471		11.65	23		34.65	48
5180	12" x 8"		16	.500		15.10	24.50		39.60	53.50
5200	14" x 4"		15	.533		11.75	26		37.75	52.50
5220	14" x 6"		14	.571		12.95	28		40.95	57
5240	14" x 8"		13	.615		16.85	30		46.85	64.50
5260	16" x 6"		12.50	.640		18.50	31.50		50	68
6000	For steel diffusers instead of aluminum, deduct					10%				

23 37 13.30 Grilles		Crew	Daily Output	Labor-Hours	Unit	Material	Labor	Equipment	Total	Total Incl O&P
0010	**GRILLES**									
0020	Aluminum									
0100	Air supply, single deflection, adjustable									
0120	8" x 4"	1 Shee	30	.267	Ea.	18.85	13.10		31.95	41
0140	8" x 8"		28	.286		23	14.05		37.05	46
0160	10" x 4"		24	.333		21	16.35		37.35	48
0180	10" x 10"		23	.348		27.50	17.10		44.60	56
0200	12" x 6"		23	.348		24.50	17.10		41.60	53
0220	12" x 12"		22	.364		32.50	17.85		50.35	62.50
0230	14" x 6"		23	.348		27.50	17.10		44.60	56
0240	14" x 8"		23	.348		29.50	17.10		46.60	58.50
0260	14" x 14"		22	.364		39.50	17.85		57.35	70.50
0270	18" x 8"		23	.348		34	17.10		51.10	63
0280	18" x 10"		23	.348		38.50	17.10		55.60	68.50
0300	18" x 18"		21	.381		60.50	18.70		79.20	95
0320	20" x 12"		22	.364		44	17.85		61.85	75.50
0340	20" x 20"		21	.381		73	18.70		91.70	109
0350	24" x 8"		20	.400		42.50	19.65		62.15	76
0360	24" x 14"		18	.444		59.50	22		81.50	98.50
0380	24" x 24"		15	.533		103	26		129	154
0400	30" x 8"		20	.400		54.50	19.65		74.15	89.50
0420	30" x 10"		19	.421		60	20.50		80.50	97.50
0440	30" x 12"		18	.444		66	22		88	106
0460	30" x 16"		17	.471		85.50	23		108.50	129
0480	30" x 18"		17	.471		96	23		119	140
0500	30" x 30"		14	.571		171	28		199	231
0520	36" x 12"		17	.471		82	23		105	125
0540	36" x 16"		16	.500		104	24.50		128.50	152
0560	36" x 18"		15	.533		116	26		142	167

23 37 13 – Diffusers, Registers, and Grilles

23 37 13.30 Grilles		Crew	Daily Output	Labor-Hours	Unit	Material	Labor	Equipment	Total	Total Incl O&P
						2010 Bare Costs				
0570	36" x 20"	1 Shee	15	.533	Ea.	127	26		153	179
0580	36" x 24"		14	.571		161	28		189	220
0600	36" x 28"		13	.615		183	30		213	247
0620	36" x 30"		12	.667		221	32.50		253.50	294
0640	36" x 32"		12	.667		238	32.50		270.50	310
0660	36" x 34"		11	.727		265	35.50		300.50	345
0680	36" x 36"		11	.727		287	35.50		322.50	370
0700	For double deflecting, add					70%				
1000	Air return, 6" x 6"	1 Shee	26	.308		17.55	15.10		32.65	42.50
1020	10" x 6"		24	.333		17.55	16.35		33.90	44.50
1040	14" x 6"		23	.348		20.50	17.10		37.60	48.50
1060	10" x 8"		23	.348		20.50	17.10		37.60	48.50
1080	16" x 8"		22	.364		22	17.85		39.85	51.50
1100	12" x 12"		22	.364		24	17.85		41.85	53
1120	24" x 12"		18	.444		33	22		55	69
1140	30" x 12"		16	.500		42.50	24.50		67	83.50
1160	14" x 14"		22	.364		28.50	17.85		46.35	58
1180	16" x 16"		22	.364		31.50	17.85		49.35	62
1200	18" x 18"		21	.381		36.50	18.70		55.20	68.50
1220	24" x 18"		16	.500		41.50	24.50		66	82.50
1240	36" x 18"		15	.533		63.50	26		89.50	110
1260	24" x 24"		15	.533		48.50	26		74.50	93
1280	36" x 24"		14	.571		71	28		99	121
1300	48" x 24"		12	.667		122	32.50		154.50	184
1320	48" x 30"		11	.727		157	35.50		192.50	227
1340	36" x 36"		13	.615		94.50	30		124.50	150
1360	48" x 36"		11	.727		195	35.50		230.50	269
1380	48" x 48"		8	1		241	49		290	340
2000	Door grilles, 12" x 12"		22	.364		54	17.85		71.85	86
2020	18" x 12"		22	.364		61	17.85		78.85	94.50
2040	24" x 12"		18	.444		66.50	22		88.50	106
2060	18" x 18"		18	.444		82	22		104	124
2080	24" x 18"		16	.500		93.50	24.50		118	140
2100	24" x 24"		15	.533		129	26		155	182
3000	Filter grille with filter, 12" x 12"		24	.333		47	16.35		63.35	76.50
3020	18" x 12"		20	.400		65.50	19.65		85.15	102
3040	24" x 18"		18	.444		77	22		99	118
3060	24" x 24"		16	.500		89	24.50		113.50	135
3080	30" x 24"		14	.571		111	28		139	165
3100	30" x 30"		13	.615		131	30		161	190
3950	Eggcrate, framed, 6" x 6" opening		26	.308		19.25	15.10		34.35	44
3954	8" x 8" opening		24	.333		20.50	16.35		36.85	47.50
3960	10" x 10" opening		23	.348		22	17.10		39.10	50.50
3970	12" x 12" opening		22	.364		26.50	17.85		44.35	56.50
3980	14" x 14" opening		22	.364		28	17.85		45.85	57.50
3984	16" x 16" opening		21	.381		41	18.70		59.70	73.50
3990	18" x 18" opening		21	.381		47	18.70		65.70	80
4020	22" x 22" opening		17	.471		61	23		84	103
4040	24" x 24" opening		15	.533		77	26		103	124
4044	28" x 28" opening		14	.571		119	28		147	174
4048	36" x 36" opening		12	.667		136	32.50		168.50	200
4050	48" x 24" opening		12	.667		173	32.50		205.50	240
4060	Eggcrate, lay-in, T-bar system									

23 37 13 – Diffusers, Registers, and Grilles

23 37 13.30 Grilles		Crew	Daily Output	Labor-Hours	Unit	Material	2010 Bare Costs Labor	Equipment	Total	Total Incl O&P
4070	48" x 24" sheet	1 Shee	40	.200	Ea.	93	9.80		102.80	117
5000	Transfer grille, vision proof, 8" x 4"		30	.267		34.50	13.10		47.60	58
5020	8" x 6"		28	.286		34.50	14.05		48.55	59
5040	8" x 8"		26	.308		39	15.10		54.10	65.50
5060	10" x 6"		24	.333		36	16.35		52.35	64.50
5080	10" x 10"		23	.348		45.50	17.10		62.60	76.50
5090	12" x 6"		23	.348		37.50	17.10		54.60	67.50
5100	12" x 10"		22	.364		50	17.85		67.85	82
5110	12" x 12"		22	.364		54	17.85		71.85	86.50
5120	14" x 10"		22	.364		54	17.85		71.85	86.50
5140	16" x 8"		21	.381		54	18.70		72.70	88
5160	18" x 12"		20	.400		70.50	19.65		90.15	108
5170	18" x 18"		21	.381		100	18.70		118.70	139
5180	20" x 12"		20	.400		76	19.65		95.65	114
5200	20" x 20"		19	.421		121	20.50		141.50	165
5220	24" x 12"		17	.471		86	23		109	130
5240	24" x 24"		16	.500		165	24.50		189.50	218
5260	30" x 6"		15	.533		76	26		102	124
5280	30" x 8"		15	.533		88.50	26		114.50	137
5300	30" x 12"		14	.571		111	28		139	165
5320	30" x 16"		13	.615		132	30		162	191
5340	30" x 20"		12	.667		166	32.50		198.50	233
5360	30" x 24"		11	.727		205	35.50		240.50	279
5380	30" x 30"		10	.800		270	39.50		309.50	355
6000	For steel grilles instead of aluminum in above, deduct					10%				
6200	Plastic, eggcrate, lay-in, T-bar system									
6210	48" x 24" sheet	1 Shee	50	.160	Ea.	23	7.85		30.85	37.50
6250	Steel door louver									
6270	With fire link, steel only									
6350	12" x 18"	1 Shee	22	.364	Ea.	187	17.85		204.85	233
6410	12" x 24"		21	.381		214	18.70		232.70	264
6470	18" x 18"		20	.400		209	19.65		228.65	260
6530	18" x 24"		19	.421		240	20.50		260.50	296
6600	18" x 36"		17	.471		279	23		302	340
6660	24" x 24"		15	.533		242	26		268	305
6720	24" x 30"		15	.533		279	26		305	345
6780	30" x 30"		13	.615		325	30		355	405
6840	30" x 36"		12	.667		380	32.50		412.50	465
6900	36" x 36"		10	.800		435	39.50		474.50	540

23 37 13.60 Registers		Crew	Daily Output	Labor-Hours	Unit	Material	Labor	Equipment	Total	Total Incl O&P
0010	**REGISTERS**									
0980	Air supply									
1000	Ceiling/wall, O.B. damper, anodized aluminum									
1010	One or two way deflection, adj. curved face bars									
1014	6" x 6"	1 Shee	24	.333	Ea.	32.50	16.35		48.85	61
1020	8" x 4"		26	.308		32.50	15.10		47.60	59
1040	8" x 8"		24	.333		40.50	16.35		56.85	69.50
1060	10" x 6"		20	.400		37.50	19.65		57.15	70.50
1080	10" x 10"		19	.421		50.50	20.50		71	87
1100	12" x 6"		19	.421		37.50	20.50		58	73
1120	12" x 12"		18	.444		59.50	22		81.50	98.50
1140	14" x 8"		17	.471		51.50	23		74.50	92

23 37 13 – Diffusers, Registers, and Grilles

23 37 13.60 Registers		Crew	Daily Output	Labor-Hours	Unit	Material	Labor	Equipment	Total	Total Incl O&P
						2010 Bare Costs				
1160	14" x 14"	1 Shee	18	.444	Ea.	69.50	22		91.50	110
1170	16" x 16"		17	.471		77	23		100	120
1180	18" x 8"		18	.444		57.50	22		79.50	96.50
1200	18" x 18"		17	.471		105	23		128	151
1220	20" x 4"		19	.421		53	20.50		73.50	90
1240	20" x 6"		18	.444		53	22		75	91.50
1260	20" x 8"		18	.444		58	22		80	97
1280	20" x 20"		17	.471		128	23		151	176
1290	22" x 22"		15	.533		147	26		173	202
1300	24" x 4"		17	.471		65.50	23		88.50	107
1320	24" x 6"		16	.500		65.50	24.50		90	109
1340	24" x 8"		13	.615		71.50	30		101.50	125
1350	24" x 18"		12	.667		133	32.50		165.50	196
1360	24" x 24"		11	.727		181	35.50		216.50	253
1380	30" x 4"		16	.500		82	24.50		106.50	127
1400	30" x 6"		15	.533		82	26		108	130
1420	30" x 8"		14	.571		94	28		122	146
1440	30" x 24"		12	.667		234	32.50		266.50	305
1460	30" x 30"		10	.800		299	39.50		338.50	390
1504	4 way deflection, adjustable curved face bars									
1510	6" x 6"	1 Shee	26	.308	Ea.	39	15.10		54.10	66
1514	8" x 8"		24	.333		48.50	16.35		64.85	78.50
1518	10" x 10"		19	.421		60.50	20.50		81	98
1522	12" x 6"		19	.421		45	20.50		65.50	81
1526	12" x 12"		18	.444		71.50	22		93.50	112
1530	14" x 14"		18	.444		83	22		105	125
1534	16" x 16"		17	.471		92.50	23		115.50	137
1538	18" x 18"		17	.471		127	23		150	174
1542	22" x 22"		16	.500		177	24.50		201.50	232
1980	One way deflection, adj. vert. or horiz. face bars									
1990	6" x 6"	1 Shee	26	.308	Ea.	34	15.10		49.10	60.50
2000	8" x 4"		26	.308		33.50	15.10		48.60	59.50
2020	8" x 8"		24	.333		39.50	16.35		55.85	68.50
2040	10" x 6"		20	.400		40	19.65		59.65	73.50
2060	10" x 10"		19	.421		46.50	20.50		67	82.50
2080	12" x 6"		19	.421		43	20.50		63.50	79
2100	12" x 12"		18	.444		56	22		78	94.50
2120	14" x 8"		17	.471		51.50	23		74.50	91.50
2140	14" x 12"		18	.444		61.50	22		83.50	101
2160	14" x 14"		18	.444		69.50	22		91.50	110
2180	16" x 6"		18	.444		51.50	22		73.50	89.50
2200	16" x 12"		18	.444		68	22		90	108
2220	16" x 16"		17	.471		86	23		109	130
2240	18" x 8"		18	.444		60	22		82	99
2260	18" x 12"		17	.471		73	23		96	115
2280	18" x 18"		16	.500		106	24.50		130.50	153
2300	20" x 10"		18	.444		71	22		93	112
2320	20" x 16"		16	.500		103	24.50		127.50	150
2340	20" x 20"		16	.500		130	24.50		154.50	180
2360	24" x 12"		15	.533		89.50	26		115.50	138
2380	24" x 16"		14	.571		119	28		147	174
2400	24" x 20"		12	.667		149	32.50		181.50	214
2420	24" x 24"		10	.800		182	39.50		221.50	261

23 37 13 – Diffusers, Registers, and Grilles

23 37 13.60 Registers	Crew	Daily Output	Labor-Hours	Unit	Material	2010 Bare Costs Labor	Equipment	Total	Total Incl O&P
2440 30" x 12"	1 Shee	13	.615	Ea.	116	30		146	174
2460 30" x 16"		13	.615		151	30		181	212
2480 30" x 24"		11	.727		235	35.50		270.50	315
2500 30" x 30"		9	.889		300	43.50		343.50	395
2520 36" x 12"		11	.727		144	35.50		179.50	213
2540 36" x 24"		10	.800		290	39.50		329.50	380
2560 36" x 36"		8	1		485	49		534	610
2600 For 2 way deflect., adj. vert. or horiz. face bars, add					40%				
2700 Above registers in steel instead of aluminum, deduct					10%				
3000 Baseboard, hand adj. damper, enameled steel									
3012 8" x 6"	1 Shee	26	.308	Ea.	13.90	15.10		29	38.50
3020 10" x 6"		24	.333		15.10	16.35		31.45	41.50
3040 12" x 5"		23	.348		16.40	17.10		33.50	44
3060 12" x 6"		23	.348		16.40	17.10		33.50	44
3080 12" x 8"		22	.364		24	17.85		41.85	53
3100 14" x 6"		20	.400		17.80	19.65		37.45	49
4000 Floor, toe operated damper, enameled steel									
4020 4" x 8"	1 Shee	32	.250	Ea.	25.50	12.30		37.80	47
4040 4" x 12"		26	.308		30	15.10		45.10	56
4060 6" x 8"		28	.286		27.50	14.05		41.55	51.50
4080 6" x 14"		22	.364		35	17.85		52.85	65.50
4100 8" x 10"		22	.364		31.50	17.85		49.35	61.50
4120 8" x 16"		20	.400		52.50	19.65		72.15	87.50
4140 10" x 10"		20	.400		37.50	19.65		57.15	71
4160 10" x 16"		18	.444		57	22		79	95.50
4180 12" x 12"		18	.444		46	22		68	84
4200 12" x 24"		16	.500		160	24.50		184.50	213
4220 14" x 14"		16	.500		132	24.50		156.50	182
4240 14" x 20"		15	.533		176	26		202	233
4300 Spiral pipe supply register									
4310 Aluminum, double deflection, w/damper extractor									
4320 4" x 12", for 6" thru 10" diameter duct	1 Shee	25	.320	Ea.	58	15.70		73.70	88
4330 4" x 18", for 6" thru 10" diameter duct		18	.444		72.50	22		94.50	113
4340 6" x 12", for 8" thru 12" diameter duct		19	.421		63.50	20.50		84	102
4350 6" x 18", for 8" thru 12" diameter duct		18	.444		82.50	22		104.50	124
4360 6" x 24", for 8" thru 12" diameter duct		16	.500		102	24.50		126.50	150
4370 6" x 30", for 8" thru 12" diameter duct		15	.533		130	26		156	183
4380 8" x 18", for 10" thru 14" diameter duct		18	.444		88	22		110	130
4390 8" x 24", for 10" thru 14" diameter duct		15	.533		110	26		136	161
4400 8" x 30", for 10" thru 14" diameter duct		14	.571		146	28		174	204
4410 10" x 24", for 12" thru 18" diameter duct		13	.615		121	30		151	179
4420 10" x 30", for 12" thru 18" diameter duct		12	.667		160	32.50		192.50	226
4430 10" x 36", for 12" thru 18" diameter duct		11	.727		198	35.50		233.50	272
4980 Air return									
5000 Ceiling or wall, fixed 45° face blades									
5010 Adjustable O.B. damper, anodized aluminum									
5020 4" x 8"	1 Shee	26	.308	Ea.	24	15.10		39.10	49
5040 6" x 8"		24	.333		27	16.35		43.35	55
5060 6" x 10"		19	.421		27	20.50		47.50	61.50
5070 6" x 12"		19	.421		31.50	20.50		52	66.50
5080 6" x 16"		18	.444		35	22		57	71.50
5100 8" x 10"		19	.421		31.50	20.50		52	66.50
5120 8" x 12"		16	.500		33.50	24.50		58	74

23 37 Air Outlets and Inlets

23 37 13 – Diffusers, Registers, and Grilles

23 37 13.60 Registers		Crew	Daily Output	Labor-Hours	Unit	2010 Bare Costs Material	Labor	Equipment	Total	Total Incl O&P
5140	10″ x 10″	1 Shee	18	.444	Ea.	33.50	22		55.50	70
5160	10″ x 16″		17	.471		41.50	23		64.50	80.50
5170	12″ x 12″		18	.444		36.50	22		58.50	73
5180	12″ x 18″		18	.444		47.50	22		69.50	85.50
5184	12″ x 24″		14	.571		54.50	28		82.50	103
5200	12″ x 30″		12	.667		72.50	32.50		105	130
5210	12″ x 36″		10	.800		99	39.50		138.50	169
5220	16″ x 16″		17	.471		50.50	23		73.50	90.50
5240	18″ x 18″		16	.500		60	24.50		84.50	103
5250	18″ x 24″		13	.615		71.50	30		101.50	125
5254	18″ x 30″		11	.727		110	35.50		145.50	175
5260	18″ x 36″		10	.800		130	39.50		169.50	203
5270	20″ x 20″		10	.800		69	39.50		108.50	136
5274	20″ x 24″		12	.667		82	32.50		114.50	140
5280	24″ x 24″		11	.727		92	35.50		127.50	155
5290	24″ x 30″		10	.800		102	39.50		141.50	172
5300	24″ x 36″		8	1		144	49		193	234
5320	24″ x 48″		6	1.333		215	65.50		280.50	335
5340	30″ x 30″	↓	6.50	1.231		169	60.50		229.50	278
5344	30″ x 120″	Q-9	6	2.667		725	118		843	980
5360	36″ x 36″	1 Shee	6	1.333		365	65.50		430.50	500
5370	40″ x 32″	″	6	1.333	↓	450	65.50		515.50	595
5400	Ceiling or wall, removable-reversible core, single deflection									
5402	Adjustable O.B. damper, radiused frame with border, aluminum									
5410	8″ x 4″	1 Shee	26	.308	Ea.	29.50	15.10		44.60	55.50
5414	8″ x 6″		24	.333		29.50	16.35		45.85	57.50
5418	10″ x 6″		19	.421		37.50	20.50		58	72.50
5422	10″ x 10″		18	.444		42.50	22		64.50	80
5426	12″ x 6″		19	.421		37.50	20.50		58	72.50
5430	12″ x 12″		18	.444		48	22		70	86
5434	16″ x 16″		16	.500		72	24.50		96.50	117
5438	18″ x 18″		17	.471		91	23		114	135
5442	20″ x 20″		16	.500		109	24.50		133.50	157
5446	24″ x 12″		15	.533		72	26		98	119
5450	24″ x 18″		13	.615		109	30		139	166
5454	24″ x 24″		11	.727		152	35.50		187.50	221
5458	30″ x 12″		13	.615		96	30		126	152
5462	30″ x 18″		12.50	.640		139	31.50		170.50	201
5466	30″ x 24″		11	.727		195	35.50		230.50	268
5470	30″ x 30″		9	.889		243	43.50		286.50	335
5474	36″ x 12″		11	.727		120	35.50		155.50	186
5478	36″ x 18″		10.50	.762		168	37.50		205.50	242
5482	36″ x 24″		10	.800		235	39.50		274.50	320
5486	36″ x 30″		9	.889		315	43.50		358.50	410
5490	36″ x 36″	↓	8	1	↓	405	49		454	520
6000	For steel construction instead of aluminum, deduct					10%				

23 37 15 – Louvers

23 37 15.40 HVAC Louvers		Crew	Daily Output	Labor-Hours	Unit	Material	Labor	Equipment	Total	Total Incl O&P
0010	**HVAC LOUVERS**									
0100	Aluminum, extruded, with screen, mill finish									
1002	Brick vent, see also Div. 04 05 23.19									
1100	Standard, 4″ deep, 8″ wide, 5″ high	1 Shee	24	.333	Ea.	34.50	16.35		50.85	62.50

23 37 Air Outlets and Inlets

23 37 15 – Louvers

23 37 15.40 HVAC Louvers		Crew	Daily Output	Labor-Hours	Unit	Material (2010 Bare Costs)	Labor (2010 Bare Costs)	Equipment (2010 Bare Costs)	Total (2010 Bare Costs)	Total Incl O&P
1200	Modular, 4" deep, 7-3/4" wide, 5" high	1 Shee	24	.333	Ea.	36	16.35		52.35	64.50
1300	Speed brick, 4" deep, 11-5/8" wide, 3-7/8" high		24	.333		36	16.35		52.35	64.50
1400	Fuel oil brick, 4" deep, 8" wide, 5" high		24	.333		61.50	16.35		77.85	93
2000	Cooling tower and mechanical equip., screens, light weight		40	.200	S.F.	15.40	9.80		25.20	32
2020	Standard weight		35	.229		41	11.20		52.20	62
2500	Dual combination, automatic, intake or exhaust		20	.400		56	19.65		75.65	91
2520	Manual operation		20	.400		41.50	19.65		61.15	75.50
2540	Electric or pneumatic operation		20	.400		41.50	19.65		61.15	75.50
2560	Motor, for electric or pneumatic		14	.571	Ea.	480	28		508	575
3000	Fixed blade, continuous line									
3100	Mullion type, stormproof	1 Shee	28	.286	S.F.	41.50	14.05		55.55	67
3200	Stormproof		28	.286		41.50	14.05		55.55	67
3300	Vertical line		28	.286		49.50	14.05		63.55	75
3500	For damper to use with above, add					50%	30%			
3520	Motor, for damper, electric or pneumatic	1 Shee	14	.571	Ea.	480	28		508	575
4000	Operating, 45°, manual, electric or pneumatic		24	.333	S.F.	50	16.35		66.35	80
4100	Motor, for electric or pneumatic		14	.571	Ea.	480	28		508	575
4200	Penthouse, roof		56	.143	S.F.	24.50	7		31.50	37.50
4300	Walls		40	.200		58	9.80		67.80	79
5000	Thinline, under 4" thick, fixed blade		40	.200		24	9.80		33.80	41.50
5010	Finishes, applied by mfr. at additional cost, available in colors									
5020	Prime coat only, add				S.F.	3.30			3.30	3.63
5040	Baked enamel finish coating, add					6.10			6.10	6.70
5060	Anodized finish, add					6.60			6.60	7.25
5080	Duranodic finish, add					12			12	13.20
5100	Fluoropolymer finish coating, add					18.90			18.90	21
9980	For small orders (under 10 pieces), add					25%				

23 37 23 – HVAC Gravity Ventilators

23 37 23.10 HVAC Gravity Air Ventilators		Crew	Daily Output	Labor-Hours	Unit	Material	Labor	Equipment	Total	Total Incl O&P
0010	**HVAC GRAVITY AIR VENTILATORS**, Incl. base and damper									
1280	Rotary ventilators, wind driven, galvanized									
1300	4" neck diameter	Q-9	20	.800	Ea.	50	35.50		85.50	109
1320	5" neck diameter		18	.889		56	39.50		95.50	121
1340	6" neck diameter		16	1		56	44		100	129
1360	8" neck diameter		14	1.143		58	50.50		108.50	141
1380	10" neck diameter		12	1.333		75	59		134	172
1400	12" neck diameter		10	1.600		81	70.50		151.50	197
1420	14" neck diameter		10	1.600		113	70.50		183.50	231
1440	16" neck diameter		9	1.778		136	78.50		214.50	269
1460	18" neck diameter, 1700 CFM		9	1.778		158	78.50		236.50	293
1480	20" neck diameter, 2100 CFM		8	2		176	88.50		264.50	325
1500	24" neck diameter, 3,100 CFM		8	2		236	88.50		324.50	395
1520	30" neck diameter, 4500 CFM		7	2.286		490	101		591	695
1540	36" neck diameter, 5,500 CFM		6	2.667		695	118		813	945
1600	For aluminum, add					300%				
1620	For stainless steel, add					600%				
1630	For copper, add					600%				
2000	Stationary, gravity, syphon, galvanized									
2100	3" neck diameter, 40 CFM	Q-9	24	.667	Ea.	45.50	29.50		75	94.50
2120	4" neck diameter, 50 CFM		20	.800		49.50	35.50		85	108
2140	5" neck diameter, 58 CFM		18	.889		50	39.50		89.50	115
2160	6" neck diameter, 66 CFM		16	1		53	44		97	126

23 37 23 – HVAC Gravity Ventilators

23 37 23.10 HVAC Gravity Air Ventilators		Crew	Daily Output	Labor-Hours	Unit	2010 Bare Costs Material	Labor	Equipment	Total	Total Incl O&P
2180	7" neck diameter, 86 CFM	Q-9	15	1.067	Ea.	54.50	47		101.50	132
2200	8" neck diameter, 110 CFM		14	1.143		55	50.50		105.50	137
2220	10" neck diameter, 140 CFM		12	1.333		71	59		130	167
2240	12" neck diameter, 160 CFM		10	1.600		91	70.50		161.50	207
2260	14" neck diameter, 250 CFM		10	1.600		114	70.50		184.50	233
2280	16" neck diameter, 380 CFM		9	1.778		153	78.50		231.50	287
2300	18" neck diameter, 500 CFM		9	1.778		203	78.50		281.50	340
2320	20" neck diameter, 625 CFM		8	2		215	88.50		303.50	370
2340	24" neck diameter, 900 CFM		8	2		248	88.50		336.50	405
2360	30" neck diameter, 1375 CFM		7	2.286		565	101		666	775
2380	36" neck diameter, 2,000 CFM		6	2.667		775	118		893	1,025
2400	42" neck diameter, 3000 CFM		4	4		1,100	177		1,277	1,475
2410	48" neck diameter, 4000 CFM		3	5.333		1,475	236		1,711	1,975
2500	For aluminum, add					300%				
2520	For stainless steel, add					600%				
2530	For copper, add					600%				
3000	Rotating chimney cap, galvanized, 4" neck diameter	Q-9	20	.800		39	35.50		74.50	96
3020	5" neck diameter		18	.889		40	39.50		79.50	103
3040	6" neck diameter		16	1		40.50	44		84.50	112
3060	7" neck diameter		15	1.067		44	47		91	120
3080	8" neck diameter		14	1.143		45.50	50.50		96	127
3100	10" neck diameter		12	1.333		58.50	59		117.50	154
3600	Stationary chimney rain cap, galvanized, 3" neck diameter		24	.667		34	29.50		63.50	82
3620	4" neck diameter		20	.800		35	35.50		70.50	92
3640	6" neck diameter		16	1		36.50	44		80.50	107
3680	8" neck diameter		14	1.143		39	50.50		89.50	120
3700	10" neck diameter		12	1.333		49.50	59		108.50	144
3720	12" neck diameter		10	1.600		82.50	70.50		153	198
3740	14" neck diameter		10	1.600		88	70.50		158.50	204
3760	16" neck diameter		9	1.778		108	78.50		186.50	238
3770	18" neck diameter		9	1.778		156	78.50		234.50	290
3780	20" neck diameter		8	2		208	88.50		296.50	365
3790	24" neck diameter		8	2		243	88.50		331.50	400
4200	Stationary mushroom, aluminum, 16" orifice diameter		10	1.600		535	70.50		605.50	695
4220	26" orifice diameter		6.15	2.602		790	115		905	1,050
4230	30" orifice diameter		5.71	2.802		1,150	124		1,274	1,475
4240	38" orifice diameter		5	3.200		1,650	141		1,791	2,050
4250	42" orifice diameter		4.70	3.404		2,200	150		2,350	2,650
4260	50" orifice diameter		4.44	3.604		2,625	159		2,784	3,125
5000	Relief vent									
5500	Rectangular, aluminum, galvanized curb									
5510	intake/exhaust, 0.033" SP									
5580	500 CFM, 12" x 12"	Q-9	8.60	1.860	Ea.	610	82		692	795
5600	600 CFM, 12" x 16"		8	2		685	88.50		773.50	890
5620	750 CFM, 12" x 20"		7.20	2.222		735	98		833	960
5640	1000 CFM, 12" x 24"		6.60	2.424		775	107		882	1,000
5660	1500 CFM, 12" x 36"		5.80	2.759		1,025	122		1,147	1,300
5680	3000 CFM, 20" x 42"		4	4		1,350	177		1,527	1,775
5700	5000 CFM, 20" x 72"		3	5.333		1,950	236		2,186	2,500
5720	6000 CFM, 24" x 72"		2.30	6.957		2,100	305		2,405	2,775
5740	10,000 CFM, 48" x 60"		1.80	8.889		2,675	395		3,070	3,525
5760	12,000 CFM, 48" x 72"		1.60	10		3,175	440		3,615	4,150
5780	13,750 CFM, 60" x 66"		1.40	11.429		3,500	505		4,005	4,625

23 37 Air Outlets and Inlets

23 37 23 – HVAC Gravity Ventilators

23 37 23.10 HVAC Gravity Air Ventilators		Crew	Daily Output	Labor-Hours	Unit	Material	2010 Bare Costs Labor	Equipment	Total	Total Incl O&P
5800	15,000 CFM, 60" x 72"	Q-9	1.30	12.308	Ea.	3,750	545		4,295	4,950
5820	18,000 CFM, 72" x 72"	↓	1.20	13.333	↓	4,250	590		4,840	5,575
5880	Size is throat area, volume is at 500 FPM									
7000	Note: sizes based on exhaust. Intake, with 0.125" SP									
7100	loss, approximately twice listed capacity.									

23 38 Ventilation Hoods

23 38 13 – Commercial-Kitchen Hoods

23 38 13.10 Hood and Ventilation Equipment

		Crew	Daily Output	Labor-Hours	Unit	Material	Labor	Equipment	Total	Total Incl O&P
0010	**HOOD AND VENTILATION EQUIPMENT**									
2970	Exhaust hood, sst, gutter on all sides, 4' x 4' x 2'	1 Carp	1.80	4.444	Ea.	4,325	185		4,510	5,050
2980	4' x 4' x 7'	"	1.60	5	"	6,800	208		7,008	7,800
7800	Vent hood, wall canopy with fire protection	L-3A	9	1.333	L.F.	375	60.50		435.50	505
7810	Without fire protection		10	1.200		261	54.50		315.50	370
7820	Island canopy with fire protection		7	1.714		395	78		473	555
7830	Without fire protection		8	1.500		286	68		354	420
7840	Back shelf with fire protection		11	1.091		375	49.50		424.50	485
7850	Without fire protection	↓	12	1	↓	261	45.50		306.50	355
7860	Range hood & CO_2 system, minimum	1 Carp	2.50	3.200	Ea.	2,750	133		2,883	3,225
7870	Maximum	"	1	8		33,000	330		33,330	36,800
7950	Hood fire protection system, minimum	Q-1	3	5.333		4,300	250		4,550	5,100
8050	Maximum	"	1	16	↓	33,800	750		34,550	38,300

23 38 13.20 Kitchen Ventilation

		Crew	Daily Output	Labor-Hours	Unit	Material	Labor	Equipment	Total	Total Incl O&P
0010	**KITCHEN VENTILATION**, Commercial									
1010	Heat reclaim unit, air to air									
1100	Heat pipe exchanger									
1110	Combined supply/exhaust air volume									
1120	2.5 to 6.0 MCFM	Q-10	2.80	8.571	MCFM	6,775	395		7,170	8,050
1130	6 to 16 MCFM		5	4.800		4,150	220		4,370	4,900
1140	16 to 22 MCFM	↓	6	4	↓	3,250	183		3,433	3,850
2010	Packaged ventilation									
2100	Combined supply/exhaust air volume									
2110	With ambient supply									
2120	3 to 7 MCFM	Q-10	5	4.800	MCFM	1,225	220		1,445	1,675
2130	7 to 12 MCFM		7.10	3.380		815	155		970	1,125
2140	12 to 22 MCFM	↓	10.70	2.243	↓	575	103		678	790
2400	With tempered supply									
2410	3 to 7 MCFM	Q-10	5	4.800	MCFM	2,125	220		2,345	2,650
2420	7 to 12 MCFM		7.10	3.380		1,375	155		1,530	1,725
2430	12 to 22 MCFM	↓	10.70	2.243	↓	1,000	103		1,103	1,250
6000	Exhaust hoods									
6100	Centrifugal grease extraction									
6110	Water wash type									
6120	Per foot of length, minimum	Q-10	7.10	3.380	L.F.	1,625	155		1,780	2,000
6130	Maximum	"	7.10	3.380	"	2,775	155		2,930	3,275
6200	Non water wash type									
6210	Per foot of length	Q-10	13.40	1.791	L.F.	1,200	82		1,282	1,425
6400	Island style ventilator									
6410	Water wash type									
6420	Per foot of length, minimum	Q-10	7.10	3.380	L.F.	2,150	155		2,305	2,600
6430	Maximum	"	7.10	3.380	"	3,600	155		3,755	4,200

23 38 Ventilation Hoods

23 38 13 – Commercial-Kitchen Hoods

23 38 13.20 Kitchen Ventilation		Crew	Daily Output	Labor-Hours	Unit	2010 Bare Costs Material	Labor	Equipment	Total	Total Incl O&P
6460	Non water wash type									
6470	Per foot of length, minimum	Q-10	13.40	1.791	L.F.	935	82		1,017	1,150
6480	Maximum	"	13.40	1.791	"	1,900	82		1,982	2,225

23 41 Particulate Air Filtration

23 41 13 – Panel Air Filters

23 41 13.10 Panel Type Air Filters

		Crew	Daily Output	Labor-Hours	Unit	Material	Labor	Equipment	Total	Total Incl O&P
0010	PANEL TYPE AIR FILTERS									
2950	Mechanical media filtration units									
3000	High efficiency type, with frame, non-supported G				MCFM	45			45	49.50
3100	Supported type G				"	60			60	66
5500	Throwaway glass or paper media type				Ea.	3.35			3.35	3.69

23 41 16 – Renewable-Media Air Filters

23 41 16.10 Disposable Media Air Filters

		Crew	Daily Output	Labor-Hours	Unit	Material	Labor	Equipment	Total	Total Incl O&P
0010	DISPOSABLE MEDIA AIR FILTERS									
5000	Renewable disposable roll				MCFM	250			250	275
5800	Filter, bag type									
5810	90-95% efficiency									
5820	24" x 12" x 29", .75-1.25 MCFM	1 Shee	1.90	4.211	Ea.	45.50	207		252.50	365
5830	24" x 24" x 29", 1.5-2.5 MCFM	"	1.90	4.211	"	83	207		290	405
5850	80-85% efficiency									
5860	24" x 12" x 29", .75-1.25 MCFM	1 Shee	1.90	4.211	Ea.	40.50	207		247.50	360
5870	24" x 24" x 29", 1.5-2.5 MCFM	"	1.90	4.211	"	74.50	207		281.50	395

23 41 19 – Washable Air Filters

23 41 19.10 Permanent Air Filters

		Crew	Daily Output	Labor-Hours	Unit	Material	Labor	Equipment	Total	Total Incl O&P
0010	PERMANENT AIR FILTERS									
4500	Permanent washable G				MCFM	20			20	22

23 41 23 – Extended Surface Filters

23 41 23.10 Expanded Surface Filters

		Crew	Daily Output	Labor-Hours	Unit	Material	Labor	Equipment	Total	Total Incl O&P
0010	EXPANDED SURFACE FILTERS									
4000	Medium efficiency, extended surface G				MCFM	5.50			5.50	6.05

23 41 33 – High-Efficiency Particulate Filtration

23 41 33.10 HEPA Filters

		Crew	Daily Output	Labor-Hours	Unit	Material	Labor	Equipment	Total	Total Incl O&P
0010	HEPA FILTERS									
6000	HEPA filter complete w/particle board,									
6010	kraft paper frame, separator material									
6020	95% DOP efficiency									
6030	12" x 12" x 6", 150 CFM	1 Shee	3.70	2.162	Ea.	44	106		150	210
6034	24" x 12" x 6", 375 CFM		1.90	4.211		73.50	207		280.50	395
6038	24" x 18" x 6", 450 CFM		1.90	4.211		90	207		297	415
6042	24" x 24" x 6", 700 CFM		1.90	4.211		101	207		308	425
6046	12" x 12" x 12", 250 CFM		3.70	2.162		74.50	106		180.50	243
6050	24" x 12" x 12", 500 CFM		1.90	4.211		105	207		312	430
6054	24" x 18" x 12", 875 CFM		1.90	4.211		146	207		353	475
6058	24" x 24" x 12", 1000 CFM	↓	1.90	4.211	↓	158	207		365	490
6100	99% DOP efficiency									
6110	12" x 12" x 6", 150 CFM	1 Shee	3.70	2.162	Ea.	46	106		152	212
6114	24" x 12" x 6", 325 CFM		1.90	4.211		76	207		283	400
6118	24" x 18" x 6", 550 CFM	↓	1.90	4.211	↓	93	207		300	415

23 41 Particulate Air Filtration

23 41 33 – High-Efficiency Particulate Filtration

23 41 33.10 HEPA Filters		Crew	Daily Output	Labor-Hours	Unit	Material	2010 Bare Costs Labor	Equipment	Total	Total Incl O&P
6122	24" x 24" x 6", 775 CFM	1 Shee	1.90	4.211	Ea.	105	207		312	430
6124	24" x 48" x 6", 775 CFM		1.30	6.154		201	300		501	680
6126	12" x 12" x 12", 250 CFM		3.60	2.222		77	109		186	250
6130	24" x 12" x 12", 500 CFM		1.90	4.211		109	207		316	435
6134	24" x 18" x 12", 775 CFM		1.90	4.211		152	207		359	480
6138	24" x 24" x 12", 1100 CFM		1.90	4.211		164	207		371	495
6500	HEPA filter housing, 14 ga. galv. sheet metal									
6510	12" x 12" x 6"	1 Shee	2.50	3.200	Ea.	660	157		817	965
6514	24" x 12" x 6"		2	4		740	196		936	1,100
6518	12" x 12" x 12"		2.40	3.333		660	164		824	975
6522	14" x 12" x 12"		2.30	3.478		660	171		831	985
6526	24" x 18" x 6"		1.90	4.211		940	207		1,147	1,350
6530	24" x 24" x 6"		1.80	4.444		855	218		1,073	1,275
6534	24" x 48" x 6"		1.70	4.706		1,150	231		1,381	1,600
6538	24" x 72" x 6"		1.60	5		1,475	246		1,721	2,000
6542	24" x 18" x 12"		1.80	4.444		940	218		1,158	1,350
6546	24" x 24" x 12"		1.70	4.706		1,050	231		1,281	1,500
6550	24" x 48" x 12"		1.60	5		1,225	246		1,471	1,725
6554	24" x 72" x 12"		1.50	5.333		1,525	262		1,787	2,100
6558	48" x 48" x 6"	Q-9	2.80	5.714		1,700	253		1,953	2,225
6562	48" x 72" x 6"		2.60	6.154		2,025	272		2,297	2,625
6566	48" x 96" x 6"		2.40	6.667		2,575	295		2,870	3,275
6570	48" x 48" x 12"		2.70	5.926		1,700	262		1,962	2,250
6574	48" x 72" x 12"		2.50	6.400		2,025	283		2,308	2,650
6578	48" x 96" x 12"		2.30	6.957		2,575	305		2,880	3,300
6582	114" x 72" x 12"		2	8		3,550	355		3,905	4,425

23 42 Gas-Phase Air Filtration

23 42 13 – Activated-Carbon Air Filtration

23 42 13.10 Charcoal Type Air Filtration		Crew	Daily Output	Labor-Hours	Unit	Material	Labor	Equipment	Total	Total Incl O&P
0010	**CHARCOAL TYPE AIR FILTRATION**									
0050	Activated charcoal type, full flow				MCFM	600			600	660
0060	Full flow, impregnated media 12" deep					225			225	248
0070	HEPA filter & frame for field erection					300			300	330
0080	HEPA filter-diffuser, ceiling install.					275			275	305

23 42 16 – Chemically-Impregnated Adsorption Air Filtration

23 42 16.10 Chemical Adsorption Air Filtration		Crew	Daily Output	Labor-Hours	Unit	Material	Labor	Equipment	Total	Total Incl O&P
0010	**CHEMICAL ADSORPTION AIR FILTRATION**									
0500	Chemical media filtration type									
1100	Industrial air fume & odor scrubber unit w/pump & motor									
1110	corrosion resistant PVC construction									
1120	Single pack filter, horizontal type									
1130	500 CFM	Q-9	14	1.143	Ea.	5,975	50.50		6,025.50	6,650
1140	1000 CFM		11	1.455		7,025	64.50		7,089.50	7,825
1150	2000 CFM		8	2		8,500	88.50		8,588.50	9,475
1160	3000 CFM		7	2.286		9,850	101		9,951	11,000
1170	5000 CFM		5	3.200		12,800	141		12,941	14,300
1180	8000 CFM		4	4		18,000	177		18,177	20,000
1190	12,000 CFM		3	5.333		23,800	236		24,036	26,600
1200	16,000 CFM		2.50	6.400		29,400	283		29,683	32,700

23 42 16 – Chemically-Impregnated Adsorption Air Filtration

23 42 16.10 Chemical Adsorption Air Filtration		Crew	Daily Output	Labor-Hours	Unit	Material	2010 Bare Costs Labor	Equipment	Total	Total Incl O&P
1210	20,000 CFM	Q-9	2	8	Ea.	36,900	355		37,255	41,000
1220	26,000 CFM		1.50	10.667		42,400	470		42,870	47,400
1230	30,000 CFM	Q-10	2	12		48,200	550		48,750	54,000
1240	40,000 CFM	"	1.50	16		63,000	735		63,735	70,500
1250	50,000 CFM	Q-11	2	16		77,000	750		77,750	85,500
1260	55,000 CFM		1.80	17.778		84,000	830		84,830	93,500
1270	60,000 CFM		1.50	21.333		90,500	1,000		91,500	101,000
1300	Double pack filter, horizontal type									
1310	500 CFM	Q-9	10	1.600	Ea.	7,775	70.50		7,845.50	8,675
1320	1000 CFM		8	2		9,350	88.50		9,438.50	10,400
1330	2000 CFM		6	2.667		11,200	118		11,318	12,600
1340	3000 CFM		5	3.200		13,400	141		13,541	15,000
1350	5000 CFM		4	4		17,100	177		17,277	19,100
1360	8000 CFM		3	5.333		22,600	236		22,836	25,200
1370	12,000 CFM		2.50	6.400		30,600	283		30,883	34,100
1380	16,000 CFM		2	8		40,000	355		40,355	44,500
1390	20,000 CFM		1.50	10.667		48,500	470		48,970	54,000
1400	26,000 CFM		1	16		56,000	705		56,705	62,500
1410	30,000 CFM	Q-10	1.50	16		66,000	735		66,735	73,500
1420	40,000 CFM	"	1.30	18.462		84,500	845		85,345	94,000
1430	50,000 CFM	Q-11	1.50	21.333		105,500	1,000		106,500	117,500
1440	55,000 CFM		1.30	24.615		115,000	1,150		116,150	129,000
1450	60,000 CFM		1	32		125,000	1,500		126,500	139,500
1500	Single pack filter, vertical type									
1510	500 CFM	Q-9	24	.667	Ea.	5,725	29.50		5,754.50	6,325
1520	1000 CFM		18	.889		6,750	39.50		6,789.50	7,475
1530	2000 CFM		12	1.333		8,100	59		8,159	9,000
1540	3000 CFM		9	1.778		9,475	78.50		9,553.50	10,500
1550	5000 CFM		6	2.667		12,600	118		12,718	14,100
1560	8000 CFM		4	4		18,000	177		18,177	20,000
1570	12,000 CFM		3	5.333		24,400	236		24,636	27,300
1580	16,000 CFM		2	8		28,600	355		28,955	31,900
1590	20,000 CFM		1.80	8.889		36,000	395		36,395	40,200
1600	24,000 CFM		1.60	10		40,300	440		40,740	45,000
1650	Double pack filter, vertical type									
1660	500 CFM	Q-9	22	.727	Ea.	7,425	32		7,457	8,200
1670	1000 CFM		16	1		8,850	44		8,894	9,800
1680	2000 CFM		10	1.600		10,600	70.50		10,670.50	11,800
1690	3000 CFM		7	2.286		12,500	101		12,601	13,900
1700	5000 CFM		5	3.200		16,400	141		16,541	18,200
1710	8000 CFM		3	5.333		22,400	236		22,636	25,000
1720	12,000 CFM		2.50	6.400		30,200	283		30,483	33,700
1730	16,000 CFM		2	8		38,000	355		38,355	42,300
1740	20,000 CFM		1.50	10.667		47,400	470		47,870	52,500
1750	24,000 CFM		1	16		53,000	705		53,705	59,000
1800	Inlet or outlet transition, horizontal									
1810	Single pack to 12,000 CFM, add					4%				
1820	Single pack to 30,000 CFM, add					6%				
1830	Single pack to 60,000 CFM, add					8%				
1840	Double pack to 12,000 CFM, add					3%				
1850	Double pack to 30,000 CFM, add					5%				
1860	Double pack to 60,000 CFM, add					6%				
1870	Inlet or outlet transition, vertical									

23 42 Gas-Phase Air Filtration

23 42 16 – Chemically-Impregnated Adsorption Air Filtration

23 42 16.10 Chemical Adsorption Air Filtration		Crew	Daily Output	Labor-Hours	Unit	2010 Bare Costs Material	Labor	Equipment	Total	Total Incl O&P
1880	Single pack to 5000 CFM, add					2%				
1890	Single pack to 24,000 CFM, add					3%				
1900	Double pack to 24,000 CFM, add					2%				

23 43 Electronic Air Cleaners

23 43 13 – Washable Electronic Air Cleaners

23 43 13.10 Electronic Air Cleaners		Crew	Daily Output	Labor-Hours	Unit	Material	Labor	Equipment	Total	Total Incl O&P
0010	**ELECTRONIC AIR CLEANERS**									
2000	Electronic air cleaner, duct mounted									
2150	400 – 1000 CFM	1 Shee	2.30	3.478	Ea.	1,675	171		1,846	2,100
2200	1000 – 1400 CFM		2.20	3.636		1,525	179		1,704	1,950
2250	1400 – 2000 CFM		2.10	3.810		1,525	187		1,712	1,950
2260	2000 – 2500 CFM		2	4		1,500	196		1,696	1,950

23 51 Breechings, Chimneys, and Stacks

23 51 13 – Draft Control Devices

23 51 13.13 Draft-Induction Fans		Crew	Daily Output	Labor-Hours	Unit	Material	Labor	Equipment	Total	Total Incl O&P
0010	**DRAFT-INDUCTION FANS**									
1000	Breeching installation									
1800	Hot gas, 600°F, variable pitch pulley and motor									
1840	6" diam. inlet, 1/4 H.P., 1 phase, 400 CFM	Q-9	6	2.667	Ea.	2,000	118		2,118	2,375
1850	7" diam. inlet, 1/4 H.P., 1 phase, 800 CFM		5	3.200		2,125	141		2,266	2,575
1860	8" diam. inlet, 1/4 H.P., 1phase, 1120 CFM		4	4		3,100	177		3,277	3,700
1870	9" diam. inlet, 3/4 H.P., 1 phase, 1440 CFM		3.60	4.444		3,675	196		3,871	4,350
1880	10" diam. inlet, 3/4 H.P., 1 phase, 2000 CFM		3.30	4.848		4,125	214		4,339	4,875
1900	12" diam. inlet, 3/4 H.P., 3 phase, 2960 CFM		3	5.333		4,250	236		4,486	5,025
1910	14" diam. inlet, 1 H.P., 3 phase, 4160 CFM		2.60	6.154		5,000	272		5,272	5,900
1920	16" diam. inlet, 2 H.P., 3 phase, 6720 CFM		2.30	6.957		5,475	305		5,780	6,500
1940	18" diam. inlet, 3 H.P., 3 phase, 9120 CFM		2	8		6,600	355		6,955	7,800
1950	20" diam. inlet, 3 H.P., 3 phase, 9760 CFM		1.50	10.667		7,200	470		7,670	8,625
1960	22" diam. inlet, 5 H.P., 3 phase, 13,360 CFM		1	16		8,425	705		9,130	10,300
1980	24" diam. inlet, 7-1/2 H.P., 3 phase, 17,760 CFM		.80	20		10,400	885		11,285	12,800
2300	For multi-blade damper at fan inlet, add					20%				
3600	Chimney-top installation									
3700	6" size	1 Shee	8	1	Ea.	785	49		834	940
3740	8" size		7	1.143		790	56		846	955
3750	10" size		6.50	1.231		1,025	60.50		1,085.50	1,250
3780	13" size		6	1.333		1,050	65.50		1,115.50	1,275
3880	For speed control switch, add					60			60	66
3920	For thermal fan control, add					60			60	66
5500	Flue shutter damper for draft control,									
5510	parallel blades									
5550	8" size	Q-9	8	2	Ea.	700	88.50		788.50	905
5560	9" size		7.50	2.133		755	94.50		849.50	975
5570	10" size		7	2.286		780	101		881	1,025
5580	12" size		6.50	2.462		855	109		964	1,100
5590	14" size		6	2.667		970	118		1,088	1,250
5600	16" size		5.50	2.909		1,050	129		1,179	1,350
5610	18" size		5	3.200		1,150	141		1,291	1,500

23 51 Breechings, Chimneys, and Stacks

23 51 13 – Draft Control Devices

23 51 13.13 Draft-Induction Fans		Crew	Daily Output	Labor-Hours	Unit	2010 Bare Costs Material	Labor	Equipment	Total	Total Incl O&P
5620	20″ size	Q-9	4.50	3.556	Ea.	1,275	157		1,432	1,650
5630	22″ size		4	4		1,400	177		1,577	1,825
5640	24″ size		3.50	4.571		1,550	202		1,752	2,000
5650	27″ size		3	5.333		1,575	236		1,811	2,100
5660	30″ size		2.50	6.400		1,650	283		1,933	2,225
5670	32″ size		2	8		1,725	355		2,080	2,425
5680	36″ size		1.50	10.667		1,850	470		2,320	2,750

23 51 13.16 Vent Dampers		Crew	Daily Output	Labor-Hours	Unit	Material	Labor	Equipment	Total	Total Incl O&P
0010	**VENT DAMPERS**									
5000	Vent damper, bi-metal, gas, 3″ diameter	Q-9	24	.667	Ea.	40	29.50		69.50	88.50
5010	4″ diameter		24	.667		40	29.50		69.50	88.50
5020	5″ diameter		23	.696		40	31		71	90.50
5030	6″ diameter		22	.727		40	32		72	92.50
5040	7″ diameter		21	.762		42.50	33.50		76	98
5050	8″ diameter		20	.800		46	35.50		81.50	105

23 51 13.19 Barometric Dampers		Crew	Daily Output	Labor-Hours	Unit	Material	Labor	Equipment	Total	Total Incl O&P
0010	**BAROMETRIC DAMPERS**									
1000	Barometric, gas fired system only, 6″ size for 5″ and 6″ pipes	1 Shee	20	.400	Ea.	59.50	19.65		79.15	95
1020	7″ size, for 6″ and 7″ pipes		19	.421		63.50	20.50		84	102
1040	8″ size, for 7″ and 8″ pipes		18	.444		82.50	22		104.50	124
1060	9″ size, for 8″ and 9″ pipes		16	.500		92.50	24.50		117	139
2000	All fuel, oil, oil/gas, coal									
2020	10″ for 9″ and 10″ pipes	1 Shee	15	.533	Ea.	138	26		164	191
2040	12″ for 11″ and 12″ pipes		15	.533		180	26		206	238
2060	14″ for 13″ and 14″ pipes		14	.571		233	28		261	300
2080	16″ for 15″ and 16″ pipes		13	.615		325	30		355	405
2100	18″ for 17″ and 18″ pipes		12	.667		435	32.50		467.50	530
2120	20″ for 19″ and 21″ pipes		10	.800		520	39.50		559.50	630
2140	24″ for 22″ and 25″ pipes	Q-9	12	1.333		635	59		694	790
2160	28″ for 26″ and 30″ pipes		10	1.600		790	70.50		860.50	975
2180	32″ for 31″ and 34″ pipes		8	2		1,000	88.50		1,088.50	1,225
3260	For thermal switch for above, add	1 Shee	24	.333		57	16.35		73.35	87.50

23 51 23 – Gas Vents

23 51 23.10 Gas Chimney Vents		Crew	Daily Output	Labor-Hours	Unit	Material	Labor	Equipment	Total	Total Incl O&P
0010	**GAS CHIMNEY VENTS,** Prefab metal, U.L. listed									
0020	Gas, double wall, galvanized steel									
0080	3″ diameter	Q-9	72	.222	V.L.F.	5.35	9.80		15.15	21
0100	4″ diameter		68	.235		6.70	10.40		17.10	23
0120	5″ diameter		64	.250		7.80	11.05		18.85	25.50
0140	6″ diameter		60	.267		9.10	11.80		20.90	28
0160	7″ diameter		56	.286		13.35	12.65		26	34
0180	8″ diameter		52	.308		14.85	13.60		28.45	37
0200	10″ diameter		48	.333		34	14.75		48.75	59.50
0220	12″ diameter		44	.364		45	16.05		61.05	74
0240	14″ diameter		42	.381		76	16.85		92.85	109
0260	16″ diameter		40	.400		103	17.70		120.70	140
0280	18″ diameter		38	.421		133	18.60		151.60	174
0300	20″ diameter	Q-10	36	.667		139	30.50		169.50	200
0320	22″ diameter		34	.706		199	32.50		231.50	268
0340	24″ diameter		32	.750		245	34.50		279.50	320
0360	26″ diameter		31	.774		294	35.50		329.50	380
0380	28″ diameter		30	.800		310	36.50		346.50	395

23 51 23.10 Gas Chimney Vents		Crew	Daily Output	Labor-Hours	Unit	Material	Labor	Equipment	Total	Total Incl O&P
						2010 Bare Costs				
0400	30" diameter	Q-10	28	.857	V.L.F.	330	39.50		369.50	420
0420	32" diameter		27	.889		380	40.50		420.50	480
0440	34" diameter		26	.923		435	42.50		477.50	540
0460	36" diameter		25	.960		460	44		504	570
0480	38" diameter		24	1		500	46		546	620
0500	40" diameter		23	1.043		560	48		608	690
0520	42" diameter		22	1.091		585	50		635	720
0540	44" diameter		21	1.143		645	52.50		697.50	790
0560	46" diameter		20	1.200		710	55		765	870
0580	48" diameter		19	1.263		790	58		848	955
0600	For 4", 5" and 6" oval, add					50%				
0650	Gas, double wall, galvanized steel, fittings									
0660	Elbow 45°, 3" diameter	Q-9	36	.444	Ea.	13.25	19.65		32.90	44.50
0670	4" diameter		34	.471		15.75	21		36.75	49
0680	5" diameter		32	.500		18.50	22		40.50	54
0690	6" diameter		30	.533		22	23.50		45.50	60
0700	7" diameter		28	.571		32	25.50		57.50	73.50
0710	8" diameter		26	.615		37	27		64	81.50
0720	10" diameter		24	.667		98	29.50		127.50	153
0730	12" diameter		22	.727		124	32		156	186
0740	14" diameter		21	.762		174	33.50		207.50	242
0750	16" diameter		20	.800		218	35.50		253.50	294
0760	18" diameter		19	.842		281	37		318	365
0770	20" diameter	Q-10	18	1.333		330	61		391	460
0780	22" diameter		17	1.412		510	64.50		574.50	660
0790	24" diameter		16	1.500		635	69		704	805
0800	26" diameter		16	1.500		1,250	69		1,319	1,475
0810	28" diameter		15	1.600		1,275	73.50		1,348.50	1,500
0820	30" diameter		14	1.714		1,350	78.50		1,428.50	1,625
0830	32" diameter		14	1.714		1,525	78.50		1,603.50	1,800
0840	34" diameter		13	1.846		1,675	84.50		1,759.50	1,975
0850	36" diameter		12	2		1,725	91.50		1,816.50	2,025
0860	38" diameter		12	2		1,850	91.50		1,941.50	2,175
0870	40" diameter		11	2.182		1,950	100		2,050	2,300
0880	42" diameter		11	2.182		2,050	100		2,150	2,400
0890	44" diameter		10	2.400		2,200	110		2,310	2,600
0900	46" diameter		10	2.400		2,350	110		2,460	2,750
0910	48" diameter		10	2.400		2,450	110		2,560	2,850
0916	Adjustable length									
0918	3" diameter, to 12"	Q-9	36	.444	Ea.	10.90	19.65		30.55	42
0920	4" diameter, to 12"		34	.471		12.90	21		33.90	45.50
0924	6" diameter, to 12"		30	.533		16.60	23.50		40.10	54
0928	8" diameter, to 12"		26	.615		24.50	27		51.50	68
0930	10" diameter, to 18"		24	.667		99	29.50		128.50	154
0932	12" diameter, to 18"		22	.727		121	32		153	182
0936	16" diameter, to 18"		20	.800		251	35.50		286.50	330
0938	18" diameter, to 18"		19	.842		277	37		314	360
0944	24" diameter, to 18"	Q-10	16	1.500		495	69		564	650
0950	Elbow 90°, adjustable, 3" diameter	Q-9	36	.444		22	19.65		41.65	54.50
0960	4" diameter		34	.471		27	21		48	61
0970	5" diameter		32	.500		31.50	22		53.50	68
0980	6" diameter		30	.533		37.50	23.50		61	77
0990	7" diameter		28	.571		59.50	25.50		85	104

23 51 23 – Gas Vents

23 51 23.10 Gas Chimney Vents		Crew	Daily Output	Labor-Hours	Unit	Material	2010 Bare Costs Labor	Equipment	Total	Total Incl O&P
1010	8″ diameter	Q-9	26	.615	Ea.	63.50	27		90.50	111
1020	Wall thimble, 4 to 7″ adjustable, 3″ diameter		36	.444		14.05	19.65		33.70	45.50
1022	4″ diameter		34	.471		15.65	21		36.65	49
1024	5″ diameter		32	.500		18.90	22		40.90	54.50
1026	6″ diameter		30	.533		21.50	23.50		45	59
1028	7″ diameter		28	.571		25	25.50		50.50	66
1030	8″ diameter		26	.615		26	27		53	69.50
1040	Roof flashing, 3″ diameter		36	.444		24.50	19.65		44.15	57
1050	4″ diameter		34	.471		28	21		49	62.50
1060	5″ diameter		32	.500		29.50	22		51.50	66
1070	6″ diameter		30	.533		31.50	23.50		55	70
1080	7″ diameter		28	.571		53	25.50		78.50	96.50
1090	8″ diameter		26	.615		54	27		81	101
1100	10″ diameter		24	.667		83.50	29.50		113	136
1110	12″ diameter		22	.727		106	32		138	166
1120	14″ diameter		20	.800		124	35.50		159.50	191
1130	16″ diameter		18	.889		154	39.50		193.50	229
1140	18″ diameter		16	1		208	44		252	296
1150	20″ diameter	Q-10	18	1.333		270	61		331	390
1160	22″ diameter		14	1.714		340	78.50		418.50	495
1170	24″ diameter		12	2		410	91.50		501.50	590
1200	Tee, 3″ diameter	Q-9	27	.593		33	26		59	75.50
1210	4″ diameter		26	.615		34	27		61	78.50
1220	5″ diameter		25	.640		37	28.50		65.50	84
1230	6″ diameter		24	.667		40	29.50		69.50	88.50
1240	7″ diameter		23	.696		44	31		75	95
1250	8″ diameter		22	.727		60.50	32		92.50	115
1260	10″ diameter		21	.762		176	33.50		209.50	245
1270	12″ diameter		20	.800		212	35.50		247.50	287
1280	14″ diameter		18	.889		345	39.50		384.50	440
1290	16″ diameter		16	1		490	44		534	605
1300	18″ diameter		14	1.143		610	50.50		660.50	750
1310	20″ diameter	Q-10	17	1.412		830	64.50		894.50	1,025
1320	22″ diameter		13	1.846		1,025	84.50		1,109.50	1,250
1330	24″ diameter		12	2		1,200	91.50		1,291.50	1,475
1340	26″ diameter		12	2		1,725	91.50		1,816.50	2,050
1350	28″ diameter		11	2.182		1,925	100		2,025	2,275
1360	30″ diameter		10	2.400		2,100	110		2,210	2,475
1370	32″ diameter		10	2.400		2,425	110		2,535	2,850
1380	34″ diameter		9	2.667		2,450	122		2,572	2,875
1390	36″ diameter		9	2.667		2,925	122		3,047	3,400
1400	38″ diameter		8	3		2,950	137		3,087	3,425
1410	40″ diameter		7	3.429		3,225	157		3,382	3,800
1420	42″ diameter		6	4		3,500	183		3,683	4,125
1430	44″ diameter		5	4.800		3,800	220		4,020	4,525
1440	46″ diameter		5	4.800		4,175	220		4,395	4,925
1450	48″ diameter		5	4.800		4,450	220		4,670	5,225
1460	Tee cap, 3″ diameter	Q-9	45	.356		2.34	15.70		18.04	26.50
1470	4″ diameter		42	.381		2.63	16.85		19.48	28.50
1480	5″ diameter		40	.400		3.48	17.70		21.18	31
1490	6″ diameter		37	.432		4.06	19.10		23.16	33.50
1500	7″ diameter		35	.457		6.55	20		26.55	38
1510	8″ diameter		34	.471		7.75	21		28.75	40

23 51 23.10 Gas Chimney Vents		Crew	Daily Output	Labor-Hours	Unit	Material	Labor	Equipment	Total	Total Incl O&P
						2010 Bare Costs				
1520	10" diameter	Q-9	32	.500	Ea.	15.95	22		37.95	51
1530	12" diameter		30	.533		21.50	23.50		45	59
1540	14" diameter		28	.571		44	25.50		69.50	86.50
1550	16" diameter		25	.640		51	28.50		79.50	99
1560	18" diameter	↓	24	.667		57.50	29.50		87	108
1570	20" diameter	Q-10	27	.889		61.50	40.50		102	129
1580	22" diameter		22	1.091		70.50	50		120.50	153
1590	24" diameter		21	1.143		79.50	52.50		132	167
1600	26" diameter		20	1.200		138	55		193	235
1610	28" diameter		19	1.263		325	58		383	445
1620	30" diameter		18	1.333		375	61		436	505
1630	32" diameter		17	1.412		435	64.50		499.50	580
1640	34" diameter		16	1.500		440	69		509	585
1650	36" diameter		15	1.600		450	73.50		523.50	605
1660	38" diameter		14	1.714		460	78.50		538.50	625
1670	40" diameter		14	1.714		470	78.50		548.50	640
1680	42" diameter		13	1.846		485	84.50		569.50	665
1690	44" diameter		12	2		490	91.50		581.50	680
1700	46" diameter		12	2		500	91.50		591.50	690
1710	48" diameter	↓	11	2.182		505	100		605	705
1750	Top, 3" diameter	Q-9	46	.348		12.30	15.35		27.65	37
1760	4" diameter		44	.364		16.20	16.05		32.25	42.50
1770	5" diameter		42	.381		18.10	16.85		34.95	45.50
1780	6" diameter		40	.400		22.50	17.70		40.20	51.50
1790	7" diameter		38	.421		33	18.60		51.60	64.50
1800	8" diameter		36	.444		43.50	19.65		63.15	77.50
1810	10" diameter		34	.471		109	21		130	152
1820	12" diameter		32	.500		165	22		187	215
1830	14" diameter		30	.533		188	23.50		211.50	243
1840	16" diameter		28	.571		340	25.50		365.50	415
1850	18" diameter	↓	26	.615		600	27		627	695
1860	20" diameter	Q-10	28	.857		665	39.50		704.50	790
1870	22" diameter		22	1.091		905	50		955	1,075
1880	24" diameter		20	1.200		1,125	55		1,180	1,325
1882	26" diameter		19	1.263		1,700	58		1,758	1,975
1884	28" diameter		18	1.333		1,950	61		2,011	2,225
1886	30" diameter		17	1.412		2,175	64.50		2,239.50	2,500
1888	32" diameter		16	1.500		2,300	69		2,369	2,625
1890	34" diameter		15	1.600		2,450	73.50		2,523.50	2,800
1892	36" diameter	↓	14	1.714	↓	2,625	78.50		2,703.50	3,025
1900	Gas, double wall, galvanized steel, oval									
1904	4" x 1'	Q-9	68	.235	V.L.F.	19.65	10.40		30.05	37.50
1906	5"		64	.250		54.50	11.05		65.55	77
1908	5"/6" x 1'	↓	60	.267	↓	54.50	11.80		66.30	78
1910	Oval fittings									
1912	Adjustable length									
1914	4" diameter to 12" lg	Q-9	34	.471	Ea.	20.50	21		41.50	54.50
1916	5" diameter to 12" lg		32	.500		38	22		60	75
1918	5"/6" diameter to 12"	↓	30	.533	↓	40	23.50		63.50	79.50
1920	Elbow 45°									
1922	4"	Q-9	34	.471	Ea.	29.50	21		50.50	64
1924	5"		32	.500		68	22		90	109
1926	5"/6"	↓	30	.533	↓	64.50	23.50		88	107

23 51 Breechings, Chimneys, and Stacks

23 51 23 – Gas Vents

23 51 23.10 Gas Chimney Vents		Crew	Daily Output	Labor-Hours	Unit	2010 Bare Costs Material	Labor	Equipment	Total	Total Incl O&P
1930	Elbow 45°, flat									
1932	4″	Q-9	34	.471	Ea.	36	21		57	71
1934	5″		32	.500		68	22		90	109
1936	5″/6″		30	.533		68	23.50		91.50	111
1940	Top									
1942	4″	Q-9	44	.364	Ea.	27	16.05		43.05	54
1944	5″		42	.381		18.65	16.85		35.50	46
1946	5″/6″		40	.400		45.50	17.70		63.20	77
1950	Adjustable flashing									
1952	4″	Q-9	34	.471	Ea.	16.10	21		37.10	49.50
1954	5″		32	.500		47.50	22		69.50	85.50
1956	5″/6″		30	.533		53.50	23.50		77	94.50
1960	Tee									
1962	4″	Q-9	26	.615	Ea.	53	27		80	99.50
1964	5″		25	.640		102	28.50		130.50	155
1966	5″/6″		24	.667		103	29.50		132.50	159
1970	Tee with short snout									
1972	4″	Q-9	26	.615	Ea.	53	27		80	99.50

23 51 26 – All-Fuel Vent Chimneys

23 51 26.10 All-Fuel Vent Chimneys, Press. Tight, Dbl. Wall		Crew	Daily Output	Labor-Hours	Unit	Material	Labor	Equipment	Total	Total Incl O&P
0010	**ALL-FUEL VENT CHIMNEYS, PRESSURE TIGHT, DOUBLE WALL**									
3200	All fuel, pressure tight, double wall, U.L. listed, 1400°F.									
3210	304 stainless steel liner, aluminized steel outer jacket									
3220	6″ diameter	Q-9	60	.267	L.F.	47	11.80		58.80	70
3221	8″ diameter		52	.308		58	13.60		71.60	84
3222	10″ diameter		48	.333		64.50	14.75		79.25	93.50
3223	12″ diameter		44	.364		73.50	16.05		89.55	105
3224	14″ diameter		42	.381		81	16.85		97.85	115
3225	16″ diameter		40	.400		89.50	17.70		107.20	126
3226	18″ diameter		38	.421		101	18.60		119.60	139
3227	20″ diameter	Q-10	36	.667		115	30.50		145.50	174
3228	24″ diameter		32	.750		148	34.50		182.50	215
3229	28″ diameter		30	.800		170	36.50		206.50	243
3230	32″ diameter		27	.889		191	40.50		231.50	272
3231	36″ diameter		25	.960		224	44		268	315
3232	42″ diameter		22	1.091		258	50		308	360
3233	48″ diameter		19	1.263		300	58		358	420
3260	For 316 stainless steel liner add					6%				
3280	All fuel, pressure tight, double wall fittings									
3284	304 stainless steel inner, aluminized steel jacket									
3288	Adjustable 18″/30″ section									
3291	5″ diameter	Q-9	32	.500	Ea.	121	22		143	167
3292	6″ diameter		30	.533		153	23.50		176.50	204
3293	8″ diameter		26	.615		176	27		203	235
3294	10″ diameter		24	.667		198	29.50		227.50	263
3295	12″ diameter		22	.727		225	32		257	297
3296	14″ diameter		21	.762		260	33.50		293.50	335
3297	16″ diameter		20	.800		270	35.50		305.50	350
3298	18″ diameter		19	.842		335	37		372	420
3299	20″ diameter	Q-10	18	1.333		385	61		446	520
3300	24″ diameter		16	1.500		495	69		564	650
3301	28″ diameter		15	1.600		555	73.50		628.50	720

23 51 26.10 All-Fuel Vent Chimneys, Press. Tight, Dbl. Wall		Crew	Daily Output	Labor-Hours	Unit	Material	Labor	Equipment	Total	Total Incl O&P
						2010 Bare Costs				
3302	32" diameter	Q-10	14	1.714	Ea.	650	78.50		728.50	835
3303	36" diameter		12	2		680	91.50		771.50	890
3304	42" diameter		11	2.182		745	100		845	970
3305	48" diameter	↓	10	2.400		875	110		985	1,125
3326	For 316 stainless steel liner, add				↓	15%				
3350	Elbow 90° fixed									
3353	5" diameter	Q-9	32	.500	Ea.	227	22		249	283
3354	6" diameter		30	.533		266	23.50		289.50	330
3355	8" diameter		26	.615		294	27		321	365
3356	10" diameter		24	.667		335	29.50		364.50	415
3357	12" diameter		22	.727		380	32		412	470
3358	14" diameter		21	.762		445	33.50		478.50	540
3359	16" diameter		20	.800		500	35.50		535.50	605
3360	18" diameter	↓	19	.842		580	37		617	695
3361	20" diameter	Q-10	18	1.333		650	61		711	810
3362	24" diameter		16	1.500		805	69		874	990
3363	28" diameter		15	1.600		1,000	73.50		1,073.50	1,200
3364	32" diameter		14	1.714		1,350	78.50		1,428.50	1,600
3365	36" diameter		12	2		1,575	91.50		1,666.50	1,900
3366	42" diameter		11	2.182		2,400	100		2,500	2,800
3367	48" diameter	↓	10	2.400		3,025	110		3,135	3,500
3380	For 316 stainless steel liner, add				↓	20%				
3400	Elbow 45°									
3404	6" diameter	Q-9	30	.533	Ea.	168	23.50		191.50	220
3405	8" diameter		26	.615		190	27		217	250
3406	10" diameter		24	.667		213	29.50		242.50	280
3407	12" diameter		22	.727		243	32		275	315
3408	14" diameter		21	.762		280	33.50		313.50	360
3409	16" diameter		20	.800		315	35.50		350.50	400
3410	18" diameter	↓	19	.842		355	37		392	445
3411	20" diameter	Q-10	18	1.333		410	61		471	545
3412	24" diameter		16	1.500		525	69		594	680
3413	28" diameter		15	1.600		595	73.50		668.50	765
3414	32" diameter		14	1.714		795	78.50		873.50	995
3415	36" diameter		12	2		1,125	91.50		1,216.50	1,375
3416	42" diameter		11	2.182		1,525	100		1,625	1,825
3417	48" diameter	↓	10	2.400		2,050	110		2,160	2,425
3430	For 316 stainless steel liner, add				↓	20%				
3450	Tee 90°									
3454	6" diameter	Q-9	24	.667	Ea.	175	29.50		204.50	238
3455	8" diameter		22	.727		202	32		234	271
3456	10" diameter		21	.762		236	33.50		269.50	310
3457	12" diameter		20	.800		275	35.50		310.50	355
3458	14" diameter		18	.889		325	39.50		364.50	420
3459	16" diameter		16	1		365	44		409	470
3460	18" diameter	↓	14	1.143		430	50.50		480.50	550
3461	20" diameter	Q-10	17	1.412		485	64.50		549.50	635
3462	24" diameter		12	2		600	91.50		691.50	795
3463	28" diameter		11	2.182		745	100		845	970
3464	32" diameter		10	2.400		1,025	110		1,135	1,300
3465	36" diameter		9	2.667		1,200	122		1,322	1,500
3466	42" diameter		6	4		1,900	183		2,083	2,375
3467	48" diameter	↓	5	4.800	↓	2,425	220		2,645	3,000

23 51 Breechings, Chimneys, and Stacks

23 51 26 – All-Fuel Vent Chimneys

23 51 26.10 All-Fuel Vent Chimneys, Press. Tight, Dbl. Wall		Crew	Daily Output	Labor-Hours	Unit	2010 Bare Costs Material	Labor	Equipment	Total	Total Incl O&P
3480	For Tee Cap, add				Ea.	35%	20%			
3500	For 316 stainless steel liner, add					22%				
3520	Plate support, galvanized									
3524	6″ diameter	Q-9	26	.615	Ea.	91	27		118	141
3525	8″ diameter		22	.727		103	32		135	163
3526	10″ diameter		20	.800		112	35.50		147.50	178
3527	12″ diameter		18	.889		117	39.50		156.50	189
3528	14″ diameter		17	.941		143	41.50		184.50	220
3529	16″ diameter		16	1		150	44		194	232
3530	18″ diameter		15	1.067		158	47		205	246
3531	20″ diameter	Q-10	16	1.500		168	69		237	288
3532	24″ diameter		14	1.714		176	78.50		254.50	315
3533	28″ diameter		13	1.846		216	84.50		300.50	365
3534	32″ diameter		12	2		281	91.50		372.50	450
3535	36″ diameter		10	2.400		350	110		460	550
3536	42″ diameter		9	2.667		620	122		742	865
3537	48″ diameter		8	3		745	137		882	1,025
3570	Bellows, lined, 316 stainless steel only									
3574	6″ diameter	Q-9	30	.533	Ea.	440	23.50		463.50	515
3575	8″ diameter		26	.615		475	27		502	565
3576	10″ diameter		24	.667		525	29.50		554.50	620
3577	12″ diameter		22	.727		580	32		612	690
3578	14″ diameter		21	.762		655	33.50		688.50	770
3579	16″ diameter		20	.800		725	35.50		760.50	850
3580	18″ diameter		19	.842		805	37		842	945
3581	20″ diameter	Q-10	18	1.333		910	61		971	1,100
3582	24″ diameter		16	1.500		1,150	69		1,219	1,350
3583	28″ diameter		15	1.600		1,275	73.50		1,348.50	1,500
3584	32″ diameter		14	1.714		1,475	78.50		1,553.50	1,750
3585	36″ diameter		12	2		1,700	91.50		1,791.50	2,025
3586	42″ diameter		11	2.182		2,525	100		2,625	2,950
3587	48″ diameter		10	2.400		2,625	110		2,735	3,075
3600	Ventilated roof thimble, 304 stainless steel									
3620	6″ diameter	Q-9	26	.615	Ea.	225	27		252	288
3624	8″ diameter		22	.727		241	32		273	315
3625	10″ diameter		20	.800		260	35.50		295.50	340
3626	12″ diameter		18	.889		281	39.50		320.50	370
3627	14″ diameter		17	.941		300	41.50		341.50	395
3628	16″ diameter		16	1		320	44		364	420
3629	18″ diameter		15	1.067		345	47		392	450
3630	20″ diameter	Q-10	16	1.500		370	69		439	515
3631	24″ diameter		14	1.714		420	78.50		498.50	580
3632	28″ diameter		13	1.846		480	84.50		564.50	660
3633	32″ diameter		12	2		540	91.50		631.50	735
3634	36″ diameter		10	2.400		620	110		730	845
3635	42″ diameter		9	2.667		820	122		942	1,100
3636	48″ diameter		8	3		1,025	137		1,162	1,350
3650	For 316 stainless steel, add					10%				
3670	Exit cone, 316 stainless steel only									
3674	6″ diameter	Q-9	46	.348	Ea.	178	15.35		193.35	220
3675	8″ diameter		42	.381		186	16.85		202.85	231
3676	10″ diameter		40	.400		202	17.70		219.70	249
3677	12″ diameter		38	.421		211	18.60		229.60	260

23 51 26 – All-Fuel Vent Chimneys

	23 51 26.10 All-Fuel Vent Chimneys, Press. Tight, Dbl. Wall	Crew	Daily Output	Labor-Hours	Unit	Material	2010 Bare Costs Labor	Equipment	Total	Total Incl O&P
3678	14″ diameter	Q-9	37	.432	Ea.	234	19.10		253.10	287
3679	16″ diameter		36	.444		258	19.65		277.65	315
3680	18″ diameter		35	.457		315	20		335	380
3681	20″ diameter	Q-10	28	.857		380	39.50		419.50	475
3682	24″ diameter		26	.923		485	42.50		527.50	600
3683	28″ diameter		25	.960		590	44		634	715
3684	32″ diameter		24	1		735	46		781	875
3685	36″ diameter		22	1.091		880	50		930	1,050
3686	42″ diameter		21	1.143		1,125	52.50		1,177.50	1,325
3687	48″ diameter		20	1.200		1,450	55		1,505	1,675
3720	Roof support assembly, incl. 30″ pipe sect, 304 stainless steel									
3724	6″ diameter	Q-9	25	.640	Ea.	259	28.50		287.50	330
3725	8″ diameter		21	.762		282	33.50		315.50	360
3726	10″ diameter		19	.842		310	37		347	395
3727	12″ diameter		17	.941		330	41.50		371.50	425
3728	14″ diameter		16	1		370	44		414	470
3729	16″ diameter		15	1.067		390	47		437	500
3730	18″ diameter		14	1.143		420	50.50		470.50	540
3731	20″ diameter	Q-10	15	1.600		450	73.50		523.50	605
3732	24″ diameter		13	1.846		500	84.50		584.50	680
3733	28″ diameter		12	2		590	91.50		681.50	785
3734	32″ diameter		11	2.182		705	100		805	925
3735	36″ diameter		9	2.667		855	122		977	1,125
3736	42″ diameter		8	3		1,350	137		1,487	1,675
3737	48″ diameter		7	3.429		1,650	157		1,807	2,075
3750	For 316 stainless steel, add					20%				
3770	Stack cap, 316 stainless steel only									
3773	5″ diameter	Q-9	48	.333	Ea.	169	14.75		183.75	209
3774	6″ diameter		46	.348		198	15.35		213.35	242
3775	8″ diameter		42	.381		229	16.85		245.85	277
3776	10″ diameter		40	.400		270	17.70		287.70	325
3777	12″ diameter		38	.421		315	18.60		333.60	375
3778	14″ diameter		37	.432		370	19.10		389.10	435
3779	16″ diameter		36	.444		420	19.65		439.65	495
3780	18″ diameter		35	.457		475	20		495	550
3781	20″ diameter	Q-10	28	.857		540	39.50		579.50	655
3782	24″ diameter		26	.923		655	42.50		697.50	785
3783	28″ diameter		25	.960		710	44		754	850
3784	32″ diameter		24	1		820	46		866	970
3785	36″ diameter		22	1.091		1,000	50		1,050	1,175
3786	42″ diameter		21	1.143		1,225	52.50		1,277.50	1,425
3787	48″ diameter		20	1.200		1,450	55		1,505	1,675
3800	Round storm collar									
3806	5″ diameter	Q-9	32	.500	Ea.	26.50	22		48.50	62.50

	23 51 26.30 All-Fuel Vent Chimneys, Double Wall, St. Stl.	Crew	Daily Output	Labor-Hours	Unit	Material	Labor	Equipment	Total	Total Incl O&P
0010	**ALL-FUEL VENT CHIMNEYS, DOUBLE WALL, STAINLESS STEEL**									
7800	All fuel, double wall, stainless steel, 6″ diameter	Q-9	60	.267	V.L.F.	56.50	11.80		68.30	80
7802	7″ diameter		56	.286		73	12.65		85.65	99.50
7804	8″ diameter		52	.308		86.50	13.60		100.10	116
7806	10″ diameter		48	.333		125	14.75		139.75	160
7808	12″ diameter		44	.364		167	16.05		183.05	209
7810	14″ diameter		42	.381		219	16.85		235.85	267

23 51 26 – All-Fuel Vent Chimneys

23 51 26.30 All-Fuel Vent Chimneys, Double Wall, St. Stl.		Crew	Daily Output	Labor-Hours	Unit	2010 Bare Costs Material	Labor	Equipment	Total	Total Incl O&P
8000	All fuel, double wall, stainless steel fittings									
8010	Roof support 6″ diameter	Q-9	30	.533	Ea.	103	23.50		126.50	150
8020	7″ diameter		28	.571		117	25.50		142.50	167
8030	8″ diameter		26	.615		130	27		157	184
8040	10″ diameter		24	.667		150	29.50		179.50	210
8050	12″ diameter		22	.727		181	32		213	248
8060	14″ diameter		21	.762		229	33.50		262.50	305
8100	Elbow 15°, 6″ diameter		30	.533		165	23.50		188.50	218
8120	7″ diameter		28	.571		186	25.50		211.50	244
8140	8″ diameter		26	.615		212	27		239	274
8160	10″ diameter		24	.667		278	29.50		307.50	350
8180	12″ diameter		22	.727		335	32		367	420
8200	14″ diameter		21	.762		405	33.50		438.50	495
8300	Insulated tee with insulated tee cap, 6″ diameter		30	.533		156	23.50		179.50	208
8340	7″ diameter		28	.571		205	25.50		230.50	264
8360	8″ diameter		26	.615		231	27		258	295
8380	10″ diameter		24	.667		325	29.50		354.50	405
8400	12″ diameter		22	.727		455	32		487	550
8420	14″ diameter		21	.762		600	33.50		633.50	710
8500	Joist shield, 6″ diameter		30	.533		47.50	23.50		71	88
8510	7″ diameter		28	.571		51.50	25.50		77	95.50
8520	8″ diameter		26	.615		63.50	27		90.50	111
8530	10″ diameter		24	.667		86	29.50		115.50	139
8540	12″ diameter		22	.727		107	32		139	167
8550	14″ diameter		21	.762		133	33.50		166.50	197
8600	Round top, 6″ diameter		30	.533		53	23.50		76.50	93.50
8620	7″ diameter		28	.571		72	25.50		97.50	118
8640	8″ diameter		26	.615		97	27		124	148
8660	10″ diameter		24	.667		180	29.50		209.50	243
8680	12″ diameter		22	.727		249	32		281	320
8700	14″ diameter		21	.762		330	33.50		363.50	415
8800	Adjustable roof flashing, 6″ diameter		30	.533		63	23.50		86.50	105
8820	7″ diameter		28	.571		72	25.50		97.50	118
8840	8″ diameter		26	.615		78	27		105	127
8860	10″ diameter		24	.667		100	29.50		129.50	155
8880	12″ diameter		22	.727		130	32		162	191
8900	14″ diameter		21	.762		162	33.50		195.50	229

23 51 33 – Insulated Sectional Chimneys

23 51 33.10 Prefabricated Insulated Sectional Chimneys		Crew	Daily Output	Labor-Hours	Unit	Material	Labor	Equipment	Total	Total Incl O&P
0010	**PREFABRICATED INSULATED SECTIONAL CHIMNEYS**									
9000	High temp. (2000°F), steel jacket, acid resistant refractory lining									
9010	11 ga. galvanized jacket, U.L. listed									
9020	Straight section, 48″ long, 10″ diameter	Q-10	13.30	1.805	Ea.	400	82.50		482.50	565
9030	12″ diameter		11.20	2.143		440	98		538	630
9040	18″ diameter		7.40	3.243		615	149		764	900
9050	24″ diameter		4.60	5.217		840	239		1,079	1,275
9060	30″ diameter		3.70	6.486		1,075	297		1,372	1,625
9070	36″ diameter		2.70	8.889		1,325	405		1,730	2,075
9080	42″ diameter	Q-11	3.10	10.323		1,625	485		2,110	2,525
9090	48″ diameter		2.70	11.852		1,925	555		2,480	2,950
9100	54″ diameter		2.20	14.545		2,325	680		3,005	3,575
9110	60″ diameter		2	16		3,050	750		3,800	4,475

23 51 33.10 Prefabricated Insulated Sectional Chimneys		Crew	Daily Output	Labor-Hours	Unit	Material	Labor	Equipment	Total	Total Incl O&P
						2010 Bare Costs				
9120	Tee section, 10" diameter	Q-10	4.40	5.455	Ea.	855	250		1,105	1,325
9130	12" diameter		3.70	6.486		900	297		1,197	1,450
9140	18" diameter		2.40	10		1,225	460		1,685	2,050
9150	24" diameter		1.50	16		1,675	735		2,410	2,950
9160	30" diameter		1.20	20		2,350	915		3,265	4,000
9170	36" diameter		.80	30		3,975	1,375		5,350	6,450
9180	42" diameter	Q-11	1	32		5,225	1,500		6,725	8,025
9190	48" diameter		.90	35.556		6,775	1,675		8,450	9,975
9200	54" diameter		.70	45.714		9,900	2,150		12,050	14,200
9210	60" diameter		.60	53.333		12,500	2,500		15,000	17,600
9220	Cleanout pier section, 10" diameter	Q-10	3.50	6.857		655	315		970	1,200
9230	12" diameter		2.50	9.600		690	440		1,130	1,425
9240	18" diameter		1.90	12.632		885	580		1,465	1,850
9250	24" diameter		1.30	18.462		1,125	845		1,970	2,500
9260	30" diameter		1	24		1,375	1,100		2,475	3,175
9270	36" diameter		.75	32		1,625	1,475		3,100	4,000
9280	42" diameter	Q-11	.90	35.556		2,050	1,675		3,725	4,800
9290	48" diameter		.75	42.667		2,475	2,000		4,475	5,750
9300	54" diameter		.60	53.333		2,950	2,500		5,450	7,025
9310	60" diameter		.50	64		3,775	3,000		6,775	8,675
9320	For drain, add					53%				
9330	Elbow, 30° and 45°, 10" diameter	Q-10	6.60	3.636		555	167		722	860
9340	12" diameter		5.60	4.286		600	196		796	950
9350	18" diameter		3.70	6.486		900	297		1,197	1,450
9360	24" diameter		2.30	10.435		1,350	480		1,830	2,200
9370	30" diameter		1.85	12.973		1,850	595		2,445	2,925
9380	36" diameter		1.30	18.462		2,225	845		3,070	3,725
9390	42" diameter	Q-11	1.60	20		3,050	935		3,985	4,800
9400	48" diameter		1.35	23.704		3,900	1,100		5,000	5,950
9410	54" diameter		1.10	29.091		5,050	1,350		6,400	7,600
9420	60" diameter		1	32		6,150	1,500		7,650	9,050
9430	For 60° and 90° elbow, add					112%				
9440	End cap, 10" diameter	Q-10	26	.923		955	42.50		997.50	1,125
9450	12" diameter		22	1.091		1,100	50		1,150	1,275
9460	18" diameter		15	1.600		1,350	73.50		1,423.50	1,575
9470	24" diameter		9	2.667		1,650	122		1,772	2,000
9480	30" diameter		7.30	3.288		2,050	151		2,201	2,475
9490	36" diameter		5	4.800		2,350	220		2,570	2,925
9500	42" diameter	Q-11	6.30	5.079		2,950	238		3,188	3,575
9510	48" diameter		5.30	6.038		3,475	282		3,757	4,225
9520	54" diameter		4.40	7.273		3,975	340		4,315	4,875
9530	60" diameter		3.90	8.205		4,675	385		5,060	5,725
9540	Increaser (1 diameter), 10" diameter	Q-10	6.60	3.636		475	167		642	770
9550	12" diameter		5.60	4.286		505	196		701	850
9560	18" diameter		3.70	6.486		740	297		1,037	1,275
9570	24" diameter		2.30	10.435		1,050	480		1,530	1,900
9580	30" diameter		1.85	12.973		1,325	595		1,920	2,350
9590	36" diameter		1.30	18.462		1,450	845		2,295	2,875
9592	42" diameter	Q-11	1.60	20		1,600	935		2,535	3,175
9594	48" diameter		1.35	23.704		1,925	1,100		3,025	3,775
9596	54" diameter		1.10	29.091		2,325	1,350		3,675	4,600
9600	For expansion joints, add to straight section					7%				
9610	For 1/4" hot rolled steel jacket, add					157%				

23 51 Breechings, Chimneys, and Stacks

23 51 33 – Insulated Sectional Chimneys

23 51 33.10	Prefabricated Insulated Sectional Chimneys	Crew	Daily Output	Labor-Hours	Unit	2010 Bare Costs Material	Labor	Equipment	Total	Total Incl O&P
9620	For 2950°F very high temperature, add				Ea.	89%				
9630	26 ga. aluminized jacket, straight section, 48″ long									
9640	10″ diameter	Q-10	15.30	1.569	V.L.F.	237	72		309	370
9650	12″ diameter		12.90	1.860		252	85.50		337.50	405
9660	18″ diameter		8.50	2.824		365	129		494	600
9670	24″ diameter		5.30	4.528		520	208		728	885
9680	30″ diameter		4.30	5.581		665	256		921	1,125
9690	36″ diameter		3.10	7.742		845	355		1,200	1,450
9700	Accessories (all models)									
9710	Guy band, 10″ diameter	Q-10	32	.750	Ea.	74	34.50		108.50	133
9720	12″ diameter		30	.800		78	36.50		114.50	141
9730	18″ diameter		26	.923		102	42.50		144.50	176
9740	24″ diameter		24	1		149	46		195	233
9750	30″ diameter		20	1.200		173	55		228	274
9760	36″ diameter		18	1.333		191	61		252	305
9770	42″ diameter	Q-11	22	1.455		345	68		413	485
9780	48″ diameter		20	1.600		380	75		455	530
9790	54″ diameter		16	2		420	93.50		513.50	600
9800	60″ diameter		12	2.667		455	125		580	690
9810	Draw band, galv. stl., 11 gauge, 10″ diameter	Q-10	32	.750		65	34.50		99.50	124
9820	12″ diameter		30	.800		69	36.50		105.50	131
9830	18″ diameter		26	.923		86	42.50		128.50	159
9840	24″ diameter		24	1		110	46		156	191
9850	30″ diameter		20	1.200		127	55		182	223
9860	36″ diameter		18	1.333		146	61		207	253
9870	42″ diameter	Q-11	22	1.455		196	68		264	320
9880	48″ diameter		20	1.600		251	75		326	390
9890	54″ diameter		16	2		291	93.50		384.50	460
9900	60″ diameter		12	2.667		370	125		495	595
9910	Draw band, aluminized stl., 26 gauge, 10″ diameter	Q-10	32	.750		18.95	34.50		53.45	73
9920	12″ diameter		30	.800		18.95	36.50		55.45	76.50
9930	18″ diameter		26	.923		24	42.50		66.50	90.50
9940	24″ diameter		24	1		28	46		74	100
9950	30″ diameter		20	1.200		32	55		87	119
9960	36″ diameter		18	1.333		43	61		104	140
9962	42″ diameter	Q-11	22	1.455		43	68		111	150
9964	48″ diameter		20	1.600		51	75		126	169
9966	54″ diameter		16	2		51	93.50		144.50	198
9968	60″ diameter		12	2.667		56	125		181	251

23 52 Heating Boilers

23 52 13 – Electric Boilers

23 52 13.10 Electric Boilers, ASME

			Crew	Daily Output	Labor-Hours	Unit	Material	Labor	Equipment	Total	Total Incl O&P
0010	**ELECTRIC BOILERS, ASME**, Standard controls and trim.	R235000-30									
1000	Steam, 6 KW, 20.5 MBH		Q-19	1.20	20	Ea.	3,500	950		4,450	5,275
1040	9 KW, 30.7 MBH	D3020–102		1.20	20		3,650	950		4,600	5,425
1060	18 KW, 61.4 MBH			1.20	20		3,750	950		4,700	5,550
1080	24 KW, 81.8 MBH	D3020–104		1.10	21.818		4,350	1,025		5,375	6,325
1120	36 KW, 123 MBH			1.10	21.818		4,825	1,025		5,850	6,850
1160	60 KW, 205 MBH			1	24		6,025	1,150		7,175	8,325
1220	112 KW, 382 MBH			.75	32		8,250	1,525		9,775	11,400

23 52 13 – Electric Boilers

23 52 13.10 Electric Boilers, ASME		Crew	Daily Output	Labor-Hours	Unit	2010 Bare Costs Material	Labor	Equipment	Total	Total Incl O&P
1240	148 KW, 505 MBH	Q-19	.65	36.923	Ea.	8,700	1,750		10,450	12,200
1260	168 KW, 573 MBH		.60	40		11,500	1,900		13,400	15,500
1280	222 KW, 758 MBH		.55	43.636		13,100	2,075		15,175	17,500
1300	296 KW, 1010 MBH		.45	53.333		13,200	2,525		15,725	18,300
1320	300 KW, 1023 MBH		.40	60		17,000	2,850		19,850	23,000
1340	370 KW, 1263 MBH		.35	68.571		18,300	3,250		21,550	25,000
1360	444 KW, 1515 MBH		.30	80		19,600	3,800		23,400	27,200
1380	518 KW, 1768 MBH	Q-21	.36	88.889		22,200	4,325		26,525	30,900
1400	592 KW, 2020 MBH		.34	94.118		25,100	4,575		29,675	34,500
1420	666 KW, 2273 MBH		.32	100		26,400	4,850		31,250	36,300
1460	740 KW, 2526 MBH		.28	114		27,700	5,550		33,250	38,700
1480	814 KW, 2778 MBH		.25	128		31,000	6,225		37,225	43,400
1500	962 KW, 3283 MBH		.22	145		31,800	7,075		38,875	45,600
1520	1036 KW, 3536 MBH		.20	160		33,400	7,775		41,175	48,300
1540	1110 KW, 3788 MBH		.19	168		35,100	8,175		43,275	51,000
1560	2070 KW, 7063 MBH		.18	177		49,100	8,625		57,725	67,000
1580	2250 KW, 7677 MBH		.17	188		57,500	9,150		66,650	76,500
1600	2,340 KW, 7984 MBH		.16	200		63,500	9,725		73,225	84,500
2000	Hot water, 7.5 KW, 25.6 MBH	Q-19	1.30	18.462		3,750	875		4,625	5,425
2020	15 KW, 51.2 MBH		1.30	18.462		3,850	875		4,725	5,550
2040	30 KW, 102 MBH		1.20	20		3,975	950		4,925	5,800
2060	45 KW, 164 MBH		1.20	20		4,475	950		5,425	6,350
2070	60 KW, 205 MBH		1.20	20		4,725	950		5,675	6,625
2080	75 KW, 256 MBH		1.10	21.818		5,075	1,025		6,100	7,150
2100	90 KW, 307 MBH		1.10	21.818		5,475	1,025		6,500	7,575
2120	105 KW, 358 MBH		1	24		5,850	1,150		7,000	8,150
2140	120 KW, 410 MBH		.90	26.667		6,150	1,275		7,425	8,650
2160	135 KW, 461 MBH		.75	32		6,675	1,525		8,200	9,600
2180	150 KW, 512 MBH		.65	36.923		6,925	1,750		8,675	10,200
2200	165 KW, 563 MBH		.60	40		7,175	1,900		9,075	10,800
2220	296 KW, 1010 MBH		.55	43.636		11,600	2,075		13,675	15,800
2280	370 KW, 1263 MBH		.40	60		12,800	2,850		15,650	18,400
2300	444 KW, 1515 MBH		.35	68.571		15,600	3,250		18,850	22,000
2340	518 KW, 1768 MBH	Q-21	.44	72.727		17,200	3,525		20,725	24,200
2360	592 KW, 2020 MBH		.43	74.419		19,000	3,625		22,625	26,300
2400	666 KW, 2273 MBH		.40	80		20,300	3,875		24,175	28,200
2420	740 KW, 2526 MBH		.39	82.051		20,800	3,975		24,775	28,900
2440	814 KW, 2778 MBH		.38	84.211		21,500	4,100		25,600	29,700
2460	888 KW, 3031 MBH		.37	86.486		22,600	4,200		26,800	31,200
2480	962 KW, 3283 MBH		.36	88.889		25,300	4,325		29,625	34,400
2500	1036 KW, 3536 MBH		.34	94.118		26,900	4,575		31,475	36,500
2520	1110 KW, 3788 MBH		.33	96.970		29,800	4,700		34,500	39,800
2540	1440 KW, 4915 MBH		.32	100		34,200	4,850		39,050	44,900
2560	1560 KW, 5323 MBH		.31	103		38,400	5,025		43,425	49,700
2580	1680 KW, 5733 MBH		.30	106		41,000	5,175		46,175	53,000
2600	1800 KW, 6143 MBH		.29	110		43,200	5,350		48,550	55,500
2620	1980 KW, 6757 MBH		.28	114		46,100	5,550		51,650	59,000
2640	2100 KW, 7167 MBH		.27	118		51,500	5,750		57,250	65,000
2660	2220 KW, 7576 MBH		.26	123		53,000	5,975		58,975	67,500
2680	2400 KW, 8191 MBH		.25	128		56,000	6,225		62,225	71,000
2700	2610 KW, 8905 MBH		.24	133		60,500	6,475		66,975	76,000
2720	2790 KW, 9519 MBH		.23	139		63,000	6,750		69,750	79,500
2740	2970 KW, 10133 MBH		.21	152		64,500	7,400		71,900	82,000

23 52 Heating Boilers

23 52 13 – Electric Boilers

23 52 13.10 Electric Boilers, ASME		Crew	Daily Output	Labor-Hours	Unit	2010 Bare Costs Material	Labor	Equipment	Total	Total Incl O&P
2760	3150 KW, 10748 MBH	Q-21	.19	168	Ea.	70,500	8,175		78,675	89,500
2780	3240 KW, 11055 MBH		.18	177		72,000	8,625		80,625	92,000
2800	3420 KW, 11669 MBH		.17	188		76,500	9,150		85,650	97,500
2820	3600 KW, 12,283 MBH		.16	200		79,000	9,725		88,725	101,500

23 52 16 – Condensing Boilers

23 52 16.24 Condensing Boilers		Crew	Daily Output	Labor-Hours	Unit	Material	Labor	Equipment	Total	Total Incl O&P
0010	**CONDENSING BOILERS**, Cast iron									
0020	Packaged with standard controls, circulator and trim									
0030	Intermittent (spark) pilot, natural or LP gas									
0040	Hot water, DOE MBH output, (AFUE)									
0100	42 MBH, (84.0%) G	Q-5	1.80	8.889	Ea.	1,375	415		1,790	2,125
0120	57 MBH, (84.3%) G		1.60	10		1,525	465		1,990	2,375
0140	85 MBH, (84.0%) G		1.40	11.429		1,675	535		2,210	2,650
0160	112 MBH, (83.7%) G		1.20	13.333		1,900	625		2,525	3,000
0180	140 MBH, (83.3%) G	Q-6	1.60	15		2,175	725		2,900	3,500
0200	167 MBH, (83.0%) G		1.40	17.143		2,525	830		3,355	4,025
0220	194 MBH, (82.7%) G		1.20	20		2,675	970		3,645	4,400

23 52 19 – Pulse Combustion Boilers

23 52 19.20 Pulse Type Combustion Boilers		Crew	Daily Output	Labor-Hours	Unit	Material	Labor	Equipment	Total	Total Incl O&P
0010	**PULSE TYPE COMBUSTION BOILERS**									
7990	Special feature gas fired boilers									
8000	Pulse combustion, standard controls/trim									
8050	88,000 BTU G	Q-5	1.40	11.429	Ea.	4,775	535		5,310	6,050
8080	134,000 BTU G	"	1.20	13.333	"	5,725	625		6,350	7,200

23 52 23 – Cast-Iron Boilers

23 52 23.20 Gas-Fired Boilers		Crew	Daily Output	Labor-Hours	Unit	Material	Labor	Equipment	Total	Total Incl O&P
0010	**GAS-FIRED BOILERS,** Natural or propane, standard controls, packaged R235000-10									
1000	Cast iron, with insulated jacket									
2000	Steam, gross output, 81 MBH R235000-20	Q-7	1.40	22.857	Ea.	1,950	1,125		3,075	3,850
2020	102 MBH		1.30	24.615		2,200	1,225		3,425	4,250
2040	122 MBH R235000-30		1	32		2,375	1,575		3,950	4,975
2060	163 MBH		.90	35.556		2,825	1,750		4,575	5,750
2080	203 MBH R235000-50		.90	35.556		3,100	1,750		4,850	6,050
2100	240 MBH		.85	37.647		3,100	1,850		4,950	6,225
2120	280 MBH R235000-70		.80	40		4,200	1,975		6,175	7,575
2140	320 MBH		.70	45.714		4,200	2,250		6,450	8,000
2160	360 MBH R235000-80		.63	51.200		4,925	2,525		7,450	9,200
2180	400 MBH		.56	56.838		5,200	2,800		8,000	9,925
2200	440 MBH		.51	62.500		5,525	3,100		8,625	10,700
2220	544 MBH		.45	71.588		8,475	3,550		12,025	14,600
2240	765 MBH		.43	74.419		10,200	3,675		13,875	16,800
2260	892 MBH		.38	84.211		11,900	4,175		16,075	19,300
2280	1275 MBH		.34	94.118		14,500	4,650		19,150	22,900
2300	1530 MBH		.32	100		15,500	4,950		20,450	24,500
2320	1,875 MBH		.30	106		18,300	5,275		23,575	28,000
2340	2170 MBH		.26	122		19,600	6,025		25,625	30,700
2360	2675 MBH		.20	163		22,500	8,075		30,575	36,900
2380	3060 MBH		.19	172		23,900	8,500		32,400	39,000
2400	3570 MBH		.18	181		26,500	8,975		35,475	42,700
2420	4207 MBH		.16	205		29,000	10,100		39,100	47,100
2440	4,720 MBH		.15	207		30,700	10,300		41,000	49,200

23 52 23.20 Gas-Fired Boilers		Crew	Daily Output	Labor-Hours	Unit	Material	Labor	Equipment	Total	Total Incl O&P
						2010 Bare Costs				
2460	5660 MBH	Q-7	.15	220	Ea.	67,500	10,900		78,400	90,500
2480	6,100 MBH		.13	246		77,000	12,200		89,200	102,500
2500	6390 MBH		.12	266		81,000	13,200		94,200	109,500
2520	6680 MBH		.11	290		82,500	14,400		96,900	112,000
2540	6,970 MBH		.10	320		86,500	15,800		102,300	118,500
3000	Hot water, gross output, 80 MBH		1.46	21.918		1,775	1,075		2,850	3,600
3020	100 MBH		1.35	23.704		2,300	1,175		3,475	4,275
3040	122 MBH		1.10	29.091		2,200	1,450		3,650	4,575
3060	163 MBH		1	32		2,800	1,575		4,375	5,475
3080	203 MBH		1	32		2,950	1,575		4,525	5,625
3100	240 MBH		.95	33.684		3,100	1,675		4,775	5,925
3120	280 MBH		.90	35.556		4,150	1,750		5,900	7,175
3140	320 MBH		.80	40		4,200	1,975		6,175	7,600
3160	360 MBH		.71	45.070		4,700	2,225		6,925	8,500
3180	400 MBH		.64	50		4,975	2,475		7,450	9,175
3200	440 MBH		.58	54.983		5,350	2,725		8,075	9,950
3220	544 MBH		.51	62.992		8,200	3,125		11,325	13,700
3240	765 MBH		.46	70.022		9,950	3,450		13,400	16,100
3260	1,088 MBH		.40	80		12,400	3,950		16,350	19,500
3280	1,275 MBH		.36	89.888		14,200	4,450		18,650	22,300
3300	1,530 MBH		.31	104		15,300	5,175		20,475	24,600
3320	2,000 MBH		.26	125		17,900	6,175		24,075	29,000
3340	2,312 MBH		.22	148		20,500	7,325		27,825	33,500
3360	2,856 MBH		.20	160		21,500	7,900		29,400	35,600
3380	3,264 MBH		.18	179		22,900	8,875		31,775	38,500
3400	3,808 MBH		.16	195		25,200	9,650		34,850	42,300
3420	4,488 MBH		.15	210		27,900	10,400		38,300	46,300
3440	4,720 MBH		.15	220		65,500	10,900		76,400	88,500
3460	5,520 MBH		.14	228		84,000	11,300		95,300	109,500
3480	6,100 MBH		.13	250		89,000	12,400		101,400	116,000
3500	6,390 MBH		.11	285		91,000	14,100		105,100	121,000
3520	6,680 MBH		.10	310		91,000	15,400		106,400	123,000
3540	6,970 MBH	▼	.09	359		94,000	17,800		111,800	130,000
7000	For tankless water heater, add					10%				
7050	For additional zone valves up to 312 MBH add				▼	150			150	165

23 52 23.30 Gas/Oil Fired Boilers		Crew	Daily Output	Labor-Hours	Unit	Material	Labor	Equipment	Total	Total Incl O&P
0010	**GAS/OIL FIRED BOILERS,** Combination with burners and controls, packaged									
1000	Cast iron with insulated jacket									
2000	Steam, gross output, 720 MBH	Q-7	.43	74.074	Ea.	13,900	3,650		17,550	20,800
2020	810 MBH		.38	83.990		13,900	4,150		18,050	21,500
2040	1,084 MBH		.34	93.023		15,200	4,600		19,800	23,700
2060	1,360 MBH		.33	98.160		17,000	4,850		21,850	26,000
2080	1,600 MBH		.30	107		17,000	5,300		22,300	26,600
2100	2,040 MBH		.25	130		23,900	6,425		30,325	36,000
2120	2,450 MBH		.21	156		26,200	7,725		33,925	40,400
2140	2,700 MBH		.19	165		27,500	8,200		35,700	42,600
2160	3,000 MBH		.18	175		29,000	8,700		37,700	44,900
2180	3,270 MBH		.17	183		30,400	9,100		39,500	47,100
2200	3,770 MBH		.17	191		62,500	9,475		71,975	83,000
2220	4,070 MBH		.16	200		66,000	9,875		75,875	87,500
2240	4,650 MBH		.15	210		69,500	10,400		79,900	92,000
2260	5,230 MBH	▼	.14	223	▼	77,000	11,100		88,100	101,000

23 52 23 – Cast-Iron Boilers

23 52 23.30 Gas/Oil Fired Boilers		Crew	Daily Output	Labor-Hours	Unit	Material	2010 Bare Costs Labor	Equipment	Total	Total Incl O&P
2280	5,520 MBH	Q-7	.14	235	Ea.	84,000	11,600		95,600	109,500
2300	5,810 MBH		.13	248		84,500	12,300		96,800	111,500
2320	6,100 MBH		.12	260		85,500	12,900		98,400	113,500
2340	6,390 MBH		.11	296		90,000	14,600		104,600	120,500
2360	6,680 MBH		.10	320		93,000	15,800		108,800	125,500
2380	6,970 MBH		.09	372		95,000	18,400		113,400	132,000
2900	Hot water, gross output									
2910	200 MBH	Q-6	.62	39.024	Ea.	7,775	1,900		9,675	11,400
2920	300 MBH		.49	49.080		7,775	2,375		10,150	12,100
2930	400 MBH		.41	57.971		9,100	2,800		11,900	14,200
2940	500 MBH		.36	67.039		9,800	3,250		13,050	15,700
3000	584 MBH	Q-7	.44	72.072		8,125	3,575		11,700	14,300
3020	876 MBH		.41	79.012		14,300	3,900		18,200	21,600
3040	1,168 MBH		.31	103		20,200	5,125		25,325	30,000
3060	1,460 MBH		.28	113		22,900	5,600		28,500	33,600
3080	2,044 MBH		.26	122		28,300	6,025		34,325	40,200
3100	2,628 MBH		.21	150		29,000	7,425		36,425	43,000
3120	3,210 MBH		.18	174		35,700	8,650		44,350	52,500
3140	3,796 MBH		.17	186		40,000	9,200		49,200	58,000
3160	4,088 MBH		.16	195		45,500	9,650		55,150	64,500
3180	4,672 MBH		.16	203		49,600	10,100		59,700	69,500
3200	5,256 MBH		.15	217		54,500	10,800		65,300	76,000
3220	6,000 MBH, 179 BHP		.13	256		79,500	12,700		92,200	106,500
3240	7,130 MBH, 213 BHP		.08	385		83,500	19,100		102,600	120,500
3260	9,800 MBH, 286 BHP		.06	533		96,000	26,400		122,400	145,000
3280	10,900 MBH, 325.6 BHP		.05	592		116,500	29,300		145,800	172,000
3290	12,200 MBH, 364.5 BHP		.05	666		127,000	33,000		160,000	189,000
3300	13,500 MBH, 403.3 BHP		.04	727		140,500	35,900		176,400	208,500

23 52 23.40 Oil-Fired Boilers		Crew	Daily Output	Labor-Hours	Unit	Material	Labor	Equipment	Total	Total Incl O&P
0010	**OIL-FIRED BOILERS,** Standard controls, flame retention burner, packaged									
1000	Cast iron, with insulated flush jacket									
2000	Steam, gross output, 109 MBH	Q-7	1.20	26.667	Ea.	1,975	1,325		3,300	4,150
2020	144 MBH		1.10	29.091		2,225	1,450		3,675	4,600
2040	173 MBH		1	32		2,500	1,575		4,075	5,125
2060	207 MBH		.90	35.556		2,700	1,750		4,450	5,600
2080	236 MBH		.85	37.647		3,225	1,850		5,075	6,325
2100	300 MBH		.70	45.714		3,825	2,250		6,075	7,600
2120	480 MBH		.50	64		4,925	3,175		8,100	10,200
2140	665 MBH		.45	71.111		6,700	3,525		10,225	12,600
2160	794 MBH		.41	78.049		7,925	3,850		11,775	14,500
2180	1,084 MBH		.38	85.106		9,400	4,200		13,600	16,600
2200	1,360 MBH		.33	98.160		10,800	4,850		15,650	19,200
2220	1,600 MBH		.26	122		12,100	6,025		18,125	22,400
2240	2,175 MBH		.24	133		17,300	6,625		23,925	29,000
2260	2,480 MBH		.21	156		18,600	7,725		26,325	32,000
2280	3,000 MBH		.19	170		21,600	8,425		30,025	36,300
2300	3,550 MBH		.17	187		23,900	9,250		33,150	40,200
2320	3,820 MBH		.16	200		34,000	9,875		43,875	52,000
2340	4,360 MBH		.15	214		37,300	10,600		47,900	57,000
2360	4,940 MBH		.14	225		58,500	11,100		69,600	80,500
2380	5,520 MBH		.14	235		68,500	11,600		80,100	93,000
2400	6,100 MBH		.13	256		78,500	12,700		91,200	105,500

23 52 23 – Cast-Iron Boilers

23 52 23.40 Oil-Fired Boilers		Crew	Daily Output	Labor-Hours	Unit	Material	2010 Bare Costs Labor	Equipment	Total	Total Incl O&P
2420	6,390 MBH	Q-7	.11	290	Ea.	83,000	14,400		97,400	112,500
2440	6,680 MBH		.10	313		85,000	15,500		100,500	116,500
2460	6,970 MBH		.09	363		88,000	18,000		106,000	124,000
3000	Hot water, same price as steam									
4000	For tankless coil in smaller sizes, add				Ea.	15%				

23 52 23.60 Solid-Fuel Boilers		Crew	Daily Output	Labor-Hours	Unit	Material	Labor	Equipment	Total	Total Incl O&P
0010	**SOLID-FUEL BOILERS**									
3000	Stoker fired (coal) cast iron with flush jacket and									
3400	insulation, steam or water, gross output, 1280 MBH	Q-6	.36	66.667	Ea.	33,800	3,225		37,025	42,100
3420	1460 MBH		.30	80		37,100	3,875		40,975	46,600
3440	1640 MBH		.28	85.714		39,600	4,150		43,750	49,700
3460	1820 MBH		.26	92.308		42,400	4,475		46,875	53,500
3480	2000 MBH		.25	96		45,000	4,650		49,650	56,500
3500	2360 MBH		.23	104		51,000	5,050		56,050	63,500
3540	2725 MBH		.20	120		56,500	5,800		62,300	71,000
3800	2950 MBH	Q-7	.16	200		94,500	9,875		104,375	119,000
3820	3210 MBH		.15	213		99,500	10,500		110,000	125,500
3840	3480 MBH		.14	228		104,000	11,300		115,300	131,500
3860	3745 MBH		.14	228		109,000	11,300		120,300	137,000
3880	4000 MBH		.13	246		113,500	12,200		125,700	143,000
3900	4200 MBH		.13	246		118,500	12,200		130,700	148,500
3920	4400 MBH		.12	266		123,000	13,200		136,200	155,000
3940	4600 MBH		.12	266		127,500	13,200		140,700	160,500

23 52 26 – Steel Boilers

23 52 26.40 Oil-Fired Boilers		Crew	Daily Output	Labor-Hours	Unit	Material	Labor	Equipment	Total	Total Incl O&P
0010	**OIL-FIRED BOILERS**, Standard controls, flame retention burner									
5000	Steel, with insulated flush jacket									
7000	Hot water, gross output, 103 MBH	Q-6	1.60	15	Ea.	1,725	725		2,450	3,000
7020	122 MBH		1.45	16.506		1,825	800		2,625	3,200
7040	137 MBH		1.36	17.595		1,925	850		2,775	3,400
7060	168 MBH		1.30	18.405		2,025	890		2,915	3,550
7080	225 MBH		1.22	19.704		2,700	955		3,655	4,375
7100	315 MBH		.96	25.105		5,525	1,225		6,750	7,925
7120	420 MBH		.70	34.483		5,975	1,675		7,650	9,075
7140	525 MBH		.57	42.403		7,175	2,050		9,225	11,000
7180	735 MBH		.48	50.104		11,700	2,425		14,125	16,600
7220	1,050 MBH		.37	65.753		13,400	3,175		16,575	19,600
7280	2,310 MBH		.21	114		21,700	5,550		27,250	32,200
7320	3,150 MBH		.13	184		31,600	8,950		40,550	48,200
7340	For tankless coil in steam or hot water, add					7%				

23 52 26.70 Packaged Water Tube Boilers		Crew	Daily Output	Labor-Hours	Unit	Material	Labor	Equipment	Total	Total Incl O&P
0010	**PACKAGED WATER TUBE BOILERS**									
2000	Packaged water tube, #2 oil, steam or hot water, gross output									
2010	200 MBH	Q-6	1.50	16	Ea.	4,650	775		5,425	6,275
2014	275 MBH		1.40	17.143		4,900	830		5,730	6,625
2018	360 MBH		1.10	21.818		5,125	1,050		6,175	7,225
2022	520 MBH		.65	36.923		5,400	1,800		7,200	8,600
2026	600 MBH		.60	40		5,650	1,925		7,575	9,100
2030	720 MBH		.55	43.636		7,650	2,125		9,775	11,600
2034	960 MBH		.48	50		9,325	2,425		11,750	13,800
2040	1200 MBH	Q-7	.50	64		18,200	3,175		21,375	24,800
2044	1440 MBH		.45	71.111		19,500	3,525		23,025	26,700

23 52 26 – Steel Boilers

23 52 26.70 Packaged Water Tube Boilers		Crew	Daily Output	Labor-Hours	Unit	Material	2010 Bare Costs Labor	Equipment	Total	Total Incl O&P
2060	1600 MBH	Q-7	.40	80	Ea.	22,900	3,950		26,850	31,100
2068	1920 MBH		.35	91.429		24,900	4,525		29,425	34,200
2072	2160 MBH		.33	96.970		27,600	4,800		32,400	37,500
2080	2400 MBH		.30	106		30,900	5,275		36,175	41,900
2100	3200 MBH		.25	128		35,300	6,325		41,625	48,300
2120	4800 MBH		.20	160		46,100	7,900		54,000	62,500
2200	Gas fired									
2204	200 MBH	Q-6	1.50	16	Ea.	5,825	775		6,600	7,575
2208	275 MBH		1.40	17.143		6,075	830		6,905	7,925
2212	360 MBH		1.10	21.818		6,325	1,050		7,375	8,525
2216	520 MBH		.65	36.923		6,675	1,800		8,475	10,000
2220	600 MBH		.60	40		6,825	1,925		8,750	10,400
2224	720 MBH		.55	43.636		8,075	2,125		10,200	12,100
2228	960 MBH		.48	50		10,200	2,425		12,625	14,800
2232	1220 MBH	Q-7	.50	64		14,900	3,175		18,075	21,100
2236	1440 MBH		.45	71.111		15,400	3,525		18,925	22,200
2240	1680 MBH		.40	80		19,800	3,950		23,750	27,700
2244	1920 MBH		.35	91.429		19,900	4,525		24,425	28,600
2248	2160 MBH		.33	96.970		20,200	4,800		25,000	29,400
2252	2400 MBH		.30	106		22,700	5,275		27,975	32,800

23 52 28 – Swimming Pool Boilers

23 52 28.10 Swimming Pool Heaters		Crew	Daily Output	Labor-Hours	Unit	Material	Labor	Equipment	Total	Total Incl O&P
0010	**SWIMMING POOL HEATERS**, Not including wiring, external									
0020	piping, base or pad,									
0160	Gas fired, input, 155 MBH	Q-6	1.50	16	Ea.	1,575	775		2,350	2,900
0200	199 MBH		1	24		1,700	1,150		2,850	3,625
0220	250 MBH		.70	34.286		1,850	1,650		3,500	4,525
0240	300 MBH		.60	40		1,925	1,925		3,850	5,025
0260	399 MBH		.50	48		2,175	2,325		4,500	5,850
0280	500 MBH		.40	60		7,225	2,900		10,125	12,300
0300	650 MBH		.35	68.571		7,700	3,325		11,025	13,400
0320	750 MBH		.33	72.727		8,400	3,525		11,925	14,500
0360	990 MBH		.22	109		11,300	5,275		16,575	20,300
0370	1,260 MBH		.21	114		13,200	5,525		18,725	22,900
0380	1,440 MBH		.19	126		14,100	6,125		20,225	24,700
0400	1,800 MBH		.14	171		15,700	8,300		24,000	29,700
0410	2,070 MBH		.13	184		18,500	8,950		27,450	33,800
2000	Electric, 12 KW, 4,800 gallon pool	Q-19	3	8		2,025	380		2,405	2,800
2020	15 KW, 7,200 gallon pool		2.80	8.571		2,050	405		2,455	2,850
2040	24 KW, 9,600 gallon pool		2.40	10		2,375	475		2,850	3,300
2060	30 KW, 12,000 gallon pool		2	12		2,400	570		2,970	3,500
2080	36 KW, 14,400 gallon pool		1.60	15		2,775	710		3,485	4,125
2100	57 KW, 24,000 gallon pool		1.20	20		3,475	950		4,425	5,250
9000	To select pool heater: 12 BTUH x S.F. pool area									
9010	X temperature differential = required output									
9050	For electric, KW = gallons x 2.5 divided by 1000									
9100	For family home type pool, double the									
9110	Rated gallon capacity = 1/2°F rise per hour									

23 52 39 – Fire-Tube Boilers

23 52 39.13 Scotch Marine Boilers	
0010	**SCOTCH MARINE BOILERS**
1000	Packaged fire tube, #2 oil, gross output

23 52 39 – Fire-Tube Boilers

23 52 39.13 Scotch Marine Boilers		Crew	Daily Output	Labor-Hours	Unit	Material (2010 Bare Costs)	Labor (2010 Bare Costs)	Equipment (2010 Bare Costs)	Total (2010 Bare Costs)	Total Incl O&P
1006	15 PSI steam									
1010	1005 MBH, 30 HP	Q-6	.33	72.072	Ea.	34,300	3,500		37,800	42,900
1014	1675 MBH, 50 HP	"	.26	93.023		40,500	4,500		45,000	51,500
1020	3348 MBH, 100 HP	Q-7	.21	152		52,500	7,525		60,025	69,500
1030	5025 MBH, 150 HP		.17	192		66,500	9,525		76,025	87,500
1040	6696 MBH, 200 HP		.14	223		75,500	11,100		86,600	99,500
1050	8375 MBH, 250 HP		.14	231		85,000	11,500		96,500	110,500
1060	10,044 MBH, 300 HP		.13	251		104,000	12,500		116,500	132,500
1080	16,740 MBH, 500 HP		.08	380		141,000	18,800		159,800	183,000
1100	23,435 MBH, 700 HP		.07	484		173,000	24,000		197,000	226,000
1140	To fire #6, and gas, add	↓	.42	76.190	↓	15,400	3,775		19,175	22,600
1180	For duplex package feed system									
1200	To 3348 MBH boiler, add	Q-7	.54	59.259	Ea.	8,275	2,925		11,200	13,500
1220	To 6696 MBH boiler, add		.41	78.049		9,800	3,850		13,650	16,600
1240	To 10,044 MBH boiler, add		.38	84.211		11,300	4,175		15,475	18,700
1260	To 16,740 MBH boiler, add		.28	114		12,800	5,650		18,450	22,600
1280	To 23,435 MBH boiler, add	↓	.25	128	↓	15,200	6,325		21,525	26,200
1300	150 PSI steam									
1310	1005 MBH, 30 HP	Q-6	.33	72.072	Ea.	36,500	3,500		40,000	45,300
1320	1675 MBH, 50 HP	"	.26	93.023		45,600	4,500		50,100	57,000
1330	3350 MBH, 100 HP	Q-7	.21	152		62,000	7,525		69,525	80,000
1340	5025 MBH, 150 HP		.17	192		79,500	9,525		89,025	102,000
1350	6700 MBH, 200 HP		.14	223		88,000	11,100		99,100	113,500
1360	10,050 MBH, 300 HP	↓	.13	251	↓	130,000	12,500		142,500	161,500
1400	Packaged scotch marine, #6 oil									
1410	15 PSI steam									
1420	1675 MBH, 50 HP	Q-6	.25	97.166	Ea.	53,000	4,700		57,700	65,000
1430	3350 MBH, 100 HP	Q-7	.20	159		65,000	7,875		72,875	83,500
1440	5025 MBH, 150 HP		.16	202		82,000	10,000		92,000	105,000
1450	6700 MBH, 200 HP		.14	233		89,500	11,500		101,000	116,000
1460	8375 MBH, 250 HP		.13	244		114,000	12,100		126,100	143,500
1470	10,050 MBH, 300 HP		.12	262		134,000	13,000		147,000	167,000
1480	13,400 MBH, 400 HP	↓	.09	351	↓	154,000	17,400		171,400	195,000
1500	150 PSI steam									
1510	1675 MBH, 50 HP	Q-6	.25	97.166	Ea.	62,500	4,700		67,200	76,000
1520	3350 MBH, 100 HP	Q-7	.20	159		72,000	7,875		79,875	91,500
1530	5025 MBH, 150 HP		.16	202		84,500	10,000		94,500	108,000
1540	6700 MBH, 200 HP		.14	233		96,000	11,500		107,500	123,000
1550	8375 MBH, 250 HP		.13	244		122,500	12,100		134,600	152,500
1560	10,050 MBH, 300 HP		.12	262		137,000	13,000		150,000	170,000
1570	13,400 MBH, 400 HP	↓	.09	351	↓	161,000	17,400		178,400	203,500

23 52 84 – Boiler Blowdown

23 52 84.10 Boiler Blowdown Systems

Line	Description	Crew	Daily Output	Labor-Hours	Unit	Material	Labor	Equipment	Total	Total Incl O&P
0010	**BOILER BLOWDOWN SYSTEMS**									
1010	Boiler blowdown, auto/manual to 2000 MBH	Q-5	3.75	4.267	Ea.	3,150	199		3,349	3,775
1020	7300 MBH	"	3	5.333	"	4,350	249		4,599	5,175

23 52 88 – Burners

23 52 88.10 Replacement Type Burners

Line	Description	Crew	Daily Output	Labor-Hours	Unit	Material	Labor	Equipment	Total	Total Incl O&P
0010	**REPLACEMENT TYPE BURNERS**									
0990	Residential, conversion, gas fired, LP or natural									
1000	Gun type, atmospheric input 50 to 225 MBH	Q-1	2.50	6.400	Ea.	650	300		950	1,150
1020	100 to 400 MBH	↓	2	8	↓	1,075	375		1,450	1,725

23 52 88 – Burners

23 52 88.10 Replacement Type Burners		Crew	Daily Output	Labor-Hours	Unit	Material	Labor	Equipment	Total	Total Incl O&P
1040	300 to 1000 MBH	Q-1	1.70	9.412	Ea.	3,300	440		3,740	4,275
2000	Commercial and industrial, gas/oil, input									
2050	400 MBH	Q-1	1.50	10.667	Ea.	3,400	500		3,900	4,500
2090	670 MBH		1.40	11.429		3,400	535		3,935	4,550
2140	1155 MBH		1.30	12.308		3,800	575		4,375	5,050
2200	1800 MBH		1.20	13.333		3,800	625		4,425	5,125
2260	3000 MBH		1.10	14.545		4,575	680		5,255	6,050
2320	4100 MBH		1	16		4,700	750		5,450	6,275
2500	Impinged jet, rectangular, burner only, 3-1/2" WC,									
2520	Gas fired, input, 420 to 640 MBH	Q-1	2	8	Ea.	825	375		1,200	1,475
2540	560 to 860 MBH		2	8		1,175	375		1,550	1,850
2560	630 to 950 MBH		1.90	8.421		1,225	395		1,620	1,950
2580	950 to 1200 MBH		1.80	8.889		1,750	415		2,165	2,575
2600	1120 to 1700 MBH		1.80	8.889		2,350	415		2,765	3,200
2620	1700 to 1900 MBH		1.70	9.412		2,650	440		3,090	3,550
2640	1400 to 2100 MBH		1.60	10		2,925	470		3,395	3,925
2660	1600 to 2400 MBH		1.60	10		3,100	470		3,570	4,100
2680	1700 to 2500 MBH		1.50	10.667		3,525	500		4,025	4,625
2700	1880 to 2880 MBH		1.40	11.429		3,700	535		4,235	4,875
2720	2260 to 3680 MBH		1.40	11.429		4,325	535		4,860	5,550
2740	2590 to 4320 MBH		1.30	12.308		5,275	575		5,850	6,700
2760	3020 to 5040 MBH		1.20	13.333		6,175	625		6,800	7,700
2780	3450 to 5760 MBH		1.20	13.333		7,050	625		7,675	8,675
2850	Burner pilot		50	.320		21	15		36	45.50
2860	Thermocouple		50	.320		21.50	15		36.50	46.50
2870	Thermocouple & pilot bracket		45	.356		84	16.65		100.65	118
2900	For 7" W.C. pressure, increase MBH 40%									
3000	Flame retention oil fired assembly, input									
3020	.50 to 2.25 GPH	Q-1	2.40	6.667	Ea.	289	310		599	790
3040	2.0 to 5.0 GPH		2	8		345	375		720	940
3060	3.0 to 7.0 GPH		1.80	8.889		620	415		1,035	1,300
3080	6.0 to 12.0 GPH		1.60	10		990	470		1,460	1,775
4600	Gas safety, shut off valve, 3/4" threaded	1 Stpi	20	.400		178	21		199	227
4610	1" threaded		19	.421		172	22		194	223
4620	1-1/4" threaded		15	.533		194	27.50		221.50	255
4630	1-1/2" threaded		13	.615		210	32		242	279
4640	2" threaded		11	.727		235	38		273	315
4650	2-1/2" threaded	Q-1	15	1.067		269	50		319	370
4660	3" threaded		13	1.231		370	57.50		427.50	490
4670	4" flanged		3	5.333		2,625	250		2,875	3,275
4680	6" flanged	Q-2	3	8		5,875	390		6,265	7,050

(Material, Labor, Equipment and Total are 2010 Bare Costs.)

23 53 Heating Boiler Feedwater Equipment

23 53 24 – Shot Chemical Feeder

23 53 24.10 Chemical Feeder		Crew	Daily Output	Labor-Hours	Unit	Material	2010 Bare Costs Labor	Equipment	Total	Total Incl O&P
0010	**CHEMICAL FEEDER**									
0200	Shot chem feeder, by pass, in line mount, 125 PSIG, 1.7 gallon	2 Stpi	1.80	8.889	Ea.	298	460		758	1,025
0220	Floor mount, 175 PSIG, 5 gallon		1.24	12.903		350	670		1,020	1,375
0230	12 gallon		.93	17.204		540	895		1,435	1,950
0300	150 lb., ASME, 5 gallon		1.24	12.903		350	670		1,020	1,375
0320	10 gallon		.93	17.204		500	895		1,395	1,900
0400	300 lb., ASME, 5 gallon		1.24	12.903		350	670		1,020	1,375
0420	10 gallon	↓	.93	17.204	↓	500	895		1,395	1,900

23 54 Furnaces

23 54 13 – Electric-Resistance Furnaces

23 54 13.10 Electric Furnaces		Crew	Daily Output	Labor-Hours	Unit	Material	Labor	Equipment	Total	Total Incl O&P
0010	**ELECTRIC FURNACES**, Hot air, blowers, std. controls									
0011	not including gas, oil or flue piping									
1000	Electric, UL listed									
1020	10.2 MBH	Q-20	5	4	Ea.	475	181		656	790
1040	17.1 MBH		4.80	4.167		495	188		683	830
1060	27.3 MBH		4.60	4.348		600	196		796	955
1100	34.1 MBH		4.40	4.545		615	205		820	985
1120	51.6 MBH		4.20	4.762		660	215		875	1,050
1140	68.3 MBH		4	5		750	226		976	1,175
1160	85.3 MBH	↓	3.80	5.263	↓	785	238		1,023	1,225

23 54 16 – Fuel-Fired Furnaces

23 54 16.13 Gas-Fired Furnaces		Crew	Daily Output	Labor-Hours	Unit	Material	Labor	Equipment	Total	Total Incl O&P
0010	**GAS-FIRED FURNACES**									
3000	Gas, AGA certified, upflow, direct drive models									
3020	45 MBH input	Q-9	4	4	Ea.	575	177		752	905
3040	60 MBH input		3.80	4.211		615	186		801	955
3060	75 MBH input		3.60	4.444		685	196		881	1,050
3100	100 MBH input		3.20	5		695	221		916	1,100
3120	125 MBH input		3	5.333		700	236		936	1,125
3130	150 MBH input		2.80	5.714		705	253		958	1,150
3140	200 MBH input		2.60	6.154		2,425	272		2,697	3,050
3160	300 MBH input		2.30	6.957		2,550	305		2,855	3,275
3180	400 MBH input	↓	2	8	↓	2,825	355		3,180	3,625
3200	Gas fired wall furnace									
3204	Horizontal flow									
3208	7.7 MBH	Q-9	7	2.286	Ea.	470	101		571	675
3212	14 MBH		6.50	2.462		470	109		579	685
3216	24 MBH		5	3.200		495	141		636	760
3220	49 MBH		4	4		705	177		882	1,050
3224	65 MBH	↓	3.60	4.444	↓	895	196		1,091	1,275
3240	Up-flow									
3244	7.7 MBH	Q-9	7	2.286	Ea.	470	101		571	675
3248	14 MBH		6.50	2.462		470	109		579	685
3252	24 MBH		5	3.200		595	141		736	865
3254	35 MBH		4.50	3.556		715	157		872	1,025
3256	49 MBH		4	4		1,050	177		1,227	1,425
3260	65 MBH	↓	3.60	4.444	↓	1,100	196		1,296	1,525
3500	Gas furnace									

23 54 Furnaces

23 54 16 – Fuel-Fired Furnaces

23 54 16.13 Gas-Fired Furnaces		Crew	Daily Output	Labor-Hours	Unit	2010 Bare Costs Material	Labor	Equipment	Total	Total Incl O&P
3510	Up-flow									
3514	1200 CFM, 45 MBH	Q-9	4	4	Ea.	1,125	177		1,302	1,525
3518	75 MBH		3.60	4.444		1,200	196		1,396	1,625
3522	100 MBH		3.20	5		1,275	221		1,496	1,725
3526	125 MBH		3	5.333		1,400	236		1,636	1,875
3530	150 MBH		2.80	5.714		1,425	253		1,678	1,925
3540	2000 CFM, 75 MBH		3.60	4.444		870	196		1,066	1,250
3544	100 MBH		3.20	5		875	221		1,096	1,300
3548	120 MBH		3	5.333		930	236		1,166	1,375
3552	150 MBH		2.80	5.714		1,000	253		1,253	1,475
3560	Horizontal flow									
3564	1200 CFM, 45 MBH	Q-9	4	4	Ea.	1,150	177		1,327	1,525
3568	60 MBH		3.80	4.211		1,200	186		1,386	1,600
3572	70 MBH		3.60	4.444		1,200	196		1,396	1,625
3576	2000 CFM, 90 MBH		3.30	4.848		1,250	214		1,464	1,700
3580	110 MBH		3.10	5.161		1,325	228		1,553	1,825

23 54 16.16 Oil-Fired Furnaces		Crew	Daily Output	Labor-Hours	Unit	Material	Labor	Equipment	Total	Total Incl O&P
0010	**OIL-FIRED FURNACES**									
6000	Oil, UL listed, atomizing gun type burner									
6020	56 MBH output	Q-9	3.60	4.444	Ea.	1,850	196		2,046	2,325
6030	84 MBH output		3.50	4.571		1,850	202		2,052	2,350
6040	95 MBH output		3.40	4.706		1,875	208		2,083	2,400
6060	134 MBH output		3.20	5		1,950	221		2,171	2,475
6080	151 MBH output		3	5.333		2,275	236		2,511	2,850
6100	200 MBH input		2.60	6.154		2,325	272		2,597	2,950
6120	300 MBH input		2.30	6.957		2,825	305		3,130	3,575
6140	400 MBH input		2	8		3,275	355		3,630	4,125
6200	Up flow									
6210	105 MBH	Q-9	3.40	4.706	Ea.	1,425	208		1,633	1,900
6220	140 MBH		3.30	4.848		1,725	214		1,939	2,225
6230	168 MBH		2.90	5.517		2,075	244		2,319	2,675
6260	Lo-boy style									
6270	105 MBH	Q-9	3.40	4.706	Ea.	1,575	208		1,783	2,050
6280	140 MBH		3.30	4.848		1,700	214		1,914	2,200
6290	189 MBH		2.80	5.714		2,300	253		2,553	2,900
6320	Down flow									
6330	105 MBH	Q-9	3.40	4.706	Ea.	1,500	208		1,708	1,975
6340	140 MBH		3.30	4.848		1,700	214		1,914	2,200
6350	168 MBH		2.90	5.517		2,150	244		2,394	2,750
6380	Horizontal									
6390	105 MBH	Q-9	3.40	4.706	Ea.	1,500	208		1,708	1,975
6400	140 MBH		3.30	4.848		1,700	214		1,914	2,200
6410	189 MBH		2.80	5.714		1,925	253		2,178	2,500

23 54 16.21 Solid Fuel-Fired Furnaces		Crew	Daily Output	Labor-Hours	Unit	Material	Labor	Equipment	Total	Total Incl O&P
0010	**SOLID FUEL-FIRED FURNACES**									
6020	Wood fired furnaces									
6030	Includes hot water coil, thermostat, and auto draft control									
6040	24" long firebox G	Q-9	4	4	Ea.	3,975	177		4,152	4,625
6050	30" long firebox G		3.60	4.444		4,675	196		4,871	5,425
6060	With fireplace glass doors G		3.20	5		6,000	221		6,221	6,925
6200	Wood/oil fired furnaces, includes two thermostats									
6210	Includes hot water coil and auto draft control									

23 54 Furnaces

23 54 16 – Fuel-Fired Furnaces

23 54 16.21 Solid Fuel-Fired Furnaces		Crew	Daily Output	Labor-Hours	Unit	2010 Bare Costs Material	Labor	Equipment	Total	Total Incl O&P
6240	24" long firebox G	Q-9	3.40	4.706	Ea.	4,825	208		5,033	5,625
6250	30" long firebox G		3	5.333		5,400	236		5,636	6,300
6260	With fireplace glass doors G		2.80	5.714		6,000	253		6,253	6,975
6400	Wood/gas fired furnaces, includes two thermostats									
6410	Includes hot water coil and auto draft control									
6440	24" long firebox G	Q-9	2.80	5.714	Ea.	5,450	253		5,703	6,375
6450	30" long firebox G		2.40	6.667		6,150	295		6,445	7,225
6460	With fireplace glass doors G		2	8		7,425	355		7,780	8,675
6600	Wood/oil/gas fired furnaces, optional accessories									
6610	Hot air plenum	Q-9	16	1	Ea.	93	44		137	169
6620	Safety heat dump		24	.667		82	29.50		111.50	135
6630	Auto air intake		18	.889		132	39.50		171.50	205
6640	Cold air return package		14	1.143		142	50.50		192.50	233
6650	Wood fork					44			44	48.50
6700	Wood fired outdoor furnace									
6740	24" long firebox G	Q-9	3.80	4.211	Ea.	4,200	186		4,386	4,875
6760	Wood fired outdoor furnace, optional accessories									
6770	Chimney section, stainless steel, 6" ID x 3' lg. G	Q-9	36	.444	Ea.	121	19.65		140.65	163
6780	Chimney cap, stainless steel G	"	40	.400	"	66	17.70		83.70	99.50
6800	Wood fired hot water furnace									
6820	Includes 200 gal. hot water storage, thermostat, and auto draft control									
6840	30" long firebox G	Q-9	2.10	7.619	Ea.	8,275	335		8,610	9,600
6850	Water to air heat exchanger									
6870	Includes mounting kit and blower relay									
6880	140 MBH, 18.75" W x 18.75" L	Q-9	7.50	2.133	Ea.	370	94.50		464.50	550
6890	200 MBH, 24" W x 24" L	"	7	2.286	"	525	101		626	735
6900	Water to water heat exchanger									
6940	100 MBH	Q-9	6.50	2.462	Ea.	370	109		479	575
6960	290 MBH	"	6	2.667	"	535	118		653	770
7000	Optional accessories									
7010	Large volume circulation pump, (2 Included)	Q-9	14	1.143	Ea.	242	50.50		292.50	345
7020	Air bleed fittings, (package)		24	.667		49	29.50		78.50	98.50
7030	Domestic water preheater		6	2.667		219	118		337	420
7040	Smoke pipe kit		4	4		74	177		251	350

23 54 24 – Furnace Components for Cooling

23 54 24.10 Furnace Components and Combinations

		Crew	Daily Output	Labor-Hours	Unit	Material	Labor	Equipment	Total	Total Incl O&P
0010	**FURNACE COMPONENTS AND COMBINATIONS**									
0080	Coils, A/C evaporator, for gas or oil furnaces									
0090	Add-on, with holding charge									
0100	Upflow									
0120	1-1/2 ton cooling	Q-5	4	4	Ea.	158	187		345	455
0130	2 ton cooling		3.70	4.324		188	202		390	510
0140	3 ton cooling		3.30	4.848		238	226		464	600
0150	4 ton cooling		3	5.333		330	249		579	735
0160	5 ton cooling		2.70	5.926		385	277		662	840
0300	Downflow									
0330	2-1/2 ton cooling	Q-5	3	5.333	Ea.	232	249		481	630
0340	3-1/2 ton cooling		2.60	6.154		276	287		563	735
0350	5 ton cooling		2.20	7.273		385	340		725	935
0600	Horizontal									
0630	2 ton cooling	Q-5	3.90	4.103	Ea.	260	192		452	575
0640	3 ton cooling		3.50	4.571		296	213		509	645

23 54 Furnaces

23 54 24 – Furnace Components for Cooling

23 54 24.10 Furnace Components and Combinations		Crew	Daily Output	Labor-Hours	Unit	2010 Bare Costs Material	Labor	Equipment	Total	Total Incl O&P
0650	4 ton cooling	Q-5	3.20	5	Ea.	375	234		609	760
0660	5 ton cooling	↓	2.90	5.517	↓	375	258		633	795
2000	Cased evaporator coils for air handlers									
2100	1-1/2 ton cooling	Q-5	4.40	3.636	Ea.	325	170		495	610
2110	2 ton cooling		4.10	3.902		325	182		507	635
2120	2-1/2 ton cooling		3.90	4.103		345	192		537	665
2130	3 ton cooling		3.70	4.324		385	202		587	725
2140	3-1/2 ton cooling		3.50	4.571		400	213		613	760
2150	4 ton cooling		3.20	5		480	234		714	880
2160	5 ton cooling	↓	2.90	5.517	↓	555	258		813	995
3010	Air handler, modular									
3100	With cased evaporator cooling coil									
3120	1-1/2 ton cooling	Q-5	3.80	4.211	Ea.	1,025	197		1,222	1,425
3130	2 ton cooling		3.50	4.571		1,175	213		1,388	1,600
3140	2-1/2 ton cooling		3.30	4.848		1,175	226		1,401	1,625
3150	3 ton cooling		3.10	5.161		1,275	241		1,516	1,750
3160	3-1/2 ton cooling		2.90	5.517		1,325	258		1,583	1,825
3170	4 ton cooling		2.50	6.400		1,475	299		1,774	2,075
3180	5 ton cooling	↓	2.10	7.619	↓	1,675	355		2,030	2,350
3500	With no cooling coil									
3520	1-1/2 ton coil size	Q-5	12	1.333	Ea.	795	62.50		857.50	970
3530	2 ton coil size		10	1.600		815	74.50		889.50	1,000
3540	2-1/2 ton coil size		10	1.600		890	74.50		964.50	1,075
3554	3 ton coil size		9	1.778		905	83		988	1,125
3560	3-1/2 ton coil size		9	1.778		970	83		1,053	1,200
3570	4 ton coil size		8.50	1.882		1,150	88		1,238	1,375
3580	5 ton coil size	↓	8	2	↓	1,300	93.50		1,393.50	1,600
4000	With heater									
4120	5 kW, 17.1 MBH	Q-5	16	1	Ea.	690	46.50		736.50	830
4130	7.5 kW, 25.6 MBH		15.60	1.026		810	48		858	960
4140	10 kW, 34.2 MBH		15.20	1.053		930	49		979	1,100
4150	12.5 KW, 42.7 MBH		14.80	1.081		1,050	50.50		1,100.50	1,225
4160	15 KW, 51.2 MBH		14.40	1.111		1,200	52		1,252	1,375
4170	25 KW, 85.4 MBH		14	1.143		1,325	53.50		1,378.50	1,525
4180	30 KW, 102 MBH	↓	13	1.231	↓	1,525	57.50		1,582.50	1,750

23 55 Fuel-Fired Heaters

23 55 13 – Fuel-Fired Duct Heaters

23 55 13.16 Gas-Fired Duct Heaters		Crew	Daily Output	Labor-Hours	Unit	Material	Labor	Equipment	Total	Total Incl O&P
0010	**GAS-FIRED DUCT HEATERS**, Includes burner, controls, stainless steel									
0020	heat exchanger. Gas fired, electric ignition									
0030	Indoor installation									
0080	100 MBH output	Q-5	5	3.200	Ea.	1,875	149		2,024	2,300
0100	120 MBH output		4	4		2,125	187		2,312	2,625
0130	200 MBH output		2.70	5.926		2,850	277		3,127	3,550
0140	240 MBH output		2.30	6.957		2,975	325		3,300	3,725
0160	280 MBH output		2	8		3,350	375		3,725	4,250
0180	320 MBH output	↓	1.60	10		3,550	465		4,015	4,625
0300	For powered venter and adapter, add				↓	460			460	510
0502	For required flue pipe, see Div. 23 51 23.10									
1000	Outdoor installation, with power venter									

23 55 13 – Fuel-Fired Duct Heaters

23 55 13.16 Gas-Fired Duct Heaters		Crew	Daily Output	Labor-Hours	Unit	Material	2010 Bare Costs Labor	Equipment	Total	Total Incl O&P
1020	75 MBH output	Q-5	4	4	Ea.	2,775	187		2,962	3,325
1040	94 MBH output		4	4		3,125	187		3,312	3,725
1060	120 MBH output		4	4		3,525	187		3,712	4,175
1080	157 MBH output		3.50	4.571		4,100	213		4,313	4,825
1100	187 MBH output		3	5.333		5,050	249		5,299	5,925
1120	225 MBH output		2.50	6.400		5,225	299		5,524	6,200
1140	300 MBH output		1.80	8.889		6,100	415		6,515	7,325
1160	375 MBH output		1.60	10		6,425	465		6,890	7,775
1180	450 MBH output		1.40	11.429		7,450	535		7,985	9,000
1200	600 MBH output		1	16		8,400	745		9,145	10,400
1300	Aluminized exchanger, subtract					15%				
1500	For two stage gas valve, add					190			190	208

23 55 33 – Fuel-Fired Unit Heaters

23 55 33.13 Oil-Fired Unit Heaters		Crew	Daily Output	Labor-Hours	Unit	Material	Labor	Equipment	Total	Total Incl O&P
0010	**OIL-FIRED UNIT HEATERS**, Cabinet, grilles, fan, ctrl, burner, no piping									
6000	Oil fired, suspension mounted, 94 MBH output	Q-5	4	4	Ea.	3,400	187		3,587	4,000
6040	140 MBH output		3	5.333		3,625	249		3,874	4,350
6060	184 MBH output		3	5.333		3,900	249		4,149	4,675

23 55 33.16 Gas-Fired Unit Heaters		Crew	Daily Output	Labor-Hours	Unit	Material	Labor	Equipment	Total	Total Incl O&P
0010	**GAS-FIRED UNIT HEATERS**, Cabinet, grilles, fan, ctrls., burner, no piping									
0022	thermostat, no piping. For flue see Div. 23 51 23.10									
1000	Gas fired, floor mounted									
1100	60 MBH output	Q-5	10	1.600	Ea.	805	74.50		879.50	995
1120	80 MBH output		9	1.778		815	83		898	1,025
1140	100 MBH output		8	2		890	93.50		983.50	1,125
1160	120 MBH output		7	2.286		1,050	107		1,157	1,300
1180	180 MBH output		6	2.667		1,275	125		1,400	1,575
1500	Rooftop mounted, power vent, stainless steel exchanger									
1520	75 MBH input	Q-6	4	6	Ea.	4,200	291		4,491	5,050
1540	100 MBH input		3.60	6.667		4,425	325		4,750	5,350
1560	125 MBH input		3.30	7.273		4,625	350		4,975	5,625
1580	150 MBH input		3	8		5,050	385		5,435	6,125
1600	175 MBH input		2.60	9.231		5,200	445		5,645	6,375
1620	225 MBH input		2.30	10.435		6,050	505		6,555	7,400
1640	300 MBH input		1.90	12.632		7,125	610		7,735	8,750
1660	350 MBH input		1.40	17.143		7,825	830		8,655	9,850
1600	450 MBH input		1.20	20		8,250	970		9,220	10,500
1720	700 MBH input		.80	30		10,300	1,450		11,750	13,500
1760	1200 MBH input		.30	80		10,700	3,875		14,575	17,500
1900	For aluminized steel exchanger, subtract					10%				
2000	Suspension mounted, propeller fan, 20 MBH output	Q-5	8.50	1.882		985	88		1,073	1,200
2020	40 MBH output		7.50	2.133		1,075	99.50		1,174.50	1,350
2040	60 MBH output		7	2.286		1,200	107		1,307	1,475
2060	80 MBH output		6	2.667		1,225	125		1,350	1,525
2080	100 MBH output		5.50	2.909		1,375	136		1,511	1,700
2100	130 MBH output		5	3.200		1,500	149		1,649	1,875
2120	140 MBH output		4.50	3.556		1,550	166		1,716	1,950
2140	160 MBH output		4	4		1,600	187		1,787	2,025
2160	180 MBH output		3.50	4.571		1,700	213		1,913	2,200
2180	200 MBH output		3	5.333		1,800	249		2,049	2,350
2200	240 MBH output		2.70	5.926		2,075	277		2,352	2,700
2220	280 MBH output		2.30	6.957		2,425	325		2,750	3,150

23 55 Fuel-Fired Heaters

23 55 33 – Fuel-Fired Unit Heaters

23 55 33.16 Gas-Fired Unit Heaters		Crew	Daily Output	Labor-Hours	Unit	2010 Bare Costs Material	Labor	Equipment	Total	Total Incl O&P
2240	320 MBH output	Q-5	2	8	Ea.	2,650	375		3,025	3,450
2500	For powered venter and adapter, add					355			355	390
3000	Suspension mounted, blower type, 40 MBH output	Q-5	6.80	2.353		605	110		715	830
3020	60 MBH output		6.60	2.424		775	113		888	1,025
3040	84 MBH output		5.80	2.759		810	129		939	1,075
3060	104 MBH output		5.20	3.077		920	144		1,064	1,225
3080	140 MBH output		4.30	3.721		1,025	174		1,199	1,375
3100	180 MBH output		3.30	4.848		1,150	226		1,376	1,600
3120	240 MBH output		2.50	6.400		1,500	299		1,799	2,100
3140	280 MBH output		2	8		1,875	375		2,250	2,625
4000	Suspension mounted, sealed combustion system,									
4020	Aluminized steel exchanger, powered vent									
4040	100 MBH output	Q-5	5	3.200	Ea.	1,800	149		1,949	2,225
4060	120 MBH output		4.70	3.404		2,025	159		2,184	2,475
4080	160 MBH output		3.70	4.324		2,225	202		2,427	2,725
4100	200 MBH output		2.90	5.517		2,650	258		2,908	3,300
4120	240 MBH output		2.50	6.400		2,850	299		3,149	3,575
4140	320 MBH output		1.70	9.412		3,400	440		3,840	4,375
5000	Wall furnace, 17.5 MBH output		6	2.667		715	125		840	970
5020	24 MBH output		5	3.200		740	149		889	1,050
5040	35 MBH output		4	4		895	187		1,082	1,275

23 56 Solar Energy Heating Equipment

23 56 16 – Packaged Solar Heating Equipment

23 56 16.40 Solar Heating Systems

Line	Description	Crew	Daily Output	Labor-Hours	Unit	Material	Labor	Equipment	Total	Total Incl O&P
0010	SOLAR HEATING SYSTEMS R235616-60									
0020	System/Package prices, not including connecting									
0030	pipe, insulation, or special heating/plumbing fixtures G									
0152	For solar ultraviolet pipe insulation see Div. 22 07 19.10									
0500	Hot water, standard package, low temperature G									
0540	1 collector, circulator, fittings, 65 gal. tank G	Q-1	.50	32	Ea.	1,500	1,500		3,000	3,900
0580	2 collectors, circulator, fittings, 120 gal. tank G		.40	40		2,325	1,875		4,200	5,350
0620	3 collectors, circulator, fittings, 120 gal. tank G		.34	47.059		3,175	2,200		5,375	6,775
0700	Medium temperature package G									
0720	1 collector, circulator, fittings, 80 gal. tank G	Q-1	.50	32	Ea.	2,350	1,500		3,850	4,825
0740	2 collectors, circulator, fittings, 120 gal. tank G		.40	40		3,300	1,875		5,175	6,450
0780	3 collectors, circulator, fittings, 120 gal. tank G		.30	53.333		4,300	2,500		6,800	8,475
0980	For each additional 120 gal. tank, add G					1,475			1,475	1,625

23 56 19 – Solar Heating Components

23 56 19.50 Solar Heating Ancillary

Line	Description	Crew	Daily Output	Labor-Hours	Unit	Material	Labor	Equipment	Total	Total Incl O&P
0010	SOLAR HEATING ANCILLARY									
2250	Controller, liquid temperature G	1 Plum	5	1.600	Ea.	96	83.50		179.50	230
2300	Circulators, air G									
2310	Blowers									
2330	100-300 S.F. system, 1/10 HP G	Q-9	16	1	Ea.	229	44		273	320
2340	300-500 S.F. system, 1/5 HP G		15	1.067		299	47		346	400
2350	Two speed, 100-300 S.F., 1/10 HP G		14	1.143		136	50.50		186.50	227
2400	Reversible fan, 20" diameter, 2 speed G		18	.889		108	39.50		147.50	179
2550	Booster fan 6" diameter, 120 CFM G		16	1		35	44		79	106
2570	6" diameter, 225 CFM G		16	1		43	44		87	115

23 56 19.50 Solar Heating Ancillary			Crew	Daily Output	Labor-Hours	Unit	2010 Bare Costs Material	Labor	Equipment	Total	Total Incl O&P
2580	8" diameter, 150 CFM	G	Q-9	16	1	Ea.	39.50	44		83.50	111
2590	8" diameter, 310 CFM	G		14	1.143		60.50	50.50		111	143
2600	8" diameter, 425 CFM	G		14	1.143		68	50.50		118.50	152
2650	Rheostat	G		32	.500		14.80	22		36.80	50
2660	Shutter/damper	G		12	1.333		49	59		108	143
2670	Shutter motor	G		16	1		119	44		163	198
2800	Circulators, liquid, 1/25 HP, 5.3 GPM	G	Q-1	14	1.143		268	53.50		321.50	375
2820	1/20 HP, 17 GPM	G		12	1.333		154	62.50		216.50	264
2850	1/20 HP, 17 GPM, stainless steel	G		12	1.333		243	62.50		305.50	360
2870	1/12 HP, 30 GPM	G		10	1.600		330	75		405	475
3000	Collector panels, air with aluminum absorber plate										
3010	Wall or roof mount										
3040	Flat black, plastic glazing										
3080	4' x 8'	G	Q-9	6	2.667	Ea.	635	118		753	875
3100	4' x 10'	G		5	3.200	"	780	141		921	1,075
3200	Flush roof mount, 10' to 16' x 22" wide	G		96	.167	L.F.	450	7.35		457.35	505
3210	Manifold, by L.F. width of collectors	G		160	.100	"	127	4.42		131.42	147
3300	Collector panels, liquid with copper absorber plate										
3320	Black chrome, tempered glass glazing										
3330	Alum. frame, 4' x 8', 5/32" single glazing	G	Q-1	9.50	1.684	Ea.	945	79		1,024	1,175
3390	Alum. frame, 4' x 10', 5/32" single glazing	G		6	2.667		1,075	125		1,200	1,375
3450	Flat black, alum. frame, 3.5' x 7.5'	G		9	1.778		690	83.50		773.50	885
3500	4' x 8'	G		5.50	2.909		865	136		1,001	1,150
3520	4' x 10'	G		10	1.600		970	75		1,045	1,175
3540	4' x 12.5'	G		5	3.200		1,175	150		1,325	1,500
3550	Liquid with fin tube absorber plate										
3560	Alum. frame 4' x 8' tempered glass	G	Q-1	10	1.600	Ea.	495	75		570	655
3580	Liquid with vacuum tubes, 4' x 6'-10"	G		9	1.778		850	83.50		933.50	1,050
3600	Liquid, full wetted, plastic, alum. frame, 3' x 10'	G		5	3.200		234	150		384	480
3650	Collector panel mounting, flat roof or ground rack	G		7	2.286		217	107		324	400
3670	Roof clamps	G		70	.229	Set	2.35	10.70		13.05	18.65
3700	Roof strap, teflon	G	1 Plum	205	.039	L.F.	19.95	2.03		21.98	25
3900	Differential controller with two sensors										
3930	Thermostat, hard wired	G	1 Plum	8	1	Ea.	79.50	52		131.50	165
3950	Line cord and receptacle	G		12	.667		555	34.50		589.50	660
4050	Pool valve system	G		2.50	3.200		264	167		431	540
4070	With 12 VAC actuator	G		2	4		276	208		484	615
4080	Pool pump system, 2" pipe size	G		6	1.333		172	69.50		241.50	294
4100	Five station with digital read-out	G		3	2.667		218	139		357	445
4150	Sensors										
4200	Brass plug, 1/2" MPT	G	1 Plum	32	.250	Ea.	17.95	13		30.95	39.50
4210	Brass plug, reversed	G		32	.250		24.50	13		37.50	46.50
4220	Freeze prevention	G		32	.250		22	13		35	43.50
4240	Screw attached	G		32	.250		8.35	13		21.35	28.50
4250	Brass, immersion	G		32	.250		26.50	13		39.50	48.50
4300	Heat exchanger										
4330	Fluid to air coil, up flow, 45 MBH	G	Q-1	4	4	Ea.	295	187		482	605
4380	70 MBH	G		3.50	4.571		335	214		549	685
4400	80 MBH	G		3	5.333		445	250		695	865
4580	Fluid to fluid package includes two circulating pumps										
4590	expansion tank, check valve, relief valve										
4600	controller, high temperature cutoff and sensors	G	Q-1	2.50	6.400	Ea.	695	300		995	1,225
4650	Heat transfer fluid										

23 56 Solar Energy Heating Equipment

23 56 19 – Solar Heating Components

23 56 19.50 Solar Heating Ancillary

	23 56 19.50 Solar Heating Ancillary		Crew	Daily Output	Labor-Hours	Unit	Material	2010 Bare Costs Labor	Equipment	Total	Total Incl O&P
4700	Propylene glycol, inhibited anti-freeze	G	1 Plum	28	.286	Gal.	12.85	14.85		27.70	36.50
4800	Solar storage tanks, knocked down										
4810	Air, galvanized steel clad, double wall, 4" fiberglass										
4820	insulation, 20 Mil PVC lining										
4870	4' high, 4' x 4', = 64 C.F./450 gallons	G	Q-9	2	8	Ea.	2,275	355		2,630	3,025
4880	4' x 8' = 128 C.F./900 gallons	G		1.50	10.667		3,225	470		3,695	4,275
4890	4' x 12' = 190 C.F./1300 gallons	G		1.30	12.308		4,300	545		4,845	5,550
4900	8' x 8' = 250 C.F./1700 gallons	G		1	16		4,800	705		5,505	6,350
5010	6'-3" high, 7' x 7' = 306 C.F./2000 gallons	G	Q-10	1.20	20		11,700	915		12,615	14,300
5020	7' x 10'-6" = 459 C.F./3000 gallons	G		.80	30		14,700	1,375		16,075	18,300
5030	7' x 14' = 613 C.F./4000 gallons	G		.60	40		17,200	1,825		19,025	21,800
5040	10'-6" x 10'-6" = 689 C.F./4500 gallons	G		.50	48		17,800	2,200		20,000	22,800
5050	10'-6" x 14' = 919 C.F./6000 gallons	G		.40	60		21,000	2,750		23,750	27,300
5060	14' x 14' = 1225 C.F./8000 gallons	G	Q-11	.40	80		24,500	3,750		28,250	32,600
5070	14' x 17'-6" = 1531 C.F./10,000 gallons	G		.30	106		27,600	5,000		32,600	37,900
5080	17'-6" x 17'-6" = 1914 C.F./12,500 gallons	G		.25	128		32,500	6,000		38,500	44,800
5090	17'-6" x 21' = 2297 C.F./15,000 gallons	G		.20	160		36,200	7,475		43,675	51,000
5100	21' x 21' = 2756 C.F./18,000 gallons	G		.18	177		41,500	8,325		49,825	58,000
5120	30 Mil reinforced Chemflex lining,										
5140	4' high, 4' x 4' = 64 C.F./450 gallons	G	Q-9	2	8	Ea.	2,525	355		2,880	3,300
5150	4' x 8' = 128 C.F./900 gallons	G		1.50	10.667		3,825	470		4,295	4,925
5160	4' x 12' = 190 C.F./1300 gallons	G		1.30	12.308		5,050	545		5,595	6,400
5170	8' x 8' = 250 C.F./1700 gallons	G		1	16		5,650	705		6,355	7,275
5190	6'-3" high, 7' x 7' = 306 C.F./2000 gallons	G	Q-10	1.20	20		12,500	915		13,415	15,100
5200	7' x 10'-6" = 459 C.F./3000 gallons	G		.80	30		15,200	1,375		16,575	18,900
5210	7' x 14' = 613 C.F./4000 gallons	G		.60	40		17,700	1,825		19,525	22,300
5220	10'-6" x 10'-6" = 689 C.F./4500 gallons	G		.50	48		18,300	2,200		20,500	23,400
5230	10'-6" x 14' = 919 C.F./6000 gallons	G		.40	60		21,500	2,750		24,250	27,800
5240	14' x 14' = 1225 C.F./8000 gallons	G	Q-11	.40	80		25,100	3,750		28,850	33,400
5250	14' x 17'-6" = 1531 C.F./10,000 gallons	G		.30	106		28,100	5,000		33,100	38,600
5260	17'-6" x 17'-6" = 1914 C.F./12,500 gallons	G		.25	128		33,100	6,000		39,100	45,500
5270	17'-6" x 21' = 2297 C.F./15,000 gallons	G		.20	160		37,300	7,475		44,775	52,500
5280	21' x 21' = 2756 C.F./18,000 gallons	G		.18	177		42,700	8,325		51,025	59,500
5290	30 Mil reinforced Hypalon lining, add						.02%				
7000	Solar control valves and vents										
7050	Air purger, 1" pipe size	G	1 Plum	12	.667	Ea.	43.50	34.50		78	99.50
7070	Air eliminator, automatic 3/4" size	G		32	.250		28.50	13		41.50	51
7090	Air vent, automatic, 1/8" fitting	G		32	.250		12.65	13		25.65	33.50
7100	Manual, 1/8" NPT	G		32	.250		2.78	13		15.78	22.50
7120	Backflow preventer, 1/2" pipe size	G		16	.500		58.50	26		84.50	103
7130	3/4" pipe size	G		16	.500		106	26		132	156
7150	Balancing valve, 3/4" pipe size	G		20	.400		43	21		64	78
7180	Draindown valve, 1/2" copper tube	G		9	.889		202	46.50		248.50	292
7200	Flow control valve, 1/2" pipe size	G		22	.364		107	18.95		125.95	147
7220	Expansion tank, up to 5 gal.	G		32	.250		61.50	13		74.50	87
7250	Hydronic controller (aquastat)	G		8	1		182	52		234	279
7400	Pressure gauge, 2" dial	G		32	.250		22	13		35	44
7450	Relief valve, temp. and pressure 3/4" pipe size	G		30	.267		15.60	13.90		29.50	38
7500	Solenoid valve, normally closed										
7520	Brass, 3/4" NPT, 24V	G	1 Plum	9	.889	Ea.	109	46.50		155.50	190
7530	1" NPT, 24V	G		9	.889		850	46.50		896.50	1,000
7750	Vacuum relief valve, 3/4" pipe size	G		32	.250		26.50	13		39.50	48.50
7800	Thermometers										

23 56 Solar Energy Heating Equipment

23 56 19 – Solar Heating Components

23 56 19.50 Solar Heating Ancillary		Crew	Daily Output	Labor-Hours	Unit	Material	2010 Bare Costs Labor	Equipment	Total	Total Incl O&P
7820	Digital temperature monitoring, 4 locations G	1 Plum	2.50	3.200	Ea.	128	167		295	390
7900	Upright, 1/2" NPT G		8	1		24	52		76	105
7970	Remote probe, 2" dial G		8	1		30.50	52		82.50	112
7990	Stem, 2" dial, 9" stem G	↓	16	.500	↓	20	26		46	61
8250	Water storage tank with heat exchanger and electric element									
8270	66 gal. with 2" x 2 lb. density insulation G	1 Plum	1.60	5	Ea.	970	260		1,230	1,475
8300	80 gal. with 2" x 2 lb. density insulation G		1.60	5		1,100	260		1,360	1,600
8380	120 gal. with 2" x 2 lb. density insulation G		1.40	5.714		1,225	297		1,522	1,800
8400	120 gal. with 2" x 2 lb. density insul., 40 S.F. heat coil G	↓	1.40	5.714	↓	1,575	297		1,872	2,175
8500	Water storage module, plastic									
8600	Tubular, 12" diameter, 4' high G	1 Carp	48	.167	Ea.	103	6.95		109.95	124
8610	12" diameter, 8' high G		40	.200		159	8.30		167.30	188
8620	18" diameter, 5' high G		38	.211		174	8.75		182.75	205
8630	18" diameter, 10' high G	↓	32	.250		229	10.40		239.40	268
8640	58" diameter, 5' high G	2 Carp	32	.500		490	21		511	565
8650	Cap, 12" diameter G					19			19	21
8660	18" diameter G				↓	24			24	26.50

23 57 Heat Exchangers for HVAC

23 57 16 – Steam-to-Water Heat Exchangers

23 57 16.10 Shell/Tube Type Steam-to-Water Heat Exch.		Crew	Daily Output	Labor-Hours	Unit	Material	Labor	Equipment	Total	Total Incl O&P
0010	**SHELL AND TUBE TYPE STEAM-TO-WATER HEAT EXCHANGERS**									
0016	Shell & tube type, 2 or 4 pass, 3/4" O.D. copper tubes,									
0020	C.I. heads, C.I. tube sheet, steel shell									
0100	Hot water 40°F to 180°F, by steam at 10 PSI									
0120	8 GPM	Q-5	6	2.667	Ea.	1,725	125		1,850	2,075
0140	10 GPM		5	3.200		2,600	149		2,749	3,100
0160	40 GPM		4	4		4,050	187		4,237	4,725
0180	64 GPM		2	8		6,200	375		6,575	7,375
0200	96 GPM	↓	1	16		8,300	745		9,045	10,300
0220	120 GPM	Q-6	1.50	16		10,900	775		11,675	13,200
0240	168 GPM		1	24		13,400	1,150		14,550	16,500
0260	240 GPM		.80	30		20,900	1,450		22,350	25,200
0300	600 GPM	↓	.70	34.286	↓	44,800	1,650		46,450	52,000
0500	For bronze head and tube sheet, add					50%				

23 57 19 – Liquid-to-Liquid Heat Exchangers

23 57 19.13 Plate-Type, Liquid-to-Liquid Heat Exchangers		Crew	Daily Output	Labor-Hours	Unit	Material	Labor	Equipment	Total	Total Incl O&P
0010	**PLATE-TYPE, LIQUID-TO-LIQUID HEAT EXCHANGERS**									
3000	Plate type,									
3100	400 GPM	Q-6	.80	30	Ea.	31,400	1,450		32,850	36,700
3120	800 GPM	"	.50	48		54,000	2,325		56,325	63,000
3140	1200 GPM	Q-7	.34	94.118		80,500	4,650		85,150	95,500
3160	1800 GPM	"	.24	133	↓	106,500	6,600		113,100	127,000

23 57 19.16 Shell-Type, Liquid-to-Liquid Heat Exchangers		Crew	Daily Output	Labor-Hours	Unit	Material	Labor	Equipment	Total	Total Incl O&P
0010	**SHELL-TYPE, LIQUID-TO-LIQUID HEAT EXCHANGERS**									
1000	Hot water 40°F to 140°F, by water at 200°F									
1020	7 GPM	Q-5	6	2.667	Ea.	2,125	125		2,250	2,525
1040	16 GPM		5	3.200		3,025	149		3,174	3,550
1060	34 GPM		4	4		4,575	187		4,762	5,300
1080	55 GPM	↓	3	5.333	↓	6,625	249		6,874	7,675

23 57 Heat Exchangers for HVAC

23 57 19 – Liquid-to-Liquid Heat Exchangers

23 57 19.16 Shell-Type, Liquid-to-Liquid Heat Exchangers		Crew	Daily Output	Labor-Hours	Unit	2010 Bare Costs Material	Labor	Equipment	Total	Total Incl O&P
1100	74 GPM	Q-5	1.50	10.667	Ea.	8,275	500		8,775	9,850
1120	86 GPM		1.40	11.429		11,100	535		11,635	13,000
1140	112 GPM	Q-6	2	12		13,800	580		14,380	16,100
1160	126 GPM		1.80	13.333		17,200	645		17,845	19,900
1180	152 GPM		1	24		22,000	1,150		23,150	26,000

23 61 Refrigerant Compressors

23 61 15 – Rotary Refrigerant Compressors

23 61 15.10 Rotary Compressors		Crew	Daily Output	Labor-Hours	Unit	Material	Labor	Equipment	Total	Total Incl O&P
0010	**ROTARY COMPRESSORS**									
0100	Refrigeration, hermetic, switches and protective devices									
0210	1.25 ton	1 Stpi	3	2.667	Ea.	221	138		359	450
0220	1.42 ton		3	2.667		223	138		361	450
0230	1.68 ton		2.80	2.857		243	148		391	490
0240	2.00 ton		2.60	3.077		250	160		410	515
0250	2.37 ton		2.50	3.200		282	166		448	560
0260	2.67 ton		2.40	3.333		296	173		469	585
0270	3.53 ton		2.30	3.478		325	181		506	625
0280	4.43 ton		2.20	3.636		415	189		604	740
0290	5.08 ton		2.10	3.810		455	198		653	795
0300	5.22 ton		2	4		640	208		848	1,025
0310	7.31 ton	Q-5	3	5.333		1,025	249		1,274	1,500
0320	9.95 ton		2.60	6.154		1,200	287		1,487	1,750
0330	11.6 ton		2.50	6.400		1,550	299		1,849	2,150
0340	15.25 ton		2.30	6.957		2,125	325		2,450	2,800
0350	17.7 ton		2.10	7.619		2,275	355		2,630	3,025

23 61 16 – Reciprocating Refrigerant Compressors

23 61 16.10 Reciprocating Compressors		Crew	Daily Output	Labor-Hours	Unit	Material	Labor	Equipment	Total	Total Incl O&P
0010	**RECIPROCATING COMPRESSORS**									
0990	Refrigeration, recip. hermetic, switches & protective devices									
1000	10 ton	Q-5	1	16	Ea.	8,850	745		9,595	10,900
1100	20 ton	Q-6	.72	33.333		13,400	1,625		15,025	17,100
1200	30 ton		.64	37.500		15,100	1,825		16,925	19,300
1300	40 ton		.44	54.545		16,600	2,650		19,250	22,300
1400	50 ton		.20	120		17,500	5,800		23,300	27,900
1500	75 ton	Q-7	.27	118		19,300	5,850		25,150	30,100
1600	130 ton	"	.21	152		22,100	7,525		29,625	35,600

23 61 19 – Scroll Refrigerant Compressors

23 61 19.10 Scroll Compressors		Crew	Daily Output	Labor-Hours	Unit	Material	Labor	Equipment	Total	Total Incl O&P
0010	**SCROLL COMPRESSORS**									
1800	Refrigeration, scroll type									
1810	1.9 ton	1 Stpi	2.70	2.963	Ea.	282	154		436	540
1820	2.35 ton		2.50	3.200		325	166		491	605
1830	2.82 ton		2.35	3.404		350	177		527	650
1840	3.3 ton		2.30	3.478		370	181		551	675
1850	3.83 ton		2.25	3.556		415	185		600	735
1860	4.1 ton		2.20	3.636		475	189		664	805
1870	4.8 ton		2.10	3.810		500	198		698	845
1880	5 ton		2.07	3.865		510	201		711	860

23 62 Packaged Compressor and Condenser Units

23 62 13 – Packaged Air-Cooled Refrigerant Compressor and Condenser Units

23 62 13.10 Packaged Air-Cooled Refrig. Condensing Units		Crew	Daily Output	Labor-Hours	Unit	2010 Bare Costs Material	Labor	Equipment	Total	Total Incl O&P
0010	**PACKAGED AIR-COOLED REFRIGERANT CONDENSING UNITS**									
0020	Condensing unit									
0030	Air cooled, compressor, standard controls									
0050	1.5 ton	Q-5	2.50	6.400	Ea.	1,175	299		1,474	1,750
0100	2 ton		2.10	7.619		1,200	355		1,555	1,850
0200	2.5 ton		1.70	9.412		1,300	440		1,740	2,100
0300	3 ton		1.30	12.308		1,325	575		1,900	2,325
0350	3.5 ton		1.10	14.545		1,575	680		2,255	2,750
0400	4 ton		.90	17.778		1,775	830		2,605	3,200
0500	5 ton		.60	26.667		2,125	1,250		3,375	4,225
0550	7.5 ton		.55	29.091		3,525	1,350		4,875	5,900
0560	8.5 ton		.53	30.189		4,350	1,400		5,750	6,925
0600	10 ton		.50	32		4,750	1,500		6,250	7,475
0620	12.5 ton		.48	33.333		7,225	1,550		8,775	10,300
0650	15 ton		.40	40		7,900	1,875		9,775	11,500
0700	20 ton	Q-6	.40	60		10,900	2,900		13,800	16,400
0720	25 ton		.35	68.571		15,500	3,325		18,825	22,000
0750	30 ton		.30	80		17,800	3,875		21,675	25,400
0800	40 ton		.20	120		22,500	5,800		28,300	33,500
0840	50 ton		.18	133		26,800	6,450		33,250	39,200
0860	60 ton		.16	150		30,900	7,275		38,175	44,900
0900	70 ton		.14	171		35,300	8,300		43,600	51,000
1000	80 ton		.12	200		39,200	9,675		48,875	57,500
1100	100 ton		.09	266		48,100	12,900		61,000	72,500

23 62 23 – Packaged Water-Cooled Refrigerant Compressor and Condenser Units

23 62 23.10 Packaged Water-Cooled Refrigerant Condensing Units		Crew	Daily Output	Labor-Hours	Unit	Material	Labor	Equipment	Total	Total Incl O&P
0010	**PACKAGED WATER-COOLED REFRIGERANT CONDENSING UNITS**									
2000	Water cooled, compressor, heat exchanger, controls									
2100	5 ton	Q-5	.70	22.857	Ea.	12,600	1,075		13,675	15,500
2110	10 ton		.60	26.667		14,700	1,250		15,950	18,100
2200	15 ton		.50	32		19,900	1,500		21,400	24,200
2300	20 ton	Q-6	.40	60		23,000	2,900		25,900	29,700
2310	30 ton		.30	80		26,400	3,875		30,275	34,800
2400	40 ton		.20	120		28,700	5,800		34,500	40,300
2410	60 ton		.18	133		40,200	6,450		46,650	54,000
2420	80 ton		.14	171		46,500	8,300		54,800	63,500
2500	100 ton		.11	218		57,000	10,600		67,600	78,500
2510	120 ton		.10	240		91,500	11,600		103,100	118,500

23 63 Refrigerant Condensers

23 63 13 – Air-Cooled Refrigerant Condensers

23 63 13.10 Air-Cooled Refrig. Condensers		Crew	Daily Output	Labor-Hours	Unit	Material	Labor	Equipment	Total	Total Incl O&P
0010	**AIR-COOLED REFRIG. CONDENSERS**, Ratings for 30°F TD, R-22.									
0080	Air cooled, belt drive, propeller fan									
0220	45 ton	Q-6	.70	34.286	Ea.	12,300	1,650		13,950	16,100
0240	50 ton		.69	34.985		12,700	1,700		14,400	16,600
0260	54 ton		.64	37.795		13,200	1,825		15,025	17,300
0280	59 ton		.58	41.308		14,200	2,000		16,200	18,600
0300	65 ton		.53	45.541		14,800	2,200		17,000	19,500
0320	73 ton		.47	51.173		16,700	2,475		19,175	22,100

23 63 13 – Air-Cooled Refrigerant Condensers

23 63 13.10 Air-Cooled Refrig. Condensers		Crew	Daily Output	Labor-Hours	Unit	Material	Labor	Equipment	Total	Total Incl O&P
						2010 Bare Costs				
0340	81 ton	Q-6	.42	56.738	Ea.	19,000	2,750		21,750	25,000
0360	86 ton		.40	60.302		19,800	2,925		22,725	26,200
0380	88 ton		.39	61.697		21,200	3,000		24,200	27,800
0400	101 ton	Q-7	.45	70.640		24,900	3,500		28,400	32,600
0500	159 ton		.31	102		37,200	5,075		42,275	48,500
0600	228 ton		.22	148		54,000	7,325		61,325	70,000
1500	May be specified single or multi-circuit									
1550	Air cooled, direct drive, propeller fan									
1590	1 ton	Q-5	3.80	4.211	Ea.	685	197		882	1,050
1600	1-1/2 ton		3.60	4.444		840	208		1,048	1,225
1620	2 ton		3.20	5		900	234		1,134	1,350
1630	3 ton		2.40	6.667		1,000	310		1,310	1,575
1640	5 ton		2	8		1,825	375		2,200	2,575
1650	8 ton		1.80	8.889		2,400	415		2,815	3,250
1660	10 ton		1.40	11.429		3,025	535		3,560	4,125
1670	12 ton		1.30	12.308		3,350	575		3,925	4,525
1680	14 ton		1.20	13.333		3,800	625		4,425	5,100
1690	16 ton		1.10	14.545		4,150	680		4,830	5,600
1700	21 ton		1	16		4,850	745		5,595	6,450
1720	26 ton		.84	19.002		5,400	885		6,285	7,275
1740	30 ton		.70	22.792		8,825	1,075		9,900	11,300
1760	41 ton	Q-6	.77	31.008		9,700	1,500		11,200	13,000
1780	52 ton		.66	36.419		13,800	1,775		15,575	17,900
1800	63 ton		.55	44.037		15,300	2,125		17,425	20,000
1820	76 ton		.45	52.980		17,300	2,575		19,875	23,000
1840	86 ton		.40	60		21,400	2,900		24,300	28,000
1860	97 ton		.35	67.989		24,400	3,300		27,700	31,700
1880	105 ton	Q-7	.44	73.563		28,100	3,625		31,725	36,400
1890	118 ton		.39	82.687		30,300	4,075		34,375	39,400
1900	126 ton		.36	88.154		35,200	4,350		39,550	45,200
1910	136 ton		.34	95.238		39,400	4,700		44,100	50,500
1920	142 ton		.32	99.379		40,400	4,900		45,300	52,000

23 63 30 – Lubricants

23 63 30.10 Lubricant Oils

		Crew	Daily Output	Labor-Hours	Unit	Material	Labor	Equipment	Total	Total Incl O&P
0010	**LUBRICANT OILS**									
8000	Oils									
8100	Lubricating									
8120	Oil, lubricating				Oz.	.25			.25	.28
8500	Refrigeration									
8520	Oil, refrigeration				Gal.	23			23	25.50
8525	Oil, refrigeration				Qt.	5.80			5.80	6.40

23 63 33 – Evaporative Refrigerant Condensers

23 63 33.10 Evaporative Condensers

		Crew	Daily Output	Labor-Hours	Unit	Material	Labor	Equipment	Total	Total Incl O&P
0010	**EVAPORATIVE CONDENSERS**									
3400	Evaporative, copper coil, pump, fan motor									
3440	10 ton	Q-5	.54	29.630	Ea.	5,550	1,375		6,925	8,175
3460	15 ton		.50	32		5,775	1,500		7,275	8,600
3480	20 ton		.47	34.043		6,000	1,600		7,600	8,975
3500	25 ton		.45	35.556		6,425	1,650		8,075	9,550
3520	30 ton		.42	38.095		6,825	1,775		8,600	10,200
3540	40 ton	Q-6	.49	48.980		9,050	2,375		11,425	13,500
3560	50 ton		.39	61.538		11,300	2,975		14,275	17,000

23 63 Refrigerant Condensers

23 63 33 – Evaporative Refrigerant Condensers

23 63 33.10 Evaporative Condensers		Crew	Daily Output	Labor-Hours	Unit	Material	Labor	Equipment	Total	Total Incl O&P
						2010 Bare Costs				
3580	65 ton	Q-6	.35	68.571	Ea.	12,600	3,325		15,925	18,800
3600	80 ton		.33	72.727		13,700	3,525		17,225	20,300
3620	90 ton	↓	.29	82.759		15,000	4,000		19,000	22,500
3640	100 ton	Q-7	.36	88.889		16,400	4,400		20,800	24,600
3660	110 ton		.33	96.970		18,100	4,800		22,900	27,100
3680	125 ton		.30	106		19,600	5,275		24,875	29,500
3700	135 ton		.28	114		21,000	5,650		26,650	31,600
3720	150 ton		.25	128		23,200	6,325		29,525	35,000
3740	165 ton		.23	139		25,500	6,875		32,375	38,300
3760	185 ton	↓	.22	145		26,900	7,200		34,100	40,400
3860	For fan damper control, add	Q-5	2	8	↓	575	375		950	1,200

23 64 Packaged Water Chillers

23 64 13 – Absorption Water Chillers

23 64 13.13 Direct-Fired Absorption Water Chillers

		Crew	Daily Output	Labor-Hours	Unit	Material	Labor	Equipment	Total	Total Incl O&P
0010	**DIRECT-FIRED ABSORPTION WATER CHILLERS**									
3000	Gas fired, air cooled									
3220	5 ton	Q-5	.60	26.667	Ea.	10,000	1,250		11,250	12,900
3270	10 ton	"	.40	40	"	25,600	1,875		27,475	30,900
4000	Water cooled, duplex									
4130	100 ton	Q-7	.13	246	Ea.	136,000	12,200		148,200	167,500
4140	200 ton		.11	283		183,000	14,000		197,000	222,500
4150	300 ton		.11	299		219,500	14,800		234,300	263,500
4160	400 ton		.10	316		285,500	15,700		301,200	337,500
4170	500 ton		.10	329		359,500	16,300		375,800	419,500
4180	600 ton		.09	340		424,500	16,800		441,300	491,500
4190	700 ton		.09	359		470,500	17,800		488,300	544,000
4200	800 ton		.08	380		568,000	18,800		586,800	653,000
4210	900 ton		.08	400		658,000	19,800		677,800	753,500
4220	1000 ton	↓	.08	421	↓	731,500	20,800		752,300	835,500

23 64 13.16 Indirect-Fired Absorption Water Chillers

		Crew	Daily Output	Labor-Hours	Unit	Material	Labor	Equipment	Total	Total Incl O&P
0010	**INDIRECT-FIRED ABSORPTION WATER CHILLERS**									
0020	Steam or hot water, water cooled									
0050	100 ton	Q-7	.13	240	Ea.	125,000	11,900		136,900	155,500
0100	148 ton		.12	258		194,500	12,000		207,300	233,000
0200	200 ton		.12	275		216,500	13,600		230,100	259,000
0240	250 ton		.11	290		238,500	14,400		252,900	283,500
0300	354 ton		.11	304		298,000	15,100		313,100	350,000
0400	420 ton		.10	323		325,500	16,000		341,500	382,000
0500	665 ton		.09	340		457,000	16,800		473,800	527,500
0600	750 ton		.09	363		534,000	18,000		552,000	614,000
0700	850 ton		.08	385		570,000	19,100		589,100	655,500
0800	955 ton		.08	410		609,500	20,300		629,800	701,000
0900	1125 ton		.08	421		703,500	20,800		724,300	805,000
1000	1250 ton		.07	444		756,500	22,000		778,500	865,000
1100	1465 ton		.07	463		897,000	22,900		919,900	1,021,000
1200	1660 ton	↓	.07	477	↓	1,055,000	23,600		1,078,600	1,196,000
2000	For two stage unit, add					80%	25%			

23 64 Packaged Water Chillers

23 64 16 – Centrifugal Water Chillers

	23 64 16.10 Centrifugal Type Water Chillers	Crew	Daily Output	Labor-Hours	Unit	2010 Bare Costs Material	Labor	Equipment	Total	Total Incl O&P
0010	**CENTRIFUGAL TYPE WATER CHILLERS**, With standard controls									
0020	Centrifugal liquid chiller, water cooled									
0030	not including water tower									
0100	2000 ton (twin 1000 ton units)	Q-7	.07	477	Ea.	832,500	23,600		856,100	951,500
0274	Centrifugal, packaged unit, water cooled, not incl. tower									
0280	400 ton	Q-7	.11	283	Ea.	129,500	14,000		143,500	163,500
0282	450 ton		.11	290		187,500	14,400		201,900	227,500
0290	500 ton		.11	296		208,000	14,600		222,600	251,000
0292	550 ton		.11	304		229,000	15,100		244,100	274,500
0300	600 ton		.10	310		249,500	15,400		264,900	297,500
0302	650 ton		.10	320		270,500	15,800		286,300	321,000
0304	700 ton		.10	326		291,500	16,100		307,600	344,500
0306	750 ton		.10	333		312,000	16,500		328,500	368,000
0310	800 ton		.09	340		333,000	16,800		349,800	391,500
0312	850 ton		.09	351		354,000	17,400		371,400	415,000
0316	900 ton		.09	359		374,500	17,800		392,300	438,500
0318	950 ton		.09	363		395,500	18,000		413,500	462,000
0320	1000 ton		.09	372		416,500	18,400		434,900	485,500
0324	1100 ton		.08	385		445,000	19,100		464,100	518,000
0330	1200 ton		.08	395		485,500	19,500		505,000	563,500
0340	1300 ton		.08	410		525,500	20,300		545,800	609,000
0350	1400 ton		.08	421		566,000	20,800		586,800	654,000
0360	1500 ton		.08	426		606,500	21,100		627,600	698,500
0370	1600 ton		.07	438		647,000	21,700		668,700	744,000
0380	1700 ton		.07	450		687,500	22,300		709,800	789,500
0390	1800 ton		.07	457		728,000	22,600		750,600	834,500
0400	1900 ton		.07	470		768,500	23,300		791,800	880,000
0430	2000 ton		.07	492		809,000	24,300		833,300	926,000
0440	2500 ton	↓	.06	524	↓	1,011,000	25,900		1,036,900	1,151,000

23 64 19 – Reciprocating Water Chillers

	23 64 19.10 Reciprocating Type Water Chillers	Crew	Daily Output	Labor-Hours	Unit	Material	Labor	Equipment	Total	Total Incl O&P
0010	**RECIPROCATING TYPE WATER CHILLERS**, With standard controls									
0494	Water chillers, integral air cooled condenser									
0538	60 ton cooling	Q-7	.28	115	Ea.	47,700	5,700		53,400	61,000
0546	70 ton cooling		.27	119		54,500	5,925		60,425	69,000
0552	80 ton cooling		.26	123		59,000	6,075		65,075	74,000
0554	90 ton cooling		.25	125		63,500	6,225		69,725	79,000
0600	100 ton cooling		.25	129		71,500	6,375		77,875	88,000
0620	110 ton cooling		.24	132		76,500	6,525		83,025	94,000
0630	130 ton cooling		.24	135		91,000	6,675		97,675	110,000
0640	150 ton cooling		.23	137		101,000	6,825		107,825	121,000
0650	175 ton cooling		.23	140		113,000	6,975		119,975	135,000
0654	190 ton cooling		.22	144		124,500	7,125		131,625	147,500
0660	210 ton cooling		.22	148		128,000	7,325		135,325	152,000
0662	250 ton cooling		.21	151		148,500	7,500		156,000	174,500
0664	275 ton cooling		.21	156		154,000	7,725		161,725	180,500
0666	300 ton cooling		.20	160		172,000	7,900		179,900	201,500
0668	330 ton cooling		.20	164		201,000	8,100		209,100	233,000
0670	360 ton cooling		.19	168		221,000	8,325		229,325	255,500
0672	390 ton cooling		.19	172		234,000	8,550		242,550	270,500
0674	420 ton cooling	↓	.18	177	↓	244,500	8,800		253,300	282,000
0980	Water cooled, multiple compressor, semi-hermetic, tower not incl.									

23 64 19 – Reciprocating Water Chillers

23 64 19.10 Reciprocating Type Water Chillers		Crew	Daily Output	Labor-Hours	Unit	Material	Labor	Equipment	Total	Total Incl O&P
1000	15 ton cooling	Q-6	.36	65.934	Ea.	16,000	3,200		19,200	22,400
1020	25 ton cooling	Q-7	.41	78.049		17,900	3,850		21,750	25,500
1040	28 ton cooling		.36	89.888		19,500	4,450		23,950	28,100
1060	35 ton cooling		.31	101		20,400	5,025		25,425	30,000
1080	40 ton cooling		.30	108		28,700	5,350		34,050	39,500
1100	50 ton cooling		.28	113		30,500	5,625		36,125	41,900
1120	60 ton cooling		.25	125		33,600	6,225		39,825	46,200
1130	75 ton cooling		.23	139		48,900	6,875		55,775	64,500
1140	85 ton cooling		.21	151		51,000	7,500		58,500	67,000
1150	95 ton cooling		.19	164		54,500	8,150		62,650	72,000
1160	100 ton cooling		.18	179		58,500	8,875		67,375	78,000
1170	115 ton cooling		.17	190		61,500	9,425		70,925	81,500
1180	125 ton cooling		.16	196		66,500	9,700		76,200	87,500
1200	145 ton cooling		.16	202		77,500	10,000		87,500	100,500
1210	155 ton cooling		.15	210		82,000	10,400		92,400	105,500
1300	Water cooled, single compressor, semi-hermetic, tower not incl.									
1320	40 ton cooling,	Q-7	.30	108	Ea.	16,000	5,350		21,350	25,600
1340	50 ton cooling		.28	113		19,800	5,625		25,425	30,200
1360	60 ton cooling		.25	125		22,000	6,225		28,225	33,500
1451	Water cooled, dual compressors, semi-hermetic, tower not incl.									
1500	80 ton cooling	Q-7	.14	222	Ea.	27,400	11,000		38,400	46,600
1520	100 ton cooling		.14	228		37,000	11,300		48,300	57,500
1540	120 ton cooling		.14	231		44,800	11,500		56,300	66,500
4000	Packaged chiller, remote air cooled condensers not incl.									
4020	15 ton cooling	Q-7	.30	108	Ea.	13,600	5,350		18,950	22,900
4030	20 ton cooling		.28	115		14,800	5,725		20,525	24,900
4040	25 ton cooling		.25	125		16,700	6,225		22,925	27,600
4050	35 ton cooling		.24	133		17,600	6,625		24,225	29,200
4060	40 ton cooling		.22	144		21,300	7,125		28,425	34,100
4070	50 ton cooling		.21	153		24,900	7,600		32,500	38,800
4080	60 ton cooling		.20	164		29,000	8,100		37,100	44,200
4090	75 ton cooling		.18	173		39,300	8,600		47,900	56,000
4100	85 ton cooling		.17	183		41,300	9,100		50,400	59,000
4110	95 ton cooling		.17	193		44,500	9,575		54,075	63,500
4120	105 ton cooling		.16	203		48,000	10,100		58,100	68,000
4130	115 ton cooling		.15	213		52,000	10,500		62,500	73,000
4140	125 ton cooling		.14	223		55,500	11,100		66,600	77,500
4150	145 ton cooling		.14	233		77,500	11,500		89,000	103,000

23 64 23 – Scroll Water Chillers

23 64 23.10 Scroll Water Chillers		Crew	Daily Output	Labor-Hours	Unit	Material	Labor	Equipment	Total	Total Incl O&P
0010	**SCROLL WATER CHILLERS,** With standard controls									
0490	Packaged w/integral air cooled condenser, 15 ton cool	Q-7	.37	86.486	Ea.	17,400	4,275		21,675	25,500
0500	20 ton cooling		.34	94.118		23,600	4,650		28,250	32,900
0510	25 ton cooling		.34	94.118		24,600	4,650		29,250	34,100
0515	30 ton cooling		.31	101		26,300	5,025		31,325	36,500
0517	35 ton cooling		.31	104		29,300	5,175		34,475	40,100
0520	40 ton cooling		.30	108		31,700	5,350		37,050	42,800
0528	45 ton cooling		.29	109		33,600	5,425		39,025	45,200
0536	50 ton cooling		.28	113		36,700	5,600		42,300	48,800
0680	Scroll water cooled, single compressor, hermetic, tower not incl.									
0700	2 ton cooling	Q-5	.57	28.070	Ea.	3,225	1,300		4,525	5,525
0710	5 ton cooling		.57	28.070		3,850	1,300		5,150	6,225

Column group "2010 Bare Costs" covers Material, Labor, Equipment, Total.

23 64 Packaged Water Chillers

23 64 23 – Scroll Water Chillers

23 64 23.10 Scroll Water Chillers		Crew	Daily Output	Labor-Hours	Unit	Material	Labor	Equipment	Total	Total Incl O&P
						2010 Bare Costs				
0720	6 ton cooling	Q-5	.42	38.005	Ea.	4,850	1,775		6,625	8,000
0740	8 ton cooling	↓	.31	52.117		5,325	2,425		7,750	9,500
0760	10 ton cooling	Q-6	.36	67.039		6,125	3,250		9,375	11,600
0780	15 ton cooling	"	.33	72.727		9,625	3,525		13,150	15,900
0800	20 ton cooling	Q-7	.38	83.990		10,900	4,150		15,050	18,200
0820	30 ton cooling	"	.33	96.096	↓	12,200	4,750		16,950	20,600

23 64 26 – Rotary-Screw Water Chillers

23 64 26.10 Rotary-Screw Type Water Chillers		Crew	Daily Output	Labor-Hours	Unit	Material	Labor	Equipment	Total	Total Incl O&P
0010	**ROTARY-SCREW TYPE WATER CHILLERS**, With standard controls									
0110	Screw, liquid chiller, air cooled, insulated evaporator									
0120	130 ton	Q-7	.14	228	Ea.	81,000	11,300		92,300	106,000
0124	160 ton		.13	246		100,000	12,200		112,200	128,000
0128	180 ton		.13	250		112,500	12,400		124,900	142,000
0132	210 ton		.12	258		123,500	12,800		136,300	155,000
0136	270 ton		.12	266		141,500	13,200		154,700	175,500
0140	320 ton	↓	.12	275	↓	177,500	13,600		191,100	215,500
0200	Packaged unit, water cooled, not incl. tower									
0210	80 ton	Q-7	.14	223	Ea.	42,100	11,100		53,200	63,000
0220	100 ton		.14	230		52,500	11,400		63,900	74,500
0230	150 ton		.13	240		69,500	11,900		81,400	94,500
0240	200 ton		.13	251		82,000	12,500		94,500	108,500
0250	250 ton		.12	260		95,000	12,900		107,900	124,000
0260	300 ton		.12	266		109,500	13,200		122,700	140,500
0270	350 ton	↓	.12	275	↓	156,000	13,600		169,600	192,000
1450	Water cooled, tower not included									
1560	135 ton cooling, screw compressors	Q-7	.14	235	Ea.	55,000	11,600		66,600	78,000
1580	150 ton cooling, screw compressors		.13	240		63,500	11,900		75,400	88,000
1620	200 ton cooling, screw compressors		.13	250		87,500	12,400		99,900	115,000
1660	291 ton cooling, screw compressors	↓	.12	260	↓	91,500	12,900		104,400	120,500

23 64 33 – Direct Expansion Water Chillers

23 64 33.10 Direct Expansion Type Water Chillers		Crew	Daily Output	Labor-Hours	Unit	Material	Labor	Equipment	Total	Total Incl O&P
0010	**DIRECT EXPANSION TYPE WATER CHILLERS**, With standard controls									
8000	Direct expansion, shell and tube type, for built up systems									
8020	1 ton	Q-5	2	8	Ea.	6,475	375		6,850	7,675
8030	5 ton		1.90	8.421		10,800	395		11,195	12,400
8040	10 ton		1.70	9.412		13,300	440		13,740	15,400
8050	20 ton		1.50	10.667		15,200	500		15,700	17,500
8060	30 ton		1	16		19,400	745		20,145	22,400
8070	50 ton	↓	.90	17.778		31,200	830		32,030	35,600
8080	100 ton	Q-6	.90	26.667	↓	55,500	1,300		56,800	63,000

23 65 13 – Forced-Draft Cooling Towers

23 65 13.10 Forced-Draft Type Cooling Towers		Crew	Daily Output	Labor-Hours	Unit	2010 Bare Costs Material	Labor	Equipment	Total	Total Incl O&P
0010	**FORCED-DRAFT TYPE COOLING TOWERS**, Packaged units D3030–310									
0070	Galvanized steel									
0080	Induced draft, crossflow									
0100	Vertical, belt drive, 61 tons	Q-6	90	.267	TonAC	163	12.90		175.90	198
0150	100 ton		100	.240		156	11.60		167.60	188
0200	115 ton		109	.220		145	10.65		155.65	176
0250	131 ton		120	.200		144	9.70		153.70	173
0260	162 ton		132	.182		142	8.80		150.80	169
1000	For higher capacities, use multiples									
1500	Induced air, double flow									
1900	Vertical, gear drive, 167 ton	Q-6	126	.190	TonAC	138	9.20		147.20	166
2000	297 ton		129	.186		86	9		95	109
2100	582 ton		132	.182		69.50	8.80		78.30	89.50
2150	849 ton		142	.169		67.50	8.20		75.70	86.50
2200	1016 ton		150	.160		62.50	7.75		70.25	80.50
2500	Blow through, centrifugal type									
2510	50 ton	Q-6	2.28	10.526	Ea.	10,000	510		10,510	11,800
2520	75 ton		1.52	15.789		13,000	765		13,765	15,500
2524	100 ton		1.22	19.672		15,300	955		16,255	18,200
2528	125 ton	Q-7	1.31	24.427		17,900	1,200		19,100	21,500
2532	200 ton		.82	39.216		23,400	1,950		25,350	28,700
2536	250 ton		.65	49.231		27,900	2,425		30,325	34,400
2540	300 ton		.54	59.259		32,600	2,925		35,525	40,300
2544	350 ton		.47	68.085		36,300	3,375		39,675	45,000
2548	400 ton		.41	78.049		37,800	3,850		41,650	47,400
2552	450 ton		.36	88.154		45,500	4,350		49,850	56,500
2556	500 ton		.32	100		48,500	4,950		53,450	61,000
2560	550 ton		.30	107		52,000	5,325		57,325	65,500
2564	600 ton		.27	117		54,500	5,825		60,325	68,500
2568	650 ton		.25	127		61,000	6,300		67,300	76,500
2572	700 ton		.23	137		71,500	6,800		78,300	88,500
2576	750 ton		.22	146		76,500	7,250		83,750	95,500
2580	800 ton		.21	152		83,000	7,525		90,525	102,500
2584	850 ton		.20	156		88,000	7,750		95,750	108,500
2588	900 ton		.20	161		93,500	8,000		101,500	114,500
2592	950 ton		.19	170		98,500	8,425		106,925	121,000
2596	1000 ton		.18	180		101,500	8,925		110,425	125,500
2700	Axial fan, induced draft									
2710	50 ton	Q-6	2.28	10.526	Ea.	8,650	510		9,160	10,300
2720	75 ton		1.52	15.789		10,200	765		10,965	12,400
2724	100 ton		1.22	19.672		10,800	955		11,755	13,300
2728	125 ton	Q-7	1.31	24.427		13,300	1,200		14,500	16,400
2732	200 ton		.82	39.216		22,500	1,950		24,450	27,700
2736	250 ton		.65	49.231		25,400	2,425		27,825	31,600
2740	300 ton		.54	59.259		29,600	2,925		32,525	37,000
2744	350 ton		.47	68.085		31,800	3,375		35,175	40,000
2748	400 ton		.41	78.049		36,400	3,850		40,250	45,800
2752	450 ton		.36	88.154		40,400	4,350		44,750	51,000
2756	500 ton		.32	100		42,900	4,950		47,850	54,500
2760	550 ton		.30	107		47,900	5,325		53,225	60,500
2764	600 ton		.27	117		50,500	5,825		56,325	64,000
2768	650 ton		.25	127		55,000	6,300		61,300	70,000
2772	700 ton		.23	137		63,000	6,800		69,800	79,000

23 65 Cooling Towers

23 65 13 – Forced-Draft Cooling Towers

23 65 13.10 Forced-Draft Type Cooling Towers		Crew	Daily Output	Labor-Hours	Unit	2010 Bare Costs Material	Labor	Equipment	Total	Total Incl O&P
2776	750 ton	Q-7	.22	146	Ea.	65,500	7,250		72,750	83,000
2780	800 ton		.21	152		68,000	7,525		75,525	86,500
2784	850 ton		.20	156		68,500	7,750		76,250	87,000
2788	900 ton		.20	161		71,000	8,000		79,000	90,000
2792	950 ton		.19	170		75,000	8,425		83,425	95,000
2796	1000 ton		.18	180		79,000	8,925		87,925	100,500
3000	For higher capacities, use multiples									
3500	For pumps and piping, add	Q-6	38	.632	TonAC	81.50	30.50		112	136
4000	For absorption systems, add				"	75%	75%			
4100	Cooling water chemical feeder	Q-5	3	5.333	Ea.	300	249		549	705
5000	Fiberglass tower on galvanized steel support structure									
5010	Draw thru									
5100	100 ton	Q-6	1.40	17.143	Ea.	10,100	830		10,930	12,400
5120	120 ton		1.20	20		11,700	970		12,670	14,400
5140	140 ton		1	24		12,700	1,150		13,850	15,800
5160	160 ton		.80	30		14,100	1,450		15,550	17,700
5180	180 ton		.65	36.923		16,000	1,800		17,800	20,300
5200	200 ton		.48	50		18,100	2,425		20,525	23,500
5300	For stainless steel support structure, add					30%				
5360	For higher capacities, use multiples of each size									
6000	Stainless steel									
6010	Induced draft, crossflow, horizontal, belt drive									
6100	57 ton	Q-6	1.50	16	Ea.	21,200	775		21,975	24,600
6120	91 ton		.99	24.242		25,700	1,175		26,875	30,000
6140	111 ton		.43	55.814		38,500	2,700		41,200	46,500
6160	126 ton		.22	109		38,500	5,275		43,775	50,500
6170	Induced draft, crossflow, vertical, gear drive									
6172	167 ton	Q-6	.75	32	Ea.	48,900	1,550		50,450	56,500
6174	297 ton		.43	55.814		54,500	2,700		57,200	64,000
6176	582 ton		.23	104		86,000	5,050		91,050	102,000
6178	849 ton		.17	141		121,500	6,825		128,325	143,500
6180	1016 ton		.15	160		135,000	7,750		142,750	160,000

23 72 Air-to-Air Energy Recovery Equipment

23 72 13 – Heat-Wheel Air-to-Air Energy-Recovery Equipment

23 72 13.10 Heat-Wheel Air-to-Air Energy Recovery Equip.		Crew	Daily Output	Labor-Hours	Unit	Material	Labor	Equipment	Total	Total Incl O&P
0010	**HEAT-WHEEL AIR-TO-AIR ENERGY RECOVERY EQUIPMENT**									
0100	Air to air									
4000	Enthalpy recovery wheel									
4010	1000 max CFM G	Q-9	1.20	13.333	Ea.	6,050	590		6,640	7,550
4020	2000 max CFM G		1	16		7,075	705		7,780	8,850
4030	4000 max CFM G		.80	20		8,200	885		9,085	10,400
4040	6000 max CFM G		.70	22.857		9,575	1,000		10,575	12,000
4050	8000 max CFM G	Q-10	1	24		10,600	1,100		11,700	13,300
4060	10,000 max CFM G		.90	26.667		12,700	1,225		13,925	15,800
4070	20,000 max CFM G		.80	30		22,900	1,375		24,275	27,300
4080	25,000 max CFM G		.70	34.286		28,000	1,575		29,575	33,200
4090	30,000 max CFM G		.50	48		31,100	2,200		33,300	37,500
4100	40,000 max CFM G		.45	53.333		42,900	2,450		45,350	51,000
4110	50,000 max CFM G		.40	60		49,900	2,750		52,650	59,000

23 72 Air-to-Air Energy Recovery Equipment

23 72 16 – Heat-Pipe Air-To-Air Energy-Recovery Equipment

23 72 16.10 Heat Pipes		Crew	Daily Output	Labor-Hours	Unit	Material	2010 Bare Costs Labor	Equipment	Total	Total Incl O&P
0010	**HEAT PIPES**									
8000	Heat pipe type, glycol, 50% efficient									
8010	100 MBH, 1700 CFM	1 Stpi	.80	10	Ea.	3,550	520		4,070	4,700
8020	160 MBH, 2700 CFM		.60	13.333		4,775	690		5,465	6,275
8030	620 MBH, 4000 CFM		.40	20		7,225	1,050		8,275	9,500

23 73 Indoor Central-Station Air-Handling Units

23 73 13 – Modular Indoor Central-Station Air-Handling Units

23 73 13.10 Air Handling Units		Crew	Daily Output	Labor-Hours	Unit	Material	Labor	Equipment	Total	Total Incl O&P
0010	**AIR HANDLING UNITS**, Built-Up									
0100	With cooling/heating coil section, filters, mixing box									
0880	Single zone, horizontal/vertical									
0890	Constant volume									
0900	1600 CFM	Q-5	1.20	13.333	Ea.	3,225	625		3,850	4,475
0906	2000 CFM		1.10	14.545		4,025	680		4,705	5,450
0910	3000 CFM		1	16		5,025	745		5,770	6,650
0916	4000 CFM		.95	16.842		6,700	785		7,485	8,550
0920	5000 CFM	Q-6	1.40	17.143		8,375	830		9,205	10,500
0926	6500 CFM		1.30	18.462		10,900	895		11,795	13,400
0930	7500 CFM		1.20	20		12,600	970		13,570	15,300
0936	9200 CFM		1.10	21.818		15,400	1,050		16,450	18,600
0940	11,500 CFM		1	24		17,000	1,150		18,150	20,500
0946	13,200 CFM		.90	26.667		19,500	1,300		20,800	23,300
0950	16,500 CFM		.80	30		24,300	1,450		25,750	29,000
0960	19,500 CFM		.70	34.286		28,800	1,650		30,450	34,100
0970	22,000 CFM		.60	40		32,500	1,925		34,425	38,600
0980	27,000 CFM		.50	48		36,200	2,325		38,525	43,300
0990	34,000 CFM		.40	60		45,600	2,900		48,500	54,500
1000	40,000 CFM		.30	80		49,700	3,875		53,575	60,500
1010	47,000 CFM		.20	120		63,000	5,800		68,800	78,000
1140	60,000 CFM	Q-7	.18	177		74,000	8,800		82,800	94,500
1160	75,000 CFM	"	.16	200		92,500	9,875		102,375	117,000
2000	Variable air volume									
2100	75,000 CFM	Q-7	.12	266	Ea.	93,500	13,200		106,700	123,000
2140	100,000 CFM		.11	290		127,000	14,400		141,400	161,000
2160	150,000 CFM		.10	320		194,500	15,800		210,300	237,500
2300	Multi-zone, horizontal or vertical, blow-thru fan									
2310	3000 CFM	Q-5	1	16	Ea.	5,875	745		6,620	7,575
2314	4000 CFM	"	.95	16.842		7,725	785		8,510	9,675
2318	5000 CFM	Q-6	1.40	17.143		9,625	830		10,455	11,900
2322	6500 CFM		1.30	18.462		12,500	895		13,395	15,200
2326	7500 CFM		1.20	20		14,400	970		15,370	17,400
2330	9200 CFM		1.10	21.818		17,800	1,050		18,850	21,200
2334	11,500 CFM		1	24		19,600	1,150		20,750	23,300
2336	13,200 CFM		.90	26.667		22,400	1,300		23,700	26,600
2340	16,500 CFM		.80	30		28,000	1,450		29,450	33,000
2344	19,500 CFM		.70	34.286		33,000	1,650		34,650	38,800
2348	22,000 CFM		.60	40		37,200	1,925		39,125	43,900
2352	27,000 CFM		.50	48		41,600	2,325		43,925	49,300
2356	34,000 CFM		.40	60		52,500	2,900		55,400	62,000

23 73 Indoor Central-Station Air-Handling Units

23 73 13 – Modular Indoor Central-Station Air-Handling Units

23 73 13.10 Air Handling Units		Crew	Daily Output	Labor-Hours	Unit	2010 Bare Costs Material	Labor	Equipment	Total	Total Incl O&P
2360	40,000 CFM	Q-6	.30	80	Ea.	57,000	3,875		60,875	68,500
2364	47,000 CFM	↓	.20	120	↓	72,500	5,800		78,300	88,000

23 73 13.20 Air Handling Units, Packaged Indoor Type

		Crew	Daily Output	Labor-Hours	Unit	Material	Labor	Equipment	Total	Total Incl O&P
0010	**AIR HANDLING UNITS, PACKAGED INDOOR TYPE**									
1090	Constant volume									
1200	2000 CFM	Q-5	1.10	14.545	Ea.	8,075	680		8,755	9,925
1400	5000 CFM	Q-6	.80	30		19,600	1,450		21,050	23,800
1550	10,000 CFM		.54	44.444		41,600	2,150		43,750	49,000
1670	15,000 CFM		.38	63.158		61,000	3,050		64,050	71,500
1700	20,000 CFM	↓	.31	77.419	↓	85,000	3,750		88,750	99,000
2300	Variable air volume									
2330	2000 CFM	Q-5	1	16	Ea.	3,900	745		4,645	5,425
2340	5000 CFM	Q-6	.70	34.286		22,100	1,650		23,750	26,800
2350	10,000 CFM		.50	48		46,600	2,325		48,925	55,000
2360	15,000 CFM		.40	60		70,000	2,900		72,900	81,500
2370	20,000 CFM		.29	82.759		89,500	4,000		93,500	104,500
2380	30,000 CFM	↓	.20	120	↓	134,500	5,800		140,300	156,500

23 73 39 – Indoor, Direct Gas-Fired Heating and Ventilating Units

23 73 39.10 Make-Up Air Unit

		Crew	Daily Output	Labor-Hours	Unit	Material	Labor	Equipment	Total	Total Incl O&P
0010	**MAKE-UP AIR UNIT**									
0020	Indoor suspension, natural/LP gas, direct fired,									
0032	standard control. For flue see Div. 23 51 23.10									
0040	70°F temperature rise, MBH is input									
0100	75 MBH input	Q-6	3.60	6.667	Ea.	2,400	325		2,725	3,125
0120	100 MBH input		3.40	7.059		2,425	340		2,765	3,175
0140	125 MBH input		3.20	7.500		2,550	365		2,915	3,350
0160	150 MBH input		3	8		2,850	385		3,235	3,700
0180	175 MBH input		2.80	8.571		2,950	415		3,365	3,850
0200	200 MBH input		2.60	9.231		3,050	445		3,495	4,025
0220	225 MBH input		2.40	10		3,300	485		3,785	4,350
0240	250 MBH input		2	12		3,925	580		4,505	5,175
0260	300 MBH input		1.90	12.632		4,025	610		4,635	5,375
0280	350 MBH input		1.75	13.714		4,825	665		5,490	6,300
0300	400 MBH input		1.60	15		7,450	725		8,175	9,275
0320	840 MBH input		1.20	20		10,400	970		11,370	13,000
0340	1200 MBH input	↓	.88	27.273		14,900	1,325		16,225	18,400
0600	For discharge louver assembly, add					5%				
0700	For filters, add					10%				
0800	For air shut-off damper section, add				↓	30%				

23 74 13 – Packaged, Outdoor, Central-Station Air-Handling Units

23 74 13.10 Packaged, Outdoor Type, Central-Station AHU		Crew	Daily Output	Labor-Hours	Unit	2010 Bare Costs Material	Labor	Equipment	Total	Total Incl O&P
0010	**PACKAGED, OUTDOOR TYPE, CENTRAL-STATION AIR-HANDLING UNITS**									
3000	Weathertight, ground or rooftop									
3010	Constant volume									
3100	2000 CFM	Q-6	1	24	Ea.	10,100	1,150		11,250	12,900
3140	5000 CFM		.79	30.380		24,500	1,475		25,975	29,200
3150	10,000 CFM		.52	46.154		52,000	2,225		54,225	60,500
3160	15,000 CFM		.41	58.537		76,500	2,825		79,325	88,500
3170	20,000 CFM		.30	80		106,000	3,875		109,875	122,500
3200	Variable air volume									
3210	2000 CFM	Q-6	.96	25	Ea.	12,200	1,200		13,400	15,200
3240	5000 CFM		.73	32.877		27,600	1,600		29,200	32,800
3250	10,000 CFM		.50	48		58,500	2,325		60,825	67,500
3260	15,000 CFM		.40	60		87,500	2,900		90,400	100,500
3270	20,000 CFM		.29	82.759		112,000	4,000		116,000	129,000
3280	30,000 CFM		.20	120		168,000	5,800		173,800	193,500

23 74 23 – Packaged, Outdoor, Heating-Only Makeup-Air Units

23 74 23.16 Packaged, Ind.-Fired, In/Outdoor

		Crew	Daily Output	Labor-Hours	Unit	Material	Labor	Equipment	Total	Total Incl O&P
0010	**PACKAGED, IND.-FIRED, IN/OUTDOOR**, Heat-Only Makeup-Air Units									
1000	Rooftop unit, natural gas, gravity vent, S.S. exchanger									
1010	70°F temperature rise, MBH is input									
1020	250 MBH	Q-6	4	6	Ea.	11,400	291		11,691	12,900
1040	400 MBH		3.60	6.667		11,400	325		11,725	13,100
1060	550 MBH		3.30	7.273		12,500	350		12,850	14,200
1080	750 MBH		3	8		13,800	385		14,185	15,800
1100	1000 MBH		2.60	9.231		16,400	445		16,845	18,700
1120	1750 MBH		2.30	10.435		20,600	505		21,105	23,500
1140	2500 MBH		1.90	12.632		26,900	610		27,510	30,500
1160	3250 MBH		1.40	17.143		32,500	830		33,330	37,100
1180	4000 MBH		1.20	20		39,700	970		40,670	45,200
1200	6000 MBH		1	24		49,400	1,150		50,550	56,500
1600	For cleanable filters, add					5%				
1700	For electric modulating gas control, add					10%				
8000	Rooftop unit, electric, with curb, controls									
8010	33 kW	Q-5	4	4	Ea.	3,550	187		3,737	4,175
8020	100 kW		3.60	4.444		5,925	208		6,133	6,825
8030	150 kW		3.20	5		6,825	234		7,059	7,850
8040	250 kW		2.80	5.714		8,950	267		9,217	10,200
8050	400 kW		2.20	7.273		13,200	340		13,540	15,000

23 74 33 – Packaged, Outdoor, Heating and Cooling Makeup Air-Conditioners

23 74 33.10 Roof Top Air Conditioners

		Crew	Daily Output	Labor-Hours	Unit	Material	Labor	Equipment	Total	Total Incl O&P
0010	**ROOF TOP AIR CONDITIONERS**, Standard controls, curb, economizer									
1000	Single zone, electric cool, gas heat									
1100	3 ton cooling, 60 MBH heating D3050–150	Q-5	.70	22.857	Ea.	3,475	1,075		4,550	5,425
1120	4 ton cooling, 95 MBH heating D3050–155		.61	26.403		4,225	1,225		5,450	6,500
1140	5 ton cooling, 112 MBH heating R236000-10		.56	28.521		4,850	1,325		6,175	7,350
1145	6 ton cooling, 140 MBH heating R236000-20		.52	30.769		5,625	1,425		7,050	8,350
1150	7.5 ton cooling, 170 MBH heating		.50	32.258		7,575	1,500		9,075	10,600
1160	10 ton cooling, 200 MBH heating	Q-6	.67	35.982		10,600	1,750		12,350	14,200
1170	12.5 ton cooling, 230 MBH heating		.63	37.975		12,100	1,850		13,950	16,100
1180	15 ton cooling, 270 MBH heating		.57	42.032		15,000	2,025		17,025	19,600
1190	17.5 ton cooling, 330 MBH heating		.52	45.889		17,900	2,225		20,125	23,000
1200	20 ton cooling, 360 MBH heating	Q-7	.67	47.976		21,900	2,375		24,275	27,700

23 74 33 – Packaged, Outdoor, Heating and Cooling Makeup Air-Conditioners

23 74 33.10 Roof Top Air Conditioners		Crew	Daily Output	Labor-Hours	Unit	2010 Bare Costs Material	Labor	Equipment	Total	Total Incl O&P
1210	25 ton cooling, 450 MBH heating	Q-7	.56	57.554	Ea.	25,000	2,850		27,850	31,800
1220	30 ton cooling, 540 MBH heating		.47	68.376		25,500	3,375		28,875	33,200
1240	40 ton cooling, 675 MBH heating		.35	91.168		33,400	4,500		37,900	43,500
1260	50 ton cooling, 810 MBH heating		.28	113		39,500	5,625		45,125	52,000
1265	60 ton cooling, 900 MBH heating		.23	136		47,800	6,750		54,550	62,500
1270	80 ton cooling, 1000 MBH heating		.20	160		79,000	7,900		86,900	99,000
1275	90 ton cooling, 1200 MBH heating		.17	188		82,000	9,300		91,300	104,000
1280	100 ton cooling, 1350 MBH heating		.14	228		100,000	11,300		111,300	127,000
1300	Electric cool, electric heat									
1310	5 ton	Q-5	.63	25.600	Ea.	3,625	1,200		4,825	5,775
1314	10 ton	Q-6	.74	32.432		6,575	1,575		8,150	9,575
1318	15 ton	"	.64	37.795		9,850	1,825		11,675	13,600
1322	20 ton	Q-7	.74	43.185		13,100	2,125		15,225	17,700
1326	25 ton		.62	51.696		16,400	2,550		18,950	21,900
1330	30 ton		.52	61.657		19,700	3,050		22,750	26,300
1334	40 ton		.39	82.051		26,300	4,050		30,350	35,000
1338	50 ton		.31	102		31,200	5,075		36,275	42,000
1342	60 ton		.26	123		37,500	6,075		43,575	50,500
1346	75 ton		.21	153		46,800	7,600		54,400	63,000
2000	Multizone, electric cool, gas heat, economizer									
2100	15 ton cooling, 360 MBH heating	Q-7	.61	52.545	Ea.	57,000	2,600		59,600	66,500
2120	20 ton cooling, 360 MBH heating		.53	60.038		61,000	2,975		63,975	72,000
2140	25 ton cooling, 450 MBH heating		.45	71.910		73,500	3,550		77,050	86,000
2160	28 ton cooling, 450 MBH heating		.41	79.012		83,500	3,900		87,400	98,000
2180	30 ton cooling, 540 MBH heating		.37	85.562		94,000	4,225		98,225	109,500
2200	40 ton cooling, 540 MBH heating		.28	113		107,000	5,625		112,625	126,500
2210	50 ton cooling, 540 MBH heating		.23	142		133,500	7,025		140,525	157,000
2220	70 ton cooling, 1500 MBH heating		.16	198		144,500	9,825		154,325	173,500
2240	80 ton cooling, 1500 MBH heating		.14	228		165,000	11,300		176,300	198,500
2260	90 ton cooling, 1500 MBH heating		.13	256		173,000	12,700		185,700	209,000
2280	105 ton cooling, 1500 MBH heating		.11	290		190,500	14,400		204,900	231,000
2400	For hot water heat coil, deduct					5%				
2500	For steam heat coil, deduct					2%				
2600	For electric heat, deduct					3%	5%			
4000	Single zone, electric cool, gas heat, variable air volume									
4100	12.5 ton cooling, 230 MBH heating	Q-6	.55	43.716	Ea.	13,700	2,125		15,825	18,300
4120	18 ton cooling, 330 MBH heating	"	.46	52.747		19,800	2,550		22,350	25,600
4140	25 ton cooling, 450 MBH heating	Q-7	.48	66.116		27,400	3,275		30,675	35,000
4160	40 ton cooling, 675 MBH heating		.31	104		37,600	5,175		42,775	49,100
4180	60 ton cooling, 900 MBH heating		.20	157		55,000	7,800		62,800	72,000
4200	80 ton cooling, 1000 MBH heating		.15	209		87,500	10,300		97,800	111,500
5000	Single zone, electric cool only									
5050	3 ton cooling	Q-5	.88	18.203	Ea.	2,325	850		3,175	3,850
5060	4 ton cooling		.76	21.108		2,700	985		3,685	4,450
5070	5 ton cooling		.70	22.792		3,350	1,075		4,425	5,275
5080	6 ton cooling		.65	24.502		3,875	1,150		5,025	6,000
5090	7.5 ton cooling		.62	25.806		5,350	1,200		6,550	7,675
5100	8.5 ton cooling		.59	26.981		6,000	1,250		7,250	8,500
5110	10 ton cooling	Q-6	.83	28.812		7,125	1,400		8,525	9,925
5120	12.5 ton cooling		.80	30		9,150	1,450		10,600	12,300
5130	15 ton cooling		.71	33.613		13,700	1,625		15,325	17,600
5140	17.5 ton cooling		.65	36.697		16,500	1,775		18,275	20,900
5150	20 ton cooling	Q-7	.83	38.415		20,700	1,900		22,600	25,700

23 74 33 – Packaged, Outdoor, Heating and Cooling Makeup Air-Conditioners

23 74 33.10 Roof Top Air Conditioners		Crew	Daily Output	Labor-Hours	Unit	Material	Labor	Equipment	Total	Total Incl O&P
						2010 Bare Costs				
5160	25 ton cooling	Q-7	.70	45.977	Ea.	22,700	2,275		24,975	28,400
5170	30 ton cooling		.59	54.701		24,200	2,700		26,900	30,700
5180	40 ton cooling		.44	73.059		32,500	3,600		36,100	41,200
5400	For low heat, add					7%				
5410	For high heat, add					10%				
6000	Single zone electric cooling with variable volume distribution									
6020	20 ton cooling	Q-7	.72	44.199	Ea.	24,600	2,175		26,775	30,300
6030	25 ton cooling		.61	52.893		28,900	2,625		31,525	35,700
6040	30 ton cooling		.51	62.868		29,700	3,100		32,800	37,400
6050	40 ton cooling		.38	83.990		39,900	4,150		44,050	50,000
6060	50 ton cooling		.31	104		44,200	5,175		49,375	56,500
6070	60 ton cooling		.25	125		52,000	6,225		58,225	67,000
6200	For low heat, add					7%				
6210	For high heat, add					10%				
7000	Multizone, cool/heat, variable volume distribution									
7100	50 ton cooling	Q-7	.20	164	Ea.	99,000	8,100		107,100	121,000
7110	70 ton cooling		.14	228		138,500	11,300		149,800	169,500
7120	90 ton cooling		.11	296		159,000	14,600		173,600	197,000
7130	105 ton cooling		.10	333		178,000	16,500		194,500	220,500
7140	120 ton cooling		.08	380		203,500	18,800		222,300	252,000
7150	140 ton cooling		.07	444		218,000	22,000		240,000	273,000
7300	Penthouse unit, cool/heat, variable volume distribution									
7400	150 ton cooling	Q-7	.10	310	Ea.	356,500	15,400		371,900	415,000
7410	170 ton cooling		.10	320		380,500	15,800		396,300	442,500
7420	200 ton cooling		.10	326		448,000	16,100		464,100	516,500
7430	225 ton cooling		.09	340		490,000	16,800		506,800	564,000
7440	250 ton cooling		.09	359		505,500	17,800		523,300	582,500
7450	270 ton cooling		.09	363		546,000	18,000		564,000	627,500
7460	300 ton cooling		.09	372		607,000	18,400		625,400	695,000
9400	Sub assemblies for assembly systems									
9410	Ductwork, per ton, rooftop 1 zone units	Q-9	.96	16.667	Ton	226	735		961	1,375
9420	Ductwork, per ton, rooftop multizone units		.50	32		296	1,425		1,721	2,475
9430	Ductwork, VAV cooling/ton, rooftop multizone, 1 zone/ton		.32	50		370	2,200		2,570	3,750
9440	Ductwork per ton, packaged water & air cooled units		1.26	12.698		49.50	560		609.50	905
9450	Ductwork per ton, split system remote condensing units		1.32	12.121		46.50	535		581.50	860
9500	Ductwork package for residential units, 1 ton		.80	20	Ea.	320	885		1,205	1,700
9520	2 ton		.40	40		640	1,775		2,415	3,375
9530	3 ton		.27	59.259		955	2,625		3,580	5,025
9540	4 ton	Q-10	.30	80		1,275	3,675		4,950	6,950
9550	5 ton		.23	104		1,600	4,775		6,375	9,000
9560	6 ton		.20	120		1,925	5,500		7,425	10,400
9570	7 ton		.17	141		2,225	6,475		8,700	12,300
9580	8 ton		.15	160		2,550	7,325		9,875	13,900

23 76 Evaporative Air-Cooling Equipment

23 76 13 – Direct Evaporative Air Coolers

23 76 13.10 Evaporative Coolers		Crew	Daily Output	Labor-Hours	Unit	Material	2010 Bare Costs Labor	Equipment	Total	Total Incl O&P
0010	**EVAPORATIVE COOLERS**, (Swamp Coolers) ducted, not incl. duct.									
0100	Side discharge style, capacities at .25" S.P.									
0120	1785 CFM, 1/3 HP, 115 V G	Q-9	5	3.200	Ea.	465	141		606	730
0140	2740 CFM, 1/3 HP, 115 V G		4.50	3.556		570	157		727	870
0160	3235 CFM, 1/2 HP, 115 V G		4	4		585	177		762	915
0180	3615 CFM, 1/2 HP, 230 V G		3.60	4.444		595	196		791	950
0200	4215 CFM, 3/4 HP, 230 V G		3.20	5		760	221		981	1,175
0220	5255 CFM. 1 HP, 115/230 V G		3	5.333		1,250	236		1,486	1,725
0240	6090 CFM, 1 HP, 230/460 V G		2.80	5.714		1,750	253		2,003	2,300
0260	8300 CFM, 1-1/2 HP, 230/460 V G		2.60	6.154		1,750	272		2,022	2,325
0280	8360 CFM, 1-1/2 HP, 230/460 V G		2.20	7.273		2,400	320		2,720	3,100
0300	9725 CFM, 2 HP, 230/460 V G		1.80	8.889		2,400	395		2,795	3,250
0320	11,715 CFM, 3 HP, 230/460 V G		1.40	11.429		2,450	505		2,955	3,475
0340	14,410 CFM, 5 HP, 230/460 V G		1	16		2,575	705		3,280	3,925
0400	For two-speed motor, add					5%				
0500	For down discharge style, add					10%				

23 76 16 – Indirect Evaporative Air Coolers

23 76 16.10 Evaporators		Crew	Daily Output	Labor-Hours	Unit	Material	Labor	Equipment	Total	Total Incl O&P
0010	**EVAPORATORS**, DX coils, remote compressors not included.									
1000	Coolers, reach-in type, above freezing temperatures									
1300	Shallow depth, wall mount, 7 fins per inch, air defrost									
1310	600 BTUH, 8" fan, rust-proof core	Q-5	3.80	4.211	Ea.	415	197		612	750
1320	900 BTUH, 8" fan, rust-proof core		3.30	4.848		515	226		741	905
1330	1200 BTUH, 8" fan, rust-proof core		3	5.333		605	249		854	1,050
1340	1800 BTUH, 8" fan, rust-proof core		2.40	6.667		700	310		1,010	1,225
1350	2500 BTUH, 10" fan, rust-proof core		2.20	7.273		760	340		1,100	1,350
1360	3500 BTUH, 10" fan		2	8		890	375		1,265	1,550
1370	4500 BTUH, 10" fan		1.90	8.421		940	395		1,335	1,625
1390	For pan drain, add					50			50	55
1600	Undercounter refrigerators, ceiling or wall mount,									
1610	8 fins per inch, air defrost, rust-proof core									
1630	800 BTUH, one 6" fan	Q-5	5	3.200	Ea.	261	149		410	510
1640	1300 BTUH, two 6" fans		4	4		320	187		507	630
1650	1700 BTUH, two 6" fans		3.60	4.444		355	208		563	700
2000	Coolers, reach-in and walk-in types, above freezing temperatures									
2600	Two-way discharge, ceiling mount, 150-4100 CFM,									
2610	above 34°F applications, air defrost									
2630	900 BTUH, 7 fins per inch, 8" fan	Q-5	4	4	Ea.	310	187		497	625
2660	2500 BTUH, 7 fins per inch, 10" fan		2.60	6.154		580	287		867	1,075
2690	5500 BTUH, 7 fins per inch, 12" fan		1.70	9.412		1,150	440		1,590	1,925
2720	8500 BTUH, 8 fins per inch, 16" fan		1.20	13.333		1,425	625		2,050	2,500
2750	15,000 BTUH, 7 fins per inch, 18" fan		1.10	14.545		1,975	680		2,655	3,200
2770	24,000 BTUH, 7 fins per inch, two 16" fans		1	16		3,000	745		3,745	4,425
2790	30,000 BTUH, 7 fins per inch, two 18" fans		.90	17.778		3,700	830		4,530	5,325
2850	Two-way discharge, low profile, ceiling mount,									
2860	8 fins per inch, 200-570 CFM, air defrost									
2880	800 BTUH, one 6" fan	Q-5	4	4	Ea.	335	187		522	650
2890	1300 BTUH, two 6" fans		3.50	4.571		400	213		613	760
2900	1800 BTUH, two 6" fans		3.30	4.848		420	226		646	805
2910	2700 BTUH, three 6" fans		2.70	5.926		495	277		772	960
3000	Coolers, walk-in type, above freezing temperatures									
3300	General use, ceiling mount, 108-2080 CFM									

23 76 16 – Indirect Evaporative Air Coolers

23 76 16.10 Evaporators		Crew	Daily Output	Labor-Hours	Unit	Material	2010 Bare Costs Labor	Equipment	Total	Total Incl O&P
3320	600 BTUH, 7 fins per inch, 6″ fan	Q-5	5.20	3.077	Ea.	246	144		390	485
3340	1200 BTUH, 7 fins per inch, 8″ fan		3.70	4.324		340	202		542	680
3360	1800 BTUH, 7 fins per inch, 10″ fan		2.70	5.926		475	277		752	940
3380	3500 BTUH, 7 fins per inch, 12″ fan		2.40	6.667		700	310		1,010	1,225
3400	5500 BTUH, 8 fins per inch, 12″ fan		2	8		780	375		1,155	1,425
3430	8500 BTUH, 8 fins per inch, 16″ fan		1.40	11.429		1,150	535		1,685	2,075
3460	15,000 BTUH, 7 fins per inch, two 16″ fans	↓	1.10	14.545	↓	1,875	680		2,555	3,075
3640	Low velocity, high latent load, ceiling mount,									
3650	1050-2420 CFM, air defrost, 6 fins per inch									
3670	6700 BTUH, two 10″ fans	Q-5	1.30	12.308	Ea.	1,650	575		2,225	2,650
3680	10,000 BTUH, three 10″ fans		1	16		1,900	745		2,645	3,225
3690	13,500 BTUH, three 10″ fans		1	16		2,300	745		3,045	3,650
3700	18,000 BTUH, four 10″ fans		1	16		3,050	745		3,795	4,475
3710	26,500 BTUH, four 10″ fans	↓	.80	20		3,750	935		4,685	5,525
3730	For electric defrost, add				↓	17%				
5000	Freezers and coolers, reach-in type, above 34°F									
5030	to sub-freezing temperature range, low latent load,									
5050	air defrost, 7 fins per inch									
5070	1200 BTUH, 8″ fan, rustproof core	Q-5	4.40	3.636	Ea.	345	170		515	635
5080	1500 BTUH, 8″ fan, rustproof core		4.20	3.810		355	178		533	655
5090	1800 BTUH, 8″ fan, rustproof core		3.70	4.324		370	202		572	715
5100	2500 BTUH, 10″ fan, rustproof core		2.90	5.517		425	258		683	850
5110	3500 BTUH, 10″ fan, rustproof core		2.50	6.400		580	299		879	1,100
5120	4500 BTUH, 12″ fan, rustproof core	↓	2.10	7.619	↓	770	355		1,125	1,375
6000	Freezers and coolers, walk-in type									
6050	1960-18,000 CFM, medium profile									
6060	Standard motor, 6 fins per inch, to -30°F, all aluminum									
6080	10,500 BTUH, air defrost, one 18″ fan	Q-5	1.40	11.429	Ea.	1,400	535		1,935	2,350
6090	12,500 BTUH, air defrost, one 18″ fan		1.40	11.429		1,525	535		2,060	2,475
6100	16,400 BTUH, air defrost, one 18″ fan		1.30	12.308		1,650	575		2,225	2,650
6110	20,900 BTUH, air defrost, one 18″ fans		1	16		1,900	745		2,645	3,225
6120	27,000 BTUH, air defrost, two 18″ fans	↓	1	16		2,825	745		3,570	4,250
6130	32,900 BTUH, air defrost, two 20″ fans	Q-6	1.30	18.462		3,150	895		4,045	4,825
6140	39,000 BTUH, air defrost, two 20″ fans		1.25	19.200		3,675	930		4,605	5,450
6150	44,100 BTUH, air defrost, three 20″ fans		1	24		4,175	1,150		5,325	6,325
6155	53,000 BTUH, air defrost, three 20″ fans		1	24		4,175	1,150		5,325	6,325
6160	66,200 BTUH, air defrost, four 20″ fans		1	24		4,750	1,150		5,900	6,975
6170	78,000 BTUH, air defrost, four 20″ fans		.80	30		5,900	1,450		7,350	8,675
6180	88,200 BTUH, air defrost, four 24″ fans		.60	40		6,275	1,925		8,200	9,800
6190	110,000 BTUH, air defrost, four 24″ fans	↓	.50	48		7,125	2,325		9,450	11,300
6330	Hot gas defrost, standard motor units, add					35%				
6370	Electric defrost, 230V, standard motor, add					16%				
6410	460V, standard or high capacity motor, add				↓	25%				
6800	Eight fins per inch, increases BTUH 75%									
6810	Standard capacity, add				Ea.	5%				
6830	Hot gas & electric defrost not recommended									
7000	800-4150 CFM, 12″ fans									
7010	Four fins per inch									
7030	3400 BTUH, 1 fan	Q-5	2.80	5.714	Ea.	475	267		742	920
7040	4200 BTUH, 1 fan		2.40	6.667		510	310		820	1,025
7050	5300 BTUH, 1 fan		1.90	8.421		555	395		950	1,200
7060	6800 BTUH, 1 fan		1.70	9.412		615	440		1,055	1,325
7070	8400 BTUH, 2 fans	↓	1.60	10	↓	890	465		1,355	1,675

23 76 Evaporative Air-Cooling Equipment

23 76 16 – Indirect Evaporative Air Coolers

23 76 16.10 Evaporators		Crew	Daily Output	Labor-Hours	Unit	2010 Bare Costs Material	Labor	Equipment	Total	Total Incl O&P
7080	10,500 BTUH, 2 fans	Q-5	1.40	11.429	Ea.	965	535		1,500	1,875
7090	13,000 BTUH, 3 fans		1.30	12.308		1,250	575		1,825	2,225
7100	17,000 BTUH, 4 fans		1.20	13.333		1,575	625		2,200	2,650
7110	21,500 BTUH, 5 fans		1	16		1,950	745		2,695	3,275
7160	Six fins per inch increases BTUH 35%, add					5%				
7180	Eight fins per inch increases BTUH 53%,									
7190	not recommended for below 35°F, add				Ea.	8%				
7210	For 12°F temperature differential, add					20%				
7230	For adjustable thermostat control, add					111			111	122
7240	For hot gas defrost, except 8 fin models									
7250	on applications below 35°F, add				Ea.	58%				
7260	For electric defrost, except 8 fin models									
7270	on applications below 35°F, add				Ea.	36%				
8000	Freezers, pass-thru door uprights, walk-in storage									
8300	Low temperature, thin profile, electric defrost									
8310	138-1800 CFM, 5 fins per inch									
8320	900 BTUH, one 6" fan	Q-5	3.30	4.848	Ea.	390	226		616	770
8340	1500 BTUH, two 6" fans		2.20	7.273		585	340		925	1,150
8360	2600 BTUH, three 6" fans		1.90	8.421		840	395		1,235	1,525
8370	3300 BTUH, four 6" fans		1.70	9.412		935	440		1,375	1,675
8380	4400 BTUH, five 6" fans		1.40	11.429		1,325	535		1,860	2,250
8400	7000 BTUH, three 10" fans		1.30	12.308		1,600	575		2,175	2,600
8410	8700 BTUH, four 10" fans		1.10	14.545		1,800	680		2,480	3,000
8450	For air defrost, deduct					193			193	212

23 81 Decentralized Unitary HVAC Equipment

23 81 13 – Packaged Terminal Air-Conditioners

23 81 13.10 Packaged Cabinet Type Air-Conditioners

		Crew	Daily Output	Labor-Hours	Unit	Material	Labor	Equipment	Total	Total Incl O&P
0010	**PACKAGED CABINET TYPE AIR-CONDITIONERS**, Cabinet, wall sleeve,									
0100	louver, electric heat, thermostat, manual changeover, 208 V									
0200	6,000 BTUH cooling, 8800 BTU heat	Q-5	6	2.667	Ea.	680	125		805	935
0220	9,000 BTUH cooling, 13,900 BTU heat		5	3.200		790	149		939	1,100
0240	12,000 BTUH cooling, 13,900 BTU heat		4	4		850	187		1,037	1,225
0260	15,000 BTUH cooling, 13,900 BTU heat		3	5.333		915	249		1,164	1,375
0280	18,000 BTUH cooling, 10 KW heat		2.40	6.667		1,675	310		1,985	2,300
0300	24,000 BTUH cooling, 10 KW heat		1.90	8.421		1,675	395		2,070	2,450
0320	30,000 BTUH cooling, 10 KW heat		1.40	11.429		1,950	535		2,485	2,950
0340	36,000 BTUH cooling, 10 KW heat		1.25	12.800		2,000	600		2,600	3,100
0360	42,000 BTUH cooling, 10 KW heat		1	16		2,500	745		3,245	3,875
0380	48,000 BTUH cooling, 10 KW heat		.90	17.778		2,700	830		3,530	4,225
0500	For hot water coil, increase heat by 10%, add					5%	10%			
1000	For steam, increase heat output by 30%, add					8%	10%			

23 81 19 – Self-Contained Air-Conditioners

23 81 19.10 Window Unit Air Conditioners

		Crew	Daily Output	Labor-Hours	Unit	Material	Labor	Equipment	Total	Total Incl O&P
0010	**WINDOW UNIT AIR CONDITIONERS**									
4000	Portable/window, 15 amp 125V grounded receptacle required									
4060	5000 BTUH	1 Carp	8	1	Ea.	199	41.50		240.50	283
4340	6000 BTUH		8	1		241	41.50		282.50	330
4480	8000 BTUH		6	1.333		405	55.50		460.50	530
4500	10,000 BTUH		6	1.333		515	55.50		570.50	650

23 81 19 – Self-Contained Air-Conditioners

23 81 19.10 Window Unit Air Conditioners		Crew	Daily Output	Labor-Hours	Unit	2010 Bare Costs Material	Labor	Equipment	Total	Total Incl O&P
4520	12,000 BTUH	L-2	8	2	Ea.	630	73		703	810
4600	Window/thru-the-wall, 15 amp 230V grounded receptacle required									
4780	17,000 BTUH	L-2	6	2.667	Ea.	845	97.50		942.50	1,075
4940	25,000 BTUH		4	4		1,075	146		1,221	1,400
4960	29,000 BTUH		4	4		1,200	146		1,346	1,550

23 81 19.20 Self-Contained Single Package		Crew	Daily Output	Labor-Hours	Unit	Material	Labor	Equipment	Total	Total Incl O&P
0010	**SELF-CONTAINED SINGLE PACKAGE**									
0100	Air cooled, for free blow or duct, not incl. remote condenser									
0110	Constant volume									
0200	3 ton cooling	Q-5	1	16	Ea.	3,300	745		4,045	4,750
0210	4 ton cooling	"	.80	20		3,575	935		4,510	5,350
0220	5 ton cooling	Q-6	1.20	20		3,925	970		4,895	5,775
0230	7.5 ton cooling	"	.90	26.667		5,125	1,300		6,425	7,550
0240	10 ton cooling	Q-7	1	32		6,775	1,575		8,350	9,825
0250	15 ton cooling		.95	33.684		9,425	1,675		11,100	12,900
0260	20 ton cooling		.90	35.556		13,500	1,750		15,250	17,500
0270	25 ton cooling		.85	37.647		15,600	1,850		17,450	19,900
0280	30 ton cooling		.80	40		21,600	1,975		23,575	26,800
0300	40 ton cooling		.60	53.333		33,200	2,625		35,825	40,500
0320	50 ton cooling		.50	64		41,700	3,175		44,875	50,500
0340	60 ton cooling	Q-8	.40	80		50,500	3,950	146	54,596	61,500
0380	With hot water heat, variable air volume									
0390	10 ton cooling	Q-7	.96	33.333	Ea.	8,800	1,650		10,450	12,200
0400	20 ton cooling		.86	37.209		17,200	1,850		19,050	21,700
0410	30 ton cooling		.76	42.105		27,300	2,075		29,375	33,200
0420	40 ton cooling		.58	55.172		40,800	2,725		43,525	48,900
0430	50 ton cooling		.48	66.667		51,000	3,300		54,300	61,000
0440	60 ton cooling	Q-8	.38	84.211		63,000	4,175	154	67,329	75,500
0490	For duct mounting no price change									
0500	For steam heating coils, add				Ea.	10%	10%			
0550	Packaged, with electric heat									
0552	Constant volume									
0560	5 ton cooling	Q-6	1.20	20	Ea.	7,150	970		8,120	9,300
0570	10 ton cooling	Q-7	1	32		11,700	1,575		13,275	15,300
0580	20 ton cooling		.90	35.556		20,600	1,750		22,350	25,200
0590	30 ton cooling		.80	40		33,500	1,975		35,475	39,800
0600	40 ton cooling		.60	53.333		46,000	2,625		48,625	54,500
0610	50 ton cooling		.50	64		59,500	3,175		62,675	70,500
0700	Variable air volume									
0710	10 ton cooling	Q-7	1	32	Ea.	13,100	1,575		14,675	16,800
0720	20 ton cooling		.90	35.556		22,900	1,750		24,650	27,800
0730	30 ton cooling		.80	40		37,000	1,975		38,975	43,700
0740	40 ton cooling		.60	53.333		50,000	2,625		52,625	59,000
0750	50 ton cooling		.50	64		64,500	3,175		67,675	76,000
0760	60 ton cooling	Q-8	.40	80		74,000	3,950	146	78,096	87,000
1000	Water cooled for free blow or duct, not including tower									
1010	Constant volume									
1100	3 ton cooling	Q-6	1	24	Ea.	3,250	1,150		4,400	5,325
1120	5 ton cooling		1	24		4,275	1,150		5,425	6,450
1130	7.5 ton cooling		.80	30		6,025	1,450		7,475	8,800
1140	10 ton cooling	Q-7	.90	35.556		8,300	1,750		10,050	11,800
1150	15 ton cooling		.85	37.647		12,000	1,850		13,850	16,000

23 81 19 – Self-Contained Air-Conditioners

23 81 19.20 Self-Contained Single Package		Crew	Daily Output	Labor-Hours	Unit	2010 Bare Costs Material	Labor	Equipment	Total	Total Incl O&P
1160	20 ton cooling	Q-7	.80	40	Ea.	25,800	1,975		27,775	31,300
1170	25 ton cooling		.75	42.667		30,400	2,100		32,500	36,600
1180	30 ton cooling		.70	45.714		34,200	2,250		36,450	41,000
1200	40 ton cooling		.40	80		43,300	3,950		47,250	53,500
1220	50 ton cooling		.30	106		54,000	5,275		59,275	67,500
1240	60 ton cooling	Q-8	.30	106		63,000	5,275	195	68,470	77,000
1300	For hot water or steam heat coils, add					12%	10%			
2000	Water cooled with electric heat, not including tower									
2010	Constant volume									
2100	5 ton cooling	Q-6	1.20	20	Ea.	7,475	970		8,445	9,675
2120	10 ton cooling	Q-7	1	32		13,200	1,575		14,775	17,000
2130	15 ton cooling	"	.90	35.556		19,100	1,750		20,850	23,600
2400	Variable air volume									
2440	10 ton cooling	Q-7	1	32	Ea.	14,600	1,575		16,175	18,500
2450	15 ton cooling	"	.90	35.556	"	21,400	1,750		23,150	26,100
2600	Water cooled, hot water coils, not including tower									
2610	Variable air volume									
2640	10 ton cooling	Q-7	1	32	Ea.	10,500	1,575		12,075	13,900
2650	15 ton cooling	"	.80	40	"	15,500	1,975		17,475	20,100

23 81 23 – Computer-Room Air-Conditioners

23 81 23.10 Computer Room Units		Crew	Daily Output	Labor-Hours	Unit	Material	Labor	Equipment	Total	Total Incl O&P
0010	**COMPUTER ROOM UNITS** D3050–185									
1000	Air cooled, includes remote condenser but not									
1020	interconnecting tubing or refrigerant									
1080	3 ton	Q-5	.50	32	Ea.	16,100	1,500		17,600	20,000
1120	5 ton		.45	35.556		17,200	1,650		18,850	21,400
1160	6 ton		.30	53.333		30,400	2,500		32,900	37,100
1200	8 ton		.27	59.259		32,300	2,775		35,075	39,800
1240	10 ton		.25	64		33,800	3,000		36,800	41,700
1280	15 ton		.22	72.727		37,200	3,400		40,600	46,000
1290	18 ton		.20	80		41,400	3,725		45,125	51,000
1300	20 ton	Q-6	.26	92.308		44,500	4,475		48,975	55,500
1320	22 ton		.24	100		45,100	4,850		49,950	57,000
1360	30 ton		.21	114		55,500	5,525		61,025	69,500
2200	Chilled water, for connection to									
2220	existing chiller system of adequate capacity									
2260	5 ton	Q-5	.74	21.622	Ea.	12,300	1,000		13,300	15,100
2280	6 ton		.52	30.769		12,400	1,425		13,825	15,800
2300	8 ton		.50	32		12,500	1,500		14,000	16,000
2320	10 ton		.49	32.653		12,600	1,525		14,125	16,100
2330	12 ton		.49	32.990		12,900	1,550		14,450	16,500
2360	15 ton		.48	33.333		13,300	1,550		14,850	16,900
2400	20 ton		.46	34.783		14,000	1,625		15,625	17,900
2440	25 ton	Q-6	.63	38.095		15,000	1,850		16,850	19,300
2460	30 ton		.57	42.105		16,200	2,050		18,250	20,900
2480	35 ton		.52	46.154		16,200	2,225		18,425	21,200
2500	45 ton		.46	52.174		17,300	2,525		19,825	22,900
2520	50 ton		.42	57.143		19,900	2,775		22,675	26,100
2540	60 ton		.38	63.158		21,200	3,050		24,250	27,900
4000	Glycol system, complete except for interconnecting tubing									
4060	3 ton	Q-5	.40	40	Ea.	20,300	1,875		22,175	25,100
4100	5 ton		.38	42.105		22,200	1,975		24,175	27,400

23 81 23 – Computer-Room Air-Conditioners

23 81 23.10 Computer Room Units		Crew	Daily Output	Labor-Hours	Unit	Material	Labor	Equipment	Total	Total Incl O&P
						2010 Bare Costs				
4120	6 ton	Q-5	.25	64	Ea.	32,500	3,000		35,500	40,300
4140	8 ton		.23	69.565		35,700	3,250		38,950	44,200
4160	10 ton	↓	.21	76.190		38,000	3,550		41,550	47,100
4200	15 ton	Q-6	.26	92.308		47,000	4,475		51,475	58,000
4240	20 ton		.24	100		52,000	4,850		56,850	64,500
4280	22 ton		.23	104		54,500	5,050		59,550	67,500
4430	30 ton	↓	.22	109	↓	67,500	5,275		72,775	82,000
8000	Water cooled system, not including condenser,									
8020	water supply or cooling tower									
8060	3 ton	Q-5	.62	25.806	Ea.	16,400	1,200		17,600	19,900
8100	5 ton		.54	29.630		17,900	1,375		19,275	21,800
8120	6 ton		.35	45.714		27,900	2,125		30,025	33,900
8140	8 ton		.33	48.485		29,300	2,275		31,575	35,700
8160	10 ton		.31	51.613		30,400	2,400		32,800	37,000
8200	15 ton	↓	.27	59.259		35,600	2,775		38,375	43,300
8240	20 ton	Q-6	.38	63.158		38,300	3,050		41,350	46,700
8280	22 ton		.35	68.571		40,100	3,325		43,425	49,100
8300	30 ton	↓	.30	80	↓	50,000	3,875		53,875	61,000

23 81 26 – Split-System Air-Conditioners

23 81 26.10 Split Ductless Systems		Crew	Daily Output	Labor-Hours	Unit	Material	Labor	Equipment	Total	Total Incl O&P
0010	**SPLIT DUCTLESS SYSTEMS**									
0100	Cooling only, single zone									
0110	Wall mount									
0120	3/4 ton cooling	Q-5	2	8	Ea.	655	375		1,030	1,275
0130	1 ton cooling		1.80	8.889		730	415		1,145	1,425
0140	1-1/2 ton cooling		1.60	10		1,150	465		1,615	1,950
0150	2 ton cooling	↓	1.40	11.429	↓	1,300	535		1,835	2,225
1000	Ceiling mount									
1020	2 ton cooling	Q-5	1.40	11.429	Ea.	2,475	535		3,010	3,525
1030	3 ton cooling	"	1.20	13.333	"	3,200	625		3,825	4,450
3000	Multizone									
3010	Wall mount									
3020	2 @ 3/4 ton cooling	Q-5	1.80	8.889	Ea.	2,100	415		2,515	2,950
5000	Cooling/Heating									
5010	Wall mount									
5110	1 ton cooling	Q-5	1.70	9.412	Ea.	765	440		1,205	1,500
5120	1-1/2 ton cooling	"	1.50	10.667	"	1,175	500		1,675	2,050
7000	Accessories for all split ductless systems									
7010	Add for ambient frost control	Q-5	8	2	Ea.	179	93.50		272.50	335
7020	Add for tube/wiring kit									
7030	15' kit	Q-5	32	.500	Ea.	40	23.50		63.50	79
7036	25' kit		28	.571		95	26.50		121.50	145
7040	35' kit		24	.667		129	31		160	189
7050	50' kit	↓	20	.800	↓	167	37.50		204.50	240

23 81 43 – Air-Source Unitary Heat Pumps

23 81 43.10 Air-Source Heat Pumps		Crew	Daily Output	Labor-Hours	Unit	Material	Labor	Equipment	Total	Total Incl O&P
0010	**AIR-SOURCE HEAT PUMPS**, Not including interconnecting tubing									
1000	Air to air, split system, not including curbs, pads, fan coil and ductwork									
1010	For curbs/pads see Div. 23 91 00									
1012	Outside condensing unit only, for fan coil see Div. 23 82 19.10									
1015	1.5 ton cooling, 7 MBH heat @ 0°F	Q-5	2.40	6.667	Ea.	2,075	310		2,385	2,750
1020	2 ton cooling, 8.5 MBH heat @ 0°F	↓	2	8	↓	2,125	375		2,500	2,875

23 81 43 – Air-Source Unitary Heat Pumps

23 81 43.10 Air-Source Heat Pumps		Crew	Daily Output	Labor-Hours	Unit	2010 Bare Costs Material	Labor	Equipment	Total	Total Incl O&P
1030	2.5 ton cooling, 10 MBH heat @ 0°F	Q-5	1.60	10	Ea.	2,350	465		2,815	3,275
1040	3 ton cooling, 13 MBH heat @ 0°F		1.20	13.333		2,650	625		3,275	3,850
1050	3.5 ton cooling, 18 MBH heat @ 0°F		1	16		2,850	745		3,595	4,250
1054	4 ton cooling, 24 MBH heat @ 0°F		.80	20		3,150	935		4,085	4,850
1060	5 ton cooling, 27 MBH heat @ 0°F		.50	32		3,425	1,500		4,925	6,025
1080	7.5 ton cooling, 33 MBH heat @ 0°F		.45	35.556		5,625	1,650		7,275	8,675
1100	10 ton cooling, 50 MBH heat @ 0°F	Q-6	.64	37.500		7,900	1,825		9,725	11,400
1120	15 ton cooling, 64 MBH heat @ 0°F		.50	48		10,400	2,325		12,725	15,000
1130	20 ton cooling, 85 MBH heat @ 0°F		.35	68.571		15,600	3,325		18,925	22,200
1140	25 ton cooling, 119 MBH heat @ 0°F		.25	96		18,800	4,650		23,450	27,700
1500	Single package, not including curbs, pads, or plenums									
1502	1/2 ton cooling, supplementary heat not incl.	Q-5	8	2	Ea.	2,125	93.50		2,218.50	2,475
1504	3/4 ton cooling, supplementary heat not incl.		6	2.667		2,300	125		2,425	2,700
1506	1 ton cooling, supplementary heat not incl.		4	4		2,525	187		2,712	3,050
1510	1.5 ton cooling, 5 MBH heat @ 0°F		1.55	10.323		2,550	480		3,030	3,525
1520	2 ton cooling, 6.5 MBH heat @ 0°F		1.50	10.667		2,750	500		3,250	3,775
1540	2.5 ton cooling, 8 MBH heat @ 0°F		1.40	11.429		3,000	535		3,535	4,100
1560	3 ton cooling, 10 MBH heat @ 0°F		1.20	13.333		3,275	625		3,900	4,525
1570	3.5 ton cooling, 11 MBH heat @ 0°F		1	16		3,575	745		4,320	5,075
1580	4 ton cooling, 13 MBH heat @ 0°F		.96	16.667		3,850	780		4,630	5,425
1620	5 ton cooling, 27 MBH heat @ 0°F		.65	24.615		4,150	1,150		5,300	6,300
1640	7.5 ton cooling, 35 MBH heat @ 0°F		.40	40		6,550	1,875		8,425	10,000
1648	10 ton cooling, 45 MBH heat @ 0°F	Q-6	.40	60		8,775	2,900		11,675	14,000
1652	12 ton cooling, 50 MBH heat @ 0°F	"	.36	66.667		10,100	3,225		13,325	16,000
1696	Supplementary electric heat coil incl., except as noted									

23 81 46 – Water-Source Unitary Heat Pumps

23 81 46.10 Water Source Heat Pumps		Crew	Daily Output	Labor-Hours	Unit	Material	Labor	Equipment	Total	Total Incl O&P
0010	**WATER SOURCE HEAT PUMPS**, Not incl. connecting tubing or water source									
2000	Water source to air, single package									
2100	1 ton cooling, 13 MBH heat @ 75°F	Q-5	2	8	Ea.	1,475	375		1,850	2,175
2120	1.5 ton cooling, 17 MBH heat @ 75°F		1.80	8.889		1,575	415		1,990	2,350
2140	2 ton cooling, 19 MBH heat @ 75°F		1.70	9.412		1,600	440		2,040	2,425
2160	2.5 ton cooling, 25 MBH heat @ 75°F		1.60	10		1,725	465		2,190	2,600
2180	3 ton cooling, 27 MBH heat @ 75°F		1.40	11.429		1,825	535		2,360	2,800
2190	3.5 ton cooling, 29 MBH heat @ 75°F		1.30	12.308		2,000	575		2,575	3,050
2200	4 ton cooling, 31 MBH heat @ 75°F		1.20	13.333		2,250	625		2,875	3,400
2220	5 ton cooling, 29 MBH heat @ 75°F		.90	17.778		2,625	830		3,455	4,125
2240	7.5 ton cooling, 35 MBH heat @ 75°F		.60	26.667		6,500	1,250		7,750	9,025
2250	8.5 ton cooling, 40 MBH heat @ 75°F		.58	27.586		7,875	1,300		9,175	10,600
2260	10 ton cooling, 50 MBH heat @ 75°F		.53	30.189		8,225	1,400		9,625	11,200
2280	15 ton cooling, 64 MBH heat @ 75°F	Q-6	.47	51.064		14,000	2,475		16,475	19,100
2300	20 ton cooling, 100 MBH heat @ 75°F		.41	58.537		15,300	2,825		18,125	21,100
2310	25 ton cooling, 100 MBH heat @ 75°F		.32	75		20,800	3,625		24,425	28,400
2320	30 ton cooling, 128 MBH heat @ 75°F		.24	102		22,800	4,950		27,750	32,500
2340	40 ton cooling, 200 MBH heat @ 75°F		.21	117		32,200	5,675		37,875	43,900
2360	50 ton cooling, 200 MBH heat @ 75°F		.15	160		36,400	7,750		44,150	51,500
3960	For supplementary heat coil, add					10%				
4000	For increase in capacity thru use									
4020	of solar collector, size boiler at 60%									

23 82 13 – Valance Heating and Cooling Units

23 82 13.16 Valance Units		Crew	Daily Output	Labor-Hours	Unit	Material	Labor	Equipment	Total	Total Incl O&P
						2010 Bare Costs				
0010	**VALANCE UNITS**									
6000	Valance units, complete with 1/2" cooling coil, enclosure									
6020	2 tube	Q-5	18	.889	L.F.	29.50	41.50		71	94.50
6040	3 tube		16	1		35	46.50		81.50	109
6060	4 tube		16	1		40	46.50		86.50	114
6080	5 tube		15	1.067		45.50	50		95.50	125
6100	6 tube		15	1.067		50.50	50		100.50	130
6120	8 tube		14	1.143		73	53.50		126.50	161
6200	For 3/4" cooling coil, add					10%				

23 82 16 – Air Coils

23 82 16.10 Flanged Coils		Crew	Daily Output	Labor-Hours	Unit	Material	Labor	Equipment	Total	Total Incl O&P
0010	**FLANGED COILS**									
0100	Basic water, DX, or condenser coils									
0110	Copper tubes, alum. fins, galv. end sheets									
0112	H is finned height, L is finned length									
0120	3/8" x .016 tube, .0065 aluminum fins									
0130	2 row, 8 fins per inch									
0140	4" H x 12" L	Q-5	48	.333	Ea.	570	15.55		585.55	650
0150	4" H x 24" L		38.38	.417		590	19.45		609.45	680
0160	4" H x 48" L		19.25	.831		625	39		664	750
0170	4" H x 72" L		12.80	1.250		810	58.50		868.50	980
0180	6" H x 12" L		48	.333		550	15.55		565.55	630
0190	6" H x 24" L		25.60	.625		575	29		604	675
0200	6" H x 48" L		12.80	1.250		620	58.50		678.50	775
0210	6" H x 72" L		8.53	1.876		820	87.50		907.50	1,025
0220	10" H x 12" L		30.73	.521		585	24.50		609.50	680
0230	10" H x 24" L		15.33	1.044		620	48.50		668.50	760
0240	10" H x 48" L		7.69	2.081		695	97		792	905
0250	10" H x 72" L		5.12	3.125		920	146		1,066	1,250
0260	12" H x 12" L		25.60	.625		600	29		629	705
0270	12" H x 24" L		12.80	1.250		640	58.50		698.50	795
0280	12" H x 48" L		6.40	2.500		725	117		842	970
0290	12" H x 72" L		4.27	3.747		970	175		1,145	1,325
0300	16" H x 16" L		14.38	1.113		655	52		707	800
0310	16" H x 24" L		9.59	1.668		690	78		768	875
0320	16" H x 48" L		4.80	3.333		870	156		1,026	1,200
0330	16" H x 72" L		3.20	5		1,075	234		1,309	1,525
0340	20" H x 24" L		7.69	2.081		765	97		862	985
0350	20" H x 48" L		3.84	4.167		975	195		1,170	1,375
0360	20" H x 72" L		2.56	6.250		1,200	292		1,492	1,775
0370	24" H x 24" L		6.40	2.500		825	117		942	1,075
0380	24" H x 48" L		3.20	5		1,075	234		1,309	1,525
0390	24" H x 72" L		2.13	7.512		1,325	350		1,675	1,975
0400	30" H x 28" L		4.39	3.645		1,025	170		1,195	1,400
0410	30" H x 48" L		2.56	6.250		1,225	292		1,517	1,800
0420	30" H x 72" L		1.71	9.357		1,525	435		1,960	2,325
0430	30" H x 84" L		1.46	10.959		1,650	510		2,160	2,600
0440	34" H x 32" L		3.39	4.720		1,150	220		1,370	1,575
0450	34" H x 48" L		2.26	7.080		1,300	330		1,630	1,950
0460	34" H x 72" L		1.51	10.596		1,625	495		2,120	2,550
0470	34" H x 84" L		1.29	12.403		1,800	580		2,380	2,850
0600	For 10 fins per inch, add					1%				

23 82 16.10 Flanged Coils		Crew	Daily Output	Labor-Hours	Unit	Material	2010 Bare Costs Labor	Equipment	Total	Total Incl O&P
0610	For 12 fins per inch, add					3%				
0620	For 4 row, add				Ea.	30%	50%			
0630	For 6 row, add					60%	100%			
0640	For 8 row, add					80%	150%			
1100	1/2" x .017 tube, .0065 aluminum fins									
1110	2 row, 8 fins per inch									
1120	5" H x 5" L	Q-5	40	.400	Ea.	565	18.70		583.70	650
1130	5" H x 20" L		37.10	.431		595	20		615	685
1140	5" H x 45" L		16.41	.975		650	45.50		695.50	785
1150	5" H x 90" L		8.21	1.949		950	91		1,041	1,175
1160	7.5" H x 5" L		38.50	.416		545	19.40		564.40	630
1170	7.5" H x 20" L		24.62	.650		590	30.50		620.50	690
1180	7.5" H x 45" L		10.94	1.463		660	68.50		728.50	825
1190	7.5" H x 90" L		5.46	2.930		1,000	137		1,137	1,300
1200	12.5" H x 10" L		29.49	.543		620	25.50		645.50	720
1210	12.5" H x 20" L		14.75	1.085		665	50.50		715.50	805
1220	12.5" H x 45" L		6.55	2.443		775	114		889	1,025
1230	12.5" H x 90" L		3.28	4.878		1,200	228		1,428	1,675
1240	15" H x 15" L		16.41	.975		665	45.50		710.50	805
1250	15" H x 30" L		8.21	1.949		745	91		836	955
1260	15" H x 45" L		5.46	2.930		820	137		957	1,100
1270	15" H x 90" L		2.73	5.861		1,300	274		1,574	1,825
1280	20" H x 20" L		9.22	1.735		760	81		841	955
1290	20" H x 45" L		4.10	3.902		1,000	182		1,182	1,375
1300	20" H x 90" L		2.05	7.805		1,475	365		1,840	2,175
1310	25" H x 25" L		5.90	2.712		905	127		1,032	1,175
1320	25" H x 45" L		3.28	4.878		1,150	228		1,378	1,625
1330	25" H x 90" L		1.64	9.756		1,700	455		2,155	2,550
1340	30" H x 30" L		4.10	3.902		1,100	182		1,282	1,500
1350	30" H x 45" L		2.73	5.861		1,275	274		1,549	1,800
1360	30" H x 90" L		1.37	11.679		1,900	545		2,445	2,925
1370	35" H x 35" L		3.01	5.316		1,275	248		1,523	1,775
1380	35" H x 45" L		2.34	6.838		1,400	320		1,720	2,000
1390	35" H x 90" L		1.17	13.675		2,100	640		2,740	3,250
1400	42.5" H x 45" L		1.93	8.290		1,625	385		2,010	2,350
1410	42.5" H x 75" L		1.16	13.793		2,150	645		2,795	3,350
1420	42.5" H x 105" L		.83	19.277		2,650	900		3,550	4,275
1600	For 10 fins per inch, add					2%				
1610	For 12 fins per inch, add					5%				
1620	For 4 row, add					35%	50%			
1630	For 6 row, add					65%	100%			
1640	For 8 row, add					90%	150%			
2010	5/8" x .020" tube, .0065 aluminum fins									
2020	2 row, 8 fins per inch									
2030	6" H x 6" L	Q-5	38	.421	Ea.	580	19.65		599.65	670
2040	6" H x 24" L		24.96	.641		630	30		660	735
2050	6" H x 48" L		12.48	1.282		690	60		750	850
2060	6" H x 96" L		6.24	2.564		1,025	120		1,145	1,300
2070	12" H x 12" L		24.96	.641		605	30		635	710
2080	12" H x 24" L		12.48	1.282		660	60		720	820
2090	12" H x 48" L		6.24	2.564		770	120		890	1,025
2100	12" H x 96" L		3.12	5.128		1,225	239		1,464	1,700
2110	18" H x 24" L		8.32	1.923		780	90		870	990

23 82 16 – Air Coils

23 82 16.10 Flanged Coils		Crew	Daily Output	Labor-Hours	Unit	Material	Labor	Equipment	Total	Total Incl O&P
						2010 Bare Costs				
2120	18" H x 48" L	Q-5	4.16	3.846	Ea.	1,025	180		1,205	1,400
2130	18" H x 96" L		2.08	7.692		1,500	360		1,860	2,200
2140	24" H x 24" L		6.24	2.564		865	120		985	1,125
2150	24" H x 48" L		3.12	5.128		1,150	239		1,389	1,625
2160	24" H x 96" L		1.56	10.256		1,750	480		2,230	2,650
2170	30" H x 30" L		3.99	4.010		1,125	187		1,312	1,525
2180	30" H x 48" L		2.50	6.400		1,350	299		1,649	1,925
2190	30" H x 96" L		1.25	12.800		2,050	600		2,650	3,150
2200	30" H x 120" L		1	16		2,400	745		3,145	3,750
2210	36" H x 36" L		2.77	5.776		1,350	270		1,620	1,900
2220	36" H x 48" L		2.08	7.692		1,500	360		1,860	2,200
2230	36" H x 90" L		1.11	14.414		2,225	675		2,900	3,450
2240	36" H x 120" L		.83	19.277		2,700	900		3,600	4,325
2250	42" H x 48" L		1.78	8.989		1,675	420		2,095	2,450
2260	42" H x 90" L		.95	16.842		2,475	785		3,260	3,900
2270	42" H x 120" L		.71	22.535		3,025	1,050		4,075	4,900
2280	51" H x 48" L		1.47	10.884		1,950	510		2,460	2,900
2290	51" H x 96" L		.73	21.918		3,025	1,025		4,050	4,850
2300	51" H x 120" L	↓	.59	27.119	↓	3,550	1,275		4,825	5,800
2500	For 10 fins per inch, add					3%				
2510	For 12 fins per inch, add					6%				
2520	For 4 row, add				Ea.	40%	50%			
2530	For 6 row, add					80%	100%			
2540	For 8 row, add				↓	110%	150%			
3000	Hot water booster coils									
3010	Copper tubes, alum. fins, galv. end sheets									
3012	H is finned height, L is finned length									
3020	1/2" x .017" tube, .0065 Al fins									
3030	1 row, 10 fins per inch									
3040	5"H x 10"L	Q-5	40	.400	Ea.	265	18.70		283.70	320
3050	5"H x 25"L		33.18	.482		290	22.50		312.50	355
3060	10"H x 10"L		39	.410		292	19.15		311.15	350
3070	10"H x 20"L		20.73	.772		315	36		351	400
3080	10"H x 30"L		13.83	1.157		340	54		394	450
3090	15"H x 15"L		18.44	.868		330	40.50		370.50	425
3110	15"H x 20"L		13.83	1.157		345	54		399	460
3120	15"H x 30"L		9.22	1.735		380	81		461	540
3130	20"H x 20"L		10.37	1.543		380	72		452	525
3140	20"H x 30"L		6.91	2.315		420	108		528	625
3150	20"H x 40"L		5.18	3.089		465	144		609	725
3160	25"H x 25"L		6.64	2.410		435	113		548	650
3170	25"H x 35"L		4.74	3.376		555	158		713	845
3180	25"H x 45"L		3.69	4.336		625	202		827	995
3190	30"H x 30"L		4.61	3.471		570	162		732	875
3200	30"H x 40"L		3.46	4.624		655	216		871	1,050
3210	30"H x 50"L	↓	2.76	5.797		730	271		1,001	1,200
3300	For 2 row, add				↓	20%	50%			
3400	1/2" x .020" tube, .008 Al fins									
3410	2 row, 10 fins per inch									
3420	20" H x 27" L	Q-5	7.27	2.201	Ea.	560	103		663	770
3430	20" H x 36" L		5.44	2.941		635	137		772	905
3440	20" H x 45" L		4.35	3.678		790	172		962	1,125
3450	25" H x 45" L	↓	3.49	4.585	↓	910	214		1,124	1,325

23 82 16.10 Flanged Coils		Crew	Daily Output	Labor-Hours	Unit	Material	2010 Bare Costs Labor	Equipment	Total	Total Incl O&P
3460	25" H x 54" L	Q-5	2.90	5.517	Ea.	1,075	258		1,333	1,575
3470	30" H x 54" L		2.42	6.612		1,225	310		1,535	1,825
3480	35" H x 54" L		2.07	7.729		1,375	360		1,735	2,050
3490	35" H x 63" L		1.78	8.989		1,500	420		1,920	2,275
3500	40" H x 72" L		1.36	11.765		1,850	550		2,400	2,850
3510	55" H x 63" L		1.13	14.159		2,175	660		2,835	3,400
3520	55" H x 72" L		.99	16.162		2,400	755		3,155	3,775
3530	55" H x 81" L		.88	18.182		2,650	850		3,500	4,200
3540	75" H x 81" L		.64	25		3,450	1,175		4,625	5,550
3550	75" H x 90" L		.58	27.586		3,775	1,300		5,075	6,075
3560	75" H x 108" L		.48	33.333		4,425	1,550		5,975	7,200
3800	5/8" x .020 tubing, .0065 aluminum fins									
3810	1 row, 10 fins per inch									
3820	6" H x 12" L	Q-5	38	.421	Ea.	296	19.65		315.65	355
3830	6" H x 24" L		28.80	.556		295	26		321	365
3840	6" H x 30" L		23.04	.694		305	32.50		337.50	385
3850	12" H x 12" L		28.80	.556		305	26		331	375
3860	12" H x 24" L		14.40	1.111		330	52		382	445
3870	12" H x 30" L		11.52	1.389		345	65		410	475
3880	18" H x 18" L		12.80	1.250		350	58.50		408.50	475
3890	18" H x 24" L		9.60	1.667		370	78		448	520
3900	18" H x 36" L		6.40	2.500		410	117		527	625
3910	24" H x 24" L		7.20	2.222		405	104		509	600
3920	24" H x 30" L		4.80	3.333		535	156		691	825
3930	24" H x 36" L		5.05	3.168		580	148		728	860
3940	30" H x 30" L		4.61	3.471		600	162		762	905
3950	30" H x 36" L		3.85	4.159		650	194		844	1,000
3960	30" H x 42" L		3.29	4.863		705	227		932	1,125
3970	36" H x 36" L		3.20	5		720	234		954	1,150
3980	36" H x 42" L		2.74	5.839		785	273		1,058	1,275
4100	For 2 row, add					30%	50%			
4400	Steam and hot water heating coils									
4410	Copper tubes, alum. fins. galv. end sheets									
4412	H is finned height, L is finned length									
4420	5/8" tubing, to 175 psig working pressure									
4430	1 row, 8 fins per inch									
4440	6"H x 6"L	Q-5	38	.421	Ea.	550	19.65		569.65	635
4450	6"H x 12"L		37	.432		565	20		585	650
4460	6"H x 24"L		24.48	.654		590	30.50		620.50	690
4470	6"H x 36"L		16.32	.980		610	46		656	740
4480	6"H x 48"L		12.24	1.307		635	61		696	790
4490	6"H x 84"L		6.99	2.289		1,225	107		1,332	1,500
4500	6" H x 108" L		5.44	2.941		1,375	137		1,512	1,700
4510	12" H x 12" L		24.48	.654		665	30.50		695.50	775
4520	12" H x 24" L		12.24	1.307		705	61		766	865
4530	12" H x 36" L		8.16	1.961		750	91.50		841.50	960
4540	12" H x 48" L		6.12	2.614		865	122		987	1,150
4550	12" H x 84" L		3.50	4.571		1,525	213		1,738	2,000
4560	12" H x 108" L		2.72	5.882		1,750	275		2,025	2,325
4570	18" H x18" L		10.88	1.471		795	68.50		863.50	980
4580	18" H x 24" L		8.16	1.961		825	91.50		916.50	1,050
4590	18" H x 48" L		4.08	3.922		1,025	183		1,208	1,400
4600	18" H x 84" L		2.33	6.867		1,825	320		2,145	2,500

23 82 16 – Air Coils

23 82 16.10 Flanged Coils		Crew	Daily Output	Labor-Hours	Unit	Material	Labor	Equipment	Total	Total Incl O&P
						2010 Bare Costs				
4610	18" H x 108" L	Q-5	1.81	8.840	Ea.	2,125	415		2,540	2,950
4620	24" H x 24" L		6.12	2.614		1,000	122		1,122	1,275
4630	24" H x 48" L		3.06	5.229		1,200	244		1,444	1,700
4640	24" H x 84" L		1.75	9.143		2,150	425		2,575	3,000
4650	24" H x 108" L		1.36	11.765		2,500	550		3,050	3,575
4660	36" H x 36" L		2.72	5.882		1,400	275		1,675	1,925
4670	36" H x 48" L		2.04	7.843		1,525	365		1,890	2,225
4680	36" H x 84" L		1.17	13.675		2,750	640		3,390	3,975
4690	36" H x 108" L		.91	17.582		3,250	820		4,070	4,800
4700	48"H x 48"L		1.53	10.458		1,875	490		2,365	2,775
4710	48"H x 84"L		.87	18.391		3,400	860		4,260	5,025
4720	48"H x 108"L		.68	23.529		4,025	1,100		5,125	6,075
4730	48"H x 126"L	↓	.58	27.586	↓	4,300	1,300		5,600	6,650
4910	For 10 fins per inch, add					2%				
4920	For 12 fins per inch, add					5%				
4930	For 2 row, add					30%	50%			
4940	For 3 row, add					50%	100%			
4950	For 4 row, add					80%	123%			
4960	For 230 PSIG heavy duty, add					20%	30%			
6000	Steam, heavy duty, 200 PSIG									
6010	Copper tubes, alum. fins, galv. end sheets									
6012	H is finned height, L is finned length									
6020	1" x .035" tubing, .010 aluminum fins									
6030	1 row, 8 fins per inch									
6040	12"H x 12"L	Q-5	24.48	.654	Ea.	795	30.50		825.50	920
6050	12"H x 24"L		12.24	1.307		900	61		961	1,075
6060	12"H x 36"L		8.16	1.961		1,000	91.50		1,091.50	1,225
6070	12"H x 48"L		6.12	2.614		1,200	122		1,322	1,475
6080	12"H x 84"L		3.50	4.571		1,675	213		1,888	2,150
6090	12"H x 108"L		2.72	5.882		1,975	275		2,250	2,575
6100	18"H x 24"L		8.16	1.961		1,125	91.50		1,216.50	1,350
6110	18"H x 36"L		5.44	2.941		1,350	137		1,487	1,675
6120	18"H x 48"L		4.08	3.922		1,525	183		1,708	1,950
6130	18"H x 84"L		2.33	6.867		2,150	320		2,470	2,825
6140	18"H x 108"L		1.81	8.840		2,575	415		2,990	3,475
6150	24"H x 24"L		6.12	2.614		1,450	122		1,572	1,775
6160	24"H x 36"L		4.08	3.922		1,625	183		1,808	2,050
6170	24"H x 48"L		3.06	5.229		1,825	244		2,069	2,400
6180	24"H x 84"L		1.75	9.143		2,625	425		3,050	3,550
6190	24"H x 108"L		1.36	11.765		3,175	550		3,725	4,325
6200	30"H x 36"L		3.26	4.908		1,875	229		2,104	2,425
6210	30"H x 48"L		2.45	6.531		2,150	305		2,455	2,825
6220	30"H x 84"L		1.40	11.429		3,100	535		3,635	4,225
6230	30"H x 108"L		1.09	14.679		3,775	685		4,460	5,175
6240	36"H x 36"L		2.72	5.882		2,150	275		2,425	2,775
6250	36"H x 48"L		2.04	7.843		2,475	365		2,840	3,275
6260	36"H x 84"L		1.17	13.675		3,600	640		4,240	4,900
6270	36"H x 108"L		.91	17.582		4,375	820		5,195	6,050
6280	36"H x 120"L		.82	19.512		4,700	910		5,610	6,550
6290	48"H x 48"L		1.53	10.458		4,525	490		5,015	5,700
6300	48"H x 84"L		.87	18.391		4,550	860		5,410	6,300
6310	48"H x 108"L		.68	23.529		5,600	1,100		6,700	7,800
6320	48"H x 120"L	↓	.61	26.230	↓	6,025	1,225		7,250	8,450

23 82 16 – Air Coils

23 82 16.10 Flanged Coils		Crew	Daily Output	Labor-Hours	Unit	Material	2010 Bare Costs Labor	Equipment	Total	Total Incl O&P
6400	For 10 fins per inch, add				Ea.	4%				
6410	For 12 fins per inch, add				↓	10%				

23 82 16.20 Duct Heaters		Crew	Daily Output	Labor-Hours	Unit	Material	Labor	Equipment	Total	Total Incl O&P
0010	**DUCT HEATERS**, Electric, 480 V, 3 Ph.									
0020	Finned tubular insert, 500°F									
0100	8" wide x 6" high, 4.0 kW	Q-20	16	1.250	Ea.	705	56.50		761.50	860
0120	12" high, 8.0 kW		15	1.333		1,175	60		1,235	1,375
0140	18" high, 12.0 kW		14	1.429		1,650	64.50		1,714.50	1,900
0160	24" high, 16.0 kW		13	1.538		2,100	69.50		2,169.50	2,425
0180	30" high, 20.0 kW		12	1.667		2,575	75.50		2,650.50	2,950
0300	12" wide x 6" high, 6.7 kW		15	1.333		750	60		810	915
0320	12" high, 13.3 kW		14	1.429		1,200	64.50		1,264.50	1,425
0340	18" high, 20.0 kW		13	1.538		1,700	69.50		1,769.50	1,975
0360	24" high, 26.7 kW		12	1.667		2,175	75.50		2,250.50	2,525
0380	30" high, 33.3 kW		11	1.818		2,650	82		2,732	3,050
0500	18" wide x 6" high, 13.3 kW		14	1.429		805	64.50		869.50	985
0520	12" high, 26.7 kW		13	1.538		1,375	69.50		1,444.50	1,625
0540	18" high, 40.0 kW		12	1.667		1,850	75.50		1,925.50	2,150
0560	24" high, 53.3 kW		11	1.818		2,450	82		2,532	2,825
0580	30" high, 66.7 kW		10	2		3,050	90.50		3,140.50	3,500
0700	24" wide x 6" high, 17.8 kW		13	1.538		890	69.50		959.50	1,075
0720	12" high, 35.6 kW		12	1.667		1,500	75.50		1,575.50	1,775
0740	18" high, 53.3 kW		11	1.818		2,100	82		2,182	2,450
0760	24" high, 71.1 kW		10	2		2,725	90.50		2,815.50	3,125
0780	30" high, 88.9 kW		9	2.222		3,325	100		3,425	3,825
0900	30" wide x 6" high, 22.2 kW		12	1.667		940	75.50		1,015.50	1,150
0920	12" high, 44.4 kW		11	1.818		1,575	82		1,657	1,875
0940	18" high, 66.7 kW		10	2		2,250	90.50		2,340.50	2,600
0960	24" high, 88.9 kW		9	2.222		2,750	100		2,850	3,175
0980	30" high, 111.0 kW	↓	8	2.500	↓	3,525	113		3,638	4,050
1400	Note decreased kW available for									
1410	each duct size at same cost									
1420	See line 5000 for modifications and accessories									
2000	Finned tubular flange with insulated									
2020	terminal box, 500°F									
2100	12" wide x 36" high, 54 kW	Q-20	10	2	Ea.	3,475	90.50		3,565.50	3,950
2120	40" high, 60 kW		9	2.222		4,000	100		4,100	4,550
2200	24" wide x 36" high, 118.8 kW		9	2.222		3,925	100		4,025	4,475
2220	40" high, 132 kW		8	2.500		4,675	113		4,788	5,300
2400	36" wide x 8" high, 40 kW		11	1.818		1,800	82		1,882	2,100
2420	16" high, 80 kW		10	2		2,350	90.50		2,440.50	2,725
2440	24" high, 120 kW		9	2.222		3,075	100		3,175	3,550
2460	32" high, 160 kW		8	2.500		3,950	113		4,063	4,525
2480	36" high, 180 kW		7	2.857		4,875	129		5,004	5,550
2500	40" high, 200 kW		6	3.333		5,400	151		5,551	6,150
2600	40" wide x 8" high, 45 kW		11	1.818		1,850	82		1,932	2,150
2620	16" high, 90 kW		10	2		2,550	90.50		2,640.50	2,925
2640	24" high, 135 kW		9	2.222		3,225	100		3,325	3,700
2660	32" high, 180 kW		8	2.500		4,300	113		4,413	4,900
2680	36" high, 202.5 kW		7	2.857		4,950	129		5,079	5,650
2700	40" high, 225 kW		6	3.333		5,625	151		5,776	6,425
2800	48" wide x 8" high, 54.8 kW	↓	10	2	↓	1,925	90.50		2,015.50	2,250

23 82 Convection Heating and Cooling Units

23 82 16 – Air Coils

23 82 16.20 Duct Heaters		Crew	Daily Output	Labor-Hours	Unit	Material	2010 Bare Costs Labor	Equipment	Total	Total Incl O&P
2820	16" high, 109.8 kW	Q-20	9	2.222	Ea.	2,675	100		2,775	3,100
2840	24" high, 164.4 kW		8	2.500		3,400	113		3,513	3,925
2860	32" high, 219.2 kW		7	2.857		4,550	129		4,679	5,200
2880	36" high, 246.6 kW		6	3.333		5,225	151		5,376	5,975
2900	40" high, 274 kW		5	4		5,925	181		6,106	6,775
3000	56" wide x 8" high, 64 kW		9	2.222		2,200	100		2,300	2,575
3020	16" high, 128 kW		8	2.500		3,050	113		3,163	3,525
3040	24" high, 192 kW		7	2.857		3,700	129		3,829	4,275
3060	32" high, 256 kW		6	3.333		5,175	151		5,326	5,900
3080	36" high, 288kW		5	4		5,925	181		6,106	6,800
3100	40" high, 320kW		4	5		6,550	226		6,776	7,575
3200	64" wide x 8" high, 74kW		8	2.500		2,275	113		2,388	2,675
3220	16" high, 148kW		7	2.857		3,150	129		3,279	3,650
3240	24" high, 222kW		6	3.333		4,000	151		4,151	4,625
3260	32" high, 296kW		5	4		5,350	181		5,531	6,175
3280	36" high, 333kW		4	5		6,375	226		6,601	7,375
3300	40" high, 370kW	↓	3	6.667	↓	7,050	300		7,350	8,200
3800	Note decreased kW available for									
3820	each duct size at same cost									
5000	Duct heater modifications and accessories									
5120	T.C.O. limit auto or manual reset	Q-20	42	.476	Ea.	116	21.50		137.50	160
5140	Thermostat		28	.714		495	32.50		527.50	595
5160	Overheat thermocouple (removable)		7	2.857		700	129		829	965
5180	Fan interlock relay		18	1.111		168	50		218	261
5200	Air flow switch		20	1		146	45		191	228
5220	Split terminal box cover	↓	100	.200	↓	48	9.05		57.05	66.50
8000	To obtain BTU multiply kW by 3413									

23 82 19 – Fan Coil Units

23 82 19.10 Fan Coil Air Conditioning		Crew	Daily Output	Labor-Hours	Unit	Material	Labor	Equipment	Total	Total Incl O&P
0010	**FAN COIL AIR CONDITIONING** Cabinet mounted, filters, controls									
0100	Chilled water, 1/2 ton cooling	Q-5	8	2	Ea.	710	93.50		803.50	920
0110	3/4 ton cooling		7	2.286		735	107		842	970
0120	1 ton cooling		6	2.667		815	125		940	1,075
0140	1.5 ton cooling		5.50	2.909		920	136		1,056	1,200
0150	2 ton cooling		5.25	3.048		1,150	142		1,292	1,500
0160	2.5 ton cooling		5	3.200		1,525	149		1,674	1,900
0180	3 ton cooling	↓	4	4	↓	1,725	187		1,912	2,175
0262	For hot water coil, add					40%	10%			
0300	Console, 2 pipe with electric heat									
0310	1/2 ton cooling	Q-5	8	2	Ea.	1,175	93.50		1,268.50	1,425
0315	3/4 ton cooling		7	2.286		1,275	107		1,382	1,550
0320	1 ton cooling		6	2.667		1,375	125		1,500	1,675
0330	1.5 ton cooling		5.50	2.909		1,675	136		1,811	2,050
0340	2 ton cooling		4.70	3.404		2,225	159		2,384	2,700
0345	2-1/2 ton cooling		4.70	3.404		3,225	159		3,384	3,800
0350	3 ton cooling		4	4		4,475	187		4,662	5,200
0360	4 ton cooling		4	4		4,650	187		4,837	5,400
0940	Direct expansion, for use w/air cooled condensing unit, 1.5 ton cooling		5	3.200		690	149		839	985
0950	2 ton cooling		4.80	3.333		750	156		906	1,050
0960	2.5 ton cooling		4.40	3.636		790	170		960	1,125
0970	3 ton cooling		3.80	4.211		985	197		1,182	1,375
0980	3.5 ton cooling	↓	3.60	4.444	↓	1,025	208		1,233	1,450

23 82 19.10 Fan Coil Air Conditioning		Crew	Daily Output	Labor-Hours	Unit	2010 Bare Costs Material	Labor	Equipment	Total	Total Incl O&P
0990	4 ton cooling	Q-5	3.40	4.706	Ea.	1,150	220		1,370	1,600
1000	5 ton cooling		3	5.333		1,325	249		1,574	1,825
1020	7.5 ton cooling		3	5.333		2,275	249		2,524	2,875
1040	10 ton cooling	Q-6	2.60	9.231		2,775	445		3,220	3,725
1042	12.5 ton cooling		2.30	10.435		3,625	505		4,130	4,750
1050	15 ton cooling		1.60	15		3,900	725		4,625	5,375
1060	20 ton cooling		.70	34.286		5,375	1,650		7,025	8,400
1070	25 ton cooling		.65	36.923		6,800	1,800		8,600	10,200
1080	30 ton cooling		.60	40		7,325	1,925		9,250	11,000
1500	For hot water coil, add					40%	10%			
1512	For condensing unit add see Div. 23 62									
3000	Chilled water, horizontal unit, housing, 2 pipe, fan control, no valves									
3100	1/2 ton cooling	Q-5	8	2	Ea.	1,075	93.50		1,168.50	1,350
3110	1 ton cooling		6	2.667		1,300	125		1,425	1,625
3120	1.5 ton cooling		5.50	2.909		1,575	136		1,711	1,925
3130	2 ton cooling		5.25	3.048		1,875	142		2,017	2,300
3140	3 ton cooling		4	4		2,175	187		2,362	2,675
3150	3.5 ton cooling		3.80	4.211		2,400	197		2,597	2,925
3160	4 ton cooling		3.80	4.211		2,400	197		2,597	2,925
3170	5 ton cooling		3.40	4.706		2,775	220		2,995	3,375
3180	6 ton cooling		3.40	4.706		2,775	220		2,995	3,375
3190	7 ton cooling		2.90	5.517		3,025	258		3,283	3,700
3200	8 ton cooling		2.90	5.517		3,025	258		3,283	3,700
3210	10 ton cooling	Q-6	2.80	8.571		3,425	415		3,840	4,400
3240	For hot water coil, add					40%	10%			
4000	With electric heat, 2 pipe									
4100	1/2 ton cooling	Q-5	8	2	Ea.	1,325	93.50		1,418.50	1,600
4105	3/4 ton cooling		7	2.286		1,450	107		1,557	1,750
4110	1 ton cooling		6	2.667		1,600	125		1,725	1,950
4120	1.5 ton cooling		5.50	2.909		1,775	136		1,911	2,150
4130	2 ton cooling		5.25	3.048		2,250	142		2,392	2,700
4135	2.5 ton cooling		4.80	3.333		3,100	156		3,256	3,625
4140	3 ton cooling		4	4		4,275	187		4,462	4,975
4150	3.5 ton cooling		3.80	4.211		5,125	197		5,322	5,925
4160	4 ton cooling		3.60	4.444		5,925	208		6,133	6,800
4170	5 ton cooling		3.40	4.706		6,725	220		6,945	7,700
4180	6 ton cooling		3.20	5		6,725	234		6,959	7,725
4190	7 ton cooling		2.90	5.517		6,975	258		7,233	8,050
4200	8 ton cooling		2.20	7.273		7,900	340		8,240	9,200
4210	10 ton cooling	Q-6	2.80	8.571		8,375	415		8,790	9,850

23 82 19.20 Heating and Ventilating Units

		Crew	Daily Output	Labor-Hours	Unit	Material	Labor	Equipment	Total	Total Incl O&P
0010	**HEATING AND VENTILATING UNITS**, Classroom units									
0020	Includes filter, heating/cooling coils, standard controls									
0080	750 CFM, 2 tons cooling	Q-6	2	12	Ea.	3,625	580		4,205	4,850
0100	1000 CFM, 2-1/2 tons cooling		1.60	15		4,275	725		5,000	5,800
0120	1250 CFM, 3 tons cooling		1.40	17.143		4,425	830		5,255	6,100
0140	1500 CFM, 4 tons cooling		.80	30		4,725	1,450		6,175	7,375
0160	2000 CFM 5 tons cooling		.50	48		5,550	2,325		7,875	9,575
0500	For electric heat, add					35%				
1000	For no cooling, deduct					25%	10%			

23 82 Convection Heating and Cooling Units

23 82 27 – Infrared Units

23 82 27.10 Infrared Type Heating Units		Crew	Daily Output	Labor-Hours	Unit	2010 Bare Costs Material	Labor	Equipment	Total	Total Incl O&P
0010	**INFRARED TYPE HEATING UNITS**									
0020	Gas fired, unvented, electric ignition, 100% shutoff.									
0030	Piping and wiring not included									
0060	Input, 15 MBH	Q-5	7	2.286	Ea.	435	107		542	640
0100	30 MBH		6	2.667		470	125		595	700
0120	45 MBH		5	3.200		500	149		649	775
0140	50 MBH		4.50	3.556		515	166		681	815
0160	60 MBH		4	4		585	187		772	925
0180	75 MBH		3	5.333		625	249		874	1,050
0200	90 MBH		2.50	6.400		700	299		999	1,225
0220	105 MBH		2	8		895	375		1,270	1,550
0240	120 MBH		2	8		970	375		1,345	1,625
2000	Electric, single or three phase									
2050	6 kW, 20,478 BTU	1 Elec	2.30	3.478	Ea.	430	170		600	730
2100	13.5 KW, 40,956 BTU		2.20	3.636		665	178		843	995
2150	24 KW, 81,912 BTU		2	4		1,325	196		1,521	1,750
3000	Oil fired, pump, controls, fusible valve, oil supply tank									
3050	91,000 BTU	Q-5	2.50	6.400	Ea.	6,000	299		6,299	7,050
3080	105,000 BTU		2.25	7.111		6,375	330		6,705	7,525
3110	119,000 BTU		2	8		6,975	375		7,350	8,225

23 82 29 – Radiators

23 82 29.10 Hydronic Heating

		Crew	Daily Output	Labor-Hours	Unit	Material	Labor	Equipment	Total	Total Incl O&P
0010	**HYDRONIC HEATING**, Terminal units, not incl. main supply pipe									
1000	Radiation									
1100	Panel, baseboard, C.I., including supports, no covers	Q-5	46	.348	L.F.	32.50	16.25		48.75	60
3000	Radiators, cast iron									
3100	Free standing or wall hung, 6 tube, 25" high	Q-5	96	.167	Section	36	7.80		43.80	51
3150	4 tube 25" high		96	.167		27	7.80		34.80	41
3200	4 tube, 19" high		96	.167		23.50	7.80		31.30	37.50
3250	Adj. brackets, 2 per wall radiator up to 30 sections	1 Stpi	32	.250	Ea.	27	13		40	49.50
3500	Recessed, 20" high x 5" deep, without grille	Q-5	60	.267	Section	30.50	12.45		42.95	52
3600	For inlet grille, add				"	2.12			2.12	2.33
9500	To convert SFR to BTU rating: Hot water, 150 x SFR									
9510	Forced hot water, 180 x SFR; steam, 240 x SFR									

23 82 33 – Convectors

23 82 33.10 Convector Units

		Crew	Daily Output	Labor-Hours	Unit	Material	Labor	Equipment	Total	Total Incl O&P
0010	**CONVECTOR UNITS,** Terminal units, not incl. main supply pipe									
2204	Convector, multifin, 2 pipe w/ cabinet									
2210	17" H x 24" L	Q-5	10	1.600	Ea.	98.50	74.50		173	220
2214	17" H x 36" L		8.60	1.860		148	87		235	292
2218	17" H x 48" L		7.40	2.162		197	101		298	365
2222	21" H x 24" L		9	1.778		96.50	83		179.50	230
2226	21" H x 36" L		8.20	1.951		145	91		236	296
2228	21" H x 48" L		6.80	2.353		193	110		303	375
2240	For knob operated damper, add					140%				
2241	For metal trim strips, add	Q-5	64	.250	Ea.	19.25	11.70		30.95	38.50
2243	For snap-on inlet grille, add					10%	10%			
2245	For hinged access door, add	Q-5	64	.250	Ea.	35.50	11.70		47.20	56.50
2246	For air chamber, auto-venting, add	"	58	.276	"	6.35	12.90		19.25	26.50

23 82 36 – Finned-Tube Radiation Heaters

23 82 36.10 Finned Tube Radiation		Crew	Daily Output	Labor-Hours	Unit	2010 Bare Costs Material	Labor	Equipment	Total	Total Incl O&P
0010	**FINNED TUBE RADIATION**, Terminal units, not incl. main supply pipe									
1150	Fin tube, wall hung, 14" slope top cover, with damper									
1200	1-1/4" copper tube, 4-1/4" alum. fin	Q-5	38	.421	L.F.	40.50	19.65		60.15	74.50
1250	1-1/4" steel tube, 4-1/4" steel fin		36	.444		38.50	21		59.50	73.50
1255	2" steel tube, 4-1/4" steel fin		32	.500		41.50	23.50		65	80.50
1260	21", two tier, slope top, w/damper									
1262	1-1/4" copper tube, 4-1/4" alum. fin	Q-5	30	.533	L.F.	48.50	25		73.50	90.50
1264	1-1/4" steel tube, 4-1/4" steel fin		28	.571		46	26.50		72.50	90.50
1266	2" steel tube, 4-1/4" steel fin		25	.640		49	30		79	99
1270	10", flat top, with damper									
1272	1-1/4" copper tube, 4-1/4" alum. fin	Q-5	38	.421	L.F.	28.50	19.65		48.15	61
1274	1-1/4" steel tube, 4-1/4" steel fin		36	.444		38.50	21		59.50	73.50
1276	2" steel tube, 4-1/4" steel fin		32	.500		41.50	23.50		65	80.50
1280	17", two tier, flat top, w/damper									
1282	1-1/4" copper tube, 4-1/4" alum. fin	Q-5	30	.533	L.F.	49	25		74	91.50
1284	1-1/4" steel tube, 4-1/4" steel fin		28	.571		46	26.50		72.50	90.50
1286	2" steel tube, 4-1/4" steel fin		25	.640		49	30		79	99
1301	Rough in wall hung, steel fin w/supply & balance valves	1 Stpi	.91	8.791	Ea.	253	455		708	965
1310	Baseboard, pkgd, 1/2" copper tube, alum. fin, 7" high	Q-5	60	.267	L.F.	7.55	12.45		20	27
1320	3/4" copper tube, alum. fin, 7" high		58	.276		8	12.90		20.90	28
1340	1" copper tube, alum. fin, 8-7/8" high		56	.286		25	13.35		38.35	47.50
1360	1-1/4" copper tube, alum. fin, 8-7/8" high		54	.296		26	13.85		39.85	49
1380	1-1/4" IPS steel tube with steel fins		52	.308		26	14.35		40.35	50
1381	Rough in baseboard panel & fin tube, supply & balance valves	1 Stpi	1.06	7.547	Ea.	201	390		591	805
1500	Note: fin tube may also require corners, caps, etc.									

23 82 39 – Unit Heaters

23 82 39.13 Cabinet Unit Heaters

		Crew	Daily Output	Labor-Hours	Unit	Material	Labor	Equipment	Total	Total Incl O&P
0010	**CABINET UNIT HEATERS**									
5400	Cabinet, horizontal, hot water, blower type									
5420	20 MBH	Q-5	10	1.600	Ea.	795	74.50		869.50	985
5430	60 MBH		7	2.286		1,375	107		1,482	1,675
5440	100 MBH		5	3.200		1,450	149		1,599	1,825
5450	120 MBH		4	4		1,500	187		1,687	1,925

23 82 39.16 Propeller Unit Heaters

		Crew	Daily Output	Labor-Hours	Unit	Material	Labor	Equipment	Total	Total Incl O&P
0010	**PROPELLER UNIT HEATERS**									
3950	Unit heaters, propeller, 115 V 2 psi steam, 60°F entering air									
4000	Horizontal, 12 MBH	Q-5	12	1.333	Ea.	335	62.50		397.50	465
4020	28.2 MBH		10	1.600		435	74.50		509.50	585
4040	36.5 MBH		8	2		505	93.50		598.50	695
4060	43.9 MBH		8	2		505	93.50		598.50	695
4080	56.5 MBH		7.50	2.133		540	99.50		639.50	745
4100	65.6 MBH		7	2.286		555	107		662	770
4120	87.6 MBH		6.50	2.462		600	115		715	830
4140	96.8 MBH		6	2.667		735	125		860	995
4160	133.3 MBH		5	3.200		800	149		949	1,100
4180	157.6 MBH		4	4		980	187		1,167	1,350
4200	197.7 MBH		3	5.333		1,125	249		1,374	1,600
4220	257.2 MBH		2.50	6.400		1,550	299		1,849	2,150
4240	286.9 MBH		2	8		1,600	375		1,975	2,300
4250	326.0 MBH		1.90	8.421		1,825	395		2,220	2,600
4260	364 MBH		1.80	8.889		1,900	415		2,315	2,725

23 82 39.16 Propeller Unit Heaters		Crew	Daily Output	Labor-Hours	Unit	Material	Labor	Equipment	Total	Total Incl O&P
						2010 Bare Costs				
4270	404 MBH	Q-5	1.60	10	Ea.	2,500	465		2,965	3,450
4300	For vertical diffuser, add					174			174	191
4310	Vertical flow, 40 MBH	Q-5	11	1.455		500	68		568	650
4314	58.5 MBH		8	2		520	93.50		613.50	710
4318	92 MBH		7	2.286		660	107		767	890
4322	109.7 MBH		6	2.667		780	125		905	1,050
4326	131.0 MBH		4	4		780	187		967	1,150
4330	160.0 MBH		3	5.333		825	249		1,074	1,275
4334	194.0 MBH		2.20	7.273		960	340		1,300	1,550
4338	212.0 MBH		2.10	7.619		1,175	355		1,530	1,825
4342	247.0 MBH		1.96	8.163		1,400	380		1,780	2,125
4346	297.0 MBH		1.80	8.889		1,500	415		1,915	2,275
4350	333.0 MBH	Q-6	1.90	12.632		1,625	610		2,235	2,700
4354	420 MBH,(460 V)		1.80	13.333		2,000	645		2,645	3,175
4358	500 MBH, (460 V)		1.71	14.035		2,650	680		3,330	3,925
4362	570 MBH, (460 V)		1.40	17.143		3,725	830		4,555	5,350
4366	620 MBH, (460 V)		1.30	18.462		4,275	895		5,170	6,050
4370	960 MBH, (460 V)		1.10	21.818		7,200	1,050		8,250	9,475
4400	Unit heaters, propeller, hot water, 115 V, 60 Deg. F entering air									
4404	Horizontal									
4408	6 MBH	Q-5	11	1.455	Ea.	335	68		403	470
4412	12 MBH		11	1.455		335	68		403	470
4416	16 MBH		10	1.600		375	74.50		449.50	525
4420	24 MBH		9	1.778		435	83		518	600
4424	29 MBH		8	2		505	93.50		598.50	695
4428	47 MBH		7	2.286		555	107		662	770
4432	63 MBH		6	2.667		600	125		725	845
4436	81 MBH		5.50	2.909		735	136		871	1,025
4440	90 MBH		5	3.200		800	149		949	1,100
4460	133 MBH		4	4		980	187		1,167	1,350
4464	139 MBH		3.50	4.571		1,125	213		1,338	1,550
4468	198 MBH		2.50	6.400		1,550	299		1,849	2,150
4472	224 MBH		2	8		1,750	375		2,125	2,475
4476	273 MBH		1.50	10.667		1,900	500		2,400	2,850
4500	Vertical									
4504	8 MBH	Q-5	13	1.231	Ea.	500	57.50		557.50	635
4508	12 MBH		13	1.231		500	57.50		557.50	635
4512	16 MBH		13	1.231		500	57.50		557.50	635
4516	30 MBH		12	1.333		500	62.50		562.50	645
4520	43 MBH		11	1.455		520	68		588	670
4524	57 MBH		8	2		635	93.50		728.50	840
4528	68 MBH		7	2.286		660	107		767	890
4540	105 MBH		6	2.667		780	125		905	1,050
4544	123 MBH		5	3.200		825	149		974	1,125
4550	140 MBH		5	3.200		1,125	149		1,274	1,450
4554	156 MBH		4	4		1,175	187		1,362	1,575
4560	210 MBH		4	4		1,500	187		1,687	1,925
4564	223 MBH		3	5.333		1,625	249		1,874	2,150
4568	257 MBH		2	8		1,625	375		2,000	2,325

23 83 Radiant Heating Units

23 83 16 – Radiant-Heating Hydronic Piping

	23 83 16.10 Radiant Floor Heating	Crew	Daily Output	Labor-Hours	Unit	2010 Bare Costs Material	Labor	Equipment	Total	Total Incl O&P
0010	**RADIANT FLOOR HEATING**									
0100	Tubing, PEX (cross-linked polyethylene)									
0110	Oxygen barrier type for systems with ferrous materials									
0120	1/2"	Q-5	800	.020	L.F.	1.01	.93		1.94	2.51
0130	3/4"		535	.030		1.29	1.40		2.69	3.51
0140	1"		400	.040		2.21	1.87		4.08	5.25
0200	Non barrier type for ferrous free systems									
0210	1/2"	Q-5	800	.020	L.F.	.55	.93		1.48	2.01
0220	3/4"		535	.030		1	1.40		2.40	3.19
0230	1"		400	.040		1.71	1.87		3.58	4.68
1000	Manifolds									
1110	Brass									
1120	With supply and return valves, flow meter, thermometer,									
1122	auto air vent and drain/fill valve.									
1130	1", 2 circuit	Q-5	14	1.143	Ea.	181	53.50		234.50	279
1140	1", 3 circuit		13.50	1.185		207	55.50		262.50	310
1150	1", 4 circuit		13	1.231		225	57.50		282.50	335
1154	1", 5 circuit		12.50	1.280		268	60		328	385
1158	1", 6 circuit		12	1.333		292	62.50		354.50	415
1162	1", 7 circuit		11.50	1.391		320	65		385	450
1166	1", 8 circuit		11	1.455		355	68		423	490
1172	1", 9 circuit		10.50	1.524		380	71		451	525
1174	1", 10 circuit		10	1.600		410	74.50		484.50	560
1178	1", 11 circuit		9.50	1.684		425	78.50		503.50	590
1182	1", 12 circuit		9	1.778		470	83		553	640
1610	Copper manifold header, (cut to size)									
1620	1" header, 12 – 1/2" sweat outlets	Q-5	3.33	4.805	Ea.	72.50	224		296.50	415
1630	1-1/4" header, 12 – 1/2" sweat outlets		3.20	5		84	234		318	445
1640	1-1/4" header, 12 – 3/4" sweat outlets		3	5.333		90.50	249		339.50	475
1650	1-1/2" header, 12 – 3/4" sweat outlets		3.10	5.161		109	241		350	480
1660	2" header, 12 – 3/4" sweat outlets		2.90	5.517		160	258		418	560
3000	Valves									
3110	Thermostatic zone valve actuator with end switch	Q-5	40	.400	Ea.	36.50	18.70		55.20	68
3114	Thermostatic zone valve actuator	"	36	.444	"	81	21		102	120
3120	Motorized straight zone valve with operator complete									
3130	3/4"	Q-5	35	.457	Ea.	125	21.50		146.50	170
3140	1"		32	.500		137	23.50		160.50	186
3150	1-1/4"		29.60	.541		165	25		190	220
3500	4 Way mixing valve, manual, brass									
3530	1"	Q-5	13.30	1.203	Ea.	126	56		182	223
3540	1-1/4"		11.40	1.404		137	65.50		202.50	250
3550	1-1/2"		11	1.455		203	68		271	325
3560	2"		10.60	1.509		288	70.50		358.50	420
3800	Mixing valve motor, 4 way for valves, 1" and 1-1/4"		34	.471		310	22		332	375
3810	Mixing valve motor, 4 way for valves, 1-1/2" and 2"		30	.533		355	25		380	430
5000	Radiant floor heating, zone control panel									
5120	4 Zone actuator valve control, expandable	Q-5	20	.800	Ea.	163	37.50		200.50	235
5130	6 Zone actuator valve control, expandable		18	.889		199	41.50		240.50	281
6070	Thermal track, straight panel for long continuous runs, 5.333 S.F.		40	.400		16.60	18.70		35.30	46.50
6080	Thermal track, utility panel, for direction reverse at run end, 5.333 S.F.		40	.400		16.60	18.70		35.30	46.50
6090	Combination panel, for direction reverse plus straight run, 5.333 S.F.		40	.400		16.60	18.70		35.30	46.50
7000	PEX tubing fittings									
7100	Compression type									

23 83 16 – Radiant-Heating Hydronic Piping

23 83 16.10 Radiant Floor Heating		Crew	Daily Output	Labor-Hours	Unit	Material	Labor	Equipment	Total	Total Incl O&P
						2010 Bare Costs				
7116	Coupling									
7120	1/2" x 1/2"	1 Stpi	27	.296	Ea.	6.40	15.40		21.80	30
7124	3/4" x 3/4"	"	23	.348	"	10.10	18.05		28.15	38
7130	Adapter									
7132	1/2" x female sweat 1/2"	1 Stpi	27	.296	Ea.	4.15	15.40		19.55	27.50
7134	1/2" x female sweat 3/4"		26	.308		4.88	15.95		20.83	29.50
7136	5/8" x female sweat 3/4"		24	.333		6.65	17.30		23.95	33.50
7140	Elbow									
7142	1/2" x female sweat 1/2"	1 Stpi	27	.296	Ea.	6.35	15.40		21.75	30
7144	1/2" x female sweat 3/4"		26	.308		7.40	15.95		23.35	32
7146	5/8" x female sweat 3/4"		24	.333		8.75	17.30		26.05	35.50
7200	Insert type									
7206	PEX x male NPT									
7210	1/2" x 1/2"	1 Stpi	29	.276	Ea.	1.43	14.30		15.73	23
7220	3/4" x 3/4"		27	.296		2.10	15.40		17.50	25.50
7230	1" x 1"		26	.308		2.66	15.95		18.61	27
7300	PEX coupling									
7310	1/2" x 1/2"	1 Stpi	30	.267	Ea.	.58	13.85		14.43	21.50
7320	3/4" x 3/4"		29	.276		.71	14.30		15.01	22.50
7330	1" x 1"		28	.286		1.41	14.85		16.26	23.50
7400	PEX stainless crimp ring									
7410	1/2"	1 Stpi	86	.093	Ea.	.25	4.83		5.08	7.55
7420	3/4" x 3/4"		84	.095		.34	4.94		5.28	7.75
7430	1" x 1"		82	.098		.40	5.05		5.45	8.05

23 83 33 – Electric Radiant Heaters

23 83 33.10 Electric Heating		Crew	Daily Output	Labor-Hours	Unit	Material	Labor	Equipment	Total	Total Incl O&P
0010	**ELECTRIC HEATING**, not incl. conduit or feed wiring									
1100	Rule of thumb: Baseboard units, including control	1 Elec	4.40	1.818	kW	92.50	89		181.50	234
1300	Baseboard heaters, 2' long, 375 watt		8	1	Ea.	40	49		89	117
1400	3' long, 500 watt		8	1		46	49		95	124
1600	4' long, 750 watt		6.70	1.194		54	58.50		112.50	147
1800	5' long, 935 watt		5.70	1.404		68	69		137	177
2000	6' long, 1125 watt		5	1.600		74	78.50		152.50	199
2200	7' long, 1310 watt		4.40	1.818		84	89		173	225
2400	8' long, 1500 watt		4	2		92	98		190	247
2600	9' long, 1680 watt		3.60	2.222		104	109		213	276
2800	10' long, 1875 watt		3.30	2.424		193	119		312	390
2950	Wall heaters with fan, 120 to 277 volt									
3160	Recessed, residential, 750 watt	1 Elec	6	1.333	Ea.	134	65.50		199.50	244
3170	1000 watt		6	1.333		134	65.50		199.50	244
3180	1250 watt		5	1.600		134	78.50		212.50	264
3190	1500 watt		4	2		134	98		232	293
3210	2000 watt		4	2		134	98		232	293
3230	2500 watt		3.50	2.286		335	112		447	535
3240	3000 watt		3	2.667		490	131		621	735
3250	4000 watt		2.70	2.963		495	145		640	760
3260	Commercial, 750 watt		6	1.333		176	65.50		241.50	290
3270	1000 watt		6	1.333		176	65.50		241.50	290
3280	1250 watt		5	1.600		176	78.50		254.50	310
3290	1500 watt		4	2		176	98		274	340
3300	2000 watt		4	2		176	98		274	340
3310	2500 watt		3.50	2.286		176	112		288	360

23 83 Radiant Heating Units

23 83 33 – Electric Radiant Heaters

23 83 33.10 Electric Heating		Crew	Daily Output	Labor-Hours	Unit	Material	2010 Bare Costs Labor	Equipment	Total	Total Incl O&P
3320	3000 watt	1 Elec	3	2.667	Ea.	320	131		451	545
3330	4000 watt		2.70	2.963		320	145		465	565
3600	Thermostats, integral		16	.500		30	24.50		54.50	69.50
3800	Line voltage, 1 pole		8	1		31	49		80	107
3810	2 pole		8	1		29.50	49		78.50	106
3820	Low voltage, 1 pole		8	1		34.50	49		83.50	111
4000	Heat trace system, 400 degree									
4020	115V, 2.5 watts per L.F.	1 Elec	530	.015	L.F.	7.35	.74		8.09	9.20
4030	5 watts per L.F.		530	.015		7.35	.74		8.09	9.20
4050	10 watts per L.F.		530	.015		7.35	.74		8.09	9.20
4060	208V, 5 watts per L.F.		530	.015		7.35	.74		8.09	9.20
4080	480V, 8 watts per L.F.		530	.015		7.35	.74		8.09	9.20
4200	Heater raceway									
4260	Heat transfer cement									
4280	1 gallon				Ea.	48			48	53
4300	5 gallon				″	180			180	198
4320	Snap band, clamp									
4340	3/4″ pipe size	1 Elec	470	.017	Ea.		.83		.83	1.24
4360	1″ pipe size		444	.018			.88		.88	1.31
4380	1-1/4″ pipe size		400	.020			.98		.98	1.46
4400	1-1/2″ pipe size		355	.023			1.10		1.10	1.64
4420	2″ pipe size		320	.025			1.23		1.23	1.82
4440	3″ pipe size		160	.050			2.45		2.45	3.64
4460	4″ pipe size		100	.080			3.92		3.92	5.85
4480	Thermostat NEMA 3R, 22 amp, 0-150 Deg, 10′ cap.		8	1		224	49		273	320
4500	Thermostat NEMA 4X, 25 amp, 40 Deg, 5-1/2′ cap.		7	1.143		222	56		278	330
4520	Thermostat NEMA 4X, 22 amp, 25-325 Deg, 10′ cap.		7	1.143		600	56		656	745
4540	Thermostat NEMA 4X, 22 amp, 15-140 Deg,		6	1.333		505	65.50		570.50	650
4580	Thermostat NEMA 4,7,9, 22 amp, 25-325 Deg, 10′ cap.		3.60	2.222		740	109		849	975
4600	Thermostat NEMA 4,7,9, 22 amp, 15-140 Deg,		3	2.667		735	131		866	1,000
4720	Fiberglass application tape, 36 yard roll		11	.727		69.50	35.50		105	130
5000	Radiant heating ceiling panels, 2′ x 4′, 500 watt		16	.500		253	24.50		277.50	315
5050	750 watt		16	.500		279	24.50		303.50	340
5200	For recessed plaster frame, add		32	.250		62	12.25		74.25	86.50
5300	Infrared quartz heaters, 120 volts, 1000 watts		6.70	1.194		222	58.50		280.50	330
5350	1500 watt		5	1.600		222	78.50		300.50	360
5400	240 volts, 1500 watt		5	1.600		222	78.50		300.50	360
5450	2000 watt		4	2		222	98		320	390
5500	3000 watt		3	2.667		246	131		377	465
5550	4000 watt		2.60	3.077		246	151		397	495
5570	Modulating control		.80	10		266	490		756	1,025
5600	Unit heaters, heavy duty, with fan & mounting bracket									
5650	Single phase, 208-240-277 volt, 3 kW	1 Elec	3.20	2.500	Ea.	405	123		528	625
5750	5 kW		2.40	3.333		420	163		583	705
5800	7 kW		1.90	4.211		645	206		851	1,025
5850	10 kW		1.30	6.154		735	300		1,035	1,250
5950	15 kW		.90	8.889		1,200	435		1,635	1,950
6000	480 volt, 3 kW		3.30	2.424		510	119		629	735
6020	4 kW		3	2.667		455	131		586	695
6040	5 kW		2.60	3.077		465	151		616	735
6060	7 kW		2	4		690	196		886	1,050
6080	10 kW		1.40	5.714		755	280		1,035	1,250
6100	13 kW		1.10	7.273		1,200	355		1,555	1,825

23 83 33 – Electric Radiant Heaters

23 83 33.10 Electric Heating		Crew	Daily Output	Labor-Hours	Unit	Material	2010 Bare Costs Labor	Equipment	Total	Total Incl O&P
6120	15 kW	1 Elec	1	8	Ea.	1,200	390		1,590	1,875
6140	20 kW		.90	8.889		1,550	435		1,985	2,350
6300	3 phase, 208-240 volt, 5 kW		2.40	3.333		395	163		558	680
6320	7 kW		1.90	4.211		610	206		816	975
6340	10 kW		1.30	6.154		660	300		960	1,175
6360	15 kW		.90	8.889		1,100	435		1,535	1,875
6380	20 kW		.70	11.429		1,575	560		2,135	2,550
6400	25 kW		.50	16		1,850	785		2,635	3,200
6500	480 volt, 5 kW		2.60	3.077		550	151		701	830
6520	7 kW		2	4		695	196		891	1,050
6540	10 kW		1.40	5.714		735	280		1,015	1,225
6560	13 kW		1.10	7.273		1,200	355		1,555	1,825
6580	15 kW		1	8		1,200	390		1,590	1,875
6600	20 kW		.90	8.889		1,550	435		1,985	2,350
6620	25 kW		.60	13.333		1,850	655		2,505	3,000
6630	30 kW		.70	11.429		2,150	560		2,710	3,175
6640	40 kW		.60	13.333		2,725	655		3,380	3,975
6650	50 kW		.50	16		3,300	785		4,085	4,800
6800	Vertical discharge heaters, with fan									
6820	Single phase, 208-240-277 volt, 10 kW	1 Elec	1.30	6.154	Ea.	680	300		980	1,200
6840	15 kW		.90	8.889		1,150	435		1,585	1,900
6900	3 phase, 208-240 volt, 10 kW		1.30	6.154		655	300		955	1,175
6920	15 kW		.90	8.889		1,100	435		1,535	1,875
6940	20 kW		.70	11.429		1,575	560		2,135	2,550
6960	25 kW		.50	16		1,850	785		2,635	3,200
6980	30 kW		.40	20		2,150	980		3,130	3,800
7000	40 kW		.36	22.222		2,750	1,100		3,850	4,650
7020	50 kW		.32	25		3,300	1,225		4,525	5,450
7100	480 volt, 10 kW		1.40	5.714		735	280		1,015	1,225
7120	15 kW		1	8		1,200	390		1,590	1,875
7140	20 kW		.90	8.889		1,550	435		1,985	2,350
7160	25 kW		.60	13.333		1,850	655		2,505	3,000
7180	30 kW		.50	16		2,150	785		2,935	3,525
7200	40 kW		.40	20		2,725	980		3,705	4,450
7220	50 kW		.35	22.857		3,300	1,125		4,425	5,300
7900	Cabinet convector heaters, 240 volt									
7920	3' long, 2000 watt	1 Elec	5.30	1.509	Ea.	1,825	74		1,899	2,100
7940	3000 watt		5.30	1.509		1,925	74		1,999	2,225
7960	4000 watt		5.30	1.509		1,975	74		2,049	2,275
7980	6000 watt		4.60	1.739		2,050	85		2,135	2,375
8000	8000 watt		4.60	1.739		2,125	85		2,210	2,475
8020	4' long, 4000 watt		4.60	1.739		2,000	85		2,085	2,325
8040	6000 watt		4	2		2,075	98		2,173	2,425
8060	8000 watt		4	2		2,150	98		2,248	2,525
8080	10,000 watt		4	2		2,175	98		2,273	2,525
8100	Available also in 208 or 277 volt									
8200	Cabinet unit heaters, 120 to 277 volt, 1 pole,									
8220	wall mounted, 2 kW	1 Elec	4.60	1.739	Ea.	1,850	85		1,935	2,150
8230	3 kW		4.60	1.739		1,925	85		2,010	2,250
8240	4 kW		4.40	1.818		1,975	89		2,064	2,300
8250	5 kW		4.40	1.818		2,050	89		2,139	2,375
8260	6 kW		4.20	1.905		2,075	93.50		2,168.50	2,450
8270	8 kW		4	2		2,125	98		2,223	2,500

23 83 33 – Electric Radiant Heaters

23 83 33.10 Electric Heating		Crew	Daily Output	Labor-Hours	Unit	Material	2010 Bare Costs Labor	Equipment	Total	Total Incl O&P
8280	10 kW	1 Elec	3.80	2.105	Ea.	2,175	103		2,278	2,525
8290	12 kW		3.50	2.286		2,200	112		2,312	2,600
8300	13.5 kW		2.90	2.759		2,400	135		2,535	2,825
8310	16 kW		2.70	2.963		2,850	145		2,995	3,350
8320	20 kW		2.30	3.478		3,350	170		3,520	3,925
8330	24 kW		1.90	4.211		3,400	206		3,606	4,050
8350	Recessed, 2 kW		4.40	1.818		1,850	89		1,939	2,150
8370	3 kW		4.40	1.818		2,000	89		2,089	2,325
8380	4 kW		4.20	1.905		2,050	93.50		2,143.50	2,400
8390	5 kW		4.20	1.905		2,100	93.50		2,193.50	2,475
8400	6 kW		4	2		2,150	98		2,248	2,525
8410	8 kW		3.80	2.105		2,225	103		2,328	2,600
8420	10 kW		3.50	2.286		2,575	112		2,687	3,025
8430	12 kW		2.90	2.759		2,725	135		2,860	3,200
8440	13.5 kW		2.70	2.963		2,750	145		2,895	3,250
8450	16 kW		2.30	3.478		2,750	170		2,920	3,275
8460	20 kW		1.90	4.211		3,275	206		3,481	3,900
8470	24 kW		1.60	5		3,325	245		3,570	4,025
8490	Ceiling mounted, 2 kW		3.20	2.500		1,850	123		1,973	2,200
8510	3 kW		3.20	2.500		2,000	123		2,123	2,375
8520	4 kW		3	2.667		2,050	131		2,181	2,450
8530	5 kW		3	2.667		2,100	131		2,231	2,525
8540	6 kW		2.80	2.857		2,150	140		2,290	2,575
8550	8 kW		2.40	3.333		2,225	163		2,388	2,700
8560	10 kW		2.20	3.636		2,250	178		2,428	2,750
8570	12 kW		2	4		2,725	196		2,921	3,300
8580	13.5 kW		1.50	5.333		2,750	261		3,011	3,425
8590	16 kW		1.30	6.154		2,750	300		3,050	3,475
8600	20 kW		.90	8.889		3,275	435		3,710	4,250
8610	24 kW		.60	13.333		3,325	655		3,980	4,625
8630	208 to 480 V, 3 pole									
8650	Wall mounted, 2 kW	1 Elec	4.60	1.739	Ea.	2,000	85		2,085	2,325
8670	3 kW		4.60	1.739		2,075	85		2,160	2,425
8680	4 kW		4.40	1.818		2,125	89		2,214	2,475
8690	5 kW		4.40	1.818		2,200	89		2,289	2,550
8700	6 kW		4.20	1.905		2,225	93.50		2,318.50	2,575
8710	8 kW		4	2		2,300	98		2,398	2,675
8720	10 kW		3.80	2.105		2,400	103		2,503	2,800
8730	12 kW		3.50	2.286		2,450	112		2,562	2,875
8740	13.5 kW		2.90	2.759		2,475	135		2,610	2,925
8750	16 kW		2.70	2.963		2,525	145		2,670	3,000
8760	20 kW		2.30	3.478		3,375	170		3,545	3,950
8770	24 kW		1.90	4.211		3,425	206		3,631	4,050
8790	Recessed, 2 kW		4.40	1.818		2,025	89		2,114	2,350
8810	3 kW		4.40	1.818		2,075	89		2,164	2,425
8820	4 kW		4.20	1.905		2,125	93.50		2,218.50	2,475
8830	5 kW		4.20	1.905		2,200	93.50		2,293.50	2,575
8840	6 kW		4	2		2,225	98		2,323	2,575
8850	8 kW		3.80	2.105		2,300	103		2,403	2,675
8860	10 kW		3.50	2.286		2,325	112		2,437	2,750
8870	12 kW		2.90	2.759		2,375	135		2,510	2,800
8880	13.5 kW		2.70	2.963		2,400	145		2,545	2,875
8890	16 kW		2.30	3.478		2,450	170		2,620	2,950

23 83 Radiant Heating Units

23 83 33 – Electric Radiant Heaters

23 83 33.10 Electric Heating		Crew	Daily Output	Labor-Hours	Unit	2010 Bare Costs Material	Labor	Equipment	Total	Total Incl O&P
8900	20 kW	1 Elec	1.90	4.211	Ea.	3,375	206		3,581	4,000
8920	24 kW		1.60	5		3,425	245		3,670	4,125
8940	Ceiling mount, 2 kW		3.20	2.500		2,000	123		2,123	2,375
8950	3 kW		3.20	2.500		2,075	123		2,198	2,475
8960	4 kW		3	2.667		2,125	131		2,256	2,550
8970	5 kW		3	2.667		2,200	131		2,331	2,625
8980	6 kW		2.80	2.857		2,225	140		2,365	2,625
8990	8 kW		2.40	3.333		2,300	163		2,463	2,775
9000	10 kW		2.20	3.636		2,325	178		2,503	2,850
9020	13.5 kW		1.50	5.333		2,400	261		2,661	3,050
9030	16 kW		1.30	6.154		3,325	300		3,625	4,100
9040	20 kW		.90	8.889		3,375	435		3,810	4,350
9060	24 kW		.60	13.333		3,425	655		4,080	4,725

23 84 Humidity Control Equipment

23 84 13 – Humidifiers

23 84 13.10 Humidifier Units

Line	Description	Crew	Daily Output	Labor-Hours	Unit	Material	Labor	Equipment	Total	Total Incl O&P
0010	**HUMIDIFIER UNITS**									
0520	Steam, room or duct, filter, regulators, auto. controls, 220 V									
0540	11 lb. per hour	Q-5	6	2.667	Ea.	2,350	125		2,475	2,750
0560	22 lb. per hour		5	3.200		2,600	149		2,749	3,075
0580	33 lb. per hour		4	4		2,650	187		2,837	3,200
0600	50 lb. per hour		4	4		3,250	187		3,437	3,850
0620	100 lb. per hour		3	5.333		3,900	249		4,149	4,675
0640	150 lb. per hour		2.50	6.400		5,175	299		5,474	6,125
0660	200 lb. per hour		2	8		6,400	375		6,775	7,575
0700	With blower									
0720	11 lb. per hour	Q-5	5.50	2.909	Ea.	3,475	136		3,611	4,025
0740	22 lb. per hour		4.75	3.368		3,700	157		3,857	4,300
0760	33 lb. per hour		3.75	4.267		3,775	199		3,974	4,450
0780	50 lb. per hour		3.50	4.571		4,325	213		4,538	5,100
0800	100 lb. per hour		2.75	5.818		4,800	272		5,072	5,700
0820	150 lb. per hour		2	8		7,200	375		7,575	8,475
0840	200 lb. per hour		1.50	10.667		8,375	500		8,875	9,950
1000	Steam for duct installation, manifold with pneum. controls									
1010	3 – 191 lb/hr, 24" x 24"	Q-5	5	3.200	Ea.	1,725	149		1,074	2,100
1020	3 – 191 lb/hr, 48" x 48"		4.60	3.478		3,500	162		3,662	4,100
1030	65 – 334 lb/hr, 36" x 36"		4.80	3.333		3,425	156		3,581	4,000
1040	65 – 334 lb/hr, 72" x 72"		4	4		3,700	187		3,887	4,350
1050	100 – 690 lb/hr, 48" x 48"		4	4		4,800	187		4,987	5,550
1060	100 – 690 lb/hr, 84" x 84"		3.60	4.444		4,950	208		5,158	5,750
1070	220 – 2000 lb/hr, 72" x 72"		3.80	4.211		5,025	197		5,222	5,825
1080	220 – 2000 lb/hr, 144" x 144"		3.40	4.706		5,875	220		6,095	6,775
1090	For electrically controlled unit, add					805			805	885
1092	For additional manifold, add					25%	15%			
1200	Electric steam, self generating, with manifold									
1210	1 – 96 lb/hr, 12" x 72"	Q-5	3.60	4.444	Ea.	7,825	208		8,033	8,925
1220	20 – 168 lb/hr, 12" x 72"		3.60	4.444		13,100	208		13,308	14,700
1290	For room mounting, add		16	1		225	46.50		271.50	315
5000	Furnace type, wheel bypass									
5020	10 GPD	1 Stpi	4	2	Ea.	221	104		325	400

23 84 Humidity Control Equipment

23 84 13 – Humidifiers

23 84 13.10 Humidifier Units		Crew	Daily Output	Labor-Hours	Unit	Material	2010 Bare Costs Labor	Equipment	Total	Total Incl O&P
5040	14 GPD	1 Stpi	3.80	2.105	Ea.	268	109		377	460
5060	19 GPD	↓	3.60	2.222	↓	310	115		425	520

23 84 16 – Dehumidifiers

23 84 16.10 Dehumidifier Units

		Crew	Daily Output	Labor-Hours	Unit	Material	Labor	Equipment	Total	Total Incl O&P
0010	**DEHUMIDIFIER UNITS**									
6000	Self contained with filters and standard controls									
6040	1.5 lb/hr, 50 cfm	1 Plum	8	1	Ea.	5,300	52		5,352	5,900
6060	3 lb/hr, 150 cfm	Q-1	12	1.333		5,525	62.50		5,587.50	6,175
6065	6 lb/hr, 150 cfm		9	1.778		9,000	83.50		9,083.50	10,100
6070	16 to 20 lb/hr, 600 cfm		5	3.200		18,300	150		18,450	20,400
6080	30 to 40 lb/hr, 1125 cfm		4	4		28,600	187		28,787	31,700
6090	60 to 75 lb/hr, 2250 cfm		3	5.333		37,300	250		37,550	41,400
6100	120 to 155 lb/hr, 4500 cfm		2	8		66,000	375		66,375	73,000
6110	240 to 310 lb/hr, 9000 cfm	↓	1.50	10.667		94,500	500		95,000	105,000
6120	400 to 515 lb/hr, 15,000 cfm	Q-2	1.60	15		128,000	730		128,730	142,000
6130	530 to 690 lb/hr, 20,000 cfm		1.40	17.143		139,000	835		139,835	154,000
6140	800 to 1030 lb/hr, 30,000 cfm		1.20	20		158,000	970		158,970	175,500
6150	1060 to 1375 lb/hr, 40,000 cfm	↓	1	24	↓	220,000	1,175		221,175	244,000

23 91 Prefabricated Equipment Supports

23 91 10 – Prefabricated Curbs, Pads and Stands

23 91 10.10 Prefabricated Pads and Stands

		Crew	Daily Output	Labor-Hours	Unit	Material	Labor	Equipment	Total	Total Incl O&P
0010	**PREFABRICATED PADS AND STANDS**									
6000	Pad, fiberglass reinforced concrete with polystyrene foam core									
6050	Condenser, 2" thick, 20" x 38"	1 Shee	8	1	Ea.	23.50	49		72.50	101
6070	24" x 24"		16	.500		18.90	24.50		43.40	58
6090	24" x 36"		12	.667		30	32.50		62.50	82.50
6110	24" x 42"	↓	8	1		35	49		84	113
6150	26" x 36"	Q-9	8	2		33	88.50		121.50	171
6170	28" x 38"	1 Shee	8	1		37.50	49		86.50	116
6190	30" x 30"		12	.667		31.50	32.50		64	84
6220	30" x 36"	↓	8	1		37.50	49		86.50	116
6240	30" x 40"	Q-9	8	2		41	88.50		129.50	179
6260	32" x 32"		7	2.286		35.50	101		136.50	192
6280	36" x 36"		8	2		43.50	88.50		132	182
6300	36" x 40"		7	2.286		49.50	101		150.50	208
6320	36" x 48"		7	2.286		58	101		159	217
6340	36" x 54"		6	2.667		70	118		188	255
6360	36" x 60" x 3"		7	2.286		83	101		184	245
6400	41" x 47" x 3"		6	2.667		77.50	118		195.50	263
6490	26" round		10	1.600		18.65	70.50		89.15	128
6550	30" round		9	1.778		24.50	78.50		103	146
6600	36" round	↓	8	2		33.50	88.50		122	171
6700	For 3" thick pad, add				↓	20%	5%			
8800	Stand, corrosion-resistant plastic									
8850	Heat pump, 6" high	1 Shee	32	.250	Ea.	5.10	12.30		17.40	24
8870	12" high	"	32	.250	"	11.35	12.30		23.65	31

Estimating Tips

26 05 00 Common Work Results for Electrical

- Conduit should be taken off in three main categories—power distribution, branch power, and branch lighting—so the estimator can concentrate on systems and components, therefore making it easier to ensure all items have been accounted for.
- For cost modifications for elevated conduit installation, add the percentages to labor according to the height of installation, and only to the quantities exceeding the different height levels, not to the total conduit quantities.
- Remember that aluminum wiring of equal ampacity is larger in diameter than copper and may require larger conduit.
- If more than three wires at a time are being pulled, deduct percentages from the labor hours of that grouping of wires.
- When taking off grounding systems, identify separately the type and size of wire, and list each unique type of ground connection.
- The estimator should take the weights of materials into consideration when completing a takeoff. Topics to consider include: How will the materials be supported? What methods of support are available? How high will the support structure have to reach? Will the final support structure be able to withstand the total burden? Is the support material included or separate from the fixture, equipment, and material specified?
- Do not overlook the costs for equipment used in the installation. If scaffolding or highlifts are available in the field, contractors may use them in lieu of the proposed ladders and rolling staging.

26 20 00 Low-Voltage Electrical Transmission

- Supports and concrete pads may be shown on drawings for the larger equipment, or the support system may be only a piece of plywood for the back of a panelboard. In either case, it must be included in the costs.

26 40 00 Electrical and Cathodic Protection

- When taking off cathodic protections systems, identify the type and size of cable, and list each unique type of anode connection.

26 50 00 Lighting

- Fixtures should be taken off room by room, using the fixture schedule, specifications, and the ceiling plan. For large concentrations of lighting fixtures in the same area, deduct the percentages from labor hours.

Reference Numbers

Reference numbers are shown in shaded boxes at the beginning of some major classifications. These numbers refer to related items in the Reference Section. The reference information may be an estimating procedure, an alternate pricing method, or technical information.

Note: Not all subdivisions listed here necessarily appear in this publication.

Note: **Trade Service,** *in part, has been used as a reference source for some of the material prices used in Division 26.*

26 05 33.95 Cutting and Drilling		Crew	Daily Output	Labor-Hours	Unit	Material	2010 Bare Costs Labor	Equipment	Total	Total Incl O&P
0010	**CUTTING AND DRILLING**									
0100	Hole drilling to 10' high, concrete wall									
0110	8" thick, 1/2" pipe size	R-31	12	.667	Ea.	5	32.50	4.45	41.95	59
0120	3/4" pipe size		12	.667		5	32.50	4.45	41.95	59
0130	1" pipe size		9.50	.842		9.80	41.50	5.65	56.95	78.50
0140	1-1/4" pipe size		9.50	.842		9.80	41.50	5.65	56.95	78.50
0150	1-1/2" pipe size		9.50	.842		9.80	41.50	5.65	56.95	78.50
0160	2" pipe size		4.40	1.818		15	89	12.15	116.15	162
0170	2-1/2" pipe size		4.40	1.818		15	89	12.15	116.15	162
0180	3" pipe size		4.40	1.818		15	89	12.15	116.15	162
0190	3-1/2" pipe size		3.30	2.424		18.90	119	16.20	154.10	216
0200	4" pipe size		3.30	2.424		18.90	119	16.20	154.10	216
0500	12" thick, 1/2" pipe size		9.40	.851		7.75	41.50	5.70	54.95	77
0520	3/4" pipe size		9.40	.851		7.75	41.50	5.70	54.95	77
0540	1" pipe size		7.30	1.096		14.05	53.50	7.30	74.85	104
0560	1-1/4" pipe size		7.30	1.096		14.05	53.50	7.30	74.85	104
0570	1-1/2" pipe size		7.30	1.096		14.05	53.50	7.30	74.85	104
0580	2" pipe size		3.60	2.222		20.50	109	14.85	144.35	201
0590	2-1/2" pipe size		3.60	2.222		20.50	109	14.85	144.35	201
0600	3" pipe size		3.60	2.222		20.50	109	14.85	144.35	201
0610	3-1/2" pipe size		2.80	2.857		26	140	19.10	185.10	258
0630	4" pipe size		2.50	3.200		26	157	21.50	204.50	285
0650	16" thick, 1/2" pipe size		7.60	1.053		10.45	51.50	7.05	69	96
0670	3/4" pipe size		7	1.143		10.45	56	7.65	74.10	103
0690	1" pipe size		6	1.333		18.35	65.50	8.90	92.75	127
0710	1-1/4" pipe size		5.50	1.455		18.35	71.50	9.70	99.55	137
0730	1-1/2" pipe size		5.50	1.455		18.35	71.50	9.70	99.55	137
0750	2" pipe size		3	2.667		26	131	17.80	174.80	243
0770	2-1/2" pipe size		2.70	2.963		26	145	19.80	190.80	267
0790	3" pipe size		2.50	3.200		26	157	21.50	204.50	286
0810	3-1/2" pipe size		2.30	3.478		32.50	170	23	225.50	315
0830	4" pipe size		2	4		32.50	196	26.50	255	355
0850	20" thick, 1/2" pipe size		6.40	1.250		13.15	61.50	8.35	83	115
0870	3/4" pipe size		6	1.333		13.15	65.50	8.90	87.55	121
0890	1" pipe size		5	1.600		22.50	78.50	10.70	111.70	154
0910	1-1/4" pipe size		4.80	1.667		22.50	81.50	11.15	115.15	158
0930	1-1/2" pipe size		4.60	1.739		22.50	85	11.60	119.10	165
0950	2" pipe size		2.70	2.963		32	145	19.80	196.80	273
0970	2-1/2" pipe size		2.40	3.333		32	163	22.50	217.50	305
0990	3" pipe size		2.20	3.636		32	178	24.50	234.50	325
1010	3-1/2" pipe size		2	4		39.50	196	26.50	262	365
1030	4" pipe size		1.70	4.706		39.50	231	31.50	302	425
1050	24" thick, 1/2" pipe size		5.50	1.455		15.90	71.50	9.70	97.10	134
1070	3/4" pipe size		5.10	1.569		15.90	77	10.50	103.40	143
1090	1" pipe size		4.30	1.860		27	91	12.45	130.45	179
1110	1-1/4" pipe size		4	2		27	98	13.35	138.35	190
1130	1-1/2" pipe size		4	2		27	98	13.35	138.35	190
1150	2" pipe size		2.40	3.333		37.50	163	22.50	223	310
1170	2-1/2" pipe size		2.20	3.636		37.50	178	24.50	240	335
1190	3" pipe size		2	4		37.50	196	26.50	260	360
1210	3-1/2" pipe size		1.80	4.444		46.50	218	29.50	294	410
1230	4" pipe size		1.50	5.333		46.50	261	35.50	343	480
1500	Brick wall, 8" thick, 1/2" pipe size		18	.444		5	22	2.97	29.97	41.50

26 05 33.95 Cutting and Drilling		Crew	Daily Output	Labor-Hours	Unit	Material	2010 Bare Costs Labor	Equipment	Total	Total Incl O&P
1520	3/4" pipe size	R-31	18	.444	Ea.	5	22	2.97	29.97	41.50
1540	1" pipe size		13.30	.602		9.80	29.50	4.02	43.32	59
1560	1-1/4" pipe size		13.30	.602		9.80	29.50	4.02	43.32	59
1580	1-1/2" pipe size		13.30	.602		9.80	29.50	4.02	43.32	59
1600	2" pipe size		5.70	1.404		15	69	9.40	93.40	129
1620	2-1/2" pipe size		5.70	1.404		15	69	9.40	93.40	129
1640	3" pipe size		5.70	1.404		15	69	9.40	93.40	129
1660	3-1/2" pipe size		4.40	1.818		18.90	89	12.15	120.05	166
1680	4" pipe size		4	2		18.90	98	13.35	130.25	182
1700	12" thick, 1/2" pipe size		14.50	.552		7.75	27	3.69	38.44	52.50
1720	3/4" pipe size		14.50	.552		7.75	27	3.69	38.44	52.50
1740	1" pipe size		11	.727		14.05	35.50	4.86	54.41	74
1760	1-1/4" pipe size		11	.727		14.05	35.50	4.86	54.41	74
1780	1-1/2" pipe size		11	.727		14.05	35.50	4.86	54.41	74
1800	2" pipe size		5	1.600		20.50	78.50	10.70	109.70	151
1820	2-1/2" pipe size		5	1.600		20.50	78.50	10.70	109.70	151
1840	3" pipe size		5	1.600		20.50	78.50	10.70	109.70	151
1860	3-1/2" pipe size		3.80	2.105		26	103	14.05	143.05	197
1880	4" pipe size		3.30	2.424		26	119	16.20	161.20	223
1900	16" thick, 1/2" pipe size		12.30	.650		10.45	32	4.34	46.79	64
1920	3/4" pipe size		12.30	.650		10.45	32	4.34	46.79	64
1940	1" pipe size		9.30	.860		18.35	42	5.75	66.10	89
1960	1-1/4" pipe size		9.30	.860		18.35	42	5.75	66.10	89
1980	1-1/2" pipe size		9.30	.860		18.35	42	5.75	66.10	89
2000	2" pipe size		4.40	1.818		26	89	12.15	127.15	174
2010	2-1/2" pipe size		4.40	1.818		26	89	12.15	127.15	174
2030	3" pipe size		4.40	1.818		26	89	12.15	127.15	174
2050	3-1/2" pipe size		3.30	2.424		32.50	119	16.20	167.70	231
2070	4" pipe size		3	2.667		32.50	131	17.80	181.30	250
2090	20" thick, 1/2" pipe size		10.70	.748		13.15	36.50	4.99	54.64	74.50
2110	3/4" pipe size		10.70	.748		13.15	36.50	4.99	54.64	74.50
2130	1" pipe size		8	1		22.50	49	6.70	78.20	105
2150	1-1/4" pipe size		8	1		22.50	49	6.70	78.20	105
2170	1-1/2" pipe size		8	1		22.50	49	6.70	78.20	105
2190	2" pipe size		4	2		32	98	13.35	143.35	196
2210	2-1/2" pipe size		4	2		32	98	13.35	143.35	196
2230	3" pipe size		4	2		32	98	13.35	143.35	196
2250	3-1/2" pipe size		3	2.667		39.50	131	17.80	188.30	257
2270	4" pipe size		2.70	2.963		39.50	145	19.80	204.30	282
2290	24" thick, 1/2" pipe size		9.40	.851		15.90	41.50	5.70	63.10	86
2310	3/4" pipe size		9.40	.851		15.90	41.50	5.70	63.10	86
2330	1" pipe size		7.10	1.127		27	55	7.55	89.55	120
2350	1-1/4" pipe size		7.10	1.127		27	55	7.55	89.55	120
2370	1-1/2" pipe size		7.10	1.127		27	55	7.55	89.55	120
2390	2" pipe size		3.60	2.222		37.50	109	14.85	161.35	219
2410	2-1/2" pipe size		3.60	2.222		37.50	109	14.85	161.35	219
2430	3" pipe size		3.60	2.222		37.50	109	14.85	161.35	219
2450	3-1/2" pipe size		2.80	2.857		46.50	140	19.10	205.60	280
2470	4" pipe size		2.50	3.200		46.50	157	21.50	225	310
3000	Knockouts to 8' high, metal boxes & enclosures									
3020	With hole saw, 1/2" pipe size	1 Elec	53	.151	Ea.		7.40		7.40	11
3040	3/4" pipe size		47	.170			8.35		8.35	12.40
3050	1" pipe size		40	.200			9.80		9.80	14.55

26 05 33 – Raceway and Boxes for Electrical Systems

26 05 33.95 Cutting and Drilling		Crew	Daily Output	Labor-Hours	Unit	Material	2010 Bare Costs Labor	Equipment	Total	Total Incl O&P
3060	1-1/4" pipe size	1 Elec	36	.222	Ea.		10.90		10.90	16.20
3070	1-1/2" pipe size		32	.250			12.25		12.25	18.20
3080	2" pipe size		27	.296			14.50		14.50	21.50
3090	2-1/2" pipe size		20	.400			19.60		19.60	29
4010	3" pipe size		16	.500			24.50		24.50	36.50
4030	3-1/2" pipe size		13	.615			30		30	45
4050	4" pipe size		11	.727			35.50		35.50	53
4070	With hand punch set, 1/2" pipe size		40	.200			9.80		9.80	14.55
4090	3/4" pipe size		32	.250			12.25		12.25	18.20
4110	1" pipe size		30	.267			13.05		13.05	19.45
4130	1-1/4" pipe size		28	.286			14		14	21
4150	1-1/2" pipe size		26	.308			15.10		15.10	22.50
4170	2" pipe size		20	.400			19.60		19.60	29
4190	2-1/2" pipe size		17	.471			23		23	34.50
4200	3" pipe size		15	.533			26		26	39
4220	3-1/2" pipe size		12	.667			32.50		32.50	48.50
4240	4" pipe size		10	.800			39		39	58.50
4260	With hydraulic punch, 1/2" pipe size		44	.182			8.90		8.90	13.25
4280	3/4" pipe size		38	.211			10.30		10.30	15.35
4300	1" pipe size		38	.211			10.30		10.30	15.35
4320	1-1/4" pipe size		38	.211			10.30		10.30	15.35
4340	1-1/2" pipe size		38	.211			10.30		10.30	15.35
4360	2" pipe size		32	.250			12.25		12.25	18.20
4380	2-1/2" pipe size		27	.296			14.50		14.50	21.50
4400	3" pipe size		23	.348			17.05		17.05	25.50
4420	3-1/2" pipe size		20	.400			19.60		19.60	29
4440	4" pipe size	↓	18	.444	↓		22		22	32.50

26 05 80 – Wiring Connections

26 05 80.10 Motor Connections

		Crew	Daily Output	Labor-Hours	Unit	Material	Labor	Equipment	Total	Total Incl O&P
0010	**MOTOR CONNECTIONS**									
0020	Flexible conduit and fittings, 115 volt, 1 phase, up to 1 HP motor	1 Elec	8	1	Ea.	9.55	49		58.55	83.50
0050	2 HP motor		6.50	1.231		9.75	60.50		70.25	100
0100	3 HP motor		5.50	1.455		14.90	71.50		86.40	122
0120	230 volt, 10 HP motor, 3 phase		4.20	1.905		15.90	93.50		109.40	156
0150	15 HP motor		3.30	2.424		28	119		147	208
0200	25 HP motor		2.70	2.963		31.50	145		176.50	251
0400	50 HP motor		2.20	3.636		80.50	178		258.50	355
0600	100 HP motor		1.50	5.333		197	261		458	605
1500	460 volt, 5 HP motor, 3 phase		8	1		10.30	49		59.30	84.50
1520	10 HP motor		8	1		10.30	49		59.30	84.50
1530	25 HP motor		6	1.333		17.70	65.50		83.20	117
1540	30 HP motor		6	1.333		17.70	65.50		83.20	117
1550	40 HP motor		5	1.600		28	78.50		106.50	148
1560	50 HP motor		5	1.600		30	78.50		108.50	150
1570	60 HP motor		3.80	2.105		33.50	103		136.50	190
1580	75 HP motor		3.50	2.286		46	112		158	218
1590	100 HP motor		2.50	3.200		68.50	157		225.50	310
1600	125 HP motor		2	4		91	196		287	390
1610	150 HP motor		1.80	4.444		95.50	218		313.50	430
1620	200 HP motor		1.50	5.333		150	261		411	555
2005	460 Volt, 5 HP motor, 3 Phase, w/sealtite		8	1		15.30	49		64.30	90
2010	10 HP motor	↓	8	1	↓	15.30	49		64.30	90

26 05 80 – Wiring Connections

26 05 80.10 Motor Connections		Crew	Daily Output	Labor-Hours	Unit	Material	Labor	Equipment	Total	Total Incl O&P
						2010 Bare Costs				
2015	25 HP motor	1 Elec	6	1.333	Ea.	27.50	65.50		93	127
2020	30 HP motor		6	1.333		27.50	65.50		93	127
2025	40 HP motor		5	1.600		50.50	78.50		129	173
2030	50 HP motor		5	1.600		50	78.50		128.50	172
2035	60 HP motor		3.80	2.105		80	103		183	241
2040	75 HP motor		3.50	2.286		83	112		195	258
2045	100 HP motor		2.50	3.200		111	157		268	355
2055	150 HP motor		1.80	4.444		116	218		334	455
2060	200 HP motor	↓	1.50	5.333	↓	820	261		1,081	1,300

26 05 90 – Residential Applications

26 05 90.10 Residential Wiring

		Crew	Daily Output	Labor-Hours	Unit	Material	Labor	Equipment	Total	Total Incl O&P
0010	**RESIDENTIAL WIRING**									
0020	20′ avg. runs and #14/2 wiring incl. unless otherwise noted									
1000	Service & panel, includes 24′ SE-AL cable, service eye, meter,									
1010	Socket, panel board, main bkr., ground rod, 15 or 20 amp									
1020	1-pole circuit breakers, and misc. hardware									
1100	100 amp, with 10 branch breakers	1 Elec	1.19	6.723	Ea.	565	330		895	1,125
1110	With PVC conduit and wire		.92	8.696		630	425		1,055	1,325
1120	With RGS conduit and wire		.73	10.959		805	535		1,340	1,700
1150	150 amp, with 14 branch breakers		1.03	7.767		880	380		1,260	1,525
1170	With PVC conduit and wire		.82	9.756		1,025	480		1,505	1,825
1180	With RGS conduit and wire	↓	.67	11.940		1,350	585		1,935	2,375
1200	200 amp, with 18 branch breakers	2 Elec	1.80	8.889		1,150	435		1,585	1,925
1220	With PVC conduit and wire		1.46	10.959		1,300	535		1,835	2,225
1230	With RGS conduit and wire	↓	1.24	12.903		1,750	630		2,380	2,875
1800	Lightning surge suppressor for above services, add	1 Elec	32	.250	↓	51.50	12.25		63.75	74.50
2000	Switch devices									
2100	Single pole, 15 amp, Ivory, with a 1-gang box, cover plate,									
2110	Type NM (Romex) cable	1 Elec	17.10	.468	Ea.	10.05	23		33.05	45
2120	Type MC (BX) cable		14.30	.559		27.50	27.50		55	71.50
2130	EMT & wire		5.71	1.401		32.50	68.50		101	138
2150	3-way, #14/3, type NM cable		14.55	.550		14.05	27		41.05	55.50
2170	Type MC cable		12.31	.650		38	32		70	89
2180	EMT & wire		5	1.600		36	78.50		114.50	157
2200	4-way, #14/3, type NM cable		14.55	.550		30.50	27		57.50	73.50
2220	Type MC cable		12.31	.650		54	32		86	107
2230	EMT & wire		5	1.600		52.50	78.50		131	175
2250	S.P., 20 amp, #12/2, type NM cable		13.33	.600		16.85	29.50		46.35	62
2270	Type MC cable		11.43	.700		32	34.50		66.50	86.50
2280	EMT & wire		4.85	1.649		41	81		122	165
2290	S.P. rotary dimmer, 600W, no wiring		17	.471		20.50	23		43.50	57.50
2300	S.P. rotary dimmer, 600W, type NM cable		14.55	.550		25.50	27		52.50	68
2320	Type MC cable		12.31	.650		43	32		75	94.50
2330	EMT & wire		5	1.600		49	78.50		127.50	171
2350	3-way rotary dimmer, type NM cable		13.33	.600		32.50	29.50		62	79.50
2370	Type MC cable		11.43	.700		50	34.50		84.50	106
2380	EMT & wire	↓	4.85	1.649	↓	56.50	81		137.50	182
2400	Interval timer wall switch, 20 amp, 1-30 min., #12/2									
2410	Type NM cable	1 Elec	14.55	.550	Ea.	50	27		77	95
2420	Type MC cable		12.31	.650		58.50	32		90.50	112
2430	EMT & wire	↓	5	1.600	↓	74	78.50		152.50	199
2500	Decorator style									

26 05 90 – Residential Applications

26 05 90.10 Residential Wiring		Crew	Daily Output	Labor-Hours	Unit	Material	2010 Bare Costs Labor	Equipment	Total	Total Incl O&P
2510	S.P., 15 amp, type NM cable	1 Elec	17.10	.468	Ea.	14.90	23		37.90	50.50
2520	Type MC cable		14.30	.559		32.50	27.50		60	76.50
2530	EMT & wire		5.71	1.401		37.50	68.50		106	143
2550	3-way, #14/3, type NM cable		14.55	.550		18.95	27		45.95	61
2570	Type MC cable		12.31	.650		43	32		75	94.50
2580	EMT & wire		5	1.600		41	78.50		119.50	162
2600	4-way, #14/3, type NM cable		14.55	.550		35	27		62	78.50
2620	Type MC cable		12.31	.650		59	32		91	113
2630	EMT & wire		5	1.600		57.50	78.50		136	180
2650	S.P., 20 amp, #12/2, type NM cable		13.33	.600		21.50	29.50		51	67.50
2670	Type MC cable		11.43	.700		37	34.50		71.50	92
2680	EMT & wire		4.85	1.649		45.50	81		126.50	170
2700	S.P., slide dimmer, type NM cable		17.10	.468		33	23		56	70.50
2720	Type MC cable		14.30	.559		50.50	27.50		78	96.50
2730	EMT & wire		5.71	1.401		57	68.50		125.50	165
2750	S.P., touch dimmer, type NM cable		17.10	.468		27	23		50	63.50
2770	Type MC cable		14.30	.559		44.50	27.50		72	90
2780	EMT & wire		5.71	1.401		51	68.50		119.50	158
2800	3-way touch dimmer, type NM cable		13.33	.600		45.50	29.50		75	93.50
2820	Type MC cable		11.43	.700		63	34.50		97.50	120
2830	EMT & wire		4.85	1.649		69.50	81		150.50	196
3000	Combination devices									
3100	S.P. switch/15 amp recpt., Ivory, 1-gang box, plate									
3110	Type NM cable	1 Elec	11.43	.700	Ea.	21.50	34.50		56	74.50
3120	Type MC cable		10	.800		39	39		78	102
3130	EMT & wire		4.40	1.818		45.50	89		134.50	182
3150	S.P. switch/pilot light, type NM cable		11.43	.700		22	34.50		56.50	75.50
3170	Type MC cable		10	.800		39.50	39		78.50	102
3180	EMT & wire		4.43	1.806		46	88.50		134.50	183
3190	2-S.P. switches, 2-#14/2, no wiring		14	.571		8.25	28		36.25	50.50
3200	2-S.P. switches, 2-#14/2, type NM cables		10	.800		25	39		64	86
3220	Type MC cable		8.89	.900		52	44		96	123
3230	EMT & wire		4.10	1.951		49	95.50		144.50	196
3250	3-way switch/15 amp recpt., #14/3, type NM cable		10	.800		29	39		68	90.50
3270	Type MC cable		8.89	.900		53	44		97	124
3280	EMT & wire		4.10	1.951		51	95.50		146.50	199
3300	2-3 way switches, 2-#14/3, type NM cables		8.89	.900		38	44		82	108
3320	Type MC cable		8	1		78	49		127	159
3330	EMT & wire		4	2		58.50	98		156.50	211
3350	S.P. switch/20 amp recpt., #12/2, type NM cable		10	.800		33.50	39		72.50	95.50
3370	Type MC cable		8.89	.900		42.50	44		86.50	112
3380	EMT & wire		4.10	1.951		57.50	95.50		153	206
3400	Decorator style									
3410	S.P. switch/15 amp recpt., type NM cable	1 Elec	11.43	.700	Ea.	26.50	34.50		61	80
3420	Type MC cable		10	.800		44	39		83	107
3430	EMT & wire		4.40	1.818		50	89		139	187
3450	S.P. switch/pilot light, type NM cable		11.43	.700		27	34.50		61.50	81
3470	Type MC cable		10	.800		44.50	39		83.50	108
3480	EMT & wire		4.40	1.818		51	89		140	188
3500	2-S.P. switches, 2-#14/2, type NM cables		10	.800		30	39		69	91.50
3520	Type MC cable		8.89	.900		57	44		101	128
3530	EMT & wire		4.10	1.951		54	95.50		149.50	202
3550	3-way/15 amp recpt., #14/3, type NM cable		10	.800		34	39		73	96

26 05 90.10 Residential Wiring		Crew	Daily Output	Labor-Hours	Unit	Material	2010 Bare Costs Labor	Equipment	Total	Total Incl O&P
3570	Type MC cable	1 Elec	8.89	.900	Ea.	58	44		102	129
3580	EMT & wire		4.10	1.951		56	95.50		151.50	204
3650	2-3 way switches, 2-#14/3, type NM cables		8.89	.900		43	44		87	113
3670	Type MC cable		8	1		83	49		132	164
3680	EMT & wire		4	2		63.50	98		161.50	216
3700	S.P. switch/20 amp recpt., #12/2, type NM cable		10	.800		38.50	39		77.50	101
3720	Type MC cable		8.89	.900		47	44		91	118
3730	EMT & wire		4.10	1.951		62.50	95.50		158	211
4000	Receptacle devices									
4010	Duplex outlet, 15 amp recpt., Ivory, 1-gang box, plate									
4015	Type NM cable	1 Elec	14.55	.550	Ea.	8.55	27		35.55	49.50
4020	Type MC cable		12.31	.650		26	32		58	76
4030	EMT & wire		5.33	1.501		31	73.50		104.50	143
4050	With #12/2, type NM cable		12.31	.650		10.80	32		42.80	59.50
4070	Type MC cable		10.67	.750		26	36.50		62.50	83.50
4080	EMT & wire		4.71	1.699		34.50	83		117.50	162
4100	20 amp recpt., #12/2, type NM cable		12.31	.650		19.85	32		51.85	69.50
4120	Type MC cable		10.67	.750		35	36.50		71.50	93.50
4130	EMT & wire		4.71	1.699		44	83		127	172
4140	For GFI see Div. 26 05 90.10 line 4300 below									
4150	Decorator style, 15 amp recpt., type NM cable	1 Elec	14.55	.550	Ea.	13.40	27		40.40	55
4170	Type MC cable		12.31	.650		31	32		63	81.50
4180	EMT & wire		5.33	1.501		36	73.50		109.50	149
4200	With #12/2, type NM cable		12.31	.650		15.70	32		47.70	65
4220	Type MC cable		10.67	.750		31	36.50		67.50	88.50
4230	EMT & wire		4.71	1.699		39.50	83		122.50	168
4250	20 amp recpt. #12/2, type NM cable		12.31	.650		24.50	32		56.50	74.50
4270	Type MC cable		10.67	.750		40	36.50		76.50	98.50
4280	EMT & wire		4.71	1.699		48.50	83		131.50	178
4300	GFI, 15 amp recpt., type NM cable		12.31	.650		43	32		75	95
4320	Type MC cable		10.67	.750		60.50	36.50		97	121
4330	EMT & wire		4.71	1.699		65.50	83		148.50	196
4350	GFI with #12/2, type NM cable		10.67	.750		45.50	36.50		82	105
4370	Type MC cable		9.20	.870		61	42.50		103.50	131
4380	EMT & wire		4.21	1.900		69.50	93		162.50	214
4400	20 amp recpt., #12/2 type NM cable		10.67	.750		47.50	36.50		84	107
4420	Type MC cable		9.20	.870		63	42.50		105.50	133
4430	EMT & wire		4.21	1.900		71.50	93		164.50	217
4500	Weather-proof cover for above receptacles, add		32	.250		5.25	12.25		17.50	24
4550	Air conditioner outlet, 20 amp-240 volt recpt.									
4560	30′ of #12/2, 2 pole circuit breaker									
4570	Type NM cable	1 Elec	10	.800	Ea.	60.50	39		99.50	125
4580	Type MC cable		9	.889		79.50	43.50		123	153
4590	EMT & wire		4	2		84.50	98		182.50	239
4600	Decorator style, type NM cable		10	.800		65	39		104	130
4620	Type MC cable		9	.889		84.50	43.50		128	158
4630	EMT & wire		4	2		89	98		187	244
4650	Dryer outlet, 30 amp-240 volt recpt., 20′ of #10/3									
4660	2 pole circuit breaker									
4670	Type NM cable	1 Elec	6.41	1.248	Ea.	66	61		127	164
4680	Type MC cable		5.71	1.401		80	68.50		148.50	190
4690	EMT & wire		3.48	2.299		81	113		194	256
4700	Range outlet, 50 amp-240 volt recpt., 30′ of #8/3									

26 05 90.10 Residential Wiring		Crew	Daily Output	Labor-Hours	Unit	Material	2010 Bare Costs Labor	Equipment	Total	Total Incl O&P
4710	Type NM cable	1 Elec	4.21	1.900	Ea.	93	93		186	240
4720	Type MC cable		4	2		163	98		261	325
4730	EMT & wire		2.96	2.703		111	132		243	320
4750	Central vacuum outlet, Type NM cable		6.40	1.250		62.50	61.50		124	160
4770	Type MC cable		5.71	1.401		90	68.50		158.50	201
4780	EMT & wire	↓	3.48	2.299	↓	88.50	113		201.50	265
4800	30 amp-110 volt locking recpt., #10/2 circ. bkr.									
4810	Type NM cable	1 Elec	6.20	1.290	Ea.	72.50	63		135.50	174
4830	EMT & wire	"	3.20	2.500	"	99	123		222	291
4900	Low voltage outlets									
4910	Telephone recpt., 20′ of 4/C phone wire	1 Elec	26	.308	Ea.	10.35	15.10		25.45	34
4920	TV recpt., 20′ of RG59U coax wire, F type connector	"	16	.500	"	19.30	24.50		43.80	57.50
4950	Door bell chime, transformer, 2 buttons, 60′ of bellwire									
4970	Economy model	1 Elec	11.50	.696	Ea.	75	34		109	133
4980	Custom model		11.50	.696		119	34		153	182
4990	Luxury model, 3 buttons	↓	9.50	.842	↓	320	41.50		361.50	415
6000	Lighting outlets									
6050	Wire only (for fixture), type NM cable	1 Elec	32	.250	Ea.	6.45	12.25		18.70	25.50
6070	Type MC cable		24	.333		17.05	16.35		33.40	43.50
6080	EMT & wire		10	.800		20.50	39		59.50	81
6100	Box (4″), and wire (for fixture), type NM cable		25	.320		16.25	15.70		31.95	41.50
6120	Type MC cable		20	.400		27	19.60		46.60	58.50
6130	EMT & wire	↓	11	.727	↓	30.50	35.50		66	86.50
6200	Fixtures (use with lines 6050 or 6100 above)									
6210	Canopy style, economy grade	1 Elec	40	.200	Ea.	34	9.80		43.80	52
6220	Custom grade		40	.200		52.50	9.80		62.30	72.50
6250	Dining room chandelier, economy grade		19	.421		85	20.50		105.50	124
6260	Custom grade		19	.421		250	20.50		270.50	305
6270	Luxury grade		15	.533		565	26		591	660
6310	Kitchen fixture (fluorescent), economy grade		30	.267		63.50	13.05		76.55	89.50
6320	Custom grade		25	.320		173	15.70		188.70	214
6350	Outdoor, wall mounted, economy grade		30	.267		32	13.05		45.05	54.50
6360	Custom grade		30	.267		111	13.05		124.05	141
6370	Luxury grade		25	.320		250	15.70		265.70	299
6410	Outdoor PAR floodlights, 1 lamp, 150 watt		20	.400		35	19.60		54.60	67.50
6420	2 lamp, 150 watt each		20	.400		58	19.60		77.60	93
6430	For infrared security sensor, add		32	.250		117	12.25		129.25	147
6450	Outdoor, quartz-halogen, 300 watt flood		20	.400		42	19.60		61.60	75.50
6600	Recessed downlight, round, pre-wired, 50 or 75 watt trim		30	.267		43.50	13.05		56.55	67
6610	With shower light trim		30	.267		53.50	13.05		66.55	78.50
6620	With wall washer trim		28	.286		64.50	14		78.50	92
6630	With eye-ball trim	↓	28	.286		64.50	14		78.50	92
6640	For direct contact with insulation, add					2.20			2.20	2.42
6700	Porcelain lamp holder	1 Elec	40	.200		4.20	9.80		14	19.15
6710	With pull switch		40	.200		5.05	9.80		14.85	20
6750	Fluorescent strip, 1-20 watt tube, wrap around diffuser, 24″		24	.333		54	16.35		70.35	84
6770	2-34 watt tubes, 48″		20	.400		82.50	19.60		102.10	120
6780	With residential ballast		20	.400		92.50	19.60		112.10	131
6800	Bathroom heat lamp, 1-250 watt		28	.286		46	14		60	71.50
6810	2-250 watt lamps	↓	28	.286	↓	71.50	14		85.50	99.50
6820	For timer switch, see Div. 26 05 90.10 line 2400									
6900	Outdoor post lamp, incl. post, fixture, 35′ of #14/2									
6910	Type NMC cable	1 Elec	3.50	2.286	Ea.	202	112		314	390

26 05 90.10 Residential Wiring		Crew	Daily Output	Labor-Hours	Unit	2010 Bare Costs Material	Labor	Equipment	Total	Total Incl O&P
6920	Photo-eye, add	1 Elec	27	.296	Ea.	35	14.50		49.50	60
6950	Clock dial time switch, 24 hr., w/enclosure, type NM cable		11.43	.700		61	34.50		95.50	118
6970	Type MC cable		11	.727		78.50	35.50		114	140
6980	EMT & wire	↓	4.85	1.649	↓	83.50	81		164.50	212
7000	Alarm systems									
7050	Smoke detectors, box, #14/3, type NM cable	1 Elec	14.55	.550	Ea.	34	27		61	77.50
7070	Type MC cable		12.31	.650		51.50	32		83.50	104
7080	EMT & wire	↓	5	1.600		49.50	78.50		128	172
7090	For relay output to security system, add				↓	13			13	14.30
8000	Residential equipment									
8050	Disposal hook-up, incl. switch, outlet box, 3′ of flex									
8060	20 amp-1 pole circ. bkr., and 25′ of #12/2									
8070	Type NM cable	1 Elec	10	.800	Ea.	30.50	39		69.50	92
8080	Type MC cable		8	1		48	49		97	126
8090	EMT & wire	↓	5	1.600	↓	56.50	78.50		135	180
8100	Trash compactor or dishwasher hook-up, incl. outlet box,									
8110	3′ of flex, 15 amp-1 pole circ. bkr., and 25′ of #14/2									
8120	Type NM cable	1 Elec	10	.800	Ea.	22.50	39		61.50	83.50
8130	Type MC cable		8	1		43	49		92	120
8140	EMT & wire	↓	5	1.600	↓	49	78.50		127.50	171
8150	Hot water sink dispensor hook-up, use line 8100									
8200	Vent/exhaust fan hook-up, type NM cable	1 Elec	32	.250	Ea.	6.45	12.25		18.70	25.50
8220	Type MC cable		24	.333		17.05	16.35		33.40	43.50
8230	EMT & wire	↓	10	.800	↓	20.50	39		59.50	81
8250	Bathroom vent fan, 50 CFM (use with above hook-up)									
8260	Economy model	1 Elec	15	.533	Ea.	25	26		51	66.50
8270	Low noise model		15	.533		32	26		58	74
8280	Custom model	↓	12	.667	↓	117	32.50		149.50	178
8300	Bathroom or kitchen vent fan, 110 CFM									
8310	Economy model	1 Elec	15	.533	Ea.	69.50	26		95.50	116
8320	Low noise model	"	15	.533	"	86	26		112	134
8350	Paddle fan, variable speed (w/o lights)									
8360	Economy model (AC motor)	1 Elec	10	.800	Ea.	106	39		145	176
8370	Custom model (AC motor)		10	.800		182	39		221	259
8380	Luxury model (DC motor)		8	1		360	49		409	470
8390	Remote speed switch for above, add	↓	12	.667	↓	30	32.50		62.50	81.50
8500	Whole house exhaust fan, ceiling mount, 36″, variable speed									
8510	Remote switch, incl. shutters, 20 amp-1 pole circ. bkr.									
8520	30′ of #12/2, type NM cable	1 Elec	4	2	Ea.	1,025	98		1,123	1,275
8530	Type MC cable		3.50	2.286		1,050	112		1,162	1,325
8540	EMT & wire	↓	3	2.667	↓	1,050	131		1,181	1,375
8600	Whirlpool tub hook-up, incl. timer switch, outlet box									
8610	3′ of flex, 20 amp-1 pole GFI circ. bkr.									
8620	30′ of #12/2, type NM cable	1 Elec	5	1.600	Ea.	124	78.50		202.50	253
8630	Type MC cable		4.20	1.905		136	93.50		229.50	288
8640	EMT & wire	↓	3.40	2.353	↓	144	115		259	330
8650	Hot water heater hook-up, incl. 1-2 pole circ. bkr., box;									
8660	3′ of flex, 20′ of #10/2, type NM cable	1 Elec	5	1.600	Ea.	32	78.50		110.50	153
8670	Type MC cable		4.20	1.905		55.50	93.50		149	200
8680	EMT & wire	↓	3.40	2.353	↓	52	115		167	229
9000	Heating/air conditioning									
9050	Furnace/boiler hook-up, incl. firestat, local on-off switch									
9060	Emergency switch, and 40′ of type NM cable	1 Elec	4	2	Ea.	54.50	98		152.50	206

26 05 Common Work Results for Electrical

26 05 90 – Residential Applications

26 05 90.10 Residential Wiring		Crew	Daily Output	Labor-Hours	Unit	2010 Bare Costs Material	Labor	Equipment	Total	Total Incl O&P
9070	Type MC cable	1 Elec	3.50	2.286	Ea.	81.50	112		193.50	257
9080	EMT & wire	↓	1.50	5.333	↓	89.50	261		350.50	490
9100	Air conditioner hook-up, incl. local 60 amp disc. switch									
9110	3′ sealtite, 40 amp, 2 pole circuit breaker									
9130	40′ of #8/2, type NM cable	1 Elec	3.50	2.286	Ea.	216	112		328	405
9140	Type MC cable		3	2.667		315	131		446	540
9150	EMT & wire	↓	1.30	6.154	↓	252	300		552	730
9200	Heat pump hook-up, 1-40 & 1-100 amp 2 pole circ. bkr.									
9210	Local disconnect switch, 3′ sealtite									
9220	40′ of #8/2 & 30′ of #3/2									
9230	Type NM cable	1 Elec	1.30	6.154	Ea.	545	300		845	1,050
9240	Type MC cable		1.08	7.407		745	365		1,110	1,350
9250	EMT & wire	↓	.94	8.511	↓	615	415		1,030	1,300
9500	Thermostat hook-up, using low voltage wire									
9520	Heating only, 25′ of #18-3	1 Elec	24	.333	Ea.	9.85	16.35		26.20	35.50
9530	Heating/cooling, 25′ of #18-4	"	20	.400	"	12.30	19.60		31.90	42.50

26 24 Switchboards and Panelboards

26 24 19 – Motor-Control Centers

26 24 19.40 Motor Starters and Controls		Crew	Daily Output	Labor-Hours	Unit	Material	Labor	Equipment	Total	Total Incl O&P
0010	**MOTOR STARTERS AND CONTROLS**									
0050	Magnetic, FVNR, with enclosure and heaters, 480 volt									
0080	2 HP, size 00	1 Elec	3.50	2.286	Ea.	215	112		327	405
0100	5 HP, size 0		2.30	3.478		260	170		430	540
0200	10 HP, size 1	↓	1.60	5		292	245		537	685
0300	25 HP, size 2	2 Elec	2.20	7.273		550	355		905	1,125
0400	50 HP, size 3		1.80	8.889		895	435		1,330	1,625
0500	100 HP, size 4		1.20	13.333		1,975	655		2,630	3,150
0600	200 HP, size 5		.90	17.778		4,650	870		5,520	6,425
0610	400 HP, size 6	↓	.80	20		18,300	980		19,280	21,600
0620	NEMA 7, 5 HP, size 0	1 Elec	1.60	5		1,350	245		1,595	1,875
0630	10 HP, size 1	"	1.10	7.273		1,425	355		1,780	2,075
0640	25 HP, size 2	2 Elec	1.80	8.889		2,300	435		2,735	3,175
0650	50 HP, size 3		1.20	13.333		3,450	655		4,105	4,775
0660	100 HP, size 4		.90	17.778		5,550	870		6,420	7,425
0670	200 HP, size 5	↓	.50	32		13,300	1,575		14,875	16,900
0700	Combination, with motor circuit protectors, 5 HP, size 0	1 Elec	1.80	4.444		845	218		1,063	1,250
0800	10 HP, size 1	"	1.30	6.154		875	300		1,175	1,425
0900	25 HP, size 2	2 Elec	2	8		1,225	390		1,615	1,925
1000	50 HP, size 3		1.32	12.121		1,775	595		2,370	2,825
1200	100 HP, size 4	↓	.80	20		3,850	980		4,830	5,700
1220	NEMA 7, 5 HP, size 0	1 Elec	1.30	6.154		2,425	300		2,725	3,125
1230	10 HP, size 1	"	1	8		2,500	390		2,890	3,325
1240	25 HP, size 2	2 Elec	1.32	12.121		3,325	595		3,920	4,525
1250	50 HP, size 3		.80	20		5,500	980		6,480	7,500
1260	100 HP, size 4		.60	26.667		8,575	1,300		9,875	11,400
1270	200 HP, size 5	↓	.40	40		18,600	1,950		20,550	23,400
1400	Combination, with fused switch, 5 HP, size 0	1 Elec	1.80	4.444		645	218		863	1,025
1600	10 HP, size 1	"	1.30	6.154		690	300		990	1,200
1800	25 HP, size 2	2 Elec	2	8		1,125	390		1,515	1,800
2000	50 HP, size 3	↓	1.32	12.121	↓	1,900	595		2,495	2,950

26 24 Switchboards and Panelboards

26 24 19 – Motor-Control Centers

26 24 19.40 Motor Starters and Controls		Crew	Daily Output	Labor-Hours	Unit	2010 Bare Costs Material	Labor	Equipment	Total	Total Incl O&P
2200	100 HP, size 4	2 Elec	.80	20	Ea.	3,300	980		4,280	5,075
3500	Magnetic FVNR with NEMA 12, enclosure & heaters, 480 volt									
3600	5 HP, size 0	1 Elec	2.20	3.636	Ea.	245	178		423	535
3700	10 HP, size 1	"	1.50	5.333		370	261		631	795
3800	25 HP, size 2	2 Elec	2	8		690	390		1,080	1,350
3900	50 HP, size 3		1.60	10		1,075	490		1,565	1,900
4000	100 HP, size 4		1	16		2,550	785		3,335	3,975
4100	200 HP, size 5	↓	.80	20		6,100	980		7,080	8,150
4200	Combination, with motor circuit protectors, 5 HP, size 0	1 Elec	1.70	4.706		810	231		1,041	1,250
4300	10 HP, size 1	"	1.20	6.667		845	325		1,170	1,425
4400	25 HP, size 2	2 Elec	1.80	8.889		1,250	435		1,685	2,050
4500	50 HP, size 3		1.20	13.333		2,050	655		2,705	3,250
4600	100 HP, size 4	↓	.74	21.622		4,650	1,050		5,700	6,675
4700	Combination, with fused switch, 5 HP, size 0	1 Elec	1.70	4.706		785	231		1,016	1,200
4800	10 HP, size 1	"	1.20	6.667		820	325		1,145	1,375
4900	25 HP, size 2	2 Elec	1.80	8.889		1,250	435		1,685	2,025
5000	50 HP, size 3		1.20	13.333		2,000	655		2,655	3,175
5100	100 HP, size 4	↓	.74	21.622	↓	4,050	1,050		5,100	6,025
5200	Factory installed controls, adders to size 0 thru 5									
5300	Start-stop push button	1 Elec	32	.250	Ea.	51.50	12.25		63.75	74.50
5400	Hand-off-auto-selector switch		32	.250		51.50	12.25		63.75	74.50
5500	Pilot light		32	.250		96.50	12.25		108.75	124
5600	Start-stop-pilot		32	.250		148	12.25		160.25	180
5700	Auxiliary contact, NO or NC		32	.250		70.50	12.25		82.75	95.50
5800	NO-NC	↓	32	.250	↓	141	12.25		153.25	173

26 29 Low-Voltage Controllers

26 29 13 – Enclosed Controllers

26 29 13.20 Control Stations		Crew	Daily Output	Labor-Hours	Unit	Material	Labor	Equipment	Total	Total Incl O&P
0010	**CONTROL STATIONS**									
0050	NEMA 1, heavy duty, stop/start	1 Elec	8	1	Ea.	148	49		197	235
0100	Stop/start, pilot light		6.20	1.290		201	63		264	315
0200	Hand/off/automatic		6.20	1.290		109	63		172	214
0400	Stop/start/reverse		5.30	1.509		199	74		273	330
0500	NEMA 7, heavy duty, stop/start		6	1.333		340	65.50		405.50	470
0600	Stop/start, pilot light	↓	4	2	↓	415	98		513	600

26 29 23 – Variable-Frequency Motor Controllers

26 29 23.10 Variable Frequency Drives/Adj. Frequency Drives			Crew	Daily Output	Labor-Hours	Unit	Material	Labor	Equipment	Total	Total Incl O&P
0010	**VARIABLE FREQUENCY DRIVES/ADJ. FREQUENCY DRIVES**										
0100	Enclosed (NEMA 1), 460 volt, for 3 HP motor size	G	1 Elec	.80	10	Ea.	1,000	490		1,490	1,825
0110	5 HP motor size	G		.80	10		1,150	490		1,640	1,975
0120	7.5 HP motor size	G		.67	11.940		1,375	585		1,960	2,375
0130	10 HP motor size	G	↓	.67	11.940		1,625	585		2,210	2,650
0140	15 HP motor size	G	2 Elec	.89	17.978		1,925	880		2,805	3,425
0150	20 HP motor size	G		.89	17.978		2,300	880		3,180	3,850
0160	25 HP motor size	G		.67	23.881		2,850	1,175		4,025	4,875
0170	30 HP motor size	G		.67	23.881		3,525	1,175		4,700	5,625
0180	40 HP motor size	G		.67	23.881		4,200	1,175		5,375	6,375
0190	50 HP motor size	G	↓	.53	30.189		5,250	1,475		6,725	7,975
0200	60 HP motor size	G	R-3	.56	35.714	↓	6,275	1,725	247	8,247	9,750

26 29 Low-Voltage Controllers

26 29 23 – Variable-Frequency Motor Controllers

26 29 23.10 Variable Frequency Drives/Adj. Frequency Drives		Crew	Daily Output	Labor-Hours	Unit	Material	2010 Bare Costs Labor	Equipment	Total	Total Incl O&P
0210	75 HP motor size G	R-3	.56	35.714	Ea.	7,250	1,725	247	9,222	10,800
0220	100 HP motor size G		.50	40		8,450	1,925	276	10,651	12,500
0230	125 HP motor size G		.50	40		9,400	1,925	276	11,601	13,600
0240	150 HP motor size G		.50	40		10,800	1,925	276	13,001	15,100
0250	200 HP motor size G		.42	47.619		14,000	2,300	330	16,630	19,200
1100	Custom-engineered, 460 volt, for 3 HP motor size G	1 Elec	.56	14.286		2,350	700		3,050	3,625
1110	5 HP motor size G		.56	14.286		2,350	700		3,050	3,625
1120	7.5 HP motor size G		.47	17.021		2,450	835		3,285	3,950
1130	10 HP motor size G		.47	17.021		2,575	835		3,410	4,075
1140	15 HP motor size G	2 Elec	.62	25.806		3,225	1,275		4,500	5,425
1150	20 HP motor size G		.62	25.806		3,625	1,275		4,900	5,875
1160	25 HP motor size G		.47	34.043		4,250	1,675		5,925	7,150
1170	30 HP motor size G		.47	34.043		5,300	1,675		6,975	8,300
1180	40 HP motor size G		.47	34.043		6,275	1,675		7,950	9,375
1190	50 HP motor size G		.37	43.243		7,200	2,125		9,325	11,100
1200	60 HP motor size G	R-3	.39	51.282		10,900	2,475	355	13,730	16,000
1210	75 HP motor size G		.39	51.282		11,600	2,475	355	14,430	16,800
1220	100 HP motor size G		.35	57.143		12,600	2,750	395	15,745	18,400
1230	125 HP motor size G		.35	57.143		13,500	2,750	395	16,645	19,400
1240	150 HP motor size G		.35	57.143		15,500	2,750	395	18,645	21,500
1250	200 HP motor size G		.29	68.966		20,200	3,325	475	24,000	27,700
2000	For complex & special design systems to meet specific									
2010	requirements, obtain quote from vendor.									

26 42 Cathodic Protection

26 42 16 – Passive Cathodic Protection for Underground Storage Tank

26 42 16.50 Cathodic Protection Wiring Methods

		Crew	Daily Output	Labor-Hours	Unit	Material	Labor	Equipment	Total	Total Incl O&P
0010	**CATHODIC PROTECTION WIRING METHODS**									
1000	Anodes, magnesium type, 9 #	R-15	18.50	2.595	Ea.	45	124	18.80	187.80	255
1010	17 #		13	3.692		83	176	26.50	285.50	385
1020	32 #		10	4.800		150	229	35	414	545
1030	48 #		7.20	6.667		228	320	48.50	596.50	780
1100	Graphite type w/ epoxy cap, 3" x 60" (32 #)	R-22	8.40	4.438		117	185		302	405
1110	4" x 80" (68 #)		6	6.213		222	258		480	635
1120	6" x 72" (80 #)		5.20	7.169		1,475	298		1,773	2,075
1130	6" x 36" (45 #)		9.60	3.883		745	162		907	1,075
2000	Rectifiers, silicon type, air cooled, 28 V/10 A	R-19	3.50	5.714		1,950	281		2,231	2,575
2010	20 V/20 A		3.50	5.714		2,025	281		2,306	2,650
2100	Oil immersed, 28 V/10 A		3	6.667		2,750	325		3,075	3,500
2110	20 V/20 A		3	6.667		2,850	325		3,175	3,625
3000	Anode backfill, coke breeze	R-22	3850	.010	Lb.	.14	.40		.54	.75
4000	Cable, HMWPE, No. 8		2.40	15.533	M.L.F.	282	645		927	1,275
4010	No. 6		2.40	15.533		405	645		1,050	1,425
4020	No. 4		2.40	15.533		580	645		1,225	1,600
4030	No. 2		2.40	15.533		925	645		1,570	2,000
4040	No. 1		2.20	16.945		1,225	705		1,930	2,400
4050	No. 1/0		2.20	16.945		1,500	705		2,205	2,700
4060	No. 2/0		2.20	16.945		1,825	705		2,530	3,050
4070	No. 4/0		2	18.640		3,475	775		4,250	5,000
5000	Test station, 7 terminal box, flush curb type w/lockable cover	R-19	12	1.667	Ea.	70.50	82		152.50	200
5010	Reference cell, 2" dia PVC conduit, cplg, plug, set flush	"	4.80	4.167	"	148	205		353	470

Estimating Tips

- When estimating material costs for electronic safety and security systems, it is always prudent to obtain manufacturers' quotations for equipment prices and special installation requirements that affect the total cost.
- Fire alarm systems consist of control panels, annunciator panels, battery with rack, charger, and fire alarm actuating and indicating devices. Some fire alarm systems include speakers, telephone lines, door closer controls, and other components. Be careful not to overlook the costs related to installation for these items. Also be aware of costs for integrated automation instrumentation and terminal devices, control equipment, control wiring, and programming.
- Security equipment includes items such as CCTV, access control, and other detection and identification systems to perform alert and alarm functions. Be sure to consider the costs related to installation for this security equipment, such as for integrated automation instrumentation and terminal devices, control equipment, control wiring, and programming.

Reference Numbers

Reference numbers are shown in shaded boxes at the beginning of some major classifications. These numbers refer to related items in the Reference Section. The reference information may be an estimating procedure, an alternate pricing method, or technical information.

Note: Not all subdivisions listed here necessarily appear in this publication.

28 13 Access Control

28 13 53 – Security Access Detection

28 13 53.13 Security Access Metal Detectors		Crew	Daily Output	Labor-Hours	Unit	2010 Bare Costs Material	Labor	Equipment	Total	Total Incl O&P
0010	**SECURITY ACCESS METAL DETECTORS**									
0240	Metal detector, hand-held, wand type, unit only				Ea.	81.50			81.50	90
0250	Metal detector, walk through portal type, single zone	1 Elec	2	4		2,750	196		2,946	3,325
0260	Multi-zone	"	2	4		3,500	196		3,696	4,150

28 13 53.16 Security Access X-Ray Equipment		Crew	Daily Output	Labor-Hours	Unit	Material	Labor	Equipment	Total	Total Incl O&P
0010	**SECURITY ACCESS X-RAY EQUIPMENT**									
0290	X-ray machine, desk top, for mail/small packages/letters	1 Elec	4	2	Ea.	3,000	98		3,098	3,450
0300	Conveyor type, incl monitor, minimum		2	4		14,000	196		14,196	15,700
0310	Maximum		2	4		25,000	196		25,196	27,800
0320	X-ray machine, large unit, for airports, incl monitor, min	2 Elec	1	16		35,000	785		35,785	39,700
0330	Maximum	"	.50	32		60,000	1,575		61,575	68,500

28 13 53.23 Security Access Explosive Detection Equipment		Crew	Daily Output	Labor-Hours	Unit	Material	Labor	Equipment	Total	Total Incl O&P
0010	**SECURITY ACCESS EXPLOSIVE DETECTION EQUIPMENT**									
0270	Explosives detector, walk through portal type	1 Elec	2	4	Ea.	3,500	196		3,696	4,150
0280	Hand-held, battery operated				"				25,500	28,100

28 16 Intrusion Detection

28 16 16 – Intrusion Detection Systems Infrastructure

28 16 16.50 Intrusion Detection		Crew	Daily Output	Labor-Hours	Unit	Material	Labor	Equipment	Total	Total Incl O&P
0010	**INTRUSION DETECTION**, not including wires & conduits									
0100	Burglar alarm, battery operated, mechanical trigger	1 Elec	4	2	Ea.	272	98		370	445
0200	Electrical trigger		4	2		325	98		423	500
0400	For outside key control, add		8	1		82	49		131	164
0600	For remote signaling circuitry, add		8	1		122	49		171	208
0800	Card reader, flush type, standard		2.70	2.963		910	145		1,055	1,225
1000	Multi-code		2.70	2.963		1,175	145		1,320	1,525

28 23 Video Surveillance

28 23 23 – Video Surveillance Systems Infrastructure

28 23 23.50 Video Surveillance Equipments		Crew	Daily Output	Labor-Hours	Unit	Material	Labor	Equipment	Total	Total Incl O&P
0010	**VIDEO SURVEILLANCE EQUIPMENTS**									
0200	Video cameras, wireless, hidden in exit signs, clocks, etc, incl receiver	1 Elec	3	2.667	Ea.	149	131		280	360
0210	Accessories for VCR, single camera		3	2.667		570	131		701	820
0220	For multiple cameras		3	2.667		1,225	131		1,356	1,550
0230	Video cameras, wireless, for under vehicle searching, complete		2	4		9,975	196		10,171	11,300

28 31 Fire Detection and Alarm

28 31 23 – Fire Detection and Alarm Annunciation Panels and Fire Stations

28 31 23.50 Alarm Panels and Devices		Crew	Daily Output	Labor-Hours	Unit	Material	Labor	Equipment	Total	Total Incl O&P
0010	**ALARM PANELS AND DEVICES**, not including wires & conduits									
3594	Fire, alarm control panel									
3600	4 zone	2 Elec	2	8	Ea.	730	390		1,120	1,400
3800	8 zone		1	16		760	785		1,545	2,000
4000	12 zone		.67	23.988		2,300	1,175		3,475	4,275
4020	Alarm device	1 Elec	8	1		229	49		278	325
4050	Actuating device		8	1		325	49		374	430
4200	Battery and rack		4	2		395	98		493	580

28 31 Fire Detection and Alarm

28 31 23 – Fire Detection and Alarm Annunciation Panels and Fire Stations

28 31 23.50 Alarm Panels and Devices		Crew	Daily Output	Labor-Hours	Unit	2010 Bare Costs Material	Labor	Equipment	Total	Total Incl O&P
4400	Automatic charger	1 Elec	8	1	Ea.	560	49		609	695
4600	Signal bell		8	1		70.50	49		119.50	151
4800	Trouble buzzer or manual station		8	1		80.50	49		129.50	162
5600	Strobe and horn		5.30	1.509		147	74		221	271
5800	Fire alarm horn		6.70	1.194		58.50	58.50		117	152
6000	Door holder, electro-magnetic		4	2		99	98		197	255
6200	Combination holder and closer		3.20	2.500		119	123		242	315
6600	Drill switch		8	1		360	49		409	470
6800	Master box		2.70	2.963		6,175	145		6,320	7,025
7000	Break glass station		8	1		65	49		114	145
7800	Remote annunciator, 8 zone lamp		1.80	4.444		201	218		419	545
8000	12 zone lamp	2 Elec	2.60	6.154		345	300		645	830
8200	16 zone lamp	"	2.20	7.273		345	355		700	910

28 31 43 – Fire Detection Sensors

28 31 43.50 Fire and Heat Detectors		Crew	Daily Output	Labor-Hours	Unit	Material	Labor	Equipment	Total	Total Incl O&P
0010	**FIRE & HEAT DETECTORS**									
5000	Detector, rate of rise	1 Elec	8	1	Ea.	49	49		98	127
5100	Fixed temperature	"	8	1	"	49	49		98	127

28 31 46 – Smoke Detection Sensors

28 31 46.50 Smoke Detectors		Crew	Daily Output	Labor-Hours	Unit	Material	Labor	Equipment	Total	Total Incl O&P
0010	**SMOKE DETECTORS**									
5200	Smoke detector, ceiling type	1 Elec	6.20	1.290	Ea.	126	63		189	233
5400	Duct type	"	3.20	2.500	"	315	123		438	525

28 33 Fuel-Gas Detection and Alarm

28 33 33 – Fuel-Gas Detection Sensors

28 33 33.50 Tank Leak Detection Systems		Crew	Daily Output	Labor-Hours	Unit	Material	Labor	Equipment	Total	Total Incl O&P
0010	**TANK LEAK DETECTION SYSTEMS** Liquid and vapor									
0100	For hydrocarbons and hazardous liquids/vapors									
0120	Controller, data acquisition, incl. printer, modem, RS232 port									
0140	24 channel, for use with all probes				Ea.	3,700			3,700	4,075
0160	9 channel, for external monitoring				"	805			805	885
0200	Probes									
0210	Well monitoring									
0220	Liquid phase detection				Ea.	425			425	470
0230	Hydrocarbon vapor, fixed position					425			425	470
0240	Hydrocarbon vapor, float mounted					425			425	470
0250	Both liquid and vapor hydrocarbon					425			425	470
0300	Secondary containment, liquid phase									
0310	Pipe trench/manway sump				Ea.	740			740	810
0320	Double wall pipe and manual sump					740			740	810
0330	Double wall fiberglass annular space					290			290	320
0340	Double wall steel tank annular space					300			300	330
0500	Accessories									
0510	Modem, non-dedicated phone line				Ea.	272			272	299
0600	Monitoring, internal									
0610	Automatic tank gauge, incl. overfill				Ea.	1,025			1,025	1,125

28 33 Fuel-Gas Detection and Alarm

28 33 33 – Fuel-Gas Detection Sensors

28 33 33.50 Tank Leak Detection Systems		Crew	Daily Output	Labor-Hours	Unit	2010 Bare Costs Material	Labor	Equipment	Total	Total Incl O&P
0620	Product line				Ea.	1,025			1,025	1,125
0700	Monitoring, special									
0710	Cathodic protection				Ea.	650			650	715
0720	Annular space chemical monitor				"	880			880	970

Estimating Tips

31 05 00 Common Work Results for Earthwork

- Estimating the actual cost of performing earthwork requires careful consideration of the variables involved. This includes items such as type of soil, whether water will be encountered, dewatering, whether banks need bracing, disposal of excavated earth, and length of haul to fill or spoil sites, etc. If the project has large quantities of cut or fill, consider raising or lowering the site to reduce costs, while paying close attention to the effect on site drainage and utilities.
- If the project has large quantities of fill, creating a borrow pit on the site can significantly lower the costs.
- It is very important to consider what time of year the project is scheduled for completion. Bad weather can create large cost overruns from dewatering, site repair, and lost productivity from cold weather.

Reference Numbers

Reference numbers are shown in shaded boxes at the beginning of some major classifications. These numbers refer to related items in the Reference Section. The reference information may be an estimating procedure, an alternate pricing method, or technical information.

Note: Not all subdivisions listed here necessarily appear in this publication.

31 05 Common Work Results for Earthwork

31 05 23 – Cement and Concrete for Earthwork

31 05 23.30 Plant Mixed Bituminous Concrete		Crew	Daily Output	Labor-Hours	Unit	Material	2010 Bare Costs Labor	Equipment	Total	Total Incl O&P
0010	**PLANT MIXED BITUMINOUS CONCRETE**									
0020	Asphaltic concrete plant mix (145 LB per C.F.)				Ton	65			65	71.50
0040	Asphaltic concrete less than 300 tons add trucking costs									
0050	See Div. 31 23 23.20 for hauling costs									
0200	All weather patching mix, hot				Ton	68			68	75
0250	Cold patch					68			68	75
0300	Berm mix				↓	68			68	75

31 23 Excavation and Fill

31 23 16 – Excavation

31 23 16.13 Excavating, Trench

31 23 16.13 Excavating, Trench		Crew	Daily Output	Labor-Hours	Unit	Material	2010 Bare Costs Labor	Equipment	Total	Total Incl O&P
0010	**EXCAVATING, TRENCH** G1030–805									
0011	Or continuous footing									
0020	Common earth with no sheeting or dewatering included									
0050	1′ to 4′ deep, 3/8 C.Y. excavator	B-11C	150	.107	B.C.Y.		4.06	2.25	6.31	8.65
0060	1/2 C.Y. excavator	B-11M	200	.080			3.04	1.94	4.98	6.75
0090	4′ to 6′ deep, 1/2 C.Y. excavator	″	200	.080			3.04	1.94	4.98	6.75
0100	5/8 C.Y. excavator	B-12Q	250	.064			2.48	2.26	4.74	6.25
0110	3/4 C.Y. excavator	B-12F	300	.053			2.07	2.26	4.33	5.60
0300	1/2 C.Y. excavator, truck mounted	B-12J	200	.080			3.10	4.06	7.16	9.15
0500	6′ to 10′ deep, 3/4 C.Y. excavator	B-12F	225	.071			2.76	3.01	5.77	7.50
0600	1 C.Y. excavator, truck mounted	B-12K	400	.040			1.55	2.40	3.95	4.99
0900	10′ to 14′ deep, 3/4 C.Y. excavator	B-12F	200	.080			3.10	3.39	6.49	8.45
1000	1-1/2 C.Y. excavator	B-12B	540	.030			1.15	1.88	3.03	3.81
1300	14′ to 20′ deep, 1 C.Y. excavator	B-12A	320	.050			1.94	2.51	4.45	5.70
1340	20′ to 24′ deep, 1 C.Y. excavator	″	288	.056			2.15	2.78	4.93	6.30
1352	4′ to 6′ deep, 1/2 C.Y. excavator w/trench box	B-13H	188	.085			3.30	4.93	8.23	10.40
1354	5/8 C.Y. excavator	″	235	.068			2.64	3.94	6.58	8.35
1356	3/4 C.Y. excavator	B-13G	282	.057			2.20	2.81	5.01	6.40
1362	6′ to 10′ deep, 3/4 C.Y. excavator w/trench box		212	.075			2.92	3.73	6.65	8.55
1374	10′ to 14′ deep, 3/4 C.Y. excavator w/trench box	↓	188	.085			3.30	4.21	7.51	9.65
1376	1-1/2 C.Y. excavator	B-13E	508	.032			1.22	2.23	3.45	4.30
1381	14′ to 20′ deep, 1 C.Y. excavator w/trench box	B-13D	301	.053			2.06	3.04	5.10	6.45
1386	20′ to 24′ deep, 1 C.Y. excavator w/trench box	″	271	.059			2.29	3.38	5.67	7.20
1400	By hand with pick and shovel 2′ to 6′ deep, light soil	1 Clab	8	1			33		33	51
1500	Heavy soil	″	4	2	↓		66		66	102
1700	For tamping backfilled trenches, air tamp, add	A-1G	100	.080	E.C.Y.		2.65	.52	3.17	4.65
1900	Vibrating plate, add	B-18	180	.133	″		4.50	.24	4.74	7.20
2100	Trim sides and bottom for concrete pours, common earth		1500	.016	S.F.		.54	.03	.57	.86
2300	Hardpan	↓	600	.040	″		1.35	.07	1.42	2.16
5020	Loam & Sandy clay with no sheeting or dewatering included									
5050	1′ to 4′ deep, 3/8 C.Y. tractor loader/backhoe	B-11C	162	.099	B.C.Y.		3.76	2.09	5.85	8
5060	1/2 C.Y. excavator	B-11M	216	.074			2.82	1.79	4.61	6.25
5080	4′ to 6′ deep, 1/2 C.Y. excavator	″	216	.074			2.82	1.79	4.61	6.25
5090	5/8 C.Y. excavator	B-12Q	276	.058			2.25	2.05	4.30	5.65
5100	3/4 C.Y. excavator	B-12F	324	.049			1.91	2.09	4	5.20
5130	1/2 C.Y. excavator, truck mounted	B-12J	216	.074			2.87	3.76	6.63	8.50
5140	6′ to 10′ deep, 3/4 C.Y. excavator	B-12F	243	.066			2.55	2.79	5.34	6.95
5160	1 C.Y. excavator, truck mounted	B-12K	432	.037			1.44	2.22	3.66	4.62
5190	10′ to 14′ deep, 3/4 C.Y. excavator	B-12F	216	.074			2.87	3.14	6.01	7.80
5210	1-1/2 C.Y. excavator	B-12B	583	.027	↓		1.06	1.74	2.80	3.53

31 23 16.13 Excavating, Trench

	31 23 16.13 Excavating, Trench	Crew	Daily Output	Labor-Hours	Unit	Material	2010 Bare Costs Labor	Equipment	Total	Total Incl O&P
5250	14' to 20' deep, 1 C.Y. excavator	B-12A	346	.046	B.C.Y.		1.79	2.32	4.11	5.25
5300	20' to 24' deep, 1 C.Y. excavator	"	311	.051			1.99	2.58	4.57	5.85
5352	4' to 6' deep, 1/2 C.Y. excavator w/trench box	B-13H	205	.078			3.02	4.52	7.54	9.55
5354	5/8 C.Y. excavator	"	257	.062			2.41	3.60	6.01	7.60
5356	3/4 C.Y. excavator	B-13G	308	.052			2.01	2.57	4.58	5.90
5362	6' to 10' deep, 3/4 C.Y. excavator w/trench box		231	.069			2.68	3.42	6.10	7.85
5370	10' to 14' deep, 3/4 C.Y. excavator w/trench box		205	.078			3.02	3.86	6.88	8.85
5374	1-1/2 C.Y. excavator	B-13E	554	.029			1.12	2.04	3.16	3.94
5382	14' to 20' deep, 1 C.Y. excavator w/trench box	B-13D	329	.049			1.88	2.78	4.66	5.90
5392	20' to 24' deep, 1 C.Y. excavator w/trench box	"	295	.054			2.10	3.10	5.20	6.60
6020	Sand & gravel with no sheeting or dewatering included									
6050	1' to 4' deep, 3/8 C.Y. excavator	B-11C	165	.097	B.C.Y.		3.69	2.05	5.74	7.85
6060	1/2 C.Y. excavator	B-11M	220	.073			2.77	1.76	4.53	6.15
6080	4' to 6' deep, 1/2 C.Y. excavator	"	220	.073			2.77	1.76	4.53	6.15
6090	5/8 C.Y. excavator	B-12Q	275	.058			2.25	2.05	4.30	5.70
6100	3/4 C.Y. excavator	B-12F	330	.048			1.88	2.05	3.93	5.10
6130	1/2 C.Y. excavator, truck mounted	B-12J	220	.073			2.82	3.69	6.51	8.35
6140	6' to 10' deep, 3/4 C.Y. excavator	B-12F	248	.065			2.50	2.73	5.23	6.80
6160	1 C.Y. excavator, truck mounted	B-12K	440	.036			1.41	2.18	3.59	4.54
6190	10' to 14' deep, 3/4 C.Y. excavator	B-12F	220	.073			2.82	3.08	5.90	7.65
6210	1-1/2 C.Y. excavator	B-12B	594	.027			1.04	1.71	2.75	3.46
6250	14' to 20' deep, 1 C.Y. excavator	B-12A	352	.045			1.76	2.28	4.04	5.15
6300	20' to 24' deep, 1 C.Y. excavator	"	317	.050			1.96	2.53	4.49	5.75
6352	4' to 6' deep, 1/2 C.Y. excavator w/trench box	B-13H	209	.077			2.97	4.43	7.40	9.40
6354	5/8 C.Y. excavator	"	261	.061			2.38	3.55	5.93	7.50
6356	3/4 C.Y. excavator	B-13G	314	.051			1.97	2.52	4.49	5.75
6362	6' to 10' deep, 3/4 C.Y. excavator w/trench box		236	.068			2.63	3.35	5.98	7.65
6370	10' to 14' deep, 3/4 C.Y. excavator w/trench box		209	.077			2.97	3.79	6.76	8.65
6374	1-1/2 C.Y. excavator	B-13E	564	.028			1.10	2	3.10	3.87
6382	14' to 20' deep, 1 C.Y. excavator w/trench box	B-13D	334	.048			1.86	2.74	4.60	5.80
6392	20' to 24' deep, 1 C.Y. excavator w/trench box	"	301	.053			2.06	3.04	5.10	6.45
7020	Dense hard clay with no sheeting or dewatering included									
7050	1' to 4' deep, 3/8 C.Y. excavator	B-11C	132	.121	B.C.Y.		4.61	2.56	7.17	9.80
7060	1/2 C.Y. excavator	B-11M	176	.091			3.46	2.20	5.66	7.65
7080	4' to 6' deep, 1/2 C.Y. excavator	"	176	.091			3.46	2.20	5.66	7.65
7090	5/8 C.Y. excavator	B-12Q	220	.073			2.82	2.57	5.39	7.10
7100	3/4 C.Y. excavator	B-12F	264	.061			2.35	2.57	4.92	6.40
7130	1/2 C.Y. oxcavator, truck mounted	B-12J	176	.091			3.52	4.62	8.14	10.45
7140	6' to 10' deep, 3/4 C.Y. excavator	B-12F	198	.081			3.13	3.42	6.55	8.50
7160	1 C.Y. excavator, truck mounted	B-12K	352	.045			1.76	2.72	4.48	5.65
7190	10' to 14' deep, 3/4 C.Y. excavator	B-12F	176	.091			3.52	3.85	7.37	9.60
7210	1-1/2 C.Y. excavator	B-12B	475	.034			1.31	2.14	3.45	4.33
7250	14' to 20' deep, 1 C.Y. excavator	B-12A	282	.057			2.20	2.84	5.04	6.45
7300	20' to 24' deep, 1 C.Y. excavator	"	254	.063			2.44	3.16	5.60	7.15

31 23 16.14 Excavating, Utility Trench

	31 23 16.14 Excavating, Utility Trench	Crew	Daily Output	Labor-Hours	Unit	Material	Labor	Equipment	Total	Total Incl O&P
0010	**EXCAVATING, UTILITY TRENCH**									
0011	Common earth									
0050	Trenching with chain trencher, 12 H.P., operator walking									
0100	4" wide trench, 12" deep	B-53	800	.010	L.F.		.41	.08	.49	.71
0150	18" deep		750	.011			.44	.09	.53	.76
0200	24" deep		700	.011			.47	.09	.56	.81
0300	6" wide trench, 12" deep		650	.012			.51	.10	.61	.87

31 23 Excavation and Fill

31 23 16 – Excavation

31 23 16.14 Excavating, Utility Trench		Crew	Daily Output	Labor-Hours	Unit	Material	2010 Bare Costs Labor	Equipment	Total	Total Incl O&P
0350	18" deep	B-53	600	.013	L.F.		.55	.11	.66	.94
0400	24" deep		550	.015			.60	.12	.72	1.03
0450	36" deep		450	.018			.73	.15	.88	1.26
0600	8" wide trench, 12" deep		475	.017			.70	.14	.84	1.19
0650	18" deep		400	.020			.83	.17	1	1.42
0700	24" deep		350	.023			.94	.19	1.13	1.62
0750	36" deep	↓	300	.027	↓		1.10	.22	1.32	1.89
1000	Backfill by hand including compaction, add									
1050	4" wide trench, 12" deep	A-1G	800	.010	L.F.		.33	.06	.39	.58
1100	18" deep		530	.015			.50	.10	.60	.88
1150	24" deep		400	.020			.66	.13	.79	1.16
1300	6" wide trench, 12" deep		540	.015			.49	.10	.59	.87
1350	18" deep		405	.020			.65	.13	.78	1.15
1400	24" deep		270	.030			.98	.19	1.17	1.72
1450	36" deep		180	.044			1.47	.29	1.76	2.59
1600	8" wide trench, 12" deep		400	.020			.66	.13	.79	1.16
1650	18" deep		265	.030			1	.20	1.20	1.76
1700	24" deep		200	.040			1.32	.26	1.58	2.32
1750	36" deep	↓	135	.059	↓		1.96	.38	2.34	3.45
2000	Chain trencher, 40 H.P. operator riding									
2050	6" wide trench and backfill, 12" deep	B-54	1200	.007	L.F.		.28	.28	.56	.72
2100	18" deep		1000	.008			.33	.33	.66	.86
2150	24" deep		975	.008			.34	.34	.68	.89
2200	36" deep		900	.009			.37	.37	.74	.96
2250	48" deep		750	.011			.44	.45	.89	1.15
2300	60" deep		650	.012			.51	.51	1.02	1.33
2400	8" wide trench and backfill, 12" deep		1000	.008			.33	.33	.66	.86
2450	18" deep		950	.008			.35	.35	.70	.91
2500	24" deep		900	.009			.37	.37	.74	.96
2550	36" deep		800	.010			.41	.42	.83	1.08
2600	48" deep		650	.012			.51	.51	1.02	1.33
2700	12" wide trench and backfill, 12" deep		975	.008			.34	.34	.68	.89
2750	18" deep		860	.009			.38	.39	.77	1.01
2800	24" deep		800	.010			.41	.42	.83	1.08
2850	36" deep		725	.011			.46	.46	.92	1.19
3000	16" wide trench and backfill, 12" deep		835	.010			.40	.40	.80	1.03
3050	18" deep		750	.011			.44	.45	.89	1.15
3100	24" deep	↓	700	.011	↓		.47	.48	.95	1.24
3200	Compaction with vibratory plate, add								35%	35%
5100	Hand excavate and trim for pipe bells after trench excavation									
5200	8" pipe	1 Clab	155	.052	L.F.		1.71		1.71	2.63
5300	18" pipe	"	130	.062	"		2.04		2.04	3.14

31 23 16.16 Structural Excavation for Minor Structures		Crew	Daily Output	Labor-Hours	Unit	Material	Labor	Equipment	Total	Total Incl O&P
0010	**STRUCTURAL EXCAVATION FOR MINOR STRUCTURES**									
0015	Hand, pits to 6' deep, sandy soil	1 Clab	8	1	B.C.Y.		33		33	51
0100	Heavy soil or clay		4	2			66		66	102
0300	Pits 6' to 12' deep, sandy soil		5	1.600			53		53	81.50
0500	Heavy soil or clay		3	2.667			88.50		88.50	136
0700	Pits 12' to 18' deep, sandy soil		4	2			66		66	102
0900	Heavy soil or clay	↓	2	4	↓		132		132	204
1500	For wet or muck hand excavation, add to above				%				50%	50%

31 23 16 – Excavation

31 23 16.30 Drilling and Blasting Rock		Crew	Daily Output	Labor-Hours	Unit	Material	2010 Bare Costs Labor	Equipment	Total	Total Incl O&P
0010	**DRILLING AND BLASTING ROCK**									
0020	Rock, open face, under 1500 C.Y.	B-47	225	.107	B.C.Y.	2.81	3.89	6.60	13.30	16.30
0100	Over 1500 C.Y.		300	.080		2.81	2.92	4.94	10.67	13
0200	Areas where blasting mats are required, under 1500 C.Y.		175	.137		2.81	5	8.45	16.26	20
0250	Over 1500 C.Y.		250	.096		2.81	3.50	5.90	12.21	14.95
2200	Trenches, up to 1500 C.Y.		22	1.091		8.15	40	67.50	115.65	143
2300	Over 1500 C.Y.		26	.923		8.15	33.50	57	98.65	123

31 23 19 – Dewatering

31 23 19.20 Dewatering Systems		Crew	Daily Output	Labor-Hours	Unit	Material	Labor	Equipment	Total	Total Incl O&P
0010	**DEWATERING SYSTEMS**									
0020	Excavate drainage trench, 2' wide, 2' deep	B-11C	90	.178	C.Y.		6.75	3.75	10.50	14.40
0100	2' wide, 3' deep, with backhoe loader	"	135	.119			4.51	2.50	7.01	9.60
0200	Excavate sump pits by hand, light soil	1 Clab	7.10	1.127			37.50		37.50	57.50
0300	Heavy soil	"	3.50	2.286			75.50		75.50	117
0500	Pumping 8 hr., attended 2 hrs. per day, including 20 L.F.									
0550	of suction hose & 100 L.F. discharge hose									
0600	2" diaphragm pump used for 8 hours	B-10H	4	3	Day		119	17.95	136.95	200
0650	4" diaphragm pump used for 8 hours	B-10I	4	3			119	25	144	207
0800	8 hrs. attended, 2" diaphragm pump	B-10H	1	12			475	72	547	800
0900	3" centrifugal pump	B-10J	1	12			475	81	556	810
1000	4" diaphragm pump	B-10I	1	12			475	99	574	830
1100	6" centrifugal pump	B-10K	1	12			475	365	840	1,125
1300	CMP, incl. excavation 3' deep, 12" diameter	B-6	115	.209	L.F.	13.25	7.50	2.94	23.69	29
1400	18" diameter		100	.240	"	16.50	8.60	3.38	28.48	35
1600	Sump hole construction, incl. excavation and gravel, pit		1250	.019	C.F.	.98	.69	.27	1.94	2.42
1700	With 12" gravel collar, 12" pipe, corrugated, 16 ga.		70	.343	L.F.	23	12.30	4.83	40.13	49.50
1800	15" pipe, corrugated, 16 ga.		55	.436		29.50	15.65	6.15	51.30	63.50
1900	18" pipe, corrugated, 16 ga.		50	.480		34.50	17.20	6.75	58.45	71.50
2000	24" pipe, corrugated, 14 ga.		40	.600		41.50	21.50	8.45	71.45	88
2200	Wood lining, up to 4' x 4', add		300	.080	SFCA	17	2.87	1.13	21	24.50
9950	See Div. 31 23 19.40 for wellpoints									
9960	See Div. 31 23 19.30 for deep well systems									

31 23 19.30 Wells		Crew	Daily Output	Labor-Hours	Unit	Material	Labor	Equipment	Total	Total Incl O&P
0010	**WELLS**									
0011	For dewatering 10' to 20' deep, 2' diameter									
0020	with steel casing, minimum	B-6	165	.145	V.L.F.	40	5.20	2.05	47.25	54
0050	Average		98	.245		40	8.75	3.45	52.20	61
0100	Maximum		49	.490		40	17.55	6.90	64.45	78.50
0300	For dewatering pumps see 01 54 33 in Reference Section									
0500	For domestic water wells, see Div. 33 21 13.10									

31 23 19.40 Wellpoints		Crew	Daily Output	Labor-Hours	Unit	Material	Labor	Equipment	Total	Total Incl O&P
0010	**WELLPOINTS** R312319-90									
0011	For equipment rental, see 01 54 33 in Reference Section									
0100	Installation and removal of single stage system									
0110	Labor only, .75 labor-hours per L.F., minimum	1 Clab	10.70	.748	LF Hdr		25		25	38
0200	2.0 labor-hours per L.F., maximum	"	4	2	"		66		66	102
0400	Pump operation, 4 @ 6 hr. shifts									
0410	Per 24 hour day	4 Eqlt	1.27	25.197	Day		1,050		1,050	1,550
0500	Per 168 hour week, 160 hr. straight, 8 hr. double time		.18	177	Week		7,350		7,350	11,000
0550	Per 4.3 week month		.04	800	Month		33,000		33,000	49,500
0600	Complete installation, operation, equipment rental, fuel &									

31 23 Excavation and Fill

31 23 19 – Dewatering

31 23 19.40 Wellpoints		Crew	Daily Output	Labor-Hours	Unit	Material	2010 Bare Costs Labor	Equipment	Total	Total Incl O&P
0610	removal of system with 2″ wellpoints 5′ O.C.									
0700	100′ long header, 6″ diameter, first month	4 Eqlt	3.23	9.907	LF Hdr	157	410		567	785
0800	Thereafter, per month		4.13	7.748		125	320		445	620
1000	200′ long header, 8″ diameter, first month		6	5.333		142	220		362	485
1100	Thereafter, per month		8.39	3.814		70.50	158		228.50	315
1300	500′ long header, 8″ diameter, first month		10.63	3.010		55	124		179	247
1400	Thereafter, per month		20.91	1.530		39	63		102	138
1600	1,000′ long header, 10″ diameter, first month		11.62	2.754		47	114		161	222
1700	Thereafter, per month	↓	41.81	.765	↓	23.50	31.50		55	73.50
1900	Note: above figures include pumping 168 hrs. per week									
1910	and include the pump operator and one stand-by pump.									

31 23 23 – Fill

31 23 23.13 Backfill

		Crew	Daily Output	Labor-Hours	Unit	Material	Labor	Equipment	Total	Total Incl O&P
0010	**BACKFILL** G1030–805									
0015	By hand, no compaction, light soil	1 Clab	14	.571	L.C.Y.		18.90		18.90	29
0100	Heavy soil		11	.727	″		24		24	37
0300	Compaction in 6″ layers, hand tamp, add to above	↓	20.60	.388	E.C.Y.		12.85		12.85	19.85
0400	Roller compaction operator walking, add	B-10A	100	.120			4.76	1.50	6.26	8.85
0500	Air tamp, add	B-9D	190	.211			7.05	1.35	8.40	12.40
0600	Vibrating plate, add	A-1D	60	.133			4.41	.57	4.98	7.45
0800	Compaction in 12″ layers, hand tamp, add to above	1 Clab	34	.235			7.80		7.80	12
0900	Roller compaction operator walking, add	B-10A	150	.080			3.17	1	4.17	5.90
1000	Air tamp, add	B-9	285	.140			4.70	.79	5.49	8.10
1100	Vibrating plate, add	A-1E	90	.089	↓		2.94	.48	3.42	5.05
1300	Dozer backfilling, bulk, up to 300′ haul, no compaction	B-10B	1200	.010	L.C.Y.		.40	.99	1.39	1.69
1400	Air tamped, add	B-11B	80	.200	E.C.Y.		7.45	3.67	11.12	15.35
1600	Compacting backfill, 6″ to 12″ lifts, vibrating roller	B-10C	800	.015			.60	2.03	2.63	3.14
1700	Sheepsfoot roller	B-10D	750	.016	↓		.63	2.24	2.87	3.43
1900	Dozer backfilling, trench, up to 300′ haul, no compaction	B-10B	900	.013	L.C.Y.		.53	1.32	1.85	2.26
2000	Air tamped, add	B-11B	80	.200	E.C.Y.		7.45	3.67	11.12	15.35
2200	Compacting backfill, 6″ to 12″ lifts, vibrating roller	B-10C	700	.017			.68	2.32	3	3.59
2300	Sheepsfoot roller	B-10D	650	.018	↓		.73	2.59	3.32	3.95

31 23 23.16 Fill By Borrow and Utility Bedding

		Crew	Daily Output	Labor-Hours	Unit	Material	Labor	Equipment	Total	Total Incl O&P
0010	**FILL BY BORROW AND UTILITY BEDDING**									
0049	Utility bedding, for pipe & conduit, not incl. compaction									
0050	Crushed or screened bank run gravel	B-6	150	.160	L.C.Y.	25	5.75	2.25	33	38.50
0100	Crushed stone 3/4″ to 1/2″		150	.160		35	5.75	2.25	43	49.50
0200	Sand, dead or bank	↓	150	.160	↓	10.55	5.75	2.25	18.55	23
0500	Compacting bedding in trench	A-1D	90	.089	E.C.Y.		2.94	.38	3.32	4.96
0600	If material source exceeds 2 miles, add for extra mileage.									
0610	See Div. 31 23 23 .20 for hauling mileage add.									

31 23 23.17 General Fill

		Crew	Daily Output	Labor-Hours	Unit	Material	Labor	Equipment	Total	Total Incl O&P
0010	**GENERAL FILL**									
0011	Spread dumped material, no compaction									
0020	By dozer, no compaction	B-10B	1000	.012	L.C.Y.		.48	1.19	1.67	2.03
0100	By hand	1 Clab	12	.667	″		22		22	34
0500	Gravel fill, compacted, under floor slabs, 4″ deep	B-37	10000	.005	S.F.	.25	.17	.02	.44	.56
0600	6″ deep		8600	.006		.38	.19	.02	.59	.74
0700	9″ deep		7200	.007		.63	.23	.02	.88	1.08
0800	12″ deep		6000	.008	↓	.89	.28	.03	1.20	1.43
1000	Alternate pricing method, 4″ deep		120	.400	E.C.Y.	19	13.90	1.27	34.17	44
1100	6″ deep	↓	160	.300	↓	19	10.45	.95	30.40	38

31 23 23.17 General Fill		Crew	Daily Output	Labor-Hours	Unit	2010 Bare Costs Material	Labor	Equipment	Total	Total Incl O&P
1200	9" deep	B-37	200	.240	E.C.Y.	19	8.35	.76	28.11	34.50
1300	12" deep	↓	220	.218	↓	19	7.60	.69	27.29	33.50
1400	Granular fill				L.C.Y.	12.65			12.65	13.90

31 23 23.20 Hauling		Crew	Daily Output	Labor-Hours	Unit	Material	Labor	Equipment	Total	Total Incl O&P
0010	**HAULING**									
0011	Excavated or borrow, loose cubic yards									
0012	no loading equipment, including hauling,waiting, loading/dumping									
0013	time per cycle (wait, load, travel, unload or dump & return)									
0014	8 CY truck, 15 MPH ave, cycle 0.5 miles, 10 min. wait/Ld./Uld.	B-34A	320	.025	L.C.Y.		.83	1.26	2.09	2.64
0016	cycle 1 mile		272	.029			.97	1.48	2.45	3.11
0018	cycle 2 miles		208	.038			1.27	1.93	3.20	4.06
0020	cycle 4 miles		144	.056			1.84	2.79	4.63	5.90
0022	cycle 6 miles		112	.071			2.37	3.59	5.96	7.55
0024	cycle 8 miles		88	.091			3.01	4.57	7.58	9.60
0026	20 MPH ave,cycle 0.5 mile		336	.024			.79	1.20	1.99	2.52
0028	cycle 1 mile		296	.027			.90	1.36	2.26	2.86
0030	cycle 2 miles		240	.033			1.10	1.67	2.77	3.52
0032	cycle 4 miles		176	.045			1.51	2.28	3.79	4.81
0034	cycle 6 miles		136	.059			1.95	2.95	4.90	6.20
0036	cycle 8 miles		112	.071			2.37	3.59	5.96	7.55
0044	25 MPH ave, cycle 4 miles		192	.042			1.38	2.09	3.47	4.41
0046	cycle 6 miles		160	.050			1.66	2.51	4.17	5.30
0048	cycle 8 miles		128	.063			2.07	3.14	5.21	6.60
0050	30 MPH ave, cycle 4 miles		216	.037			1.23	1.86	3.09	3.92
0052	cycle 6 miles		176	.045			1.51	2.28	3.79	4.81
0054	cycle 8 miles		144	.056			1.84	2.79	4.63	5.90
0114	15 MPH ave, cycle 0.5 mile, 15 min. wait/Ld./Uld.		224	.036			1.18	1.79	2.97	3.78
0116	cycle 1 mile		200	.040			1.33	2.01	3.34	4.23
0118	cycle 2 miles		168	.048			1.58	2.39	3.97	5.05
0120	cycle 4 miles		120	.067			2.21	3.35	5.56	7.05
0122	cycle 6 miles		96	.083			2.76	4.19	6.95	8.80
0124	cycle 8 miles		80	.100			3.32	5	8.32	10.60
0126	20 MPH ave, cycle 0.5 mile		232	.034			1.14	1.73	2.87	3.65
0128	cycle 1 mile		208	.038			1.27	1.93	3.20	4.06
0130	cycle 2 miles		184	.043			1.44	2.18	3.62	4.60
0132	cycle 4 miles		144	.056			1.84	2.79	4.63	5.90
0134	cycle 6 miles		112	.071			2.37	3.59	5.96	7.55
0136	cycle 8 miles		96	.083			2.76	4.19	6.95	8.80
0144	25 MPH ave, cycle 4 miles		152	.053			1.74	2.64	4.38	5.55
0146	cycle 6 miles		128	.063			2.07	3.14	5.21	6.60
0148	cycle 8 miles		112	.071			2.37	3.59	5.96	7.55
0150	30 MPH ave, cycle 4 miles		168	.048			1.58	2.39	3.97	5.05
0152	cycle 6 miles		144	.056			1.84	2.79	4.63	5.90
0154	cycle 8 miles		120	.067			2.21	3.35	5.56	7.05
0214	15 MPH ave, cycle 0.5 mile, 20 min wait/Ld./Uld.		176	.045			1.51	2.28	3.79	4.81
0216	cycle 1 mile		160	.050			1.66	2.51	4.17	5.30
0218	cycle 2 miles		136	.059			1.95	2.95	4.90	6.20
0220	cycle 4 miles		104	.077			2.55	3.86	6.41	8.15
0222	cycle 6 miles		88	.091			3.01	4.57	7.58	9.60
0224	cycle 8 miles		72	.111			3.68	5.60	9.28	11.75
0226	20 MPH ave, cycle 0.5 mile		176	.045			1.51	2.28	3.79	4.81
0228	cycle 1 mile	↓	168	.048	↓		1.58	2.39	3.97	5.05

31 23 23.20 Hauling		Crew	Daily Output	Labor-Hours	Unit	Material	2010 Bare Costs Labor	Equipment	Total	Total Incl O&P
0230	cycle 2 miles	B-34A	144	.056	L.C.Y.		1.84	2.79	4.63	5.90
0232	cycle 4 miles		120	.067			2.21	3.35	5.56	7.05
0234	cycle 6 miles		96	.083			2.76	4.19	6.95	8.80
0236	cycle 8 miles		88	.091			3.01	4.57	7.58	9.60
0244	25 MPH ave, cycle 4 miles		128	.063			2.07	3.14	5.21	6.60
0246	cycle 6 miles		112	.071			2.37	3.59	5.96	7.55
0248	cycle 8 miles		96	.083			2.76	4.19	6.95	8.80
0250	30 MPH ave, cycle 4 miles		136	.059			1.95	2.95	4.90	6.20
0252	cycle 6 miles		120	.067			2.21	3.35	5.56	7.05
0254	cycle 8 miles		104	.077			2.55	3.86	6.41	8.15
0314	15 MPH ave, cycle 0.5 mile, 25 min wait/Ld./Uld.		144	.056			1.84	2.79	4.63	5.90
0316	cycle 1 mile		128	.063			2.07	3.14	5.21	6.60
0318	cycle 2 miles		112	.071			2.37	3.59	5.96	7.55
0320	cycle 4 miles		96	.083			2.76	4.19	6.95	8.80
0322	cycle 6 miles		80	.100			3.32	5	8.32	10.60
0324	cycle 8 miles		64	.125			4.14	6.30	10.44	13.20
0326	20 MPH ave, cycle 0.5 mile		144	.056			1.84	2.79	4.63	5.90
0328	cycle 1 mile		136	.059			1.95	2.95	4.90	6.20
0330	cycle 2 miles		120	.067			2.21	3.35	5.56	7.05
0332	cycle 4 miles		104	.077			2.55	3.86	6.41	8.15
0334	cycle 6 miles		88	.091			3.01	4.57	7.58	9.60
0336	cycle 8 miles		80	.100			3.32	5	8.32	10.60
0344	25 MPH ave, cycle 4 miles		112	.071			2.37	3.59	5.96	7.55
0346	cycle 6 miles		96	.083			2.76	4.19	6.95	8.80
0348	cycle 8 miles		88	.091			3.01	4.57	7.58	9.60
0350	30 MPH ave, cycle 4 miles		112	.071			2.37	3.59	5.96	7.55
0352	cycle 6 miles		104	.077			2.55	3.86	6.41	8.15
0354	cycle 8 miles		96	.083			2.76	4.19	6.95	8.80
0414	15 MPH ave, cycle 0.5 mile, 30 min wait/Ld./Uld.		120	.067			2.21	3.35	5.56	7.05
0416	cycle 1 mile		112	.071			2.37	3.59	5.96	7.55
0418	cycle 2 miles		96	.083			2.76	4.19	6.95	8.80
0420	cycle 4 miles		80	.100			3.32	5	8.32	10.60
0422	cycle 6 miles		72	.111			3.68	5.60	9.28	11.75
0424	cycle 8 miles		64	.125			4.14	6.30	10.44	13.20
0426	20 MPH ave, cycle 0.5 mile		120	.067			2.21	3.35	5.56	7.05
0428	cycle 1 mile		112	.071			2.37	3.59	5.96	7.55
0430	cycle 2 miles		104	.077			2.55	3.86	6.41	8.15
0432	cycle 4 miles		88	.091			3.01	4.57	7.58	9.60
0434	cycle 6 miles		80	.100			3.32	5	8.32	10.60
0436	cycle 8 miles		72	.111			3.68	5.60	9.28	11.75
0444	25 MPH ave, cycle 4 miles		96	.083			2.76	4.19	6.95	8.80
0446	cycle 6 miles		88	.091			3.01	4.57	7.58	9.60
0448	cycle 8 miles		80	.100			3.32	5	8.32	10.60
0450	30 MPH ave, cycle 4 miles		96	.083			2.76	4.19	6.95	8.80
0452	cycle 6 miles		88	.091			3.01	4.57	7.58	9.60
0454	cycle 8 miles		80	.100			3.32	5	8.32	10.60
0514	15 MPH ave, cycle 0.5 mile, 35 min wait/Ld./Uld.		104	.077			2.55	3.86	6.41	8.15
0516	cycle 1 mile		96	.083			2.76	4.19	6.95	8.80
0518	cycle 2 miles		88	.091			3.01	4.57	7.58	9.60
0520	cycle 4 miles		72	.111			3.68	5.60	9.28	11.75
0522	cycle 6 miles		64	.125			4.14	6.30	10.44	13.20
0524	cycle 8 miles		56	.143			4.74	7.20	11.94	15.10
0526	20 MPH ave, cycle 0.5 mile		104	.077			2.55	3.86	6.41	8.15

31 23 23.20 Hauling		Crew	Daily Output	Labor-Hours	Unit	Material	2010 Bare Costs Labor	Equipment	Total	Total Incl O&P
0528	cycle 1 mile	B-34A	96	.083	L.C.Y.		2.76	4.19	6.95	8.80
0530	cycle 2 miles		96	.083			2.76	4.19	6.95	8.80
0532	cycle 4 miles		80	.100			3.32	5	8.32	10.60
0534	cycle 6 miles		72	.111			3.68	5.60	9.28	11.75
0536	cycle 8 miles		64	.125			4.14	6.30	10.44	13.20
0544	25 MPH ave, cycle 4 miles		88	.091			3.01	4.57	7.58	9.60
0546	cycle 6 miles		80	.100			3.32	5	8.32	10.60
0548	cycle 8 miles		72	.111			3.68	5.60	9.28	11.75
0550	30 MPH ave, cycle 4 miles		88	.091			3.01	4.57	7.58	9.60
0552	cycle 6 miles		80	.100			3.32	5	8.32	10.60
0554	cycle 8 miles		72	.111			3.68	5.60	9.28	11.75
1014	12 CY truck, cycle 0.5 mile, 15 MPH ave, 15 min. wait/Ld./Uld.	B-34B	336	.024			.79	1.99	2.78	3.39
1016	cycle 1 mile		300	.027			.88	2.23	3.11	3.80
1018	cycle 2 miles		252	.032			1.05	2.65	3.70	4.53
1020	cycle 4 miles		180	.044			1.47	3.71	5.18	6.35
1022	cycle 6 miles		144	.056			1.84	4.64	6.48	7.90
1024	cycle 8 miles		120	.067			2.21	5.55	7.76	9.45
1025	cycle 10 miles		96	.083			2.76	6.95	9.71	11.85
1026	20 MPH ave, cycle 0.5 mile		348	.023			.76	1.92	2.68	3.27
1028	cycle 1 mile		312	.026			.85	2.14	2.99	3.65
1030	cycle 2 miles		276	.029			.96	2.42	3.38	4.13
1032	cycle 4 miles		216	.037			1.23	3.09	4.32	5.25
1034	cycle 6 miles		168	.048			1.58	3.98	5.56	6.80
1036	cycle 8 miles		144	.056			1.84	4.64	6.48	7.90
1038	cycle 10 miles		120	.067			2.21	5.55	7.76	9.45
1040	25 MPH ave, cycle 4 miles		228	.035			1.16	2.93	4.09	4.99
1042	cycle 6 miles		192	.042			1.38	3.48	4.86	5.95
1044	cycle 8 miles		168	.048			1.58	3.98	5.56	6.80
1046	cycle 10 miles		144	.056			1.84	4.64	6.48	7.90
1050	30 MPH ave, cycle 4 miles		252	.032			1.05	2.65	3.70	4.53
1052	cycle 6 miles		216	.037			1.23	3.09	4.32	5.25
1054	cycle 8 miles		180	.044			1.47	3.71	5.18	6.35
1056	cycle 10 miles		156	.051			1.70	4.28	5.98	7.30
1060	35 MPH ave, cycle 4 miles		264	.030			1	2.53	3.53	4.31
1062	cycle 6 miles		228	.035			1.16	2.93	4.09	4.99
1064	cycle 8 miles		204	.039			1.30	3.27	4.57	5.60
1066	cycle 10 miles		180	.044			1.47	3.71	5.18	6.35
1068	cycle 20 miles		120	.067			2.21	5.55	7.76	9.45
1069	cycle 30 miles		84	.095			3.16	7.95	11.11	13.55
1070	cycle 40 miles		72	.111			3.68	9.30	12.98	15.80
1072	40 MPH ave, cycle 6 miles		240	.033			1.10	2.78	3.88	4.74
1074	cycle 8 miles		216	.037			1.23	3.09	4.32	5.25
1076	cycle 10 miles		192	.042			1.38	3.48	4.86	5.95
1078	cycle 20 miles		120	.067			2.21	5.55	7.76	9.45
1080	cycle 30 miles		96	.083			2.76	6.95	9.71	11.85
1082	cycle 40 miles		72	.111			3.68	9.30	12.98	15.80
1084	cycle 50 miles		60	.133			4.42	11.15	15.57	19
1094	45 MPH ave, cycle 8 miles		216	.037			1.23	3.09	4.32	5.25
1096	cycle 10 miles		204	.039			1.30	3.27	4.57	5.60
1098	cycle 20 miles		132	.061			2.01	5.05	7.06	8.60
1100	cycle 30 miles		108	.074			2.46	6.20	8.66	10.55
1102	cycle 40 miles		84	.095			3.16	7.95	11.11	13.55
1104	cycle 50 miles		72	.111			3.68	9.30	12.98	15.80

31 23 23.20 Hauling		Crew	Daily Output	Labor-Hours	Unit	Material	2010 Bare Costs Labor	Equipment	Total	Total Incl O&P
1106	50 MPH ave, cycle 10 miles	B-34B	216	.037	L.C.Y.		1.23	3.09	4.32	5.25
1108	cycle 20 miles		144	.056			1.84	4.64	6.48	7.90
1110	cycle 30 miles		108	.074			2.46	6.20	8.66	10.55
1112	cycle 40 miles		84	.095			3.16	7.95	11.11	13.55
1114	cycle 50 miles		72	.111			3.68	9.30	12.98	15.80
1214	15 MPH ave, cycle 0.5 mile, 20 min. wait/Ld./Uld.		264	.030			1	2.53	3.53	4.31
1216	cycle 1 mile		240	.033			1.10	2.78	3.88	4.74
1218	cycle 2 miles		204	.039			1.30	3.27	4.57	5.60
1220	cycle 4 miles		156	.051			1.70	4.28	5.98	7.30
1222	cycle 6 miles		132	.061			2.01	5.05	7.06	8.60
1224	cycle 8 miles		108	.074			2.46	6.20	8.66	10.55
1225	cycle 10 miles		96	.083			2.76	6.95	9.71	11.85
1226	20 MPH ave, cycle 0.5 mile		264	.030			1	2.53	3.53	4.31
1228	cycle 1 mile		252	.032			1.05	2.65	3.70	4.53
1230	cycle 2 miles		216	.037			1.23	3.09	4.32	5.25
1232	cycle 4 miles		180	.044			1.47	3.71	5.18	6.35
1234	cycle 6 miles		144	.056			1.84	4.64	6.48	7.90
1236	cycle 8 miles		132	.061			2.01	5.05	7.06	8.60
1238	cycle 10 miles		108	.074			2.46	6.20	8.66	10.55
1240	25 MPH ave, cycle 4 miles		192	.042			1.38	3.48	4.86	5.95
1242	cycle 6 miles		168	.048			1.58	3.98	5.56	6.80
1244	cycle 8 miles		144	.056			1.84	4.64	6.48	7.90
1246	cycle 10 miles		132	.061			2.01	5.05	7.06	8.60
1250	30 MPH ave, cycle 4 miles		204	.039			1.30	3.27	4.57	5.60
1252	cycle 6 miles		180	.044			1.47	3.71	5.18	6.35
1254	cycle 8 miles		156	.051			1.70	4.28	5.98	7.30
1256	cycle 10 miles		144	.056			1.84	4.64	6.48	7.90
1260	35 MPH ave, cycle 4 miles		216	.037			1.23	3.09	4.32	5.25
1262	cycle 6 miles		192	.042			1.38	3.48	4.86	5.95
1264	cycle 8 miles		168	.048			1.58	3.98	5.56	6.80
1266	cycle 10 miles		156	.051			1.70	4.28	5.98	7.30
1268	cycle 20 miles		108	.074			2.46	6.20	8.66	10.55
1269	cycle 30 miles		72	.111			3.68	9.30	12.98	15.80
1270	cycle 40 miles		60	.133			4.42	11.15	15.57	19
1272	40 MPH ave, cycle 6 miles		192	.042			1.38	3.48	4.86	5.95
1274	cycle 8 miles		180	.044			1.47	3.71	5.18	6.35
1276	cycle 10 miles		156	.051			1.70	4.28	5.98	7.30
1278	cycle 20 miles		108	.074			2.46	6.20	8.66	10.55
1280	cycle 30 miles		84	.095			3.16	7.95	11.11	13.55
1282	cycle 40 miles		72	.111			3.68	9.30	12.98	15.80
1284	cycle 50 miles		60	.133			4.42	11.15	15.57	19
1294	45 MPH ave, cycle 8 miles		180	.044			1.47	3.71	5.18	6.35
1296	cycle 10 miles		168	.048			1.58	3.98	5.56	6.80
1298	cycle 20 miles		120	.067			2.21	5.55	7.76	9.45
1300	cycle 30 miles		96	.083			2.76	6.95	9.71	11.85
1302	cycle 40 miles		72	.111			3.68	9.30	12.98	15.80
1304	cycle 50 miles		60	.133			4.42	11.15	15.57	19
1306	50 MPH ave, cycle 10 miles		180	.044			1.47	3.71	5.18	6.35
1308	cycle 20 miles		132	.061			2.01	5.05	7.06	8.60
1310	cycle 30 miles		96	.083			2.76	6.95	9.71	11.85
1312	cycle 40 miles		84	.095			3.16	7.95	11.11	13.55
1314	cycle 50 miles		72	.111			3.68	9.30	12.98	15.80
1414	15 MPH ave, cycle 0.5 mile, 25 min. wait/Ld./Uld.		204	.039			1.30	3.27	4.57	5.60

31 23 23.20 Hauling		Crew	Daily Output	Labor-Hours	Unit	Material	2010 Bare Costs Labor	Equipment	Total	Total Incl O&P
1416	cycle 1 mile	B-34B	192	.042	L.C.Y.		1.38	3.48	4.86	5.95
1418	cycle 2 miles		168	.048			1.58	3.98	5.56	6.80
1420	cycle 4 miles		132	.061			2.01	5.05	7.06	8.60
1422	cycle 6 miles		120	.067			2.21	5.55	7.76	9.45
1424	cycle 8 miles		96	.083			2.76	6.95	9.71	11.85
1425	cycle 10 miles		84	.095			3.16	7.95	11.11	13.55
1426	20 MPH ave, cycle 0.5 mile		216	.037			1.23	3.09	4.32	5.25
1428	cycle 1 mile		204	.039			1.30	3.27	4.57	5.60
1430	cycle 2 miles		180	.044			1.47	3.71	5.18	6.35
1432	cycle 4 miles		156	.051			1.70	4.28	5.98	7.30
1434	cycle 6 miles		132	.061			2.01	5.05	7.06	8.60
1436	cycle 8 miles		120	.067			2.21	5.55	7.76	9.45
1438	cycle 10 miles		96	.083			2.76	6.95	9.71	11.85
1440	25 MPH ave, cycle 4 miles		168	.048			1.58	3.98	5.56	6.80
1442	cycle 6 miles		144	.056			1.84	4.64	6.48	7.90
1444	cycle 8 miles		132	.061			2.01	5.05	7.06	8.60
1446	cycle 10 miles		108	.074			2.46	6.20	8.66	10.55
1450	30 MPH ave, cycle 4 miles		168	.048			1.58	3.98	5.56	6.80
1452	cycle 6 miles		156	.051			1.70	4.28	5.98	7.30
1454	cycle 8 miles		132	.061			2.01	5.05	7.06	8.60
1456	cycle 10 miles		120	.067			2.21	5.55	7.76	9.45
1460	35 MPH ave, cycle 4 miles		180	.044			1.47	3.71	5.18	6.35
1462	cycle 6 miles		156	.051			1.70	4.28	5.98	7.30
1464	cycle 8 miles		144	.056			1.84	4.64	6.48	7.90
1466	cycle 10 miles		132	.061			2.01	5.05	7.06	8.60
1468	cycle 20 miles		96	.083			2.76	6.95	9.71	11.85
1469	cycle 30 miles		72	.111			3.68	9.30	12.98	15.80
1470	cycle 40 miles		60	.133			4.42	11.15	15.57	19
1472	40 MPH ave, cycle 6 miles		168	.048			1.58	3.98	5.56	6.80
1474	cycle 8 miles		156	.051			1.70	4.28	5.98	7.30
1476	cycle 10 miles		144	.056			1.84	4.64	6.48	7.90
1478	cycle 20 miles		96	.083			2.76	6.95	9.71	11.85
1480	cycle 30 miles		84	.095			3.16	7.95	11.11	13.55
1482	cycle 40 miles		60	.133			4.42	11.15	15.57	19
1484	cycle 50 miles		60	.133			4.42	11.15	15.57	19
1494	45 MPH ave, cycle 8 miles		156	.051			1.70	4.28	5.98	7.30
1496	cycle 10 miles		144	.056			1.84	4.64	6.48	7.90
1498	cycle 20 miles		108	.074			2.46	6.20	8.66	10.55
1500	cycle 30 miles		84	.095			3.16	7.95	11.11	13.55
1502	cycle 40 miles		72	.111			3.68	9.30	12.98	15.80
1504	cycle 50 miles		60	.133			4.42	11.15	15.57	19
1506	50 MPH ave, cycle 10 miles		156	.051			1.70	4.28	5.98	7.30
1508	cycle 20 miles		120	.067			2.21	5.55	7.76	9.45
1510	cycle 30 miles		96	.083			2.76	6.95	9.71	11.85
1512	cycle 40 miles		72	.111			3.68	9.30	12.98	15.80
1514	cycle 50 miles		60	.133			4.42	11.15	15.57	19
1614	15 MPH, cycle 0.5 mile, 30 min. wait/Ld./Uld.		180	.044			1.47	3.71	5.18	6.35
1616	cycle 1 mile		168	.048			1.58	3.98	5.56	6.80
1618	cycle 2 miles		144	.056			1.84	4.64	6.48	7.90
1620	cycle 4 miles		120	.067			2.21	5.55	7.76	9.45
1622	cycle 6 miles		108	.074			2.46	6.20	8.66	10.55
1624	cycle 8 miles		84	.095			3.16	7.95	11.11	13.55
1625	cycle 10 miles		84	.095			3.16	7.95	11.11	13.55

31 23 23.20 Hauling		Crew	Daily Output	Labor-Hours	Unit	Material	2010 Bare Costs Labor	Equipment	Total	Total Incl O&P
1626	20 MPH ave, cycle 0.5 mile	B-34B	180	.044	L.C.Y.		1.47	3.71	5.18	6.35
1628	cycle 1 mile		168	.048			1.58	3.98	5.56	6.80
1630	cycle 2 miles		156	.051			1.70	4.28	5.98	7.30
1632	cycle 4 miles		132	.061			2.01	5.05	7.06	8.60
1634	cycle 6 miles		120	.067			2.21	5.55	7.76	9.45
1636	cycle 8 miles		108	.074			2.46	6.20	8.66	10.55
1638	cycle 10 miles		96	.083			2.76	6.95	9.71	11.85
1640	25 MPH ave, cycle 4 miles		144	.056			1.84	4.64	6.48	7.90
1642	cycle 6 miles		132	.061			2.01	5.05	7.06	8.60
1644	cycle 8 miles		108	.074			2.46	6.20	8.66	10.55
1646	cycle 10 miles		108	.074			2.46	6.20	8.66	10.55
1650	30 MPH ave, cycle 4 miles		144	.056			1.84	4.64	6.48	7.90
1652	cycle 6 miles		132	.061			2.01	5.05	7.06	8.60
1654	cycle 8 miles		120	.067			2.21	5.55	7.76	9.45
1656	cycle 10 miles		108	.074			2.46	6.20	8.66	10.55
1660	35 MPH ave, cycle 4 miles		156	.051			1.70	4.28	5.98	7.30
1662	cycle 6 miles		144	.056			1.84	4.64	6.48	7.90
1664	cycle 8 miles		132	.061			2.01	5.05	7.06	8.60
1666	cycle 10 miles		120	.067			2.21	5.55	7.76	9.45
1668	cycle 20 miles		84	.095			3.16	7.95	11.11	13.55
1669	cycle 30 miles		72	.111			3.68	9.30	12.98	15.80
1670	cycle 40 miles		60	.133			4.42	11.15	15.57	19
1672	40 MPH, cycle 6 miles		144	.056			1.84	4.64	6.48	7.90
1674	cycle 8 miles		132	.061			2.01	5.05	7.06	8.60
1676	cycle 10 miles		120	.067			2.21	5.55	7.76	9.45
1678	cycle 20 miles		96	.083			2.76	6.95	9.71	11.85
1680	cycle 30 miles		72	.111			3.68	9.30	12.98	15.80
1682	cycle 40 miles		60	.133			4.42	11.15	15.57	19
1684	cycle 50 miles		48	.167			5.55	13.90	19.45	24
1694	45 MPH ave, cycle 8 miles		144	.056			1.84	4.64	6.48	7.90
1696	cycle 10 miles		132	.061			2.01	5.05	7.06	8.60
1698	cycle 20 miles		96	.083			2.76	6.95	9.71	11.85
1700	cycle 30 miles		84	.095			3.16	7.95	11.11	13.55
1702	cycle 40 miles		60	.133			4.42	11.15	15.57	19
1704	cycle 50 miles		60	.133			4.42	11.15	15.57	19
1706	50 MPH ave, cycle 10 miles		132	.061			2.01	5.05	7.06	8.60
1708	cycle 20 miles		108	.074			2.46	6.20	8.66	10.55
1710	cycle 30 miles		84	.095			3.16	7.95	11.11	13.55
1712	cycle 40 miles		72	.111			3.68	9.30	12.98	15.80
1714	cycle 50 miles		60	.133			4.42	11.15	15.57	19
2000	Hauling, 8 CY truck, small project cost per hour	B-34A	8	1	Hr.		33	50	83	106
2100	12 CY Truck	B-34B	8	1			33	83.50	116.50	143
2150	16.5 CY Truck	B-34C	8	1			33	89	122	149
2175	18 CY 8 wheel Truck	B-34I	8	1			33	105	138	166
2200	20 CY Truck	B-34D	8	1			33	91	124	151
2300	Grading at dump, or embankment if required, by dozer	B-10B	1000	.012	L.C.Y.		.48	1.19	1.67	2.03
2310	Spotter at fill or cut, if required	1 Clab	8	1	Hr.		33		33	51
9014	18 CY truck, 8 wheels,15 min. wait/Ld./Uld.,15 MPH, cycle 0.5 mi.	B-34I	504	.016	L.C.Y.		.53	1.66	2.19	2.63
9016	cycle 1 mile		450	.018			.59	1.86	2.45	2.95
9018	cycle 2 miles		378	.021			.70	2.22	2.92	3.51
9020	cycle 4 miles		270	.030			.98	3.11	4.09	4.92
9022	cycle 6 miles		216	.037			1.23	3.88	5.11	6.15
9024	cycle 8 miles		180	.044			1.47	4.66	6.13	7.35

31 23 23.20 Hauling		Crew	Daily Output	Labor-Hours	Unit	2010 Bare Costs Material	Labor	Equipment	Total	Total Incl O&P
9025	cycle 10 miles	B-34I	144	.056	L.C.Y.		1.84	5.80	7.64	9.20
9026	20 MPH ave, cycle 0.5 mile		522	.015			.51	1.61	2.12	2.54
9028	cycle 1 mile		468	.017			.57	1.79	2.36	2.83
9030	cycle 2 miles		414	.019			.64	2.02	2.66	3.21
9032	cycle 4 miles		324	.025			.82	2.59	3.41	4.10
9034	cycle 6 miles		252	.032			1.05	3.33	4.38	5.25
9036	cycle 8 miles		216	.037			1.23	3.88	5.11	6.15
9038	cycle 10 miles		180	.044			1.47	4.66	6.13	7.35
9040	25 MPH ave, cycle 4 miles		342	.023			.78	2.45	3.23	3.88
9042	cycle 6 miles		288	.028			.92	2.91	3.83	4.60
9044	cycle 8 miles		252	.032			1.05	3.33	4.38	5.25
9046	cycle 10 miles		216	.037			1.23	3.88	5.11	6.15
9050	30 MPH ave, cycle 4 miles		378	.021			.70	2.22	2.92	3.51
9052	cycle 6 miles		324	.025			.82	2.59	3.41	4.10
9054	cycle 8 miles		270	.030			.98	3.11	4.09	4.92
9056	cycle 10 miles		234	.034			1.13	3.58	4.71	5.65
9060	35 MPH ave, cycle 4 miles		396	.020			.67	2.12	2.79	3.35
9062	cycle 6 miles		342	.023			.78	2.45	3.23	3.88
9064	cycle 8 miles		288	.028			.92	2.91	3.83	4.60
9066	cycle 10 miles		270	.030			.98	3.11	4.09	4.92
9068	cycle 20 miles		162	.049			1.64	5.20	6.84	8.20
9070	cycle 30 miles		126	.063			2.10	6.65	8.75	10.50
9072	cycle 40 miles		90	.089			2.95	9.30	12.25	14.75
9074	40 MPH ave, cycle 6 miles		360	.022			.74	2.33	3.07	3.68
9076	cycle 8 miles		324	.025			.82	2.59	3.41	4.10
9078	cycle 10 miles		288	.028			.92	2.91	3.83	4.60
9080	cycle 20 miles		180	.044			1.47	4.66	6.13	7.35
9082	cycle 30 miles		144	.056			1.84	5.80	7.64	9.20
9084	cycle 40 miles		108	.074			2.46	7.75	10.21	12.30
9086	cycle 50 miles		90	.089			2.95	9.30	12.25	14.75
9094	45 MPH ave, cycle 8 miles		324	.025			.82	2.59	3.41	4.10
9096	cycle 10 miles		306	.026			.87	2.74	3.61	4.33
9098	cycle 20 miles		198	.040			1.34	4.23	5.57	6.70
9100	cycle 30 miles		144	.056			1.84	5.80	7.64	9.20
9102	cycle 40 miles		126	.063			2.10	6.65	8.75	10.50
9104	cycle 50 miles		108	.074			2.46	7.75	10.21	12.30
9106	50 MPH ave, cycle 10 miles		324	.025			.82	2.59	3.41	4.10
9108	cycle 20 miles		216	.037			1.23	3.88	5.11	6.15
9110	cycle 30 miles		162	.049			1.64	5.20	6.84	8.20
9112	cycle 40 miles		126	.063			2.10	6.65	8.75	10.50
9114	cycle 50 miles		108	.074			2.46	7.75	10.21	12.30
9214	20 min. wait/Ld./Uld.,15 MPH, cycle 0.5 mi.		396	.020			.67	2.12	2.79	3.35
9216	cycle 1 mile		360	.022			.74	2.33	3.07	3.68
9218	cycle 2 miles		306	.026			.87	2.74	3.61	4.33
9220	cycle 4 miles		234	.034			1.13	3.58	4.71	5.65
9222	cycle 6 miles		198	.040			1.34	4.23	5.57	6.70
9224	cycle 8 miles		162	.049			1.64	5.20	6.84	8.20
9225	cycle 10 miles		144	.056			1.84	5.80	7.64	9.20
9226	20 MPH ave, cycle 0.5 mile		396	.020			.67	2.12	2.79	3.35
9228	cycle 1 mile		378	.021			.70	2.22	2.92	3.51
9230	cycle 2 miles		324	.025			.82	2.59	3.41	4.10
9232	cycle 4 miles		270	.030			.98	3.11	4.09	4.92
9234	cycle 6 miles		216	.037			1.23	3.88	5.11	6.15

31 23 23.20 Hauling		Crew	Daily Output	Labor-Hours	Unit	Material	2010 Bare Costs Labor	Equipment	Total	Total Incl O&P
9236	cycle 8 miles	B-34I	198	.040	L.C.Y.		1.34	4.23	5.57	6.70
9238	cycle 10 miles		162	.049			1.64	5.20	6.84	8.20
9240	25 MPH ave, cycle 4 miles		288	.028			.92	2.91	3.83	4.60
9242	cycle 6 miles		252	.032			1.05	3.33	4.38	5.25
9244	cycle 8 miles		216	.037			1.23	3.88	5.11	6.15
9246	cycle 10 miles		198	.040			1.34	4.23	5.57	6.70
9250	30 MPH ave, cycle 4 miles		306	.026			.87	2.74	3.61	4.33
9252	cycle 6 miles		270	.030			.98	3.11	4.09	4.92
9254	cycle 8 miles		234	.034			1.13	3.58	4.71	5.65
9256	cycle 10 miles		216	.037			1.23	3.88	5.11	6.15
9260	35 MPH ave, cycle 4 miles		324	.025			.82	2.59	3.41	4.10
9262	cycle 6 miles		288	.028			.92	2.91	3.83	4.60
9264	cycle 8 miles		252	.032			1.05	3.33	4.38	5.25
9266	cycle 10 miles		234	.034			1.13	3.58	4.71	5.65
9268	cycle 20 miles		162	.049			1.64	5.20	6.84	8.20
9270	cycle 30 miles		108	.074			2.46	7.75	10.21	12.30
9272	cycle 40 miles		90	.089			2.95	9.30	12.25	14.75
9274	40 MPH ave, cycle 6 miles		288	.028			.92	2.91	3.83	4.60
9276	cycle 8 miles		270	.030			.98	3.11	4.09	4.92
9278	cycle 10 miles		234	.034			1.13	3.58	4.71	5.65
9280	cycle 20 miles		162	.049			1.64	5.20	6.84	8.20
9282	cycle 30 miles		126	.063			2.10	6.65	8.75	10.50
9284	cycle 40 miles		108	.074			2.46	7.75	10.21	12.30
9286	cycle 50 miles		90	.089			2.95	9.30	12.25	14.75
9294	45 MPH ave, cycle 8 miles		270	.030			.98	3.11	4.09	4.92
9296	cycle 10 miles		252	.032			1.05	3.33	4.38	5.25
9298	cycle 20 miles		180	.044			1.47	4.66	6.13	7.35
9300	cycle 30 miles		144	.056			1.84	5.80	7.64	9.20
9302	cycle 40 miles		108	.074			2.46	7.75	10.21	12.30
9304	cycle 50 miles		90	.089			2.95	9.30	12.25	14.75
9306	50 MPH ave, cycle 10 miles		270	.030			.98	3.11	4.09	4.92
9308	cycle 20 miles		198	.040			1.34	4.23	5.57	6.70
9310	cycle 30 miles		144	.056			1.84	5.80	7.64	9.20
9312	cycle 40 miles		126	.063			2.10	6.65	8.75	10.50
9314	cycle 50 miles		108	.074			2.46	7.75	10.21	12.30
9414	25 min. wait/Ld./Uld.,15 MPH, cycle 0.5 mi.		306	.026			.87	2.74	3.61	4.33
9416	cycle 1 mile		288	.028			.92	2.91	3.83	4.60
9418	cycle 2 miles		252	.032			1.05	3.33	4.38	5.25
9420	cycle 4 miles		198	.040			1.34	4.23	5.57	6.70
9422	cycle 6 miles		180	.044			1.47	4.66	6.13	7.35
9424	cycle 8 miles		144	.056			1.84	5.80	7.64	9.20
9425	cycle 10 miles		126	.063			2.10	6.65	8.75	10.50
9426	20 MPH ave, cycle 0.5 mile		324	.025			.82	2.59	3.41	4.10
9428	cycle 1 mile		306	.026			.87	2.74	3.61	4.33
9430	cycle 2 miles		270	.030			.98	3.11	4.09	4.92
9432	cycle 4 miles		234	.034			1.13	3.58	4.71	5.65
9434	cycle 6 miles		198	.040			1.34	4.23	5.57	6.70
9436	cycle 8 miles		180	.044			1.47	4.66	6.13	7.35
9438	cycle 10 miles		144	.056			1.84	5.80	7.64	9.20
9440	25 MPH ave, cycle 4 miles		252	.032			1.05	3.33	4.38	5.25
9442	cycle 6 miles		216	.037			1.23	3.88	5.11	6.15
9444	cycle 8 miles		198	.040			1.34	4.23	5.57	6.70
9446	cycle 10 miles		180	.044			1.47	4.66	6.13	7.35

31 23 23.20 Hauling		Crew	Daily Output	Labor-Hours	Unit	Material	2010 Bare Costs Labor	Equipment	Total	Total Incl O&P
9450	30 MPH ave, cycle 4 miles	B-34I	252	.032	L.C.Y.		1.05	3.33	4.38	5.25
9452	cycle 6 miles		234	.034			1.13	3.58	4.71	5.65
9454	cycle 8 miles		198	.040			1.34	4.23	5.57	6.70
9456	cycle 10 miles		180	.044			1.47	4.66	6.13	7.35
9460	35 MPH ave, cycle 4 miles		270	.030			.98	3.11	4.09	4.92
9462	cycle 6 miles		234	.034			1.13	3.58	4.71	5.65
9464	cycle 8 miles		216	.037			1.23	3.88	5.11	6.15
9466	cycle 10 miles		198	.040			1.34	4.23	5.57	6.70
9468	cycle 20 miles		144	.056			1.84	5.80	7.64	9.20
9470	cycle 30 miles		108	.074			2.46	7.75	10.21	12.30
9472	cycle 40 miles		90	.089			2.95	9.30	12.25	14.75
9474	40 MPH ave, cycle 6 miles		252	.032			1.05	3.33	4.38	5.25
9476	cycle 8 miles		234	.034			1.13	3.58	4.71	5.65
9478	cycle 10 miles		216	.037			1.23	3.88	5.11	6.15
9480	cycle 20 miles		144	.056			1.84	5.80	7.64	9.20
9482	cycle 30 miles		126	.063			2.10	6.65	8.75	10.50
9484	cycle 40 miles		90	.089			2.95	9.30	12.25	14.75
9486	cycle 50 miles		90	.089			2.95	9.30	12.25	14.75
9494	45 MPH ave, cycle 8 miles		234	.034			1.13	3.58	4.71	5.65
9496	cycle 10 miles		216	.037			1.23	3.88	5.11	6.15
9498	cycle 20 miles		162	.049			1.64	5.20	6.84	8.20
9500	cycle 30 miles		126	.063			2.10	6.65	8.75	10.50
9502	cycle 40 miles		108	.074			2.46	7.75	10.21	12.30
9504	cycle 50 miles		90	.089			2.95	9.30	12.25	14.75
9506	50 MPH ave, cycle 10 miles		234	.034			1.13	3.58	4.71	5.65
9508	cycle 20 miles		180	.044			1.47	4.66	6.13	7.35
9510	cycle 30 miles		144	.056			1.84	5.80	7.64	9.20
9512	cycle 40 miles		108	.074			2.46	7.75	10.21	12.30
9514	cycle 50 miles		90	.089			2.95	9.30	12.25	14.75
9614	25 min. wait/Ld./Uld.,15 MPH, cycle 0.5 mi.		270	.030			.98	3.11	4.09	4.92
9616	cycle 1 mile		252	.032			1.05	3.33	4.38	5.25
9618	cycle 2 miles		216	.037			1.23	3.88	5.11	6.15
9620	cycle 4 miles		180	.044			1.47	4.66	6.13	7.35
9622	cycle 6 miles		162	.049			1.64	5.20	6.84	8.20
9624	cycle 8 miles		126	.063			2.10	6.65	8.75	10.50
9625	cycle 10 miles		126	.063			2.10	6.65	8.75	10.50
9626	20 MPH ave, cycle 0.5 mile		270	.030			.98	3.11	4.09	4.92
9628	cycle 1 mile		252	.032			1.05	3.33	4.38	5.25
9630	cycle 2 miles		234	.034			1.13	3.58	4.71	5.65
9632	cycle 4 miles		198	.040			1.34	4.23	5.57	6.70
9634	cycle 6 miles		180	.044			1.47	4.66	6.13	7.35
9636	cycle 8 miles		162	.049			1.64	5.20	6.84	8.20
9638	cycle 10 miles		144	.056			1.84	5.80	7.64	9.20
9640	25 MPH ave, cycle 4 miles		216	.037			1.23	3.88	5.11	6.15
9642	cycle 6 miles		198	.040			1.34	4.23	5.57	6.70
9644	cycle 8 miles		180	.044			1.47	4.66	6.13	7.35
9646	cycle 10 miles		162	.049			1.64	5.20	6.84	8.20
9650	30 MPH ave, cycle 4 miles		216	.037			1.23	3.88	5.11	6.15
9652	cycle 6 miles		198	.040			1.34	4.23	5.57	6.70
9654	cycle 8 miles		180	.044			1.47	4.66	6.13	7.35
9656	cycle 10 miles		162	.049			1.64	5.20	6.84	8.20
9660	35 MPH ave, cycle 4 miles		234	.034			1.13	3.58	4.71	5.65
9662	cycle 6 miles	▼	216	.037	▼		1.23	3.88	5.11	6.15

31 23 Excavation and Fill

31 23 23 – Fill

31 23 23.20 Hauling		Crew	Daily Output	Labor-Hours	Unit	Material	2010 Bare Costs Labor	Equipment	Total	Total Incl O&P
9664	cycle 8 miles	B-34I	198	.040	L.C.Y.		1.34	4.23	5.57	6.70
9666	cycle 10 miles		180	.044			1.47	4.66	6.13	7.35
9668	cycle 20 miles		126	.063			2.10	6.65	8.75	10.50
9670	cycle 30 miles		108	.074			2.46	7.75	10.21	12.30
9672	cycle 40 miles		90	.089			2.95	9.30	12.25	14.75
9674	40 MPH ave, cycle 6 miles		216	.037			1.23	3.88	5.11	6.15
9676	cycle 8 miles		198	.040			1.34	4.23	5.57	6.70
9678	cycle 10 miles		180	.044			1.47	4.66	6.13	7.35
9680	cycle 20 miles		144	.056			1.84	5.80	7.64	9.20
9682	cycle 30 miles		108	.074			2.46	7.75	10.21	12.30
9684	cycle 40 miles		90	.089			2.95	9.30	12.25	14.75
9686	cycle 50 miles		72	.111			3.68	11.65	15.33	18.40
9694	45 MPH ave, cycle 8 miles		216	.037			1.23	3.88	5.11	6.15
9696	cycle 10 miles		198	.040			1.34	4.23	5.57	6.70
9698	cycle 20 miles		144	.056			1.84	5.80	7.64	9.20
9700	cycle 30 miles		126	.063			2.10	6.65	8.75	10.50
9702	cycle 40 miles		108	.074			2.46	7.75	10.21	12.30
9704	cycle 50 miles		90	.089			2.95	9.30	12.25	14.75
9706	50 MPH ave, cycle 10 miles		198	.040			1.34	4.23	5.57	6.70
9708	cycle 20 miles		162	.049			1.64	5.20	6.84	8.20
9710	cycle 30 miles		126	.063			2.10	6.65	8.75	10.50
9712	cycle 40 miles		108	.074			2.46	7.75	10.21	12.30
9714	cycle 50 miles	↓	90	.089	↓		2.95	9.30	12.25	14.75

31 41 Shoring

31 41 16 – Sheet Piling

31 41 16.10 Sheet Piling Systems		Crew	Daily Output	Labor-Hours	Unit	Material	Labor	Equipment	Total	Total Incl O&P
0010	**SHEET PILING SYSTEMS**									
0020	Sheet piling steel, not incl. wales, 22 psf, 15' excav., left in place	B-40	10.81	5.920	Ton	1,250	245	355	1,850	2,150
0100	Drive, extract & salvage R314116-45		6	10.667		515	440	640	1,595	1,950
0300	20' deep excavation, 27 psf, left in place		12.95	4.942		1,250	204	296	1,750	2,025
0400	Drive, extract & salvage		6.55	9.771	↓	515	405	585	1,505	1,850
1200	15' deep excavation, 22 psf, left in place		983	.065	S.F.	14.65	2.69	3.90	21.24	24.50
1300	Drive, extract & salvage		545	.117		5.75	4.85	7.05	17.65	21.50
1500	20' deep excavation, 27 psf, left in place		960	.067		18.40	2.76	3.99	25.15	28.50
1600	Drive, extract & salvage	↓	485	.132	↓	7.50	5.45	7.90	20.85	25.50
2100	Rent steel sheet piling and wales, first month				Ton	268			268	295
2200	Per added month					27			27	29.50
2300	Rental piling left in place, add to rental					1,000			1,000	1,100
2500	Wales, connections & struts, 2/3 salvage				↓	280			280	310
3900	Wood, solid sheeting, incl. wales, braces and spacers, R314116-40									
3910	drive, extract & salvage, 8' deep excavation	B-31	330	.121	S.F.	1.16	4.27	.64	6.07	8.60
4000	10' deep, 50 S.F./hr. in & 150 S.F./hr. out		300	.133		1.20	4.69	.71	6.60	9.35
4100	12' deep, 45 S.F./hr. in & 135 S.F./hr. out		270	.148		1.23	5.20	.78	7.21	10.25
4200	14' deep, 42 S.F./hr. in & 126 S.F./hr. out		250	.160		1.27	5.65	.85	7.77	11
4300	16' deep, 40 S.F./hr. in & 120 S.F./hr. out		240	.167		1.31	5.85	.88	8.04	11.45
4400	18' deep, 38 S.F./hr. in & 114 S.F./hr. out		230	.174		1.35	6.10	.92	8.37	11.95
4520	Left in place, 8' deep, 55 S.F./hr.		440	.091		2.09	3.20	.48	5.77	7.75
4540	10' deep, 50 S.F./hr.		400	.100		2.20	3.52	.53	6.25	8.45
4560	12' deep, 45 S.F./hr.		360	.111		2.32	3.91	.59	6.82	9.25
4565	14' deep, 42 S.F./hr.	↓	335	.119	↓	2.46	4.20	.63	7.29	9.90

31 41 Shoring

31 41 16 – Sheet Piling

31 41 16.10 Sheet Piling Systems		Crew	Daily Output	Labor-Hours	Unit	Material	2010 Bare Costs Labor	Equipment	Total	Total Incl O&P
4570	16′ deep, 40 S.F./hr.	B-31	320	.125	S.F.	2.61	4.40	.66	7.67	10.40
4700	Alternate pricing, left in place, 8′ deep		1.76	22.727	M.B.F.	470	800	120	1,390	1,875
4800	Drive, extract and salvage, 8′ deep	↓	1.32	30.303	"	420	1,075	160	1,655	2,275

31 52 Cofferdams

31 52 16 – Timber Cofferdams

31 52 16.10 Cofferdams

		Crew	Daily Output	Labor-Hours	Unit	Material	Labor	Equipment	Total	Total Incl O&P
0010	**COFFERDAMS**									
0011	Incl. mobilization and temporary sheeting									
0080	Soldier beams & lagging H piles with 3″ wood sheeting									
0090	horizontal between piles, including removal of wales & braces									
0100	No hydrostatic head, 15′ deep, 1 line of braces, minimum	B-50	545	.206	S.F.	5.30	8.10	4.48	17.88	23.50
0200	Maximum		495	.226		5.90	8.95	4.94	19.79	26
1300	No hydrostatic head, left in place, 15′ dp., 1 line of braces, min.		635	.176		7.05	6.95	3.85	17.85	23
1400	Maximum	↓	575	.195		7.55	7.70	4.25	19.50	25
2350	Lagging only, 3″ thick wood between piles 8′ O.C., minimum	B-46	400	.120		1.18	4.44	.10	5.72	8.40
2370	Maximum		250	.192		1.77	7.10	.16	9.03	13.30
2400	Open sheeting no bracing, for trenches to 10′ deep, min.		1736	.028		.53	1.02	.02	1.57	2.22
2450	Maximum	↓	1510	.032		.59	1.18	.03	1.80	2.54
2500	Tie-back method, add to open sheeting, add, minimum								20%	20%
2550	Maximum				↓				60%	60%
2700	Tie-backs only, based on tie-backs total length, minimum	B-46	86.80	.553	L.F.	13.20	20.50	.45	34.15	47.50
2750	Maximum		38.50	1.247	"	23	46	1.02	70.02	99.50
3500	Tie-backs only, typical average, 25′ long		2	24	Ea.	580	890	19.70	1,489.70	2,050
3600	35′ long	↓	1.58	30.380	"	775	1,125	25	1,925	2,650

Division Notes

		CREW	DAILY OUTPUT	LABOR-HOURS	UNIT	2010 BARE COSTS MAT.	LABOR	EQUIP.	TOTAL	TOTAL INCL O&P

Estimating Tips

32 01 00 Operations and Maintenance of Exterior Improvements

- Recycling of asphalt pavement is becoming very popular and is an alternative to removal and replacement of asphalt pavement. It can be a good value engineering proposal if removed pavement can be recycled either at the site or another site that is reasonably close to the project site.

32 10 00 Bases, Ballasts, and Paving

- When estimating paving, keep in mind the project schedule. If an asphaltic paving project is in a colder climate and runs through to the spring, consider placing the base course in the autumn and then topping it in the spring, just prior to completion. This could save considerable costs in spring repair. Keep in mind that prices for asphalt and concrete are generally higher in the cold seasons.

32 90 00 Planting

- The timing of planting and guarantee specifications often dictate the costs for establishing tree and shrub growth and a stand of grass or ground cover. Establish the work performance schedule to coincide with the local planting season. Maintenance and growth guarantees can add from 20%–100% to the total landscaping cost. The cost to replace trees and shrubs can be as high as 5% of the total cost, depending on the planting zone, soil conditions, and time of year.

Reference Numbers

Reference numbers are shown in shaded boxes at the beginning of some major classifications. These numbers refer to related items in the Reference Section. The reference information may be an estimating procedure, an alternate pricing method, or technical information.

Note: Not all subdivisions listed here necessarily appear in this publication.

32 12 16.13 Plant-Mix Asphalt Paving

	32 12 16.13 Plant-Mix Asphalt Paving	Crew	Daily Output	Labor-Hours	Unit	2010 Bare Costs Material	Labor	Equipment	Total	Total Incl O&P
0010	**PLANT-MIX ASPHALT PAVING**									
0020	And large paved areas with no hauling included									
0025	See Div. 31 23 23.20 for hauling costs									
0080	Binder course, 1-1/2" thick	B-25	7725	.011	S.Y.	5.30	.41	.33	6.04	6.85
0120	2" thick		6345	.014		7.05	.50	.40	7.95	9
0130	2-1/2" thick		5620	.016		8.85	.56	.46	9.87	11.05
0160	3" thick		4905	.018		10.60	.65	.52	11.77	13.20
0170	3-1/2 " thick		4520	.019		12.35	.70	.57	13.62	15.30
0200	4" thick		4140	.021		14.15	.76	.62	15.53	17.40
0300	Wearing course, 1" thick	B-25B	10575	.009		3.51	.33	.27	4.11	4.66
0340	1-1/2" thick		7725	.012		5.90	.45	.36	6.71	7.60
0380	2" thick		6345	.015		7.90	.55	.44	8.89	10.05
0420	2-1/2" thick		5480	.018		9.75	.64	.51	10.90	12.30
0460	3" thick		4900	.020		11.65	.72	.57	12.94	14.50
0470	3-1/2" thick		4520	.021		13.65	.78	.62	15.05	16.85
0480	4 " thick		4140	.023		15.60	.85	.68	17.13	19.20
0500	Open graded friction course	B-25C	5000	.010		3.30	.35	.45	4.10	4.66
0800	Alternate method of figuring paving costs									
0810	Binder course, 1-1/2" thick	B-25	630	.140	Ton	65	5	4.07	74.07	83.50
0811	2" thick		690	.128		65	4.59	3.72	73.31	82.50
0812	3" thick		800	.110		65	3.96	3.21	72.17	81
0813	4" thick		900	.098		65	3.52	2.85	71.37	80
0850	Wearing course, 1" thick	B-25B	575	.167		68	6.10	4.87	78.97	89.50
0851	1-1/2" thick		630	.152		68	5.55	4.45	78	88.50
0852	2" thick		690	.139		68	5.10	4.06	77.16	87
0853	2-1/2" thick		765	.125		68	4.59	3.66	76.25	86
0854	3" thick		800	.120		68	4.39	3.50	75.89	85.50
1000	Pavement replacement over trench, 2" thick	B-37	90	.533	S.Y.	7.30	18.55	1.69	27.54	38.50
1050	4" thick		70	.686		14.45	24	2.17	40.62	55
1080	6" thick		55	.873		23	30.50	2.77	56.27	75

32 12 16.14 Asphaltic Concrete Paving

	32 12 16.14 Asphaltic Concrete Paving	Crew	Daily Output	Labor-Hours	Unit	Material	Labor	Equipment	Total	Total Incl O&P
0011	**ASPHALTIC CONCRETE PAVING**, parking lots & driveways									
0015	No asphalt hauling included									
0018	Use 6.05 C. Y. per inch per MSF for hauling									
0020	6" stone base, 2" binder course, 1" topping	B-25C	9000	.005	S.F.	1.82	.20	.25	2.27	2.58
0025	2" binder course, 2" topping		9000	.005		2.24	.20	.25	2.69	3.03
0030	3" binder course, 2" topping		9000	.005		2.63	.20	.25	3.08	3.46
0035	4" binder course, 2" topping		9000	.005		3.02	.20	.25	3.47	3.89
0040	1.5" binder course, 1" topping		9000	.005		1.63	.20	.25	2.08	2.36
0042	3" binder course, 1" topping		9000	.005		2.22	.20	.25	2.67	3.01
0045	3" binder course, 3" topping		9000	.005		3.05	.20	.25	3.50	3.92
0050	4" binder course, 3" topping		9000	.005		3.44	.20	.25	3.89	4.35
0055	4" binder course, 4" topping		9000	.005		3.84	.20	.25	4.29	4.80
0300	Binder course, 1-1/2" thick		35000	.001		.59	.05	.06	.70	.80
0400	2" thick		25000	.002		.76	.07	.09	.92	1.05
0500	3" thick		15000	.003		1.18	.12	.15	1.45	1.64
0600	4" thick		10800	.004		1.55	.16	.21	1.92	2.18
0800	Sand finish course, 3/4" thick		41000	.001		.32	.04	.05	.41	.49
0900	1" thick		34000	.001		.40	.05	.07	.52	.59
1000	Fill pot holes, hot mix, 2" thick	B-16	4200	.008		.84	.26	.16	1.26	1.49
1100	4" thick		3500	.009		1.23	.31	.19	1.73	2.03
1120	6" thick		3100	.010		1.65	.35	.22	2.22	2.59

32 12 Flexible Paving

32 12 16 – Asphalt Paving

32 12 16.14 Asphaltic Concrete Paving		Crew	Daily Output	Labor-Hours	Unit	2010 Bare Costs Material	Labor	Equipment	Total	Total Incl O&P
1140	Cold patch, 2" thick	B-51	3000	.016	S.F.	.85	.53	.08	1.46	1.85
1160	4" thick		2700	.018		1.62	.59	.09	2.30	2.79
1180	6" thick		1900	.025		2.52	.84	.13	3.49	4.20

32 84 Planting Irrigation

32 84 23 – Underground Sprinklers

32 84 23.10 Sprinkler Irrigation System		Crew	Daily Output	Labor-Hours	Unit	Material	Labor	Equipment	Total	Total Incl O&P
0010	**SPRINKLER IRRIGATION SYSTEM**									
0011	For lawns									
0100	Golf course with fully automatic system	C-17	.05	1600	9 holes	99,000	69,000		168,000	215,000
0200	24' diam. head at 15' O.C incl. piping, auto oper., minimum	B-20	70	.343	Head	20	12.65		32.65	41.50
0300	Maximum		40	.600		50	22		72	89
0500	60' diam. head at 40' O.C. incl. piping, auto oper., minimum		28	.857		65	31.50		96.50	121
0600	Maximum		23	1.043		180	38.50		218.50	258
0800	Residential system, custom, 1" supply		2000	.012	S.F.	.30	.44		.74	1.01
0900	1-1/2" supply		1800	.013	"	.35	.49		.84	1.15

Division Notes

	CREW	DAILY OUTPUT	LABOR-HOURS	UNIT	2010 BARE COSTS MAT.	LABOR	EQUIP.	TOTAL	TOTAL INCL O&P

Estimating Tips

33 10 00 Water Utilities
33 30 00 Sanitary Sewerage Utilities
33 40 00 Storm Drainage Utilities

- Never assume that the water, sewer, and drainage lines will go in at the early stages of the project. Consider the site access needs before dividing the site in half with open trenches, loose pipe, and machinery obstructions. Always inspect the site to establish that the site drawings are complete. Check off all existing utilities on your drawings as you locate them. Be especially careful with underground because appurtenances are sometimes buried during regrading or repaving operations. If you find any discrepancies, mark up the site plan for further research. Differing site conditions can be very costly if discovered later in the project.
- See also Section 33 01 00 for restoration of pipe where removal/replacement may be undesirable. Use of new types of piping materials can reduce the overall project cost. Owners/design engineers should consider the installing contractor as a valuable source of current information on utility products and local conditions that could lead to significant cost savings.

Reference Numbers

Reference numbers are shown in shaded boxes at the beginning of some major classifications. These numbers refer to related items in the Reference Section. The reference information may be an estimating procedure, an alternate pricing method, or technical information.

Note: Not all subdivisions listed here necessarily appear in this publication.

Note: **Trade Service,** *in part, has been used as a reference source for some of the material prices used in Division 33.*

33 01 10 – Operation and Maintenance of Water Utilities

33 01 10.10 Corrosion Resistance		Crew	Daily Output	Labor-Hours	Unit	2010 Bare Costs Material	Labor	Equipment	Total	Total Incl O&P
0010	**CORROSION RESISTANCE**									
0012	Wrap & coat, add to pipe, 4" diameter				L.F.	2.02			2.02	2.22
0020	5" diameter					2.10			2.10	2.31
0040	6" diameter					2.10			2.10	2.31
0060	8" diameter					3.25			3.25	3.58
0080	10" diameter					4.86			4.86	5.35
0100	12" diameter					4.86			4.86	5.35
0120	14" diameter					7.10			7.10	7.85
0140	16" diameter					7.10			7.10	7.85
0160	18" diameter					7.20			7.20	7.90
0180	20" diameter					7.55			7.55	8.30
0200	24" diameter					9.20			9.20	10.15
0220	Small diameter pipe, 1" diameter, add					1.52			1.52	1.67
0240	2" diameter					1.67			1.67	1.84
0260	2-1/2" diameter					1.84			1.84	2.02
0280	3" diameter				▼	1.84			1.84	2.02
0300	Fittings, field covered, add				S.F.	9.60			9.60	10.60
0500	Coating, bituminous, per diameter inch, 1 coat, add				L.F.	.39			.39	.43
0540	3 coat					.58			.58	.64
0560	Coal tar epoxy, per diameter inch, 1 coat, add					.23			.23	.25
0600	3 coat					.48			.48	.53
1000	Polyethylene H.D. extruded, .025" thk., 1/2" diameter add					.35			.35	.39
1020	3/4" diameter					.37			.37	.41
1040	1" diameter					.42			.42	.46
1060	1-1/4" diameter					.52			.52	.57
1080	1-1/2" diameter					.54			.54	.59
1100	.030" thk., 2" diameter					.58			.58	.64
1120	2-1/2" diameter					.68			.68	.75
1140	.035" thk., 3" diameter					.85			.85	.94
1160	3-1/2" diameter					.96			.96	1.06
1180	4" diameter					1.07			1.07	1.18
1200	.040" thk, 5" diameter					1.47			1.47	1.62
1220	6" diameter					1.54			1.54	1.69
1240	8" diameter					2			2	2.20
1260	10" diameter					2.45			2.45	2.70
1280	12" diameter					2.94			2.94	3.23
1300	.060" thk., 14" diameter					4.04			4.04	4.44
1320	16" diameter					4.73			4.73	5.20
1340	18" diameter					5.60			5.60	6.15
1360	20" diameter				▼	6.30			6.30	6.90
1380	Fittings, field wrapped, add				S.F.	6.50			6.50	7.15

33 01 10.20 Pipe Repair		Crew	Daily Output	Labor-Hours	Unit	Material	Labor	Equipment	Total	Total Incl O&P
0010	**PIPE REPAIR**									
0020	Not including excavation or backfill									
0100	Clamp, stainless steel, lightweight, for steel pipe									
0110	3" long, 1/2" diameter pipe	1 Plum	34	.235	Ea.	12.65	12.25		24.90	32.50
0120	3/4" diameter pipe		32	.250		13.10	13		26.10	34
0130	1" diameter pipe		30	.267		13.80	13.90		27.70	36
0140	1-1/4" diameter pipe		28	.286		14.60	14.85		29.45	38.50
0150	1-1/2" diameter pipe		26	.308		15.25	16		31.25	41
0160	2" diameter pipe		24	.333		16.70	17.35		34.05	44.50
0170	2-1/2" diameter pipe	▼	23	.348	▼	20	18.10		38.10	49

33 01 10 – Operation and Maintenance of Water Utilities

33 01 10.20 Pipe Repair		Crew	Daily Output	Labor-Hours	Unit	Material	Labor	Equipment	Total	Total Incl O&P
						2010 Bare Costs				
0180	3" diameter pipe	1 Plum	22	.364	Ea.	20.50	18.95		39.45	51
0190	3-1/2" diameter pipe		21	.381		22.50	19.85		42.35	54.50
0200	4" diameter pipe	B-20	56	.429		24.50	15.85		40.35	51.50
0210	5" diameter pipe		53	.453		27.50	16.70		44.20	56
0220	6" diameter pipe		48	.500		28.50	18.45		46.95	60
0230	8" diameter pipe		30	.800		34	29.50		63.50	83
0240	10" diameter pipe		28	.857		63	31.50		94.50	118
0250	12" diameter pipe		24	1		72	37		109	137
0260	14" diameter pipe		22	1.091		230	40.50		270.50	315
0270	16" diameter pipe		20	1.200		247	44.50		291.50	340
0280	18" diameter pipe		18	1.333		261	49		310	365
0290	20" diameter pipe		16	1.500		276	55.50		331.50	390
0300	24" diameter pipe		14	1.714		305	63.50		368.50	435
0360	For 6" long, add					100%	40%			
0370	For 9" long, add					200%	100%			
0380	For 12" long, add					300%	150%			
0390	For 18" long, add					500%	200%			
0400	Pipe Freezing for live repairs of systems 3/8 inch to 6 inch									
0410	Note: Pipe Freezing can also be used to install a valve into a live system									
0420	Pipe Freezing each side 3/8 inch	2 Skwk	8	2	Ea.	490	85		575	670
0425	Pipe Freezing each side 3/8 inch, second location same kit		8	2		19.75	85		104.75	153
0430	Pipe Freezing each side 3/4 inch		8	2		490	85		575	670
0435	Pipe Freezing each side 3/4 inch, second location same kit		8	2		19.75	85		104.75	153
0440	Pipe Freezing each side 1 1/2 inch		6	2.667		490	114		604	715
0445	Pipe Freezing each side 1 1/2 inch , second location same kit		6	2.667		19.75	114		133.75	197
0450	Pipe Freezing each side 2 inch		6	2.667		725	114		839	975
0455	Pipe Freezing each side 2 inch, second location same kit		6	2.667		19.75	114		133.75	197
0460	Pipe Freezing each side 2 1/2 inch-3 inch		6	2.667		1,025	114		1,139	1,300
0465	Pipe Freeze each side 2 1/2-3 inch, second location same kit		6	2.667		39.50	114		153.50	219
0470	Pipe Freezing each side 4 inch		4	4		1,650	170		1,820	2,050
0475	Pipe Freezing each side 4 inch, second location same kit		4	4		64.50	170		234.50	335
0480	Pipe Freezing each side 5-6 inch		4	4		3,975	170		4,145	4,625
0485	Pipe Freezing each side 5-6 inch, second location same kit		4	4		193	170		363	475
0490	Pipe Frz extra 20 lb CO_2 cylinders (3/8 to 2 " - 1 ea, 3 " -2 ea)					216			216	238
0500	Pipe Freezing extra 50 lb CO_2 cylinders (4 " - 2 ea, 5-6 " -6 ea)					485			485	530
1000	Clamp, stainless steel, with threaded service tap									
1040	Full seal for iron, steel, PVC pipe									
1100	6" long, 2" diameter pipe	1 Plum	17	.471	Ea.	163	24.50		187.50	216
1110	2-1/2" diameter pipe		16	.500		173	26		199	229
1120	3" diameter pipe		15.60	.513		184	26.50		210.50	243
1130	3-1/2" diameter pipe		15	.533		192	28		220	253
1140	4" diameter pipe	B-20	40	.600		211	22		233	266
1150	6" diameter pipe		34	.706		244	26		270	310
1160	8" diameter pipe		21	1.143		305	42		347	400
1170	10" diameter pipe		20	1.200		335	44.50		379.50	440
1180	12" diameter pipe		17	1.412		385	52		437	500
1205	For 9" long, add					20%	45%			
1210	For 12" long, add					40%	80%			
1220	For 18" long, add					70%	110%			
1600	Clamp, stainless steel, single section									
1640	Full seal for iron, steel, PVC pipe									
1700	6" long, 2" diameter pipe	1 Plum	17	.471	Ea.	84.50	24.50		109	130
1710	2-1/2" diameter pipe		16	.500		92	26		118	140

33 01 Operation and Maintenance of Utilities

33 01 10 – Operation and Maintenance of Water Utilities

33 01 10.20 Pipe Repair		Crew	Daily Output	Labor-Hours	Unit	Material	2010 Bare Costs Labor	Equipment	Total	Total Incl O&P
1720	3" diameter pipe	1 Plum	15.60	.513	Ea.	96	26.50		122.50	146
1730	3-1/2" diameter pipe	"	15	.533		106	28		134	158
1740	4" diameter pipe	B-20	40	.600		115	22		137	161
1750	6" diameter pipe		34	.706		144	26		170	198
1760	8" diameter pipe		21	1.143		192	42		234	276
1770	10" diameter pipe		20	1.200		221	44.50		265.50	310
1780	12" diameter pipe		17	1.412		240	52		292	345
1800	For 9" long, add					40%	45%			
1810	For 12" long, add					60%	80%			
1820	For 18" long, add					120%	110%			
2000	Clamp, stainless steel, two section									
2040	Full seal, for iron, steel, PVC pipe									
2100	6" long, 4" diameter pipe	B-20	24	1	Ea.	182	37		219	257
2110	6" diameter pipe		20	1.200		210	44.50		254.50	300
2120	8" diameter pipe		13	1.846		256	68		324	385
2130	10" diameter pipe		12	2		270	74		344	410
2140	12" diameter pipe		10	2.400		380	88.50		468.50	550
2200	9" long, 4" diameter pipe		16	1.500		241	55.50		296.50	350
2210	6" diameter pipe		13	1.846		271	68		339	405
2220	8" diameter pipe		9	2.667		299	98.50		397.50	480
2230	10" diameter pipe		8	3		400	111		511	610
2240	12" diameter pipe		7	3.429		450	127		577	690
2250	14" diameter pipe		6.40	3.750		635	138		773	910
2260	16" diameter pipe		6	4		680	148		828	980
2270	18" diameter pipe		5	4.800		720	177		897	1,075
2280	20" diameter pipe		4.60	5.217		780	193		973	1,150
2290	24" diameter pipe		4	6		1,275	222		1,497	1,750
2320	For 12" long, add to 9"					15%	25%			
2330	For 18" long, add to 9"					70%	55%			
8000	For internal cleaning and inspection, see Div. 33 01 30.16									
8100	For pipe testing, see Div. 23 05 93.50									

33 01 30 – Operation and Maintenance of Sewer Utilities

33 01 30.16 TV Inspection of Sewer Pipelines		Crew	Daily Output	Labor-Hours	Unit	Material	Labor	Equipment	Total	Total Incl O&P
0010	**TV INSPECTION OF SEWER PIPELINES**									
0100	Pipe internal cleaning & inspection, cleaning, pressure pipe systems									
0120	Pig method, lengths 1000' to 10,000'									
0140	4" diameter thru 24" diameter, minimum				L.F.				3.60	4.17
0160	Maximum				"				8.75	10.13
6000	Sewage/sanitary systems									
6100	Power rodder with header & cutters									
6110	Mobilization charge, minimum				Total				435	500
6120	Mobilization charge, maximum				"				1,030	1,180
6140	Cleaning 4-12 " diameter				L.F.				3.60	4.15
6190	14-24" diameter								4.24	4.88
6240	30" diameter								6.18	7.11
6250	36" diameter								7.21	8.29
6260	48" diameter								8.24	9.48
6270	60" diameter								9.27	10.66
6280	72" diameter								10.30	11.85
9000	Inspection, television camera with video									
9060	up to 500 linear feet				Total				750	865

33 01 Operation and Maintenance of Utilities

33 01 30 – Operation and Maintenance of Sewer Utilities

33 01 30.71 Pipebursting		Crew	Daily Output	Labor-Hours	Unit	Material	Labor	Equipment	Total	Total Incl O&P
						2010 Bare Costs				
0010	**PIPEBURSTING**									
0011	300′ runs, replace with HDPE pipe									
0020	Not including excavation, backfill, shoring, or dewatering									
0100	6″ to 15″ diameter, minimum				L.F.				102	112
0200	Maximum								204	224
0300	18″ to 36″ diameter, minimum								230	253
0400	Maximum				↓				408	449
0500	Mobilize and demobilize, minimum				Job				3,000	3,300
0600	Maximum				″				30,000	33,000

33 01 30.74 HDPE Pipe Lining		Crew	Daily Output	Labor-Hours	Unit	Material	Labor	Equipment	Total	Total Incl O&P
0010	**HDPE PIPE LINING**, excludes cleaning and video inspection									
0020	Pipe relined with one pipe size smaller than original (4″ for 6 ″)									
0100	6″ diameter, original size	B-6B	600	.080	L.F.	3.65	2.70	1.08	7.43	9.40
0150	8″ diameter, original size		600	.080		10.70	2.70	1.08	14.48	17.10
0200	10″ diameter, original size		600	.080		12.35	2.70	1.08	16.13	18.90
0250	12″ diameter, original size		400	.120		19.10	4.05	1.63	24.78	29
0300	14″ diameter, original size		400	.120		27	4.05	1.63	32.68	37.50
0350	16″ diameter, original size		300	.160		32.50	5.40	2.17	40.07	46.50
0400	18″ diameter, original size	↓	300	.160	↓	42	5.40	2.17	49.57	57
1000	Pipe HDPE lining, make service line taps	B-6	4	6	Ea.	111	215	84.50	410.50	545

33 05 Common Work Results for Utilities

33 05 23 – Trenchless Utility Installation

33 05 23.19 Microtunneling		Crew	Daily Output	Labor-Hours	Unit	Material	Labor	Equipment	Total	Total Incl O&P
0010	**MICROTUNNELING**									
0011	Not including excavation, backfill, shoring,									
0020	or dewatering, average 50′/day, slurry method									
0100	24″ to 48″ outside diameter, minimum				L.F.				800	880
0110	Adverse conditions, add				%				50%	50%
1000	Rent microtunneling machine, average monthly lease				Month				90,000	99,000
1010	Operating technician				Day				560	620
1100	Mobilization and demobilization, minimum				Job				40,000	44,000
1110	Maximum				″				410,000	451,000

33 05 23.20 Horizontal Boring		Crew	Daily Output	Labor-Hours	Unit	Material	Labor	Equipment	Total	Total Incl O&P
0010	**HORIZONTAL BORING**									
0011	Casing only, 100′ minimum,									
0020	not incl. jacking pits or dewatering									
0100	Roadwork, 1/2″ thick wall, 24″ diameter casing	B-42	20	3.200	L.F.	117	119	68.50	304.50	390
0200	36″ diameter		16	4		186	149	85.50	420.50	530
0300	48″ diameter		15	4.267		273	158	91	522	650
0500	Railroad work, 24″ diameter		15	4.267		117	158	91	366	475
0600	36″ diameter		14	4.571		186	170	97.50	453.50	575
0700	48″ diameter	↓	12	5.333		273	198	114	585	735
0900	For ledge, add									20%
1000	Small diameter boring, 3″, sandy soil	B-82	900	.018		28	.66	.09	28.75	32
1040	Rocky soil	″	500	.032	↓	28	1.19	.16	29.35	33
1100	Prepare jacking pits, incl. mobilization & demobilization, minimum				Ea.				3,000	3,450
1101	Maximum				″				20,000	23,000

33 05 Common Work Results for Utilities

33 05 26 – Utility Line Signs, Markers, and Flags

33 05 26.10 Utility Accessories		Crew	Daily Output	Labor-Hours	Unit	Material	2010 Bare Costs Labor	Equipment	Total	Total Incl O&P
0010	UTILITY ACCESSORIES G1030–805									
0400	Underground tape, detectable, reinforced, alum. foil core, 2"	1 Clab	150	.053	C.L.F.	2.36	1.77		4.13	5.30
0500	6"	"	140	.057	"	7.20	1.89		9.09	10.80

33 11 Water Utility Distribution Piping

33 11 13 – Public Water Utility Distribution Piping

33 11 13.15 Water Supply, Ductile Iron Pipe

		Crew	Daily Output	Labor-Hours	Unit	Material	2010 Bare Costs Labor	Equipment	Total	Total Incl O&P
0010	WATER SUPPLY, DUCTILE IRON PIPE R331113-80									
0020	Not including excavation or backfill									
2000	Pipe, class 50 water piping, 18' lengths									
2020	Mechanical joint, 4" diameter	B-21A	200	.200	L.F.	16.55	8.25	3.28	28.08	34.50
2040	6" diameter		160	.250		19.20	10.30	4.10	33.60	41
2060	8" diameter		133.33	.300		21	12.40	4.92	38.32	47.50
2080	10" diameter		114.29	.350		28.50	14.45	5.75	48.70	60
2100	12" diameter		105.26	.380		35.50	15.70	6.25	57.45	69.50
2120	14" diameter		100	.400		45	16.50	6.55	68.05	81.50
2140	16" diameter		72.73	.550		49	22.50	9	80.50	98.50
2160	18" diameter		68.97	.580		61.50	24	9.50	95	114
2170	20" diameter		57.14	.700		72.50	29	11.45	112.95	136
2180	24" diameter		47.06	.850		92.50	35	13.95	141.45	170
3000	Tyton, push-on joint, 4" diameter		400	.100		12.40	4.13	1.64	18.17	21.50
3020	6" diameter		333.33	.120		15.85	4.95	1.97	22.77	27
3040	8" diameter		200	.200		22	8.25	3.28	33.53	40.50
3060	10" diameter		181.82	.220		31	9.10	3.61	43.71	51.50
3080	12" diameter		160	.250		33	10.30	4.10	47.40	56
3100	14" diameter		133.33	.300		36	12.40	4.92	53.32	63.50
3120	16" diameter		114.29	.350		58	14.45	5.75	78.20	92.50
3140	18" diameter		100	.400		64.50	16.50	6.55	87.55	103
3160	20" diameter		88.89	.450		71	18.55	7.40	96.95	115
3180	24" diameter		76.92	.520		84.50	21.50	8.50	114.50	134
8000	Fittings, mechanical joint									
8006	90° bend, 4" diameter	B-20A	16	2	Ea.	230	81		311	375
8020	6" diameter		12.80	2.500		315	101		416	500
8040	8" diameter		10.67	2.999		455	121		576	685
8060	10" diameter	B-21A	11.43	3.500		500	144	57.50	701.50	830
8080	12" diameter		10.53	3.799		905	157	62.50	1,124.50	1,300
8100	14" diameter		10	4		1,375	165	65.50	1,605.50	1,825
8120	16" diameter		7.27	5.502		2,050	227	90	2,367	2,700
8140	18" diameter		6.90	5.797		3,400	239	95	3,734	4,225
8160	20" diameter		5.71	7.005		4,575	289	115	4,979	5,575
8180	24" diameter		4.70	8.511		5,025	350	139	5,514	6,225
8200	Wye or tee, 4" diameter	B-20A	10.67	2.999		350	121		471	570
8220	6" diameter		8.53	3.751		465	152		617	745
8240	8" diameter		7.11	4.501		665	182		847	1,000
8260	10" diameter	B-21A	7.62	5.249		1,375	217	86	1,678	1,950
8280	12" diameter		7.02	5.698		1,750	235	93.50	2,078.50	2,375
8300	14" diameter		6.67	5.997		1,950	247	98.50	2,295.50	2,625
8320	16" diameter		4.85	8.247		2,025	340	135	2,500	2,900
8340	18" diameter		4.60	8.696		5,300	360	143	5,803	6,525
8360	20" diameter		3.81	10.499		8,275	435	172	8,882	9,950

33 11 13 – Public Water Utility Distribution Piping

33 11 13.15 Water Supply, Ductile Iron Pipe		Crew	Daily Output	Labor-Hours	Unit	2010 Bare Costs Material	Labor	Equipment	Total	Total Incl O&P
8380	24" diameter	B-21A	3.14	12.739	Ea.	9,800	525	209	10,534	11,800
8398	45° bends, 4" diameter	B-20A	16	2		180	81		261	320
8400	6" diameter	"	12.80	2.500		249	101		350	430
8410	12" diameter	B-21A	10.53	3.799		545	157	62.50	764.50	905
8420	16" diameter		7.27	5.502		1,300	227	90	1,617	1,875
8430	20" diameter		5.71	7.005		2,850	289	115	3,254	3,675
8440	24" diameter		4.70	8.511		3,925	350	139	4,414	5,000
8450	Decreaser, 6" x 4" diameter	B-20A	14.22	2.250		193	91		284	350
8460	8" x 6" diameter	"	11.64	2.749		247	111		358	440
8470	10" x 6 " diameter	B-21A	13.33	3.001		310	124	49	483	585
8480	12" x 6" diameter		12.70	3.150		385	130	51.50	566.50	680
8490	16" x 6" diameter		10	4		885	165	65.50	1,115.50	1,300
8500	20" x 6" diameter		8.42	4.751		2,175	196	78	2,449	2,775
8552	For water utility valves see Div. 33 12 16									
8600	12" diameter	B-21	3	9.333	Ea.	2,850	355	46	3,251	3,725
8610	14" diameter		2	14		3,475	530	69	4,074	4,725
8620	16" diameter		2	14		5,375	530	69	5,974	6,825
8700	Joint restraint, ductile iron mechanical joints									
8710	4" diameter	B-20A	32	1	Ea.	34	40.50		74.50	99
8720	6" diameter		25.60	1.250		41.50	50.50		92	123
8730	8" diameter		21.33	1.500		59	60.50		119.50	157
8740	10" diameter		18.28	1.751		86.50	71		157.50	202
8750	12" diameter		16.84	1.900		121	77		198	250
8760	14" diameter		16	2		165	81		246	305
8770	16" diameter		11.64	2.749		219	111		330	410
8780	18" diameter		11.03	2.901		305	117		422	515
8785	20" diameter		9.14	3.501		375	142		517	625
8790	24" diameter		7.53	4.250		445	172		617	745
9600	Steel sleeve with tap, 4" diameter	B-20	3	8		415	295		710	910
9620	6" diameter		2	12		450	445		895	1,175
9630	8" diameter		2	12		495	445		940	1,225

33 11 13.25 Water Supply, Polyvinyl Chloride Pipe

Code	Description	Crew	Daily Output	Labor-Hours	Unit	Material	Labor	Equipment	Total	Total Incl O&P
0010	**WATER SUPPLY, POLYVINYL CHLORIDE PIPE** R331113-80									
0020	Not including excavation or backfill, unless specified									
2100	PVC pipe, Class 150, 1-1/2" diameter	Q-1A	750	.013	L.F.	.30	.70		1	1.38
2120	2" diameter		686	.015		.46	.76		1.22	1.66
2140	2-1/2" diameter		500	.020		.68	1.05		1.73	2.32
2160	3" diameter	B-20	430	.056		1.04	2.06		3.10	4.31
3010	AWWA C905, PR 100, DR 25									
3030	14" diameter	B-20A	213	.150	L.F.	10	6.10		16.10	20
3040	16" diameter		200	.160		13	6.50		19.50	24
3050	18" diameter		160	.200		19.50	8.10		27.60	34
3060	20" diameter		133	.241		19.50	9.75		29.25	36.50
3070	24" diameter		107	.299		27.50	12.10		39.60	49
3080	30" diameter		80	.400		55	16.20		71.20	85
3090	36" diameter		80	.400		85	16.20		101.20	118
3100	42" diameter		60	.533		115	21.50		136.50	160
3200	48" diameter		60	.533		150	21.50		171.50	198
4520	Pressure pipe Class 150, SDR 18, AWWA C900, 4" diameter		380	.084		2.34	3.41		5.75	7.70
4530	6" diameter		316	.101		4.63	4.10		8.73	11.30
4540	8" diameter		264	.121		8.15	4.91		13.06	16.40
4550	10" diameter		220	.145		12.35	5.90		18.25	22.50

33 11 13.25 Water Supply, Polyvinyl Chloride Pipe		Crew	Daily Output	Labor-Hours	Unit	Material	2010 Bare Costs Labor	Equipment	Total	Total Incl O&P
4560	12″ diameter	B-20A	186	.172	L.F.	17.45	6.95		24.40	29.50
8000	Fittings with rubber gasket									
8003	Class 150, D.R. 18									
8006	90° Bend , 4″ diameter	B-20	100	.240	Ea.	51.50	8.85		60.35	70
8020	6″ diameter		90	.267		93	9.85		102.85	117
8040	8″ diameter		80	.300		176	11.10		187.10	210
8060	10″ diameter		50	.480		268	17.75		285.75	325
8080	12″ diameter		30	.800		410	29.50		439.50	495
8100	Tee, 4″ diameter		90	.267		54.50	9.85		64.35	75
8120	6″ diameter		80	.300		146	11.10		157.10	178
8140	8″ diameter		70	.343		205	12.65		217.65	245
8160	10″ diameter		40	.600		430	22		452	510
8180	12″ diameter		20	1.200		605	44.50		649.50	735
8200	45° Bend, 4″ diameter		100	.240		51	8.85		59.85	69.50
8220	6″ diameter		90	.267		92	9.85		101.85	116
8240	8″ diameter		50	.480		173	17.75		190.75	218
8260	10″ diameter		50	.480		273	17.75		290.75	330
8280	12″ diameter		30	.800		365	29.50		394.50	445
8300	Reducing tee 6″ x 4″		100	.240		120	8.85		128.85	146
8320	8″ x 6″		90	.267		290	9.85		299.85	335
8330	10″ x 6″		90	.267		335	9.85		344.85	385
8340	10″ x 8″		90	.267		360	9.85		369.85	410
8350	12″ x 6″		90	.267		450	9.85		459.85	510
8360	12″ x 8″		90	.267		500	9.85		509.85	565
8400	Tapped service tee (threaded type) 6″ x 6″ x 3/4″		100	.240		78	8.85		86.85	99.50
8430	6″ x 6″ x 1″		90	.267		78	9.85		87.85	101
8440	6″ x 6″ x 1 1/2″		90	.267		78	9.85		87.85	101
8450	6″ x 6″ x 2″		90	.267		78	9.85		87.85	101
8460	8″ x 8″ x 3/4″		90	.267		123	9.85		132.85	151
8470	8″ x 8″ x 1″		90	.267		123	9.85		132.85	151
8480	8″ x 8″ x 1 1/2″		90	.267		123	9.85		132.85	151
8490	8″ x 8″ x 2″		90	.267		123	9.85		132.85	151
8500	Repair coupling 4″		100	.240		108	8.85		116.85	133
8520	6″ diameter		90	.267		131	9.85		140.85	159
8540	8″ diameter		50	.480		155	17.75		172.75	198
8560	10″ diameter		50	.480		198	17.75		215.75	246
8580	12″ diameter		50	.480		234	17.75		251.75	285
8600	Plug end 4″		100	.240		38	8.85		46.85	55.50
8620	6″ diameter		90	.267		73.50	9.85		83.35	96
8640	8″ diameter		50	.480		79	17.75		96.75	115
8660	10″ diameter		50	.480		126	17.75		143.75	167
8680	12″ diameter		50	.480		156	17.75		173.75	200
8700	PVC pipe, joint restraint									
8710	4″ diameter	B-20A	32	1	Ea.	38.50	40.50		79	104
8720	6″ diameter		25.60	1.250		46.50	50.50		97	128
8730	8″ diameter		21.33	1.500		67	60.50		127.50	166
8740	10″ diameter		18.28	1.751		124	71		195	243
8750	12″ diameter		16.84	1.900		131	77		208	261
8760	14″ diameter		16	2		206	81		287	350
8770	16″ diameter		11.64	2.749		263	111		374	460
8780	18″ diameter		11.03	2.901		325	117		442	540
8785	20″ diameter		9.14	3.501		405	142		547	660
8790	24″ diameter		7.53	4.250		470	172		642	775

33 11 Water Utility Distribution Piping

33 11 13 – Public Water Utility Distribution Piping

33 11 13.35 Water Supply, HDPE		Crew	Daily Output	Labor-Hours	Unit	Material	2010 Bare Costs Labor	Equipment	Total	Total Incl O&P
0010	**WATER SUPPLY, HDPE**									
0011	Butt fusion joints, SDR 21,40′ lengths not including excavation or backfill									
0100	4″ diameter	B-22A	400	.095	L.F.	3.65	3.54	.97	8.16	10.55
0200	6″ diameter		380	.100		10.70	3.73	1.02	15.45	18.55
0300	8″ diameter		320	.119		12.35	4.43	1.21	17.99	21.50
0400	10″ diameter		300	.127		19.10	4.73	1.29	25.12	29.50
0500	12″ diameter		260	.146		27	5.45	1.49	33.94	39.50
0600	14″ diameter		220	.173		32.50	6.45	1.76	40.71	48
0700	16″ diameter		180	.211		42	7.90	2.15	52.05	61
0800	18″ diameter		140	.271		53.50	10.15	2.77	66.42	77.50
0900	24″ diameter		100	.380		95.50	14.20	3.88	113.58	131
1000	Fittings									
1100	Elbows, 90 degrees									
1200	4″ diameter	B-22B	32	.500	Ea.	33.50	18.95	3.08	55.53	69.50
1300	6″ diameter		28	.571		104	21.50	3.52	129.02	151
1400	8″ diameter		24	.667		220	25	4.11	249.11	286
1500	10″ diameter		18	.889		350	33.50	5.50	389	445
1600	12″ diameter		12	1.333		530	50.50	8.20	588.70	670
1700	14″ diameter		9	1.778		635	67.50	10.95	713.45	815
1800	16″ diameter		6	2.667		770	101	16.45	887.45	1,025
1900	18″ diameter		4	4		1,050	151	24.50	1,225.50	1,400
2000	24″ diameter		3	5.333		2,125	202	33	2,360	2,675
2100	Tees									
2200	4″ diameter	B-22B	30	.533	Ea.	41.50	20	3.29	64.79	80.50
2300	6″ diameter		26	.615		123	23.50	3.79	150.29	175
2400	8″ diameter		22	.727		310	27.50	4.48	341.98	385
2500	10″ diameter		15	1.067		405	40.50	6.55	452.05	515
2600	12″ diameter		10	1.600		565	60.50	9.85	635.35	725
2700	14″ diameter		8	2		670	75.50	12.30	757.80	865
2800	16″ diameter		6	2.667		785	101	16.45	902.45	1,050
2900	18″ diameter		4	4		1,050	151	24.50	1,225.50	1,400
3000	24″ diameter		2	8		2,175	305	49.50	2,529.50	2,900

33 12 Water Utility Distribution Equipment

33 12 13 – Water Service Connections

33 12 13.15 Tapping, Crosses and Sleeves

		Crew	Daily Output	Labor-Hours	Unit	Material	Labor	Equipment	Total	Total Incl O&P
0010	**TAPPING, CROSSES AND SLEEVES**									
4000	Drill and tap pressurized main (labor only)									
4100	6″ main, 1″ to 2″ service	Q-1	3	5.333	Ea.		250		250	375
4150	8″ main, 1″ to 2″ service	″	2.75	5.818	″		273		273	410
4500	Tap and insert gate valve									
4600	8″ main, 4″ branch	B-21	3.20	8.750	Ea.		335	43	378	560
4650	6″ branch		2.70	10.370			395	51	446	660
4700	10″ Main, 4″ branch		2.70	10.370			395	51	446	660
4750	6″ branch		2.35	11.915			455	58.50	513.50	760
4800	12″ main, 6″ branch		2.35	11.915			455	58.50	513.50	760
4850	8″ branch		2.35	11.915			455	58.50	513.50	760
7020	Crosses, 4″ x 4″		37	.757		610	29	3.73	642.73	725
7030	6″ x 4″		25	1.120		730	42.50	5.50	778	870
7040	6″ x 6″		25	1.120		730	42.50	5.50	778	870

33 12 13 – Water Service Connections

33 12 13.15 Tapping, Crosses and Sleeves		Crew	Daily Output	Labor-Hours	Unit	Material	2010 Bare Costs Labor	Equipment	Total	Total Incl O&P
7060	8″ x 6″	B-21	21	1.333	Ea.	895	50.50	6.55	952.05	1,075
7080	8″ x 8″		21	1.333		970	50.50	6.55	1,027.05	1,150
7100	10″ x 6″		21	1.333		1,775	50.50	6.55	1,832.05	2,025
7120	10″ x 10″		21	1.333		1,925	50.50	6.55	1,982.05	2,175
7140	12″ x 6″		18	1.556		1,775	59	7.65	1,841.65	2,050
7160	12″ x 12″		18	1.556		2,225	59	7.65	2,291.65	2,550
7180	14″ x 6″		16	1.750		4,425	66.50	8.65	4,500.15	4,975
7200	14″ x 14″		16	1.750		4,675	66.50	8.65	4,750.15	5,250
7220	16″ x 6″		14	2		4,750	76	9.85	4,835.85	5,350
7240	16″ x 10″		14	2		4,825	76	9.85	4,910.85	5,425
7260	16″ x 16″		14	2		5,050	76	9.85	5,135.85	5,700
7280	18″ x 6″		10	2.800		7,100	106	13.80	7,219.80	7,975
7300	18″ x 12″		10	2.800		7,150	106	13.80	7,269.80	8,050
7320	18″ x 18″		10	2.800		6,550	106	13.80	6,669.80	7,375
7340	20″ x 6″		8	3.500		5,750	133	17.25	5,900.25	6,550
7360	20″ x 12″		8	3.500		6,150	133	17.25	6,300.25	7,000
7380	20″ x 20″		8	3.500		9,000	133	17.25	9,150.25	10,100
7400	24″ x 6″		6	4.667		7,425	177	23	7,625	8,450
7420	24″ x 12″		6	4.667		7,525	177	23	7,725	8,575
7440	24″ x 18″		6	4.667		11,200	177	23	11,400	12,700
7460	24″ x 24″		6	4.667		11,700	177	23	11,900	13,200
7600	Cut-in sleeves with rubber gaskets, 4″		18	1.556		193	59	7.65	259.65	310
7620	6″		12	2.333		251	88.50	11.50	351	425
7640	8″		10	2.800		330	106	13.80	449.80	545
7660	10″		10	2.800		410	106	13.80	529.80	635
7680	12″		9	3.111		545	118	15.35	678.35	800
7800	Cut-in valves with rubber gaskets, 4″		18	1.556		440	59	7.65	506.65	585
7820	6″		12	2.333		600	88.50	11.50	700	810
7840	8″		10	2.800		830	106	13.80	949.80	1,100
7860	10″		10	2.800		990	106	13.80	1,109.80	1,250
7880	12″		9	3.111		1,225	118	15.35	1,358.35	1,550
7900	Tapping Valve 4 inch, MJ, ductile iron		18	1.556		585	59	7.65	651.65	740
7920	Tapping Valve 6 inch, MJ, ductile iron		12	2.333		795	88.50	11.50	895	1,025
8000	Sleeves with rubber gaskets, 4″ x 4″		37	.757		570	29	3.73	602.73	675
8010	6″ x 4″		25	1.120		640	42.50	5.50	688	775
8020	6″ x 6″		25	1.120		665	42.50	5.50	713	805
8030	8″ x 4″		21	1.333		745	50.50	6.55	802.05	905
8040	8″ x 6″		21	1.333		760	50.50	6.55	817.05	920
8060	8″ x 8″		21	1.333		870	50.50	6.55	927.05	1,050
8070	10″ x 4″		21	1.333		1,200	50.50	6.55	1,257.05	1,400
8080	10″ x 6″		21	1.333		1,250	50.50	6.55	1,307.05	1,450
8090	10″ x 8″		21	1.333		1,375	50.50	6.55	1,432.05	1,600
8100	10″ x 10″		21	1.333		1,550	50.50	6.55	1,607.05	1,775
8110	12″ x 4″		18	1.556		1,125	59	7.65	1,191.65	1,350
8120	12″ x 6″		18	1.556		1,275	59	7.65	1,341.65	1,500
8130	12″ x 8″		18	1.556		1,400	59	7.65	1,466.65	1,650
8135	12″ x 10″		18	1.556		1,675	59	7.65	1,741.65	1,950
8140	12″ x 12″		18	1.556		1,700	59	7.65	1,766.65	1,975
8160	14″ x 6″		16	1.750		1,100	66.50	8.65	1,175.15	1,300
8180	14″ x 14″		16	1.750		1,725	66.50	8.65	1,800.15	2,000
8200	16″ x 6″		14	2		2,500	76	9.85	2,585.85	2,875
8220	16″ x 10″		14	2		2,800	76	9.85	2,885.85	3,200
8240	16″ x 16″		14	2		2,925	76	9.85	3,010.85	3,350

33 12 13 – Water Service Connections

33 12 13.15 Tapping, Crosses and Sleeves		Crew	Daily Output	Labor-Hours	Unit	Material	Labor	Equipment	Total	Total Incl O&P
						2010 Bare Costs				
8260	18″ x 6″	B-21	10	2.800	Ea.	1,800	106	13.80	1,919.80	2,150
8280	18″ x 12″		10	2.800		2,275	106	13.80	2,394.80	2,675
8300	18″ x 18″		10	2.800		3,100	106	13.80	3,219.80	3,575
8320	20″ x 6″		8	3.500		2,400	133	17.25	2,550.25	2,875
8340	20″ x 12″		8	3.500		2,525	133	17.25	2,675.25	3,025
8360	20″ x 20″		8	3.500		3,000	133	17.25	3,150.25	3,525
8380	24″ x 6″		6	4.667		1,550	177	23	1,750	2,000
8400	24″ x 12″		6	4.667		2,650	177	23	2,850	3,225
8420	24″ x 18″		6	4.667		3,200	177	23	3,400	3,825
8440	24″ x 24″		6	4.667		4,900	177	23	5,100	5,700
8800	Hydrant valve box, 6′ long	B-20	20	1.200		296	44.50		340.50	395
8820	8′ long		18	1.333		405	49		454	520
8830	Valve box w/lid 4′ deep		14	1.714		225	63.50		288.50	345
8840	Valve box and large base w/lid		14	1.714		266	63.50		329.50	390

33 12 16 – Water Utility Distribution Valves

33 12 16.10 Valves		Crew	Daily Output	Labor-Hours	Unit	Material	Labor	Equipment	Total	Total Incl O&P
0010	**VALVES**, water distribution									
0011	See Div. 22 05 23.20 and 22 05 23.60									
3000	Butterfly valves with boxes, cast iron, mech. jt.									
3100	4″ diameter	B-6	6	4	Ea.	1,125	143	56.50	1,324.50	1,500
3140	6″ diameter		6	4		1,325	143	56.50	1,524.50	1,725
3180	8″ diameter		6	4		1,725	143	56.50	1,924.50	2,175
3300	10″ diameter		6	4		2,225	143	56.50	2,424.50	2,725
3340	12″ diameter		6	4		3,075	143	56.50	3,274.50	3,650
3400	14″ diameter		4	6		3,750	215	84.50	4,049.50	4,550
3440	16″ diameter		4	6		5,650	215	84.50	5,949.50	6,625
3460	18″ diameter		4	6		7,600	215	84.50	7,899.50	8,800
3480	20″ diameter		4	6		8,775	215	84.50	9,074.50	10,100
3500	24″ diameter		4	6		12,200	215	84.50	12,499.50	13,800
3600	With lever operator									
3610	4″ diameter	B-6	6	4	Ea.	900	143	56.50	1,099.50	1,275
3614	6″ diameter		6	4		1,100	143	56.50	1,299.50	1,475
3616	8″ diameter		6	4		1,500	143	56.50	1,699.50	1,925
3618	10″ diameter		6	4		2,000	143	56.50	2,199.50	2,475
3620	12″ diameter		6	4		2,850	143	56.50	3,049.50	3,400
3622	14″ diameter		4	6		3,475	215	84.50	3,774.50	4,250
3624	16″ diameter		4	6		5,375	215	84.50	5,674.50	6,350
3626	18″ diameter		4	6		7,350	215	84.50	7,649.50	8,500
3628	20″ diameter		4	6		8,500	215	84.50	8,799.50	9,775
3630	24″ diameter		4	6		11,900	215	84.50	12,199.50	13,500
3700	Check valves, flanged									
3710	4″ diameter	B-6	6	4	Ea.	565	143	56.50	764.50	900
3714	6″ diameter		6	4		880	143	56.50	1,079.50	1,250
3716	8″ diameter		6	4		1,475	143	56.50	1,674.50	1,900
3718	10″ diameter		6	4		2,325	143	56.50	2,524.50	2,825
3720	12″ diameter		6	4		2,700	143	56.50	2,899.50	3,250
3722	14″ diameter		4	6		4,600	215	84.50	4,899.50	5,475
3724	16″ diameter		4	6		3,650	215	84.50	3,949.50	4,425
3726	18″ diameter		4	6		5,950	215	84.50	6,249.50	6,975
3728	20″ diameter		4	6		11,600	215	84.50	11,899.50	13,100
3730	24″ diameter		4	6		17,500	215	84.50	17,799.50	19,700
3800	Gate valves, C.I., 250 PSI, mechanical joint, w/boxes									

33 12 16 – Water Utility Distribution Valves

33 12 16.10 Valves		Crew	Daily Output	Labor-Hours	Unit	Material	2010 Bare Costs Labor	Equipment	Total	Total Incl O&P
3810	4" diameter	B-6	6	4	Ea.	625	143	56.50	824.50	965
3814	6" diameter		6	4		800	143	56.50	999.50	1,150
3816	8" diameter		6	4		1,025	143	56.50	1,224.50	1,425
3818	10" diameter		6	4		1,450	143	56.50	1,649.50	1,875
3820	12" diameter		6	4		1,825	143	56.50	2,024.50	2,275
3822	14" diameter		4	6		5,075	215	84.50	5,374.50	6,000
3824	16" diameter		4	6		5,475	215	84.50	5,774.50	6,450
3826	18" diameter		4	6		8,775	215	84.50	9,074.50	10,100
3828	20" diameter		4	6		11,800	215	84.50	12,099.50	13,400
3830	24" diameter		4	6		17,800	215	84.50	18,099.50	19,900
3831	30" diameter		4	6		39,500	215	84.50	39,799.50	43,800
3832	36" diameter	↓	4	6	↓	66,000	215	84.50	66,299.50	73,000

33 12 16.20 Valves		Crew	Daily Output	Labor-Hours	Unit	Material	Labor	Equipment	Total	Total Incl O&P
0010	**VALVES**									
0011	Special trim or use									
9000	Valves, gate valve, N.R.S. post type, 4" diameter	B-6	6	4	Ea.	625	143	56.50	824.50	965
9020	6" diameter		6	4		800	143	56.50	999.50	1,150
9040	8" diameter		6	4		1,025	143	56.50	1,224.50	1,425
9060	10" diameter		6	4		1,450	143	56.50	1,649.50	1,875
9080	12" diameter		6	4		1,825	143	56.50	2,024.50	2,275
9100	14" diameter		6	4		5,025	143	56.50	5,224.50	5,800
9120	O.S.&Y., 4" diameter		6	4		920	143	56.50	1,119.50	1,275
9140	6" diameter		6	4		1,275	143	56.50	1,474.50	1,675
9160	8" diameter		6	4		1,925	143	56.50	2,124.50	2,400
9180	10" diameter		6	4		2,950	143	56.50	3,149.50	3,500
9200	12" diameter		6	4		4,000	143	56.50	4,199.50	4,675
9220	14" diameter		4	6		9,475	215	84.50	9,774.50	10,800
9400	Check valves, rubber disc, 2-1/2" diameter		6	4		380	143	56.50	579.50	700
9420	3" diameter		6	4		420	143	56.50	619.50	740
9440	4" diameter		6	4		565	143	56.50	764.50	900
9480	6" diameter		6	4		880	143	56.50	1,079.50	1,250
9500	8" diameter		6	4		1,475	143	56.50	1,674.50	1,900
9520	10" diameter		6	4		2,325	143	56.50	2,524.50	2,825
9540	12" diameter		6	4		2,700	143	56.50	2,899.50	3,250
9542	14" diameter		4	6		4,600	215	84.50	4,899.50	5,475
9700	Detector check valves, reducing, 4" diameter		6	4		1,300	143	56.50	1,499.50	1,725
9720	6" diameter		6	4		1,750	143	56.50	1,949.50	2,200
9740	8" diameter		6	4		2,550	143	56.50	2,749.50	3,075
9760	10" diameter		6	4		4,500	143	56.50	4,699.50	5,225
9800	Galvanized, 4" diameter		6	4		1,625	143	56.50	1,824.50	2,075
9820	6" diameter		6	4		2,375	143	56.50	2,574.50	2,875
9840	8" diameter		6	4		3,800	143	56.50	3,999.50	4,450
9860	10" diameter	↓	6	4	↓	7,500	143	56.50	7,699.50	8,525

33 16 Water Utility Storage Tanks

33 16 13 – Aboveground Water Utility Storage Tanks

33 16 13.13 Steel Water Storage Tanks

	33 16 13.13 Steel Water Storage Tanks	Crew	Daily Output	Labor-Hours	Unit	Material	2010 Bare Costs Labor	Equipment	Total	Total Incl O&P
0010	**STEEL WATER STORAGE TANKS**									
0910	Steel, ground level, ht/diam. less than 1, not incl. fdn., 100,000 gallons				Ea.				168,000	184,000
1000	250,000 gallons								178,500	196,000
1200	500,000 gallons								272,000	299,000
1250	750,000 gallons								324,000	356,000
1300	1,000,000 gallons								494,000	543,000
1500	2,000,000 gallons								797,000	877,000
1600	4,000,000 gallons								1,239,000	1,363,000
1800	6,000,000 gallons								1,719,000	1,891,000
1850	8,000,000 gallons								2,250,000	2,474,000
1910	10,000,000 gallons								3,356,000	3,691,000
2100	Steel standpipes, hgt/diam. more than 1, 100′ to overflow, no fdn									
2200	500,000 gallons				Ea.				356,000	392,000
2400	750,000 gallons								435,000	478,000
2500	1,000,000 gallons								529,000	582,000
2700	1,500,000 gallons								739,000	813,000
2800	2,000,000 gallons								908,000	999,000

33 16 13.16 Prestressed Conc. Aboveground Water Utility Storage Tanks

	33 16 13.16 Prestressed Conc. Aboveground Water Utility Storage Tanks	Crew	Daily Output	Labor-Hours	Unit	Material	Labor	Equipment	Total	Total Incl O&P
0010	**PRESTRESSED CONC. ABOVEGROUND WATER UTILITY STORAGE TANKS**									
0020	Not including fdn., pipe or pumps, 250,000 gallons				Ea.				270,000	297,000
0100	500,000 gallons								412,000	453,000
0300	1,000,000 gallons								597,000	657,000
0400	2,000,000 gallons								906,000	997,000
0600	4,000,000 gallons								1,442,000	1,586,000
0700	6,000,000 gallons								1,916,000	2,107,000
0750	8,000,000 gallons								2,472,000	2,719,000
0800	10,000,000 gallons								2,987,000	3,286,000

33 16 13.23 Plastic-Coated Fabric Pillow Water Tanks

	33 16 13.23 Plastic-Coated Fabric Pillow Water Tanks	Crew	Daily Output	Labor-Hours	Unit	Material	Labor	Equipment	Total	Total Incl O&P
0010	**PLASTIC-COATED FABRIC PILLOW WATER TANKS**									
7000	Water tanks, vinyl coated fabric pillow tanks, freestanding, 5,000 gallons	4 Clab	4	8	Ea.	9,500	265		9,765	10,900
7100	Supporting embankment not included, 25,000 gallons	6 Clab	2	24		14,900	795		15,695	17,600
7200	50,000 gallons	8 Clab	1.50	42.667		27,600	1,400		29,000	32,600
7300	100,000 gallons	9 Clab	.90	80		60,000	2,650		62,650	70,000
7400	150,000 gallons		.50	144		82,000	4,775		86,775	97,500
7500	200,000 gallons		.40	180		100,500	5,950		106,450	119,500
7600	250,000 gallons		.30	240		143,000	7,950		150,950	170,000

33 16 19 – Elevated Water Utility Storage Tanks

33 16 19.50 Elevated Water Storage Tanks

	33 16 19.50 Elevated Water Storage Tanks	Crew	Daily Output	Labor-Hours	Unit	Material	Labor	Equipment	Total	Total Incl O&P
0010	**ELEVATED WATER STORAGE TANKS**									
0011	Not incl. pipe, pumps or foundation									
3000	Elevated water tanks, 100′ to bottom capacity line, incl. painting									
3010	50,000 gallons				Ea.				213,000	233,000
3300	100,000 gallons								297,000	327,000
3400	250,000 gallons								409,000	450,000
3600	500,000 gallons								660,000	726,000
3700	750,000 gallons								912,000	1,003,000
3900	1,000,000 gallons								916,000	1,007,000

33 21 13.10 Wells and Accessories		Crew	Daily Output	Labor-Hours	Unit	Material	2010 Bare Costs Labor	Equipment	Total	Total Incl O&P
0010	**WELLS & ACCESSORIES**									
0011	Domestic									
0100	Drilled, 4" to 6" diameter	B-23	120	.333	L.F.		11.15	24	35.15	43.50
0200	8" diameter	"	95.20	.420	"		14.10	30.50	44.60	55
0400	Gravel pack well, 40' deep, incl. gravel & casing, complete									
0500	24" diameter casing x 18" diameter screen	B-23	.13	307	Total	31,800	10,300	22,200	64,300	75,500
0600	36" diameter casing x 18" diameter screen		.12	333	"	34,200	11,200	24,100	69,500	81,500
0800	Observation wells, 1-1/4" riser pipe		163	.245	V.L.F.	17.50	8.20	17.70	43.40	51.50
0900	For flush Buffalo roadway box, add	1 Skwk	16.60	.482	Ea.	48	20.50		68.50	84
1200	Test well, 2-1/2" diameter, up to 50' deep (15 to 50 GPM)	B-23	1.51	26.490	"	715	885	1,900	3,500	4,275
1300	Over 50' deep, add	"	121.80	.328	L.F.	19.10	11	23.50	53.60	64
1500	Pumps, installed in wells to 100' deep, 4" submersible									
1510	1/2 H.P.	Q-1	3.22	4.969	Ea.	455	233		688	850
1520	3/4 H.P.		2.66	6.015		620	282		902	1,100
1600	1 H.P.		2.29	6.987		840	325		1,165	1,425
1700	1-1/2 H.P.	Q-22	1.60	10		1,050	470	410	1,930	2,325
1800	2 H.P.		1.33	12.030		1,275	565	495	2,335	2,775
1900	3 H.P.		1.14	14.035		1,725	660	575	2,960	3,525
2000	5 H.P.		1.14	14.035		2,350	660	575	3,585	4,225
2050	Remove and install motor only, 4 H.P.		1.14	14.035		2,450	660	575	3,685	4,325
3000	Pump, 6" submersible, 25' to 150' deep, 25 H.P., 249 to 297 GPM		.89	17.978		5,275	840	735	6,850	7,875
3100	25' to 500' deep, 30 H.P., 100 to 300 GPM		.73	21.918		5,600	1,025	900	7,525	8,700
8110	Well screen assembly, stainless steel, 2" diameter	B-23A	273	.088	L.F.	74	3.26	10	87.26	97.50
8120	3" diameter		253	.095		102	3.51	10.80	116.31	130
8130	4" diameter		200	.120		117	4.45	13.65	135.10	151
8140	5" diameter		168	.143		137	5.30	16.25	158.55	177
8150	6" diameter		126	.190		164	7.05	21.50	192.55	216
8160	8" diameter		98.50	.244		214	9.05	27.50	250.55	279
8170	10" diameter		73	.329		269	12.20	37.50	318.70	355
8180	12" diameter		62.50	.384		315	14.25	43.50	372.75	420
8190	14" diameter		54.30	.442		360	16.40	50.50	426.90	475
8200	16" diameter		48.30	.497		395	18.40	56.50	469.90	525
8210	18" diameter		39.20	.612		490	22.50	69.50	582	650
8220	20" diameter		31.20	.769		560	28.50	87.50	676	755
8230	24" diameter		23.80	1.008		695	37.50	115	847.50	950
8240	26" diameter		21	1.143		780	42.50	130	952.50	1,075
8300	Slotted PVC, 1-1/4" diameter		521	.046		1.83	1.71	5.25	8.79	10.35
8310	1-1/2" diameter		488	.049		2.65	1.82	5.60	10.07	11.85
8320	2" diameter		273	.088		3.67	3.26	10	16.93	20
8330	3" diameter		253	.095		3.90	3.51	10.80	18.21	21.50
8340	4" diameter		200	.120		4.52	4.45	13.65	22.62	27
8350	5" diameter		168	.143		4.73	5.30	16.25	26.28	31
8360	6" diameter		126	.190		6.55	7.05	21.50	35.10	42
8370	8" diameter		98.50	.244		9.90	9.05	27.50	46.45	55
8400	Artificial gravel pack, 2" screen, 6" casing	B-23B	174	.138		3.41	5.10	17.65	26.16	31
8405	8" casing		111	.216		4.66	8	27.50	40.16	48
8410	10" casing		74.50	.322		5.85	11.95	41	58.80	70
8415	12" casing		60	.400		7.90	14.80	51	73.70	87.50
8420	14" casing		50.20	.478		10.25	17.70	61	88.95	106
8425	16" casing		40.70	.590		14.10	22	75.50	111.60	132
8430	18" casing		36	.667		16.40	24.50	85.50	126.40	150
8435	20" casing		29.50	.814		18.90	30	104	152.90	182
8440	24" casing		25.70	.934		21	34.50	120	175.50	207

33 21 Water Supply Wells

33 21 13 – Public Water Supply Wells

33 21 13.10 Wells and Accessories		Crew	Daily Output	Labor-Hours	Unit	Material	Labor (2010 Bare Costs)	Equipment (2010 Bare Costs)	Total (2010 Bare Costs)	Total Incl O&P
8445	26″ casing	B-23B	24.60	.976	L.F.	23.50	36	125	184.50	218
8450	30″ casing		20	1.200		27	44.50	154	225.50	267
8455	36″ casing		16.40	1.463		29	54	187	270	320
8500	Develop well		8	3	Hr.	283	111	385	779	900
8550	Pump test well		8	3		75	111	385	571	670
8560	Standby well	B-23A	8	3		73	111	340	524	625
8570	Standby, drill rig		8	3			111	340	451	545
8580	Surface seal well, concrete filled		1	24	Ea.	765	890	2,725	4,380	5,200
8590	Well test pump, install & remove	B-23	1	40			1,350	2,875	4,225	5,250
8600	Well sterilization, chlorine	2 Clab	1	16		150	530		680	980
8610	Well water pressure switch	1 Clab	12	.667		67	22		89	108
9950	See Div. 31 23 19.40 for wellpoints									
9960	See Div. 31 23 19.30 for drainage wells									

33 21 13.20 Water Supply Wells, Pumps		Crew	Daily Output	Labor-Hours	Unit	Material	Labor	Equipment	Total	Total Incl O&P
0010	**WATER SUPPLY WELLS, PUMPS**									
0011	With pressure control									
1000	Deep well, jet, 42 gal. galvanized tank									
1040	3/4 HP	1 Plum	.80	10	Ea.	1,450	520		1,970	2,375
3000	Shallow well, jet, 30 gal. galvanized tank									
3040	1/2 HP	1 Plum	2	4	Ea.	1,200	208		1,408	1,625

33 31 Sanitary Utility Sewerage Piping

33 31 13 – Public Sanitary Utility Sewerage Piping

33 31 13.15 Sewage Collection, Concrete Pipe		Crew	Daily Output	Labor-Hours	Unit	Material	Labor	Equipment	Total	Total Incl O&P
0010	**SEWAGE COLLECTION, CONCRETE PIPE**									
0020	See Div. 33 41 13.60 for sewage/drainage collection, concrete pipe									

33 31 13.25 Sewage Collection, Polyvinyl Chloride Pipe		Crew	Daily Output	Labor-Hours	Unit	Material	Labor	Equipment	Total	Total Incl O&P
0010	**SEWAGE COLLECTION, POLYVINYL CHLORIDE PIPE**									
0020	Not including excavation or backfill									
2000	20′ lengths, S.D.R. 35, B&S, 4″ diameter	B-20	375	.064	L.F.	1.68	2.36		4.04	5.50
2040	6″ diameter		350	.069		3.64	2.53		6.17	7.90
2080	13′ lengths , S.D.R. 35, B&S, 8″ diameter		335	.072		7.85	2.65		10.50	12.70
2120	10″ diameter	B-21	330	.085		11.90	3.22	.42	15.54	18.50
2160	12″ diameter		320	.088		13.60	3.33	.43	17.36	20.50
2200	15″ diameter		240	.117		15.90	4.43	.58	20.91	25
4000	Piping, DWV PVC, no exc/bkfill, 10′ L, Sch 40, 4″ diameter	B-20	375	.064		3.71	2.36		6.07	7.75
4010	6″ diameter		350	.069		7.25	2.53		9.78	11.85
4020	8″ diameter		335	.072		12.35	2.65		15	17.60

33 41 13 – Public Storm Utility Drainage Piping

33 41 13.40 Piping, Storm Drainage, Corrugated Metal		Crew	Daily Output	Labor-Hours	Unit	Material	2010 Bare Costs Labor	Equipment	Total	Total Incl O&P
0010	PIPING, STORM DRAINAGE, CORRUGATED METAL									
0020	Not including excavation or backfill									
2000	Corrugated metal pipe, galvanized									
2020	Bituminous coated with paved invert, 20′ lengths									
2040	8″ diameter, 16 ga.	B-14	330	.145	L.F.	12.30	5.05	1.02	18.37	22.50
2060	10″ diameter, 16 ga.		260	.185		14.75	6.40	1.30	22.45	27.50
2080	12″ diameter, 16 ga.		210	.229		17.65	7.95	1.61	27.21	33.50
2100	15″ diameter, 16 ga.		200	.240		21.50	8.35	1.69	31.54	38
2120	18″ diameter, 16 ga.		190	.253		27.50	8.80	1.78	38.08	45.50
2140	24″ diameter, 14 ga.		160	.300		33.50	10.45	2.11	46.06	55.50
2160	30″ diameter, 14 ga.	B-13	120	.467		44.50	16.70	6.25	67.45	81.50
2180	36″ diameter, 12 ga.		120	.467		65	16.70	6.25	87.95	104
2200	48″ diameter, 12 ga.		100	.560		99	20	7.50	126.50	148
2220	60″ diameter, 10 ga.	B-13B	75	.747		128	26.50	16.10	170.60	200
2240	72″ diameter, 8 ga.	″	45	1.244		192	44.50	27	263.50	310
2500	Galvanized, uncoated, 20′ lengths									
2520	8″ diameter, 16 ga.	B-14	355	.135	L.F.	10.05	4.71	.95	15.71	19.30
2540	10″ diameter, 16 ga.		280	.171		11	5.95	1.21	18.16	22.50
2560	12″ diameter, 16 ga.		220	.218		12.55	7.60	1.54	21.69	27
2580	15″ diameter, 16 ga.		220	.218		16	7.60	1.54	25.14	31
2600	18″ diameter, 16 ga.		205	.234		21	8.15	1.65	30.80	38
2620	24″ diameter, 14 ga.		175	.274		30.50	9.55	1.93	41.98	51
2640	30″ diameter, 14 ga.	B-13	130	.431		39	15.40	5.75	60.15	73
2660	36″ diameter, 12 ga.		130	.431		64	15.40	5.75	85.15	100
2680	48″ diameter, 12 ga.		110	.509		85.50	18.20	6.80	110.50	130
2690	60″ diameter, 10 ga.	B-13B	78	.718		135	25.50	15.50	176	204
2780	End sections, 8″ diameter	B-14	24	2	Ea.	98	69.50	14.10	181.60	231
2785	10″ diameter		22	2.182		101	76	15.35	192.35	244
2790	12″ diameter		35	1.371		113	47.50	9.65	170.15	209
2800	18″ diameter		30	1.600		135	55.50	11.25	201.75	246
2810	24″ diameter	B-13	25	2.240		196	80	30	306	370
2820	30″ diameter		25	2.240		375	80	30	485	565
2825	36″ diameter		20	2.800		495	100	37.50	632.50	740
2830	48″ diameter		10	5.600		1,025	200	75	1,300	1,525
2835	60″ diameter	B-13B	5	11.200		1,325	400	242	1,967	2,350
2840	72″ diameter	″	4	14		2,400	500	300	3,200	3,750

33 41 13.60 Sewage/Drainage Collection, Concrete Pipe		Crew	Daily Output	Labor-Hours	Unit	Material	Labor	Equipment	Total	Total Incl O&P
0010	SEWAGE/DRAINAGE COLLECTION, CONCRETE PIPE									
0020	Not including excavation or backfill									
1000	Non-reinforced pipe, extra strength, B&S or T&G joints									
1010	6″ diameter	B-14	265.04	.181	L.F.	6.25	6.30	1.27	13.82	17.95
1020	8″ diameter		224	.214		6.90	7.45	1.51	15.86	20.50
1030	10″ diameter		216	.222		7.65	7.75	1.56	16.96	22
1040	12″ diameter		200	.240		9.35	8.35	1.69	19.39	25
1050	15″ diameter		180	.267		11.15	9.30	1.88	22.33	28.50
1060	18″ diameter		144	.333		13.15	11.60	2.35	27.10	35
1070	21″ diameter		112	.429		16.20	14.90	3.02	34.12	44
1080	24″ diameter		100	.480		21.50	16.70	3.38	41.58	53
2000	Reinforced culvert, class 3, no gaskets									
2010	12″ diameter	B-14	150	.320	L.F.	10.65	11.15	2.25	24.05	31.50
2020	15″ diameter		150	.320		13.95	11.15	2.25	27.35	35
2030	18″ diameter		132	.364		18.45	12.65	2.56	33.66	42.50

33 41 13 – Public Storm Utility Drainage Piping

33 41 13.60 Sewage/Drainage Collection, Concrete Pipe		Crew	Daily Output	Labor-Hours	Unit	Material	Labor	Equipment	Total	Total Incl O&P
						2010 Bare Costs				
2035	21" diameter	B-14	120	.400	L.F.	20	13.90	2.82	36.72	46.50
2040	24" diameter		100	.480		26.50	16.70	3.38	46.58	58
2045	27" diameter	B-13	92	.609		33.50	22	8.15	63.65	79
2050	30" diameter		88	.636		40	22.50	8.50	71	87.50
2060	36" diameter		72	.778		53.50	28	10.40	91.90	113
2070	42" diameter	B-13B	72	.778		67.50	28	16.80	112.30	135
2080	48" diameter		64	.875		82	31.50	18.90	132.40	159
2090	60" diameter		48	1.167		121	41.50	25	187.50	224
2100	72" diameter		40	1.400		175	50	30	255	305
2120	84" diameter		32	1.750		266	62.50	38	366.50	430
2140	96" diameter		24	2.333		320	83.50	50.50	454	535
2200	With gaskets, class 3, 12" diameter	B-21	168	.167		11.70	6.35	.82	18.87	23.50
2220	15" diameter		160	.175		15.35	6.65	.86	22.86	28
2230	18" diameter		152	.184		20.50	7	.91	28.41	34.50
2240	24" diameter		136	.206		33.50	7.80	1.02	42.32	50
2260	30" diameter	B-13	88	.636		48.50	22.50	8.50	79.50	97
2270	36" diameter	"	72	.778		64	28	10.40	102.40	124
2290	48" diameter	B-13B	64	.875		98.50	31.50	18.90	148.90	177
2310	72" diameter	"	40	1.400		202	50	30	282	330
2330	Flared ends, 6'-1" long, 12" diameter	B-21	190	.147		43.50	5.60	.73	49.83	57
2340	15" diameter		155	.181		48.50	6.85	.89	56.24	64.50
2400	6'-2" long, 18" diameter		122	.230		56.50	8.70	1.13	66.33	77
2420	24" diameter		88	.318		76	12.10	1.57	89.67	104
2440	36" diameter	B-13	60	.933		134	33.50	12.45	179.95	212
3080	Radius pipe, add to pipe prices, 12" to 60" diameter					50%				
3090	Over 60" diameter, add					20%				
3500	Reinforced elliptical, 8' lengths, C507 class 3									
3520	14" x 23" inside, round equivalent 18" diameter	B-21	82	.341	L.F.	52.50	13	1.68	67.18	80
3530	24" x 38" inside, round equivalent 30" diameter	B-13	58	.966		78	34.50	12.90	125.40	152
3540	29" x 45" inside, round equivalent 36" diameter		52	1.077		93.50	38.50	14.40	146.40	178
3550	38" x 60" inside, round equivalent 48" diameter		38	1.474		155	52.50	19.65	227.15	272
3560	48" x 76" inside, round equivalent 60" diameter		26	2.154		196	77	29	302	365
3570	58" x 91" inside, round equivalent 72" diameter		22	2.545		296	91	34	421	500
3780	Concrete slotted pipe, class 4 mortar joint									
3800	12" diameter	B-21	168	.167	L.F.	12	6.35	.82	19.17	24
3840	18" diameter	"	152	.184	"	18	7	.91	25.91	31.50
3900	Class 4 O-ring									
3940	12" diameter	B-21	168	.167	L.F.	11	6.35	.82	18.17	22.50
3960	18" diameter	"	152	.184	"	17.75	7	.91	25.66	31.50

33 42 Culverts

33 42 16 – Concrete Culverts

33 42 16.15 Oval Arch Culverts		Crew	Daily Output	Labor-Hours	Unit	Material	Labor	Equipment	Total	Total Incl O&P
0010	**OVAL ARCH CULVERTS**									
3000	Corrugated galvanized or aluminum, coated & paved									
3020	17" x 13", 16 ga., 15" equivalent	B-14	200	.240	L.F.	38.50	8.35	1.69	48.54	56.50
3040	21" x 15", 16 ga., 18" equivalent		150	.320		49.50	11.15	2.25	62.90	74
3060	28" x 20", 14 ga., 24" equivalent		125	.384		71	13.35	2.70	87.05	101
3080	35" x 24", 14 ga., 30" equivalent		100	.480		86.50	16.70	3.38	106.58	124
3100	42" x 29", 12 ga., 36" equivalent	B-13	100	.560		129	20	7.50	156.50	181
3120	49" x 33", 12 ga., 42" equivalent		90	.622		155	22	8.30	185.30	214

33 42 Culverts

33 42 16 – Concrete Culverts

33 42 16.15 Oval Arch Culverts		Crew	Daily Output	Labor-Hours	Unit	Material	2010 Bare Costs Labor	Equipment	Total	Total Incl O&P
3140	57" x 38", 12 ga., 48" equivalent	B-13	75	.747	L.F.	172	26.50	9.95	208.45	242
3160	Steel, plain oval arch culverts, plain									
3180	17" x 13", 16 ga., 15" equivalent	B-14	225	.213	L.F.	21	7.40	1.50	29.90	36
3200	21" x 15", 16 ga., 18" equivalent		175	.274		24.50	9.55	1.93	35.98	44
3220	28" x 20", 14 ga., 24" equivalent	↓	150	.320		39.50	11.15	2.25	52.90	63
3240	35" x 24", 14 ga., 30" equivalent	B-13	108	.519		49.50	18.55	6.90	74.95	90.50
3260	42" x 29", 12 ga., 36" equivalent		108	.519		82	18.55	6.90	107.45	126
3280	49" x 33", 12 ga., 42" equivalent		92	.609		96.50	22	8.15	126.65	148
3300	57" x 38", 12 ga., 48" equivalent		75	.747	↓	114	26.50	9.95	150.45	177
3320	End sections, 17" x 13"		22	2.545	Ea.	114	91	34	239	300
3340	42" x 29"	↓	17	3.294	"	465	118	44	627	745
3360	Multi-plate arch, steel	B-20	1690	.014	Lb.	1.30	.52		1.82	2.24

33 44 Storm Utility Water Drains

33 44 13 – Utility Area Drains

33 44 13.13 Catchbasins

		Crew	Daily Output	Labor-Hours	Unit	Material	Labor	Equipment	Total	Total Incl O&P
0010	**CATCHBASINS**									
0011	Not including footing & excavation									
1600	Frames & covers, C.I., 24" square, 500 lb.	B-6	7.80	3.077	Ea.	300	110	43.50	453.50	545
1700	26" D shape, 600 lb.		7	3.429		515	123	48.50	686.50	810
1800	Light traffic, 18" diameter, 100 lb.		10	2.400		169	86	34	289	355
1900	24" diameter, 300 lb.		8.70	2.759		261	99	39	399	480
2000	36" diameter, 900 lb.		5.80	4.138		525	148	58.50	731.50	865
2100	Heavy traffic, 24" diameter, 400 lb.		7.80	3.077		252	110	43.50	405.50	495
2200	36" diameter, 1150 lb.		3	8		835	287	113	1,235	1,475
2300	Mass. State standard, 26" diameter, 475 lb.		7	3.429		630	123	48.50	801.50	935
2400	30" diameter, 620 lb.		7	3.429		400	123	48.50	571.50	680
2500	Watertight, 24" diameter, 350 lb.		7.80	3.077		430	110	43.50	583.50	685
2600	26" diameter, 500 lb.		7	3.429		405	123	48.50	576.50	685
2700	32" diameter, 575 lb.	↓	6	4	↓	905	143	56.50	1,104.50	1,275
2800	3 piece cover & frame, 10" deep,									
2900	1200 lbs., for heavy equipment	B-6	3	8	Ea.	1,300	287	113	1,700	1,975
3000	Raised for paving 1-1/4" to 2" high									
3100	4 piece expansion ring									
3200	20" to 26" diameter	1 Clab	3	2.667	Ea.	141	88.50		229.50	291
3300	30" to 36" diameter	"	3	2.667	"	197	88.50		285.50	355
3320	Frames and covers, existing, raised for paving, 2", including									
3340	row of brick, concrete collar, up to 12" wide frame	B-6	18	1.333	Ea.	44	48	18.75	110.75	142
3360	20" to 26" wide frame		11	2.182		67	78	30.50	175.50	227
3380	30" to 36" wide frame	↓	9	2.667		82.50	95.50	37.50	215.50	279
3400	Inverts, single channel brick	D-1	3	5.333		94.50	201		295.50	410
3500	Concrete		5	3.200		74	121		195	264
3600	Triple channel, brick		2	8		144	300		444	615
3700	Concrete	↓	3	5.333	↓	127	201		328	445

33 46 Subdrainage

33 46 16 – Subdrainage Piping

33 46 16.20 Piping, Subdrainage, Concrete		Crew	Daily Output	Labor-Hours	Unit	Material	Labor	Equipment	Total	Total Incl O&P
						2010 Bare Costs				
0010	**PIPING, SUBDRAINAGE, CONCRETE**									
0021	Not including excavation and backfill									
3000	Porous wall concrete underdrain, std. strength, 4" diameter	B-20	335	.072	L.F.	3.40	2.65		6.05	7.80
3020	6" diameter	"	315	.076		4.42	2.81		7.23	9.20
3040	8" diameter	B-21	310	.090		5.45	3.43	.45	9.33	11.75
3100	18" diameter	"	165	.170		17.50	6.45	.84	24.79	30
4000	Extra strength, 6" diameter	B-20	315	.076		4.09	2.81		6.90	8.85
4020	8" diameter	B-21	310	.090		6.15	3.43	.45	10.03	12.50
4040	10" diameter		295	.095		12.25	3.61	.47	16.33	19.55
4060	12" diameter		285	.098		13.30	3.73	.48	17.51	21
4080	15" diameter		230	.122		14.70	4.63	.60	19.93	24
4100	18" diameter	↓	165	.170	↓	21.50	6.45	.84	28.79	34.50

33 46 16.25 Piping, Subdrainage, Corrugated Metal		Crew	Daily Output	Labor-Hours	Unit	Material	Labor	Equipment	Total	Total Incl O&P
0010	**PIPING, SUBDRAINAGE, CORRUGATED METAL**									
0021	Not including excavation and backfill									
2010	Aluminum, perforated									
2020	6" diameter, 18 ga.	B-14	380	.126	L.F.	7.25	4.40	.89	12.54	15.75
2200	8" diameter, 16 ga.		370	.130		10.50	4.51	.91	15.92	19.45
2220	10" diameter, 16 ga.		360	.133		13.15	4.64	.94	18.73	22.50
2240	12" diameter, 16 ga.		285	.168		14.70	5.85	1.19	21.74	26.50
2260	18" diameter, 16 ga.	↓	205	.234	↓	22	8.15	1.65	31.80	39
3000	Uncoated galvanized, perforated									
3020	6" diameter, 18 ga.	B-20	380	.063	L.F.	7.25	2.33		9.58	11.60
3200	8" diameter, 16 ga.	"	370	.065		10.50	2.40		12.90	15.25
3220	10" diameter, 16 ga.	B-21	360	.078		15.75	2.96	.38	19.09	22.50
3240	12" diameter, 16 ga.		285	.098		16.50	3.73	.48	20.71	24.50
3260	18" diameter, 16 ga.	↓	205	.137	↓	25	5.20	.67	30.87	36
4000	Steel, perforated, asphalt coated									
4020	6" diameter 18 ga.	B-20	380	.063	L.F.	7.25	2.33		9.58	11.60
4030	8" diameter 18 ga.	"	370	.065		10.50	2.40		12.90	15.25
4040	10" diameter 16 ga.	B-21	360	.078		12.75	2.96	.38	16.09	19
4050	12" diameter 16 ga.		285	.098		14.50	3.73	.48	18.71	22
4060	18" diameter 16 ga.	↓	205	.137	↓	20.50	5.20	.67	26.37	31

33 46 16.30 Piping, Subdrainage, Plastic		Crew	Daily Output	Labor-Hours	Unit	Material	Labor	Equipment	Total	Total Incl O&P
0010	**PIPING, SUBDRAINAGE, PLASTIC**									
0020	Not including excavation and backfill									
2100	Perforated PVC, 4" diameter	B-14	314	.153	L.F.	1.68	5.30	1.08	8.06	11.20
2110	6" diameter		300	.160		3.64	5.55	1.13	10.32	13.80
2120	8" diameter		290	.166		7.15	5.75	1.17	14.07	18
2130	10" diameter		280	.171		9.50	5.95	1.21	16.66	21
2140	12" diameter	↓	270	.178	↓	13.60	6.20	1.25	21.05	26

33 49 Storm Drainage Structures

33 49 13 – Storm Drainage Manholes, Frames, and Covers

33 49 13.10 Storm Drainage Manholes, Frames and Covers		Crew	Daily Output	Labor-Hours	Unit	Material	2010 Bare Costs Labor	Equipment	Total	Total Incl O&P
0010	**STORM DRAINAGE MANHOLES, FRAMES & COVERS**									
0020	Excludes footing, excavation, backfill (See line items for frame & cover)									
0050	Brick, 4' inside diameter, 4' deep	D-1	1	16	Ea.	440	605		1,045	1,400
0100	6' deep		.70	22.857		625	860		1,485	1,975
0150	8' deep		.50	32		800	1,200		2,000	2,700
0200	For depths over 8', add		4	4	V.L.F.	167	151		318	410
0400	Concrete blocks (radial), 4' I.D., 4' deep		1.50	10.667	Ea.	365	400		765	1,000
0500	6' deep		1	16		490	605		1,095	1,450
0600	8' deep		.70	22.857		615	860		1,475	1,975
0700	For depths over 8', add		5.50	2.909	V.L.F.	64.50	110		174.50	236
0800	Concrete, cast in place, 4' x 4', 8" thick, 4' deep	C-14H	2	24	Ea.	475	985	15.10	1,475.10	2,075
0900	6' deep		1.50	32		690	1,325	20	2,035	2,800
1000	8' deep		1	48		990	1,975	30	2,995	4,150
1100	For depths over 8', add		8	6	V.L.F.	108	246	3.78	357.78	505
1110	Precast, 4' I.D., 4' deep	B-22	4.10	7.317	Ea.	450	281	50.50	781.50	980
1120	6' deep		3	10		800	385	69	1,254	1,550
1130	8' deep		2	15		925	575	104	1,604	2,025
1140	For depths over 8', add		16	1.875	V.L.F.	75	72	12.95	159.95	207
1150	5' I.D., 4' deep	B-6	3	8	Ea.	1,100	287	113	1,500	1,750
1160	6' deep		2	12		1,200	430	169	1,799	2,175
1170	8' deep		1.50	16		1,300	575	225	2,100	2,550
1180	For depths over 8', add		12	2	V.L.F.	184	71.50	28	283.50	340
1190	6' I.D., 4' deep		2	12	Ea.	1,825	430	169	2,424	2,850
1200	6' deep		1.50	16		1,975	575	225	2,775	3,300
1210	8' deep		1	24		2,700	860	340	3,900	4,650
1220	For depths over 8', add		8	3	V.L.F.	310	107	42	459	550
1250	Slab tops, precast, 8" thick									
1300	4' diameter manhole	B-6	8	3	Ea.	195	107	42	344	425
1400	5' diameter manhole		7.50	3.200		400	115	45	560	665
1500	6' diameter manhole		7	3.429		555	123	48.50	726.50	850
3800	Steps, heavyweight cast iron, 7" x 9"	1 Bric	40	.200		13.10	8.35		21.45	27
3900	8" x 9"		40	.200		19.60	8.35		27.95	34
3928	12" x 10-1/2"		40	.200		22.50	8.35		30.85	37
4000	Standard sizes, galvanized steel		40	.200		18.90	8.35		27.25	33.50
4100	Aluminum		40	.200		25.50	8.35		33.85	41

33 51 Natural-Gas Distribution

33 51 13 – Natural-Gas Piping

33 51 13.10 Piping, Gas Service and Distribution, P.E.		Crew	Daily Output	Labor-Hours	Unit	Material	Labor	Equipment	Total	Total Incl O&P
0010	**PIPING, GAS SERVICE AND DISTRIBUTION, POLYETHYLENE**									
0020	Not including excavation or backfill									
1000	60 psi coils, compression coupling @ 100', 1/2" diameter, SDR 11	B-20A	608	.053	L.F.	1.66	2.13		3.79	5.05
1010	1" diameter, SDR 11		544	.059		1.83	2.38		4.21	5.60
1040	1-1/4" diameter, SDR 11		544	.059		2.57	2.38		4.95	6.45
1100	2" diameter, SDR 11		488	.066		3.18	2.65		5.83	7.55
1160	3" diameter, SDR 11		408	.078		6.65	3.17		9.82	12.10
1500	60 PSI 40' joints with coupling, 3" diameter, SDR 11	B-21A	408	.098		6.65	4.05	1.61	12.31	15.15
1540	4" diameter, SDR 11		352	.114		15.20	4.69	1.86	21.75	26
1600	6" diameter, SDR 11		328	.122		47.50	5.05	2	54.55	62
1640	8" diameter, SDR 11		272	.147		64.50	6.05	2.41	72.96	83

33 52 16 – Gasoline Distribution

33 52 16.13 Gasoline Piping		Crew	Daily Output	Labor-Hours	Unit	Material	2010 Bare Costs Labor	Equipment	Total	Total Incl O&P
0010	**GASOLINE PIPING**									
0020	Primary containment pipe, fiberglass-reinforced									
0030	Plastic pipe 15′ & 30′ lengths									
0040	2″ diameter	Q-6	425	.056	L.F.	5.10	2.73		7.83	9.70
0050	3″ diameter		400	.060		6.65	2.91		9.56	11.70
0060	4″ diameter		375	.064		8.60	3.10		11.70	14.10
0100	Fittings									
0110	Elbows, 90° & 45°, bell-ends, 2″	Q-6	24	1	Ea.	51.50	48.50		100	129
0120	3″ diameter		22	1.091		54	53		107	139
0130	4″ diameter		20	1.200		71.50	58		129.50	166
0200	Tees, bell ends, 2″		21	1.143		62.50	55.50		118	152
0210	3″ diameter		18	1.333		63	64.50		127.50	166
0220	4″ diameter		15	1.600		86	77.50		163.50	211
0230	Flanges bell ends, 2″		24	1		20.50	48.50		69	95
0240	3″ diameter		22	1.091		26	53		79	108
0250	4″ diameter		20	1.200		35.50	58		93.50	126
0260	Sleeve couplings, 2″		21	1.143		13.15	55.50		68.65	97.50
0270	3″ diameter		18	1.333		18.60	64.50		83.10	118
0280	4″ diameter		15	1.600		25.50	77.50		103	145
0290	Threaded adapters 2″		21	1.143		17.25	55.50		72.75	102
0300	3″ diameter		18	1.333		30.50	64.50		95	131
0310	4″ diameter		15	1.600		41	77.50		118.50	161
0320	Reducers, 2″		27	.889		23	43		66	89.50
0330	3″ diameter		22	1.091		26.50	53		79.50	108
0340	4″ diameter		20	1.200		34.50	58		92.50	125
1010	Gas station product line for secondary containment (double wall)									
1100	Fiberglass reinforced plastic pipe 25′ lengths									
1120	Pipe, plain end, 3″ diameter	Q-6	375	.064	L.F.	8.75	3.10		11.85	14.25
1130	4″ diameter		350	.069		14.35	3.32		17.67	20.50
1140	5″ diameter		325	.074		18.60	3.58		22.18	26
1150	6″ diameter		300	.080		19.15	3.87		23.02	27
1200	Fittings									
1230	Elbows, 90° & 45°, 3″ diameter	Q-6	18	1.333	Ea.	63	64.50		127.50	166
1240	4″ diameter		16	1.500		110	72.50		182.50	230
1250	5″ diameter		14	1.714		254	83		337	405
1260	6″ diameter		12	2		257	97		354	430
1270	Tees, 3″ diameter		15	1.600		93	77.50		170.50	218
1280	4″ diameter		12	2		137	97		234	296
1290	5″ diameter		9	2.667		277	129		406	500
1300	6″ diameter		6	4		290	194		484	610
1310	Couplings, 3″ diameter		18	1.333		44	64.50		108.50	146
1320	4″ diameter		16	1.500		114	72.50		186.50	234
1330	5″ diameter		14	1.714		237	83		320	385
1340	6″ diameter		12	2		245	97		342	415
1350	Cross-over nipples, 3″ diameter		18	1.333		10.10	64.50		74.60	108
1360	4″ diameter		16	1.500		11.85	72.50		84.35	122
1370	5″ diameter		14	1.714		17.55	83		100.55	143
1380	6″ diameter		12	2		18.40	97		115.40	166
1400	Telescoping, reducers, concentric 4″ x 3″		18	1.333		33.50	64.50		98	134
1410	5″ x 4″		17	1.412		87.50	68.50		156	199
1420	6″ x 5″		16	1.500		210	72.50		282.50	340

	33 61 13.20 Pipe Conduit, Prefabricated/Preinsulated	Crew	Daily Output	Labor-Hours	Unit	2010 Bare Costs Material	Labor	Equipment	Total	Total Incl O&P
0010	**PIPE CONDUIT, PREFABRICATED/PREINSULATED**									
0020	Does not include trenching, fittings or crane.									
0300	For cathodic protection, add 12 to 14%									
0310	of total built-up price (casing plus service pipe)									
0580	Polyurethane insulated system, 250°F. max. temp.									
0620	Black steel service pipe, standard wt., 1/2" insulation									
0660	3/4" diam. pipe size	Q-17	54	.296	L.F.	47.50	13.85	1.08	62.43	73.50
0670	1" diam. pipe size		50	.320		52	14.95	1.17	68.12	81
0680	1-1/4" diam. pipe size		47	.340		58	15.90	1.25	75.15	89.50
0690	1-1/2" diam. pipe size		45	.356		63	16.60	1.30	80.90	95.50
0700	2" diam. pipe size		42	.381		65.50	17.80	1.39	84.69	100
0710	2-1/2" diam. pipe size		34	.471		66.50	22	1.72	90.22	108
0720	3" diam. pipe size		28	.571		77	26.50	2.09	105.59	127
0730	4" diam. pipe size		22	.727		97	34	2.66	133.66	161
0740	5" diam. pipe size		18	.889		124	41.50	3.25	168.75	202
0750	6" diam. pipe size	Q-18	23	1.043		144	50.50	2.55	197.05	237
0760	8" diam. pipe size		19	1.263		210	61	3.08	274.08	325
0770	10" diam. pipe size		16	1.500		277	72.50	3.66	353.16	420
0780	12" diam. pipe size		13	1.846		345	89.50	4.50	439	520
0790	14" diam. pipe size		11	2.182		385	106	5.30	496.30	585
0800	16" diam. pipe size		10	2.400		440	116	5.85	561.85	665
0810	18" diam. pipe size		8	3		510	145	7.30	662.30	785
0820	20" diam. pipe size		7	3.429		565	166	8.35	739.35	880
0830	24" diam. pipe size		6	4		690	194	9.75	893.75	1,050
0900	For 1" thick insulation, add					10%				
0940	For 1-1/2" thick insulation, add					13%				
0980	For 2" thick insulation, add					20%				
1500	Gland seal for system, 3/4" diam. pipe size	Q-17	32	.500	Ea.	640	23.50	1.83	665.33	740
1510	1" diam. pipe size		32	.500		640	23.50	1.83	665.33	740
1540	1-1/4" diam. pipe size		30	.533		685	25	1.95	711.95	795
1550	1-1/2" diam. pipe size		30	.533		685	25	1.95	711.95	795
1560	2" diam. pipe size		28	.571		815	26.50	2.09	843.59	935
1570	2-1/2" diam. pipe size		26	.615		875	28.50	2.25	905.75	1,000
1580	3" diam. pipe size		24	.667		940	31	2.44	973.44	1,075
1590	4" diam. pipe size		22	.727		1,125	34	2.66	1,161.66	1,275
1600	5" diam. pipe size		19	.842		1,375	39.50	3.08	1,417.58	1,550
1610	6" diam. pipe size	Q-18	26	.923		1,450	44.50	2.25	1,496.75	1,675
1620	8" diam. pipe size		25	.960		1,700	46.50	2.34	1,748.84	1,925
1630	10" diam. pipe size		23	1.043		2,050	50.50	2.55	2,103.05	2,325
1640	12" diam. pipe size		21	1.143		2,275	55.50	2.79	2,333.29	2,575
1650	14" diam. pipe size		19	1.263		2,550	61	3.08	2,614.08	2,900
1660	16" diam. pipe size		18	1.333		2,975	64.50	3.25	3,042.75	3,375
1670	18" diam. pipe size		16	1.500		3,175	72.50	3.66	3,251.16	3,625
1680	20" diam. pipe size		14	1.714		3,625	83	4.18	3,712.18	4,100
1690	24" diam. pipe size		12	2		4,025	97	4.88	4,126.88	4,575
2000	Elbow, 45° for system									
2020	3/4" diam. pipe size	Q-17	14	1.143	Ea.	405	53.50	4.18	462.68	530
2040	1" diam. pipe size		13	1.231		415	57.50	4.50	477	545
2050	1-1/4" diam. pipe size		11	1.455		470	68	5.30	543.30	625
2060	1-1/2" diam. pipe size		9	1.778		490	83	6.50	579.50	670
2070	2" diam. pipe size		6	2.667		515	125	9.75	649.75	765
2080	2-1/2" diam. pipe size		4	4		555	187	14.65	756.65	905
2090	3" diam. pipe size		3.50	4.571		645	213	16.75	874.75	1,050

33 61 Hydronic Energy Distribution

33 61 13 – Underground Hydronic Energy Distribution

33 61 13.20 Pipe Conduit, Prefabricated/Preinsulated		Crew	Daily Output	Labor-Hours	Unit	2010 Bare Costs Material	Labor	Equipment	Total	Total Incl O&P
2100	4" diam. pipe size	Q-17	3	5.333	Ea.	750	249	19.50	1,018.50	1,225
2110	5" diam. pipe size		2.80	5.714		965	267	21	1,253	1,475
2120	6" diam. pipe size	Q-18	4	6		1,100	291	14.65	1,405.65	1,650
2130	8" diam. pipe size		3	8		1,575	385	19.50	1,979.50	2,350
2140	10" diam. pipe size		2.40	10		2,025	485	24.50	2,534.50	2,975
2150	12" diam. pipe size		2	12		2,650	580	29.50	3,259.50	3,825
2160	14" diam. pipe size		1.80	13.333		3,300	645	32.50	3,977.50	4,625
2170	16" diam. pipe size		1.60	15		3,925	725	36.50	4,686.50	5,475
2180	18" diam. pipe size		1.30	18.462		4,925	895	45	5,865	6,825
2190	20" diam. pipe size		1	24		6,200	1,150	58.50	7,408.50	8,650
2200	24" diam. pipe size		.70	34.286		7,800	1,650	83.50	9,533.50	11,200
2260	For elbow, 90°, add					25%				
2300	For tee, straight, add					85%	30%			
2340	For tee, reducing, add					170%	30%			
2380	For weldolet, straight, add					50%				
2400	Polyurethane insulation, 1"									
2410	FRP carrier and casing									
2420	4"	Q-5	18	.889	L.F.	51	41.50		92.50	119
2422	6"	Q-6	23	1.043		89	50.50		139.50	174
2424	8"		19	1.263		145	61		206	251
2426	10"		16	1.500		195	72.50		267.50	325
2428	12"		13	1.846		255	89.50		344.50	415
2430	FRP carrier and PVC casing									
2440	4"	Q-5	18	.889	L.F.	20	41.50		61.50	84
2444	8"	Q-6	19	1.263		42	61		103	138
2446	10"		16	1.500		60.50	72.50		133	176
2448	12"		13	1.846		77.50	89.50		167	220
2450	PVC carrier and casing									
2460	4"	Q-1	36	.444	L.F.	10	21		31	42
2462	6"	"	29	.552		13.95	26		39.95	54.50
2464	8"	Q-2	36	.667		19.45	32.50		51.95	70
2466	10"		32	.750		26.50	36.50		63	84
2468	12"		31	.774		31	37.50		68.50	90.50

33 63 Steam Energy Distribution

33 63 13 – Underground Steam and Condensate Distribution Piping

33 63 13.10 Calcium Silicate Insulated System		Crew	Daily Output	Labor-Hours	Unit	Material	Labor	Equipment	Total	Total Incl O&P
0010	**CALCIUM SILICATE INSULATED SYSTEM**									
0011	High temp. (1200 degrees F)									
2840	Steel casing with protective exterior coating									
2850	6-5/8" diameter	Q-18	52	.462	L.F.	91	22.50	1.13	114.63	135
2860	8-5/8" diameter		50	.480		99.50	23.50	1.17	124.17	145
2870	10-3/4" diameter		47	.511		116	24.50	1.25	141.75	166
2880	12-3/4" diameter		44	.545		127	26.50	1.33	154.83	180
2890	14" diameter		41	.585		143	28.50	1.43	172.93	201
2900	16" diameter		39	.615		153	30	1.50	184.50	214
2910	18" diameter		36	.667		169	32.50	1.63	203.13	236
2920	20" diameter		34	.706		190	34	1.72	225.72	262
2930	22" diameter		32	.750		264	36.50	1.83	302.33	345
2940	24" diameter		29	.828		300	40	2.02	342.02	390
2950	26" diameter		26	.923		345	44.50	2.25	391.75	450

33 63 13.10 Calcium Silicate Insulated System		Crew	Daily Output	Labor-Hours	Unit	Material	2010 Bare Costs Labor	Equipment	Total	Total Incl O&P
2960	28" diameter	Q-18	23	1.043	L.F.	430	50.50	2.55	483.05	550
2970	30" diameter		21	1.143		455	55.50	2.79	513.29	585
2980	32" diameter		19	1.263		510	61	3.08	574.08	660
2990	34" diameter		18	1.333		520	64.50	3.25	587.75	670
3000	36" diameter		16	1.500		555	72.50	3.66	631.16	725
3040	For multi-pipe casings, add					10%				
3060	For oversize casings, add					2%				
3400	Steel casing gland seal, single pipe									
3420	6-5/8" diameter	Q-18	25	.960	Ea.	1,275	46.50	2.34	1,323.84	1,475
3440	8-5/8" diameter		23	1.043		1,475	50.50	2.55	1,528.05	1,700
3450	10-3/4" diameter		21	1.143		1,675	55.50	2.79	1,733.29	1,900
3460	12-3/4" diameter		19	1.263		1,975	61	3.08	2,039.08	2,275
3470	14" diameter		17	1.412		2,175	68.50	3.44	2,246.94	2,475
3480	16" diameter		16	1.500		2,525	72.50	3.66	2,601.16	2,925
3490	18" diameter		15	1.600		2,825	77.50	3.90	2,906.40	3,225
3500	20" diameter		13	1.846		3,125	89.50	4.50	3,219	3,575
3510	22" diameter		12	2		3,500	97	4.88	3,601.88	4,025
3520	24" diameter		11	2.182		3,925	106	5.30	4,036.30	4,500
3530	26" diameter		10	2.400		4,500	116	5.85	4,621.85	5,100
3540	28" diameter		9.50	2.526		5,075	122	6.15	5,203.15	5,775
3550	30" diameter		9	2.667		5,175	129	6.50	5,310.50	5,900
3560	32" diameter		8.50	2.824		5,800	137	6.90	5,943.90	6,625
3570	34" diameter		8	3		6,325	145	7.30	6,477.30	7,200
3580	36" diameter		7	3.429		6,725	166	8.35	6,899.35	7,650
3620	For multi-pipe casings, add					5%				
4000	Steel casing anchors, single pipe									
4020	6-5/8" diameter	Q-18	8	3	Ea.	1,125	145	7.30	1,277.30	1,475
4040	8-5/8" diameter		7.50	3.200		1,200	155	7.80	1,362.80	1,550
4050	10-3/4" diameter		7	3.429		1,550	166	8.35	1,724.35	1,975
4060	12-3/4" diameter		6.50	3.692		1,675	179	9	1,863	2,100
4070	14" diameter		6	4		1,950	194	9.75	2,153.75	2,450
4080	16" diameter		5.50	4.364		2,275	211	10.65	2,496.65	2,825
4090	18" diameter		5	4.800		2,550	232	11.70	2,793.70	3,200
4100	20" diameter		4.50	5.333		2,825	258	13	3,096	3,500
4110	22" diameter		4	6		3,150	291	14.65	3,455.65	3,900
4120	24" diameter		3.50	6.857		3,425	330	16.75	3,771.75	4,300
4130	26" diameter		3	8		3,875	385	19.50	4,279.50	4,875
4140	28" diameter		2.50	9.600		4,250	465	23.50	4,738.50	5,400
4150	30" diameter		2	12		4,525	580	29.50	5,134.50	5,875
4160	32" diameter		1.50	16		5,425	775	39	6,239	7,150
4170	34" diameter		1	24		6,075	1,150	58.50	7,283.50	8,500
4180	36" diameter		1	24		6,600	1,150	58.50	7,808.50	9,075
4220	For multi-pipe, add					5%	20%			
4800	Steel casing elbow									
4820	6-5/8" diameter	Q-18	15	1.600	Ea.	1,550	77.50	3.90	1,631.40	1,850
4830	8-5/8" diameter		15	1.600		1,675	77.50	3.90	1,756.40	1,950
4850	10-3/4" diameter		14	1.714		2,000	83	4.18	2,087.18	2,325
4860	12-3/4" diameter		13	1.846		2,375	89.50	4.50	2,469	2,775
4870	14" diameter		12	2		2,525	97	4.88	2,626.88	2,950
4880	16" diameter		11	2.182		2,775	106	5.30	2,886.30	3,225
4890	18" diameter		10	2.400		3,175	116	5.85	3,296.85	3,650
4900	20" diameter		9	2.667		3,375	129	6.50	3,510.50	3,925
4910	22" diameter		8	3		3,650	145	7.30	3,802.30	4,225

33 63 13.10 Calcium Silicate Insulated System		Crew	Daily Output	Labor-Hours	Unit	Material	2010 Bare Costs Labor	Equipment	Total	Total Incl O&P
4920	24″ diameter	Q-18	7	3.429	Ea.	4,025	166	8.35	4,199.35	4,675
4930	26″ diameter		6	4		4,375	194	9.75	4,578.75	5,125
4940	28″ diameter		5	4.800		4,725	232	11.70	4,968.70	5,575
4950	30″ diameter		4	6		4,750	291	14.65	5,055.65	5,675
4960	32″ diameter		3	8		5,425	385	19.50	5,829.50	6,550
4970	34″ diameter		2	12		5,950	580	29.50	6,559.50	7,425
4980	36″ diameter		2	12		6,325	580	29.50	6,934.50	7,875
5500	Black steel service pipe, std. wt., 1″ thick insulation									
5510	3/4″ diameter pipe size	Q-17	54	.296	L.F.	39.50	13.85	1.08	54.43	65
5540	1″ diameter pipe size		50	.320		41	14.95	1.17	57.12	69.50
5550	1-1/4″ diameter pipe size		47	.340		46	15.90	1.25	63.15	76.50
5560	1-1/2″ diameter pipe size		45	.356		50.50	16.60	1.30	68.40	82
5570	2″ diameter pipe size		42	.381		56	17.80	1.39	75.19	89.50
5580	2-1/2″ diameter pipe size		34	.471		59	22	1.72	82.72	100
5590	3″ diameter pipe size		28	.571		67	26.50	2.09	95.59	116
5600	4″ diameter pipe size		22	.727		86	34	2.66	122.66	148
5610	5″ diameter pipe size		18	.889		116	41.50	3.25	160.75	194
5620	6″ diameter pipe size	Q-18	23	1.043		127	50.50	2.55	180.05	218
6000	Black steel service pipe, std. wt., 1-1/2″ thick insul.									
6010	3/4″ diameter pipe size	Q-17	54	.296	L.F.	40	13.85	1.08	54.93	65.50
6040	1″ diameter pipe size		50	.320		45	14.95	1.17	61.12	73.50
6050	1-1/4″ diameter pipe size		47	.340		50.50	15.90	1.25	67.65	81
6060	1-1/2″ diameter pipe size		45	.356		55	16.60	1.30	72.90	87
6070	2″ diameter pipe size		42	.381		59.50	17.80	1.39	78.69	93.50
6080	2-1/2″ diameter pipe size		34	.471		63.50	22	1.72	87.22	105
6090	3″ diameter pipe size		28	.571		71.50	26.50	2.09	100.09	121
6100	4″ diameter pipe size		22	.727		91	34	2.66	127.66	154
6110	5″ diameter pipe size		18	.889		116	41.50	3.25	160.75	194
6120	6″ diameter pipe size	Q-18	23	1.043		132	50.50	2.55	185.05	224
6130	8″ diameter pipe size		19	1.263		190	61	3.08	254.08	305
6140	10″ diameter pipe size		16	1.500		248	72.50	3.66	324.16	385
6150	12″ diameter pipe size		13	1.846		296	89.50	4.50	390	465
6190	For 2″ thick insulation, add					15%				
6220	For 2-1/2″ thick insulation, add					25%				
6260	For 3″ thick insulation, add					30%				
6800	Black steel service pipe, ex. hvy. wt., 1″ thick insul.									
6820	3/4″ diameter pipe size	Q-17	50	.320	L.F.	41.50	14.95	1.17	57.62	70
6040	1″ diameter pipe size		47	.340		44.50	15.90	1.25	61.65	74.50
6850	1-1/4″ diameter pipe size		44	.364		51	17	1.33	69.33	83.50
6860	1-1/2″ diameter pipe size		42	.381		54.50	17.80	1.39	73.69	88
6870	2″ diameter pipe size		40	.400		57.50	18.70	1.46	77.66	93
6880	2-1/2″ diameter pipe size		31	.516		71.50	24	1.89	97.39	117
6890	3″ diameter pipe size		27	.593		80.50	27.50	2.17	110.17	132
6900	4″ diameter pipe size		21	.762		106	35.50	2.79	144.29	173
6910	5″ diameter pipe size		17	.941		148	44	3.44	195.44	233
6920	6″ diameter pipe size	Q-18	22	1.091		136	53	2.66	191.66	232
7400	Black steel service pipe, ex. hvy. wt., 1-1/2″ thick insul.									
7420	3/4″ diameter pipe size	Q-17	50	.320	L.F.	34.50	14.95	1.17	50.62	62
7440	1″ diameter pipe size		47	.340		39.50	15.90	1.25	56.65	69
7450	1-1/4″ diameter pipe size		44	.364		46	17	1.33	64.33	77.50
7460	1-1/2″ diameter pipe size		42	.381		50.50	17.80	1.39	69.69	83.50
7470	2″ diameter pipe size		40	.400		51	18.70	1.46	71.16	85.50
7480	2-1/2″ diameter pipe size		31	.516		60.50	24	1.89	86.39	105

33 63 13 – Underground Steam and Condensate Distribution Piping

33 63 13.10 Calcium Silicate Insulated System		Crew	Daily Output	Labor-Hours	Unit	2010 Bare Costs Material	Labor	Equipment	Total	Total Incl O&P
7490	3″ diameter pipe size	Q-17	27	.593	L.F.	71	27.50	2.17	100.67	122
7500	4″ diameter pipe size		21	.762		92.50	35.50	2.79	130.79	159
7510	5″ diameter pipe size		17	.941		128	44	3.44	175.44	210
7520	6″ diameter pipe size	Q-18	22	1.091		141	53	2.66	196.66	237
7530	8″ diameter pipe size		18	1.333		211	64.50	3.25	278.75	335
7540	10″ diameter pipe size		15	1.600		251	77.50	3.90	332.40	395
7550	12″ diameter pipe size		13	1.846		310	89.50	4.50	404	480
7590	For 2″ thick insulation, add					13%				
7640	For 2-1/2″ thick insulation, add					18%				
7680	For 3″ thick insulation, add					24%				
33 63 13.20 Combined Steam Pipe With Condensate Return										
9000	**COMBINED STEAM PIPE WITH CONDENSATE RETURN**									
9010	8″ and 4″ in 24″ case	Q-18	100	.240	L.F.	445	11.60	.59	457.19	510
9020	6″ and 3″ in 20″ case		104	.231		305	11.20	.56	316.76	350
9030	3″ and 1-1/2″ in 16″ case		110	.218		228	10.55	.53	239.08	266
9040	2″ and 1-1/4″ in 12-3/4″ case		114	.211		198	10.20	.51	208.71	234
9050	1-1/2″ and 1-1/4″ in 10-3/4″ case		116	.207		173	10	.50	183.50	206
9100	Steam pipe only (no return)									
9110	6″ in 18″ case	Q-18	108	.222	L.F.	250	10.75	.54	261.29	292
9120	4″ in 14″ case		112	.214		198	10.40	.52	208.92	234
9130	3″ in 14″ case		114	.211		183	10.20	.51	193.71	217
9140	2-1/2″ in 12-3/4″ case		118	.203		171	9.85	.50	181.35	203
9150	2″ in 12-3/4″ case		122	.197		163	9.55	.48	173.03	195

Estimating Tips

There are four basic types of pollution: Air, Noise, Water, and Solid.

These systems may be interrelated and care must be taken that the complete systems are estimated. For example, Air Pollution Equipment may include dust and air-entrained particles that have to be collected. The vacuum systems could be noisy, requiring silencers to reduce noise pollution, and the collected solids have to be disposed of to prevent Solid Pollution.

Water Treatment may be accomplished as comparatively small packaged plants providing Chemical, Biological or Thermal Treatment. These are usually priced by GPH or GPD.

Large plants and municipal treatment facilities are sized MGD. These facilities are usually priced in two steps. First, the large concrete structures used as settling ponds, etc.; and second, the equipment required. Storm water does not need as much treatment as sanitary wastewater.

The total cost of the system is greatly affected by the level of purification required by specifications and codes before release.

An additional cost may be encountered if there is a possibility that the water being treated may contain petroleum products, which have to be isolated and disposed of in an environmentally acceptable manner.

Reference Numbers

Reference numbers are shown in shaded boxes at the beginning of some major classifications. These numbers refer to related items in the Reference Section. The reference information may be an estimating procedure, an alternate pricing method, or technical information.

Note: Not all subdivisions listed here necessarily appear in this publication.

44 11 16 – Industrial Dust Collectors

44 11 16.10 Dust Collection Systems		Crew	Daily Output	Labor-Hours	Unit	2010 Bare Costs Material	Labor	Equipment	Total	Total Incl O&P
0010	**DUST COLLECTION SYSTEMS** Commercial/Industrial									
0120	Central vacuum units									
0130	Includes stand, filters and motorized shaker									
0200	500 CFM, 10" inlet, 2 HP	Q-20	2.40	8.333	Ea.	3,900	375		4,275	4,875
0220	1000 CFM, 10" inlet, 3 HP		2.20	9.091		4,100	410		4,510	5,150
0240	1500 CFM, 10" inlet, 5 HP		2	10		4,325	450		4,775	5,425
0260	3000 CFM, 13" inlet, 10 HP		1.50	13.333		12,200	600		12,800	14,300
0280	5000 CFM, 16" inlet, 2 @ 10 HP		1	20		12,900	905		13,805	15,600
1000	Vacuum tubing, galvanized									
1100	2-1/8" OD, 16 ga.	Q-9	440	.036	L.F.	2.55	1.61		4.16	5.25
1110	2-1/2" OD, 16 ga.		420	.038		2.93	1.68		4.61	5.75
1120	3" OD, 16 ga.		400	.040		3.35	1.77		5.12	6.35
1130	3-1/2" OD, 16 ga.		380	.042		4.86	1.86		6.72	8.15
1140	4" OD, 16 ga.		360	.044		5.45	1.96		7.41	8.95
1150	5" OD, 14 ga.		320	.050		9.80	2.21		12.01	14.10
1160	6" OD, 14 ga.		280	.057		11.30	2.53		13.83	16.25
1170	8" OD, 14 ga.		200	.080		17.05	3.54		20.59	24
1180	10" OD, 12 ga.		160	.100		17.05	4.42		21.47	25.50
1190	12" OD, 12 ga.		120	.133		29.50	5.90		35.40	41.50
1200	14" OD, 12 ga.		80	.200		36	8.85		44.85	53
1940	Hose, flexible wire reinforced rubber									
1956	3" dia.	Q-9	400	.040	L.F.	7.60	1.77		9.37	11.05
1960	4" dia.		360	.044		8.80	1.96		10.76	12.60
1970	5" dia.		320	.050		11.05	2.21		13.26	15.50
1980	6" dia.		280	.057		11.95	2.53		14.48	16.95
2000	90° Elbow, slip fit									
2110	2-1/8" dia.	Q-9	70	.229	Ea.	10.65	10.10		20.75	27
2120	2-1/2" dia.		65	.246		14.90	10.90		25.80	33
2130	3" dia.		60	.267		20	11.80		31.80	40
2140	3-1/2" dia.		55	.291		25	12.85		37.85	47
2150	4" dia.		50	.320		31	14.15		45.15	55.50
2160	5" dia.		45	.356		58.50	15.70		74.20	88.50
2170	6" dia.		40	.400		81	17.70		98.70	116
2180	8" dia.		30	.533		154	23.50		177.50	205
2400	45° Elbow, slip fit									
2410	2-1/8" dia.	Q-9	70	.229	Ea.	9.30	10.10		19.40	25.50
2420	2-1/2" dia.		65	.246		13.70	10.90		24.60	31.50
2430	3" dia.		60	.267		16.55	11.80		28.35	36
2440	3-1/2" dia.		55	.291		21	12.85		33.85	42.50
2450	4" dia.		50	.320		27	14.15		41.15	51
2460	5" dia.		45	.356		45.50	15.70		61.20	74
2470	6" dia.		40	.400		61.50	17.70		79.20	95
2480	8" dia.		35	.457		121	20		141	164
2800	90° TY, slip fit thru 6" dia.									
2810	2-1/8" dia.	Q-9	42	.381	Ea.	21	16.85		37.85	48.50
2820	2-1/2" dia.		39	.410		27	18.15		45.15	57.50
2830	3" dia.		36	.444		35.50	19.65		55.15	69.50
2840	3-1/2" dia.		33	.485		45.50	21.50		67	82.50
2850	4" dia.		30	.533		65.50	23.50		89	108
2860	5" dia.		27	.593		124	26		150	177
2870	6" dia.		24	.667		169	29.50		198.50	231
2880	8" dia., butt end					360			360	395
2890	10" dia., butt end					650			650	715

44 11 16 – Industrial Dust Collectors

44 11 16.10 Dust Collection Systems		Crew	Daily Output	Labor-Hours	Unit	2010 Bare Costs Material	Labor	Equipment	Total	Total Incl O&P
2900	12" dia., butt end				Ea.	650			650	715
2910	14" dia., butt end					1,175			1,175	1,300
2920	6" x 4" dia., butt end					103			103	114
2930	8" x 4" dia., butt end					196			196	215
2940	10" x 4" dia., butt end					256			256	282
2950	12" x 4" dia., butt end					294			294	325
3100	90° Elbow, butt end segmented									
3110	8" dia., butt end, segmented				Ea.	154			154	169
3120	10" dia., butt end, segmented					380			380	420
3130	12" dia., butt end, segmented					485			485	530
3140	14" dia., butt end, segmented					605			605	665
3200	45° Elbow, butt end segmented									
3210	8" dia., butt end, segmented				Ea.	121			121	133
3220	10" dia., butt end, segmented					276			276	305
3230	12" dia., butt end, segmented					282			282	310
3240	14" dia., butt end, segmented					345			345	380
3400	All butt end fittings require one coupling per joint.									
3410	Labor for fitting included with couplings.									
3460	Compression coupling, galvanized, neoprene gasket									
3470	2-1/8" dia.	Q-9	44	.364	Ea.	12.35	16.05		28.40	38
3480	2-1/2" dia.		44	.364		12.35	16.05		28.40	38
3490	3" dia.		38	.421		17.80	18.60		36.40	47.50
3500	3-1/2" dia.		35	.457		20	20		40	52.50
3510	4" dia.		33	.485		22	21.50		43.50	56.50
3520	5" dia.		29	.552		25	24.50		49.50	64.50
3530	6" dia.		26	.615		25	27		52	68.50
3540	8" dia.		22	.727		52	32		84	106
3550	10" dia.		20	.800		70.50	35.50		106	131
3560	12" dia.		18	.889		87	39.50		126.50	156
3570	14" dia.		16	1		111	44		155	189
3800	Air gate valves, galvanized									
3810	2-1/8" dia.	Q-9	30	.533	Ea.	180	23.50		203.50	234
3820	2-1/2" dia.		28	.571		199	25.50		224.50	258
3830	3" dia.		26	.615		217	27		244	280
3840	4" dia.		23	.696		273	31		304	345
3850	6" dia.		18	.889		370	39.50		409.50	465

44 11 26 – Electrostatic Precipitators

44 11 26.20 Electrostatic Precipitators		Crew	Daily Output	Labor-Hours	Unit	Material	Labor	Equipment	Total	Total Incl O&P
0010	**ELECTROSTATIC PRECIPITATORS**									
4100	Smoke pollution control									
4110	Exhaust air volume									
4120	1 to 2.4 MCFM	Q-10	2.15	11.163	MCFM	7,975	510		8,485	9,525
4130	2.4 to 4.8 MCFM		2.70	8.889		9,050	405		9,455	10,600
4140	4.8 to 7.2 MCFM		3.60	6.667		9,225	305		9,530	10,600
4150	7.2 to 12 MCFM		5.40	4.444		9,650	204		9,854	10,900
4400	Electrostatic precipitators with built-in exhaust fans									
4410	Smoke pollution control									
4420	Exhaust air volume									
4440	1 to 2.4 MCFM	Q-10	2.15	11.163	MCFM	9,575	510		10,085	11,300
4450	2.4 to 4.8 MCFM		2.70	8.889		7,600	405		8,005	8,975
4460	4.8 to 7.2 MCFM		3.60	6.667		6,500	305		6,805	7,625
4470	7.2 to 12 MCFM		5.40	4.444		5,600	204		5,804	6,475

44 42 56 – Water Treatment Pumps

44 42 56.10 Pumps, Pneumatic Ejector		Crew	Daily Output	Labor-Hours	Unit	2010 Bare Costs Material	Labor	Equipment	Total	Total Incl O&P
0010	**PUMPS, PNEUMATIC EJECTOR**									
0020	With steel receiver, level controls, inlet/outlet gate/check valves									
0030	Cross connect. not incl. compressor, fittings or piping									
0040	Duplex									
0050	30 GPM	Q-2	1.70	14.118	Ea.	22,500	685		23,185	25,800
0060	50 GPM		1.56	15.385		26,400	745		27,145	30,200
0070	100 GPM		1.30	18.462		36,500	895		37,395	41,600
0080	150 GPM		1.10	21.818		67,000	1,050		68,050	75,000
0090	200 GPM	↓	.85	28.235		94,500	1,375		95,875	106,000
0100	250 GPM	Q-3	.91	35.165		108,000	1,750		109,750	121,000
0110	300 GPM	"	.57	56.140	Ea.	146,500	2,775		149,275	165,000

Assembles Section

Table of Contents

How to Use the Assemblies Cost Tables

The following is a detailed explanation of a sample Assemblies Cost Table. Most Assembly Tables are separated into three parts: (1) an illustration of the system to be estimated, (2) the components and related costs of a typical system, and (3) the costs for similar systems with dimensional and/or size variations. For costs of the components that comprise these systems, or assemblies, refer to the Unit Price Section. Next to each bold number below is the described item with the appropriate component of the sample entry following in parentheses. In most cases, if the work is to be subcontracted, the general contractor will need to add an additional markup (RSMeans suggests using 10%) to the "Total" figures.

1 *System/Line Numbers (D3010 510 1760)*

Each Assemblies Cost Line has been assigned a unique identification number based on the UniFormat classification sytem.

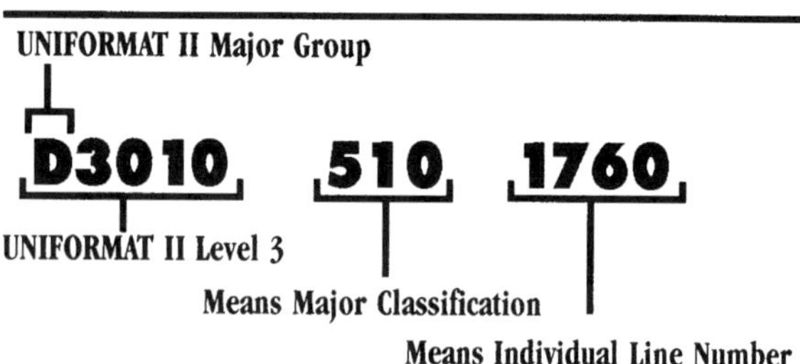

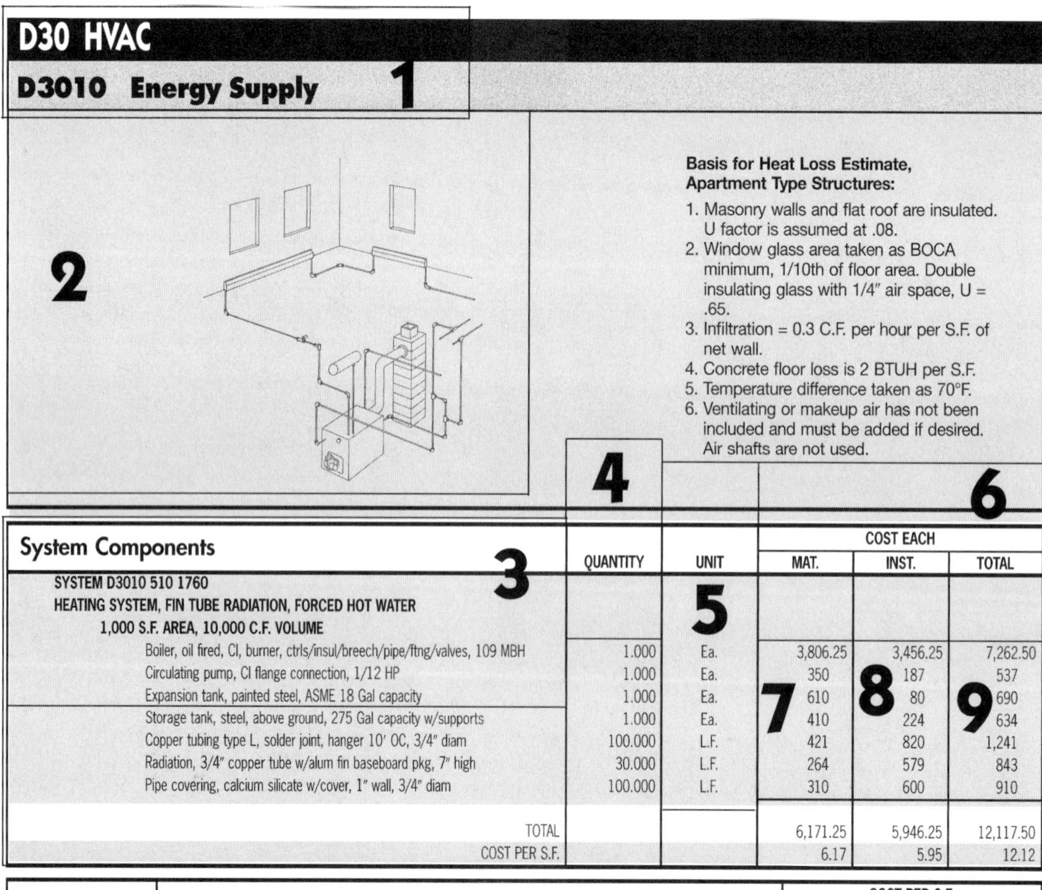

D30 HVAC

D3010 Energy Supply 1

2

Basis for Heat Loss Estimate, Apartment Type Structures:

1. Masonry walls and flat roof are insulated. U factor is assumed at .08.
2. Window glass area taken as BOCA minimum, 1/10th of floor area. Double insulating glass with 1/4" air space, U = .65.
3. Infiltration = 0.3 C.F. per hour per S.F. of net wall.
4. Concrete floor loss is 2 BTUH per S.F.
5. Temperature difference taken as 70°F.
6. Ventilating or makeup air has not been included and must be added if desired. Air shafts are not used.

System Components 3	QUANTITY 4	UNIT 5	COST EACH MAT. 7	COST EACH INST. 8	COST EACH TOTAL 9 (6)
SYSTEM D3010 510 1760 HEATING SYSTEM, FIN TUBE RADIATION, FORCED HOT WATER 1,000 S.F. AREA, 10,000 C.F. VOLUME					
Boiler, oil fired, CI, burner, ctrls/insul/breech/pipe/ftng/valves, 109 MBH	1.000	Ea.	3,806.25	3,456.25	7,262.50
Circulating pump, CI flange connection, 1/12 HP	1.000	Ea.	350	187	537
Expansion tank, painted steel, ASME 18 Gal capacity	1.000	Ea.	610	80	690
Storage tank, steel, above ground, 275 Gal capacity w/supports	1.000	Ea.	410	224	634
Copper tubing type L, solder joint, hanger 10' OC, 3/4" diam	100.000	L.F.	421	820	1,241
Radiation, 3/4" copper tube w/alum fin baseboard pkg, 7" high	30.000	L.F.	264	579	843
Pipe covering, calcium silicate w/cover, 1" wall, 3/4" diam	100.000	L.F.	310	600	910
TOTAL			6,171.25	5,946.25	12,117.50
COST PER S.F.			6.17	5.95	12.12

D3010 510	**Apartment Building Heating - Fin Tube Radiation**		COST PER S.F. MAT.	COST PER S.F. INST.	COST PER S.F. TOTAL
1740	Heating systems, fin tube radiation, forced hot water				
1760	1,000 S.F. area, 10,000 C.F. volume		6.18	5.95	12.13
1800	10,000 S.F. area, 100,000 C.F. volume	R235000 -10	2.51	3.59	6.10
1840	20,000 S.F. area, 200,000 C.F. volume		2.83	4.03	6.86
1880	30,000 S.F. area, 300,000 C.F. volume	R235000 -20	2.73	3.92	6.65
1890					

2 *Illustration*

At the top of most assembly pages are an illustration, a brief description, and the design criteria used to develop the cost.

3 *System Components*

The components of a typical system are listed separately to show what has been included in the development of the total system price. The table below contains prices for other similar systems with dimensional and/or size variations.

4 *Quantity*

This is the number of line item units required for one system unit. For example, we assume that it will take 30 linear feet of radiation for the 1,000 S.F. area shown.

5 *Unit of Measure for Each Item*

The abbreviated designation indicates the unit of measure, as defined by industry standards, upon which the price of the component is based. For example, baseboard radiation is priced by the linear foot. For a complete listing of abbreviations, see the Reference Section.

6 *Unit of Measure for Each System (Each)*

Costs shown in the three right-hand columns have been adjusted by the component quantity and unit of measure for the entire system. In this example, "Cost Each" is the unit of measure for this system, or assembly.

7 *Materials (6,171.25)*

This column contains the Materials Cost of each component. These cost figures are bare costs plus 10% for profit.

8 *Installation (5,946.25)*

Installation includes labor and equipment plus the installing contractor's overhead and profit. Equipment costs are the bare rental costs plus 10% for profit. The labor overhead and profit is defined on the inside back cover of this book.

9 *Total (12,117.50)*

The figure in this column is the sum of the material and installation costs.

Material Cost	+	Installation Cost	=	Total
$6,171.25	+	$5,946.25	=	$12,117.50

A Substructure

A2020 Basement Walls

A2020 220	Subdrainage Piping	COST PER L.F.		
		MAT.	INST.	TOTAL
2000	Piping, excavation & backfill excluded, PVC, perforated			
2110	3″ diameter	1.85	3.64	5.49
2130	4″ diameter	1.85	3.64	5.49
2140	5″ diameter	4	3.90	7.90
2150	6″ diameter	4	3.90	7.90
3000	Metal alum. or steel, perforated asphalt coated			
3150	6″ diameter	8	7.75	15.75
3160	8″ diameter	11.55	7.90	19.45
3170	10″ diameter	14.45	8.15	22.60
3180	12″ diameter	16.15	10.30	26.45
3220	18″ diameter	24.50	14.30	38.80
4000	Porous wall concrete			
4130	4″ diameter	3.74	4.07	7.81
4150	6″ diameter	4.86	4.33	9.19
4160	8″ diameter	6	5.75	11.75
4180	12″ diameter	12.70	6.25	18.95
4200	15″ diameter	14.60	7.75	22.35
4220	18″ diameter	19.25	10.80	30.05

D Services

D2020 Domestic Water Distribution

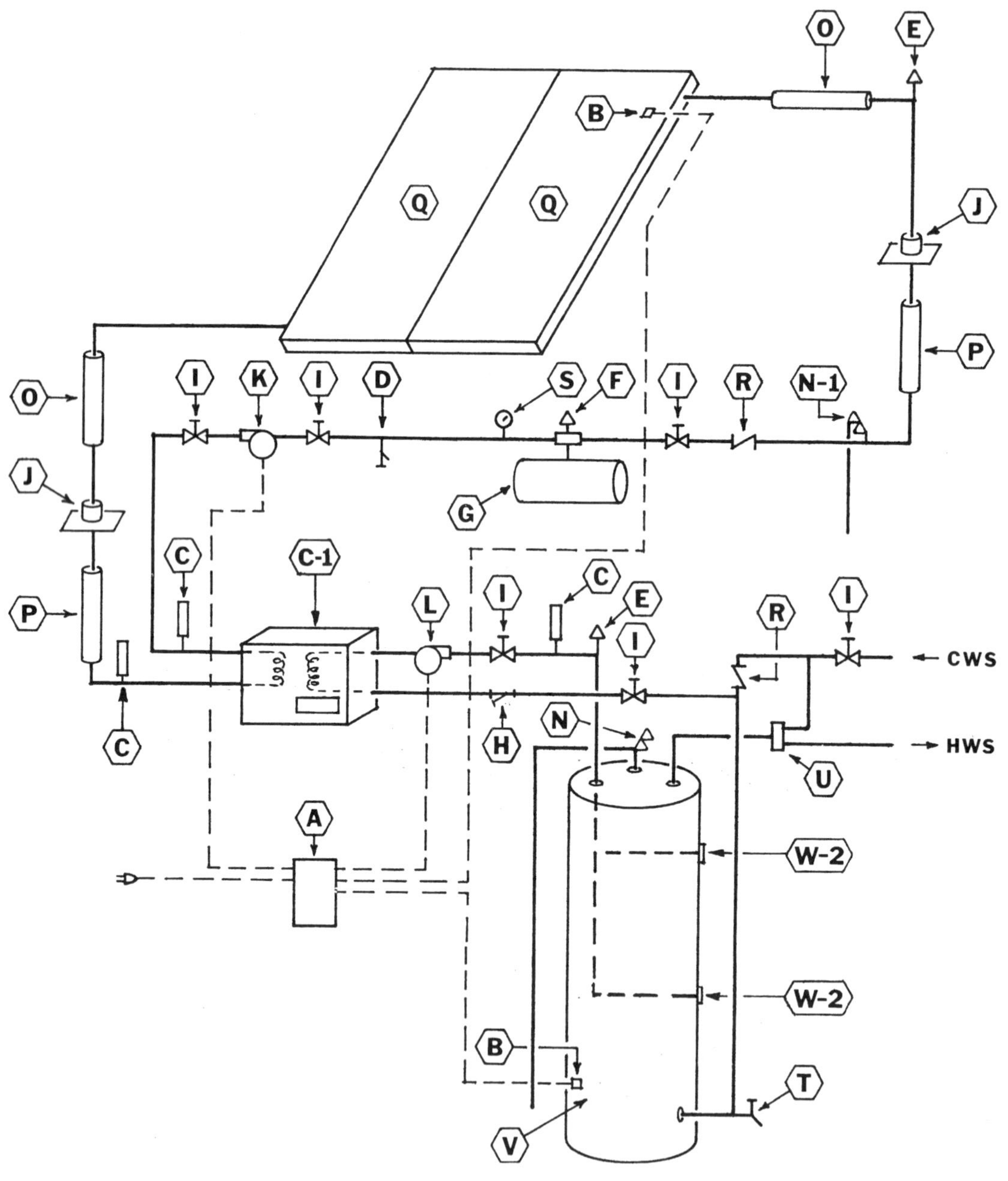

In this closed-loop indirect collection system, fluid with a low freezing temperature, such as propylene glycol, transports heat from the collectors to water storage. The transfer fluid is contained in a closed-loop consisting of collectors, supply and return piping, and a remote heat exchanger. The heat exchanger transfers heat energy from the fluid in the collector loop to potable water circulated in a storage loop. A typical two-or-three panel system contains 5 to 6 gallons of heat transfer fluid.

When the collectors become approximately 20°F warmer than the storage temperature, a controller activates the circulator on the collector and storage loops. The circulators will move the fluid and potable water through the heat exchanger until heat collection no longer occurs. At that point, the system shuts down. Since the heat transfer medium is a fluid with a very low freezing temperature, there is no need for it to be drained from the system between periods of collection.

D2020 Domestic Water Distribution

System Components	QUANTITY	UNIT	COST EACH MAT.	INST.	TOTAL
SYSTEM D2020 265 2760					
SOLAR, CLOSED LOOP, ADD-ON HOT WATER SYS., EXTERNAL HEAT EXCHANGER					
3/4" TUBING, TWO 3'X7' BLACK CHROME COLLECTORS					
A,B,G,L,K,M Heat exchanger fluid-fluid pkg incl 2 circulators, expansion tank,					
Check valve, relief valve, controller, hi temp cutoff, & 2 sensors	1.000	Ea.	765	450	1,215
C Thermometer, 2" dial	3.000	Ea.	66	117	183
D, T Fill & drain valve, brass, 3/4" connection	1.000	Ea.	8.85	26	34.85
E Air vent, manual, 1/8" fitting	2.000	Ea.	6.12	39	45.12
F Air purger	1.000	Ea.	47.50	52	99.50
H Strainer, Y type, bronze body, 3/4" IPS	1.000	Ea.	28.50	33	61.50
I Valve, gate, bronze, NRS, soldered 3/4" diam	6.000	Ea.	240	186	426
J Neoprene vent flashing	2.000	Ea.	26.80	62	88.80
N-1, N Relief valve temp & press, 150 psi 210°F self-closing 3/4" IPS	1.000	Ea.	17.15	21	38.15
O Pipe covering, urethane, ultraviolet cover, 1" wall 3/4" diam	20.000	L.F.	48	111	159
P Pipe covering, fiberglass, all service jacket, 1" wall, 3/4" diam	50.000	L.F.	57	221.50	278.50
Q Collector panel solar energy blk chrome on copper, 1/8" temp glass 3'x7'	2.000	Ea.	2,100	236	2,336
Roof clamps for solar energy collector panels	2.000	Set	5.18	32.10	37.28
R Valve, swing check, bronze, regrinding disc, 3/4" diam	2.000	Ea.	144	62	206
S Pressure gauge, 60 psi, 2" dial	1.000	Ea.	24.50	19.50	44
U Valve, water tempering, bronze, sweat connections, 3/4" diam	1.000	Ea.	100	31	131
W-2, V Tank water storage w/heating element, drain, relief valve, existing	1.000	Ea.			
Copper tubing type L, solder joint, hanger 10' OC 3/4" diam	20.000	L.F.	84.20	164	248.20
Copper tubing, type M, solder joint, hanger 10' OC 3/4" diam	70.000	L.F.	225.40	560	785.40
Sensor wire, #22-2 conductor multistranded	.500	C.L.F.	7.53	29.25	36.78
Solar energy heat transfer fluid, propylene glycol anti-freeze	6.000	Gal.	84.90	135	219.90
Wrought copper fittings & solder, 3/4" diam	76.000	Ea.	273.60	2,508	2,781.60
TOTAL			4,360.23	5,095.35	9,455.58

D2020 265	Solar, Closed Loop, Add-On Hot Water Systems	COST EACH MAT.	INST.	TOTAL
2550	Solar, closed loop, add-on hot water system, external heat exchanger			
2570	3/8" tubing, 3 ea. 4' x 4'-4" vacuum tube collectors	4,975	4,775	9,750
2580	1/2" tubing, 4 ea. 4 x 4'-4" vacuum tube collectors R235616-60	5,625	5,125	10,750
2600	2 ea. 3'x7' black chrome collectors	4,000	4,825	8,825
2620	3 ea. 3'x7' black chrome collectors	5,050	4,975	10,025
2640	2 ea. 3'x7' flat black collectors	3,400	4,850	8,250
2660	3 ea. 3'x7' flat black collectors	4,175	5,000	9,175
2700	3/4" tubing, 3 ea. 3'x7' black chrome collectors	5,425	5,225	10,650
2720	3 ea. 3'x7' flat black absorber plate collectors	4,550	5,250	9,800
2740	2 ea. 4'x9' flat black w/plastic glazing collectors	4,150	5,275	9,425
2760	2 ea. 3'x7' black chrome collectors	4,350	5,100	9,450
2780	1" tubing, 4 ea 2'x9' plastic absorber & glazing collectors	5,200	5,900	11,100
2800	4 ea. 3'x7' black chrome absorber collectors	7,225	5,925	13,150
2820	4 ea. 3'x7' flat black absorber collectors	6,075	5,950	12,025

D2020 Domestic Water Distribution

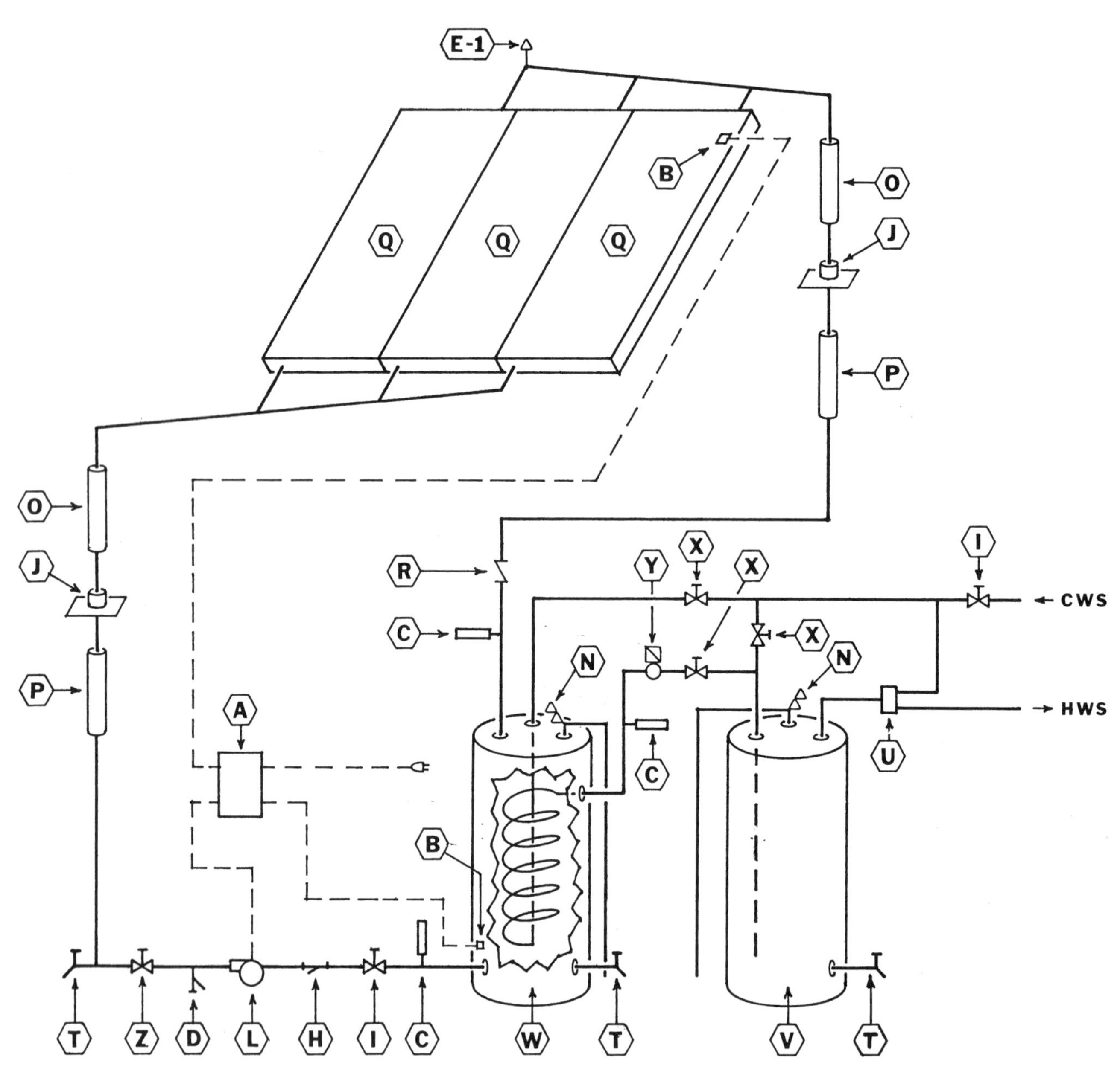

In the drainback indirect-collection system, the heat transfer fluid is distilled water contained in a loop consisting of collectors, supply and return piping, and an unpressurized holding tank. A large heat exchanger containing incoming potable water is immersed in the holding tank. When a controller activates solar collection, the distilled water is pumped through the collectors and heated and pumped back down to the holding tank. When the temperature differential between the water in the collectors and water in storage is such that collection no longer occurs, the pump turns off and gravity causes the distilled water in the collector loop to drain back to the holding tank. All the loop piping is pitched so that the water can drain out of the collectors and piping and not freeze there. As hot water is needed in the home, incoming water first flows through the holding tank with the immersed heat exchanger and is warmed and then flows through a conventional heater for any supplemental heating that is necessary.

D2020 Domestic Water Distribution

System Components		QUANTITY	UNIT	COST EACH MAT.	INST.	TOTAL
	SYSTEM D2020 270 2760					
	SOLAR, DRAINBACK, ADD ON, HOT WATER, IMMERSED HEAT EXCHANGER					
	3/4″ TUBING, THREE EA 3′X7′ BLACK CHROME COLLECTOR					
A, B	Differential controller 2 sensors, thermostat, solar energy system	1.000	Ea.	610	52	662
C	Thermometer 2″ dial	3.000	Ea.	66	117	183
D, T	Fill & drain valve, brass, 3/4″ connection	1.000	Ea.	8.85	26	34.85
E-1	Automatic air vent 1/8″ fitting	1.000	Ea.	13.95	19.50	33.45
H	Strainer, Y type, bronze body, 3/4″ IPS	1.000	Ea.	28.50	33	61.50
I	Valve, gate, bronze, NRS, soldered 3/4″ diam	2.000	Ea.	80	62	142
J	Neoprene vent flashing	2.000	Ea.	26.80	62	88.80
L	Circulator, solar heated liquid, 1/20 HP	1.000	Ea.	170	93.50	263.50
N	Relief valve temp. & press. 150 psi 210°F self-closing 3/4″ IPS	1.000	Ea.	17.15	21	38.15
O	Pipe covering, urethane, ultraviolet cover, 1″ wall, 3/4″ diam	20.000	L.F.	48	111	159
P	Pipe covering, fiberglass, all service jacket, 1″ wall, 3/4″ diam	50.000	L.F.	57	221.50	278.50
Q	Collector panel solar energy blk chrome on copper, 1/8″ temp glas 3′x7′	3.000	Ea.	3,150	354	3,504
	Roof clamps for solar energy collector panels	3.000	Set	7.77	48.15	55.92
R	Valve, swing check, bronze, regrinding disc, 3/4″ diam	1.000	Ea.	72	31	103
U	Valve, water tempering, bronze sweat connections, 3/4″ diam	1.000	Ea.	100	31	131
V	Tank, water storage w/heating element, drain, relief valve, existing	1.000	Ea.			
W	Tank, water storage immersed heat exchr elec 2″x1/2# insul 120 gal	1.000	Ea.	1,350	445	1,795
X	Valve, globe, bronze, rising stem, 3/4″ diam, soldered	3.000	Ea.	306	93	399
Y	Flow control valve	1.000	Ea.	118	28.50	146.50
Z	Valve, ball, bronze, solder 3/4″ diam, solar loop flow control	1.000	Ea.	19.60	31	50.60
	Copper tubing, type L, solder joint, hanger 10′ OC 3/4″ diam	20.000	L.F.	84.20	164	248.20
	Copper tubing, type M, solder joint, hanger 10′ OC 3/4″ diam	70.000	L.F.	225.40	560	785.40
	Sensor wire, #22-2 conductor, multistranded	.500	C.L.F.	7.53	29.25	36.78
	Wrought copper fittings & solder, 3/4″ diam	76.000	Ea.	273.60	2,508	2,781.60
	TOTAL			6,840.35	5,141.40	11,981.75

D2020 270	Solar, Drainback, Hot Water Systems	COST EACH MAT.	INST.	TOTAL
2550	Solar, drainback, hot water, immersed heat exchanger R235616-60			
2560	3/8″ tubing, 3 ea. 4′ x 4′-4″ vacuum tube collectors	6,175	4,625	10,800
2580	1/2″ tubing, 4 ea 4′ x 4′-4″ vacuum tube collectors, 80 gal tank	6,875	5,000	11,875
2600	120 gal tank	7,025	5,050	12,075
2640	2 ea. 3′x7′ blk chrome collectors, 80 gal tank	5,225	4,700	9,925
2660	3 ea. 3′x7′ blk chrome collectors, 120 gal tank	6,425	4,900	11,325
2700	2 ea. 3′x7′ flat blk collectors, 120 gal tank	4,800	4,775	9,575
2720	3 ea. 3′x7′ flat blk collectors, 120 gal tank	5,550	4,900	10,450
2760	3/4″ tubing, 3 ea 3′x7′ black chrome collectors, 120 gal tank	6,850	5,150	12,000
2780	3 ea. 3′x7′ flat black absorber collectors, 120 gal tank	5,975	5,150	11,125
2800	2 ea. 4′x9′ flat blk w/plastic glazing collectors 120 gal tank	5,600	5,175	10,775
2840	1″ tubing, 4 ea. 2′x9′ plastic absorber & glazing collectors, 120 gal tank	6,700	5,825	12,525
2860	4 ea. 3′x7′ black chrome absorber collectors, 120 gal tank	8,725	5,850	14,575
2880	4 ea. 3′x7′ flat black absorber collectors, 120 gal tank	7,575	5,875	13,450

D2020 Domestic Water Distribution

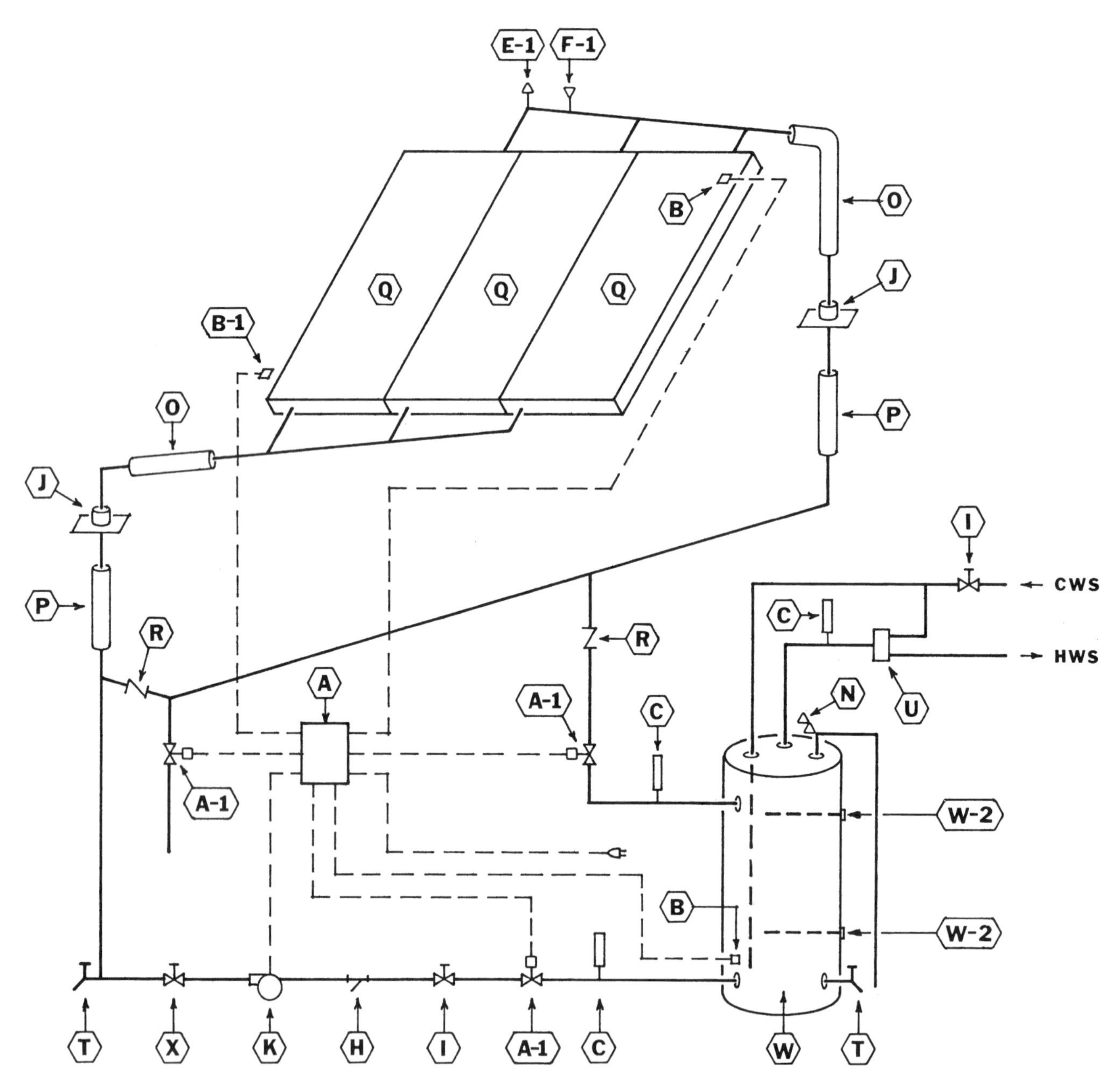

In the draindown direct-collection system, incoming domestic water is heated in the collectors. When the controller activates solar collection, domestic water is first heated as it flows through the collectors and is then pumped to storage. When conditions are no longer suitable for heat collection, the pump shuts off and the water in the loop drains down and out of the system by means of solenoid valves and properly pitched piping.

D2020 Domestic Water Distribution

System Components		QUANTITY	UNIT	COST EACH MAT.	INST.	TOTAL
	SYSTEM D2020 275 2760					
	SOLAR, DRAINDOWN, HOT WATER, DIRECT COLLECTION					
	3/4″ TUBING, THREE 3′X7′ BLACK CHROME COLLECTORS					
A, B	Differential controller, 2 sensors, thermostat, solar energy system	1.000	Ea.	610	52	662
A-1	Solenoid valve, solar heating loop, brass, 3/4″ diam, 24 volts	3.000	Ea.	360	208.50	568.50
B-1	Solar energy sensor, freeze prevention	1.000	Ea.	24	19.50	43.50
C	Thermometer, 2″ dial	3.000	Ea.	66	117	183
E-1	Vacuum relief valve, 3/4″ diam	1.000	Ea.	29	19.50	48.50
F-1	Air vent, automatic, 1/8″ fitting	1.000	Ea.	13.95	19.50	33.45
H	Strainer, Y type, bronze body, 3/4″ IPS	1.000	Ea.	28.50	33	61.50
I	Valve, gate, bronze, NRS, soldered, 3/4″ diam	2.000	Ea.	80	62	142
J	Vent flashing neoprene	2.000	Ea.	26.80	62	88.80
K	Circulator, solar heated liquid, 1/25 HP	1.000	Ea.	295	80.50	375.50
N	Relief valve temp & press 150 psi 210°F self-closing 3/4″ IPS	1.000	Ea.	17.15	21	38.15
O	Pipe covering, urethane, ultraviolet cover, 1″ wall, 3/4″ diam	20.000	L.F.	48	111	159
P	Pipe covering, fiberglass, all service jacket, 1″ wall, 3/4″ diam	50.000	L.F.	57	221.50	278.50
	Roof clamps for solar energy collector panels	3.000	Set	7.77	48.15	55.92
Q	Collector panel solar energy blk chrome on copper, 1/8″ temp glass 3′x7′	3.000	Ea.	3,150	354	3,504
R	Valve, swing check, bronze, regrinding disc, 3/4″ diam, soldered	2.000	Ea.	144	62	206
T	Drain valve, brass, 3/4″ connection	2.000	Ea.	17.70	52	69.70
U	Valve, water tempering, bronze, sweat connections, 3/4″ diam	1.000	Ea.	100	31	131
W-2, W	Tank, water storage elec elem 2″x1/2# insul 120 gal	1.000	Ea.	1,350	445	1,795
X	Valve, globe, bronze, rising stem, 3/4″ diam, soldered	1.000	Ea.	102	31	133
	Copper tubing, type L, solder joints, hangers 10′ OC 3/4″ diam	20.000	L.F.	84.20	164	248.20
	Copper tubing, type M, solder joints, hangers 10′ OC 3/4″ diam	70.000	L.F.	225.40	560	785.40
	Sensor wire, #22-2 conductor, multistranded	.500	C.L.F.	7.53	29.25	36.78
	Wrought copper fittings & solder, 3/4″ diam	76.000	Ea.	273.60	2,508	2,781.60
	TOTAL			7,117.60	5,311.40	12,429

D2020 275	Solar, Draindown, Hot Water Systems	COST EACH MAT.	INST.	TOTAL
2550	Solar, draindown, hot water			
2560	3/8″ tubing, 3 ea. 4′ x 4′-4″ vacuum tube collectors, 80 gal tank	6,250	5,025	11,275
2580	1/2″ tubing, 4 ea. 4′ x 4′-4″ vacuum tube collectors, 80 gal tank R235616 -60	7,175	5,175	12,350
2600	120 gal tank	7,325	5,225	12,550
2640	2 ea. 3′x7′ black chrome collectors, 80 gal tank	5,550	4,875	10,425
2660	3 ea. 3′x7′ black chrome collectors, 120 gal tank	6,750	5,075	11,825
2700	2 ea. 3′x7′ flat black collectors, 120 gal tank	5,100	4,950	10,050
2720	3 ea. 3′x7′ flat black collectors, 120 gal tank	5,825	5,075	10,900
2760	3/4″ tubing, 3 ea. 3′x7′ black chrome collectors, 120 gal tank	7,125	5,300	12,425
2780	3 ea. 3′x7′ flat collectors, 120 gal tank	6,250	5,325	11,575
2800	2 ea. 4′x9′ flat black & plastic glazing collectors, 120 gal tank	5,875	5,350	11,225
2840	1″ tubing, 4 ea. 2′x9′ plastic absorber & glazing collectors, 120 gal tank	9,350	5,975	15,325
2860	4 ea. 3′x7′ black chrome absorber collectors, 120 gal tank	11,400	6,000	17,400
2880	4 ea. 3′x7′ flat black absorber collectors, 120 gal tank	10,200	6,050	16,250

D2020 Domestic Water Distribution

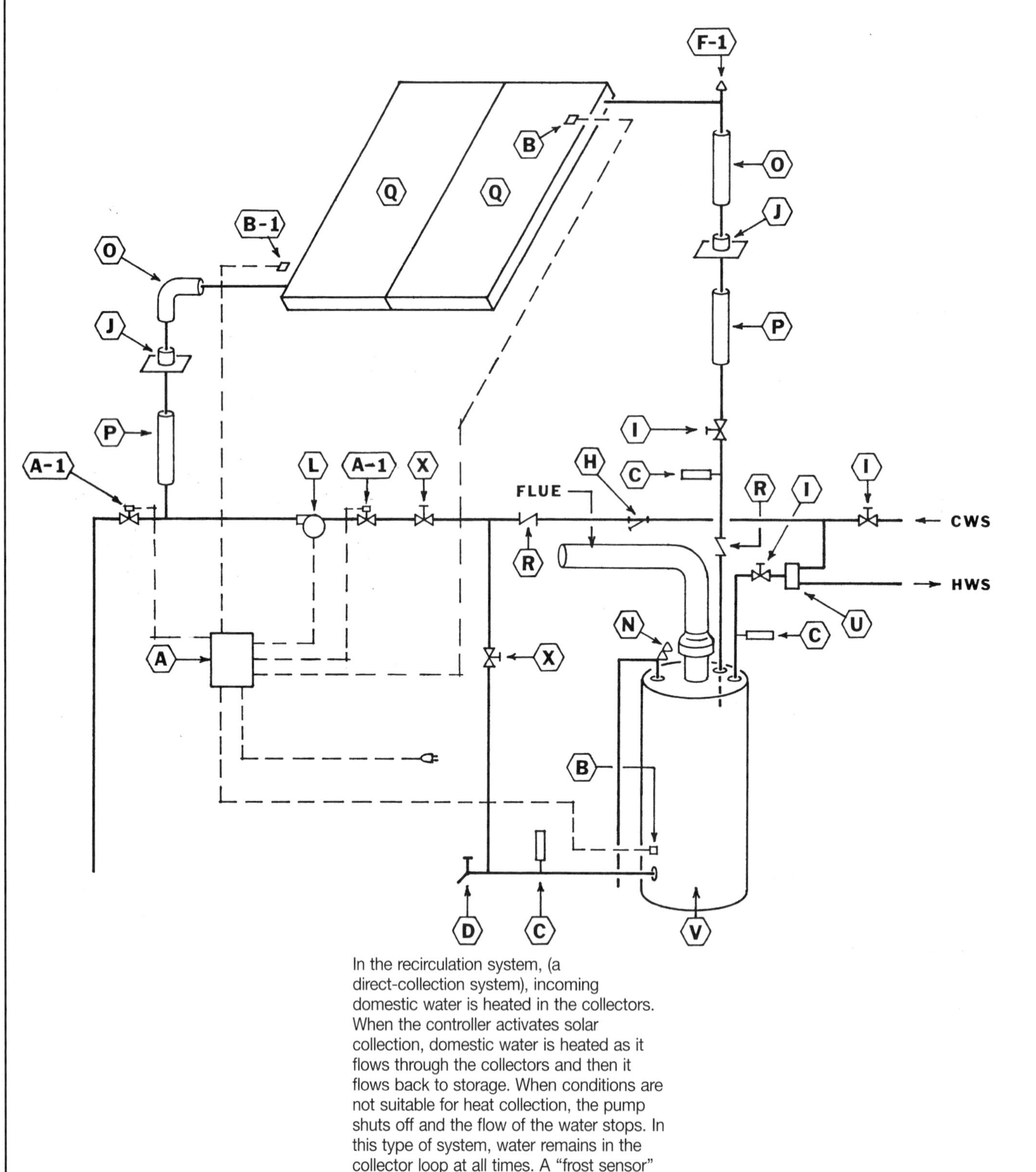

In the recirculation system, (a direct-collection system), incoming domestic water is heated in the collectors. When the controller activates solar collection, domestic water is heated as it flows through the collectors and then it flows back to storage. When conditions are not suitable for heat collection, the pump shuts off and the flow of the water stops. In this type of system, water remains in the collector loop at all times. A "frost sensor" at the collector activates the circulation of warm water from storage through the collectors when protection from freezing is required.

D2020 Domestic Water Distribution

System Components		QUANTITY	UNIT	COST EACH MAT.	COST EACH INST.	COST EACH TOTAL
	SYSTEM D2020 280 2820					
	SOLAR, RECIRCULATION, HOT WATER					
	3/4" TUBING, TWO 3'X7' BLACK CHROME COLLECTORS					
A, B	Differential controller 2 sensors, thermostat, for solar energy system	1.000	Ea.	610	52	662
A-1	Solenoid valve, solar heating loop, brass, 3/4" IPS, 24 volts	2.000	Ea.	240	139	379
B-1	Solar energy sensor freeze prevention	1.000	Ea.	24	19.50	43.50
C	Thermometer, 2" dial	3.000	Ea.	66	117	183
D	Drain valve, brass, 3/4" connection	1.000	Ea.	8.85	26	34.85
F-1	Air vent, automatic, 1/8" fitting	1.000	Ea.	13.95	19.50	33.45
H	Strainer, Y type, bronze body, 3/4" IPS	1.000	Ea.	28.50	33	61.50
I	Valve, gate, bronze, 125 lb, soldered 3/4" diam	3.000	Ea.	120	93	213
J	Vent flashing, neoprene	2.000	Ea.	26.80	62	88.80
L	Circulator, solar heated liquid, 1/20 HP	1.000	Ea.	170	93.50	263.50
N	Relief valve, temp & press 150 psi 210°F self-closing 3/4" IPS	2.000	Ea.	34.30	42	76.30
O	Pipe covering, urethane, ultraviolet cover, 1" wall, 3/4" diam	20.000	L.F.	48	111	159
P	Pipe covering, fiberglass, all service jacket, 1" wall 3/4" diam	50.000	L.F.	57	221.50	278.50
Q	Collector panel solar energy blk chrome on copper, 1/8" temp glass 3'x7'	2.000	Ea.	2,100	236	2,336
	Roof clamps for solar energy collector panels	2.000	Set	5.18	32.10	37.28
R	Valve, swing check, bronze, 125 lb, regrinding disc, soldered 3/4" diam	2.000	Ea.	144	62	206
U	Valve, water tempering, bronze, sweat connections, 3/4" diam	1.000	Ea.	100	31	131
V	Tank, water storage, w/heating element, drain, relief valve, existing	1.000	Ea.			
X	Valve, globe, bronze, 125 lb, 3/4" diam	2.000	Ea.	204	62	266
	Copper tubing, type L, solder joints, hangers 10' OC 3/4" diam	20.000	L.F.	84.20	164	248.20
	Copper tubing, type M, solder joints, hangers 10' OC 3/4" diam	70.000	L.F.	225.40	560	785.40
	Wrought copper fittings & solder, 3/4" diam	76.000	Ea.	273.60	2,508	2,781.60
	Sensor wire, #22-2 conductor, multistranded	.500	C.L.F.	7.53	29.25	36.78
	TOTAL			4,591.31	4,713.35	9,304.66

D2020 280	Solar Recirculation, Domestic Hot Water Systems	COST EACH MAT.	COST EACH INST.	COST EACH TOTAL
2550	Solar recirculation, hot water			
2560	3/8" tubing, 3 ea. 4' x 4'-4" vacuum tube collectors	5,150	4,400	9,550
2580	1/2" tubing, 4 ea. 4' x 4'-4" vacuum tube collectors (R235616 -60)	5,825	4,750	10,575
2640	2 ea. 3'x7' black chrome collectors	4,175	4,450	8,625
2660	3 ea. 3'x7' black chrome collectors	5,250	4,600	9,850
2700	2 ea. 3'x7' flat black collectors	3,600	4,475	8,075
2720	3 ea. 3'x7' flat black collectors	4,375	4,625	9,000
2760	3/4" tubing, 3 ea. 3'x7' black chrome collectors	5,650	4,850	10,500
2780	3 ea. 3'x7' flat black absorber plate collectors	4,775	4,875	9,650
2800	2 ea. 4'x9' flat black w/plastic glazing collectors	4,400	4,875	9,275
2820	2 ea. 3'x7' black chrome collectors	4,600	4,725	9,325
2840	1" tubing, 4 ea. 2'x9' black plastic absorber & glazing collectors	7,125	5,525	12,650
2860	4 ea. 3'x7' black chrome absorber collectors	9,150	5,550	14,700
2880	4 ea. 3'x7' flat black absorber collectors	8,000	5,575	13,575

D2020 Domestic Water Distribution

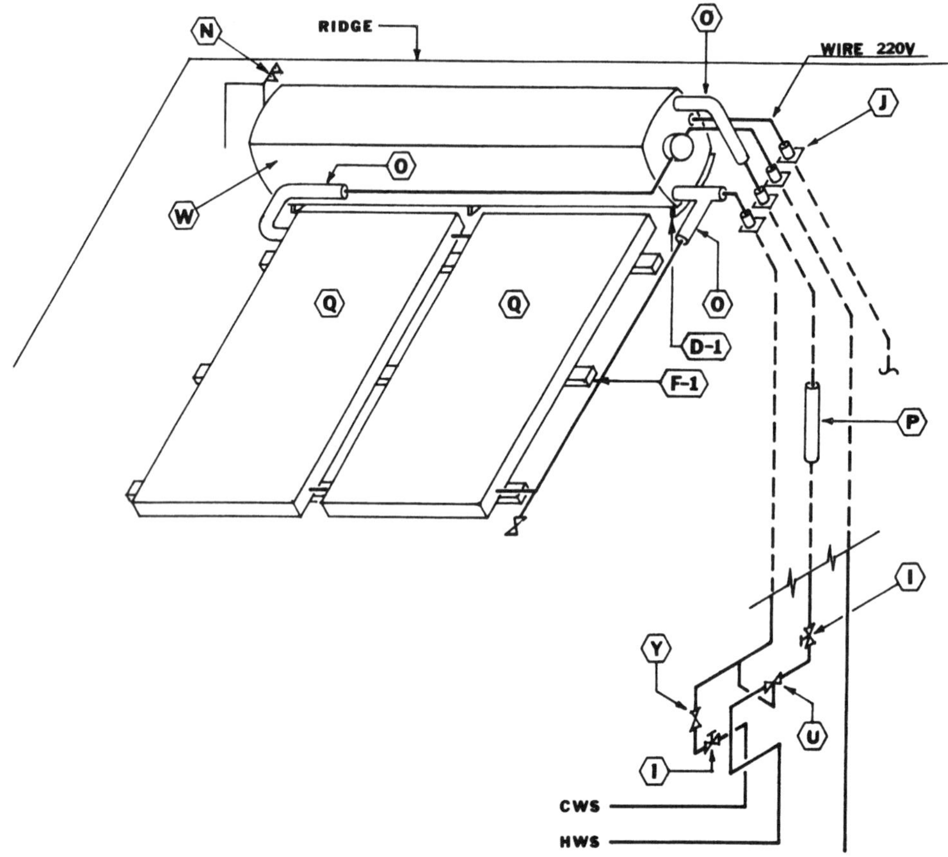

The thermosyphon domestic hot water system, a direct collection system, operates under city water pressure and does not require pumps for system operation. An insulated water storage tank is located above the collectors. As the sun heats the collectors, warm water in them rises by means of natural convection; the colder water in the storage tank flows into the collectors by means of gravity. As long as the sun is shining the water continues to flow through the collectors and to become warmer.

To prevent freezing, the system must be drained or the collectors covered with an insulated lid when the temperature drops below 32°F.

D2020 Domestic Water Distribution

System Components		QUANTITY	UNIT	COST EACH MAT.	INST.	TOTAL
SYSTEM D2020 285 0960						
SOLAR, THERMOSYPHON, WATER HEATER						
3/4″ TUBING, TWO 3′X7′ BLACK CHROME COLLECTORS						
D-1	Framing lumber, fir, 2″ x 6″ x 8′, tank cradle	.008	M.B.F.	3.32	6.56	9.88
F-1	Framing lumber, fir, 2″ x 4″ x 24′, sleepers	.016	M.B.F.	6.56	19.60	26.16
I	Valve, gate, bronze, 125 lb, soldered 1/2″ diam	2.000	Ea.	70	52	122
J	Vent flashing, neoprene	4.000	Ea.	53.60	124	177.60
O	Pipe covering, urethane, ultraviolet cover, 1″ wall, 1/2″ diam	40.000	L.F.	86.40	218	304.40
P	Pipe covering fiberglass all service jacket 1″ wall 1/2″ diam	160.000	L.F.	168	680	848
Q	Collector panel solar, blk chrome on copper, 3/16″ temp glass 3′-6″ x 7.5′	2.000	Ea.	1,520	250	1,770
Y	Flow control valve, globe, bronze, 125#, soldered, 1/2″ diam	1.000	Ea.	75.50	26	101.50
U	Valve, water tempering, bronze, sweat connections, 1/2″ diam	1.000	Ea.	82	26	108
W	Tank, water storage, solar energy system, 80 Gal, 2″ x 1/2 lb insul	1.000	Ea.	1,200	390	1,590
	Copper tubing type L, solder joints, hangers 10′ OC 1/2″ diam	150.000	L.F.	411	1,155	1,566
	Copper tubing type M, solder joints, hangers 10′ OC 1/2″ diam	50.000	L.F.	102.50	372.50	475
	Sensor wire, #22-2 gauge multistranded	.500	C.L.F.	7.53	29.25	36.78
	Wrought copper fittings & solder, 1/2″ diam	75.000	Ea.	120	2,325	2,445
	TOTAL			3,906.41	5,673.91	9,580.32

D2020 285	Thermosyphon, Hot Water		COST EACH MAT.	INST.	TOTAL
0960	Solar, thermosyphon, hot water, two collector system	R235616-60	3,900	5,675	9,575
0970					

D2020 Domestic Water Distribution

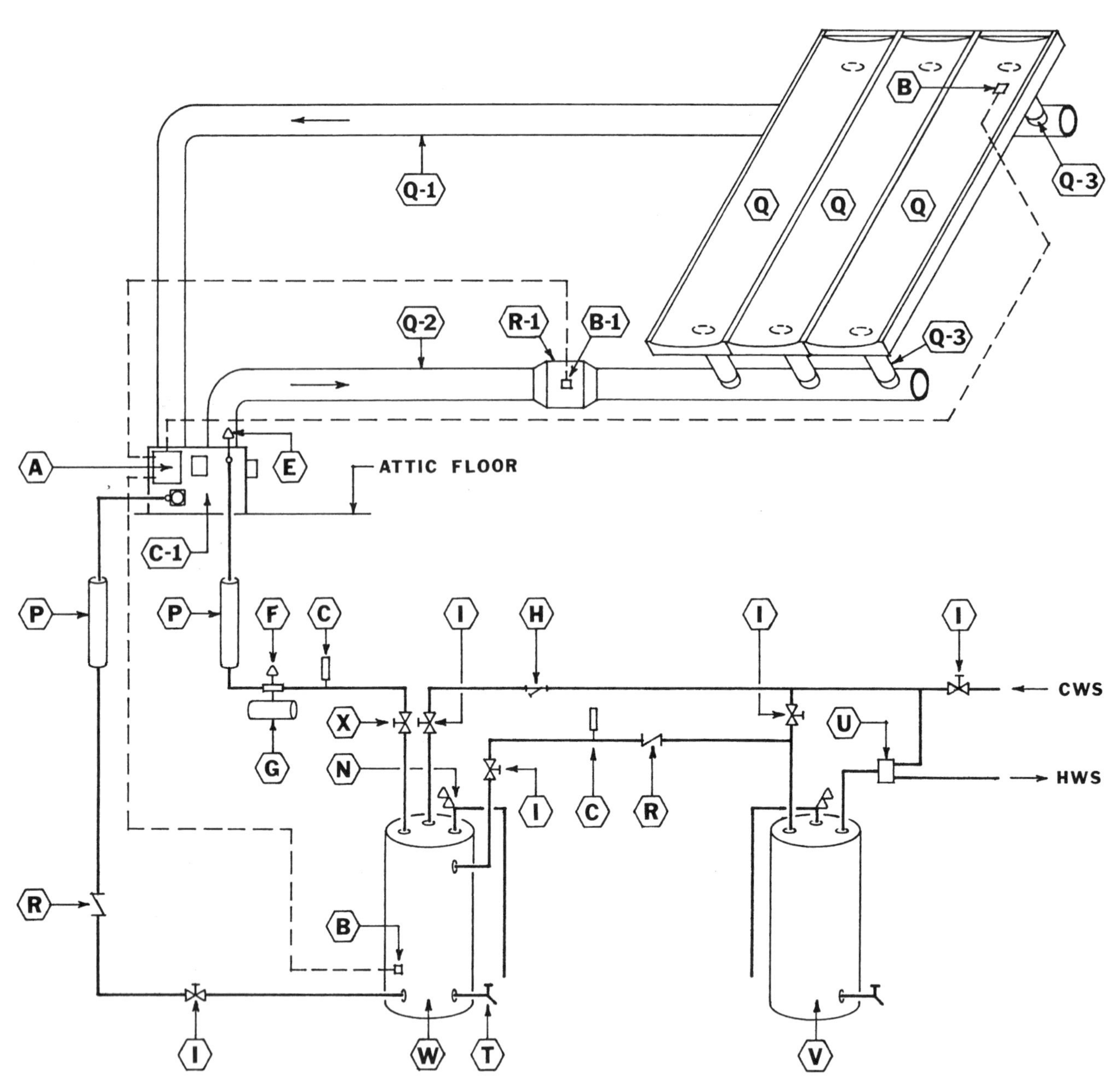

This domestic hot water pre-heat system includes heat exchanger with a circulating pump, blower, air-to-water, coil and controls, mounted in the upper collector manifold. Heat from the hot air coming out of the collectors is transferred through the heat exchanger. For each degree of DHW preheating gained, one degree less heating is needed from the fuel fired water heater. The system is simple, inexpensive to operate and can provide a substantial portion of DHW requirements for modest additional cost.

D2020 Domestic Water Distribution

System Components		QUANTITY	UNIT	COST EACH MAT.	INST.	TOTAL
	SYSTEM D2020 290 2560					
	SOLAR, HOT WATER, AIR TO WATER HEAT EXCHANGE					
	THREE COLLECTORS, OPTICAL BLACK ON ALUMINUM, 10'X2',80 GAL TANK					
A, B	Differential controller, 2 sensors, thermostat, solar energy system	1.000	Ea.	610	52	662
C	Thermometer, 2" dial	2.000	Ea.	44	78	122
C-1	Heat exchanger, air to fluid, up flow 70 MBH	1.000	Ea.	365	320	685
E	Air vent, manual, for solar energy system 1/8" fitting	1.000	Ea.	3.06	19.50	22.56
F	Air purger	1.000	Ea.	47.50	52	99.50
G	Expansion tank	1.000	Ea.	67.50	19.50	87
H	Strainer, Y type, bronze body, 1/2" IPS	1.000	Ea.	25	31	56
I	Valve, gate, bronze, 125 lb, NRS, soldered 1/2" diam	5.000	Ea.	175	130	305
N	Relief valve, temp & pressure solar 150 psi 210°F self-closing	1.000	Ea.	17.15	21	38.15
P	Pipe covering, fiberglass, all service jacket, 1" wall, 1/2" diam	60.000	L.F.	63	255	318
Q	Collector panel solar energy, air, black on alum plate, flush mount, 10'x2'	30.000	L.F.	14,850	334.50	15,184.50
B-1, R-1	Shutter damper for solar heater circulator	1.000	Ea.	54	89	143
B-1, R-1	Shutter motor for solar heater circulator	1.000	Ea.	131	67	198
R	Backflow preventer for solar energy system, 1/2" diam	2.000	Ea.	128	78	206
T	Drain valve, brass, 3/4" connection	1.000	Ea.	8.85	26	34.85
U	Valve, water tempering, bronze, sweat connections, 1/2" diam	1.000	Ea.	82	26	108
W	Tank, water storage, solar energy system, 80 Gal, 2" x 2 lb insul	1.000	Ea.	1,200	390	1,590
V	Tank, water storage, w/heating element, drain, relief valve, existing	1.000	System			
X	Valve, globe, bronze, 125 lb, rising stem, 1/2" diam	1.000	Ea.	75.50	26	101.50
	Copper tubing type M, solder joints, hangers 10' OC 1/2" diam	50.000	L.F.	102.50	372.50	475
	Copper tubing type L, solder joints, hangers 10' OC 1/2" diam	10.000	L.F.	27.40	77	104.40
	Wrought copper fittings & solder, 1/2" diam	10.000	Ea.	16	310	326
	Sensor wire, #22-2 conductor multistranded	.500	C.L.F.	7.53	29.25	36.78
Q-1, Q-2	Ductwork, fiberglass, aluminzed jacket, 1-1/2" thick, 8" diam	32.000	S.F.	184	190.40	374.40
Q-3	Manifold for flush mount solar energy collector panels, air	6.000	L.F.	840	40.20	880.20
	TOTAL			19,123.99	3,033.85	22,157.84

D2020 290	Solar Hot Water, Air To Water Heat Exchange		COST EACH MAT.	INST.	TOTAL
2550	Solar hot water, air to water heat exchange				
2560	Three collectors, optical black on aluminum, 10' x 2', 80 Gal tank		19,100	3,025	22,125
2580	Four collectors, optical black on aluminum, 10' x 2', 80 Gal tank	R235616 -60	24,400	3,200	27,600
2600	Four collectors, optical black on aluminum, 10' x 2', 120 Gal tank		24,600	3,250	27,850

D2020 Domestic Water Distribution

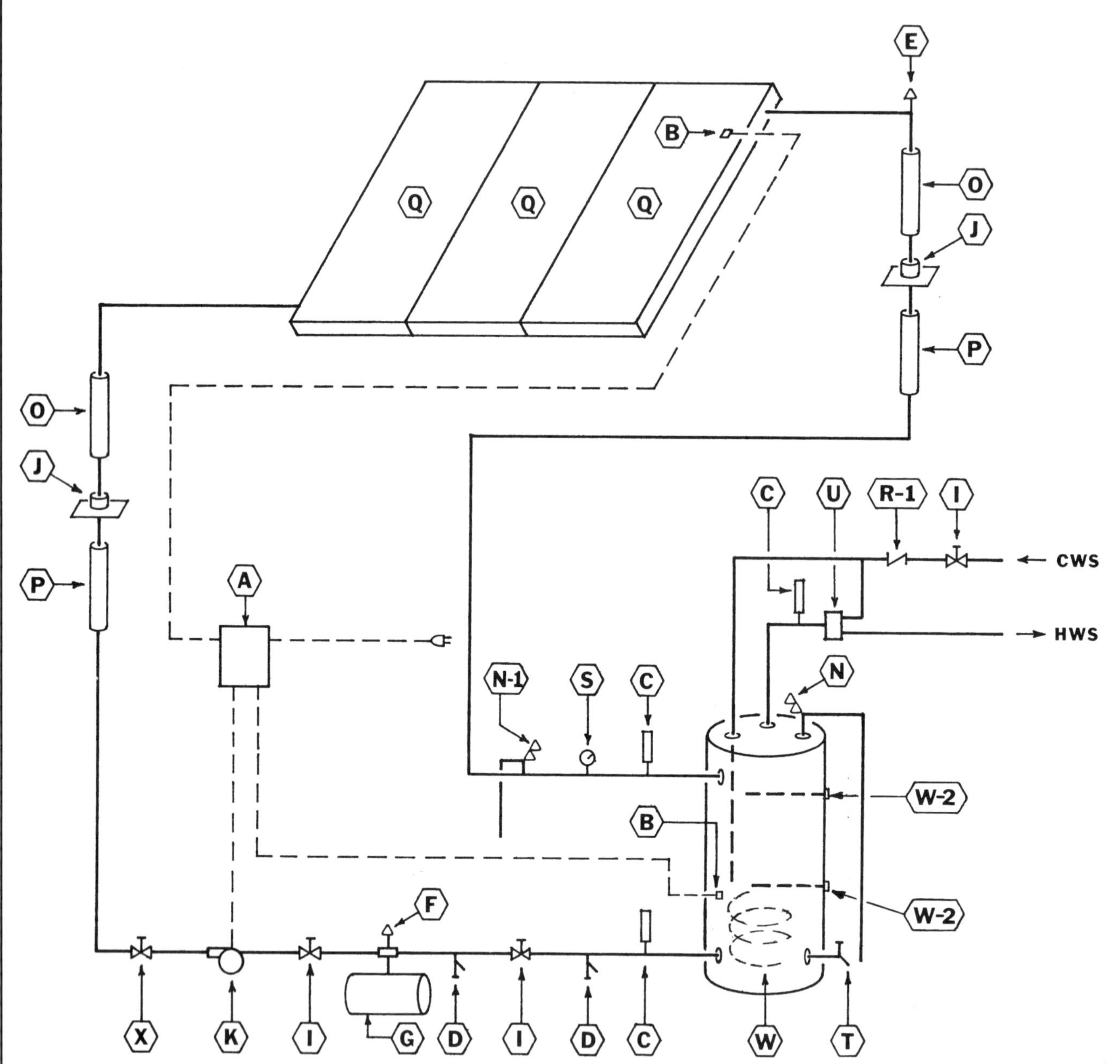

In this closed-loop indirect collection system, fluid with a low freezing temperature, such as propylene glycol, transports heat from the collectors to water storage. The transfer fluid is contained in a closed-loop consisting of collectors, supply and return piping, and a heat exchanger immersed in the storage tank. A typical two-or-three panel system contains 5 to 6 gallons of heat transfer fluid.

When the collectors become approximately 20°F warmer than the storage temperature, a controller activates the circulator. The circulator moves the fluid continuously through the collectors until the temperature difference between the collectors and storage is such that heat collection no longer occurs; at that point, the circulator shuts off. Since the heat transfer fluid has a very low freezing temperature, there is no need for it to be drained from the collectors between periods of collection.

D2020 Domestic Water Distribution

System Components		QUANTITY	UNIT	COST EACH MAT.	INST.	TOTAL
	SYSTEM D2020 295 2760 SOLAR, CLOSED LOOP, HOT WATER SYSTEM, IMMERSED HEAT EXCHANGER 3/4″ TUBING, THREE 3′ X 7′ BLACK CHROME COLLECTORS					
A, B	Differential controller, 2 sensors, thermostat, solar energy system	1.000	Ea.	610	52	662
C	Thermometer 2″ dial	3.000	Ea.	66	117	183
D, T	Fill & drain valves, brass, 3/4″ connection	3.000	Ea.	26.55	78	104.55
E	Air vent, manual, 1/8″ fitting	1.000	Ea.	3.06	19.50	22.56
F	Air purger	1.000	Ea.	47.50	52	99.50
G	Expansion tank	1.000	Ea.	67.50	19.50	87
I	Valve, gate, bronze, NRS, soldered 3/4″ diam	3.000	Ea.	120	93	213
J	Neoprene vent flashing	2.000	Ea.	26.80	62	88.80
K	Circulator, solar heated liquid, 1/25 HP	1.000	Ea.	295	80.50	375.50
N-1, N	Relief valve, temp & press 150 psi 210°F self-closing 3/4″ IPS	2.000	Ea.	34.30	42	76.30
O	Pipe covering, urethane ultraviolet cover, 1″ wall, 3/4″ diam	20.000	L.F.	48	111	159
P	Pipe covering, fiberglass, all service jacket, 1″ wall, 3/4″ diam	50.000	L.F.	57	221.50	278.50
	Roof clamps for solar energy collector panel	3.000	Set	7.77	48.15	55.92
Q	Collector panel solar blk chrome on copper, 1/8″ temp glass, 3′x7′	3.000	Ea.	3,150	354	3,504
R-1	Valve, swing check, bronze, regrinding disc, 3/4″ diam, soldered	1.000	Ea.	72	31	103
S	Pressure gauge, 60 psi, 2-1/2″ dial	1.000	Ea.	24.50	19.50	44
U	Valve, water tempering, bronze, sweat connections, 3/4″ diam	1.000	Ea.	100	31	131
W-2, W	Tank, water storage immersed heat exchr elec elem 2″x2# insul 120 Gal	1.000	Ea.	1,350	445	1,795
X	Valve, globe, bronze, rising stem, 3/4″ diam, soldered	1.000	Ea.	102	31	133
	Copper tubing type L, solder joint, hanger 10′ OC 3/4″ diam	20.000	L.F.	84.20	164	248.20
	Copper tubing, type M, solder joint, hanger 10′ OC 3/4″ diam	70.000	L.F.	225.40	560	785.40
	Sensor wire, #22-2 conductor multistranded	.500	C.L.F.	7.53	29.25	36.78
	Solar energy heat transfer fluid, propylene glycol, anti-freeze	6.000	Gal.	84.90	135	219.90
	Wrought copper fittings & solder, 3/4″ diam	76.000	Ea.	273.60	2,508	2,781.60
	TOTAL			6,883.61	5,303.90	12,187.51

D2020 295	Solar, Closed Loop, Hot Water Systems	COST EACH MAT.	INST.	TOTAL
2550	Solar, closed loop, hot water system, immersed heat exchanger			
2560	3/8″ tubing, 3 ea. 4′ x 4′-4″ vacuum tube collectors, 80 gal. tank	6,300	4,800	11,100
2580	1/2″ tubing, 4 ea. 4′ x 4′-4″ vacuum tube collectors, 80 gal. tank	6,975	5,175	12,150
2600	120 gal. tank	7,125	5,225	12,350
2640	2 ea. 3′x7′ black chrome collectors, 80 gal. tank (R235616-60)	5,375	4,925	10,300
2660	120 gal. tank	5,475	4,925	10,400
2700	2 ea. 3′x7′ flat black collectors, 120 gal. tank	4,900	4,950	9,850
2720	3 ea. 3′x7′ flat black collectors, 120 gal. tank	5,650	5,075	10,725
2760	3/4″ tubing, 3 ea. 3′x7′ black chrome collectors, 120 gal. tank	6,875	5,300	12,175
2780	3 ea. 3′x7′ flat black collectors, 120 gal. tank	6,025	5,325	11,350
2800	2 ea. 4′x9′ flat black w/plastic glazing collectors 120 gal. tank	5,625	5,350	10,975
2840	1″ tubing, 4 ea. 2′x9′ plastic absorber & glazing collectors 120 gal. tank	6,600	5,950	12,550
2860	4 ea. 3′x7′ black chrome collectors, 120 gal. tank	8,625	5,975	14,600
2880	4 ea. 3′x7′ flat black absorber collectors, 120 gal. tank	7,450	6,000	13,450

D3010 Energy Supply

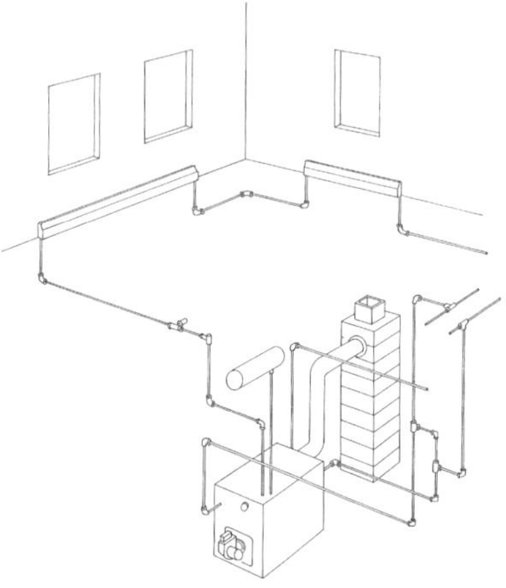

Basis for Heat Loss Estimate, Apartment Type Structures:

1. Masonry walls and flat roof are insulated. U factor is assumed at .08.
2. Window glass area taken as BOCA minimum, 1/10th of floor area. Double insulating glass with 1/4″ air space, U = .65.
3. Infiltration = 0.3 C.F. per hour per S.F. of net wall.
4. Concrete floor loss is 2 BTUH per S.F.
5. Temperature difference taken as 70°F.
6. Ventilating or makeup air has not been included and must be added if desired. Air shafts are not used.

System Components	QUANTITY	UNIT	COST EACH		
			MAT.	INST.	TOTAL
SYSTEM D3010 510 1760					
HEATING SYSTEM, FIN TUBE RADIATION, FORCED HOT WATER					
1,000 S.F. AREA, 10,000 C.F. VOLUME					
Boiler, oil fired, CI, burner, ctrls/insul/breech/pipe/ftng/valves, 109 MBH	1.000	Ea.	3,806.25	3,456.25	7,262.50
Circulating pump, CI flange connection, 1/12 HP	1.000	Ea.	350	187	537
Expansion tank, painted steel, ASME 18 Gal capacity	1.000	Ea.	610	80	690
Storage tank, steel, above ground, 275 Gal capacity w/supports	1.000	Ea.	410	224	634
Copper tubing type L, solder joint, hanger 10′ OC, 3/4″ diam	100.000	L.F.	421	820	1,241
Radiation, 3/4″ copper tube w/alum fin baseboard pkg, 7″ high	30.000	L.F.	264	579	843
Pipe covering, calcium silicate w/cover, 1″ wall, 3/4″ diam	100.000	L.F.	310	600	910
TOTAL			6,171.25	5,946.25	12,117.50
COST PER S.F.			6.17	5.95	12.12

D3010 510	Apartment Building Heating - Fin Tube Radiation		COST PER S.F.		
			MAT.	INST.	TOTAL
1740	Heating systems, fin tube radiation, forced hot water				
1760	1,000 S.F. area, 10,000 C.F. volume		6.18	5.95	12.13
1800	10,000 S.F. area, 100,000 C.F. volume	R235000 -10	2.51	3.59	6.10
1840	20,000 S.F. area, 200,000 C.F. volume		2.83	4.03	6.86
1880	30,000 S.F. area, 300,000 C.F. volume	R235000 -20	2.73	3.92	6.65
1890					

D3010 Energy Supply

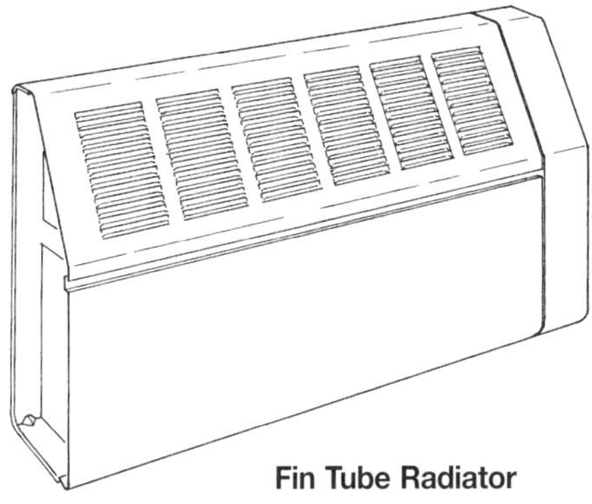
Fin Tube Radiator

Basis for Heat Loss Estimate, Factory or Commercial Type Structures:

1. Walls and flat roof are of lightly insulated concrete block or metal. U factor is assumed at .17.
2. Windows and doors are figured at 1-1/2 S.F. per linear foot of building perimeter. The U factor for flat single glass is 1.13.
3. Infiltration is approximately 5 C.F. per hour per S.F. of wall.
4. Concrete floor loss is 2 BTUH per S.F. of floor.
5. Temperature difference is assumed at 70°F.
6. Ventilation or makeup air has not been included and must be added if desired.

System Components	QUANTITY	UNIT	COST EACH MAT.	COST EACH INST.	COST EACH TOTAL
SYSTEM D3010 520 1960					
HEATING SYSTEM, FIN TUBE RADIATION, FORCED HOT WATER					
1,000 S.F. BLDG., ONE FLOOR					
Boiler, oil fired, CI, burner/ctrls/insul/breech/pipe/ftngs/valves, 109 MBH	1.000	Ea.	3,806.25	3,456.25	7,262.50
Expansion tank, painted steel, ASME 18 Gal capacity	1.000	Ea.	2,375	93.50	2,468.50
Storage tank, steel, above ground, 550 Gal capacity w/supports	1.000	Ea.	2,475	415	2,890
Circulating pump, CI flanged, 1/8 HP	1.000	Ea.	590	187	777
Pipe, steel, black, sch. 40, threaded, cplg & hngr 10'OC, 1-1/2" diam.	260.000	L.F.	2,704	3,653	6,357
Pipe covering, calcium silicate w/cover , 1" wall, 1-1/2" diam	260.000	L.F.	819	1,612	2,431
Radiation, steel 1-1/4" tube & 4-1/4" fin w/cover & damper, wall hung	42.000	L.F.	1,785	1,302	3,087
Rough in, steel fin tube radiation w/supply & balance valves	4.000	Set	1,112	2,740	3,852
TOTAL			15,666.25	13,458.75	29,125
COST PER S.F.			15.67	13.46	29.13

D3010 520	Commercial Building Heating - Fin Tube Radiation		COST PER S.F. MAT.	COST PER S.F. INST.	COST PER S.F. TOTAL
1940	Heating systems, fin tube radiation, forced hot water				
1960	1,000 S.F. bldg, one floor		15.70	13.50	29.20
2000	10,000 S.F., 100,000 C.F., total two floors	R235000 -10	4.56	5.10	9.66
2040	100,000 S.F., 1,000,000 C.F., total three floors		1.63	2.27	3.90
2080	1,000,000 S.F., 10,000,000 C.F., total five floors	R235000 -20	.86	1.20	2.06
2090					

D3010 Energy Supply

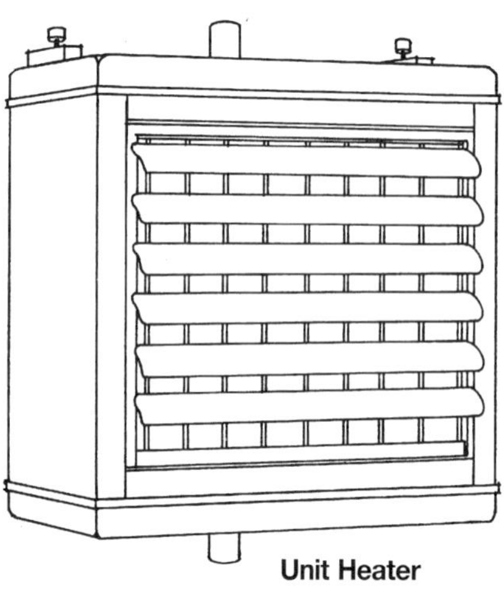

Unit Heater

Basis for Heat Loss Estimate, Factory or Commercial Type Structures:

1. Walls and flat roof are of lightly insulated concrete block or metal. U factor is assumed at .17.
2. Windows and doors are figured at 1-1/2 S.F. per linear foot of building perimeter. The U factor for flat single glass is 1.13.
3. Infiltration is approximately 5 C.F. per hour per S.F. of wall.
4. Concrete floor loss is 2 BTUH per S.F. of floor.
5. Temperature difference is assumed at 70°F.
6. Ventilation or makeup air has not been included and must be added if desired.

System Components	QUANTITY	UNIT	COST EACH MAT.	COST EACH INST.	COST EACH TOTAL
SYSTEM D3010 530 1880					
HEATING SYSTEM, TERMINAL UNIT HEATERS, FORCED HOT WATER					
1,000 S.F. BLDG., ONE FLOOR					
Boiler oil fired, CI, burner/ctrls/insul/breech/pipe/ftngs/valves, 109 MBH	1.000	Ea.	3,806.25	3,456.25	7,262.50
Expansion tank, painted steel, ASME 18 Gal capacity	1.000	Ea.	2,375	93.50	2,468.50
Storage tank, steel, above ground, 550 Gal capacity w/supports	1.000	Ea.	2,475	415	2,890
Circulating pump, CI, flanged, 1/8 HP	1.000	Ea.	590	187	777
Pipe, steel, black, schedule 40, threaded, cplg & hngr 10′ OC 1-1/2″ diam	260.000	L.F.	2,704	3,653	6,357
Pipe covering, calcium silicate w/ cover, 1″ wall, 1-1/2″ diam	260.000	L.F.	819	1,612	2,431
Unit heater, 1 speed propeller, horizontal, 200° EWT, 26.9 MBH	2.000	Ea.	1,110	280	1,390
Unit heater piping hookup with controls	2.000	Set	1,310	2,650	3,960
TOTAL			15,189.25	12,346.75	27,536
COST PER S.F.			15.19	12.35	27.54

D3010 530	Commercial Bldg. Heating - Terminal Unit Heaters		COST PER S.F. MAT.	COST PER S.F. INST.	COST PER S.F. TOTAL
1860	Heating systems, terminal unit heaters, forced hot water				
1880	1,000 S.F. bldg., one floor		15.20	12.30	27.50
1920	10,000 S.F. bldg., 100,000 C.F. total two floors	R235000 -10	4.08	4.25	8.33
1960	100,000 S.F. bldg., 1,000,000 C.F. total three floors		1.63	2.07	3.70
2000	1,000,000 S.F. bldg., 10,000,000 C.F. total five floors	R235000 -20	1.10	1.35	2.45
2010					

D3010 Energy Supply

Many styles of active solar energy systems exist. Those shown on the following page represent the majority of systems now being installed in different regions of the country.

The five active domestic hot water (DHW) systems which follow are typically specified as two or three panel systems with additional variations being the type of glazing and size of the storage tanks. Various combinations have been costed for the user's evaluation and comparison. The basic specifications from which the following systems were developed satisfy the construction detail requirements specified in the HUD Intermediate Minimum Property Standards (IMPS). If these standards are not complied with, the renewable energy system's costs could be significantly lower than shown.

To develop the system's specifications and costs it was necessary to make a number of assumptions about the systems. Certain systems are more appropriate to one climatic region than another or the systems may require modifications to be usable in particular locations. Specific instances in which the systems are not appropriate throughout the country as specified, or in which modification will be needed include the following:

- The freeze protection mechanisms provided in the DHW systems vary greatly. In harsh climates, where freeze is a major concern, a closed-loop indirect collection system may be more appropriate than a direct collection system.
- The thermosyphon water heater system described cannot be used when temperatures drop below 32°F.
- In warm climates it may be necessary to modify the systems installed to prevent overheating.

For each renewable resource (solar) system a schematic diagram and descriptive summary of the system is provided along with a list of all the components priced as part of the system. The costs were developed based on these specifications.

Considerations affecting costs which may increase or decrease beyond the estimates presented here include the following:

- Special structural qualities (allowance for earthquake, future expansion, high winds, and unusual spans or shapes);
- Isolated building site or rough terrain that would affect the transportation of personnel, material, or equipment;
- Unusual climatic conditions during the construction process;
- Substitution of other materials or system components for those used in the system specifications.

D3010 Energy Supply

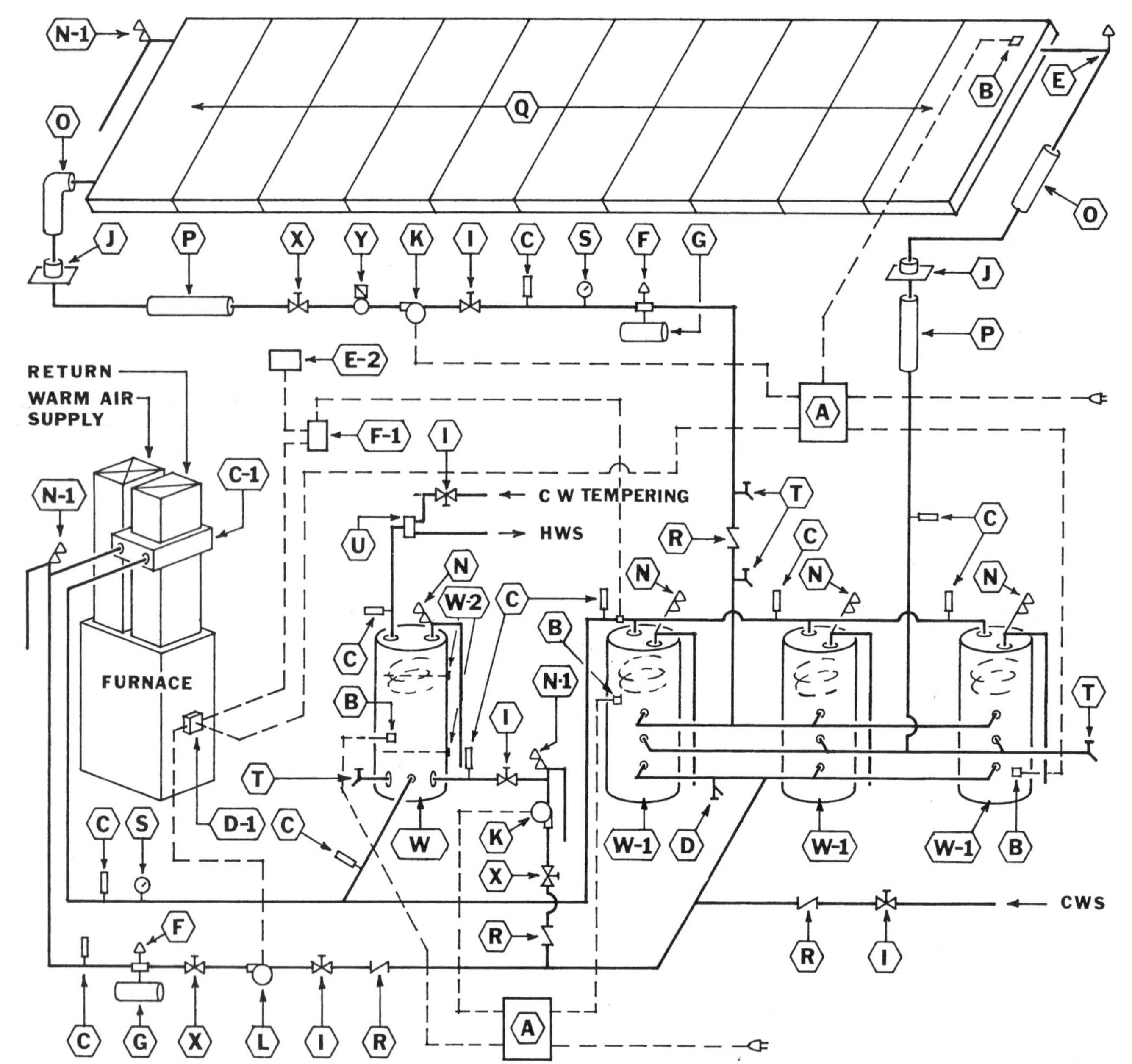

In this closed-loop indirect collection system, fluid with a low freezing temperature, propylene glycol, transports heat from the collectors to water storage. The transfer fluid is contained in a closed loop consisting of collectors, supply and return piping, and a heat exchanger immersed in the storage tank.

When the collectors become approximately 20°F warmer than the storage temperature, the controller activates the circulator. The circulator moves the fluid continuously until the temperature difference between fluid in the collectors and storage is such that the collection will no longer occur and then the circulator turns off. Since the heat transfer fluid has a very low freezing temperature, there is no need for it to be drained from the collectors between periods of collection.

D3010 Energy Supply

System Components

System Components		QUANTITY	UNIT	COST EACH MAT.	COST EACH INST.	COST EACH TOTAL
SYSTEM D3010 650 2750						
SOLAR, CLOSED LOOP, SPACE/HOT WATER						
1″ TUBING, TEN 3′X7′ BLK CHROME ON COPPER ABSORBER COLLECTORS						
A, B	Differential controller 2 sensors, thermostat, solar energy system	2.000	Ea.	1,220	104	1,324
C	Thermometer, 2″ dial	10.000	Ea.	220	390	610
C-1	Heat exchanger, solar energy system, fluid to air, up flow, 80 MBH	1.000	Ea.	490	375	865
D, T	Fill & drain valves, brass, 3/4″ connection	5.000	Ea.	44.25	130	174.25
D-1	Fan center	1.000	Ea.	141	130	271
E	Air vent, manual, 1/8″ fitting	1.000	Ea.	3.06	19.50	22.56
E-2	Thermostat, 2 stage for sensing room temperature	1.000	Ea.	157	74.50	231.50
F	Air purger	2.000	Ea.	95	104	199
F-1	Controller, liquid temperature, solar energy system	1.000	Ea.	105	125	230
G	Expansion tank	2.000	Ea.	135	39	174
I	Valve, gate, bronze, 125 lb, soldered, 1″ diam	5.000	Ea.	285	165	450
J	Vent flashing, neoprene	2.000	Ea.	26.80	62	88.80
K	Circulator, solar heated liquid, 1/25 HP	2.000	Ea.	590	161	751
L	Circulator, solar heated liquid, 1/20 HP	1.000	Ea.	170	93.50	263.50
N	Relief valve, temp & pressure 150 psi 210° F self-closing	4.000	Ea.	68.60	84	152.60
N-1	Relief valve, pressure poppet, bronze, 30 psi, 3/4″ IPS	3.000	Ea.	115.50	67.50	183
O	Pipe covering, urethane, ultraviolet cover, 1″ wall, 1″ diam	50.000	L.F.	120	277.50	397.50
P	Pipe covering, fiberglass, all service jacket, 1″ wall, 1″ diam	60.000	L.F.	68.40	265.80	334.20
Q	Collector panel solar energy blk chrome on copper 1/8″ temp glass 3′x7′	10.000	Ea.	10,500	1,180	11,680
	Roof clamps for solar energy collector panel	10.000	Set	25.90	160.50	186.40
R	Valve, swing check, bronze, 125 lb, regrinding disc, 3/4″ & 1″ diam	4.000	Ea.	416	132	548
S	Pressure gage, 0-60 psi, for solar energy system	2.000	Ea.	49	39	88
U	Valve, water tempering, bronze, sweat connections, 3/4″ diam	1.000	Ea.	100	31	131
W-2, W-1, W	Tank, water storage immersed heat xchr elec elem 2″x1/2# ins 120 gal	4.000	Ea.	5,400	1,780	7,180
X	Valve, globe, bronze, 125 lb, rising stem, 1″ diam	3.000	Ea.	480	99	579
Y	Valve, flow control	1.000	Ea.	118	28.50	146.50
	Copper tubing type M, solder joint, hanger 10′ OC, 1″ diam	110.000	L.F.	572	979	1,551
	Copper tubing, type L, solder joint, hanger 10′ OC 3/4″ diam	20.000	L.F.	84.20	164	248.20
	Wrought copper fittings & solder, 3/4″ & 1″ diam	121.000	Ea.	1,064.80	4,719	5,783.80
	Sensor, wire, #22-2 conductor, multistranded	.700	C.L.F.	10.54	40.95	51.49
	Ductwork, galvanized steel, for heat exchanger	8.000	Lb.	4.16	56.80	60.96
	Solar energy heat transfer fluid propylene glycol, anti-freeze	25.000	Gal.	353.75	562.50	916.25
	TOTAL			23,232.96	12,639.55	35,872.51

D3010 650 Solar, Closed Loop, Space/Hot Water Systems

D3010 650	Solar, Closed Loop, Space/Hot Water Systems	COST EACH MAT.	COST EACH INST.	COST EACH TOTAL
2540	Solar, closed loop, space/hot water			
2550	1/2″ tubing, 12 ea. 4′x4′4″ vacuum tube collectors	22,100	11,800	33,900
2600	3/4″ tubing, 12 ea. 4′x4′4″ vacuum tube collectors (R235616 -60)	23,100	12,200	35,300
2650	10 ea. 3′ x 7′ black chrome absorber collectors	22,300	11,800	34,100
2700	10 ea. 3′ x 7′ flat black absorber collectors	19,400	11,900	31,300
2750	1″ tubing, 10 ea. 3′ x 7′ black chrome absorber collectors	23,200	12,600	35,800
2800	10 ea. 3′ x 7′ flat black absorber collectors	20,400	12,700	33,100
2850	6 ea. 4′ x 9′ flat black w/plastic glazing collectors	18,500	12,600	31,100
2900	12 ea. 2′ x 9′ plastic absorber and glazing collectors	19,300	12,900	32,200

D3010 Energy Supply

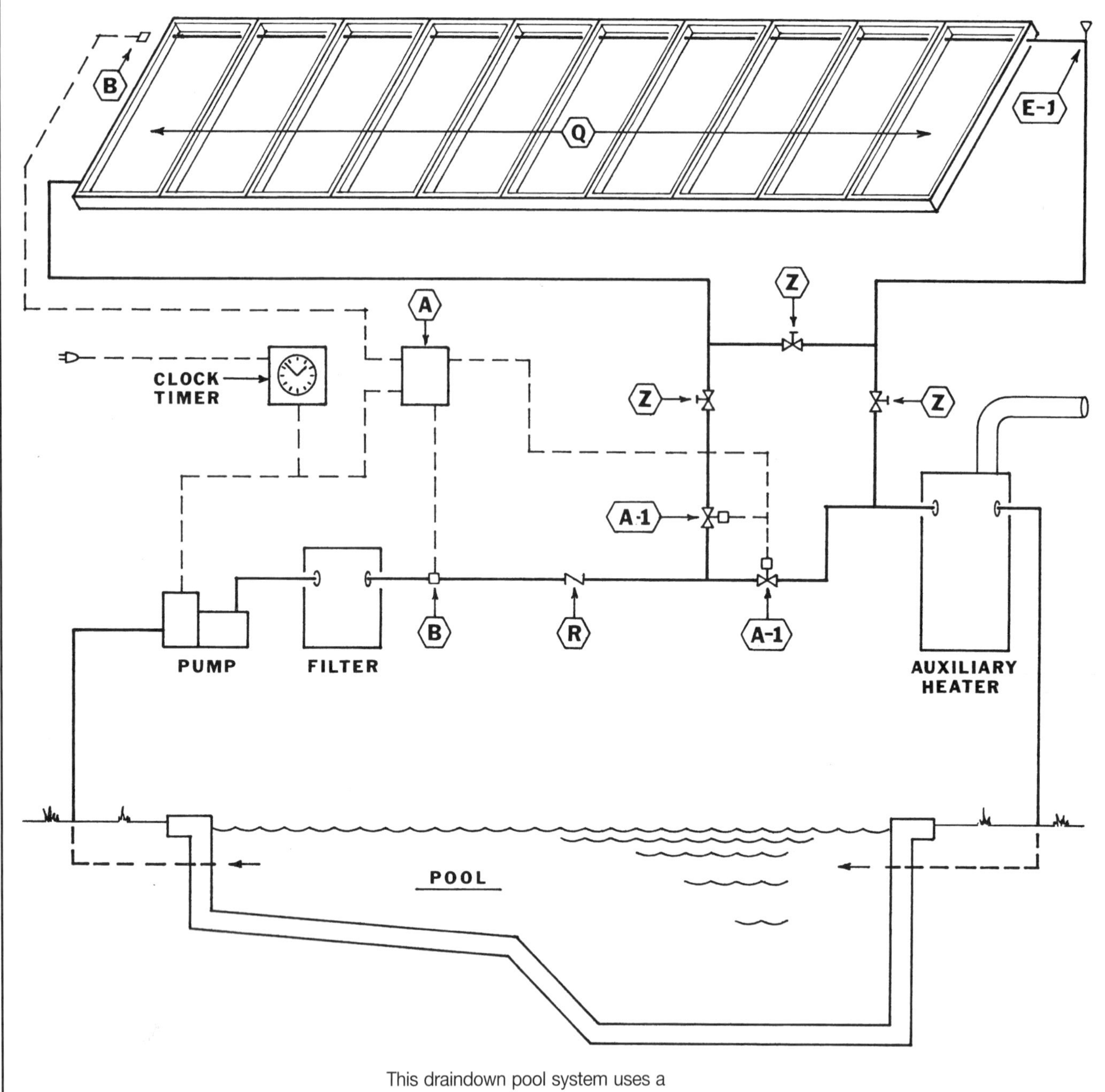

This draindown pool system uses a differential thermostat similar to those used in solar domestic hot water and space heating applications. To heat the pool, the pool water passes through the conventional pump-filter loop and then flows through the collectors. When collection is not possible, or when the pool temperature is reached, all water drains from the solar loop back to the pool through the existing piping. The modes are controlled by solenoid valves or other automatic valves in conjunction with a vacuum breaker relief valve, which facilitates draindown.

D3010 Energy Supply

System Components

System Components	QUANTITY	UNIT	COST EACH MAT.	COST EACH INST.	COST EACH TOTAL
SYSTEM D3010 660 2640					
SOLAR SWIMMING POOL HEATER, ROOF MOUNTED COLLECTORS					
TEN 4′ X 10′ FULLY WETTED UNGLAZED PLASTIC ABSORBERS					
A Differential thermostat/controller, 110V, adj pool pump system	1.000	Ea.	305	310	615
A-1 Solenoid valve, PVC, normally 1 open 1 closed (included)	2.000	Ea.			
B Sensor, thermistor type (included)	2.000	Ea.			
E-1 Valve, vacuum relief	1.000	Ea.	29	19.50	48.50
Q Collector panel, solar energy, plastic, liquid full wetted, 4′ x 10′	10.000	Ea.	2,570	2,250	4,820
R Valve, ball check, PVC, socket, 1-1/2″ diam	1.000	Ea.	85.50	31	116.50
Z Valve, ball, PVC, socket, 1-1/2″ diam	3.000	Ea.	160.50	93	253.50
Pipe, PVC, sch 40, 1-1/2″ diam	80.000	L.F.	256.80	1,388	1,644.80
Pipe fittings, PVC sch 40, socket joint, 1-1/2″ diam	10.000	Ea.	17.40	310	327.40
Sensor wire, #22-2 conductor, multistranded	.500	C.L.F.	7.53	29.25	36.78
Roof clamps for solar energy collector panels	10.000	Set	25.90	160.50	186.40
Roof strap, teflon for solar energy collector panels	26.000	L.F.	572	79.04	651.04
TOTAL			4,029.63	4,670.29	8,699.92

D3010 660 Solar Swimming Pool Heater Systems

		COST EACH MAT.	COST EACH INST.	COST EACH TOTAL
2530	Solar swimming pool heater systems, roof mounted collectors			
2540	10 ea. 3′x7′ black chrome absorber, 1/8″ temp. glass	12,000	3,600	15,600
2560	10 ea. 4′x8′ black chrome absorber, 3/16″ temp. glass R235616-60	13,500	4,300	17,800
2580	10 ea. 3′8″x6′ flat black absorber, 3/16″ temp. glass	9,050	3,675	12,725
2600	10 ea. 4′x9′ flat black absorber, plastic glazing	11,000	4,450	15,450
2620	10 ea. 2′x9′ rubber absorber, plastic glazing	12,400	4,825	17,225
2640	10 ea. 4′x10′ fully wetted unglazed plastic absorber	4,025	4,675	8,700
2660	Ground mounted collectors			
2680	10 ea. 3′x7′ black chrome absorber, 1/8″ temp. glass	12,000	4,150	16,150
2700	10 ea. 4′x8′ black chrome absorber, 3/16″ temp. glass	13,500	4,825	18,325
2720	10 ea. 3′8″x6′ flat blk absorber, 3/16″ temp. glass	9,150	4,225	13,375
2740	10 ea. 4′x9′ flat blk absorber, plastic glazing	11,000	5,000	16,000
2760	10 ea. 2′x9′ rubber absorber, plastic glazing	12,400	5,200	17,600
2780	10 ea. 4′x10′ fully wetted unglazed plastic absorber	4,100	5,225	9,325

D3010 Energy Supply

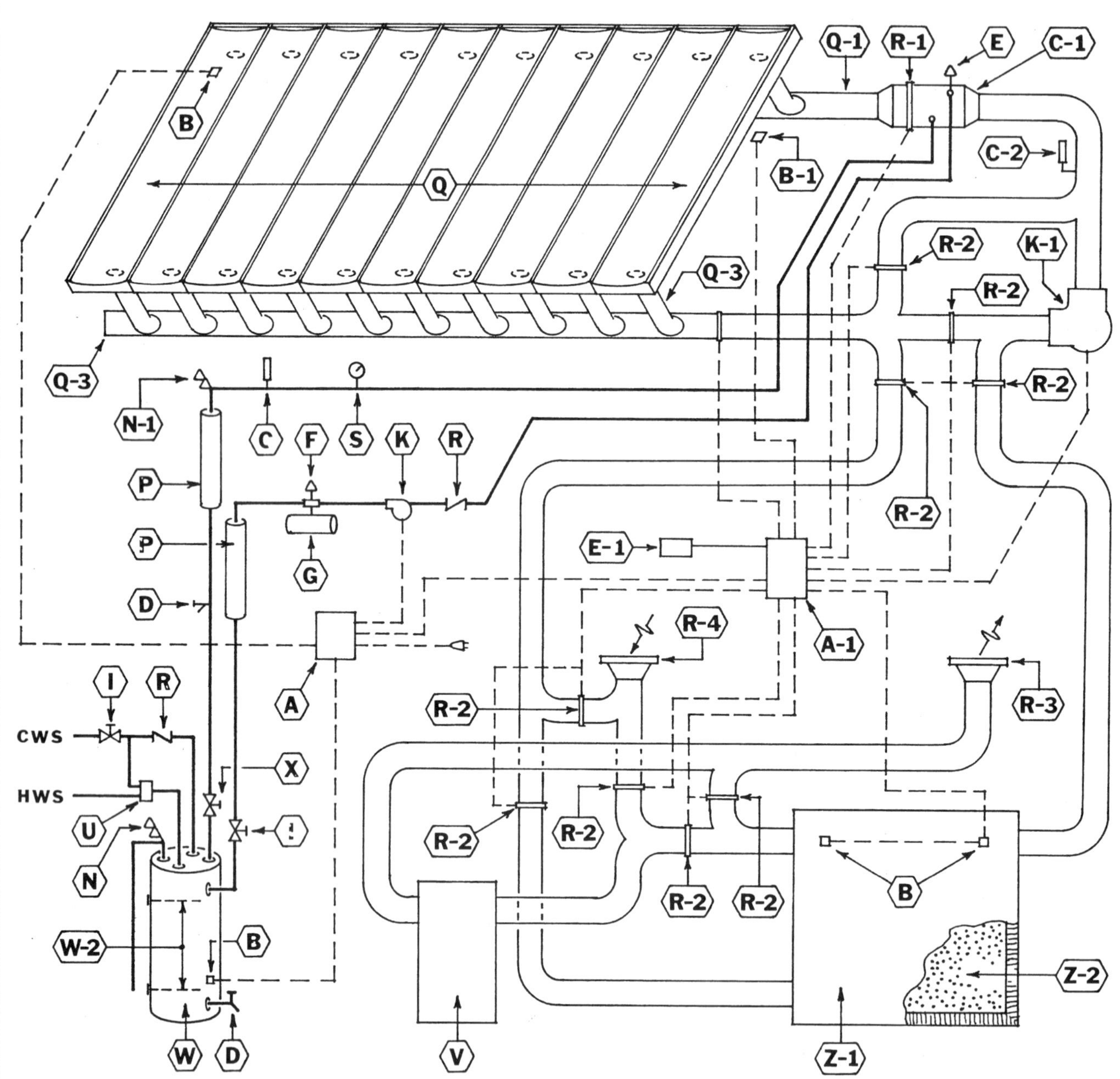

The complete Solar Air Heating System provides maximum savings of conventional fuel with both space heating and year-round domestic hot water heating. It allows for the home air conditioning to operate simultaneously and independently from the solar domestic water heating in summer. The system's modes of operation are:

Mode 1: The building is heated directly from the collectors with air circulated by the Solar Air Mover.

Mode 2: When heat is not needed in the building, the dampers change within the air mover to circulate the air from the collectors to the rock storage bin.

Mode 3: When heat is not available from the collector array and is available in rock storage, the air mover draws heated air from rock storage and directs it into the building. When heat is not available from the collectors or the rock storage bin, the auxiliary heating unit will provide heat for the building. The size of the collector array is typically 25% the size of the main floor area.

D3010 Energy Supply

System Components		QUANTITY	UNIT	COST EACH MAT.	COST EACH INST.	COST EACH TOTAL
	SYSTEM D3010 675 1210					
	SOLAR, SPACE/HOT WATER, AIR TO WATER HEAT EXCHANGE					
A, B	Differential controller 2 sensors thermos., solar energy sys liquid loop	1.000	Ea.	610	52	662
A-1, B	Differential controller 2 sensors 6 station solar energy sys air loop	1.000	Ea.	239	208	447
B-1	Solar energy sensor, freeze prevention	1.000	Ea.	24	19.50	43.50
C	Thermometer for solar energy system, 2″ dial	2.000	Ea.	44	78	122
C-1	Heat exchanger, solar energy system, air to fluid, up flow, 70 MBH	1.000	Ea.	365	320	685
D	Drain valve, brass, 3/4″ connection	2.000	Ea.	17.70	52	69.70
E	Air vent, manual, for solar energy system 1/8″ fitting	1.000	Ea.	3.06	19.50	22.56
E-1	Thermostat, 2 stage for sensing room temperature	1.000	Ea.	157	74.50	231.50
F	Air purger	1.000	Ea.	47.50	52	99.50
G	Expansion tank, for solar energy system	1.000	Ea.	67.50	19.50	87
I	Valve, gate, bronze, 125 lb, NRS, soldered 3/4″ diam	2.000	Ea.	80	62	142
K	Circulator, solar heated liquid, 1/25 HP	1.000	Ea.	295	80.50	375.50
N	Relief valve temp & press 150 psi 210°F self-closing, 3/4″ IPS	1.000	Ea.	17.15	21	38.15
N-1	Relief valve, pressure, poppet, bronze, 30 psi, 3/4″ IPS	1.000	Ea.	38.50	22.50	61
P	Pipe covering, fiberglass, all service jacket, 1″ wall, 3/4″ diam	60.000	L.F.	68.40	265.80	334.20
Q-3	Manifold for flush mount solar energy collector panels	20.000	L.F.	2,800	134	2,934
Q	Collector panel solar energy, air, black on alum. plate, flush mount 10′x2′	100.000	L.F.	49,500	1,115	50,615
R	Valve, swing check, bronze, 125 lb, regrinding disc, 3/4″ diam	2.000	Ea.	144	62	206
S	Pressure gage, 2″ dial, for solar energy system	1.000	Ea.	24.50	19.50	44
U	Valve, water tempering, bronze, sweat connections, 3/4″ diam	1.000	Ea.	100	31	131
W-2, W	Tank, water storage, solar, elec element 2″x1/2# insul, 80 Gal	1.000	Ea.	1,200	390	1,590
X	Valve, globe, bronze, 125 lb, soldered, 3/4″ diam	1.000	Ea.	102	31	133
	Copper tubing type L, solder joints, hangers 10′ OC 3/4″ diam	10.000	L.F.	42.10	82	124.10
	Copper tubing type M, solder joints, hangers 10′ OC 3/4″ diam	60.000	L.F.	193.20	480	673.20
	Wrought copper fittings & solder, 3/4″ diam	26.000	Ea.	93.60	858	951.60
	Sensor wire, #22-2 conductor multistranded	1.200	C.L.F.	18.06	70.20	88.26
Q-1	Duct work, rigid fiberglass, rectangular	400.000	S.F.	340	1,904	2,244
	12″ x 10″	8.000	Ea.	164	356	520
R-2	Shutter/damper for solar heater circulator	9.000	Ea.	486	801	1,287
K-1	Shutter motor for solar heater circulator blower	9.000	Ea.	1,179	603	1,782
K-1	Fan, solar energy heated air circulator, space & DHW system	1.000	Ea.	1,625	2,150	3,775
Z-1	Tank, solar energy air storage, 6′-3″H 7′x7′ = 306 CF/2000 Gal	1.000	Ea.	12,900	1,400	14,300
Z-2	Crushed stone 1-1/2″	11.000	C.Y.	423.50	78.98	502.48
C-2	Themometer, remote probe, 2″ dial	1.000	Ea.	33.50	78	111.50
R-1	Solenoid valve	1.000	Ea.	120	69.50	189.50
V, R-4, R-3	Furnace, supply diffusers, return grilles, existing	1.000	Ea.	12,900	1,400	14,300
	TOTAL			86,462.27	13,459.98	99,922.25

D3010 675	Air To Water Heat Exchange		COST EACH MAT.	COST EACH INST.	COST EACH TOTAL
1210	Solar, air to water heat exchange, for space/hot water heating	R235616 -60	86,500	13,500	100,000
1220					

D3020 Heat Generating Systems

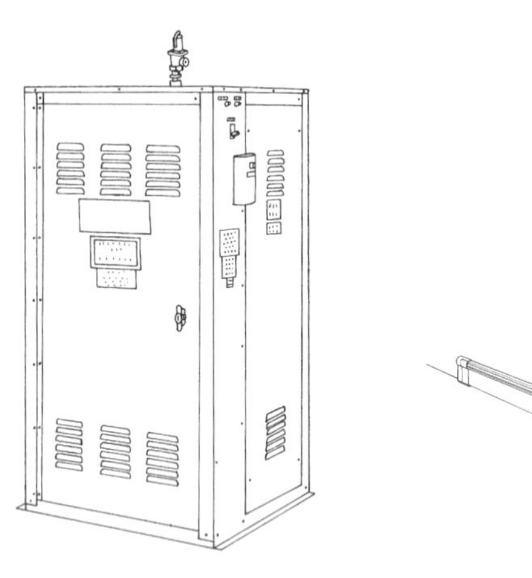

Boiler

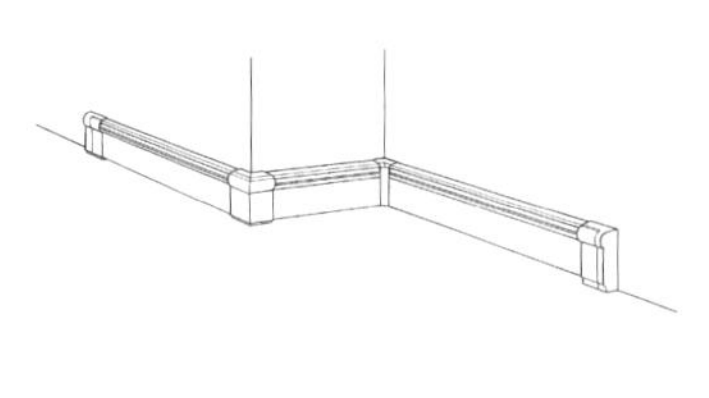

Baseboard Radiation

Small Electric Boiler System Considerations:

1. Terminal units are fin tube baseboard radiation rated at 720 BTU/hr with 200° water temperature or 820 BTU/hr steam.
2. Primary use being for residential or smaller supplementary areas, the floor levels are based on 7-1/2′ ceiling heights.
3. All distribution piping is copper for boilers through 205 MBH. All piping for larger systems is steel pipe.

System Components	QUANTITY	UNIT	COST EACH MAT.	COST EACH INST.	COST EACH TOTAL
SYSTEM D3020 102 1120					
SMALL HEATING SYSTEM, HYDRONIC, ELECTRIC BOILER					
1,480 S.F., 61 MBH, STEAM, 1 FLOOR					
Boiler, electric steam, std cntrls, trim, ftngs and valves, 18 KW, 61.4 MBH	1.000	Ea.	4,537.50	1,567.50	6,105
Copper tubing type L, solder joint, hanger 10′OC, 1-1/4″ diam	160.000	L.F.	1,472	1,720	3,192
Radiation, 3/4″ copper tube w/alum fin baseboard pkg 7″ high	60.000	L.F.	528	1,158	1,686
Rough in baseboard panel or fin tube with valves & traps	10.000	Set	2,210	5,850	8,060
Pipe covering, calcium silicate w/cover 1″ wall 1-1/4″ diam	160.000	L.F.	496	992	1,488
Low water cut-off, quick hookup, in gage glass tappings	1.000	Ea.	231	39	270
TOTAL			9,474.50	11,326.50	20,801
COST PER S.F.			6.40	7.65	14.05

D3020 102	Small Heating Systems, Hydronic, Electric Boilers		COST PER S.F. MAT.	COST PER S.F. INST.	COST PER S.F. TOTAL
1100	Small heating systems, hydronic, electric boilers				
1120	Steam, 1 floor, 1480 S.F., 61 M.B.H.		6.40	7.64	14.04
1160	3,000 S.F., 123 M.B.H.	R235000 -10	4.90	6.70	11.60
1200	5,000 S.F., 205 M.B.H.		4.28	6.20	10.48
1240	2 floors, 12,400 S.F., 512 M.B.H.	R235000 -20	3.64	6.15	9.79
1280	3 floors, 24,800 S.F., 1023 M.B.H.		3.79	6.05	9.84
1320	34,750 S.F., 1,433 M.B.H.	R235000 -30	3.44	5.90	9.34
1360	Hot water, 1 floor, 1,000 S.F., 41 M.B.H.		10	4.26	14.26
1400	2,500 S.F., 103 M.B.H.		7.45	7.65	15.10
1440	2 floors, 4,850 S.F., 205 M.B.H.		7.35	9.20	16.55
1480	3 floors, 9,700 S.F., 410 M.B.H.		7.70	9.50	17.20

D3020 Heat Generating Systems

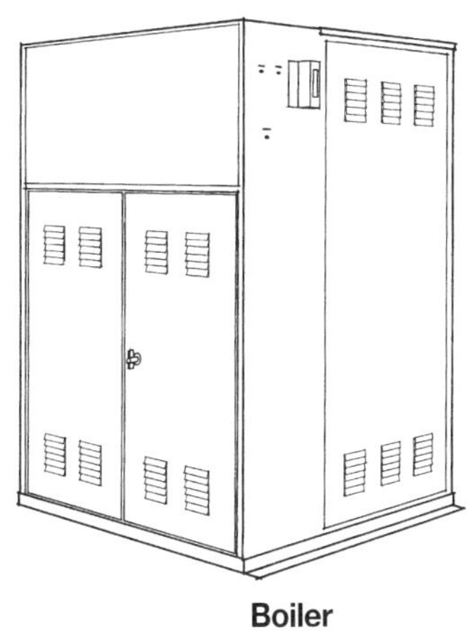

Boiler

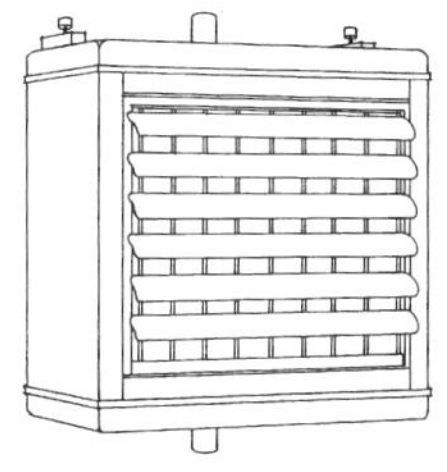

Unit Heater

Large Electric Boiler System Considerations:

1. Terminal units are all unit heaters of the same size. Quantities are varied to accommodate total requirements.
2. All air is circulated through the heaters a minimum of three times per hour.
3. As the capacities are adequate for commercial use, floor levels are based on 10′ ceiling heights.
4. All distribution piping is black steel pipe.

System Components	QUANTITY	UNIT	COST EACH MAT.	COST EACH INST.	COST EACH TOTAL
SYSTEM D3020 104 1240					
LARGE HEATING SYSTEM, HYDRONIC, ELECTRIC BOILER					
9,280 S.F., 135 KW, 461 MBH, 1 FLOOR					
Boiler, electric hot water, std ctrls, trim, ftngs, valves, 135 KW, 461 MBH	1.000	Ea.	10,987.50	3,412.50	14,400
Expansion tank, painted steel, 60 Gal capacity ASME	1.000	Ea.	3,700	187	3,887
Circulating pump, CI, close cpld, 50 GPM, 2 HP, 2″ pipe conn	1.000	Ea.	2,350	375	2,725
Unit heater, 1 speed propeller, horizontal, 200° EWT, 72.7 MBH	7.000	Ea.	5,670	1,428	7,098
Unit heater piping hookup with controls	7.000	Set	4,585	9,275	13,860
Pipe, steel, black, schedule 40, welded, 2-1/2″ diam	380.000	L.F.	3,914	9,640.60	13,554.60
Pipe covering, calcium silicate w/cover, 1″ wall, 2-1/2″ diam	380.000	L.F.	1,459.20	2,413	3,872.20
TOTAL			32,665.70	26,731.10	59,396.80
COST PER S.F.			3.52	2.88	6.40

D3020 104	Large Heating Systems, Hydronic, Electric Boilers		COST PER S.F. MAT.	COST PER S.F. INST.	COST PER S.F. TOTAL
1230	Large heating systems, hydronic, electric boilers				
1240	9,280 S.F., 135 K.W., 461 M.B.H., 1 floor		3.52	2.88	6.40
1280	14,900 S.F., 240 K.W., 820 M.B.H., 2 floors	R235000 -10	4.28	4.68	8.96
1320	18,600 S.F., 296 K.W., 1,010 M.B.H., 3 floors		4.22	5.10	9.32
1360	26,100 S.F., 420 K.W., 1,432 M.B.H., 4 floors	R235000 -20	4.19	4.99	9.18
1400	39,100 S.F., 666 K.W., 2,273 M.B.H., 4 floors		3.59	4.16	7.75
1440	57,700 S.F., 900 K.W., 3,071 M.B.H., 5 floors	R235000 -30	3.42	4.08	7.50
1480	111,700 S.F., 1,800 K.W., 6,148 M.B.H., 6 floors		3.14	3.52	6.66
1520	149,000 S.F., 2,400 K.W., 8,191 M.B.H., 8 floors		3.11	3.49	6.60
1560	223,300 S.F., 3,600 K.W., 12,283 M.B.H., 14 floors		3.30	3.98	7.28

D3020 Heat Generating Systems

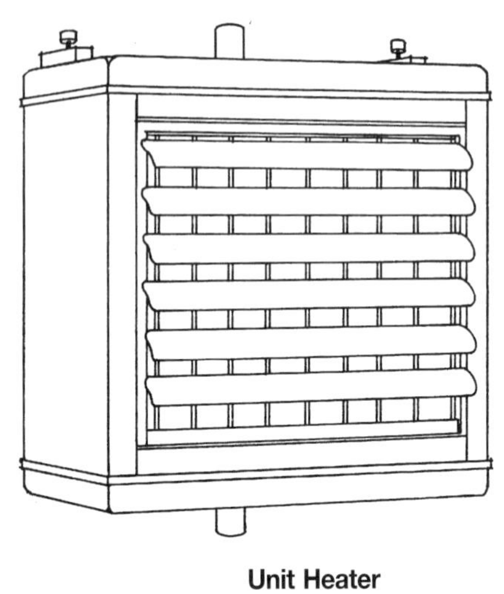

Unit Heater

Fossil Fuel Boiler System Considerations:

1. Terminal units are horizontal unit heaters. Quantities are varied to accommodate total heat loss per building.
2. Unit heater selection was determined by their capacity to circulate the building volume a minimum of three times per hour in addition to the BTU output.
3. Systems shown are forced hot water. Steam boilers cost slightly more than hot water boilers. However, this is compensated for by the smaller size or fewer terminal units required with steam.
4. Floor levels are based on 10′ story heights.
5. MBH requirements are gross boiler output.

System Components	QUANTITY	UNIT	COST EACH MAT.	COST EACH INST.	COST EACH TOTAL
SYSTEM D3020 108 1280					
HEATING SYSTEM, HYDRONIC, FOSSIL FUEL, TERMINAL UNIT HEATERS					
CAST IRON BOILER, GAS, 80 MBH, 1,070 S.F. BUILDING					
Boiler, gas, hot water, CI, burner, controls, insulation, breeching, 80 MBH	1.000	Ea.	2,073.75	1,706.25	3,780
Pipe, steel, black, schedule 40, threaded, cplg & hngr 10′OC, 2″ diam	200.000	L.F.	2,770	3,510	6,280
Unit heater, 1 speed propeller, horizontal, 200° EWT, 72.7 MBH	2.000	Ea.	1,620	408	2,028
Unit heater piping hookup with controls	2.000	Set	1,310	2,650	3,960
Expansion tank, painted steel, ASME, 18 Gal capacity	1.000	Ea.	2,375	93.50	2,468.50
Circulating pump, CI, flange connection, 1/12 HP	1.000	Ea.	350	187	537
Pipe covering, calcium silicate w/cover, 1″ wall, 2″ diam	200.000	L.F.	722	1,270	1,992
TOTAL			11,220.75	9,824.75	21,045.50
COST PER S.F.			10.49	9.18	19.67

D3020 108	Heating Systems, Unit Heaters		COST PER S.F. MAT.	COST PER S.F. INST.	COST PER S.F. TOTAL
1260	Heating systems, hydronic, fossil fuel, terminal unit heaters,				
1280	Cast iron boiler, gas, 80 M.B.H., 1,070 S.F. bldg.		10.47	9.18	19.65
1320	163 M.B.H., 2,140 S.F. bldg.	R235000 -10	7	6.25	13.25
1360	544 M.B.H., 7,250 S.F. bldg.		4.52	4.43	8.95
1400	1,088 M.B.H., 14,500 S.F. bldg.	R235000 -20	3.80	4.20	8
1440	3,264 M.B.H., 43,500 S.F. bldg.		2.95	3.14	6.09
1480	5,032 M.B.H., 67,100 S.F. bldg.	R235000 -30	3.58	3.26	6.84
1520	Oil, 109 M.B.H., 1,420 S.F. bldg.		10.40	8.20	18.60
1560	235 M.B.H., 3,150 S.F. bldg.		6.30	5.85	12.15
1600	940 M.B.H., 12,500 S.F. bldg.		4.42	3.82	8.24
1640	1,600 M.B.H., 21,300 S.F. bldg.		3.98	3.66	7.64
1680	2,480 M.B.H., 33,100 S.F. bldg.		3.98	3.29	7.27
1720	3,350 M.B.H., 44,500 S.F. bldg.		3.65	3.38	7.03
1760	Coal, 148 M.B.H., 1,975 S.F. bldg.		9.15	5.30	14.45
1800	300 M.B.H., 4,000 S.F. bldg.		6.90	4.20	11.10
1840	2,360 M.B.H., 31,500 S.F. bldg.		4.42	3.36	7.78

D3020 Heat Generating Systems

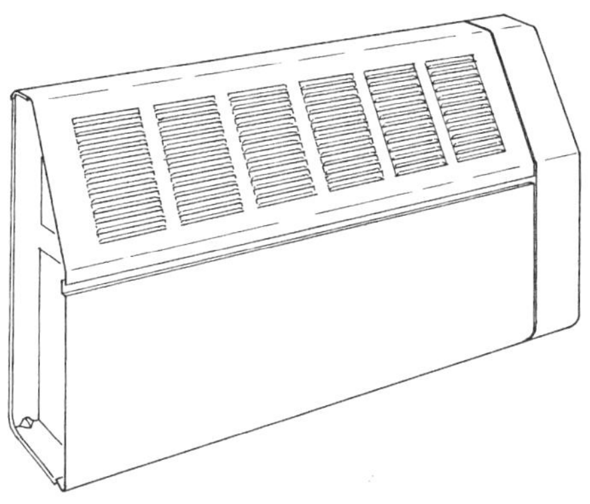

Fin Tube Radiator

Fossil Fuel Boiler System Considerations:

1. Terminal units are commercial steel fin tube radiation. Quantities are varied to accommodate total heat loss per building.
2. Systems shown are forced hot water. Steam boilers cost slightly more than hot water boilers. However, this is compensated for by the smaller size or fewer terminal units required with steam.
3. Floor levels are based on 10′ story heights.
4. MBH requirements are gross boiler output.

System Components	QUANTITY	UNIT	COST EACH MAT.	COST EACH INST.	COST EACH TOTAL
SYSTEM D3020 110 3240					
HEATING SYSTEM, HYDRONIC, FOSSIL FUEL, FIN TUBE RADIATION					
CAST IRON BOILER, GAS, 80 MBH, 1,070 S.F. BLDG.					
Boiler, gas, hot water, CI, burner, controls, insulation, breeching, 80 MBH	1.000	Ea.	2,073.75	1,706.25	3,780
Pipe, steel, black, schedule 40, threaded, 2″ diam	200.000	L.F.	2,770	3,510	6,280
Radiation,steel 1-1/4″ tube & 4-1/4″ fin w/cover & damper, wall hung	80.000	L.F.	3,400	2,480	5,880
Rough-in, wall hung steel fin radiation with supply & balance valves	8.000	Set	2,224	5,480	7,704
Expansion tank, painted steel, ASME, 18 Gal capacity	1.000	Ea.	2,375	93.50	2,468.50
Circulating pump, CI flanged, 1/12 HP	1.000	Ea.	350	187	537
Pipe covering, calcium silicate w/cover, 1″ wall, 2″ diam	200.000	L.F.	722	1,270	1,992
TOTAL			13,914.75	14,726.75	28,641.50
COST PER S.F.			13	13.76	26.76

D3020 110	Heating System, Fin Tube Radiation		COST PER S.F. MAT.	COST PER S.F. INST.	COST PER S.F. TOTAL
3230	Heating systems, hydronic, fossil fuel, fin tube radiation				
3240	Cast iron boiler, gas, 80 MBH, 1,070 S.F. bldg.		12.99	13.74	26.73
3280	169 M.B.H., 2,140 S.F. bldg.	R235000 -10	8.20	8.70	16.90
3320	544 M.B.H., 7,250 S.F. bldg.		6.30	7.50	13.80
3360	1,088 M.B.H., 14,500 S.F. bldg.	R235000 -20	5.70	7.35	13.05
3400	3,264 M.B.H., 43,500 S.F. bldg.		4.97	6.45	11.42
3440	5,032 M.B.H., 67,100 S.F. bldg.	R235000 30	5.60	6.55	12.15
3480	Oil, 109 M.B.H., 1,420 S.F. bldg.		14.60	15	29.60
3520	235 M.B.H., 3,150 S.F. bldg.		8.10	8.85	16.95
3560	940 M.B.H., 12,500 S.F. bldg.		6.35	7	13.35
3600	1,600 M.B.H., 21,300 S.F. bldg.		6	6.95	12.95
3640	2,480 M.B.H., 33,100 S.F. bldg.		6	6.60	12.60
3680	3,350 M.B.H., 44,500 S.F. bldg.		5.65	6.65	12.30
3720	Coal, 148 M.B.H., 1,975 S.F. bldg.		10.85	8.30	19.15
3760	300 M.B.H., 4,000 S.F. bldg.		8.60	7.15	15.75
3800	2,360 M.B.H., 31,500 S.F. bldg.		6.40	6.60	13
4080	Steel boiler, oil, 97 M.B.H., 1,300 S.F. bldg.		10.90	12.35	23.25
4120	315 M.B.H., 4,550 S.F. bldg.		6.05	6.25	12.30
4160	525 M.B.H., 7,000 S.F. bldg.		7.65	7.25	14.90
4200	1,050 M.B.H., 14,000 S.F. bldg.		6.60	7.25	13.85
4240	2,310 M.B.H., 30,800 S.F. bldg.		6.15	6.65	12.80
4280	3,150 M.B.H., 42,000 S.F. bldg.		5.95	6.75	12.70

D3020 Heat Generating Systems

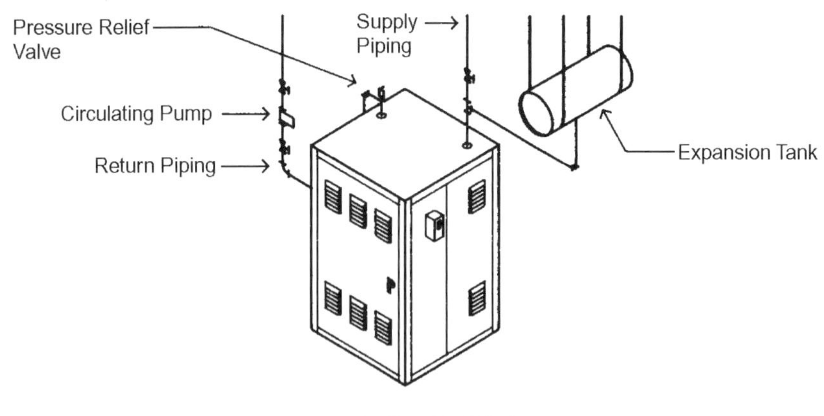

Electric Boiler, Hot Water

System Components	QUANTITY	UNIT	COST EACH MAT.	COST EACH INST.	COST EACH TOTAL
SYSTEM D3020 126 1010					
BOILER, ELECTRIC, HOT WATER, 15 kW, 52 MBH					
Boilers, electric, ASME, hot water, 15 KW, 52 MBH	1.000	Ea.	4,250	1,300	5,550
Pipe, black steel, Sch 40, threaded, W/coupling & hangers, 10′ OC, 3/4″ dia	10.000	L.F.	54.17	117.88	172.05
Pipe, black steel, Sch 40, threaded, W/coupling & hangers, 10′ OC, 1″ dia	20.000	L.F.	148.35	252.63	400.98
Elbow, 90°, black, straight, 3/4″ dia.	6.000	Ea.	18.96	267	285.96
Elbow, 90°, black, straight, 1″ dia.	6.000	Ea.	33	288	321
Tee, black, straight, 3/4″ dia.	1.000	Ea.	5	69.50	74.50
Tee, black, straight, 1″ dia.	2.000	Ea.	17.10	156	173.10
Tee, black, reducing, 1″ dia.	2.000	Ea.	27.50	156	183.50
Union, black with brass seat, 3/4″ dia.	1.000	Ea.	13.25	48	61.25
Union, black with brass seat, 1″ dia.	2.000	Ea.	34.60	104	138.60
Pipe nipples, 3/4″ diam.	3.000	Ea.	7.07	15.38	22.45
Pipe nipples, 1″ diam.	3.000	Ea.	10.35	17.63	27.98
Valves, bronze, gate, N.R.S., threaded, class 150, 3/4″ size	2.000	Ea.	127	62	189
Valves, bronze, gate, N.R.S., threaded, class 150, 1″ size	2.000	Ea.	162	66	228
Thermometer, stem type, 9″ case, 8″ stem, 3/4″ NPT	1.000	Ea.	372	44	416
Tank, steel, liquid expansion, ASME, painted, 15 gallon capacity	1.000	Ea.	575	66	641
Pump, circulating, bronze, flange connection, 3/4″ to 1-1/2″ size, 1/8 HP	1.000	Ea.	940	187	1,127
Insulation, fiberglass pipe covering, 1″ wall, 1″ IPS	20.000	L.F.	24.60	92.80	117.40
TOTAL			6,819.95	3,309.82	10,129.77

D3020 126	Electric Boiler, Hot Water	COST EACH MAT.	COST EACH INST.	COST EACH TOTAL
1010	Boiler, electric, hot water, 15 kW, 52 MBH	6,825	3,300	10,125
1020	30 kW, 103 MBH	7,650	3,500	11,150
1030	60 kW, 205 MBH	7,825	3,625	11,450
1040	120 kW, 410 MBH	10,100	4,350	14,450
1050	150 kW, 510 MBH	15,600	8,375	23,975
1060	210 kW, 716 MBH	18,400	9,200	27,600
1070	296 kW, 1010 MBH	23,500	9,475	32,975
1080	444 kW, 1515 MBH	28,300	12,800	41,100
1090	666 kW, 2273 MBH	33,900	13,800	47,700
1100	900 kW, 3071 MBH	42,000	16,000	58,000
1110	1320 kW, 4505 MBH	49,800	16,800	66,600
1120	2100 kW, 7167 MBH	75,000	20,500	95,500
1130	2610 kW, 8905 MBH	93,000	24,600	117,600
1140	3600 kW, 12,283 MBH	113,500	29,400	142,900

D3020 Heat Generating Systems

D3020 128	Electric Boiler, Steam	COST EACH		
		MAT.	INST.	TOTAL
1010	Boiler, electric, steam, 18 kW, 61.4 MBH	5,225	3,200	8,425
1020	36 kW, 123 MBH	6,550	3,375	9,925
1030	60 kW, 205 MBH	13,500	6,875	20,375
1040	120 kW, 409 MBH	16,000	7,575	23,575
1050	150 kW, 512 MBH	19,100	11,400	30,500
1060	210 kW, 716 MBH	24,600	12,300	36,900
1070	300 kW, 1023 MBH	28,900	13,400	42,300
1080	510 kW, 1740 MBH	36,200	17,500	53,700
1090	720 kW, 2456 MBH	40,800	18,300	59,100
1100	1080 kW, 3685 MBH	47,900	21,700	69,600
1110	1260 kW, 4300 MBH	51,000	23,200	74,200
1120	1620 kW, 5527 MBH	53,000	26,100	79,100
1130	2070 kW, 7063 MBH	70,500	27,400	97,900
1140	2340 kW, 7984 MBH	92,000	31,100	123,100

D3020 Heat Generating Systems

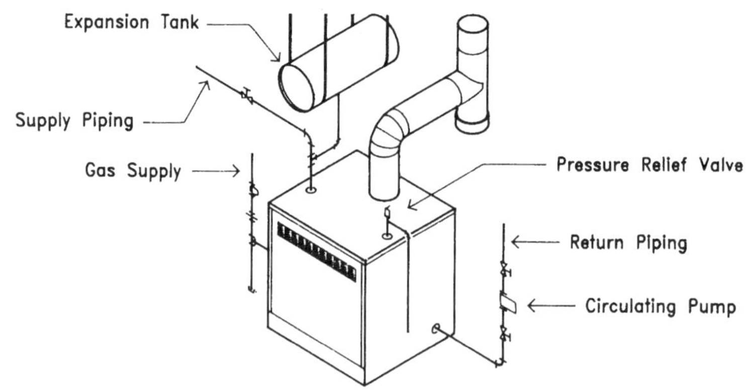

Cast Iron Boiler, Hot Water, Gas Fired

System Components	QUANTITY	UNIT	COST EACH MAT.	COST EACH INST.	COST EACH TOTAL
SYSTEM D3020 130 1010					
BOILER, CAST IRON, GAS, HOT WATER, 100 MBH					
Boilers, gas fired, std controls, CI, insulated, HW, gross output 100 MBH	1.000	Ea.	2,525	1,750	4,275
Pipe, black steel, Sch 40, threaded, W/coupling & hangers, 10′ OC, 3/4″ dia	20.000	L.F.	105.98	230.63	336.61
Pipe, black steel, Sch 40, threaded, W/coupling & hangers, 10′ OC, 1″ dia.	20.000	L.F.	148.35	252.63	400.98
Elbow, 90°, black, straight, 3/4″ dia.	9.000	Ea.	28.44	400.50	428.94
Elbow, 90°, black, straight, 1″ dia.	6.000	Ea.	33	288	321
Tee, black, straight, 3/4″ dia.	2.000	Ea.	10	139	149
Tee, black, straight, 1″ dia.	2.000	Ea.	17.10	156	173.10
Tee, black, reducing, 1″ dia.	2.000	Ea.	27.50	156	183.50
Pipe cap, black, 3/4″ dia.	1.000	Ea.	3.58	19.50	23.08
Union, black with brass seat, 3/4″ dia.	2.000	Ea.	26.50	96	122.50
Union, black with brass seat, 1″ dia.	2.000	Ea.	34.60	104	138.60
Pipe nipples, black, 3/4″ dia.	5.000	Ea.	11.78	25.63	37.41
Pipe nipples, black, 1″ dia.	3.000	Ea.	10.35	17.63	27.98
Valves, bronze, gate, N.R.S., threaded, class 150, 3/4″ size	2.000	Ea.	127	62	189
Valves, bronze, gate, N.R.S., threaded, class 150, 1″ size	2.000	Ea.	162	66	228
Gas cock, brass, 3/4″ size	1.000	Ea.	12.30	28.50	40.80
Thermometer, stem type, 9″ case, 8″ stem, 3/4″ NPT	2.000	Ea.	372	44	416
Tank, steel, liquid expansion, ASME, painted, 15 gallon capacity	1.000	Ea.	575	66	641
Pump, circulating, bronze, flange connection, 3/4″ to 1-1/2″ size, 1/8 HP	1.000	Ea.	940	187	1,127
Vent chimney, all fuel, pressure tight, double wall, SS, 6″ dia.	20.000	L.F.	1,040	357	1,397
Vent chimney, elbow, 90° fixed, 6″ dia.	2.000	Ea.	586	71	657
Vent chimney, Tee, 6″ dia.	2.000	Ea.	386	89	475
Vent chimney, ventilated roof thimble, 6″ dia.	1.000	Ea.	247	41	288
Vent chimney, adjustable roof flashing, 6″ dia.	1.000	Ea.	69	35.50	104.50
Vent chimney, stack cap, 6″ diameter	1.000	Ea.	218	23.50	241.50
Insulation, fiberglass pipe covering, 1″ wall, 1″ IPS	20.000	L.F.	24.60	92.80	117.40
TOTAL			7,741.08	4,798.82	12,539.90

D3020 130	Boiler, Cast Iron, Hot Water, Gas	COST EACH MAT.	COST EACH INST.	COST EACH TOTAL
1010	Boiler, cast iron, gas, hot water, 100 MBH	7,750	4,800	12,550
1020	200 MBH	8,650	5,500	14,150
1030	320 MBH	9,300	5,975	15,275
1040	440 MBH	12,300	7,700	20,000
1050	544 MBH	20,600	12,100	32,700
1060	765 MBH	25,400	13,000	38,400

D3020 Heat Generating Systems

D3020 130	Boiler, Cast Iron, Hot Water, Gas	COST EACH		
		MAT.	INST.	TOTAL
1070	1088 MBH	28,200	13,900	42,100
1080	1530 MBH	32,800	18,400	51,200
1090	2312 MBH	38,500	21,700	60,200
1100	2856 MBH	45,200	23,300	68,500
1110	3808 MBH	50,500	27,200	77,700
1120	4720 MBH	95,500	30,600	126,100
1130	6100 MBH	122,000	34,100	156,100
1140	6970 MBH	128,000	41,900	169,900

D3020 134	Boiler, Cast Iron, Steam, Gas	COST EACH		
		MAT.	INST.	TOTAL
1010	Boiler, cast iron, gas, steam, 100 MBH	6,250	4,650	10,900
1020	200 MBH	12,900	8,775	21,675
1030	320 MBH	14,600	9,850	24,450
1040	544 MBH	22,100	15,400	37,500
1050	765 MBH	25,100	16,100	41,200
1060	1275 MBH	29,900	17,700	47,600
1070	2675 MBH	41,100	25,800	66,900
1080	3570 MBH	48,200	28,900	77,100
1090	4720 MBH	53,000	30,800	83,800
1100	6100 MBH	106,500	35,800	142,300
1110	6970 MBH	117,000	41,900	158,900

D3020 136	Boiler, Cast Iron, Hot Water, Gas/Oil	COST EACH		
		MAT.	INST.	TOTAL
1010	Boiler, cast iron, gas & oil, hot water, 584 MBH	20,700	13,400	34,100
1020	876 MBH	30,500	14,400	44,900
1030	1168 MBH	37,100	16,300	53,400
1040	1460 MBH	41,300	19,700	61,000
1050	2044 MBH	47,200	20,300	67,500
1060	2628 MBH	54,500	24,500	79,000
1070	3210 MBH	62,000	26,400	88,400
1080	3796 MBH	66,500	27,200	93,700
1090	4672 MBH	77,000	28,500	105,500
1100	5256 MBH	93,000	34,500	127,500
1110	6000 MBH	112,000	35,200	147,200
1120	9800 MBH	140,000	60,000	200,000
1130	12,200 MBH	174,000	69,500	243,500
1140	13,500 MBH	189,000	74,500	263,500

D3020 138	Boiler, Cast Iron, Steam, Gas/Oil	COST EACH		
		MAT.	INST.	TOTAL
1010	Boiler, cast iron, gas & oil, steam, 810 MBH	29,400	17,500	46,900
1020	1360 MBH	32,900	18,600	51,500
1030	2040 MBH	42,900	24,000	66,900
1040	2700 MBH	46,900	26,600	73,500
1050	3270 MBH	53,000	29,600	82,600
1060	3770 MBH	88,500	30,200	118,700
1070	4650 MBH	99,000	34,400	133,400
1080	5520 MBH	114,500	36,200	150,700
1090	6100 MBH	116,500	38,100	154,600
1100	6970 MBH	127,000	46,400	173,400

D3020 Heat Generating Systems

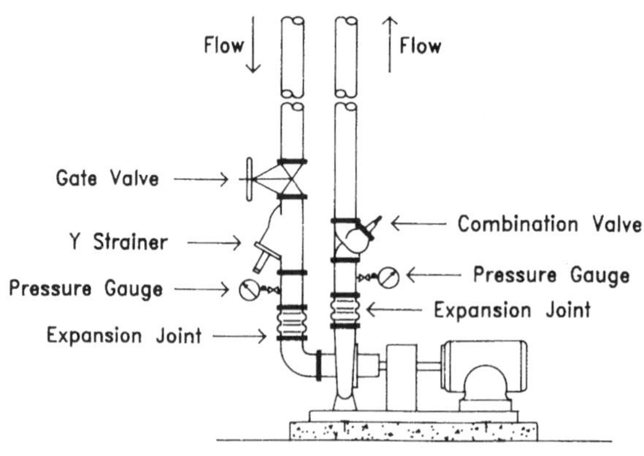

Base Mounted End–Suction Pump

System Components	QUANTITY	UNIT	COST EACH MAT.	COST EACH INST.	COST EACH TOTAL
SYSTEM D3020 330 1010					
PUMP, BASE MTD. WITH MOTOR, END-SUCTION, 2-1/2″ SIZE, 3 HP, TO 150 GPM					
Pump, circulating, CI, base mounted, 2-1/2″ size, 3 HP, to 150 GPM	1.000	Ea.	6,775	625	7,400
Pipe, black steel, Sch. 40, on yoke & roll hangers, 10′ O.C., 2-1/2″ dia	12.000	L.F.	123.60	304.44	428.04
Elbow, 90°, weld joint, steel, 2-1/2″ pipe size	1.000	Ea.	30.50	148.05	178.55
Flange, weld neck, 150 LB, 2-1/2″ pipe size	8.000	Ea.	280	592.24	872.24
Valve, iron body, gate, 125 lb., N.R.S., flanged, 2-1/2″ size	1.000	Ea.	595	225	820
Strainer, Y type, iron body, flanged, 125 lb., 2-1/2″ pipe size	1.000	Ea.	113	224	337
Multipurpose valve, CI body, 2″ size	1.000	Ea.	445	78	523
Expansion joint, flanged spool, 6″ F to F, 2-1/2″ dia	2.000	Ea.	572	181	753
T-O-L, weld joint, socket, 1/4″ pipe size, nozzle	2.000	Ea.	14.70	103.60	118.30
T-O-L, weld joint, socket, 1/2″ pipe size, nozzle	2.000	Ea.	14.70	107.86	122.56
Control gauges, pressure or vacuum, 3-1/2″ diameter dial	2.000	Ea.	79	38.90	117.90
Pressure/temperature relief plug, 316 SS, 3/4″ OD, 7-1/2″ insertion	2.000	Ea.	188	38.90	226.90
Insulation, fiberglass pipe covering, 1-1/2″ wall, 2-1/2″ IPS	12.000	L.F.	34.56	67.80	102.36
Pump control system	1.000	Ea.	1,200	570	1,770
Pump balancing	1.000	Ea.		274	274
TOTAL			10,465.06	3,578.79	14,043.85

D3020 330	Circulating Pump Systems, End Suction	COST EACH MAT.	COST EACH INST.	COST EACH TOTAL
1010	Pump, base mtd with motor, end-suction, 2-1/2″ size, 3 HP, to 150 GPM	10,500	3,575	14,075
1020	3″ size, 5 HP, to 225 GPM	11,700	4,100	15,800
1030	4″ size, 7-1/2 HP, to 350 GPM	12,900	4,925	17,825
1040	5″ size, 15 HP, to 1000 GPM	17,800	7,500	25,300
1050	6″ size, 25 HP, to 1550 GPM	24,700	9,100	33,800

D3020 340	Circulating Pump Systems, Double Suction	COST EACH MAT.	COST EACH INST.	COST EACH TOTAL
1010	Pump, base mtd w/ motor, double suction, 6″ size, 50 HP, to 1200 GPM	19,200	13,200	32,400
1020	8″ size, 75 HP, to 2500 GPM	35,200	17,800	53,000
1030	8″ size, 100 HP, to 3000 GPM	51,000	19,800	70,800
1040	10″ size, 150 HP, to 4000 GPM	76,500	22,500	99,000

D3030 Cooling Generating Systems

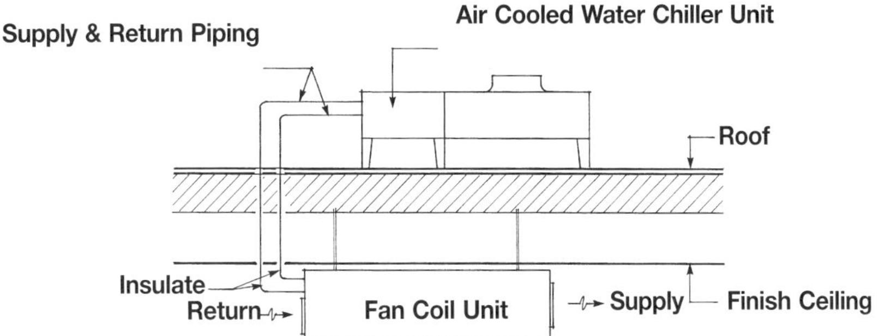

Design Assumptions: The chilled water, air cooled systems priced, utilize reciprocating hermetic compressors and propeller-type condenser fans. Piping with pumps and expansion tanks is included based on a two pipe system. No ducting is included and the fan-coil units are cooling only. Water treatment and balancing are not included. Chilled water piping is insulated. Area distribution is through the use of multiple fan coil units. Fewer but larger fan coil units with duct distribution would be approximately the same S.F. cost.

System Components	QUANTITY	UNIT	COST EACH		
			MAT.	INST.	TOTAL
SYSTEM D3030 110 1200					
PACKAGED CHILLER, AIR COOLED, WITH FAN COIL UNIT					
APARTMENT CORRIDORS, 3,000 S.F., 5.50 TON					
Fan coil air conditioning unit, cabinet mounted & filters chilled water	1.000	Ea.	3,482.70	513.24	3,995.94
Water chiller, air conditioning unit, air cooled	1.000	Ea.	7,122.50	1,918.13	9,040.63
Chilled water unit coil connections	1.000	Ea.	1,300	1,400	2,700
Chilled water distribution piping	440.000	L.F.	7,854	18,920	26,774
TOTAL			19,759.20	22,751.37	42,510.57
COST PER S.F.			6.59	7.58	14.17
*Cooling requirements would lead to choosing a water cooled unit.					

D3030 110	Chilled Water, Air Cooled Condenser Systems	COST PER S.F.		
		MAT.	INST.	TOTAL
1180	Packaged chiller, air cooled, with fan coil unit			
1200	Apartment corridors, 3,000 S.F., 5.50 ton	6.60	7.60	14.20
1240	6,000 S.F., 11.00 ton	5.50	6.05	11.55
1280	10,000 S.F., 18.33 ton	4.83	4.61	9.44
1320	20,000 S.F., 36.66 ton	3.56	3.34	6.90
1360	40,000 S.F., 73.33 ton	4.16	3.35	7.51
1440	Banks and libraries, 3,000 S.F., 12.50 ton	10.40	8.95	19.35
1480	6,000 S.F., 25.00 ton	9.55	7.25	16.80
1520	10,000 S.F., 41.66 ton	7.05	5.05	12.10
1560	20,000 S.F., 83.33 ton	7.50	4.56	12.06
1600	40,000 S.F., 167 ton*			
1680	Bars and taverns, 3,000 S.F., 33.25 ton	17.40	10.60	28
1720	6,000 S.F., 66.50 ton	16.45	8.10	24.55
1760	10,000 S.F., 110.83 ton	13.25	2.63	15.88
1800	20,000 S.F., 220 ton*			
1840	40,000 S.F., 440 ton*			
1920	Bowling alleys, 3,000 S.F., 17.00 ton	13.25	10.05	23.30
1960	6,000 S.F., 34.00 ton	9.55	7.05	16.60
2000	10,000 S.F., 56.66 ton	8.75	5.05	13.80
2040	20,000 S.F., 113.33 ton	8.45	4.63	13.08
2080	40,000 S.F., 227 ton*			
2160	Department stores, 3,000 S.F., 8.75 ton	9.30	8.45	17.75
2200	6,000 S.F., 17.50 ton	7.25	6.55	13.80
2240	10,000 S.F., 29.17 ton	5.60	4.72	10.32
2280	20,000 S.F., 58.33 ton	4.93	3.47	8.40
2320	40,000 S.F., 116.66 ton	5.40	3.47	8.87

D3030 Cooling Generating Systems

D3030 110	Chilled Water, Air Cooled Condenser Systems	COST PER S.F.		
		MAT.	INST.	TOTAL
2400	Drug stores, 3,000 S.F., 20.00 ton	15.05	10.35	25.40
2440	6,000 S.F., 40.00 ton	11.40	7.70	19.10
2480	10,000 S.F., 66.66 ton	11.65	6.55	18.20
2520	20,000 S.F., 133.33 ton	10.45	4.89	15.34
2560	40,000 S.F., 267 ton*			
2640	Factories, 2,000 S.F., 10.00 ton	9.05	8.60	17.65
2680	6,000 S.F., 20.00 ton	8.25	6.90	15.15
2720	10,000 S.F., 33.33 ton	6.05	4.85	10.90
2760	20,000 S.F., 66.66 ton	6.45	4.34	10.79
2800	40,000 S.F., 133.33 ton	5.95	3.58	9.53
2880	Food supermarkets, 3,000 S.F., 8.50 ton	9.15	8.40	17.55
2920	6,000 S.F., 17.00 ton	7.10	6.50	13.60
2960	10,000 S.F., 28.33 ton	5.30	4.60	9.90
3000	20,000 S.F., 56.66 ton	4.74	3.40	8.14
3040	40,000 S.F., 113.33 ton	5.10	3.48	8.58
3120	Medical centers, 3,000 S.F., 7.00 ton	8.20	8.20	16.40
3160	6,000 S.F., 14.00 ton	6.30	6.30	12.60
3200	10,000 S.F., 23.33 ton	5.70	4.80	10.50
3240	20,000 S.F., 46.66 ton	4.28	3.53	7.81
3280	40,000 S.F., 93.33 ton	4.62	3.41	8.03
3360	Offices, 3,000 S.F., 9.50 ton	8.40	8.35	16.75
3400	6,000 S.F., 19.00 ton	7.95	6.80	14.75
3440	10,000 S.F., 31.66 ton	5.90	4.79	10.69
3480	20,000 S.F., 63.33 ton	6.45	4.41	10.86
3520	40,000 S.F., 126.66 ton	5.85	3.59	9.44
3600	Restaurants, 3,000 S.F., 15.00 ton	11.70	9.30	21
3640	6,000 S.F., 30.00 ton	9.10	7.05	16.15
3680	10,000 S.F., 50.00 ton	8.35	5.15	13.50
3720	20,000 S.F., 100.00 ton	8.60	4.85	13.45
3760	40,000 S.F., 200 ton*			
3840	Schools and colleges, 3,000 S.F., 11.50 ton	9.85	8.80	18.65
3880	6,000 S.F., 23.00 ton	9.05	7.10	16.15
3920	10,000 S.F., 38.33 ton	6.65	4.98	11.63
3960	20,000 S.F., 76.66 ton	7.30	4.60	11.90
4000	40,000 S.F., 153 ton*			

D3030 Cooling Generating Systems

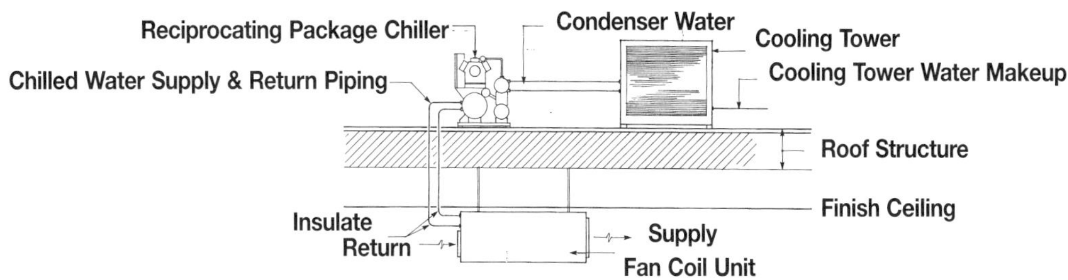

General: Water cooled chillers are available in the same sizes as air cooled units. They are also available in larger capacities.

Design Assumptions: The chilled water systems with water cooled condenser include reciprocating hermetic compressors, water cooling tower, pumps, piping and expansion tanks and are based on a two pipe system. Chilled water piping is insulated. No ducts are included and fan-coil units are cooling only. Area distribution is through use of multiple fan coil units. Fewer but larger fan coil units with duct distribution would be approximately the same S.F. cost. Water treatment and balancing are not included.

System Components	QUANTITY	UNIT	COST EACH MAT.	COST EACH INST.	COST EACH TOTAL
SYSTEM D3030 115 1320					
PACKAGED CHILLER, WATER COOLED, WITH FAN COIL UNIT					
APARTMENT CORRIDORS, 4,000 S.F., 7.33 TON					
Fan coil air conditioner unit, cabinet mounted & filters, chilled water	2.000	Ea.	4,641.70	684.04	5,325.74
Water chiller, water cooled, 1 compressor, hermetic scroll,	1.000	Ea.	5,358.60	3,343.40	8,702
Cooling tower, draw thru single flow, belt drive	1.000	Ea.	1,312.07	141.84	1,453.91
Cooling tower pumps & piping	1.000	System	656.04	337.18	993.22
Chilled water unit coil connections	2.000	Ea.	2,600	2,800	5,400
Chilled water distribution piping	520.000	L.F.	9,282	22,360	31,642
TOTAL			23,850.41	29,666.46	53,516.87
COST PER S.F.			5.96	7.42	13.38
*Cooling requirements would lead to choosing a water cooled unit.					

D3030 115	Chilled Water, Cooling Tower Systems	COST PER S.F. MAT.	COST PER S.F. INST.	COST PER S.F. TOTAL
1300	Packaged chiller, water cooled, with fan coil unit			
1320	Apartment corridors, 4,000 S.F., 7.33 ton	5.98	7.43	13.41
1360	6,000 S.F., 11.00 ton (R236000-20)	4.95	6.40	11.35
1400	10,000 S.F., 18.33 ton	4.94	4.74	9.68
1440	20,000 S.F., 26.66 ton	3.86	3.54	7.40
1480	40,000 S.F., 73.33 ton	3.77	3.62	7.39
1520	60,000 S.F., 110.00 ton	3.94	3.71	7.65
1600	Banks and libraries, 4,000 S.F., 16.66 ton	10.40	8.15	18.55
1640	6,000 S.F., 25.00 ton	8.85	7.20	16.05
1680	10,000 S.F., 41.66 ton	7.85	5.35	13.20
1720	20,000 S.F., 83.33 ton	6.55	5.10	11.65
1760	40,000 S.F., 166.66 ton	7.25	6.25	13.50
1800	60,000 S.F., 250.00 ton	7.40	6.55	13.95
1880	Bars and taverns, 4,000 S.F., 44.33 ton	18.75	10.30	29.05
1920	6,000 S.F., 66.50 ton	18.70	10.60	29.30
1960	10,000 S.F., 110.83 ton	15.10	8.45	23.55
2000	20,000 S.F., 221.66 ton	18.50	8.55	27.05
2040	40,000 S.F., 440 ton*			
2080	60,000 S.F., 660 ton*			
2160	Bowling alleys, 4,000 S.F., 22.66 ton	12.45	8.95	21.40
2200	6,000 S.F., 34.00 ton	10.70	7.90	18.60
2240	10,000 S.F., 56.66 ton	9.40	5.80	15.20
2280	20,000 S.F., 113.33 ton	8.35	5.40	13.75
2320	40,000 S.F., 226.66 ton	10.30	6.25	16.55

D3030 Cooling Generating Systems

D3030 115	Chilled Water, Cooling Tower Systems	COST PER S.F.		
		MAT.	INST.	TOTAL
2360	60,000 S.F., 340 ton			
2440	Department stores, 4,000 S.F., 11.66 ton	6.75	8.05	14.80
2480	6,000 S.F., 17.50 ton	7.65	6.65	14.30
2520	10,000 S.F., 29.17 ton	5.80	4.95	10.75
2560	20,000 S.F., 58.33 ton	4.90	3.68	8.58
2600	40,000 S.F., 116.66 ton	4.90	3.81	8.71
2640	60,000 S.F., 175.00 ton	6.60	6	12.60
2720	Drug stores, 4,000 S.F., 26.66 ton	13.20	9.20	22.40
2760	6,000 S.F., 40.00 ton	12.40	7.90	20.30
2800	10,000 S.F., 66.66 ton	11.85	7.15	19
2840	20,000 S.F., 133.33 ton	9.70	5.65	15.35
2880	40,000 S.F., 266.67 ton	10.20	7.05	17.25
2920	60,000 S.F., 400 ton*			
3000	Factories, 4,000 S.F., 13.33 ton	8.95	7.75	16.70
3040	6,000 S.F., 20.00 ton	7.85	6.70	14.55
3080	10,000 S.F., 33.33 ton	6.35	5.10	11.45
3120	20,000 S.F., 66.66 ton	6.45	4.59	11.04
3160	40,000 S.F., 133.33 ton	5.55	3.94	9.49
3200	60,000 S.F., 200.00 ton	6.60	6.30	12.90
3280	Food supermarkets, 4,000 S.F., 11.33 ton	6.65	8	14.65
3320	6,000 S.F., 17.00 ton	7	6.50	13.50
3360	10,000 S.F., 28.33 ton	5.70	4.92	10.62
3400	20,000 S.F., 56.66 ton	4.97	3.70	8.67
3440	40,000 S.F., 113.33 ton	4.87	3.82	8.69
3480	60,000 S.F., 170.00 ton	6.55	6	12.55
3560	Medical centers, 4.000 S.F., 9.33 ton	5.70	7.35	13.05
3600	6,000 S.F., 14.00 ton	6.65	6.35	13
3640	10,000 S.F., 23.33 ton	5.30	4.77	10.07
3680	20,000 S.F., 46.66 ton	4.44	3.61	8.05
3720	40,000 S.F., 93.33 ton	4.37	3.74	8.11
3760	60,000 S.F., 140.00 ton	5.30	6.05	11.35
3840	Offices, 4,000 S.F., 12.66 ton	8.65	7.70	16.35
3880	6,000 S.F., 19.00 ton	7.85	6.85	14.70
3920	10,000 S.F., 31.66 ton	6.35	5.15	11.50
3960	20,000 S.F., 63.33 ton	6.35	4.60	10.95
4000	40,000 S.F., 126.66 ton	5.90	5.95	11.85
4040	60,000 S.F., 190.00 ton	6.35	6.25	12.60
4120	Restaurants, 4,000 S.F., 20.00 ton	11.10	8.30	19.40
4160	6,000 S.F., 30.00 ton	9.60	7.40	17
4200	10,000 S.F., 50.00 ton	8.65	5.55	14.20
4240	20,000 S.F., 100.00 ton	7.80	5.35	13.15
4280	40,000 S.F., 200.00 ton	7.75	6.05	13.80
4320	60,000 S.F., 300.00 ton	8.30	6.80	15.10
4400	Schools and colleges, 4,000 S.F., 15.33 ton	9.85	8	17.85
4440	6,000 S.F., 23.00 ton	8.40	7.05	15.45
4480	10,000 S.F., 38.33 ton	7.40	5.25	12.65
4520	20,000 S.F., 76.66 ton	6.30	5.05	11.35
4560	40,000 S.F., 153.33 ton	6.85	6.10	12.95
4600	60,000 S.F., 230.00 ton	6.80	6.35	13.15

D3030 Cooling Generating Systems

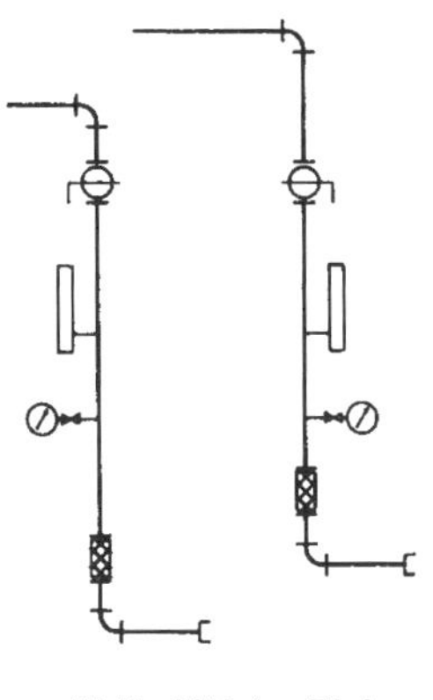
Chilled Water Piping

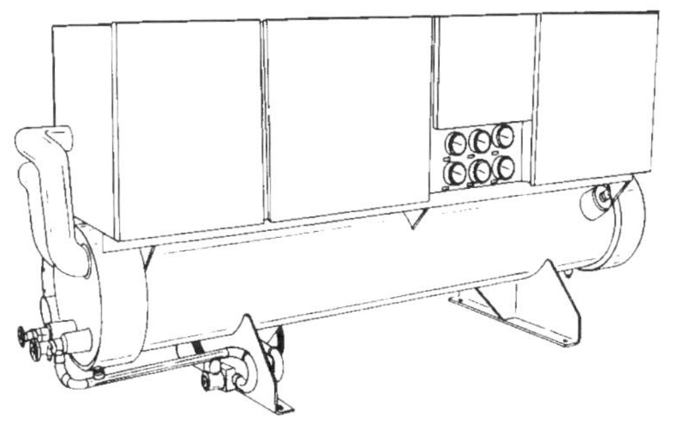
Reciprocating Chiller, Water Cooled

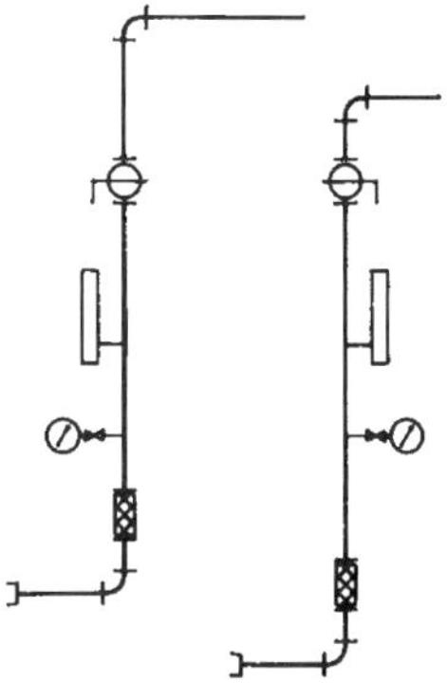
Condenser Water Piping

System Components	QUANTITY	UNIT	COST EACH MAT.	COST EACH INST.	COST EACH TOTAL
SYSTEM D3030 130 1010					
CHILLER, RECIPROCATING, WATER COOLED, STD. CONTROLS, 60 TON					
Water chiller, reciprocating, water cooled, 60 ton cooling	1.000	Ea.	36,900	9,325	46,225
Pipe, black steel, welded, Sch. 40, on yoke & roll hgrs, 10′ O.C., 4″ dia	40.000	L.F.	702	1,289.60	1,991.60
Elbow, 90°, weld joint, butt, 4″ pipe size	8.000	Ea.	424	1,903.20	2,327.20
Flange, welding neck, 150 lb., 4″ pipe size	16.000	Ea.	744	1,895.20	2,639.20
Valves, iron body, butterfly, lug type, lever actuator, 4″ size	4.000	Ea.	676	900	1,576
Expansion joints, bellows type, flange spool, 6″ F to F, 4″ dia	4.000	Ea.	1,380	532	1,912
T-O-L socket, weld joint, 1/4″ pipe size, nozzle	4.000	Ea.	29.40	207.20	236.60
T-O-L socket, weld joint, 3/4″ pipe size, nozzle	4.000	Ea.	34	226.28	260.28
Thermometers, stem type, 9″ case, 8″ stem, 3/4″ NPT	4.000	Ea.	744	88	832
Gauges, pressure or vacuum, 3-1/2″ diameter dial	4.000	Ea.	158	77.80	235.80
Insulation, fiberglass pipe covering, 1-1/2″ wall, 4″ IPS	20.000	L.F.	68.40	146	214.40
Chiller balancing	1.000	Ea.		530	530
TOTAL			41,859.80	17,120.28	58,980.08

D3030 130	Reciprocating Chiller, Water Cooled	COST EACH MAT.	COST EACH INST.	COST EACH TOTAL
1010	Chiller, reciprocating, water cooled, std. controls, 60 Ton	41,900	17,100	59,000
1020	100 Ton	46,400	25,700	72,100
1030	150 Ton	77,000	28,300	105,300
1040	200 Ton	103,500	29,900	133,400

D3030 135	Reciprocating Chiller, Air Cooled	COST EACH MAT.	COST EACH INST.	COST EACH TOTAL
1010	Chiller, reciprocating, air cooled, std. controls, 20 Ton	27,900	10,000	37,900
1020	30 Ton	30,900	10,600	41,500
1030	40 Ton	37,000	11,300	48,300
1040	60 Ton	55,000	12,800	67,800
1050	70 Ton	62,500	13,200	75,700
1060	80 Ton	67,500	13,400	80,900
1070	90 Ton	72,000	13,600	85,600
1080	100 Ton	81,000	13,800	94,800
1090	110 Ton	87,500	14,900	102,400
1100	130 Ton	103,500	15,100	118,600

D3030 Cooling Generating Systems

D3030 135	Reciprocating Chiller, Air Cooled	COST EACH		
		MAT.	INST.	TOTAL
1110	150 Ton	114,500	15,300	129,800
1120	175 Ton	128,000	15,500	143,500
1130	210 Ton	144,500	17,000	161,500

D3030 Cooling Generating Systems

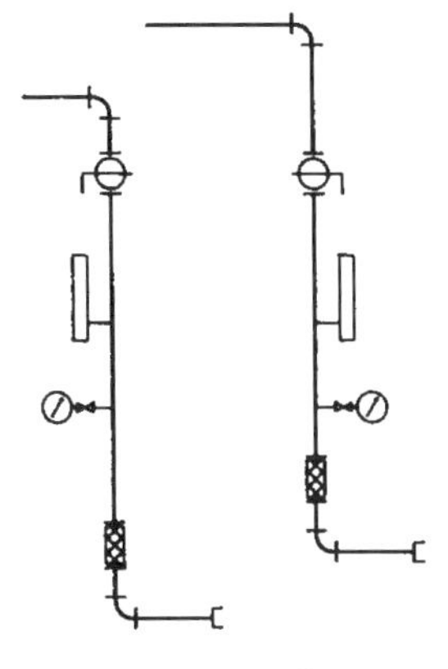

Chilled Water Piping

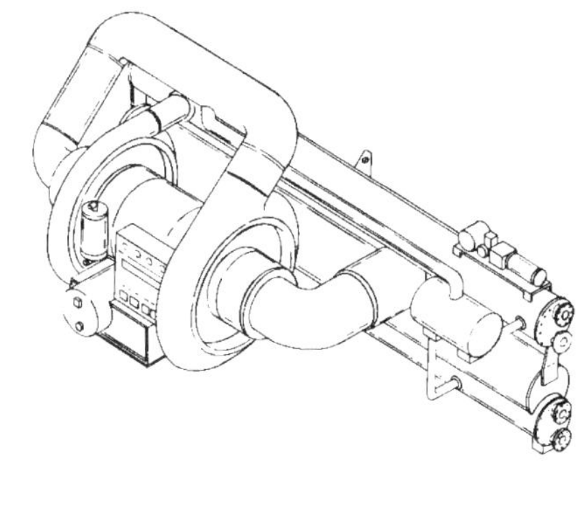

Centrifugal Chiller, Water Cooled, Hermetic

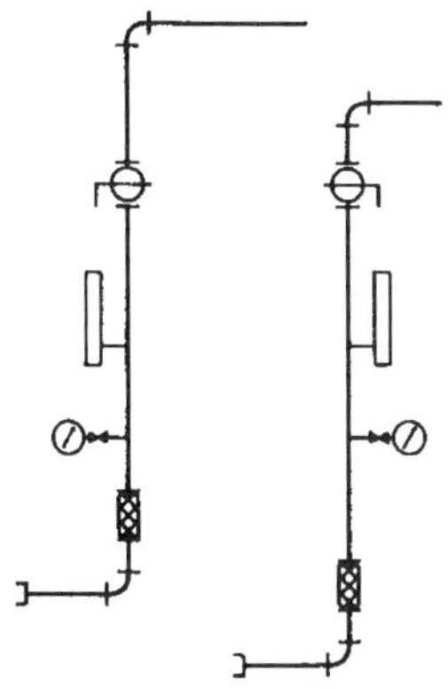

Condenser Water Piping

System Components	QUANTITY	UNIT	COST EACH MAT.	COST EACH INST.	COST EACH TOTAL
SYSTEM D3030 140 1010					
CHILLER, CENTRIF., WATER COOLED, PKGD. HERMETIC, STD. CONTROLS, 200 TON					
Water chiller, screw, water cooled, not incl tower, 200 ton	1.000	Ea.	90,000	18,700	108,700
Pipe, black steel, welded, Sch. 40, on yoke & roll hgrs, 10′ o.c., 6″ dia	40.000	L.F.	1,160	2,011.60	3,171.60
Elbow, 90°, weld joint, 6″ pipe size	8.000	Ea.	928	2,902.80	3,830.80
Flange, welding neck, 150 lb., 6″ pipe size	16.000	Ea.	1,120	2,903.20	4,023.20
Valves, iron body, butterfly, lug type, lever actuator, 6″ size	4.000	Ea.	1,132	1,400	2,532
Expansion joints, bellows type, flange spool, 6″ F to F, 6″ dia	4.000	Ea.	1,740	660	2,400
T-O-L socket, weld jiont, 1/4″ pipe size, nozzle	4.000	Ea.	29.40	207.20	236.60
T-O-L socket, weld joint, 3/4″ pipe size, nozzle	4.000	Ea.	34	226.28	260.28
Thermometers, stem type, 9″ case, 8″ stem, 3/4″ NPT	4.000	Ea.	744	88	832
Gauges, pressure or vacuum, 3-1/2″ diameter dial	4.000	Ea.	158	77.80	235.80
Insulation, fiberglass pipe covering, 1-1/2″ wall, 6″ IPS	40.000	L.F.	162	370	532
Chiller balancing	1.000	Ea.		530	530
TOTAL			97,207.40	30,076.88	127,284.28

D3030 140	Centrifugal Chiller, Water Cooled	COST EACH MAT.	COST EACH INST.	COST EACH TOTAL
1010	Chiller, centrif., water cooled, pkgd. hermetic, std controls, 200 Ton	97,000	30,100	127,100
1020	400 Ton	156,000	41,300	197,300
1030	1,000 Ton	492,500	62,500	555,000
1040	1,500 Ton	721,500	76,000	797,500

D3030 Cooling Generating Systems

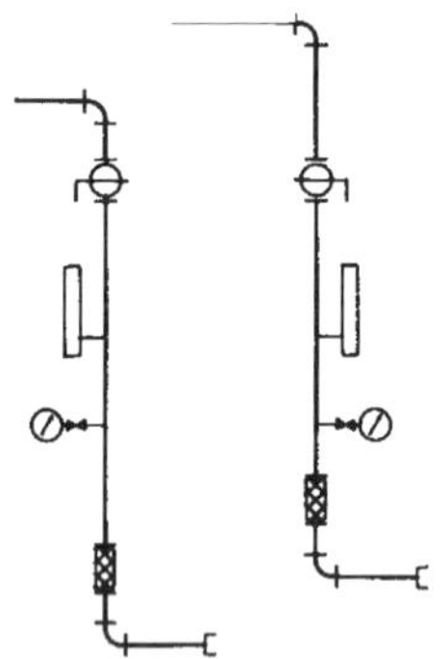
Chilled Water Piping
(Condenser Water Piping Similar)

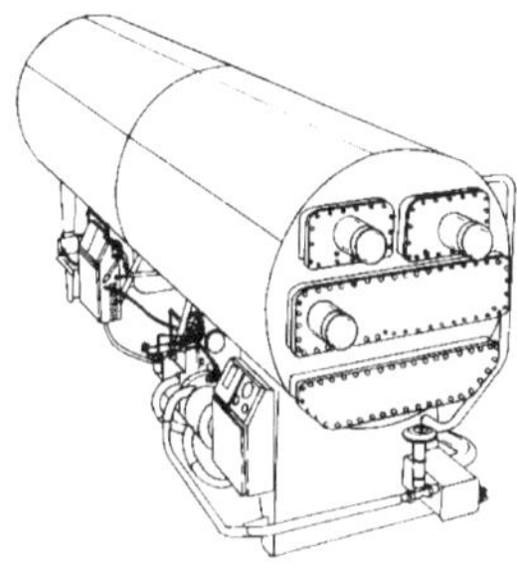
Gas Absorption Chiller, Water Cooled

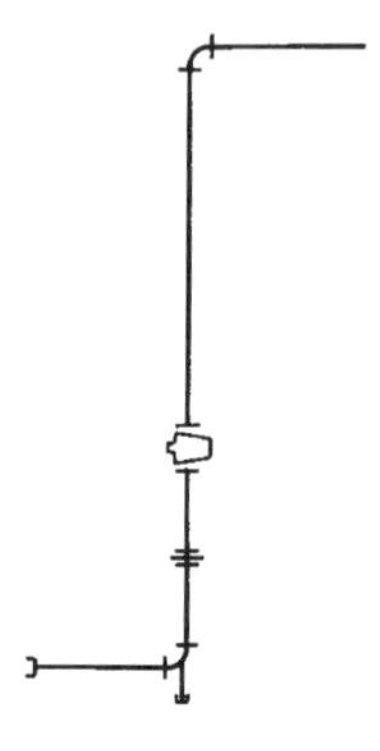
Gas Piping

System Components	QUANTITY	UNIT	COST EACH MAT.	COST EACH INST.	COST EACH TOTAL
SYSTEM D3030 145 1010					
CHILLER, GAS FIRED ABSORPTION, WATER COOLED, STD. CONTROLS, 800 TON					
Absorption water chiller, gas fired, water cooled, 800 ton	1.000	Ea.	625,000	28,200	653,200
Pipe, black steel, weld, Sch. 40, on yoke & roll hangers, 10′ O.C., 10″ dia	20.000	L.F.	1,210	1,513.60	2,723.60
Pipe, black steel, welded, Sch. 40, 14″ dia., (two rod roll type hanger)	44.000	L.F.	4,180	5,336.76	9,516.76
Pipe, black steel, welded, Sch. 40, on yoke & roll hangers, 10′ O.C., 4″ dia	20.000	L.F.	351	644.80	995.80
Elbow, 90°, weld joint, 10″ pipe size	4.000	Ea.	1,740	2,426	4,166
Elbow, 90°, weld joint, 14″ pipe size	6.000	Ea.	6,450	5,442	11,892
Elbow, 90°, weld joint, 4″ pipe size	3.000	Ea.	159	713.70	872.70
Flange, welding neck, 150 lb., 10″ pipe size	8.000	Ea.	1,592	2,413.60	4,005.60
Flange, welding neck, 150 lb., 14″ pipe size	14.000	Ea.	5,810	5,660.20	11,470.20
Flange, welding neck, 150 lb., 4″ pipe size	3.000	Ea.	139.50	355.35	494.85
Valves, iron body, butterfly, lug type, gear operated, 10″ size	2.000	Ea.	1,620	870	2,490
Valves, iron body, butterfly, lug type, gear operated, 14″ size	5.000	Ea.	11,250	3,800	15,050
Valves, semi-steel, lubricated plug valve, flanged, 4″ pipe size	1.000	Ea.	420	375	795
Expansion joints, bellows type, flange spool, 6″ F to F, 10″ dia	2.000	Ea.	1,400	448	1,848
Expansion joints, bellows type, flange spool, 10″ F to F, 14″ dia	2.000	Ea.	2,400	590	2,990
T-O-L socket, weld joint, 1/4″ pipe size, nozzle	4.000	Ea.	29.40	207.20	236.60
T-O-L socket, weld joint, 3/4″ pipe size, nozzle	4.000	Ea.	34	226.28	260.28
Thermometers, stem type, 9″ case, 8″ stem, 3/4″ NPT	4.000	Ea.	744	88	832
Gauges, pressure or vacuum, 3-1/2″ diameter dial	4.000	Ea.	158	77.80	235.80
Insulation, fiberglass pipe covering, 2″ wall, 10″ IPS	20.000	L.F.	180	291	471
Vent chimney, all fuel, pressure tight, double wall, SS, 12″ dia	20.000	L.F.	1,610	490	2,100
Chiller balancing	1.000	Ea.		530	530
TOTAL			666,476.90	60,699.29	727,176.19

D3030 145	Gas Absorption Chiller, Water Cooled	COST EACH MAT.	COST EACH INST.	COST EACH TOTAL
1010	Chiller, gas fired absorption, water cooled, std. ctrls, 800 Ton	666,500	61,000	727,500
1020	1000 Ton	874,000	78,000	952,000

D3030 150	Steam Absorption Chiller, Water Cooled	COST EACH MAT.	COST EACH INST.	COST EACH TOTAL
1010	Chiller, steam absorption, water cooled, std. ctrls, 750 Ton	643,500	72,000	715,500
1020	955 Ton	752,000	86,000	838,000
1030	1465 Ton	1,087,500	98,500	1,186,000
1040	1660 Ton	1,313,500	112,000	1,425,500

D3030 Cooling Generating Systems

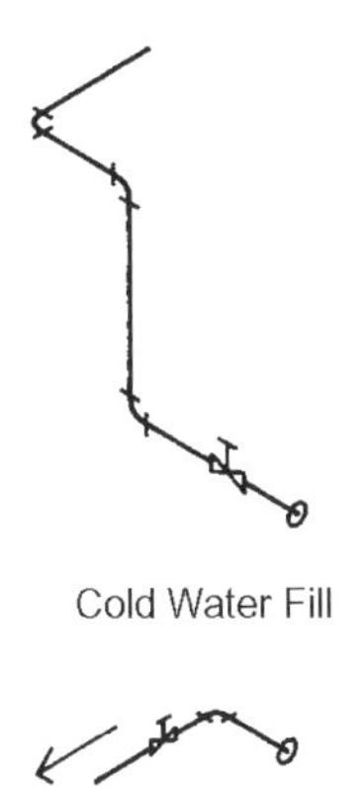

Cold Water Fill

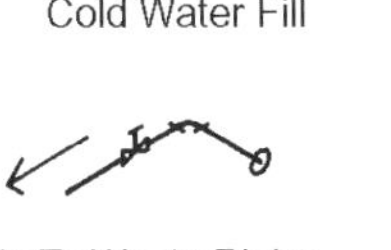

Drain To Waste Piping

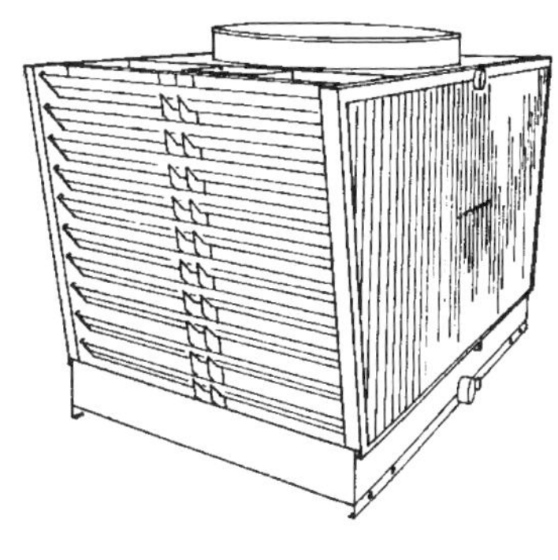

Cooling Tower

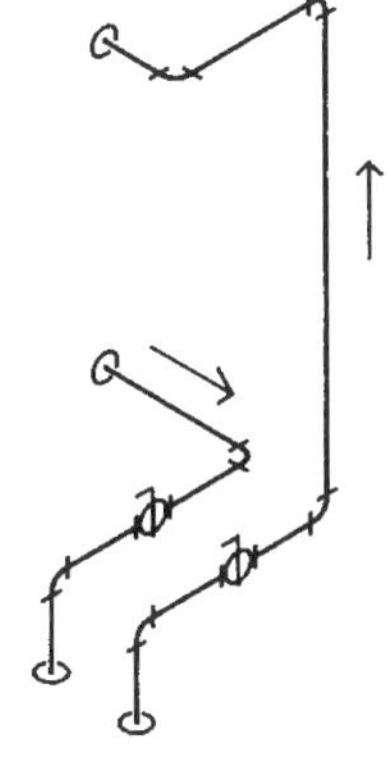

Condenser Water Piping

System Components	QUANTITY	UNIT	COST EACH MAT.	COST EACH INST.	COST EACH TOTAL
SYSTEM D3030 310 1010					
COOLING TOWER, GALVANIZED STEEL, PACKAGED UNIT, DRAW THRU, 60 TON					
Cooling towers, draw thru, single flow, belt drive, 60 tons	1.000	Ea.	10,740	1,161	11,901
Pipe, black steel, welded, Sch. 40, on yoke & roll hangers, 10′ OC, 4″ dia.	40.000	L.F.	702	1,289.60	1,991.60
Tubing, copper, Type L, couplings & hangers 10′ OC, 1-1/2″ dia.	20.000	L.F.	236	240	476
Elbow, 90°, copper, cu x cu, 1-1/2″ dia.	3.000	Ea.	63	144	207
Elbow, 90°, weld joint, 4″ pipe size	6.000	Ea.	318	1,427.40	1,745.40
Flange, welding neck, 150 lb., 4″ pipe size	6.000	Ea.	279	710.70	989.70
Valves, bronze, gate, N.R.S., soldered, 125 psi, 1-1/2″ size	1.000	Ea.	97	48	145
Valves, iron body, butterfly, lug type, lever actuator, 4″ size	2.000	Ea.	338	450	788
Electric heat trace system, 400 degree, 115v, 5 watts per L.F.	160.000	L.F.	1,296	176	1,472
Insulation, fiberglass pipe covering, , 1-1/2″ wall, 4″ IPS	40.000	L.F.	136.80	292	428.80
Insulation, fiberglass pipe covering, 1″ wall, 1-1/2″ IPS	20.000	L.F.	28.80	97.20	126
Cooling tower condenser control system	1.000	Ea.	5,175	2,550	7,725
Cooling tower balancing	1.000	Ea.		405	405
TOTAL			19,409.60	8,990.90	28,400.50

D3030 310	Galvanized Draw Through Cooling Tower	COST EACH MAT.	COST EACH INST.	COST EACH TOTAL
1010	Cooling tower, galvanized steel, packaged unit, draw thru, 60 Ton	19,400	8,975	28,375
1020	110 Ton	27,000	10,600	37,600
1030	300 Ton	40,300	15,900	56,200
1040	600 Ton	65,500	23,800	89,300
1050	1000 Ton	100,000	35,100	135,100

D3030 320	Fiberglass Draw Through Cooling Tower	COST EACH MAT.	COST EACH INST.	COST EACH TOTAL
1010	Cooling tower, fiberglass, packaged unit, draw thru, 100 Ton	19,800	9,075	28,875
1020	120 Ton	22,300	10,300	32,600
1030	140 Ton	25,800	13,600	39,400
1040	160 Ton	35,300	18,100	53,400
1050	180 Ton	48,400	26,200	74,600

D3030 330	Stainless Steel Draw Through Cooling Tower	COST EACH MAT.	COST EACH INST.	COST EACH TOTAL
1010	Cooling tower, stainless steel, packaged unit, draw thru, 60 Ton	32,100	8,975	41,075
1020	110 Ton	37,600	10,600	48,200
1030	300 Ton	72,000	15,900	87,900
1040	600 Ton	114,500	23,500	138,000
1050	1000 Ton	179,500	35,100	214,600

D3040 Distribution Systems

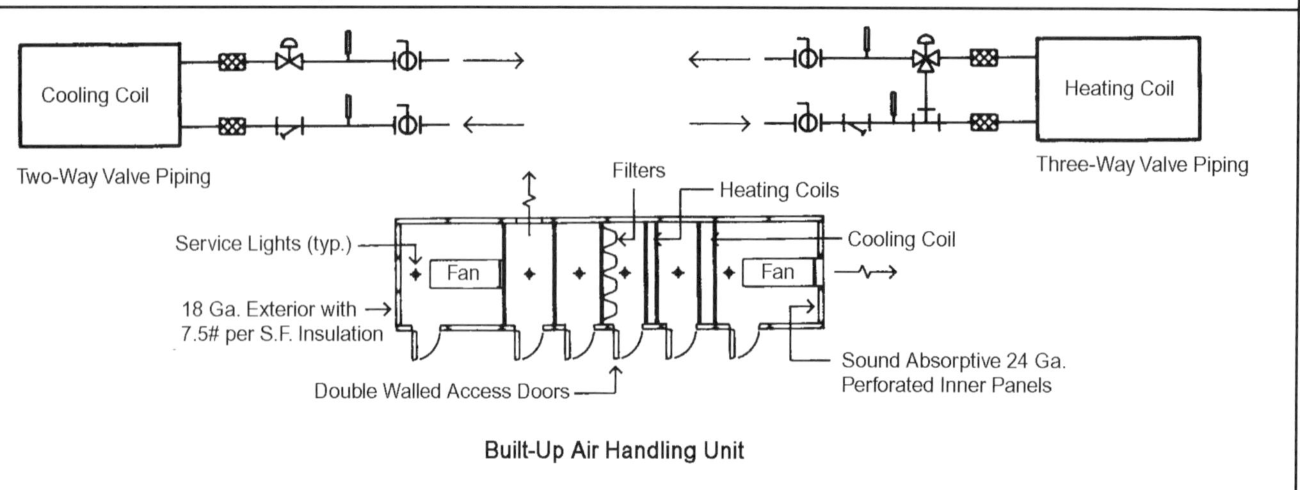

Built-Up Air Handling Unit

System Components	QUANTITY	UNIT	COST EACH MAT.	COST EACH INST.	COST EACH TOTAL
SYSTEM D3040 106 1010					
AHU, FIELD FAB., BLT. UP, COOL/HEAT COILS, FLTRS., CONST. VOL., 40,000 CFM					
AHU, Built-up, cool/heat coils, filters, mix box, constant volume	1.000	Ea.	54,500	5,800	60,300
Pipe, black steel, welded, Sch. 40, on yoke & roll hgrs, 10′ OC, 3″ dia.	26.000	L.F.	326.30	715	1,041.30
Pipe, black steel, welded, Sch. 40, on yoke & roll hgrs, 10′ OC, 4″ dia	20.000	L.F.	351	644.80	995.80
Elbow, 90°, weld joint, 3″ pipe size	5.000	Ea.	185	851	1,036
Elbow, 90°, weld joint, 4″ pipe size	4.000	Ea.	212	951.60	1,163.60
Tee, welded, reducing on outlet, 3″ by 2-1/2″ pipe	3.000	Ea.	390	891.30	1,281.30
Tee, welded, reducing on outlet, 4″ by 3″ pipe	2.000	Ea.	274	793	1,067
Reducer, eccentric, weld joint, 3″ pipe size	4.000	Ea.	260	592.20	852.20
Reducer, eccentric, weld joint, 4″ pipe size	4.000	Ea.	332	791	1,123
Flange, weld neck, 150 LB, 2-1/2″ pipe size	3.000	Ea.	105	222.09	327.09
Flange, weld neck, 150 lb., 3″ pipe size	14.000	Ea.	539	1,191.54	1,730.54
Flange, weld neck, 150 lb., 4″ pipe size	10.000	Ea.	465	1,184.50	1,649.50
Valves, iron body, butterfly, lug type, lever actuator, 3″ size	2.000	Ea.	278	280	558
Valves, iron body, butterfly, lug type, lever actuator, 4″ size	2.000	Ea.	338	450	788
Strainer, Y type, iron body, flanged, 125 lb., 3″ pipe size	1.000	Ea.	146	249	395
Strainer, Y type, iron body, flanged, 125 lb., 4″ pipe size	1.000	Ea.	267	375	642
Valve, electric motor actuated, iron body, 3 way, flanged, 2-1/2″ pipe	1.000	Ea.	1,225	305	1,530
Valve, electric motor actuated, iron body, two way, flanged, 3″ pipe size	1.000	Ea.	1,275	305	1,580
Expansion joints, bellows type, flange spool, 6″ F to F, 3″ dia	2.000	Ea.	640	197	837
Expansion joints, bellows type, flange spool, 6″ F to F, 4″ dia	2.000	Ea.	690	266	956
T-O-L socket, weld joint, 1/4″ pipe size, nozzle	4.000	Ea.	29.40	207.20	236.60
T-O-L socket, weld joint, 1/2″ pipe size, nozzle	12.000	Ea.	88.20	647.16	735.36
Thermometers, stem type, 9″ case, 8″ stem, 3/4″ NPT	4.000	Ea.	744	88	832
Pressure/temperature relief plug, 316 SS, 3/4″OD, 7-1/2″ insertion	4.000	Ea.	376	77.80	453.80
Insulation, fiberglass pipe covering, 1-1/2″ wall, 3″ IPS	26.000	L.F.	78.26	156	234.26
Insulation, fiberglass pipe covering, 1-1/2″ wall, 4″ IPS	20.000	L.F.	68.40	146	214.40
Fan coil control system	1.000	Ea.	1,175	570	1,745
Re-heat coil balancing	2.000	Ea.		230	230
Air conditioner balancing	1.000	Ea.		715	715
TOTAL			65,357.56	19,892.19	85,249.75

D3040 106	Field Fabricated Air Handling Unit	COST EACH MAT.	COST EACH INST.	COST EACH TOTAL
1010	AHU, Field fab, blt up, cool/heat coils, fltrs, const vol, 40,000 CFM	65,500	19,900	85,400
1020	60,000 CFM	96,000	30,800	126,800
1030	75,000 CFM	116,500	32,400	148,900

D3040 Distribution Systems

D3040 108	Field Fabricated VAV Air Handling Unit	COST EACH		
		MAT.	INST.	TOTAL
1010	AHU, Field fab, blt up, cool/heat coils, fltrs, VAV, 75,000 CFM	117,500	37,300	154,800
1020	100,000 CFM	177,000	49,300	226,300
1030	150,000 CFM	293,000	79,000	372,000

D3040 Distribution Systems

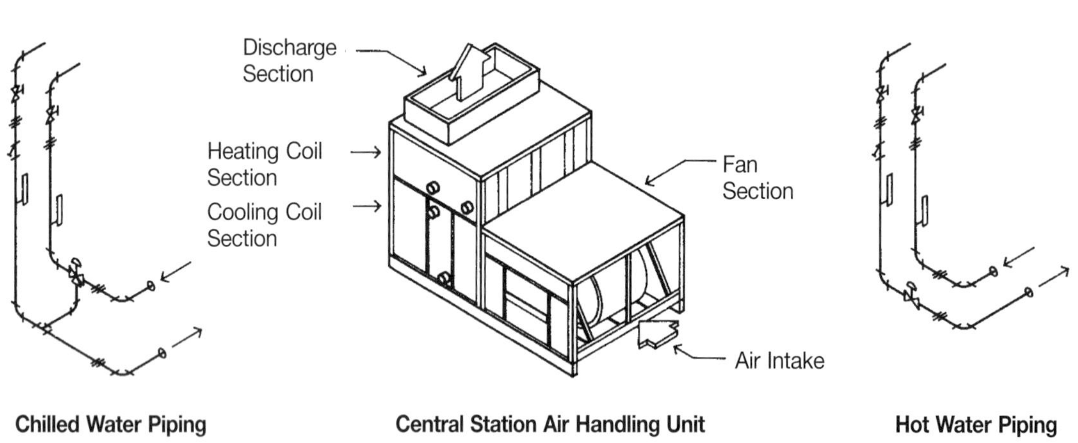

System Components	QUANTITY	UNIT	COST EACH MAT.	COST EACH INST.	COST EACH TOTAL
SYSTEM D3040 110 1010					
AHU, CENTRAL STATION, COOL/HEAT COILS, CONST. VOL., FLTRS., 2,000 CFM					
Central station air handling unit, chilled water, 2000 CFM	1.000	Ea.	8,900	1,025	9,925
Pipe, black steel, Sch 40, threaded, W/cplgs & hangers, 10′ OC, 1″ dia.	26.000	L.F.	224.25	381.88	606.13
Pipe, black steel, Sch 40, threaded, W/cplg & hangers, 10′ OC, 1-1/4″ dia.	20.000	L.F.	227.50	328.90	556.40
Elbow, 90°, black, straight, 1″ dia.	4.000	Ea.	22	192	214
Elbow, 90°, black, straight, 1-1/4″ dia.	4.000	Ea.	36	204	240
Tee, black, reducing, 1″ dia.	8.000	Ea.	110	624	734
Tee, black, reducing, 1-1/4″ dia.	8.000	Ea.	188	644	832
Reducer, concentric, black, 1″ dia.	4.000	Ea.	39.80	166	205.80
Reducer, concentric, black, 1-1/4″ dia.	4.000	Ea.	44.60	172	216.60
Union, black with brass seat, 1″ dia.	4.000	Ea.	69.20	208	277.20
Union, black with brass seat, 1-1/4″ dia.	2.000	Ea.	50	107	157
Pipe nipple, black, 1″ dia	13.000	Ea.	44.85	76.38	121.23
Pipe nipple, black, 1-1/4″ dia	12.000	Ea.	52.50	75.90	128.40
Valves, bronze, gate, N.R.S., threaded, class 150, 1″ size	2.000	Ea.	162	66	228
Valves, bronze, gate, N.R.S., threaded, class 150, 1-1/4″ size	2.000	Ea.	216	83	299
Strainer, Y type, bronze body, screwed, 150 lb., 1″ pipe size	1.000	Ea.	34	36.50	70.50
Strainer, Y type, bronze body, screwed, 150 lb., 1-1/4″ pipe size	1.000	Ea.	68	41.50	109.50
Valve, electric motor actuated, brass, 3 way, screwed, 3/4″ pipe	1.000	Ea.	325	34	359
Valve, electric motor actuated, brass, 2 way, screwed, 1″ pipe size	1.000	Ea.	345	32.50	377.50
Thermometers, stem type, 9″ case, 8″ stem, 3/4″ NPT	4.000	Ea.	744	88	832
Pressure/temperature relief plug, 316 SS, 3/4″OD, 7-1/2″ insertion	4.000	Ea.	376	77.80	453.80
Insulation, fiberglass pipe covering, 1″ wall, 1″ IPS	26.000	L.F.	31.98	120.64	152.62
Insulation, fiberglass pipe covering, 1″ wall, 1-1/4″ IPS	20.000	L.F.	26.60	97.20	123.80
Fan coil control system	1.000	Ea.	1,175	570	1,745
Re-heat coil balancing	2.000	Ea.		230	230
Multizone A/C balancing	1.000	Ea.		535	535
TOTAL			13,512.28	6,217.20	19,729.48

D3040 110	Central Station Air Handling Unit	COST EACH MAT.	COST EACH INST.	COST EACH TOTAL
1010	AHU, Central station, cool/heat coils, constant vol, fltrs, 2,000 CFM	13,500	6,225	19,725
1020	5,000 CFM	27,600	7,975	35,575
1030	10,000 CFM	54,500	11,200	65,700
1040	15,000 CFM	78,500	15,100	93,600
1050	20,000 CFM	105,000	17,700	122,700

D3040 Distribution Systems

D3040 112	Central Station Air Handling Unit, VAV	COST EACH		
		MAT.	INST.	TOTAL
1010	AHU, Central station, cool/heat coils, VAV, fltrs, 5,000 CFM	30,300	8,300	38,600
1020	10,000 CFM	60,000	11,500	71,500
1030	15,000 CFM	88,500	14,900	103,400
1040	20,000 CFM	110,000	18,100	128,100
1050	30,000 CFM	159,500	20,800	180,300

D3040 Distribution Systems

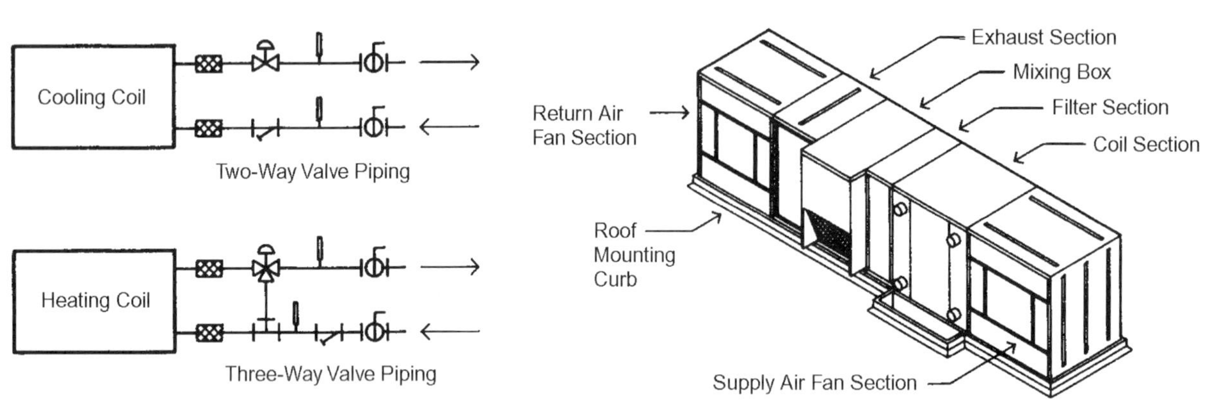

Rooftop Air Handling Unit

System Components	QUANTITY	UNIT	COST EACH MAT.	INST.	TOTAL
SYSTEM D3040 114 1010					
AHU, ROOFTOP, COOL/HEAT COILS, FLTRS., CONST. VOL., 2,000 CFM					
AHU, Built-up, cool/heat coils, filter, mix box, rooftop, constant volume	1.000	Ea.	11,100	1,750	12,850
Pipe, black steel, Sch 40, threaded, W/cplgs & hangers, 10′ OC, 1″ dia.	26.000	L.F.	224.25	381.88	606.13
Pipe, black steel, Sch 40, threaded, W/cplg & hangers, 10′ OC, 1-1/4″ dia.	20.000	L.F.	227.50	328.90	556.40
Elbow, 90°, black, straight, 1″ dia.	4.000	Ea.	22	192	214
Elbow, 90°, black, straight, 1-1/4″ dia.	4.000	Ea.	36	204	240
Tee, black, reducing, 1″ dia.	8.000	Ea.	110	624	734
Tee, black, reducing, 1-1/4″ dia.	8.000	Ea.	188	644	832
Reducer, black, concentric, 1″ dia.	4.000	Ea.	39.80	166	205.80
Reducer, black, concentric, 1-1/4″ dia.	4.000	Ea.	44.60	172	216.60
Union, black with brass seat, 1″ dia.	4.000	Ea.	69.20	208	277.20
Union, black with brass seat, 1-1/4″ dia.	2.000	Ea.	50	107	157
Pipe nipples, 1″ dia	13.000	Ea.	44.85	76.38	121.23
Pipe nipples, 1-1/4″ dia	12.000	Ea.	52.50	75.90	128.40
Valves, bronze, gate, N.R.S., threaded, class 150, 1″ size	2.000	Ea.	162	66	228
Valves, bronze, gate, N.R.S., threaded, class 150, 1-1/4″ size	2.000	Ea.	216	83	299
Strainer, Y type, bronze body, screwed, 150 lb., 1″ pipe size	1.000	Ea.	34	36.50	70.50
Strainer, Y type, bronze body, screwed, 150 lb., 1-1/4″ pipe size	1.000	Ea.	68	41.50	109.50
Valve, electric motor actuated, brass, 3 way, screwed, 3/4″ pipe	1.000	Ea.	325	34	359
Valve, electric motor actuated, brass, 2 way, screwed, 1″ pipe size	1.000	Ea.	345	32.50	377.50
Thermometers, stem type, 9″ case, 8″ stem, 3/4″ NPT	4.000	Ea.	744	88	832
Pressure/temperature relief plug, 316 SS, 3/4″OD, 7-1/2″ insertion	4.000	Ea.	376	77.80	453.80
Insulation, fiberglass pipe covering, 1″ wall, 1″ IPS	26.000	L.F.	31.98	120.64	152.62
Insulation, fiberglass pipe covering, 1″ wall, 1-1/4″ IPS	20.000	L.F.	26.60	97.20	123.80
Insulation pipe covering .016″ aluminum jacket finish, add	45.000	S.F.	55.35	229.50	284.85
Fan coil control system	1.000	Ea.	1,175	570	1,745
Re-heat coil balancing	2.000	Ea.		230	230
Rooftop unit heat/cool balancing	1.000	Ea.		415	415
TOTAL			15,767.63	7,051.70	22,819.33

D3040 114	Rooftop Air Handling Unit	COST EACH MAT.	INST.	TOTAL
1010	AHU, Rooftop, cool/heat coils, constant vol., fltrs, 2,000 CFM	15,800	7,050	22,850
1020	5,000 CFM	33,000	8,175	41,175
1030	10,000 CFM	65,500	11,400	76,900
1040	15,000 CFM	95,500	15,100	110,600

D3040 Distribution Systems

D3040 114	Rooftop Air Handling Unit	COST EACH		
		MAT.	INST.	TOTAL
1050	20,000 CFM	128,500	18,300	146,800

D3040 116	Rooftop Air Handling Unit, VAV	COST EACH		
		MAT.	INST.	TOTAL
1010	AHU, Rooftop, cool/heat coils, VAV, fltrs, 5,000 CFM	36,400	8,350	44,750
1020	10,000 CFM	72,500	11,600	84,100
1030	15,000 CFM	107,500	15,200	122,700
1040	20,000 CFM	135,000	18,500	153,500
1050	30,000 CFM	197,000	21,200	218,200

D3040 Distribution Systems

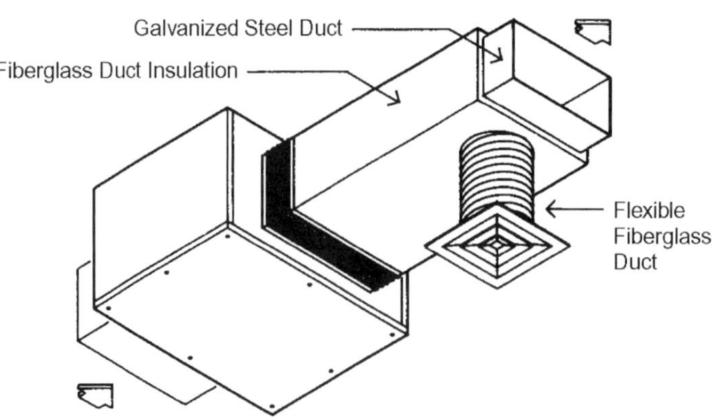

Horizontal Fan Coil Air Conditioning System

System Components	QUANTITY	UNIT	COST EACH MAT.	COST EACH INST.	COST EACH TOTAL
SYSTEM D3040 124 1010					
FAN COIL A/C SYSTEM, HORIZONTAL, W/HOUSING, CONTROLS, 2 PIPE, 1/2 TON					
Fan coil A/C, horizontal housing, filters, chilled water, 1/2 ton cooling	1.000	Ea.	1,200	140	1,340
Pipe, black steel, Sch 40, threaded, W/cplgs & hangers, 10′ OC, 3/4″ dia.	20.000	L.F.	101.27	220.38	321.65
Elbow, 90°, black, straight, 3/4″ dia.	6.000	Ea.	18.96	267	285.96
Tee, black, straight, 3/4″ dia.	2.000	Ea.	10	139	149
Union, black with brass seat, 3/4″ dia.	2.000	Ea.	26.50	96	122.50
Pipe nipples, 3/4″ diam	3.000	Ea.	7.07	15.38	22.45
Valves, bronze, gate, N.R.S., threaded, class 150, 3/4″ size	2.000	Ea.	127	62	189
Circuit setter, balance valve, bronze body, threaded, 3/4″ pipe size	1.000	Ea.	78.50	31	109.50
Insulation, fiberglass pipe covering, 1″ wall, 3/4″ IPS	20.000	L.F.	22.80	88.60	111.40
Ductwork, 12″ x 8″ fabricated, galvanized steel, 12 LF	55.000	Lb.	28.60	390.50	419.10
Insulation, ductwork, blanket type, fiberglass, 1″ thk, 1-1/2 LB density	40.000	S.F.	30	116.40	146.40
Diffusers, aluminum, OB damper, ceiling, perf, 24″x24″ panel size, 6″x6″	2.000	Ea.	127	74	201
Ductwork, flexible, fiberglass fabric, insulated, 1″thk, PE jacket, 6″ dia	16.000	L.F.	41.92	65.92	107.84
Round volume control damper 6″ dia.	2.000	Ea.	51	54	105
Fan coil unit balancing	1.000	Ea.		77.50	77.50
Re-heat coil balancing	1.000	Ea.		115	115
Diffuser/register, high, balancing	2.000	Ea.		214	214
TOTAL			1,870.62	2,166.68	4,037.30

D3040 118	Fan Coil A/C Unit, Two Pipe	COST EACH MAT.	COST EACH INST.	COST EACH TOTAL
1010	Fan coil A/C system, cabinet mounted, controls, 2 pipe, 1/2 Ton	1,175	1,175	2,350
1020	1 Ton	1,425	1,300	2,725
1030	1-1/2 Ton	1,525	1,325	2,850
1040	2 Ton	2,425	1,625	4,050
1050	3 Ton	3,050	1,675	4,725

D3040 120	Fan Coil A/C Unit, Two Pipe, Electric Heat	COST EACH MAT.	COST EACH INST.	COST EACH TOTAL
1010	Fan coil A/C system, cabinet mntd, elect. ht, controls, 2 pipe, 1/2 Ton	1,675	1,175	2,850
1020	1 Ton	2,025	1,300	3,325
1030	1-1/2 Ton	2,375	1,325	3,700
1040	2 Ton	3,600	1,650	5,250
1050	3 Ton	6,075	1,675	7,750

D3040 Distribution Systems

D3040 122	Fan Coil A/C Unit, Four Pipe	COST EACH		
		MAT.	INST.	TOTAL
1010	Fan coil A/C system, cabinet mounted, controls, 4 pipe, 1/2 Ton	1,900	2,275	4,175
1020	1 Ton	2,175	2,425	4,600
1030	1-1/2 Ton	2,325	2,450	4,775
1040	2 Ton	2,900	2,525	5,425
1050	3 Ton	3,925	2,725	6,650

D3040 124	Fan Coil A/C, Horizontal, Duct Mount, 2 Pipe	COST EACH		
		MAT.	INST.	TOTAL
1010	Fan coil A/C system, horizontal w/housing, controls, 2 pipe, 1/2 Ton	1,875	2,175	4,050
1020	1 Ton	2,425	3,150	5,575
1030	1-1/2 Ton	2,925	4,075	7,000
1040	2 Ton	3,975	4,900	8,875
1050	3 Ton	4,675	6,475	11,150
1060	3-1/2 Ton	4,900	6,700	11,600
1070	4 Ton	5,025	7,600	12,625
1080	5 Ton	5,875	9,650	15,525
1090	6 Ton	5,925	10,300	16,225
1100	7 Ton	6,000	10,700	16,700
1110	8 Ton	6,000	10,900	16,900
1120	10 Ton	6,750	11,400	18,150

D3040 126	Fan Coil A/C, Horiz., Duct Mount, 2 Pipe, Elec. Ht.	COST EACH		
		MAT.	INST.	TOTAL
1010	Fan coil A/C system, horiz. hsng, elect. ht, ctrls, 2 pipe, 1/2 Ton	2,125	2,175	4,300
1020	1 Ton	2,750	3,150	5,900
1030	1-1/2 Ton	3,150	4,075	7,225
1040	2 Ton	4,375	4,900	9,275
1050	3 Ton	6,975	6,675	13,650
1060	3-1/2 Ton	7,900	6,700	14,600
1070	4 Ton	8,900	7,600	16,500
1080	5 Ton	10,200	9,650	19,850
1090	6 Ton	10,300	10,400	20,700
1100	7 Ton	10,300	10,300	20,600
1110	8 Ton	11,400	10,400	21,800
1120	10 Ton	12,200	10,700	22,900

D3040 128	Fan Coil A/C, Horizontal, Duct Mount, 4 Pipe	COST EACH		
		MAT.	INST.	TOTAL
1010	Fan coil A/C system, horiz. w/cabinet, controls, 4 pipe, 1/2 Ton	2,750	3,250	6,000
1020	1 Ton	3,400	4,350	7,750
1030	1-1/2 Ton	4,025	5,175	9,200
1040	2 Ton	4,775	5,800	10,575
1050	3 Ton	5,750	7,725	13,475
1060	3.5 Ton	6,475	9,425	15,900
1070	4 Ton	6,425	9,525	15,950
1080	5 Ton	7,650	13,100	20,750
1090	6 Ton	8,150	15,400	23,550
1100	7 Ton	9,300	17,400	26,700
1110	8 Ton	9,425	18,300	27,725
1120	10 Ton	11,300	22,500	33,800

D3040 Distribution Systems

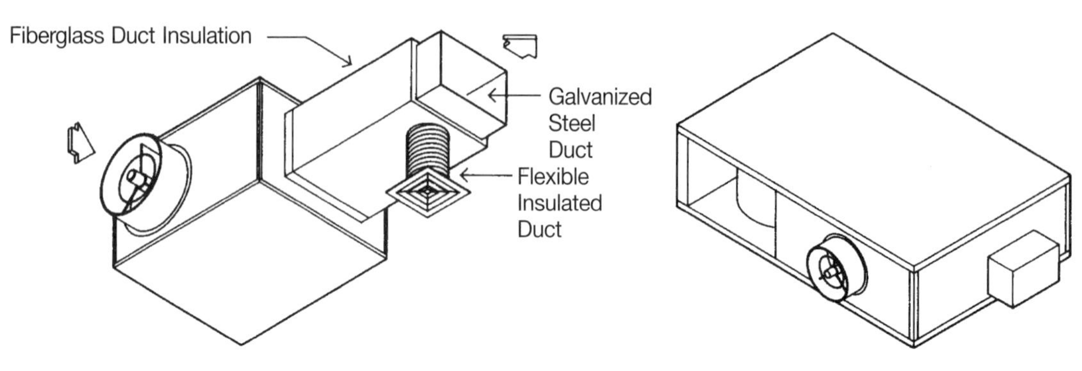

VAV Terminal

Fan Powered VAV Terminal

System Components	QUANTITY	UNIT	COST EACH MAT.	COST EACH INST.	COST EACH TOTAL
SYSTEM D3040 132 1010					
VAV TERMINAL, COOLING ONLY, WITH ACTUATOR/CONTROLS, 200 CFM					
Mixing box variable volume 300 to 600 CFM, cool only	1.000	Ea.	500.50	107.25	607.75
Ductwork, 12″ x 8″ fabricated, galvanized steel, 12 LF	55.000	Lb.	28.60	390.50	419.10
Insulation, ductwork, blanket type, fiberglass, 1″ thk, 1-1/2 LB density	40.000	S.F.	30	116.40	146.40
Diffusers, aluminum, OB damper, ceiling, perf, 24″x24″ panel size, 6″x6″	2.000	Ea.	127	74	201
Ductwork, flexible, fiberglass fabric, insulated, 1″thk, PE jacket, 6″ dia	16.000	L.F.	41.92	65.92	107.84
Round volume control damper 6″ dia.	2.000	Ea.	51	54	105
VAV box balancing	1.000	Ea.		71.50	71.50
Diffuser/register, high, balancing	2.000	Ea.		214	214
TOTAL			779.02	1,093.57	1,872.59

D3040 132	VAV Terminal, Cooling Only	COST EACH MAT.	COST EACH INST.	COST EACH TOTAL
1010	VAV Terminal, cooling only, with actuator/controls, 200 CFM	780	1,100	1,880
1020	400 CFM	940	1,925	2,865
1030	600 CFM	1,200	2,875	4,075
1040	800 CFM	1,275	3,475	4,750
1050	1000 CFM	1,450	4,025	5,475
1060	1250 CFM	1,675	5,125	6,800
1070	1500 CFM	1,825	6,050	7,875
1080	2000 CFM	2,325	8,575	10,900

D3040 134	VAV Terminal, Hot Water Reheat	COST EACH MAT.	COST EACH INST.	COST EACH TOTAL
1010	VAV Terminal, cooling, HW reheat, with actuator/controls, 200 CFM	1,775	2,050	3,825
1020	400 CFM	1,950	2,800	4,750
1030	600 CFM	2,200	3,825	6,025
1040	800 CFM	2,300	4,450	6,750
1050	1000 CFM	2,450	5,025	7,475
1060	1250 CFM	2,750	6,150	8,900
1070	1500 CFM	3,100	7,300	10,400
1080	2000 CFM	3,675	9,875	13,550

D3040 136	Fan Powered VAV Terminal, Cooling Only	COST EACH MAT.	COST EACH INST.	COST EACH TOTAL
1010	VAV Terminal, cooling, fan powrd, with actuator/controls, 200 CFM	1,350	1,100	2,450
1020	400 CFM	1,575	1,800	3,375

D3040 Distribution Systems

D3040 136	Fan Powered VAV Terminal, Cooling Only	COST EACH		
		MAT.	INST.	TOTAL
1030	600 CFM	1,800	2,850	4,650
1040	800 CFM	1,950	3,425	5,375
1050	1000 CFM	2,100	3,950	6,050
1060	1250 CFM	2,425	5,200	7,625
1070	1500 CFM	2,525	6,075	8,600
1080	2000 CFM	3,175	8,850	12,025

D3040 138	Fan Powered VAV Terminal, Hot Water Reheat	COST EACH		
		MAT.	INST.	TOTAL
1010	VAV Terminal, cool, HW reht, fan powrd, with actuator/ctrls, 200 CFM	2,300	2,050	4,350
1020	400 CFM	2,500	2,825	5,325
1030	600 CFM	2,725	3,825	6,550
1040	800 CFM	2,875	4,400	7,275
1050	1000 CFM	3,025	4,950	7,975
1060	1250 CFM	3,475	6,175	9,650
1070	1500 CFM	3,825	7,350	11,175
1080	2000 CFM	4,450	10,200	14,650

D3040 Distribution Systems

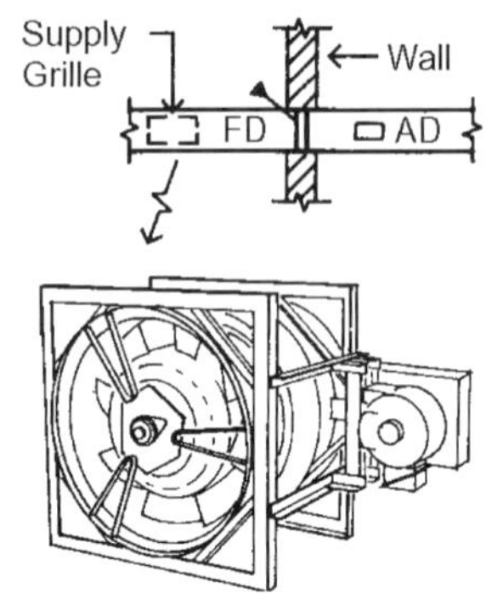

Axial Flow Centrifugal Fan Belt Drive

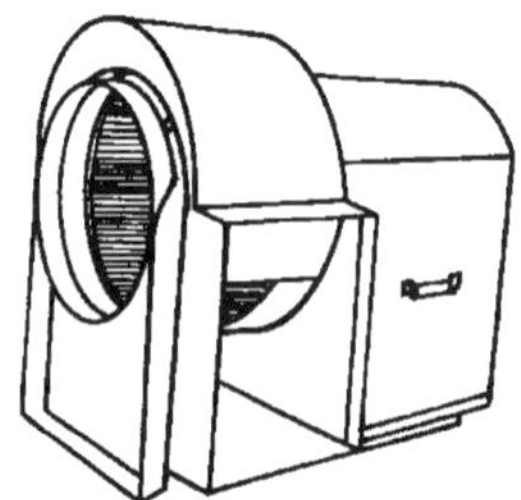

Belt Drive Utility Set

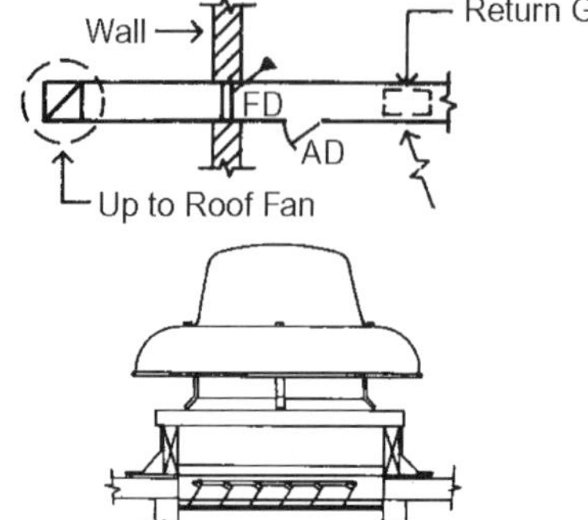

Centrifugal Roof Exhaust Fan Direct Drive

System Components	QUANTITY	UNIT	COST EACH MAT.	COST EACH INST.	COST EACH TOTAL
SYSTEM D3040 220 1010					
FAN SYSTEM, IN-LINE CENTRIFUGAL, 500 CFM					
Fans, in-line centrifugal, supply/exhaust, 500 CFM, 10″ dia conn	1.000	Ea.	1,275	455	1,730
Louver, alum., W/screen, damper, mill fin, fxd blade, cont line, stormproof	4.050	S.F.	186.30	85.05	271.35
Motor for louver damper, electric or pneumatic	1.000	Ea.	530	42.50	572.50
Ductwork, 12″ x 8″ fabricated, galvanized steel, 60 LF	200.000	Lb.	132.60	1,768	1,900.60
Grilles, aluminum, air supply, single deflection, adjustable, 12″ x 6″	1.000	Ea.	81	78	159
Fire damper, curtain type, vertical, 12″ x 8″	1.000	Ea.	27.50	27	54.50
Duct access door, insulated, 10″ x 10″	1.000	Ea.	19.95	54	73.95
Duct access door, insulated, 16″ x 12″	1.000	Ea.	30	66	96
Pressure controller/switch, air flow controller	1.000	Ea.	180	52	232
Fan control switch and mounting base	1.000	Ea.	107	39	146
Fan balancing	1.000	Ea.		355	355
Diffuser/register balancing	3.000	Ea.		214.50	214.50
TOTAL			2,569.35	3,236.05	5,805.40

D3040 220	Centrifugal In Line Fan Systems	COST EACH MAT.	COST EACH INST.	COST EACH TOTAL
1010	Fan system, in-line centrifugal, 500 CFM	2,575	3,225	5,800
1020	1300 CFM	3,025	5,600	8,625
1030	1500 CFM	3,400	6,475	9,875
1040	2500 CFM	5,000	19,700	24,700
1050	3500 CFM	6,200	26,600	32,800
1060	5000 CFM	7,950	39,600	47,550
1070	7500 CFM	9,525	40,400	49,925
1080	10,000 CFM	11,600	47,400	59,000

D3040 230	Utility Set Fan Systems	COST EACH MAT.	COST EACH INST.	COST EACH TOTAL
1010	Utility fan set system, belt drive, 2000 CFM	2,175	10,500	12,675
1020	3500 CFM	3,575	18,500	22,075
1030	5000 CFM	4,525	25,700	30,225
1040	7500 CFM	6,400	29,800	36,200
1050	10,000 CFM	8,525	39,200	47,725
1060	15,000 CFM	9,100	40,000	49,100
1070	20,000 CFM	10,500	45,300	55,800

D3040 Distribution Systems

D3040 240	Roof Exhaust Fan Systems	COST EACH		
		MAT.	INST.	TOTAL
1010	Roof vent. system, centrifugal, alum., galv curb, BDD, 500 CFM	1,150	2,125	3,275
1020	800 CFM	1,375	4,100	5,475
1030	1500 CFM	1,925	6,075	8,000
1040	2750 CFM	2,850	13,800	16,650
1050	3500 CFM	3,425	17,700	21,125
1060	5000 CFM	4,725	29,100	33,825
1070	8500 CFM	6,200	37,500	43,700
1080	13,800 CFM	8,625	54,500	63,125

D3040 Controls & Instrumentation

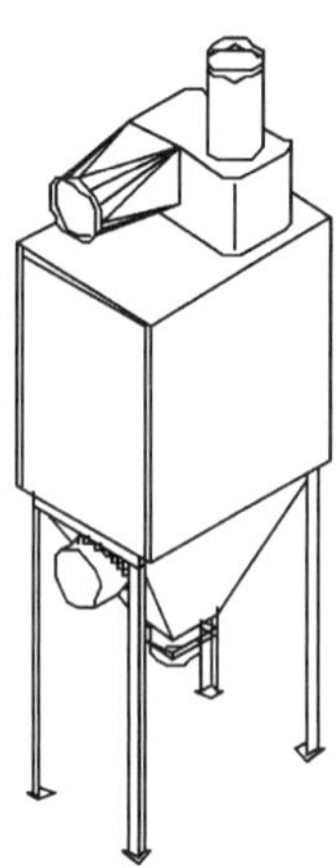

Commercial/Industrial Dust Collection System

System Components	QUANTITY	UNIT	COST EACH MAT.	COST EACH INST.	COST EACH TOTAL
SYSTEM D3040 250 1010					
COMMERCIAL/INDUSTRIAL VACUUM DUST COLLECTION SYSTEM, 500 CFM					
Dust collection central vac unit inc. stand, filter & shaker, 500 CFM 2 HP	1.000	Ea.	4,300	570	4,870
Galvanized tubing, 16 ga., 4″ OD	80.000	L.F.	480	237.60	717.60
Galvanized tubing, 90° ell, 4″ dia.	3.000	Ea.	102	64.50	166.50
Galvanized tubing, 45° ell, 4″ dia.	4.000	Ea.	118	86	204
Galvanized tubing, TY slip fit, 4″ dia.	4.000	Ea.	288	142	430
Flexible rubber hose, 4″ dia.	32.000	L.F.	308.80	95.04	403.84
Galvanized air gate valve, 4″ dia.	4.000	Ea.	1,200	186	1,386
Galvanized compression coupling, 4″ dia.	36.000	Ea.	864	1,170	2,034
Pipe hangers, steel, 4″ pipe size	16.000	Ea.	60.16	163.20	223.36
TOTAL			7,720.96	2,714.34	10,435.30

D3040 250	Commercial/Industrial Vacuum Dust Collection	COST EACH MAT.	COST EACH INST.	COST EACH TOTAL
1010	Commercial/industrial, vacuum dust collection system, 500 CFM	7,725	2,725	10,450
1020	1000 CFM	10,100	3,100	13,200
1030	1500 CFM	12,700	3,425	16,125
1040	3000 CFM	26,200	4,775	30,975
1050	5000 CFM	34,900	7,175	42,075

D3040 Controls & Instrumentation

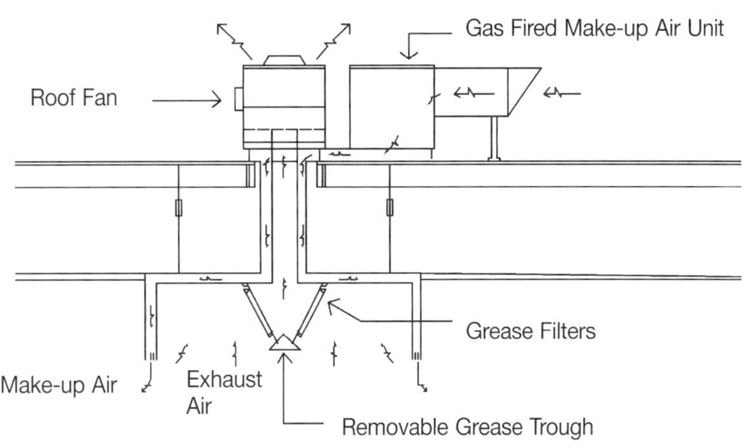

Commercial Kitchen Exhaust/Make-up Air Rooftop System

System Components	QUANTITY	UNIT	COST EACH MAT.	COST EACH INST.	COST EACH TOTAL
SYSTEM D3040 260 1010					
COMMERCIAL KITCHEN EXHAUST/MAKE-UP AIR SYSTEM, ROOFTOP, GAS, 2000 CFM					
Make-up air rooftop unit, gas, 750 MBH	1.000	Ea.	3,480.80	132.82	3,613.62
Pipe, black steel, Sch 40, thrded, nipples, W/cplngs/hngrs, 10′ OC, 1″ dia.	20.000	L.F.	151.80	258.50	410.30
Elbow, 90°, black steel, straight, 3/4″ dia.	2.000	Ea.	6.32	89	95.32
Elbow, 90°, black steel, straight, 1″ dia.	3.000	Ea.	16.50	144	160.50
Tee, black steel, reducing, 1″ dia.	1.000	Ea.	13.75	78	91.75
Union, black with brass seat, 1″ dia.	1.000	Ea.	17.30	52	69.30
Pipe nipples, black, 3/4″ dia	2.000	Ea.	9.42	20.50	29.92
Pipe cap, malleable iron, black, 1″ dia.	1.000	Ea.	4.32	21	25.32
Gas cock, brass, 1″ size	1.000	Ea.	19.45	33	52.45
Ductwork, 20″ x 14″ fabricated, galv. steel, 30 LF	220.000	Lb.	110	1,441	1,551
Ductwork, 18″ x 10″ fabricated, galv. steel, 12 LF	72.000	Lb.	36	471.60	507.60
Ductwork, 30″ x 10″ fabricated, galv. steel, 30 LF	260.000	Lb.	130	1,703	1,833
Ductwork, 16″ x 10″ fabricated, galv. steel, 12 LF	70.000	Lb.	35	458.50	493.50
Kitchen ventilation, island style, water wash, Stainless Steel	5.000	L.F.	11,875	1,175	13,050
Control system, Make-up Air Unit/Exhauster	1.000	Ea.	2,575	2,200	4,775
Rooftop unit heat/cool balancing	1.000	Ea.		415	415
Roof fan balancing	1.000	Ea.		238	238
TOTAL			18,480.66	8,930.92	27,411.58

D3040 260	Kitchen Exhaust/Make-Up Air	COST EACH MAT.	COST EACH INST.	COST EACH TOTAL
1010	Commercial kitchen exhaust/make-up air system, rooftop, gas, 2000 CFM	18,500	8,925	27,425
1020	3000 CFM	27,700	12,400	40,100
1030	5000 CFM	30,500	14,700	45,200
1040	8000 CFM	45,700	21,700	67,400
1050	12,000 CFM	60,500	25,500	86,000
1060	16,000 CFM	74,500	38,900	113,400

D3040 Distribution Systems

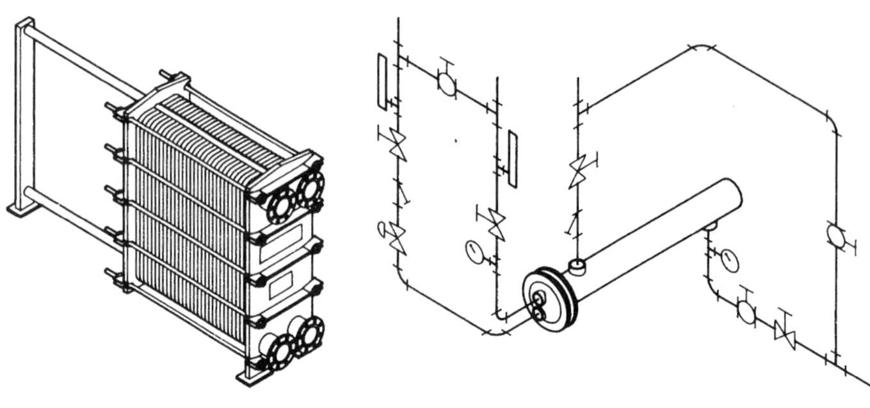

Plate Heat Exchanger

Shell and Tube Heat Exchanger

System Components	QUANTITY	UNIT	COST EACH MAT.	COST EACH INST.	COST EACH TOTAL
SYSTEM D3040 620 1010					
SHELL & TUBE HEAT EXCHANGER, 40 GPM					
Heat exchanger, 4 pass, 3/4" O.D. copper tubes, by steam at 10 psi, 40 GPM	1.000	Ea.	4,450	280	4,730
Pipe, black steel, Sch 40, threaded, W/couplings & hangers, 10' OC, 2" dia.	40.000	L.F.	699.43	886.28	1,585.71
Elbow, 90°, straight, 2" dia.	16.000	Ea.	328	1,000	1,328
Reducer, black steel, concentric, 2" dia.	2.000	Ea.	46	107	153
Valves, bronze, gate, rising stem, threaded, class 150, 2" size	6.000	Ea.	1,074	339	1,413
Valves, bronze, globe, class 150, rising stem, threaded, 2" size	2.000	Ea.	930	113	1,043
Strainers, Y type, bronze body, screwed, 150 lb., 2" pipe size	2.000	Ea.	194	96	290
Valve, electric motor actuated, brass, 2 way, screwed, 1-1/2" pipe size	1.000	Ea.	450	48	498
Union, black with brass seat, 2" dia.	8.000	Ea.	288	528	816
Tee, black, straight, 2" dia.	9.000	Ea.	234	918	1,152
Tee, black, reducing run and outlet, 2" dia.	4.000	Ea.	162	408	570
Pipe nipple, black, 2" dia.	21.000	Ea.	6.93	8.78	15.71
Thermometers, stem type, 9" case, 8" stem, 3/4" NPT	2.000	Ea.	372	44	416
Gauges, pressure or vacuum, 3-1/2" diameter dial	2.000	Ea.	79	38.90	117.90
Insulation, fiberglass pipe covering, 1" wall, 2" IPS	40.000	L.F.	62	204	266
Heat exchanger control system	1.000	Ea.	2,500	1,975	4,475
Coil balancing	1.000	Ea.		115	115
TOTAL			11,875.36	7,108.96	18,984.32

D3040 610	Heat Exchanger, Plate Type	COST EACH MAT.	COST EACH INST.	COST EACH TOTAL
1010	Plate heat exchanger, 400 GPM	47,600	14,400	62,000
1020	800 GPM	76,500	18,400	94,900
1030	1200 GPM	115,000	24,800	139,800
1040	1800 GPM	156,000	30,900	186,900

D3040 620	Heat Exchanger, Shell & Tube	COST EACH MAT.	COST EACH INST.	COST EACH TOTAL
1010	Shell & tube heat exchanger, 40 GPM	11,900	7,100	19,000
1020	96 GPM	21,400	12,800	34,200
1030	240 GPM	38,900	17,200	56,100
1040	600 GPM	75,000	24,300	99,300

D3050 Terminal & Package Units

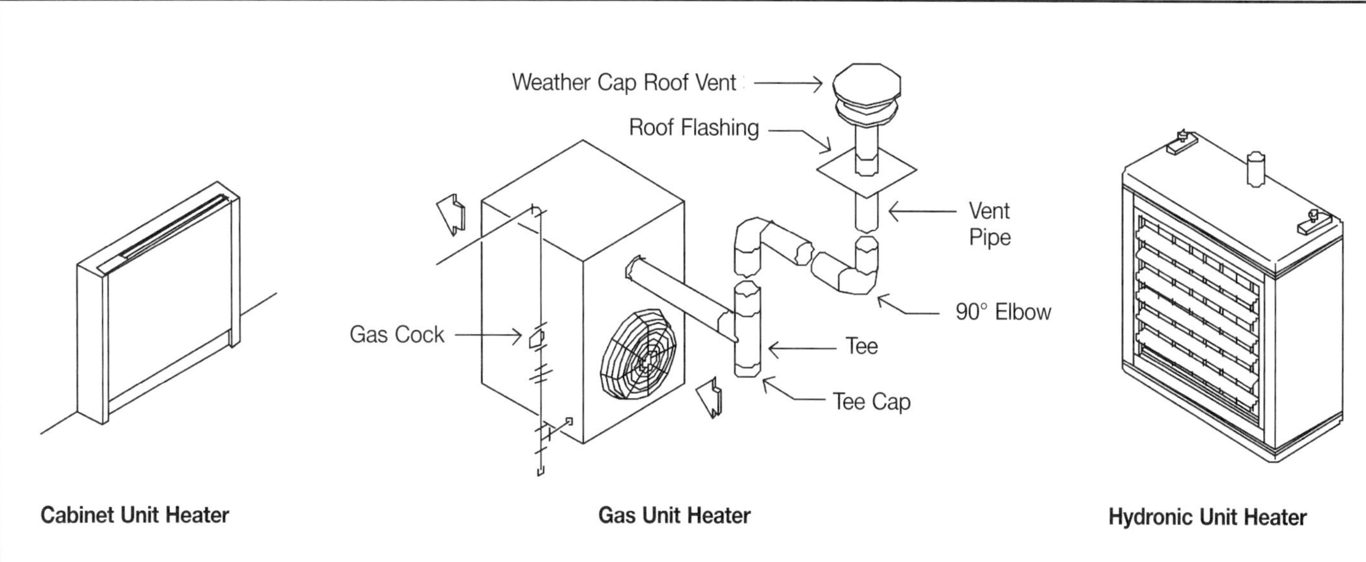

System Components	QUANTITY	UNIT	COST EACH MAT.	COST EACH INST.	COST EACH TOTAL
SYSTEM D3050 120 1010					
SPACE HEATER, SUSPENDED, GAS FIRED, PROPELLER FAN, 20 MBH					
Space heater, propeller fan, 20 MBH output	1.000	Ea.	1,075	132	1,207
Pipe, black steel, Sch 40, threaded, W/coupling & hangers, 10′ OC, 3/4″ dia	20.000	L.F.	98.91	215.25	314.16
Elbow, 90°, black steel, straight, 1/2″ dia.	3.000	Ea.	7.83	124.50	132.33
Elbow, 90°, black steel, straight, 3/4″ dia.	3.000	Ea.	9.48	133.50	142.98
Tee, black steel, reducing, 3/4″ dia.	1.000	Ea.	7.40	69.50	76.90
Union, black with brass seat, 3/4″ dia.	1.000	Ea.	13.25	48	61.25
Pipe nipples, black, 1/2″ dia	2.000	Ea.	7.94	19.80	27.74
Pipe nipples, black, 3/4″ dia	2.000	Ea.	4.71	10.25	14.96
Pipe cap, black, 3/4″ dia.	1.000	Ea.	3.58	19.50	23.08
Gas cock, brass, 3/4″ size	1.000	Ea.	12.30	28.50	40.80
Thermostat, 1 set back, electric, timed	1.000	Ea.	36.50	74.50	111
Wiring, thermostat hook-up, 25′ of #18-3	1.000	Ea.	10.80	24.50	35.30
Vent chimney, prefab metal, U.L. listed, gas, double wall, galv. st, 4″ dia	12.000	L.F.	88.20	189	277.20
Vent chimney, gas, double wall, galv. steel, elbow 90°, 4″ dia.	2.000	Ea.	59	63	122
Vent chimney, gas, double wall, galv. steel, Tee, 4″ dia.	1.000	Ea.	37.50	41	78.50
Vent chimney, gas, double wall, galv. steel, T cap, 4″ dia.	1.000	Ea.	2.89	25.50	28.39
Vent chimney, gas, double wall, galv. steel, roof flashing, 4″ dia.	1.000	Ea.	31	31.50	62.50
Vent chimney, gas, double wall, galv. steel, top, 4″ dia.	1.000	Ea.	17.80	24.50	42.30
TOTAL			1,524.09	1,274.30	2,798.39

D3050 120	Unit Heaters, Gas	COST EACH MAT.	COST EACH INST.	COST EACH TOTAL
1010	Space heater, suspended, gas fired, propeller fan, 20 MBH	1,525	1,275	2,800
1020	60 MBH	1,800	1,325	3,125
1030	100 MBH	2,025	1,400	3,425
1040	160 MBH	2,350	1,550	3,900
1050	200 MBH	2,800	1,700	4,500
1060	280 MBH	3,500	1,825	5,325
1070	320 MBH	4,600	2,075	6,675

D3050 Terminal & Package Units

D3050 130	Unit Heaters, Hydronic	COST EACH		
		MAT.	INST.	TOTAL
1010	Space heater, suspended, horiz. mount, HW, prop. fan, 20 MBH	1,800	1,375	3,175
1020	60 MBH	2,350	1,550	3,900
1030	100 MBH	2,800	1,675	4,475
1040	150 MBH	3,425	1,850	5,275
1050	200 MBH	3,975	2,075	6,050
1060	300 MBH	4,450	2,450	6,900

D3050 140	Cabinet Unit Heaters, Hydronic	COST EACH		
		MAT.	INST.	TOTAL
1010	Unit heater, cabinet type, horizontal blower, hot water, 20 MBH	2,125	1,350	3,475
1020	60 MBH	2,975	1,475	4,450
1030	100 MBH	3,325	1,625	4,950
1040	120 MBH	3,375	1,675	5,050

D3050 Terminal & Package Units

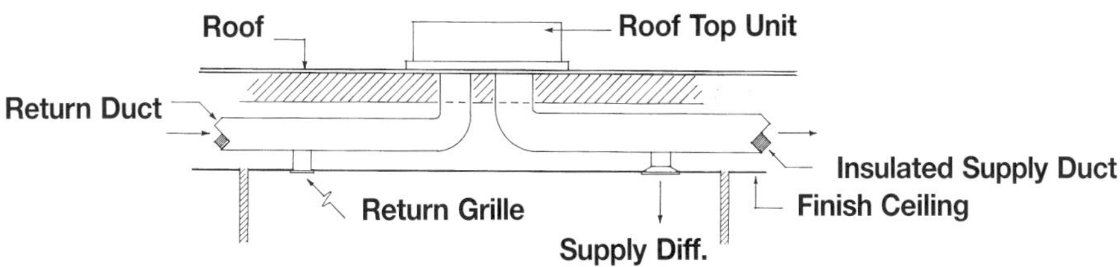

System Description: Rooftop single zone units are electric cooling and gas heat. Duct systems are low velocity, galvanized steel supply and return. Price variations between sizes are due to several factors. Jumps in the cost of the rooftop unit occur when the manufacturer shifts from the largest capacity unit on a small frame to the smallest capacity on the next larger frame, or changes from one compressor to two. As the unit capacity increases for larger areas the duct distribution grows in proportion. For most applications there is a tradeoff point where it is less expensive and more efficient to utilize smaller units with short simple distribution systems. Larger units also require larger initial supply and return ducts which can create a space problem. Supplemental heat may be desired in colder locations. The table below is based on one unit supplying the area listed. The 10,000 S.F. unit for bars and taverns is not listed because a nominal 110 ton unit would be required and this is above the normal single zone rooftop capacity.

System Components	QUANTITY	UNIT	COST EACH MAT.	COST EACH INST.	COST EACH TOTAL
SYSTEM D3050 150 1280					
ROOFTOP, SINGLE ZONE, AIR CONDITIONER					
APARTMENT CORRIDORS, 500 S.F., .92 TON					
Rooftop air conditioner, 1 zone, electric cool, standard controls, curb	1.000	Ea.	1,759.50	736	2,495.50
Ductwork package for rooftop single zone units	1.000	System	229.08	1,035	1,264.08
TOTAL			1,988.58	1,771	3,759.58
COST PER S.F.			3.98	3.54	7.52
*Size would suggest multiple units					

D3050 150	Rooftop Single Zone Unit Systems	COST PER S.F. MAT.	COST PER S.F. INST.	COST PER S.F. TOTAL
1260	Rooftop, single zone, air conditioner			
1280	Apartment corridors, 500 S.F., .92 ton	4	3.55	7.55
1320	1,000 S.F., 1.83 ton (R236000-20)	3.96	3.52	7.48
1360	1500 S.F., 2.75 ton	2.59	2.91	5.50
1400	3,000 S.F., 5.50 ton	2.42	2.79	5.21
1440	5,000 S.F., 9.17 ton	2.59	2.54	5.13
1480	10,000 S.F., 18.33 ton	2.67	2.39	5.06
1560	Banks or libraries, 500 S.F., 2.08 ton	9	8	17
1600	1,000 S.F., 4.17 ton	5.90	6.60	12.50
1640	1,500 S.F., 6.25 ton	5.50	6.35	11.85
1680	3,000 S.F., 12.50 ton	5.85	5.75	11.60
1720	5,000 S.F., 20.80 ton	6.05	5.40	11.45
1760	10,000 S.F., 41.67 ton	4.86	5.40	10.26
1840	Bars and taverns, 500 S.F. 5.54 ton	13.25	10.65	23.90
1880	1,000 S.F., 11.08 ton	14.25	9.10	23.35
1920	1,500 S.F., 16.62 ton	13.55	8.50	22.05
1960	3,000 S.F., 33.25 ton	11.75	8.10	19.85
2000	5,000 S.F., 55.42 ton	11	8.10	19.10
2040	10,000 S.F., 110.83 ton*			
2080	Bowling alleys, 500 S.F., 2.83 ton	8	9	17
2120	1,000 S.F., 5.67 ton	7.45	8.65	16.10
2160	1,500 S.F., 8.50 ton	8	7.85	15.85
2200	3,000 S.F., 17.00 ton	7.65	7.55	15.20
2240	5,000 S.F., 28.33 ton	6.70	7.35	14.05
2280	10,000 S.F., 56.67 ton	6.35	7.35	13.70
2360	Department stores, 500 S.F., 1.46 ton	6.30	5.60	11.90

D3050 Terminal & Package Units

D3050 150	Rooftop Single Zone Unit Systems	COST PER S.F.		
		MAT.	INST.	TOTAL
2400	1,000 S.F., 2.92 ton	4.12	4.64	8.76
2440	1,500 S.F., 4.37 ton	3.85	4.45	8.30
2480	3,000 S.F., 8.75 ton	4.11	4.04	8.15
2520	5,000 S.F., 14.58 ton	3.94	3.87	7.81
2560	10,000 S.F., 29.17 ton	3.46	3.77	7.23
2640	Drug stores, 500 S.F., 3.33 ton	9.40	10.55	19.95
2680	1,000 S.F., 6.67 ton	8.80	10.15	18.95
2720	1,500 S.F., 10.00 ton	9.40	9.25	18.65
2760	3,000 S.F., 20.00 ton	9.70	8.70	18.40
2800	5,000 S.F., 33.33 ton	7.90	8.65	16.55
2840	10,000 S.F., 66.67 ton	7.45	8.60	16.05
2920	Factories, 500 S.F., 1.67 ton	7.20	6.40	13.60
2960	1,000 S.F., 3.33 ton	4.70	5.30	10
3000	1,500 S.F., 5.00 ton	4.40	5.10	9.50
3040	3,000 S.F., 10.00 ton	4.70	4.62	9.32
3080	5,000 S.F., 16.67 ton	4.50	4.43	8.93
3120	10,000 S.F., 33.33 ton	3.95	4.31	8.26
3200	Food supermarkets, 500 S.F., 1.42 ton	6.15	5.45	11.60
3240	1,000 S.F., 2.83 ton	3.96	4.48	8.44
3280	1,500 S.F., 4.25 ton	3.74	4.32	8.06
3320	3,000 S.F., 8.50 ton	4	3.93	7.93
3360	5,000 S.F., 14.17 ton	3.83	3.77	7.60
3400	10,000 S.F., 28.33 ton	3.36	3.67	7.03
3480	Medical centers, 500 S.F., 1.17 ton	5.05	4.50	9.55
3520	1,000 S.F., 2.33 ton	5.05	4.49	9.54
3560	1,500 S.F., 3.50 ton	3.29	3.71	7
3600	3,000 S.F., 7.00 ton	3.08	3.56	6.64
3640	5,000 S.F., 11.67 ton	3.29	3.24	6.53
3680	10,000 S.F., 23.33 ton	3.39	3.03	6.42
3760	Offices, 500 S.F., 1.58 ton	6.85	6.10	12.95
3800	1,000 S.F., 3.17 ton	4.47	5.05	9.52
3840	1,500 S.F., 4.75 ton	4.18	4.83	9.01
3880	3,000 S.F., 9.50 ton	4.46	4.38	8.84
3920	5,000 S.F., 15.83 ton	4.27	4.20	8.47
3960	10,000 S.F., 31.67 ton	3.76	4.10	7.86
4000	Restaurants, 500 S.F., 2.50 ton	10.80	9.65	20.45
4040	1,000 S.F., 5.00 ton	6.60	7.65	14.25
4080	1,500 S.F., 7.50 ton	7.05	6.95	14
4120	3,000 S.F., 15.00 ton	6.75	6.65	13.40
4160	5,000 S.F., 25.00 ton	7.30	6.50	13.80
4200	10,000 S.F., 50.00 ton	5.60	6.45	12.05
4240	Schools and colleges, 500 S.F., 1.92 ton	8.30	7.40	15.70
4280	1,000 S.F., 3.83 ton	5.40	6.10	11.50
4320	1,500 S.F., 5.75 ton	5.05	5.85	10.90
4360	3,000 S.F., 11.50 ton	5.40	5.30	10.70
4400	5,000 S.F., 19.17 ton	5.55	4.99	10.54
4440	10,000 S.F., 38.33 ton	4.47	4.96	9.43
5000				
5100				
5200				
9000	Components of ductwork packages for above systems, per ton of cooling:			
9010	Ductwork; galvanized steel, 120 pounds			
9020	Insulation; fiberglass 2" thick, FRK faced, 52 SF			
9030	Diffusers; aluminum, 24" x 12", one			
9040	Registers; aluminum, return with dampers, one			

D3050 Terminal & Package Units

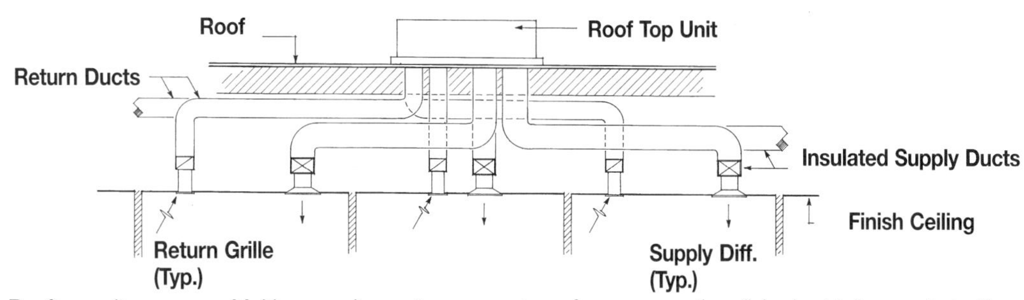

System Description: Rooftop units are multizone with up to 12 zones, and include electric cooling, gas heat, thermostats, filters, supply and return fans complete. Duct systems are low velocity, galvanized steel supply and return with insulated supplies.

Multizone units cost more per ton of cooling than single zone. However, they offer flexibility where load conditions are varied due to heat generating areas or exposure to radiational heating. For example, perimeter offices on the "sunny side" may require cooling at the same time "shady side" or central offices may require heating. It is possible to accomplish similar results using duct heaters in branches of the single zone unit. However, heater location could be a problem and total system operating energy efficiency could be lower.

System Components	QUANTITY	UNIT	COST EACH MAT.	COST EACH INST.	COST EACH TOTAL
SYSTEM D3050 155 1280					
ROOFTOP, MULTIZONE, AIR CONDITIONER					
APARTMENT CORRIDORS, 3,000 S.F., 5.50 TON					
Rooftop multizone unit, standard controls, curb	1.000	Ea.	27,500	1,716	29,216
Ductwork package for rooftop multizone units	1.000	System	1,787.50	11,825	13,612.50
TOTAL			29,287.50	13,541	42,828.50
COST PER S.F.			9.76	4.51	14.27
Note A: Small single zone unit recommended.					
Note B: A combination of multizone units recommended.					

D3050 155	Rooftop Multizone Unit Systems	COST PER S.F. MAT.	COST PER S.F. INST.	COST PER S.F. TOTAL
1240	Rooftop, multizone, air conditioner			
1260	Apartment corridors, 1,500 S.F., 2.75 ton. See Note A.			
1280	3,000 S.F., 5.50 ton (R236000-20)	9.77	4.50	14.27
1320	10,000 S.F., 18.30 ton	6.80	4.34	11.14
1360	15,000 S.F., 27.50 ton	6.60	4.32	10.92
1400	20,000 S.F., 36.70 ton	6.90	4.34	11.24
1440	25,000 S.F., 45.80 ton	6.45	4.36	10.81
1520	Banks or libraries, 1,500 S.F., 6.25 ton	22	10.25	32.25
1560	3,000 S.F., 12.50 ton	18.70	10.05	28.75
1600	10,000 S.F., 41.67 ton	14.65	9.90	24.55
1640	15,000 S.F., 62.50 ton	10.80	9.85	20.65
1680	20,000 S.F., 83.33 ton	10.80	9.85	20.65
1720	25,000 S.F., 104.00 ton	9.65	9.80	19.45
1800	Bars and taverns, 1,500 S.F., 16.62 ton	48	14.80	62.80
1840	3,000 S.F., 33.24 ton	40	14.25	54.25
1880	10,000 S.F., 110.83 ton	24	14.20	38.20
1920	15,000 S.F., 165 ton, See Note B.			
1960	20,000 S.F., 220 ton, See Note B.			
2000	25,000 S.F., 275 ton, See Note B.			
2080	Bowling alleys, 1,500 S.F., 8.50 ton	30	13.95	43.95
2120	3,000 S.F., 17.00 ton	25.50	13.65	39.15
2160	10,000 S.F., 56.70 ton	19.90	13.50	33.40
2200	15,000 S.F., 85.00 ton	14.70	13.40	28.10
2240	20,000 S.F., 113.00 ton	13.10	13.30	26.40
2280	25,000 S.F., 140.00 ton see Note B.			
2360	Department stores, 1,500 S.F., 4.37 ton, See Note A.			

D3050 Terminal & Package Units

D3050 155	Rooftop Multizone Unit Systems	COST PER S.F. MAT.	INST.	TOTAL
2400	3,000 S.F., 8.75 ton	15.55	7.20	22.75
2440	10,000 S.F., 29.17 ton	10.95	6.90	17.85
2480	15,000 S.F., 43.75 ton	10.25	6.95	17.20
2520	20,000 S.F., 58.33 ton	7.55	6.90	14.45
2560	25,000 S.F., 72.92 ton	7.55	6.90	14.45
2640	Drug stores, 1,500 S.F., 10.00 ton	35.50	16.40	51.90
2680	3,000 S.F., 20.00 ton	24.50	15.80	40.30
2720	10,000 S.F., 66.66 ton	17.30	15.75	33.05
2760	15,000 S.F., 100.00 ton	15.45	15.70	31.15
2800	20,000 S.F., 135 ton, See Note B.			
2840	25,000 S.F., 165 ton, See Note B.			
2920	Factories, 1,500 S.F., 5 ton, See Note A.			
2960	3,000 S.F., 10.00 ton	17.75	8.20	25.95
3000	10,000 S.F., 33.33 ton	12.50	7.90	20.40
3040	15,000 S.F., 50.00 ton	11.70	7.95	19.65
3080	20,000 S.F., 66.66 ton	8.65	7.85	16.50
3120	25,000 S.F., 83.33 ton	8.65	7.85	16.50
3200	Food supermarkets, 1,500 S.F., 4.25 ton, See Note A.			
3240	3,000 S.F., 8.50 ton	15.10	6.95	22.05
3280	10,000 S.F., 28.33 ton	10.25	6.70	16.95
3320	15,000 S.F., 42.50 ton	9.95	6.75	16.70
3360	20,000 S.F., 56.67 ton	7.35	6.70	14.05
3400	25,000 S.F., 70.83 ton	7.35	6.70	14.05
3480	Medical centers, 1,500 S.F., 3.5 ton, See Note A.			
3520	3,000 S.F., 7.00 ton	12.45	5.75	18.20
3560	10,000 S.F., 23.33 ton	8.25	5.50	13.75
3600	15,000 S.F., 35.00 ton	8.75	5.50	14.25
3640	20,000 S.F., 46.66 ton	8.20	5.55	13.75
3680	25,000 S.F., 58.33 ton	6.05	5.50	11.55
3760	Offices, 1,500 S.F., 4.75 ton, See Note A.			
3800	3,000 S.F., 9.50 ton	16.85	7.80	24.65
3840	10,000 S.F., 31.66 ton	11.90	7.50	19.40
3880	15,000 S.F., 47.50 ton	11.15	7.55	18.70
3920	20,000 S.F., 63.33 ton	8.20	7.50	15.70
3960	25,000 S.F., 79.16 ton	8.20	7.50	15.70
4000	Restaurants, 1,500 S.F., 7.50 ton	26.50	12.30	38.80
4040	3,000 S.F., 15.00 ton	22.50	12.05	34.55
4080	10,000 S.F., 50.00 ton	17.55	11.90	29.45
4120	15,000 S.F., 75.00 ton	13	11.80	24.80
4160	20,000 S.F., 100.00 ton	11.60	11.80	23.40
4200	25,000 S.F., 125 ton, See Note B.			
4240	Schools and colleges, 1,500 S.F., 5.75 ton	20.50	9.45	29.95
4280	3,000 S.F., 11.50 ton	17.20	9.25	26.45
4320	10,000 S.F., 38.33 ton	13.45	9.10	22.55
4360	15,000 S.F., 57.50 ton	9.95	9.05	19
4400	20,000 S.F.,76.66 ton	9.95	9.05	19
4440	25,000 S.F., 95.83 ton	9.35	9.05	18.40
4450				
9000	Components of ductwork packages for above systems, per ton of cooling:			
9010	Ductwork; galvanized steel, 240 pounds			
9020	Insulation; fiberglass, 2″ thick, FRK faced, 104 SF			
9030	Diffusers; aluminum, 24″ x 12″, two			
9040	Registers; aluminum, return with dampers, one			

D3050 Terminal & Package Units

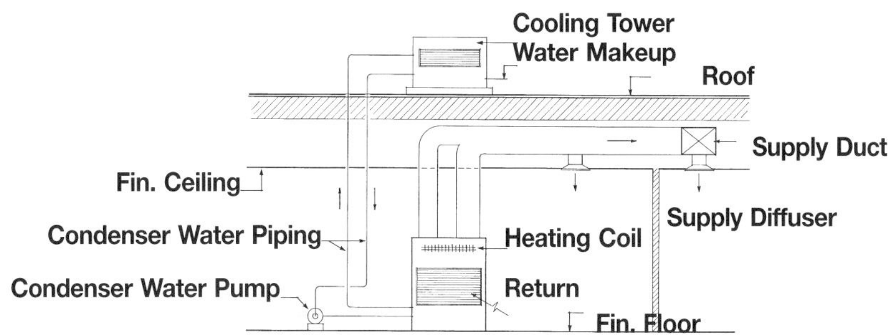

System Description: Self-contained, single package water cooled units include cooling tower, pump, piping allowance. Systems for 1000 S.F. and up include duct and diffusers to provide for even distribution of air. Smaller units distribute air through a supply air plenum, which is integral with the unit.

Returns are not ducted and supplies are not insulated.

Hot water or steam heating coils are included but piping to boiler and the boiler itself is not included.

Where local codes or conditions permit single pass cooling for the smaller units, deduct 10%.

System Components	QUANTITY	UNIT	COST EACH		
			MAT.	INST.	TOTAL
SYSTEM D3050 160 1300					
SELF-CONTAINED, WATER COOLED UNIT					
APARTMENT CORRIDORS, 500 S.F., .92 TON					
Self contained, water cooled, single package air conditioner unit	1.000	Ea.	1,201.20	588	1,789.20
Ductwork package for water or air cooled packaged units	1.000	System	50.14	782	832.14
Cooling tower, draw thru single flow, belt drive	1.000	Ea.	164.68	17.80	182.48
Cooling tower pumps & piping	1.000	System	82.34	42.32	124.66
TOTAL			1,498.36	1,430.12	2,928.48
COST PER S.F.			3	2.86	5.86

D3050 160	Self-contained, Water Cooled Unit Systems	COST PER S.F.		
		MAT.	INST.	TOTAL
1280	Self-contained, water cooled unit	2.99	2.86	5.85
1300	Apartment corridors, 500 S.F., .92 ton	3	2.85	5.85
1320	1,000 S.F., 1.83 ton (R236000-20)	2.99	2.85	5.84
1360	3,000 S.F., 5.50 ton	2.49	2.39	4.88
1400	5,000 S.F., 9.17 ton	2.43	2.21	4.64
1440	10,000 S.F., 18.33 ton	3.45	1.98	5.43
1520	Banks or libraries, 500 S.F., 2.08 ton	6.55	2.94	9.49
1560	1,000 S.F., 4.17 ton	5.65	5.40	11.05
1600	3,000 S.F., 12.50 ton	5.55	5	10.55
1640	5,000 S.F., 20.80 ton	7.80	4.49	12.29
1680	10,000 S.F., 41.66 ton	6.80	4.49	11.29
1760	Bars and taverns, 500 S.F., 5.54 ton	14.35	4.96	19.31
1800	1,000 S.F., 11.08 ton	14.30	8.60	22.90
1840	3,000 S.F., 33.25 ton	18.45	6.80	25.25
1880	5,000 S.F., 55.42 ton	17.30	7.10	24.40
1920	10,000 S.F., 110.00 ton	16.95	7	23.95
2000	Bowling alleys, 500 S.F., 2.83 ton	8.95	4	12.95
2040	1,000 S.F., 5.66 ton	7.70	7.35	15.05
2080	3,000 S.F., 17.00 ton	10.65	6.10	16.75
2120	5,000 S.F., 28.33 ton	9.65	5.90	15.55
2160	10,000 S.F., 56.66 ton	9	6	15
2200	Department stores, 500 S.F., 1.46 ton	4.61	2.06	6.67
2240	1,000 S.F., 2.92 ton	3.96	3.79	7.75
2280	3,000 S.F., 8.75 ton	3.87	3.51	7.38
2320	5,000 S.F., 14.58 ton	5.50	3.15	8.65
2360	10,000 S.F., 29.17 ton	4.96	3.03	7.99

D3050 Terminal & Package Units

D3050 160	Self-contained, Water Cooled Unit Systems	COST PER S.F.		
		MAT.	INST.	TOTAL
2440	Drug stores, 500 S.F., 3.33 ton	10.55	4.72	15.27
2480	1,000 S.F., 6.66 ton	9.05	8.70	17.75
2520	3,000 S.F., 20.00 ton	12.55	7.20	19.75
2560	5,000 S.F., 33.33 ton	13.15	7.10	20.25
2600	10,000 S.F., 66.66 ton	10.60	7.15	17.75
2680	Factories, 500 S.F., 1.66 ton	5.25	2.35	7.60
2720	1,000 S.F. 3.37 ton	4.56	4.39	8.95
2760	3,000 S.F., 10.00 ton	4.43	4	8.43
2800	5,000 S.F., 16.66 ton	6.25	3.59	9.84
2840	10,000 S.F., 33.33 ton	5.65	3.45	9.10
2920	Food supermarkets, 500 S.F., 1.42 ton	4.47	2	6.47
2960	1,000 S.F., 2.83 ton	4.62	4.41	9.03
3000	3,000 S.F., 8.50 ton	3.84	3.68	7.52
3040	5,000 S.F., 14.17 ton	3.75	3.41	7.16
3080	10,000 S.F., 28.33 ton	4.82	2.94	7.76
3160	Medical centers, 500 S.F., 1.17 ton	3.68	1.65	5.33
3200	1,000 S.F., 2.33 ton	3.82	3.64	7.46
3240	3,000 S.F., 7.00 ton	3.17	3.04	6.21
3280	5,000 S.F., 11.66 ton	3.10	2.81	5.91
3320	10,000 S.F., 23.33 ton	4.39	2.52	6.91
3400	Offices, 500 S.F., 1.58 ton	5	2.24	7.24
3440	1,000 S.F., 3.17 ton	5.15	4.93	10.08
3480	3,000 S.F., 9.50 ton	4.20	3.81	8.01
3520	5,000 S.F., 15.83 ton	5.95	3.42	9.37
3560	10,000 S.F., 31.67 ton	5.40	3.29	8.69
3640	Restaurants, 500 S.F., 2.50 ton	7.90	3.54	11.44
3680	1,000 S.F., 5.00 ton	6.80	6.50	13.30
3720	3,000 S.F., 15.00 ton	6.65	6	12.65
3760	5,000 S.F., 25.00 ton	9.40	5.40	14.80
3800	10,000 S.F., 50.00 ton	5.75	4.95	10.70
3880	Schools and colleges, 500 S.F., 1.92 ton	6.05	2.71	8.76
3920	1,000 S.F., 3.83 ton	5.20	4.99	10.19
3960	3,000 S.F., 11.50 ton	5.10	4.62	9.72
4000	5,000 S.F., 19.17 ton	7.20	4.14	11.34
4040	10,000 S.F., 38.33 ton	6.25	4.13	10.38
4050				
9000	Components of ductwork packages for above systems, per ton of cooling:			
9010	Ductwork; galvanized steel, 108 pounds			
9020	Diffusers, aluminum, 24" x 12", two			

D3050 Terminal & Package Units

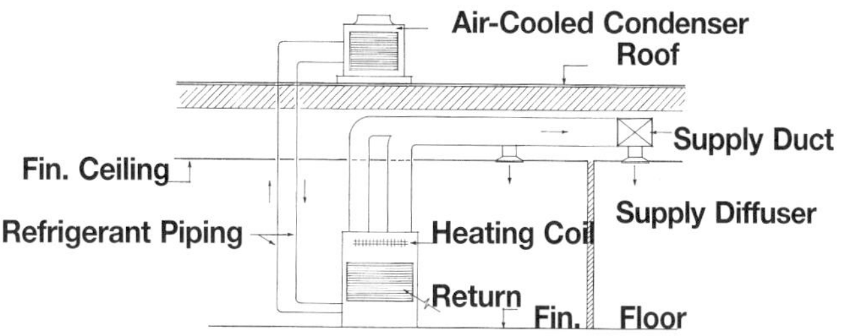

System Description: Self-contained air cooled units with remote air cooled condenser and interconnecting tubing. Systems for 1000 S.F. and up include duct and diffusers. Smaller units distribute air directly.

Returns are not ducted and supplies are not insulated.

Potential savings may be realized by using a single zone rooftop system or through-the-wall unit, especially in the smaller capacities, if the application permits.

Hot water or steam heating coils are included but piping to boiler and the boiler itself is not included.

Condenserless models are available for 15% less where remote refrigerant source is available.

System Components	QUANTITY	UNIT	COST EACH MAT.	COST EACH INST.	COST EACH TOTAL
SYSTEM D3050 165 1320					
SELF-CONTAINED, AIR COOLED UNIT					
APARTMENT CORRIDORS, 500 S.F., .92 TON					
Air cooled, package unit	1.000	Ea.	1,232.50	382.50	1,615
Ductwork package for water or air cooled packaged units	1.000	System	50.14	782	832.14
Refrigerant piping	1.000	System	340.40	515.20	855.60
Air cooled condenser, direct drive, propeller fan	1.000	Ea.	341	144.15	485.15
TOTAL			1,964.04	1,823.85	3,787.89
COST PER S.F.			3.93	3.65	7.58

D3050 165	Self-contained, Air Cooled Unit Systems	COST PER S.F. MAT.	COST PER S.F. INST.	COST PER S.F. TOTAL
1300	Self-contained, air cooled unit			
1320	Apartment corridors, 500 S.F., .92 ton	3.95	3.65	7.60
1360	1,000 S.F., 1.83 ton (R236000-20)	3.88	3.61	7.49
1400	3,000 S.F., 5.50 ton	3.26	3.38	6.64
1440	5,000 S.F., 9.17 ton	2.89	3.22	6.11
1480	10,000 S.F., 18.33 ton	2.74	2.96	5.70
1560	Banks or libraries, 500 S.F., 2.08 ton	8.55	4.67	13.22
1600	1,000 S.F., 4.17 ton	7.40	7.65	15.05
1640	3,000 S.F., 12.50 ton	6.60	7.30	13.90
1680	5,000 S.F., 20.80 ton	6.30	6.70	13
1720	10,000 S.F., 41.66 ton	7.05	6.55	13.60
1800	Bars and taverns, 500 S.F., 5.54 ton	17.45	10.20	27.65
1840	1,000 S.F., 11.08 ton	17.15	14.70	31.85
1880	3,000 S.F., 33.25 ton	17.65	12.70	30.35
1920	5,000 S.F., 55.42 ton	18.65	12.75	31.40
1960	10,000 S.F., 110.00 ton	18.80	12.65	31.45
2040	Bowling alleys, 500 S.F., 2.83 ton	11.70	6.40	18.10
2080	1,000 S.F., 5.66 ton	10.10	10.45	20.55
2120	3,000 S.F., 17.00 ton	8.50	9.10	17.60
2160	5,000 S.F., 28.33 ton	9.20	8.90	18.10
2200	10,000 S.F., 56.66 ton	9.70	8.90	18.60
2240	Department stores, 500 S.F., 1.46 ton	6	3.28	9.28
2280	1,000 S.F., 2.92 ton	5.20	5.35	10.55
2320	3,000 S.F., 8.75 ton	4.60	5.10	9.70

D3050 Terminal & Package Units

D3050 165	Self-contained, Air Cooled Unit Systems	COST PER S.F.		
		MAT.	INST.	TOTAL
2360	5,000 S.F., 14.58 ton	4.60	5.10	9.70
2400	10,000 S.F., 29.17 ton	4.72	4.59	9.31
2480	Drug stores, 500 S.F., 3.33 ton	13.75	7.50	21.25
2520	1,000 S.F., 6.66 ton	11.85	12.25	24.10
2560	3,000 S.F., 20.00 ton	10.05	10.75	20.80
2600	5,000 S.F., 33.33 ton	10.80	10.50	21.30
2640	10,000 S.F., 66.66 ton	11.45	10.50	21.95
2720	Factories, 500 S.F., 1.66 ton	7	3.80	10.80
2760	1,000 S.F., 3.33 ton	5.95	6.15	12.10
2800	3,000 S.F., 10.00 ton	5.25	5.85	11.10
2840	5,000 S.F., 16.66 ton	4.96	5.35	10.31
2880	10,000 S.F., 33.33 ton	5.40	5.25	10.65
2960	Food supermarkets, 500 S.F., 1.42 ton	5.85	3.20	9.05
3000	1,000 S.F., 2.83 ton	5.95	5.60	11.55
3040	3,000 S.F., 8.50 ton	5.05	5.20	10.25
3080	5,000 S.F., 14.17 ton	4.46	4.97	9.43
3120	10,000 S.F., 28.33 ton	4.59	4.46	9.05
3200	Medical centers, 500 S.F., 1.17 ton	4.84	2.64	7.48
3240	1,000 S.F., 2.33 ton	4.97	4.62	9.59
3280	3,000 S.F., 7.00 ton	4.16	4.29	8.45
3320	5,000 S.F., 16.66 ton	3.69	4.09	7.78
3360	10,000 S.F., 23.33 ton	3.50	3.75	7.25
3440	Offices, 500 S.F., 1.58 ton	6.55	3.57	10.12
3480	1,000 S.F., 3.16 ton	6.75	6.25	13
3520	3,000 S.F., 9.50 ton	5	5.55	10.55
3560	5,000 S.F., 15.83 ton	4.77	5.10	9.87
3600	10,000 S.F., 31.66 ton	5.15	4.98	10.13
3680	Restaurants, 500 S.F., 2.50 ton	10.30	5.60	15.90
3720	1,000 S.F., 5.00 ton	8.90	9.20	18.10
3760	3,000 S.F., 15.00 ton	7.90	8.75	16.65
3800	5,000 S.F., 25.00 ton	8.05	7.85	15.90
3840	10,000 S.F., 50.00 ton	8.60	7.80	16.40
3920	Schools and colleges, 500 S.F., 1.92 ton	7.90	4.33	12.23
3960	1,000 S.F., 3.83 ton	6.85	7.05	13.90
4000	3,000 S.F., 11.50 ton	6.05	6.75	12.80
4040	5,000 S.F., 19.17 ton	5.75	6.15	11.90
4080	10,000 S.F., 38.33 ton	6.50	6.05	12.55
4090				
9000	Components of ductwork packages for above systems, per ton of cooling:			
9010	Ductwork; galvanized steel, 102 pounds			
9020	Diffusers; aluminum, 24" x 12", two			

D3050 Terminal & Package Units

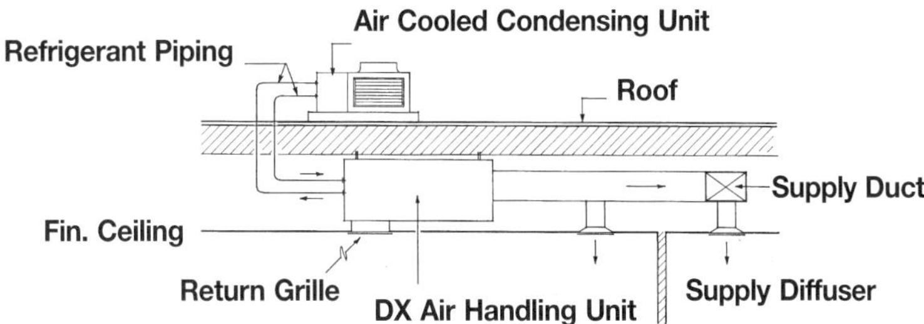

General: Split systems offer several important advantages which should be evaluated when a selection is to be made. They provide a greater degree of flexibility in component selection which permits an accurate match-up of the proper equipment size and type with the particular needs of the building. This allows for maximum use of modern energy saving concepts in heating and cooling. Outdoor installation of the air cooled condensing unit allows space savings in the building and also isolates the equipment operating sounds from building occupants.

Design Assumptions: The systems below are comprised of a direct expansion air handling unit and air cooled condensing unit with interconnecting copper tubing. Ducts and diffusers are also included for distribution of air. Systems are priced for cooling only. Heat can be added as desired either by putting hot water/steam coils into the air unit or into the duct supplying the particular area of need. Gas fired duct furnaces are also available. Refrigerant liquid line is insulated.

System Components	QUANTITY	UNIT	COST EACH MAT.	COST EACH INST.	COST EACH TOTAL
SYSTEM D3050 170 1280					
SPLIT SYSTEM, AIR COOLED CONDENSING UNIT					
APARTMENT CORRIDORS, 1,000 S.F., 1.80 TON					
Fan coil AC unit, cabinet mntd & filters direct expansion air cool	1.000	Ea.	530.70	137.25	667.95
Ductwork package, for split system, remote condensing unit	1.000	System	47.12	741.15	788.27
Refrigeration piping	1.000	System	276.33	850.95	1,127.28
Condensing unit, air cooled, incls compressor & standard controls	1.000	Ea.	1,586	549	2,135
TOTAL			2,440.15	2,278.35	4,718.50
COST PER S.F.			2.44	2.28	4.72
*Cooling requirements would lead to choosing a water cooled unit.					

D3050 170	Split Systems With Air Cooled Condensing Units	COST PER S.F. MAT.	COST PER S.F. INST.	COST PER S.F. TOTAL
1260	Split system, air cooled condensing unit			
1280	Apartment corridors, 1,000 S.F., 1.83 ton	2.45	2.28	4.73
1320	2,000 S.F., 3.66 ton (R236000-20)	1.75	2.30	4.05
1360	5,000 S.F., 9.17 ton	1.89	2.86	4.75
1400	10,000 S.F., 18.33 ton	2.15	3.14	5.29
1440	20,000 S.F., 36.66 ton	2.34	3.19	5.53
1520	Banks and libraries, 1,000 S.F., 4.17 ton	3.98	5.25	9.23
1560	2,000 S.F., 8.33 ton	4.29	6.55	10.84
1600	5,000 S.F., 20.80 ton	4.89	7.10	11.99
1640	10,000 S.F., 41.66 ton	5.30	7.25	12.55
1680	20,000 S.F., 83.32 ton	6.20	7.55	13.75
1760	Bars and taverns, 1,000 S.F., 11.08 ton	10.30	10.30	20.60
1800	2,000 S.F., 22.16 ton	13.95	12.35	26.30
1840	5,000 S.F., 55.42 ton	12.15	11.55	23.70
1880	10,000 S.F., 110.84 ton	14.15	12.05	26.20
1920	20,000 S.F., 220 ton*			
2000	Bowling alleys, 1,000 S.F., 5.66 ton	5.45	9.75	15.20
2040	2,000 S.F., 11.33 ton	5.85	8.85	14.70
2080	5,000 S.F., 28.33 ton	6.65	9.70	16.35
2120	10,000 S.F., 56.66 ton	7.25	9.85	17.10
2160	20,000 S.F., 113.32 ton	9.10	10.70	19.80
2320	Department stores, 1,000 S.F., 2.92 ton	2.80	3.60	6.40
2360	2,000 S.F., 5.83 ton	2.81	5.05	7.86
2400	5,000 S.F., 14.58 ton	3	4.58	7.58
2440	10,000 S.F., 29.17 ton	3.43	4.98	8.41
2480	20,000 S.F., 58.33 ton	3.72	5.05	8.77

D3050 Terminal & Package Units

D3050 170	Split Systems With Air Cooled Condensing Units	COST PER S.F.		
		MAT.	INST.	TOTAL
2560	Drug stores, 1,000 S.F., 6.66 ton	6.40	11.50	17.90
2600	2,000 S.F., 13.32 ton	6.85	10.45	17.30
2640	5,000 S.F., 33.33 ton	8.50	11.60	20.10
2680	10,000 S.F., 66.66 ton	8.85	12.10	20.95
2720	20,000 S.F., 133.32 ton*			
2800	Factories, 1,000 S.F., 3.33 ton	3.20	4.09	7.29
2840	2,000 S.F., 6.66 ton	3.21	5.75	8.96
2880	5,000 S.F., 16.66 ton	3.91	5.70	9.61
2920	10,000 S.F., 33.33 ton	4.25	5.80	10.05
2960	20,000 S.F., 66.66 ton	4.41	6.05	10.46
3040	Food supermarkets, 1,000 S.F., 2.83 ton	2.71	3.49	6.20
3080	2,000 S.F., 5.66 ton	2.73	4.88	7.61
3120	5,000 S.F., 14.66 ton	2.92	4.44	7.36
3160	10,000 S.F., 28.33 ton	3.34	4.85	8.19
3200	20,000 S.F., 56.66 ton	3.63	4.93	8.56
3280	Medical centers, 1,000 S.F., 2.33 ton	2.44	2.82	5.26
3320	2,000 S.F., 4.66 ton	2.25	4.01	6.26
3360	5,000 S.F., 11.66 ton	2.40	3.66	6.06
3400	10,000 S.F., 23.33 ton	2.74	4	6.74
3440	20,000 S.F., 46.66 ton	2.98	4.07	7.05
3520	Offices, 1,000 S.F., 3.17 ton	3.04	3.90	6.94
3560	2,000 S.F., 6.33 ton	3.05	5.45	8.50
3600	5,000 S.F., 15.83 ton	3.26	4.95	8.21
3640	10,000 S.F., 31.66 ton	3.72	5.40	9.12
3680	20,000 S.F., 63.32 ton	4.20	5.75	9.95
3760	Restaurants, 1,000 S.F., 5.00 ton	4.82	8.65	13.47
3800	2,000 S.F., 10.00 ton	5.15	7.85	13
3840	5,000 S.F., 25.00 ton	5.90	8.55	14.45
3880	10,000 S.F., 50.00 ton	6.40	8.70	15.10
3920	20,000 S.F., 100.00 ton	8.05	9.45	17.50
4000	Schools and colleges, 1,000 S.F., 3.83 ton	3.66	4.82	8.48
4040	2,000 S.F., 7.66 ton	3.95	6	9.95
4080	5,000 S.F., 19.17 ton	4.51	6.55	11.06
4120	10,000 S.F., 38.33 ton	4.90	6.65	11.55
4160	20,000 S.F., 76.66 ton	5.10	6.95	12.05
5000				
5100				
5200				
9000	Components of ductwork packages for above systems, per ton of cooling:			
9010	Ductwork; galvanized steel, 102 pounds			
9020	Diffusers; aluminum, 24″ x 12″, two			

D3050 Terminal & Package Units

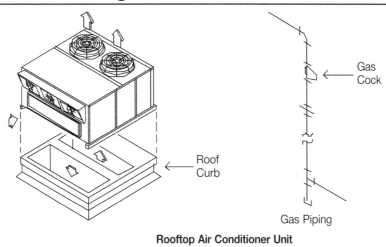

Gas Cock

Roof Curb

Gas Piping

Rooftop Air Conditioner Unit

System Components	QUANTITY	UNIT	COST EACH MAT.	COST EACH INST.	COST EACH TOTAL
SYSTEM D3050 175 1010					
A/C, ROOFTOP, DX COOL, GAS HEAT, CURB, ECONOMIZER, FILTERS., 5 TON					
Roof top A/C, curb, economizer, sgl zone, elec cool, gas ht, 5 ton, 112 MBH	1.000	Ea.	5,350	2,000	7,350
Pipe, black steel, Sch 40, threaded, W/cplgs & hangers, 10′ OC, 1″ dia	20.000	L.F.	144.90	246.75	391.65
Elbow, 90°, black, straight, 3/4″ dia.	3.000	Ea.	9.48	133.50	142.98
Elbow, 90°, black, straight, 1″ dia.	3.000	Ea.	16.50	144	160.50
Tee, black, reducing, 1″ dia.	1.000	Ea.	13.75	78	91.75
Union, black with brass seat, 1″ dia.	1.000	Ea.	17.30	52	69.30
Pipe nipple, black, 3/4″ dia	2.000	Ea.	9.42	20.50	29.92
Pipe nipple, black, 1″ dia	2.000	Ea.	6.90	11.75	18.65
Cap, black, 1″ dia.	1.000	Ea.	4.32	21	25.32
Gas cock, brass, 1″ size	1.000	Ea.	19.45	33	52.45
Control system, pneumatic, Rooftop A/C unit	1.000	Ea.	2,575	2,200	4,775
Rooftop unit heat/cool balancing	1.000	Ea.		415	415
TOTAL			8,167.02	5,355.50	13,522.52

D3050 175	Rooftop Air Conditioner, Const. Volume	COST EACH MAT.	COST EACH INST.	COST EACH TOTAL
1010	A/C, Rooftop, DX cool, gas heat, curb, economizer, fltrs, 5 Ton	8,175	5,350	13,525
1020	7-1/2 Ton	11,200	5,550	16,750
1030	12-1/2 Ton	16,200	6,125	22,325
1040	18 Ton	22,600	6,700	29,300
1050	25 Ton	30,400	7,650	38,050
1060	40 Ton	39,700	10,200	49,900

D3050 180	Rooftop Air Conditioner, Variable Air Volume	COST EACH MAT.	COST EACH INST.	COST EACH TOTAL
1010	A/C, Rooftop, DX cool, gas heat, curb, ecmizr, fltrs, VAV, 12-1/2 Ton	18,000	6,550	24,550
1020	18 Ton	24,700	7,200	31,900
1030	25 Ton	33,000	8,275	41,275
1040	40 Ton	44,300	11,200	55,500
1050	60 Ton	63,500	15,300	78,800
1060	80 Ton	99,000	19,100	118,100

D3050 Terminal & Package Units

Computer rooms impose special requirements on air conditioning systems. A prime requirement is reliability, due to the potential monetary loss that could be incurred by a system failure. A second basic requirement is the tolerance of control with which temperature and humidity are regulated, and dust eliminated. As the air conditioning system reliability is so vital, the additional cost of reserve capacity and redundant components is often justified.

System Descriptions: Computer areas may be environmentally controlled by one of three methods as follows:

1. **Self-contained Units** These are units built to higher standards of performance and reliability. They usuallycontain alarms and controls to indicate component operation failure, filter change, etc. It should be remembered that these units in the room will occupy space that is relatively expensive to build and that all alterations and service of the equipment will also have to be accomplished within the computer area.
2. **Decentralized Air Handling Units** In operation these are similar to the self-contained units except that their cooling capability comes from remotely located refrigeration equipment as refrigerant or chilled water. As no compressors or refrigerating equipment are required in the air units, they are smaller and require less service than self-contained units. An added plus for this type of system occurs if some of the computer components themselves also require chilled water for cooling.
3. **Central System Supply** Cooling is obtained from a central source which, since it is not located within the computer room, may have excess capacity and permit greater flexibility without interfering with the computer components. System performance criteria must still be met.

Note: The costs shown below do not include an allowance for ductwork or piping.

D3050 185	Computer Room Cooling Units	COST EACH		
		MAT.	INST.	TOTAL
0560	Computer room unit, air cooled, includes remote condenser			
0580	3 ton	17,700	2,250	19,950
0600	5 ton	18,900	2,500	21,400
0620	8 ton	35,600	4,150	39,750
0640	10 ton	37,200	4,475	41,675
0660	15 ton	40,900	5,100	46,000
0680	20 ton	49,600	7,250	56,850
0700	23 ton	61,000	8,300	69,300
0800	Chilled water, for connection to existing chiller system			
0820	5 ton	13,600	1,525	15,125
0840	8 ton	13,700	2,250	15,950
0860	10 ton	13,800	2,275	16,075
0880	15 ton	14,600	2,325	16,925
0900	20 ton	15,500	2,425	17,925
0920	23 ton	16,500	2,775	19,275
1000	Glycol system, complete except for interconnecting tubing			
1020	3 ton	22,300	2,800	25,100
1040	5 ton	24,400	2,950	27,350
1060	8 ton	39,300	4,875	44,175
1080	10 ton	41,800	5,325	47,125
1100	15 ton	51,500	6,700	58,200
1120	20 ton	57,000	7,250	64,250
1140	23 ton	60,000	7,575	67,575
1240	Water cooled, not including condenser water supply or cooling tower			
1260	3 ton	18,100	1,800	19,900
1280	5 ton	19,700	2,075	21,775
1300	8 ton	32,300	3,400	35,700
1320	15 ton	39,100	4,150	43,250
1340	20 ton	42,100	4,575	46,675
1360	23 ton	44,100	4,975	49,075

D3050 201	Packaged A/C, Elec. Ht., Const. Volume	COST EACH		
		MAT.	INST.	TOTAL
1010	A/C Packaged, DX, air cooled, electric heat, constant vol., 5 Ton	8,075	3,800	11,875
1020	10 Ton	13,200	5,250	18,450

D3050 Terminal & Package Units

D3050 201	Packaged A/C, Elec. Ht., Const. Volume	COST EACH		
		MAT.	INST.	TOTAL
1030	20 Ton	23,000	6,850	29,850
1040	30 Ton	37,300	7,625	44,925
1050	40 Ton	51,000	8,775	59,775
1060	50 Ton	66,000	9,850	75,850

D3050 202	Packaged Air Conditioner, Elect. Heat, VAV	COST EACH		
		MAT.	INST.	TOTAL
1010	A/C Packaged, DX, air cooled, electric heat, VAV, 10 Ton	14,800	6,525	21,325
1020	20 Ton	25,800	8,600	34,400
1030	30 Ton	41,400	9,375	50,775
1040	40 Ton	55,500	10,800	66,300
1050	50 Ton	71,500	12,000	83,500
1060	60 Ton	82,000	14,000	96,000

D3050 203	Packaged A/C, Hot Wtr. Heat, Const. Volume	COST EACH		
		MAT.	INST.	TOTAL
1010	A/C Packaged, DX, air cooled, H/W heat, constant vol., 5 Ton	6,825	5,625	12,450
1020	10 Ton	10,300	7,375	17,675
1030	20 Ton	19,700	9,925	29,625
1040	30 Ton	30,600	12,400	43,000
1050	40 Ton	45,000	13,900	58,900
1060	50 Ton	56,000	15,100	71,100

D3050 204	Packaged A/C, Hot Wtr. Heat, VAV	COST EACH		
		MAT.	INST.	TOTAL
1010	A/C Packaged, DX, air cooled, H/W heat, VAV, 10 Ton	12,000	8,550	20,550
1020	20 Ton	22,400	11,200	33,600
1030	30 Ton	34,700	14,000	48,700
1040	40 Ton	49,800	15,100	64,900
1050	50 Ton	61,500	17,500	79,000
1060	60 Ton	74,500	19,700	94,200

D3050 Terminal & Package Units

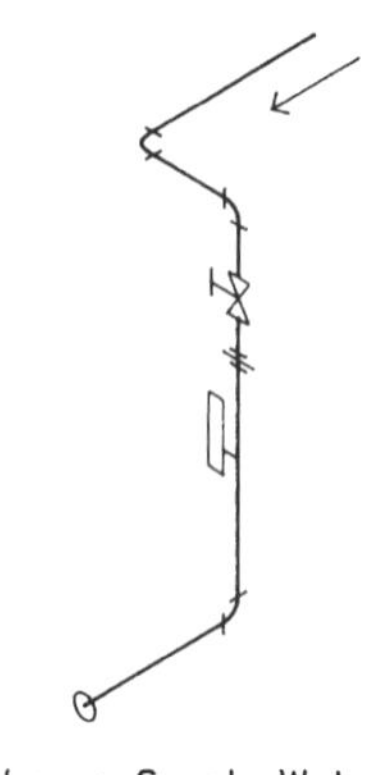

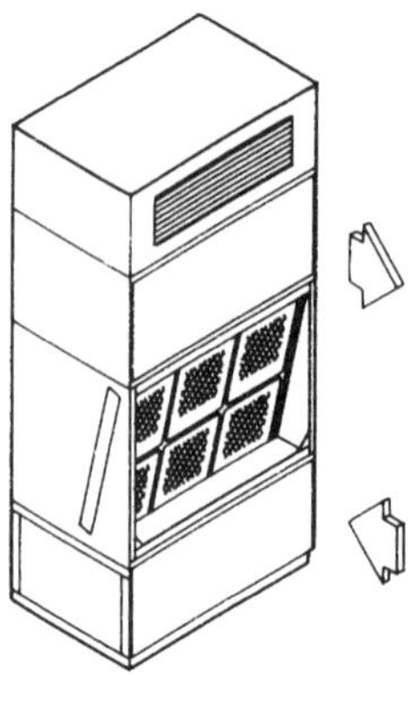

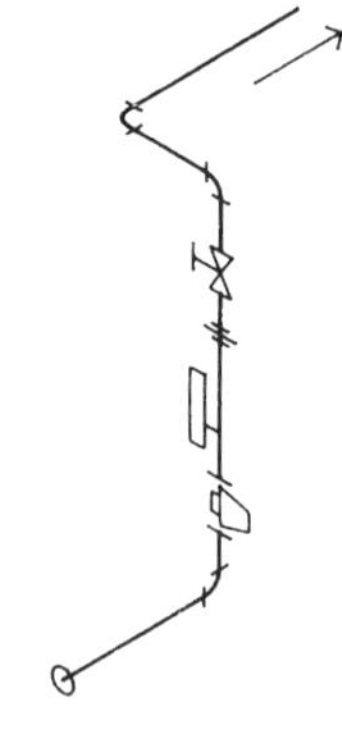

Self–Contained Air Conditioner Unit, Water Cooled

System Components	QUANTITY	UNIT	COST EACH MAT.	COST EACH INST.	COST EACH TOTAL
SYSTEM D3050 210 1010					
A/C, SELF CONTAINED, SINGLE PKG., WATER COOLED, ELECT. HEAT, 5 TON					
Self-contained, water cooled, elect. heat, not inc tower, 5 ton, const vol	1.000	Ea.	8,225	1,450	9,675
Pipe, black steel, Sch 40, threaded, W/cplg & hangers, 10′ OC, 1-1/4″ dia.	20.000	L.F.	188.13	271.98	460.11
Elbow, 90°, black, straight, 1-1/4″ dia.	6.000	Ea.	54	306	360
Tee, black, straight, 1-1/4″ dia.	2.000	Ea.	27.80	161	188.80
Tee, black, reducing, 1-1/4″ dia.	2.000	Ea.	47	161	208
Thermometers, stem type, 9″ case, 8″ stem, 3/4″ NPT	2.000	Ea.	372	44	416
Union, black with brass seat, 1-1/4″ dia.	2.000	Ea.	50	107	157
Pipe nipple, black, 1-1/4″ dia	3.000	Ea.	13.13	18.98	32.11
Valves, bronze, gate, N.R.S., threaded, class 150, 1-1/4″ size	2.000	Ea.	216	83	299
Circuit setter, bal valve, bronze body, threaded, 1-1/4″ pipe size	1.000	Ea.	149	41.50	190.50
Insulation, fiberglass pipe covering, 1″ wall, 1-1/4″ IPS	20.000	L.F.	26.60	97.20	123.80
Control system, pneumatic, A/C Unit with heat	1.000	Ea.	2,575	2,200	4,775
Re-heat coil balancing	1.000	Ea.		115	115
Rooftop unit heat/cool balancing	1.000	Ea.		415	415
TOTAL			11,943.66	5,471.66	17,415.32

D3050 210	AC Unit, Package, Elec. Ht., Water Cooled	COST EACH MAT.	COST EACH INST.	COST EACH TOTAL
1010	A/C, Self contained, single pkg., water cooled, elect. heat, 5 Ton	11,900	5,475	17,375
1020	10 Ton	18,800	6,750	25,550
1030	20 Ton	25,600	8,600	34,200
1040	30 Ton	41,500	9,350	50,850
1050	40 Ton	55,000	10,400	65,400
1060	50 Ton	70,500	12,400	82,900

D3050 215	AC Unit, Package, Elec. Ht., Water Cooled, VAV	COST EACH MAT.	COST EACH INST.	COST EACH TOTAL
1010	A/C, Self contained, single pkg., water cooled, elect. ht, VAV, 10 Ton	20,300	6,750	27,050
1020	20 Ton	28,100	8,600	36,700
1030	30 Ton	45,400	9,350	54,750
1040	40 Ton	59,500	10,400	69,900
1050	50 Ton	76,000	12,400	88,400
1060	60 Ton	87,000	15,000	102,000

D3050 Terminal & Package Units

D3050 220	AC Unit, Package, Hot Water Coil, Water Cooled	COST EACH		
		MAT.	INST.	TOTAL
1010	A/C, Self contn'd, single pkg., water cool, H/W ht, const. vol, 5 Ton	10,500	7,600	18,100
1020	10 Ton	15,700	9,050	24,750
1030	20 Ton	38,700	11,300	50,000
1040	30 Ton	50,500	14,500	65,000
1050	40 Ton	62,000	17,500	79,500
1060	50 Ton	76,500	21,500	98,000

D3050 225	AC Unit, Package, HW Coil, Water Cooled, VAV	COST EACH		
		MAT.	INST.	TOTAL
1010	A/C, Self contn'd, single pkg., water cool, H/W ht, VAV, 10 Ton	17,100	8,525	25,625
1020	20 Ton	24,700	11,000	35,700
1030	30 Ton	39,100	13,900	53,000
1040	40 Ton	54,000	15,100	69,100
1050	50 Ton	67,000	17,800	84,800
1060	60 Ton	80,000	19,300	99,300

D3050 Terminal & Package Units

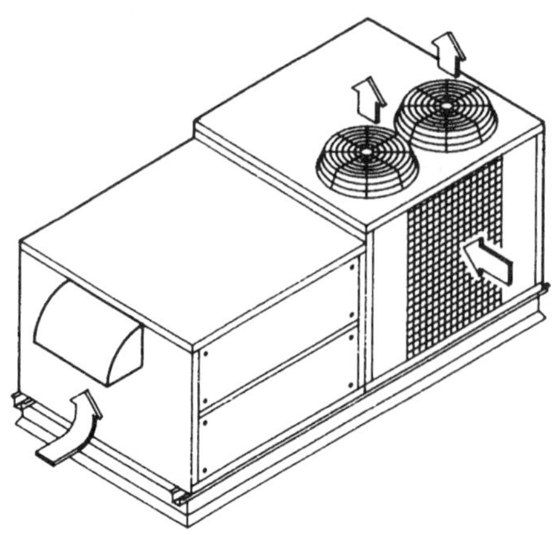

Heat Pump, Air/Air, Rooftop w/ Economizer

System Components	QUANTITY	UNIT	COST EACH MAT.	COST EACH INST.	COST EACH TOTAL
SYSTEM D3050 245 1010					
HEAT PUMP, ROOF TOP, AIR/AIR, CURB., ECONOMIZER., SUP. ELECT. HEAT, 10 TON					
Heat pump, air to air, rooftop, curb, economizer, elect ht, 10 ton cooling	1.000	Ea.	55,500	4,350	59,850
Ductwork, 32" x 12" fabricated, galvanized steel, 16 LF	176.000	Lb.	91.52	1,249.60	1,341.12
Insulation, ductwork, blanket type, fiberglass, 1" thk, 1-1/2 LB density	117.000	S.F.	87.75	340.47	428.22
Rooftop unit heat/cool balancing	1.000	Ea.		415	415
TOTAL			55,679.27	6,355.07	62,034.34

D3050 230	Heat Pump, Water Source, Central Station	COST EACH MAT.	COST EACH INST.	COST EACH TOTAL
1010	Heat pump, central station, water source, constant vol., 5 Ton	4,025	3,075	7,100
1020	10 Ton	10,700	4,300	15,000
1030	20 Ton	18,800	8,025	26,825
1040	30 Ton	27,300	11,700	39,000
1050	40 Ton	37,600	12,700	50,300
1060	50 Ton	42,600	17,100	59,700

D3050 235	Heat Pump, Water Source, Console	COST EACH MAT.	COST EACH INST.	COST EACH TOTAL
1020	Heat pump, console, water source, 1 Ton	2,150	1,675	3,825
1030	1-1/2 Ton	2,250	1,750	4,000
1040	2 Ton	2,925	2,075	5,000
1050	3 Ton	3,150	2,200	5,350
1060	3-1/2 Ton	3,350	2,275	5,625

D3050 240	Heat Pump, Horizontal, Duct Mounted	COST EACH MAT.	COST EACH INST.	COST EACH TOTAL
1020	Heat pump, horizontal, ducted, water source, 1 Ton	2,500	3,550	6,050
1030	1-1/2 Ton	2,750	4,475	7,225
1040	2 Ton	3,450	5,350	8,800
1050	3 Ton	3,925	7,000	10,925
1060	3-1/2 Ton	4,150	7,225	11,375

D3050 Terminal & Package Units

D3050 245	Heat Pump, Roof Top, Air/Air	COST EACH		
		MAT.	INST.	TOTAL
1010	Heat pump, roof top, air/air, curb, economizer, sup. elect heat, 10 Ton	55,500	6,350	61,850
1020	20 Ton	79,500	10,900	90,400
1030	30 Ton	108,000	15,200	123,200
1040	40 Ton	120,000	17,400	137,400
1050	50 Ton	140,000	19,600	159,600
1060	60 Ton	147,500	22,300	169,800

D3050 Terminal & Package Units

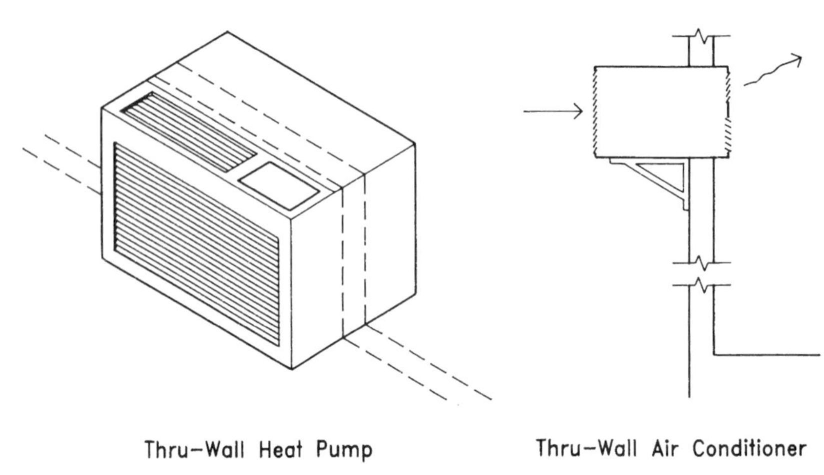
Thru–Wall Heat Pump
Thru–Wall Air Conditioner

System Components	QUANTITY	UNIT	COST EACH MAT.	COST EACH INST.	COST EACH TOTAL
SYSTEM D3050 255 1010					
A/C UNIT, THRU-THE-WALL, SUPPLEMENTAL ELECT. HEAT, CABINET, LOUVER, 1/2 TON					
A/C unit, thru-wall, electric heat., cabinet, louver, 1/2 ton	1.000	Ea.	750	187	937
TOTAL			750	187	937

D3050 255	Thru-Wall A/C Unit	COST EACH MAT.	COST EACH INST.	COST EACH TOTAL
1010	A/C Unit, thru-the-wall, supplemental elect. heat, cabinet, louver, 1/2 Ton	750	187	937
1020	3/4 Ton	865	224	1,089
1030	1 Ton	935	280	1,215
1040	1-1/2 Ton	1,825	465	2,290
1050	2 Ton	1,850	590	2,440

D3050 260	Thru-Wall Heat Pump	COST EACH MAT.	COST EACH INST.	COST EACH TOTAL
1010	Heat pump, thru-the-wall, cabinet, louver, 1/2 Ton	2,325	140	2,465
1020	3/4 Ton	2,525	187	2,712
1030	1 Ton	2,775	280	3,055
1040	Supplemental. elect. heat, 1-1/2 Ton	2,800	725	3,525
1050	2 Ton	3,025	745	3,770

D3090 Fume Exhaust Systems

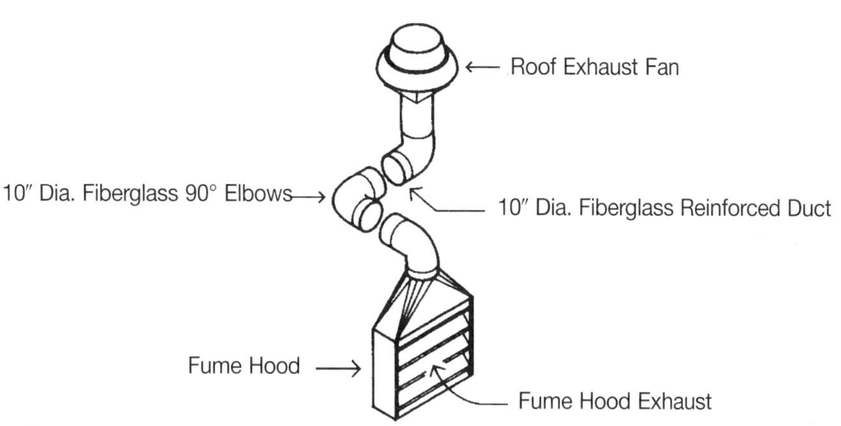

System Components	QUANTITY	UNIT	COST EACH MAT.	COST EACH INST.	COST EACH TOTAL
SYSTEM D3090 310 1010					
FUME HOOD EXHAUST SYSTEM, 3′ LONG, 1000 CFM					
Fume hood	3.000	L.F.	5,925	1,275	7,200
Ductwork, plastic, FM approved for acid fume & smoke, straight, 10″ dia	60.000	L.F.	1,320	756	2,076
Ductwork, plastic, FM approved for acid fume & smoke, elbow 90°, 10″ dia	3.000	Ea.	342	411	753
Fan, corrosive fume resistant plastic, 1630 CFM, 1/2 HP	1.000	Ea.	5,475	340	5,815
Fan control switch & mounting base	1.000	Ea.	107	39	146
Roof fan balancing	1.000	Ea.		238	238
TOTAL			13,169	3,059	16,228

D3090 310	Fume Hood Exhaust Systems	COST EACH MAT.	COST EACH INST.	COST EACH TOTAL
1010	Fume hood exhaust system, 3′ long, 1000 CFM	13,200	3,050	16,250
1020	4′ long, 2000 CFM	15,700	3,875	19,575
1030	6′ long, 3500 CFM	21,500	5,800	27,300
1040	6′ long, 5000 CFM	29,700	6,800	36,500
1050	10′ long, 8000 CFM	40,100	9,300	49,400

D3090 Other HVAC Systems/Equip

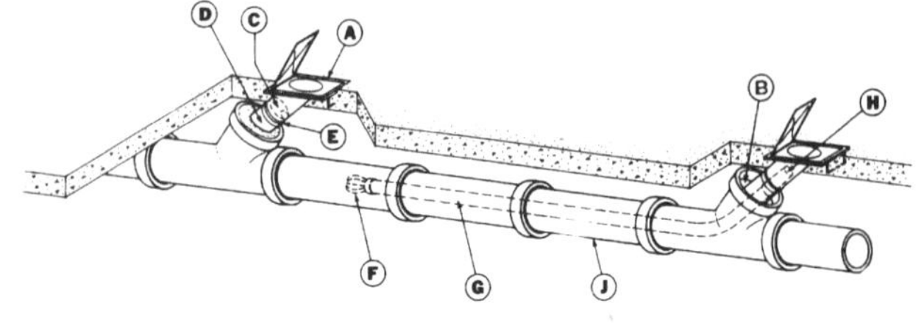

Cast Iron Garage Exhaust System

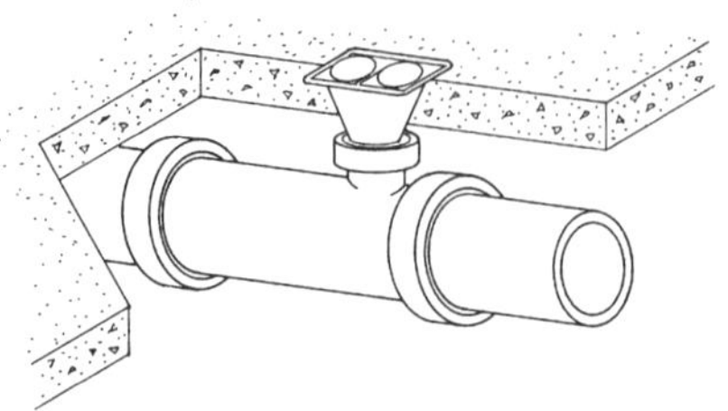
Dual Exhaust System

System Components

System Components		QUANTITY	UNIT	COST EACH MAT.	COST EACH INST.	COST EACH TOTAL
SYSTEM D3090 320 1040						
GARAGE, EXHAUST, SINGLE OUTLET, 3″ EXHAUST, CARS & LIGHT TRUCKS						
A	Outlet top assy. for engine exhaust system with adapters and ftngs, 3″ diam	1.000	Ea.	288	37	325
F	Bullnose (guide) for engine exhaust system, 3″ diam	1.000	Ea.	33		33
G	Galvanized flexible tubing for engine exhaust system, 3″ diam	8.000	L.F.	111.60		111.60
H	Adapter for metal tubing end of engine exhaust system, 3″ tail pipe	1.000	Ea.	59.50		59.50
J	Pipe, sewer, cast iron, push-on joint, 8″ diam.	18.000	L.F.	738	693	1,431
	24″ deep	18.000	L.F.		29.16	29.16
	Backfill utility trench by hand, incl. compaction, 8″ wide 24″ deep	18.000	L.F.		41.76	41.76
	Stand for blower, concrete over polystyrene core, 6″ high	1.000	Ea.	5.60	18.60	24.20
	AC&V duct spiral reducer 10″x8″	1.000	Ea.	14.30	33.50	47.80
	AC&V duct spiral reducer 12″x10″	1.000	Ea.	16.25	44.50	60.75
	AC&V duct spiral preformed 45° elbow 8″ diam	1.000	Ea.	8.90	38.50	47.40
	AC&V utility fan, belt drive, 3 phase, 2000 CFM, 1 HP	2.000	Ea.	2,100	592	2,692
	Safety switch, heavy duty fused, 240V, 3 pole, 30 amp	1.000	Ea.	201	182	383
	TOTAL			3,576.15	1,710.02	5,286.17

D3090 320	Garage Exhaust Systems	COST PER BAY MAT.	COST PER BAY INST.	COST PER BAY TOTAL
1040	Garage, single exhaust, 3″ outlet, cars & light trucks, one bay	3,575	1,700	5,275
1060	Additional bays up to seven bays	900	460	1,360
1500	4″ outlet, trucks, one bay	3,600	1,700	5,300
1520	Additional bays up to six bays	935	460	1,395
1600	5″ outlet, diesel trucks, one bay	3,950	1,700	5,650
1650	Additional single bays up to six	1,375	545	1,920
1700	Two adjoining bays	3,950	1,700	5,650
2000	Dual exhaust, 3″ outlets, pair of adjoining bays	4,475	2,150	6,625
2100	Additional pairs of adjoining bays	1,450	545	1,995

D3090 Other HVAC Systems/Equip

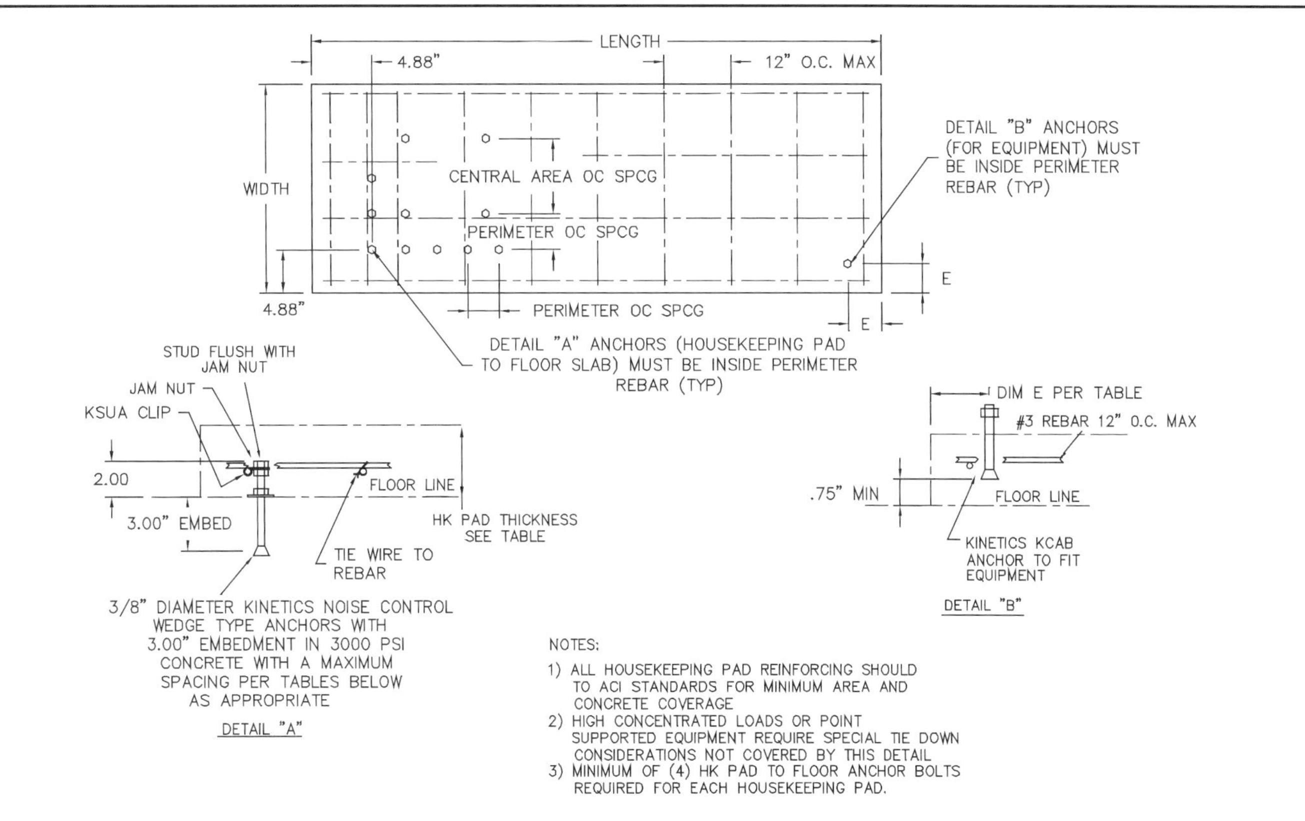

System Components	QUANTITY	UNIT	COST EACH MAT.	COST EACH INST.	COST EACH TOTAL
SYSTEM D3090 750 1000					
SEISMIC PAD FOR 20-38 SELF-CONTAINED UNIT					
Form equipment pad	28.000	L.F.	19.88	106.96	126.84
Reinforcing in hold down pad, #3 to #7	60.000	Lb.	26.40	34.20	60.60
Concrete 3000 PSI	.866	C.Y.	92.66		92.66
Place concrete pad	.866	C.Y.		28.80	28.80
Finish concrete equipment pad	47.000	S.F.		33.84	33.84
Drill for 3/8 inch slab anchors	10.000	Ea.	.70	81.50	82.20
Exisiting slab anchors 3/8 inch	10.000	Ea.	6.60	36.60	43.20
Equipment anchors 3/8 inch for plate/foot	8.000	Ea.	13.44	126	139.44
TOTAL			159.68	447.90	607.58

D3090 750	Seismic Concrete Equipment Pads	COST PER EACH MAT.	COST PER EACH INST.	COST PER EACH TOTAL
0990	Seismic pad for self-contained AC units for plate/foot attachment			
1000	Seismic pad for 20-38 tons Self-contained AC unit without economizer .25g	160	445	605
1010	0.5 g seismic acceleration	160	445	605
1020	0.75 g seismic acceleration	160	445	605
1030	1.0 g seismic acceleration	160	445	605
1040	1.5 g seismic acceleration	163	495	658
1050	2.0 g seismic acceleration	175	670	845
1060	2.5 g seismic acceleration	183	810	993
1100	Seismic pad for 40-80 tons Self contained AC unit without economizer .25g	252	630	882
1110	0.5 g seismic acceleration	252	630	882
1120	0.75 g seismic acceleration	252	630	882
1130	1.0 g seismic acceleration	252	630	882
1140	1.5 g seismic acceleration	274	845	1,119
1150	2.0 g seismic acceleration	320	1,275	1,595

D3090 Other HVAC Systems/Equip

D3090 750	Seismic Concrete Equipment Pads	COST PER EACH		
		MAT.	INST.	TOTAL
1160	2.5 g seismic acceleration	370	1,800	2,170
1200	Seismic pad for 80-110 tons Self contained AC unit without economizer .25g	278	670	948
1210	0.5 g seismic acceleration	278	670	948
1220	0.75 g seismic acceleration	278	670	948
1230	1.0 g seismic acceleration	278	670	948
1240	1.5 g seismic acceleration	300	900	1,200
1250	2.0 g seismic acceleration	350	1,375	1,725
1260	2.5 g seismic acceleration	405	1,925	2,330
1990	Seismic pad for self-contained AC units for isolator/snubber bolts			
2000	Seismic pad for 20-38 tons Self contained AC unit without economizer .25g	173	575	748
2010	0.5 g seismic acceleration	173	575	748
2020	0.75 g seismic acceleration	173	575	748
2030	1.0 g seismic acceleration	173	575	748
2040	1.5 g seismic acceleration	176	620	796
2050	2.0 g seismic acceleration	192	825	1,017
2060	2.5 g seismic acceleration	200	965	1,165
2100	Seismic pad for 40-80 tons Self contained AC unit without economizer .25g	269	785	1,054
2110	0.5 g seismic acceleration	269	785	1,054
2120	0.75 g seismic acceleration	269	785	1,054
2130	1.0 g seismic acceleration	269	785	1,054
2140	1.5 g seismic acceleration	290	1,000	1,290
2150	2.0 g seismic acceleration	345	1,500	1,845
2160	2.5 g seismic acceleration	395	2,050	2,445
2200	Seismic pad for 80-110 tons Self contained AC unit without economizer .25g	295	825	1,120
2210	0.5 g seismic acceleration	295	825	1,120
2220	0.75 g seismic acceleration	295	825	1,120
2230	1.0 g seismic acceleration	295	825	1,120
2240	1.5 g seismic acceleration	320	1,050	1,370
2250	2.0 g seismic acceleration	375	1,600	1,975
2260	2.5 g seismic acceleration	430	2,175	2,605
2990	Seismic pad for self-contained AC economizer units for plate/foot attachment			
3000	Seismic pad for 20-38 tons Self contained AC unit economizer .25 g-2.0 g	106	242	348
3010	2.5 g seismic acceleration	113	335	448
3100	Seismic pad for 40-80 tons Self contained AC unit economizer .25-1.5 g	143	345	488
3110	2.0 g seismic acceleration	148	385	533
3120	2.5 g seismic acceleration	153	445	598
3190	Seismic pad for self-contained AC unit economizer for isolator/snubber bolts			
3200	Seismic pad for 20-38 tons Self contained AC unit economizer .25 g-2.0 g	111	289	400
3210	2.5 g seismic acceleration	121	410	531
3300	Seismic pad for 40-80 tons Self contained AC unit economizer .25-1.5 g	153	440	593
3310	2.0 g seismic acceleration	158	480	638
3320	2.5 g seismic acceleration	164	540	704

G Building Sitework

G1030 Site Earthwork

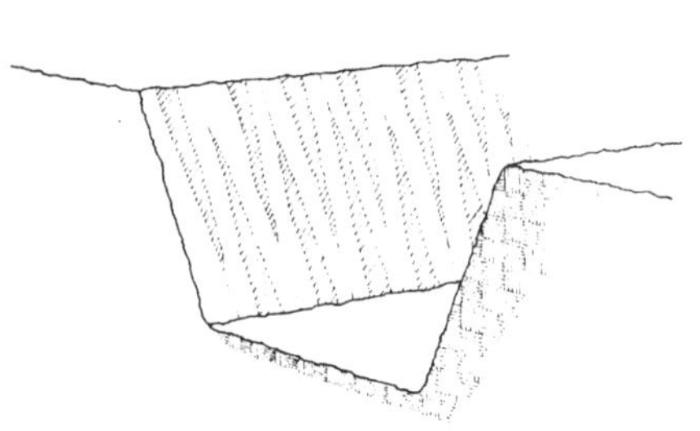

Trenching Systems are shown on a cost per linear foot basis. The systems include: excavation; backfill and removal of spoil; and compaction for various depths and trench bottom widths. The backfill has been reduced to accommodate a pipe of suitable diameter and bedding.

The slope for trench sides varies from none to 1:1.

The Expanded System Listing shows Trenching Systems that range from 2′ to 12′ in width. Depths range from 2′ to 25′.

System Components	QUANTITY	UNIT	COST PER L.F. EQUIP.	LABOR	TOTAL
SYSTEM G1030 805 1310					
TRENCHING COMMON EARTH, NO SLOPE, 2′ WIDE, 2′ DP, 3/8 C.Y. BUCKET					
Excavation, trench, hyd. backhoe, track mtd., 3/8 C.Y. bucket	.148	B.C.Y.	.37	.91	1.28
Backfill and load spoil, from stockpile	.153	L.C.Y.	.11	.28	.39
Compaction by vibrating plate, 6″ lifts, 4 passes	.118	E.C.Y.	.03	.34	.37
Remove excess spoil, 8 C.Y. dump truck, 2 mile roundtrip	.040	L.C.Y.	.15	.13	.28
TOTAL			.66	1.66	2.32

G1030 805	Trenching Common Earth	COST PER L.F. EQUIP.	LABOR	TOTAL
1310	Trenching, common earth, no slope, 2′ wide, 2′ deep, 3/8 C.Y. bucket	.66	1.66	2.32
1320	3′ deep, 3/8 C.Y. bucket	.94	2.51	3.45
1330	4′ deep, 3/8 C.Y. bucket	1.21	3.35	4.56
1340	6′ deep, 3/8 C.Y. bucket	1.61	4.36	5.97
1350	8′ deep, 1/2 C.Y. bucket	2.11	5.80	7.91
1360	10′ deep, 1 C.Y. bucket	3.45	6.90	10.35
1400	4′ wide, 2′ deep, 3/8 C.Y. bucket	1.51	3.28	4.79
1410	3′ deep, 3/8 C.Y. bucket	2.07	4.97	7.04
1420	4′ deep, 1/2 C.Y. bucket	2.47	5.60	8.07
1430	6′ deep, 1/2 C.Y. bucket	3.86	8.90	12.76
1440	8′ deep, 1/2 C.Y. bucket	6.45	11.50	17.95
1450	10′ deep, 1 C.Y. bucket	7.75	14.30	22.05
1460	12′ deep, 1 C.Y. bucket	10.05	18.20	28.25
1470	15′ deep, 1-1/2 C.Y. bucket	8.75	16.25	25
1480	18′ deep, 2-1/2 C.Y. bucket	12.20	23	35.20
1520	6′ wide, 6′ deep, 5/8 C.Y. bucket w/trench box	8.60	13.20	21.80
1530	8′ deep, 3/4 C.Y. bucket	11.35	17.40	28.75
1540	10′ deep, 1 C.Y. bucket	11.10	18	29.10
1550	12′ deep, 1-1/2 C.Y. bucket	11.80	19.35	31.15
1560	16′ deep, 2-1/2 C.Y. bucket	16.10	24	40.10
1570	20′ deep, 3-1/2 C.Y. bucket	20.50	29	49.50
1580	24′ deep, 3-1/2 C.Y. bucket	24.50	34.50	59
1640	8′ wide, 12′ deep, 1-1/2 C.Y. bucket w/trench box	16.65	24.50	41.15
1650	15′ deep, 1-1/2 C.Y. bucket	21.50	32	53.50
1660	18′ deep, 2-1/2 C.Y. bucket	23.50	31.50	55
1680	24′ deep, 3-1/2 C.Y. bucket	33.50	45	78.50
1730	10′ wide, 20′ deep, 3-1/2 C.Y. bucket w/trench box	26.50	42.50	69
1740	24′ deep, 3-1/2 C.Y. bucket	40	51	91
1780	12′ wide, 20′ deep, 3-1/2 C.Y. bucket w/trench box	42.50	54	96.50
1790	25′ deep, bucket	52.50	69	121.50
1800	1/2 to 1 slope, 2′ wide, 2′ deep, 3/8 C.Y. bucket	.94	2.51	3.45
1810	3′ deep, 3/8 C.Y. bucket	1.56	4.41	5.97

G1030 Site Earthwork

G1030 805	Trenching Common Earth	COST PER L.F.		
		EQUIP.	LABOR	TOTAL
1820	4′ deep, 3/8 C.Y. bucket	2.31	6.75	9.06
1840	6′ deep, 3/8 C.Y. bucket	3.85	10.95	14.80
1860	8′ deep, 1/2 C.Y. bucket	6.10	17.50	23.60
1880	10′ deep, 1 C.Y. bucket	11.85	24.50	36.35
2300	4′ wide, 2′ deep, 3/8 C.Y. bucket	1.79	4.13	5.92
2310	3′ deep, 3/8 C.Y. bucket	2.69	6.90	9.59
2320	4′ deep, 1/2 C.Y. bucket	3.47	8.50	11.97
2340	6′ deep, 1/2 C.Y. bucket	6.45	15.80	22.25
2360	8′ deep, 1/2 C.Y. bucket	12.40	23	35.40
2380	10′ deep, 1 C.Y. bucket	17.10	32.50	49.60
2400	12′ deep, 1 C.Y. bucket	22.50	44	66.50
2430	15′ deep, 1-1/2 C.Y. bucket	24.50	47.50	72
2460	18′ deep, 2-1/2 C.Y. bucket	43.50	73	116.50
2840	6′ wide, 6′ deep, 5/8 C.Y. bucket w/trench box	12.65	19.65	32.30
2860	8′ deep, 3/4 C.Y. bucket	18.30	29.50	47.80
2880	10′ deep, 1 C.Y. bucket	17.70	30	47.70
2900	12′ deep, 1-1/2 C.Y. bucket	22.50	39.50	62
2940	16′ deep, 2-1/2 C.Y. bucket	36.50	56.50	93
2980	20′ deep, 3-1/2 C.Y. bucket	51	77.50	128.50
3020	24′ deep, 3-1/2 C.Y. bucket	72.50	106	178.50
3100	8′ wide, 12′ deep, 1-1/2 C.Y. bucket w/trench box	28	44.50	72.50
3120	15′ deep, 1-1/2 C.Y. bucket	40.50	65	105.50
3140	18′ deep, 2-1/2 C.Y. bucket	50.50	76	126.50
3180	24′ deep, 3-1/2 C.Y. bucket	81.50	116	197.50
3270	10′ wide, 20′ deep, 3-1/2 C.Y. bucket w/trench box	51.50	89.50	141
3280	24′ deep, 3-1/2 C.Y. bucket	90	127	217
3370	12′ wide, 20′ deep, 3-1/2 C.Y. bucket w/ trench box	76	104	180
3380	25′ deep, 3-1/2 C.Y. bucket	105	148	253
3500	1 to 1 slope, 2′ wide, 2′ deep, 3/8 C.Y. bucket	1.21	3.35	4.56
3520	3′ deep, 3/8 C.Y. bucket	3.50	7.55	11.05
3540	4′ deep, 3/8 C.Y. bucket	3.41	10.15	13.56
3560	6′ deep, 3/8 C.Y. bucket	3.84	10.95	14.79
3580	8′ deep, 1/2 C.Y. bucket	7.60	22	29.60
3600	10′ deep, 1 C.Y. bucket	20.50	42	62.50
3800	4′ wide, 2′ deep, 3/8 C.Y. bucket	2.07	4.97	7.04
3820	3′ deep, 3/8 C.Y. bucket	3.31	8.75	12.06
3840	4′ deep, 1/2 C.Y. bucket	4.46	11.40	15.86
3860	6′ deep, 1/2 C.Y. bucket	9.05	22.50	31.55
3880	8′ deep, 1/2 C.Y. bucket	18.40	35	53.40
3900	10′ deep, 1 C.Y. bucket	26.50	51	77.50
3920	12′ deep, 1 C.Y. bucket	39	74	113
3940	15′ deep, 1-1/2 C.Y. bucket	40.50	79	119.50
3960	18′ deep, 2-1/2 C.Y. bucket	59.50	100	159.50
4030	6′ wide, 6′ deep, 5/8 C.Y. bucket w/trench box	16.55	26.50	43.05
4040	8′ deep, 3/4 C.Y. bucket	24	36.50	60.50
4050	10′ deep, 1 C.Y. bucket	25.50	44	69.50
4060	12′ deep, 1-1/2 C.Y. bucket	34	60	94
4070	16′ deep, 2-1/2 C.Y. bucket	57.50	89.50	147
4080	20′ deep, 3-1/2 C.Y. bucket	82.50	127	209.50
4090	24′ deep, 3-1/2 C.Y. bucket	121	178	299
4500	8′ wide, 12′ deep, 1-1/2 C.Y. bucket w/trench box	39	65	104
4550	15′ deep, 1-1/2 C.Y. bucket	59.50	97.50	157
4600	18′ deep, 2-1/2 C.Y. bucket	76.50	117	193.50
4650	24′ deep, 3-1/2 C.Y. bucket	129	188	317
4800	10′ wide, 20′ deep, 3-1/2 C.Y. bucket w/trench box	76.50	136	212.50
4850	24′ deep, 3-1/2 C.Y. bucket	138	199	337
4950	12′ wide, 20′ deep, 3-1/2 C.Y. bucket w/ trench box	109	153	262
4980	25′ deep, 3-1/2 C.Y. bucket	157	224	381

G1030 Site Earthwork

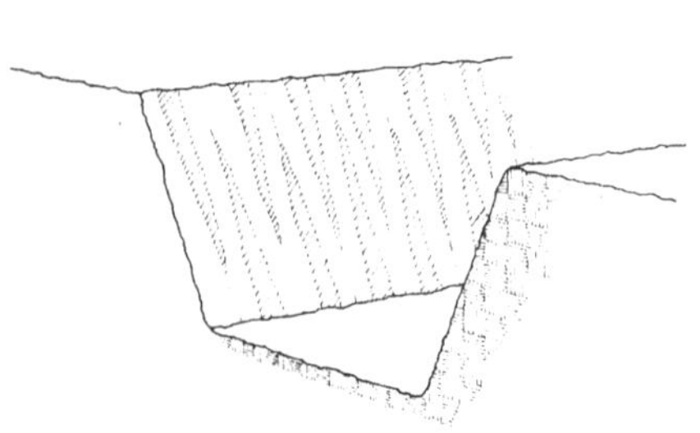

Trenching Systems are shown on a cost per linear foot basis. The systems include: excavation; backfill and removal of spoil; and compaction for various depths and trench bottom widths. The backfill has been reduced to accommodate a pipe of suitable diameter and bedding.

The slope for trench sides varies from none to 1:1.

The Expanded System Listing shows Trenching Systems that range from 2′ to 12′ in width. Depths range from 2′ to 25′.

System Components	QUANTITY	UNIT	COST PER L.F. EQUIP.	LABOR	TOTAL
SYSTEM G1030 806 1310					
TRENCHING LOAM & SANDY CLAY, NO SLOPE, 2′ WIDE, 2′ DP, 3/8 C.Y. BUCKET					
Excavation, trench, hyd. backhoe, track mtd., 3/8 C.Y. bucket	.148	B.C.Y.	.34	.84	1.18
Backfill and load spoil, from stockpile	.165	L.C.Y.	.12	.30	.42
Compaction by vibrating plate 18″ wide, 6″ lifts, 4 passes	.118	E.C.Y.	.03	.34	.37
Remove excess spoil, 8 C.Y. dump truck, 2 mile roundtrip	.042	L.C.Y.	.15	.14	.29
TOTAL			.66	1.66	2.32

G1030 806	Trenching Loam & Sandy Clay	COST PER L.F. EQUIP.	LABOR	TOTAL
1310	Trenching, loam & sandy clay, no slope, 2′ wide, 2′ deep, 3/8 C.Y. bucket	.64	1.62	2.26
1320	3′ deep, 3/8 C.Y. bucket	1.01	2.68	3.69
1330	4′ deep, 3/8 C.Y. bucket	1.18	3.28	4.46
1340	6′ deep, 3/8 C.Y. bucket	1.69	3.90	5.60
1350	8′ deep, 1/2 C.Y. bucket	2.23	5.15	7.40
1360	10′ deep, 1 C.Y. bucket	2.57	5.55	8.10
1400	4′ wide, 2′ deep, 3/8 C.Y. bucket	1.52	3.22	4.74
1410	3′ deep, 3/8 C.Y. bucket	2.05	4.87	6.90
1420	4′ deep, 1/2 C.Y. bucket	2.45	5.50	7.95
1430	6′ deep, 1/2 C.Y. bucket	4.06	8	12.05
1440	8′ deep, 1/2 C.Y. bucket	6.25	11.35	17.60
1450	10′ deep, 1 C.Y. bucket	6	11.60	17.60
1460	12′ deep, 1 C.Y. bucket	7.55	14.45	22
1470	15′ deep, 1-1/2 C.Y. bucket	9.15	16.80	26
1480	18′ deep, 2-1/2 C.Y. bucket	10.75	18.10	29
1520	6′ wide, 6′ deep, 5/8 C.Y. bucket w/trench box	8.25	12.95	21
1530	8′ deep, 3/4 C.Y. bucket	10.90	17.05	28
1540	10′ deep, 1 C.Y. bucket	10.10	17.30	27.50
1550	12′ deep, 1-1/2 C.Y. bucket	11.45	19.40	31
1560	16′ deep, 2-1/2 C.Y. bucket	15.60	24	39.50
1570	20′ deep, 3-1/2 C.Y. bucket	18.80	29	48
1580	24′ deep, 3-1/2 C.Y. bucket	23.50	35	58.50
1640	8′ wide, 12′ deep, 1-1/4 C.Y. bucket w/trench box	16.35	24.50	41
1650	15′ deep, 1-1/2 C.Y. bucket	19.90	31.50	51.50
1660	18′ deep, 2-1/2 C.Y. bucket	24	34.50	58.50
1680	24′ deep, 3-1/2 C.Y. bucket	32.50	45.50	78
1730	10′ wide, 20′ deep, 3-1/2 C.Y. bucket w/trench box	32.50	45.50	78
1740	24′ deep, 3-1/2 C.Y. bucket	40.50	56.50	97
1780	12′ wide, 20′ deep, 3-1/2 C.Y. bucket w/trench box	39.50	54	93.50
1790	25′ deep, bucket	51	70	121
1800	1/2:1 slope, 2′ wide, 2′ deep, 3/8 C.Y. BK	.91	2.45	3.36
1810	3′ deep, 3/8 C.Y. bucket	1.51	4.31	5.80
1820	4′ deep, 3/8 C.Y. bucket	2.24	6.60	8.85
1840	6′ deep, 3/8 C.Y. bucket	4.06	9.80	13.85

G1030 Site Earthwork

G1030 806	Trenching Loam & Sandy Clay	COST PER L.F.		
		EQUIP.	LABOR	TOTAL
1860	8′ deep, 1/2 C.Y. bucket	6.45	15.65	22
1880	10′ deep, 1 C.Y. bucket	8.75	19.65	28.50
2300	4′ wide, 2′ deep, 3/8 C.Y. bucket	1.78	4.04	5.80
2310	3′ deep, 3/8 C.Y. bucket	2.64	6.75	9.40
2320	4′ deep, 1/2 C.Y. bucket	3.42	8.35	11.75
2340	6′ deep, 1/2 C.Y. bucket	6.75	14.20	21
2360	8′ deep, 1/2 C.Y. bucket	12.05	23	35
2380	10′ deep, 1 C.Y. bucket	13.15	26.50	39.50
2400	12′ deep, 1 C.Y. bucket	22	44	66
2430	15′ deep, 1-1/2 C.Y. bucket	25.50	49	74.50
2460	18′ deep, 2-1/2 C.Y. bucket	42.50	74	117
2840	6′ wide, 6′ deep, 5/8 C.Y. bucket w/trench box	12	19.65	31.50
2860	8′ deep, 3/4 C.Y. bucket	17.55	29	46.50
2880	10′ deep, 1 C.Y. bucket	18	32.50	50.50
2900	12′ deep, 1-1/2 C.Y. bucket	22.50	40	62.50
2940	16′ deep, 2-1/2 C.Y. bucket	35.50	57.50	93
2980	20′ deep, 3-1/2 C.Y. bucket	49.50	79	129
3020	24′ deep, 3-1/2 C.Y. bucket	70	108	178
3100	8′ wide, 12′ deep, 1-1/2 C.Y. bucket w/trench box	27	45	72
3120	15′ deep, 1-1/2 C.Y. bucket	37	63	100
3140	18′ deep, 2-1/2 C.Y. bucket	49	76.50	126
3180	24′ deep, 3-1/2 C.Y. bucket	78.50	118	197
3270	10′ wide, 20′ deep, 3-1/2 C.Y. bucket w/trench box	63	95.50	159
3280	24′ deep, 3-1/2 C.Y. bucket	87	129	216
3320	12′ wide, 20′ deep, 3-1/2 C.Y. bucket w/trench box	70	104	174
3380	25′ deep, 3-1/2 C.Y. bucket w/trench box	95.50	139	235
3500	1:1 slope, 2′ wide, 2′ deep, 3/8 C.Y. bucket	1.18	3.28	4.46
3520	3′ deep, 3/8 C.Y. bucket	2.10	6.15	8.25
3540	4′ deep, 3/8 C.Y. bucket	3.31	9.90	13.20
3560	6′ deep, 3/8 C.Y. bucket	4.06	9.80	13.85
3580	8′ deep, 1/2 C.Y. bucket	10.65	26	36.50
3600	10′ deep, 1 C.Y. bucket	14.95	34	49
3800	2′ deep, 3/8 C.Y. bucket	2.05	4.87	6.90
3820	4′ wide, 3′ deep, 3/8 C.Y. bucket	3.25	8.60	11.85
3840	4′ deep, 1/2 C.Y. bucket	4.39	11.25	15.65
3860	6′ deep, 1/2 C.Y. bucket	9.50	20.50	30
3880	8′ deep, 1/2 C.Y. bucket	17.85	34.50	52.50
3900	10′ deep, 1 C.Y. bucket	20.50	41.50	62
3920	12′ deep, 1 C.Y. bucket	29.50	59	88.50
3940	15′ deep, 1-1/2 C.Y. bucket	42.50	81.50	124
3960	18′ deep, 2-1/2 C.Y. bucket	58	102	160
4030	6′ wide, 6′ deep, 5/8 C.Y. bucket w/trench box	15.75	26.50	42.50
4040	8′ deep, 3/4 C.Y. bucket	24	41	65
4050	10′ deep, 1 C.Y. bucket	26	48	74
4060	12′ deep, 1-1/2 C.Y. bucket	33	60	93
4070	16′ deep, 2-1/2 C.Y. bucket	55.50	90.50	146
4080	20′ deep, 3-1/2 C.Y. bucket	80	129	209
4090	24′ deep, 3-1/2 C.Y. bucket	116	181	297
4500	8′ wide, 12′ deep, 1-1/4 C.Y. bucket w/trench box	38	65.50	104
4550	15′ deep, 1-1/2 C.Y. bucket	53.50	95	149
4600	18′ deep, 2-1/2 C.Y. bucket	74.50	119	194
4650	24′ deep, 3-1/2 C.Y. bucket	125	191	315
4800	10′ wide, 20′ deep, 3-1/2 C.Y. bucket w/trench box	93.50	146	240
4850	24′ deep, 3-1/2 C.Y. bucket	133	202	335
4950	12′ wide, 20′ deep, 3-1/2 C.Y. bucket w/trench box	100	154	254
4980	25′ deep, bucket	152	228	380

G1030 Site Earthwork

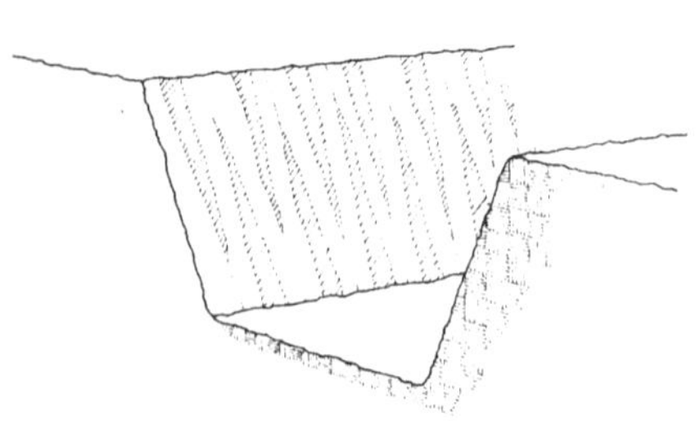

Trenching Systems are shown on a cost per linear foot basis. The systems include: excavation; backfill and removal of spoil; and compaction for various depths and trench bottom widths. The backfill has been reduced to accommodate a pipe of suitable diameter and bedding.

The slope for trench sides varies from none to 1:1.

The Expanded System Listing shows Trenching Systems that range from 2′ to 12′ in width. Depths range from 2′ to 25′.

System Components	QUANTITY	UNIT	COST PER L.F. EQUIP.	LABOR	TOTAL
SYSTEM G1030 807 1310					
TRENCHING SAND & GRAVEL, NO SLOPE, 2′ WIDE, 2′ DEEP, 3/8 C.Y. BUCKET					
Excavation, trench, hyd. backhoe, track mtd., 3/8 C.Y. bucket	.148	B.C.Y.	.33	.83	1.16
Backfill and load spoil, from stockpile	.140	L.C.Y.	.10	.25	.35
Compaction by vibrating plate 18″ wide, 6″ lifts, 4 passes	.118	E.C.Y.	.03	.34	.37
Remove excess spoil, 8 C.Y. dump truck, 2 mile roundtrip	.035	L.C.Y.	.13	.12	.25
TOTAL			.59	1.54	2.13

G1030 807	Trenching Sand & Gravel	COST PER L.F. EQUIP.	LABOR	TOTAL
1310	Trenching, sand & gravel, no slope, 2′ wide, 2′ deep, 3/8 C.Y. bucket	.59	1.54	2.13
1320	3′ deep, 3/8 C.Y. bucket	.95	2.59	3.54
1330	4′ deep, 3/8 C.Y. bucket	1.10	3.13	4.23
1340	6′ deep, 3/8 C.Y. bucket	1.60	3.73	5.35
1350	8′ deep, 1/2 C.Y. bucket	2.12	4.95	7.05
1360	10′ deep, 1 C.Y. bucket	2.40	5.25	7.65
1400	4′ wide, 2′ deep, 3/8 C.Y. bucket	1.38	3.03	4.41
1410	3′ deep, 3/8 C.Y. bucket	1.88	4.61	6.50
1420	4′ deep, 1/2 C.Y. bucket	2.26	5.20	7.45
1430	6′ deep, 1/2 C.Y. bucket	3.84	7.65	11.50
1440	8′ deep, 1/2 C.Y. bucket	5.90	10.75	16.65
1450	10′ deep, 1 C.Y. bucket	5.65	10.95	16.60
1460	12′ deep, 1 C.Y. bucket	7.15	13.65	21
1470	15′ deep, 1-1/2 C.Y. bucket	8.60	15.85	24.50
1480	18′ deep, 2-1/2 C.Y. bucket	10.20	17.05	27.50
1520	6′ wide, 6′ deep, 5/8 C.Y. bucket w/trench box	7.80	12.25	20
1530	8′ deep, 3/4 C.Y. bucket	10.35	16.20	26.50
1540	10′ deep, 1 C.Y. bucket	9.55	16.35	26
1550	12′ deep, 1-1/2 C.Y. bucket	10.85	18.30	29
1560	16′ deep, 2 C.Y. bucket	14.95	22.50	37.50
1570	20′ deep, 3-1/2 C.Y. bucket	17.80	27	45
1580	24′ deep, 3-1/2 C.Y. bucket	22.50	33	55.50
1640	8′ wide, 12′ deep, 1-1/2 C.Y. bucket w/trench box	15.35	23	38.50
1650	15′ deep, 1-1/2 C.Y. bucket	18.65	29.50	48
1660	18′ deep, 2-1/2 C.Y. bucket	22.50	32.50	55
1680	24′ deep, 3-1/2 C.Y. bucket	30.50	42.50	73
1730	10′ wide, 20′ deep, 3-1/2 C.Y. bucket w/trench box	30.50	42.50	73
1740	24′ deep, 3-1/2 C.Y. bucket	38.50	53	91.50
1780	12′ wide, 20′ deep, 3-1/2 C.Y. bucket w/trench box	37	50.50	87.50
1790	25′ deep, bucket	48	65.50	114
1800	1/2:1 slope, 2′ wide, 2′ deep, 3/8 CY bk	.85	2.33	3.18
1810	3′ deep, 3/8 C.Y. bucket	1.42	4.11	5.55
1820	4′ deep, 3/8 C.Y. bucket	2.10	6.30	8.40
1840	6′ deep, 3/8 C.Y. bucket	3.87	9.40	13.25

G1030 Site Earthwork

G1030 807	Trenching Sand & Gravel	COST PER L.F.		
		EQUIP.	LABOR	TOTAL
1860	8' deep, 1/2 C.Y. bucket	6.15	14.95	21
1880	10' deep, 1 C.Y. bucket	8.25	18.55	27
2300	4' wide, 2' deep, 3/8 C.Y. bucket	1.62	3.81	5.45
2310	3' deep, 3/8 C.Y. bucket	2.43	6.40	8.85
2320	4' deep, 1/2 C.Y. bucket	3.16	7.95	11.10
2340	6' deep, 1/2 C.Y. bucket	6.45	13.60	20
2360	8' deep, 1/2 C.Y. bucket	11.45	22	33.50
2380	10' deep, 1 C.Y. bucket	12.40	25	37.50
2400	12' deep, 1 C.Y. bucket	21	42	63
2430	15' deep, 1-1/2 C.Y. bucket	24.50	46.50	71
2460	18' deep, 2-1/2 C.Y. bucket	40	69.50	110
2840	6' wide, 6' deep, 5/8 C.Y. bucket w/trench box	11.40	18.65	30
2860	8' deep, 3/4 C.Y. bucket	16.65	27.50	44
2880	10' deep, 1 C.Y. bucket	17.10	31	48
2900	12' deep, 1-1/2 C.Y. bucket	21	37.50	58.50
2940	16' deep, 2 C.Y. bucket	34	54	88
2980	20' deep, 3-1/2 C.Y. bucket	47	74	121
3020	24' deep, 3-1/2 C.Y. bucket	66.50	101	168
3100	8' wide, 12' deep, 1-1/4 C.Y. bucket w/trench box	25.50	42.50	68
3120	15' deep, 1-1/2 C.Y. bucket	34.50	59.50	94
3140	18' deep, 2-1/2 C.Y. bucket	46.50	72	119
3180	24' deep, 3-1/2 C.Y. bucket	74.50	111	186
3270	10' wide, 20' deep, 3-1/2 C.Y. bucket w/trench box	59.50	89.50	149
3280	24' deep, 3-1/2 C.Y. bucket	82.50	121	204
3370	12' wide, 20' deep, 3-1/2 C.Y. bucket w/trench box	66.50	99.50	166
3380	25' deep, bucket	96	140	236
3500	1:1 slope, 2' wide, 2' deep, 3/8 CY bk	1.94	3.82	5.75
3520	3' deep, 3/8 C.Y. bucket	1.98	5.90	7.90
3540	4' deep, 3/8 C.Y. bucket	3.11	9.45	12.55
3560	6' deep, 3/8 C.Y. bucket	3.87	9.40	13.25
3580	8' deep, 1/2 C.Y. bucket	10.20	25	35
3600	10' deep, 1 C.Y. bucket	14.05	32	46
3800	4' wide, 2' deep, 3/8 C.Y. bucket	1.88	4.61	6.50
3820	3' deep, 3/8 C.Y. bucket	3.01	8.15	11.15
3840	4' deep, 1/2 C.Y. bucket	4.08	10.70	14.80
3860	6' deep, 1/2 C.Y. bucket	9.05	19.60	28.50
3880	8' deep, 1/2 C.Y. bucket	16.95	33	50
3900	10' deep, 1 C.Y. bucket	19.20	39.50	58.50
3920	12' deep, 1 C.Y. bucket	28	56	84
3940	15' deep, 1-1/2 C.Y. bucket	40	77	117
3960	18' deep, 2-1/2 C.Y. bucket	55	95.50	151
4030	6' wide, 6' deep, 5/8 C.Y. bucket w/trench box	14.95	25	40
4040	8' deep, 3/4 C.Y. bucket	23	39	62
4050	10' deep, 1 C.Y. bucket	24.50	45	69.50
4060	12' deep, 1-1/2 C.Y. bucket	31.50	57	88.50
4070	16' deep, 2 C.Y. bucket	53.50	85.50	139
4080	20' deep, 3-1/2 C.Y. bucket	76	121	197
4090	24' deep, 3-1/2 C.Y. bucket	111	170	281
4500	8' wide, 12' deep, 1-1/2 C.Y. bucket w/trench box	36	62	98
4550	15' deep, 1-1/2 C.Y. bucket	50.50	89.50	140
4600	18' deep, 2-1/2 C.Y. bucket	70.50	112	183
4650	24' deep, 3-1/2 C.Y. bucket	119	179	298
4800	10' wide, 20' deep, 3-1/2 C.Y. bucket w/trench box	88.50	136	225
4850	24' deep, 3-1/2 C.Y. bucket	126	189	315
4950	12' wide, 20' deep, 3-1/2 C.Y. bucket w/trench box	95	144	239
4980	25' deep, bucket	144	214	360

G1030 Site Earthwork

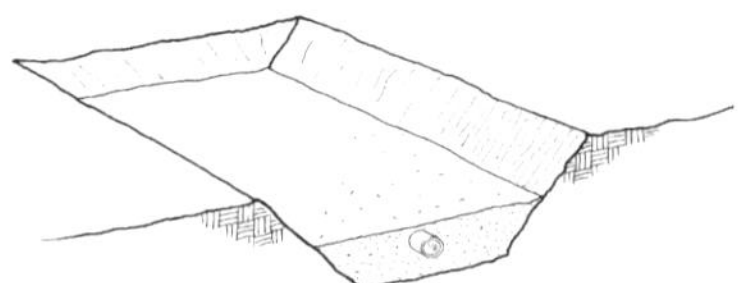

The Pipe Bedding System is shown for various pipe diameters. Compacted bank sand is used for pipe bedding and to fill 12″ over the pipe. No backfill is included. Various side slopes are shown to accommodate different soil conditions. Pipe sizes vary from 6″ to 84″ diameter.

System Components	QUANTITY	UNIT	COST PER L.F. MAT.	COST PER L.F. INST.	COST PER L.F. TOTAL
SYSTEM G1030 815 1440					
PIPE BEDDING, SIDE SLOPE 0 TO 1, 1′ WIDE, PIPE SIZE 6″ DIAMETER					
Borrow, bank sand, 2 mile haul, machine spread	.067	C.Y.	.78	.48	1.26
Compaction, vibrating plate	.067	C.Y.		.15	.15
TOTAL			.78	.63	1.41

G1030 815	Pipe Bedding	COST PER L.F. MAT.	COST PER L.F. INST.	COST PER L.F. TOTAL
1440	Pipe bedding, side slope 0 to 1, 1′ wide, pipe size 6″ diameter	.78	.63	1.41
1460	2′ wide, pipe size 8″ diameter	1.72	1.40	3.12
1480	Pipe size 10″ diameter	1.76	1.43	3.19
1500	Pipe size 12″ diameter	1.81	1.47	3.28
1520	3′ wide, pipe size 14″ diameter	2.98	2.42	5.40
1540	Pipe size 15″ diameter	3.02	2.44	5.46
1560	Pipe size 16″ diameter	3.05	2.48	5.53
1580	Pipe size 18″ diameter	3.12	2.54	5.66
1600	4′ wide, pipe size 20″ diameter	4.50	3.65	8.15
1620	Pipe size 21″ diameter	4.55	3.68	8.23
1640	Pipe size 24″ diameter	4.66	3.79	8.45
1660	Pipe size 30″ diameter	4.77	3.87	8.64
1680	6′ wide, pipe size 32″ diameter	8.35	6.75	15.10
1700	Pipe size 36″ diameter	8.55	6.95	15.50
1720	7′ wide, pipe size 48″ diameter	11.15	9.05	20.20
1740	8′ wide, pipe size 60″ diameter	13.90	11.30	25.20
1760	10′ wide, pipe size 72″ diameter	20	16.30	36.30
1780	12′ wide, pipe size 84″ diameter	27.50	22	49.50
2140	Side slope 1/2 to 1, 1′ wide, pipe size 6″ diameter	1.64	1.33	2.97
2160	2′ wide, pipe size 8″ diameter	2.71	2.20	4.91
2180	Pipe size 10″ diameter	2.95	2.40	5.35
2200	Pipe size 12″ diameter	3.14	2.54	5.68
2220	3′ wide, pipe size 14″ diameter	4.51	3.65	8.16
2240	Pipe size 15″ diameter	4.63	3.76	8.39
2260	Pipe size 16″ diameter	4.77	3.87	8.64
2280	Pipe size 18″ diameter	5.05	4.09	9.14
2300	4′ wide, pipe size 20″ diameter	6.65	5.40	12.05
2320	Pipe size 21″ diameter	6.80	5.55	12.35
2340	Pipe size 24″ diameter	7.30	5.95	13.25
2360	Pipe size 30″ diameter	8.20	6.65	14.85
2380	6′ wide, pipe size 32″ diameter	12.10	9.75	21.85
2400	Pipe size 36″ diameter	12.90	10.45	23.35
2420	7′ wide, pipe size 48″ diameter	17.65	14.35	32
2440	8′ wide, pipe size 60″ diameter	23	18.65	41.65
2460	10′ wide, pipe size 72″ diameter	32	26	58
2480	12′ wide, pipe size 84″ diameter	43	35	78
2620	Side slope 1 to 1, 1′ wide, pipe size 6″ diameter	2.49	2.03	4.52
2640	2′ wide, pipe size 8″ diameter	3.75	3.04	6.79

G1030 Site Earthwork

G1030 815	Pipe Bedding	COST PER L.F.		
		MAT.	INST.	TOTAL
2660	Pipe size 10" diameter	4.11	3.33	7.44
2680	Pipe size 12" diameter	4.51	3.65	8.16
2700	3' wide, pipe size 14" diameter	6.05	4.90	10.95
2720	Pipe size 15" diameter	6.25	5.05	11.30
2740	Pipe size 16" diameter	6.50	5.25	11.75
2760	Pipe size 18" diameter	6.95	5.65	12.60
2780	4' wide, pipe size 20" diameter	8.80	7.15	15.95
2800	Pipe size 21" diameter	9.10	7.40	16.50
2820	Pipe size 24" diameter	9.95	8.05	18
2840	Pipe size 30" diameter	11.60	9.40	21
2860	6' wide, pipe size 32" diameter	15.80	12.85	28.65
2880	Pipe size 36" diameter	17.25	14	31.25
2900	7' wide, pipe size 48" diameter	24	19.60	43.60
2920	8' wide, pipe size 60" diameter	32	26	58
2940	10' wide, pipe size 72" diameter	44.50	36	80.50
2960	12' wide, pipe size 84" diameter	58.50	47	105.50

G3010 Water Supply

G3010 110	Water Distribution Piping	COST PER L.F.		
		MAT.	INST.	TOTAL
2000	Piping, excav. & backfill excl., ductile iron class 250, mech. joint			
2130	4″ diameter	18.20	16.10	34.30
2150	6″ diameter	21	20	41
2160	8″ diameter	23.50	24	47.50
2170	10″ diameter	31.50	28.50	60
2180	12″ diameter	39	30.50	69.50
2210	16″ diameter	54	44.50	98.50
2220	18″ diameter	68	46.50	114.50
3000	Tyton joint			
3130	4″ diameter	13.65	8.05	21.70
3150	6″ diameter	17.45	9.65	27.10
3160	8″ diameter	24.50	16.10	40.60
3170	10″ diameter	34	17.70	51.70
3180	12″ diameter	36	20	56
3210	16″ diameter	64	28.50	92.50
3220	18″ diameter	71	32	103
3230	20″ diameter	78.50	36	114.50
3250	24″ diameter	92.50	42	134.50
4000	Copper tubing, type K			
4050	3/4″ diameter	6.05	2.81	8.86
4060	1″ diameter	7.90	3.51	11.41
4080	1-1/2″ diameter	12.80	4.24	17.04
4090	2″ diameter	19.70	4.89	24.59
4110	3″ diameter	40.50	8.40	48.90
4130	4″ diameter	66.50	11.85	78.35
4150	6″ diameter	169	22	191
5000	Polyvinyl chloride class 160, S.D.R. 26			
5130	1-1/2″ diameter	.33	1.05	1.38
5150	2″ diameter	.51	1.15	1.66
5160	4″ diameter	2.57	5.15	7.72
5170	6″ diameter	5.10	6.20	11.30
5180	8″ diameter	8.95	7.45	16.40
6000	Polyethylene 160 psi, S.D.R. 7			
6050	3/4″ diameter	.65	1.50	2.15
6060	1″ diameter	.89	1.62	2.51
6080	1-1/2″ diameter	1.02	1.75	2.77
6090	2″ diameter	1.71	2.15	3.86

G3020 Sanitary Sewer

G3020 110	Drainage & Sewage Piping	COST PER L.F.		
		MAT.	INST.	TOTAL
2000	Piping, excavation & backfill excluded, PVC, plain			
2130	4″ diameter	1.85	3.64	5.49
2150	6″ diameter	4	3.90	7.90
2160	8″ diameter	8.65	4.07	12.72
2900	Box culvert, precast, 8′ long			
3000	6′ x 3′	325	31	356
3020	6′ x 7′	490	34.50	524.50
3040	8′ x 3′	445	32.50	477.50
3060	8′ x 8′	600	43.50	643.50
3080	10′ x 3′	660	39.50	699.50
3100	10′ x 8′	750	54.50	804.50
3120	12′ x 3′	640	43.50	683.50
3140	12′ x 8′	1,100	65	1,165
4000	Concrete, nonreinforced			
4150	6″ diameter	6.90	11.05	17.95
4160	8″ diameter	7.55	13.10	20.65
4170	10″ diameter	8.40	13.55	21.95
4180	12″ diameter	10.30	14.65	24.95
4200	15″ diameter	12.25	16.30	28.55
4220	18″ diameter	14.45	20.50	34.95
4250	24″ diameter	24	29	53
4400	Reinforced, no gasket			
4580	12″ diameter	11.70	19.60	31.30
4600	15″ diameter	15.35	19.60	34.95
4620	18″ diameter	20.50	22	42.50
4650	24″ diameter	29	29	58
4670	30″ diameter	43.50	44	87.50
4680	36″ diameter	59	54	113
4690	42″ diameter	74	61	135
4700	48″ diameter	90	69	159
4720	60″ diameter	133	91	224
4730	72″ diameter	193	110	303
4740	84″ diameter	293	137	430
4800	With gasket			
4980	12″ diameter	12.90	10.60	23.50
5000	15″ diameter	16.90	11.15	28.05
5020	18″ diameter	22.50	11.75	34.25
5050	24″ diameter	37	13.10	50.10
5070	30″ diameter	53	44	97
5080	36″ diameter	70.50	54	124.50
5090	42″ diameter	86.50	55.50	142
5100	48″ diameter	108	69	177
5120	60″ diameter	154	88	242
5130	72″ diameter	222	110	332
5140	84″ diameter	350	183	533
5700	Corrugated metal, alum. or galv. bit. coated			
5760	8″ diameter	13.50	8.90	22.40
5770	10″ diameter	16.20	11.30	27.50
5780	12″ diameter	19.45	13.95	33.40
5800	15″ diameter	23.50	14.65	38.15
5820	18″ diameter	30	15.45	45.45
5850	24″ diameter	37	18.30	55.30
5870	30″ diameter	49	32.50	81.50
5880	36″ diameter	72	32.50	104.50
5900	48″ diameter	109	39	148
5920	60″ diameter	141	59	200
5930	72″ diameter	211	97.50	308.50
6000	Plain			

G3020 Sanitary Sewer

G3020 110	Drainage & Sewage Piping	COST PER L.F.		
		MAT.	INST.	TOTAL
6060	8″ diameter	11.05	8.25	19.30
6070	10″ diameter	12.10	10.50	22.60
6080	12″ diameter	13.85	13.35	27.20
6100	15″ diameter	17.60	13.35	30.95
6120	18″ diameter	23.50	14.30	37.80
6140	24″ diameter	34	16.75	50.75
6170	30″ diameter	43	30	73
6180	36″ diameter	70.50	30	100.50
6200	48″ diameter	94.50	35.50	130
6220	60″ diameter	148	56	204
6230	72″ diameter	207	137	344
6300	Steel or alum. oval arch, coated & paved invert			
6400	15″ equivalent diameter	42	14.65	56.65
6420	18″ equivalent diameter	54.50	19.60	74.10
6450	24″ equivalent diameter	78	23.50	101.50
6470	30″ equivalent diameter	95	29	124
6480	36″ equivalent diameter	142	39	181
6490	42″ equivalent diameter	171	43	214
6500	48″ equivalent diameter	190	52	242
6600	Plain			
6700	15″ equivalent diameter	23	13.05	36.05
6720	18″ equivalent diameter	27	16.75	43.75
6750	24″ equivalent diameter	43.50	19.60	63.10
6770	30″ equivalent diameter	54.50	36	90.50
6780	36″ equivalent diameter	90	36	126
6790	42″ equivalent diameter	106	42	148
6800	48″ equivalent diameter	125	52	177
8000	Polyvinyl chloride SDR 35			
8130	4″ diameter	1.85	3.64	5.49
8150	6″ diameter	4	3.90	7.90
8160	8″ diameter	8.65	4.07	12.72
8170	10″ diameter	13.10	5.40	18.50
8180	12″ diameter	14.95	5.60	20.55
8200	15″ diameter	17.45	7.45	24.90

G3030 Storm Sewer

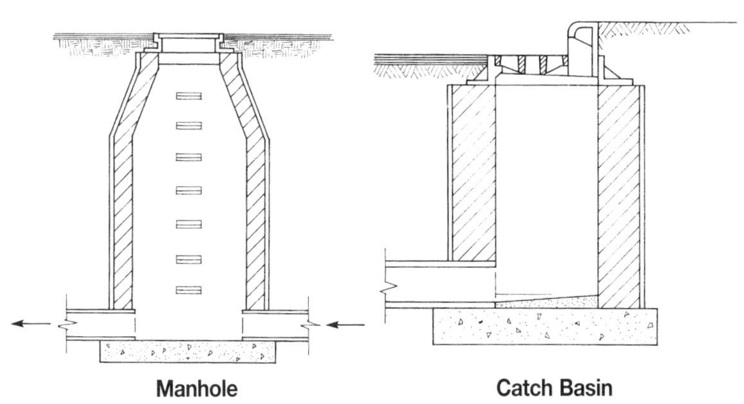

The Manhole and Catch Basin System includes: excavation with a backhoe; a formed concrete footing; frame and cover; cast iron steps and compacted backfill.

The Expanded System Listing shows manholes that have a 4′, 5′ and 6′ inside diameter riser. Depths range from 4′ to 14′. Construction material shown is either concrete, concrete block, precast concrete, or brick.

System Components	QUANTITY	UNIT	COST PER EACH MAT.	COST PER EACH INST.	COST PER EACH TOTAL
SYSTEM G3030 210 1920					
MANHOLE/CATCH BASIN, BRICK, 4′ I.D. RISER, 4′ DEEP					
Excavation, hydraulic backhoe, 3/8 C.Y. bucket	14.815	B.C.Y.		110.97	110.97
Trim sides and bottom of excavation	64.000	S.F.		55.04	55.04
Forms in place, manhole base, 4 uses	20.000	SFCA	11	94	105
Reinforcing in place footings, #4 to #7	.019	Ton	15.87	21.85	37.72
Concrete, 3000 psi	.925	C.Y.	98.98		98.98
Place and vibrate concrete, footing, direct chute	.925	C.Y.		43.67	43.67
Catch basin or MH, brick, 4′ ID, 4′ deep	1.000	Ea.	485	910	1,395
Catch basin or MH steps; heavy galvanized cast iron	1.000	Ea.	14.40	12.60	27
Catch basin or MH frame and cover	1.000	Ea.	277	215.50	492.50
Fill, granular	12.954	L.C.Y.	180.06		180.06
Backfill, spread with wheeled front end loader	12.954	L.C.Y.		29.27	29.27
Backfill compaction, 12″ lifts, air tamp	12.954	E.C.Y.		105.19	105.19
TOTAL			1,082.31	1,598.09	2,680.40

G3030 210	Manholes & Catch Basins	COST PER EACH MAT.	COST PER EACH INST.	COST PER EACH TOTAL
1920	Manhole/catch basin, brick, 4′ I.D. riser, 4′ deep	1,075	1,600	2,675
1940	6′ deep	1,475	2,225	3,700
1960	8′ deep	1,900	3,025	4,925
1980	10′ deep	2,425	3,775	6,200
3000	12′ deep	3,125	4,075	7,200
3020	14′ deep	3,875	5,725	9,600
3200	Block, 4′ I.D. riser, 4′ deep	995	1,300	2,295
3220	6′ deep	1,325	1,825	3,150
3240	8′ deep	1,700	2,500	4,200
3260	10′ deep	2,000	3,125	5,125
3280	12′ deep	2,450	3,975	6,425
3300	14′ deep	3,000	4,825	7,825
4620	Concrete, cast-in-place, 4′ I.D. riser, 4′ deep	1,125	2,225	3,350
4640	6′ deep	1,550	2,975	4,525
4660	8′ deep	2,125	4,275	6,400
4680	10′ deep	2,500	5,300	7,800
4700	12′ deep	3,075	6,600	9,675
4720	14′ deep	3,725	7,925	11,650
5820	Concrete, precast, 4′ I.D. riser, 4′ deep	1,100	1,175	2,275
5840	6′ deep	1,650	1,600	3,250

G3030 Storm Sewer

G3030 210	Manholes & Catch Basins	COST PER EACH		
		MAT.	INST.	TOTAL
5860	8′ deep	2,050	2,225	4,275
5880	10′ deep	2,350	2,725	5,075
5900	12′ deep	2,850	3,350	6,200
5920	14′ deep	3,425	4,275	7,700
6000	5′ I.D. riser, 4′ deep	1,850	1,350	3,200
6020	6′ deep	2,175	1,875	4,050
6040	8′ deep	2,525	2,475	5,000
6060	10′ deep	3,225	3,175	6,400
6080	12′ deep	4,025	4,025	8,050
6100	14′ deep	4,875	4,925	9,800
6200	6′ I.D. riser, 4′ deep	2,775	1,750	4,525
6220	6′ deep	3,175	2,325	5,500
6240	8′ deep	4,250	3,225	7,475
6260	10′ deep	5,275	4,100	9,375
6280	12′ deep	6,375	5,200	11,575
6300	14′ deep	7,575	6,300	13,875

G3060 Fuel Distribution

G3060 110	Gas Service Piping	COST PER L.F.		
		MAT.	INST.	TOTAL
2000	Piping, excavation & backfill excluded, polyethylene			
2070	1-1/4" diam, SDR 10	2.82	3.61	6.43
2090	2" diam, SDR 11	3.50	4.03	7.53
2110	3" diam, SDR 11.5	7.30	4.82	12.12
2130	4" diam,SDR 11	16.70	9.15	25.85
2150	6" diam, SDR 21	52	9.80	61.80
2160	8" diam, SDR 21	71	11.85	82.85
3000	Steel, schedule 40, plain end, tarred & wrapped			
3060	1" diameter	8.55	8.15	16.70
3090	2" diameter	13.40	8.75	22.15
3110	3" diameter	22	9.40	31.40
3130	4" diameter	29	14.90	43.90
3140	5" diameter	42	17.25	59.25
3150	6" diameter	51.50	21	72.50
3160	8" diameter	81	27.50	108.50

G3060 Fuel Distribution

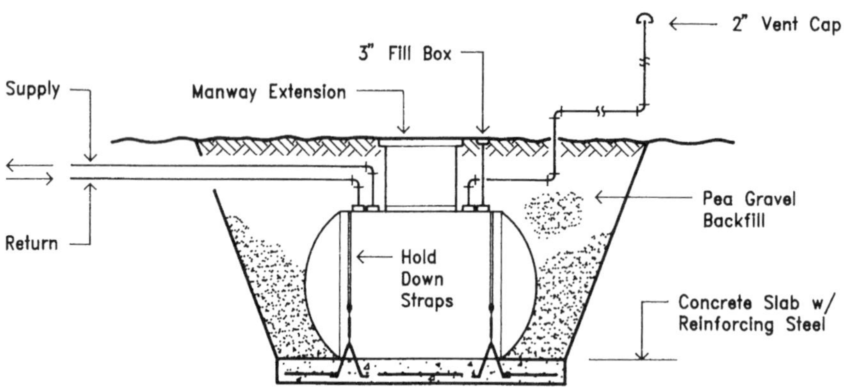

Fiberglass Underground Fuel Storage Tank, Single Wall

System Components	QUANTITY	UNIT	COST EACH MAT.	COST EACH INST.	COST EACH TOTAL
SYSTEM G3060 310 1010					
STORAGE TANK, FUEL, UNDERGROUND, DOUBLE WALL FIBERGLASS, 550 GAL.		Ea.			
Excavating trench, no sheeting or dewatering, 6'-10 ' D, 3/4 CY hyd backhoe	18.000	Ea.		134.82	134.82
Corrosion resistance wrap & coat, small diam pipe, 1" diam, add	120.000	Ea.	200.40		200.40
Fiberglass-reinforced elbows 90 & 45 degree, 2 inch	7.000	Ea.	395.50	507.50	903
Fiberglass-reinforced threaded adapters, 2 inch	2.000	Ea.	37.90	166	203.90
Remote tank gauging system, 30', 5" pointer travel	1.000	Ea.	5,050	310	5,360
Corrosion resistance wrap & coat, small diam pipe. 4 " diam, add	30.000	Ea.	66.60		66.60
Pea gravel	9.000	Ea.	679.50	400.50	1,080
Reinforcing in hold down pad, #3 to #7	120.000	Ea.	52.80	68.40	121.20
Tank leak detector system, 8 channel, external monitoring	1.000	Ea.	885		885
Tubbing copper, type L, 3/4" diam	120.000	Ea.	505.20	984	1,489.20
Elbow 90 degree ,wrought cu x cu 3/4 " diam	4.000	Ea.	14.40	132	146.40
Pipe, black steel welded, sch 40, 3" diam	30.000	Ea.	376.50	825	1,201.50
Elbow 90 degree, steel welded butt joint, 3 " diam	3.000	Ea.	111	510.60	621.60
Fiberglass-reinforced plastic pipe, 2 inch	156.000	Ea.	873.60	639.60	1,513.20
Fiberglass-reinforced plastic pipe, 3 inch	36.000	Ea.	345.60	167.40	513
Fiberglass-reinforced elbows 90 & 45 degree, 3 inch	3.000	Ea.	207	291	498
Tank leak detection system, probes, well monitoring, liquid phase detection	2.000	Ea.	940		940
Double wall fiberglass annular space	1.000	Ea.	320		320
Flange, weld neck, 150 lb, 3 " pipe size	1.000	Ea.	38.50	85.11	123.61
Vent protector/breather, 2" dia.	1.000	Ea.	23	19.45	42.45
Corrosion resistance wrap & coat, small diam pipe, 2" diam, add	36.000	Ea.	66.24		66.24
Concrete hold down pad, 6' x 5' x 8" thick	30.000	Ea.	84.90	40.50	125.40
600 gallon capacity	1.000	Ea.	7,375	465	7,840
Tank, for hold-downs 500-4000 gal., add	1.000	Ea.	470	148	618
Foot valve, single poppet, 3/4" dia	1.000	Ea.	62.50	34.50	97
Fuel fill box, locking inner cover, 3" dia.	1.000	Ea.	130	124	254
Fuel oil specialties, valve,ball chk, globe type fusible, 3/4" dia	2.000	Ea.	147	62	209
TOTAL			19,458.14	6,115.38	25,573.52

G3060 310	Fiberglass Fuel Tank, Double Wall	COST EACH MAT.	COST EACH INST.	COST EACH TOTAL
1010	Storage tank, fuel, underground, double wall fiberglass, 550 Gal.	19,500	6,100	25,600
1020	2500 Gal.	29,600	8,450	38,050
1030	4000 Gal.	36,100	10,500	46,600
1040	6000 Gal.	37,100	11,800	48,900

G3060 Fuel Distribution

G3060 310	Fiberglass Fuel Tank, Double Wall	COST EACH		
		MAT.	INST.	TOTAL
1050	8000 Gal.	45,600	14,300	59,900
1060	10,000 Gal.	51,500	16,400	67,900
1070	15,000 Gal.	69,000	18,900	87,900
1080	20,000 Gal.	86,000	22,800	108,800
1090	25,000 Gal.	107,500	26,600	134,100
1100	30,000 Gal.	122,500	28,200	150,700

G3060 Fuel Distribution

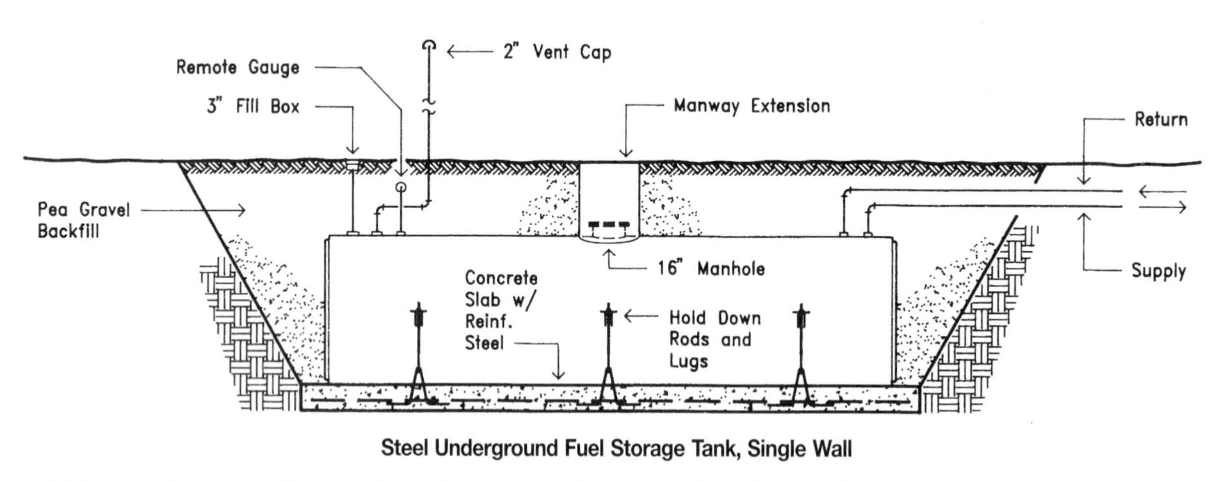

Steel Underground Fuel Storage Tank, Single Wall

System Components	QUANTITY	UNIT	COST EACH MAT.	INST.	TOTAL
SYSTEM G3060 320 1010					
STORAGE TANK, FUEL, UNDERGROUND, DOUBLE WALL STEEL, 550 GAL.		Ea.			
Fiberglass-reinforced plastic pipe, 2 inch	156.000	Ea.	873.60	639.60	1,513.20
Fiberglass-reinforced plastic pipe, 3 inch	36.000	Ea.	345.60	167.40	513
Fiberglass-reinforced elbows 90 & 45 degree, 3 inch	3.000	Ea.	207	291	498
Tank leak detection system, probes, well monitoring, liquid phase detection	2.000	Ea.	940		940
Flange, weld neck, 150 lb, 3" pipe size	1.000	Ea.	38.50	85.11	123.61
Vent protector/breather, 2" dia.	1.000	Ea.	23	19.45	42.45
Concrete hold down pad, 6′ x 5′ x 8" thick	30.000	Ea.	84.90	40.50	125.40
Tanks, for hold-downs, 500-2000 gal, add	1.000	Ea.	166	148	314
Double wall steel tank annular space	1.000	Ea.	330		330
Foot valve, single poppet, 3/4" dia	1.000	Ea.	62.50	34.50	97
Fuel fill box, locking inner cover, 3" dia.	1.000	Ea.	130	124	254
Fuel oil specialties, valve, ball chk, globe type fusible, 3/4" dia	2.000	Ea.	147	62	209
Pea gravel	9.000	Ea.	679.50	400.50	1,080
Reinforcing in hold down pad, #3 to #7	120.000	Ea.	52.80	68.40	121.20
Tank, for manways, add	1.000	Ea.	1,650		1,650
Tank leak detector system, 8 channel, external monitoring	1.000	Ea.	885		885
Tubbing copper, type L, 3/4" diam	120.000	Ea.	505.20	984	1,489.20
Elbow 90 degree, wrought cu x cu 3/4" diam	8.000	Ea.	28.80	264	292.80
Pipe, black steel welded, sch 40, 3" diam	30.000	Ea.	376.50	825	1,201.50
Elbow 90 degree, steel welded butt joint, 3" diam	3.000	Ea.	111	510.60	621.60
Excavating trench, no sheeting or dewatering, 6′-10′ D, 3/4 CY hyd backhoe	18.000	Ea.		134.82	134.82
Fiberglass-reinforced elbows 90 & 45 degree, 2 inch	7.000	Ea.	395.50	507.50	903
Fiberglass-reinforced threaded adapters, 2 inch	2.000	Ea.	37.90	166	203.90
500 gallon capacity	1.000	Ea.	4,300	465	4,765
Remote tank gauging system, 30′, 5" pointer travel	1.000	Ea.	5,050	310	5,360
TOTAL			17,420.30	6,247.38	23,667.68

G3060 320	Steel Fuel Tank, Double Wall	COST EACH MAT.	INST.	TOTAL
1010	Storage tank, fuel, underground, double wall steel, 500 Gal.	17,400	6,225	23,625
1020	2000 Gal.	24,100	8,175	32,275
1030	4000 Gal.	33,800	10,600	44,400
1040	6000 Gal.	38,500	11,700	50,200
1050	8000 Gal.	44,500	14,500	59,000
1060	10,000 Gal.	51,000	16,600	67,600

G3060 Fuel Distribution

G3060 320	Steel Fuel Tank, Double Wall	COST EACH		
		MAT.	INST.	TOTAL
1070	15,000 Gal.	61,500	19,300	80,800
1080	20,000 Gal.	80,500	23,200	103,700
1090	25,000 Gal.	107,000	26,500	133,500
1100	30,000 Gal.	122,500	28,300	150,800
1110	40,000 Gal.	153,500	35,700	189,200
1120	50,000 Gal.	185,000	42,700	227,700

G3060 325	Steel Tank, Above Ground, Single Wall	COST EACH		
		MAT.	INST.	TOTAL
1010	Storage tank, fuel, above ground, single wall steel, 550 Gal.	11,600	4,400	16,000
1020	2000 Gal.	14,500	4,500	19,000
1030	5000 Gal.	17,400	6,900	24,300
1040	10,000 Gal.	30,700	7,325	38,025
1050	15,000 Gal.	33,500	7,550	41,050
1060	20,000 Gal.	39,300	7,775	47,075
1070	25,000 Gal.	45,800	7,975	53,775
1080	30,000 Gal.	50,000	8,300	58,300

G3060 330	Steel Tank, Above Ground, Double Wall	COST EACH		
		MAT.	INST.	TOTAL
1010	Storage tank, fuel, above ground, double wall steel, 500 Gal.	13,000	4,450	17,450
1020	2000 Gal.	19,500	4,550	24,050
1030	4000 Gal.	26,900	4,650	31,550
1040	6000 Gal.	31,300	7,150	38,450
1050	8000 Gal.	37,100	7,325	44,425
1060	10,000 Gal.	40,100	7,475	47,575
1070	15,000 Gal.	54,500	7,725	62,225
1080	20,000 Gal.	60,000	7,975	67,975
1090	25,000 Gal.	70,000	8,200	78,200
1100	30,000 Gal.	76,000	8,525	84,525

Reference Section

All the reference information is in one section, making it easy to find what you need to know . . . and easy to use the book on a daily basis. This section is visually identified by a vertical gray bar on the page edges.

In this Reference Section, we've included Equipment Rental Costs, a listing of rental and operating costs; Crew Listings, a full listing of all crews and equipment, and their costs; Historical Cost Indexes for cost comparisons over time; City Cost Indexes and Location Factors for adjusting costs to the region you are in; Reference Tables, where you will find explanations, estimating information and procedures, or technical data; Change Orders, information on pricing changes to contract documents; Square Foot Costs that allow you to make a rough estimate for the overall cost of a project; and an explanation of all the Abbreviations in the book.

Table of Contents

Estimating Tips

- This section contains the average costs to rent and operate hundreds of pieces of construction equipment. This is useful information when estimating the time and material requirements of any particular operation in order to establish a unit or total cost. Equipment costs include not only rental, but also operating costs for equipment under normal use.

Rental Costs

- Equipment rental rates are obtained from industry sources throughout North America–contractors, suppliers, dealers, manufacturers, and distributors.
- Rental rates vary throughout the country, with larger cities generally having lower rates. Lease plans for new equipment are available for periods in excess of six months, with a percentage of payments applying toward purchase.
- Monthly rental rates vary from 2% to 5% of the purchase price of the equipment depending on the anticipated life of the equipment and its wearing parts.
- Weekly rental rates are about 1/3 the monthly rates, and daily rental rates are about 1/3 the weekly rate.

Operating Costs

- The operating costs include parts and labor for routine servicing, such as repair and replacement of pumps, filters and worn lines. Normal operating expendables, such as fuel, lubricants, tires and electricity (where applicable), are also included.
- Extraordinary operating expendables with highly variable wear patterns, such as diamond bits and blades, are excluded. These costs can be found as material costs in the Unit Price section.
- The hourly operating costs listed do not include the operator's wages.

Equipment Cost/Day

- Any power equipment required by a crew is shown in the Crew Listings with a daily cost.
- The daily cost of equipment needed by a crew is based on dividing the weekly rental rate by 5 (number of working days in the week), and then adding the hourly operating cost times 8 (the number of hours in a day). This "Equipment Cost/ Day" is shown in the far right column of the Equipment Rental pages.
- If equipment is needed for only one or two days, it is best to develop your own cost by including components for daily rent and hourly operating cost. This is important when the listed Crew for a task does not contain the equipment needed, such as a crane for lifting mechanical heating/cooling equipment up onto a roof.
- If the quantity of work is less than the crew's Daily Output shown for a Unit Price line item that includes a bare unit equipment cost, it is recommended to estimate one day's rental cost and operating cost for equipment shown in the Crew Listing for that line item.

Mobilization/ Demobilization

- The cost to move construction equipment from an equipment yard or rental company to the jobsite and back again is not included in equipment rental costs listed in the Reference section, nor in the bare equipment cost of any Unit Price line item, nor in any equipment costs shown in the Crew listings.
- Mobilization (to the site) and demobilization (from the site) costs can be found in the Unit Price section.
- If a piece of equipment is already at the jobsite, it is not appropriate to utilize mobil./ demob. costs again in an estimate.

01 54 33 | Equipment Rental

		01 54 33 | Equipment Rental	UNIT	HOURLY OPER. COST	RENT PER DAY	RENT PER WEEK	RENT PER MONTH	EQUIPMENT COST/DAY	
10	0010	**CONCRETE EQUIPMENT RENTAL** without operators R015433 -10							10
	0200	Bucket, concrete lightweight, 1/2 C.Y.	Ea.	.70	20	60	180	17.60	
	0300	1 C.Y.		.75	24.50	73	219	20.60	
	0400	1-1/2 C.Y.		.95	33	99	297	27.40	
	0500	2 C.Y.		1.05	40	120	360	32.40	
	0580	8 C.Y.		5.65	260	780	2,350	201.20	
	0600	Cart, concrete, self-propelled, operator walking, 10 C.F.		3.05	56.50	170	510	58.40	
	0700	Operator riding, 18 C.F.		5.00	90	270	810	94	
	0800	Conveyer for concrete, portable, gas, 16″ wide, 26′ long		11.95	123	370	1,100	169.60	
	0900	46′ long		12.30	150	450	1,350	188.40	
	1000	56′ long		12.45	158	475	1,425	194.60	
	1100	Core drill, electric, 2-1/2 H.P., 1″ to 8″ bit diameter		1.75	65.50	197	590	53.40	
	1150	11 H.P., 8″ to 18″ cores		5.60	115	345	1,025	113.80	
	1200	Finisher, concrete floor, gas, riding trowel, 96″ wide		12.35	145	435	1,300	185.80	
	1300	Gas, walk-behind, 3 blade, 36″ trowel		2.05	19.65	59	177	28.20	
	1400	4 blade, 48″ trowel		4.10	27	81	243	49	
	1500	Float, hand-operated (Bull float) 48″ wide		.08	13.65	41	123	8.85	
	1570	Curb builder, 14 H.P., gas, single screw		13.55	247	740	2,225	256.40	
	1590	Double screw		14.20	287	860	2,575	285.60	
	1600	Grinder, concrete and terrazzo, electric, floor		2.48	104	312	935	82.25	
	1700	Wall grinder		1.24	52	156	470	41.10	
	1800	Mixer, powered, mortar and concrete, gas, 6 C.F., 18 H.P.		8.20	118	355	1,075	136.60	
	1900	10 C.F., 25 H.P.		10.30	143	430	1,300	168.40	
	2000	16 C.F.		10.60	167	500	1,500	184.80	
	2100	Concrete, stationary, tilt drum, 2 C.Y.		6.30	232	695	2,075	189.40	
	2120	Pump, concrete, truck mounted 4″ line 80′ boom		23.35	920	2,760	8,275	738.80	
	2140	5″ line, 110′ boom		30.30	1,225	3,670	11,000	976.40	
	2160	Mud jack, 50 C.F. per hr		6.95	128	385	1,150	132.60	
	2180	225 C.F. per hr		9.00	147	440	1,325	160	
	2190	Shotcrete pump rig, 12 C.Y./hr		14.15	227	680	2,050	249.20	
	2200	35 C.Y./hr		16.75	243	730	2,200	280	
	2600	Saw, concrete, manual, gas, 18 H.P.		6.35	38.50	115	345	73.80	
	2650	Self-propelled, gas, 30 H.P.		12.80	98.50	295	885	161.40	
	2700	Vibrators, concrete, electric, 60 cycle, 2 H.P.		.45	7.35	22	66	8	
	2800	3 H.P.		.67	10.35	31	93	11.55	
	2900	Gas engine, 5 H.P.		1.95	14.35	43	129	24.20	
	3000	8 H.P.		2.65	15.35	46	138	30.40	
	3050	Vibrating screed, gas engine, 8 H.P.		2.72	72.50	217	650	65.15	
	3120	Concrete transit mixer, 6 x 4, 250 H.P., 8 C.Y., rear discharge		59.20	575	1,725	5,175	818.60	
	3200	Front discharge		69.95	710	2,125	6,375	984.60	
	3300	6 x 6, 285 H.P., 12 C.Y., rear discharge		69.00	665	2,000	6,000	952	
	3400	Front discharge	↓	72.20	715	2,150	6,450	1,008	
20	0010	**EARTHWORK EQUIPMENT RENTAL** without operators R015433 -10							20
	0040	Aggregate spreader, push type 8′ to 12′ wide	Ea.	2.90	25.50	77	231	38.60	
	0045	Tailgate type, 8′ wide		2.75	33.50	100	300	42	
	0055	Earth auger, truck-mounted, for fence & sign posts, utility poles		18.90	495	1,490	4,475	449.20	
	0060	For borings and monitoring wells		43.85	665	1,995	5,975	749.80	
	0070	Portable, trailer mounted		3.05	25.50	77	231	39.80	
	0075	Truck-mounted, for caissons, water wells		96.05	3,050	9,115	27,300	2,591	
	0080	Horizontal boring machine, 12″ to 36″ diameter, 45 H.P.		22.80	200	600	1,800	302.40	
	0090	12″ to 48″ diameter, 65 H.P.		32.65	355	1,060	3,175	473.20	
	0095	Auger, for fence posts, gas engine, hand held		.55	5.65	17	51	7.80	
	0100	Excavator, diesel hydraulic, crawler mounted, 1/2 C.Y. cap.		20.20	375	1,125	3,375	386.60	
	0120	5/8 C.Y. capacity		30.30	535	1,610	4,825	564.40	
	0140	3/4 C.Y. capacity		37.85	625	1,875	5,625	677.80	
	0150	1 C.Y. capacity		47.70	700	2,100	6,300	801.60	
	0200	1-1/2 C.Y. capacity		57.00	935	2,805	8,425	1,017	
	0300	2 C.Y. capacity	↓	76.90	1,175	3,530	10,600	1,321	

01 54 33 | Equipment Rental

		01 54 33 Equipment Rental	UNIT	HOURLY OPER. COST	RENT PER DAY	RENT PER WEEK	RENT PER MONTH	EQUIPMENT COST/DAY	
20	0320	2-1/2 C.Y. capacity	Ea.	104.45	1,575	4,740	14,200	1,784	20
	0325	3-1/2 C.Y. capacity		138.85	2,075	6,240	18,700	2,359	
	0330	4-1/2 C.Y. capacity		166.15	2,575	7,705	23,100	2,870	
	0335	6 C.Y. capacity		211.30	2,825	8,500	25,500	3,390	
	0340	7 C.Y. capacity		216.05	3,050	9,125	27,400	3,553	
	0342	Excavator attachments, bucket thumbs		2.95	238	715	2,150	166.60	
	0345	Grapples		2.65	207	620	1,850	145.20	
	0347	Hydraulic hammer for boom mounting, 5000 ft lbs		12.45	410	1,225	3,675	344.60	
	0349	11,000 ft lbs		21.05	735	2,200	6,600	608.40	
	0350	Gradall type, truck mounted, 3 ton @ 15' radius, 5/8 C.Y.		38.60	840	2,520	7,550	812.80	
	0370	1 C.Y. capacity		43.55	1,025	3,050	9,150	958.40	
	0400	Backhoe-loader, 40 to 45 H.P., 5/8 C.Y. capacity		12.80	193	580	1,750	218.40	
	0450	45 H.P. to 60 H.P., 3/4 C.Y. capacity		20.35	292	875	2,625	337.80	
	0460	80 H.P., 1-1/4 C.Y. capacity		24.45	320	960	2,875	387.60	
	0470	112 H.P., 1-1/2 C.Y. capacity		38.55	525	1,570	4,700	622.40	
	0482	Backhoe-loader attachment, compactor, 20,000 lb.		5.25	142	425	1,275	127	
	0485	Hydraulic hammer, 750 ft lbs		2.95	91.50	275	825	78.60	
	0486	Hydraulic hammer, 1200 ft lbs		5.40	185	555	1,675	154.20	
	0500	Brush chipper, gas engine, 6" cutter head, 35 H.P.		10.65	108	325	975	150.20	
	0550	12" cutter head, 130 H.P.		18.15	180	540	1,625	253.20	
	0600	15" cutter head, 165 H.P.		27.75	217	650	1,950	352	
	0750	Bucket, clamshell, general purpose, 3/8 C.Y.		1.15	36.50	110	330	31.20	
	0800	1/2 C.Y.		1.25	45	135	405	37	
	0850	3/4 C.Y.		1.40	55	165	495	44.20	
	0900	1 C.Y.		1.45	58.50	175	525	46.60	
	0950	1-1/2 C.Y.		2.35	80	240	720	66.80	
	1000	2 C.Y.		2.45	90	270	810	73.60	
	1010	Bucket, dragline, medium duty, 1/2 C.Y.		.65	23.50	71	213	19.40	
	1020	3/4 C.Y.		.70	24.50	74	222	20.40	
	1030	1 C.Y.		.70	26.50	80	240	21.60	
	1040	1-1/2 C.Y.		1.10	40	120	360	32.80	
	1050	2 C.Y.		1.20	45	135	405	36.60	
	1070	3 C.Y.		1.80	63.50	190	570	52.40	
	1200	Compactor, manually guided 2-drum vibratory smooth roller, 7.5 H.P.		6.20	167	500	1,500	149.60	
	1250	Rammer/tamper, gas, 8"		2.60	43.50	130	390	46.80	
	1260	15"		2.85	48.50	145	435	51.80	
	1300	Vibratory plate, gas, 18" plate, 3000 lb. blow		2.55	23	69	207	34.20	
	1350	21" plate, 5000 lb. blow		3.20	30	90	270	43.60	
	1370	Curb builder/extruder, 14 H.P., gas, single screw		13.55	247	740	2,225	256.40	
	1390	Double screw		14.20	287	860	2,575	285.60	
	1500	Disc harrow attachment, for tractor		.40	67	201	605	43.40	
	1810	Feller buncher, shearing & accumulating trees, 100 H.P.		34.20	560	1,680	5,050	609.60	
	1860	Grader, self-propelled, 25,000 lb.		31.75	475	1,420	4,250	538	
	1910	30,000 lb.		36.65	550	1,645	4,925	622.20	
	1920	40,000 lb.		55.90	1,050	3,120	9,350	1,071	
	1930	55,000 lb.		72.55	1,375	4,155	12,500	1,411	
	1950	Hammer, pavement breaker, self-propelled, diesel, 1000 to 1250 lb.		28.25	345	1,040	3,125	434	
	2000	1300 to 1500 lb.		42.38	695	2,080	6,250	755.05	
	2050	Pile driving hammer, steam or air, 4150 ft lbs @ 225 bpm		9.65	470	1,405	4,225	358.20	
	2100	8750 ft lbs @ 145 bpm		11.65	650	1,945	5,825	482.20	
	2150	15,000 ft lbs @ 60 bpm		13.20	785	2,360	7,075	577.60	
	2200	24,450 ft lbs @ 111 bpm		14.25	880	2,635	7,900	641	
	2250	Leads, 60' high for pile driving hammers up to 20,000 ft lbs		3.00	78	234	700	70.80	
	2300	90' high for hammers over 20,000 ft lbs		4.50	141	422	1,275	120.40	
	2350	Diesel type hammer, 22,400 ft lbs		29.80	635	1,905	5,725	619.40	
	2400	41,300 ft lbs		38.60	690	2,070	6,200	722.80	
	2450	141,000 ft lbs		56.40	1,175	3,540	10,600	1,159	
	2500	Vib. elec. hammer/extractor, 200 kW diesel generator, 34 H.P.	↓	53.70	660	1,980	5,950	825.60	

01 54 33 | Equipment Rental

			UNIT	HOURLY OPER. COST	RENT PER DAY	RENT PER WEEK	RENT PER MONTH	EQUIPMENT COST/DAY	
20	2550	80 H.P.	Ea.	99.00	970	2,910	8,725	1,374	20
	2600	150 H.P.		188.90	1,875	5,650	17,000	2,641	
	2700	Extractor, steam or air, 700 ft lbs		17.50	510	1,525	4,575	445	
	2750	1000 ft lbs		19.95	625	1,870	5,600	533.60	
	2800	Log chipper, up to 22" diameter, 600 H.P.		52.70	445	1,340	4,025	689.60	
	2850	Logger, for skidding & stacking logs, 150 H.P.		50.90	835	2,500	7,500	907.20	
	2860	Mulcher, diesel powered, trailer mounted		23.50	218	655	1,975	319	
	2900	Rake, spring tooth, with tractor		14.34	285	855	2,575	285.70	
	3000	Roller, vibratory, tandem, smooth drum, 20 H.P.		8.25	143	430	1,300	152	
	3050	35 H.P.		10.40	253	760	2,275	235.20	
	3100	Towed type vibratory compactor, smooth drum, 50 H.P.		25.90	380	1,135	3,400	434.20	
	3150	Sheepsfoot, 50 H.P.		28.15	440	1,320	3,950	489.20	
	3170	Landfill compactor, 220 H.P.		82.85	1,475	4,390	13,200	1,541	
	3200	Pneumatic tire roller, 80 H.P.		16.15	345	1,040	3,125	337.20	
	3250	120 H.P.		24.20	610	1,835	5,500	560.60	
	3300	Sheepsfoot vibratory roller, 240 H.P.		69.85	1,150	3,420	10,300	1,243	
	3320	340 H.P.		94.00	1,400	4,220	12,700	1,596	
	3350	Smooth drum vibratory roller, 75 H.P.		23.20	550	1,655	4,975	516.60	
	3400	125 H.P.		29.90	655	1,970	5,900	633.20	
	3410	Rotary mower, brush, 60", with tractor		18.25	260	780	2,350	302	
	3420	Rototiller, walk-behind, gas, 5 H.P.		2.12	65.50	196	590	56.15	
	3422	8 H.P.		3.30	100	300	900	86.40	
	3440	Scrapers, towed type, 7 C.Y. capacity		5.15	107	320	960	105.20	
	3450	10 C.Y. capacity		6.45	175	525	1,575	156.60	
	3500	15 C.Y. capacity		7.15	192	575	1,725	172.20	
	3525	Self-propelled, single engine, 14 C.Y. capacity		103.95	1,475	4,425	13,300	1,717	
	3550	Dual engine, 21 C.Y. capacity		153.65	1,700	5,065	15,200	2,242	
	3600	31 C.Y. capacity		208.65	2,575	7,750	23,300	3,219	
	3640	44 C.Y. capacity		264.30	3,625	10,910	32,700	4,296	
	3650	Elevating type, single engine, 11 C.Y. capacity		66.70	1,075	3,255	9,775	1,185	
	3700	22 C.Y. capacity		126.05	2,125	6,375	19,100	2,283	
	3710	Screening plant 110 H.P. w/5' x 10' screen		36.90	415	1,240	3,725	543.20	
	3720	5' x 16' screen		39.05	515	1,550	4,650	622.40	
	3850	Shovel, crawler-mounted, front-loading, 7 C.Y. capacity		228.10	2,850	8,555	25,700	3,536	
	3855	12 C.Y. capacity		305.30	3,625	10,860	32,600	4,614	
	3860	Shovel/backhoe bucket, 1/2 C.Y.		2.20	63.50	190	570	55.60	
	3870	3/4 C.Y.		2.25	70	210	630	60	
	3880	1 C.Y.		2.35	78.50	235	705	65.80	
	3890	1-1/2 C.Y.		2.50	93.50	280	840	76	
	3910	3 C.Y.		2.80	127	380	1,150	98.40	
	3950	Stump chipper, 18" deep, 30 H.P.		7.56	217	652	1,950	190.90	
	4110	Tractor, crawler, with bulldozer, torque converter, diesel 80 H.P.		26.45	375	1,125	3,375	436.60	
	4150	105 H.P.		39.30	575	1,730	5,200	660.40	
	4200	140 H.P.		42.45	715	2,150	6,450	769.60	
	4260	200 H.P.		68.60	1,075	3,215	9,650	1,192	
	4310	300 H.P.		82.50	1,550	4,660	14,000	1,592	
	4360	410 H.P.		109.70	2,000	6,005	18,000	2,079	
	4370	500 H.P.		147.85	2,700	8,135	24,400	2,810	
	4380	700 H.P.		245.45	4,400	13,170	39,500	4,598	
	4400	Loader, crawler, torque conv., diesel, 1-1/2 C.Y., 80 H.P.		24.10	400	1,205	3,625	433.80	
	4450	1-1/2 to 1-3/4 C.Y., 95 H.P.		28.20	515	1,545	4,625	534.60	
	4510	1-3/4 to 2-1/4 C.Y., 130 H.P.		42.30	835	2,505	7,525	839.40	
	4530	2-1/2 to 3-1/4 C.Y., 190 H.P.		56.50	1,050	3,125	9,375	1,077	
	4560	3-1/2 to 5 C.Y., 275 H.P.		81.25	1,500	4,490	13,500	1,548	
	4610	Front end loader, 4WD, articulated frame, 1 to 1-1/4 C.Y., 70 H.P.		18.05	208	625	1,875	269.40	
	4620	1-1/2 to 1-3/4 C.Y., 95 H.P.		23.95	293	880	2,650	367.60	
	4650	1-3/4 to 2 C.Y., 130 H.P.		27.65	345	1,035	3,100	428.20	
	4710	2-1/2 to 3-1/2 C.Y., 145 H.P.	↓	26.35	375	1,125	3,375	435.80	

01 54 33 | Equipment Rental

		01 54 33 Equipment Rental	UNIT	HOURLY OPER. COST	RENT PER DAY	RENT PER WEEK	RENT PER MONTH	EQUIPMENT COST/DAY
20	4730	3 to 4-1/2 C.Y., 185 H.P.	Ea.	35.85	570	1,705	5,125	627.80
	4760	5-1/4 to 5-3/4 C.Y., 270 H.P.		56.30	905	2,720	8,150	994.40
	4810	7 to 9 C.Y., 475 H.P.		97.15	1,625	4,900	14,700	1,757
	4870	9 - 11 C.Y., 620 H.P.		129.55	2,600	7,825	23,500	2,601
	4880	Skid steer loader, wheeled, 10 C.F., 30 H.P. gas		9.30	145	435	1,300	161.40
	4890	1 C.Y., 78 H.P., diesel		19.00	232	695	2,075	291
	4892	Skid-steer attachment, auger		.55	91.50	275	825	59.40
	4893	Backhoe		.67	112	336	1,000	72.55
	4894	Broom		.73	121	364	1,100	78.65
	4895	Forks		.24	40	120	360	25.90
	4896	Grapple		.57	95.50	286	860	61.75
	4897	Concrete hammer		1.06	177	532	1,600	114.90
	4898	Tree spade		.82	136	408	1,225	88.15
	4899	Trencher		.79	131	393	1,175	84.90
	4900	Trencher, chain, boom type, gas, operator walking, 12 H.P.		4.75	46.50	140	420	66
	4910	Operator riding, 40 H.P.		19.55	297	890	2,675	334.40
	5000	Wheel type, diesel, 4' deep, 12" wide		82.05	825	2,480	7,450	1,152
	5100	6' deep, 20" wide		86.95	1,900	5,735	17,200	1,843
	5150	Chain type, diesel, 5' deep, 8" wide		36.25	580	1,745	5,225	639
	5200	Diesel, 8' deep, 16" wide		88.95	2,300	6,890	20,700	2,090
	5202	Rock trencher, wheel type, 6" wide x 18" deep		19.55	315	940	2,825	344.40
	5206	Chain type, 18" wide x 7' deep		106.25	2,875	8,605	25,800	2,571
	5210	Tree spade, self-propelled		13.10	267	800	2,400	264.80
	5250	Truck, dump, 2-axle, 12 ton, 8 C.Y. payload, 220 H.P.		33.35	225	675	2,025	401.80
	5300	Three axle dump, 16 ton, 12 C.Y. payload, 400 H.P.		59.35	320	965	2,900	667.80
	5310	Four axle dump, 25 ton, 18 C.Y. payload, 450 H.P.		69.55	470	1,410	4,225	838.40
	5350	Dump trailer only, rear dump, 16-1/2 C.Y.		4.90	135	405	1,225	120.20
	5400	20 C.Y.		5.35	153	460	1,375	134.80
	5450	Flatbed, single axle, 1-1/2 ton rating		24.85	68.50	205	615	239.80
	5500	3 ton rating		29.75	96.50	290	870	296
	5550	Off highway rear dump, 25 ton capacity		74.80	1,325	4,005	12,000	1,399
	5600	35 ton capacity		70.50	1,175	3,515	10,500	1,267
	5610	50 ton capacity		95.65	1,625	4,895	14,700	1,744
	5620	65 ton capacity		99.70	1,625	4,860	14,600	1,770
	5630	100 ton capacity		128.25	2,125	6,350	19,100	2,296
	6000	Vibratory plow, 25 H.P., walking		8.15	61.50	185	555	102.20
40	0010	**GENERAL EQUIPMENT RENTAL** without operators (R015433-10)						
	0150	Aerial lift, scissor type, to 15' high, 1000 lb. cap., electric	Ea.	2.80	55	165	495	55.40
	0160	To 25' high, 2000 lb. capacity		3.20	70	210	630	67.60
	0170	Telescoping boom to 40' high, 500 lb. capacity, gas		19.30	320	965	2,900	347.40
	0180	To 45' high, 500 lb. capacity		20.35	370	1,115	3,350	385.80
	0190	To 60' high, 600 lb. capacity		22.55	490	1,465	4,400	473.40
	0195	Air compressor, portable, 6.5 cfm, electric		.50	13	39	117	11.80
	0196	Gasoline		.75	19.65	59	177	17.80
	0200	Towed type, gas engine, 60 cfm		13.70	48.50	145	435	138.60
	0300	160 cfm		16.00	50	150	450	158
	0400	Diesel engine, rotary screw, 250 cfm		16.30	105	315	945	193.40
	0500	365 cfm		22.20	130	390	1,175	255.60
	0550	450 cfm		28.25	163	490	1,475	324
	0600	600 cfm		49.20	228	685	2,050	530.60
	0700	750 cfm		50.10	237	710	2,125	542.80
	0800	For silenced models, small sizes, add to rent		3%	5%	5%	5%	
	0900	Large sizes, add to rent		5%	7%	7%	7%	
	0930	Air tools, breaker, pavement, 60 lb.	Ea.	.45	9.65	29	87	9.40
	0940	80 lb.		.45	10.35	31	93	9.80
	0950	Drills, hand (jackhammer) 65 lb.		.55	17.35	52	156	14.80
	0960	Track or wagon, swing boom, 4" drifter		57.35	765	2,295	6,875	917.80
	0970	5" drifter		71.65	825	2,480	7,450	1,069

01 54 33 | Equipment Rental

			UNIT	HOURLY OPER. COST	RENT PER DAY	RENT PER WEEK	RENT PER MONTH	EQUIPMENT COST/DAY
40	0975	Track mounted quarry drill, 6" diameter drill	Ea.	96.10	1,275	3,790	11,400	1,527
	0980	Dust control per drill		1.04	19	57	171	19.70
	0990	Hammer, chipping, 12 lb.		.50	23.50	70	210	18
	1000	Hose, air with couplings, 50' long, 3/4" diameter		.03	4.33	13	39	2.85
	1100	1" diameter		.04	6.35	19	57	4.10
	1200	1-1/2" diameter		.06	9.65	29	87	6.30
	1300	2" diameter		.11	19	57	171	12.30
	1400	2-1/2" diameter		.12	20	60	180	12.95
	1410	3" diameter		.15	25	75	225	16.20
	1450	Drill, steel, 7/8" x 2'		.05	8	24	72	5.20
	1460	7/8" x 6'		.05	9	27	81	5.80
	1520	Moil points		.02	4	12	36	2.55
	1525	Pneumatic nailer w/accessories		.46	30.50	91	273	21.90
	1530	Sheeting driver for 60 lb. breaker		.04	6	18	54	3.90
	1540	For 90 lb. breaker		.12	8	24	72	5.75
	1550	Spade, 25 lb.		.40	6	18	54	6.80
	1560	Tamper, single, 35 lb.		.53	35	105	315	25.25
	1570	Triple, 140 lb.		.79	52.50	158	475	37.90
	1580	Wrenches, impact, air powered, up to 3/4" bolt		.40	9	27	81	8.60
	1590	Up to 1-1/4" bolt		.50	23	69	207	17.80
	1600	Barricades, barrels, reflectorized, 1 to 50 barrels		.03	4.60	13.80	41.50	3
	1610	100 to 200 barrels		.02	3.53	10.60	32	2.30
	1620	Barrels with flashers, 1 to 50 barrels		.03	5.25	15.80	47.50	3.40
	1630	100 to 200 barrels		.03	4.20	12.60	38	2.75
	1640	Barrels with steady burn type C lights		.04	7	21	63	4.50
	1650	Illuminated board, trailer mounted, with generator		.75	118	355	1,075	77
	1670	Portable barricade, stock, with flashers, 1 to 6 units		.03	5.25	15.80	47.50	3.40
	1680	25 to 50 units		.03	4.90	14.70	44	3.20
	1690	Butt fusion machine, electric		12.80	53.50	160	480	134.40
	1695	Electro fusion machine		9.45	38.50	115	345	98.60
	1700	Carts, brick, hand powered, 1000 lb. capacity		.36	60.50	182	545	39.30
	1800	Gas engine, 1500 lb., 7-1/2' lift		4.01	120	359	1,075	103.90
	1822	Dehumidifier, medium, 6 lb/hr, 150 cfm		.98	60	180	540	43.85
	1824	Large, 18 lb/hr, 600 cfm		1.98	122	366	1,100	89.05
	1830	Distributor, asphalt, trailer mounted, 2000 gal., 38 H.P. diesel		9.40	340	1,015	3,050	278.20
	1840	3000 gal., 38 H.P. diesel		10.75	370	1,105	3,325	307
	1850	Drill, rotary hammer, electric		.88	25	75	225	22.05
	1860	Carbide bit, 1-1/2" diameter, add to electric rotary hammer		.04	7.60	22.75	68.50	4.85
	1865	Rotary, crawler, 250 H.P.		141.75	2,075	6,245	18,700	2,383
	1870	Emulsion sprayer, 65 gal., 5 H.P. gas engine		2.83	99	297	890	82.05
	1880	200 gal., 5 H.P. engine		7.45	165	495	1,475	158.60
	1930	Floodlight, mercury vapor, or quartz, on tripod, 1000 watt		.36	14	42	126	11.30
	1940	2000 watt		.63	26	78	234	20.65
	1950	Floodlights, trailer mounted with generator, 1 - 300 watt light		3.35	71.50	215	645	69.80
	1960	2 - 1000 watt lights		4.45	90	270	810	89.60
	2000	4 - 300 watt lights		4.20	88.50	265	795	86.60
	2005	Foam spray rig, incl. box trailer, compressor, generator, proportioner		32.33	490	1,465	4,400	551.65
	2020	Forklift, straight mast, 12' lift, 5000 lb., 2 wheel drive, gas		25.20	200	600	1,800	321.60
	2040	21' lift, 5000 lb., 4 wheel drive, diesel		20.20	247	740	2,225	309.60
	2050	For rough terrain, 42' lift, 35' reach, 9000 lb., 110 H.P.		26.40	445	1,335	4,000	478.20
	2060	For plant, 4 ton capacity, 80 H.P., 2 wheel drive, gas		15.50	102	305	915	185
	2080	10 ton capacity, 120 H.P., 2 wheel drive, diesel		23.40	178	535	1,600	294.20
	2100	Generator, electric, gas engine, 1.5 kW to 3 kW		3.55	11.35	34	102	35.20
	2200	5 kW		4.60	15.35	46	138	46
	2300	10 kW		8.70	36.50	110	330	91.60
	2400	25 kW		10.20	75	225	675	126.60
	2500	Diesel engine, 20 kW		11.50	75	225	675	137
	2600	50 kW	↓	21.55	103	310	930	234.40

01 54 33 | Equipment Rental

			UNIT	HOURLY OPER. COST	RENT PER DAY	RENT PER WEEK	RENT PER MONTH	EQUIPMENT COST/DAY	
40	2700	100 kW	Ea.	42.45	132	395	1,175	418.60	40
	2800	250 kW		82.90	233	700	2,100	803.20	
	2850	Hammer, hydraulic, for mounting on boom, to 500 ft lbs		2.25	68.50	205	615	59	
	2860	1000 ft lbs		3.80	112	335	1,000	97.40	
	2900	Heaters, space, oil or electric, 50 MBH		1.69	7	21	63	17.70	
	3000	100 MBH		3.06	9.35	28	84	30.10	
	3100	300 MBH		9.79	31.50	94	282	97.10	
	3150	500 MBH		19.75	41.50	125	375	183	
	3200	Hose, water, suction with coupling, 20' long, 2" diameter		.02	3	9	27	1.95	
	3210	3" diameter		.03	4.67	14	42	3.05	
	3220	4" diameter		.03	5.35	16	48	3.45	
	3230	6" diameter		.11	18.35	55	165	11.90	
	3240	8" diameter		.20	33.50	100	300	21.60	
	3250	Discharge hose with coupling, 50' long, 2" diameter		.01	1.33	4	12	.90	
	3260	3" diameter		.02	2.67	8	24	1.75	
	3270	4" diameter		.02	3.67	11	33	2.35	
	3280	6" diameter		.06	9.65	29	87	6.30	
	3290	8" diameter		.20	33.50	100	300	21.60	
	3295	Insulation blower		.22	6	18	54	5.35	
	3300	Ladders, extension type, 16' to 36' long		.14	23	69	207	14.90	
	3400	40' to 60' long		.20	33.50	100	300	21.60	
	3405	Lance for cutting concrete		3.07	104	313	940	87.15	
	3407	Lawn mower, rotary, 22", 5 H.P.		2.12	65	195	585	55.95	
	3408	48" self propelled		3.28	114	341	1,025	94.45	
	3410	Level, laser type, for pipe and sewer leveling		1.46	97	291	875	69.90	
	3430	Electronic		.75	50	150	450	36	
	3440	Laser type, rotating beam for grade control		1.17	78	234	700	56.15	
	3460	Builders level with tripod and rod		.09	14.65	44	132	9.50	
	3500	Light towers, towable, with diesel generator, 2000 watt		4.20	88.50	265	795	86.60	
	3600	4000 watt		4.45	98.50	295	885	94.60	
	3700	Mixer, powered, plaster and mortar, 6 C.F., 7 H.P.		2.55	18.65	56	168	31.60	
	3800	10 C.F., 9 H.P.		2.70	30	90	270	39.60	
	3850	Nailer, pneumatic		.46	30.50	91	273	21.90	
	3900	Paint sprayers complete, 8 cfm		.69	45.50	137	410	32.90	
	4000	17 cfm		1.28	85	255	765	61.25	
	4020	Pavers, bituminous, rubber tires, 8' wide, 50 H.P., diesel		39.45	1,050	3,165	9,500	948.60	
	4030	10' wide, 150 H.P.		85.85	1,700	5,070	15,200	1,701	
	4050	Crawler, 8' wide, 100 H.P., diesel		83.90	1,800	5,420	16,300	1,755	
	4060	10' wide, 150 H.P.		96.80	2,125	6,365	19,100	2,047	
	4070	Concrete paver, 12' to 24' wide, 250 H.P.		96.25	1,575	4,750	14,300	1,720	
	4080	Placer-spreader-trimmer, 24' wide, 300 H.P.		140.15	2,600	7,785	23,400	2,678	
	4100	Pump, centrifugal gas pump, 1-1/2" diam., 65 GPM		3.60	45	135	405	55.80	
	4200	2" diameter, 130 GPM		5.00	51.50	155	465	71	
	4300	3" diameter, 250 GPM		5.30	53.50	160	480	74.40	
	4400	6" diameter, 1500 GPM		29.25	177	530	1,600	340	
	4500	Submersible electric pump, 1-1/4" diameter, 55 GPM		.41	16	48	144	12.90	
	4600	1-1/2" diameter, 83 GPM		.44	18.35	55	165	14.50	
	4700	2" diameter, 120 GPM		1.25	23	69	207	23.80	
	4800	3" diameter, 300 GPM		2.15	41.50	125	375	42.20	
	4900	4" diameter, 560 GPM		9.55	163	490	1,475	174.40	
	5000	6" diameter, 1590 GPM		14.05	220	660	1,975	244.40	
	5100	Diaphragm pump, gas, single, 1-1/2" diameter		1.10	48	144	430	37.60	
	5200	2" diameter		4.00	60	180	540	68	
	5300	3" diameter		4.00	60	180	540	68	
	5400	Double, 4" diameter		5.35	80	240	720	90.80	
	5450	Pressure Washer 5 GPM, 3000 psi		4.60	40	120	360	60.80	
	5500	Trash pump, self-priming, gas, 2" diameter		4.35	20.50	62	186	47.20	
	5600	Diesel, 4" diameter		11.90	58.50	175	525	130.20	

01 54 33 | Equipment Rental

			UNIT	HOURLY OPER. COST	RENT PER DAY	RENT PER WEEK	RENT PER MONTH	EQUIPMENT COST/DAY	
40	5650	Diesel, 6″ diameter	Ea.	40.15	130	390	1,175	399.20	40
	5655	Grout Pump		18.10	95	285	855	201.80	
	5700	Salamanders, L.P. gas fired, 100,000 Btu		3.10	16	48	144	34.40	
	5705	50,000 Btu		2.31	9.35	28	84	24.10	
	5720	Sandblaster, portable, open top, 3 C.F. capacity		.55	27	81	243	20.60	
	5730	6 C.F. capacity		.85	40	120	360	30.80	
	5740	Accessories for above		.12	20	60	180	12.95	
	5750	Sander, floor		.85	19	57	171	18.20	
	5760	Edger		.81	28	84	252	23.30	
	5800	Saw, chain, gas engine, 18″ long		2.15	18.35	55	165	28.20	
	5900	Hydraulic powered, 36″ long		.65	56.50	170	510	39.20	
	5950	60″ long		.65	56.50	170	510	39.20	
	6000	Masonry, table mounted, 14″ diameter, 5 H.P.		1.34	56.50	170	510	44.70	
	6050	Portable cut-off, 8 H.P.		2.40	32.50	98	294	38.80	
	6100	Circular, hand held, electric, 7-1/4″ diameter		.21	4.67	14	42	4.50	
	6200	12″ diameter		.27	7.65	23	69	6.75	
	6250	Wall saw, w/hydraulic power, 10 H.P		9.30	60	180	540	110.40	
	6275	Shot blaster, walk-behind, 20″ wide		4.60	298	895	2,675	215.80	
	6280	Sidewalk broom, walk-behind		2.05	61.50	185	555	53.40	
	6300	Steam cleaner, 100 gallons per hour		3.50	53.50	161	485	60.20	
	6310	200 gallons per hour		5.05	63.50	190	570	78.40	
	6340	Tar Kettle/Pot, 400 gallons		4.68	80	240	720	85.45	
	6350	Torch, cutting, acetylene-oxygen, 150′ hose		.50	25	75	225	19	
	6360	Hourly operating cost includes tips and gas		9.90				79.20	
	6410	Toilet, portable chemical		.12	19.35	58	174	12.55	
	6420	Recycle flush type		.14	23.50	70	210	15.10	
	6430	Toilet, fresh water flush, garden hose,		.16	27	81	243	17.50	
	6440	Hoisted, non-flush, for high rise		.14	23	69	207	14.90	
	6450	Toilet, trailers, minimum		.24	40	120	360	25.90	
	6460	Maximum		.73	122	365	1,100	78.85	
	6465	Tractor, farm with attachment		17.10	267	800	2,400	296.80	
	6500	Trailers, platform, flush deck, 2 axle, 25 ton capacity		4.95	110	330	990	105.60	
	6600	40 ton capacity		6.40	153	460	1,375	143.20	
	6700	3 axle, 50 ton capacity		6.95	170	510	1,525	157.60	
	6800	75 ton capacity		8.65	223	670	2,000	203.20	
	6810	Trailer mounted cable reel for high voltage line work		4.94	235	706	2,125	180.70	
	6820	Trailer mounted cable tensioning rig		9.80	465	1,400	4,200	358.40	
	6830	Cable pulling rig		65.52	2,625	7,860	23,600	2,096	
	6900	Water tank trailer, engine driven discharge, 5000 gallons		6.45	147	440	1,325	139.60	
	6925	10,000 gallons		8.80	203	610	1,825	192.40	
	6950	Water truck, off highway, 6000 gallons		84.65	805	2,420	7,250	1,161	
	7010	Tram car for high voltage line work, powered, 2 conductor		6.81	128	383	1,150	131.10	
	7020	Transit (builder's level) with tripod		.09	14.65	44	132	9.50	
	7030	Trench box, 3000 lbs. 6′ x 8′		.56	93	279	835	60.30	
	7040	7200 lbs. 6′ x 20′		1.05	175	525	1,575	113.40	
	7050	8000 lbs., 8′ x 16′		.95	158	475	1,425	102.60	
	7060	9500 lbs., 8′ x 20′		1.19	199	596	1,800	128.70	
	7065	11,000 lbs., 8′ x 24′		1.27	212	637	1,900	137.55	
	7070	12,000 lbs., 10′ x 20′		1.71	285	855	2,575	184.70	
	7100	Truck, pickup, 3/4 ton, 2 wheel drive		13.25	56.50	170	510	140	
	7200	4 wheel drive		13.50	71.50	215	645	151	
	7250	Crew carrier, 9 passenger		18.80	86.50	260	780	202.40	
	7290	Flat bed truck, 20,000 lbs. GVW		21.20	127	380	1,150	245.60	
	7300	Tractor, 4 x 2, 220 H.P.		29.50	197	590	1,775	354	
	7410	330 H.P.		43.80	270	810	2,425	512.40	
	7500	6 x 4, 380 H.P.		50.30	315	950	2,850	592.40	
	7600	450 H.P.		60.85	380	1,145	3,425	715.80	
	7620	Vacuum truck, hazardous material, 2500 gallons		11.15	305	920	2,750	273.20	

01 54 33 | Equipment Rental

		01 54 33 | Equipment Rental		UNIT	HOURLY OPER. COST	RENT PER DAY	RENT PER WEEK	RENT PER MONTH	EQUIPMENT COST/DAY	
40	7625	5,000 gallons		Ea.	13.23	430	1,290	3,875	363.85	40
	7640	Tractor, with A frame, boom and winch, 225 H.P.			32.15	273	820	2,450	421.20	
	7650	Vacuum, HEPA, 16 gallon, wet/dry			.84	18	54	162	17.50	
	7655	55 gallon, wet/dry			.86	27	81	243	23.10	
	7660	Water tank, portable			.17	28.50	85.50	257	18.45	
	7690	Sewer/catch basin vacuum, 14 C.Y., 1500 gallons			17.78	645	1,940	5,825	530.25	
	7700	Welder, electric, 200 amp			3.86	17.35	52	156	41.30	
	7800	300 amp			5.72	21.50	64	192	58.55	
	7900	Gas engine, 200 amp			14.15	25.50	76	228	128.40	
	8000	300 amp			16.20	27	81	243	145.80	
	8100	Wheelbarrow, any size			.07	11.65	35	105	7.55	
	8200	Wrecking ball, 4000 lbs.		↓	2.15	71.50	215	645	60.20	
50	0010	**HIGHWAY EQUIPMENT RENTAL** without operators	R015433 -10							50
	0050	Asphalt batch plant, portable drum mixer, 100 ton/hr		Ea.	66.15	1,450	4,355	13,100	1,400	
	0060	200 ton/hr			73.90	1,525	4,600	13,800	1,511	
	0070	300 ton/hr			85.75	1,800	5,425	16,300	1,771	
	0100	Backhoe attachment, long stick, up to 185 H.P., 10.5' long			.33	21.50	65	195	15.65	
	0140	Up to 250 H.P., 12' long			.35	23.50	70	210	16.80	
	0180	Over 250 H.P., 15' long			.48	31.50	95	285	22.85	
	0200	Special dipper arm, up to 100 H.P., 32' long			.98	65	195	585	46.85	
	0240	Over 100 H.P., 33' long			1.22	81	243	730	58.35	
	0280	Catch basin/sewer cleaning truck, 3 ton, 9 C.Y., 1000 gal			40.90	405	1,210	3,625	569.20	
	0300	Concrete batch plant, portable, electric, 200 C.Y./hr			17.80	515	1,540	4,625	450.40	
	0520	Grader/dozer attachment, ripper/scarifier, rear mounted, up to 135 H.P.			3.20	65	195	585	64.60	
	0540	Up to 180 H.P.			3.75	83.50	250	750	80	
	0580	Up to 250 H.P.			4.15	95	285	855	90.20	
	0700	Pvmt. removal bucket, for hyd. excavator, up to 90 H.P.			1.70	50	150	450	43.60	
	0740	Up to 200 H.P.			1.90	71.50	215	645	58.20	
	0780	Over 200 H.P.			2.05	83.50	250	750	66.40	
	0900	Aggregate spreader, self-propelled, 187 H.P.			50.35	695	2,090	6,275	820.80	
	1000	Chemical spreader, 3 C.Y.			3.05	43.50	130	390	50.40	
	1900	Hammermill, traveling, 250 H.P.			63.25	1,875	5,650	17,000	1,636	
	2000	Horizontal borer, 3" diameter, 13 H.P. gas driven			6.00	56.50	170	510	82	
	2150	Horizontal directional drill, 20,000 lb. thrust, 78 H.P. diesel			29.70	690	2,070	6,200	651.60	
	2160	30,000 lb. thrust, 115 H.P.			37.15	1,050	3,155	9,475	928.20	
	2170	50,000 lb. thrust, 170 H.P.			53.00	1,350	4,040	12,100	1,232	
	2190	Mud trailer for HDD, 1500 gallons, 175 H.P., gas			30.75	153	460	1,375	338	
	2200	Hydromulchers, gas power, 3000 gallons, for truck mounting			22.90	258	775	2,325	338.20	
	2400	Joint & crack cleaner, walk behind, 25 H.P.			3.80	50	150	450	60.40	
	2500	Filler, trailer mounted, 400 gallons, 20 H.P.			9.15	217	650	1,950	203.20	
	3000	Paint striper, self-propelled, double line, 30 H.P.			6.95	162	485	1,450	152.60	
	3200	Post drivers, 6" I-Beam frame, for truck mounting			18.75	425	1,280	3,850	406	
	3400	Road sweeper, self-propelled, 8' wide, 90 H.P.			37.80	580	1,740	5,225	650.40	
	3450	Road sweeper, vacuum assisted, 4 C.Y., 220 gallons			73.55	635	1,905	5,725	969.40	
	4000	Road mixer, self-propelled, 130 H.P.			44.05	775	2,325	6,975	817.40	
	4100	310 H.P.			79.75	2,200	6,585	19,800	1,955	
	4220	Cold mix paver, incl. pug mill and bitumen tank, 165 H.P.			96.80	2,150	6,440	19,300	2,062	
	4250	Paver, asphalt, wheel or crawler, 130 H.P., diesel			95.35	2,050	6,155	18,500	1,994	
	4300	Paver, road widener, gas 1' to 6', 67 H.P.			44.65	835	2,500	7,500	857.20	
	4400	Diesel, 2' to 14', 88 H.P.			59.80	1,075	3,190	9,575	1,116	
	4600	Slipform pavers, curb and gutter, 2 track, 75 H.P.			44.25	785	2,355	7,075	825	
	4700	4 track, 165 H.P.			54.95	825	2,475	7,425	934.60	
	4800	Median barrier, 215 H.P.			55.65	855	2,570	7,700	959.20	
	4901	Trailer, low bed, 75 ton capacity			9.35	218	655	1,975	205.80	
	5000	Road planer, walk behind, 10" cutting width, 10 H.P.			3.45	30.50	91	273	45.80	
	5100	Self-propelled, 12" cutting width, 64 H.P.			10.20	127	380	1,150	157.60	
	5120	Traffic line remover, metal ball blaster, truck mounted, 115 H.P.			52.80	770	2,305	6,925	883.40	
	5140	Grinder, truck mounted, 115 H.P.		↓	56.60	835	2,505	7,525	953.80	

01 54 33 | Equipment Rental

		01 54 33 Equipment Rental	UNIT	HOURLY OPER. COST	RENT PER DAY	RENT PER WEEK	RENT PER MONTH	EQUIPMENT COST/DAY	
50	5160	Walk-behind, 11 H.P.	Ea.	3.90	46.50	140	420	59.20	50
	5200	Pavement profiler, 4' to 6' wide, 450 H.P.		244.25	3,400	10,220	30,700	3,998	
	5300	8' to 10' wide, 750 H.P.		391.35	4,950	14,865	44,600	6,104	
	5400	Roadway plate, steel, 1" x 8' x 20'		.07	11.65	35	105	7.55	
	5600	Stabilizer, self-propelled, 150 H.P.		47.20	615	1,850	5,550	747.60	
	5700	310 H.P.		81.20	1,350	4,070	12,200	1,464	
	5800	Striper, thermal, truck mounted 120 gallon paint, 150 H.P.		62.30	505	1,520	4,550	802.40	
	6000	Tar kettle, 330 gallon, trailer mounted		4.30	61.50	185	555	71.40	
	7000	Tunnel locomotive, diesel, 8 to 12 ton		30.00	580	1,745	5,225	589	
	7005	Electric, 10 ton		24.00	665	1,995	5,975	591	
	7010	Muck cars, 1/2 C.Y. capacity		1.85	23.50	70	210	28.80	
	7020	1 C.Y. capacity		2.05	31.50	95	285	35.40	
	7030	2 C.Y. capacity		2.15	36.50	110	330	39.20	
	7040	Side dump, 2 C.Y. capacity		2.35	45	135	405	45.80	
	7050	3 C.Y. capacity		3.20	51.50	155	465	56.60	
	7060	5 C.Y. capacity		4.55	65	195	585	75.40	
	7100	Ventilating blower for tunnel, 7-1/2 H.P.		1.39	51.50	155	465	42.10	
	7110	10 H.P.		.83	.52	1.55	4.65	6.95	
	7120	20 H.P.		1.66	.67	2	6	13.70	
	7140	40 H.P.		3.31	.97	2.90	8.70	27.05	
	7160	60 H.P.		7.23	152	455	1,375	148.85	
	7175	75 H.P.		9.22	202	607	1,825	195.15	
	7180	200 H.P.		21.05	305	910	2,725	350.40	
	7800	Windrow loader, elevating	↓	52.00	1,350	4,060	12,200	1,228	
60	0010	**LIFTING AND HOISTING EQUIPMENT RENTAL** without operators R015433 -10							60
	0120	Aerial lift truck, 2 person, to 80'	Ea.	28.05	735	2,205	6,625	665.40	
	0140	Boom work platform, 40' snorkel R015433 -15		17.60	265	795	2,375	299.80	
	0150	Crane, flatbed mounted, 3 ton capacity		15.70	195	585	1,750	242.60	
	0200	Crane, climbing, 106' jib, 6000 lb. capacity, 410 fpm R312316 -45		38.95	1,500	4,490	13,500	1,210	
	0300	101' jib, 10,250 lb. capacity, 270 fpm		44.95	1,900	5,690	17,100	1,498	
	0500	Tower, static, 130' high, 106' jib, 6200 lb. capacity at 400 fpm		42.45	1,725	5,190	15,600	1,378	
	0600	Crawler mounted, lattice boom, 1/2 C.Y., 15 tons at 12' radius		35.47	685	2,050	6,150	693.75	
	0700	3/4 C.Y., 20 tons at 12' radius		47.29	855	2,570	7,700	892.30	
	0800	1 C.Y., 25 tons at 12' radius		63.05	1,150	3,420	10,300	1,188	
	0900	1-1/2 C.Y., 40 tons at 12' radius		63.05	1,150	3,445	10,300	1,193	
	1000	2 C.Y., 50 tons at 12' radius		66.90	1,350	4,055	12,200	1,346	
	1100	3 C.Y., 75 tons at 12' radius		71.80	1,575	4,710	14,100	1,516	
	1200	100 ton capacity, 60' boom		81.40	1,800	5,405	16,200	1,732	
	1300	165 ton capacity, 60' boom		103.75	2,100	6,320	19,000	2,094	
	1400	200 ton capacity, 70' boom		125.80	2,675	7,990	24,000	2,604	
	1500	350 ton capacity, 80' boom		177.05	4,075	12,195	36,600	3,855	
	1600	Truck mounted, lattice boom, 6 x 4, 20 tons at 10' radius		36.33	1,225	3,690	11,100	1,029	
	1700	25 tons at 10' radius		39.34	1,350	4,020	12,100	1,119	
	1800	8 x 4, 30 tons at 10' radius		42.56	1,425	4,280	12,800	1,196	
	1900	40 tons at 12' radius		45.29	1,500	4,470	13,400	1,256	
	2000	60 tons at 15' radius		50.61	1,575	4,730	14,200	1,351	
	2050	82 tons at 15' radius		56.35	1,675	5,050	15,200	1,461	
	2100	90 tons at 15' radius		63.07	1,825	5,510	16,500	1,607	
	2200	115 tons at 15' radius		71.12	2,050	6,160	18,500	1,801	
	2300	150 tons at 18' radius		88.35	2,150	6,480	19,400	2,003	
	2350	165 tons at 18' radius		83.09	2,300	6,870	20,600	2,039	
	2400	Truck mounted, hydraulic, 12 ton capacity		41.45	540	1,620	4,850	655.60	
	2500	25 ton capacity		43.70	665	1,990	5,975	747.60	
	2550	33 ton capacity		44.25	680	2,045	6,125	763	
	2560	40 ton capacity		57.20	800	2,405	7,225	938.60	
	2600	55 ton capacity		75.80	1,000	3,015	9,050	1,209	
	2700	80 ton capacity		97.75	1,275	3,815	11,400	1,545	
	2720	100 ton capacity	↓	91.65	1,525	4,600	13,800	1,653	

01 54 33 | Equipment Rental

			UNIT	HOURLY OPER. COST	RENT PER DAY	RENT PER WEEK	RENT PER MONTH	EQUIPMENT COST/DAY	
60	2740	120 ton capacity	Ea.	106.15	1,650	4,920	14,800	1,833	60
	2760	150 ton capacity		124.30	2,250	6,735	20,200	2,341	
	2800	Self-propelled, 4 x 4, with telescoping boom, 5 ton		16.90	235	705	2,125	276.20	
	2900	12-1/2 ton capacity		41.45	540	1,620	4,850	655.60	
	3000	15 ton capacity		39.70	585	1,755	5,275	668.60	
	3050	20 ton capacity		35.20	575	1,720	5,150	625.60	
	3100	25 ton capacity		36.45	605	1,820	5,450	655.60	
	3150	40 ton capacity		57.95	915	2,750	8,250	1,014	
	3200	Derricks, guy, 20 ton capacity, 60' boom, 75' mast		21.70	365	1,100	3,300	393.60	
	3300	100' boom, 115' mast		34.23	630	1,890	5,675	651.85	
	3400	Stiffleg, 20 ton capacity, 70' boom, 37' mast		24.01	475	1,430	4,300	478.10	
	3500	100' boom, 47' mast		36.96	760	2,280	6,850	751.70	
	3550	Helicopter, small, lift to 1250 lbs. maximum, w/pilot		92.16	2,950	8,880	26,600	2,513	
	3600	Hoists, chain type, overhead, manual, 3/4 ton		.10	1	3	9	1.40	
	3900	10 ton		.75	9.35	28	84	11.60	
	4000	Hoist and tower, 5000 lb. cap., portable electric, 40' high		4.82	211	633	1,900	165.15	
	4100	For each added 10' section, add		.10	16.65	50	150	10.80	
	4200	Hoist and single tubular tower, 5000 lb. electric, 100' high		6.48	295	884	2,650	228.65	
	4300	For each added 6'-6" section, add		.17	28.50	85	255	18.35	
	4400	Hoist and double tubular tower, 5000 lb., 100' high		6.93	325	974	2,925	250.25	
	4500	For each added 6'-6" section, add		.19	31.50	95	285	20.50	
	4550	Hoist and tower, mast type, 6000 lb., 100' high		7.53	335	1,010	3,025	262.25	
	4570	For each added 10' section, add		.12	20	60	180	12.95	
	4600	Hoist and tower, personnel, electric, 2000 lb., 100' @ 125 fpm		15.10	895	2,690	8,075	658.80	
	4700	3000 lb., 100' @ 200 fpm		17.31	1,025	3,050	9,150	748.50	
	4800	3000 lb., 150' @ 300 fpm		19.11	1,125	3,410	10,200	834.90	
	4900	4000 lb., 100' @ 300 fpm		19.88	1,150	3,480	10,400	855.05	
	5000	6000 lb., 100' @ 275 fpm		21.55	1,225	3,650	11,000	902.40	
	5100	For added heights up to 500', add	L.F.	.01	1.67	5	15	1.10	
	5200	Jacks, hydraulic, 20 ton	Ea.	.05	2	6	18	1.60	
	5500	100 ton		.35	11.35	34	102	9.60	
	6100	Jacks, hydraulic, climbing w/ 50' jackrods, control console, 30 ton cap.		1.82	121	364	1,100	87.35	
	6150	For each added 10' jackrod section, add		.05	3.33	10	30	2.40	
	6300	50 ton capacity		2.93	195	585	1,750	140.45	
	6350	For each added 10' jackrod section, add		.06	4	12	36	2.90	
	6500	125 ton capacity		7.70	515	1,540	4,625	369.60	
	6550	For each added 10' jackrod section, add		.53	35	105	315	25.25	
	6600	Cable jack, 10 ton capacity with 200' cable		1.53	102	305	915	73.25	
	6650	For each added 50' of cable, add		.18	11.65	35	105	8.45	
70	0010	**WELLPOINT EQUIPMENT RENTAL** without operators R015433-10							70
	0020	Based on 2 months rental							
	0100	Combination jetting & wellpoint pump, 60 H.P. diesel	Ea.	14.71	300	902	2,700	298.10	
	0200	High pressure gas jet pump, 200 H.P., 300 psi	"	33.40	257	771	2,325	421.40	
	0300	Discharge pipe, 8" diameter	L.F.	.01	.49	1.47	4.41	.35	
	0350	12" diameter		.01	.72	2.16	6.50	.50	
	0400	Header pipe, flows up to 150 GPM, 4" diameter		.01	.44	1.32	3.96	.35	
	0500	400 GPM, 6" diameter		.01	.52	1.57	4.71	.40	
	0600	800 GPM, 8" diameter		.01	.72	2.16	6.50	.50	
	0700	1500 GPM, 10" diameter		.01	.75	2.26	6.80	.55	
	0800	2500 GPM, 12" diameter		.02	1.43	4.29	12.85	1	
	0900	4500 GPM, 16" diameter		.03	1.83	5.50	16.50	1.35	
	0950	For quick coupling aluminum and plastic pipe, add		.03	1.90	5.69	17.05	1.40	
	1100	Wellpoint, 25' long, with fittings & riser pipe, 1-1/2" or 2" diameter	Ea.	.06	3.78	11.35	34	2.75	
	1200	Wellpoint pump, diesel powered, 4" suction, 20 H.P.		6.44	173	520	1,550	155.50	
	1300	6" suction, 30 H.P.		8.72	215	645	1,925	198.75	
	1400	8" suction, 40 H.P.		11.79	295	884	2,650	271.10	
	1500	10" suction, 75 H.P.		17.73	345	1,033	3,100	348.45	

01 54 33 | Equipment Rental

			UNIT	HOURLY OPER. COST	RENT PER DAY	RENT PER WEEK	RENT PER MONTH	EQUIPMENT COST/DAY	
70	1600	12″ suction, 100 H.P.	Ea.	25.55	550	1,650	4,950	534.40	70
	1700	12″ suction, 175 H.P.	″	37.17	605	1,810	5,425	659.35	
80	0010	**MARINE EQUIPMENT RENTAL** without operators R015433-10							80
	0200	Barge, 400 Ton, 30′ wide x 90′ long	Ea.	15.50	1,075	3,220	9,650	768	
	0240	800 Ton, 45′ wide x 90′ long		18.80	1,300	3,900	11,700	930.40	
	2000	Tugboat, diesel, 100 H.P.		36.45	203	610	1,825	413.60	
	2040	250 H.P.		77.10	375	1,120	3,350	840.80	
	2080	380 H.P.	↓	149.05	1,100	3,325	9,975	1,857	

Crew A-1	Bare Costs Hr.	Bare Costs Daily	Incl. Subs O&P Hr.	Incl. Subs O&P Daily	Bare Costs	Incl. O&P
1 Building Laborer	$33.10	$264.80	$51.05	$408.40	$33.10	$51.05
1 Concrete saw, gas manual		73.80		81.18	9.22	10.15
8 L.H., Daily Totals		$338.60		$489.58	$42.33	$61.20

Crew A-1A	Hr.	Daily	Hr.	Daily	Bare Costs	Incl. O&P
1 Skilled Worker	$42.60	$340.80	$65.50	$524.00	$42.60	$65.50
1 Shot Blaster, 20″		215.80		237.38	26.98	29.67
8 L.H., Daily Totals		$556.60		$761.38	$69.58	$95.17

Crew A-1B	Hr.	Daily	Hr.	Daily	Bare Costs	Incl. O&P
1 Building Laborer	$33.10	$264.80	$51.05	$408.40	$33.10	$51.05
1 Concrete Saw		161.40		177.54	20.18	22.19
8 L.H., Daily Totals		$426.20		$585.94	$53.27	$73.24

Crew A-1C	Hr.	Daily	Hr.	Daily	Bare Costs	Incl. O&P
1 Building Laborer	$33.10	$264.80	$51.05	$408.40	$33.10	$51.05
1 Chain Saw, gas, 18″		28.20		31.02	3.52	3.88
8 L.H., Daily Totals		$293.00		$439.42	$36.63	$54.93

Crew A-1D	Hr.	Daily	Hr.	Daily	Bare Costs	Incl. O&P
1 Building Laborer	$33.10	$264.80	$51.05	$408.40	$33.10	$51.05
1 Vibrating plate, gas, 18″		34.20		37.62	4.28	4.70
8 L.H., Daily Totals		$299.00		$446.02	$37.38	$55.75

Crew A-1E	Hr.	Daily	Hr.	Daily	Bare Costs	Incl. O&P
1 Building Laborer	$33.10	$264.80	$51.05	$408.40	$33.10	$51.05
1 Vibratory Plate, gas, 21″		43.60		47.96	5.45	6.00
8 L.H., Daily Totals		$308.40		$456.36	$38.55	$57.05

Crew A-1F	Hr.	Daily	Hr.	Daily	Bare Costs	Incl. O&P
1 Building Laborer	$33.10	$264.80	$51.05	$408.40	$33.10	$51.05
1 Rammer/tamper, gas, 8″		46.80		51.48	5.85	6.43
8 L.H., Daily Totals		$311.60		$459.88	$38.95	$57.48

Crew A-1G	Hr.	Daily	Hr.	Daily	Bare Costs	Incl. O&P
1 Building Laborer	$33.10	$264.80	$51.05	$408.40	$33.10	$51.05
1 Rammer/tamper, gas, 15″		51.80		56.98	6.47	7.12
8 L.H., Daily Totals		$316.60		$465.38	$39.58	$58.17

Crew A-1H	Hr.	Daily	Hr.	Daily	Bare Costs	Incl. O&P
1 Building Laborer	$33.10	$264.80	$51.05	$408.40	$33.10	$51.05
1 Exterior Steam Cleaner		60.20		66.22	7.53	8.28
8 L.H., Daily Totals		$325.00		$474.62	$40.63	$59.33

Crew A-1J	Hr.	Daily	Hr.	Daily	Bare Costs	Incl. O&P
1 Building Laborer	$33.10	$264.80	$51.05	$408.40	$33.10	$51.05
1 Cultivator, Walk-Behind, 5 H.P.		56.15		61.77	7.02	7.72
8 L.H., Daily Totals		$320.95		$470.17	$40.12	$58.77

Crew A-1K	Hr.	Daily	Hr.	Daily	Bare Costs	Incl. O&P
1 Building Laborer	$33.10	$264.80	$51.05	$408.40	$33.10	$51.05
1 Cultivator, Walk-Behind, 8 H.P.		86.40		95.04	10.80	11.88
8 L.H., Daily Totals		$351.20		$503.44	$43.90	$62.93

Crew A-1M	Hr.	Daily	Hr.	Daily	Bare Costs	Incl. O&P
1 Building Laborer	$33.10	$264.80	$51.05	$408.40	$33.10	$51.05
1 Snow Blower, Walk-Behind		53.40		58.74	6.67	7.34
8 L.H., Daily Totals		$318.20		$467.14	$39.77	$58.39

Crew A-2	Hr.	Daily	Hr.	Daily	Bare Costs	Incl. O&P
2 Laborers	$33.10	$529.60	$51.05	$816.80	$32.82	$50.43
1 Truck Driver (light)	32.25	258.00	49.20	393.60		
1 Flatbed Truck, Gas, 1.5 Ton		239.80		263.78	9.99	10.99
24 L.H., Daily Totals		$1027.40		$1474.18	$42.81	$61.42

Crew A-2A	Hr.	Daily	Hr.	Daily	Bare Costs	Incl. O&P
2 Laborers	$33.10	$529.60	$51.05	$816.80	$32.82	$50.43
1 Truck Driver (light)	32.25	258.00	49.20	393.60		
1 Flatbed Truck, Gas, 1.5 Ton		239.80		263.78		
1 Concrete Saw		161.40		177.54	16.72	18.39
24 L.H., Daily Totals		$1188.80		$1651.72	$49.53	$68.82

Crew A-2B	Hr.	Daily	Hr.	Daily	Bare Costs	Incl. O&P
1 Truck Driver (light)	$32.25	$258.00	$49.20	$393.60	$32.25	$49.20
1 Flatbed Truck, Gas, 1.5 Ton		239.80		263.78	29.98	32.97
8 L.H., Daily Totals		$497.80		$657.38	$62.23	$82.17

Crew A-3A	Hr.	Daily	Hr.	Daily	Bare Costs	Incl. O&P
1 Truck Driver (light)	$32.25	$258.00	$49.20	$393.60	$32.25	$49.20
1 Pickup truck, 4 x 4, 3/4 ton		151.00		166.10	18.88	20.76
8 L.H., Daily Totals		$409.00		$559.70	$51.13	$69.96

Crew A-3B	Hr.	Daily	Hr.	Daily	Bare Costs	Incl. O&P
1 Equip. Oper. (med.)	$42.95	$343.60	$64.30	$514.40	$38.05	$57.42
1 Truck Driver (heav.)	33.15	265.20	50.55	404.40		
1 Dump Truck, 12 C.Y., 400 H.P.		667.80		734.58		
1 F.E. Loader, W.M.,2.5 C.Y.		435.80		479.38	68.97	75.87
16 L.H., Daily Totals		$1712.40		$2132.76	$107.03	$133.30

Crew A-3C	Hr.	Daily	Hr.	Daily	Bare Costs	Incl. O&P
1 Equip. Oper. (light)	$41.30	$330.40	$61.85	$494.80	$41.30	$61.85
1 Loader, Skid Steer, 78 H.P.		291.00		320.10	36.38	40.01
8 L.H., Daily Totals		$621.40		$814.90	$77.67	$101.86

Crew A-3D	Hr.	Daily	Hr.	Daily	Bare Costs	Incl. O&P
1 Truck Driver, Light	$32.25	$258.00	$49.20	$393.60	$32.25	$49.20
1 Pickup truck, 4 x 4, 3/4 ton		151.00		166.10		
1 Flatbed Trailer, 25 Ton		105.60		116.16	32.08	35.28
8 L.H., Daily Totals		$514.60		$675.86	$64.33	$84.48

Crew A-3E	Hr.	Daily	Hr.	Daily	Bare Costs	Incl. O&P
1 Equip. Oper. (crane)	$44.40	$355.20	$66.45	$531.60	$38.77	$58.50
1 Truck Driver (heavy)	33.15	265.20	50.55	404.40		
1 Pickup truck, 4 x 4, 3/4 ton		151.00		166.10	9.44	10.38
16 L.H., Daily Totals		$771.40		$1102.10	$48.21	$68.88

Crew A-3F	Hr.	Daily	Hr.	Daily	Bare Costs	Incl. O&P
1 Equip. Oper. (crane)	$44.40	$355.20	$66.45	$531.60	$38.77	$58.50
1 Truck Driver (heavy)	33.15	265.20	50.55	404.40		
1 Pickup truck, 4 x 4, 3/4 ton		151.00		166.10		
1 Truck Tractor, 6x4, 380 H.P.		592.40		651.64		
1 Lowbed Trailer, 75 Ton		205.80		226.38	59.33	65.26
16 L.H., Daily Totals		$1569.60		$1980.12	$98.10	$123.76

Crew No.	Bare Costs		Incl. Subs O&P		Cost Per Labor-Hour	

Crew A-3G	Hr.	Daily	Hr.	Daily	Bare Costs	Incl. O&P
1 Equip. Oper. (crane)	$44.40	$355.20	$66.45	$531.60	$38.77	$58.50
1 Truck Driver (heavy)	33.15	265.20	50.55	404.40		
1 Pickup truck, 4 x 4, 3/4 ton		151.00		166.10		
1 Truck Tractor, 6x4, 450 H.P.		715.80		787.38		
1 Lowbed Trailer, 75 Ton		205.80		226.38	67.04	73.74
16 L.H., Daily Totals		$1693.00		$2115.86	$105.81	$132.24

Crew A-3H	Hr.	Daily	Hr.	Daily	Bare Costs	Incl. O&P
1 Equip. Oper. (crane)	$44.40	$355.20	$66.45	$531.60	$44.40	$66.45
1 Hyd. crane, 12 Ton (daily)		871.60		958.76	108.95	119.85
8 L.H., Daily Totals		$1226.80		$1490.36	$153.35	$186.29

Crew A-3I	Hr.	Daily	Hr.	Daily	Bare Costs	Incl. O&P
1 Equip. Oper. (crane)	$44.40	$355.20	$66.45	$531.60	$44.40	$66.45
1 Hyd. crane, 25 Ton (daily)		1015.00		1116.50	126.88	139.56
8 L.H., Daily Totals		$1370.20		$1648.10	$171.28	$206.01

Crew A-3J	Hr.	Daily	Hr.	Daily	Bare Costs	Incl. O&P
1 Equip. Oper. (crane)	$44.40	$355.20	$66.45	$531.60	$44.40	$66.45
1 Hyd. crane, 40 Ton (daily)		1258.00		1383.80	157.25	172.97
8 L.H., Daily Totals		$1613.20		$1915.40	$201.65	$239.43

Crew A-3K	Hr.	Daily	Hr.	Daily	Bare Costs	Incl. O&P
1 Equip. Oper. (crane)	$44.40	$355.20	$66.45	$531.60	$41.35	$61.90
1 Equip. Oper. Oiler	38.30	306.40	57.35	458.80		
1 Hyd. crane, 55 Ton (daily)		1611.00		1772.10		
1 P/U Truck, 3/4 Ton (daily)		161.00		177.10	110.75	121.83
16 L.H., Daily Totals		$2433.60		$2939.60	$152.10	$183.72

Crew A-3L	Hr.	Daily	Hr.	Daily	Bare Costs	Incl. O&P
1 Equip. Oper. (crane)	$44.40	$355.20	$66.45	$531.60	$41.35	$61.90
1 Equip. Oper. Oiler	38.30	306.40	57.35	458.80		
1 Hyd. crane, 80 Ton (daily)		2052.00		2257.20		
1 P/U Truck, 3/4 Ton (daily)		161.00		177.10	138.31	152.14
16 L.H., Daily Totals		$2874.60		$3424.70	$179.66	$214.04

Crew A-3M	Hr.	Daily	Hr.	Daily	Bare Costs	Incl. O&P
1 Equip. Oper. (crane)	$44.40	$355.20	$66.45	$531.60	$41.35	$61.90
1 Equip. Oper. Oiler	38.30	306.40	57.35	458.80		
1 Hyd. crane, 100 Ton (daily)		2268.00		2494.80		
1 P/U Truck, 3/4 Ton (daily)		161.00		177.10	151.81	166.99
16 L.H., Daily Totals		$3090.60		$3662.30	$193.16	$228.89

Crew A-3N	Hr.	Daily	Hr.	Daily	Bare Costs	Incl. O&P
1 Equip. Oper. (crane)	$44.40	$355.20	$66.45	$531.60	$44.40	$66.45
1 Tower crane (monthly)		1049.00		1153.90	131.13	144.24
8 L.H., Daily Totals		$1404.20		$1685.50	$175.53	$210.69

Crew A-3P	Hr.	Daily	Hr.	Daily	Bare Costs	Incl. O&P
1 Equip. Oper., Light	$41.30	$330.40	$61.85	$494.80	$41.30	$61.85
1 A.T. Forklift, 42' lift		478.20		526.02	59.77	65.75
8 L.H., Daily Totals		$808.60		$1020.82	$101.08	$127.60

Crew A-4	Hr.	Daily	Hr.	Daily	Bare Costs	Incl. O&P
2 Carpenters	$41.55	$664.80	$64.05	$1024.80	$39.82	$60.73
1 Painter, Ordinary	36.35	290.80	54.10	432.80		
24 L.H., Daily Totals		$955.60		$1457.60	$39.82	$60.73

Crew A-5	Hr.	Daily	Hr.	Daily	Bare Costs	Incl. O&P
2 Laborers	$33.10	$529.60	$51.05	$816.80	$33.01	$50.84
.25 Truck Driver (light)	32.25	64.50	49.20	98.40		
.25 Flatbed Truck, Gas, 1.5 Ton		59.95		65.94	3.33	3.66
18 L.H., Daily Totals		$654.05		$981.14	$36.34	$54.51

Crew A-6	Hr.	Daily	Hr.	Daily	Bare Costs	Incl. O&P
1 Instrument Man	$42.60	$340.80	$65.50	$524.00	$41.40	$63.02
1 Rodman/Chainman	40.20	321.60	60.55	484.40		
1 Laser Transit/Level		69.90		76.89	4.37	4.81
16 L.H., Daily Totals		$732.30		$1085.29	$45.77	$67.83

Crew A-7	Hr.	Daily	Hr.	Daily	Bare Costs	Incl. O&P
1 Chief Of Party	$52.25	$418.00	$79.90	$639.20	$45.02	$68.65
1 Instrument Man	42.60	340.80	65.50	524.00		
1 Rodman/Chainman	40.20	321.60	60.55	484.40		
1 Laser Transit/Level		69.90		76.89	2.91	3.20
24 L.H., Daily Totals		$1150.30		$1724.49	$47.93	$71.85

Crew A-8	Hr.	Daily	Hr.	Daily	Bare Costs	Incl. O&P
1 Chief of Party	$52.25	$418.00	$79.90	$639.20	$43.81	$66.63
1 Instrument Man	42.60	340.80	65.50	524.00		
2 Rodmen/Chainmen	40.20	643.20	60.55	968.80		
1 Laser Transit/Level		69.90		76.89	2.18	2.40
32 L.H., Daily Totals		$1471.90		$2208.89	$46.00	$69.03

Crew A-9	Hr.	Daily	Hr.	Daily	Bare Costs	Incl. O&P
1 Asbestos Foreman	$46.05	$368.40	$71.55	$572.40	$45.61	$70.89
7 Asbestos Workers	45.55	2550.80	70.80	3964.80		
64 L.H., Daily Totals		$2919.20		$4537.20	$45.61	$70.89

Crew A-10A	Hr.	Daily	Hr.	Daily	Bare Costs	Incl. O&P
1 Asbestos Foreman	$46.05	$368.40	$71.55	$572.40	$45.72	$71.05
2 Asbestos Workers	45.55	728.80	70.80	1132.80		
24 L.H., Daily Totals		$1097.20		$1705.20	$45.72	$71.05

Crew A-10B	Hr.	Daily	Hr.	Daily	Bare Costs	Incl. O&P
1 Asbestos Foreman	$46.05	$368.40	$71.55	$572.40	$45.67	$70.99
3 Asbestos Workers	45.55	1093.20	70.80	1699.20		
32 L.H., Daily Totals		$1461.60		$2271.60	$45.67	$70.99

Crew A-10C	Hr.	Daily	Hr.	Daily	Bare Costs	Incl. O&P
3 Asbestos Workers	$45.55	$1093.20	$70.80	$1699.20	$45.55	$70.80
1 Flatbed Truck, Gas, 1.5 Ton		239.80		263.78	9.99	10.99
24 L.H., Daily Totals		$1333.00		$1962.98	$55.54	$81.79

Crew A-10D	Hr.	Daily	Hr.	Daily	Bare Costs	Incl. O&P
2 Asbestos Workers	$45.55	$728.80	$70.80	$1132.80	$43.45	$66.35
1 Equip. Oper. (crane)	44.40	355.20	66.45	531.60		
1 Equip. Oper. Oiler	38.30	306.40	57.35	458.80		
1 Hydraulic Crane, 33 Ton		763.00		839.30	23.84	26.23
32 L.H., Daily Totals		$2153.40		$2962.50	$67.29	$92.58

Crew A-11	Hr.	Daily	Hr.	Daily	Bare Costs	Incl. O&P
1 Asbestos Foreman	$46.05	$368.40	$71.55	$572.40	$45.61	$70.89
7 Asbestos Workers	45.55	2550.80	70.80	3964.80		
2 Chip. Hammers, 12 Lb., Elec.		36.00		39.60	.56	.62
64 L.H., Daily Totals		$2955.20		$4576.80	$46.17	$71.51

Crew No.	Bare Costs Hr.	Bare Costs Daily	Incl. Subs O&P Hr.	Incl. Subs O&P Daily	Cost Per Labor-Hour Bare Costs	Cost Per Labor-Hour Incl. O&P
Crew A-12	**Hr.**	**Daily**	**Hr.**	**Daily**	**Bare Costs**	**Incl. O&P**
1 Asbestos Foreman	$46.05	$368.40	$71.55	$572.40	$45.61	$70.89
7 Asbestos Workers	45.55	2550.80	70.80	3964.80		
1 Trk-mtd vac, 14 CY, 1500 Gal.		530.25		583.27		
1 Flatbed Truck, 20,000 GVW		245.60		270.16	12.12	13.33
64 L.H., Daily Totals		$3695.05		$5390.64	$57.74	$84.23
Crew A-13	**Hr.**	**Daily**	**Hr.**	**Daily**	**Bare Costs**	**Incl. O&P**
1 Equip. Oper. (light)	$41.30	$330.40	$61.85	$494.80	$41.30	$61.85
1 Trk-mtd vac, 14 CY, 1500 Gal.		530.25		583.27		
1 Flatbed Truck, 20,000 GVW		245.60		270.16	96.98	106.68
8 L.H., Daily Totals		$1106.25		$1348.23	$138.28	$168.53
Crew B-1	**Hr.**	**Daily**	**Hr.**	**Daily**	**Bare Costs**	**Incl. O&P**
1 Labor Foreman (outside)	$35.10	$280.80	$54.10	$432.80	$33.77	$52.07
2 Laborers	33.10	529.60	51.05	816.80		
24 L.H., Daily Totals		$810.40		$1249.60	$33.77	$52.07
Crew B-1A	**Hr.**	**Daily**	**Hr.**	**Daily**	**Bare Costs**	**Incl. O&P**
1 Laborer Foreman	$35.10	$280.80	$54.10	$432.80	$33.77	$52.07
2 Laborers	33.10	529.60	51.05	816.80		
2 Cutting Torches		38.00		41.80		
2 Set of Gases		158.40		174.24	8.18	9.00
24 L.H., Daily Totals		$1006.80		$1465.64	$41.95	$61.07
Crew B-1B	**Hr.**	**Daily**	**Hr.**	**Daily**	**Bare Costs**	**Incl. O&P**
1 Laborer Foreman	$35.10	$280.80	$54.10	$432.80	$36.42	$55.66
2 Laborers	33.10	529.60	51.05	816.80		
1 Equip. Oper. (crane)	44.40	355.20	66.45	531.60		
2 Cutting Torches		38.00		41.80		
2 Set of Gases		158.40		174.24		
1 Hyd. Crane, 12 Ton		655.60		721.16	26.63	29.29
32 L.H., Daily Totals		$2017.60		$2718.40	$63.05	$84.95
Crew B-2	**Hr.**	**Daily**	**Hr.**	**Daily**	**Bare Costs**	**Incl. O&P**
1 Labor Foreman (outside)	$35.10	$280.80	$54.10	$432.80	$33.50	$51.66
4 Laborers	33.10	1059.20	51.05	1633.60		
40 L.H., Daily Totals		$1340.00		$2066.40	$33.50	$51.66
Crew B-3	**Hr.**	**Daily**	**Hr.**	**Daily**	**Bare Costs**	**Incl. O&P**
1 Labor Foreman (outside)	$35.10	$280.80	$54.10	$432.80	$35.09	$53.60
2 Laborers	33.10	529.60	51.05	816.80		
1 Equip. Oper. (med.)	42.95	343.60	64.30	514.40		
2 Truck Drivers (heavy)	33.15	530.40	50.55	808.80		
1 Crawler Loader, 3 C.Y.		1077.00		1184.70		
2 Dump Trucks 12 C.Y., 400 H.P.		1335.60		1469.16	50.26	55.29
48 L.H., Daily Totals		$4097.00		$5226.66	$85.35	$108.89
Crew B-3A	**Hr.**	**Daily**	**Hr.**	**Daily**	**Bare Costs**	**Incl. O&P**
4 Laborers	$33.10	$1059.20	$51.05	$1633.60	$35.07	$53.70
1 Equip. Oper. (med.)	42.95	343.60	64.30	514.40		
1 Hyd. Excavator, 1.5 C.Y.		1017.00		1118.70	25.43	27.97
40 L.H., Daily Totals		$2419.80		$3266.70	$60.49	$81.67
Crew B-3B	**Hr.**	**Daily**	**Hr.**	**Daily**	**Bare Costs**	**Incl. O&P**
2 Laborers	$33.10	$529.60	$51.05	$816.80	$35.58	$54.24
1 Equip. Oper. (med.)	42.95	343.60	64.30	514.40		
1 Truck Driver (heavy)	33.15	265.20	50.55	404.40		
1 Backhoe Loader, 80 H.P.		387.60		426.36		
1 Dump Truck, 12 C.Y., 400 H.P.		667.80		734.58	32.98	36.28
32 L.H., Daily Totals		$2193.80		$2896.54	$68.56	$90.52

Crew No.	Bare Costs Hr.	Bare Costs Daily	Incl. Subs O&P Hr.	Incl. Subs O&P Daily	Cost Per Labor-Hour Bare Costs	Cost Per Labor-Hour Incl. O&P
Crew B-3C	**Hr.**	**Daily**	**Hr.**	**Daily**	**Bare Costs**	**Incl. O&P**
3 Laborers	$33.10	$794.40	$51.05	$1225.20	$35.56	$54.36
1 Equip. Oper. (med.)	42.95	343.60	64.30	514.40		
1 Crawler Loader, 4 C.Y.		1548.00		1702.80	48.38	53.21
32 L.H., Daily Totals		$2686.00		$3442.40	$83.94	$107.58
Crew B-4	**Hr.**	**Daily**	**Hr.**	**Daily**	**Bare Costs**	**Incl. O&P**
1 Labor Foreman (outside)	$35.10	$280.80	$54.10	$432.80	$33.44	$51.48
4 Laborers	33.10	1059.20	51.05	1633.60		
1 Truck Driver (heavy)	33.15	265.20	50.55	404.40		
1 Truck Tractor, 220 H.P.		354.00		389.40		
1 Flatbed Trailer, 40 Ton		143.20		157.52	10.36	11.39
48 L.H., Daily Totals		$2102.40		$3017.72	$43.80	$62.87
Crew B-5	**Hr.**	**Daily**	**Hr.**	**Daily**	**Bare Costs**	**Incl. O&P**
1 Labor Foreman (outside)	$35.10	$280.80	$54.10	$432.80	$36.20	$55.27
4 Laborers	33.10	1059.20	51.05	1633.60		
2 Equip. Oper. (med.)	42.95	687.20	64.30	1028.80		
1 Air Compressor, 250 cfm		193.40		212.74		
2 Breakers, Pavement, 60 lb.		18.80		20.68		
2 -50' Air Hoses, 1.5"		12.60		13.86		
1 Crawler Loader, 3 C.Y.		1077.00		1184.70	23.25	25.57
56 L.H., Daily Totals		$3329.00		$4527.18	$59.45	$80.84
Crew B-5A	**Hr.**	**Daily**	**Hr.**	**Daily**	**Bare Costs**	**Incl. O&P**
1 Foreman	$35.10	$280.80	$54.10	$432.80	$35.60	$54.33
6 Laborers	33.10	1588.80	51.05	2450.40		
2 Equip. Oper. (med.)	42.95	687.20	64.30	1028.80		
1 Equip. Oper. (light)	41.30	330.40	61.85	494.80		
2 Truck Drivers (heavy)	33.15	530.40	50.55	808.80		
1 Air Compressor, 365 cfm		255.60		281.16		
2 Breakers, Pavement, 60 lb.		18.80		20.68		
8 -50' Air Hoses, 1"		32.80		36.08		
2 Dump Trucks, 8 C.Y., 220 H.P.		803.60		883.96	11.57	12.73
96 L.H., Daily Totals		$4528.40		$6437.48	$47.17	$67.06
Crew B-5B	**Hr.**	**Daily**	**Hr.**	**Daily**	**Bare Costs**	**Incl. O&P**
1 Powderman	$42.60	$340.80	$65.50	$524.00	$37.99	$57.63
2 Equip. Oper. (med.)	42.95	687.20	64.30	1028.80		
3 Truck Drivers (heavy)	33.15	795.60	50.55	1213.20		
1 F.E. Loader, W.M.,2.5 C.Y.		435.80		479.38		
3 Dump Trucks, 12 C.Y., 400 H.P.		2003.40		2203.74		
1 Air Compressor, 365 cfm		255.60		281.16	56.14	61.76
48 L.H., Daily Totals		$4518.40		$5730.28	$94.13	$119.38
Crew B-5C	**Hr.**	**Daily**	**Hr.**	**Daily**	**Bare Costs**	**Incl. O&P**
3 Laborers	$33.10	$794.40	$51.05	$1225.20	$36.41	$55.29
1 Equip. Oper. (med.)	42.95	343.60	64.30	514.40		
2 Truck Drivers (heavy)	33.15	530.40	50.55	808.80		
1 Equip. Oper. (crane)	44.40	355.20	66.45	531.60		
1 Equip. Oper. Oiler	38.30	306.40	57.35	458.80		
2 Dump Trucks, 12 C.Y., 400 H.P.		1335.60		1469.16		
1 Crawler Loader, 4 C.Y.		1548.00		1702.80		
1 S.P. Crane, 4x4, 25 Ton		655.60		721.16	55.30	60.83
64 L.H., Daily Totals		$5869.20		$7431.92	$91.71	$116.12
Crew B-6	**Hr.**	**Daily**	**Hr.**	**Daily**	**Bare Costs**	**Incl. O&P**
2 Laborers	$33.10	$529.60	$51.05	$816.80	$35.83	$54.65
1 Equip. Oper. (light)	41.30	330.40	61.85	494.80		
1 Backhoe Loader, 48 H.P.		337.80		371.58	14.07	15.48
24 L.H., Daily Totals		$1197.80		$1683.18	$49.91	$70.13

Crew No.	Bare Costs		Incl. Subs O&P		Cost Per Labor-Hour	
Crew B-6A	**Hr.**	**Daily**	**Hr.**	**Daily**	**Bare Costs**	**Incl. O&P**
.5 Labor Foreman (outside)	$35.10	$140.40	$54.10	$216.40	$37.44	$56.96
1 Laborer	33.10	264.80	51.05	408.40		
1 Equip. Oper. (med.)	42.95	343.60	64.30	514.40		
1 Vacuum Trk.,5000 Gal.		363.85		400.24	18.19	20.01
20 L.H., Daily Totals		$1112.65		$1539.43	$55.63	$76.97
Crew B-6B	**Hr.**	**Daily**	**Hr.**	**Daily**	**Bare Costs**	**Incl. O&P**
2 Labor Foremen (out)	$35.10	$561.60	$54.10	$865.60	$33.77	$52.07
4 Laborers	33.10	1059.20	51.05	1633.60		
1 S.P. Crane, 4x4, 5 Ton		276.20		303.82		
1 Flatbed Truck, Gas, 1.5 Ton		239.80		263.78		
1 Butt Fusion Machine		134.40		147.84	13.55	14.90
48 L.H., Daily Totals		$2271.20		$3214.64	$47.32	$66.97
Crew B-7	**Hr.**	**Daily**	**Hr.**	**Daily**	**Bare Costs**	**Incl. O&P**
1 Labor Foreman (outside)	$35.10	$280.80	$54.10	$432.80	$35.08	$53.77
4 Laborers	33.10	1059.20	51.05	1633.60		
1 Equip. Oper. (med.)	42.95	343.60	64.30	514.40		
1 Brush Chipper, 12", 130 H.P.		253.20		278.52		
1 Crawler Loader, 3 C.Y.		1077.00		1184.70		
2 Chain Saws, gas, 36" Long		78.40		86.24	29.35	32.28
48 L.H., Daily Totals		$3092.20		$4130.26	$64.42	$86.05
Crew B-7A	**Hr.**	**Daily**	**Hr.**	**Daily**	**Bare Costs**	**Incl. O&P**
2 Laborers	$33.10	$529.60	$51.05	$816.80	$35.83	$54.65
1 Equip. Oper. (light)	41.30	330.40	61.85	494.80		
1 Rake w/Tractor		285.70		314.27		
2 Chain Saws, gas, 18"		56.40		62.04	14.25	15.68
24 L.H., Daily Totals		$1202.10		$1687.91	$50.09	$70.33
Crew B-8	**Hr.**	**Daily**	**Hr.**	**Daily**	**Bare Costs**	**Incl. O&P**
1 Labor Foreman (outside)	$35.10	$280.80	$54.10	$432.80	$36.48	$55.41
2 Laborers	33.10	529.60	51.05	816.80		
2 Equip. Oper. (med.)	42.95	687.20	64.30	1028.80		
1 Equip. Oper. Oiler	38.30	306.40	57.35	458.80		
2 Truck Drivers (heavy)	33.15	530.40	50.55	808.80		
1 Hyd. Crane, 25 Ton		747.60		822.36		
1 Crawler Loader, 3 C.Y.		1077.00		1184.70		
2 Dump Trucks, 12 C.Y., 400 H.P.		1335.60		1469.16	49.38	54.32
64 L.H., Daily Totals		$5494.60		$7022.22	$85.85	$109.72
Crew B-9	**Hr.**	**Daily**	**Hr.**	**Daily**	**Bare Costs**	**Incl. O&P**
1 Labor Foreman (outside)	$35.10	$280.80	$54.10	$432.80	$33.50	$51.66
4 Laborers	33.10	1059.20	51.05	1633.60		
1 Air Compressor, 250 cfm		193.40		212.74		
2 Breakers, Pavement, 60 lb.		18.80		20.68		
2 -50' Air Hoses, 1.5"		12.60		13.86	5.62	6.18
40 L.H., Daily Totals		$1564.80		$2313.68	$39.12	$57.84
Crew B-9A	**Hr.**	**Daily**	**Hr.**	**Daily**	**Bare Costs**	**Incl. O&P**
2 Laborers	$33.10	$529.60	$51.05	$816.80	$33.12	$50.88
1 Truck Driver (heavy)	33.15	265.20	50.55	404.40		
1 Water Tank Trailer, 5000 Gal.		139.60		153.56		
1 Truck Tractor, 220 H.P.		354.00		389.40		
2 -50' Discharge Hoses, 3"		3.50		3.85	20.71	22.78
24 L.H., Daily Totals		$1291.90		$1768.01	$53.83	$73.67
Crew B-9B	**Hr.**	**Daily**	**Hr.**	**Daily**	**Bare Costs**	**Incl. O&P**
2 Laborers	$33.10	$529.60	$51.05	$816.80	$33.12	$50.88
1 Truck Driver (heavy)	33.15	265.20	50.55	404.40		
2 -50' Discharge Hoses, 3"		3.50		3.85		
1 Water Tank Trailer, 5000 Gal.		139.60		153.56		
1 Truck Tractor, 220 H.P.		354.00		389.40		
1 Pressure Washer		60.80		66.88	23.25	25.57
24 L.H., Daily Totals		$1352.70		$1834.89	$56.36	$76.45
Crew B-9D	**Hr.**	**Daily**	**Hr.**	**Daily**	**Bare Costs**	**Incl. O&P**
1 Labor Foreman (Outside)	$35.10	$280.80	$54.10	$432.80	$33.50	$51.66
4 Common Laborers	33.10	1059.20	51.05	1633.60		
1 Air Compressor, 250 cfm		193.40		212.74		
2 -50' Air Hoses, 1.5"		12.60		13.86		
2 Air Powered Tampers		50.50		55.55	6.41	7.05
40 L.H., Daily Totals		$1596.50		$2348.55	$39.91	$58.71
Crew B-10	**Hr.**	**Daily**	**Hr.**	**Daily**	**Bare Costs**	**Incl. O&P**
1 Equip. Oper. (med.)	$42.95	$343.60	$64.30	$514.40	$39.67	$59.88
.5 Laborer	33.10	132.40	51.05	204.20		
12 L.H., Daily Totals		$476.00		$718.60	$39.67	$59.88
Crew B-10A	**Hr.**	**Daily**	**Hr.**	**Daily**	**Bare Costs**	**Incl. O&P**
1 Equip. Oper. (med.)	$42.95	$343.60	$64.30	$514.40	$39.67	$59.88
.5 Laborer	33.10	132.40	51.05	204.20		
1 Roller, 2-Drum, W.B., 7.5 H.P.		149.60		164.56	12.47	13.71
12 L.H., Daily Totals		$625.60		$883.16	$52.13	$73.60
Crew B-10B	**Hr.**	**Daily**	**Hr.**	**Daily**	**Bare Costs**	**Incl. O&P**
1 Equip. Oper. (med.)	$42.95	$343.60	$64.30	$514.40	$39.67	$59.88
.5 Laborer	33.10	132.40	51.05	204.20		
1 Dozer, 200 H.P.		1192.00		1311.20	99.33	109.27
12 L.H., Daily Totals		$1668.00		$2029.80	$139.00	$169.15
Crew B-10C	**Hr.**	**Daily**	**Hr.**	**Daily**	**Bare Costs**	**Incl. O&P**
1 Equip. Oper. (med.)	$42.95	$343.60	$64.30	$514.40	$39.67	$59.88
.5 Laborer	33.10	132.40	51.05	204.20		
1 Dozer, 200 H.P.		1192.00		1311.20		
1 Vibratory Roller, Towed, 23 Ton		434.20		477.62	135.52	149.07
12 L.H., Daily Totals		$2102.20		$2507.42	$175.18	$208.95
Crew B-10D	**Hr.**	**Daily**	**Hr.**	**Daily**	**Bare Costs**	**Incl. O&P**
1 Equip. Oper. (med.)	$42.95	$343.60	$64.30	$514.40	$39.67	$59.88
.5 Laborer	33.10	132.40	51.05	204.20		
1 Dozer, 200 H.P.		1192.00		1311.20		
1 Sheepsft. Roller, Towed		489.20		538.12	140.10	154.11
12 L.H., Daily Totals		$2157.20		$2567.92	$179.77	$213.99
Crew B-10E	**Hr.**	**Daily**	**Hr.**	**Daily**	**Bare Costs**	**Incl. O&P**
1 Equip. Oper. (med.)	$42.95	$343.60	$64.30	$514.40	$39.67	$59.88
.5 Laborer	33.10	132.40	51.05	204.20		
1 Tandem Roller, 5 Ton		152.00		167.20	12.67	13.93
12 L.H., Daily Totals		$628.00		$885.80	$52.33	$73.82
Crew B-10F	**Hr.**	**Daily**	**Hr.**	**Daily**	**Bare Costs**	**Incl. O&P**
1 Equip. Oper. (med.)	$42.95	$343.60	$64.30	$514.40	$39.67	$59.88
.5 Laborer	33.10	132.40	51.05	204.20		
1 Tandem Roller, 10 Ton		235.20		258.72	19.60	21.56
12 L.H., Daily Totals		$711.20		$977.32	$59.27	$81.44

Crew No.	Bare Costs		Incl. Subs O&P		Cost Per Labor-Hour	
Crew B-10G	**Hr.**	**Daily**	**Hr.**	**Daily**	**Bare Costs**	**Incl. O&P**
1 Equip. Oper. (med.)	$42.95	$343.60	$64.30	$514.40	$39.67	$59.88
.5 Laborer	33.10	132.40	51.05	204.20		
1 Sheepsft. Roll., 240 H.P.		1243.00		1367.30	103.58	113.94
12 L.H., Daily Totals		$1719.00		$2085.90	$143.25	$173.82
Crew B-10H	**Hr.**	**Daily**	**Hr.**	**Daily**	**Bare Costs**	**Incl. O&P**
1 Equip. Oper. (med.)	$42.95	$343.60	$64.30	$514.40	$39.67	$59.88
.5 Laborer	33.10	132.40	51.05	204.20		
1 Diaphragm Water Pump, 2″		68.00		74.80		
1 -20' Suction Hose, 2″		1.95		2.15		
2 -50' Discharge Hoses, 2″		1.80		1.98	5.98	6.58
12 L.H., Daily Totals		$547.75		$797.52	$45.65	$66.46
Crew B-10I	**Hr.**	**Daily**	**Hr.**	**Daily**	**Bare Costs**	**Incl. O&P**
1 Equip. Oper. (med.)	$42.95	$343.60	$64.30	$514.40	$39.67	$59.88
.5 Laborer	33.10	132.40	51.05	204.20		
1 Diaphragm Water Pump, 4″		90.80		99.88		
1 -20' Suction Hose, 4″		3.45		3.79		
2 -50' Discharge Hoses, 4″		4.70		5.17	8.25	9.07
12 L.H., Daily Totals		$574.95		$827.45	$47.91	$68.95
Crew B-10J	**Hr.**	**Daily**	**Hr.**	**Daily**	**Bare Costs**	**Incl. O&P**
1 Equip. Oper. (med.)	$42.95	$343.60	$64.30	$514.40	$39.67	$59.88
.5 Laborer	33.10	132.40	51.05	204.20		
1 Centrifugal Water Pump, 3″		74.40		81.84		
1 -20' Suction Hose, 3″		3.05		3.36		
2 -50' Discharge Hoses, 3″		3.50		3.85	6.75	7.42
12 L.H., Daily Totals		$556.95		$807.64	$46.41	$67.30
Crew B-10K	**Hr.**	**Daily**	**Hr.**	**Daily**	**Bare Costs**	**Incl. O&P**
1 Equip. Oper. (med.)	$42.95	$343.60	$64.30	$514.40	$39.67	$59.88
.5 Laborer	33.10	132.40	51.05	204.20		
1 Centr. Water Pump, 6″		340.00		374.00		
1 -20' Suction Hose, 6″		11.90		13.09		
2 -50' Discharge Hoses, 6″		12.60		13.86	30.38	33.41
12 L.H., Daily Totals		$840.50		$1119.55	$70.04	$93.30
Crew B-10L	**Hr.**	**Daily**	**Hr.**	**Daily**	**Bare Costs**	**Incl. O&P**
1 Equip. Oper. (med.)	$42.95	$343.60	$64.30	$514.40	$39.67	$59.88
.5 Laborer	33.10	132.40	51.05	204.20		
1 Dozer, 80 H.P.		436.60		480.26	36.38	40.02
12 L.H., Daily Totals		$912.60		$1198.86	$76.05	$99.91
Crew B-10M	**Hr.**	**Daily**	**Hr.**	**Daily**	**Bare Costs**	**Incl. O&P**
1 Equip. Oper. (med.)	$42.95	$343.60	$64.30	$514.40	$39.67	$59.88
.5 Laborer	33.10	132.40	51.05	204.20		
1 Dozer, 300 H.P.		1592.00		1751.20	132.67	145.93
12 L.H., Daily Totals		$2068.00		$2469.80	$172.33	$205.82
Crew B-10N	**Hr.**	**Daily**	**Hr.**	**Daily**	**Bare Costs**	**Incl. O&P**
1 Equip. Oper. (med.)	$42.95	$343.60	$64.30	$514.40	$39.67	$59.88
.5 Laborer	33.10	132.40	51.05	204.20		
1 F.E. Loader, T.M., 1.5 C.Y.		433.80		477.18	36.15	39.77
12 L.H., Daily Totals		$909.80		$1195.78	$75.82	$99.65
Crew B-10O	**Hr.**	**Daily**	**Hr.**	**Daily**	**Bare Costs**	**Incl. O&P**
1 Equip. Oper. (med.)	$42.95	$343.60	$64.30	$514.40	$39.67	$59.88
.5 Laborer	33.10	132.40	51.05	204.20		
1 F.E. Loader, T.M., 2.25 C.Y.		839.40		923.34	69.95	76.94
12 L.H., Daily Totals		$1315.40		$1641.94	$109.62	$136.83

Crew No.	Bare Costs		Incl. Subs O&P		Cost Per Labor-Hour	
Crew B-10P	**Hr.**	**Daily**	**Hr.**	**Daily**	**Bare Costs**	**Incl. O&P**
1 Equip. Oper. (med.)	$42.95	$343.60	$64.30	$514.40	$39.67	$59.88
.5 Laborer	33.10	132.40	51.05	204.20		
1 Crawler Loader, 3 C.Y.		1077.00		1184.70	89.75	98.72
12 L.H., Daily Totals		$1553.00		$1903.30	$129.42	$158.61
Crew B-10Q	**Hr.**	**Daily**	**Hr.**	**Daily**	**Bare Costs**	**Incl. O&P**
1 Equip. Oper. (med.)	$42.95	$343.60	$64.30	$514.40	$39.67	$59.88
.5 Laborer	33.10	132.40	51.05	204.20		
1 Crawler Loader, 4 C.Y.		1548.00		1702.80	129.00	141.90
12 L.H., Daily Totals		$2024.00		$2421.40	$168.67	$201.78
Crew B-10R	**Hr.**	**Daily**	**Hr.**	**Daily**	**Bare Costs**	**Incl. O&P**
1 Equip. Oper. (med.)	$42.95	$343.60	$64.30	$514.40	$39.67	$59.88
.5 Laborer	33.10	132.40	51.05	204.20		
1 F.E. Loader, W.M., 1 C.Y.		269.40		296.34	22.45	24.70
12 L.H., Daily Totals		$745.40		$1014.94	$62.12	$84.58
Crew B-10S	**Hr.**	**Daily**	**Hr.**	**Daily**	**Bare Costs**	**Incl. O&P**
1 Equip. Oper. (med.)	$42.95	$343.60	$64.30	$514.40	$39.67	$59.88
.5 Laborer	33.10	132.40	51.05	204.20		
1 F.E. Loader, W.M., 1.5 C.Y.		367.60		404.36	30.63	33.70
12 L.H., Daily Totals		$843.60		$1122.96	$70.30	$93.58
Crew B-10T	**Hr.**	**Daily**	**Hr.**	**Daily**	**Bare Costs**	**Incl. O&P**
1 Equip. Oper. (med.)	$42.95	$343.60	$64.30	$514.40	$39.67	$59.88
.5 Laborer	33.10	132.40	51.05	204.20		
1 F.E. Loader, W.M.,2.5 C.Y.		435.80		479.38	36.32	39.95
12 L.H., Daily Totals		$911.80		$1197.98	$75.98	$99.83
Crew B-10U	**Hr.**	**Daily**	**Hr.**	**Daily**	**Bare Costs**	**Incl. O&P**
1 Equip. Oper. (med.)	$42.95	$343.60	$64.30	$514.40	$39.67	$59.88
.5 Laborer	33.10	132.40	51.05	204.20		
1 F.E. Loader, W.M., 5.5 C.Y.		994.40		1093.84	82.87	91.15
12 L.H., Daily Totals		$1470.40		$1812.44	$122.53	$151.04
Crew B-10V	**Hr.**	**Daily**	**Hr.**	**Daily**	**Bare Costs**	**Incl. O&P**
1 Equip. Oper. (med.)	$42.95	$343.60	$64.30	$514.40	$39.67	$59.88
.5 Laborer	33.10	132.40	51.05	204.20		
1 Dozer, 700 H.P.		4598.00		5057.80	383.17	421.48
12 L.H., Daily Totals		$5074.00		$5776.40	$422.83	$481.37
Crew B-10W	**Hr.**	**Daily**	**Hr.**	**Daily**	**Bare Costs**	**Incl. O&P**
1 Equip. Oper. (med.)	$42.95	$343.60	$64.30	$514.40	$39.67	$59.88
.5 Laborer	33.10	132.40	51.05	204.20		
1 Dozer, 105 H.P.		660.40		726.44	55.03	60.54
12 L.H., Daily Totals		$1136.40		$1445.04	$94.70	$120.42
Crew B-10X	**Hr.**	**Daily**	**Hr.**	**Daily**	**Bare Costs**	**Incl. O&P**
1 Equip. Oper. (med.)	$42.95	$343.60	$64.30	$514.40	$39.67	$59.88
.5 Laborer	33.10	132.40	51.05	204.20		
1 Dozer, 410 H.P.		2079.00		2286.90	173.25	190.57
12 L.H., Daily Totals		$2555.00		$3005.50	$212.92	$250.46
Crew B-10Y	**Hr.**	**Daily**	**Hr.**	**Daily**	**Bare Costs**	**Incl. O&P**
1 Equip. Oper. (med.)	$42.95	$343.60	$64.30	$514.40	$39.67	$59.88
.5 Laborer	33.10	132.40	51.05	204.20		
1 Vibr. Roller, Towed, 12 Ton		516.60		568.26	43.05	47.35
12 L.H., Daily Totals		$992.60		$1286.86	$82.72	$107.24

Crew No.	Bare Costs		Incl. Subs O&P		Cost Per Labor-Hour	
Crew B-11A	**Hr.**	**Daily**	**Hr.**	**Daily**	**Bare Costs**	**Incl. O&P**
1 Equipment Oper. (med.)	$42.95	$343.60	$64.30	$514.40	$38.02	$57.67
1 Laborer	33.10	264.80	51.05	408.40		
1 Dozer, 200 H.P.		1192.00		1311.20	74.50	81.95
16 L.H., Daily Totals		$1800.40		$2234.00	$112.53	$139.63

Crew B-11B	Hr.	Daily	Hr.	Daily	Bare Costs	Incl. O&P
1 Equipment Oper. (light)	$41.30	$330.40	$61.85	$494.80	$37.20	$56.45
1 Laborer	33.10	264.80	51.05	408.40		
1 Air Powered Tamper		25.25		27.77		
1 Air Compressor, 365 cfm		255.60		281.16		
2 -50' Air Hoses, 1.5"		12.60		13.86	18.34	20.17
16 L.H., Daily Totals		$888.65		$1225.99	$55.54	$76.62

Crew B-11C	Hr.	Daily	Hr.	Daily	Bare Costs	Incl. O&P
1 Equipment Oper. (med.)	$42.95	$343.60	$64.30	$514.40	$38.02	$57.67
1 Laborer	33.10	264.80	51.05	408.40		
1 Backhoe Loader, 48 H.P.		337.80		371.58	21.11	23.22
16 L.H., Daily Totals		$946.20		$1294.38	$59.14	$80.90

Crew B-11J	Hr.	Daily	Hr.	Daily	Bare Costs	Incl. O&P
1 Equipment Oper. (med.)	$42.95	$343.60	$64.30	$514.40	$38.02	$57.67
1 Laborer	33.10	264.80	51.05	408.40		
1 Grader, 30,000 Lbs.		622.20		684.42		
1 Ripper, beam & 1 shank		80.00		88.00	43.89	48.28
16 L.H., Daily Totals		$1310.60		$1695.22	$81.91	$105.95

Crew B-11K	Hr.	Daily	Hr.	Daily	Bare Costs	Incl. O&P
1 Equipment Oper. (med.)	$42.95	$343.60	$64.30	$514.40	$38.02	$57.67
1 Laborer	33.10	264.80	51.05	408.40		
1 Trencher, Chain Type, 8' D		2090.00		2299.00	130.63	143.69
16 L.H., Daily Totals		$2698.40		$3221.80	$168.65	$201.36

Crew B-11L	Hr.	Daily	Hr.	Daily	Bare Costs	Incl. O&P
1 Equipment Oper. (med.)	$42.95	$343.60	$64.30	$514.40	$38.02	$57.67
1 Laborer	33.10	264.80	51.05	408.40		
1 Grader, 30,000 Lbs.		622.20		684.42	38.89	42.78
16 L.H., Daily Totals		$1230.60		$1607.22	$76.91	$100.45

Crew B-11M	Hr.	Daily	Hr.	Daily	Bare Costs	Incl. O&P
1 Equipment Oper. (med.)	$42.95	$343.60	$64.30	$514.40	$38.02	$57.67
1 Laborer	33.10	264.80	51.05	408.40		
1 Backhoe Loader, 80 H.P.		387.60		426.36	24.23	26.65
16 L.H., Daily Totals		$996.00		$1349.16	$62.25	$84.32

Crew B-11N	Hr.	Daily	Hr.	Daily	Bare Costs	Incl. O&P
1 Labor Foreman	$35.10	$280.80	$54.10	$432.80	$35.54	$54.00
2 Equipment Operators (med.)	42.95	687.20	64.30	1028.80		
6 Truck Drivers (hvy.)	33.15	1591.20	50.55	2426.40		
1 F.E. Loader, W.M., 5.5 C.Y.		994.40		1093.84		
1 Dozer, 410 H.P.		2079.00		2286.90		
6 Dump Trucks, Off Hwy., 50 Ton		10464.00		11510.40	188.02	206.82
72 L.H., Daily Totals		$16096.60		$18779.14	$223.56	$260.82

Crew B-11Q	Hr.	Daily	Hr.	Daily	Bare Costs	Incl. O&P
1 Equipment Operator (med.)	$42.95	$343.60	$64.30	$514.40	$39.67	$59.88
.5 Laborer	33.10	132.40	51.05	204.20		
1 Dozer, 140 H.P.		769.60		846.56	64.13	70.55
12 L.H., Daily Totals		$1245.60		$1565.16	$103.80	$130.43

Crew No.	Bare Costs		Incl. Subs O&P		Cost Per Labor-Hour	
Crew B-11R	**Hr.**	**Daily**	**Hr.**	**Daily**	**Bare Costs**	**Incl. O&P**
1 Equipment Operator (med.)	$42.95	$343.60	$64.30	$514.40	$39.67	$59.88
.5 Laborer	33.10	132.40	51.05	204.20		
1 Dozer, 200 H.P.		1192.00		1311.20	99.33	109.27
12 L.H., Daily Totals		$1668.00		$2029.80	$139.00	$169.15

Crew B-11S	Hr.	Daily	Hr.	Daily	Bare Costs	Incl. O&P
1 Equipment Operator (med.)	$42.95	$343.60	$64.30	$514.40	$39.67	$59.88
.5 Laborer	33.10	132.40	51.05	204.20		
1 Dozer, 300 H.P.		1592.00		1751.20		
1 Ripper, beam & 1 shank		80.00		88.00	139.33	153.27
12 L.H., Daily Totals		$2148.00		$2557.80	$179.00	$213.15

Crew B-11T	Hr.	Daily	Hr.	Daily	Bare Costs	Incl. O&P
1 Equipment Operator (med.)	$42.95	$343.60	$64.30	$514.40	$39.67	$59.88
.5 Laborer	33.10	132.40	51.05	204.20		
1 Dozer, 410 H.P.		2079.00		2286.90		
1 Ripper, beam & 2 shanks		90.20		99.22	180.77	198.84
12 L.H., Daily Totals		$2645.20		$3104.72	$220.43	$258.73

Crew B-11U	Hr.	Daily	Hr.	Daily	Bare Costs	Incl. O&P
1 Equipment Operator (med.)	$42.95	$343.60	$64.30	$514.40	$39.67	$59.88
.5 Laborer	33.10	132.40	51.05	204.20		
1 Dozer, 520 H.P.		2810.00		3091.00	234.17	257.58
12 L.H., Daily Totals		$3286.00		$3809.60	$273.83	$317.47

Crew B-11V	Hr.	Daily	Hr.	Daily	Bare Costs	Incl. O&P
3 Laborers	$33.10	$794.40	$51.05	$1225.20	$33.10	$51.05
1 Roller, 2-Drum, W.B., 7.5 H.P.		149.60		164.56	6.23	6.86
24 L.H., Daily Totals		$944.00		$1389.76	$39.33	$57.91

Crew B-11W	Hr.	Daily	Hr.	Daily	Bare Costs	Incl. O&P
1 Equipment Operator (med.)	$42.95	$343.60	$64.30	$514.40	$33.96	$51.74
1 Common Laborer	33.10	264.80	51.05	408.40		
10 Truck Drivers (hvy.)	33.15	2652.00	50.55	4044.00		
1 Dozer, 200 H.P.		1192.00		1311.20		
1 Vibratory Roller, Towed, 23 Ton		434.20		477.62		
10 Dump Trucks, 8 C.Y., 220 H.P.		4018.00		4419.80	58.79	64.67
96 L.H., Daily Totals		$8904.60		$11175.42	$92.76	$116.41

Crew B-11Y	Hr.	Daily	Hr.	Daily	Bare Costs	Incl. O&P
1 Labor Foreman (Outside)	$35.10	$280.80	$54.10	$432.80	$36.61	$55.81
5 Common Laborers	33.10	1324.00	51.05	2042.00		
3 Equipment Operators (med.)	42.95	1030.80	64.30	1543.20		
1 Dozer, 80 H.P.		436.60		480.26		
2 Rollers, 2-Drum, W.B., 7.5 H.P.		299.20		329.12		
4 Vibratory Plates, gas, 21"		174.40		191.84	12.64	13.91
72 L.H., Daily Totals		$3545.80		$5019.22	$49.25	$69.71

Crew B-12A	Hr.	Daily	Hr.	Daily	Bare Costs	Incl. O&P
1 Equip. Oper. (crane)	$44.40	$355.20	$66.45	$531.60	$38.75	$58.75
1 Laborer	33.10	264.80	51.05	408.40		
1 Hyd. Excavator, 1 C.Y.		801.60		881.76	50.10	55.11
16 L.H., Daily Totals		$1421.60		$1821.76	$88.85	$113.86

Crew B-12B	Hr.	Daily	Hr.	Daily	Bare Costs	Incl. O&P
1 Equip. Oper. (crane)	$44.40	$355.20	$66.45	$531.60	$38.75	$58.75
1 Laborer	33.10	264.80	51.05	408.40		
1 Hyd. Excavator, 1.5 C.Y.		1017.00		1118.70	63.56	69.92
16 L.H., Daily Totals		$1637.00		$2058.70	$102.31	$128.67

Crew No.	Bare Costs		Incl. Subs O&P		Cost Per Labor-Hour	
Crew B-12C	**Hr.**	**Daily**	**Hr.**	**Daily**	**Bare Costs**	**Incl. O&P**
1 Equip. Oper. (crane)	$44.40	$355.20	$66.45	$531.60	$38.75	$58.75
1 Laborer	33.10	264.80	51.05	408.40		
1 Hyd. Excavator, 2 C.Y.		1321.00		1453.10	82.56	90.82
16 L.H., Daily Totals		$1941.00		$2393.10	$121.31	$149.57

Crew B-12D	Hr.	Daily	Hr.	Daily	Bare Costs	Incl. O&P
1 Equip. Oper. (crane)	$44.40	$355.20	$66.45	$531.60	$38.75	$58.75
1 Laborer	33.10	264.80	51.05	408.40		
1 Hyd. Excavator, 3.5 C.Y.		2359.00		2594.90	147.44	162.18
16 L.H., Daily Totals		$2979.00		$3534.90	$186.19	$220.93

Crew B-12E	Hr.	Daily	Hr.	Daily	Bare Costs	Incl. O&P
1 Equip. Oper. (crane)	$44.40	$355.20	$66.45	$531.60	$38.75	$58.75
1 Laborer	33.10	264.80	51.05	408.40		
1 Hyd. Excavator, .5 C.Y.		386.60		425.26	24.16	26.58
16 L.H., Daily Totals		$1006.60		$1365.26	$62.91	$85.33

Crew B-12F	Hr.	Daily	Hr.	Daily	Bare Costs	Incl. O&P
1 Equip. Oper. (crane)	$44.40	$355.20	$66.45	$531.60	$38.75	$58.75
1 Laborer	33.10	264.80	51.05	408.40		
1 Hyd. Excavator, .75 C.Y.		677.80		745.58	42.36	46.60
16 L.H., Daily Totals		$1297.80		$1685.58	$81.11	$105.35

Crew B-12G	Hr.	Daily	Hr.	Daily	Bare Costs	Incl. O&P
1 Equip. Oper. (crane)	$44.40	$355.20	$66.45	$531.60	$38.75	$58.75
1 Laborer	33.10	264.80	51.05	408.40		
1 Crawler Crane, 15 Ton		693.75		763.13		
1 Clamshell Bucket, .5 C.Y.		37.00		40.70	45.67	50.24
16 L.H., Daily Totals		$1350.75		$1743.83	$84.42	$108.99

Crew B-12H	Hr.	Daily	Hr.	Daily	Bare Costs	Incl. O&P
1 Equip. Oper. (crane)	$44.40	$355.20	$66.45	$531.60	$38.75	$58.75
1 Laborer	33.10	264.80	51.05	408.40		
1 Crawler Crane, 25 Ton		1188.00		1306.80		
1 Clamshell Bucket, 1 C.Y.		46.60		51.26	77.16	84.88
16 L.H., Daily Totals		$1854.60		$2298.06	$115.91	$143.63

Crew B-12I	Hr.	Daily	Hr.	Daily	Bare Costs	Incl. O&P
1 Equip. Oper. (crane)	$44.40	$355.20	$66.45	$531.60	$38.75	$58.75
1 Laborer	33.10	264.80	51.05	408.40		
1 Crawler Crane, 20 Ton		892.30		981.53		
1 Dragline Bucket, .75 C.Y.		20.40		22.44	57.04	62.75
16 L.H., Daily Totals		$1532.70		$1943.97	$95.79	$121.50

Crew B-12J	Hr.	Daily	Hr.	Daily	Bare Costs	Incl. O&P
1 Equip. Oper. (crane)	$44.40	$355.20	$66.45	$531.60	$38.75	$58.75
1 Laborer	33.10	264.80	51.05	408.40		
1 Gradall, 5/8 C.Y.		812.80		894.08	50.80	55.88
16 L.H., Daily Totals		$1432.80		$1834.08	$89.55	$114.63

Crew B-12K	Hr.	Daily	Hr.	Daily	Bare Costs	Incl. O&P
1 Equip. Oper. (crane)	$44.40	$355.20	$66.45	$531.60	$38.75	$58.75
1 Laborer	33.10	264.80	51.05	408.40		
1 Gradall, 3 Ton, 1 C.Y.		958.40		1054.24	59.90	65.89
16 L.H., Daily Totals		$1578.40		$1994.24	$98.65	$124.64

Crew B-12L	Hr.	Daily	Hr.	Daily	Bare Costs	Incl. O&P
1 Equip. Oper. (crane)	$44.40	$355.20	$66.45	$531.60	$38.75	$58.75
1 Laborer	33.10	264.80	51.05	408.40		
1 Crawler Crane, 15 Ton		693.75		763.13		
1 F.E. Attachment, .5 C.Y.		55.60		61.16	46.83	51.52
16 L.H., Daily Totals		$1369.35		$1764.29	$85.58	$110.27

Crew B-12M	Hr.	Daily	Hr.	Daily	Bare Costs	Incl. O&P
1 Equip. Oper. (crane)	$44.40	$355.20	$66.45	$531.60	$38.75	$58.75
1 Laborer	33.10	264.80	51.05	408.40		
1 Crawler Crane, 20 Ton		892.30		981.53		
1 F.E. Attachment, .75 C.Y.		60.00		66.00	59.52	65.47
16 L.H., Daily Totals		$1572.30		$1987.53	$98.27	$124.22

Crew B-12N	Hr.	Daily	Hr.	Daily	Bare Costs	Incl. O&P
1 Equip. Oper. (crane)	$44.40	$355.20	$66.45	$531.60	$38.75	$58.75
1 Laborer	33.10	264.80	51.05	408.40		
1 Crawler Crane, 25 Ton		1188.00		1306.80		
1 F.E. Attachment, 1 C.Y.		65.80		72.38	78.36	86.20
16 L.H., Daily Totals		$1873.80		$2319.18	$117.11	$144.95

Crew B-12O	Hr.	Daily	Hr.	Daily	Bare Costs	Incl. O&P
1 Equip. Oper. (crane)	$44.40	$355.20	$66.45	$531.60	$38.75	$58.75
1 Laborer	33.10	264.80	51.05	408.40		
1 Crawler Crane, 40 Ton		1193.00		1312.30		
1 F.E. Attachment, 1.5 C.Y.		76.00		83.60	79.31	87.24
16 L.H., Daily Totals		$1889.00		$2335.90	$118.06	$145.99

Crew B-12P	Hr.	Daily	Hr.	Daily	Bare Costs	Incl. O&P
1 Equip. Oper. (crane)	$44.40	$355.20	$66.45	$531.60	$38.75	$58.75
1 Laborer	33.10	264.80	51.05	408.40		
1 Crawler Crane, 40 Ton		1193.00		1312.30		
1 Dragline Bucket, 1.5 C.Y.		32.80		36.08	76.61	84.27
16 L.H., Daily Totals		$1845.80		$2288.38	$115.36	$143.02

Crew B-12Q	Hr.	Daily	Hr.	Daily	Bare Costs	Incl. O&P
1 Equip. Oper. (crane)	$44.40	$355.20	$66.45	$531.60	$38.75	$58.75
1 Laborer	33.10	264.80	51.05	408.40		
1 Hyd. Excavator, 5/8 C.Y.		564.40		620.84	35.27	38.80
16 L.H., Daily Totals		$1184.40		$1560.84	$74.03	$97.55

Crew B-12S	Hr.	Daily	Hr.	Daily	Bare Costs	Incl. O&P
1 Equip. Oper. (crane)	$44.40	$355.20	$66.45	$531.60	$38.75	$58.75
1 Laborer	33.10	264.80	51.05	408.40		
1 Hyd. Excavator, 2.5 C.Y.		1784.00		1962.40	111.50	122.65
16 L.H., Daily Totals		$2404.00		$2902.40	$150.25	$181.40

Crew B-12T	Hr.	Daily	Hr.	Daily	Bare Costs	Incl. O&P
1 Equip. Oper. (crane)	$44.40	$355.20	$66.45	$531.60	$38.75	$58.75
1 Laborer	33.10	264.80	51.05	408.40		
1 Crawler Crane, 75 Ton		1516.00		1667.60		
1 F.E. Attachment, 3 C.Y.		98.40		108.24	100.90	110.99
16 L.H., Daily Totals		$2234.40		$2715.84	$139.65	$169.74

Crew B-12V	Hr.	Daily	Hr.	Daily	Bare Costs	Incl. O&P
1 Equip. Oper. (crane)	$44.40	$355.20	$66.45	$531.60	$38.75	$58.75
1 Laborer	33.10	264.80	51.05	408.40		
1 Crawler Crane, 75 Ton		1516.00		1667.60		
1 Dragline Bucket, 3 C.Y.		52.40		57.64	98.03	107.83
16 L.H., Daily Totals		$2188.40		$2665.24	$136.78	$166.58

Crew B-13	Hr.	Daily	Hr.	Daily	Bare Costs	Incl. O&P
1 Labor Foreman (outside)	$35.10	$280.80	$54.10	$432.80	$35.74	$54.59
4 Laborers	33.10	1059.20	51.05	1633.60		
1 Equip. Oper. (crane)	44.40	355.20	66.45	531.60		
1 Equip. Oper. Oiler	38.30	306.40	57.35	458.80		
1 Hyd. Crane, 25 Ton		747.60		822.36	13.35	14.69
56 L.H., Daily Totals		$2749.20		$3879.16	$49.09	$69.27

Crew B-13A	Hr.	Daily	Hr.	Daily	Bare Costs	Incl. O&P
1 Foreman	$35.10	$280.80	$54.10	$432.80	$36.21	$55.13
2 Laborers	33.10	529.60	51.05	816.80		
2 Equipment Operators(med.)	42.95	687.20	64.30	1028.80		
2 Truck Drivers (heavy)	33.15	530.40	50.55	808.80		
1 Crawler Crane, 75 Ton		1516.00		1667.60		
1 Crawler Loader, 4 C.Y.		1548.00		1702.80		
2 Dump Trucks, 8 C.Y., 220 H.P.		803.60		883.96	69.06	75.97
56 L.H., Daily Totals		$5895.60		$7341.56	$105.28	$131.10

Crew B-13B	Hr.	Daily	Hr.	Daily	Bare Costs	Incl. O&P
1 Labor Foreman (outside)	$35.10	$280.80	$54.10	$432.80	$35.74	$54.59
4 Laborers	33.10	1059.20	51.05	1633.60		
1 Equip. Oper. (crane)	44.40	355.20	66.45	531.60		
1 Equip. Oper. Oiler	38.30	306.40	57.35	458.80		
1 Hyd. Crane, 55 Ton		1209.00		1329.90	21.59	23.75
56 L.H., Daily Totals		$3210.60		$4386.70	$57.33	$78.33

Crew B-13C	Hr.	Daily	Hr.	Daily	Bare Costs	Incl. O&P
1 Labor Foreman (outside)	$35.10	$280.80	$54.10	$432.80	$35.74	$54.59
4 Laborers	33.10	1059.20	51.05	1633.60		
1 Equip. Oper. (crane)	44.40	355.20	66.45	531.60		
1 Equip. Oper. Oiler	38.30	306.40	57.35	458.80		
1 Crawler Crane, 100 Ton		1732.00		1905.20	30.93	34.02
56 L.H., Daily Totals		$3733.60		$4962.00	$66.67	$88.61

Crew B-13D	Hr.	Daily	Hr.	Daily	Bare Costs	Incl. O&P
1 Laborer	$33.10	$264.80	$51.05	$408.40	$38.75	$58.75
1 Equip. Oper. (crane)	44.40	355.20	66.45	531.60		
1 Hyd. Excavator, 1 C.Y.		801.60		881.76		
1 Trench Box		113.40		124.74	57.19	62.91
16 L.H., Daily Totals		$1535.00		$1946.50	$95.94	$121.66

Crew B-13E	Hr.	Daily	Hr.	Daily	Bare Costs	Incl. O&P
1 Laborer	$33.10	$264.80	$51.05	$408.40	$38.75	$58.75
1 Equip. Oper. (crane)	44.40	355.20	66.45	531.60		
1 Hyd. Excavator, 1.5 C.Y.		1017.00		1118.70		
1 Trench Box		113.40		124.74	70.65	77.72
16 L.H., Daily Totals		$1750.40		$2183.44	$109.40	$136.47

Crew B-13F	Hr.	Daily	Hr.	Daily	Bare Costs	Incl. O&P
1 Laborer	$33.10	$264.80	$51.05	$408.40	$38.75	$58.75
1 Equip. Oper. (crane)	44.40	355.20	66.45	531.60		
1 Hyd. Excavator, 3.5 C.Y.		2359.00		2594.90		
1 Trench Box		113.40		124.74	154.53	169.98
16 L.H., Daily Totals		$3092.40		$3659.64	$193.28	$228.73

Crew B-13G	Hr.	Daily	Hr.	Daily	Bare Costs	Incl. O&P
1 Laborer	$33.10	$264.80	$51.05	$408.40	$38.75	$58.75
1 Equip. Oper. (crane)	44.40	355.20	66.45	531.60		
1 Hyd. Excavator, .75 C.Y.		677.80		745.58		
1 Trench Box		113.40		124.74	49.45	54.40
16 L.H., Daily Totals		$1411.20		$1810.32	$88.20	$113.15

Crew B-13H	Hr.	Daily	Hr.	Daily	Bare Costs	Incl. O&P
1 Laborer	$33.10	$264.80	$51.05	$408.40	$38.75	$58.75
1 Equip. Oper. (crane)	44.40	355.20	66.45	531.60		
1 Gradall, 5/8 C.Y.		812.80		894.08		
1 Trench Box		113.40		124.74	57.89	63.68
16 L.H., Daily Totals		$1546.20		$1958.82	$96.64	$122.43

Crew B-13I	Hr.	Daily	Hr.	Daily	Bare Costs	Incl. O&P
1 Laborer	$33.10	$264.80	$51.05	$408.40	$38.75	$58.75
1 Equip. Oper. (crane)	44.40	355.20	66.45	531.60		
1 Gradall, 3 Ton, 1 C.Y.		958.40		1054.24		
1 Trench Box		113.40		124.74	66.99	73.69
16 L.H., Daily Totals		$1691.80		$2118.98	$105.74	$132.44

Crew B-13J	Hr.	Daily	Hr.	Daily	Bare Costs	Incl. O&P
1 Laborer	$33.10	$264.80	$51.05	$408.40	$38.75	$58.75
1 Equip. Oper. (crane)	44.40	355.20	66.45	531.60		
1 Hyd. Excavator, 2.5 C.Y.		1784.00		1962.40		
1 Trench Box		113.40		124.74	118.59	130.45
16 L.H., Daily Totals		$2517.40		$3027.14	$157.34	$189.20

Crew B-14	Hr.	Daily	Hr.	Daily	Bare Costs	Incl. O&P
1 Labor Foreman (outside)	$35.10	$280.80	$54.10	$432.80	$34.80	$53.36
4 Laborers	33.10	1059.20	51.05	1633.60		
1 Equip. Oper. (light)	41.30	330.40	61.85	494.80		
1 Backhoe Loader, 48 H.P.		337.80		371.58	7.04	7.74
48 L.H., Daily Totals		$2008.20		$2932.78	$41.84	$61.10

Crew B-14A	Hr.	Daily	Hr.	Daily	Bare Costs	Incl. O&P
1 Equip. Oper. (crane)	$44.40	$355.20	$66.45	$531.60	$40.63	$61.32
.5 Laborer	33.10	132.40	51.05	204.20		
1 Hyd. Excavator, 4.5 C.Y.		2870.00		3157.00	239.17	263.08
12 L.H., Daily Totals		$3357.60		$3892.80	$279.80	$324.40

Crew B-14B	Hr.	Daily	Hr.	Daily	Bare Costs	Incl. O&P
1 Equip. Oper. (crane)	$44.40	$355.20	$66.45	$531.60	$40.63	$61.32
.5 Laborer	33.10	132.40	51.05	204.20		
1 Hyd. Excavator, 6 C.Y.		3390.00		3729.00	282.50	310.75
12 L.H., Daily Totals		$3877.60		$4464.80	$323.13	$372.07

Crew B-14C	Hr.	Daily	Hr.	Daily	Bare Costs	Incl. O&P
1 Equip. Oper. (crane)	$44.40	$355.20	$66.45	$531.60	$40.63	$61.32
.5 Laborer	33.10	132.40	51.05	204.20		
1 Hyd. Excavator, 7 C.Y.		3553.00		3908.30	296.08	325.69
12 L.H., Daily Totals		$4040.60		$4644.10	$336.72	$387.01

Crew B-14F	Hr.	Daily	Hr.	Daily	Bare Costs	Incl. O&P
1 Equip. Oper. (crane)	$44.40	$355.20	$66.45	$531.60	$40.63	$61.32
.5 Laborer	33.10	132.40	51.05	204.20		
1 Hyd. Shovel, 7 C.Y.		3536.00		3889.60	294.67	324.13
12 L.H., Daily Totals		$4023.60		$4625.40	$335.30	$385.45

Crew B-14G	Hr.	Daily	Hr.	Daily	Bare Costs	Incl. O&P
1 Equip. Oper. (crane)	$44.40	$355.20	$66.45	$531.60	$40.63	$61.32
.5 Laborer	33.10	132.40	51.05	204.20		
1 Hyd. Shovel, 12 C.Y.		4614.00		5075.40	384.50	422.95
12 L.H., Daily Totals		$5101.60		$5811.20	$425.13	$484.27

Crew No.	Bare Costs		Incl. Subs O&P		Cost Per Labor-Hour	
Crew B-14J	**Hr.**	**Daily**	**Hr.**	**Daily**	**Bare Costs**	**Incl. O&P**
1 Equip. Oper. (med.)	$42.95	$343.60	$64.30	$514.40	$39.67	$59.88
.5 Laborer	33.10	132.40	51.05	204.20		
1 F.E. Loader, 8 C.Y.		1757.00		1932.70	146.42	161.06
12 L.H., Daily Totals		$2233.00		$2651.30	$186.08	$220.94
Crew B-14K	**Hr.**	**Daily**	**Hr.**	**Daily**	**Bare Costs**	**Incl. O&P**
1 Equip. Oper. (med.)	$42.95	$343.60	$64.30	$514.40	$39.67	$59.88
.5 Laborer	33.10	132.40	51.05	204.20		
1 F.E. Loader, 10 C.Y.		2601.00		2861.10	216.75	238.43
12 L.H., Daily Totals		$3077.00		$3579.70	$256.42	$298.31
Crew B-15	**Hr.**	**Daily**	**Hr.**	**Daily**	**Bare Costs**	**Incl. O&P**
1 Equipment Oper. (med.)	$42.95	$343.60	$64.30	$514.40	$35.94	$54.55
.5 Laborer	33.10	132.40	51.05	204.20		
2 Truck Drivers (heavy)	33.15	530.40	50.55	808.80		
2 Dump Trucks, 12 C.Y., 400 H.P.		1335.60		1469.16		
1 Dozer, 200 H.P.		1192.00		1311.20	90.27	99.30
28 L.H., Daily Totals		$3534.00		$4307.76	$126.21	$153.85
Crew B-16	**Hr.**	**Daily**	**Hr.**	**Daily**	**Bare Costs**	**Incl. O&P**
1 Labor Foreman (outside)	$35.10	$280.80	$54.10	$432.80	$33.61	$51.69
2 Laborers	33.10	529.60	51.05	816.80		
1 Truck Driver (heavy)	33.15	265.20	50.55	404.40		
1 Dump Truck, 12 C.Y., 400 H.P.		667.80		734.58	20.87	22.96
32 L.H., Daily Totals		$1743.40		$2388.58	$54.48	$74.64
Crew B-17	**Hr.**	**Daily**	**Hr.**	**Daily**	**Bare Costs**	**Incl. O&P**
2 Laborers	$33.10	$529.60	$51.05	$816.80	$35.16	$53.63
1 Equip. Oper. (light)	41.30	330.40	61.85	494.80		
1 Truck Driver (heavy)	33.15	265.20	50.55	404.40		
1 Backhoe Loader, 48 H.P.		337.80		371.58		
1 Dump Truck, 8 C.Y., 220 H.P.		401.80		441.98	23.11	25.42
32 L.H., Daily Totals		$1864.80		$2529.56	$58.27	$79.05
Crew B-17A	**Hr.**	**Daily**	**Hr.**	**Daily**	**Bare Costs**	**Incl. O&P**
2 Laborer Foremen	$35.10	$561.60	$54.10	$865.60	$35.60	$54.85
6 Laborers	33.10	1588.80	51.05	2450.40		
1 Skilled Worker Foreman	44.60	356.80	68.55	548.40		
1 Skilled Worker	42.60	340.80	65.50	524.00		
80 L.H., Daily Totals		$2848.00		$4388.40	$35.60	$54.85
Crew B-18	**Hr.**	**Daily**	**Hr.**	**Daily**	**Bare Costs**	**Incl. O&P**
1 Labor Foreman (outside)	$35.10	$280.80	$54.10	$432.80	$33.77	$52.07
2 Laborers	33.10	529.60	51.05	816.80		
1 Vibratory Plate, gas, 21"		43.60		47.96	1.82	2.00
24 L.H., Daily Totals		$854.00		$1297.56	$35.58	$54.06
Crew B-19	**Hr.**	**Daily**	**Hr.**	**Daily**	**Bare Costs**	**Incl. O&P**
1 Pile Driver Foreman	$42.30	$338.40	$68.00	$544.00	$41.33	$64.68
4 Pile Drivers	40.30	1289.60	64.80	2073.60		
2 Equip. Oper. (crane)	44.40	710.40	66.45	1063.20		
1 Equip. Oper. Oiler	38.30	306.40	57.35	458.80		
1 Crawler Crane, 40 Ton		1193.00		1312.30		
1 Lead, 90' high		120.40		132.44		
1 Hammer, Diesel, 22k ft-lb		619.40		681.34	30.20	33.22
64 L.H., Daily Totals		$4577.60		$6265.68	$71.53	$97.90

Crew No.	Bare Costs		Incl. Subs O&P		Cost Per Labor-Hour	
Crew B-19A	**Hr.**	**Daily**	**Hr.**	**Daily**	**Bare Costs**	**Incl. O&P**
1 Pile Driver Foreman	$42.30	$338.40	$68.00	$544.00	$41.33	$64.68
4 Pile Drivers	40.30	1289.60	64.80	2073.60		
2 Equip. Oper. (crane)	44.40	710.40	66.45	1063.20		
1 Equip. Oper. Oiler	38.30	306.40	57.35	458.80		
1 Crawler Crane, 75 Ton		1516.00		1667.60		
1 Lead, 90' high		120.40		132.44		
1 Hammer, Diesel, 41k ft-lb		722.80		795.08	36.86	40.55
64 L.H., Daily Totals		$5004.00		$6734.72	$78.19	$105.23
Crew B-20	**Hr.**	**Daily**	**Hr.**	**Daily**	**Bare Costs**	**Incl. O&P**
1 Labor Foreman (out)	$35.10	$280.80	$54.10	$432.80	$36.93	$56.88
1 Skilled Worker	42.60	340.80	65.50	524.00		
1 Laborer	33.10	264.80	51.05	408.40		
24 L.H., Daily Totals		$886.40		$1365.20	$36.93	$56.88
Crew B-20A	**Hr.**	**Daily**	**Hr.**	**Daily**	**Bare Costs**	**Incl. O&P**
1 Labor Foreman	$35.10	$280.80	$54.10	$432.80	$40.48	$61.40
1 Laborer	33.10	264.80	51.05	408.40		
1 Plumber	52.05	416.40	78.00	624.00		
1 Plumber Apprentice	41.65	333.20	62.45	499.60		
32 L.H., Daily Totals		$1295.20		$1964.80	$40.48	$61.40
Crew B-21	**Hr.**	**Daily**	**Hr.**	**Daily**	**Bare Costs**	**Incl. O&P**
1 Labor Foreman (out)	$35.10	$280.80	$54.10	$432.80	$38.00	$58.25
1 Skilled Worker	42.60	340.80	65.50	524.00		
1 Laborer	33.10	264.80	51.05	408.40		
.5 Equip. Oper. (crane)	44.40	177.60	66.45	265.80		
.5 S.P. Crane, 4x4, 5 Ton		138.10		151.91	4.93	5.43
28 L.H., Daily Totals		$1202.10		$1782.91	$42.93	$63.68
Crew B-21A	**Hr.**	**Daily**	**Hr.**	**Daily**	**Bare Costs**	**Incl. O&P**
1 Labor Foreman	$35.10	$280.80	$54.10	$432.80	$41.26	$62.41
1 Laborer	33.10	264.80	51.05	408.40		
1 Plumber	52.05	416.40	78.00	624.00		
1 Plumber Apprentice	41.65	333.20	62.45	499.60		
1 Equip. Oper. (crane)	44.40	355.20	66.45	531.60		
1 S.P. Crane, 4x4, 12 Ton		655.60		721.16	16.39	18.03
40 L.H., Daily Totals		$2306.00		$3217.56	$57.65	$80.44
Crew B-21B	**Hr.**	**Daily**	**Hr.**	**Daily**	**Bare Costs**	**Incl. O&P**
1 Laborer Foreman	$35.10	$280.80	$54.10	$432.80	$35.76	$54.74
3 Laborers	33.10	794.40	51.05	1225.20		
1 Equip. Oper. (crane)	44.40	355.20	66.45	531.60		
1 Hyd. Crane, 12 Ton		655.60		721.16	16.39	18.03
40 L.H., Daily Totals		$2086.00		$2910.76	$52.15	$72.77
Crew B-21C	**Hr.**	**Daily**	**Hr.**	**Daily**	**Bare Costs**	**Incl. O&P**
1 Laborer Foreman	$35.10	$280.80	$54.10	$432.80	$35.74	$54.59
4 Laborers	33.10	1059.20	51.05	1633.60		
1 Equip. Oper. (crane)	44.40	355.20	66.45	531.60		
1 Equip. Oper. Oiler	38.30	306.40	57.35	458.80		
2 Cutting Torches		38.00		41.80		
2 Set of Gases		158.40		174.24		
1 Lattice Boom Crane, 90 Ton		1607.00		1767.70	32.20	35.42
56 L.H., Daily Totals		$3805.00		$5040.54	$67.95	$90.01

Crew No.	Bare Costs		Incl. Subs O&P		Cost Per Labor-Hour	
Crew B-22	**Hr.**	**Daily**	**Hr.**	**Daily**	**Bare Costs**	**Incl. O&P**
1 Labor Foreman (out)	$35.10	$280.80	$54.10	$432.80	$38.43	$58.80
1 Skilled Worker	42.60	340.80	65.50	524.00		
1 Laborer	33.10	264.80	51.05	408.40		
.75 Equip. Oper. (crane)	44.40	266.40	66.45	398.70		
.75 S.P. Crane, 4x4, 5 Ton		207.15		227.87	6.91	7.60
30 L.H., Daily Totals		$1359.95		$1991.77	$45.33	$66.39
Crew B-22A	**Hr.**	**Daily**	**Hr.**	**Daily**	**Bare Costs**	**Incl. O&P**
1 Labor Foreman (out)	$35.10	$280.80	$54.10	$432.80	$37.31	$57.17
1 Skilled Worker	42.60	340.80	65.50	524.00		
2 Laborers	33.10	529.60	51.05	816.80		
.75 Equipment Oper. (crane)	44.40	266.40	66.45	398.70		
.75 S.P. Crane, 4x4, 5 Ton		207.15		227.87		
1 Generator, 5 kW		46.00		50.60		
1 Butt Fusion Machine		134.40		147.84	10.20	11.22
38 L.H., Daily Totals		$1805.15		$2598.61	$47.50	$68.38
Crew B-22B	**Hr.**	**Daily**	**Hr.**	**Daily**	**Bare Costs**	**Incl. O&P**
1 Skilled Worker	$42.60	$340.80	$65.50	$524.00	$37.85	$58.27
1 Laborer	33.10	264.80	51.05	408.40		
1 Electro Fusion Machine		98.60		108.46	6.16	6.78
16 L.H., Daily Totals		$704.20		$1040.86	$44.01	$65.05
Crew B-23	**Hr.**	**Daily**	**Hr.**	**Daily**	**Bare Costs**	**Incl. O&P**
1 Labor Foreman (outside)	$35.10	$280.80	$54.10	$432.80	$33.50	$51.66
4 Laborers	33.10	1059.20	51.05	1633.60		
1 Drill Rig, Truck-Mounted		2591.00		2850.10		
1 Flatbed Truck, Gas, 3 Ton		296.00		325.60	72.17	79.39
40 L.H., Daily Totals		$4227.00		$5242.10	$105.68	$131.05
Crew B-23A	**Hr.**	**Daily**	**Hr.**	**Daily**	**Bare Costs**	**Incl. O&P**
1 Labor Foreman (outside)	$35.10	$280.80	$54.10	$432.80	$37.05	$56.48
1 Laborer	33.10	264.80	51.05	408.40		
1 Equip. Operator (med.)	42.95	343.60	64.30	514.40		
1 Drill Rig, Truck-Mounted		2591.00		2850.10		
1 Pickup Truck, 3/4 Ton		140.00		154.00	113.79	125.17
24 L.H., Daily Totals		$3620.20		$4359.70	$150.84	$181.65
Crew B-23B	**Hr.**	**Daily**	**Hr.**	**Daily**	**Bare Costs**	**Incl. O&P**
1 Labor Foreman (outside)	$35.10	$280.80	$54.10	$432.80	$37.05	$56.48
1 Laborer	33.10	264.80	51.05	408.40		
1 Equip. Operator (med.)	42.95	343.60	64.30	514.40		
1 Drill Rig, Truck-Mounted		2591.00		2850.10		
1 Pickup Truck, 3/4 Ton		140.00		154.00		
1 Centr. Water Pump, 6"		340.00		374.00	127.96	140.75
24 L.H., Daily Totals		$3960.20		$4733.70	$165.01	$197.24
Crew B-24	**Hr.**	**Daily**	**Hr.**	**Daily**	**Bare Costs**	**Incl. O&P**
1 Cement Finisher	$39.70	$317.60	$57.95	$463.60	$38.12	$57.68
1 Laborer	33.10	264.80	51.05	408.40		
1 Carpenter	41.55	332.40	64.05	512.40		
24 L.H., Daily Totals		$914.80		$1384.40	$38.12	$57.68
Crew B-25	**Hr.**	**Daily**	**Hr.**	**Daily**	**Bare Costs**	**Incl. O&P**
1 Labor Foreman	$35.10	$280.80	$54.10	$432.80	$35.97	$54.94
7 Laborers	33.10	1853.60	51.05	2858.80		
3 Equip. Oper. (med.)	42.95	1030.80	64.30	1543.20		
1 Asphalt Paver, 130 H.P.		1994.00		2193.40		
1 Tandem Roller, 10 Ton		235.20		258.72		
1 Roller, Pneum. Whl, 12 Ton		337.20		370.92	29.16	32.08
88 L.H., Daily Totals		$5731.60		$7657.84	$65.13	$87.02

Crew No.	Bare Costs		Incl. Subs O&P		Cost Per Labor-Hour	
Crew B-25B	**Hr.**	**Daily**	**Hr.**	**Daily**	**Bare Costs**	**Incl. O&P**
1 Labor Foreman	$35.10	$280.80	$54.10	$432.80	$36.55	$55.72
7 Laborers	33.10	1853.60	51.05	2858.80		
4 Equip. Oper. (med.)	42.95	1374.40	64.30	2057.60		
1 Asphalt Paver, 130 H.P.		1994.00		2193.40		
2 Tandem Rollers, 10 Ton		470.40		517.44		
1 Roller, Pneum. Whl, 12 Ton		337.20		370.92	29.18	32.10
96 L.H., Daily Totals		$6310.40		$8430.96	$65.73	$87.82
Crew B-25C	**Hr.**	**Daily**	**Hr.**	**Daily**	**Bare Costs**	**Incl. O&P**
1 Labor Foreman	$35.10	$280.80	$54.10	$432.80	$36.72	$55.98
3 Laborers	33.10	794.40	51.05	1225.20		
2 Equip. Oper. (med.)	42.95	687.20	64.30	1028.80		
1 Asphalt Paver, 130 H.P.		1994.00		2193.40		
1 Tandem Roller, 10 Ton		235.20		258.72	46.44	51.09
48 L.H., Daily Totals		$3991.60		$5138.92	$83.16	$107.06
Crew B-26	**Hr.**	**Daily**	**Hr.**	**Daily**	**Bare Costs**	**Incl. O&P**
1 Labor Foreman (outside)	$35.10	$280.80	$54.10	$432.80	$36.92	$56.58
6 Laborers	33.10	1588.80	51.05	2450.40		
2 Equip. Oper. (med.)	42.95	687.20	64.30	1028.80		
1 Rodman (reinf.)	46.80	374.40	75.40	603.20		
1 Cement Finisher	39.70	317.60	57.95	463.60		
1 Grader, 30,000 Lbs.		622.20		684.42		
1 Paving Mach. & Equip.		2678.00		2945.80	37.50	41.25
88 L.H., Daily Totals		$6549.00		$8609.02	$74.42	$97.83
Crew B-26A	**Hr.**	**Daily**	**Hr.**	**Daily**	**Bare Costs**	**Incl. O&P**
1 Labor Foreman (outside)	$35.10	$280.80	$54.10	$432.80	$36.92	$56.58
6 Laborers	33.10	1588.80	51.05	2450.40		
2 Equip. Oper. (med.)	42.95	687.20	64.30	1028.80		
1 Rodman (reinf.)	46.80	374.40	75.40	603.20		
1 Cement Finisher	39.70	317.60	57.95	463.60		
1 Grader, 30,000 Lbs.		622.20		684.42		
1 Paving Mach. & Equip.		2678.00		2945.80		
1 Concrete Saw		161.40		177.54	39.34	43.27
88 L.H., Daily Totals		$6710.40		$8786.56	$76.25	$99.85
Crew B-26B	**Hr.**	**Daily**	**Hr.**	**Daily**	**Bare Costs**	**Incl. O&P**
1 Labor Foreman (outside)	$35.10	$280.80	$54.10	$432.80	$37.42	$57.22
6 Laborers	33.10	1588.80	51.05	2450.40		
3 Equip. Oper. (med.)	42.95	1030.80	64.30	1543.20		
1 Rodman (reinf.)	46.80	374.40	75.40	603.20		
1 Cement Finisher	39.70	317.60	57.95	463.60		
1 Grader, 30,000 Lbs.		622.20		684.42		
1 Paving Mach. & Equip.		2678.00		2945.80		
1 Concrete Pump, 110' Boom		976.40		1074.04	44.55	49.00
96 L.H., Daily Totals		$7869.00		$10197.46	$81.97	$106.22
Crew B-27	**Hr.**	**Daily**	**Hr.**	**Daily**	**Bare Costs**	**Incl. O&P**
1 Labor Foreman (outside)	$35.10	$280.80	$54.10	$432.80	$33.60	$51.81
3 Laborers	33.10	794.40	51.05	1225.20		
1 Berm Machine		285.60		314.16	8.93	9.82
32 L.H., Daily Totals		$1360.80		$1972.16	$42.52	$61.63
Crew B-28	**Hr.**	**Daily**	**Hr.**	**Daily**	**Bare Costs**	**Incl. O&P**
2 Carpenters	$41.55	$664.80	$64.05	$1024.80	$38.73	$59.72
1 Laborer	33.10	264.80	51.05	408.40		
24 L.H., Daily Totals		$929.60		$1433.20	$38.73	$59.72

Crew No.	Bare Costs		Incl. Subs O&P		Cost Per Labor-Hour	
Crew B-29	**Hr.**	**Daily**	**Hr.**	**Daily**	**Bare Costs**	**Incl. O&P**
1 Labor Foreman (outside)	$35.10	$280.80	$54.10	$432.80	$35.74	$54.59
4 Laborers	33.10	1059.20	51.05	1633.60		
1 Equip. Oper. (crane)	44.40	355.20	66.45	531.60		
1 Equip. Oper. Oiler	38.30	306.40	57.35	458.80		
1 Gradall, 5/8 C.Y.		812.80		894.08	14.51	15.97
56 L.H., Daily Totals		$2814.40		$3950.88	$50.26	$70.55
Crew B-30	**Hr.**	**Daily**	**Hr.**	**Daily**	**Bare Costs**	**Incl. O&P**
1 Equip. Oper. (med.)	$42.95	$343.60	$64.30	$514.40	$36.42	$55.13
2 Truck Drivers (heavy)	33.15	530.40	50.55	808.80		
1 Hyd. Excavator, 1.5 C.Y.		1017.00		1118.70		
2 Dump Trucks, 12 C.Y., 400 H.P.		1335.60		1469.16	98.03	107.83
24 L.H., Daily Totals		$3226.60		$3911.06	$134.44	$162.96
Crew B-31	**Hr.**	**Daily**	**Hr.**	**Daily**	**Bare Costs**	**Incl. O&P**
1 Labor Foreman (outside)	$35.10	$280.80	$54.10	$432.80	$35.19	$54.26
3 Laborers	33.10	794.40	51.05	1225.20		
1 Carpenter	41.55	332.40	64.05	512.40		
1 Air Compressor, 250 cfm		193.40		212.74		
1 Sheeting Driver		5.75		6.33		
2 -50' Air Hoses, 1.5"		12.60		13.86	5.29	5.82
40 L.H., Daily Totals		$1619.35		$2403.32	$40.48	$60.08
Crew B-32	**Hr.**	**Daily**	**Hr.**	**Daily**	**Bare Costs**	**Incl. O&P**
1 Laborer	$33.10	$264.80	$51.05	$408.40	$40.49	$60.99
3 Equip. Oper. (med.)	42.95	1030.80	64.30	1543.20		
1 Grader, 30,000 Lbs.		622.20		684.42		
1 Tandem Roller, 10 Ton		235.20		258.72		
1 Dozer, 200 H.P.		1192.00		1311.20	64.04	70.45
32 L.H., Daily Totals		$3345.00		$4205.94	$104.53	$131.44
Crew B-32A	**Hr.**	**Daily**	**Hr.**	**Daily**	**Bare Costs**	**Incl. O&P**
1 Laborer	$33.10	$264.80	$51.05	$408.40	$39.67	$59.88
2 Equip. Oper. (med.)	42.95	687.20	64.30	1028.80		
1 Grader, 30,000 Lbs.		622.20		684.42		
1 Roller, Vibratory, 25 Ton		633.20		696.52	52.31	57.54
24 L.H., Daily Totals		$2207.40		$2818.14	$91.97	$117.42
Crew B-32B	**Hr.**	**Daily**	**Hr.**	**Daily**	**Bare Costs**	**Incl. O&P**
1 Laborer	$33.10	$264.80	$51.05	$408.40	$39.67	$59.88
2 Equip. Oper. (med.)	42.95	687.20	64.30	1028.80		
1 Dozer, 200 H.P.		1192.00		1311.20		
1 Roller, Vibratory, 25 Ton		633.20		696.52	76.05	83.66
24 L.H., Daily Totals		$2777.20		$3444.92	$115.72	$143.54
Crew B-32C	**Hr.**	**Daily**	**Hr.**	**Daily**	**Bare Costs**	**Incl. O&P**
1 Labor Foreman	$35.10	$280.80	$54.10	$432.80	$38.36	$58.18
2 Laborers	33.10	529.60	51.05	816.80		
3 Equip. Oper. (med.)	42.95	1030.80	64.30	1543.20		
1 Grader, 30,000 Lbs.		622.20		684.42		
1 Tandem Roller, 10 Ton		235.20		258.72		
1 Dozer, 200 H.P.		1192.00		1311.20	42.70	46.97
48 L.H., Daily Totals		$3890.60		$5047.14	$81.05	$105.15
Crew B-33A	**Hr.**	**Daily**	**Hr.**	**Daily**	**Bare Costs**	**Incl. O&P**
1 Equip. Oper. (med.)	$42.95	$343.60	$64.30	$514.40	$40.14	$60.51
.5 Laborer	33.10	132.40	51.05	204.20		
.25 Equip. Oper. (med.)	42.95	85.90	64.30	128.60		
1 Scraper, Towed, 7 C.Y.		105.20		115.72		
1.25 Dozers, 300 H.P.		1990.00		2189.00	149.66	164.62
14 L.H., Daily Totals		$2657.10		$3151.92	$189.79	$225.14

Crew No.	Bare Costs		Incl. Subs O&P		Cost Per Labor-Hour	
Crew B-33B	**Hr.**	**Daily**	**Hr.**	**Daily**	**Bare Costs**	**Incl. O&P**
1 Equip. Oper. (med.)	$42.95	$343.60	$64.30	$514.40	$40.14	$60.51
.5 Laborer	33.10	132.40	51.05	204.20		
.25 Equip. Oper. (med.)	42.95	85.90	64.30	128.60		
1 Scraper, Towed, 10 C.Y.		156.60		172.26		
1.25 Dozers, 300 H.P.		1990.00		2189.00	153.33	168.66
14 L.H., Daily Totals		$2708.50		$3208.46	$193.46	$229.18
Crew B-33C	**Hr.**	**Daily**	**Hr.**	**Daily**	**Bare Costs**	**Incl. O&P**
1 Equip. Oper. (med.)	$42.95	$343.60	$64.30	$514.40	$40.14	$60.51
.5 Laborer	33.10	132.40	51.05	204.20		
.25 Equip. Oper. (med.)	42.95	85.90	64.30	128.60		
1 Scraper, Towed, 15 C.Y.		172.20		189.42		
1.25 Dozers, 300 H.P.		1990.00		2189.00	154.44	169.89
14 L.H., Daily Totals		$2724.10		$3225.62	$194.58	$230.40
Crew B-33D	**Hr.**	**Daily**	**Hr.**	**Daily**	**Bare Costs**	**Incl. O&P**
1 Equip. Oper. (med.)	$42.95	$343.60	$64.30	$514.40	$40.14	$60.51
.5 Laborer	33.10	132.40	51.05	204.20		
.25 Equip. Oper. (med.)	42.95	85.90	64.30	128.60		
1 S.P. Scraper, 14 C.Y.		1717.00		1888.70		
.25 Dozer, 300 H.P.		398.00		437.80	151.07	166.18
14 L.H., Daily Totals		$2676.90		$3173.70	$191.21	$226.69
Crew B-33E	**Hr.**	**Daily**	**Hr.**	**Daily**	**Bare Costs**	**Incl. O&P**
1 Equip. Oper. (med.)	$42.95	$343.60	$64.30	$514.40	$40.14	$60.51
.5 Laborer	33.10	132.40	51.05	204.20		
.25 Equip. Oper. (med.)	42.95	85.90	64.30	128.60		
1 S.P. Scraper, 21 C.Y.		2242.00		2466.20		
.25 Dozer, 300 H.P.		398.00		437.80	188.57	207.43
14 L.H., Daily Totals		$3201.90		$3751.20	$228.71	$267.94
Crew B-33F	**Hr.**	**Daily**	**Hr.**	**Daily**	**Bare Costs**	**Incl. O&P**
1 Equip. Oper. (med.)	$42.95	$343.60	$64.30	$514.40	$40.14	$60.51
.5 Laborer	33.10	132.40	51.05	204.20		
.25 Equip. Oper. (med.)	42.95	85.90	64.30	128.60		
1 Elev. Scraper, 11 C.Y.		1185.00		1303.50		
.25 Dozer, 300 H.P.		398.00		437.80	113.07	124.38
14 L.H., Daily Totals		$2144.90		$2588.50	$153.21	$184.89
Crew B-33G	**Hr.**	**Daily**	**Hr.**	**Daily**	**Bare Costs**	**Incl. O&P**
1 Equip. Oper. (med.)	$42.95	$343.60	$64.30	$514.40	$40.14	$60.51
.5 Laborer	33.10	132.40	51.05	204.20		
.25 Equip. Oper. (med.)	42.95	85.90	64.30	128.60		
1 Elev. Scraper, 22 C.Y.		2283.00		2511.30		
.25 Dozer, 300 H.P.		398.00		437.80	191.50	210.65
14 L.H., Daily Totals		$3242.90		$3796.30	$231.64	$271.16
Crew B-33H	**Hr.**	**Daily**	**Hr.**	**Daily**	**Bare Costs**	**Incl. O&P**
.5 Laborer	$33.10	$132.40	$51.05	$204.20	$40.14	$60.51
1 Equipment Operator (med.)	42.95	343.60	64.30	514.40		
.25 Equipment Operator (med.)	42.95	85.90	64.30	128.60		
1 S.P. Scraper, 44 C.Y.		4296.00		4725.60		
.25 Dozer, 410 H.P.		519.75		571.73	343.98	378.38
14 L.H., Daily Totals		$5377.65		$6144.52	$384.12	$438.89
Crew B-33J	**Hr.**	**Daily**	**Hr.**	**Daily**	**Bare Costs**	**Incl. O&P**
1 Equipment Operator (med.)	$42.95	$343.60	$64.30	$514.40	$42.95	$64.30
1 S.P. Scraper, 14 C.Y.		1717.00		1888.70	214.63	236.09
8 L.H., Daily Totals		$2060.60		$2403.10	$257.57	$300.39

Crew No.	Bare Costs		Incl. Subs O&P		Cost Per Labor-Hour	
Crew B-33K	**Hr.**	**Daily**	**Hr.**	**Daily**	**Bare Costs**	**Incl. O&P**
1 Equipment Operator (med.)	$42.95	$343.60	$64.30	$514.40	$40.14	$60.51
.25 Equipment Operator (med.)	42.95	85.90	64.30	128.60		
.5 Laborer	33.10	132.40	51.05	204.20		
1 S.P. Scraper, 31 C.Y.		3219.00		3540.90		
.25 Dozer, 410 H.P.		519.75		571.73	267.05	293.76
14 L.H., Daily Totals		$4300.65		$4959.82	$307.19	$354.27

Crew B-34A	Hr.	Daily	Hr.	Daily	Bare Costs	Incl. O&P
1 Truck Driver (heavy)	$33.15	$265.20	$50.55	$404.40	$33.15	$50.55
1 Dump Truck, 8 C.Y., 220 H.P.		401.80		441.98	50.23	55.25
8 L.H., Daily Totals		$667.00		$846.38	$83.38	$105.80

Crew B-34B	Hr.	Daily	Hr.	Daily	Bare Costs	Incl. O&P
1 Truck Driver (heavy)	$33.15	$265.20	$50.55	$404.40	$33.15	$50.55
1 Dump Truck, 12 C.Y., 400 H.P.		667.80		734.58	83.47	91.82
8 L.H., Daily Totals		$933.00		$1138.98	$116.63	$142.37

Crew B-34C	Hr.	Daily	Hr.	Daily	Bare Costs	Incl. O&P
1 Truck Driver (heavy)	$33.15	$265.20	$50.55	$404.40	$33.15	$50.55
1 Truck Tractor, 6x4, 380 H.P.		592.40		651.64		
1 Dump Trailer, 16.5 C.Y.		120.20		132.22	89.08	97.98
8 L.H., Daily Totals		$977.80		$1188.26	$122.22	$148.53

Crew B-34D	Hr.	Daily	Hr.	Daily	Bare Costs	Incl. O&P
1 Truck Driver (heavy)	$33.15	$265.20	$50.55	$404.40	$33.15	$50.55
1 Truck Tractor, 6x4, 380 H.P.		592.40		651.64		
1 Dump Trailer, 20 C.Y.		134.80		148.28	90.90	99.99
8 L.H., Daily Totals		$992.40		$1204.32	$124.05	$150.54

Crew B-34E	Hr.	Daily	Hr.	Daily	Bare Costs	Incl. O&P
1 Truck Driver (heavy)	$33.15	$265.20	$50.55	$404.40	$33.15	$50.55
1 Dump Truck, Off Hwy., 25 Ton		1399.00		1538.90	174.88	192.36
8 L.H., Daily Totals		$1664.20		$1943.30	$208.03	$242.91

Crew B-34F	Hr.	Daily	Hr.	Daily	Bare Costs	Incl. O&P
1 Truck Driver (heavy)	$33.15	$265.20	$50.55	$404.40	$33.15	$50.55
1 Dump Truck, Off Hwy., 35 Ton		1267.00		1393.70	158.38	174.21
8 L.H., Daily Totals		$1532.20		$1798.10	$191.53	$224.76

Crew B-34G	Hr.	Daily	Hr.	Daily	Bare Costs	Incl. O&P
1 Truck Driver (heavy)	$33.15	$265.20	$50.55	$404.40	$33.15	$50.55
1 Dump Truck, Off Hwy., 50 Ton		1744.00		1918.40	218.00	239.80
8 L.H., Daily Totals		$2009.20		$2322.80	$251.15	$290.35

Crew B-34H	Hr.	Daily	Hr.	Daily	Bare Costs	Incl. O&P
1 Truck Driver (heavy)	$33.15	$265.20	$50.55	$404.40	$33.15	$50.55
1 Dump Truck, Off Hwy., 65 Ton		1770.00		1947.00	221.25	243.38
8 L.H., Daily Totals		$2035.20		$2351.40	$254.40	$293.93

Crew B-34I	Hr.	Daily	Hr.	Daily	Bare Costs	Incl. O&P
1 Truck Driver (heavy)	$33.15	$265.20	$50.55	$404.40	$33.15	$50.55
1 Dump Truck, 18 C.Y., 450 H.P.		838.40		922.24	104.80	115.28
8 L.H., Daily Totals		$1103.60		$1326.64	$137.95	$165.83

Crew B-34J	Hr.	Daily	Hr.	Daily	Bare Costs	Incl. O&P
1 Truck Driver (heavy)	$33.15	$265.20	$50.55	$404.40	$33.15	$50.55
1 Dump Truck, Off Hwy., 100 Ton		2296.00		2525.60	287.00	315.70
8 L.H., Daily Totals		$2561.20		$2930.00	$320.15	$366.25

Crew B-34K	Hr.	Daily	Hr.	Daily	Bare Costs	Incl. O&P
1 Truck Driver (heavy)	$33.15	$265.20	$50.55	$404.40	$33.15	$50.55
1 Truck Tractor, 6x4, 450 H.P.		715.80		787.38		
1 Lowbed Trailer, 75 Ton		205.80		226.38	115.20	126.72
8 L.H., Daily Totals		$1186.80		$1418.16	$148.35	$177.27

Crew B-34L	Hr.	Daily	Hr.	Daily	Bare Costs	Incl. O&P
1 Equip. Oper. (light)	$41.30	$330.40	$61.85	$494.80	$41.30	$61.85
1 Flatbed Truck, Gas, 1.5 Ton		239.80		263.78	29.98	32.97
8 L.H., Daily Totals		$570.20		$758.58	$71.28	$94.82

Crew B-34N	Hr.	Daily	Hr.	Daily	Bare Costs	Incl. O&P
1 Truck Driver (heavy)	$33.15	$265.20	$50.55	$404.40	$33.15	$50.55
1 Dump Truck, 8 C.Y., 220 H.P.		401.80		441.98		
1 Flatbed Trailer, 40 Ton		143.20		157.52	68.13	74.94
8 L.H., Daily Totals		$810.20		$1003.90	$101.28	$125.49

Crew B-34P	Hr.	Daily	Hr.	Daily	Bare Costs	Incl. O&P
1 Pipe Fitter	$51.90	$415.20	$77.80	$622.40	$42.37	$63.77
1 Truck Driver (light)	32.25	258.00	49.20	393.60		
1 Equip. Oper. (med.)	42.95	343.60	64.30	514.40		
1 Flatbed Truck, Gas, 3 Ton		296.00		325.60		
1 Backhoe Loader, 48 H.P.		337.80		371.58	26.41	29.05
24 L.H., Daily Totals		$1650.60		$2227.58	$68.78	$92.82

Crew B-34Q	Hr.	Daily	Hr.	Daily	Bare Costs	Incl. O&P
1 Pipe Fitter	$51.90	$415.20	$77.80	$622.40	$42.85	$64.48
1 Truck Driver (light)	32.25	258.00	49.20	393.60		
1 Eqip. Oper. (crane)	44.40	355.20	66.45	531.60		
1 Flatbed Trailer, 25 Ton		105.60		116.16		
1 Dump Truck, 8 C.Y., 220 H.P.		401.80		441.98		
1 Hyd. Crane, 25 Ton		747.60		822.36	52.29	57.52
24 L.H., Daily Totals		$2283.40		$2928.10	$95.14	$122.00

Crew B-34R	Hr.	Daily	Hr.	Daily	Bare Costs	Incl. O&P
1 Pipe Fitter	$51.90	$415.20	$77.80	$622.40	$42.85	$64.48
1 Truck Driver (light)	32.25	258.00	49.20	393.60		
1 Eqip. Oper. (crane)	44.40	355.20	66.45	531.60		
1 Flatbed Trailer, 25 Ton		105.60		116.16		
1 Dump Truck, 8 C.Y., 220 H.P.		401.80		441.98		
1 Hyd. Crane, 25 Ton		747.60		822.36		
1 Hyd. Excavator, 1 C.Y.		801.60		881.76	85.69	94.26
24 L.H., Daily Totals		$3085.00		$3809.86	$128.54	$158.74

Crew B-34S	Hr.	Daily	Hr.	Daily	Bare Costs	Incl. O&P
2 Pipe Fitters	$51.90	$830.40	$77.80	$1244.80	$45.34	$68.15
1 Truck Driver (heavy)	33.15	265.20	50.55	404.40		
1 Eqip. Oper. (crane)	44.40	355.20	66.45	531.60		
1 Flatbed Trailer, 40 Ton		143.20		157.52		
1 Truck Tractor, 6x4, 380 H.P.		592.40		651.64		
1 Hyd. Crane, 80 Ton		1545.00		1699.50		
1 Hyd. Excavator, 2 C.Y.		1321.00		1453.10	112.55	123.81
32 L.H., Daily Totals		$5052.40		$6142.56	$157.89	$191.96

Crew B-34T	Hr.	Daily	Hr.	Daily	Bare Costs	Incl. O&P
2 Pipe Fitters	$51.90	$830.40	$77.80	$1244.80	$45.34	$68.15
1 Truck Driver (heavy)	33.15	265.20	50.55	404.40		
1 Eqip. Oper. (crane)	44.40	355.20	66.45	531.60		
1 Flatbed Trailer, 40 Ton		143.20		157.52		
1 Truck Tractor, 6x4, 380 H.P.		592.40		651.64		
1 Hyd. Crane, 80 Ton		1545.00		1699.50	71.27	78.40
32 L.H., Daily Totals		$3731.40		$4689.46	$116.61	$146.55

Crew No.	Bare Costs		Incl. Subs O&P		Cost Per Labor-Hour	

Crew B-35	Hr.	Daily	Hr.	Daily	Bare Costs	Incl. O&P
1 Laborer Foreman (out)	$35.10	$280.80	$54.10	$432.80	$40.92	$62.08
1 Skilled Worker	42.60	340.80	65.50	524.00		
1 Welder (plumber)	52.05	416.40	78.00	624.00		
1 Laborer	33.10	264.80	51.05	408.40		
1 Equip. Oper. (crane)	44.40	355.20	66.45	531.60		
1 Equip. Oper. Oiler	38.30	306.40	57.35	458.80		
1 Welder, electric, 300 amp		58.55		64.41		
1 Hyd. Excavator, .75 C.Y.		677.80		745.58	15.34	16.87
48 L.H., Daily Totals		$2700.75		$3789.59	$56.27	$78.95

Crew B-35A	Hr.	Daily	Hr.	Daily	Bare Costs	Incl. O&P
1 Laborer Foreman (out)	$35.10	$280.80	$54.10	$432.80	$39.81	$60.50
2 Laborers	33.10	529.60	51.05	816.80		
1 Skilled Worker	42.60	340.80	65.50	524.00		
1 Welder (plumber)	52.05	416.40	78.00	624.00		
1 Equip. Oper. (crane)	44.40	355.20	66.45	531.60		
1 Equip. Oper. Oiler	38.30	306.40	57.35	458.80		
1 Welder, gas engine, 300 amp		145.80		160.38		
1 Crawler Crane, 75 Ton		1516.00		1667.60	29.68	32.64
56 L.H., Daily Totals		$3891.00		$5215.98	$69.48	$93.14

Crew B-36	Hr.	Daily	Hr.	Daily	Bare Costs	Incl. O&P
1 Labor Foreman (outside)	$35.10	$280.80	$54.10	$432.80	$37.44	$56.96
2 Laborers	33.10	529.60	51.05	816.80		
2 Equip. Oper. (med.)	42.95	687.20	64.30	1028.80		
1 Dozer, 200 H.P.		1192.00		1311.20		
1 Aggregate Spreader		38.60		42.46		
1 Tandem Roller, 10 Ton		235.20		258.72	36.65	40.31
40 L.H., Daily Totals		$2963.40		$3890.78	$74.08	$97.27

Crew B-36A	Hr.	Daily	Hr.	Daily	Bare Costs	Incl. O&P
1 Labor Foreman (outside)	$35.10	$280.80	$54.10	$432.80	$39.01	$59.06
2 Laborers	33.10	529.60	51.05	816.80		
4 Equip. Oper. (med.)	42.95	1374.40	64.30	2057.60		
1 Dozer, 200 H.P.		1192.00		1311.20		
1 Aggregate Spreader		38.60		42.46		
1 Tandem Roller, 10 Ton		235.20		258.72		
1 Roller, Pneum. Whl, 12 Ton		337.20		370.92	32.20	35.42
56 L.H., Daily Totals		$3987.80		$5290.50	$71.21	$94.47

Crew B-36B	Hr.	Daily	Hr.	Daily	Bare Costs	Incl. O&P
1 Labor Foreman (outside)	$35.10	$280.80	$54.10	$432.80	$38.28	$57.99
2 Laborers	33.10	529.60	51.05	816.80		
4 Equip. Oper. (med.)	42.95	1374.40	64.30	2057.60		
1 Truck Driver, Heavy	33.15	265.20	50.55	404.40		
1 Grader, 30,000 Lbs.		622.20		684.42		
1 F.E. Loader, crl, 1.5 C.Y.		534.60		588.06		
1 Dozer, 300 H.P.		1592.00		1751.20		
1 Roller, Vibratory, 25 Ton		633.20		696.52		
1 Truck Tractor, 6x4, 450 H.P.		715.80		787.38		
1 Water Tank Trailer, 5000 Gal.		139.60		153.56	66.21	72.83
64 L.H., Daily Totals		$6687.40		$8372.74	$104.49	$130.82

Crew B-36C	Hr.	Daily	Hr.	Daily	Bare Costs	Incl. O&P
1 Labor Foreman (outside)	$35.10	$280.80	$54.10	$432.80	$39.42	$59.51
3 Equip. Oper. (med.)	42.95	1030.80	64.30	1543.20		
1 Truck Driver, Heavy	33.15	265.20	50.55	404.40		
1 Grader, 30,000 Lbs.		622.20		684.42		
1 Dozer, 300 H.P.		1592.00		1751.20		
1 Roller, Vibratory, 25 Ton		633.20		696.52		
1 Truck Tractor, 6x4, 450 H.P.		715.80		787.38		
1 Water Tank Trailer, 5000 Gal.		139.60		153.56	92.57	101.83
40 L.H., Daily Totals		$5279.60		$6453.48	$131.99	$161.34

Crew B-37	Hr.	Daily	Hr.	Daily	Bare Costs	Incl. O&P
1 Labor Foreman (outside)	$35.10	$280.80	$54.10	$432.80	$34.80	$53.36
4 Laborers	33.10	1059.20	51.05	1633.60		
1 Equip. Oper. (light)	41.30	330.40	61.85	494.80		
1 Tandem Roller, 5 Ton		152.00		167.20	3.17	3.48
48 L.H., Daily Totals		$1822.40		$2728.40	$37.97	$56.84

Crew B-38	Hr.	Daily	Hr.	Daily	Bare Costs	Incl. O&P
1 Labor Foreman (outside)	$35.10	$280.80	$54.10	$432.80	$37.11	$56.47
2 Laborers	33.10	529.60	51.05	816.80		
1 Equip. Oper. (light)	41.30	330.40	61.85	494.80		
1 Equip. Oper. (med.)	42.95	343.60	64.30	514.40		
1 Backhoe Loader, 48 H.P.		337.80		371.58		
1 Hyd.Hammer, (1200 lb.)		154.20		169.62		
1 F.E. Loader, W.M., 4 C.Y.		627.80		690.58		
1 Pvmt. Rem. Bucket		58.20		64.02	29.45	32.40
40 L.H., Daily Totals		$2662.40		$3554.60	$66.56	$88.86

Crew B-39	Hr.	Daily	Hr.	Daily	Bare Costs	Incl. O&P
1 Labor Foreman (outside)	$35.10	$280.80	$54.10	$432.80	$34.80	$53.36
4 Laborers	33.10	1059.20	51.05	1633.60		
1 Equip. Oper. (light)	41.30	330.40	61.85	494.80		
1 Air Compressor, 250 cfm		193.40		212.74		
2 Breakers, Pavement, 60 lb.		18.80		20.68		
2 -50' Air Hoses, 1.5"		12.60		13.86	4.68	5.15
48 L.H., Daily Totals		$1895.20		$2808.48	$39.48	$58.51

Crew B-40	Hr.	Daily	Hr.	Daily	Bare Costs	Incl. O&P
1 Pile Driver Foreman (out)	$42.30	$338.40	$68.00	$544.00	$41.33	$64.68
4 Pile Drivers	40.30	1289.60	64.80	2073.60		
2 Equip. Oper. (crane)	44.40	710.40	66.45	1063.20		
1 Equip. Oper. Oiler	38.30	306.40	57.35	458.80		
1 Crawler Crane, 40 Ton		1193.00		1312.30		
1 Vibratory Hammer & Gen.		2641.00		2905.10	59.91	65.90
64 L.H., Daily Totals		$6478.80		$8357.00	$101.23	$130.58

Crew B-40B	Hr.	Daily	Hr.	Daily	Bare Costs	Incl. O&P
1 Laborer Foreman	$35.10	$280.80	$54.10	$432.80	$36.18	$55.17
3 Laborers	33.10	794.40	51.05	1225.20		
1 Equip. Oper. (crane)	44.40	355.20	66.45	531.60		
1 Equip. Oper. Oiler	38.30	306.40	57.35	458.80		
1 Lattice Boom Crane, 40 Ton		1256.00		1381.60	26.17	28.78
48 L.H., Daily Totals		$2992.80		$4030.00	$62.35	$83.96

Crew B-41	Hr.	Daily	Hr.	Daily	Bare Costs	Incl. O&P
1 Labor Foreman (outside)	$35.10	$280.80	$54.10	$432.80	$34.21	$52.59
4 Laborers	33.10	1059.20	51.05	1633.60		
.25 Equip. Oper. (crane)	44.40	88.80	66.45	132.90		
.25 Equip. Oper. Oiler	38.30	76.60	57.35	114.70		
.25 Crawler Crane, 40 Ton		298.25		328.07	6.78	7.46
44 L.H., Daily Totals		$1803.65		$2642.07	$40.99	$60.05

Crew No.	Bare Costs		Incl. Subs O&P		Cost Per Labor-Hour	

Crew B-42	Hr.	Daily	Hr.	Daily	Bare Costs	Incl. O&P
1 Labor Foreman (outside)	$35.10	$280.80	$54.10	$432.80	$37.14	$58.07
4 Laborers	33.10	1059.20	51.05	1633.60		
1 Equip. Oper. (crane)	44.40	355.20	66.45	531.60		
1 Equip. Oper. Oiler	38.30	306.40	57.35	458.80		
1 Welder	46.90	375.20	82.45	659.60		
1 Hyd. Crane, 25 Ton		747.60		822.36		
1 Welder, gas engine, 300 amp		145.80		160.38		
1 Horz. Boring Csg. Mch.		473.20		520.52	21.35	23.49
64 L.H., Daily Totals		$3743.40		$5219.66	$58.49	$81.56

Crew B-43	Hr.	Daily	Hr.	Daily	Bare Costs	Incl. O&P
1 Labor Foreman (outside)	$35.10	$280.80	$54.10	$432.80	$36.18	$55.17
3 Laborers	33.10	794.40	51.05	1225.20		
1 Equip. Oper. (crane)	44.40	355.20	66.45	531.60		
1 Equip. Oper. Oiler	38.30	306.40	57.35	458.80		
1 Drill Rig, Truck-Mounted		2591.00		2850.10	53.98	59.38
48 L.H., Daily Totals		$4327.80		$5498.50	$90.16	$114.55

Crew B-44	Hr.	Daily	Hr.	Daily	Bare Costs	Incl. O&P
1 Pile Driver Foreman	$42.30	$338.40	$68.00	$544.00	$40.67	$63.89
4 Pile Drivers	40.30	1289.60	64.80	2073.60		
2 Equip. Oper. (crane)	44.40	710.40	66.45	1063.20		
1 Laborer	33.10	264.80	51.05	408.40		
1 Crawler Crane, 40 Ton		1193.00		1312.30		
1 Lead, 60' high		70.80		77.88		
1 Hammer, diesel, 15K ft.-lbs.		577.60		635.36	28.77	31.65
64 L.H., Daily Totals		$4444.60		$6114.74	$69.45	$95.54

Crew B-45	Hr.	Daily	Hr.	Daily	Bare Costs	Incl. O&P
1 Equip. Oper. (med.)	$42.95	$343.60	$64.30	$514.40	$38.05	$57.42
1 Truck Driver (heavy)	33.15	265.20	50.55	404.40		
1 Dist. Tanker, 3000 Gallon		307.00		337.70		
1 Truck Tractor, 6x4, 380 H.P.		592.40		651.64	56.21	61.83
16 L.H., Daily Totals		$1508.20		$1908.14	$94.26	$119.26

Crew B-46	Hr.	Daily	Hr.	Daily	Bare Costs	Incl. O&P
1 Pile Driver Foreman	$42.30	$338.40	$68.00	$544.00	$37.03	$58.46
2 Pile Drivers	40.30	644.80	64.80	1036.80		
3 Laborers	33.10	794.40	51.05	1225.20		
1 Chain Saw, gas, 36" Long		39.20		43.12	.82	.90
48 L.H., Daily Totals		$1816.80		$2849.12	$37.85	$59.36

Crew B-47	Hr.	Daily	Hr.	Daily	Bare Costs	Incl. O&P
1 Blast Foreman	$35.10	$280.80	$54.10	$432.80	$36.50	$55.67
1 Driller	33.10	264.80	51.05	408.40		
1 Equip. Oper. (light)	41.30	330.40	61.85	494.80		
1 Air Track Drill, 4"		917.80		1009.58		
1 Air Compressor, 600 cfm		530.60		583.66		
2 -50' Air Hoses, 3"		32.40		35.64	61.70	67.87
24 L.H., Daily Totals		$2356.80		$2964.88	$98.20	$123.54

Crew B-47A	Hr.	Daily	Hr.	Daily	Bare Costs	Incl. O&P
1 Drilling Foreman	$35.10	$280.80	$54.10	$432.80	$39.27	$59.30
1 Equip. Oper. (heavy)	44.40	355.20	66.45	531.60		
1 Oiler	38.30	306.40	57.35	458.80		
1 Air Track Drill, 5"		1069.00		1175.90	44.54	49.00
24 L.H., Daily Totals		$2011.40		$2599.10	$83.81	$108.30

Crew B-47C	Hr.	Daily	Hr.	Daily	Bare Costs	Incl. O&P
1 Laborer	$33.10	$264.80	$51.05	$408.40	$37.20	$56.45
1 Equip. Oper. (light)	41.30	330.40	61.85	494.80		
1 Air Compressor, 750 cfm		542.80		597.08		
2 -50' Air Hoses, 3"		32.40		35.64		
1 Air Track Drill, 4"		917.80		1009.58	93.31	102.64
16 L.H., Daily Totals		$2088.20		$2545.50	$130.51	$159.09

Crew B-47E	Hr.	Daily	Hr.	Daily	Bare Costs	Incl. O&P
1 Laborer Foreman	$35.10	$280.80	$54.10	$432.80	$33.60	$51.81
3 Laborers	33.10	794.40	51.05	1225.20		
1 Flatbed Truck, Gas, 3 Ton		296.00		325.60	9.25	10.18
32 L.H., Daily Totals		$1371.20		$1983.60	$42.85	$61.99

Crew B-47G	Hr.	Daily	Hr.	Daily	Bare Costs	Incl. O&P
1 Laborer Foreman	$35.10	$280.80	$54.10	$432.80	$35.65	$54.51
2 Laborers	33.10	529.60	51.05	816.80		
1 Equip. Oper. (light)	41.30	330.40	61.85	494.80		
1 Air Track Drill, 4"		917.80		1009.58		
1 Air Compressor, 600 cfm		530.60		583.66		
2 -50' Air Hoses, 3"		32.40		35.64		
1 Grout Pump		201.80		221.98	52.58	57.84
32 L.H., Daily Totals		$2823.40		$3595.26	$88.23	$112.35

Crew B-47H	Hr.	Daily	Hr.	Daily	Bare Costs	Incl. O&P
1 Skilled Worker Foreman	$44.60	$356.80	$68.55	$548.40	$43.10	$66.26
3 Skilled Workers	42.60	1022.40	65.50	1572.00		
1 Flatbed Truck, Gas, 3 Ton		296.00		325.60	9.25	10.18
32 L.H., Daily Totals		$1675.20		$2446.00	$52.35	$76.44

Crew B-48	Hr.	Daily	Hr.	Daily	Bare Costs	Incl. O&P
1 Labor Foreman (outside)	$35.10	$280.80	$54.10	$432.80	$36.91	$56.13
3 Laborers	33.10	794.40	51.05	1225.20		
1 Equip. Oper. (crane)	44.40	355.20	66.45	531.60		
1 Equip. Oper. Oiler	38.30	306.40	57.35	458.80		
1 Equip. Oper. (light)	41.30	330.40	61.85	494.80		
1 Centr. Water Pump, 6"		340.00		374.00		
1 -20' Suction Hose, 6"		11.90		13.09		
1 -50' Discharge Hose, 6"		6.30		6.93		
1 Drill Rig, Truck-Mounted		2591.00		2850.10	52.66	57.93
56 L.H., Daily Totals		$5016.40		$6387.32	$89.58	$114.06

Crew B-49	Hr.	Daily	Hr.	Daily	Bare Costs	Incl. O&P
1 Labor Foreman (outside)	$35.10	$280.80	$54.10	$432.80	$38.34	$58.75
3 Laborers	33.10	794.40	51.05	1225.20		
2 Equip. Oper. (crane)	44.40	710.40	66.45	1063.20		
2 Equip. Oper. Oilers	38.30	612.80	57.35	917.60		
1 Equip. Oper. (light)	41.30	330.40	61.85	494.80		
2 Pile Drivers	40.30	644.80	64.80	1036.80		
1 Hyd. Crane, 25 Ton		747.60		822.36		
1 Centr. Water Pump, 6"		340.00		374.00		
1 -20' Suction Hose, 6"		11.90		13.09		
1 -50' Discharge Hose, 6"		6.30		6.93		
1 Drill Rig, Truck-Mounted		2591.00		2850.10	42.01	46.21
88 L.H., Daily Totals		$7070.40		$9236.88	$80.35	$104.96

Crew B-50	Bare Costs Hr.	Bare Costs Daily	Incl. Subs O&P Hr.	Incl. Subs O&P Daily	Cost Per Labor-Hour Bare Costs	Cost Per Labor-Hour Incl. O&P
2 Pile Driver Foremen	$42.30	$676.80	$68.00	$1088.00	$39.49	$62.01
6 Pile Drivers	40.30	1934.40	64.80	3110.40		
2 Equip. Oper. (crane)	44.40	710.40	66.45	1063.20		
1 Equip. Oper. Oiler	38.30	306.40	57.35	458.80		
3 Laborers	33.10	794.40	51.05	1225.20		
1 Crawler Crane, 40 Ton		1193.00		1312.30		
1 Lead, 60' high		70.80		77.88		
1 Hammer, diesel, 15K ft.-lbs.		577.60		635.36		
1 Air Compressor, 600 cfm		530.60		583.66		
2 -50' Air Hoses, 3"		32.40		35.64		
1 Chain Saw, gas, 36" Long		39.20		43.12	21.82	24.00
112 L.H., Daily Totals		$6866.00		$9633.56	$61.30	$86.01

Crew B-51	Hr.	Daily	Hr.	Daily	Bare Costs	Incl. O&P
1 Labor Foreman (outside)	$35.10	$280.80	$54.10	$432.80	$33.29	$51.25
4 Laborers	33.10	1059.20	51.05	1633.60		
1 Truck Driver (light)	32.25	258.00	49.20	393.60		
1 Flatbed Truck, Gas, 1.5 Ton		239.80		263.78	5.00	5.50
48 L.H., Daily Totals		$1837.80		$2723.78	$38.29	$56.75

Crew B-52	Hr.	Daily	Hr.	Daily	Bare Costs	Incl. O&P
1 Carpenter Foreman	$43.55	$348.40	$67.15	$537.20	$38.42	$58.88
1 Carpenter	41.55	332.40	64.05	512.40		
3 Laborers	33.10	794.40	51.05	1225.20		
1 Cement Finisher	39.70	317.60	57.95	463.60		
.5 Rodman (reinf.)	46.80	187.20	75.40	301.60		
.5 Equip. Oper. (med.)	42.95	171.80	64.30	257.20		
.5 Crawler Loader, 3 C.Y.		538.50		592.35	9.62	10.58
56 L.H., Daily Totals		$2690.30		$3889.55	$48.04	$69.46

Crew B-53	Hr.	Daily	Hr.	Daily	Bare Costs	Incl. O&P
1 Equip. Oper. (light)	$41.30	$330.40	$61.85	$494.80	$41.30	$61.85
1 Trencher, Chain, 12 H.P.		66.00		72.60	8.25	9.07
8 L.H., Daily Totals		$396.40		$567.40	$49.55	$70.92

Crew B-54	Hr.	Daily	Hr.	Daily	Bare Costs	Incl. O&P
1 Equip. Oper. (light)	$41.30	$330.40	$61.85	$494.80	$41.30	$61.85
1 Trencher, Chain, 40 H.P.		334.40		367.84	41.80	45.98
8 L.H., Daily Totals		$664.80		$862.64	$83.10	$107.83

Crew B-54A	Hr.	Daily	Hr.	Daily	Bare Costs	Incl. O&P
.17 Labor Foreman (outside)	$35.10	$47.74	$54.10	$73.58	$41.81	$62.82
1 Equipment Operator (med.)	42.95	343.60	64.30	514.40		
1 Wheel Trencher, 67 H.P.		1152.00		1267.20	123.08	135.38
9.36 L.H., Daily Totals		$1543.34		$1855.18	$164.89	$198.20

Crew B-54B	Hr.	Daily	Hr.	Daily	Bare Costs	Incl. O&P
.25 Labor Foreman (outside)	$35.10	$70.20	$54.10	$108.20	$41.38	$62.26
1 Equipment Operator (med.)	42.95	343.60	64.30	514.40		
1 Wheel Trencher, 150 H.P.		1843.00		2027.30	184.30	202.73
10 L.H., Daily Totals		$2256.80		$2649.90	$225.68	$264.99

Crew B-54C	Hr.	Daily	Hr.	Daily	Bare Costs	Incl. O&P
1 Laborer	$33.10	$264.80	$51.05	$408.40	$38.02	$57.67
1 Equipment Operator (med.)	42.95	343.60	64.30	514.40		
1 Wheel Trencher, 67 H.P.		1152.00		1267.20	72.00	79.20
16 L.H., Daily Totals		$1760.40		$2190.00	$110.03	$136.88

Crew B-54D	Hr.	Daily	Hr.	Daily	Bare Costs	Incl. O&P
1 Laborer	$33.10	$264.80	$51.05	$408.40	$38.02	$57.67
1 Equipment Operator (med.)	42.95	343.60	64.30	514.40		
1 Rock trencher, 6" width		344.40		378.84	21.52	23.68
16 L.H., Daily Totals		$952.80		$1301.64	$59.55	$81.35

Crew B-54E	Hr.	Daily	Hr.	Daily	Bare Costs	Incl. O&P
1 Laborer	$33.10	$264.80	$51.05	$408.40	$38.02	$57.67
1 Equipment Operator (med.)	42.95	343.60	64.30	514.40		
1 Rock trencher, 18" width		2571.00		2828.10	160.69	176.76
16 L.H., Daily Totals		$3179.40		$3750.90	$198.71	$234.43

Crew B-55	Hr.	Daily	Hr.	Daily	Bare Costs	Incl. O&P
2 Laborers	$33.10	$529.60	$51.05	$816.80	$32.82	$50.43
1 Truck Driver (light)	32.25	258.00	49.20	393.60		
1 Truck-mounted earth auger		749.80		824.78		
1 Flatbed Truck, Gas, 3 Ton		296.00		325.60	43.58	47.93
24 L.H., Daily Totals		$1833.40		$2360.78	$76.39	$98.37

Crew B-56	Hr.	Daily	Hr.	Daily	Bare Costs	Incl. O&P
1 Laborer	$33.10	$264.80	$51.05	$408.40	$37.20	$56.45
1 Equip. Oper. (light)	41.30	330.40	61.85	494.80		
1 Air Track Drill, 4"		917.80		1009.58		
1 Air Compressor, 600 cfm		530.60		583.66		
1 -50' Air Hose, 3"		16.20		17.82	91.54	100.69
16 L.H., Daily Totals		$2059.80		$2514.26	$128.74	$157.14

Crew B-57	Hr.	Daily	Hr.	Daily	Bare Costs	Incl. O&P
1 Labor Foreman (outside)	$35.10	$280.80	$54.10	$432.80	$37.55	$56.98
2 Laborers	33.10	529.60	51.05	816.80		
1 Equip. Oper. (crane)	44.40	355.20	66.45	531.60		
1 Equip. Oper. (light)	41.30	330.40	61.85	494.80		
1 Equip. Oper. Oiler	38.30	306.40	57.35	458.80		
1 Crawler Crane, 25 Ton		1188.00		1306.80		
1 Clamshell Bucket, 1 C.Y.		46.60		51.26		
1 Centr. Water Pump, 6"		340.00		374.00		
1 -20' Suction Hose, 6"		11.90		13.09		
20 -50' Discharge Hoses, 6"		126.00		138.60	35.68	39.24
48 L.H., Daily Totals		$3514.90		$4618.55	$73.23	$96.22

Crew B-58	Hr.	Daily	Hr.	Daily	Bare Costs	Incl. O&P
2 Laborers	$33.10	$529.60	$51.05	$816.80	$35.83	$54.65
1 Equip. Oper. (light)	41.30	330.40	61.85	494.80		
1 Backhoe Loader, 48 H.P.		337.80		371.58		
1 Small Helicopter, w/pilot		2513.00		2764.30	118.78	130.66
24 L.H., Daily Totals		$3710.80		$4447.48	$154.62	$185.31

Crew B-59	Hr.	Daily	Hr.	Daily	Bare Costs	Incl. O&P
1 Truck Driver (heavy)	$33.15	$265.20	$50.55	$404.40	$33.15	$50.55
1 Truck Tractor, 220 H.P.		354.00		389.40		
1 Water Tank Trailer, 5000 Gal.		139.60		153.56	61.70	67.87
8 L.H., Daily Totals		$758.80		$947.36	$94.85	$118.42

Crew B-59A	Hr.	Daily	Hr.	Daily	Bare Costs	Incl. O&P
2 Laborers	$33.10	$529.60	$51.05	$816.80	$33.12	$50.88
1 Truck Driver (heavy)	33.15	265.20	50.55	404.40		
1 Water Tank Trailer, 5000 Gal.		139.60		153.56		
1 Truck Tractor, 220 H.P.		354.00		389.40	20.57	22.62
24 L.H., Daily Totals		$1288.40		$1764.16	$53.68	$73.51

Crew No.	Bare Costs		Incl. Subs O&P		Cost Per Labor-Hour	

Crew B-60	Hr.	Daily	Hr.	Daily	Bare Costs	Incl. O&P
1 Labor Foreman (outside)	$35.10	$280.80	$54.10	$432.80	$38.09	$57.67
2 Laborers	33.10	529.60	51.05	816.80		
1 Equip. Oper. (crane)	44.40	355.20	66.45	531.60		
2 Equip. Oper. (light)	41.30	660.80	61.85	989.60		
1 Equip. Oper. Oiler	38.30	306.40	57.35	458.80		
1 Crawler Crane, 40 Ton		1193.00		1312.30		
1 Lead, 60' high		70.80		77.88		
1 Hammer, diesel, 15K ft.-lbs.		577.60		635.36		
1 Backhoe Loader, 48 H.P.		337.80		371.58	38.91	42.81
56 L.H., Daily Totals		$4312.00		$5626.72	$77.00	$100.48

Crew B-61	Hr.	Daily	Hr.	Daily	Bare Costs	Incl. O&P
1 Labor Foreman (outside)	$35.10	$280.80	$54.10	$432.80	$35.14	$53.82
3 Laborers	33.10	794.40	51.05	1225.20		
1 Equip. Oper. (light)	41.30	330.40	61.85	494.80		
1 Cement Mixer, 2 C.Y.		189.40		208.34		
1 Air Compressor, 160 cfm		158.00		173.80	8.69	9.55
40 L.H., Daily Totals		$1753.00		$2534.94	$43.83	$63.37

Crew B-62	Hr.	Daily	Hr.	Daily	Bare Costs	Incl. O&P
2 Laborers	$33.10	$529.60	$51.05	$816.80	$35.83	$54.65
1 Equip. Oper. (light)	41.30	330.40	61.85	494.80		
1 Loader, Skid Steer, 30 H.P., gas		161.40		177.54	6.72	7.40
24 L.H., Daily Totals		$1021.40		$1489.14	$42.56	$62.05

Crew B-63	Hr.	Daily	Hr.	Daily	Bare Costs	Incl. O&P
4 Laborers	$33.10	$1059.20	$51.05	$1633.60	$34.74	$53.21
1 Equip. Oper. (light)	41.30	330.40	61.85	494.80		
1 Loader, Skid Steer, 30 H.P., gas		161.40		177.54	4.04	4.44
40 L.H., Daily Totals		$1551.00		$2305.94	$38.77	$57.65

Crew B-64	Hr.	Daily	Hr.	Daily	Bare Costs	Incl. O&P
1 Laborer	$33.10	$264.80	$51.05	$408.40	$32.67	$50.13
1 Truck Driver (light)	32.25	258.00	49.20	393.60		
1 Power Mulcher (small)		150.20		165.22		
1 Flatbed Truck, Gas, 1.5 Ton		239.80		263.78	24.38	26.81
16 L.H., Daily Totals		$912.80		$1231.00	$57.05	$76.94

Crew B-65	Hr.	Daily	Hr.	Daily	Bare Costs	Incl. O&P
1 Laborer	$33.10	$264.80	$51.05	$408.40	$32.67	$50.13
1 Truck Driver (light)	32.25	258.00	49.20	393.60		
1 Power Mulcher (large)		319.00		350.90		
1 Flatbed Truck, Gas, 1.5 Ton		239.80		263.78	34.92	38.42
16 L.H., Daily Totals		$1081.60		$1416.68	$67.60	$88.54

Crew B-66	Hr.	Daily	Hr.	Daily	Bare Costs	Incl. O&P
1 Equip. Oper. (light)	$41.30	$330.40	$61.85	$494.80	$41.30	$61.85
1 Loader-Backhoe		218.40		240.24	27.30	30.03
8 L.H., Daily Totals		$548.80		$735.04	$68.60	$91.88

Crew B-67	Hr.	Daily	Hr.	Daily	Bare Costs	Incl. O&P
1 Millwright	$42.95	$343.60	$62.70	$501.60	$42.13	$62.27
1 Equip. Oper. (light)	41.30	330.40	61.85	494.80		
1 Forklift, R/T, 4,000 Lb.		309.60		340.56	19.35	21.29
16 L.H., Daily Totals		$983.60		$1336.96	$61.48	$83.56

Crew B-68	Hr.	Daily	Hr.	Daily	Bare Costs	Incl. O&P
2 Millwrights	$42.95	$687.20	$62.70	$1003.20	$42.40	$62.42
1 Equip. Oper. (light)	41.30	330.40	61.85	494.80		
1 Forklift, R/T, 4,000 Lb.		309.60		340.56	12.90	14.19
24 L.H., Daily Totals		$1327.20		$1838.56	$55.30	$76.61

Crew B-69	Hr.	Daily	Hr.	Daily	Bare Costs	Incl. O&P
1 Labor Foreman (outside)	$35.10	$280.80	$54.10	$432.80	$36.18	$55.17
3 Laborers	33.10	794.40	51.05	1225.20		
1 Equip Oper. (crane)	44.40	355.20	66.45	531.60		
1 Equip Oper. Oiler	38.30	306.40	57.35	458.80		
1 Hyd. Crane, 80 Ton		1545.00		1699.50	32.19	35.41
48 L.H., Daily Totals		$3281.80		$4347.90	$68.37	$90.58

Crew B-69A	Hr.	Daily	Hr.	Daily	Bare Costs	Incl. O&P
1 Labor Foreman	$35.10	$280.80	$54.10	$432.80	$36.17	$54.92
3 Laborers	33.10	794.40	51.05	1225.20		
1 Equip. Oper. (med.)	42.95	343.60	64.30	514.40		
1 Concrete Finisher	39.70	317.60	57.95	463.60		
1 Curb/Gutter Paver, 2-Track		825.00		907.50	17.19	18.91
48 L.H., Daily Totals		$2561.40		$3543.50	$53.36	$73.82

Crew B-69B	Hr.	Daily	Hr.	Daily	Bare Costs	Incl. O&P
1 Labor Foreman	$35.10	$280.80	$54.10	$432.80	$36.17	$54.92
3 Laborers	33.10	794.40	51.05	1225.20		
1 Equip. Oper. (med.)	42.95	343.60	64.30	514.40		
1 Cement Finisher	39.70	317.60	57.95	463.60		
1 Curb/Gutter Paver, 4-Track		934.60		1028.06	19.47	21.42
48 L.H., Daily Totals		$2671.00		$3664.06	$55.65	$76.33

Crew B-70	Hr.	Daily	Hr.	Daily	Bare Costs	Incl. O&P
1 Labor Foreman (outside)	$35.10	$280.80	$54.10	$432.80	$37.61	$57.16
3 Laborers	33.10	794.40	51.05	1225.20		
3 Equip. Oper. (med.)	42.95	1030.80	64.30	1543.20		
1 Grader, 30,000 Lbs.		622.20		684.42		
1 Ripper, beam & 1 shank		80.00		88.00		
1 Road Sweeper, S.P., 8' wide		650.40		715.44		
1 F.E. Loader, W.M., 1.5 C.Y.		367.60		404.36	30.72	33.79
56 L.H., Daily Totals		$3826.20		$5093.42	$68.33	$90.95

Crew B-70A	Hr.	Daily	Hr.	Daily	Bare Costs	Incl. O&P
1 Laborer	$33.10	$264.80	$51.05	$408.40	$40.98	$61.65
4 Equip. Oper. (med.)	42.95	1374.40	64.30	2057.60		
1 Grader, 40,000 Lbs.		1071.00		1178.10		
1 F.E. Loader, W.M.,2.5 C.Y.		435.80		479.38		
1 Dozer, 80 H.P.		436.60		480.26		
1 Roller, Pneum. Whl, 12 Ton		337.20		370.92	57.02	62.72
40 L.H., Daily Totals		$3919.80		$4974.66	$98.00	$124.37

Crew B-71	Hr.	Daily	Hr.	Daily	Bare Costs	Incl. O&P
1 Labor Foreman (outside)	$35.10	$280.80	$54.10	$432.80	$37.61	$57.16
3 Laborers	33.10	794.40	51.05	1225.20		
3 Equip. Oper. (med.)	42.95	1030.80	64.30	1543.20		
1 Pvmt. Profiler, 750 H.P.		6104.00		6714.40		
1 Road Sweeper, S.P., 8' wide		650.40		715.44		
1 F.E. Loader, W.M., 1.5 C.Y.		367.60		404.36	127.18	139.90
56 L.H., Daily Totals		$9228.00		$11035.40	$164.79	$197.06

Crew No.	Bare Costs		Incl. Subs O&P		Cost Per Labor-Hour	
Crew B-72	**Hr.**	**Daily**	**Hr.**	**Daily**	**Bare Costs**	**Incl. O&P**
1 Labor Foreman (outside)	$35.10	$280.80	$54.10	$432.80	$38.27	$58.06
3 Laborers	33.10	794.40	51.05	1225.20		
4 Equip. Oper. (med.)	42.95	1374.40	64.30	2057.60		
1 Pvmt. Profiler, 750 H.P.		6104.00		6714.40		
1 Hammermill, 250 H.P.		1636.00		1799.60		
1 Windrow Loader		1228.00		1350.80		
1 Mix Paver 165 H.P.		2062.00		2268.20		
1 Roller, Pneum. Whl, 12 Ton		337.20		370.92	177.61	195.37
64 L.H., Daily Totals		$13816.80		$16219.52	$215.89	$253.43

Crew B-73	**Hr.**	**Daily**	**Hr.**	**Daily**	**Bare Costs**	**Incl. O&P**
1 Labor Foreman (outside)	$35.10	$280.80	$54.10	$432.80	$39.51	$59.71
2 Laborers	33.10	529.60	51.05	816.80		
5 Equip. Oper. (med.)	42.95	1718.00	64.30	2572.00		
1 Road Mixer, 310 H.P.		1955.00		2150.50		
1 Tandem Roller, 10 Ton		235.20		258.72		
1 Hammermill, 250 H.P.		1636.00		1799.60		
1 Grader, 30,000 Lbs.		622.20		684.42		
.5 F.E. Loader, W.M., 1.5 C.Y.		183.80		202.18		
.5 Truck Tractor, 220 H.P.		177.00		194.70		
.5 Water Tank Trailer, 5000 Gal.		69.80		76.78	76.23	83.86
64 L.H., Daily Totals		$7407.40		$9188.50	$115.74	$143.57

Crew B-74	**Hr.**	**Daily**	**Hr.**	**Daily**	**Bare Costs**	**Incl. O&P**
1 Labor Foreman (outside)	$35.10	$280.80	$54.10	$432.80	$38.29	$57.93
1 Laborer	33.10	264.80	51.05	408.40		
4 Equip. Oper. (med.)	42.95	1374.40	64.30	2057.60		
2 Truck Drivers (heavy)	33.15	530.40	50.55	808.80		
1 Grader, 30,000 Lbs.		622.20		684.42		
1 Ripper, beam & 1 shank		80.00		88.00		
2 Stabilizers, 310 H.P.		2928.00		3220.80		
1 Flatbed Truck, Gas, 3 Ton		296.00		325.60		
1 Chem. Spreader, Towed		50.40		55.44		
1 Roller, Vibratory, 25 Ton		633.20		696.52		
1 Water Tank Trailer, 5000 Gal.		139.60		153.56		
1 Truck Tractor, 220 H.P.		354.00		389.40	79.74	87.71
64 L.H., Daily Totals		$7553.80		$9321.34	$118.03	$145.65

Crew B-75	**Hr.**	**Daily**	**Hr.**	**Daily**	**Bare Costs**	**Incl. O&P**
1 Labor Foreman (outside)	$35.10	$280.80	$54.10	$432.80	$39.02	$58.99
1 Laborer	33.10	264.80	51.05	408.40		
4 Equip. Oper. (med.)	42.95	1374.40	64.30	2057.60		
1 Truck Driver (heavy)	33.15	265.20	50.55	404.40		
1 Grader, 30,000 Lbs.		622.20		684.42		
1 Ripper, beam & 1 shank		80.00		88.00		
2 Stabilizers, 310 H.P.		2928.00		3220.80		
1 Dist. Tanker, 3000 Gallon		307.00		337.70		
1 Truck Tractor, 6x4, 380 H.P.		592.40		651.64		
1 Roller, Vibratory, 25 Ton		633.20		696.52	92.19	101.41
56 L.H., Daily Totals		$7348.00		$8982.28	$131.21	$160.40

Crew No.	Bare Costs		Incl. Subs O&P		Cost Per Labor-Hour	
Crew B-76	**Hr.**	**Daily**	**Hr.**	**Daily**	**Bare Costs**	**Incl. O&P**
1 Dock Builder Foreman	$42.30	$338.40	$68.00	$544.00	$41.21	$64.69
5 Dock Builders	40.30	1612.00	64.80	2592.00		
2 Equip. Oper. (crane)	44.40	710.40	66.45	1063.20		
1 Equip. Oper. Oiler	38.30	306.40	57.35	458.80		
1 Crawler Crane, 50 Ton		1346.00		1480.60		
1 Barge, 400 Ton		768.00		844.80		
1 Hammer, diesel, 15K ft.-lbs.		577.60		635.36		
1 Lead, 60' high		70.80		77.88		
1 Air Compressor, 600 cfm		530.60		583.66		
2 -50' Air Hoses, 3"		32.40		35.64	46.19	50.80
72 L.H., Daily Totals		$6292.60		$8315.94	$87.40	$115.50

Crew B-76A	**Hr.**	**Daily**	**Hr.**	**Daily**	**Bare Costs**	**Incl. O&P**
1 Laborer Foreman	$35.10	$280.80	$54.10	$432.80	$35.41	$54.14
5 Laborers	33.10	1324.00	51.05	2042.00		
1 Equip. Oper. (crane)	44.40	355.20	66.45	531.60		
1 Equip. Oper. Oiler	38.30	306.40	57.35	458.80		
1 Crawler Crane, 50 Ton		1346.00		1480.60		
1 Barge, 400 Ton		768.00		844.80	33.03	36.33
64 L.H., Daily Totals		$4380.40		$5790.60	$68.44	$90.48

Crew B-77	**Hr.**	**Daily**	**Hr.**	**Daily**	**Bare Costs**	**Incl. O&P**
1 Labor Foreman	$35.10	$280.80	$54.10	$432.80	$33.33	$51.29
3 Laborers	33.10	794.40	51.05	1225.20		
1 Truck Driver (light)	32.25	258.00	49.20	393.60		
1 Crack Cleaner, 25 H.P.		60.40		66.44		
1 Crack Filler, Trailer Mtd.		203.20		223.52		
1 Flatbed Truck, Gas, 3 Ton		296.00		325.60	13.99	15.39
40 L.H., Daily Totals		$1892.80		$2667.16	$47.32	$66.68

Crew B-78	**Hr.**	**Daily**	**Hr.**	**Daily**	**Bare Costs**	**Incl. O&P**
1 Labor Foreman	$35.10	$280.80	$54.10	$432.80	$33.29	$51.25
4 Laborers	33.10	1059.20	51.05	1633.60		
1 Truck Driver (light)	32.25	258.00	49.20	393.60		
1 Paint Striper, S.P.		152.60		167.86		
1 Flatbed Truck, Gas, 3 Ton		296.00		325.60		
1 Pickup Truck, 3/4 Ton		140.00		154.00	12.26	13.49
48 L.H., Daily Totals		$2186.60		$3107.46	$45.55	$64.74

Crew B-78A	**Hr.**	**Daily**	**Hr.**	**Daily**	**Bare Costs**	**Incl. O&P**
1 Equip. Oper. (light)	$41.30	$330.40	$61.85	$494.80	$41.30	$61.85
1 Line Rem. (metal balls) 115 H.P.		883.40		971.74	110.43	121.47
8 L.H., Daily Totals		$1213.80		$1466.54	$151.72	$183.32

Crew B-78B	**Hr.**	**Daily**	**Hr.**	**Daily**	**Bare Costs**	**Incl. O&P**
2 Laborers	$33.10	$529.60	$51.05	$816.80	$34.01	$52.25
.25 Equip. Oper. (light)	41.30	82.60	61.85	123.70		
1 Pickup Truck, 3/4 Ton		140.00		154.00		
1 Line Rem., 11 H.P., walk behind		59.20		65.12		
.25 Road Sweeper, S.P., 8' wide		162.60		178.86	20.10	22.11
18 L.H., Daily Totals		$974.00		$1338.48	$54.11	$74.36

Crew B-79	**Hr.**	**Daily**	**Hr.**	**Daily**	**Bare Costs**	**Incl. O&P**
1 Labor Foreman	$35.10	$280.80	$54.10	$432.80	$33.33	$51.29
3 Laborers	33.10	794.40	51.05	1225.20		
1 Truck Driver (light)	32.25	258.00	49.20	393.60		
1 Thermo. Striper, T.M.		802.40		882.64		
1 Flatbed Truck, Gas, 3 Ton		296.00		325.60		
2 Pickup Trucks, 3/4 Ton		280.00		308.00	34.46	37.91
40 L.H., Daily Totals		$2711.60		$3567.84	$67.79	$89.20

Crew No.	Bare Costs		Incl. Subs O&P		Cost Per Labor-Hour	
Crew B-79A	**Hr.**	**Daily**	**Hr.**	**Daily**	**Bare Costs**	**Incl. O&P**
1.5 Equip. Oper. (light)	$41.30	$495.60	$61.85	$742.20	$41.30	$61.85
.5 Line Remov. (grinder) 115 H.P.		476.90		524.59		
1 Line Rem. (metal balls) 115 H.P.		883.40		971.74	113.36	124.69
12 L.H., Daily Totals		$1855.90		$2238.53	$154.66	$186.54

Crew B-80	Hr.	Daily	Hr.	Daily	Bare Costs	Incl. O&P
1 Labor Foreman	$35.10	$280.80	$54.10	$432.80	$35.44	$54.05
1 Laborer	33.10	264.80	51.05	408.40		
1 Truck Driver (light)	32.25	258.00	49.20	393.60		
1 Equip. Oper. (light)	41.30	330.40	61.85	494.80		
1 Flatbed Truck, Gas, 3 Ton		296.00		325.60		
1 Earth Auger, Truck-Mtd.		449.20		494.12	23.29	25.62
32 L.H., Daily Totals		$1879.20		$2549.32	$58.73	$79.67

Crew B-80A	Hr.	Daily	Hr.	Daily	Bare Costs	Incl. O&P
3 Laborers	$33.10	$794.40	$51.05	$1225.20	$33.10	$51.05
1 Flatbed Truck, Gas, 3 Ton		296.00		325.60	12.33	13.57
24 L.H., Daily Totals		$1090.40		$1550.80	$45.43	$64.62

Crew B-80B	Hr.	Daily	Hr.	Daily	Bare Costs	Incl. O&P
3 Laborers	$33.10	$794.40	$51.05	$1225.20	$35.15	$53.75
1 Equip. Oper. (light)	41.30	330.40	61.85	494.80		
1 Crane, Flatbed Mounted, 3 Ton		242.60		266.86	7.58	8.34
32 L.H., Daily Totals		$1367.40		$1986.86	$42.73	$62.09

Crew B-80C	Hr.	Daily	Hr.	Daily	Bare Costs	Incl. O&P
2 Laborers	$33.10	$529.60	$51.05	$816.80	$32.82	$50.43
1 Truck Driver (light)	32.25	258.00	49.20	393.60		
1 Flatbed Truck, Gas, 1.5 Ton		239.80		263.78		
1 Manual fence post auger, gas		7.80		8.58	10.32	11.35
24 L.H., Daily Totals		$1035.20		$1482.76	$43.13	$61.78

Crew B-81	Hr.	Daily	Hr.	Daily	Bare Costs	Incl. O&P
1 Laborer	$33.10	$264.80	$51.05	$408.40	$36.40	$55.30
1 Equip. Oper. (med.)	42.95	343.60	64.30	514.40		
1 Truck Driver (heavy)	33.15	265.20	50.55	404.40		
1 Hydromulcher, T.M.		338.20		372.02		
1 Truck Tractor, 220 H.P.		354.00		389.40	28.84	31.73
24 L.H., Daily Totals		$1565.80		$2088.62	$65.24	$87.03

Crew B-82	Hr.	Daily	Hr.	Daily	Bare Costs	Incl. O&P
1 Laborer	$33.10	$264.80	$51.05	$408.40	$37.20	$56.45
1 Equip. Oper. (light)	41.30	330.40	61.85	494.80		
1 Horiz. Borer, 6 H.P.		82.00		90.20	5.13	5.64
16 L.H., Daily Totals		$677.20		$993.40	$42.33	$62.09

Crew B-82A	Hr.	Daily	Hr.	Daily	Bare Costs	Incl. O&P
1 Laborer	$33.10	$264.80	$51.05	$408.40	$37.20	$56.45
1 Equip. Oper. (light)	41.30	330.40	61.85	494.80		
1 Flatbed Truck, Gas, 3 Ton		296.00		325.60		
1 Flatbed Trailer, 25 Ton		105.60		116.16		
1 Horiz. Dir. Drill, 20k lb. thrust		651.60		716.76	65.83	72.41
16 L.H., Daily Totals		$1648.40		$2061.72	$103.03	$128.86

Crew B-82B	Hr.	Daily	Hr.	Daily	Bare Costs	Incl. O&P
2 Laborers	$33.10	$529.60	$51.05	$816.80	$35.83	$54.65
1 Equip. Oper. (light)	41.30	330.40	61.85	494.80		
1 Flatbed Truck, Gas, 3 Ton		296.00		325.60		
1 Flatbed Trailer, 25 Ton		105.60		116.16		
1 Horiz. Dir. Drill, 30k lb. thrust		928.20		1021.02	55.41	60.95
24 L.H., Daily Totals		$2189.80		$2774.38	$91.24	$115.60

Crew B-82C	Hr.	Daily	Hr.	Daily	Bare Costs	Incl. O&P
2 Laborers	$33.10	$529.60	$51.05	$816.80	$35.83	$54.65
1 Equip. Oper. (light)	41.30	330.40	61.85	494.80		
1 Flatbed Truck, Gas, 3 Ton		296.00		325.60		
1 Flatbed Trailer, 25 Ton		105.60		116.16		
1 Horiz. Dir. Drill, 50k lb. thrust		1232.00		1355.20	68.07	74.87
24 L.H., Daily Totals		$2493.60		$3108.56	$103.90	$129.52

Crew B-82D	Hr.	Daily	Hr.	Daily	Bare Costs	Incl. O&P
1 Equip. Oper. (light)	$41.30	$330.40	$61.85	$494.80	$41.30	$61.85
1 Mud Trailer for HDD, 1500 gallon		338.00		371.80	42.25	46.48
8 L.H., Daily Totals		$668.40		$866.60	$83.55	$108.33

Crew B-83	Hr.	Daily	Hr.	Daily	Bare Costs	Incl. O&P
1 Tugboat Captain	$42.95	$343.60	$64.30	$514.40	$38.02	$57.67
1 Tugboat Hand	33.10	264.80	51.05	408.40		
1 Tugboat, 250 H.P.		840.80		924.88	52.55	57.81
16 L.H., Daily Totals		$1449.20		$1847.68	$90.58	$115.48

Crew B-84	Hr.	Daily	Hr.	Daily	Bare Costs	Incl. O&P
1 Equip. Oper. (med.)	$42.95	$343.60	$64.30	$514.40	$42.95	$64.30
1 Rotary Mower/Tractor		302.00		332.20	37.75	41.52
8 L.H., Daily Totals		$645.60		$846.60	$80.70	$105.83

Crew B-85	Hr.	Daily	Hr.	Daily	Bare Costs	Incl. O&P
3 Laborers	$33.10	$794.40	$51.05	$1225.20	$35.08	$53.60
1 Equip. Oper. (med.)	42.95	343.60	64.30	514.40		
1 Truck Driver (heavy)	33.15	265.20	50.55	404.40		
1 Aerial Lift Truck, 80'		665.40		731.94		
1 Brush Chipper, 12", 130 H.P.		253.20		278.52		
1 Pruning Saw, Rotary		6.75		7.42	23.13	25.45
40 L.H., Daily Totals		$2328.55		$3161.89	$58.21	$79.05

Crew B-86	Hr.	Daily	Hr.	Daily	Bare Costs	Incl. O&P
1 Equip. Oper. (med.)	$42.95	$343.60	$64.30	$514.40	$42.95	$64.30
1 Stump Chipper, S.P.		190.90		209.99	23.86	26.25
8 L.H., Daily Totals		$534.50		$724.39	$66.81	$90.55

Crew B-86A	Hr.	Daily	Hr.	Daily	Bare Costs	Incl. O&P
1 Equip. Oper. (med.)	$42.95	$343.60	$64.30	$514.40	$42.95	$64.30
1 Grader, 30,000 Lbs.		622.20		684.42	77.78	85.55
8 L.H., Daily Totals		$965.80		$1198.82	$120.72	$149.85

Crew B-86B	Hr.	Daily	Hr.	Daily	Bare Costs	Incl. O&P
1 Equip. Oper. (med.)	$42.95	$343.60	$64.30	$514.40	$42.95	$64.30
1 Dozer, 200 H.P.		1192.00		1311.20	149.00	163.90
8 L.H., Daily Totals		$1535.60		$1825.60	$191.95	$228.20

Crew B-87	Bare Costs Hr.	Bare Costs Daily	Incl. Subs O&P Hr.	Incl. Subs O&P Daily	Bare Costs (Cost Per Labor-Hour)	Incl. O&P (Cost Per Labor-Hour)
1 Laborer	$33.10	$264.80	$51.05	$408.40	$40.98	$61.65
4 Equip. Oper. (med.)	42.95	1374.40	64.30	2057.60		
2 Feller Bunchers, 100 H.P.		1219.20		1341.12		
1 Log Chipper, 22" Tree		689.60		758.56		
1 Dozer, 105 H.P.		660.40		726.44		
1 Chain Saw, gas, 36" Long		39.20		43.12	65.21	71.73
40 L.H., Daily Totals		$4247.60		$5335.24	$106.19	$133.38

Crew B-88	Hr.	Daily	Hr.	Daily	Bare Costs	Incl. O&P
1 Laborer	$33.10	$264.80	$51.05	$408.40	$41.54	$62.41
6 Equip. Oper. (med.)	42.95	2061.60	64.30	3086.40		
2 Feller Bunchers, 100 H.P.		1219.20		1341.12		
1 Log Chipper, 22" Tree		689.60		758.56		
2 Log Skidders, 50 H.P.		1814.40		1995.84		
1 Dozer, 105 H.P.		660.40		726.44		
1 Chain Saw, gas, 36" Long		39.20		43.12	78.98	86.88
56 L.H., Daily Totals		$6749.20		$8359.88	$120.52	$149.28

Crew B-89	Hr.	Daily	Hr.	Daily	Bare Costs	Incl. O&P
1 Equip. Oper. (light)	$41.30	$330.40	$61.85	$494.80	$36.77	$55.52
1 Truck Driver (light)	32.25	258.00	49.20	393.60		
1 Flatbed Truck, Gas, 3 Ton		296.00		325.60		
1 Concrete Saw		161.40		177.54		
1 Water Tank, 65 Gal.		18.45		20.30	29.74	32.71
16 L.H., Daily Totals		$1064.25		$1411.84	$66.52	$88.24

Crew B-89A	Hr.	Daily	Hr.	Daily	Bare Costs	Incl. O&P
1 Skilled Worker	$42.60	$340.80	$65.50	$524.00	$37.85	$58.27
1 Laborer	33.10	264.80	51.05	408.40		
1 Core Drill (large)		113.80		125.18	7.11	7.82
16 L.H., Daily Totals		$719.40		$1057.58	$44.96	$66.10

Crew B-89B	Hr.	Daily	Hr.	Daily	Bare Costs	Incl. O&P
1 Equip. Oper. (light)	$41.30	$330.40	$61.85	$494.80	$36.77	$55.52
1 Truck Driver, Light	32.25	258.00	49.20	393.60		
1 Wall Saw, Hydraulic, 10 H.P.		110.40		121.44		
1 Generator, Diesel, 100 kW		418.60		460.46		
1 Water Tank, 65 Gal.		18.45		20.30		
1 Flatbed Truck, Gas, 3 Ton		296.00		325.60	52.72	57.99
16 L.H., Daily Totals		$1431.85		$1816.19	$89.49	$113.51

Crew B-90	Hr.	Daily	Hr.	Daily	Bare Costs	Incl. O&P
1 Labor Foreman (outside)	$35.10	$280.80	$54.10	$432.80	$35.41	$54.01
3 Laborers	33.10	794.40	51.05	1225.20		
2 Equip. Oper. (light)	41.30	660.80	61.85	989.60		
2 Truck Drivers (heavy)	33.15	530.40	50.55	808.80		
1 Road Mixer, 310 H.P.		1955.00		2150.50		
1 Dist. Truck, 2000 Gal.		278.20		306.02	34.89	38.38
64 L.H., Daily Totals		$4499.60		$5912.92	$70.31	$92.39

Crew B-90A	Hr.	Daily	Hr.	Daily	Bare Costs	Incl. O&P
1 Labor Foreman	$35.10	$280.80	$54.10	$432.80	$39.01	$59.06
2 Laborers	33.10	529.60	51.05	816.80		
4 Equip. Oper. (med.)	42.95	1374.40	64.30	2057.60		
2 Graders, 30,000 Lbs.		1244.40		1368.84		
1 Tandem Roller, 10 Ton		235.20		258.72		
1 Roller, Pneum. Whl, 12 Ton		337.20		370.92	32.44	35.69
56 L.H., Daily Totals		$4001.60		$5305.68	$71.46	$94.74

Crew B-90B	Hr.	Daily	Hr.	Daily	Bare Costs	Incl. O&P
1 Labor Foreman	$35.10	$280.80	$54.10	$432.80	$38.36	$58.18
2 Laborers	33.10	529.60	51.05	816.80		
3 Equip. Oper. (med.)	42.95	1030.80	64.30	1543.20		
1 Tandem Roller, 10 Ton		235.20		258.72		
1 Roller, Pneum. Whl, 12 Ton		337.20		370.92		
1 Road Mixer, 310 H.P.		1955.00		2150.50	52.65	57.92
48 L.H., Daily Totals		$4368.60		$5572.94	$91.01	$116.10

Crew B-91	Hr.	Daily	Hr.	Daily	Bare Costs	Incl. O&P
1 Labor Foreman (outside)	$35.10	$280.80	$54.10	$432.80	$38.28	$57.99
2 Laborers	33.10	529.60	51.05	816.80		
4 Equip. Oper. (med.)	42.95	1374.40	64.30	2057.60		
1 Truck Driver (heavy)	33.15	265.20	50.55	404.40		
1 Dist. Tanker, 3000 Gallon		307.00		337.70		
1 Truck Tractor, 6x4, 380 H.P.		592.40		651.64		
1 Aggreg. Spreader, S.P.		820.80		902.88		
1 Roller, Pneum. Whl, 12 Ton		337.20		370.92		
1 Tandem Roller, 10 Ton		235.20		258.72	35.82	39.40
64 L.H., Daily Totals		$4742.60		$6233.46	$74.10	$97.40

Crew B-92	Hr.	Daily	Hr.	Daily	Bare Costs	Incl. O&P
1 Labor Foreman (outside)	$35.10	$280.80	$54.10	$432.80	$33.60	$51.81
3 Laborers	33.10	794.40	51.05	1225.20		
1 Crack Cleaner, 25 H.P.		60.40		66.44		
1 Air Compressor, 60 cfm		138.60		152.46		
1 Tar Kettle, T.M.		71.40		78.54		
1 Flatbed Truck, Gas, 3 Ton		296.00		325.60	17.70	19.47
32 L.H., Daily Totals		$1641.60		$2281.04	$51.30	$71.28

Crew B-93	Hr.	Daily	Hr.	Daily	Bare Costs	Incl. O&P
1 Equip. Oper. (med.)	$42.95	$343.60	$64.30	$514.40	$42.95	$64.30
1 Feller Buncher, 100 H.P.		609.60		670.56	76.20	83.82
8 L.H., Daily Totals		$953.20		$1184.96	$119.15	$148.12

Crew B-94A	Hr.	Daily	Hr.	Daily	Bare Costs	Incl. O&P
1 Laborer	$33.10	$264.80	$51.05	$408.40	$33.10	$51.05
1 Diaphragm Water Pump, 2"		68.00		74.80		
1 -20' Suction Hose, 2"		1.95		2.15		
2 -50' Discharge Hoses, 2"		1.80		1.98	8.97	9.87
8 L.H., Daily Totals		$336.55		$487.32	$42.07	$60.92

Crew B-94B	Hr.	Daily	Hr.	Daily	Bare Costs	Incl. O&P
1 Laborer	$33.10	$264.80	$51.05	$408.40	$33.10	$51.05
1 Diaphragm Water Pump, 4"		90.80		99.88		
1 -20' Suction Hose, 4"		3.45		3.79		
2 -50' Discharge Hoses, 4"		4.70		5.17	12.37	13.61
8 L.H., Daily Totals		$363.75		$517.25	$45.47	$64.66

Crew B-94C	Hr.	Daily	Hr.	Daily	Bare Costs	Incl. O&P
1 Laborer	$33.10	$264.80	$51.05	$408.40	$33.10	$51.05
1 Centrifugal Water Pump, 3"		74.40		81.84		
1 -20' Suction Hose, 3"		3.05		3.36		
2 -50' Discharge Hoses, 3"		3.50		3.85	10.12	11.13
8 L.H., Daily Totals		$345.75		$497.44	$43.22	$62.18

Crew B-94D	Hr.	Daily	Hr.	Daily	Bare Costs	Incl. O&P
1 Laborer	$33.10	$264.80	$51.05	$408.40	$33.10	$51.05
1 Centr. Water Pump, 6"		340.00		374.00		
1 -20' Suction Hose, 6"		11.90		13.09		
2 -50' Discharge Hoses, 6"		12.60		13.86	45.56	50.12
8 L.H., Daily Totals		$629.30		$809.35	$78.66	$101.17

Crew No.	Bare Costs		Incl. Subs O&P		Cost Per Labor-Hour	
Crew C-1	**Hr.**	**Daily**	**Hr.**	**Daily**	**Bare Costs**	**Incl. O&P**
3 Carpenters	$41.55	$997.20	$64.05	$1537.20	$39.44	$60.80
1 Laborer	33.10	264.80	51.05	408.40		
32 L.H., Daily Totals		$1262.00		$1945.60	$39.44	$60.80

Crew C-2	Hr.	Daily	Hr.	Daily	Bare Costs	Incl. O&P
1 Carpenter Foreman (out)	$43.55	$348.40	$67.15	$537.20	$40.48	$62.40
4 Carpenters	41.55	1329.60	64.05	2049.60		
1 Laborer	33.10	264.80	51.05	408.40		
48 L.H., Daily Totals		$1942.80		$2995.20	$40.48	$62.40

Crew C-2A	Hr.	Daily	Hr.	Daily	Bare Costs	Incl. O&P
1 Carpenter Foreman (out)	$43.55	$348.40	$67.15	$537.20	$40.17	$61.38
3 Carpenters	41.55	997.20	64.05	1537.20		
1 Cement Finisher	39.70	317.60	57.95	463.60		
1 Laborer	33.10	264.80	51.05	408.40		
48 L.H., Daily Totals		$1928.00		$2946.40	$40.17	$61.38

Crew C-3	Hr.	Daily	Hr.	Daily	Bare Costs	Incl. O&P
1 Rodman Foreman	$48.80	$390.40	$78.60	$628.80	$42.94	$68.02
4 Rodmen (reinf.)	46.80	1497.60	75.40	2412.80		
1 Equip. Oper. (light)	41.30	330.40	61.85	494.80		
2 Laborers	33.10	529.60	51.05	816.80		
3 Stressing Equipment		28.80		31.68		
.5 Grouting Equipment		80.00		88.00	1.70	1.87
64 L.H., Daily Totals		$2856.80		$4472.88	$44.64	$69.89

Crew C-4	Hr.	Daily	Hr.	Daily	Bare Costs	Incl. O&P
1 Rodman Foreman	$48.80	$390.40	$78.60	$628.80	$47.30	$76.20
3 Rodmen (reinf.)	46.80	1123.20	75.40	1809.60		
3 Stressing Equipment		28.80		31.68	.90	.99
32 L.H., Daily Totals		$1542.40		$2470.08	$48.20	$77.19

Crew C-4A	Hr.	Daily	Hr.	Daily	Bare Costs	Incl. O&P
2 Rodmen (reinf.)	$46.80	$748.80	$75.40	$1206.40	$46.80	$75.40
4 Stressing Equipment		38.40		42.24	2.40	2.64
16 L.H., Daily Totals		$787.20		$1248.64	$49.20	$78.04

Crew C-5	Hr.	Daily	Hr.	Daily	Bare Costs	Incl. O&P
1 Rodman Foreman	$48.80	$390.40	$78.60	$628.80	$45.53	$72.00
4 Rodmen (reinf.)	46.80	1497.60	75.40	2412.80		
1 Equip. Oper. (crane)	44.40	355.20	66.45	531.60		
1 Equip. Oper. Oiler	38.30	306.40	57.35	458.80		
1 Hyd. Crane, 25 Ton		747.60		822.36	13.35	14.69
56 L.H., Daily Totals		$3297.20		$4854.36	$58.88	$86.69

Crew C-6	Hr.	Daily	Hr.	Daily	Bare Costs	Incl. O&P
1 Labor Foreman (outside)	$35.10	$280.80	$54.10	$432.80	$34.53	$52.71
4 Laborers	33.10	1059.20	51.05	1633.60		
1 Cement Finisher	39.70	317.60	57.95	463.60		
2 Gas Engine Vibrators		60.80		66.88	1.27	1.39
48 L.H., Daily Totals		$1718.40		$2596.88	$35.80	$54.10

Crew No.	Bare Costs		Incl. Subs O&P		Cost Per Labor-Hour	
Crew C-7	**Hr.**	**Daily**	**Hr.**	**Daily**	**Bare Costs**	**Incl. O&P**
1 Labor Foreman (outside)	$35.10	$280.80	$54.10	$432.80	$35.73	$54.33
5 Laborers	33.10	1324.00	51.05	2042.00		
1 Cement Finisher	39.70	317.60	57.95	463.60		
1 Equip. Oper. (med.)	42.95	343.60	64.30	514.40		
1 Equip. Oper. (oiler)	38.30	306.40	57.35	458.80		
2 Gas Engine Vibrators		60.80		66.88		
1 Concrete Bucket, 1 C.Y.		20.60		22.66		
1 Hyd. Crane, 55 Ton		1209.00		1329.90	17.92	19.71
72 L.H., Daily Totals		$3862.80		$5331.04	$53.65	$74.04

Crew C-7A	Hr.	Daily	Hr.	Daily	Bare Costs	Incl. O&P
1 Labor Foreman (outside)	$35.10	$280.80	$54.10	$432.80	$33.36	$51.31
5 Laborers	33.10	1324.00	51.05	2042.00		
2 Truck Drivers (Heavy)	33.15	530.40	50.55	808.80		
2 Conc. Transit Mixers		2016.00		2217.60	31.50	34.65
64 L.H., Daily Totals		$4151.20		$5501.20	$64.86	$85.96

Crew C-7B	Hr.	Daily	Hr.	Daily	Bare Costs	Incl. O&P
1 Labor Foreman (outside)	$35.10	$280.80	$54.10	$432.80	$35.41	$54.14
5 Laborers	33.10	1324.00	51.05	2042.00		
1 Equipment Oper. (crane)	44.40	355.20	66.45	531.60		
1 Equipment Oiler	38.30	306.40	57.35	458.80		
1 Conc. Bucket, 2 C.Y.		32.40		35.64		
1 Lattice Boom Crane, 165 Ton		2039.00		2242.90	32.37	35.60
64 L.H., Daily Totals		$4337.80		$5743.74	$67.78	$89.75

Crew C-7C	Hr.	Daily	Hr.	Daily	Bare Costs	Incl. O&P
1 Labor Foreman (outside)	$35.10	$280.80	$54.10	$432.80	$35.81	$54.74
5 Laborers	33.10	1324.00	51.05	2042.00		
2 Equipment Operators (med.)	42.95	687.20	64.30	1028.80		
2 F.E. Loaders, W.M., 4 C.Y.		1255.60		1381.16	19.62	21.58
64 L.H., Daily Totals		$3547.60		$4884.76	$55.43	$76.32

Crew C-7D	Hr.	Daily	Hr.	Daily	Bare Costs	Incl. O&P
1 Labor Foreman (outside)	$35.10	$280.80	$54.10	$432.80	$34.79	$53.38
5 Laborers	33.10	1324.00	51.05	2042.00		
1 Equip. Oper. (med.)	42.95	343.60	64.30	514.40		
1 Concrete Conveyer		194.60		214.06	3.48	3.82
56 L.H., Daily Totals		$2143.00		$3203.26	$38.27	$57.20

Crew C-8	Hr.	Daily	Hr.	Daily	Bare Costs	Incl. O&P
1 Labor Foreman (outside)	$35.10	$280.80	$54.10	$432.80	$36.68	$55.35
3 Laborers	33.10	794.40	51.05	1225.20		
2 Cement Finishers	39.70	635.20	57.95	927.20		
1 Equip. Oper. (med.)	42.95	343.60	64.30	514.40		
1 Concrete Pump (small)		738.80		812.68	13.19	14.51
56 L.H., Daily Totals		$2792.80		$3912.28	$49.87	$69.86

Crew C-8A	Hr.	Daily	Hr.	Daily	Bare Costs	Incl. O&P
1 Labor Foreman (outside)	$35.10	$280.80	$54.10	$432.80	$35.63	$53.86
3 Laborers	33.10	794.40	51.05	1225.20		
2 Cement Finishers	39.70	635.20	57.95	927.20		
48 L.H., Daily Totals		$1710.40		$2585.20	$35.63	$53.86

Crew No.	Bare Costs		Incl. Subs O&P		Cost Per Labor-Hour	

Crew C-8B	Hr.	Daily	Hr.	Daily	Bare Costs	Incl. O&P
1 Labor Foreman (outside)	$35.10	$280.80	$54.10	$432.80	$35.47	$54.31
3 Laborers	33.10	794.40	51.05	1225.20		
1 Equip. Oper. (med.)	42.95	343.60	64.30	514.40		
1 Vibrating Power Screed		65.15		71.67		
1 Roller, Vibratory, 25 Ton		633.20		696.52		
1 Dozer, 200 H.P.		1192.00		1311.20	47.26	51.98
40 L.H., Daily Totals		$3309.15		$4251.78	$82.73	$106.29

Crew C-8C	Hr.	Daily	Hr.	Daily	Bare Costs	Incl. O&P
1 Labor Foreman (outside)	$35.10	$280.80	$54.10	$432.80	$36.17	$54.92
3 Laborers	33.10	794.40	51.05	1225.20		
1 Cement Finisher	39.70	317.60	57.95	463.60		
1 Equip. Oper. (med.)	42.95	343.60	64.30	514.40		
1 Shotcrete Rig, 12 C.Y./hr		249.20		274.12	5.19	5.71
48 L.H., Daily Totals		$1985.60		$2910.12	$41.37	$60.63

Crew C-8D	Hr.	Daily	Hr.	Daily	Bare Costs	Incl. O&P
1 Labor Foreman (outside)	$35.10	$280.80	$54.10	$432.80	$37.30	$56.24
1 Laborer	33.10	264.80	51.05	408.40		
1 Cement Finisher	39.70	317.60	57.95	463.60		
1 Equipment Oper. (light)	41.30	330.40	61.85	494.80		
1 Air Compressor, 250 cfm		193.40		212.74		
2 -50' Air Hoses, 1"		8.20		9.02	6.30	6.93
32 L.H., Daily Totals		$1395.20		$2021.36	$43.60	$63.17

Crew C-8E	Hr.	Daily	Hr.	Daily	Bare Costs	Incl. O&P
1 Labor Foreman (outside)	$35.10	$280.80	$54.10	$432.80	$37.30	$56.24
1 Laborer	33.10	264.80	51.05	408.40		
1 Cement Finisher	39.70	317.60	57.95	463.60		
1 Equipment Oper. (light)	41.30	330.40	61.85	494.80		
1 Air Compressor, 250 cfm		193.40		212.74		
2 -50' Air Hoses, 1"		8.20		9.02		
1 Concrete Pump (small)		738.80		812.68	29.39	32.33
32 L.H., Daily Totals		$2134.00		$2834.04	$66.69	$88.56

Crew C-10	Hr.	Daily	Hr.	Daily	Bare Costs	Incl. O&P
1 Laborer	$33.10	$264.80	$51.05	$408.40	$37.50	$55.65
2 Cement Finishers	39.70	635.20	57.95	927.20		
24 L.H., Daily Totals		$900.00		$1335.60	$37.50	$55.65

Crew C-10B	Hr.	Daily	Hr.	Daily	Bare Costs	Incl. O&P
3 Laborers	$33.10	$794.40	$51.05	$1225.20	$35.74	$53.81
2 Cement Finishers	39.70	635.20	57.95	927.20		
1 Concrete Mixer, 10 C.F.		168.40		185.24		
2 Trowels, 48" Walk-Behind		98.00		107.80	6.66	7.33
40 L.H., Daily Totals		$1696.00		$2445.44	$42.40	$61.14

Crew C-10C	Hr.	Daily	Hr.	Daily	Bare Costs	Incl. O&P
1 Laborer	$33.10	$264.80	$51.05	$408.40	$37.50	$55.65
2 Cement Finishers	39.70	635.20	57.95	927.20		
1 Trowel, 48" Walk-Behind		49.00		53.90	2.04	2.25
24 L.H., Daily Totals		$949.00		$1389.50	$39.54	$57.90

Crew C-10D	Hr.	Daily	Hr.	Daily	Bare Costs	Incl. O&P
1 Laborer	$33.10	$264.80	$51.05	$408.40	$37.50	$55.65
2 Cement Finishers	39.70	635.20	57.95	927.20		
1 Vibrating Power Screed		65.15		71.67		
1 Trowel, 48" Walk-Behind		49.00		53.90	4.76	5.23
24 L.H., Daily Totals		$1014.15		$1461.17	$42.26	$60.88

Crew C-10E	Hr.	Daily	Hr.	Daily	Bare Costs	Incl. O&P
1 Laborer	$33.10	$264.80	$51.05	$408.40	$37.50	$55.65
2 Cement Finishers	39.70	635.20	57.95	927.20		
1 Vibrating Power Screed		65.15		71.67		
1 Cement Trowel, 96" Ride-On		185.80		204.38	10.46	11.50
24 L.H., Daily Totals		$1150.95		$1611.65	$47.96	$67.15

Crew C-11	Hr.	Daily	Hr.	Daily	Bare Costs	Incl. O&P
1 Struc. Steel Foreman	$48.90	$391.20	$85.95	$687.60	$45.89	$78.27
6 Struc. Steel Workers	46.90	2251.20	82.45	3957.60		
1 Equip. Oper. (crane)	44.40	355.20	66.45	531.60		
1 Equip. Oper. Oiler	38.30	306.40	57.35	458.80		
1 Lattice Boom Crane, 150 Ton		2003.00		2203.30	27.82	30.60
72 L.H., Daily Totals		$5307.00		$7838.90	$73.71	$108.87

Crew C-12	Hr.	Daily	Hr.	Daily	Bare Costs	Incl. O&P
1 Carpenter Foreman (out)	$43.55	$348.40	$67.15	$537.20	$40.95	$62.80
3 Carpenters	41.55	997.20	64.05	1537.20		
1 Laborer	33.10	264.80	51.05	408.40		
1 Equip. Oper. (crane)	44.40	355.20	66.45	531.60		
1 Hyd. Crane, 12 Ton		655.60		721.16	13.66	15.02
48 L.H., Daily Totals		$2621.20		$3735.56	$54.61	$77.82

Crew C-13	Hr.	Daily	Hr.	Daily	Bare Costs	Incl. O&P
1 Struc. Steel Worker	$46.90	$375.20	$82.45	$659.60	$45.12	$76.32
1 Welder	46.90	375.20	82.45	659.60		
1 Carpenter	41.55	332.40	64.05	512.40		
1 Welder, gas engine, 300 amp		145.80		160.38	6.08	6.68
24 L.H., Daily Totals		$1228.60		$1991.98	$51.19	$83.00

Crew C-14	Hr.	Daily	Hr.	Daily	Bare Costs	Incl. O&P
1 Carpenter Foreman (out)	$43.55	$348.40	$67.15	$537.20	$40.72	$62.94
5 Carpenters	41.55	1662.00	64.05	2562.00		
4 Laborers	33.10	1059.20	51.05	1633.60		
4 Rodmen (reinf.)	46.80	1497.60	75.40	2412.80		
2 Cement Finishers	39.70	635.20	57.95	927.20		
1 Equip. Oper. (crane)	44.40	355.20	66.45	531.60		
1 Equip. Oper. Oiler	38.30	306.40	57.35	458.80		
1 Hyd. Crane, 80 Ton		1545.00		1699.50	10.73	11.80
144 L.H., Daily Totals		$7409.00		$10762.70	$51.45	$74.74

Crew C-14A	Hr.	Daily	Hr.	Daily	Bare Costs	Incl. O&P
1 Carpenter Foreman (out)	$43.55	$348.40	$67.15	$537.20	$41.78	$64.72
16 Carpenters	41.55	5318.40	64.05	8198.40		
4 Rodmen (reinf.)	46.80	1497.60	75.40	2412.80		
2 Laborers	33.10	529.60	51.05	816.80		
1 Cement Finisher	39.70	317.60	57.95	463.60		
1 Equip. Oper. (med.)	42.95	343.60	64.30	514.40		
1 Gas Engine Vibrator		30.40		33.44		
1 Concrete Pump (small)		738.80		812.68	3.85	4.23
200 L.H., Daily Totals		$9124.40		$13789.32	$45.62	$68.95

Crew C-14B	Hr.	Daily	Hr.	Daily	Bare Costs	Incl. O&P
1 Carpenter Foreman (out)	$43.55	$348.40	$67.15	$537.20	$41.70	$64.46
16 Carpenters	41.55	5318.40	64.05	8198.40		
4 Rodmen (reinf.)	46.80	1497.60	75.40	2412.80		
2 Laborers	33.10	529.60	51.05	816.80		
2 Cement Finishers	39.70	635.20	57.95	927.20		
1 Equip. Oper. (med.)	42.95	343.60	64.30	514.40		
1 Gas Engine Vibrator		30.40		33.44		
1 Concrete Pump (small)		738.80		812.68	3.70	4.07
208 L.H., Daily Totals		$9442.00		$14252.92	$45.39	$68.52

Crew No.	Bare Costs		Incl. Subs O&P		Cost Per Labor-Hour	
Crew C-14C	**Hr.**	**Daily**	**Hr.**	**Daily**	**Bare Costs**	**Incl. O&P**
1 Carpenter Foreman (out)	$43.55	$348.40	$67.15	$537.20	$39.90	$61.74
6 Carpenters	41.55	1994.40	64.05	3074.40		
2 Rodmen (reinf.)	46.80	748.80	75.40	1206.40		
4 Laborers	33.10	1059.20	51.05	1633.60		
1 Cement Finisher	39.70	317.60	57.95	463.60		
1 Gas Engine Vibrator		30.40		33.44	.27	.30
112 L.H., Daily Totals		$4498.80		$6948.64	$40.17	$62.04
Crew C-14D	**Hr.**	**Daily**	**Hr.**	**Daily**	**Bare Costs**	**Incl. O&P**
1 Carpenter Foreman (out)	$43.55	$348.40	$67.15	$537.20	$41.36	$63.81
18 Carpenters	41.55	5983.20	64.05	9223.20		
2 Rodmen (reinf.)	46.80	748.80	75.40	1206.40		
2 Laborers	33.10	529.60	51.05	816.80		
1 Cement Finisher	39.70	317.60	57.95	463.60		
1 Equip. Oper. (med.)	42.95	343.60	64.30	514.40		
1 Gas Engine Vibrator		30.40		33.44		
1 Concrete Pump (small)		738.80		812.68	3.85	4.23
200 L.H., Daily Totals		$9040.40		$13607.72	$45.20	$68.04
Crew C-14E	**Hr.**	**Daily**	**Hr.**	**Daily**	**Bare Costs**	**Incl. O&P**
1 Carpenter Foreman (out)	$43.55	$348.40	$67.15	$537.20	$41.17	$64.36
2 Carpenters	41.55	664.80	64.05	1024.80		
4 Rodmen (reinf.)	46.80	1497.60	75.40	2412.80		
3 Laborers	33.10	794.40	51.05	1225.20		
1 Cement Finisher	39.70	317.60	57.95	463.60		
1 Gas Engine Vibrator		30.40		33.44	.35	.38
88 L.H., Daily Totals		$3653.20		$5697.04	$41.51	$64.74
Crew C-14F	**Hr.**	**Daily**	**Hr.**	**Daily**	**Bare Costs**	**Incl. O&P**
1 Laborer Foreman (out)	$35.10	$280.80	$54.10	$432.80	$37.72	$55.99
2 Laborers	33.10	529.60	51.05	816.80		
6 Cement Finishers	39.70	1905.60	57.95	2781.60		
1 Gas Engine Vibrator		30.40		33.44	.42	.46
72 L.H., Daily Totals		$2746.40		$4064.64	$38.14	$56.45
Crew C-14G	**Hr.**	**Daily**	**Hr.**	**Daily**	**Bare Costs**	**Incl. O&P**
1 Laborer Foreman (out)	$35.10	$280.80	$54.10	$432.80	$37.16	$55.43
2 Laborers	33.10	529.60	51.05	816.80		
4 Cement Finishers	39.70	1270.40	57.95	1854.40		
1 Gas Engine Vibrator		30.40		33.44	.54	.60
56 L.H., Daily Totals		$2111.20		$3137.44	$37.70	$56.03
Crew C-14H	**Hr.**	**Daily**	**Hr.**	**Daily**	**Bare Costs**	**Incl. O&P**
1 Carpenter Foreman (out)	$43.55	$348.40	$67.15	$537.20	$41.04	$63.27
2 Carpenters	41.55	664.80	64.05	1024.80		
1 Rodman (reinf.)	46.80	374.40	75.40	603.20		
1 Laborer	33.10	264.80	51.05	408.40		
1 Cement Finisher	39.70	317.60	57.95	463.60		
1 Gas Engine Vibrator		30.40		33.44	.63	.70
48 L.H., Daily Totals		$2000.40		$3070.64	$41.67	$63.97
Crew C-14L	**Hr.**	**Daily**	**Hr.**	**Daily**	**Bare Costs**	**Incl. O&P**
1 Carpenter Foreman (out)	$43.55	$348.40	$67.15	$537.20	$38.75	$59.47
6 Carpenters	41.55	1994.40	64.05	3074.40		
4 Laborers	33.10	1059.20	51.05	1633.60		
1 Cement Finisher	39.70	317.60	57.95	463.60		
1 Gas Engine Vibrator		30.40		33.44	.32	.35
96 L.H., Daily Totals		$3750.00		$5742.24	$39.06	$59.81

Crew No.	Bare Costs		Incl. Subs O&P		Cost Per Labor-Hour	
Crew C-15	**Hr.**	**Daily**	**Hr.**	**Daily**	**Bare Costs**	**Incl. O&P**
1 Carpenter Foreman (out)	$43.55	$348.40	$67.15	$537.20	$39.13	$59.97
2 Carpenters	41.55	664.80	64.05	1024.80		
3 Laborers	33.10	794.40	51.05	1225.20		
2 Cement Finishers	39.70	635.20	57.95	927.20		
1 Rodman (reinf.)	46.80	374.40	75.40	603.20		
72 L.H., Daily Totals		$2817.20		$4317.60	$39.13	$59.97
Crew C-16	**Hr.**	**Daily**	**Hr.**	**Daily**	**Bare Costs**	**Incl. O&P**
1 Labor Foreman (outside)	$35.10	$280.80	$54.10	$432.80	$38.93	$59.81
3 Laborers	33.10	794.40	51.05	1225.20		
2 Cement Finishers	39.70	635.20	57.95	927.20		
1 Equip. Oper. (med.)	42.95	343.60	64.30	514.40		
2 Rodmen (reinf.)	46.80	748.80	75.40	1206.40		
1 Concrete Pump (small)		738.80		812.68	10.26	11.29
72 L.H., Daily Totals		$3541.60		$5118.68	$49.19	$71.09
Crew C-17	**Hr.**	**Daily**	**Hr.**	**Daily**	**Bare Costs**	**Incl. O&P**
2 Skilled Worker Foremen	$44.60	$713.60	$68.55	$1096.80	$43.00	$66.11
8 Skilled Workers	42.60	2726.40	65.50	4192.00		
80 L.H., Daily Totals		$3440.00		$5288.80	$43.00	$66.11
Crew C-17A	**Hr.**	**Daily**	**Hr.**	**Daily**	**Bare Costs**	**Incl. O&P**
2 Skilled Worker Foremen	$44.60	$713.60	$68.55	$1096.80	$43.02	$66.11
8 Skilled Workers	42.60	2726.40	65.50	4192.00		
.125 Equip. Oper. (crane)	44.40	44.40	66.45	66.45		
.125 Hyd. Crane, 80 Ton		193.13		212.44	2.38	2.62
81 L.H., Daily Totals		$3677.53		$5567.69	$45.40	$68.74
Crew C-17B	**Hr.**	**Daily**	**Hr.**	**Daily**	**Bare Costs**	**Incl. O&P**
2 Skilled Worker Foremen	$44.60	$713.60	$68.55	$1096.80	$43.03	$66.12
8 Skilled Workers	42.60	2726.40	65.50	4192.00		
.25 Equip. Oper. (crane)	44.40	88.80	66.45	132.90		
.25 Hyd. Crane, 80 Ton		386.25		424.88		
.25 Trowel, 48" Walk-Behind		12.25		13.48	4.86	5.35
82 L.H., Daily Totals		$3927.30		$5860.05	$47.89	$71.46
Crew C-17C	**Hr.**	**Daily**	**Hr.**	**Daily**	**Bare Costs**	**Incl. O&P**
2 Skilled Worker Foremen	$44.60	$713.60	$68.55	$1096.80	$43.05	$66.12
8 Skilled Workers	42.60	2726.40	65.50	4192.00		
.375 Equip. Oper. (crane)	44.40	133.20	66.45	199.35		
.375 Hyd. Crane, 80 Ton		579.38		637.31	6.98	7.68
83 L.H., Daily Totals		$4152.57		$6125.46	$50.03	$73.80
Crew C-17D	**Hr.**	**Daily**	**Hr.**	**Daily**	**Bare Costs**	**Incl. O&P**
2 Skilled Worker Foremen	$44.60	$713.60	$68.55	$1096.80	$43.07	$66.13
8 Skilled Workers	42.60	2726.40	65.50	4192.00		
.5 Equip. Oper. (crane)	44.40	177.60	66.45	265.80		
.5 Hyd. Crane, 80 Ton		772.50		849.75	9.20	10.12
84 L.H., Daily Totals		$4390.10		$6404.35	$52.26	$76.24
Crew C-17E	**Hr.**	**Daily**	**Hr.**	**Daily**	**Bare Costs**	**Incl. O&P**
2 Skilled Worker Foremen	$44.60	$713.60	$68.55	$1096.80	$43.00	$66.11
8 Skilled Workers	42.60	2726.40	65.50	4192.00		
1 Hyd. Jack with Rods		87.35		96.08	1.09	1.20
80 L.H., Daily Totals		$3527.35		$5384.89	$44.09	$67.31

Crew No.	Bare Costs		Incl. Subs O&P		Cost Per Labor-Hour	
Crew C-18	**Hr.**	**Daily**	**Hr.**	**Daily**	**Bare Costs**	**Incl. O&P**
.125 Labor Foreman (out)	$35.10	$35.10	$54.10	$54.10	$33.32	$51.39
1 Laborer	33.10	264.80	51.05	408.40		
1 Concrete Cart, 10 C.F.		58.40		64.24	6.49	7.14
9 L.H., Daily Totals		$358.30		$526.74	$39.81	$58.53
Crew C-19	**Hr.**	**Daily**	**Hr.**	**Daily**	**Bare Costs**	**Incl. O&P**
.125 Labor Foreman (out)	$35.10	$35.10	$54.10	$54.10	$33.32	$51.39
1 Laborer	33.10	264.80	51.05	408.40		
1 Concrete Cart, 18 C.F.		94.00		103.40	10.44	11.49
9 L.H., Daily Totals		$393.90		$565.90	$43.77	$62.88
Crew C-20	**Hr.**	**Daily**	**Hr.**	**Daily**	**Bare Costs**	**Incl. O&P**
1 Labor Foreman (outside)	$35.10	$280.80	$54.10	$432.80	$35.41	$53.95
5 Laborers	33.10	1324.00	51.05	2042.00		
1 Cement Finisher	39.70	317.60	57.95	463.60		
1 Equip. Oper. (med.)	42.95	343.60	64.30	514.40		
2 Gas Engine Vibrators		60.80		66.88		
1 Concrete Pump (small)		738.80		812.68	12.49	13.74
64 L.H., Daily Totals		$3065.60		$4332.36	$47.90	$67.69
Crew C-21	**Hr.**	**Daily**	**Hr.**	**Daily**	**Bare Costs**	**Incl. O&P**
1 Labor Foreman (outside)	$35.10	$280.80	$54.10	$432.80	$35.41	$53.95
5 Laborers	33.10	1324.00	51.05	2042.00		
1 Cement Finisher	39.70	317.60	57.95	463.60		
1 Equip. Oper. (med.)	42.95	343.60	64.30	514.40		
2 Gas Engine Vibrators		60.80		66.88		
1 Concrete Conveyer		194.60		214.06	3.99	4.39
64 L.H., Daily Totals		$2521.40		$3733.74	$39.40	$58.34
Crew C-22	**Hr.**	**Daily**	**Hr.**	**Daily**	**Bare Costs**	**Incl. O&P**
1 Rodman Foreman	$48.80	$390.40	$78.60	$628.80	$46.92	$75.37
4 Rodmen (reinf.)	46.80	1497.60	75.40	2412.80		
.125 Equip. Oper. (crane)	44.40	44.40	66.45	66.45		
.125 Equip. Oper. Oiler	38.30	38.30	57.35	57.35		
.125 Hyd. Crane, 25 Ton		93.45		102.80	2.23	2.45
42 L.H., Daily Totals		$2064.15		$3268.20	$49.15	$77.81
Crew C-23	**Hr.**	**Daily**	**Hr.**	**Daily**	**Bare Costs**	**Incl. O&P**
2 Skilled Worker Foremen	$44.60	$713.60	$68.55	$1096.80	$42.75	$65.39
6 Skilled Workers	42.60	2044.80	65.50	3144.00		
1 Equip. Oper. (crane)	44.40	355.20	66.45	531.60		
1 Equip. Oper. Oiler	38.30	306.40	57.35	458.80		
1 Lattice Boom Crane, 90 Ton		1607.00		1767.70	20.09	22.10
80 L.H., Daily Totals		$5027.00		$6998.90	$62.84	$87.49
Crew C-23A	**Hr.**	**Daily**	**Hr.**	**Daily**	**Bare Costs**	**Incl. O&P**
1 Labor Foreman (outside)	$35.10	$280.80	$54.10	$432.80	$36.80	$56.00
2 Laborers	33.10	529.60	51.05	816.80		
1 Equip. Oper. (crane)	44.40	355.20	66.45	531.60		
1 Equip. Oper. Oiler	38.30	306.40	57.35	458.80		
1 Crawler Crane, 100 Ton		1732.00		1905.20		
3 Conc. Buckets, 8 C.Y.		603.60		663.96	58.39	64.23
40 L.H., Daily Totals		$3807.60		$4809.16	$95.19	$120.23
Crew C-24	**Hr.**	**Daily**	**Hr.**	**Daily**	**Bare Costs**	**Incl. O&P**
2 Skilled Worker Foremen	$44.60	$713.60	$68.55	$1096.80	$42.75	$65.39
6 Skilled Workers	42.60	2044.80	65.50	3144.00		
1 Equip. Oper. (crane)	44.40	355.20	66.45	531.60		
1 Equip. Oper. Oiler	38.30	306.40	57.35	458.80		
1 Lattice Boom Crane, 150 Ton		2003.00		2203.30	25.04	27.54
80 L.H., Daily Totals		$5423.00		$7434.50	$67.79	$92.93

Crew No.	Bare Costs		Incl. Subs O&P		Cost Per Labor-Hour	
Crew C-25	**Hr.**	**Daily**	**Hr.**	**Daily**	**Bare Costs**	**Incl. O&P**
2 Rodmen (reinf.)	$46.80	$748.80	$75.40	$1206.40	$36.50	$59.48
2 Rodmen Helpers	26.20	419.20	43.55	696.80		
32 L.H., Daily Totals		$1168.00		$1903.20	$36.50	$59.48
Crew C-27	**Hr.**	**Daily**	**Hr.**	**Daily**	**Bare Costs**	**Incl. O&P**
2 Cement Finishers	$39.70	$635.20	$57.95	$927.20	$39.70	$57.95
1 Concrete Saw		161.40		177.54	10.09	11.10
16 L.H., Daily Totals		$796.60		$1104.74	$49.79	$69.05
Crew C-28	**Hr.**	**Daily**	**Hr.**	**Daily**	**Bare Costs**	**Incl. O&P**
1 Cement Finisher	$39.70	$317.60	$57.95	$463.60	$39.70	$57.95
1 Portable Air Compressor, Gas		17.80		19.58	2.23	2.45
8 L.H., Daily Totals		$335.40		$483.18	$41.92	$60.40
Crew D-1	**Hr.**	**Daily**	**Hr.**	**Daily**	**Bare Costs**	**Incl. O&P**
1 Bricklayer	$41.75	$334.00	$62.95	$503.60	$37.70	$56.85
1 Bricklayer Helper	33.65	269.20	50.75	406.00		
16 L.H., Daily Totals		$603.20		$909.60	$37.70	$56.85
Crew D-2	**Hr.**	**Daily**	**Hr.**	**Daily**	**Bare Costs**	**Incl. O&P**
3 Bricklayers	$41.75	$1002.00	$62.95	$1510.80	$38.79	$58.61
2 Bricklayer Helpers	33.65	538.40	50.75	812.00		
.5 Carpenter	41.55	166.20	64.05	256.20		
44 L.H., Daily Totals		$1706.60		$2579.00	$38.79	$58.61
Crew D-3	**Hr.**	**Daily**	**Hr.**	**Daily**	**Bare Costs**	**Incl. O&P**
3 Bricklayers	$41.75	$1002.00	$62.95	$1510.80	$38.65	$58.35
2 Bricklayer Helpers	33.65	538.40	50.75	812.00		
.25 Carpenter	41.55	83.10	64.05	128.10		
42 L.H., Daily Totals		$1623.50		$2450.90	$38.65	$58.35
Crew D-4	**Hr.**	**Daily**	**Hr.**	**Daily**	**Bare Costs**	**Incl. O&P**
1 Bricklayer	$41.75	$334.00	$62.95	$503.60	$37.59	$56.58
2 Bricklayer Helpers	33.65	538.40	50.75	812.00		
1 Equip. Oper. (light)	41.30	330.40	61.85	494.80		
1 Grout Pump, 50 C.F./hr.		132.60		145.86	4.14	4.56
32 L.H., Daily Totals		$1335.40		$1956.26	$41.73	$61.13
Crew D-5	**Hr.**	**Daily**	**Hr.**	**Daily**	**Bare Costs**	**Incl. O&P**
1 Bricklayer	$41.75	$334.00	$62.95	$503.60	$41.75	$62.95
8 L.H., Daily Totals		$334.00		$503.60	$41.75	$62.95
Crew D-6	**Hr.**	**Daily**	**Hr.**	**Daily**	**Bare Costs**	**Incl. O&P**
3 Bricklayers	$41.75	$1002.00	$62.95	$1510.80	$37.85	$57.14
3 Bricklayer Helpers	33.65	807.60	50.75	1218.00		
.25 Carpenter	41.55	83.10	64.05	128.10		
50 L.H., Daily Totals		$1892.70		$2856.90	$37.85	$57.14
Crew D-7	**Hr.**	**Daily**	**Hr.**	**Daily**	**Bare Costs**	**Incl. O&P**
1 Tile Layer	$39.35	$314.80	$57.50	$460.00	$35.27	$51.55
1 Tile Layer Helper	31.20	249.60	45.60	364.80		
16 L.H., Daily Totals		$564.40		$824.80	$35.27	$51.55
Crew D-8	**Hr.**	**Daily**	**Hr.**	**Daily**	**Bare Costs**	**Incl. O&P**
3 Bricklayers	$41.75	$1002.00	$62.95	$1510.80	$38.51	$58.07
2 Bricklayer Helpers	33.65	538.40	50.75	812.00		
40 L.H., Daily Totals		$1540.40		$2322.80	$38.51	$58.07

Crew No.	Bare Costs		Incl. Subs O&P		Cost Per Labor-Hour	
Crew D-9	**Hr.**	**Daily**	**Hr.**	**Daily**	**Bare Costs**	**Incl. O&P**
3 Bricklayers	$41.75	$1002.00	$62.95	$1510.80	$37.70	$56.85
3 Bricklayer Helpers	33.65	807.60	50.75	1218.00		
48 L.H., Daily Totals		$1809.60		$2728.80	$37.70	$56.85
Crew D-10	**Hr.**	**Daily**	**Hr.**	**Daily**	**Bare Costs**	**Incl. O&P**
1 Bricklayer Foreman	$43.75	$350.00	$66.00	$528.00	$40.89	$61.54
1 Bricklayer	41.75	334.00	62.95	503.60		
1 Bricklayer Helper	33.65	269.20	50.75	406.00		
1 Equip. Oper. (crane)	44.40	355.20	66.45	531.60		
1 S.P. Crane, 4x4, 12 Ton		655.60		721.16	20.49	22.54
32 L.H., Daily Totals		$1964.00		$2690.36	$61.38	$84.07
Crew D-11	**Hr.**	**Daily**	**Hr.**	**Daily**	**Bare Costs**	**Incl. O&P**
1 Bricklayer Foreman	$43.75	$350.00	$66.00	$528.00	$39.72	$59.90
1 Bricklayer	41.75	334.00	62.95	503.60		
1 Bricklayer Helper	33.65	269.20	50.75	406.00		
24 L.H., Daily Totals		$953.20		$1437.60	$39.72	$59.90
Crew D-12	**Hr.**	**Daily**	**Hr.**	**Daily**	**Bare Costs**	**Incl. O&P**
1 Bricklayer Foreman	$43.75	$350.00	$66.00	$528.00	$38.20	$57.61
1 Bricklayer	41.75	334.00	62.95	503.60		
2 Bricklayer Helpers	33.65	538.40	50.75	812.00		
32 L.H., Daily Totals		$1222.40		$1843.60	$38.20	$57.61
Crew D-13	**Hr.**	**Daily**	**Hr.**	**Daily**	**Bare Costs**	**Incl. O&P**
1 Bricklayer Foreman	$43.75	$350.00	$66.00	$528.00	$39.79	$60.16
1 Bricklayer	41.75	334.00	62.95	503.60		
2 Bricklayer Helpers	33.65	538.40	50.75	812.00		
1 Carpenter	41.55	332.40	64.05	512.40		
1 Equip. Oper. (crane)	44.40	355.20	66.45	531.60		
1 S.P. Crane, 4x4, 12 Ton		655.60		721.16	13.66	15.02
48 L.H., Daily Totals		$2565.60		$3608.76	$53.45	$75.18
Crew E-1	**Hr.**	**Daily**	**Hr.**	**Daily**	**Bare Costs**	**Incl. O&P**
1 Welder Foreman	$48.90	$391.20	$85.95	$687.60	$45.70	$76.75
1 Welder	46.90	375.20	82.45	659.60		
1 Equip. Oper. (light)	41.30	330.40	61.85	494.80		
1 Welder, gas engine, 300 amp		145.80		160.38	6.08	6.68
24 L.H., Daily Totals		$1242.60		$2002.38	$51.77	$83.43
Crew E-2	**Hr.**	**Daily**	**Hr.**	**Daily**	**Bare Costs**	**Incl. O&P**
1 Struc. Steel Foreman	$48.90	$391.20	$85.95	$687.60	$45.60	$77.08
4 Struc. Steel Workers	46.90	1500.80	82.45	2638.40		
1 Equip. Oper. (crane)	44.40	355.20	66.45	531.60		
1 Equip. Oper. Oiler	38.30	306.40	57.35	458.80		
1 Lattice Boom Crane, 90 Ton		1607.00		1767.70	28.70	31.57
56 L.H., Daily Totals		$4160.60		$6084.10	$74.30	$108.64
Crew E-3	**Hr.**	**Daily**	**Hr.**	**Daily**	**Bare Costs**	**Incl. O&P**
1 Struc. Steel Foreman	$48.90	$391.20	$85.95	$687.60	$47.57	$83.62
1 Struc. Steel Worker	46.90	375.20	82.45	659.60		
1 Welder	46.90	375.20	82.45	659.60		
1 Welder, gas engine, 300 amp		145.80		160.38	6.08	6.68
24 L.H., Daily Totals		$1287.40		$2167.18	$53.64	$90.30
Crew E-4	**Hr.**	**Daily**	**Hr.**	**Daily**	**Bare Costs**	**Incl. O&P**
1 Struc. Steel Foreman	$48.90	$391.20	$85.95	$687.60	$47.40	$83.33
3 Struc. Steel Workers	46.90	1125.60	82.45	1978.80		
1 Welder, gas engine, 300 amp		145.80		160.38	4.56	5.01
32 L.H., Daily Totals		$1662.60		$2826.78	$51.96	$88.34

Crew No.	Bare Costs		Incl. Subs O&P		Cost Per Labor-Hour	
Crew E-5	**Hr.**	**Daily**	**Hr.**	**Daily**	**Bare Costs**	**Incl. O&P**
2 Struc. Steel Foremen	$48.90	$782.40	$85.95	$1375.20	$46.19	$79.04
5 Struc. Steel Workers	46.90	1876.00	82.45	3298.00		
1 Equip. Oper. (crane)	44.40	355.20	66.45	531.60		
1 Welder	46.90	375.20	82.45	659.60		
1 Equip. Oper. Oiler	38.30	306.40	57.35	458.80		
1 Lattice Boom Crane, 90 Ton		1607.00		1767.70		
1 Welder, gas engine, 300 amp		145.80		160.38	21.91	24.10
80 L.H., Daily Totals		$5448.00		$8251.28	$68.10	$103.14
Crew E-6	**Hr.**	**Daily**	**Hr.**	**Daily**	**Bare Costs**	**Incl. O&P**
3 Struc. Steel Foremen	$48.90	$1173.60	$85.95	$2062.80	$46.23	$79.25
9 Struc. Steel Workers	46.90	3376.80	82.45	5936.40		
1 Equip. Oper. (crane)	44.40	355.20	66.45	531.60		
1 Welder	46.90	375.20	82.45	659.60		
1 Equip. Oper. Oiler	38.30	306.40	57.35	458.80		
1 Equip. Oper. (light)	41.30	330.40	61.85	494.80		
1 Lattice Boom Crane, 90 Ton		1607.00		1767.70		
1 Welder, gas engine, 300 amp		145.80		160.38		
1 Air Compressor, 160 cfm		158.00		173.80		
2 Impact Wrenches		35.60		39.16	15.21	16.73
128 L.H., Daily Totals		$7864.00		$12285.04	$61.44	$95.98
Crew E-7	**Hr.**	**Daily**	**Hr.**	**Daily**	**Bare Costs**	**Incl. O&P**
1 Struc. Steel Foreman	$48.90	$391.20	$85.95	$687.60	$46.19	$79.04
4 Struc. Steel Workers	46.90	1500.80	82.45	2638.40		
1 Equip. Oper. (crane)	44.40	355.20	66.45	531.60		
1 Equip. Oper. Oiler	38.30	306.40	57.35	458.80		
1 Welder Foreman	48.90	391.20	85.95	687.60		
2 Welders	46.90	750.40	82.45	1319.20		
1 Lattice Boom Crane, 90 Ton		1607.00		1767.70		
2 Welders, gas engine, 300 amp		291.60		320.76	23.73	26.11
80 L.H., Daily Totals		$5593.80		$8411.66	$69.92	$105.15
Crew E-8	**Hr.**	**Daily**	**Hr.**	**Daily**	**Bare Costs**	**Incl. O&P**
1 Struc. Steel Foreman	$48.90	$391.20	$85.95	$687.60	$45.92	$78.24
4 Struc. Steel Workers	46.90	1500.80	82.45	2638.40		
1 Welder Foreman	48.90	391.20	85.95	687.60		
4 Welders	46.90	1500.80	82.45	2638.40		
1 Equip. Oper. (crane)	44.40	355.20	66.45	531.60		
1 Equip. Oper. Oiler	38.30	306.40	57.35	458.80		
1 Equip. Oper. (light)	41.30	330.40	61.85	494.80		
1 Lattice Boom Crane, 90 Ton		1607.00		1767.70		
4 Welders, gas engine, 300 amp		583.20		641.52	21.06	23.17
104 L.H., Daily Totals		$6966.20		$10546.42	$66.98	$101.41
Crew E-9	**Hr.**	**Daily**	**Hr.**	**Daily**	**Bare Costs**	**Incl. O&P**
2 Struc. Steel Foremen	$48.90	$782.40	$85.95	$1375.20	$46.23	$79.25
5 Struc. Steel Workers	46.90	1876.00	82.45	3298.00		
1 Welder Foreman	48.90	391.20	85.95	687.60		
5 Welders	46.90	1876.00	82.45	3298.00		
1 Equip. Oper. (crane)	44.40	355.20	66.45	531.60		
1 Equip. Oper. Oiler	38.30	306.40	57.35	458.80		
1 Equip. Oper. (light)	41.30	330.40	61.85	494.80		
1 Lattice Boom Crane, 90 Ton		1607.00		1767.70		
5 Welders, gas engine, 300 amp		729.00		801.90	18.25	20.07
128 L.H., Daily Totals		$8253.60		$12713.60	$64.48	$99.33

Crew E-10	Bare Costs Hr.	Bare Costs Daily	Incl. Subs O&P Hr.	Incl. Subs O&P Daily	Cost Per Labor-Hour Bare Costs	Cost Per Labor-Hour Incl. O&P
1 Welder Foreman	$48.90	$391.20	$85.95	$687.60	$47.90	$84.20
1 Welder	46.90	375.20	82.45	659.60		
1 Welder, gas engine, 300 amp		145.80		160.38		
1 Flatbed Truck, Gas, 3 Ton		296.00		325.60	27.61	30.37
16 L.H., Daily Totals		$1208.20		$1833.18	$75.51	$114.57

Crew E-11	Hr.	Daily	Hr.	Daily	Bare Costs	Incl. O&P
2 Painters, Struc. Steel	$37.40	$598.40	$66.90	$1070.40	$37.30	$61.67
1 Building Laborer	33.10	264.80	51.05	408.40		
1 Equip. Oper. (light)	41.30	330.40	61.85	494.80		
1 Air Compressor, 250 cfm		193.40		212.74		
1 Sandblaster, portable, 3 C.F.		20.60		22.66		
1 Set Sand Blasting Accessories		12.95		14.24	7.09	7.80
32 L.H., Daily Totals		$1420.55		$2223.24	$44.39	$69.48

Crew E-12	Hr.	Daily	Hr.	Daily	Bare Costs	Incl. O&P
1 Welder Foreman	$48.90	$391.20	$85.95	$687.60	$45.10	$73.90
1 Equip. Oper. (light)	41.30	330.40	61.85	494.80		
1 Welder, gas engine, 300 amp		145.80		160.38	9.11	10.02
16 L.H., Daily Totals		$867.40		$1342.78	$54.21	$83.92

Crew E-13	Hr.	Daily	Hr.	Daily	Bare Costs	Incl. O&P
1 Welder Foreman	$48.90	$391.20	$85.95	$687.60	$46.37	$77.92
.5 Equip. Oper. (light)	41.30	165.20	61.85	247.40		
1 Welder, gas engine, 300 amp		145.80		160.38	12.15	13.37
12 L.H., Daily Totals		$702.20		$1095.38	$58.52	$91.28

Crew E-14	Hr.	Daily	Hr.	Daily	Bare Costs	Incl. O&P
1 Welder Foreman	$48.90	$391.20	$85.95	$687.60	$48.90	$85.95
1 Welder, gas engine, 300 amp		145.80		160.38	18.23	20.05
8 L.H., Daily Totals		$537.00		$847.98	$67.13	$106.00

Crew E-16	Hr.	Daily	Hr.	Daily	Bare Costs	Incl. O&P
1 Welder Foreman	$48.90	$391.20	$85.95	$687.60	$47.90	$84.20
1 Welder	46.90	375.20	82.45	659.60		
1 Welder, gas engine, 300 amp		145.80		160.38	9.11	10.02
16 L.H., Daily Totals		$912.20		$1507.58	$57.01	$94.22

Crew E-17	Hr.	Daily	Hr.	Daily	Bare Costs	Incl. O&P
1 Structural Steel Foreman	$48.90	$391.20	$85.95	$687.60	$47.90	$84.20
1 Structural Steel Worker	46.90	375.20	82.45	659.60		
16 L.H., Daily Totals		$766.40		$1347.20	$47.90	$84.20

Crew E-18	Hr.	Daily	Hr.	Daily	Bare Costs	Incl. O&P
1 Structural Steel Foreman	$48.90	$391.20	$85.95	$687.60	$46.51	$79.52
3 Structural Steel Workers	46.90	1125.60	82.45	1978.80		
1 Equipment Operator (med.)	42.95	343.60	64.30	514.40		
1 Lattice Boom Crane, 20 Ton		1029.00		1131.90	25.73	28.30
40 L.H., Daily Totals		$2889.40		$4312.70	$72.23	$107.82

Crew E-19	Hr.	Daily	Hr.	Daily	Bare Costs	Incl. O&P
1 Structural Steel Worker	$46.90	$375.20	$82.45	$659.60	$45.70	$76.75
1 Structural Steel Foreman	48.90	391.20	85.95	687.60		
1 Equip. Oper. (light)	41.30	330.40	61.85	494.80		
1 Lattice Boom Crane, 20 Ton		1029.00		1131.90	42.88	47.16
24 L.H., Daily Totals		$2125.80		$2973.90	$88.58	$123.91

Crew E-20	Hr.	Daily	Hr.	Daily	Bare Costs	Incl. O&P
1 Structural Steel Foreman	$48.90	$391.20	$85.95	$687.60	$45.76	$77.75
5 Structural Steel Workers	46.90	1876.00	82.45	3298.00		
1 Equip. Oper. (crane)	44.40	355.20	66.45	531.60		
1 Oiler	38.30	306.40	57.35	458.80		
1 Lattice Boom Crane, 40 Ton		1256.00		1381.60	19.63	21.59
64 L.H., Daily Totals		$4184.80		$6357.60	$65.39	$99.34

Crew E-22	Hr.	Daily	Hr.	Daily	Bare Costs	Incl. O&P
1 Skilled Worker Foreman	$44.60	$356.80	$68.55	$548.40	$43.27	$66.52
2 Skilled Workers	42.60	681.60	65.50	1048.00		
24 L.H., Daily Totals		$1038.40		$1596.40	$43.27	$66.52

Crew E-24	Hr.	Daily	Hr.	Daily	Bare Costs	Incl. O&P
3 Structural Steel Workers	$46.90	$1125.60	$82.45	$1978.80	$45.91	$77.91
1 Equipment Operator (med.)	42.95	343.60	64.30	514.40		
1 Hyd. Crane, 25 Ton		747.60		822.36	23.36	25.70
32 L.H., Daily Totals		$2216.80		$3315.56	$69.28	$103.61

Crew E-25	Hr.	Daily	Hr.	Daily	Bare Costs	Incl. O&P
1 Welder Foreman	$48.90	$391.20	$85.95	$687.60	$48.90	$85.95
1 Cutting Torch		19.00		20.90		
1 Set of Gases		79.20		87.12	12.28	13.50
8 L.H., Daily Totals		$489.40		$795.62	$61.17	$99.45

Crew F-3	Hr.	Daily	Hr.	Daily	Bare Costs	Incl. O&P
4 Carpenters	$41.55	$1329.60	$64.05	$2049.60	$42.12	$64.53
1 Equip. Oper. (crane)	44.40	355.20	66.45	531.60		
1 Hyd. Crane, 12 Ton		655.60		721.16	16.39	18.03
40 L.H., Daily Totals		$2340.40		$3302.36	$58.51	$82.56

Crew F-4	Hr.	Daily	Hr.	Daily	Bare Costs	Incl. O&P
4 Carpenters	$41.55	$1329.60	$64.05	$2049.60	$41.48	$63.33
1 Equip. Oper. (crane)	44.40	355.20	66.45	531.60		
1 Equip. Oper. Oiler	38.30	306.40	57.35	458.80		
1 Hyd. Crane, 55 Ton		1209.00		1329.90	25.19	27.71
48 L.H., Daily Totals		$3200.20		$4369.90	$66.67	$91.04

Crew F-5	Hr.	Daily	Hr.	Daily	Bare Costs	Incl. O&P
1 Carpenter Foreman	$43.55	$348.40	$67.15	$537.20	$42.05	$64.83
3 Carpenters	41.55	997.20	64.05	1537.20		
32 L.H., Daily Totals		$1345.60		$2074.40	$42.05	$64.83

Crew F-6	Hr.	Daily	Hr.	Daily	Bare Costs	Incl. O&P
2 Carpenters	$41.55	$664.80	$64.05	$1024.80	$38.74	$59.33
2 Building Laborers	33.10	529.60	51.05	816.80		
1 Equip. Oper. (crane)	44.40	355.20	66.45	531.60		
1 Hyd. Crane, 12 Ton		655.60		721.16	16.39	18.03
40 L.H., Daily Totals		$2205.20		$3094.36	$55.13	$77.36

Crew F-7	Hr.	Daily	Hr.	Daily	Bare Costs	Incl. O&P
2 Carpenters	$41.55	$664.80	$64.05	$1024.80	$37.33	$57.55
2 Building Laborers	33.10	529.60	51.05	816.80		
32 L.H., Daily Totals		$1194.40		$1841.60	$37.33	$57.55

Crew No.	Bare Costs		Incl. Subs O&P		Cost Per Labor-Hour	
Crew G-1	**Hr.**	**Daily**	**Hr.**	**Daily**	**Bare Costs**	**Incl. O&P**
1 Roofer Foreman	$37.40	$299.20	$62.15	$497.20	$33.06	$54.95
4 Roofers, Composition	35.40	1132.80	58.85	1883.20		
2 Roofer Helpers	26.20	419.20	43.55	696.80		
1 Application Equipment		188.40		207.24		
1 Tar Kettle/Pot		85.45		94.00		
1 Crew Truck		202.40		222.64	8.50	9.35
56 L.H., Daily Totals		$2327.45		$3601.07	$41.56	$64.30
Crew G-2	**Hr.**	**Daily**	**Hr.**	**Daily**	**Bare Costs**	**Incl. O&P**
1 Plasterer	$37.30	$298.40	$55.65	$445.20	$34.77	$52.43
1 Plasterer Helper	33.90	271.20	50.60	404.80		
1 Building Laborer	33.10	264.80	51.05	408.40		
1 Grout Pump, 50 C.F./hr		132.60		145.86	5.53	6.08
24 L.H., Daily Totals		$967.00		$1404.26	$40.29	$58.51
Crew G-2A	**Hr.**	**Daily**	**Hr.**	**Daily**	**Bare Costs**	**Incl. O&P**
1 Roofer, composition	$35.40	$283.20	$58.85	$470.80	$31.57	$51.15
1 Roofer Helper	26.20	209.60	43.55	348.40		
1 Building Laborer	33.10	264.80	51.05	408.40		
1 Foam spray rig, trailer-mtd.		551.65		606.82		
1 Pickup Truck, 3/4 Ton		140.00		154.00	28.82	31.70
24 L.H., Daily Totals		$1449.25		$1988.42	$60.39	$82.85
Crew G-3	**Hr.**	**Daily**	**Hr.**	**Daily**	**Bare Costs**	**Incl. O&P**
2 Sheet Metal Workers	$49.10	$785.60	$74.35	$1189.60	$41.10	$62.70
2 Building Laborers	33.10	529.60	51.05	816.80		
32 L.H., Daily Totals		$1315.20		$2006.40	$41.10	$62.70
Crew G-4	**Hr.**	**Daily**	**Hr.**	**Daily**	**Bare Costs**	**Incl. O&P**
1 Labor Foreman (outside)	$35.10	$280.80	$54.10	$432.80	$33.77	$52.07
2 Building Laborers	33.10	529.60	51.05	816.80		
1 Flatbed Truck, Gas, 1.5 Ton		239.80		263.78		
1 Air Compressor, 160 cfm		158.00		173.80	16.57	18.23
24 L.H., Daily Totals		$1208.20		$1687.18	$50.34	$70.30
Crew G-5	**Hr.**	**Daily**	**Hr.**	**Daily**	**Bare Costs**	**Incl. O&P**
1 Roofer Foreman	$37.40	$299.20	$62.15	$497.20	$32.12	$53.39
2 Roofers, Composition	35.40	566.40	58.85	941.60		
2 Roofer Helpers	26.20	419.20	43.55	696.80		
1 Application Equipment		188.40		207.24	4.71	5.18
40 L.H., Daily Totals		$1473.20		$2342.84	$36.83	$58.57
Crew G-6A	**Hr.**	**Daily**	**Hr.**	**Daily**	**Bare Costs**	**Incl. O&P**
2 Roofers Composition	$35.40	$566.40	$58.85	$941.60	$35.40	$58.85
1 Small Compressor, Electric		11.80		12.98		
2 Pneumatic Nailers		43.80		48.18	3.48	3.82
16 L.H., Daily Totals		$622.00		$1002.76	$38.88	$62.67
Crew G-7	**Hr.**	**Daily**	**Hr.**	**Daily**	**Bare Costs**	**Incl. O&P**
1 Carpenter	$41.55	$332.40	$64.05	$512.40	$41.55	$64.05
1 Small Compressor, Electric		11.80		12.98		
1 Pneumatic Nailer		21.90		24.09	4.21	4.63
8 L.H., Daily Totals		$366.10		$549.47	$45.76	$68.68
Crew H-1	**Hr.**	**Daily**	**Hr.**	**Daily**	**Bare Costs**	**Incl. O&P**
2 Glaziers	$40.20	$643.20	$60.55	$968.80	$43.55	$71.50
2 Struc. Steel Workers	46.90	750.40	82.45	1319.20		
32 L.H., Daily Totals		$1393.60		$2288.00	$43.55	$71.50

Crew No.	Bare Costs		Incl. Subs O&P		Cost Per Labor-Hour	
Crew H-2	**Hr.**	**Daily**	**Hr.**	**Daily**	**Bare Costs**	**Incl. O&P**
2 Glaziers	$40.20	$643.20	$60.55	$968.80	$37.83	$57.38
1 Building Laborer	33.10	264.80	51.05	408.40		
24 L.H., Daily Totals		$908.00		$1377.20	$37.83	$57.38
Crew H-3	**Hr.**	**Daily**	**Hr.**	**Daily**	**Bare Costs**	**Incl. O&P**
1 Glazier	$40.20	$321.60	$60.55	$484.40	$35.90	$54.50
1 Helper	31.60	252.80	48.45	387.60		
16 L.H., Daily Totals		$574.40		$872.00	$35.90	$54.50
Crew J-1	**Hr.**	**Daily**	**Hr.**	**Daily**	**Bare Costs**	**Incl. O&P**
3 Plasterers	$37.30	$895.20	$55.65	$1335.60	$35.94	$53.63
2 Plasterer Helpers	33.90	542.40	50.60	809.60		
1 Mixing Machine, 6 C.F.		136.60		150.26	3.42	3.76
40 L.H., Daily Totals		$1574.20		$2295.46	$39.35	$57.39
Crew J-2	**Hr.**	**Daily**	**Hr.**	**Daily**	**Bare Costs**	**Incl. O&P**
3 Plasterers	$37.30	$895.20	$55.65	$1335.60	$36.11	$53.75
2 Plasterer Helpers	33.90	542.40	50.60	809.60		
1 Lather	36.95	295.60	54.35	434.80		
1 Mixing Machine, 6 C.F.		136.60		150.26	2.85	3.13
48 L.H., Daily Totals		$1869.80		$2730.26	$38.95	$56.88
Crew J-3	**Hr.**	**Daily**	**Hr.**	**Daily**	**Bare Costs**	**Incl. O&P**
1 Terrazzo Worker	$39.00	$312.00	$57.00	$456.00	$35.45	$51.80
1 Terrazzo Helper	31.90	255.20	46.60	372.80		
1 Terrazzo Grinder, Electric		82.25		90.47		
1 Terrazzo Mixer		184.80		203.28	16.69	18.36
16 L.H., Daily Totals		$834.25		$1122.56	$52.14	$70.16
Crew K-1	**Hr.**	**Daily**	**Hr.**	**Daily**	**Bare Costs**	**Incl. O&P**
1 Carpenter	$41.55	$332.40	$64.05	$512.40	$36.90	$56.63
1 Truck Driver (light)	32.25	258.00	49.20	393.60		
1 Flatbed Truck, Gas, 3 Ton		296.00		325.60	18.50	20.35
16 L.H., Daily Totals		$886.40		$1231.60	$55.40	$76.97
Crew K-2	**Hr.**	**Daily**	**Hr.**	**Daily**	**Bare Costs**	**Incl. O&P**
1 Struc. Steel Foreman	$48.90	$391.20	$85.95	$687.60	$42.68	$72.53
1 Struc. Steel Worker	46.90	375.20	82.45	659.60		
1 Truck Driver (light)	32.25	258.00	49.20	393.60		
1 Flatbed Truck, Gas, 3 Ton		296.00		325.60	12.33	13.57
24 L.H., Daily Totals		$1320.40		$2066.40	$55.02	$86.10
Crew L-1	**Hr.**	**Daily**	**Hr.**	**Daily**	**Bare Costs**	**Incl. O&P**
1 Electrician	$49.00	$392.00	$72.85	$582.80	$50.52	$75.42
1 Plumber	52.05	416.40	78.00	624.00		
16 L.H., Daily Totals		$808.40		$1206.80	$50.52	$75.42
Crew L-2	**Hr.**	**Daily**	**Hr.**	**Daily**	**Bare Costs**	**Incl. O&P**
1 Carpenter	$41.55	$332.40	$64.05	$512.40	$36.58	$56.25
1 Carpenter Helper	31.60	252.80	48.45	387.60		
16 L.H., Daily Totals		$585.20		$900.00	$36.58	$56.25
Crew L-3	**Hr.**	**Daily**	**Hr.**	**Daily**	**Bare Costs**	**Incl. O&P**
1 Carpenter	$41.55	$332.40	$64.05	$512.40	$45.30	$68.83
.5 Electrician	49.00	196.00	72.85	291.40		
.5 Sheet Metal Worker	49.10	196.40	74.35	297.40		
16 L.H., Daily Totals		$724.80		$1101.20	$45.30	$68.83

Crew No.	Bare Costs		Incl. Subs O&P		Cost Per Labor-Hour	
Crew L-3A	**Hr.**	**Daily**	**Hr.**	**Daily**	**Bare Costs**	**Incl. O&P**
1 Carpenter Foreman (outside)	$43.55	$348.40	$67.15	$537.20	$45.40	$69.55
.5 Sheet Metal Worker	49.10	196.40	74.35	297.40		
12 L.H., Daily Totals		$544.80		$834.60	$45.40	$69.55

Crew L-4	Hr.	Daily	Hr.	Daily	Bare Costs	Incl. O&P
2 Skilled Workers	$42.60	$681.60	$65.50	$1048.00	$38.93	$59.82
1 Helper	31.60	252.80	48.45	387.60		
24 L.H., Daily Totals		$934.40		$1435.60	$38.93	$59.82

Crew L-5	Hr.	Daily	Hr.	Daily	Bare Costs	Incl. O&P
1 Struc. Steel Foreman	$48.90	$391.20	$85.95	$687.60	$46.83	$80.66
5 Struc. Steel Workers	46.90	1876.00	82.45	3298.00		
1 Equip. Oper. (crane)	44.40	355.20	66.45	531.60		
1 Hyd. Crane, 25 Ton		747.60		822.36	13.35	14.69
56 L.H., Daily Totals		$3370.00		$5339.56	$60.18	$95.35

Crew L-5A	Hr.	Daily	Hr.	Daily	Bare Costs	Incl. O&P
1 Structural Steel Foreman	$48.90	$391.20	$85.95	$687.60	$46.77	$79.33
2 Structural Steel Workers	46.90	750.40	82.45	1319.20		
1 Equip. Oper. (crane)	44.40	355.20	66.45	531.60		
1 S.P. Crane, 4x4, 25 Ton		655.60		721.16	20.49	22.54
32 L.H., Daily Totals		$2152.40		$3259.56	$67.26	$101.86

Crew L-5B	Hr.	Daily	Hr.	Daily	Bare Costs	Incl. O&P
1 Structural Steel Foreman	$48.90	$391.20	$85.95	$687.60	$47.47	$75.11
2 Structural Steel Workers	46.90	750.40	82.45	1319.20		
2 Electricians	49.00	784.00	72.85	1165.60		
2 Steamfitters/Pipefitters	51.90	830.40	77.80	1244.80		
1 Equip. Oper. (crane)	44.40	355.20	66.45	531.60		
1 Equip. Oper. Oiler	38.30	306.40	57.35	458.80		
1 Hyd. Crane, 80 Ton		1545.00		1699.50	21.46	23.60
72 L.H., Daily Totals		$4962.60		$7107.10	$68.92	$98.71

Crew L-6	Hr.	Daily	Hr.	Daily	Bare Costs	Incl. O&P
1 Plumber	$52.05	$416.40	$78.00	$624.00	$51.03	$76.28
.5 Electrician	49.00	196.00	72.85	291.40		
12 L.H., Daily Totals		$612.40		$915.40	$51.03	$76.28

Crew L-7	Hr.	Daily	Hr.	Daily	Bare Costs	Incl. O&P
2 Carpenters	$41.55	$664.80	$64.05	$1024.80	$40.20	$61.59
1 Building Laborer	33.10	264.80	51.05	408.40		
.5 Electrician	49.00	196.00	72.85	291.40		
28 L.H., Daily Totals		$1125.60		$1724.60	$40.20	$61.59

Crew L-8	Hr.	Daily	Hr.	Daily	Bare Costs	Incl. O&P
2 Carpenters	$41.55	$664.80	$64.05	$1024.80	$43.65	$66.84
.5 Plumber	52.05	208.20	78.00	312.00		
20 L.H., Daily Totals		$873.00		$1336.80	$43.65	$66.84

Crew L-9	Hr.	Daily	Hr.	Daily	Bare Costs	Incl. O&P
1 Labor Foreman (inside)	$33.60	$268.80	$51.80	$414.40	$38.04	$60.62
2 Building Laborers	33.10	529.60	51.05	816.80		
1 Struc. Steel Worker	46.90	375.20	82.45	659.60		
.5 Electrician	49.00	196.00	72.85	291.40		
36 L.H., Daily Totals		$1369.60		$2182.20	$38.04	$60.62

Crew L-10	Hr.	Daily	Hr.	Daily	Bare Costs	Incl. O&P
1 Structural Steel Foreman	$48.90	$391.20	$85.95	$687.60	$46.73	$78.28
1 Structural Steel Worker	46.90	375.20	82.45	659.60		
1 Equip. Oper. (crane)	44.40	355.20	66.45	531.60		
1 Hyd. Crane, 12 Ton		655.60		721.16	27.32	30.05
24 L.H., Daily Totals		$1777.20		$2599.96	$74.05	$108.33

Crew L-11	Hr.	Daily	Hr.	Daily	Bare Costs	Incl. O&P
2 Wreckers	$33.10	$529.60	$55.50	$888.00	$37.98	$59.83
1 Equip. Oper. (crane)	44.40	355.20	66.45	531.60		
1 Equip. Oper. (light)	41.30	330.40	61.85	494.80		
1 Hyd. Excavator, 2.5 C.Y.		1784.00		1962.40		
1 Loader, Skid Steer, 78 H.P.		291.00		320.10	64.84	71.33
32 L.H., Daily Totals		$3290.20		$4196.90	$102.82	$131.15

Crew M-1	Hr.	Daily	Hr.	Daily	Bare Costs	Incl. O&P
3 Elevator Constructors	$61.70	$1480.80	$91.70	$2200.80	$58.61	$87.11
1 Elevator Apprentice	49.35	394.80	73.35	586.80		
5 Hand Tools		58.00		63.80	1.81	1.99
32 L.H., Daily Totals		$1933.60		$2851.40	$60.42	$89.11

Crew M-3	Hr.	Daily	Hr.	Daily	Bare Costs	Incl. O&P
1 Electrician Foreman (out)	$51.00	$408.00	$75.85	$606.80	$48.44	$72.48
1 Common Laborer	33.10	264.80	51.05	408.40		
.25 Equipment Operator, Med.	42.95	85.90	64.30	128.60		
1 Elevator Constructor	61.70	493.60	91.70	733.60		
1 Elevator Apprentice	49.35	394.80	73.35	586.80		
.25 S.P. Crane, 4x4, 20 Ton		156.40		172.04	4.60	5.06
34 L.H., Daily Totals		$1803.50		$2636.24	$53.04	$77.54

Crew M-4	Hr.	Daily	Hr.	Daily	Bare Costs	Incl. O&P
1 Electrician Foreman (out)	$51.00	$408.00	$75.85	$606.80	$47.96	$71.76
1 Common Laborer	33.10	264.80	51.05	408.40		
.25 Equipment Operator, Crane	44.40	88.80	66.45	132.90		
.25 Equipment Operator, Oiler	38.30	76.60	57.35	114.70		
1 Elevator Constructor	61.70	493.60	91.70	733.60		
1 Elevator Apprentice	49.35	394.80	73.35	586.80		
.25 S.P. Crane, 4x4, 40 Ton		253.50		278.85	7.04	7.75
36 L.H., Daily Totals		$1980.10		$2862.05	$55.00	$79.50

Crew Q-1	Hr.	Daily	Hr.	Daily	Bare Costs	Incl. O&P
1 Plumber	$52.05	$416.40	$78.00	$624.00	$46.85	$70.22
1 Plumber Apprentice	41.65	333.20	62.45	499.60		
16 L.H., Daily Totals		$749.60		$1123.60	$46.85	$70.22

Crew Q-1A	Hr.	Daily	Hr.	Daily	Bare Costs	Incl. O&P
.25 Plumber Foreman (out)	$54.05	$108.10	$81.00	$162.00	$52.45	$78.60
1 Plumber	52.05	416.40	78.00	624.00		
10 L.H., Daily Totals		$524.50		$786.00	$52.45	$78.60

Crew Q-1C	Hr.	Daily	Hr.	Daily	Bare Costs	Incl. O&P
1 Plumber	$52.05	$416.40	$78.00	$624.00	$45.55	$68.25
1 Plumber Apprentice	41.65	333.20	62.45	499.60		
1 Equip. Oper. (med.)	42.95	343.60	64.30	514.40		
1 Trencher, Chain Type, 8' D		2090.00		2299.00	87.08	95.79
24 L.H., Daily Totals		$3183.20		$3937.00	$132.63	$164.04

Crew Q-2	Hr.	Daily	Hr.	Daily	Bare Costs	Incl. O&P
2 Plumbers	$52.05	$832.80	$78.00	$1248.00	$48.58	$72.82
1 Plumber Apprentice	41.65	333.20	62.45	499.60		
24 L.H., Daily Totals		$1166.00		$1747.60	$48.58	$72.82

Crew No.	Bare Costs		Incl. Subs O&P		Cost Per Labor-Hour	
Crew Q-3	**Hr.**	**Daily**	**Hr.**	**Daily**	**Bare Costs**	**Incl. O&P**
1 Plumber Foreman (inside)	$52.55	$420.40	$78.75	$630.00	$49.58	$74.30
2 Plumbers	52.05	832.80	78.00	1248.00		
1 Plumber Apprentice	41.65	333.20	62.45	499.60		
32 L.H., Daily Totals		$1586.40		$2377.60	$49.58	$74.30

Crew Q-4	Hr.	Daily	Hr.	Daily	Bare Costs	Incl. O&P
1 Plumber Foreman (inside)	$52.55	$420.40	$78.75	$630.00	$49.58	$74.30
1 Plumber	52.05	416.40	78.00	624.00		
1 Welder (plumber)	52.05	416.40	78.00	624.00		
1 Plumber Apprentice	41.65	333.20	62.45	499.60		
1 Welder, electric, 300 amp		58.55		64.41	1.83	2.01
32 L.H., Daily Totals		$1644.95		$2442.01	$51.40	$76.31

Crew Q-5	Hr.	Daily	Hr.	Daily	Bare Costs	Incl. O&P
1 Steamfitter	$51.90	$415.20	$77.80	$622.40	$46.70	$70.00
1 Steamfitter Apprentice	41.50	332.00	62.20	497.60		
16 L.H., Daily Totals		$747.20		$1120.00	$46.70	$70.00

Crew Q-6	Hr.	Daily	Hr.	Daily	Bare Costs	Incl. O&P
2 Steamfitters	$51.90	$830.40	$77.80	$1244.80	$48.43	$72.60
1 Steamfitter Apprentice	41.50	332.00	62.20	497.60		
24 L.H., Daily Totals		$1162.40		$1742.40	$48.43	$72.60

Crew Q-7	Hr.	Daily	Hr.	Daily	Bare Costs	Incl. O&P
1 Steamfitter Foreman (inside)	$52.40	$419.20	$78.55	$628.40	$49.42	$74.09
2 Steamfitters	51.90	830.40	77.80	1244.80		
1 Steamfitter Apprentice	41.50	332.00	62.20	497.60		
32 L.H., Daily Totals		$1581.60		$2370.80	$49.42	$74.09

Crew Q-8	Hr.	Daily	Hr.	Daily	Bare Costs	Incl. O&P
1 Steamfitter Foreman (inside)	$52.40	$419.20	$78.55	$628.40	$49.42	$74.09
1 Steamfitter	51.90	415.20	77.80	622.40		
1 Welder (steamfitter)	51.90	415.20	77.80	622.40		
1 Steamfitter Apprentice	41.50	332.00	62.20	497.60		
1 Welder, electric, 300 amp		58.55		64.41	1.83	2.01
32 L.H., Daily Totals		$1640.15		$2435.20	$51.25	$76.10

Crew Q-9	Hr.	Daily	Hr.	Daily	Bare Costs	Incl. O&P
1 Sheet Metal Worker	$49.10	$392.80	$74.35	$594.80	$44.20	$66.92
1 Sheet Metal Apprentice	39.30	314.40	59.50	476.00		
16 L.H., Daily Totals		$707.20		$1070.80	$44.20	$66.92

Crew Q-10	Hr.	Daily	Hr.	Daily	Bare Costs	Incl. O&P
2 Sheet Metal Workers	$49.10	$785.60	$74.35	$1189.60	$45.83	$69.40
1 Sheet Metal Apprentice	39.30	314.40	59.50	476.00		
24 L.H., Daily Totals		$1100.00		$1665.60	$45.83	$69.40

Crew Q-11	Hr.	Daily	Hr.	Daily	Bare Costs	Incl. O&P
1 Sheet Metal Foreman (inside)	$49.60	$396.80	$75.10	$600.80	$46.77	$70.83
2 Sheet Metal Workers	49.10	785.60	74.35	1189.60		
1 Sheet Metal Apprentice	39.30	314.40	59.50	476.00		
32 L.H., Daily Totals		$1496.80		$2266.40	$46.77	$70.83

Crew Q-12	Hr.	Daily	Hr.	Daily	Bare Costs	Incl. O&P
1 Sprinkler Installer	$50.40	$403.20	$75.45	$603.60	$45.35	$67.90
1 Sprinkler Apprentice	40.30	322.40	60.35	482.80		
16 L.H., Daily Totals		$725.60		$1086.40	$45.35	$67.90

Crew Q-13	Hr.	Daily	Hr.	Daily	Bare Costs	Incl. O&P
1 Sprinkler Foreman (inside)	$50.90	$407.20	$76.20	$609.60	$48.00	$71.86
2 Sprinkler Installers	50.40	806.40	75.45	1207.20		
1 Sprinkler Apprentice	40.30	322.40	60.35	482.80		
32 L.H., Daily Totals		$1536.00		$2299.60	$48.00	$71.86

Crew Q-14	Hr.	Daily	Hr.	Daily	Bare Costs	Incl. O&P
1 Asbestos Worker	$45.55	$364.40	$70.80	$566.40	$41.00	$63.73
1 Asbestos Apprentice	36.45	291.60	56.65	453.20		
16 L.H., Daily Totals		$656.00		$1019.60	$41.00	$63.73

Crew Q-15	Hr.	Daily	Hr.	Daily	Bare Costs	Incl. O&P
1 Plumber	$52.05	$416.40	$78.00	$624.00	$46.85	$70.22
1 Plumber Apprentice	41.65	333.20	62.45	499.60		
1 Welder, electric, 300 amp		58.55		64.41	3.66	4.03
16 L.H., Daily Totals		$808.15		$1188.01	$50.51	$74.25

Crew Q-16	Hr.	Daily	Hr.	Daily	Bare Costs	Incl. O&P
2 Plumbers	$52.05	$832.80	$78.00	$1248.00	$48.58	$72.82
1 Plumber Apprentice	41.65	333.20	62.45	499.60		
1 Welder, electric, 300 amp		58.55		64.41	2.44	2.68
24 L.H., Daily Totals		$1224.55		$1812.01	$51.02	$75.50

Crew Q-17	Hr.	Daily	Hr.	Daily	Bare Costs	Incl. O&P
1 Steamfitter	$51.90	$415.20	$77.80	$622.40	$46.70	$70.00
1 Steamfitter Apprentice	41.50	332.00	62.20	497.60		
1 Welder, electric, 300 amp		58.55		64.41	3.66	4.03
16 L.H., Daily Totals		$805.75		$1184.41	$50.36	$74.03

Crew Q-17A	Hr.	Daily	Hr.	Daily	Bare Costs	Incl. O&P
1 Steamfitter	$51.90	$415.20	$77.80	$622.40	$45.93	$68.82
1 Steamfitter Apprentice	41.50	332.00	62.20	497.60		
1 Equip. Oper. (crane)	44.40	355.20	66.45	531.60		
1 Hyd. Crane, 12 Ton		655.60		721.16		
1 Welder, electric, 300 amp		58.55		64.41	29.76	32.73
24 L.H., Daily Totals		$1816.55		$2437.17	$75.69	$101.55

Crew Q-18	Hr.	Daily	Hr.	Daily	Bare Costs	Incl. O&P
2 Steamfitters	$51.90	$830.40	$77.80	$1244.80	$48.43	$72.60
1 Steamfitter Apprentice	41.50	332.00	62.20	497.60		
1 Welder, electric, 300 amp		58.55		64.41	2.44	2.68
24 L.H., Daily Totals		$1220.95		$1806.81	$50.87	$75.28

Crew Q-19	Hr.	Daily	Hr.	Daily	Bare Costs	Incl. O&P
1 Steamfitter	$51.90	$415.20	$77.80	$622.40	$47.47	$70.95
1 Steamfitter Apprentice	41.50	332.00	62.20	497.60		
1 Electrician	49.00	392.00	72.85	582.80		
24 L.H., Daily Totals		$1139.20		$1702.80	$47.47	$70.95

Crew Q-20	Hr.	Daily	Hr.	Daily	Bare Costs	Incl. O&P
1 Sheet Metal Worker	$49.10	$392.80	$74.35	$594.80	$45.16	$68.11
1 Sheet Metal Apprentice	39.30	314.40	59.50	476.00		
.5 Electrician	49.00	196.00	72.85	291.40		
20 L.H., Daily Totals		$903.20		$1362.20	$45.16	$68.11

Crew Q-21	Hr.	Daily	Hr.	Daily	Bare Costs	Incl. O&P
2 Steamfitters	$51.90	$830.40	$77.80	$1244.80	$48.58	$72.66
1 Steamfitter Apprentice	41.50	332.00	62.20	497.60		
1 Electrician	49.00	392.00	72.85	582.80		
32 L.H., Daily Totals		$1554.40		$2325.20	$48.58	$72.66

Crew Q-22	Bare Costs Hr.	Bare Costs Daily	Incl. Subs O&P Hr.	Incl. Subs O&P Daily	Cost Per Labor-Hour Bare Costs	Cost Per Labor-Hour Incl. O&P
1 Plumber	$52.05	$416.40	$78.00	$624.00	$46.85	$70.22
1 Plumber Apprentice	41.65	333.20	62.45	499.60		
1 Hyd. Crane, 12 Ton		655.60		721.16	40.98	45.07
16 L.H., Daily Totals		$1405.20		$1844.76	$87.83	$115.30

Crew Q-22A	Hr.	Daily	Hr.	Daily	Bare Costs	Incl. O&P
1 Plumber	$52.05	$416.40	$78.00	$624.00	$42.80	$64.49
1 Plumber Apprentice	41.65	333.20	62.45	499.60		
1 Laborer	33.10	264.80	51.05	408.40		
1 Equip. Oper. (crane)	44.40	355.20	66.45	531.60		
1 Hyd. Crane, 12 Ton		655.60		721.16	20.49	22.54
32 L.H., Daily Totals		$2025.20		$2784.76	$63.29	$87.02

Crew Q-23	Hr.	Daily	Hr.	Daily	Bare Costs	Incl. O&P
1 Plumber Foreman	$54.05	$432.40	$81.00	$648.00	$49.68	$74.43
1 Plumber	52.05	416.40	78.00	624.00		
1 Equip. Oper. (med.)	42.95	343.60	64.30	514.40		
1 Lattice Boom Crane, 20 Ton		1029.00		1131.90	42.88	47.16
24 L.H., Daily Totals		$2221.40		$2918.30	$92.56	$121.60

Crew R-1	Hr.	Daily	Hr.	Daily	Bare Costs	Incl. O&P
1 Electrician Foreman	$49.50	$396.00	$73.60	$588.80	$43.28	$64.84
3 Electricians	49.00	1176.00	72.85	1748.40		
2 Helpers	31.60	505.60	48.45	775.20		
48 L.H., Daily Totals		$2077.60		$3112.40	$43.28	$64.84

Crew R-1A	Hr.	Daily	Hr.	Daily	Bare Costs	Incl. O&P
1 Electrician	$49.00	$392.00	$72.85	$582.80	$40.30	$60.65
1 Helper	31.60	252.80	48.45	387.60		
16 L.H., Daily Totals		$644.80		$970.40	$40.30	$60.65

Crew R-2	Hr.	Daily	Hr.	Daily	Bare Costs	Incl. O&P
1 Electrician Foreman	$49.50	$396.00	$73.60	$588.80	$43.44	$65.07
3 Electricians	49.00	1176.00	72.85	1748.40		
2 Helpers	31.60	505.60	48.45	775.20		
1 Equip. Oper. (crane)	44.40	355.20	66.45	531.60		
1 S.P. Crane, 4x4, 5 Ton		276.20		303.82	4.93	5.43
56 L.H., Daily Totals		$2709.00		$3947.82	$48.38	$70.50

Crew R-3	Hr.	Daily	Hr.	Daily	Bare Costs	Incl. O&P
1 Electrician Foreman	$49.50	$396.00	$73.60	$588.80	$48.28	$71.87
1 Electrician	49.00	392.00	72.85	582.80		
.5 Equip. Oper. (crane)	44.40	177.60	66.45	265.80		
.5 S.P. Crane, 4x4, 5 Ton		138.10		151.91	6.91	7.60
20 L.H., Daily Totals		$1103.70		$1589.31	$55.19	$79.47

Crew R-4	Hr.	Daily	Hr.	Daily	Bare Costs	Incl. O&P
1 Struc. Steel Foreman	$48.90	$391.20	$85.95	$687.60	$47.72	$81.23
3 Struc. Steel Workers	46.90	1125.60	82.45	1978.80		
1 Electrician	49.00	392.00	72.85	582.80		
1 Welder, gas engine, 300 amp		145.80		160.38	3.65	4.01
40 L.H., Daily Totals		$2054.60		$3409.58	$51.37	$85.24

Crew R-5	Hr.	Daily	Hr.	Daily	Bare Costs	Incl. O&P
1 Electrician Foreman	$49.50	$396.00	$73.60	$588.80	$42.72	$64.05
4 Electrician Linemen	49.00	1568.00	72.85	2331.20		
2 Electrician Operators	49.00	784.00	72.85	1165.60		
4 Electrician Groundmen	31.60	1011.20	48.45	1550.40		
1 Crew Truck		202.40		222.64		
1 Flatbed Truck, 20,000 GVW		245.60		270.16		
1 Pickup Truck, 3/4 Ton		140.00		154.00		
.2 Hyd. Crane, 55 Ton		241.80		265.98		
.2 Hyd. Crane, 12 Ton		131.12		144.23		
.2 Earth Auger, Truck-Mtd.		89.84		98.82		
1 Tractor w/Winch		421.20		463.32	16.73	18.40
88 L.H., Daily Totals		$5231.16		$7255.16	$59.45	$82.44

Crew R-6	Hr.	Daily	Hr.	Daily	Bare Costs	Incl. O&P
1 Electrician Foreman	$49.50	$396.00	$73.60	$588.80	$42.72	$64.05
4 Electrician Linemen	49.00	1568.00	72.85	2331.20		
2 Electrician Operators	49.00	784.00	72.85	1165.60		
4 Electrician Groundmen	31.60	1011.20	48.45	1550.40		
1 Crew Truck		202.40		222.64		
1 Flatbed Truck, 20,000 GVW		245.60		270.16		
1 Pickup Truck, 3/4 Ton		140.00		154.00		
.2 Hyd. Crane, 55 Ton		241.80		265.98		
.2 Hyd. Crane, 12 Ton		131.12		144.23		
.2 Earth Auger, Truck-Mtd.		89.84		98.82		
1 Tractor w/Winch		421.20		463.32		
3 Cable Trailers		542.10		596.31		
.5 Tensioning Rig		179.20		197.12		
.5 Cable Pulling Rig		1048.00		1152.80	36.83	40.52
88 L.H., Daily Totals		$7000.46		$9201.39	$79.55	$104.56

Crew R-7	Hr.	Daily	Hr.	Daily	Bare Costs	Incl. O&P
1 Electrician Foreman	$49.50	$396.00	$73.60	$588.80	$34.58	$52.64
5 Electrician Groundmen	31.60	1264.00	48.45	1938.00		
1 Crew Truck		202.40		222.64	4.22	4.64
48 L.H., Daily Totals		$1862.40		$2749.44	$38.80	$57.28

Crew R-8	Hr.	Daily	Hr.	Daily	Bare Costs	Incl. O&P
1 Electrician Foreman	$49.50	$396.00	$73.60	$588.80	$43.28	$64.84
3 Electrician Linemen	49.00	1176.00	72.85	1748.40		
2 Electrician Groundmen	31.60	505.60	48.45	775.20		
1 Pickup Truck, 3/4 Ton		140.00		154.00		
1 Crew Truck		202.40		222.64	7.13	7.85
48 L.H., Daily Totals		$2420.00		$3489.04	$50.42	$72.69

Crew R-9	Hr.	Daily	Hr.	Daily	Bare Costs	Incl. O&P
1 Electrician Foreman	$49.50	$396.00	$73.60	$588.80	$40.36	$60.74
1 Electrician Lineman	49.00	392.00	72.85	582.80		
2 Electrician Operators	49.00	784.00	72.85	1165.60		
4 Electrician Groundmen	31.60	1011.20	48.45	1550.40		
1 Pickup Truck, 3/4 Ton		140.00		154.00		
1 Crew Truck		202.40		222.64	5.35	5.88
64 L.H., Daily Totals		$2925.60		$4264.24	$45.71	$66.63

Crew R-10	Hr.	Daily	Hr.	Daily	Bare Costs	Incl. O&P
1 Electrician Foreman	$49.50	$396.00	$73.60	$588.80	$46.18	$68.91
4 Electrician Linemen	49.00	1568.00	72.85	2331.20		
1 Electrician Groundman	31.60	252.80	48.45	387.60		
1 Crew Truck		202.40		222.64		
3 Tram Cars		393.30		432.63	12.41	13.65
48 L.H., Daily Totals		$2812.50		$3962.87	$58.59	$82.56

Crew No.	Bare Costs		Incl. Subs O&P		Cost Per Labor-Hour	
Crew R-11	**Hr.**	**Daily**	**Hr.**	**Daily**	**Bare Costs**	**Incl. O&P**
1 Electrician Foreman	$49.50	$396.00	$73.60	$588.80	$46.14	$68.93
4 Electricians	49.00	1568.00	72.85	2331.20		
1 Equip. Oper. (crane)	44.40	355.20	66.45	531.60		
1 Common Laborer	33.10	264.80	51.05	408.40		
1 Crew Truck		202.40		222.64		
1 Hyd. Crane, 12 Ton		655.60		721.16	15.32	16.85
56 L.H., Daily Totals		$3442.00		$4803.80	$61.46	$85.78
Crew R-12	**Hr.**	**Daily**	**Hr.**	**Daily**	**Bare Costs**	**Incl. O&P**
1 Carpenter Foreman	$42.05	$336.40	$64.85	$518.80	$39.14	$61.09
4 Carpenters	41.55	1329.60	64.05	2049.60		
4 Common Laborers	33.10	1059.20	51.05	1633.60		
1 Equip. Oper. (med.)	42.95	343.60	64.30	514.40		
1 Steel Worker	46.90	375.20	82.45	659.60		
1 Dozer, 200 H.P.		1192.00		1311.20		
1 Pickup Truck, 3/4 Ton		140.00		154.00	15.14	16.65
88 L.H., Daily Totals		$4776.00		$6841.20	$54.27	$77.74
Crew R-13	**Hr.**	**Daily**	**Hr.**	**Daily**	**Bare Costs**	**Incl. O&P**
1 Electrician Foreman	$49.50	$396.00	$73.60	$588.80	$46.84	$69.74
3 Electricians	49.00	1176.00	72.85	1748.40		
.25 Equip. Oper. (crane)	44.40	88.80	66.45	132.90		
1 Equipment Oiler	38.30	306.40	57.35	458.80		
.25 Hydraulic Crane, 33 Ton		190.75		209.82	4.54	5.00
42 L.H., Daily Totals		$2157.95		$3138.72	$51.38	$74.73
Crew R-15	**Hr.**	**Daily**	**Hr.**	**Daily**	**Bare Costs**	**Incl. O&P**
1 Electrician Foreman	$49.50	$396.00	$73.60	$588.80	$47.80	$71.14
4 Electricians	49.00	1568.00	72.85	2331.20		
1 Equipment Oper. (light)	41.30	330.40	61.85	494.80		
1 Aerial Lift Truck		347.40		382.14	7.24	7.96
48 L.H., Daily Totals		$2641.80		$3796.94	$55.04	$79.10
Crew R-18	**Hr.**	**Daily**	**Hr.**	**Daily**	**Bare Costs**	**Incl. O&P**
.25 Electrician Foreman	$49.50	$99.00	$73.60	$147.20	$38.33	$57.89
1 Electrician	49.00	392.00	72.85	582.80		
2 Helpers	31.60	505.60	48.45	775.20		
26 L.H., Daily Totals		$996.60		$1505.20	$38.33	$57.89
Crew R-19	**Hr.**	**Daily**	**Hr.**	**Daily**	**Bare Costs**	**Incl. O&P**
.5 Electrician Foreman	$49.50	$198.00	$73.60	$294.40	$49.10	$73.00
2 Electricians	49.00	784.00	72.85	1165.60		
20 L.H., Daily Totals		$982.00		$1460.00	$49.10	$73.00
Crew R-21	**Hr.**	**Daily**	**Hr.**	**Daily**	**Bare Costs**	**Incl. O&P**
1 Electrician Foreman	$49.50	$396.00	$73.60	$588.80	$48.97	$72.82
3 Electricians	49.00	1176.00	72.85	1748.40		
.1 Equip. Oper. (med.)	42.95	34.36	64.30	51.44		
.1 S.P. Crane, 4x4, 25 Ton		65.56		72.12	2.00	2.20
32.8 L.H., Daily Totals		$1671.92		$2460.76	$50.97	$75.02
Crew R-22	**Hr.**	**Daily**	**Hr.**	**Daily**	**Bare Costs**	**Incl. O&P**
.66 Electrician Foreman	$49.50	$261.36	$73.60	$388.61	$41.60	$62.48
2 Helpers	31.60	505.60	48.45	775.20		
2 Electricians	49.00	784.00	72.85	1165.60		
37.28 L.H., Daily Totals		$1550.96		$2329.41	$41.60	$62.48

Crew No.	Bare Costs		Incl. Subs O&P		Cost Per Labor-Hour	
Crew R-30	**Hr.**	**Daily**	**Hr.**	**Daily**	**Bare Costs**	**Incl. O&P**
.25 Electrician Foreman (out)	$51.00	$102.00	$75.85	$151.70	$39.37	$59.67
1 Electrician	49.00	392.00	72.85	582.80		
2 Laborers, (Semi-Skilled)	33.10	529.60	51.05	816.80		
26 L.H., Daily Totals		$1023.60		$1551.30	$39.37	$59.67
Crew R-31	**Hr.**	**Daily**	**Hr.**	**Daily**	**Bare Costs**	**Incl. O&P**
1 Electrician	$49.00	$392.00	$72.85	$582.80	$49.00	$72.85
1 Core Drill, Electric, 2.5 H.P.		53.40		58.74	6.67	7.34
8 L.H., Daily Totals		$445.40		$641.54	$55.67	$80.19
Crew W-41E	**Hr.**	**Daily**	**Hr.**	**Daily**	**Bare Costs**	**Incl. O&P**
.5 Plumber Foreman (out)	$54.05	$216.20	$81.00	$324.00	$44.87	$67.82
1 Plumber	52.05	416.40	78.00	624.00		
1 Laborer	33.10	264.80	51.05	408.40		
20 L.H., Daily Totals		$897.40		$1356.40	$44.87	$67.82

Historical Cost Indexes

The table below lists both the RSMeans Historical Cost Index based on Jan. 1, 1993 = 100 as well as the computed value of an index based on Jan. 1, 2010 costs. Since the Jan. 1, 2010 figure is estimated, space is left to write in the actual index figures as they become available through either the quarterly "RSMeans Construction Cost Indexes" or as printed in the "Engineering News-Record." To compute the actual index based on Jan. 1, 2010 = 100, divide the Historical Cost Index for a particular year by the actual Jan. 1, 2010 Construction Cost Index. Space has been left to advance the index figures as the year progresses.

Year	Historical Cost Index Jan. 1, 1993 = 100		Current Index Based on Jan. 1, 2010 = 100		Year	Historical Cost Index Jan. 1, 1993 = 100	Current Index Based on Jan. 1, 2010 = 100		Year	Historical Cost Index Jan. 1, 1993 = 100	Current Index Based on Jan. 1, 2010 = 100	
	Est.	Actual	Est.	Actual		Actual	Est.	Actual		Actual	Est.	Actual
Oct 2010					July 1995	107.6	58.9		July 1977	49.5	27.1	
July 2010					1994	104.4	57.1		1976	46.9	25.7	
April 2010					1993	101.7	55.6		1975	44.8	24.5	
Jan 2010	182.8		100.0	100.0	1992	99.4	54.4		1974	41.4	22.6	
July 2009		180.1	98.5		1991	96.8	53.0		1973	37.7	20.6	
2008		180.4	98.7		1990	94.3	51.6		1972	34.8	19.0	
2007		169.4	92.7		1989	92.1	50.4		1971	32.1	17.6	
2006		162.0	88.6		1988	89.9	49.2		1970	28.7	15.7	
2005		151.6	82.9		1987	87.7	48.0		1969	26.9	14.7	
2004		143.7	78.6		1986	84.2	46.1		1968	24.9	13.6	
2003		132.0	72.2		1985	82.6	45.2		1967	23.5	12.9	
2002		128.7	70.4		1984	82.0	44.8		1966	22.7	12.4	
2001		125.1	68.4		1983	80.2	43.9		1965	21.7	11.9	
2000		120.9	66.1		1982	76.1	41.7		1964	21.2	11.6	
1999		117.6	64.3		1981	70.0	38.3		1963	20.7	11.3	
1998		115.1	63.0		1980	62.9	34.4		1962	20.2	11.1	
1997		112.8	61.7		1979	57.8	31.6		1961	19.8	10.8	
1996		110.2	60.3		1978	53.5	29.3		1960	19.7	10.8	

Adjustments to Costs

The Historical Cost Index can be used to convert National Average building costs at a particular time to the approximate building costs for some other time.

Time Adjustment using the Historical Cost Indexes:

$$\frac{\text{Index for Year A}}{\text{Index for Year B}} \times \text{Cost in Year B} = \text{Cost in Year A}$$

Example:

Estimate and compare construction costs for different years in the same city.

To estimate the National Average construction cost of a building in 1970, knowing that it cost $900,000 in 2010:

INDEX in 1970 = 28.7

INDEX in 2010 = 182.8

$$\frac{\text{INDEX 1970}}{\text{INDEX 2010}} \times \text{Cost 2010} = \text{Cost 1970}$$

$$\frac{28.7}{182.8} \times \$900{,}000 = .157 \times \$900{,}000 = \$141{,}300$$

The construction cost of the building in 1970 is $141,300.

Note: The City Cost Indexes for Canada can be used to convert U.S. National averages to local costs in Canadian dollars.

How to Use the City Cost Indexes

What you should know before you begin

RSMeans City Cost Indexes (CCI) are an extremely useful tool to use when you want to compare costs from city to city and region to region.

This publication contains average construction cost indexes for 316 U.S. and Canadian cities covering over 930 three-digit zip code locations.

Keep in mind that a City Cost Index number is a percentage ratio of a specific city's cost to the national average cost of the same item at a stated time period.

In other words, these index figures represent relative construction factors (or, if you prefer, multipliers) for Material and Installation costs, as well as the weighted average for Total In Place costs for each CSI MasterFormat division. Installation costs include both labor and equipment rental costs. When estimating equipment rental rates only, for a specific location, use 01543 CONTRACTOR EQUIPMENT index.

The 30 City Average Index is the average of 30 major U.S. cities and serves as a National Average.

Index figures for both material and installation are based on the 30 major city average of 100 and represent the cost relationship as of July 1, 2008. The index for each division is computed from representative material and labor quantities for that division. The weighted average for each city is a weighted total of the components listed above it, but does not include relative productivity between trades or cities.

As changes occur in local material prices, labor rates, and equipment rental rates, (including fuel costs) the impact of these changes should be accurately measured by the change in the City Cost Index for each particular city (as compared to the 30 City Average).

Therefore, if you know (or have estimated) building costs in one city today, you can easily convert those costs to expected building costs in another city.

In addition, by using the Historical Cost Index, you can easily convert National Average building costs at a particular time to the approximate building costs for some other time. The City Cost Indexes can then be applied to calculate the costs for a particular city.

Quick Calculations

Location Adjustment Using the City Cost Indexes:

$$\frac{\text{Index for City A}}{\text{Index for City B}} \text{ x Cost in City B = Cost in City A}$$

Time Adjustment for the National Average Using the Historical Cost Index:

$$\frac{\text{Index for Year A}}{\text{Index for Year B}} \text{ x Cost in Year B = Cost in Year A}$$

Adjustment from the National Average:

$$\frac{\text{Index for City A}}{100} \text{ x National Average Cost = Cost in City A}$$

Since each of the other RSMeans publications contains many different items, any *one* item multiplied by the particular city index may give incorrect results. However, the larger the number of items compiled, the closer the results should be to actual costs for that particular city.

The City Cost Indexes for Canadian cities are calculated using Canadian material and equipment prices and labor rates, in Canadian dollars. Therefore, indexes for Canadian cities can be used to convert U.S. National Average prices to local costs in Canadian dollars.

How to use this section

1. Compare costs from city to city.

In using the RSMeans Indexes, remember that an index number is not a fixed number but a ratio: It's a percentage ratio of a building component's cost at any stated time to the National Average cost of that same component at the same time period. Put in the form of an equation:

$$\frac{\text{Specific City Cost}}{\text{National Average Cost}} \text{ x 100 = City Index Number}$$

Therefore, when making cost comparisons between cities, do not subtract one city's index number from the index number of another city and read the result as a percentage difference. Instead, divide one city's index number by that of the other city. The resulting number may then be used as a multiplier to calculate cost differences from city to city.

The formula used to find cost differences between cities for the purpose of comparison is as follows:

$$\frac{\text{City A Index}}{\text{City B Index}} \text{ x City B Cost (Known) = City A Cost (Unknown)}$$

In addition, you can use RSMeans CCI to calculate and compare costs division by division between cities using the same basic formula. (Just be sure that you're comparing similar divisions.)

2. Compare a specific city's construction costs with the National Average.

When you're studying construction location feasibility, it's advisable to compare a prospective project's cost index with an index of the National Average cost.

For example, divide the weighted average index of construction costs of a specific city by that of the 30 City Average, which = 100.

$$\frac{\text{City Index}}{100} = \text{\% of National Average}$$

As a result, you get a ratio that indicates the relative cost of construction in that city in comparison with the National Average.

3. Convert U.S. National Average to actual costs in Canadian City.

$$\frac{\text{Index for Canadian City}}{100} \times \text{National Average Cost = Cost in Canadian City in \$ CAN}$$

4. **Adjust construction cost data based on a National Average.**

When you use a source of construction cost data which is based on a National Average (such as RSMeans cost data publications), it is necessary to adjust those costs to a specific location.

$$\frac{\text{City Index}}{100} \times \text{"Book" Cost Based on National Average Costs} = \text{City Cost (Unknown)}$$

5. **When applying the City Cost Indexes to demolition projects, use the appropriate division installation index.** For example, for removal of existing doors and windows, use Division 8 (Openings) index.

What you might like to know about how we developed the Indexes

The information presented in the CCI is organized according to the Construction Specifications Institute (CSI) MasterFormat 2004.

To create a reliable index, RSMeans researched the building type most often constructed in the United States and Canada. Because it was concluded that no one type of building completely represented the building construction industry, nine different types of buildings were combined to create a composite model.

The exact material, labor and equipment quantities are based on detailed analysis of these nine building types, then each quantity is weighted in proportion to expected usage. These various material items, labor hours, and equipment rental rates are thus combined to form a composite building representing as closely as possible the actual usage of materials, labor and equipment used in the North American Building Construction Industry.

The following structures were chosen to make up that composite model:

1. Factory, 1 story
2. Office, 2-4 story
3. Store, Retail
4. Town Hall, 2-3 story
5. High School, 2-3 story
6. Hospital, 4-8 story
7. Garage, Parking
8. Apartment, 1-3 story
9. Hotel/Motel, 2-3 story

For the purposes of ensuring the timeliness of the data, the components of the index for the composite model have been streamlined. They currently consist of:

- specific quantities of 66 commonly used construction materials;
- specific labor-hours for 21 building construction trades; and
- specific days of equipment rental for 6 types of construction equipment (normally used to install the 66 material items by the 21 trades.) Fuel costs and routine maintenance costs are included in the equipment costs.

A sophisticated computer program handles the updating of all costs for each city on a quarterly basis. Material and equipment price quotations are gathered quarterly from 316 cities in the United States and Canada. These prices and the latest negotiated labor wage rates for 21 different building trades are used to compile the quarterly update of the City Cost Index.

The 30 major U.S. cities used to calculate the National Average are:

Atlanta, GA	Memphis, TN
Baltimore, MD	Milwaukee, WI
Boston, MA	Minneapolis, MN
Buffalo, NY	Nashville, TN
Chicago, IL	New Orleans, LA
Cincinnati, OH	New York, NY
Cleveland, OH	Philadelphia, PA
Columbus, OH	Phoenix, AZ
Dallas, TX	Pittsburgh, PA
Denver, CO	St. Louis, MO
Detroit, MI	San Antonio, TX
Houston, TX	San Diego, CA
Indianapolis, IN	San Francisco, CA
Kansas City, MO	Seattle, WA
Los Angeles, CA	Washington, DC

What the CCI does not indicate

The weighted average for each city is a total of the divisional components weighted to reflect typical usage, but it does not include the productivity variations between trades or cities.

In addition, the CCI does not take into consideration factors such as the following:

- managerial efficiency
- competitive conditions
- automation
- restrictive union practices
- unique local requirements
- regional variations due to specific building codes

City Cost Indexes

	DIVISION	UNITED STATES			ALABAMA														
		ANYTOWN			BIRMINGHAM			HUNTSVILLE			MOBILE			MONTGOMERY			TUSCALOOSA		
		MAT.	INST.	TOTAL	MAT.	INST.	TOTAL	MAT.	INST.	TOTAL	MAT.	INST.	TOTAL	MAT.	INST.	TOTAL	MAT.	INST.	TOTAL
015433	**CONTRACTOR EQUIPMENT**		100.0	100.0		101.0	101.0		100.9	100.9		97.8	97.8		97.8	97.8		100.9	100.9
0241, 31 - 34	**SITE & INFRASTRUCTURE, DEMOLITION**	100.0	100.0	100.0	87.0	93.8	91.8	82.7	92.9	89.8	96.1	88.3	90.7	94.1	87.7	89.6	83.1	92.9	89.9
0310	Concrete Forming & Accessories	100.0	100.0	100.0	90.6	70.2	72.9	94.1	67.4	71.0	94.2	59.6	64.3	92.2	47.9	53.9	94.0	51.7	57.4
0320	Concrete Reinforcing	100.0	100.0	100.0	83.4	87.7	85.6	83.4	77.2	80.2	86.0	85.7	85.8	86.0	86.3	86.1	83.4	86.4	84.9
0330	Cast-in-Place Concrete	100.0	100.0	100.0	100.8	68.8	88.7	91.2	65.4	81.4	96.2	72.0	87.1	98.6	49.7	80.1	94.9	52.2	78.8
03	**CONCRETE**	100.0	100.0	100.0	92.3	74.1	83.8	87.7	69.7	79.3	90.6	70.2	81.1	91.6	57.5	75.7	89.6	60.0	75.8
04	**MASONRY**	100.0	100.0	100.0	87.6	83.4	85.0	87.1	63.5	72.9	87.8	58.7	70.3	86.2	43.7	60.7	86.2	51.1	65.1
05	**METALS**	100.0	100.0	100.0	96.6	94.7	95.9	97.8	89.3	94.9	96.7	93.8	95.7	96.5	91.6	94.9	96.9	92.0	95.2
06	**WOOD, PLASTICS & COMPOSITES**	100.0	100.0	100.0	92.0	67.1	77.9	94.7	68.7	80.0	94.9	59.1	74.6	91.6	47.3	66.5	94.7	49.6	69.2
07	**THERMAL & MOISTURE PROTECTION**	100.0	100.0	100.0	94.0	83.9	90.2	91.6	75.7	85.6	91.6	75.5	85.5	90.8	66.5	81.6	91.7	69.4	83.3
08	**OPENINGS**	100.0	100.0	100.0	97.2	73.0	91.2	97.5	65.3	89.5	97.5	64.0	89.2	97.5	57.2	87.5	97.5	63.6	89.1
0920	Plaster & Gypsum Board	100.0	100.0	100.0	98.5	66.6	76.4	95.3	68.2	76.6	95.3	58.4	69.8	98.5	46.2	62.3	95.3	48.6	63.0
0950, 0980	Ceilings & Acoustic Treatment	100.0	100.0	100.0	94.1	66.6	76.6	98.4	68.2	79.2	98.4	58.4	72.9	94.1	46.2	63.6	98.4	48.6	66.7
0960	Flooring	100.0	100.0	100.0	103.1	64.9	92.6	103.1	58.2	90.7	112.8	60.3	98.3	111.1	31.9	89.2	103.1	40.4	85.8
0970, 0990	Wall Finishes & Painting/Coating	100.0	100.0	100.0	96.1	61.7	75.4	96.1	64.9	77.4	100.1	60.3	76.1	96.1	59.2	73.9	96.1	51.0	69.0
09	**FINISHES**	100.0	100.0	100.0	96.5	67.6	80.8	96.7	65.2	79.6	101.6	58.8	78.4	99.9	44.8	70.0	96.7	48.5	70.5
COVERS	**DIVS. 10 - 14, 25, 28, 41, 43, 44**	100.0	100.0	100.0	100.0	85.2	96.9	100.0	82.2	96.2	100.0	82.0	96.2	100.0	76.8	95.1	100.0	79.0	95.5
21, 22, 23	**FIRE SUPPRESSION, PLUMBING & HVAC**	100.0	100.0	100.0	99.8	66.0	85.9	99.7	58.6	82.8	99.7	60.4	83.5	99.8	38.4	74.5	99.8	38.2	74.4
26, 27, 3370	**ELECTRICAL, COMMUNICATIONS & UTIL.**	100.0	100.0	100.0	108.0	64.3	85.2	99.6	68.4	83.3	99.6	56.2	76.9	102.7	68.2	84.7	99.0	64.3	80.8
MF2004	**WEIGHTED AVERAGE**	100.0	100.0	100.0	97.4	75.2	87.6	96.0	70.1	84.6	97.1	67.4	84.0	97.1	58.3	80.0	96.0	60.2	80.2

	DIVISION	ALASKA									ARIZONA								
		ANCHORAGE			FAIRBANKS			JUNEAU			FLAGSTAFF			MESA/TEMPE			PHOENIX		
		MAT.	INST.	TOTAL	MAT.	INST.	TOTAL	MAT.	INST.	TOTAL	MAT.	INST.	TOTAL	MAT.	INST.	TOTAL	MAT.	INST.	TOTAL
015433	**CONTRACTOR EQUIPMENT**		116.2	116.2		116.2	116.2		116.2	116.2		94.9	94.9		98.0	98.0		98.6	98.6
0241, 31 - 34	**SITE & INFRASTRUCTURE, DEMOLITION**	137.3	131.5	133.2	123.9	131.5	129.2	136.4	131.5	132.9	80.1	99.5	93.6	83.9	101.9	96.5	84.4	103.3	97.5
0310	Concrete Forming & Accessories	135.6	118.5	120.8	141.4	119.1	122.1	137.8	118.5	121.1	100.0	63.3	68.3	100.5	57.4	63.2	101.4	71.4	75.4
0320	Concrete Reinforcing	146.4	106.5	126.0	150.6	106.5	128.1	112.1	106.5	109.3	96.6	79.7	88.0	97.2	79.6	88.2	95.6	79.9	87.6
0330	Cast-in-Place Concrete	162.5	114.5	144.4	142.6	117.1	132.9	163.4	114.5	144.9	91.9	75.7	85.8	98.1	62.6	84.6	98.1	76.3	89.8
03	**CONCRETE**	136.5	114.2	126.1	123.1	115.3	119.5	132.2	114.2	123.8	119.0	70.8	96.4	105.3	63.7	85.8	105.0	74.6	90.8
04	**MASONRY**	202.7	119.0	152.4	211.4	118.8	155.7	194.2	119.0	149.0	95.2	60.8	74.5	110.0	45.8	71.4	96.6	64.6	77.3
05	**METALS**	120.2	101.2	113.8	120.3	101.4	113.9	120.2	101.1	113.8	97.3	73.6	89.3	93.7	73.8	87.0	95.4	75.4	88.7
06	**WOOD, PLASTICS & COMPOSITES**	114.1	118.6	116.7	117.8	119.3	118.7	114.1	118.6	116.7	105.0	62.3	80.8	98.8	62.5	78.2	99.7	72.8	84.4
07	**THERMAL & MOISTURE PROTECTION**	185.0	114.3	158.3	183.1	114.4	157.2	183.7	114.3	157.5	95.6	66.2	84.5	102.4	56.0	84.9	102.3	68.1	89.4
08	**OPENINGS**	121.0	113.5	119.1	118.1	113.6	117.0	118.1	113.5	117.0	101.1	67.9	92.8	96.8	64.8	88.9	97.8	73.6	91.8
0920	Plaster & Gypsum Board	137.8	119.0	124.8	157.2	119.7	131.3	137.8	119.0	124.8	95.4	61.2	71.7	102.4	61.2	73.9	105.2	71.9	82.1
0950, 0980	Ceilings & Acoustic Treatment	121.3	119.0	119.8	126.5	119.7	122.2	121.3	119.0	119.8	126.2	61.2	84.8	116.2	61.2	81.2	123.9	71.9	90.8
0960	Flooring	159.3	122.5	149.1	159.3	122.5	149.1	159.3	122.5	149.1	97.4	42.8	82.3	102.1	53.3	88.7	102.4	55.7	89.5
0970, 0990	Wall Finishes & Painting/Coating	154.8	112.8	129.5	154.8	125.7	137.3	154.8	112.8	129.5	90.4	55.1	69.2	95.0	55.1	71.0	95.0	60.9	74.5
09	**FINISHES**	147.2	119.2	132.0	148.7	121.0	133.7	146.0	119.2	131.4	105.1	57.7	79.4	103.9	55.8	77.8	106.1	66.9	84.8
COVERS	**DIVS. 10 - 14, 25, 28, 41, 43, 44**	100.0	110.3	102.2	100.0	110.3	102.2	100.0	110.3	102.2	100.0	82.4	96.3	100.0	77.3	95.2	100.0	84.0	96.6
21, 22, 23	**FIRE SUPPRESSION, PLUMBING & HVAC**	100.4	101.7	100.9	100.3	104.5	102.1	100.4	99.0	99.8	100.2	76.8	90.6	100.3	62.7	84.8	100.3	76.9	90.6
26, 27, 3370	**ELECTRICAL, COMMUNICATIONS & UTIL.**	141.6	110.0	125.1	147.9	110.0	128.1	142.0	110.0	125.3	100.8	58.9	78.9	94.2	58.9	75.7	101.4	67.0	83.4
MF2004	**WEIGHTED AVERAGE**	127.4	112.3	120.7	126.3	113.3	120.5	126.0	111.7	119.7	101.6	70.4	87.9	99.4	64.4	83.9	99.9	74.6	88.7

	DIVISION	ARIZONA						ARKANSAS											
		PRESCOTT			TUCSON			FORT SMITH			JONESBORO			LITTLE ROCK			PINE BLUFF		
		MAT.	INST.	TOTAL	MAT.	INST.	TOTAL	MAT.	INST.	TOTAL	MAT.	INST.	TOTAL	MAT.	INST.	TOTAL	MAT.	INST.	TOTAL
015433	**CONTRACTOR EQUIPMENT**		94.9	94.9		98.0	98.0		87.4	87.4		106.6	106.6		87.4	87.4		87.4	87.4
0241, 31 - 34	**SITE & INFRASTRUCTURE, DEMOLITION**	68.9	97.9	89.1	79.9	103.0	96.0	78.6	84.1	82.4	100.1	98.0	98.6	86.5	84.1	84.8	80.5	84.1	83.0
0310	Concrete Forming & Accessories	96.6	47.3	53.9	100.7	70.8	74.9	97.4	39.1	47.0	87.8	53.2	57.9	89.6	65.6	68.9	81.5	65.4	67.5
0320	Concrete Reinforcing	96.6	70.7	83.4	80.4	79.7	80.0	99.4	71.9	85.4	95.0	68.3	81.4	99.6	66.9	82.9	99.5	66.9	82.8
0330	Cast-in-Place Concrete	91.8	56.1	78.3	100.9	76.1	91.5	95.4	64.9	83.9	90.9	59.2	78.9	95.9	65.1	84.2	87.5	65.0	79.0
03	**CONCRETE**	103.7	55.1	81.0	104.1	74.2	90.2	91.3	55.1	74.4	88.1	59.4	74.7	91.1	66.0	79.3	89.1	65.8	78.2
04	**MASONRY**	95.5	48.9	67.5	94.6	60.8	74.3	97.8	54.8	71.9	93.3	46.7	65.3	94.6	54.8	70.7	117.9	54.8	80.0
05	**METALS**	97.3	65.6	86.6	94.6	74.1	87.6	100.8	70.7	90.6	95.2	80.3	90.2	96.9	69.4	87.6	99.2	69.1	89.0
06	**WOOD, PLASTICS & COMPOSITES**	100.9	45.6	69.6	99.0	72.8	84.2	105.1	34.6	65.2	92.8	54.0	70.8	99.7	70.0	82.9	85.8	70.0	76.9
07	**THERMAL & MOISTURE PROTECTION**	94.4	52.8	78.7	103.6	63.6	88.5	98.1	54.7	81.7	101.7	54.0	83.7	97.4	58.4	82.7	97.0	58.4	82.5
08	**OPENINGS**	101.1	50.6	88.6	93.2	73.6	88.3	97.5	43.8	84.2	99.5	57.6	89.1	97.5	63.1	88.9	92.7	63.1	85.4
0920	Plaster & Gypsum Board	92.3	44.0	58.9	108.8	71.9	83.2	84.1	33.2	48.9	94.6	52.8	65.7	79.9	69.6	72.8	77.0	69.6	71.9
0950, 0980	Ceilings & Acoustic Treatment	124.4	44.0	73.2	117.0	71.9	88.3	93.0	33.2	54.9	92.7	52.8	67.3	91.3	69.6	77.5	89.6	69.6	76.9
0960	Flooring	95.8	42.6	81.1	93.0	42.8	79.2	111.6	65.5	98.8	74.7	57.5	70.0	112.9	65.5	99.8	101.3	65.5	91.4
0970, 0990	Wall Finishes & Painting/Coating	90.4	41.0	60.7	95.2	55.1	71.1	98.6	59.8	75.2	87.6	53.7	67.2	98.6	61.3	76.2	98.6	61.3	76.2
09	**FINISHES**	102.6	44.9	71.3	101.9	63.9	81.3	94.9	45.0	67.8	87.1	53.8	69.0	94.5	65.9	79.0	90.3	65.9	77.1
COVERS	**DIVS. 10 - 14, 25, 28, 41, 43, 44**	100.0	78.4	95.4	100.0	83.9	96.6	100.0	71.7	94.0	100.0	52.9	90.0	100.0	75.8	94.9	100.0	75.8	94.9
21, 22, 23	**FIRE SUPPRESSION, PLUMBING & HVAC**	100.2	67.1	86.6	100.2	65.6	86.0	100.2	45.4	77.6	100.4	45.3	77.7	100.1	62.6	84.7	100.2	47.8	78.6
26, 27, 3370	**ELECTRICAL, COMMUNICATIONS & UTIL.**	100.5	58.8	78.7	96.5	58.4	76.6	97.0	72.4	84.1	104.0	50.2	75.8	102.4	73.6	87.3	96.3	73.6	84.4
MF2004	**WEIGHTED AVERAGE**	99.1	61.1	82.4	98.2	69.9	85.7	97.3	57.8	79.9	96.9	57.9	79.7	97.2	67.0	83.9	96.8	63.9	82.3

DIVISION		ARKANSAS			CALIFORNIA																	
		TEXARKANA			ANAHEIM			BAKERSFIELD			FRESNO			LOS ANGELES			OAKLAND					
		MAT.	INST.	TOTAL	MAT.	INST.	TOTAL	MAT.	INST.	TOTAL	MAT.	INST.	TOTAL	MAT.	INST.	TOTAL	MAT.	INST.	TOTAL			
015433	CONTRACTOR EQUIPMENT		88.0	88.0		101.6	101.6		99.3	99.3		99.3	99.3		100.6	100.6		98.7	98.7			
0241, 31 - 34	SITE & INFRASTRUCTURE, DEMOLITION	94.8	84.6	87.7	100.6	109.4	106.7	106.3	106.7	106.6	106.0	106.3	106.2	97.8	109.0	105.6	133.5	103.2	112.4			
0310	Concrete Forming & Accessories	87.4	43.9	49.7	105.8	125.9	123.2	104.3	125.2	122.3	101.9	127.5	124.0	105.6	126.2	123.4	111.1	143.2	138.8			
0320	Concrete Reinforcing	99.1	66.3	82.4	100.6	118.0	109.5	113.0	118.2	115.6	96.0	118.2	107.4	111.3	118.6	115.0	111.5	118.9	115.3			
0330	Cast-in-Place Concrete	95.4	46.0	76.7	97.5	119.8	105.9	104.3	118.6	109.7	102.8	114.2	107.1	103.9	118.5	109.4	126.3	117.7	123.1			
03	CONCRETE	88.1	49.7	70.1	105.2	121.3	112.7	107.7	120.5	113.7	106.9	120.0	113.0	110.3	121.1	115.3	122.7	128.1	125.2			
04	MASONRY	97.8	31.9	58.2	86.3	111.0	101.2	107.0	116.1	112.5	110.0	115.8	113.5	93.1	120.4	109.5	153.5	124.9	136.3			
05	METALS	92.5	66.0	83.5	103.7	102.1	103.2	99.0	101.4	99.8	103.6	100.4	102.5	102.2	103.7	102.7	98.2	100.2	98.9			
06	WOOD, PLASTICS & COMPOSITES	94.0	46.9	67.3	93.6	127.0	112.5	87.9	127.1	110.1	99.4	129.1	116.2	88.9	126.8	110.4	106.6	147.9	130.0			
07	THERMAL & MOISTURE PROTECTION	97.8	43.9	77.4	103.4	115.2	107.8	102.0	107.2	104.0	96.9	111.1	102.3	102.4	118.9	108.7	113.9	122.8	117.3			
08	OPENINGS	98.0	49.6	86.0	104.4	120.8	108.5	101.9	118.6	106.0	103.7	118.2	107.3	98.6	123.9	104.9	103.5	132.9	110.7			
0920	Plaster & Gypsum Board	81.6	45.9	56.8	107.3	127.7	121.4	102.1	127.7	119.8	102.3	129.8	121.3	100.9	127.7	119.5	106.0	148.9	135.7			
0950, 0980	Ceilings & Acoustic Treatment	94.7	45.9	63.6	112.2	127.7	122.1	113.6	127.7	122.6	115.1	129.8	124.5	114.7	127.7	123.0	114.2	148.9	136.3			
0960	Flooring	104.0	54.9	90.4	114.5	114.5	114.5	114.4	84.4	106.1	121.1	128.5	123.1	102.4	114.5	105.8	118.2	111.6	116.4			
0970, 0990	Wall Finishes & Painting/Coating	98.6	31.5	58.3	104.6	102.2	103.2	106.3	98.2	101.4	124.0	97.1	107.8	100.0	110.0	106.0	108.3	139.6	127.2			
09	FINISHES	93.4	44.5	66.9	108.4	121.9	115.7	110.0	115.8	113.1	113.9	125.7	120.3	105.8	122.6	114.9	113.5	138.8	127.3			
COVERS	DIVS. 10 - 14, 25, 28, 41, 43, 44	100.0	38.0	86.8	100.0	112.6	102.7	100.0	109.6	102.0	100.0	124.6	105.2	100.0	112.2	102.6	100.0	128.2	106.0			
21, 22, 23	FIRE SUPPRESSION, PLUMBING & HVAC	100.2	33.4	72.7	100.1	112.9	105.4	100.1	94.2	97.7	100.2	107.8	103.4	100.2	112.9	105.4	100.2	136.1	115.0			
26, 27, 3370	ELECTRICAL, COMMUNICATIONS & UTIL.	98.4	40.3	68.0	93.5	101.1	97.5	105.5	93.8	99.4	92.8	90.8	91.8	101.9	115.8	109.1	100.9	125.0	113.5			
MF2004	WEIGHTED AVERAGE	96.2	46.5	74.2	101.1	112.6	106.2	102.9	106.6	104.6	102.8	110.6	106.3	101.7	116.0	108.0	108.5	126.0	116.2			

DIVISION		CALIFORNIA																	
		OXNARD			REDDING			RIVERSIDE			SACRAMENTO			SAN DIEGO			SAN FRANCISCO		
		MAT.	INST.	TOTAL	MAT.	INST.	TOTAL	MAT.	INST.	TOTAL	MAT.	INST.	TOTAL	MAT.	INST.	TOTAL	MAT.	INST.	TOTAL
015433	CONTRACTOR EQUIPMENT		98.2	98.2		99.0	99.0		100.3	100.3		98.3	98.3		98.6	98.6		107.5	107.5
0241, 31 - 34	SITE & INFRASTRUCTURE, DEMOLITION	106.4	104.7	105.2	110.4	105.0	106.6	98.9	107.3	104.7	106.4	110.0	108.9	105.1	102.8	103.5	135.8	110.2	118.0
0310	Concrete Forming & Accessories	107.0	125.9	123.3	106.5	127.0	124.2	105.9	125.7	123.0	109.4	128.8	126.2	108.8	110.4	110.2	110.6	145.2	140.5
0320	Concrete Reinforcing	113.0	117.8	115.4	109.6	117.8	113.8	112.0	117.8	114.9	103.4	118.3	111.0	115.0	118.1	116.6	127.0	119.6	123.2
0330	Cast-in-Place Concrete	103.5	119.0	109.4	117.3	113.9	116.0	101.1	119.7	108.1	106.8	115.2	110.0	100.2	106.7	102.7	129.5	120.0	125.9
03	CONCRETE	107.5	120.9	113.8	117.1	119.6	118.2	106.1	121.1	113.1	111.8	120.6	116.0	108.5	110.0	109.2	126.5	130.4	128.3
04	MASONRY	111.8	105.7	108.2	113.7	108.0	110.3	85.0	106.6	98.0	122.8	115.9	118.6	95.4	109.4	103.8	154.1	133.0	141.4
05	METALS	98.8	101.4	99.7	103.4	99.2	102.0	103.9	101.6	103.1	94.0	96.9	95.0	101.8	102.7	102.1	104.8	110.0	106.6
06	WOOD, PLASTICS & COMPOSITES	94.1	127.1	112.8	94.1	129.1	113.9	93.6	127.0	112.5	100.0	130.4	117.2	94.9	108.1	102.4	106.6	148.1	130.1
07	THERMAL & MOISTURE PROTECTION	106.8	111.2	108.5	107.3	106.6	107.0	103.6	108.8	105.5	120.6	112.0	117.4	108.8	104.0	107.0	116.5	136.4	124.0
08	OPENINGS	100.8	120.8	105.8	103.5	119.0	107.3	102.9	120.8	107.4	116.3	120.9	117.4	100.1	109.0	102.3	107.8	136.8	115.0
0920	Plaster & Gypsum Board	107.3	127.7	121.4	105.3	129.8	122.3	106.8	127.7	121.3	100.3	131.0	121.5	115.8	108.1	110.5	108.8	148.9	136.5
0950, 0980	Ceilings & Acoustic Treatment	117.0	127.7	123.8	122.2	129.8	127.0	115.3	127.7	123.2	112.3	131.0	124.2	108.4	108.1	108.2	121.7	148.9	139.0
0960	Flooring	112.9	114.5	113.3	113.7	105.3	111.3	118.5	114.5	117.4	117.6	109.8	115.4	105.7	105.9	105.7	118.2	125.5	120.2
0970, 0990	Wall Finishes & Painting/Coating	105.6	93.2	98.1	105.6	101.2	102.9	102.7	102.2	102.4	105.6	110.5	108.5	104.3	108.1	106.6	108.3	151.6	134.4
09	FINISHES	110.8	120.9	116.3	112.5	122.0	117.7	109.9	121.9	116.4	110.1	124.7	118.1	107.6	109.1	108.4	115.8	143.3	130.7
COVERS	DIVS. 10 - 14, 25, 28, 41, 43, 44	100.0	112.8	102.7	100.0	124.6	105.2	100.0	112.5	102.7	100.0	125.2	105.4	100.0	109.5	102.0	100.0	128.8	106.1
21, 22, 23	FIRE SUPPRESSION, PLUMBING & HVAC	100.1	112.9	105.4	100.1	98.7	99.5	100.1	112.9	105.3	100.1	110.8	104.5	100.2	111.3	104.8	100.2	165.1	126.9
26, 27, 3370	ELECTRICAL, COMMUNICATIONS & UTIL.	97.6	95.6	96.6	101.0	94.3	97.5	93.6	92.6	93.1	96.3	99.7	98.1	101.7	102.5	102.1	101.0	157.9	130.7
MF2004	WEIGHTED AVERAGE	102.5	110.6	106.1	105.4	107.6	106.3	101.1	110.6	105.3	104.9	112.6	108.3	102.4	107.5	104.6	110.7	140.4	123.8

DIVISION		CALIFORNIA												COLORADO					
		SAN JOSE			SANTA BARBARA			STOCKTON			VALLEJO			COLORADO SPRINGS			DENVER		
		MAT.	INST.	TOTAL	MAT.	INST.	TOTAL	MAT.	INST.	TOTAL	MAT.	INST.	TOTAL	MAT.	INST.	TOTAL	MAT.	INST.	TOTAL
015433	CONTRACTOR EQUIPMENT		99.8	99.8		99.3	99.3		99.0	99.0		98.8	98.8		94.7	94.7		99.2	99.2
0241, 31 - 34	SITE & INFRASTRUCTURE, DEMOLITION	140.9	100.9	113.0	106.3	106.7	106.6	104.1	106.1	105.5	106.0	110.0	108.8	90.2	94.7	93.3	89.4	103.2	99.0
0310	Concrete Forming & Accessories	109.1	144.1	139.4	108.0	125.6	123.2	107.1	127.4	124.6	107.8	142.3	137.6	94.9	77.5	79.8	99.9	81.6	84.1
0320	Concrete Reinforcing	100.0	119.4	109.9	113.0	118.0	115.5	113.5	117.8	115.7	111.1	118.6	114.9	101.0	82.2	91.4	101.0	82.3	91.5
0330	Cast-in-Place Concrete	123.1	118.2	121.2	103.1	118.8	109.1	103.1	114.5	107.4	116.3	116.3	116.3	99.8	80.0	92.3	94.5	82.0	89.8
03	CONCRETE	115.4	129.2	121.9	107.4	120.7	113.6	107.4	120.0	113.3	117.6	127.0	122.0	104.6	79.6	92.9	98.8	82.1	91.0
04	MASONRY	143.9	132.7	137.1	108.3	110.4	109.5	111.4	108.2	109.5	87.3	118.9	106.3	97.4	73.1	82.8	99.3	79.1	87.2
05	METALS	98.6	108.5	102.0	99.2	101.4	99.9	100.4	100.8	100.5	97.1	98.0	97.4	105.4	87.0	99.2	108.0	87.4	101.1
06	WOOD, PLASTICS & COMPOSITES	102.0	147.8	127.9	94.1	127.1	112.8	95.7	129.1	114.6	97.3	147.7	125.8	93.1	80.8	86.2	99.7	83.6	90.6
07	THERMAL & MOISTURE PROTECTION	103.2	135.7	115.5	103.2	106.2	104.3	106.8	107.7	107.2	116.2	119.4	117.4	99.6	77.5	91.3	98.5	76.2	90.1
08	OPENINGS	94.1	137.3	104.8	101.9	120.8	106.6	101.1	120.2	105.9	115.9	136.5	121.0	99.7	83.3	95.6	99.5	85.4	96.0
0920	Plaster & Gypsum Board	102.3	148.9	134.5	107.3	127.7	121.4	107.7	129.8	123.0	105.9	148.9	135.6	90.2	80.3	83.3	104.5	83.4	89.9
0950, 0980	Ceilings & Acoustic Treatment	107.5	148.9	133.8	117.0	127.7	123.8	117.0	129.8	125.2	121.9	148.9	139.1	118.1	80.3	94.0	117.3	83.4	95.7
0960	Flooring	106.4	125.5	111.7	114.4	84.4	106.1	114.4	101.2	110.7	126.5	125.5	126.2	103.5	73.5	95.2	108.6	91.3	103.8
0970, 0990	Wall Finishes & Painting/Coating	106.2	141.3	127.3	105.6	93.2	98.1	105.6	101.5	103.2	105.4	123.9	116.5	104.2	45.0	68.6	104.5	74.0	86.1
09	FINISHES	108.1	141.9	126.4	111.5	115.7	113.8	111.4	121.3	116.8	113.6	139.2	127.5	105.3	73.5	88.0	108.2	82.7	94.4
COVERS	DIVS. 10 - 14, 25, 28, 41, 43, 44	100.0	127.8	105.9	100.0	112.8	102.7	100.0	124.7	105.2	100.0	127.0	105.7	100.0	88.1	97.5	100.0	89.6	97.8
21, 22, 23	FIRE SUPPRESSION, PLUMBING & HVAC	100.1	138.5	115.9	100.1	112.9	105.4	100.1	98.8	99.6	100.1	112.0	105.0	100.1	70.8	88.1	100.0	83.2	93.1
26, 27, 3370	ELECTRICAL, COMMUNICATIONS & UTIL.	104.2	142.1	124.0	90.6	96.4	93.7	100.5	107.4	104.1	93.3	111.9	103.0	102.2	80.0	90.6	103.9	85.4	94.2
MF2004	WEIGHTED AVERAGE	105.8	131.3	117.1	101.7	110.5	105.6	103.0	109.7	106.0	103.9	118.8	110.5	101.5	78.7	91.4	101.6	85.1	94.3

DIVISION		COLORADO												CONNECTICUT					
		FORT COLLINS			GRAND JUNCTION			GREELEY			PUEBLO			BRIDGEPORT			BRISTOL		
		MAT.	INST.	TOTAL	MAT.	INST.	TOTAL	MAT.	INST.	TOTAL	MAT.	INST.	TOTAL	MAT.	INST.	TOTAL	MAT.	INST.	TOTAL
015433	CONTRACTOR EQUIPMENT		96.1	96.1		99.2	99.2		96.1	96.1		96.4	96.4		100.6	100.6		100.6	100.6
0241, 31 - 34	SITE & INFRASTRUCTURE, DEMOLITION	100.2	96.6	97.7	118.1	99.3	105.0	87.6	95.4	93.1	110.0	94.5	99.2	97.7	104.4	102.3	96.9	104.3	102.1
0310	Concrete Forming & Accessories	101.0	71.4	75.4	109.5	71.7	76.8	99.2	45.0	52.3	107.1	77.8	81.7	97.7	121.9	118.6	97.7	121.6	118.3
0320	Concrete Reinforcing	102.1	74.0	87.7	103.5	73.9	88.4	102.0	73.5	87.4	99.7	82.1	90.7	110.6	127.2	119.1	110.6	127.1	119.1
0330	Cast-in-Place Concrete	109.7	74.8	96.5	113.4	75.7	99.1	91.7	53.1	77.1	101.6	81.1	93.9	101.0	117.5	107.3	94.7	117.4	103.3
03	CONCRETE	111.6	73.4	93.8	116.7	73.8	96.6	96.4	54.2	76.7	106.3	80.1	94.0	109.2	121.2	114.8	106.1	121.0	113.0
04	MASONRY	109.6	50.7	74.2	136.9	52.7	86.3	104.1	33.6	61.7	100.8	71.5	83.2	107.1	132.0	122.1	99.6	132.0	119.1
05	METALS	103.5	78.4	95.0	100.0	77.6	92.5	103.4	76.9	94.5	101.7	87.8	97.0	100.8	125.0	109.0	100.8	124.7	108.9
06	WOOD, PLASTICS & COMPOSITES	99.5	75.2	85.8	104.0	75.4	87.8	97.2	44.7	67.4	101.5	81.1	90.0	98.8	119.9	110.8	98.8	119.9	110.8
07	THERMAL & MOISTURE PROTECTION	99.4	61.9	85.2	101.9	58.7	85.6	98.8	51.9	81.1	101.5	76.9	92.2	95.9	127.1	107.6	96.0	123.9	106.5
08	OPENINGS	95.4	78.2	91.2	99.7	78.3	94.4	95.4	61.7	87.1	94.2	83.4	91.5	104.3	129.7	110.6	104.3	126.7	109.9
0920	Plaster & Gypsum Board	102.4	74.7	83.3	123.6	74.7	89.8	101.2	43.3	61.1	89.1	80.3	83.0	99.6	119.8	113.6	99.6	119.8	113.6
0950, 0980	Ceilings & Acoustic Treatment	110.4	74.7	87.7	120.1	74.7	91.2	110.4	43.3	67.6	131.3	80.3	98.8	98.7	119.8	112.1	98.7	119.8	112.1
0960	Flooring	108.5	59.0	94.8	118.9	59.0	102.3	107.2	59.0	93.9	115.3	91.3	108.7	97.6	123.1	104.6	97.6	107.5	100.3
0970, 0990	Wall Finishes & Painting/Coating	104.5	44.7	68.5	108.4	74.0	87.7	104.5	27.8	58.4	108.4	41.7	68.3	94.6	118.2	108.8	94.6	118.2	108.8
09	FINISHES	107.0	66.4	85.0	116.0	70.3	91.2	105.7	44.4	72.4	112.9	76.8	93.3	101.1	121.5	112.2	101.1	118.8	110.8
COVERS	DIVS. 10 - 14, 25, 28, 41, 43, 44	100.0	86.3	97.1	100.0	87.3	97.3	100.0	79.5	95.6	100.0	88.8	97.6	100.0	114.1	103.0	100.0	114.1	103.0
21, 22, 23	FIRE SUPPRESSION, PLUMBING & HVAC	100.0	74.5	89.5	100.0	61.4	84.1	100.0	69.8	87.6	100.0	64.1	85.2	100.1	112.8	105.4	100.1	112.8	105.3
26, 27, 3370	ELECTRICAL, COMMUNICATIONS & UTIL.	98.3	82.0	89.8	95.1	59.3	76.4	98.3	82.0	89.8	96.1	74.6	84.9	97.9	109.4	103.9	97.9	109.1	103.7
MF2004	WEIGHTED AVERAGE	102.4	74.4	90.0	105.4	69.3	89.5	99.7	64.5	84.1	101.5	76.9	90.7	101.7	118.2	109.0	100.9	117.5	108.2

DIVISION		CONNECTICUT																	
		HARTFORD			NEW BRITAIN			NEW HAVEN			NORWALK			STAMFORD			WATERBURY		
		MAT.	INST.	TOTAL	MAT.	INST.	TOTAL	MAT.	INST.	TOTAL	MAT.	INST.	TOTAL	MAT.	INST.	TOTAL	MAT.	INST.	TOTAL
015433	CONTRACTOR EQUIPMENT		100.6	100.6		100.6	100.6		101.0	101.0		100.6	100.6		100.6	100.6		100.6	100.6
0241, 31 - 34	SITE & INFRASTRUCTURE, DEMOLITION	94.4	104.3	101.3	97.1	104.3	102.1	96.9	105.1	102.6	97.5	104.4	102.3	98.1	104.4	102.5	97.3	104.4	102.2
0310	Concrete Forming & Accessories	95.6	121.6	118.1	97.9	121.6	118.4	97.4	121.5	118.3	97.7	122.1	118.8	97.6	122.4	119.1	97.7	121.7	118.5
0320	Concrete Reinforcing	110.6	127.1	119.1	110.6	127.1	119.1	110.6	127.1	119.0	110.6	127.3	119.1	110.6	127.4	119.2	110.6	127.2	119.1
0330	Cast-in-Place Concrete	97.1	117.4	104.8	96.2	117.4	104.2	97.9	108.5	101.9	99.4	128.8	110.5	101.0	129.0	111.6	101.0	117.4	107.2
03	CONCRETE	107.1	121.0	113.6	106.8	121.0	113.4	123.6	117.9	121.0	108.4	125.2	116.2	109.2	125.4	116.8	109.2	121.1	114.8
04	MASONRY	99.8	132.0	119.2	101.6	132.0	119.9	99.8	132.0	119.2	99.3	133.6	119.9	100.1	133.6	120.3	100.1	132.0	119.3
05	METALS	105.5	124.7	112.0	97.1	124.7	106.4	97.3	124.7	106.6	100.8	125.4	109.1	100.8	125.7	109.2	100.8	124.9	108.9
06	WOOD, PLASTICS & COMPOSITES	97.4	119.9	110.1	98.8	119.9	110.8	98.8	119.9	110.8	98.8	119.9	110.8	98.8	119.9	110.8	98.8	119.9	110.8
07	THERMAL & MOISTURE PROTECTION	96.7	123.9	106.9	96.0	123.9	106.5	96.0	122.5	106.0	96.0	129.2	108.6	96.0	129.3	108.5	96.0	123.9	106.5
08	OPENINGS	105.0	126.7	110.4	104.3	126.7	109.9	104.3	129.7	110.6	104.3	129.7	110.6	104.3	129.7	110.6	104.3	129.7	110.6
0920	Plaster & Gypsum Board	101.1	119.8	114.1	99.6	119.8	113.6	99.6	119.8	113.6	99.6	119.8	113.6	99.6	119.8	113.6	99.6	119.8	113.6
0950, 0980	Ceilings & Acoustic Treatment	98.7	119.8	112.1	98.7	119.8	112.1	98.7	119.8	112.1	98.7	119.8	112.1	98.7	119.8	112.1	98.7	119.8	112.1
0960	Flooring	97.6	107.5	100.3	97.6	107.5	100.3	97.6	123.1	104.6	97.6	123.1	104.6	97.6	123.1	104.6	97.6	123.1	104.6
0970, 0990	Wall Finishes & Painting/Coating	94.6	118.2	108.8	94.6	118.2	108.8	94.6	118.2	108.8	94.6	118.2	108.8	94.6	118.2	108.8	94.6	118.2	108.8
09	FINISHES	100.8	118.8	110.6	101.1	118.8	110.8	101.2	121.5	112.2	101.2	121.5	112.2	101.2	121.5	112.2	101.0	121.5	112.2
COVERS	DIVS. 10 - 14, 25, 28, 41, 43, 44	100.0	114.1	103.0	100.0	114.1	103.0	100.0	114.1	103.0	100.0	114.1	103.0	100.0	114.3	103.0	100.0	114.1	103.0
21, 22, 23	FIRE SUPPRESSION, PLUMBING & HVAC	100.1	112.8	105.3	100.1	112.8	105.3	100.1	112.8	105.3	100.1	112.8	105.4	100.1	112.9	105.4	100.1	112.8	105.3
26, 27, 3370	ELECTRICAL, COMMUNICATIONS & UTIL.	98.0	110.0	104.3	98.0	109.1	103.8	97.8	109.1	103.7	97.9	107.8	103.1	97.9	156.0	128.3	97.4	109.4	103.7
MF2004	WEIGHTED AVERAGE	101.7	117.6	108.7	100.6	117.5	108.1	102.6	117.6	109.2	101.2	118.7	109.0	101.4	125.5	112.0	101.3	118.0	108.7

DIVISION		D.C.			DELAWARE			FLORIDA											
		WASHINGTON			WILMINGTON			DAYTONA BEACH			FORT LAUDERDALE			JACKSONVILLE			MELBOURNE		
		MAT.	INST.	TOTAL	MAT.	INST.	TOTAL	MAT.	INST.	TOTAL	MAT.	INST.	TOTAL	MAT.	INST.	TOTAL	MAT.	INST.	TOTAL
015433	CONTRACTOR EQUIPMENT		104.4	104.4		117.8	117.8		97.8	97.8		90.4	90.4		97.8	97.8		97.8	97.8
0241, 31 - 34	SITE & INFRASTRUCTURE, DEMOLITION	112.3	92.8	98.7	90.9	112.3	105.8	116.3	89.2	97.5	101.6	77.3	84.7	116.4	88.5	97.0	124.8	88.5	99.5
0310	Concrete Forming & Accessories	101.3	82.1	84.7	100.6	101.9	101.7	94.8	75.7	78.2	93.5	76.5	78.8	94.6	60.1	64.8	91.5	77.5	79.4
0320	Concrete Reinforcing	103.0	89.0	95.8	93.9	102.4	98.2	86.0	80.7	83.3	86.0	80.4	83.1	86.0	62.8	74.1	86.9	80.7	83.8
0330	Cast-in-Place Concrete	126.8	88.3	112.2	89.5	102.7	94.5	93.4	76.7	87.1	98.0	84.6	92.9	94.3	73.8	86.5	112.6	79.3	100.0
03	CONCRETE	117.1	86.8	102.9	95.5	103.2	99.1	89.3	78.0	84.0	91.5	81.1	86.7	89.8	66.7	79.0	101.1	79.7	91.1
04	MASONRY	104.8	83.0	91.7	108.7	98.7	102.7	86.6	73.1	78.4	86.8	76.2	80.4	86.4	68.1	75.4	84.3	77.6	80.3
05	METALS	94.3	108.4	99.0	104.7	117.1	108.9	98.1	93.8	96.7	98.1	95.7	97.3	96.8	85.2	92.8	106.1	93.9	102.0
06	WOOD, PLASTICS & COMPOSITES	99.9	81.8	89.6	101.1	101.2	101.2	95.6	77.5	85.4	92.0	75.2	82.5	95.6	57.9	74.3	91.6	77.5	83.6
07	THERMAL & MOISTURE PROTECTION	102.4	86.1	96.3	99.4	115.3	105.4	93.9	81.8	89.3	93.9	89.9	92.4	94.1	68.1	84.2	94.2	84.4	90.5
08	OPENINGS	107.8	91.0	103.6	95.9	108.8	99.1	100.1	74.2	93.7	97.5	73.3	91.5	100.1	57.2	89.5	99.3	77.9	94.0
0920	Plaster & Gypsum Board	107.5	81.2	89.3	111.7	101.1	104.4	95.3	77.4	82.9	94.9	75.0	81.1	95.3	57.1	68.9	90.8	77.4	81.5
0950, 0980	Ceilings & Acoustic Treatment	107.9	81.2	90.9	92.2	101.1	97.9	98.4	77.4	85.0	98.4	75.0	83.5	98.4	57.1	72.1	93.2	77.4	83.1
0960	Flooring	127.0	97.0	118.7	83.3	108.2	90.2	116.6	79.3	106.3	116.6	74.7	105.0	116.6	69.5	103.6	113.4	79.3	104.0
0970, 0990	Wall Finishes & Painting/Coating	125.0	84.5	100.6	92.5	102.7	98.7	111.0	82.3	93.7	106.1	58.2	77.3	111.0	51.6	75.3	111.0	102.1	105.6
09	FINISHES	108.9	84.6	95.7	97.3	102.5	100.2	105.7	77.0	90.1	103.1	73.5	87.0	105.7	60.3	81.1	103.6	80.4	91.0
COVERS	DIVS. 10 - 14, 25, 28, 41, 43, 44	100.0	98.3	99.6	100.0	101.2	100.2	100.0	85.7	97.0	100.0	89.1	97.7	100.0	81.6	96.1	100.0	87.4	97.3
21, 22, 23	FIRE SUPPRESSION, PLUMBING & HVAC	100.2	93.1	97.3	100.1	115.7	106.5	99.8	66.8	86.2	99.8	68.7	87.0	99.8	49.2	79.0	99.8	69.1	87.2
26, 27, 3370	ELECTRICAL, COMMUNICATIONS & UTIL.	104.1	99.9	101.9	103.7	109.4	106.7	98.4	61.6	79.1	98.4	76.0	86.7	98.1	69.8	83.3	99.6	71.7	85.0
MF2004	WEIGHTED AVERAGE	104.0	92.2	98.8	100.1	108.7	103.9	98.2	75.3	88.1	97.5	77.4	88.7	98.0	66.3	84.0	100.8	78.5	91.0

	DIVISION	FLORIDA																	
		MIAMI			ORLANDO			PANAMA CITY			PENSACOLA			ST. PETERSBURG			TALLAHASSEE		
		MAT.	INST.	TOTAL	MAT.	INST.	TOTAL	MAT.	INST.	TOTAL	MAT.	INST.	TOTAL	MAT.	INST.	TOTAL	MAT.	INST.	TOTAL
015433	CONTRACTOR EQUIPMENT		90.4	90.4		97.8	97.8		97.8	97.8		97.8	97.8		97.8	97.8		97.8	97.8
0241, 31 - 34	SITE & INFRASTRUCTURE, DEMOLITION	105.1	77.1	85.6	119.0	88.3	97.6	131.2	86.5	100.1	130.8	87.8	100.9	117.2	88.1	96.9	114.5	87.5	95.7
0310	Concrete Forming & Accessories	97.7	77.0	79.8	94.5	80.2	82.1	94.0	47.4	53.7	92.0	57.4	62.0	93.0	52.2	57.7	96.9	48.7	55.2
0320	Concrete Reinforcing	86.0	80.5	83.2	86.0	77.7	81.7	89.7	58.9	74.0	92.1	58.8	75.1	89.1	66.2	77.4	86.0	62.4	73.9
0330	Cast-in-Place Concrete	106.3	85.7	98.5	123.1	76.5	105.5	99.1	52.8	81.6	122.4	71.3	103.1	105.7	69.5	92.0	103.9	60.5	87.5
03	CONCRETE	95.9	81.7	89.2	105.3	79.4	93.2	98.0	52.7	76.8	108.9	63.9	87.9	95.7	62.4	80.2	94.6	57.1	77.1
04	MASONRY	86.1	80.2	82.6	90.6	73.1	80.1	90.8	52.2	67.6	109.1	60.5	79.9	132.7	54.5	85.7	89.4	56.1	69.4
05	METALS	102.8	94.4	99.9	103.7	92.2	99.8	97.9	71.4	89.0	99.1	83.5	93.8	101.0	85.8	95.9	91.9	84.5	89.4
06	WOOD, PLASTICS & COMPOSITES	97.3	75.2	84.8	95.4	84.3	89.1	94.6	46.6	67.5	93.0	57.6	73.0	93.3	50.7	69.2	93.2	46.1	66.5
07	THERMAL & MOISTURE PROTECTION	102.7	83.6	95.5	95.2	82.5	90.4	94.3	55.2	79.5	94.2	67.2	84.0	93.8	62.4	82.0	99.8	78.4	91.7
08	OPENINGS	97.5	73.2	91.5	100.1	74.7	93.8	97.7	41.2	83.7	97.7	57.9	87.8	98.7	54.9	87.9	100.6	50.5	88.2
0920	Plaster & Gypsum Board	92.8	75.0	80.5	105.8	84.3	90.9	93.4	45.5	60.3	94.3	56.9	68.4	92.5	49.8	62.9	111.1	45.0	65.4
0950, 0980	Ceilings & Acoustic Treatment	94.1	75.0	81.9	98.4	84.3	89.4	93.2	45.5	62.8	93.2	56.9	70.1	93.2	49.8	65.5	98.4	45.0	64.4
0960	Flooring	124.0	78.6	111.5	116.8	79.3	106.4	116.0	46.3	96.8	110.4	67.8	98.6	115.1	59.4	99.7	116.6	60.6	101.2
0970, 0990	Wall Finishes & Painting/Coating	106.1	58.2	77.3	111.0	61.1	81.0	111.0	52.3	75.7	111.0	50.9	74.9	111.0	53.7	76.5	111.0	48.1	73.2
09	FINISHES	104.4	74.7	88.3	106.9	78.6	91.6	105.5	47.5	74.0	103.5	58.1	78.9	103.6	52.8	76.0	107.2	49.6	75.9
COVERS	DIVS. 10 - 14, 25, 28, 41, 43, 44	100.0	89.8	97.8	100.0	86.4	97.1	100.0	54.2	90.3	100.0	54.4	90.3	100.0	61.5	91.8	100.0	77.8	95.3
21, 22, 23	FIRE SUPPRESSION, PLUMBING & HVAC	99.8	74.4	89.3	99.7	64.1	85.0	99.8	42.1	76.1	99.8	59.1	83.0	99.8	57.2	82.2	99.8	44.0	76.8
26, 27, 3370	ELECTRICAL, COMMUNICATIONS & UTIL.	106.3	78.8	91.9	106.5	48.4	76.1	97.0	46.4	70.6	101.1	58.6	78.9	98.7	51.5	74.0	106.4	46.0	74.8
MF2004	WEIGHTED AVERAGE	100.1	79.4	90.9	102.2	73.2	89.4	99.5	53.1	79.0	102.2	64.4	85.5	101.6	61.6	83.9	99.2	57.7	80.9

	DIVISION	FLORIDA			GEORGIA														
		TAMPA			ALBANY			ATLANTA			AUGUSTA			COLUMBUS			MACON		
		MAT.	INST.	TOTAL	MAT.	INST.	TOTAL	MAT.	INST.	TOTAL	MAT.	INST.	TOTAL	MAT.	INST.	TOTAL	MAT.	INST.	TOTAL
015433	CONTRACTOR EQUIPMENT		97.8	97.8		91.2	91.2		93.1	93.1		92.6	92.6		91.2	91.2		102.1	102.1
0241, 31 - 34	SITE & INFRASTRUCTURE, DEMOLITION	117.6	88.8	97.5	101.9	78.4	85.6	103.6	94.4	97.2	100.0	92.1	94.5	101.8	78.7	85.7	102.3	94.2	96.6
0310	Concrete Forming & Accessories	95.6	84.3	85.8	94.1	47.7	54.0	96.2	75.0	77.9	94.0	68.3	71.8	94.0	59.7	64.4	93.3	56.9	61.8
0320	Concrete Reinforcing	86.0	104.3	95.3	85.8	87.1	86.5	90.5	88.8	89.6	91.5	75.8	83.5	86.0	88.1	87.0	87.1	87.4	87.2
0330	Cast-in-Place Concrete	103.2	75.8	92.9	99.8	49.0	80.6	110.5	73.0	96.3	104.4	51.2	84.2	99.3	58.1	83.7	98.0	51.6	80.5
03	CONCRETE	94.2	86.0	90.4	92.3	57.2	75.9	99.8	77.0	89.1	95.2	64.1	80.7	92.1	65.8	79.8	91.6	62.2	77.9
04	MASONRY	87.8	84.4	85.8	89.5	43.9	62.1	89.5	71.4	78.6	89.6	43.5	61.9	88.7	61.6	72.4	101.2	41.6	65.4
05	METALS	100.0	103.8	101.3	97.5	88.2	94.3	90.9	81.0	87.6	89.8	72.7	84.0	96.9	90.6	94.7	93.2	89.8	92.1
06	WOOD, PLASTICS & COMPOSITES	96.8	86.8	91.1	94.7	42.4	65.1	94.0	76.7	84.2	91.4	73.8	81.4	94.7	58.3	74.1	104.5	57.2	77.7
07	THERMAL & MOISTURE PROTECTION	94.0	94.1	94.0	91.6	62.7	80.7	91.8	76.4	86.0	91.4	58.5	79.0	91.5	63.8	81.0	90.1	65.3	80.7
08	OPENINGS	100.1	83.8	96.1	97.5	49.0	85.5	98.7	73.8	92.5	93.0	65.6	86.2	97.5	59.6	88.1	95.7	58.1	86.4
0920	Plaster & Gypsum Board	95.3	86.9	89.5	95.1	41.2	57.8	113.6	76.4	87.9	112.3	73.4	85.4	95.1	57.5	69.1	104.4	56.4	71.2
0950, 0980	Ceilings & Acoustic Treatment	98.4	86.9	91.1	97.5	41.2	61.6	106.8	76.4	87.4	107.7	73.4	85.9	97.5	57.5	72.1	92.6	56.4	69.6
0960	Flooring	116.6	60.9	101.2	116.6	33.2	93.6	89.4	70.2	84.1	88.3	42.3	75.6	116.6	54.3	99.4	90.9	39.1	76.5
0970, 0990	Wall Finishes & Painting/Coating	111.0	66.2	84.0	106.1	55.5	75.6	97.5	86.2	90.7	97.5	40.6	63.3	106.1	67.5	82.9	108.2	55.5	76.5
09	FINISHES	105.7	77.9	90.6	102.8	43.9	70.8	97.7	75.2	85.5	97.2	61.1	77.6	102.8	58.7	78.8	91.0	52.3	70.0
COVERS	DIVS. 10 - 14, 25, 28, 41, 43, 44	100.0	86.5	97.1	100.0	77.2	95.2	100.0	81.8	96.1	100.0	58.1	91.1	100.0	79.2	95.6	100.0	77.0	95.1
21, 22, 23	FIRE SUPPRESSION, PLUMBING & HVAC	99.8	91.7	96.4	99.7	69.9	87.4	100.0	76.0	90.1	100.0	69.4	87.4	99.8	52.6	80.4	99.8	67.7	86.6
26, 27, 3370	ELECTRICAL, COMMUNICATIONS & UTIL.	98.4	51.5	73.9	93.8	57.0	74.6	98.4	77.1	87.3	99.7	69.3	83.8	94.0	74.3	83.7	92.8	63.6	77.5
MF2004	WEIGHTED AVERAGE	99.2	83.3	92.2	97.1	61.7	81.5	97.4	77.8	88.8	96.0	66.5	83.0	97.0	66.2	83.4	95.8	65.8	82.6

	DIVISION	GEORGIA						HAWAII						IDAHO					
		SAVANNAH			VALDOSTA			HONOLULU			STATES & POSS., GUAM			BOISE			LEWISTON		
		MAT.	INST.	TOTAL	MAT.	INST.	TOTAL	MAT.	INST.	TOTAL	MAT.	INST.	TOTAL	MAT.	INST.	TOTAL	MAT.	INST.	TOTAL
015433	CONTRACTOR EQUIPMENT		92.2	92.2		91.2	91.2		99.8	99.8		166.6	166.6		100.7	100.7		94.6	94.6
0241, 31 - 34	SITE & INFRASTRUCTURE, DEMOLITION	104.4	79.3	86.9	112.2	78.5	88.7	134.6	106.3	114.9	172.3	105.9	126.0	76.7	99.8	92.8	83.2	95.3	91.6
0310	Concrete Forming & Accessories	94.1	53.0	58.6	84.7	46.8	51.9	116.4	136.8	134.0	111.4	62.8	69.3	99.9	72.2	75.9	122.7	60.8	69.1
0320	Concrete Reinforcing	86.8	76.0	81.3	87.8	73.8	80.7	113.0	107.1	110.0	205.9	33.5	117.9	98.3	69.2	83.5	106.5	91.0	98.6
0330	Cast-in-Place Concrete	115.9	50.5	91.2	97.6	53.1	80.8	168.1	121.2	150.3	171.1	113.1	149.1	92.7	77.9	87.1	104.2	86.6	97.5
03	CONCRETE	100.4	57.0	80.5	96.3	55.7	77.3	139.9	124.6	132.7	154.2	75.1	117.2	102.5	73.8	89.1	113.7	75.8	96.0
04	MASONRY	88.6	54.3	68.0	94.3	49.3	67.3	122.9	124.9	124.1	187.2	48.4	103.8	126.2	62.0	87.6	128.1	82.2	100.5
05	METALS	93.6	84.7	90.6	96.5	83.8	92.2	131.3	102.1	121.5	142.2	80.0	121.2	102.5	72.9	92.5	96.4	84.2	92.3
06	WOOD, PLASTICS & COMPOSITES	110.4	50.0	76.2	83.3	41.3	59.5	108.1	141.5	127.0	108.4	64.2	83.3	93.5	72.9	81.8	104.9	53.2	75.6
07	THERMAL & MOISTURE PROTECTION	93.9	58.2	80.4	91.7	63.5	81.1	116.5	123.2	119.0	122.9	70.2	103.0	93.4	68.3	83.9	144.8	75.1	118.5
08	OPENINGS	98.7	50.3	86.7	93.0	44.7	81.0	106.0	129.9	111.9	105.5	54.3	92.8	93.0	68.3	86.9	112.6	60.0	99.6
0920	Plaster & Gypsum Board	94.1	49.0	62.9	87.3	40.1	54.6	133.6	142.5	139.8	211.9	51.6	100.9	89.8	72.0	77.5	157.4	51.8	84.3
0950, 0980	Ceilings & Acoustic Treatment	98.4	49.0	66.9	93.2	40.1	59.4	116.2	142.5	132.9	233.1	51.6	117.5	125.5	72.0	91.4	161.2	51.8	91.5
0960	Flooring	117.4	49.9	98.8	108.8	41.7	90.3	165.2	128.4	155.0	164.9	50.0	133.2	99.5	68.4	90.9	141.8	86.1	126.4
0970, 0990	Wall Finishes & Painting/Coating	108.0	59.1	78.6	106.1	55.5	75.6	104.1	138.8	124.9	105.6	39.9	66.1	97.9	43.8	65.4	118.7	65.2	86.5
09	FINISHES	103.5	52.3	75.7	99.3	45.1	69.9	132.7	137.1	135.1	197.0	58.7	121.9	104.7	69.2	85.4	174.1	64.6	114.7
COVERS	DIVS. 10 - 14, 25, 28, 41, 43, 44	100.0	57.3	90.9	100.0	56.3	90.7	100.0	115.1	103.2	100.0	86.0	97.0	100.0	67.2	93.0	100.0	70.2	93.7
21, 22, 23	FIRE SUPPRESSION, PLUMBING & HVAC	99.8	60.9	83.8	99.8	69.8	87.4	100.1	107.0	102.9	102.4	42.2	77.7	100.0	65.2	85.7	101.0	78.7	91.8
26, 27, 3370	ELECTRICAL, COMMUNICATIONS & UTIL.	97.2	58.1	76.8	92.0	57.7	74.1	112.8	116.6	114.7	161.4	46.2	101.2	96.1	74.1	84.6	85.9	75.0	80.2
MF2004	WEIGHTED AVERAGE	98.4	61.4	82.0	97.1	61.1	81.2	116.9	117.9	117.3	135.6	61.7	103.0	100.4	71.8	87.7	110.1	76.8	95.4

	DIVISION	IDAHO						ILLINOIS											
		POCATELLO			TWIN FALLS			CHICAGO			DECATUR			EAST ST. LOUIS			JOLIET		
		MAT.	INST.	TOTAL	MAT.	INST.	TOTAL	MAT.	INST.	TOTAL	MAT.	INST.	TOTAL	MAT.	INST.	TOTAL	MAT.	INST.	TOTAL
015433	CONTRACTOR EQUIPMENT		100.7	100.7		100.7	100.7		93.4	93.4		101.9	101.9		107.7	107.7		91.4	91.4
0241, 31 - 34	SITE & INFRASTRUCTURE, DEMOLITION	77.6	99.8	93.0	83.9	98.0	93.7	103.4	93.6	96.6	89.5	97.6	95.1	105.2	98.4	100.5	103.2	93.1	96.2
0310	Concrete Forming & Accessories	100.1	71.9	75.7	100.8	30.6	40.0	101.7	154.8	147.6	100.5	114.4	112.6	97.0	111.0	109.1	103.9	158.6	151.2
0320	Concrete Reinforcing	98.7	69.5	83.8	100.7	67.8	83.9	98.8	151.1	125.5	96.7	104.7	100.8	98.6	105.2	102.0	98.8	148.6	124.2
0330	Cast-in-Place Concrete	95.3	77.8	88.7	97.8	40.6	76.1	109.7	143.5	122.5	100.2	107.4	102.9	93.8	112.2	100.7	109.6	147.2	123.8
03	CONCRETE	101.7	73.6	88.6	109.8	42.2	78.2	103.6	149.0	124.8	97.2	110.2	103.3	88.2	111.0	98.9	103.7	151.3	126.0
04	MASONRY	123.5	67.9	90.1	126.3	32.5	69.9	97.7	153.4	131.2	71.8	109.2	94.3	75.3	116.5	100.1	100.7	149.8	130.2
05	METALS	110.4	73.0	97.8	110.5	69.4	96.6	92.0	130.3	105.0	94.8	108.2	99.3	92.4	119.1	101.4	89.8	127.1	102.4
06	WOOD, PLASTICS & COMPOSITES	93.5	72.9	81.8	94.1	29.6	57.6	106.0	154.3	133.3	101.0	114.6	108.7	99.0	108.5	104.4	107.8	159.9	137.3
07	THERMAL & MOISTURE PROTECTION	93.4	63.7	82.2	94.1	39.9	73.7	102.2	143.2	117.7	100.2	103.8	101.6	94.7	110.2	100.6	102.2	141.7	117.1
08	OPENINGS	93.7	62.7	86.0	97.0	36.9	82.1	101.8	153.2	114.6	96.6	112.0	100.4	86.4	114.1	93.3	99.8	155.7	113.6
0920	Plaster & Gypsum Board	87.0	72.0	76.6	86.9	27.4	45.7	95.0	155.8	137.0	99.6	115.0	110.2	95.7	108.7	104.7	92.0	161.7	140.2
0950, 0980	Ceilings & Acoustic Treatment	131.3	72.0	93.5	125.3	27.4	62.9	108.2	155.8	138.5	98.1	115.0	108.8	92.0	108.7	102.6	108.2	161.7	142.2
0960	Flooring	102.7	68.4	93.2	104.0	46.4	88.1	91.3	152.8	108.3	100.9	106.8	102.6	110.4	118.0	112.5	90.9	142.0	105.0
0970, 0990	Wall Finishes & Painting/Coating	97.8	45.6	66.4	97.9	26.9	55.2	93.0	152.1	128.6	96.2	107.7	103.1	103.9	109.2	107.1	91.2	142.2	121.9
09	FINISHES	106.6	69.4	86.4	106.3	32.2	66.1	97.6	155.2	128.9	99.9	113.3	107.1	98.7	112.2	106.1	97.1	155.5	128.8
COVERS	DIVS. 10 - 14, 25, 28, 41, 43, 44	100.0	67.2	93.0	100.0	38.3	86.9	100.0	126.3	105.6	100.0	98.1	99.6	100.0	93.7	98.7	100.0	126.7	105.7
21, 22, 23	FIRE SUPPRESSION, PLUMBING & HVAC	100.0	65.2	85.6	100.0	54.2	81.1	100.1	134.5	114.2	100.0	100.1	100.0	100.0	95.4	98.1	100.2	129.5	112.2
26, 27, 3370	ELECTRICAL, COMMUNICATIONS & UTIL.	93.1	69.9	81.0	89.1	61.1	74.5	95.8	130.5	113.9	96.7	94.3	95.4	93.6	103.5	98.8	95.1	126.5	111.5
MF2004	WEIGHTED AVERAGE	101.2	71.4	88.1	102.5	51.8	80.1	99.0	137.6	116.0	96.5	104.5	100.0	94.0	106.6	99.6	98.5	135.8	115.0

	DIVISION	ILLINOIS									INDIANA								
		PEORIA			ROCKFORD			SPRINGFIELD			ANDERSON			BLOOMINGTON			EVANSVILLE		
		MAT.	INST.	TOTAL	MAT.	INST.	TOTAL	MAT.	INST.	TOTAL	MAT.	INST.	TOTAL	MAT.	INST.	TOTAL	MAT.	INST.	TOTAL
015433	CONTRACTOR EQUIPMENT		101.2	101.2		101.2	101.2		101.9	101.9		96.3	96.3		86.8	86.8		117.5	117.5
0241, 31 - 34	SITE & INFRASTRUCTURE, DEMOLITION	98.2	96.5	97.0	97.4	97.6	97.5	96.6	97.7	97.3	87.7	94.6	92.5	76.4	94.7	89.2	81.0	125.6	112.0
0310	Concrete Forming & Accessories	98.9	114.5	112.4	102.9	132.7	128.7	101.0	112.4	110.8	94.7	79.5	81.6	98.6	74.9	78.1	93.4	79.9	81.7
0320	Concrete Reinforcing	96.7	103.1	100.0	91.5	138.5	115.5	96.7	104.8	100.8	95.7	79.9	87.6	91.2	79.7	85.4	100.0	78.0	88.7
0330	Cast-in-Place Concrete	97.9	112.0	103.2	100.2	118.0	107.0	89.1	106.0	95.5	96.2	81.8	90.7	95.2	78.2	88.8	91.0	87.7	89.7
03	CONCRETE	96.0	111.6	103.3	96.6	128.5	111.5	91.7	108.8	99.7	91.6	80.9	86.6	99.0	76.9	88.7	99.5	82.4	91.5
04	MASONRY	113.4	108.1	110.2	88.0	132.5	114.7	76.6	110.0	96.7	82.6	83.2	83.0	86.0	80.9	83.0	82.2	84.6	83.7
05	METALS	94.8	109.5	99.7	94.7	127.3	105.7	96.9	108.6	100.9	96.4	88.0	93.5	94.8	77.0	88.8	88.7	84.9	87.4
06	WOOD, PLASTICS & COMPOSITES	101.0	111.5	106.9	101.0	131.5	118.2	102.3	110.9	107.2	103.7	78.8	89.6	117.9	72.2	92.0	96.7	77.9	86.1
07	THERMAL & MOISTURE PROTECTION	100.3	110.2	104.1	102.8	125.8	111.5	100.1	109.2	103.5	96.5	80.3	90.4	93.3	80.6	88.5	97.3	85.4	92.8
08	OPENINGS	96.7	110.2	100.1	96.7	131.7	105.4	97.7	110.0	100.8	95.7	80.6	91.9	103.5	77.0	96.9	95.7	79.6	91.7
0920	Plaster & Gypsum Board	99.6	111.7	108.0	99.6	132.3	122.2	95.4	111.1	106.3	92.7	78.6	82.9	87.6	72.0	76.8	83.1	76.6	78.6
0950, 0980	Ceilings & Acoustic Treatment	98.1	111.7	106.8	98.1	132.3	119.9	96.4	111.1	105.8	87.5	78.6	81.8	76.7	72.0	73.7	81.4	76.6	78.4
0960	Flooring	100.9	112.1	104.0	100.9	115.0	104.8	100.9	106.4	102.4	93.2	83.6	90.5	104.5	80.8	98.0	98.0	86.9	94.9
0970, 0990	Wall Finishes & Painting/Coating	96.2	126.2	114.3	96.2	133.7	118.8	96.2	107.3	102.9	93.8	73.8	81.8	90.4	87.8	88.8	96.4	87.2	90.9
09	FINISHES	100.0	115.5	108.4	100.0	130.2	116.4	98.8	111.2	105.5	92.0	79.6	85.3	93.5	76.7	84.4	92.0	81.8	86.5
COVERS	DIVS. 10 - 14, 25, 28, 41, 43, 44	100.0	105.6	101.2	100.0	101.9	100.4	100.0	97.9	99.6	100.0	91.4	98.2	100.0	90.2	97.9	100.0	81.3	96.0
21, 22, 23	FIRE SUPPRESSION, PLUMBING & HVAC	100.0	107.4	103.0	100.1	117.6	107.3	99.9	104.5	101.8	99.8	77.0	90.4	99.7	82.0	92.4	99.9	82.9	92.9
26, 27, 3370	ELECTRICAL, COMMUNICATIONS & UTIL.	95.8	98.9	97.4	96.1	123.9	110.7	101.6	96.2	98.8	86.1	90.4	88.4	100.1	86.5	93.0	95.8	83.9	89.6
MF2004	WEIGHTED AVERAGE	98.7	107.3	102.5	97.5	122.8	108.7	97.0	105.4	100.7	94.5	83.5	89.7	97.4	81.6	90.4	95.1	86.6	91.4

	DIVISION	INDIANA																	
		FORT WAYNE			GARY			INDIANAPOLIS			MUNCIE			SOUTH BEND			TERRE HAUTE		
		MAT.	INST.	TOTAL	MAT.	INST.	TOTAL	MAT.	INST.	TOTAL	MAT.	INST.	TOTAL	MAT.	INST.	TOTAL	MAT.	INST.	TOTAL
015433	CONTRACTOR EQUIPMENT		96.3	96.3		96.3	96.3		91.7	91.7		95.8	95.8		105.1	105.1		117.5	117.5
0241, 31 - 34	SITE & INFRASTRUCTURE, DEMOLITION	88.7	94.6	92.8	88.2	98.0	95.0	87.4	97.9	94.7	76.4	94.8	89.2	90.5	95.0	93.6	82.9	125.8	112.8
0310	Concrete Forming & Accessories	93.5	75.9	78.3	94.8	113.2	110.7	95.4	83.4	85.0	92.2	79.2	81.0	96.2	79.6	81.8	94.3	81.6	83.3
0320	Concrete Reinforcing	95.7	79.0	87.2	95.7	103.7	99.8	95.1	80.0	87.4	101.0	79.9	90.2	95.7	77.4	86.3	100.0	80.1	89.8
0330	Cast-in-Place Concrete	102.2	77.8	93.0	100.5	111.0	104.5	92.5	83.7	89.2	100.1	81.0	92.9	97.4	81.7	91.5	88.1	90.3	88.9
03	CONCRETE	94.5	77.8	86.7	93.7	110.5	101.6	93.6	82.6	88.4	97.9	80.5	89.8	88.3	81.2	85.0	102.2	84.5	93.9
04	MASONRY	86.2	81.2	83.2	83.8	113.0	101.4	89.9	86.0	87.6	88.0	83.2	85.1	82.9	81.4	82.0	88.9	83.4	85.6
05	METALS	96.4	87.9	93.5	96.4	105.2	99.4	98.6	78.7	91.9	96.7	88.1	93.8	96.4	99.5	97.4	89.3	86.9	88.5
06	WOOD, PLASTICS & COMPOSITES	103.4	74.9	87.3	101.7	112.3	107.7	100.5	82.7	90.4	113.3	78.5	93.6	101.1	79.0	88.6	98.5	80.4	88.3
07	THERMAL & MOISTURE PROTECTION	96.3	83.9	91.6	95.4	109.7	100.8	94.7	83.7	90.6	96.0	80.3	90.0	91.1	85.3	88.9	97.4	84.1	92.4
08	OPENINGS	95.7	74.9	90.5	95.7	114.2	100.3	103.3	82.7	98.2	98.5	80.5	94.0	89.2	79.1	86.7	96.3	81.5	92.6
0920	Plaster & Gypsum Board	91.8	74.6	79.9	87.1	113.1	105.1	90.4	82.4	84.9	83.1	78.6	80.0	84.4	78.8	80.6	83.1	79.2	80.4
0950, 0980	Ceilings & Acoustic Treatment	87.5	74.6	79.3	87.5	113.1	103.8	91.0	82.4	85.5	77.5	78.6	78.2	83.2	78.8	80.4	81.4	79.2	80.0
0960	Flooring	93.2	80.7	89.8	93.2	119.0	100.3	93.0	90.0	92.2	97.5	83.6	93.6	91.6	85.7	90.0	98.0	90.0	95.8
0970, 0990	Wall Finishes & Painting/Coating	93.8	78.4	84.5	93.8	115.4	106.8	93.8	87.8	90.2	90.4	73.8	80.4	93.3	85.1	88.4	96.4	88.4	91.5
09	FINISHES	91.8	77.1	83.8	91.3	114.7	104.0	92.9	85.2	88.7	90.7	79.4	84.6	89.5	81.3	85.0	92.0	83.7	87.5
COVERS	DIVS. 10 - 14, 25, 28, 41, 43, 44	100.0	91.9	98.3	100.0	108.7	101.9	100.0	92.8	98.5	100.0	90.8	98.0	100.0	76.9	95.1	100.0	94.2	98.8
21, 22, 23	FIRE SUPPRESSION, PLUMBING & HVAC	99.8	72.5	88.6	99.8	107.6	103.0	99.8	82.9	92.8	99.7	76.9	90.3	99.7	79.7	91.5	99.9	81.9	92.5
26, 27, 3370	ELECTRICAL, COMMUNICATIONS & UTIL.	86.8	76.9	81.6	99.0	103.9	101.5	101.0	90.4	95.5	91.1	76.6	83.5	106.3	88.8	97.2	93.9	86.6	90.1
MF2004	WEIGHTED AVERAGE	95.2	79.6	88.3	96.1	108.3	101.5	97.7	85.7	92.4	96.0	81.5	89.6	95.1	84.7	90.5	95.9	87.9	92.3

	DIVISION	IOWA																	
		CEDAR RAPIDS			COUNCIL BLUFFS			DAVENPORT			DES MOINES			DUBUQUE			SIOUX CITY		
		MAT.	INST.	TOTAL	MAT.	INST.	TOTAL	MAT.	INST.	TOTAL	MAT.	INST.	TOTAL	MAT.	INST.	TOTAL	MAT.	INST.	TOTAL
015433	CONTRACTOR EQUIPMENT		96.5	96.5		96.1	96.1		100.0	100.0		101.7	101.7		95.2	95.2		100.0	100.0
0241, 31 - 34	SITE & INFRASTRUCTURE, DEMOLITION	97.7	95.3	96.1	105.4	92.3	96.3	95.5	99.3	98.1	87.9	99.6	96.0	95.5	92.2	93.2	106.8	95.5	99.0
0310	Concrete Forming & Accessories	102.3	73.9	77.7	86.8	51.9	56.6	102.0	84.3	86.7	103.2	69.3	73.9	88.0	65.7	68.7	102.3	59.3	65.1
0320	Concrete Reinforcing	93.8	76.7	85.1	95.7	71.8	83.5	93.8	88.1	90.9	93.8	73.4	83.4	92.6	76.6	84.4	93.8	57.8	75.5
0330	Cast-in-Place Concrete	107.1	75.8	95.3	111.4	73.8	97.2	103.2	88.7	97.7	98.9	89.1	95.2	104.9	85.7	97.6	106.2	51.1	85.4
03	CONCRETE	101.1	75.8	89.3	103.7	64.5	85.4	99.1	87.2	93.6	95.9	77.8	87.4	98.0	75.6	87.5	100.4	57.4	80.3
04	MASONRY	106.9	72.9	86.5	107.7	73.3	87.0	102.8	81.8	90.1	97.9	66.8	79.2	107.9	64.7	81.9	100.3	51.0	70.6
05	METALS	88.3	88.2	88.3	92.9	85.6	90.4	88.3	98.2	91.7	88.5	87.8	88.2	86.9	87.7	87.2	88.3	77.8	84.8
06	WOOD, PLASTICS & COMPOSITES	106.2	73.1	87.5	87.3	45.6	63.7	106.2	83.0	93.0	106.0	69.0	85.1	88.9	65.5	75.6	106.2	59.6	79.8
07	THERMAL & MOISTURE PROTECTION	102.6	74.7	92.0	102.1	61.9	86.9	102.1	83.5	95.1	103.6	71.6	91.5	102.3	65.8	88.5	102.1	50.1	82.5
08	OPENINGS	99.3	76.2	93.6	98.2	60.2	88.8	99.3	85.5	95.9	99.3	73.6	92.9	98.3	74.0	92.2	99.3	57.2	88.8
0920	Plaster & Gypsum Board	101.4	72.6	81.5	92.8	44.2	59.2	101.4	82.5	88.3	90.4	68.2	75.0	93.3	64.7	73.5	101.4	58.5	71.7
0950, 0980	Ceilings & Acoustic Treatment	111.3	72.6	86.6	107.0	44.2	67.0	111.3	82.5	93.0	107.0	68.2	82.3	107.0	64.7	80.1	111.3	58.5	77.7
0960	Flooring	131.3	48.0	108.3	105.1	39.2	86.9	116.0	89.4	108.7	115.7	35.8	93.6	120.4	33.4	96.4	116.5	50.6	98.3
0970, 0990	Wall Finishes & Painting/Coating	105.6	70.8	84.7	98.6	67.9	80.1	102.1	85.5	92.1	102.1	73.8	85.1	104.6	70.8	84.3	103.1	58.9	76.5
09	FINISHES	113.8	68.0	88.9	103.2	49.2	73.9	109.0	85.1	96.0	105.3	62.8	82.2	108.3	59.3	81.7	110.8	57.6	81.9
COVERS	DIVS. 10 - 14, 25, 28, 41, 43, 44	100.0	79.9	95.7	100.0	78.7	95.5	100.0	83.0	96.4	100.0	78.5	95.4	100.0	76.8	95.1	100.0	75.9	94.9
21, 22, 23	FIRE SUPPRESSION, PLUMBING & HVAC	100.1	76.6	90.4	100.1	73.3	89.1	100.1	82.9	93.0	100.0	66.4	86.2	100.1	68.7	87.2	100.1	68.1	87.0
26, 27, 3370	ELECTRICAL, COMMUNICATIONS & UTIL.	95.3	75.5	84.9	100.9	79.2	89.6	93.3	83.9	88.4	108.1	69.9	88.2	99.4	72.1	85.1	95.3	68.7	81.4
MF2004	WEIGHTED AVERAGE	99.5	77.5	89.8	100.2	71.6	87.5	98.4	86.7	93.2	98.7	73.6	87.7	98.6	72.6	87.2	99.1	66.1	84.5

	DIVISION	IOWA			KANSAS														
		WATERLOO			DODGE CITY			KANSAS CITY			SALINA			TOPEKA			WICHITA		
		MAT.	INST.	TOTAL	MAT.	INST.	TOTAL	MAT.	INST.	TOTAL	MAT.	INST.	TOTAL	MAT.	INST.	TOTAL	MAT.	INST.	TOTAL
015433	CONTRACTOR EQUIPMENT		100.0	100.0		102.9	102.9		99.7	99.7		102.9	102.9		101.2	101.2		102.9	102.9
0241, 31 - 34	SITE & INFRASTRUCTURE, DEMOLITION	96.0	94.7	95.1	112.8	92.5	98.7	93.0	92.1	92.4	102.5	92.9	95.8	96.7	90.5	92.4	95.2	92.7	93.5
0310	Concrete Forming & Accessories	102.7	41.6	49.8	98.3	58.3	63.7	102.1	98.8	99.2	95.2	47.1	53.6	102.4	44.0	51.9	101.3	48.6	55.7
0320	Concrete Reinforcing	93.8	76.3	84.9	104.7	52.1	77.9	99.5	92.0	95.6	104.1	67.2	85.3	96.7	94.1	95.4	96.7	77.2	86.7
0330	Cast-in-Place Concrete	107.1	43.1	82.9	116.3	50.9	91.5	91.2	98.0	93.8	100.9	44.0	79.3	92.3	48.3	75.6	88.0	52.0	74.4
03	CONCRETE	101.1	50.3	77.4	114.5	55.9	87.1	95.4	97.5	96.4	101.5	51.4	78.1	93.4	56.6	76.2	91.2	56.6	75.0
04	MASONRY	101.8	53.5	72.8	105.5	49.4	71.7	105.5	97.0	100.4	120.7	43.3	74.2	100.0	56.6	73.9	93.6	56.0	71.0
05	METALS	88.3	86.1	87.6	92.0	76.0	86.6	97.7	98.5	98.0	91.8	82.5	88.6	96.9	94.6	96.1	96.9	85.3	93.0
06	WOOD, PLASTICS & COMPOSITES	106.7	41.3	69.7	94.8	64.8	77.8	100.4	100.2	100.3	91.6	49.6	67.8	96.7	41.6	65.5	98.1	47.5	69.4
07	THERMAL & MOISTURE PROTECTION	101.9	46.4	80.9	100.9	55.2	83.6	99.8	97.5	98.9	100.5	50.9	81.8	101.7	62.1	86.7	100.1	57.0	83.8
08	OPENINGS	94.9	53.6	84.7	96.4	55.3	86.2	95.5	92.5	94.7	96.3	49.5	84.7	96.7	57.8	87.1	96.7	55.3	86.5
0920	Plaster & Gypsum Board	101.4	39.6	58.7	96.6	63.6	73.8	92.2	100.1	97.7	95.7	48.0	62.7	94.1	39.7	56.5	91.0	45.8	59.7
0950, 0980	Ceilings & Acoustic Treatment	111.3	39.6	65.7	92.0	63.6	74.0	92.0	100.1	97.2	92.0	48.0	64.0	92.0	39.7	58.7	91.2	45.8	62.3
0960	Flooring	117.9	50.8	99.3	100.3	42.2	84.2	89.8	97.7	92.0	98.3	30.8	79.7	101.9	39.2	84.5	100.7	67.9	91.7
0970, 0990	Wall Finishes & Painting/Coating	103.1	30.9	59.7	96.2	43.2	64.3	103.9	75.2	86.7	96.2	43.2	64.3	96.2	58.4	73.4	96.2	53.3	70.4
09	FINISHES	109.6	41.3	72.6	99.2	54.5	74.9	96.7	95.3	96.0	97.7	43.1	68.1	98.1	42.8	68.1	97.2	52.3	72.8
COVERS	DIVS. 10 - 14, 25, 28, 41, 43, 44	100.0	70.2	93.7	100.0	42.8	87.9	100.0	71.0	93.8	100.0	52.5	89.9	100.0	55.5	90.6	100.0	54.7	90.4
21, 22, 23	FIRE SUPPRESSION, PLUMBING & HVAC	100.1	37.5	74.4	100.0	55.9	81.9	99.9	96.5	98.5	100.0	55.9	81.8	100.0	60.3	83.7	99.8	54.2	81.0
26, 27, 3370	ELECTRICAL, COMMUNICATIONS & UTIL.	95.3	62.4	78.1	96.0	69.0	81.9	101.8	94.9	98.2	95.7	69.0	81.7	107.7	69.3	87.6	103.3	69.0	85.3
MF2004	WEIGHTED AVERAGE	98.4	56.0	79.7	100.4	61.5	83.3	98.6	95.3	97.1	99.1	59.2	81.5	99.0	63.8	83.4	97.7	62.7	82.2

	DIVISION	KENTUCKY												LOUISIANA					
		BOWLING GREEN			LEXINGTON			LOUISVILLE			OWENSBORO			ALEXANDRIA			BATON ROUGE		
		MAT.	INST.	TOTAL	MAT.	INST.	TOTAL	MAT.	INST.	TOTAL	MAT.	INST.	TOTAL	MAT.	INST.	TOTAL	MAT.	INST.	TOTAL
015433	CONTRACTOR EQUIPMENT		95.5	95.5		102.1	102.1		95.5	95.5		117.5	117.5		88.0	88.0		88.0	88.0
0241, 31 - 34	SITE & INFRASTRUCTURE, DEMOLITION	69.1	98.7	89.7	76.8	101.1	93.7	66.9	98.5	88.9	81.1	125.8	112.2	101.9	85.4	90.4	116.4	85.1	94.6
0310	Concrete Forming & Accessories	83.8	82.3	82.5	96.7	74.5	77.5	91.8	80.7	82.2	90.5	71.9	74.4	83.5	37.9	44.0	94.9	60.9	65.5
0320	Concrete Reinforcing	90.6	84.0	87.2	100.0	91.5	95.6	100.0	92.0	95.9	90.7	90.6	90.6	101.0	60.2	80.1	104.9	60.3	82.1
0330	Cast-in-Place Concrete	83.1	103.2	90.7	89.7	81.6	86.6	93.6	75.5	86.8	85.7	99.4	90.9	99.7	40.0	77.1	102.9	59.4	86.4
03	CONCRETE	91.3	89.9	90.6	93.4	80.5	87.3	95.0	81.2	88.5	100.4	85.2	93.3	93.4	44.2	70.4	98.0	60.8	80.6
04	MASONRY	89.2	74.0	80.1	86.2	65.7	73.0	87.4	79.0	82.4	85.9	71.5	77.2	115.3	50.4	76.3	106.8	52.2	74.0
05	METALS	92.8	84.2	89.9	94.3	88.5	92.3	101.0	87.9	96.6	85.4	88.5	86.4	91.1	72.8	84.9	103.1	70.0	91.9
06	WOOD, PLASTICS & COMPOSITES	93.0	81.8	86.7	102.0	74.5	86.4	99.7	81.8	89.5	93.4	68.2	79.1	89.1	36.8	59.5	102.7	66.1	82.0
07	THERMAL & MOISTURE PROTECTION	84.9	84.1	84.6	97.6	85.9	93.2	91.5	78.8	86.7	97.4	75.7	89.2	98.2	49.3	79.7	98.0	57.8	82.8
08	OPENINGS	95.3	77.7	90.9	96.3	78.5	91.9	96.3	84.4	93.3	93.4	78.6	89.7	99.6	43.7	85.8	102.6	59.4	91.9
0920	Plaster & Gypsum Board	78.8	81.6	80.7	85.9	73.1	77.1	82.0	81.6	81.7	78.9	66.6	70.4	78.2	35.5	48.7	94.2	65.5	74.4
0950, 0980	Ceilings & Acoustic Treatment	81.4	81.6	81.5	86.6	73.1	78.0	82.3	81.6	81.8	74.5	66.6	69.5	92.2	35.5	56.1	99.9	65.5	78.0
0960	Flooring	93.9	81.6	90.5	98.8	59.8	88.0	97.3	70.3	89.8	96.4	81.6	92.3	101.6	66.0	91.8	107.0	66.0	95.7
0970, 0990	Wall Finishes & Painting/Coating	96.4	71.5	81.4	96.4	69.7	80.3	96.4	79.0	85.9	96.4	98.7	97.7	98.6	35.4	60.6	103.2	40.6	65.5
09	FINISHES	90.3	81.0	85.3	93.9	70.6	81.2	91.5	78.5	84.4	89.5	75.7	82.0	92.2	42.0	64.9	101.7	60.3	79.3
COVERS	DIVS. 10 - 14, 25, 28, 41, 43, 44	100.0	66.1	92.8	100.0	90.9	98.1	100.0	90.4	98.0	100.0	84.2	96.6	100.0	55.5	90.5	100.0	78.6	95.5
21, 22, 23	FIRE SUPPRESSION, PLUMBING & HVAC	99.9	83.1	93.0	99.9	65.0	85.5	99.9	81.0	92.1	99.9	68.8	87.1	100.2	52.1	80.4	99.9	51.2	79.9
26, 27, 3370	ELECTRICAL, COMMUNICATIONS & UTIL.	94.4	76.3	84.9	95.0	73.6	83.8	106.2	76.3	90.6	93.9	82.3	87.8	94.9	53.4	73.2	116.1	60.4	87.0
MF2004	WEIGHTED AVERAGE	93.9	82.6	88.9	95.5	76.3	87.0	97.1	82.4	90.6	94.3	81.7	88.8	97.5	54.0	78.3	103.0	61.2	84.5

DIVISION		LOUISIANA												MAINE					
		LAKE CHARLES			MONROE			NEW ORLEANS			SHREVEPORT			AUGUSTA			BANGOR		
		MAT.	INST.	TOTAL	MAT.	INST.	TOTAL	MAT.	INST.	TOTAL	MAT.	INST.	TOTAL	MAT.	INST.	TOTAL	MAT.	INST.	TOTAL
015433	CONTRACTOR EQUIPMENT		88.0	88.0		88.0	88.0		89.0	89.0		88.0	88.0		100.6	100.6		100.6	100.6
0241, 31 - 34	SITE & INFRASTRUCTURE, DEMOLITION	118.0	84.9	95.0	101.9	84.7	89.9	121.0	87.7	97.8	103.5	84.6	90.4	81.0	101.1	95.0	80.2	98.8	93.1
0310	Concrete Forming & Accessories	97.8	52.1	58.3	83.0	38.4	44.4	100.1	63.7	68.6	96.9	40.7	48.3	96.5	102.2	101.4	92.3	91.5	91.6
0320	Concrete Reinforcing	104.9	58.0	81.0	99.8	60.1	79.5	104.9	61.8	82.9	99.4	60.5	79.5	90.4	103.1	96.9	90.1	102.8	96.6
0330	Cast-in-Place Concrete	105.0	68.8	91.3	99.7	57.6	83.8	104.0	69.0	90.7	99.1	46.0	79.0	84.3	63.7	76.5	78.3	62.3	72.3
03	CONCRETE	99.2	59.7	80.7	93.2	50.2	73.1	98.8	65.8	83.4	93.1	47.4	71.7	103.0	88.5	96.2	99.3	82.9	91.7
04	MASONRY	104.9	59.8	77.8	110.2	48.3	73.0	106.2	57.5	76.9	101.3	49.9	70.4	98.0	63.2	77.1	112.7	59.0	80.4
05	METALS	97.9	69.6	88.4	91.0	70.6	84.1	111.0	74.0	98.5	87.8	70.7	82.0	88.3	85.1	87.2	87.9	80.1	85.3
06	WOOD, PLASTICS & COMPOSITES	107.1	53.7	76.9	88.4	37.2	59.4	103.7	66.1	82.4	104.4	40.4	68.2	97.5	114.0	106.8	92.7	99.9	96.8
07	THERMAL & MOISTURE PROTECTION	99.6	59.3	84.4	98.2	52.2	80.8	100.4	61.1	85.6	97.5	51.1	79.9	95.8	64.5	83.9	96.1	60.1	82.5
08	OPENINGS	102.6	53.2	90.4	99.6	47.8	86.8	103.4	66.4	94.2	97.5	45.8	84.7	101.1	91.4	98.7	101.0	85.1	97.1
0920	Plaster & Gypsum Board	100.9	52.8	67.6	77.8	35.9	48.8	99.6	65.5	76.0	83.6	39.2	52.9	102.3	113.7	110.2	94.7	99.2	97.8
0950, 0980	Ceilings & Acoustic Treatment	99.1	52.8	69.6	92.2	35.9	56.3	98.1	65.5	77.3	91.3	39.2	58.1	90.1	113.7	105.1	90.9	99.2	96.2
0960	Flooring	107.2	78.5	99.3	101.1	61.5	90.2	107.6	66.0	96.1	111.0	66.0	98.6	97.6	59.5	87.1	95.3	59.2	85.3
0970, 0990	Wall Finishes & Painting/Coating	103.2	37.4	63.6	98.6	42.3	64.7	104.7	64.0	80.2	98.6	35.4	60.6	94.6	38.3	60.7	94.6	35.9	59.3
09	FINISHES	102.3	55.3	76.8	92.0	42.2	65.0	102.6	64.2	81.8	95.8	44.1	67.7	98.1	89.1	93.2	96.7	81.2	88.3
COVERS	DIVS. 10 - 14, 25, 28, 41, 43, 44	100.0	77.2	95.2	100.0	50.8	89.6	100.0	80.5	95.9	100.0	70.9	93.8	100.0	91.1	98.1	100.0	105.3	101.1
21, 22, 23	FIRE SUPPRESSION, PLUMBING & HVAC	100.0	52.7	80.5	100.2	46.2	78.0	100.0	64.1	85.2	100.1	54.8	81.5	100.0	69.0	87.3	100.1	64.7	85.5
26, 27, 3370	ELECTRICAL, COMMUNICATIONS & UTIL.	102.2	58.7	79.5	96.9	54.5	74.8	102.8	71.7	86.6	104.4	68.9	85.8	98.9	81.5	89.8	95.8	80.7	87.9
MF2004	WEIGHTED AVERAGE	101.1	60.7	83.3	97.4	53.4	77.9	103.2	68.2	87.8	97.4	57.7	79.9	97.7	81.5	90.6	97.5	77.5	88.7

DIVISION		MAINE						MARYLAND						MASSACHUSETTS					
		LEWISTON			PORTLAND			BALTIMORE			HAGERSTOWN			BOSTON			BROCKTON		
		MAT.	INST.	TOTAL	MAT.	INST.	TOTAL	MAT.	INST.	TOTAL	MAT.	INST.	TOTAL	MAT.	INST.	TOTAL	MAT.	INST.	TOTAL
015433	CONTRACTOR EQUIPMENT		100.6	100.6		100.6	100.6		102.9	102.9		99.2	99.2		106.5	106.5		102.5	102.5
0241, 31 - 34	SITE & INFRASTRUCTURE, DEMOLITION	77.8	98.8	92.4	79.3	98.8	92.9	95.2	94.4	94.6	85.4	90.9	89.2	85.3	108.1	101.2	82.7	104.3	97.8
0310	Concrete Forming & Accessories	96.7	91.5	92.2	95.5	91.5	92.0	98.3	75.9	78.9	89.1	78.1	79.5	100.0	139.7	134.4	100.5	127.1	123.6
0320	Concrete Reinforcing	110.6	102.8	106.6	110.6	102.8	106.6	109.9	84.8	97.1	97.1	75.7	86.2	109.4	147.7	129.0	110.6	147.5	129.5
0330	Cast-in-Place Concrete	79.9	62.3	73.3	101.6	62.3	86.8	103.9	79.1	94.5	88.4	58.6	77.1	101.1	150.6	119.8	96.2	146.0	115.0
03	CONCRETE	98.7	82.9	91.4	109.4	82.9	97.0	105.1	79.7	93.2	90.0	71.9	81.6	112.4	144.0	127.2	109.7	136.7	122.3
04	MASONRY	96.9	59.0	74.1	97.2	59.0	74.2	101.1	70.5	82.7	105.0	81.3	90.7	115.7	156.6	140.3	111.9	149.4	134.4
05	METALS	91.1	80.1	87.4	92.3	80.1	88.2	93.3	96.4	94.3	90.1	91.7	90.6	103.0	128.5	111.6	99.6	125.8	108.5
06	WOOD, PLASTICS & COMPOSITES	98.1	99.9	99.2	97.3	99.9	98.8	95.9	77.5	85.5	85.5	77.3	80.9	101.4	138.9	122.7	100.2	125.9	114.8
07	THERMAL & MOISTURE PROTECTION	96.0	60.1	82.4	99.1	60.1	84.4	94.4	79.6	88.8	93.7	69.9	84.7	96.1	150.2	116.5	96.2	144.2	114.3
08	OPENINGS	104.3	85.1	99.6	104.3	85.1	99.6	93.1	83.7	90.8	90.4	77.9	87.3	100.5	142.6	111.0	98.6	132.7	107.0
0920	Plaster & Gypsum Board	100.4	99.2	99.6	106.9	99.2	101.6	100.5	77.0	84.2	95.3	77.0	82.7	103.3	139.5	128.3	97.6	126.0	117.2
0950, 0980	Ceilings & Acoustic Treatment	98.7	99.2	99.0	96.1	99.2	98.1	93.8	77.0	83.1	94.7	77.0	83.4	97.7	139.5	124.3	99.7	126.0	116.4
0960	Flooring	98.3	59.2	87.5	97.6	59.2	87.0	89.2	75.1	85.3	84.7	84.5	84.7	98.0	168.9	117.6	98.5	168.9	118.0
0970, 0990	Wall Finishes & Painting/Coating	94.6	35.9	59.3	94.6	35.9	59.3	97.0	81.4	87.6	97.0	34.5	59.4	97.2	159.0	134.4	96.7	143.5	124.9
09	FINISHES	99.7	81.2	89.7	99.6	81.2	89.6	92.7	75.7	83.5	90.2	74.7	81.8	99.7	147.6	125.7	99.6	136.7	119.7
COVERS	DIVS. 10 - 14, 25, 28, 41, 43, 44	100.0	105.3	101.1	100.0	105.3	101.1	100.0	88.1	97.5	100.0	91.1	98.1	100.0	120.6	104.4	100.0	116.8	103.6
21, 22, 23	FIRE SUPPRESSION, PLUMBING & HVAC	100.1	64.7	85.5	100.0	64.7	85.5	100.0	87.7	94.9	100.0	87.8	94.9	100.1	129.1	112.0	100.1	111.5	104.8
26, 27, 3370	ELECTRICAL, COMMUNICATIONS & UTIL.	98.0	81.3	89.2	104.5	81.3	92.4	97.5	95.0	96.2	94.8	79.5	86.8	97.1	136.8	117.8	96.2	100.5	98.4
MF2004	WEIGHTED AVERAGE	97.8	77.6	88.9	100.1	77.6	90.2	97.8	85.2	92.2	94.5	81.9	88.9	102.0	136.5	117.2	100.6	123.2	110.6

DIVISION		MASSACHUSETTS																	
		FALL RIVER			HYANNIS			LAWRENCE			LOWELL			NEW BEDFORD			PITTSFIELD		
		MAT.	INST.	TOTAL	MAT.	INST.	TOTAL	MAT.	INST.	TOTAL	MAT.	INST.	TOTAL	MAT.	INST.	TOTAL	MAT.	INST.	TOTAL
015433	CONTRACTOR EQUIPMENT		103.5	103.5		102.5	102.5		102.5	102.5		100.6	100.6		103.5	103.5		100.6	100.6
0241, 31 - 34	SITE & INFRASTRUCTURE, DEMOLITION	81.8	104.4	97.5	79.9	104.2	96.8	83.5	104.3	98.0	82.3	104.3	97.6	80.4	104.4	97.1	83.3	102.7	96.8
0310	Concrete Forming & Accessories	100.5	126.8	123.3	93.4	126.5	122.0	100.3	127.3	123.6	97.7	127.5	123.5	100.5	126.8	123.3	97.7	108.4	106.9
0320	Concrete Reinforcing	110.6	123.3	117.1	88.7	123.2	106.3	109.7	137.2	123.7	110.6	137.2	124.2	110.6	123.3	117.1	92.2	109.7	101.1
0330	Cast-in-Place Concrete	93.0	146.4	113.2	87.8	145.8	109.7	97.0	146.1	115.6	88.2	146.2	110.1	82.0	146.4	106.3	96.2	122.7	106.2
03	CONCRETE	108.2	132.0	119.3	99.8	131.7	114.7	109.9	134.8	121.6	100.3	134.7	116.4	102.7	132.0	116.4	101.6	112.9	106.9
04	MASONRY	112.0	149.3	134.4	110.8	149.4	134.0	111.4	149.4	134.2	97.5	149.5	128.7	110.8	149.3	133.9	98.2	122.1	112.5
05	METALS	99.6	115.8	105.1	96.0	115.3	102.5	96.9	121.7	105.3	96.9	119.5	104.5	99.6	115.8	105.1	96.7	102.3	98.6
06	WOOD, PLASTICS & COMPOSITES	100.2	126.2	114.9	91.6	125.9	111.0	100.2	125.9	114.8	99.8	125.9	114.6	100.2	126.2	114.9	99.8	108.5	104.7
07	THERMAL & MOISTURE PROTECTION	96.2	138.1	112.0	95.9	138.8	112.1	96.2	144.2	114.3	96.0	144.2	114.2	96.2	138.1	112.0	96.0	117.6	104.2
08	OPENINGS	98.6	122.1	104.4	94.7	121.9	101.4	98.6	129.9	106.3	104.3	129.9	110.6	98.6	122.1	104.4	104.3	109.2	105.5
0920	Plaster & Gypsum Board	97.6	126.0	117.2	90.1	126.0	114.9	100.4	126.0	118.1	100.4	126.0	118.1	97.6	126.0	117.2	100.4	108.0	105.7
0950, 0980	Ceilings & Acoustic Treatment	99.7	126.0	116.4	91.9	126.0	113.6	98.7	126.0	116.1	98.7	126.0	116.1	99.7	126.0	116.4	98.7	108.0	104.6
0960	Flooring	98.4	168.9	117.8	94.7	168.9	115.2	97.6	168.9	117.3	97.6	168.9	117.3	98.4	168.9	117.8	97.8	138.1	108.9
0970, 0990	Wall Finishes & Painting/Coating	96.7	143.5	124.9	96.7	143.5	124.9	94.7	143.5	124.1	94.6	143.5	124.0	96.7	143.5	124.9	94.6	108.0	102.7
09	FINISHES	99.6	136.9	119.8	95.5	136.7	117.9	99.3	136.7	119.6	99.3	136.7	119.6	99.5	136.9	119.8	99.3	114.1	107.3
COVERS	DIVS. 10 - 14, 25, 28, 41, 43, 44	100.0	117.5	103.7	100.0	116.8	103.6	100.0	116.9	103.6	100.0	116.9	103.6	100.0	117.5	103.7	100.0	105.6	101.2
21, 22, 23	FIRE SUPPRESSION, PLUMBING & HVAC	100.1	111.3	104.7	100.1	111.2	104.7	100.1	114.4	106.0	100.1	124.9	110.3	100.1	111.3	104.7	100.1	98.7	99.6
26, 27, 3370	ELECTRICAL, COMMUNICATIONS & UTIL.	96.1	100.4	98.4	93.8	100.4	97.3	97.4	128.5	113.7	98.0	128.5	113.9	97.0	100.4	98.8	97.9	93.4	95.6
MF2004	WEIGHTED AVERAGE	100.4	121.1	109.5	97.7	120.9	107.9	100.3	127.0	112.1	99.0	129.0	112.2	99.7	121.1	109.1	99.2	106.1	102.3

	DIVISION	MASSACHUSETTS						MICHIGAN											
		SPRINGFIELD			WORCESTER			ANN ARBOR			DEARBORN			DETROIT			FLINT		
		MAT.	INST.	TOTAL	MAT.	INST.	TOTAL	MAT.	INST.	TOTAL	MAT.	INST.	TOTAL	MAT.	INST.	TOTAL	MAT.	INST.	TOTAL
015433	CONTRACTOR EQUIPMENT		100.6	100.6		100.6	100.6		110.3	110.3		110.3	110.3		97.6	97.6		110.3	110.3
0241, 31 - 34	SITE & INFRASTRUCTURE, DEMOLITION	82.8	102.9	96.8	82.7	104.2	97.7	75.3	95.7	89.5	75.1	95.8	89.5	89.0	97.4	94.9	65.7	95.0	86.1
0310	Concrete Forming & Accessories	97.9	109.4	107.9	98.2	122.6	119.3	95.4	106.7	105.2	95.2	115.6	112.8	96.0	115.7	113.0	97.9	93.4	94.0
0320	Concrete Reinforcing	110.6	115.1	112.9	110.6	147.5	129.4	116.3	126.4	121.5	116.3	126.6	121.6	115.5	126.6	121.2	116.3	125.9	121.2
0330	Cast-in-Place Concrete	91.8	124.2	104.0	91.2	146.0	111.9	80.9	117.1	94.6	79.1	114.8	92.6	86.8	114.8	97.4	81.5	98.5	87.9
03	CONCRETE	102.1	114.9	108.1	101.8	134.5	117.1	91.9	114.4	102.4	91.0	117.6	103.4	94.7	116.4	104.8	92.3	102.0	96.8
04	MASONRY	97.8	124.7	114.0	97.3	149.5	128.7	92.3	113.2	104.9	92.2	117.3	107.3	91.5	117.3	107.0	92.4	100.4	97.2
05	METALS	99.5	104.5	101.2	99.6	123.0	107.5	91.2	122.0	101.6	91.3	122.6	101.9	95.0	104.2	98.1	91.3	119.5	100.8
06	WOOD, PLASTICS & COMPOSITES	99.8	108.5	104.7	100.1	120.3	111.5	98.9	105.0	102.3	98.9	115.7	108.4	98.9	115.7	108.4	101.8	91.3	95.9
07	THERMAL & MOISTURE PROTECTION	96.0	118.7	104.6	96.0	138.5	112.1	102.3	114.5	106.9	100.8	119.5	107.9	99.1	119.5	106.8	100.0	101.0	100.4
08	OPENINGS	104.3	110.6	105.9	104.3	129.6	110.6	94.0	108.4	97.6	94.0	114.2	99.0	95.7	114.2	100.3	94.0	98.7	95.2
0920	Plaster & Gypsum Board	100.4	108.0	105.7	100.4	120.2	114.1	97.2	104.1	101.9	97.2	115.2	109.6	97.2	115.2	109.6	98.9	90.1	92.8
0950, 0980	Ceilings & Acoustic Treatment	98.7	108.0	104.6	98.7	120.2	112.4	91.1	104.1	99.4	91.1	115.2	106.4	92.0	115.2	106.8	91.1	90.1	90.4
0960	Flooring	97.5	138.1	108.7	97.6	168.9	117.3	92.7	120.4	100.4	92.4	121.2	100.3	92.4	121.2	100.4	92.5	84.9	90.4
0970, 0990	Wall Finishes & Painting/Coating	96.1	108.0	103.3	94.6	143.5	124.0	93.5	107.5	101.9	93.5	107.6	102.0	95.0	107.6	102.6	93.5	94.9	94.3
09	FINISHES	99.3	114.7	107.7	99.3	133.4	117.8	92.3	109.1	101.4	92.2	116.0	105.1	93.4	116.0	105.7	91.6	91.1	91.3
COVERS	DIVS. 10 - 14, 25, 28, 41, 43, 44	100.0	106.5	101.4	100.0	110.1	102.2	100.0	107.0	101.5	100.0	108.9	101.9	100.0	108.9	101.9	100.0	100.0	100.0
21, 22, 23	FIRE SUPPRESSION, PLUMBING & HVAC	100.1	101.6	100.7	100.1	111.8	104.9	100.1	100.2	100.1	100.1	110.5	104.3	100.1	112.7	105.2	100.1	98.9	99.6
26, 27, 3370	ELECTRICAL, COMMUNICATIONS & UTIL.	98.0	93.4	95.6	98.0	102.9	100.5	94.6	92.4	93.5	94.6	112.2	103.8	95.7	112.1	104.3	94.6	100.2	97.5
MF2004	WEIGHTED AVERAGE	99.6	107.7	103.2	99.6	122.1	109.5	94.9	106.1	99.9	94.7	113.4	103.0	96.4	112.1	103.3	94.6	100.2	97.1

	DIVISION	MICHIGAN															MINNESOTA		
		GRAND RAPIDS			KALAMAZOO			LANSING			MUSKEGON			SAGINAW			DULUTH		
		MAT.	INST.	TOTAL	MAT.	INST.	TOTAL	MAT.	INST.	TOTAL	MAT.	INST.	TOTAL	MAT.	INST.	TOTAL	MAT.	INST.	TOTAL
015433	CONTRACTOR EQUIPMENT		103.5	103.5		103.5	103.5		110.3	110.3		103.5	103.5		110.3	110.3		101.0	101.0
0241, 31 - 34	SITE & INFRASTRUCTURE, DEMOLITION	82.9	87.6	86.2	81.7	87.8	85.9	83.5	94.5	91.1	79.4	87.7	85.2	68.4	94.3	86.5	95.2	102.1	100.0
0310	Concrete Forming & Accessories	97.8	82.8	84.8	97.6	93.7	94.3	98.7	92.0	92.9	98.1	92.1	92.9	95.4	91.5	92.0	100.9	121.7	118.9
0320	Concrete Reinforcing	100.0	85.9	92.8	100.0	88.0	93.8	116.3	125.7	121.1	100.6	87.0	93.7	116.3	125.4	121.0	92.2	114.4	103.5
0330	Cast-in-Place Concrete	90.7	98.6	93.7	91.4	103.1	95.9	83.1	96.3	88.1	89.3	99.8	93.3	79.9	95.7	85.9	108.3	109.5	108.7
03	CONCRETE	93.9	88.3	91.3	96.6	95.1	95.9	93.2	100.6	96.6	91.7	93.1	92.3	91.4	100.1	95.5	101.5	116.6	108.6
04	MASONRY	88.6	56.3	69.2	91.5	90.9	91.1	86.3	98.1	93.4	90.3	74.2	80.6	93.7	91.0	92.1	109.6	122.8	117.5
05	METALS	94.3	81.0	89.8	93.2	85.7	90.7	89.9	119.0	99.7	90.9	84.0	88.6	91.3	118.5	100.5	89.4	128.6	102.7
06	WOOD, PLASTICS & COMPOSITES	95.7	82.7	88.3	98.5	94.5	96.3	101.1	90.3	95.0	94.6	93.3	93.8	94.3	91.0	92.5	110.6	122.8	117.5
07	THERMAL & MOISTURE PROTECTION	95.4	60.7	82.3	93.5	91.2	92.6	101.3	95.9	99.3	92.7	79.7	87.8	100.8	97.5	99.5	103.6	121.5	110.4
08	OPENINGS	94.2	71.2	88.5	90.7	86.5	89.7	94.0	98.1	95.0	89.9	87.6	89.4	91.9	98.5	93.6	95.2	126.7	103.0
0920	Plaster & Gypsum Board	82.8	78.1	79.5	84.6	90.3	88.5	95.4	89.0	91.0	69.4	89.0	82.9	97.2	89.7	92.0	93.2	123.9	114.5
0950, 0980	Ceilings & Acoustic Treatment	84.8	78.1	80.5	86.6	90.3	88.9	89.4	89.0	89.1	87.4	89.0	88.4	91.1	89.7	90.2	92.2	123.9	112.4
0960	Flooring	97.8	36.5	80.9	97.7	78.0	92.2	102.5	84.9	97.6	96.9	67.2	88.7	92.7	80.3	89.3	107.4	131.5	114.0
0970, 0990	Wall Finishes & Painting/Coating	96.4	35.8	60.0	96.4	83.8	88.8	105.5	88.6	95.3	95.7	68.5	79.3	93.5	87.8	90.1	89.6	113.4	103.9
09	FINISHES	92.9	69.0	79.9	93.3	89.9	91.5	96.3	89.7	92.7	90.8	85.1	87.7	92.0	88.4	90.0	98.6	123.4	112.1
COVERS	DIVS. 10 - 14, 25, 28, 41, 43, 44	100.0	107.6	101.6	100.0	109.3	102.0	100.0	103.1	100.7	100.0	108.9	101.9	100.0	98.8	99.7	100.0	106.5	101.4
21, 22, 23	FIRE SUPPRESSION, PLUMBING & HVAC	99.9	83.2	93.0	99.9	81.8	92.4	100.0	96.9	98.7	99.7	85.9	94.0	100.1	88.6	95.3	99.8	100.0	99.9
26, 27, 3370	ELECTRICAL, COMMUNICATIONS & UTIL.	104.6	51.5	76.9	94.3	84.5	89.2	100.6	98.2	99.3	94.8	81.4	87.8	92.5	93.5	93.0	99.7	114.9	107.6
MF2004	WEIGHTED AVERAGE	96.4	74.8	86.8	95.3	88.2	92.2	95.7	98.7	97.0	93.9	85.6	90.2	94.2	95.3	94.7	98.6	114.6	105.6

	DIVISION	MINNESOTA												MISSISSIPPI					
		MINNEAPOLIS			ROCHESTER			SAINT PAUL			ST. CLOUD			BILOXI			GREENVILLE		
		MAT.	INST.	TOTAL	MAT.	INST.	TOTAL	MAT.	INST.	TOTAL	MAT.	INST.	TOTAL	MAT.	INST.	TOTAL	MAT.	INST.	TOTAL
015433	CONTRACTOR EQUIPMENT		104.7	104.7		101.0	101.0		101.0	101.0		101.6	101.6		98.3	98.3		98.3	98.3
0241, 31 - 34	SITE & INFRASTRUCTURE, DEMOLITION	93.9	107.7	103.5	92.9	101.9	99.2	97.7	102.7	101.2	86.1	105.8	99.8	104.0	88.7	93.4	107.8	88.8	94.6
0310	Concrete Forming & Accessories	102.9	135.3	130.9	103.7	112.4	111.2	98.1	131.8	127.2	90.8	130.6	125.3	93.0	59.1	63.7	81.9	71.4	72.8
0320	Concrete Reinforcing	92.4	125.6	109.4	92.2	125.2	109.0	89.0	125.6	107.7	94.7	125.4	110.4	86.0	60.2	72.8	93.2	67.1	79.9
0330	Cast-in-Place Concrete	112.4	122.9	116.4	110.3	105.0	108.3	114.4	122.2	117.4	97.2	121.9	106.6	108.5	57.2	89.1	109.0	60.5	90.6
03	CONCRETE	105.5	129.4	116.7	102.7	113.1	107.6	108.2	127.7	117.3	91.6	126.0	100.1	96.6	60.2	79.6	99.8	68.1	85.0
04	MASONRY	108.4	133.6	123.5	107.8	120.4	115.3	118.3	133.5	127.5	109.5	131.1	122.5	88.9	48.2	64.4	129.8	53.9	84.2
05	METALS	92.2	136.2	107.1	89.3	134.1	104.4	88.7	135.8	104.6	90.6	134.0	105.3	94.7	84.7	91.3	92.8	87.9	91.2
06	WOOD, PLASTICS & COMPOSITES	113.7	134.9	125.7	113.7	111.0	112.2	106.9	130.3	120.2	95.0	129.1	114.3	95.1	62.8	76.8	80.6	76.3	78.2
07	THERMAL & MOISTURE PROTECTION	103.8	133.1	114.8	104.0	107.9	105.4	105.1	132.4	115.4	103.0	117.5	108.5	91.6	59.7	79.6	91.8	66.1	82.1
08	OPENINGS	98.4	142.2	109.2	95.2	129.3	103.7	92.8	139.7	104.4	91.7	139.0	103.4	97.5	63.2	89.0	96.9	68.6	89.9
0920	Plaster & Gypsum Board	98.4	136.2	124.6	97.9	111.8	107.5	90.5	131.7	119.0	86.3	130.5	116.9	99.2	62.2	73.6	87.3	76.1	79.6
0950, 0980	Ceilings & Acoustic Treatment	95.6	136.2	121.5	93.9	111.8	105.3	90.4	131.7	116.7	69.0	130.5	108.2	98.4	62.2	75.3	93.2	76.1	82.3
0960	Flooring	105.1	130.1	112.0	107.5	93.6	103.7	101.1	130.1	109.1	104.3	130.1	111.5	116.6	56.9	100.1	107.6	56.9	93.6
0970, 0990	Wall Finishes & Painting/Coating	97.3	126.8	115.1	92.5	105.0	100.1	97.3	132.2	118.3	103.7	126.8	117.6	106.1	48.3	71.3	106.1	73.2	86.3
09	FINISHES	99.6	134.1	118.4	99.6	108.9	104.7	96.7	132.0	115.9	93.2	130.7	113.6	103.5	57.7	78.6	98.5	69.8	82.9
COVERS	DIVS. 10 - 14, 25, 28, 41, 43, 44	100.0	111.7	102.5	100.0	104.9	101.0	100.0	110.8	102.3	100.0	110.5	102.2	100.0	65.4	92.6	100.0	69.0	93.4
21, 22, 23	FIRE SUPPRESSION, PLUMBING & HVAC	99.8	124.1	109.8	99.8	99.3	99.6	99.7	115.9	106.4	99.5	124.0	109.6	99.8	53.0	80.5	99.8	54.0	80.9
26, 27, 3370	ELECTRICAL, COMMUNICATIONS & UTIL.	99.2	115.0	107.5	98.2	84.9	91.3	98.1	115.0	106.9	99.2	115.0	107.4	98.9	60.5	78.8	98.2	70.3	83.6
MF2004	WEIGHTED AVERAGE	99.7	126.3	111.5	98.5	107.8	102.6	99.3	123.4	109.9	96.2	124.2	108.6	97.9	62.2	82.1	99.6	67.9	85.6

	DIVISION	MISSISSIPPI						MISSOURI											
		JACKSON			MERIDIAN			CAPE GIRARDEAU			COLUMBIA			JOPLIN			KANSAS CITY		
		MAT.	INST.	TOTAL	MAT.	INST.	TOTAL	MAT.	INST.	TOTAL	MAT.	INST.	TOTAL	MAT.	INST.	TOTAL	MAT.	INST.	TOTAL
015433	CONTRACTOR EQUIPMENT		98.3	98.3		98.3	98.3		106.7	106.7		107.7	107.7		105.3	105.3		103.2	103.2
0241, 31 - 34	SITE & INFRASTRUCTURE, DEMOLITION	99.2	88.8	91.9	99.0	89.1	92.1	95.3	93.2	93.8	103.4	95.8	98.1	98.7	97.8	98.1	92.4	97.9	96.2
0310	Concrete Forming & Accessories	92.3	64.3	68.1	80.9	70.8	72.2	93.2	70.9	73.9	92.5	75.4	77.7	103.9	66.1	71.2	102.8	107.2	106.6
0320	Concrete Reinforcing	86.0	59.2	72.3	92.1	59.2	75.3	104.5	79.1	91.5	100.4	112.4	106.5	102.1	67.5	84.4	96.9	113.6	105.4
0330	Cast-in-Place Concrete	100.8	61.3	85.9	102.8	63.9	88.1	94.5	80.4	89.2	89.7	85.2	88.0	98.5	66.6	86.4	91.6	107.5	97.6
03	CONCRETE	92.7	63.7	79.2	92.4	67.5	80.8	93.7	77.4	86.1	85.4	87.4	86.3	97.6	67.7	83.6	93.8	108.8	100.8
04	MASONRY	90.0	53.9	68.3	88.5	58.3	70.4	116.7	74.3	91.2	132.4	82.3	102.3	93.8	56.6	71.4	98.0	108.1	104.1
05	METALS	94.7	84.5	91.2	92.8	84.7	90.0	94.1	102.3	96.9	90.0	118.4	99.6	93.0	84.2	90.1	100.7	113.9	105.1
06	WOOD, PLASTICS & COMPOSITES	94.0	66.7	78.5	79.6	73.0	75.9	87.3	69.0	76.9	94.3	72.1	81.7	99.3	65.2	80.0	98.6	106.5	103.1
07	THERMAL & MOISTURE PROTECTION	93.6	65.1	82.8	91.4	67.7	82.4	100.9	75.1	91.1	94.8	80.4	89.4	98.8	62.3	85.0	98.7	109.8	102.9
08	OPENINGS	97.9	61.2	88.8	96.8	70.8	90.4	98.4	69.3	91.2	91.2	92.3	91.5	90.6	67.3	84.8	96.5	109.0	99.6
0920	Plaster & Gypsum Board	93.4	66.2	74.6	87.3	72.7	77.2	99.9	68.0	77.8	94.8	71.2	78.4	100.4	64.0	75.2	96.5	106.6	103.5
0950, 0980	Ceilings & Acoustic Treatment	94.1	66.2	76.3	93.2	72.7	80.1	87.6	68.0	75.1	92.0	71.2	78.7	89.4	64.0	73.2	95.4	106.6	102.5
0960	Flooring	117.0	56.9	100.4	106.4	56.9	92.7	89.9	64.5	82.9	107.3	86.1	101.4	122.0	43.4	100.3	98.5	107.9	101.1
0970, 0990	Wall Finishes & Painting/Coating	106.1	73.2	86.3	106.1	84.6	93.1	101.0	75.2	85.4	103.9	72.9	85.3	95.5	39.0	61.5	100.5	117.6	110.8
09	FINISHES	102.0	64.2	81.5	97.4	70.3	82.7	94.2	69.3	80.7	97.5	74.4	85.0	104.7	59.1	80.0	99.9	108.2	104.4
COVERS	DIVS. 10 - 14, 25, 28, 41, 43, 44	100.0	67.9	93.2	100.0	70.2	93.7	100.0	56.7	90.8	100.0	96.7	99.3	100.0	65.3	92.6	100.0	101.4	100.3
21, 22, 23	FIRE SUPPRESSION, PLUMBING & HVAC	99.8	59.7	83.3	99.8	62.1	84.3	99.8	84.7	93.6	100.0	86.4	94.4	100.1	49.5	79.3	100.1	112.9	105.3
26, 27, 3370	ELECTRICAL, COMMUNICATIONS & UTIL.	103.3	70.3	86.0	97.8	61.3	78.7	100.5	104.1	102.4	95.1	83.7	89.1	90.5	66.1	77.8	101.4	104.4	103.0
MF2004	WEIGHTED AVERAGE	97.8	67.0	84.2	96.1	68.6	84.0	98.4	83.8	92.0	96.8	88.2	93.0	96.8	65.2	82.9	98.7	108.2	102.9

	DIVISION	MISSOURI									MONTANA								
		SPRINGFIELD			ST. JOSEPH			ST. LOUIS			BILLINGS			BUTTE			GREAT FALLS		
		MAT.	INST.	TOTAL	MAT.	INST.	TOTAL	MAT.	INST.	TOTAL	MAT.	INST.	TOTAL	MAT.	INST.	TOTAL	MAT.	INST.	TOTAL
015433	CONTRACTOR EQUIPMENT		102.1	102.1		101.9	101.9		107.8	107.8		99.0	99.0		98.8	98.8		98.8	98.8
0241, 31 - 34	SITE & INFRASTRUCTURE, DEMOLITION	96.9	93.6	94.6	93.9	93.3	93.5	95.2	97.2	96.6	93.7	96.9	95.9	103.4	96.2	98.4	107.3	96.9	100.0
0310	Concrete Forming & Accessories	103.6	71.2	75.6	102.6	76.0	79.6	104.4	103.9	103.9	99.8	63.2	68.2	89.6	66.8	69.9	99.7	63.9	68.7
0320	Concrete Reinforcing	96.7	112.0	104.5	95.8	99.7	97.8	95.6	108.7	102.3	93.9	67.3	80.3	101.8	67.3	84.2	93.8	67.2	80.2
0330	Cast-in-Place Concrete	97.4	74.2	88.6	91.5	101.4	95.3	94.5	110.2	100.4	110.6	66.9	94.0	121.7	67.8	101.3	128.6	65.4	104.7
03	CONCRETE	96.0	80.9	89.0	93.6	90.2	92.0	93.1	107.8	100.0	103.0	66.3	85.8	106.4	68.2	88.5	111.5	66.0	90.3
04	MASONRY	88.9	81.5	84.5	97.7	86.9	91.2	97.6	110.0	105.0	126.2	69.4	92.0	121.6	71.6	91.6	126.0	70.4	92.6
05	METALS	94.0	104.8	97.6	97.0	104.6	99.6	98.8	119.8	105.9	105.6	84.0	98.3	99.0	84.1	94.0	102.0	83.7	95.8
06	WOOD, PLASTICS & COMPOSITES	101.2	69.5	83.3	99.3	71.4	83.5	100.3	101.8	101.1	101.0	62.2	79.0	90.2	67.0	77.1	102.1	62.2	79.5
07	THERMAL & MOISTURE PROTECTION	99.1	75.1	90.0	99.0	89.1	95.3	100.8	107.3	103.2	102.2	70.2	90.1	102.1	71.0	90.4	102.6	65.6	88.6
08	OPENINGS	96.7	78.5	92.2	95.3	86.5	93.1	97.2	108.8	100.1	97.9	61.0	88.7	95.8	62.9	87.7	99.3	60.2	89.6
0920	Plaster & Gypsum Board	97.8	68.5	77.5	101.9	70.4	80.1	107.9	101.8	103.7	95.8	61.3	71.9	94.0	66.3	74.8	101.4	61.3	73.7
0950, 0980	Ceilings & Acoustic Treatment	92.0	68.5	77.0	94.5	70.4	79.2	92.8	101.8	98.5	104.5	61.3	77.0	109.6	66.3	82.0	111.3	61.3	79.5
0960	Flooring	110.0	43.4	91.6	101.5	93.2	99.2	97.3	98.6	97.6	109.4	75.7	100.1	106.1	79.9	98.9	114.7	79.9	105.1
0970, 0990	Wall Finishes & Painting/Coating	98.2	55.8	72.7	96.0	82.5	87.9	101.0	109.0	105.8	100.7	52.0	71.4	98.6	56.9	73.5	98.6	54.0	71.8
09	FINISHES	101.3	63.7	80.9	101.0	78.4	88.7	98.8	102.7	100.9	104.4	63.8	82.4	104.0	68.1	84.5	108.2	65.3	84.9
COVERS	DIVS. 10 - 14, 25, 28, 41, 43, 44	100.0	92.8	98.5	100.0	94.0	98.7	100.0	104.5	101.0	100.0	81.1	96.0	100.0	81.8	96.1	100.0	81.9	96.2
21, 22, 23	FIRE SUPPRESSION, PLUMBING & HVAC	100.0	67.7	86.7	100.1	85.0	93.9	99.8	105.0	102.0	100.0	72.4	88.6	100.1	66.0	86.1	100.1	70.1	87.8
26, 27, 3370	ELECTRICAL, COMMUNICATIONS & UTIL.	99.3	70.3	84.1	99.6	80.7	89.7	103.3	110.5	107.0	93.9	72.6	82.7	101.8	69.7	85.0	93.9	69.7	81.2
MF2004	WEIGHTED AVERAGE	97.7	77.8	88.9	98.0	87.2	93.3	98.6	107.2	102.4	102.0	72.9	89.2	102.0	72.3	88.9	103.4	72.2	89.6

	DIVISION	MONTANA						NEBRASKA											
		HELENA			MISSOULA			GRAND ISLAND			LINCOLN			NORTH PLATTE			OMAHA		
		MAT.	INST.	TOTAL	MAT.	INST.	TOTAL	MAT.	INST.	TOTAL	MAT.	INST.	TOTAL	MAT.	INST.	TOTAL	MAT.	INST.	TOTAL
015433	CONTRACTOR EQUIPMENT		98.8	98.8		98.8	98.8		101.2	101.2		101.2	101.2		101.2	101.2		91.8	91.8
0241, 31 - 34	SITE & INFRASTRUCTURE, DEMOLITION	104.4	96.1	98.6	84.1	96.3	92.6	100.9	91.9	94.7	90.6	91.9	91.5	101.9	91.5	94.7	80.8	90.8	87.7
0310	Concrete Forming & Accessories	99.9	63.1	68.1	92.1	65.4	69.0	100.4	79.1	82.0	105.2	62.5	68.3	100.4	78.0	81.0	95.4	69.2	72.7
0320	Concrete Reinforcing	97.1	67.1	81.8	103.5	76.4	89.7	105.3	70.0	87.3	96.7	69.5	82.8	106.7	70.8	88.4	101.3	69.7	85.1
0330	Cast-in-Place Concrete	108.5	63.9	91.6	89.7	65.0	80.3	117.0	71.4	99.7	93.3	74.0	86.0	117.0	68.0	98.4	100.3	75.9	91.1
03	CONCRETE	102.1	65.2	84.9	84.1	68.4	76.7	108.2	75.4	92.8	94.1	68.8	82.3	108.4	73.8	92.2	96.7	71.9	85.1
04	MASONRY	118.2	68.9	88.5	146.2	70.2	100.5	108.1	81.6	92.2	96.9	71.7	81.8	92.3	93.4	93.0	102.4	77.2	87.2
05	METALS	101.1	83.4	95.1	95.6	88.0	93.0	89.9	84.7	88.2	94.8	83.7	91.0	90.2	85.3	88.5	94.1	75.4	87.8
06	WOOD, PLASTICS & COMPOSITES	102.2	62.2	79.5	93.3	65.7	77.7	97.9	81.0	88.4	100.6	58.8	76.9	97.8	81.0	88.3	91.3	68.2	78.2
07	THERMAL & MOISTURE PROTECTION	102.3	70.9	90.5	101.5	70.5	89.8	100.0	77.4	91.5	99.6	72.5	89.3	100.0	84.6	94.2	95.2	79.8	89.4
08	OPENINGS	98.7	59.0	88.9	95.9	64.5	88.1	90.4	74.5	86.5	95.9	58.2	86.6	89.7	73.8	85.8	98.9	68.0	91.2
0920	Plaster & Gypsum Board	93.2	61.3	71.1	94.5	65.0	74.1	95.7	80.3	85.1	91.4	57.4	67.9	95.7	80.3	85.1	94.1	67.7	75.8
0950, 0980	Ceilings & Acoustic Treatment	107.0	61.3	77.9	109.6	65.0	81.2	92.0	80.3	84.6	93.8	57.4	70.6	92.0	80.3	84.6	107.1	67.7	82.0
0960	Flooring	114.7	63.6	100.6	108.3	79.9	100.5	99.2	115.5	103.7	100.9	90.0	97.9	99.1	108.1	101.6	129.0	52.3	107.8
0970, 0990	Wall Finishes & Painting/Coating	98.6	48.0	68.2	98.6	52.9	71.1	96.2	75.1	83.5	96.2	86.5	90.4	96.2	70.8	80.9	167.0	69.3	108.2
09	FINISHES	106.0	61.5	81.8	103.5	66.6	83.5	98.3	85.0	91.1	98.3	69.6	82.7	98.2	82.5	89.7	119.9	65.4	90.3
COVERS	DIVS. 10 - 14, 25, 28, 41, 43, 44	100.0	81.4	96.1	100.0	81.1	96.0	100.0	83.2	96.4	100.0	80.6	95.9	100.0	77.4	95.2	100.0	80.4	95.8
21, 22, 23	FIRE SUPPRESSION, PLUMBING & HVAC	100.0	66.2	86.1	100.1	64.6	85.5	100.0	75.4	89.9	99.9	75.4	89.8	100.0	73.7	89.2	99.7	75.3	89.7
26, 27, 3370	ELECTRICAL, COMMUNICATIONS & UTIL.	99.7	69.7	84.0	99.5	73.7	86.0	89.7	71.1	79.9	103.1	71.1	86.3	91.8	86.4	89.0	102.9	80.0	90.9
MF2004	WEIGHTED AVERAGE	101.9	70.6	88.1	99.2	72.7	87.5	97.9	79.2	89.6	97.8	74.1	87.3	97.2	81.6	90.4	99.8	75.6	89.2

DIVISION		NEVADA									NEW HAMPSHIRE								
		CARSON CITY			LAS VEGAS			RENO			MANCHESTER			NASHUA			PORTSMOUTH		
		MAT.	INST.	TOTAL	MAT.	INST.	TOTAL	MAT.	INST.	TOTAL	MAT.	INST.	TOTAL	MAT.	INST.	TOTAL	MAT.	INST.	TOTAL
015433	CONTRACTOR EQUIPMENT		100.7	100.7		100.7	100.7		100.7	100.7		100.6	100.6		100.6	100.6		100.6	100.6
0241, 31 - 34	SITE & INFRASTRUCTURE, DEMOLITION	68.0	101.5	91.3	62.6	105.1	92.2	60.8	101.5	89.2	81.9	98.8	93.7	84.4	98.8	94.5	78.8	100.2	93.7
0310	Concrete Forming & Accessories	100.7	88.9	90.5	103.0	116.4	114.6	98.1	89.0	90.2	96.0	83.0	84.7	97.9	83.0	85.0	88.3	87.1	87.2
0320	Concrete Reinforcing	103.5	107.8	105.7	95.5	125.7	110.9	98.2	124.2	111.5	110.6	93.4	101.8	110.6	93.4	101.8	88.5	93.5	91.0
0330	Cast-in-Place Concrete	104.8	87.7	98.4	102.0	116.3	107.4	111.5	87.8	102.5	102.0	113.3	106.3	89.1	113.3	98.3	84.5	119.6	97.8
03	CONCRETE	107.1	92.2	100.1	104.7	117.6	110.8	109.5	95.3	102.9	109.6	95.3	102.9	103.3	95.3	99.6	94.7	99.4	96.9
04	MASONRY	120.3	75.0	93.1	114.6	112.5	113.4	121.1	75.0	93.4	98.3	84.5	90.0	98.1	84.5	89.9	93.7	94.3	94.0
05	METALS	96.9	98.6	97.5	103.2	111.1	105.9	97.3	105.4	100.1	99.6	89.5	96.2	99.5	89.5	96.2	95.9	92.5	94.8
06	WOOD, PLASTICS & COMPOSITES	93.3	89.7	91.3	93.3	115.7	106.0	89.8	89.7	89.8	96.5	83.7	89.3	99.8	83.7	90.7	88.2	83.7	85.7
07	THERMAL & MOISTURE PROTECTION	97.0	84.4	92.2	112.1	104.9	109.4	98.3	84.4	93.0	95.5	90.1	93.5	96.1	90.1	93.9	95.8	114.5	102.9
08	OPENINGS	93.6	103.8	96.1	95.2	122.3	101.9	94.0	103.6	96.4	104.3	78.7	98.0	104.3	78.7	98.0	105.0	73.8	97.3
0920	Plaster & Gypsum Board	86.8	89.3	88.5	93.9	116.0	109.2	87.0	89.3	88.6	104.1	82.5	89.1	100.4	82.5	88.0	92.1	82.5	85.5
0950, 0980	Ceilings & Acoustic Treatment	121.0	89.3	100.8	135.6	116.0	123.1	131.3	89.3	104.5	96.1	82.5	87.4	98.7	82.5	88.4	90.9	82.5	85.6
0960	Flooring	104.0	55.1	90.5	97.9	106.4	100.2	103.6	55.1	90.2	98.6	96.4	98.0	97.6	96.4	97.3	92.4	96.4	93.5
0970, 0990	Wall Finishes & Painting/Coating	98.9	84.8	90.4	100.3	122.4	113.6	97.9	84.8	90.0	94.6	101.8	98.9	94.6	101.8	98.9	94.6	101.8	98.9
09	FINISHES	104.8	81.6	92.2	106.4	115.0	111.1	106.2	81.6	92.8	99.3	87.3	92.8	99.7	87.3	93.0	94.9	89.6	92.0
COVERS	DIVS. 10 - 14, 25, 28, 41, 43, 44	100.0	103.7	100.8	100.0	101.7	100.4	100.0	103.7	100.8	100.0	66.4	92.9	100.0	66.4	92.9	100.0	69.8	93.6
21, 22, 23	FIRE SUPPRESSION, PLUMBING & HVAC	99.9	82.3	92.7	100.1	111.1	104.6	100.0	82.5	92.8	100.0	85.6	94.1	100.1	85.6	94.2	100.1	90.5	96.2
26, 27, 3370	ELECTRICAL, COMMUNICATIONS & UTIL.	98.1	95.5	96.7	98.4	123.0	111.3	94.5	95.5	95.0	97.4	80.3	88.4	97.9	80.3	88.7	95.8	80.3	87.7
MF2004	WEIGHTED AVERAGE	99.9	89.5	95.3	101.0	113.8	106.7	100.0	90.5	95.8	100.4	87.0	94.5	99.9	87.0	94.2	97.3	90.9	94.4

DIVISION		NEW JERSEY																	
		CAMDEN			ELIZABETH			JERSEY CITY			NEWARK			PATERSON			TRENTON		
		MAT.	INST.	TOTAL	MAT.	INST.	TOTAL	MAT.	INST.	TOTAL	MAT.	INST.	TOTAL	MAT.	INST.	TOTAL	MAT.	INST.	TOTAL
015433	CONTRACTOR EQUIPMENT		98.7	98.7		100.6	100.6		98.7	98.7		100.6	100.6		100.6	100.6		98.2	98.2
0241, 31 - 34	SITE & INFRASTRUCTURE, DEMOLITION	87.3	104.4	99.2	99.9	105.0	103.4	87.3	105.4	99.9	102.6	105.0	104.2	98.5	105.5	103.4	87.1	105.3	99.8
0310	Concrete Forming & Accessories	97.9	118.8	115.9	104.9	119.4	117.4	97.9	119.5	116.6	94.6	119.4	116.0	96.7	119.4	116.3	95.1	119.0	115.7
0320	Concrete Reinforcing	110.6	121.9	116.4	85.2	123.8	104.9	110.6	123.8	117.4	110.6	123.8	117.4	110.6	123.8	117.4	110.6	119.3	115.0
0330	Cast-in-Place Concrete	77.5	133.4	98.7	85.3	130.4	102.4	77.5	130.5	97.5	99.1	130.4	110.9	98.8	130.4	110.8	88.2	131.4	104.6
03	CONCRETE	97.6	123.4	109.7	100.2	123.3	111.0	97.6	123.2	109.6	108.1	123.3	115.2	108.0	123.3	115.2	102.7	122.4	111.9
04	MASONRY	91.7	128.4	113.8	111.4	129.0	122.0	89.2	129.0	113.1	100.4	129.0	117.6	95.9	129.0	115.8	87.8	128.4	112.2
05	METALS	99.4	106.6	101.8	95.6	112.1	101.2	99.5	109.7	102.9	99.4	112.1	103.7	94.7	112.1	100.6	94.6	107.1	98.8
06	WOOD, PLASTICS & COMPOSITES	99.8	115.9	108.9	112.6	115.9	114.5	99.8	115.9	108.9	98.5	115.9	108.4	101.5	115.9	109.7	96.9	115.9	107.6
07	THERMAL & MOISTURE PROTECTION	95.7	125.0	106.8	96.3	128.6	108.5	95.7	128.6	108.1	95.8	128.6	108.2	96.3	125.3	107.2	91.9	125.4	104.5
08	OPENINGS	104.3	118.0	107.7	105.1	119.7	108.7	104.3	119.7	108.1	110.1	119.7	112.5	110.1	119.7	112.5	104.9	118.2	108.2
0920	Plaster & Gypsum Board	100.4	115.7	111.0	104.2	115.7	112.1	100.4	115.7	111.0	104.1	115.7	112.1	100.4	115.7	111.0	104.1	115.7	112.1
0950, 0980	Ceilings & Acoustic Treatment	98.7	115.7	109.5	90.9	115.7	106.7	98.7	115.7	109.5	96.1	115.7	108.6	98.7	115.7	109.5	96.1	115.7	108.6
0960	Flooring	97.6	136.9	108.4	102.1	145.8	114.1	97.6	145.8	110.9	97.8	145.8	111.0	97.6	145.8	110.9	98.0	145.8	111.2
0970, 0990	Wall Finishes & Painting/Coating	94.6	127.6	114.5	94.5	129.7	115.7	94.6	129.7	115.7	94.5	129.7	115.7	94.5	129.7	115.7	94.6	127.6	114.5
09	FINISHES	100.1	122.8	112.4	100.9	124.7	113.8	100.1	124.7	113.5	100.2	124.7	113.5	100.3	124.7	113.5	99.9	124.3	113.2
COVERS	DIVS. 10 - 14, 25, 28, 41, 43, 44	100.0	111.6	102.5	100.0	118.4	103.9	100.0	118.4	103.9	100.0	118.4	103.9	100.0	118.4	103.9	100.0	111.5	102.4
21, 22, 23	FIRE SUPPRESSION, PLUMBING & HVAC	100.1	126.3	110.9	100.1	127.1	111.2	100.1	129.6	112.3	100.2	127.1	111.3	100.1	129.6	112.2	100.2	126.1	110.9
26, 27, 3370	ELECTRICAL, COMMUNICATIONS & UTIL.	98.3	139.6	119.9	95.0	136.2	116.5	99.4	140.0	120.6	99.4	140.0	120.7	99.4	136.2	118.6	102.5	142.2	123.3
MF2004	WEIGHTED AVERAGE	99.0	122.9	109.5	100.1	123.8	110.6	99.0	124.7	110.3	101.9	124.3	111.8	100.9	124.3	111.2	99.0	123.4	109.8

DIVISION		NEW MEXICO															NEW YORK		
		ALBUQUERQUE			FARMINGTON			LAS CRUCES			ROSWELL			SANTA FE			ALBANY		
		MAT.	INST.	TOTAL	MAT.	INST.	TOTAL	MAT.	INST.	TOTAL	MAT.	INST.	TOTAL	MAT.	INST.	TOTAL	MAT.	INST.	TOTAL
015433	CONTRACTOR EQUIPMENT		113.4	113.4		113.4	113.4		87.8	87.8		113.4	113.4		113.4	113.4		114.6	114.6
0241, 31 - 34	SITE & INFRASTRUCTURE, DEMOLITION	77.2	108.6	99.1	83.6	108.6	101.0	87.8	86.5	86.9	88.2	108.6	102.4	82.9	108.6	100.8	73.0	107.5	97.0
0310	Concrete Forming & Accessories	98.8	68.8	72.8	98.9	68.8	72.9	95.6	67.1	70.9	98.9	68.5	72.6	98.1	68.8	72.7	98.1	96.2	96.5
0320	Concrete Reinforcing	98.9	71.5	84.9	108.1	71.5	89.4	101.6	48.2	74.3	107.2	48.5	77.2	106.1	71.5	88.4	94.7	92.9	93.7
0330	Cast-in-Place Concrete	104.2	76.1	93.6	105.2	76.1	94.2	91.5	67.9	82.6	96.9	76.0	89.0	120.2	76.1	103.5	80.9	103.4	89.4
03	CONCRETE	106.1	72.8	90.5	109.9	72.8	92.6	86.5	64.5	76.2	110.5	68.2	90.7	114.0	72.8	95.2	98.7	98.8	98.8
04	MASONRY	111.2	63.5	82.6	117.6	63.5	85.1	104.5	61.1	78.4	118.9	63.5	85.6	113.7	63.5	83.6	92.1	102.0	98.0
05	METALS	103.1	88.8	98.3	100.9	88.8	96.8	100.8	70.3	90.5	100.9	77.5	93.0	100.9	88.8	96.8	99.0	105.6	101.2
06	WOOD, PLASTICS & COMPOSITES	93.4	69.7	80.0	93.5	69.7	80.0	83.8	68.5	75.1	93.5	69.7	80.0	92.9	69.7	79.8	97.5	94.7	95.9
07	THERMAL & MOISTURE PROTECTION	97.8	72.8	88.4	98.1	72.8	88.5	84.5	67.4	78.1	98.3	72.8	88.7	97.7	72.8	88.3	90.2	96.0	92.4
08	OPENINGS	93.8	71.6	88.3	96.6	71.6	90.4	86.8	63.7	81.0	92.5	64.4	85.6	92.7	71.6	87.5	96.1	88.3	94.2
0920	Plaster & Gypsum Board	92.2	68.4	75.7	83.9	68.4	73.2	86.5	68.4	73.9	83.9	68.4	73.2	93.7	68.4	76.2	106.0	94.3	97.9
0950, 0980	Ceilings & Acoustic Treatment	125.7	68.4	89.2	121.0	68.4	87.5	113.2	68.4	84.7	121.0	68.4	87.5	121.0	68.4	87.5	93.6	94.3	94.0
0960	Flooring	101.2	66.3	91.6	103.6	66.3	93.3	137.0	62.6	116.4	103.6	66.3	93.3	103.6	66.3	93.3	85.0	102.6	89.8
0970, 0990	Wall Finishes & Painting/Coating	102.4	53.1	72.7	97.9	53.1	70.9	90.4	53.1	67.9	97.9	53.1	70.9	97.9	53.1	70.9	82.9	82.7	82.8
09	FINISHES	105.6	66.7	84.5	104.2	66.7	83.8	118.9	65.0	89.7	104.7	66.7	84.1	105.5	66.7	84.4	94.4	95.9	95.2
COVERS	DIVS. 10 - 14, 25, 28, 41, 43, 44	100.0	71.3	93.9	100.0	71.3	93.9	100.0	68.1	93.2	100.0	71.3	93.9	100.0	71.3	93.9	100.0	97.9	99.6
21, 22, 23	FIRE SUPPRESSION, PLUMBING & HVAC	100.1	70.8	88.0	100.0	70.8	88.0	100.3	70.2	87.9	100.0	70.6	87.9	100.0	70.8	88.0	100.1	93.1	97.2
26, 27, 3370	ELECTRICAL, COMMUNICATIONS & UTIL.	89.5	70.9	79.8	87.6	70.9	78.9	88.4	54.9	70.9	88.4	70.9	79.3	100.9	70.9	85.3	104.7	93.4	98.8
MF2004	WEIGHTED AVERAGE	99.8	74.8	88.8	100.4	74.8	89.1	96.8	66.7	83.5	100.4	72.8	88.2	101.8	74.8	89.9	97.7	97.6	97.7

DIVISION		NEW YORK																	
		BINGHAMTON			BUFFALO			HICKSVILLE			NEW YORK			RIVERHEAD			ROCHESTER		
		MAT.	INST.	TOTAL	MAT.	INST.	TOTAL	MAT.	INST.	TOTAL	MAT.	INST.	TOTAL	MAT.	INST.	TOTAL	MAT.	INST.	TOTAL
015433	CONTRACTOR EQUIPMENT		114.9	114.9		96.3	96.3		115.8	115.8		115.5	115.5		115.8	115.8		117.1	117.1
0241, 31 - 34	SITE & INFRASTRUCTURE, DEMOLITION	95.1	91.5	92.6	97.2	96.7	96.9	114.2	128.9	124.5	128.2	126.8	127.2	115.4	128.9	124.8	73.0	109.1	98.1
0310	Concrete Forming & Accessories	102.4	82.5	85.2	98.3	112.5	110.6	93.2	146.8	139.6	111.6	186.1	176.0	97.4	146.7	140.1	99.6	96.6	97.0
0320	Concrete Reinforcing	93.8	93.1	93.4	93.7	97.5	95.7	95.4	187.7	142.5	102.0	188.0	145.9	97.2	184.6	141.8	94.5	89.8	92.1
0330	Cast-in-Place Concrete	99.8	93.9	97.6	107.6	116.9	111.2	93.4	156.6	117.3	107.6	171.3	131.7	94.9	156.5	118.2	96.8	99.8	97.9
03	CONCRETE	98.5	90.1	94.6	109.2	110.5	109.8	103.3	156.6	128.2	115.3	179.1	145.1	103.9	155.9	128.2	114.6	97.3	106.5
04	MASONRY	107.7	87.3	95.5	107.0	117.0	113.0	111.4	158.5	139.7	107.4	170.5	145.3	117.1	158.5	142.0	103.1	101.6	102.2
05	METALS	95.7	114.8	102.2	99.4	93.2	97.3	105.0	140.0	116.8	115.5	143.4	125.0	105.4	138.8	116.7	96.0	106.4	99.5
06	WOOD, PLASTICS & COMPOSITES	107.3	82.3	93.1	101.2	112.3	107.5	90.1	145.4	121.4	112.8	190.4	156.8	95.2	145.4	123.6	98.1	96.7	97.3
07	THERMAL & MOISTURE PROTECTION	104.9	84.1	97.1	101.7	108.4	104.2	107.4	150.5	123.7	110.9	166.2	131.8	108.0	150.5	124.0	97.2	96.7	97.0
08	OPENINGS	91.0	80.7	88.4	94.5	100.4	96.0	88.9	153.1	104.8	97.4	179.4	117.7	88.9	150.8	104.2	98.3	89.3	96.0
0920	Plaster & Gypsum Board	119.4	81.3	93.0	99.4	112.4	108.4	106.0	146.7	134.2	128.6	192.7	172.9	108.3	146.7	134.9	98.4	96.6	97.2
0950, 0980	Ceilings & Acoustic Treatment	98.7	81.3	87.6	103.9	112.4	109.3	82.4	146.7	123.3	105.3	192.7	161.0	84.1	146.7	123.9	103.1	96.6	99.0
0960	Flooring	95.7	89.1	93.9	91.4	115.0	97.9	95.4	153.7	111.5	93.6	178.4	117.1	96.6	153.7	112.4	81.6	103.5	87.7
0970, 0990	Wall Finishes & Painting/Coating	86.9	86.2	86.5	91.2	111.8	103.6	112.3	149.3	134.5	91.2	156.8	130.7	112.3	149.3	134.5	90.4	96.7	94.2
09	FINISHES	97.6	83.8	90.1	95.0	113.7	105.1	105.8	147.7	128.5	108.5	183.1	149.0	106.8	147.7	129.0	94.6	98.2	96.5
COVERS	DIVS. 10 - 14, 25, 28, 41, 43, 44	100.0	94.1	98.7	100.0	105.2	101.1	100.0	127.3	105.8	100.0	138.7	108.2	100.0	127.3	105.8	100.0	99.7	99.9
21, 22, 23	FIRE SUPPRESSION, PLUMBING & HVAC	100.4	82.6	93.1	100.0	94.7	97.8	99.8	142.6	117.4	100.3	163.8	126.4	99.9	141.2	116.9	100.0	91.9	96.7
26, 27, 3370	ELECTRICAL, COMMUNICATIONS & UTIL.	100.9	91.6	96.0	98.3	95.5	96.8	99.8	148.8	125.4	107.2	178.5	144.5	101.6	148.8	126.3	99.8	90.9	95.1
MF2004	WEIGHTED AVERAGE	98.8	89.5	94.7	100.3	102.8	101.4	101.6	146.4	121.4	107.0	166.3	133.2	102.4	145.9	121.6	99.8	97.4	98.7

DIVISION		NEW YORK																	
		SCHENECTADY			SYRACUSE			UTICA			WATERTOWN			WHITE PLAINS			YONKERS		
		MAT.	INST.	TOTAL	MAT.	INST.	TOTAL	MAT.	INST.	TOTAL	MAT.	INST.	TOTAL	MAT.	INST.	TOTAL	MAT.	INST.	TOTAL
015433	CONTRACTOR EQUIPMENT		114.6	114.6		114.6	114.6		114.6	114.6		114.6	114.6		115.2	115.2		115.2	115.2
0241, 31 - 34	SITE & INFRASTRUCTURE, DEMOLITION	72.9	107.5	97.0	94.0	106.5	102.7	71.1	104.6	94.4	79.2	107.0	98.5	118.5	121.0	120.3	127.6	120.8	122.9
0310	Concrete Forming & Accessories	102.1	95.9	96.7	101.1	88.4	90.1	102.8	82.4	85.2	89.4	91.7	91.4	111.0	132.7	129.7	111.3	132.7	129.8
0320	Concrete Reinforcing	93.4	92.9	93.1	94.7	90.5	92.5	94.7	90.8	92.7	95.3	90.1	92.6	94.9	187.1	142.0	98.7	187.2	143.9
0330	Cast-in-Place Concrete	91.6	102.8	95.9	92.6	97.7	94.5	84.7	90.9	87.0	98.6	98.4	98.5	94.8	136.0	110.4	106.1	136.0	117.4
03	CONCRETE	104.1	98.4	101.4	102.6	92.9	98.0	100.9	87.9	94.8	114.7	94.5	105.2	103.4	143.0	121.9	114.2	143.0	127.7
04	MASONRY	91.6	101.0	97.2	98.4	98.9	98.7	90.2	83.5	86.2	91.3	97.5	95.0	101.9	134.2	121.3	107.0	134.2	123.4
05	METALS	99.1	105.6	101.3	99.0	103.0	100.3	96.9	102.7	98.9	97.0	102.6	98.9	101.3	136.9	113.3	111.1	137.0	119.9
06	WOOD, PLASTICS & COMPOSITES	103.4	94.7	98.5	103.4	86.8	94.0	103.4	84.0	92.4	86.9	90.6	89.0	114.0	135.4	126.1	113.8	135.4	126.0
07	THERMAL & MOISTURE PROTECTION	90.6	95.5	92.5	101.2	96.4	99.4	90.4	90.3	90.4	90.6	94.6	92.1	111.3	140.5	122.3	111.5	140.5	122.5
08	OPENINGS	96.1	88.3	94.2	94.1	80.2	90.7	96.1	81.0	92.3	96.1	86.1	93.6	91.9	147.7	105.7	95.5	149.5	108.9
0920	Plaster & Gypsum Board	113.7	94.3	100.2	113.7	86.1	94.6	113.7	83.3	92.6	106.3	90.1	95.1	121.9	136.0	131.7	128.1	136.0	133.6
0950, 0980	Ceilings & Acoustic Treatment	98.7	94.3	95.9	98.7	86.1	90.7	98.7	83.3	88.9	98.7	90.1	93.2	81.1	136.0	116.1	103.6	136.0	124.2
0960	Flooring	84.6	102.6	89.6	86.3	89.1	87.1	84.6	93.0	86.9	77.0	93.0	81.5	90.1	162.9	110.2	89.6	162.9	109.9
0970, 0990	Wall Finishes & Painting/Coating	82.9	82.7	82.8	89.1	93.2	91.5	82.9	82.5	82.7	82.9	89.2	86.7	89.2	149.3	125.3	89.2	149.3	125.3
09	FINISHES	96.2	95.7	95.9	97.8	88.8	92.9	96.2	84.3	89.7	93.7	91.8	92.7	100.1	139.7	121.6	106.6	139.7	124.6
COVERS	DIVS. 10 - 14, 25, 28, 41, 43, 44	100.0	97.6	99.5	100.0	96.5	99.2	100.0	91.7	98.2	100.0	90.5	98.0	100.0	121.3	104.5	100.0	122.1	104.7
21, 22, 23	FIRE SUPPRESSION, PLUMBING & HVAC	100.2	92.6	97.1	100.2	85.9	94.3	100.2	82.8	93.1	100.2	78.9	91.5	100.5	124.5	110.4	100.5	124.5	110.4
26, 27, 3370	ELECTRICAL, COMMUNICATIONS & UTIL.	100.9	93.4	97.0	100.9	91.3	95.9	98.8	90.4	94.4	100.9	90.4	95.4	96.6	152.0	125.5	105.1	152.0	129.6
MF2004	WEIGHTED AVERAGE	98.2	97.3	97.8	99.3	93.0	96.5	97.2	88.9	93.5	99.1	91.9	95.9	100.7	136.0	116.3	105.7	136.1	119.2

DIVISION		NORTH CAROLINA																	
		ASHEVILLE			CHARLOTTE			DURHAM			FAYETTEVILLE			GREENSBORO			RALEIGH		
		MAT.	INST.	TOTAL	MAT.	INST.	TOTAL	MAT.	INST.	TOTAL	MAT.	INST.	TOTAL	MAT.	INST.	TOTAL	MAT.	INST.	TOTAL
015433	CONTRACTOR EQUIPMENT		95.2	95.2		95.2	95.2		100.8	100.8		100.8	100.8		100.8	100.8		100.8	100.8
0241, 31 - 34	SITE & INFRASTRUCTURE, DEMOLITION	106.6	75.1	84.6	110.9	75.1	86.0	107.0	84.0	91.0	105.7	84.1	90.7	106.9	84.1	91.0	111.1	84.2	92.3
0310	Concrete Forming & Accessories	95.4	39.3	46.9	99.4	41.3	49.1	97.6	39.9	47.7	95.0	40.7	48.0	97.4	40.6	48.3	97.8	41.5	49.1
0320	Concrete Reinforcing	100.5	43.0	71.2	101.0	40.1	69.9	102.2	50.3	75.7	104.4	50.4	76.8	101.0	59.0	79.5	101.0	50.4	75.1
0330	Cast-in-Place Concrete	106.9	46.7	84.1	120.1	48.7	93.1	108.6	46.9	85.3	112.1	50.0	88.6	107.8	48.2	85.3	123.2	53.5	96.8
03	CONCRETE	106.1	44.5	77.3	112.2	45.5	81.0	106.6	46.2	78.4	108.1	47.6	79.8	106.0	48.6	79.2	113.6	49.2	83.5
04	MASONRY	88.3	40.3	59.4	95.1	44.1	64.5	93.9	36.3	59.2	92.0	41.3	61.5	91.9	36.9	58.9	94.2	42.2	62.9
05	METALS	85.0	74.3	81.4	88.9	73.4	83.6	97.7	77.5	90.9	101.3	77.6	93.3	92.1	80.9	88.3	89.9	77.6	85.7
06	WOOD, PLASTICS & COMPOSITES	95.3	38.8	63.3	100.3	41.0	66.7	97.5	40.3	65.1	94.2	40.3	63.7	97.2	40.2	64.9	97.5	40.9	65.5
07	THERMAL & MOISTURE PROTECTION	99.3	43.3	78.1	97.0	44.9	77.3	100.0	45.4	79.3	99.2	43.9	78.3	99.8	42.9	78.3	99.1	44.1	78.3
08	OPENINGS	94.3	38.6	80.5	98.5	40.1	84.0	98.5	42.7	84.6	94.4	42.7	81.6	98.5	45.0	85.2	95.2	42.8	82.2
0920	Plaster & Gypsum Board	103.3	36.8	57.2	105.1	39.0	59.3	106.6	38.3	59.3	107.2	38.3	59.5	108.6	38.2	59.9	105.1	39.0	59.3
0950, 0980	Ceilings & Acoustic Treatment	92.6	36.8	57.0	95.2	39.0	59.4	96.1	38.3	59.3	93.5	38.3	58.3	96.1	38.2	59.2	95.2	39.0	59.4
0960	Flooring	106.8	44.9	89.7	110.4	45.0	92.3	110.9	44.9	92.7	106.9	44.9	89.8	110.9	41.5	91.7	110.9	44.9	92.7
0970, 0990	Wall Finishes & Painting/Coating	116.0	35.6	67.6	116.0	40.6	70.6	116.0	37.3	68.6	116.0	35.6	67.6	116.0	34.6	67.0	116.0	39.2	69.8
09	FINISHES	99.4	39.7	67.0	101.2	41.7	68.9	101.9	40.1	68.4	100.1	40.5	67.8	102.2	39.8	68.3	101.6	41.6	69.0
COVERS	DIVS. 10 - 14, 25, 28, 41, 43, 44	100.0	70.6	93.8	100.0	71.1	93.9	100.0	75.6	94.8	100.0	76.4	95.0	100.0	70.8	93.8	100.0	76.8	95.1
21, 22, 23	FIRE SUPPRESSION, PLUMBING & HVAC	100.2	39.0	75.1	100.0	39.2	75.0	100.3	37.8	74.6	100.1	39.0	75.0	100.2	39.1	75.1	100.1	37.3	74.2
26, 27, 3370	ELECTRICAL, COMMUNICATIONS & UTIL.	98.3	38.6	67.1	100.5	50.3	74.2	99.0	48.9	72.8	97.1	44.4	69.6	98.0	38.8	67.1	104.7	38.5	70.1
MF2004	WEIGHTED AVERAGE	97.4	47.2	75.3	99.8	49.7	77.7	100.3	49.7	78.0	100.0	50.1	78.0	99.2	49.2	77.1	100.3	49.4	77.8

	DIVISION	NORTH CAROLINA						NORTH DAKOTA											
		WILMINGTON			WINSTON-SALEM			BISMARCK			FARGO			GRAND FORKS			MINOT		
		MAT.	INST.	TOTAL	MAT.	INST.	TOTAL	MAT.	INST.	TOTAL	MAT.	INST.	TOTAL	MAT.	INST.	TOTAL	MAT.	INST.	TOTAL
015433	CONTRACTOR EQUIPMENT		95.2	95.2		100.8	100.8		98.8	98.8		98.8	98.8		98.8	98.8		98.8	98.8
0241, 31 - 34	SITE & INFRASTRUCTURE, DEMOLITION	108.0	75.0	85.1	107.2	84.1	91.1	99.3	96.7	97.5	98.8	96.7	97.3	108.1	94.1	98.3	105.3	96.7	99.3
0310	Concrete Forming & Accessories	96.6	39.9	47.5	98.6	39.6	47.6	97.2	43.3	50.6	98.5	43.6	51.0	94.6	37.5	45.2	91.1	57.8	62.3
0320	Concrete Reinforcing	101.3	48.6	74.4	101.0	39.9	69.8	101.8	66.3	83.7	93.8	65.6	79.4	100.3	66.7	83.1	103.7	66.7	84.8
0330	Cast-in-Place Concrete	106.4	47.9	84.3	110.5	45.6	86.0	102.7	50.1	82.8	98.9	51.9	81.1	106.7	47.6	84.3	106.7	49.0	84.9
03	CONCRETE	106.0	46.2	78.0	107.4	43.6	77.6	99.8	51.4	77.1	107.1	52.0	81.3	107.2	47.8	79.4	105.7	57.5	83.2
04	MASONRY	77.6	38.2	53.9	92.1	38.0	59.6	104.8	55.8	75.3	106.4	41.8	67.6	107.4	62.7	80.5	107.3	62.7	80.5
05	METALS	84.9	76.6	82.1	89.9	72.8	84.1	92.5	77.7	87.5	95.0	76.6	88.8	92.5	73.7	86.1	92.8	78.4	88.0
06	WOOD, PLASTICS & COMPOSITES	96.9	39.6	64.4	97.2	39.5	64.6	84.7	38.7	58.7	84.6	39.3	58.9	81.7	35.6	55.6	77.6	58.1	66.6
07	THERMAL & MOISTURE PROTECTION	99.3	43.3	78.1	99.8	42.6	78.1	102.4	48.5	82.0	100.8	46.3	80.2	103.2	50.7	83.4	103.0	52.2	83.8
08	OPENINGS	94.5	41.2	81.3	98.5	39.4	83.8	99.3	44.3	85.7	99.2	44.6	85.7	99.3	39.0	84.4	99.4	54.8	88.4
0920	Plaster & Gypsum Board	104.4	37.6	58.1	108.6	37.5	59.4	99.4	37.2	56.3	99.4	37.8	56.7	105.0	34.0	55.9	103.2	57.1	71.3
0950, 0980	Ceilings & Acoustic Treatment	93.5	37.6	57.9	96.1	37.5	58.8	139.6	37.2	74.4	139.6	37.8	74.7	142.2	34.0	73.3	142.2	57.1	88.0
0960	Flooring	107.7	44.9	90.4	110.9	44.9	92.7	115.2	64.7	101.2	114.8	38.7	93.7	113.4	38.7	92.8	110.8	77.6	101.6
0970, 0990	Wall Finishes & Painting/Coating	116.0	34.5	67.0	116.0	36.3	68.0	98.6	32.7	58.9	98.6	67.3	79.8	98.6	25.0	54.3	98.6	27.4	55.7
09	FINISHES	100.0	39.9	67.4	102.2	40.0	68.4	114.9	44.8	76.8	114.8	43.7	76.2	116.3	35.6	72.5	115.0	58.2	84.2
COVERS	DIVS. 10 - 14, 25, 28, 41, 43, 44	100.0	76.1	94.9	100.0	70.5	93.7	100.0	75.3	94.8	100.0	75.4	94.8	100.0	37.9	86.8	100.0	77.6	95.2
21, 22, 23	FIRE SUPPRESSION, PLUMBING & HVAC	100.2	38.7	74.9	100.2	37.7	74.5	100.1	57.4	82.6	100.1	60.9	84.0	100.2	37.4	74.4	100.2	56.9	82.4
26, 27, 3370	ELECTRICAL, COMMUNICATIONS & UTIL.	98.5	37.3	66.5	98.0	42.3	68.9	101.3	63.7	81.6	102.8	63.7	82.3	97.0	60.0	77.7	100.5	64.8	81.8
MF2004	WEIGHTED AVERAGE	96.9	47.6	75.1	99.1	47.8	76.5	100.4	60.4	82.8	101.8	59.5	83.2	101.5	52.8	80.0	101.5	64.6	85.2

	DIVISION	OHIO																	
		AKRON			CANTON			CINCINNATI			CLEVELAND			COLUMBUS			DAYTON		
		MAT.	INST.	TOTAL	MAT.	INST.	TOTAL	MAT.	INST.	TOTAL	MAT.	INST.	TOTAL	MAT.	INST.	TOTAL	MAT.	INST.	TOTAL
015433	CONTRACTOR EQUIPMENT		96.6	96.6		96.6	96.6		101.1	101.1		96.9	96.9		95.6	95.6		96.1	96.1
0241, 31 - 34	SITE & INFRASTRUCTURE, DEMOLITION	91.4	104.0	100.2	91.5	103.9	100.1	75.7	105.7	96.6	91.3	104.7	100.6	90.8	100.9	97.8	74.9	105.8	96.4
0310	Concrete Forming & Accessories	99.7	93.6	94.5	99.7	82.1	84.5	97.5	81.9	84.0	99.8	99.2	99.3	97.1	87.0	88.3	97.5	77.7	80.3
0320	Concrete Reinforcing	101.4	91.4	96.3	101.4	78.8	89.9	93.2	84.5	88.7	102.0	91.8	96.8	93.2	84.5	88.8	93.2	76.6	84.7
0330	Cast-in-Place Concrete	98.2	100.6	99.1	99.2	96.5	98.2	80.2	84.9	82.0	96.4	107.8	100.7	85.9	94.9	89.3	74.6	89.1	80.1
03	CONCRETE	96.5	94.8	95.7	97.0	85.9	91.8	86.4	83.6	85.1	95.7	99.9	97.7	89.4	88.9	89.1	83.6	81.2	82.5
04	MASONRY	90.1	97.0	94.2	90.8	86.5	88.2	76.5	89.2	84.1	94.8	105.0	100.9	94.2	98.3	96.7	76.0	86.1	82.1
05	METALS	93.6	80.2	89.1	93.6	74.4	87.1	95.0	86.2	92.0	95.2	83.8	91.4	95.2	81.4	90.5	94.2	75.8	88.0
06	WOOD, PLASTICS & COMPOSITES	94.0	92.8	93.3	94.3	81.9	87.3	97.2	79.3	87.1	93.0	96.7	95.1	103.3	83.4	92.0	98.6	74.8	85.1
07	THERMAL & MOISTURE PROTECTION	105.0	96.7	101.9	106.0	91.1	100.4	101.5	91.8	97.9	103.9	108.6	105.7	100.1	94.9	98.2	106.2	86.6	98.8
08	OPENINGS	107.4	92.8	103.8	101.6	73.1	94.5	98.4	80.0	93.8	97.8	94.9	97.1	100.9	82.1	96.2	98.7	74.8	92.8
0920	Plaster & Gypsum Board	87.3	92.2	90.7	88.2	81.0	83.2	88.6	79.0	82.0	86.4	96.3	93.2	86.4	82.9	84.0	88.6	74.3	78.7
0950, 0980	Ceilings & Acoustic Treatment	86.0	92.2	90.0	86.0	81.0	82.8	92.0	79.0	83.7	84.3	96.3	91.9	83.6	82.9	83.1	93.0	74.3	81.1
0960	Flooring	102.1	84.7	97.3	102.3	77.7	95.5	106.4	92.1	102.5	102.0	102.8	102.2	92.7	87.9	91.3	109.1	82.8	101.9
0970, 0990	Wall Finishes & Painting/Coating	104.2	104.4	104.3	104.2	84.2	92.1	105.2	87.8	94.7	104.2	111.8	108.8	94.2	93.8	93.9	105.2	82.6	91.6
09	FINISHES	98.1	93.2	95.5	98.3	81.5	89.2	97.5	83.9	90.1	97.6	101.0	99.5	93.0	87.4	90.0	98.6	78.6	87.7
COVERS	DIVS. 10 - 14, 25, 28, 41, 43, 44	100.0	85.0	96.8	100.0	64.2	92.4	100.0	88.5	97.6	100.0	97.8	99.5	100.0	94.9	98.9	100.0	87.7	97.4
21, 22, 23	FIRE SUPPRESSION, PLUMBING & HVAC	99.9	91.9	96.6	99.9	77.1	90.5	100.0	84.6	93.6	99.9	99.4	99.7	99.8	90.9	96.1	100.8	81.0	92.6
26, 27, 3370	ELECTRICAL, COMMUNICATIONS & UTIL.	101.3	84.3	92.4	100.4	91.7	95.9	95.3	81.2	87.9	100.9	106.5	103.8	99.4	84.5	91.6	93.9	81.4	87.4
MF2004	WEIGHTED AVERAGE	98.7	91.9	95.7	98.2	83.7	91.8	94.9	86.4	91.1	98.0	100.2	99.0	96.9	89.8	93.7	94.7	82.9	89.5

	DIVISION	OHIO												OKLAHOMA					
		LORAIN			SPRINGFIELD			TOLEDO			YOUNGSTOWN			ENID			LAWTON		
		MAT.	INST.	TOTAL	MAT.	INST.	TOTAL	MAT.	INST.	TOTAL	MAT.	INST.	TOTAL	MAT.	INST.	TOTAL	MAT.	INST.	TOTAL
015433	CONTRACTOR EQUIPMENT		96.6	96.6		96.1	96.1		98.3	98.3		96.6	96.6		79.5	79.5		80.4	80.4
0241, 31 - 34	SITE & INFRASTRUCTURE, DEMOLITION	90.8	103.5	99.6	75.1	104.3	95.5	90.0	100.6	97.4	91.2	104.5	100.5	107.9	88.8	94.6	103.2	90.3	94.2
0310	Concrete Forming & Accessories	99.8	91.6	92.7	97.5	83.7	85.6	97.1	99.0	98.7	99.7	86.8	88.5	94.2	35.1	43.1	97.3	47.3	54.1
0320	Concrete Reinforcing	101.4	91.7	96.4	93.2	84.0	88.5	93.2	91.3	92.2	101.4	87.1	94.1	99.2	79.7	89.2	99.4	79.7	89.4
0330	Cast-in-Place Concrete	93.5	104.3	97.6	76.6	89.1	81.4	85.9	103.1	92.4	97.3	100.7	98.6	98.4	46.5	78.8	95.2	46.5	76.8
03	CONCRETE	94.2	95.2	94.7	84.6	85.3	84.9	89.3	98.4	93.6	96.1	91.0	93.7	94.9	48.0	73.0	91.2	53.5	73.5
04	MASONRY	86.6	99.9	94.0	76.2	86.1	82.2	103.1	99.8	101.1	90.4	92.9	91.9	103.7	57.8	76.1	97.8	57.8	73.7
05	METALS	94.2	81.9	90.0	94.2	78.4	88.8	95.0	87.2	92.3	93.6	78.9	88.6	94.0	69.8	85.8	97.6	69.8	88.2
06	WOOD, PLASTICS & COMPOSITES	94.0	89.9	91.7	100.2	83.1	90.5	103.3	99.3	101.0	94.0	84.9	88.9	102.4	31.6	62.3	105.1	48.2	72.9
07	THERMAL & MOISTURE PROTECTION	106.0	103.7	105.1	106.1	87.5	99.1	101.6	105.6	103.1	106.2	93.7	101.5	98.4	59.7	83.7	98.2	61.4	84.3
08	OPENINGS	101.6	91.3	99.1	96.6	81.1	92.7	98.9	94.8	97.9	101.6	86.5	97.9	95.7	46.0	83.4	97.5	54.9	86.9
0920	Plaster & Gypsum Board	87.3	89.3	88.7	88.6	82.9	84.7	86.4	99.3	95.3	87.3	84.2	85.1	81.8	30.3	46.1	84.1	47.3	58.6
0950, 0980	Ceilings & Acoustic Treatment	86.0	89.3	88.1	93.0	82.9	86.6	83.6	99.3	93.6	86.0	84.2	84.8	85.3	30.3	50.2	93.0	47.3	63.9
0960	Flooring	102.3	107.5	103.8	109.1	82.8	101.9	91.9	94.9	92.7	102.3	89.5	98.8	108.8	44.3	91.0	111.6	44.3	93.0
0970, 0990	Wall Finishes & Painting/Coating	104.2	111.8	108.8	105.2	82.6	91.6	94.2	102.8	99.4	104.2	93.8	97.9	98.6	56.9	73.6	98.6	56.9	73.6
09	FINISHES	98.2	96.6	97.3	98.6	83.4	90.4	92.7	98.9	96.1	98.2	87.6	92.4	94.2	36.8	63.0	96.5	46.5	69.4
COVERS	DIVS. 10 - 14, 25, 28, 41, 43, 44	100.0	95.1	99.0	100.0	88.6	97.6	100.0	86.5	97.1	100.0	83.0	96.4	100.0	72.8	94.2	100.0	74.7	94.6
21, 22, 23	FIRE SUPPRESSION, PLUMBING & HVAC	99.9	79.7	91.6	100.8	81.0	92.7	99.8	95.1	97.9	99.9	93.1	97.1	100.2	62.0	84.5	100.2	62.0	84.5
26, 27, 3370	ELECTRICAL, COMMUNICATIONS & UTIL.	100.6	91.0	95.6	93.9	79.0	86.1	99.6	100.9	100.3	100.6	82.6	91.2	96.7	69.9	82.7	98.5	64.0	80.5
MF2004	WEIGHTED AVERAGE	97.7	91.5	95.0	94.6	84.3	90.1	97.1	97.1	97.1	98.1	89.6	94.3	97.7	59.8	80.9	97.9	61.7	81.9

	DIVISION	OKLAHOMA									OREGON								
		MUSKOGEE			OKLAHOMA CITY			TULSA			EUGENE			MEDFORD			PORTLAND		
		MAT.	INST.	TOTAL	MAT.	INST.	TOTAL	MAT.	INST.	TOTAL	MAT.	INST.	TOTAL	MAT.	INST.	TOTAL	MAT.	INST.	TOTAL
015433	CONTRACTOR EQUIPMENT		88.0	88.0		80.7	80.7		88.0	88.0		99.3	99.3		99.3	99.3		99.3	99.3
0241, 31 - 34	SITE & INFRASTRUCTURE, DEMOLITION	93.7	84.7	87.4	102.6	90.8	94.4	100.3	86.1	90.4	105.3	104.0	104.4	114.2	104.0	107.0	108.2	104.0	105.3
0310	Concrete Forming & Accessories	98.4	31.0	40.1	96.4	41.1	48.5	98.2	40.8	48.6	106.9	95.7	97.2	103.1	95.7	96.7	108.2	96.1	97.7
0320	Concrete Reinforcing	99.2	32.6	65.2	99.4	79.7	89.4	99.4	79.6	89.3	110.1	90.9	100.3	107.2	90.9	98.9	110.9	91.2	100.8
0330	Cast-in-Place Concrete	88.5	35.9	68.6	98.0	49.3	79.6	96.4	45.3	77.1	101.0	99.0	100.3	104.3	99.0	102.3	103.8	99.2	102.1
03	CONCRETE	86.6	34.3	62.1	92.5	51.6	73.4	91.8	51.0	72.7	105.9	95.7	101.1	112.8	95.7	104.8	107.4	96.0	102.1
04	MASONRY	113.6	47.0	73.6	97.5	59.6	74.7	96.9	59.6	74.5	111.6	94.4	101.3	108.7	94.4	100.1	112.5	99.2	104.5
05	METALS	93.9	56.2	81.1	99.4	69.8	89.4	97.2	81.3	91.8	90.3	91.2	90.6	89.9	91.2	90.3	91.1	91.9	91.4
06	WOOD, PLASTICS & COMPOSITES	106.6	31.3	63.9	104.3	38.7	67.2	105.7	39.9	68.4	94.0	95.8	95.0	88.5	95.8	92.6	95.1	95.8	95.5
07	THERMAL & MOISTURE PROTECTION	98.0	41.8	76.8	95.0	61.2	82.2	98.0	56.3	82.3	106.5	88.0	99.5	107.0	90.0	100.6	106.5	96.5	102.7
08	OPENINGS	95.7	30.4	79.5	97.5	49.8	85.7	97.5	50.0	85.7	99.2	96.8	98.6	101.8	96.8	100.5	96.8	96.8	96.8
0920	Plaster & Gypsum Board	84.5	29.8	46.6	79.9	37.6	50.6	84.1	38.7	52.6	104.0	95.5	98.1	102.0	95.5	97.5	107.5	95.5	99.2
0950, 0980	Ceilings & Acoustic Treatment	93.0	29.8	52.7	91.3	37.6	57.1	93.0	38.7	58.4	101.9	95.5	97.8	115.3	95.5	102.7	103.9	95.5	98.5
0960	Flooring	112.7	35.7	91.4	111.6	44.3	93.0	111.4	46.1	93.4	112.6	95.1	107.8	110.1	95.1	105.9	110.0	95.1	105.9
0970, 0990	Wall Finishes & Painting/Coating	98.6	30.5	57.6	98.6	56.9	73.6	98.6	45.0	66.4	110.1	65.5	83.2	110.1	60.0	80.0	109.6	70.3	86.0
09	FINISHES	96.2	32.1	61.4	95.4	41.4	66.1	95.9	41.1	66.2	107.2	92.5	99.2	110.1	91.9	100.2	107.3	93.1	99.6
COVERS	DIVS. 10 - 14, 25, 28, 41, 43, 44	100.0	60.2	91.5	100.0	74.3	94.5	100.0	62.8	92.1	100.0	95.6	99.1	100.0	95.6	99.1	100.0	95.7	99.1
21, 22, 23	FIRE SUPPRESSION, PLUMBING & HVAC	100.2	24.2	68.9	100.1	62.9	84.8	100.2	50.0	79.5	100.1	93.8	97.5	100.1	93.8	97.5	100.1	98.3	99.4
26, 27, 3370	ELECTRICAL, COMMUNICATIONS & UTIL.	96.3	29.5	61.4	103.7	69.9	86.0	98.5	38.5	67.1	100.8	91.4	95.9	104.7	81.7	92.7	101.0	95.9	98.3
MF2004	WEIGHTED AVERAGE	96.9	39.6	71.6	98.6	61.7	82.3	97.7	54.8	78.8	100.9	94.3	98.0	102.7	92.9	98.4	101.1	96.8	99.2

	DIVISION	OREGON			PENNSYLVANIA														
		SALEM			ALLENTOWN			ALTOONA			ERIE			HARRISBURG			PHILADELPHIA		
		MAT.	INST.	TOTAL	MAT.	INST.	TOTAL	MAT.	INST.	TOTAL	MAT.	INST.	TOTAL	MAT.	INST.	TOTAL	MAT.	INST.	TOTAL
015433	CONTRACTOR EQUIPMENT		99.3	99.3		114.6	114.6		114.6	114.6		114.6	114.6		113.9	113.9		96.9	96.9
0241, 31 - 34	SITE & INFRASTRUCTURE, DEMOLITION	101.9	104.0	103.3	92.9	106.4	102.3	97.1	106.2	103.5	93.6	106.8	102.8	83.5	105.3	98.7	101.0	98.1	98.9
0310	Concrete Forming & Accessories	105.8	95.9	97.2	100.5	117.6	115.3	87.5	81.0	81.8	100.4	87.4	89.2	93.4	88.5	89.2	102.1	138.8	133.8
0320	Concrete Reinforcing	111.2	91.2	101.0	94.7	111.1	103.1	91.8	91.2	91.5	93.8	93.1	93.4	94.7	101.2	98.0	95.6	153.4	125.1
0330	Cast-in-Place Concrete	91.3	99.1	94.2	83.9	104.7	91.7	93.5	88.1	91.4	91.9	86.2	89.8	93.2	96.3	94.4	99.2	133.2	112.1
03	CONCRETE	101.2	95.9	98.7	97.1	112.6	104.4	91.7	86.8	89.4	90.3	89.4	89.9	99.3	95.0	97.3	100.6	139.2	118.6
04	MASONRY	116.8	99.2	106.3	95.4	104.3	100.8	97.5	71.8	82.0	87.5	90.8	89.5	95.3	92.5	93.6	97.1	133.8	119.2
05	METALS	90.7	91.7	91.0	99.2	124.6	107.8	93.2	112.0	99.6	93.4	112.7	99.9	101.2	119.2	107.3	107.5	138.1	117.8
06	WOOD, PLASTICS & COMPOSITES	91.5	95.8	93.9	102.8	121.8	113.6	82.6	83.9	83.3	98.7	86.3	91.7	97.3	87.7	91.8	101.3	139.6	123.0
07	THERMAL & MOISTURE PROTECTION	104.1	92.4	99.7	101.2	119.9	108.3	100.6	88.6	96.1	101.0	93.4	98.1	107.2	110.6	108.5	99.3	138.5	114.1
08	OPENINGS	98.7	96.8	98.2	94.1	117.9	100.0	88.6	87.8	88.4	88.7	88.8	88.8	94.1	92.8	93.8	98.3	148.5	110.7
0920	Plaster & Gypsum Board	96.2	95.5	95.7	111.1	122.2	118.8	104.4	83.1	89.7	111.1	85.7	93.5	108.4	87.0	93.6	103.5	140.6	129.2
0950, 0980	Ceilings & Acoustic Treatment	102.4	95.5	98.0	90.1	122.2	110.6	93.6	83.1	86.9	90.1	85.7	87.3	91.8	87.0	88.8	94.0	140.6	123.7
0960	Flooring	112.8	95.1	107.9	86.3	94.9	88.6	80.2	49.1	71.6	87.8	73.1	83.7	86.8	86.4	86.7	85.6	143.3	101.5
0970, 0990	Wall Finishes & Painting/Coating	110.1	65.5	83.2	89.1	78.4	82.6	85.1	105.5	97.4	95.1	92.2	93.4	89.1	87.7	88.3	94.0	146.9	125.8
09	FINISHES	106.1	92.5	98.7	95.5	108.9	102.8	94.0	76.7	84.6	96.9	84.6	90.2	94.4	87.7	90.8	97.2	140.4	120.7
COVERS	DIVS. 10 - 14, 25, 28, 41, 43, 44	100.0	95.6	99.1	100.0	107.5	101.6	100.0	96.1	99.2	100.0	100.7	100.2	100.0	98.0	99.6	100.0	122.1	104.7
21, 22, 23	FIRE SUPPRESSION, PLUMBING & HVAC	100.0	93.8	97.5	100.2	113.3	105.6	99.8	85.5	93.9	99.8	93.5	97.2	100.1	91.1	96.4	100.0	131.7	113.0
26, 27, 3370	ELECTRICAL, COMMUNICATIONS & UTIL.	108.2	91.4	99.4	100.1	98.4	99.2	89.5	102.7	96.4	91.4	84.1	87.6	98.8	85.1	91.6	97.1	144.4	121.8
MF2004	WEIGHTED AVERAGE	101.0	95.0	98.4	98.2	110.5	103.6	94.9	90.3	92.9	94.7	93.1	94.0	98.5	95.1	97.0	100.3	134.2	115.3

	DIVISION	PENNSYLVANIA												PUERTO RICO			RHODE ISLAND		
		PITTSBURGH			READING			SCRANTON			YORK			SAN JUAN			PROVIDENCE		
		MAT.	INST.	TOTAL	MAT.	INST.	TOTAL	MAT.	INST.	TOTAL	MAT.	INST.	TOTAL	MAT.	INST.	TOTAL	MAT.	INST.	TOTAL
015433	CONTRACTOR EQUIPMENT		112.9	112.9		117.7	117.7		114.6	114.6		113.9	113.9		89.7	89.7		102.2	102.2
0241, 31 - 34	SITE & INFRASTRUCTURE, DEMOLITION	109.4	107.4	108.0	98.4	111.4	107.5	93.3	106.6	102.6	82.1	105.3	98.3	119.8	90.8	99.6	81.2	104.0	97.1
0310	Concrete Forming & Accessories	98.1	96.6	96.8	101.6	89.5	91.1	100.7	90.5	91.9	83.1	88.8	88.0	92.6	18.1	28.1	99.1	117.1	114.7
0320	Concrete Reinforcing	93.7	103.5	98.7	93.9	101.3	97.7	94.7	107.6	101.3	93.8	101.2	97.5	198.7	12.8	103.8	110.6	114.7	112.7
0330	Cast-in-Place Concrete	97.3	98.6	97.8	72.5	95.4	81.2	87.6	93.1	89.7	82.6	96.7	87.9	101.1	31.1	74.6	95.1	115.4	102.8
03	CONCRETE	97.5	99.7	98.5	87.2	95.2	90.9	99.0	96.0	97.6	97.7	95.2	96.5	112.2	22.5	70.2	109.1	115.6	112.1
04	MASONRY	90.3	99.1	95.6	98.7	95.9	97.0	95.8	99.3	97.9	95.4	93.1	94.0	87.9	17.1	45.4	108.0	126.4	119.1
05	METALS	94.2	118.6	102.5	103.7	119.7	109.1	101.3	122.4	108.4	98.7	119.2	105.6	113.3	33.7	86.4	99.6	110.3	103.2
06	WOOD, PLASTICS & COMPOSITES	96.6	94.1	95.2	101.5	87.3	93.5	102.8	88.0	94.4	89.5	87.7	88.5	93.8	17.8	50.8	99.3	116.0	108.7
07	THERMAL & MOISTURE PROTECTION	103.5	101.5	102.7	98.9	112.4	104.0	101.1	100.8	101.0	100.0	110.9	104.1	125.2	21.7	86.1	94.0	117.6	102.9
08	OPENINGS	92.6	103.0	95.2	95.9	96.3	96.0	94.1	94.8	94.3	90.9	92.8	91.3	149.0	15.2	115.8	99.2	115.5	103.3
0920	Plaster & Gypsum Board	96.1	93.7	94.4	107.9	86.8	93.3	113.7	87.4	95.5	107.5	87.0	93.3	116.3	15.3	46.4	98.1	115.8	110.3
0950, 0980	Ceilings & Acoustic Treatment	97.1	93.7	94.9	84.4	86.8	85.9	98.7	87.4	91.5	91.0	87.0	88.5	229.1	15.3	92.9	96.2	115.8	108.7
0960	Flooring	98.4	99.2	98.6	83.7	90.4	85.5	86.3	109.7	92.7	82.1	90.4	84.4	202.3	14.1	150.3	98.0	131.7	107.3
0970, 0990	Wall Finishes & Painting/Coating	97.5	107.2	103.4	92.5	98.6	96.2	89.1	97.6	94.2	89.1	87.7	88.3	203.7	13.3	89.1	96.7	124.2	113.2
09	FINISHES	100.6	97.2	98.8	95.0	89.4	92.0	97.8	94.1	95.8	92.8	88.6	90.5	205.4	17.0	103.1	98.8	120.8	110.8
COVERS	DIVS. 10 - 14, 25, 28, 41, 43, 44	100.0	103.7	100.8	100.0	100.8	100.2	100.0	101.9	100.4	100.0	98.2	99.6	100.0	17.9	82.6	100.0	108.7	101.8
21, 22, 23	FIRE SUPPRESSION, PLUMBING & HVAC	99.9	98.6	99.4	100.1	109.9	104.1	100.2	101.7	100.8	100.2	91.4	96.6	103.6	14.0	66.8	100.0	111.5	104.7
26, 27, 3370	ELECTRICAL, COMMUNICATIONS & UTIL.	96.3	102.7	99.7	99.2	92.6	95.7	100.2	96.7	98.4	93.3	85.1	89.0	124.0	13.3	66.1	97.0	100.5	98.8
MF2004	WEIGHTED AVERAGE	97.7	102.1	99.6	97.9	101.5	99.5	99.0	100.9	99.8	96.5	95.4	96.0	121.2	24.5	78.5	100.2	112.8	105.8

	DIVISION	SOUTH CAROLINA															SOUTH DAKOTA		
		CHARLESTON			COLUMBIA			FLORENCE			GREENVILLE			SPARTANBURG			ABERDEEN		
		MAT.	INST.	TOTAL	MAT.	INST.	TOTAL	MAT.	INST.	TOTAL	MAT.	INST.	TOTAL	MAT.	INST.	TOTAL	MAT.	INST.	TOTAL
015433	CONTRACTOR EQUIPMENT		100.5	100.5		100.5	100.5		100.5	100.5		100.5	100.5		100.5	100.5		98.8	98.8
0241, 31 - 34	SITE & INFRASTRUCTURE, DEMOLITION	101.1	84.2	89.3	102.4	84.1	89.7	112.0	84.1	92.6	106.5	83.8	90.7	106.2	83.8	90.6	93.7	94.9	94.5
0310	Concrete Forming & Accessories	97.2	43.2	50.5	98.8	42.2	49.9	86.2	42.4	48.3	96.8	42.2	49.6	99.7	42.2	50.0	98.1	40.9	48.6
0320	Concrete Reinforcing	101.0	61.8	81.0	101.0	61.6	80.8	100.6	61.7	80.7	100.5	49.2	74.3	100.5	49.2	74.3	100.4	42.2	70.7
0330	Cast-in-Place Concrete	99.0	51.2	80.9	106.0	51.8	85.5	84.3	51.0	71.7	84.3	50.9	71.7	84.3	50.9	71.7	101.7	47.1	81.0
03	CONCRETE	101.6	51.1	78.0	105.2	50.8	79.8	101.7	50.6	77.8	99.9	48.2	75.8	100.1	48.2	75.9	99.8	44.7	74.0
04	MASONRY	100.2	41.0	64.6	94.6	38.8	61.1	82.3	41.0	57.5	80.2	41.0	56.6	82.3	41.0	57.5	108.8	58.1	78.3
05	METALS	85.8	79.2	83.5	85.8	78.3	83.3	84.7	78.6	82.7	84.7	74.1	81.1	84.7	74.1	81.1	98.1	67.1	87.6
06	WOOD, PLASTICS & COMPOSITES	97.2	42.2	66.0	99.6	41.7	66.8	83.7	41.7	59.9	97.2	41.7	65.8	101.0	41.7	67.4	100.1	40.1	66.2
07	THERMAL & MOISTURE PROTECTION	99.4	44.4	78.6	98.6	43.1	77.6	99.7	44.4	78.8	99.6	44.4	78.7	99.7	44.4	78.8	102.0	50.8	82.7
08	OPENINGS	98.5	46.1	85.5	98.5	45.8	85.4	94.4	45.8	82.4	94.3	42.9	81.6	94.3	42.9	81.6	95.0	41.0	81.6
0920	Plaster & Gypsum Board	111.4	40.3	62.2	105.1	39.8	59.9	99.2	39.8	58.1	105.0	39.8	59.8	107.1	39.8	60.5	103.5	38.6	58.6
0950, 0980	Ceilings & Acoustic Treatment	96.1	40.3	60.5	95.2	39.8	59.9	93.5	39.8	59.3	92.6	39.8	58.9	92.6	39.8	58.9	110.0	38.6	64.6
0960	Flooring	110.9	62.7	97.6	107.4	47.0	90.7	100.9	47.0	86.0	108.1	61.6	95.3	109.8	61.6	96.5	115.4	54.2	98.5
0970, 0990	Wall Finishes & Painting/Coating	116.0	43.5	72.4	113.5	43.5	71.4	116.0	43.5	72.4	116.0	43.5	72.4	116.0	43.5	72.4	98.6	41.2	64.1
09	FINISHES	102.9	46.5	72.2	100.5	43.0	69.3	98.9	43.5	68.8	101.2	46.0	71.2	102.0	46.0	71.6	108.0	43.0	72.7
COVERS	DIVS. 10 - 14, 25, 28, 41, 43, 44	100.0	70.6	93.8	100.0	70.3	93.7	100.0	70.3	93.7	100.0	70.3	93.7	100.0	70.3	93.7	100.0	47.6	88.9
21, 22, 23	FIRE SUPPRESSION, PLUMBING & HVAC	100.2	42.1	76.3	100.0	37.3	74.2	100.2	37.3	74.4	100.2	35.8	73.7	100.2	35.8	73.7	100.0	45.1	77.4
26, 27, 3370	ELECTRICAL, COMMUNICATIONS & UTIL.	97.7	99.6	98.7	101.0	50.9	74.8	95.5	48.4	70.9	97.8	52.4	74.0	97.8	52.4	74.0	97.6	57.4	76.6
MF2004	WEIGHTED AVERAGE	98.1	59.7	81.2	98.3	51.2	77.5	96.3	51.2	76.4	96.3	50.8	76.2	96.5	50.8	76.3	100.0	54.0	79.7

	DIVISION	SOUTH DAKOTA									TENNESSEE								
		PIERRE			RAPID CITY			SIOUX FALLS			CHATTANOOGA			JACKSON			JOHNSON CITY		
		MAT.	INST.	TOTAL	MAT.	INST.	TOTAL	MAT.	INST.	TOTAL	MAT.	INST.	TOTAL	MAT.	INST.	TOTAL	MAT.	INST.	TOTAL
015433	CONTRACTOR EQUIPMENT		98.8	98.8		98.8	98.8		99.8	99.8		104.0	104.0		104.1	104.1		97.7	97.7
0241, 31 - 34	SITE & INFRASTRUCTURE, DEMOLITION	92.9	95.0	94.3	91.8	95.0	94.0	93.5	96.7	95.7	102.6	98.7	99.9	100.9	95.7	97.3	111.3	87.1	94.5
0310	Concrete Forming & Accessories	96.9	41.8	49.2	103.8	39.6	48.3	97.6	44.5	51.6	94.0	57.2	62.1	88.4	44.3	50.3	82.9	41.1	46.8
0320	Concrete Reinforcing	100.0	53.2	76.1	93.8	53.0	73.0	93.8	53.2	73.1	79.3	65.2	72.1	81.3	61.1	71.0	79.9	61.5	70.5
0330	Cast-in-Place Concrete	93.5	45.2	75.2	98.0	45.8	78.3	93.5	46.2	75.6	100.7	64.4	87.0	104.4	42.0	80.8	81.0	48.8	68.8
03	CONCRETE	95.0	46.5	72.3	96.7	45.7	72.9	92.9	48.1	72.0	90.9	62.9	77.8	93.7	48.7	72.7	96.3	49.8	74.5
04	MASONRY	104.6	57.8	76.4	105.8	53.7	74.5	101.6	55.9	74.1	96.7	54.2	71.2	110.5	38.2	67.0	108.4	37.2	65.6
05	METALS	98.1	72.4	89.4	100.0	72.4	90.7	100.2	72.9	91.0	100.0	89.4	96.4	95.0	84.1	91.3	97.2	86.4	93.5
06	WOOD, PLASTICS & COMPOSITES	98.0	41.0	65.7	103.4	37.2	65.9	98.8	43.9	67.7	101.7	58.2	77.1	86.8	45.8	63.6	76.1	41.2	56.3
07	THERMAL & MOISTURE PROTECTION	102.0	49.2	82.0	102.4	50.7	82.9	104.0	52.5	84.5	95.0	61.2	82.2	91.4	46.0	74.3	90.4	54.3	76.7
08	OPENINGS	98.2	44.8	85.0	99.3	42.8	85.3	99.5	46.4	86.4	100.4	58.2	89.9	101.0	52.8	89.1	96.1	48.8	84.4
0920	Plaster & Gypsum Board	96.5	39.5	57.1	103.2	35.6	56.4	100.2	42.5	60.2	78.3	57.5	63.9	86.7	44.7	57.6	91.7	40.0	55.9
0950, 0980	Ceilings & Acoustic Treatment	104.8	39.5	63.2	113.5	35.6	63.9	107.4	42.5	66.1	95.7	57.5	71.3	93.7	44.7	62.5	91.9	40.0	58.8
0960	Flooring	114.7	38.6	93.7	114.7	68.2	101.8	114.7	73.7	103.4	98.7	61.0	88.3	91.4	20.6	71.8	93.0	43.8	79.4
0970, 0990	Wall Finishes & Painting/Coating	98.6	49.5	69.1	98.6	49.5	69.1	98.6	49.5	69.1	105.0	58.6	77.1	96.3	55.1	71.5	102.6	46.8	69.0
09	FINISHES	105.6	41.3	70.7	108.4	45.3	74.1	106.8	50.1	76.0	94.0	57.9	74.4	93.6	41.1	65.1	95.7	41.3	66.2
COVERS	DIVS. 10 - 14, 25, 28, 41, 43, 44	100.0	71.4	93.9	100.0	71.2	93.9	100.0	71.8	94.0	100.0	47.7	88.9	100.0	49.2	89.2	100.0	49.4	89.3
21, 22, 23	FIRE SUPPRESSION, PLUMBING & HVAC	100.0	44.7	77.2	100.0	45.2	77.5	100.0	43.5	76.7	99.9	61.9	84.3	99.9	47.1	78.2	99.7	49.2	78.9
26, 27, 3370	ELECTRICAL, COMMUNICATIONS & UTIL.	98.7	49.8	73.1	93.9	49.8	70.8	101.1	66.4	83.0	104.5	70.0	86.4	102.1	65.6	83.0	93.8	52.0	71.9
MF2004	WEIGHTED AVERAGE	99.3	54.2	79.4	99.8	54.2	79.7	99.8	57.8	81.2	98.6	66.9	84.6	98.4	56.0	79.7	97.9	54.2	78.6

	DIVISION	TENNESSEE									TEXAS								
		KNOXVILLE			MEMPHIS			NASHVILLE			ABILENE			AMARILLO			AUSTIN		
		MAT.	INST.	TOTAL	MAT.	INST.	TOTAL	MAT.	INST.	TOTAL	MAT.	INST.	TOTAL	MAT.	INST.	TOTAL	MAT.	INST.	TOTAL
015433	CONTRACTOR EQUIPMENT		97.7	97.7		101.6	101.6		105.2	105.2		88.0	88.0		88.0	88.0		87.2	87.2
0241, 31 - 34	SITE & INFRASTRUCTURE, DEMOLITION	88.6	87.8	88.1	91.5	91.9	91.8	97.1	99.7	98.9	100.7	85.6	90.2	101.9	86.8	91.4	91.1	85.8	87.4
0310	Concrete Forming & Accessories	93.0	45.8	52.2	93.4	63.2	67.3	94.0	66.8	70.5	95.8	41.7	49.0	96.4	55.3	60.8	90.4	53.8	58.7
0320	Concrete Reinforcing	79.3	63.1	71.1	84.9	66.3	75.4	84.9	65.5	75.0	96.6	53.9	74.8	96.6	50.3	72.9	93.7	50.2	71.5
0330	Cast-in-Place Concrete	94.6	51.7	78.4	90.7	64.3	80.7	92.6	68.4	83.4	100.0	45.3	79.3	99.1	53.4	81.8	87.9	50.4	73.7
03	CONCRETE	88.1	53.1	71.7	87.0	65.7	77.0	88.4	68.4	79.1	93.1	46.1	71.2	92.7	54.5	74.8	82.1	52.7	68.4
04	MASONRY	74.5	57.0	64.0	87.9	67.0	75.3	81.7	62.1	70.0	100.5	53.7	72.3	100.5	51.1	70.8	92.8	50.2	67.2
05	METALS	100.8	87.9	96.4	101.5	90.0	97.6	102.1	88.3	97.5	100.8	68.7	89.9	100.8	67.5	89.5	96.8	64.9	86.0
06	WOOD, PLASTICS & COMPOSITES	88.4	41.2	61.7	92.8	64.5	76.8	99.3	68.5	81.9	102.1	40.6	67.3	100.6	58.0	76.5	93.8	56.6	72.7
07	THERMAL & MOISTURE PROTECTION	88.6	61.1	78.2	89.2	67.2	80.9	94.3	64.1	82.9	98.1	50.8	80.2	98.4	50.6	80.3	87.6	52.7	74.4
08	OPENINGS	93.0	49.1	82.1	99.2	65.0	90.8	98.4	67.4	90.7	93.7	44.3	81.5	93.7	52.6	83.6	100.5	55.2	89.3
0920	Plaster & Gypsum Board	98.9	40.0	58.1	87.2	63.8	71.0	95.2	68.0	76.3	84.1	39.4	53.2	79.4	57.3	64.1	85.2	55.9	64.9
0950, 0980	Ceilings & Acoustic Treatment	92.8	40.0	59.1	92.9	63.8	74.4	95.4	68.0	77.9	93.0	39.4	58.9	91.3	57.3	69.6	85.7	55.9	66.7
0960	Flooring	98.4	58.2	87.3	93.0	38.9	78.1	101.3	69.2	92.4	111.6	67.5	99.4	111.4	61.7	97.7	94.2	42.6	79.9
0970, 0990	Wall Finishes & Painting/Coating	102.6	46.8	69.0	97.5	58.2	73.9	105.2	72.6	85.6	97.2	54.6	71.6	97.2	43.7	65.0	95.3	44.4	64.7
09	FINISHES	90.2	47.1	66.8	89.6	57.7	72.3	99.9	68.1	82.6	95.9	47.3	69.5	95.2	55.3	73.6	87.6	50.5	67.4
COVERS	DIVS. 10 - 14, 25, 28, 41, 43, 44	100.0	53.6	90.1	100.0	81.0	96.0	100.0	76.6	95.0	100.0	76.2	95.0	100.0	77.7	95.3	100.0	77.9	95.3
21, 22, 23	FIRE SUPPRESSION, PLUMBING & HVAC	99.7	64.1	85.0	99.8	72.5	88.6	99.8	78.4	91.0	100.2	39.6	75.3	100.1	52.4	80.5	100.1	55.6	81.8
26, 27, 3370	ELECTRICAL, COMMUNICATIONS & UTIL.	100.0	57.6	77.8	101.6	68.6	84.4	101.4	64.4	82.0	98.6	52.5	74.5	103.3	70.8	86.3	102.7	65.8	83.4
MF2004	WEIGHTED AVERAGE	94.9	61.8	80.2	96.4	71.6	85.5	97.5	73.9	87.1	98.2	52.8	78.2	98.6	60.5	81.8	95.5	59.4	79.6

	DIVISION	TEXAS																	
		BEAUMONT			CORPUS CHRISTI			DALLAS			EL PASO			FORT WORTH			HOUSTON		
		MAT.	INST.	TOTAL	MAT.	INST.	TOTAL	MAT.	INST.	TOTAL	MAT.	INST.	TOTAL	MAT.	INST.	TOTAL	MAT.	INST.	TOTAL
015433	CONTRACTOR EQUIPMENT		89.1	89.1		94.9	94.9		97.9	97.9		88.0	88.0		88.0	88.0		98.4	98.4
0241, 31 - 34	SITE & INFRASTRUCTURE, DEMOLITION	96.9	86.0	89.3	123.4	81.4	94.2	115.9	87.1	95.8	98.9	85.1	89.3	103.3	86.0	91.3	120.3	84.5	95.4
0310	Concrete Forming & Accessories	102.2	52.5	59.2	97.0	39.5	47.3	94.6	58.7	63.6	95.1	47.2	53.6	91.9	58.1	62.7	91.8	64.8	68.4
0320	Concrete Reinforcing	98.3	44.6	70.9	92.9	49.5	70.8	94.6	53.9	73.8	96.6	49.4	72.5	96.6	53.8	74.7	100.1	64.7	82.0
0330	Cast-in-Place Concrete	98.0	54.7	81.6	95.0	47.0	76.8	102.0	57.8	85.3	91.8	42.6	73.2	104.3	53.5	85.0	101.1	69.3	89.0
03	CONCRETE	94.8	52.7	75.1	88.0	45.9	68.3	94.0	59.0	77.6	89.0	46.9	69.3	94.9	56.4	76.9	94.2	67.8	81.8
04	MASONRY	99.6	59.4	75.4	82.9	51.9	64.3	106.1	60.9	79.0	98.1	50.5	69.5	95.2	60.9	74.6	94.4	61.0	74.3
05	METALS	99.5	66.0	88.2	96.4	76.2	89.6	100.4	80.7	93.8	100.5	64.1	88.2	97.8	68.9	88.0	103.7	88.6	98.6
06	WOOD, PLASTICS & COMPOSITES	115.8	52.9	80.2	112.7	39.2	71.1	103.0	59.1	78.1	101.7	49.9	72.4	103.5	58.9	78.2	101.4	65.1	80.9
07	THERMAL & MOISTURE PROTECTION	106.5	58.4	88.3	96.7	47.7	78.2	89.8	63.3	79.7	94.4	54.5	79.3	99.1	55.1	82.5	97.8	66.7	86.1
08	OPENINGS	95.7	49.3	84.2	107.6	40.6	91.0	103.3	54.8	91.3	93.7	45.9	81.9	88.2	54.8	79.9	104.7	64.0	94.6
0920	Plaster & Gypsum Board	98.0	52.1	66.2	89.9	37.8	53.8	89.1	58.3	67.8	82.3	49.0	59.3	83.2	58.3	66.0	96.0	64.6	74.3
0950, 0980	Ceilings & Acoustic Treatment	103.0	52.1	70.6	87.4	37.8	55.8	94.2	58.3	71.3	91.3	49.0	64.4	93.0	58.3	70.9	103.6	64.6	78.7
0960	Flooring	110.7	72.7	100.2	106.6	43.0	89.0	100.1	51.4	86.7	111.6	67.2	99.3	144.9	43.8	117.0	98.2	56.8	86.8
0970, 0990	Wall Finishes & Painting/Coating	94.5	52.9	69.4	108.1	62.8	80.8	104.2	55.5	74.9	97.2	38.0	61.6	98.6	55.5	72.7	102.0	61.5	77.6
09	FINISHES	93.5	56.0	73.2	95.2	41.9	66.3	98.4	56.8	75.8	95.2	50.2	70.7	106.6	55.1	78.7	97.8	62.7	78.8
COVERS	DIVS. 10 - 14, 25, 28, 41, 43, 44	100.0	80.0	95.7	100.0	77.1	95.1	100.0	81.4	96.0	100.0	78.1	95.4	100.0	81.0	96.0	100.0	84.2	96.7
21, 22, 23	FIRE SUPPRESSION, PLUMBING & HVAC	100.1	62.6	84.7	100.1	42.3	76.3	100.0	64.8	85.5	100.1	37.4	74.3	100.1	60.0	83.6	100.0	69.2	87.3
26, 27, 3370	ELECTRICAL, COMMUNICATIONS & UTIL.	95.2	62.5	78.1	96.2	53.3	73.8	97.0	71.4	83.6	99.7	54.6	76.1	99.6	63.5	80.7	97.0	67.6	81.7
MF2004	WEIGHTED AVERAGE	98.1	62.1	82.2	97.8	52.7	77.9	99.6	66.8	85.2	97.4	52.7	77.6	98.2	62.7	82.6	100.0	70.3	86.9

	DIVISION	TEXAS																	
		LAREDO			LUBBOCK			ODESSA			SAN ANTONIO			WACO			WICHITA FALLS		
		MAT.	INST.	TOTAL	MAT.	INST.	TOTAL	MAT.	INST.	TOTAL	MAT.	INST.	TOTAL	MAT.	INST.	TOTAL	MAT.	INST.	TOTAL
015433	CONTRACTOR EQUIPMENT		87.2	87.2		97.0	97.0		88.0	88.0		89.8	89.8		88.0	88.0		88.0	88.0
0241, 31 - 34	SITE & INFRASTRUCTURE, DEMOLITION	90.6	85.5	87.1	127.9	83.9	97.3	100.9	86.0	90.5	90.3	89.9	90.0	99.5	86.0	90.1	100.3	85.6	90.0
0310	Concrete Forming & Accessories	91.5	39.7	46.7	93.7	44.5	51.1	95.7	41.0	48.4	91.5	53.3	58.4	96.2	42.3	49.6	96.2	40.2	47.8
0320	Concrete Reinforcing	93.7	49.4	71.1	97.9	53.7	75.3	96.6	53.5	74.6	100.1	48.9	74.0	96.6	50.2	72.9	96.6	53.4	74.5
0330	Cast-in-Place Concrete	73.1	59.5	68.0	100.2	50.0	81.2	100.0	46.0	79.6	71.7	65.8	69.5	90.2	54.7	76.8	96.4	46.8	77.6
03	CONCRETE	79.5	49.4	65.4	92.0	50.0	72.3	93.1	46.3	71.2	79.7	57.6	69.4	88.3	49.2	70.0	91.3	46.2	70.2
04	MASONRY	90.6	49.7	66.0	99.8	48.5	69.0	100.5	45.9	67.7	90.5	58.3	71.1	97.4	58.8	74.2	97.9	59.7	74.9
05	METALS	97.8	63.9	86.4	104.1	80.9	96.2	100.0	67.7	89.1	98.6	66.8	87.8	100.7	66.8	89.2	100.6	69.5	90.1
06	WOOD, PLASTICS & COMPOSITES	93.1	39.1	62.5	101.1	45.4	69.6	102.1	41.1	67.6	93.1	51.6	69.6	108.8	38.2	68.8	108.8	38.1	68.8
07	THERMAL & MOISTURE PROTECTION	92.4	50.6	76.6	88.6	50.8	74.3	98.1	45.7	78.3	92.4	63.3	81.4	98.6	50.6	80.4	98.6	53.9	81.7
08	OPENINGS	101.2	41.2	86.3	104.0	45.7	89.5	93.7	42.9	81.2	103.1	51.7	90.4	88.2	39.2	76.0	88.2	43.5	77.1
0920	Plaster & Gypsum Board	86.4	37.8	52.8	84.6	44.2	56.7	84.1	39.9	53.5	86.4	50.6	61.6	84.1	36.9	51.5	84.1	36.8	51.4
0950, 0980	Ceilings & Acoustic Treatment	87.4	37.8	55.8	94.7	44.2	62.6	93.0	39.9	59.2	87.4	50.6	64.0	93.0	36.9	57.3	93.0	36.8	57.2
0960	Flooring	91.2	43.0	77.9	103.6	41.6	86.5	111.6	41.1	92.1	91.2	68.1	84.8	144.9	38.2	115.4	146.1	77.9	127.2
0970, 0990	Wall Finishes & Painting/Coating	95.3	52.1	69.3	108.4	35.5	64.5	97.2	35.5	60.1	95.3	52.1	69.3	98.6	36.6	61.3	102.1	49.8	70.6
09	FINISHES	87.3	40.6	62.0	98.1	42.3	67.8	95.9	39.7	65.4	87.3	54.8	69.7	106.4	39.7	70.2	107.0	47.4	74.6
COVERS	DIVS. 10 - 14, 25, 28, 41, 43, 44	100.0	74.8	94.6	100.0	76.5	95.0	100.0	77.5	95.2	100.0	79.4	95.6	100.0	78.6	95.5	100.0	75.0	94.7
21, 22, 23	FIRE SUPPRESSION, PLUMBING & HVAC	100.0	40.9	75.7	99.7	45.7	77.5	100.2	37.3	74.3	100.0	65.4	85.8	100.2	54.0	81.2	100.2	48.1	78.8
26, 27, 3370	ELECTRICAL, COMMUNICATIONS & UTIL.	98.1	62.4	79.4	97.2	46.9	70.9	98.7	44.1	70.2	98.5	62.4	79.6	98.7	71.3	84.4	100.7	61.1	80.0
MF2004	WEIGHTED AVERAGE	94.9	53.0	76.4	99.9	53.8	79.6	98.1	49.3	76.5	95.3	63.7	81.3	97.8	58.1	80.3	98.4	56.5	79.9

	DIVISION	UTAH												VERMONT					
		LOGAN			OGDEN			PROVO			SALT LAKE CITY			BURLINGTON			RUTLAND		
		MAT.	INST.	TOTAL	MAT.	INST.	TOTAL	MAT.	INST.	TOTAL	MAT.	INST.	TOTAL	MAT.	INST.	TOTAL	MAT.	INST.	TOTAL
015433	CONTRACTOR EQUIPMENT		99.9	99.9		99.9	99.9		98.8	98.8		99.9	99.9		100.6	100.6		100.6	100.6
0241, 31 - 34	SITE & INFRASTRUCTURE, DEMOLITION	87.6	98.4	95.2	76.6	98.4	91.8	84.2	96.7	92.9	76.3	98.4	91.7	76.5	97.3	91.0	76.6	97.3	91.1
0310	Concrete Forming & Accessories	104.8	52.9	59.9	104.9	52.9	59.9	106.0	53.0	60.1	107.4	53.0	60.3	95.2	53.4	59.1	98.2	54.0	59.9
0320	Concrete Reinforcing	98.7	70.4	84.3	98.3	70.4	84.1	106.7	70.4	88.2	100.5	70.4	85.1	110.6	83.9	97.0	110.6	83.9	97.0
0330	Cast-in-Place Concrete	88.4	68.2	80.7	89.7	68.2	81.6	88.5	68.2	80.8	98.1	68.2	86.8	95.1	103.3	98.2	89.0	103.5	94.5
03	CONCRETE	110.3	62.1	87.8	99.2	62.1	81.8	109.8	62.1	87.5	119.8	62.1	92.8	106.1	76.8	92.4	103.3	77.1	91.1
04	MASONRY	110.5	55.7	77.6	104.3	55.7	75.1	115.8	55.7	79.7	118.0	55.7	80.6	107.6	51.7	74.0	86.6	51.7	65.6
05	METALS	102.7	72.2	92.4	103.2	72.2	92.7	100.8	72.3	91.2	108.1	72.3	96.0	101.4	78.9	93.8	99.5	79.5	92.8
06	WOOD, PLASTICS & COMPOSITES	84.4	49.9	64.9	84.4	49.9	64.9	85.7	49.9	65.4	86.0	49.9	65.6	95.0	53.4	71.4	100.1	53.4	73.7
07	THERMAL & MOISTURE PROTECTION	96.9	61.4	83.5	95.8	61.4	82.8	98.5	61.4	84.5	102.3	61.4	86.8	95.5	53.5	79.6	95.5	61.6	82.7
08	OPENINGS	88.6	52.2	79.6	88.6	52.2	79.6	92.9	52.2	82.8	90.5	52.2	81.0	104.3	54.6	92.0	104.3	54.6	92.0
0920	Plaster & Gypsum Board	84.2	48.2	59.3	84.2	48.2	59.3	84.6	48.2	59.4	87.0	48.2	60.2	100.2	51.3	66.3	98.9	51.3	65.9
0950, 0980	Ceilings & Acoustic Treatment	121.8	48.2	75.0	121.8	48.2	75.0	121.8	48.2	75.0	116.1	48.2	72.9	92.7	51.3	66.3	93.5	51.3	66.6
0960	Flooring	103.6	51.4	89.2	101.3	51.4	87.5	104.4	51.4	89.7	104.4	51.4	89.8	97.8	59.4	87.2	97.6	59.4	87.0
0970, 0990	Wall Finishes & Painting/Coating	97.9	41.7	64.1	97.9	41.7	64.1	97.9	53.3	71.1	100.6	53.3	72.1	94.6	34.5	58.4	94.6	34.5	58.4
09	FINISHES	105.4	50.6	75.6	103.5	50.6	74.8	106.1	51.9	76.7	104.2	51.9	75.8	97.2	51.6	72.5	97.1	51.6	72.4
COVERS	DIVS. 10 - 14, 25, 28, 41, 43, 44	100.0	60.0	91.5	100.0	60.0	91.5	100.0	60.0	91.5	100.0	60.0	91.5	100.0	77.1	95.1	100.0	77.1	95.1
21, 22, 23	FIRE SUPPRESSION, PLUMBING & HVAC	100.0	68.3	86.9	100.0	68.3	86.9	100.0	68.3	86.9	100.1	68.3	87.0	100.0	65.5	85.8	100.1	65.6	85.9
26, 27, 3370	ELECTRICAL, COMMUNICATIONS & UTIL.	94.5	64.2	78.7	94.9	64.2	78.9	95.2	67.0	80.5	97.9	67.0	81.8	99.7	66.1	82.1	97.8	66.1	81.2
MF2004	WEIGHTED AVERAGE	100.3	65.0	84.7	98.2	65.0	83.6	100.8	65.4	85.2	103.0	65.5	86.5	100.6	67.3	85.9	98.8	67.7	85.0

	DIVISION	VIRGINIA																	
		ALEXANDRIA			ARLINGTON			NEWPORT NEWS			NORFOLK			PORTSMOUTH			RICHMOND		
		MAT.	INST.	TOTAL	MAT.	INST.	TOTAL	MAT.	INST.	TOTAL	MAT.	INST.	TOTAL	MAT.	INST.	TOTAL	MAT.	INST.	TOTAL
015433	CONTRACTOR EQUIPMENT		102.2	102.2		100.8	100.8		105.4	105.4		106.0	106.0		105.3	105.3		105.3	105.3
0241, 31 - 34	SITE & INFRASTRUCTURE, DEMOLITION	117.5	88.4	97.2	128.2	85.2	98.3	111.0	87.6	94.7	112.1	88.6	95.7	109.4	86.8	93.6	111.9	88.0	95.2
0310	Concrete Forming & Accessories	92.6	76.2	78.4	92.5	73.8	76.3	97.4	69.5	73.3	98.7	69.6	73.6	88.1	57.0	61.2	96.8	62.8	67.4
0320	Concrete Reinforcing	89.7	81.1	85.3	101.3	75.0	87.9	101.0	73.2	86.8	101.0	73.2	86.8	100.6	70.5	85.2	101.0	73.6	87.0
0330	Cast-in-Place Concrete	110.6	82.3	99.9	107.6	79.0	96.8	108.3	62.7	91.1	119.1	62.8	97.8	107.3	62.0	90.1	110.0	57.4	90.1
03	CONCRETE	108.9	80.3	95.5	113.8	76.5	96.4	106.3	69.2	88.9	111.7	69.2	91.8	105.1	62.7	85.3	107.0	64.4	87.1
04	MASONRY	95.0	72.6	81.5	108.4	68.5	84.4	101.1	57.0	74.6	107.0	57.0	77.0	106.4	56.8	76.6	99.3	58.7	74.9
05	METALS	93.3	95.7	94.1	92.0	84.0	89.3	93.3	91.6	92.7	92.3	91.7	92.1	92.3	88.1	90.9	95.5	92.2	94.4
06	WOOD, PLASTICS & COMPOSITES	96.4	76.1	84.9	93.6	76.1	83.7	97.2	74.1	84.1	99.2	74.1	85.0	86.1	57.4	69.9	96.3	65.5	78.8
07	THERMAL & MOISTURE PROTECTION	98.3	81.7	92.0	100.1	71.2	89.2	99.4	59.1	84.2	97.1	59.1	82.7	99.5	56.7	83.3	98.4	59.4	83.7
08	OPENINGS	98.5	77.4	93.3	96.3	71.8	90.3	98.5	67.7	90.8	98.5	67.7	90.8	98.5	58.0	88.5	98.5	64.0	89.9
0920	Plaster & Gypsum Board	108.6	75.2	85.5	106.1	75.2	84.7	108.6	72.4	83.5	105.3	72.4	82.5	101.7	55.3	69.6	114.9	63.5	79.3
0950, 0980	Ceilings & Acoustic Treatment	96.1	75.2	82.8	93.5	75.2	81.9	96.1	72.4	81.0	96.1	72.4	81.0	96.1	55.3	70.1	99.5	63.5	76.6
0960	Flooring	110.9	86.0	104.0	109.0	59.3	95.2	110.9	54.4	95.3	110.4	54.4	94.9	102.0	68.8	92.8	110.4	76.7	101.1
0970, 0990	Wall Finishes & Painting/Coating	130.0	85.5	103.2	130.0	85.5	103.2	116.0	46.7	74.3	116.0	65.7	85.7	116.0	65.7	85.7	116.0	65.3	85.5
09	FINISHES	103.1	78.7	89.8	102.8	73.0	86.6	102.3	64.8	81.9	101.6	66.9	82.7	98.7	59.6	77.5	103.4	65.8	83.0
COVERS	DIVS. 10 - 14, 25, 28, 41, 43, 44	100.0	88.4	97.5	100.0	76.6	95.0	100.0	82.5	96.3	100.0	82.5	96.3	100.0	80.5	95.9	100.0	80.3	95.8
21, 22, 23	FIRE SUPPRESSION, PLUMBING & HVAC	100.2	87.9	95.2	100.2	85.5	94.2	100.2	65.0	85.7	100.1	64.1	85.3	100.2	64.0	85.3	100.1	58.2	82.9
26, 27, 3370	ELECTRICAL, COMMUNICATIONS & UTIL.	98.1	96.6	97.3	95.4	96.3	95.9	97.7	61.3	78.7	104.8	60.9	81.9	95.8	60.9	77.6	104.2	72.5	87.6
MF2004	WEIGHTED AVERAGE	100.3	85.4	93.7	101.2	81.0	92.3	100.0	69.2	86.4	101.4	69.3	87.2	99.4	66.2	84.8	101.0	68.7	86.7

	DIVISION	VIRGINIA			WASHINGTON														
		ROANOKE			EVERETT			RICHLAND			SEATTLE			SPOKANE			TACOMA		
		MAT.	INST.	TOTAL	MAT.	INST.	TOTAL	MAT.	INST.	TOTAL	MAT.	INST.	TOTAL	MAT.	INST.	TOTAL	MAT.	INST.	TOTAL
015433	CONTRACTOR EQUIPMENT		100.8	100.8		103.9	103.9		91.1	91.1		103.8	103.8		91.1	91.1		103.9	103.9
0241, 31 - 34	SITE & INFRASTRUCTURE, DEMOLITION	108.6	84.9	92.1	100.3	113.4	109.4	102.4	90.4	94.0	104.6	112.3	110.0	101.7	90.4	93.8	103.7	114.3	111.1
0310	Concrete Forming & Accessories	97.1	66.2	70.4	109.4	101.1	102.2	124.1	77.1	83.4	101.8	104.7	104.3	128.3	77.1	84.0	101.8	104.1	103.8
0320	Concrete Reinforcing	101.0	70.5	85.4	107.5	99.3	103.3	107.2	82.0	94.3	106.0	99.7	102.8	107.9	82.1	94.7	106.0	99.4	102.7
0330	Cast-in-Place Concrete	121.8	62.3	99.3	102.6	97.4	100.6	107.1	83.2	98.1	107.7	110.9	108.9	111.3	83.2	100.7	105.5	110.6	107.5
03	CONCRETE	112.9	67.0	91.4	95.9	98.9	97.3	103.1	80.1	92.3	98.9	105.3	101.9	105.5	80.1	93.6	97.8	104.9	101.1
04	MASONRY	101.7	58.9	76.0	133.3	92.7	108.9	112.6	78.2	91.9	128.4	104.2	113.9	113.3	80.5	93.6	128.3	99.5	110.9
05	METALS	93.1	88.5	91.5	106.3	89.6	100.6	90.3	78.8	86.4	108.1	93.8	103.3	92.7	79.0	88.1	108.1	91.5	102.5
06	WOOD, PLASTICS & COMPOSITES	97.2	70.1	81.9	99.5	104.1	102.1	97.5	76.5	85.6	91.5	104.1	98.7	105.9	76.5	89.2	90.3	104.1	98.1
07	THERMAL & MOISTURE PROTECTION	99.4	57.7	83.6	103.6	92.5	99.4	155.8	76.9	126.0	103.6	98.3	101.6	152.2	78.4	124.3	103.3	94.5	100.0
08	OPENINGS	98.5	64.3	90.0	103.3	99.5	102.4	110.5	71.4	100.8	105.5	100.9	104.3	111.1	71.4	101.3	104.0	100.9	103.2
0920	Plaster & Gypsum Board	108.6	69.0	81.2	119.7	104.2	109.0	139.3	75.6	95.2	114.5	104.2	107.3	136.0	75.6	94.2	117.3	104.2	108.2
0950, 0980	Ceilings & Acoustic Treatment	96.1	69.0	78.8	110.4	104.2	106.4	108.4	75.6	87.5	112.8	104.2	107.3	105.5	75.6	86.4	113.0	104.2	107.4
0960	Flooring	110.9	42.2	91.9	121.4	98.6	115.1	107.7	37.3	88.3	113.8	106.8	111.9	108.2	67.5	96.9	114.3	98.8	110.0
0970, 0990	Wall Finishes & Painting/Coating	116.0	46.2	74.0	106.3	78.8	89.8	109.4	65.0	82.7	106.3	88.6	95.7	109.3	65.0	82.7	106.3	88.6	95.7
09	FINISHES	102.1	60.2	79.4	112.2	99.0	105.0	121.1	67.8	92.2	110.0	103.4	106.4	120.2	74.0	95.1	110.5	101.7	105.7
COVERS	DIVS. 10 - 14, 25, 28, 41, 43, 44	100.0	77.2	95.2	100.0	93.5	98.6	100.0	71.5	93.9	100.0	102.9	100.6	100.0	71.4	93.9	100.0	102.8	100.6
21, 22, 23	FIRE SUPPRESSION, PLUMBING & HVAC	100.2	50.1	79.6	100.1	90.9	96.3	100.5	91.0	96.6	100.1	111.7	104.9	100.4	77.4	90.9	100.1	94.5	97.8
26, 27, 3370	ELECTRICAL, COMMUNICATIONS & UTIL.	97.7	42.7	68.9	105.0	86.8	95.5	99.3	87.2	93.0	104.8	100.4	102.5	96.4	73.4	84.4	104.8	95.2	99.8
MF2004	WEIGHTED AVERAGE	100.7	61.9	83.6	104.2	95.1	100.1	104.6	81.6	94.5	104.6	104.7	104.6	104.9	77.9	93.0	104.3	99.4	102.2

	DIVISION	WASHINGTON						WEST VIRGINIA											
		VANCOUVER			YAKIMA			CHARLESTON			HUNTINGTON			PARKERSBURG			WHEELING		
		MAT.	INST.	TOTAL	MAT.	INST.	TOTAL	MAT.	INST.	TOTAL	MAT.	INST.	TOTAL	MAT.	INST.	TOTAL	MAT.	INST.	TOTAL
015433	CONTRACTOR EQUIPMENT		98.1	98.1		103.9	103.9		100.8	100.8		100.8	100.8		100.8	100.8		100.8	100.8
0241, 31 - 34	SITE & INFRASTRUCTURE, DEMOLITION	115.8	99.9	104.7	106.7	112.2	110.5	104.3	88.2	93.1	107.7	89.4	95.0	112.3	88.2	95.5	113.1	87.4	95.2
0310	Concrete Forming & Accessories	103.2	87.8	89.9	102.2	95.5	96.4	102.5	82.5	85.2	98.5	94.0	94.6	90.1	81.1	82.3	91.5	81.3	82.7
0320	Concrete Reinforcing	107.1	99.0	102.9	106.6	81.6	93.8	101.0	85.2	92.9	101.0	87.9	94.3	99.5	84.1	91.6	98.9	85.8	92.2
0330	Cast-in-Place Concrete	117.9	94.6	109.0	112.8	79.1	100.0	102.9	104.2	103.4	114.3	105.8	111.1	106.8	95.2	102.4	106.8	102.4	105.2
03	CONCRETE	107.9	92.2	100.0	102.7	86.9	95.3	103.9	91.3	98.0	109.3	97.4	103.7	109.3	87.4	99.1	109.3	90.3	100.4
04	MASONRY	128.7	89.4	105.1	120.7	61.0	84.8	98.0	92.3	94.5	101.1	85.6	91.8	84.7	86.2	85.6	109.5	85.9	95.3
05	METALS	105.3	90.5	100.3	106.2	81.2	97.7	93.3	97.8	94.8	93.4	98.3	95.0	92.0	96.9	93.6	92.1	97.9	94.1
06	WOOD, PLASTICS & COMPOSITES	82.8	87.8	85.6	90.6	104.1	98.3	103.2	79.3	89.7	97.2	93.4	95.1	87.8	77.8	82.2	89.4	79.3	83.7
07	THERMAL & MOISTURE PROTECTION	103.4	83.8	96.0	103.4	74.9	92.6	96.9	88.1	93.6	99.6	88.6	95.5	99.4	86.5	94.6	99.8	87.4	95.1
08	OPENINGS	101.9	89.9	99.0	103.4	84.2	98.6	99.7	77.2	94.1	98.5	85.6	95.3	99.1	76.2	93.4	100.0	79.7	95.0
0920	Plaster & Gypsum Board	115.3	87.6	96.1	116.9	104.2	108.1	104.8	78.5	86.6	107.6	93.0	97.5	101.7	77.0	84.6	102.1	78.5	85.8
0950, 0980	Ceilings & Acoustic Treatment	108.8	87.6	95.3	108.7	104.2	105.8	94.3	78.5	84.2	92.6	93.0	92.9	91.7	77.0	82.3	91.7	78.5	83.3
0960	Flooring	120.1	68.6	105.9	115.5	56.0	99.1	110.9	108.1	110.1	110.7	89.9	105.0	105.2	95.7	102.6	106.3	102.1	105.2
0970, 0990	Wall Finishes & Painting/Coating	112.6	61.9	82.1	106.3	65.0	81.5	116.0	87.6	98.9	116.0	87.6	98.9	116.0	87.7	99.0	116.0	88.2	99.3
09	FINISHES	109.2	81.2	94.0	110.0	86.1	97.0	101.1	87.3	93.6	101.1	92.6	96.5	99.1	83.8	90.8	99.5	85.3	91.8
COVERS	DIVS. 10 - 14, 25, 28, 41, 43, 44	100.0	68.0	93.2	100.0	96.5	99.3	100.0	99.2	99.8	100.0	101.9	100.4	100.0	98.9	99.8	100.0	96.1	99.2
21, 22, 23	FIRE SUPPRESSION, PLUMBING & HVAC	100.2	95.2	98.2	100.1	87.7	95.0	100.1	85.6	94.1	100.2	83.7	93.4	100.1	86.2	94.4	100.2	88.9	95.6
26, 27, 3370	ELECTRICAL, COMMUNICATIONS & UTIL.	110.8	93.8	101.9	108.1	87.2	97.2	102.1	88.8	95.2	97.7	99.1	98.4	98.2	91.7	94.8	95.1	92.7	93.8
MF2004	WEIGHTED AVERAGE	105.8	90.8	99.2	104.6	86.1	96.4	99.8	89.1	95.0	100.2	91.7	96.4	99.1	87.8	94.1	100.3	89.2	95.4

DIVISION		WISCONSIN																	
		EAU CLAIRE			GREEN BAY			KENOSHA			LA CROSSE			MADISON			MILWAUKEE		
		MAT.	INST.	TOTAL	MAT.	INST.	TOTAL	MAT.	INST.	TOTAL	MAT.	INST.	TOTAL	MAT.	INST.	TOTAL	MAT.	INST.	TOTAL
015433	CONTRACTOR EQUIPMENT		100.9	100.9		98.8	98.8		99.5	99.5		100.9	100.9		101.5	101.5		89.2	89.2
0241, 31 - 34	SITE & INFRASTRUCTURE, DEMOLITION	93.7	102.7	100.0	97.3	99.0	98.4	101.0	103.6	102.8	87.1	102.7	98.0	94.5	106.5	102.8	95.5	96.8	96.4
0310	Concrete Forming & Accessories	100.7	93.8	94.8	104.9	93.1	94.6	104.1	100.3	100.8	89.8	93.4	92.9	100.2	94.4	95.2	102.9	115.3	113.6
0320	Concrete Reinforcing	92.8	95.4	94.1	90.8	81.5	86.1	93.5	101.1	97.4	92.5	86.4	89.4	93.7	86.8	90.2	93.7	101.8	97.8
0330	Cast-in-Place Concrete	99.5	97.3	98.6	102.9	94.9	99.9	111.5	97.2	106.1	89.4	91.8	90.3	100.3	97.3	99.1	100.1	109.8	103.8
03	CONCRETE	98.2	95.6	97.0	101.1	91.8	96.8	103.1	99.5	101.4	89.6	91.8	90.6	97.3	94.1	95.8	97.5	110.0	103.4
04	MASONRY	93.0	97.5	95.7	125.2	94.7	106.8	102.3	103.7	103.1	92.2	97.6	95.5	103.0	101.2	101.9	105.0	119.4	113.7
05	METALS	91.8	99.4	94.3	93.9	93.0	93.6	98.6	102.5	99.9	91.7	95.1	92.8	100.7	94.5	98.6	101.5	96.4	99.8
06	WOOD, PLASTICS & COMPOSITES	112.4	93.5	101.7	112.5	93.5	101.7	108.8	99.6	103.6	98.9	93.5	95.8	106.1	93.5	99.0	110.8	114.9	113.1
07	THERMAL & MOISTURE PROTECTION	101.2	84.6	94.9	102.5	82.7	95.0	100.9	93.8	98.2	100.8	83.4	94.2	97.5	92.0	95.4	99.3	113.4	104.6
08	OPENINGS	100.7	90.8	98.2	99.0	89.1	96.6	98.7	103.4	99.9	100.6	81.3	95.8	103.4	93.1	100.8	105.6	111.7	107.1
0920	Plaster & Gypsum Board	98.6	93.6	95.1	95.2	93.6	94.1	84.5	100.0	95.2	93.3	93.6	93.5	92.4	93.6	93.3	97.6	115.6	110.0
0950, 0980	Ceilings & Acoustic Treatment	99.2	93.6	95.6	91.5	93.6	92.8	84.7	100.0	94.4	97.5	93.6	95.0	86.7	93.6	91.1	88.2	115.6	105.6
0960	Flooring	96.4	102.5	98.1	115.9	102.5	112.2	121.4	107.7	117.6	89.0	109.1	94.5	97.6	100.9	98.5	106.1	119.3	109.8
0970, 0990	Wall Finishes & Painting/Coating	89.6	88.7	89.1	101.1	72.0	83.6	105.0	98.4	101.0	89.6	69.1	77.3	96.0	90.3	92.6	98.0	114.6	108.0
09	FINISHES	98.4	95.3	96.7	103.6	93.3	98.0	103.2	102.1	102.6	94.7	94.5	94.6	96.3	95.4	95.8	100.2	116.5	109.1
COVERS	DIVS. 10 - 14, 25, 28, 41, 43, 44	100.0	84.0	96.6	100.0	83.4	96.5	100.0	100.2	100.0	100.0	84.1	96.6	100.0	83.8	96.6	100.0	104.8	101.0
21, 22, 23	FIRE SUPPRESSION, PLUMBING & HVAC	100.0	77.6	90.8	100.2	81.0	92.3	100.0	83.5	93.2	100.0	77.4	90.7	99.6	98.8	99.3	99.8	102.9	101.1
26, 27, 3370	ELECTRICAL, COMMUNICATIONS & UTIL.	101.4	84.2	92.4	95.4	80.1	87.4	97.9	99.9	98.9	101.7	84.2	92.6	100.1	85.5	92.5	98.9	100.2	99.6
MF2004	WEIGHTED AVERAGE	98.3	90.5	94.8	100.5	88.5	95.2	100.4	97.6	99.1	96.6	89.0	93.2	99.7	95.5	97.8	100.5	106.6	103.2

DIVISION		WISCONSIN			WYOMING									CANADA					
		RACINE			CASPER			CHEYENNE			ROCK SPRINGS			CALGARY, ALBERTA			EDMONTON, ALBERTA		
		MAT.	INST.	TOTAL	MAT.	INST.	TOTAL	MAT.	INST.	TOTAL	MAT.	INST.	TOTAL	MAT.	INST.	TOTAL	MAT.	INST.	TOTAL
015433	CONTRACTOR EQUIPMENT		101.5	101.5		100.7	100.7		100.7	100.7		100.7	100.7		106.7	106.7		106.7	106.7
0241, 31 - 34	SITE & INFRASTRUCTURE, DEMOLITION	94.6	107.2	103.4	91.7	98.9	96.7	85.1	98.9	94.7	81.5	97.8	92.8	124.3	105.3	111.0	154.1	105.3	120.1
0310	Concrete Forming & Accessories	101.1	100.5	100.6	102.2	42.8	50.8	103.7	58.5	64.6	100.5	38.3	46.7	126.5	86.3	91.7	129.0	86.2	92.0
0320	Concrete Reinforcing	93.7	101.1	97.5	105.2	43.8	73.8	99.4	44.6	71.4	107.3	44.4	75.2	148.3	60.1	103.3	148.3	60.1	103.3
0330	Cast-in-Place Concrete	100.5	97.0	99.2	98.3	72.9	88.7	93.8	73.1	86.0	94.8	53.8	79.3	186.7	97.2	152.8	208.8	97.2	166.6
03	CONCRETE	97.5	99.5	98.4	104.2	54.2	80.9	101.8	61.4	82.9	102.7	45.8	76.1	159.4	85.5	124.8	170.4	85.5	130.7
04	MASONRY	105.0	103.7	104.2	104.8	37.8	64.5	101.5	52.1	71.8	158.7	46.8	91.4	213.9	87.9	138.2	202.5	87.9	133.6
05	METALS	99.7	102.6	100.7	100.4	61.0	87.1	101.7	62.5	88.5	97.5	60.3	84.9	147.7	84.3	126.3	148.3	84.2	126.6
06	WOOD, PLASTICS & COMPOSITES	109.2	99.6	103.8	97.3	40.4	65.1	98.9	60.9	77.4	96.2	37.2	62.8	113.3	85.7	97.7	110.1	85.7	96.3
07	THERMAL & MOISTURE PROTECTION	101.1	95.0	98.8	95.3	49.8	78.1	96.7	55.8	81.2	98.1	47.8	79.1	123.1	84.6	108.5	124.2	84.6	109.2
08	OPENINGS	103.4	103.4	103.4	93.1	39.9	79.9	95.2	51.1	84.3	99.5	38.2	84.3	92.8	76.3	88.7	92.8	76.3	88.7
0920	Plaster & Gypsum Board	96.6	100.0	98.9	102.2	38.4	58.1	90.9	59.6	69.2	93.2	35.1	53.0	172.8	84.8	111.9	165.3	84.8	109.6
0950, 0980	Ceilings & Acoustic Treatment	84.7	100.0	94.4	126.3	38.4	70.3	119.1	59.6	81.2	121.9	35.1	66.7	142.4	84.8	105.7	157.4	84.8	111.2
0960	Flooring	103.3	107.7	104.5	102.6	35.5	84.1	107.0	57.1	93.2	106.0	50.2	90.6	134.5	85.3	120.9	135.6	85.3	121.7
0970, 0990	Wall Finishes & Painting/Coating	96.0	98.4	97.4	97.8	47.8	67.7	101.0	47.8	69.0	97.1	30.6	57.1	110.0	97.3	102.4	110.1	89.1	97.5
09	FINISHES	98.3	102.1	100.3	109.7	41.0	72.4	108.8	57.4	80.9	106.2	38.6	69.5	132.8	87.8	108.4	137.6	86.9	110.1
COVERS	DIVS. 10 - 14, 25, 28, 41, 43, 44	100.0	100.2	100.0	100.0	74.7	94.6	100.0	77.2	95.2	100.0	61.3	91.8	140.0	96.6	130.8	140.0	96.6	130.8
21, 22, 23	FIRE SUPPRESSION, PLUMBING & HVAC	99.8	88.5	95.2	100.1	57.1	82.4	100.0	56.4	82.1	100.0	54.0	81.0	98.2	82.5	91.7	98.2	82.5	91.8
26, 27, 3370	ELECTRICAL, COMMUNICATIONS & UTIL.	97.2	101.1	99.2	100.3	57.0	77.7	97.4	69.9	83.0	93.5	58.0	74.9	126.9	81.4	103.1	117.8	81.4	98.8
MF2004	WEIGHTED AVERAGE	99.7	99.1	99.5	100.5	56.0	80.9	100.0	63.2	83.7	102.2	54.2	81.0	129.4	86.3	110.4	130.6	86.2	111.0

DIVISION		CANADA																	
		HALIFAX, NOVA SCOTIA			HAMILTON, ONTARIO			KITCHENER, ONTARIO			LAVAL, QUEBEC			LONDON, ONTARIO			MONTREAL, QUEBEC		
		MAT.	INST.	TOTAL	MAT.	INST.	TOTAL	MAT.	INST.	TOTAL	MAT.	INST.	TOTAL	MAT.	INST.	TOTAL	MAT.	INST.	TOTAL
015433	CONTRACTOR EQUIPMENT		100.9	100.9		108.6	108.6		103.8	103.8		101.8	101.8		104.0	104.0		103.4	103.4
0241, 31 - 34	SITE & INFRASTRUCTURE, DEMOLITION	99.2	97.9	98.3	115.0	112.9	113.5	101.5	105.0	103.9	94.6	99.5	98.0	114.6	105.2	108.0	101.2	99.4	99.9
0310	Concrete Forming & Accessories	97.4	72.9	76.2	130.1	91.2	96.5	116.3	83.9	88.3	126.7	82.2	88.2	130.2	85.1	91.2	134.9	89.6	95.7
0320	Concrete Reinforcing	155.9	58.2	106.0	161.3	86.0	122.8	110.8	85.9	98.1	158.1	75.8	116.1	131.1	84.6	107.4	146.6	89.4	117.4
0330	Cast-in-Place Concrete	171.3	71.9	133.7	151.9	97.5	131.3	140.8	78.9	117.4	134.5	91.5	118.2	149.3	95.5	129.0	174.6	97.8	145.6
03	CONCRETE	150.3	70.4	112.9	142.1	92.5	118.9	122.0	82.9	103.7	132.8	84.5	110.2	136.5	88.7	114.2	151.4	92.4	123.8
04	MASONRY	173.1	79.7	116.9	180.9	97.7	130.9	161.5	94.0	120.9	160.2	82.1	113.3	182.2	95.3	130.0	167.0	90.1	120.8
05	METALS	118.9	76.7	104.7	125.8	90.3	113.9	115.3	89.9	106.8	105.1	84.6	98.2	123.4	88.2	111.5	122.1	90.4	111.4
06	WOOD, PLASTICS & COMPOSITES	87.4	72.1	78.7	114.3	90.6	100.9	108.7	82.4	93.8	125.9	82.0	101.0	114.3	83.3	96.7	125.9	89.9	105.5
07	THERMAL & MOISTURE PROTECTION	106.5	73.2	93.9	121.6	92.9	110.7	110.5	89.7	102.7	105.3	84.9	97.6	122.4	89.7	110.1	118.4	93.3	108.9
08	OPENINGS	83.5	65.7	79.1	92.8	87.4	91.4	84.0	81.4	83.3	92.8	71.3	87.4	93.9	81.6	90.8	92.8	78.5	89.2
0920	Plaster & Gypsum Board	159.5	71.2	98.4	192.1	90.3	121.6	154.3	81.9	104.2	148.2	81.3	101.9	195.9	82.7	117.6	165.5	89.2	112.7
0950, 0980	Ceilings & Acoustic Treatment	105.1	71.2	83.5	124.8	90.3	102.8	105.1	81.9	90.3	95.6	81.3	86.5	137.7	82.7	102.7	114.5	89.2	98.4
0960	Flooring	110.0	62.0	96.7	134.5	91.0	122.5	127.8	91.0	117.7	132.1	91.7	120.9	134.7	91.0	122.6	134.0	91.7	122.3
0970, 0990	Wall Finishes & Painting/Coating	110.1	71.9	87.1	110.1	97.1	102.3	110.1	86.7	96.0	110.1	85.6	95.3	110.1	93.7	100.2	110.1	100.7	104.4
09	FINISHES	112.8	71.0	90.1	130.1	91.9	109.3	117.6	85.2	100.0	115.9	84.5	98.9	133.6	86.9	108.2	123.2	91.8	106.1
COVERS	DIVS. 10 - 14, 25, 28, 41, 43, 44	140.0	69.2	125.0	140.0	100.6	131.6	140.0	98.7	131.2	140.0	86.1	128.6	140.0	99.2	131.3	140.0	88.5	129.1
21, 22, 23	FIRE SUPPRESSION, PLUMBING & HVAC	97.3	71.6	86.8	98.3	86.6	93.5	97.3	83.7	91.7	97.3	87.3	93.2	98.3	83.8	92.3	98.3	88.9	94.4
26, 27, 3370	ELECTRICAL, COMMUNICATIONS & UTIL.	124.6	71.6	96.9	127.2	92.3	109.0	121.1	89.8	104.7	120.5	71.5	94.9	119.1	89.8	103.8	127.6	82.8	104.2
MF2004	WEIGHTED AVERAGE	117.4	74.6	98.5	121.9	93.2	109.2	113.3	88.5	102.4	113.5	83.8	100.4	120.6	89.6	106.9	120.8	89.7	107.1

DIVISION		CANADA																	
		OSHAWA, ONTARIO			OTTAWA, ONTARIO			QUEBEC, QUEBEC			REGINA, SASKATCHEWAN			SASKATOON, SASKATCHEWAN			ST CATHARINES, ONTARIO		
		MAT.	INST.	TOTAL	MAT.	INST.	TOTAL	MAT.	INST.	TOTAL	MAT.	INST.	TOTAL	MAT.	INST.	TOTAL	MAT.	INST.	TOTAL
015433	CONTRACTOR EQUIPMENT		103.8	103.8		103.8	103.8		103.8	103.8		99.9	99.9		99.9	99.9		101.5	101.5
0241, 31 - 34	SITE & INFRASTRUCTURE, DEMOLITION	114.1	104.3	107.3	111.9	104.9	107.0	99.8	99.4	99.6	113.0	96.0	101.2	107.3	96.2	99.5	102.4	101.5	101.7
0310	Concrete Forming & Accessories	121.7	83.8	89.0	131.2	87.7	93.6	136.5	89.8	96.1	106.2	56.7	63.4	106.2	56.6	63.3	114.5	88.6	92.1
0320	Concrete Reinforcing	175.4	81.2	127.3	160.0	84.6	121.5	140.7	89.4	114.5	127.1	59.6	92.6	119.7	59.5	89.0	111.7	85.9	98.5
0330	Cast-in-Place Concrete	162.9	83.3	132.8	153.0	94.6	130.9	148.3	98.2	129.3	156.3	66.9	122.5	142.0	66.8	113.6	134.5	97.3	120.4
03	CONCRETE	149.0	83.5	118.3	142.5	89.7	117.8	137.7	92.6	116.7	131.2	61.5	98.6	123.1	61.5	94.3	118.9	91.2	105.9
04	MASONRY	164.6	90.2	119.9	179.7	93.5	127.9	174.6	90.1	123.8	165.9	61.1	102.9	165.5	61.1	102.8	161.0	97.7	122.9
05	METALS	106.9	88.8	100.8	124.3	89.8	112.6	123.6	90.6	112.5	104.8	72.0	93.7	104.9	71.9	93.7	106.0	89.8	100.5
06	WOOD, PLASTICS & COMPOSITES	115.2	82.2	96.5	114.1	87.1	98.8	126.5	90.0	105.8	96.2	55.1	73.0	94.5	55.1	72.2	106.4	87.6	95.8
07	THERMAL & MOISTURE PROTECTION	111.4	83.1	100.7	124.6	89.2	111.3	120.6	93.4	110.4	105.9	61.0	88.9	105.0	59.9	88.0	110.5	89.9	102.7
08	OPENINGS	91.6	82.0	89.2	92.8	84.5	90.7	92.8	85.6	91.0	87.6	53.0	79.1	86.7	53.0	78.4	83.4	84.5	83.7
0920	Plaster & Gypsum Board	157.5	81.6	105.0	234.2	86.7	132.1	193.0	89.2	121.2	166.7	53.8	88.5	146.9	53.8	82.4	136.8	87.2	102.5
0950, 0980	Ceilings & Acoustic Treatment	99.9	81.6	88.3	135.1	86.7	104.3	120.2	89.2	100.4	118.9	53.8	77.4	118.9	53.8	77.4	99.9	87.2	91.8
0960	Flooring	132.1	93.4	121.4	134.5	89.7	122.1	134.5	91.7	122.7	119.8	58.3	102.8	119.8	58.3	102.8	126.1	91.0	116.4
0970, 0990	Wall Finishes & Painting/Coating	110.1	101.4	104.8	110.1	88.4	97.0	110.5	100.7	104.6	110.1	62.4	81.4	110.1	53.2	75.8	110.1	97.1	102.3
09	FINISHES	118.8	87.3	101.7	137.1	88.1	110.5	128.3	91.8	108.5	121.2	57.2	86.4	118.5	56.1	84.7	113.7	90.1	100.9
COVERS	DIVS. 10 - 14, 25, 28, 41, 43, 44	140.0	98.8	131.2	140.0	97.2	130.9	140.0	88.7	129.1	140.0	65.3	124.1	140.0	65.3	124.1	140.0	76.3	126.5
21, 22, 23	FIRE SUPPRESSION, PLUMBING & HVAC	97.3	97.2	97.3	98.3	84.2	92.5	98.2	88.9	94.4	97.5	73.1	87.4	97.4	73.1	87.4	97.3	85.3	92.4
26, 27, 3370	ELECTRICAL, COMMUNICATIONS & UTIL.	122.4	89.1	105.0	118.9	90.7	104.2	118.7	82.8	99.9	123.6	61.8	91.3	123.8	61.8	91.4	123.0	90.8	106.2
MF2004	WEIGHTED AVERAGE	117.0	90.9	105.5	121.5	90.1	107.6	119.4	90.1	106.4	114.2	67.0	93.4	112.7	66.8	92.4	111.4	90.4	102.1

DIVISION		CANADA																	
		ST JOHNS, NEWFOUNDLAND			THUNDER BAY, ONTARIO			TORONTO, ONTARIO			VANCOUVER, BRITISH COLUMBIA			WINDSOR, ONTARIO			WINNIPEG, MANITOBA		
		MAT.	INST.	TOTAL	MAT.	INST.	TOTAL	MAT.	INST.	TOTAL	MAT.	INST.	TOTAL	MAT.	INST.	TOTAL	MAT.	INST.	TOTAL
015433	CONTRACTOR EQUIPMENT		102.1	102.1		101.5	101.5		103.9	103.9		112.1	112.1		101.5	101.5		105.9	105.9
0241, 31 - 34	SITE & INFRASTRUCTURE, DEMOLITION	115.4	97.8	103.2	107.8	101.4	103.3	138.6	105.7	115.7	121.7	107.4	111.7	97.2	101.1	99.9	119.3	101.0	106.5
0310	Concrete Forming & Accessories	104.1	61.2	67.0	121.6	88.9	93.3	131.4	96.3	101.1	126.0	74.4	81.4	121.6	86.0	90.8	130.7	62.4	71.7
0320	Concrete Reinforcing	167.7	54.9	110.1	99.9	85.3	92.4	157.1	86.6	121.1	158.3	68.8	112.6	109.5	84.6	96.8	148.3	53.3	99.8
0330	Cast-in-Place Concrete	172.4	73.6	135.0	148.0	95.9	128.3	147.4	105.5	131.5	162.9	88.2	134.7	137.7	96.9	122.3	183.4	68.8	140.1
03	CONCRETE	165.2	65.1	118.4	127.8	90.7	110.5	139.3	97.5	119.8	153.1	78.8	118.3	120.6	89.7	106.2	158.0	63.7	114.0
04	MASONRY	163.3	62.6	102.8	161.8	97.5	123.1	193.3	102.7	138.8	177.0	78.0	117.5	161.2	96.1	122.0	179.8	59.8	107.7
05	METALS	108.1	73.8	96.5	105.8	88.8	100.0	126.0	91.1	114.2	162.4	84.7	136.1	105.8	89.5	100.3	155.8	74.2	128.2
06	WOOD, PLASTICS & COMPOSITES	95.8	61.3	76.3	115.2	87.7	99.7	115.2	95.2	103.9	114.9	71.8	90.5	115.2	84.5	97.8	113.2	63.0	84.8
07	THERMAL & MOISTURE PROTECTION	109.3	60.8	91.0	110.7	91.3	103.4	122.6	97.4	113.1	133.0	78.9	112.5	110.5	90.1	102.8	115.7	64.6	96.4
08	OPENINGS	99.0	57.4	88.7	82.4	84.0	82.8	91.6	91.8	91.6	94.5	72.4	89.0	82.2	82.7	82.3	92.8	58.0	84.1
0920	Plaster & Gypsum Board	165.2	60.1	92.4	170.4	87.4	112.9	179.3	95.1	121.0	155.7	70.4	96.7	159.0	84.0	107.1	163.4	61.4	92.8
0950, 0980	Ceilings & Acoustic Treatment	104.2	60.1	76.1	95.6	87.4	90.4	133.5	95.1	109.0	148.3	70.4	98.7	95.6	84.0	88.2	136.1	61.4	88.5
0960	Flooring	114.7	52.7	97.6	132.1	52.4	110.1	135.6	96.5	124.8	134.9	88.4	122.0	132.1	91.7	121.0	133.4	64.7	114.4
0970, 0990	Wall Finishes & Painting/Coating	110.1	62.4	81.4	110.1	90.1	98.0	112.0	101.4	105.6	110.0	85.7	95.4	110.1	89.7	97.8	110.1	46.4	71.8
09	FINISHES	116.0	59.7	85.4	119.1	83.0	99.5	132.8	97.0	113.4	131.9	77.7	102.5	117.3	87.5	101.1	128.9	61.4	92.2
COVERS	DIVS. 10 - 14, 25, 28, 41, 43, 44	140.0	66.5	124.4	140.0	76.7	126.6	140.0	102.4	132.0	140.0	94.1	130.3	140.0	75.8	126.4	140.0	67.2	124.5
21, 22, 23	FIRE SUPPRESSION, PLUMBING & HVAC	97.6	60.4	82.3	97.3	85.4	92.4	98.2	90.3	95.0	98.2	69.0	86.2	97.3	85.0	92.3	98.3	61.8	83.3
26, 27, 3370	ELECTRICAL, COMMUNICATIONS & UTIL.	119.2	63.2	89.9	121.0	89.1	104.4	121.1	93.0	106.4	121.7	72.1	95.8	126.3	90.8	107.7	125.1	64.4	93.4
MF2004	WEIGHTED AVERAGE	119.2	66.1	95.7	112.9	89.1	102.4	122.5	95.8	110.7	128.7	78.6	106.6	112.0	89.4	102.0	127.7	66.7	100.8

Location Factors

Costs shown in RSMeans cost data publications are based on national averages for materials and installation. To adjust these costs to a specific location, simply multiply the base cost by the factor and divide by 100 for that city. The data is arranged alphabetically by state and postal zip code numbers. For a city not listed, use the factor for a nearby city with similar economic characteristics.

STATE/ZIP	CITY	MAT.	INST.	TOTAL
ALABAMA				
350-352	Birmingham	97.4	75.2	87.6
354	Tuscaloosa	96.0	60.2	80.2
355	Jasper	96.3	58.5	79.6
356	Decatur	96.0	61.8	80.9
357-358	Huntsville	96.0	70.1	84.6
359	Gadsden	95.9	59.2	79.7
360-361	Montgomery	97.1	58.3	80.0
362	Anniston	95.2	67.0	82.8
363	Dothan	95.9	53.7	77.3
364	Evergreen	95.4	55.6	77.8
365-366	Mobile	97.1	67.4	84.0
367	Selma	95.6	53.5	77.0
368	Phenix City	96.4	57.2	79.1
369	Butler	95.8	53.9	77.3
ALASKA				
995-996	Anchorage	127.4	112.3	120.7
997	Fairbanks	126.3	113.3	120.5
998	Juneau	126.0	111.7	119.7
999	Ketchikan	139.3	111.7	127.1
ARIZONA				
850,853	Phoenix	99.9	74.6	88.7
851,852	Mesa/Tempe	99.4	64.4	83.9
855	Globe	99.5	60.5	82.3
856-857	Tucson	98.2	69.9	85.7
859	Show Low	99.6	61.6	82.8
860	Flagstaff	101.6	70.4	87.9
863	Prescott	99.1	61.1	82.4
864	Kingman	97.2	67.9	84.3
865	Chambers	97.3	61.8	81.6
ARKANSAS				
716	Pine Bluff	96.8	63.9	82.3
717	Camden	94.1	42.2	71.2
718	Texarkana	96.2	46.5	74.2
719	Hot Springs	93.3	45.3	72.1
720-722	Little Rock	97.2	67.0	83.9
723	West Memphis	95.8	57.4	78.9
724	Jonesboro	96.9	57.9	79.7
725	Batesville	94.2	51.4	75.3
726	Harrison	95.5	52.2	76.4
727	Fayetteville	92.8	51.8	74.7
728	Russellville	94.1	53.8	76.3
729	Fort Smith	97.3	57.8	79.9
CALIFORNIA				
900-902	Los Angeles	101.7	116.0	108.0
903-905	Inglewood	96.8	104.2	100.1
906-908	Long Beach	98.6	104.3	101.1
910-912	Pasadena	98.0	104.2	100.7
913-916	Van Nuys	101.4	104.2	102.6
917-918	Alhambra	100.2	104.2	102.0
919-921	San Diego	102.4	107.5	104.6
922	Palm Springs	98.3	103.0	100.4
923-924	San Bernardino	95.7	103.6	99.2
925	Riverside	101.1	110.6	105.3
926-927	Santa Ana	97.9	104.0	100.6
928	Anaheim	101.1	112.6	106.2
930	Oxnard	102.5	110.6	106.1
931	Santa Barbara	101.7	110.5	105.6
932-933	Bakersfield	102.9	106.6	104.6
934	San Luis Obispo	102.1	101.4	101.8
935	Mojave	98.9	100.4	99.6
936-938	Fresno	102.8	110.6	106.3
939	Salinas	102.9	114.3	107.9
940-941	San Francisco	110.7	140.4	123.8
942,956-958	Sacramento	104.9	112.6	108.3
943	Palo Alto	102.8	119.2	110.0
944	San Mateo	105.8	125.7	114.6
945	Vallejo	103.9	118.8	110.5
946	Oakland	108.5	126.0	116.2
947	Berkeley	107.5	119.2	112.7
948	Richmond	106.8	120.6	112.9
949	San Rafael	108.5	121.4	114.2
950	Santa Cruz	107.7	114.4	110.7

STATE/ZIP	CITY	MAT.	INST.	TOTAL
CALIFORNIA (CONT'D)				
951	San Jose	105.8	131.3	117.1
952	Stockton	103.0	109.7	106.0
953	Modesto	103.0	109.9	106.1
954	Santa Rosa	102.6	123.5	111.9
955	Eureka	104.1	107.3	105.5
959	Marysville	103.3	108.8	105.7
960	Redding	105.4	107.6	106.3
961	Susanville	104.0	107.7	105.6
COLORADO				
800-802	Denver	101.6	85.1	94.3
803	Boulder	98.2	78.9	89.7
804	Golden	100.5	77.8	90.5
805	Fort Collins	102.4	74.4	90.0
806	Greeley	99.7	64.5	84.1
807	Fort Morgan	98.9	77.5	89.5
808-809	Colorado Springs	101.5	78.7	91.4
810	Pueblo	101.5	76.9	90.7
811	Alamosa	102.6	72.5	89.3
812	Salida	102.4	73.4	89.6
813	Durango	102.9	73.2	89.8
814	Montrose	101.6	71.7	88.4
815	Grand Junction	105.4	69.3	89.5
816	Glenwood Springs	102.7	74.8	90.4
CONNECTICUT				
060	New Britain	100.6	117.5	108.1
061	Hartford	101.7	117.6	108.7
062	Willimantic	101.2	117.1	108.2
063	New London	96.9	116.6	105.6
064	Meriden	99.0	118.1	107.4
065	New Haven	102.6	117.6	109.2
066	Bridgeport	101.7	118.2	109.0
067	Waterbury	101.3	118.0	108.7
068	Norwalk	101.2	118.7	109.0
069	Stamford	101.4	125.5	112.0
D.C.				
200-205	Washington	104.0	92.2	98.8
DELAWARE				
197	Newark	99.8	108.7	103.7
198	Wilmington	100.1	108.7	103.9
199	Dover	100.5	108.7	104.1
FLORIDA				
320,322	Jacksonville	98.0	66.3	84.0
321	Daytona Beach	98.2	75.3	88.1
323	Tallahassee	99.2	57.7	80.9
324	Panama City	99.5	53.1	79.0
325	Pensacola	102.2	64.4	85.5
326,344	Gainesville	99.8	69.1	86.2
327-328,347	Orlando	102.2	73.2	89.4
329	Melbourne	100.8	78.5	91.0
330-332,340	Miami	100.1	79.4	90.9
333	Fort Lauderdale	97.5	77.4	88.7
334,349	West Palm Beach	96.0	74.2	86.4
335-336,346	Tampa	99.2	83.3	92.2
337	St. Petersburg	101.6	61.6	83.9
338	Lakeland	98.1	83.0	91.4
339,341	Fort Myers	97.3	73.9	87.0
342	Sarasota	99.4	76.1	89.1
GEORGIA				
300-303,399	Atlanta	97.4	77.8	88.8
304	Statesboro	96.6	54.7	78.1
305	Gainesville	95.4	63.4	81.3
306	Athens	94.8	65.6	81.9
307	Dalton	96.6	61.8	81.2
308-309	Augusta	96.0	66.5	83.0
310-312	Macon	95.8	65.8	82.6
313-314	Savannah	98.4	61.4	82.0
315	Waycross	96.8	61.4	81.2
316	Valdosta	97.1	61.1	81.2
317,398	Albany	97.1	61.7	81.5
318-319	Columbus	97.0	66.2	83.4

Location Factors

STATE/ZIP	CITY	MAT.	INST.	TOTAL
HAWAII				
967	Hilo	113.5	117.9	115.4
968	Honolulu	116.9	117.9	117.3
STATES & POSS.				
969	Guam	135.6	61.7	103.0
IDAHO				
832	Pocatello	101.2	71.4	88.1
833	Twin Falls	102.5	51.8	80.1
834	Idaho Falls	99.7	57.3	81.0
835	Lewiston	110.1	76.8	95.4
836-837	Boise	100.4	71.8	87.7
838	Coeur d'Alene	109.4	73.9	93.7
ILLINOIS				
600-603	North Suburban	98.5	133.3	113.9
604	Joliet	98.5	135.8	115.0
605	South Suburban	98.5	133.3	113.9
606-608	Chicago	99.0	137.6	116.0
609	Kankakee	94.3	124.3	107.5
610-611	Rockford	97.5	122.8	108.7
612	Rock Island	94.8	99.7	96.9
613	La Salle	96.0	119.4	106.3
614	Galesburg	95.8	104.8	99.8
615-616	Peoria	98.7	107.3	102.5
617	Bloomington	95.2	108.5	101.0
618-619	Champaign	98.8	107.3	102.6
620-622	East St. Louis	94.0	106.6	99.6
623	Quincy	94.9	100.0	97.2
624	Effingham	94.2	104.1	98.6
625	Decatur	96.5	104.5	100.0
626-627	Springfield	97.0	105.4	100.7
628	Centralia	92.2	106.1	98.3
629	Carbondale	91.9	101.4	96.1
INDIANA				
460	Anderson	94.5	83.5	89.7
461-462	Indianapolis	97.7	85.7	92.4
463-464	Gary	96.1	108.3	101.5
465-466	South Bend	95.1	84.7	90.5
467-468	Fort Wayne	95.2	79.6	88.3
469	Kokomo	92.0	82.7	87.9
470	Lawrenceburg	91.3	80.4	86.5
471	New Albany	92.5	76.1	85.3
472	Columbus	94.8	82.4	89.3
473	Muncie	96.0	81.5	89.6
474	Bloomington	97.4	81.6	90.4
475	Washington	93.5	83.3	89.0
476-477	Evansville	95.1	86.6	91.4
478	Terre Haute	95.9	87.9	92.3
479	Lafayette	94.5	83.4	89.6
IOWA				
500-503,509	Des Moines	98.7	73.6	87.7
504	Mason City	96.1	59.0	79.7
505	Fort Dodge	96.4	54.9	78.1
506-507	Waterloo	98.4	56.0	79.7
508	Creston	96.7	58.8	80.0
510-511	Sioux City	99.1	66.1	84.5
512	Sibley	97.5	45.7	74.6
513	Spencer	99.2	44.5	75.0
514	Carroll	96.0	49.3	75.4
515	Council Bluffs	100.2	71.6	87.5
516	Shenandoah	96.4	47.4	74.8
520	Dubuque	98.6	72.6	87.2
521	Decorah	97.0	46.8	74.9
522-524	Cedar Rapids	99.5	77.5	89.8
525	Ottumwa	97.0	66.4	83.5
526	Burlington	96.1	67.9	83.6
527-528	Davenport	98.4	86.7	93.2
KANSAS				
660-662	Kansas City	98.6	95.3	97.1
664-666	Topeka	99.0	63.8	83.4
667	Fort Scott	96.7	68.6	84.3
668	Emporia	96.9	57.3	79.4
669	Belleville	98.7	58.7	81.0
670-672	Wichita	97.7	62.7	82.2
673	Independence	98.3	60.9	81.8
674	Salina	99.1	59.2	81.5
675	Hutchinson	93.6	59.0	78.3
676	Hays	97.9	60.1	81.2
677	Colby	98.6	60.0	81.5

STATE/ZIP	CITY	MAT.	INST.	TOTAL
KANSAS (CONT'D)				
678	Dodge City	100.4	61.5	83.3
679	Liberal	97.7	59.9	81.0
KENTUCKY				
400-402	Louisville	97.1	82.4	90.6
403-405	Lexington	95.5	76.3	87.0
406	Frankfort	96.3	76.1	87.4
407-409	Corbin	92.4	63.3	79.6
410	Covington	93.2	97.4	95.1
411-412	Ashland	92.0	98.2	94.8
413-414	Campton	93.3	63.7	80.2
415-416	Pikeville	94.5	78.0	87.2
417-418	Hazard	92.6	56.5	76.7
420	Paducah	91.6	86.7	89.4
421-422	Bowling Green	93.9	82.6	88.9
423	Owensboro	94.3	81.7	88.8
424	Henderson	91.3	87.3	89.5
425-426	Somerset	90.6	67.7	80.5
427	Elizabethtown	90.0	82.3	86.6
LOUISIANA				
700-701	New Orleans	103.2	68.2	87.8
703	Thibodaux	101.1	61.0	83.4
704	Hammond	98.2	55.8	79.5
705	Lafayette	100.9	56.1	81.1
706	Lake Charles	101.1	60.7	83.3
707-708	Baton Rouge	103.0	61.2	84.5
710-711	Shreveport	97.4	57.7	79.9
712	Monroe	97.4	53.4	77.9
713-714	Alexandria	97.5	54.0	78.3
MAINE				
039	Kittery	94.3	81.9	88.9
040-041	Portland	100.1	77.6	90.2
042	Lewiston	97.8	77.6	88.9
043	Augusta	97.7	81.5	90.6
044	Bangor	97.5	77.5	88.7
045	Bath	95.6	81.3	89.3
046	Machias	95.1	80.5	88.7
047	Houlton	95.2	80.7	88.8
048	Rockland	94.2	81.1	88.4
049	Waterville	95.6	81.6	89.4
MARYLAND				
206	Waldorf	100.0	69.2	86.4
207-208	College Park	100.0	83.7	92.8
209	Silver Spring	99.2	75.2	88.6
210-212	Baltimore	97.8	85.2	92.2
214	Annapolis	98.5	79.3	90.0
215	Cumberland	93.3	81.0	87.9
216	Easton	95.0	41.8	71.5
217	Hagerstown	94.5	81.9	88.9
218	Salisbury	95.4	48.8	74.8
219	Elkton	92.3	66.7	81.0
MASSACHUSETTS				
010-011	Springfield	99.6	107.7	103.2
012	Pittsfield	99.2	106.1	102.3
013	Greenfield	96.8	107.3	101.4
014	Fitchburg	95.5	120.6	106.5
015-016	Worcester	99.6	122.1	109.5
017	Framingham	95.0	129.5	110.2
018	Lowell	99.0	129.0	112.2
019	Lawrence	100.3	127.0	112.1
020-022, 024	Boston	102.0	136.5	117.2
023	Brockton	100.6	123.2	110.6
025	Buzzards Bay	94.3	120.9	106.0
026	Hyannis	97.7	120.9	107.9
027	New Bedford	99.7	121.1	109.1
MICHIGAN				
480,483	Royal Oak	92.1	105.7	98.1
481	Ann Arbor	94.9	106.1	99.9
482	Detroit	96.4	112.1	103.3
484-485	Flint	94.6	100.2	97.1
486	Saginaw	94.2	95.3	94.7
487	Bay City	94.4	94.3	94.3
488-489	Lansing	95.7	98.7	97.0
490	Battle Creek	95.0	91.0	93.2
491	Kalamazoo	95.3	88.2	92.2
492	Jackson	92.7	95.5	94.0
493,495	Grand Rapids	96.4	74.8	86.8
494	Muskegon	93.9	85.6	90.2

STATE/ZIP	CITY	MAT.	INST.	TOTAL
MICHIGAN (CONT'D)				
496	Traverse City	92.2	79.8	86.8
497	Gaylord	93.4	74.2	84.9
498-499	Iron mountain	95.5	88.8	92.5
MINNESOTA				
550-551	Saint Paul	99.3	123.4	109.9
553-555	Minneapolis	99.7	126.3	111.5
556-558	Duluth	98.6	114.6	105.6
MINNESOTA (CONT'd)				
559	Rochester	98.5	107.8	102.6
560	Mankato	96.2	105.5	100.3
561	Windom	94.8	77.8	87.3
562	Willmar	94.4	84.4	90.0
563	St. Cloud	96.2	124.2	108.6
564	Brainerd	96.1	107.4	101.1
565	Detroit Lakes	97.9	97.7	97.8
566	Bemidji	97.2	100.1	98.5
567	Thief River Falls	96.7	96.4	96.6
MISSISSIPPI				
386	Clarksdale	95.8	57.8	79.0
387	Greenville	99.6	67.9	85.6
388	Tupelo	97.3	60.4	81.0
389	Greenwood	97.1	57.3	79.5
390-392	Jackson	97.8	67.0	84.2
393	Meridian	96.1	68.6	84.0
394	Laurel	97.1	60.9	81.2
395	Biloxi	97.9	62.2	82.1
396	Mccomb	95.5	56.8	78.4
397	Columbus	97.0	59.2	80.4
MISSOURI				
630-631	St. Louis	98.6	107.2	102.4
633	Bowling Green	97.0	86.6	92.4
634	Hannibal	95.8	76.8	87.4
635	Kirksville	98.9	70.7	86.5
636	Flat River	98.0	87.4	93.3
637	Cape Girardeau	98.4	83.8	92.0
638	Sikeston	96.0	74.2	86.4
639	Poplar Bluff	95.4	74.7	86.3
640-641	Kansas City	98.7	108.2	102.9
644-645	St. Joseph	98.0	87.2	93.3
646	Chillicothe	94.7	66.1	82.1
647	Harrisonville	94.2	94.8	94.5
648	Joplin	96.8	65.2	82.9
650-651	Jefferson City	95.6	85.0	90.9
652	Columbia	96.8	88.2	93.0
653	Sedalia	95.7	78.8	88.2
654-655	Rolla	94.0	75.8	86.0
656-658	Springfield	97.7	77.8	88.9
MONTANA				
590-591	Billings	102.0	72.9	89.2
592	Wolf Point	101.5	68.6	87.0
593	Miles City	99.0	70.4	86.4
594	Great Falls	103.4	72.2	89.6
595	Havre	100.3	71.3	87.5
596	Helena	101.9	70.6	88.1
597	Butte	102.0	72.3	88.9
598	Missoula	99.2	72.7	87.5
599	Kalispell	97.8	70.9	86.0
NEBRASKA				
680-681	Omaha	99.8	75.6	89.2
683-685	Lincoln	97.8	74.1	87.3
686	Columbus	95.8	73.4	85.9
687	Norfolk	97.5	77.1	88.5
688	Grand Island	97.9	79.2	89.6
689	Hastings	96.8	81.5	90.0
690	Mccook	96.5	72.2	85.8
691	North Platte	97.2	81.6	90.4
692	Valentine	98.5	70.6	86.2
693	Alliance	98.7	68.7	85.5
NEVADA				
889-891	Las Vegas	101.0	113.8	106.7
893	Ely	99.6	69.1	86.1
894-895	Reno	100.0	90.5	95.8
897	Carson City	99.9	89.5	95.3
898	Elko	98.2	74.4	87.7
NEW HAMPSHIRE				
030	Nashua	99.9	87.0	94.2
031	Manchester	100.4	87.0	94.5

STATE/ZIP	CITY	MAT.	INST.	TOTAL
NEW HAMPSHIRE (CONT'D)				
032-033	Concord	98.5	83.3	91.8
034	Keene	96.3	51.3	76.4
035	Littleton	96.1	58.2	79.3
036	Charleston	95.8	48.8	75.0
037	Claremont	94.7	48.8	74.5
038	Portsmouth	97.3	90.9	94.4
NEW JERSEY				
070-071	Newark	101.9	124.3	111.8
072	Elizabeth	100.1	123.8	110.6
073	Jersey City	99.0	124.7	110.3
074-075	Paterson	100.9	124.3	111.2
076	Hackensack	98.8	124.8	110.3
077	Long Branch	98.5	122.4	109.1
078	Dover	99.1	124.3	110.2
079	Summit	99.0	123.8	110.0
080,083	Vineland	97.1	122.2	108.2
081	Camden	99.0	122.9	109.5
082,084	Atlantic City	97.6	122.2	108.5
085-086	Trenton	99.0	123.4	109.8
087	Point Pleasant	99.2	122.2	109.3
088-089	New Brunswick	99.7	123.6	110.2
NEW MEXICO				
870-872	Albuquerque	99.8	74.8	88.8
873	Gallup	100.0	74.8	88.9
874	Farmington	100.4	74.8	89.1
875	Santa Fe	101.8	74.8	89.9
877	Las Vegas	98.2	74.6	87.8
878	Socorro	97.8	74.8	87.6
879	Truth/Consequences	97.9	68.9	85.1
880	Las Cruces	96.8	66.7	83.5
881	Clovis	98.5	72.7	87.1
882	Roswell	100.4	72.8	88.2
883	Carrizozo	100.8	74.8	89.3
884	Tucumcari	99.3	72.7	87.6
NEW YORK				
100-102	New York	107.0	166.3	133.2
103	Staten Island	103.1	162.5	129.3
104	Bronx	100.8	162.5	128.0
105	Mount Vernon	100.6	136.0	116.2
106	White Plains	100.7	136.0	116.3
107	Yonkers	105.7	136.1	119.2
108	New Rochelle	100.9	136.0	116.4
109	Suffern	100.7	126.7	112.2
110	Queens	101.8	162.4	128.6
111	Long Island City	103.5	162.4	129.5
112	Brooklyn	103.9	162.4	129.7
113	Flushing	103.4	162.4	129.4
114	Jamaica	101.5	162.4	128.4
115,117,118	Hicksville	101.6	146.4	121.4
116	Far Rockaway	103.6	162.4	129.5
119	Riverhead	102.4	145.9	121.6
120-122	Albany	97.7	97.6	97.7
123	Schenectady	98.2	97.3	97.8
124	Kingston	101.2	117.9	108.6
125-126	Poughkeepsie	100.3	134.3	115.3
127	Monticello	99.6	122.5	109.7
128	Glens Falls	92.2	90.5	91.5
129	Plattsburgh	97.1	87.3	92.8
130-132	Syracuse	99.3	93.0	96.5
133-135	Utica	97.2	88.9	93.5
136	Watertown	99.1	91.9	95.9
137-139	Binghamton	98.8	89.5	94.7
140-142	Buffalo	100.3	102.8	101.4
143	Niagara Falls	97.7	100.8	99.1
144-146	Rochester	99.8	97.4	98.7
147	Jamestown	96.6	86.9	92.3
148-149	Elmira	96.3	87.8	92.6
NORTH CAROLINA				
270,272-274	Greensboro	99.2	49.2	77.1
271	Winston-Salem	99.1	47.8	76.5
275-276	Raleigh	100.3	49.4	77.8
277	Durham	100.3	49.7	78.0
278	Rocky Mount	96.3	40.3	71.6
279	Elizabeth City	97.3	41.6	72.7
280	Gastonia	98.2	47.2	75.7
281-282	Charlotte	99.8	49.7	77.7
283	Fayetteville	100.0	50.1	78.0
284	Wilmington	96.9	47.6	75.1
285	Kinston	94.6	41.6	71.2

STATE/ZIP	CITY	MAT.	INST.	TOTAL
NORTH CAROLINA (CONT'D)				
286	Hickory	94.9	43.9	72.4
287-288	Asheville	97.4	47.2	75.3
289	Murphy	95.8	34.0	68.5
NORTH DAKOTA				
580-581	Fargo	101.8	59.5	83.2
582	Grand Forks	101.5	52.8	80.0
583	Devils Lake	100.3	55.0	80.3
584	Jamestown	100.4	47.1	76.9
585	Bismarck	100.4	60.4	82.8
586	Dickinson	101.3	57.7	82.0
587	Minot	101.5	64.6	85.2
588	Williston	99.6	57.7	81.1
OHIO				
430-432	Columbus	96.9	89.8	93.7
433	Marion	92.9	80.6	87.5
434-436	Toledo	97.1	97.1	97.1
437-438	Zanesville	93.3	80.0	87.5
439	Steubenville	94.9	89.6	92.6
440	Lorain	97.7	91.5	95.0
441	Cleveland	98.0	100.2	99.0
442-443	Akron	98.7	91.9	95.7
444-445	Youngstown	98.1	89.6	94.3
446-447	Canton	98.2	83.7	91.8
448-449	Mansfield	94.9	87.1	91.5
450	Hamilton	94.6	83.6	89.7
451-452	Cincinnati	94.9	86.4	91.1
453-454	Dayton	94.7	82.9	89.5
455	Springfield	94.6	84.3	90.1
456	Chillicothe	93.1	89.1	91.3
457	Athens	95.6	78.4	88.0
458	Lima	96.0	83.3	90.4
OKLAHOMA				
730-731	Oklahoma City	98.6	61.7	82.3
734	Ardmore	95.1	61.2	80.2
735	Lawton	97.9	61.7	81.9
736	Clinton	96.6	59.8	80.3
737	Enid	97.7	59.8	80.9
738	Woodward	95.3	59.9	79.6
739	Guymon	96.4	30.8	67.4
740-741	Tulsa	97.7	54.8	78.8
743	Miami	93.8	62.6	80.1
744	Muskogee	96.9	39.6	71.6
745	Mcalester	93.5	51.0	74.7
746	Ponca City	94.2	59.2	78.8
747	Durant	94.2	58.6	78.5
748	Shawnee	95.9	57.4	78.9
749	Poteau	93.2	61.1	79.0
OREGON				
970-972	Portland	101.1	96.8	99.2
973	Salem	101.0	95.0	98.4
974	Eugene	100.9	94.3	98.0
975	Medford	102.7	92.9	98.4
976	Klamath Falls	102.4	92.7	98.1
977	Bend	101.2	94.4	98.2
978	Pendleton	95.5	94.7	95.1
979	Vale	93.1	86.3	90.1
PENNSYLVANIA				
150-152	Pittsburgh	97.7	102.1	99.6
153	Washington	94.3	102.1	97.7
154	Uniontown	94.6	100.4	97.2
155	Bedford	95.6	91.0	93.6
156	Greensburg	95.6	100.6	97.8
157	Indiana	94.4	99.4	96.6
158	Dubois	96.1	94.7	95.5
159	Johnstown	95.6	94.9	95.3
160	Butler	92.2	101.6	96.3
161	New Castle	92.2	98.9	95.2
162	Kittanning	92.7	102.6	97.1
163	Oil City	92.2	95.6	93.7
164-165	Erie	94.7	93.1	94.0
166	Altoona	94.9	90.3	92.9
167	Bradford	95.6	93.8	94.8
168	State College	95.2	90.7	93.2
169	Wellsboro	96.3	92.8	94.8
170-171	Harrisburg	98.5	95.1	97.0
172	Chambersburg	95.8	89.7	93.1
173-174	York	96.5	95.4	96.0
175-176	Lancaster	94.4	89.5	92.2

STATE/ZIP	CITY	MAT.	INST.	TOTAL
PENNSYLVANIA (CONT'D)				
177	Williamsport	92.9	81.1	87.7
178	Sunbury	95.2	93.8	94.6
179	Pottsville	94.3	97.7	95.8
180	Lehigh Valley	95.7	114.3	103.9
181	Allentown	98.2	110.5	103.6
182	Hazleton	95.1	96.8	95.9
183	Stroudsburg	95.0	105.1	99.5
184-185	Scranton	99.0	100.9	99.8
186-187	Wilkes-Barre	94.8	97.4	95.9
188	Montrose	94.5	98.4	96.2
189	Doylestown	94.8	122.6	107.1
190-191	Philadelphia	100.3	134.2	115.3
193	Westchester	96.4	127.0	109.9
194	Norristown	95.4	132.6	111.8
195-196	Reading	97.9	101.5	99.5
PUERTO RICO				
009	San Juan	121.2	24.5	78.5
RHODE ISLAND				
028	Newport	99.0	112.8	105.1
029	Providence	100.2	112.8	105.8
SOUTH CAROLINA				
290-292	Columbia	98.3	51.2	77.5
293	Spartanburg	96.5	50.8	76.3
294	Charleston	98.1	59.7	81.2
295	Florence	96.3	51.2	76.4
296	Greenville	96.3	50.8	76.2
297	Rock Hill	95.6	48.9	74.9
298	Aiken	96.5	72.0	85.7
299	Beaufort	97.3	44.1	73.8
SOUTH DAKOTA				
570-571	Sioux Falls	99.8	57.8	81.2
572	Watertown	98.0	53.1	78.2
573	Mitchell	96.8	53.0	77.4
574	Aberdeen	100.0	54.0	79.7
575	Pierre	99.3	54.2	79.4
576	Mobridge	97.5	53.2	77.9
577	Rapid City	99.8	54.2	79.7
TENNESSEE				
370-372	Nashville	97.5	73.9	87.1
373-374	Chattanooga	98.6	66.9	84.6
375,380-381	Memphis	96.4	71.6	85.5
376	Johnson City	97.9	54.2	78.6
377-379	Knoxville	94.9	61.8	80.2
382	Mckenzie	96.4	55.6	78.4
383	Jackson	98.4	56.0	79.7
384	Columbia	94.9	62.3	80.5
385	Cookeville	96.3	60.5	80.5
TEXAS				
750	Mckinney	98.8	51.2	77.8
751	Waxahackie	98.7	56.4	80.0
752-753	Dallas	99.6	66.8	85.2
754	Greenville	98.8	40.0	72.9
755	Texarkana	98.3	52.6	78.1
756	Longview	98.9	41.0	73.4
757	Tyler	99.3	55.6	80.0
758	Palestine	95.1	41.5	71.4
759	Lufkin	95.8	45.6	73.6
760-761	Fort Worth	98.2	62.7	82.6
762	Denton	97.1	49.2	76.0
763	Wichita Falls	98.4	56.5	79.9
764	Eastland	96.8	40.8	72.1
765	Temple	95.1	51.0	75.6
766-767	Waco	97.8	58.1	80.3
768	Brownwood	97.9	38.6	71.7
769	San Angelo	97.4	49.3	76.2
770-772	Houston	100.0	70.3	86.9
773	Huntsville	98.1	39.3	72.2
774	Wharton	99.6	42.9	74.6
775	Galveston	97.4	70.1	85.3
776-777	Beaumont	98.1	62.1	82.2
778	Bryan	94.4	63.0	80.6
779	Victoria	99.6	45.2	75.6
780	Laredo	94.9	53.0	76.4
781-782	San Antonio	95.3	63.7	81.3
783-784	Corpus Christi	97.8	52.7	77.9
785	Mc Allen	97.2	48.0	75.5
786-787	Austin	95.5	59.4	79.6

STATE/ZIP	CITY	MAT.	INST.	TOTAL
TEXAS (CONT'D)				
788	Del Rio	96.9	33.7	69.0
789	Giddings	93.9	41.6	70.8
790-791	Amarillo	98.6	60.5	81.8
792	Childress	97.4	53.4	78.0
793-794	Lubbock	99.9	53.8	79.6
795-796	Abilene	98.2	52.8	78.2
797	Midland	99.6	50.5	77.9
798-799,885	El Paso	97.4	52.7	77.6
UTAH				
840-841	Salt Lake City	103.0	65.5	86.5
842,844	Ogden	98.2	65.0	83.6
843	Logan	100.3	65.0	84.7
845	Price	100.7	44.4	75.8
846-847	Provo	100.8	65.4	85.2
VERMONT				
050	White River Jct.	97.6	56.1	79.3
051	Bellows Falls	95.9	67.2	83.2
052	Bennington	96.3	67.0	83.4
053	Brattleboro	96.7	71.5	85.6
054	Burlington	100.6	67.3	85.9
056	Montpelier	97.1	67.7	84.1
057	Rutland	98.8	67.7	85.0
058	St. Johnsbury	97.6	56.6	79.5
059	Guildhall	96.2	56.4	78.6
VIRGINIA				
220-221	Fairfax	99.6	82.8	92.2
222	Arlington	101.2	81.0	92.3
223	Alexandria	100.3	85.4	93.7
224-225	Fredericksburg	98.2	73.9	87.5
226	Winchester	98.9	64.2	83.5
227	Culpeper	98.8	69.6	85.9
228	Harrisonburg	99.0	65.7	84.3
229	Charlottesville	99.4	65.2	84.3
230-232	Richmond	101.0	68.7	86.7
233-235	Norfolk	101.4	69.3	87.2
236	Newport News	100.0	69.2	86.4
237	Portsmouth	99.4	66.2	84.8
238	Petersburg	99.3	68.7	85.8
239	Farmville	98.4	54.5	79.0
240-241	Roanoke	100.7	61.9	83.6
242	Bristol	98.0	56.3	79.6
243	Pulaski	97.7	52.3	77.7
244	Staunton	98.6	60.7	81.9
245	Lynchburg	98.8	66.1	84.3
246	Grundy	98.1	51.7	77.6
WASHINGTON				
980-981,987	Seattle	104.6	104.7	104.6
982	Everett	104.2	95.1	100.1
983-984	Tacoma	104.3	99.4	102.2
985	Olympia	102.5	99.4	101.1
986	Vancouver	105.8	90.8	99.2
988	Wenatchee	103.9	78.5	92.7
989	Yakima	104.6	86.1	96.4
990-992	Spokane	104.9	77.9	93.0
993	Richland	104.6	81.6	94.5
994	Clarkston	102.8	78.3	92.0
WEST VIRGINIA				
247-248	Bluefield	96.7	77.1	88.1
249	Lewisburg	98.6	82.8	91.6
250-253	Charleston	99.8	89.1	95.0
254	Martinsburg	98.3	80.3	90.3
255-257	Huntington	100.2	91.7	96.4
258-259	Beckley	96.6	86.7	92.2
260	Wheeling	100.3	89.2	95.4
261	Parkersburg	99.1	87.8	94.1
262	Buckhannon	98.2	89.0	94.1
263-264	Clarksburg	98.8	87.7	93.9
265	Morgantown	98.9	88.3	94.2
266	Gassaway	98.1	89.3	94.2
267	Romney	98.1	82.7	91.3
268	Petersburg	97.9	85.4	92.4
WISCONSIN				
530,532	Milwaukee	100.5	106.6	103.2
531	Kenosha	100.4	97.6	99.1
534	Racine	99.7	99.1	99.5
535	Beloit	99.6	92.1	96.3
537	Madison	99.7	95.5	97.8

STATE/ZIP	CITY	MAT.	INST.	TOTAL
WISCONSIN (CONT'D)				
538	Lancaster	96.9	86.8	92.4
539	Portage	95.4	90.6	93.3
540	New Richmond	95.7	89.9	93.2
541-543	Green Bay	100.5	88.5	95.2
544	Wausau	94.8	89.4	92.4
545	Rhinelander	98.3	85.2	92.5
546	La Crosse	96.6	89.0	93.2
547	Eau Claire	98.3	90.5	94.8
548	Superior	95.5	93.6	94.7
549	Oshkosh	95.5	86.6	91.6
WYOMING				
820	Cheyenne	100.0	63.2	83.7
821	Yellowstone Nat'l Park	97.4	56.2	79.2
822	Wheatland	98.5	56.2	79.8
823	Rawlins	100.3	56.0	80.7
824	Worland	98.0	54.2	78.7
825	Riverton	99.2	54.2	79.3
826	Casper	100.5	56.0	80.9
827	Newcastle	97.8	56.0	79.3
828	Sheridan	100.7	58.0	81.9
829-831	Rock Springs	102.2	54.2	81.0
CANADIAN FACTORS (reflect Canadian currency)				
ALBERTA				
	Calgary	129.4	86.3	110.4
	Edmonton	130.6	86.2	111.0
	Fort McMurray	126.0	88.8	109.6
	Lethbridge	119.9	88.3	105.9
	Lloydminster	115.1	85.2	101.9
	Medicine Hat	115.3	84.5	101.7
	Red Deer	115.9	84.5	102.0
BRITISH COLUMBIA				
	Kamloops	116.4	86.9	103.4
	Prince George	117.6	86.9	104.1
	Vancouver	128.7	78.6	106.6
	Victoria	117.7	75.6	99.1
MANITOBA				
	Brandon	115.0	72.9	96.4
	Portage la Prairie	115.0	71.6	95.9
	Winnipeg	127.7	66.7	100.8
NEW BRUNSWICK				
	Bathurst	113.2	64.8	91.8
	Dalhousie	113.2	64.8	91.8
	Fredericton	115.8	69.2	95.2
	Moncton	113.5	65.6	92.3
	Newcastle	113.2	64.8	91.8
	St. John	115.9	69.2	95.3
NEWFOUNDLAND				
	Corner Brook	118.7	65.0	95.0
	St Johns	119.2	66.1	95.7
NORTHWEST TERRITORIES				
	Yellowknife	120.5	82.0	103.5
NOVA SCOTIA				
	Bridgewater	114.8	71.7	95.8
	Dartmouth	116.2	71.7	96.5
	Halifax	117.4	74.6	98.5
	New Glasgow	114.2	71.7	95.5
	Sydney	111.6	71.7	94.0
	Truro	114.2	71.7	95.5
	Yarmouth	114.1	71.7	95.4
ONTARIO				
	Barrie	118.0	89.2	105.3
	Brantford	117.1	92.9	106.4
	Cornwall	116.9	89.5	104.8
	Hamilton	121.9	93.2	109.2
	Kingston	117.8	89.6	105.4
	Kitchener	113.3	88.5	102.4
	London	120.6	89.6	106.9
	North Bay	117.1	87.5	104.0
	Oshawa	117.0	90.9	105.5
	Ottawa	121.5	90.1	107.6
	Owen Sound	118.1	87.5	104.6
	Peterborough	117.1	89.3	104.8
	Sarnia	116.8	93.5	106.5
	Sault Ste Marie	111.7	88.8	101.6

STATE/ZIP	CITY	MAT.	INST.	TOTAL
	St. Catharines	111.4	90.4	102.1
	Sudbury	111.4	89.1	101.6
	Thunder Bay	112.9	89.1	102.4
	Timmins	117.3	87.5	104.1
	Toronto	122.5	95.8	110.7
	Windsor	112.0	89.4	102.0
PRINCE EDWARD ISLAND				
	Charlottetown	116.6	60.5	91.8
	Summerside	116.1	60.5	91.6
QUEBEC				
	Cap-de-la-Madeleine	113.8	84.0	100.7
	Charlesbourg	113.8	84.0	100.7
	Chicoutimi	112.8	89.4	102.5
	Gatineau	113.4	83.8	100.3
	Granby	113.7	83.8	100.5
	Hull	113.6	83.8	100.4
	Joliette	114.1	84.0	100.8
	Laval	113.5	83.8	100.4
	Montreal	120.8	89.7	107.1
	Quebec	119.4	90.1	106.4
	Rimouski	113.3	89.4	102.8
	Rouyn-Noranda	113.4	83.8	100.3
	Saint Hyacinthe	113.2	83.8	100.2
	Sherbrooke	113.8	83.8	100.5
	Sorel	114.0	84.0	100.8
	St Jerome	113.4	83.8	100.4
	Trois Rivieres	114.0	84.0	100.8
SASKATCHEWAN				
	Moose Jaw	112.2	66.9	92.2
	Prince Albert	111.1	65.2	90.9
	Regina	114.2	67.0	93.4
	Saskatoon	112.7	66.8	92.4
YUKON				
	Whitehorse	112.0	65.9	91.7

R011105-05 Tips for Accurate Estimating

1. Use pre-printed or columnar forms for orderly sequence of dimensions and locations and for recording telephone quotations.
2. Use only the front side of each paper or form except for certain pre-printed summary forms.
3. Be consistent in listing dimensions: For example, length x width x height. This helps in rechecking to ensure that, the total length of partitions is appropriate for the building area.
4. Use printed (rather than measured) dimensions where given.
5. Add up multiple printed dimensions for a single entry where possible.
6. Measure all other dimensions carefully.
7. Use each set of dimensions to calculate multiple related quantities.
8. Convert foot and inch measurements to decimal feet when listing. Memorize decimal equivalents to .01 parts of a foot (1/8″ equals approximately .01′).
9. Do not "round off" quantities until the final summary.
10. Mark drawings with different colors as items are taken off.
11. Keep similar items together, different items separate.
12. Identify location and drawing numbers to aid in future checking for completeness.
13. Measure or list everything on the drawings or mentioned in the specifications.
14. It may be necessary to list items not called for to make the job complete.
15. Be alert for: Notes on plans such as N.T.S. (not to scale); changes in scale throughout the drawings; reduced size drawings; discrepancies between the specifications and the drawings.
16. Develop a consistent pattern of performing an estimate. For example:
    a. Start the quantity takeoff at the lower floor and move to the next higher floor.
    b. Proceed from the main section of the building to the wings.
    c. Proceed from south to north or vice versa, clockwise or counterclockwise.
    d. Take off floor plan quantities first, elevations next, then detail drawings.
17. List all gross dimensions that can be either used again for different quantities, or used as a rough check of other quantities for verification (exterior perimeter, gross floor area, individual floor areas, etc.).
18. Utilize design symmetry or repetition (repetitive floors, repetitive wings, symmetrical design around a center line, similar room layouts, etc.). Note: Extreme caution is needed here so as not to omit or duplicate an area.
19. Do not convert units until the final total is obtained. For instance, when estimating concrete work, keep all units to the nearest cubic foot, then summarize and convert to cubic yards.
20. When figuring alternatives, it is best to total all items involved in the basic system, then total all items involved in the alternates. Therefore you work with positive numbers in all cases. When adds and deducts are used, it is often confusing whether to add or subtract a portion of an item; especially on a complicated or involved alternate.

R011105-10 Unit Gross Area Requirements

The figures in the table below indicate typical ranges in square feet as a function of the "occupant" unit. This table is best used in the preliminary design stages to help determine the probable size requirement for the total project.

Building Type	Unit	Gross Area in S.F.		
		1/4	Median	3/4
Apartments	Unit	660	860	1,100
Auditorium & Play Theaters	Seat	18	25	38
Bowling Alleys	Lane		940	
Churches & Synagogues	Seat	20	28	39
Dormitories	Bed	200	230	275
Fraternity & Sorority Houses	Bed	220	315	370
Garages, Parking	Car	325	355	385
Hospitals	Bed	685	850	1,075
Hotels	Rental Unit	475	600	710
Housing for the elderly	Unit	515	635	755
Housing, Public	Unit	700	875	1,030
Ice Skating Rinks	Total	27,000	30,000	36,000
Motels	Rental Unit	360	465	620
Nursing Homes	Bed	290	350	450
Restaurants	Seat	23	29	39
Schools, Elementary	Pupil	65	77	90
Junior High & Middle	↓	85	110	129
Senior High	↓	102	130	145
Vocational	↓	110	135	195
Shooting Ranges	Point		450	
Theaters & Movies	Seat		15	

R011105-20 Floor Area Ratios

Table below lists commonly used gross to net area and net to gross area ratios expressed in % for various building types.

Building Type	Gross to Net Ratio	Net to Gross Ratio	Building Type	Gross to Net Ratio	Net to Gross Ratio
Apartment	156	64	School Buildings (campus type)		
Bank	140	72	Administrative	150	67
Church	142	70	Auditorium	142	70
Courthouse	162	61	Biology	161	62
Department Store	123	81	Chemistry	170	59
Garage	118	85	Classroom	152	66
Hospital	183	55	Dining Hall	138	72
Hotel	158	63	Dormitory	154	65
Laboratory	171	58	Engineering	164	61
Library	132	76	Fraternity	160	63
Office	135	75	Gymnasium	142	70
Restaurant	141	70	Science	167	60
Warehouse	108	93	Service	120	83
			Student Union	172	59

The gross area of a building is the total floor area based on outside dimensions.

The net area of a building is the usable floor area for the function intended and excludes such items as stairways, corridors and mechanical rooms. In the case of a commercial building, it might be considered as the "leasable area."

R011105-30 Occupancy Determinations

Description		S.F. Required per Person		
		BOCA	SBC	UBC
Assembly Areas	Fixed Seats	**	6	7
	Movable Seats		15	15
	Concentrated	7		
	Unconcentrated	15		
	Standing Space	3		
Educational	Unclassified			
	Classrooms	20	40	20
	Shop Areas	50	100	50
Institutional	Unclassified		125	
	In-Patient Areas	240		
	Sleeping Areas	120		
Mercantile	Basement	30	30	20
	Ground Floor	30	30	30
	Upper Floors	60	60	50
Office		100	100	100

BOCA=Building Officials & Code Administrators
SBC=Southern Building Code
UBC=Uniform Building Code

** The occupancy load for assembly area with fixed seats shall be determined by the number of fixed seats installed.

R011110-10 Architectural Fees

Tabulated below are typical percentage fees by project size, for good professional architectural service. Fees may vary from those listed depending upon degree of design difficulty and economic conditions in any particular area.

Rates can be interpolated horizontally and vertically. Various portions of the same project requiring different rates should be adjusted proportionately. For alterations, add 50% to the fee for the first $500,000 of project cost and add 25% to the fee for project cost over $500,000.

Architectural fees tabulated below include Structural, Mechanical and Electrical Engineering Fees. They do not include the fees for special consultants such as kitchen planning, security, acoustical, interior design, etc.

Civil Engineering fees are included in the Architectural fee for project sites requiring minimal design such as city sites. However, separate Civil Engineering fees must be added when utility connections require design, drainage calculations are needed, stepped foundations are required, or provisions are required to protect adjacent wetlands.

	Total Project Size in Thousands of Dollars						
Building Types	100	250	500	1,000	5,000	10,000	50,000
Factories, garages, warehouses, repetitive housing	9.0%	8.0%	7.0%	6.2%	5.3%	4.9%	4.5%
Apartments, banks, schools, libraries, offices, municipal buildings	12.2	12.3	9.2	8.0	7.0	6.6	6.2
Churches, hospitals, homes, laboratories, museums, research	15.0	13.6	12.7	11.9	9.5	8.8	8.0
Memorials, monumental work, decorative furnishings	—	16.0	14.5	13.1	10.0	9.0	8.3

R011110-30 Engineering Fees

Typical **Structural Engineering Fees** based on type of construction and total project size. These fees are included in Architectural Fees.

	Total Project Size (in thousands of dollars)			
Type of Construction	$500	$500-$1,000	$1,000-$5,000	Over $5000
Industrial buildings, factories & warehouses	Technical payroll times 2.0 to 2.5	1.60%	1.25%	1.00%
Hotels, apartments, offices, dormitories, hospitals, public buildings, food stores		2.00%	1.70%	1.20%
Museums, banks, churches and cathedrals		2.00%	1.75%	1.25%
Thin shells, prestressed concrete, earthquake resistive		2.00%	1.75%	1.50%
Parking ramps, auditoriums, stadiums, convention halls, hangars & boiler houses		2.50%	2.00%	1.75%
Special buildings, major alterations, underpinning & future expansion	▼	Add to above 0.5%	Add to above 0.5%	Add to above 0.5%

For complex reinforced concrete or unusually complicated structures, add 20% to 50%.

Typical **Mechanical and Electrical Engineering Fees** are based on the size of the subcontract. The fee structure for both are shown below. These fees are included in Architectural Fees.

	Subcontract Size							
Type of Construction	$25,000	$50,000	$100,000	$225,000	$350,000	$500,000	$750,000	$1,000,000
Simple structures	6.4%	5.7%	4.8%	4.5%	4.4%	4.3%	4.2%	4.1%
Intermediate structures	8.0	7.3	6.5	5.6	5.1	5.0	4.9	4.8
Complex structures	10.1	9.0	9.0	8.0	7.5	7.5	7.0	7.0

For renovations, add 15% to 25% to applicable fee.

R012153-10 Repair and Remodeling

Cost figures are based on new construction utilizing the most cost-effective combination of labor, equipment and material with the work scheduled in proper sequence to allow the various trades to accomplish their work in an efficient manner.

The costs for repair and remodeling work must be modified due to the following factors that may be present in any given repair and remodeling project.

1. Equipment usage curtailment due to the physical limitations of the project, with only hand-operated equipment being used.
2. Increased requirement for shoring and bracing to hold up the building while structural changes are being made and to allow for temporary storage of construction materials on above-grade floors.
3. Material handling becomes more costly due to having to move within the confines of an enclosed building. For multi-story construction, low capacity elevators and stairwells may be the only access to the upper floors.
4. Large amount of cutting and patching and attempting to match the existing construction is required. It is often more economical to remove entire walls rather than create many new door and window openings. This sort of trade-off has to be carefully analyzed.
5. Cost of protection of completed work is increased since the usual sequence of construction usually cannot be accomplished.
6. Economies of scale usually associated with new construction may not be present. If small quantities of components must be custom fabricated due to job requirements, unit costs will naturally increase. Also, if only small work areas are available at a given time, job scheduling between trades becomes difficult and subcontractor quotations may reflect the excessive start-up and shut-down phases of the job.
7. Work may have to be done on other than normal shifts and may have to be done around an existing production facility which has to stay in production during the course of the repair and remodeling.
8. Dust and noise protection of adjoining non-construction areas can involve substantial special protection and alter usual construction methods.
9. Job may be delayed due to unexpected conditions discovered during demolition or removal. These delays ultimately increase construction costs.
10. Piping and ductwork runs may not be as simple as for new construction. Wiring may have to be snaked through walls and floors.
11. Matching "existing construction" may be impossible because materials may no longer be manufactured. Substitutions may be expensive.
12. Weather protection of existing structure requires additional temporary structures to protect building at openings.
13. On small projects, because of local conditions, it may be necessary to pay a tradesman for a minimum of four hours for a task that is completed in one hour.

All of the above areas can contribute to increased costs for a repair and remodeling project. Each of the above factors should be considered in the planning, bidding and construction stage in order to minimize the increased costs associated with repair and remodeling jobs.

R012909-80 Sales Tax by State

State sales tax on materials is tabulated below (5 states have no sales tax). Many states allow local jurisdictions, such as a county or city, to levy additional sales tax.

Some projects may be sales tax exempt, particularly those constructed with public funds.

State	Tax (%)	State	Tax (%)	State	Tax (%)	State	Tax (%)
Alabama	4	Illinois	6.25	Montana	0	Rhode Island	7
Alaska	0	Indiana	7	Nebraska	5.5	South Carolina	6
Arizona	5.6	Iowa	6	Nevada	6.5	South Dakota	4
Arkansas	6	Kansas	5.3	New Hampshire	0	Tennessee	7
California	8.25	Kentucky	6	New Jersey	7	Texas	6.25
Colorado	2.9	Louisiana	4	New Mexico	5	Utah	4.75
Connecticut	6	Maine	5	New York	4	Vermont	6
Delaware	0	Maryland	6	North Carolina	4.5	Virginia	4
District of Columbia	5.75	Massachusetts	6.25	North Dakota	5	Washington	6.5
Florida	6	Michigan	6	Ohio	7	West Virginia	6
Georgia	4	Minnesota	6.5	Oklahoma	4.5	Wisconsin	5
Hawaii	4	Mississippi	7	Oregon	0	Wyoming	4
Idaho	6	Missouri	4.225	Pennsylvania	6	Average	5.01 %

Sales Tax by Province (Canada)

GST - a value-added tax, which the government imposes on most goods and services provided in or imported into Canada. PST - a retail sales tax, which five of the provinces impose on the price of most goods and some services. QST - a value-added tax, similar to the federal GST, which Quebec imposes. HST - Three provinces have combined their retail sales tax with the federal GST into one harmonized tax.

Province	PST (%)	QST (%)	GST(%)	HST(%)
Alberta	0	0	5	0
British Columbia	7	0	5	0
Manitoba	7	0	5	0
New Brunswick	0	0	0	13
Newfoundland	0	0	0	13
Northwest Territories	0	0	5	0
Nova Scotia	0	0	0	13
Ontario	8	0	5	0
Prince Edward Island	10	0	5	0
Quebec	0	7.5	5	0
Saskatchewan	5	0	5	0
Yukon	0	0	5	0

R012909-85 Unemployment Taxes and Social Security Taxes

State Unemployment Tax rates vary not only from state to state, but also with the experience rating of the contractor. The Federal Unemployment Tax rate is 6.2% of the first $7,000 of wages. This is reduced by a credit of up to 5.4% for timely payment to the state. The minimum Federal Unemployment Tax is 0.8% after all credits.

Social Security (FICA) for 2010 is estimated at time of publication to be 7.65% of wages up to $106,800.

R012909-90 Overtime

One way to improve the completion date of a project or eliminate negative float from a schedule is to compress activity duration times. This can be achieved by increasing the crew size or working overtime with the proposed crew.

To determine the costs of working overtime to compress activity duration times, consider the following examples. Below is an overtime efficiency and cost chart based on a five, six, or seven day week with an eight through twelve hour day. Payroll percentage increases for time and one half and double time are shown for the various working days.

Days per Week	Hours per Day	Production Efficiency					Payroll Cost Factors	
		1st Week	2nd Week	3rd Week	4th Week	Average 4 Weeks	@ 1-1/2 Times	@ 2 Times
5	8	100%	100%	100%	100%	100 %	100 %	100 %
	9	100	100	95	90	96.25	105.6	111.1
	10	100	95	90	85	91.25	110.0	120.0
	11	95	90	75	65	81.25	113.6	127.3
	12	90	85	70	60	76.25	116.7	133.3
6	8	100	100	95	90	96.25	108.3	116.7
	9	100	95	90	85	92.50	113.0	125.9
	10	95	90	85	80	87.50	116.7	133.3
	11	95	85	70	65	78.75	119.7	139.4
	12	90	80	65	60	73.75	122.2	144.4
7	8	100	95	85	75	88.75	114.3	128.6
	9	95	90	80	70	83.75	118.3	136.5
	10	90	85	75	65	78.75	121.4	142.9
	11	85	80	65	60	72.50	124.0	148.1
	12	85	75	60	55	68.75	126.2	152.4

R013113-40 Builder's Risk Insurance

Builder's Risk Insurance is insurance on a building during construction. Premiums are paid by the owner or the contractor. Blasting, collapse and underground insurance would raise total insurance costs above those listed. Floater policy for materials delivered to the job runs $.75 to $1.25 per $100 value. Contractor equipment insurance runs $.50 to $1.50 per $100 value. Insurance for miscellaneous tools to $1,500 value runs from $3.00 to $7.50 per $100 value.

Tabulated below are New England Builder's Risk insurance rates in dollars per $100 value for $1,000 deductible. For $25,000 deductible, rates can be reduced 13% to 34%. On contracts over $1,000,000, rates may be lower than those tabulated. Policies are written annually for the total completed value in place. For "all risk" insurance (excluding flood, earthquake and certain other perils) add $.025 to total rates below.

Coverage	Frame Construction (Class 1)		Brick Construction (Class 4)		Fire Resistive (Class 6)	
	Range	Average	Range	Average	Range	Average
Fire Insurance	$.350 to $ .850	$.600	$.158 to $.189	$.174	$.052 to $.080	$.070
Extended Coverage	.115 to .200	.158	.080 to .105	.101	.081 to .105	.100
Vandalism	.012 to .016	.014	.008 to .011	.011	.008 to .011	.010
Total Annual Rate	$.477 to $1.066	$.772	$.246 to $.305	$.286	$.141 to $.196	$.180

R013113-50 General Contractor's Overhead

There are two distinct types of overhead on a construction project: Project Overhead and Main Office Overhead. Project Overhead includes those costs at a construction site not directly associated with the installation of construction materials. Examples of Project Overhead costs include the following:

1. Superintendent
2. Construction office and storage trailers
3. Temporary sanitary facilities
4. Temporary utilities
5. Security fencing
6. Photographs
7. Clean up
8. Performance and payment bonds

The above Project Overhead items are also referred to as General Requirements and therefore are estimated in Division 1. Division 1 is the first division listed in the CSI MasterFormat but it is usually the last division estimated. The sum of the costs in Divisions 1 through 49 is referred to as the sum of the direct costs.

All construction projects also include indirect costs. The primary components of indirect costs are the contractor's Main Office Overhead and profit. The amount of the Main Office Overhead expense varies depending on the following:

1. Owner's compensation
2. Project managers and estimator's wages
3. Clerical support wages
4. Office rent and utilities
5. Corporate legal and accounting costs
6. Advertising
7. Automobile expenses
8. Association dues
9. Travel and entertainment expenses

These costs are usually calculated as a percentage of annual sales volume. This percentage can range from 35% for a small contractor doing less than $500,000 to 5% for a large contractor with sales in excess of $100 million.

R013113-60 Workers' Compensation Insurance Rates by Trade

The table below tabulates the national averages for Workers' Compensation insurance rates by trade and type of building. The average "Insurance Rate" is multiplied by the "% of Building Cost" for each trade. This produces the "Workers' Compensation Cost" by % of total labor cost, to be added for each trade by building type to determine the weighted average Workers' Compensation rate for the building types analyzed.

Trade	Insurance Rate (% Labor Cost) Range	Insurance Rate Average	% of Building Cost Office Bldgs.	% of Building Cost Schools & Apts.	% of Building Cost Mfg.	Workers' Compensation Office Bldgs.	Workers' Compensation Schools & Apts.	Workers' Compensation Mfg.
Excavation, Grading, etc.	4.2 % to 17.5%	9.4%	4.8%	4.9%	4.5%	0.45%	0.46%	0.42%
Piles & Foundations	4.8 to 57.1	18.5	7.1	5.2	8.7	1.31	0.96	1.61
Concrete	4.8 to 29.9	13.7	5.0	14.8	3.7	0.69	2.03	0.51
Masonry	4.6 to 29.8	13.5	6.9	7.5	1.9	0.93	1.01	0.26
Structural Steel	4.8 to 170.8	36.0	10.7	3.9	17.6	3.85	1.40	6.34
Miscellaneous & Ornamental Metals	4 to 20.3	10.1	2.8	4.0	3.6	0.28	0.40	0.36
Carpentry & Millwork	4.8 to 53.2	16.9	3.7	4.0	0.5	0.63	0.68	0.08
Metal or Composition Siding	4.8 to 35.7	15.5	2.3	0.3	4.3	0.36	0.05	0.67
Roofing	4.8 to 77.1	28.9	2.3	2.6	3.1	0.66	0.75	0.90
Doors & Hardware	3.4 to 34.5	11.3	0.9	1.4	0.4	0.10	0.16	0.05
Sash & Glazing	4.8 to 36	13.3	3.5	4.0	1.0	0.47	0.53	0.13
Lath & Plaster	3.1 to 30.3	11.9	3.3	6.9	0.8	0.39	0.82	0.10
Tile, Marble & Floors	3.1 to 16.8	8.8	2.6	3.0	0.5	0.23	0.26	0.04
Acoustical Ceilings	4.4 to 32	9.8	2.4	0.2	0.3	0.24	0.02	0.03
Painting	4.8 to 29.6	11.5	1.5	1.6	1.6	0.17	0.18	0.18
Interior Partitions	4.8 to 53.2	16.9	3.9	4.3	4.4	0.66	0.73	0.74
Miscellaneous Items	2.4 to 134.5	14.9	5.2	3.7	9.7	0.78	0.55	1.45
Elevators	2.6 to 13.5	6.3	2.1	1.1	2.2	0.13	0.07	0.14
Sprinklers	2.5 to 13.6	7.4	0.5	—	2.0	0.04	—	0.15
Plumbing	3.9 to 13.3	7.6	4.9	7.2	5.2	0.37	0.55	0.40
Heat., Vent., Air Conditioning	3.9 to 21.2	9.1	13.5	11.0	12.9	1.23	1.00	1.17
Electrical	2.9 to 15.4	6.4	10.1	8.4	11.1	0.65	0.54	0.71
Total	2.4 % to 170.8%	—	100.0%	100.0%	100.0%	14.62%	13.15%	16.44%
	Overall Weighted Average 14.74%							

Workers' Compensation Insurance Rates by States

The table below lists the weighted average Workers' Compensation base rate for each state with a factor comparing this with the national average of 14.4%.

State	Weighted Average	Factor	State	Weighted Average	Factor	State	Weighted Average	Factor
Alabama	19.8%	138	Kentucky	16.3%	113	North Dakota	12.9%	90
Alaska	19.6	136	Louisiana	28.2	196	Ohio	13.0	90
Arizona	9.7	67	Maine	14.2	99	Oklahoma	14.5	101
Arkansas	10.4	72	Maryland	14.5	101	Oregon	12.4	86
California	21.4	149	Massachusetts	12.1	84	Pennsylvania	14.7	102
Colorado	9.0	63	Michigan	19.0	132	Rhode Island	11.6	81
Connecticut	20.0	139	Minnesota	26.0	181	South Carolina	19.8	138
Delaware	11.5	80	Mississippi	15.3	106	South Dakota	17.2	119
District of Columbia	11.3	78	Missouri	17.5	122	Tennessee	14.2	99
Florida	12.6	88	Montana	15.0	104	Texas	11.2	78
Georgia	26.8	186	Nebraska	21.2	147	Utah	9.1	63
Hawaii	16.9	117	Nevada	10.3	72	Vermont	15.0	104
Idaho	9.9	69	New Hampshire	19.2	133	Virginia	10.9	76
Illinois	23.0	160	New Jersey	13.3	92	Washington	9.8	68
Indiana	7.2	50	New Mexico	18.4	128	West Virginia	12.0	83
Iowa	11.4	79	New York	6.9	48	Wisconsin	15.0	104
Kansas	8.6	60	North Carolina	16.9	117	Wyoming	5.3	37
			Weighted Average for U.S. is 14.7% of payroll = 100%					

Rates in the following table are the base or manual costs per $100 of payroll for Workers' Compensation in each state. Rates are usually applied to straight time wages only and not to premium time wages and bonuses.

The weighted average skilled worker rate for 35 trades is 14.4%. For bidding purposes, apply the full value of Workers' Compensation directly to total labor costs, or if labor is 38%, materials 42% and overhead and profit 20% of total cost, carry 38/80 x 14.4% =6.8% of cost (before overhead and profit) into overhead. Rates vary not only from state to state but also with the experience rating of the contractor.

Rates are the most current available at the time of publication.

R013113-60 Workers' Compensation Insurance Rates by Trade and State (cont.)

State	Carpentry — 3 stories or less 5651	Carpentry — interior cab. work 5437	Carpentry — general 5403	Concrete Work — NOC 5213	Concrete Work — flat (flr., sdwk.) 5221	Electrical Wiring — inside 5190	Excavation — earth NOC 6217	Excavation — rock 6217	Glaziers 5462	Insulation Work 5479	Lathing 5443	Masonry 5022	Painting & Decorating 5474	Pile Driving 6003	Plastering 5480	Plumbing 5183	Roofing 5551	Sheet Metal Work (HVAC) 5538	Steel Erection — door & sash 5102	Steel Erection — inter., ornam. 5102	Steel Erection — structure 5040	Steel Erection — NOC 5057	Tile Work — (interior ceramic) 5348	Waterproofing 9014	Wrecking 5701
AL	28.11	16.91	33.75	15.19	9.51	10.24	10.24	10.24	22.09	16.53	10.78	23.87	21.44	24.16	15.68	10.18	43.60	10.03	8.28	8.28	41.21	18.79	13.84	7.13	41.21
AK	11.41	14.35	15.47	11.20	10.66	8.55	12.54	12.54	35.96	31.60	8.46	29.81	16.71	31.00	30.26	10.01	35.75	8.86	8.04	8.04	29.65	32.73	6.25	6.29	29.65
AZ	11.55	6.95	17.61	9.58	5.16	4.96	5.32	5.32	7.21	14.50	5.56	7.35	6.53	11.22	5.24	5.77	15.08	6.92	9.96	9.96	17.31	14.27	3.56	3.13	17.31
AR	9.83	6.03	13.66	8.16	6.58	4.73	6.47	6.47	7.41	9.31	5.61	7.02	9.22	12.87	8.87	5.43	17.31	5.37	5.41	5.41	25.61	31.50	5.46	3.15	25.61
CA	34.52	34.52	34.52	12.92	12.92	9.65	15.56	15.56	18.87	14.65	12.69	21.99	16.27	16.32	24.72	13.26	47.80	16.92	13.77	13.77	22.36	17.89	8.82	16.27	17.89
CO	10.43	5.56	7.98	7.19	5.24	3.48	7.24	7.24	6.53	11.47	4.36	9.05	5.77	10.08	5.59	4.64	17.78	4.59	6.83	6.83	27.27	10.66	5.43	3.46	10.66
CT	14.94	15.49	24.99	28.69	11.11	8.50	13.15	13.15	20.92	17.38	17.82	23.28	16.39	20.36	16.27	10.08	34.95	12.68	15.41	15.41	41.41	23.21	13.64	5.64	41.41
DE	11.96	11.96	8.80	8.99	9.49	4.26	7.16	7.16	10.17	8.80	10.97	10.39	11.94	14.39	10.17	6.10	21.13	7.75	9.69	9.69	21.36	9.69	7.19	10.39	21.36
DC	10.70	10.28	7.97	9.28	7.58	6.22	9.12	9.12	16.60	6.06	6.77	8.27	6.06	12.61	10.50	8.51	14.67	8.61	11.03	11.03	24.14	13.86	13.64	3.40	24.14
FL	12.62	7.98	13.59	13.18	5.99	15.44	7.01	7.01	8.77	8.12	5.53	9.08	8.76	38.24	19.38	5.47	21.29	7.98	8.12	8.12	22.59	10.37	5.16	4.53	22.59
GA	35.73	21.35	25.92	17.90	15.31	10.45	16.30	16.30	19.14	22.15	14.61	24.16	24.99	26.75	24.86	12.18	63.28	17.45	20.26	20.26	71.30	39.53	13.97	8.92	71.30
HI	17.38	11.62	28.20	13.74	12.27	7.10	7.73	7.73	20.55	21.19	10.94	17.87	11.33	20.15	15.61	6.06	33.24	8.02	11.11	11.11	31.71	21.70	9.78	11.54	31.71
ID	9.41	6.65	10.29	10.24	5.84	3.48	5.84	5.84	8.87	8.06	9.22	7.35	7.88	11.56	5.48	5.01	26.07	6.56	7.00	7.00	24.33	9.22	9.07	4.91	24.33
IL	25.30	15.57	23.56	29.89	12.90	10.43	10.77	10.77	18.01	20.90	17.31	21.46	11.94	31.27	17.49	12.01	33.86	14.59	19.11	19.11	79.28	21.91	16.76	5.22	79.28
IN	11.86	4.44	6.93	5.02	3.12	2.85	4.38	4.38	6.11	7.10	22.54	4.56	4.86	5.84	3.09	12.76	10.06	4.48	3.95	3.95	12.69	6.07	3.12	2.39	12.69
IA	13.14	10.98	12.06	14.63	7.56	4.65	7.06	7.06	9.01	6.62	5.84	9.42	7.53	9.92	7.28	6.74	22.71	6.08	5.56	5.56	30.86	26.76	8.01	3.61	30.86
KS	12.20	7.16	9.07	6.95	5.58	3.39	5.22	5.22	9.12	7.26	4.46	7.09	7.12	9.74	5.27	4.53	12.45	5.13	4.53	4.53	24.78	16.44	6.24	4.25	16.44
KY	14.98	12.29	25.40	10.80	7.22	5.36	9.95	9.95	13.90	14.09	9.23	7.30	11.88	22.46	12.14	5.65	31.57	11.42	9.86	9.86	60.00	18.72	16.40	4.59	60.00
LA	22.02	24.93	53.17	26.43	15.61	9.88	17.49	17.49	20.14	21.43	24.51	28.33	29.58	31.25	22.37	8.64	77.12	21.20	18.98	18.98	51.76	24.21	13.77	13.78	66.41
ME	15.11	10.18	25.32	18.95	9.40	6.61	8.79	8.79	14.50	11.15	7.69	13.82	13.31	15.89	12.02	9.12	18.69	8.40	8.34	8.34	32.51	19.53	6.58	5.84	32.51
MD	14.33	11.61	13.75	15.69	5.67	5.60	7.82	7.82	11.43	13.56	8.01	9.92	7.12	20.15	12.77	5.69	29.53	6.93	8.31	8.31	52.61	26.15	7.99	5.44	26.15
MA	7.50	5.93	11.92	21.45	6.62	3.17	4.17	4.17	8.93	8.94	6.53	11.62	5.01	4.99	4.89	3.88	30.35	5.74	6.76	6.76	44.61	39.97	6.45	2.50	29.46
MI	20.53	10.95	22.88	15.72	9.41	5.27	10.95	10.95	14.98	13.04	13.59	14.45	13.57	57.09	12.36	7.21	32.64	11.89	11.07	11.07	57.09	19.42	10.83	6.26	57.09
MN	22.68	23.13	32.18	10.25	13.85	6.68	11.88	11.88	27.48	16.33	12.68	19.35	14.58	22.85	12.68	11.28	54.88	16.13	11.70	11.70	170.80	13.03	11.23	6.85	13.03
MS	16.49	10.30	17.28	10.71	7.75	6.48	11.17	11.17	10.34	10.46	6.84	13.96	12.78	30.91	19.04	7.94	34.45	7.79	12.45	12.45	33.05	15.81	8.83	3.94	33.05
MO	20.17	12.95	15.43	19.32	10.66	7.57	10.57	10.57	11.63	15.03	9.23	16.87	12.93	18.60	16.85	10.79	34.32	9.57	11.61	11.61	33.95	41.29	11.92	6.56	33.95
MT	13.10	10.20	23.32	12.63	10.84	5.93	14.51	14.51	8.57	24.07	7.98	12.12	8.43	22.50	10.60	9.71	37.56	10.63	9.09	9.09	22.59	11.57	9.08	7.91	11.57
NE	23.73	18.63	20.88	26.18	14.23	9.80	16.00	16.00	16.90	23.33	11.75	21.90	17.52	20.73	17.43	12.53	36.95	12.48	15.00	15.00	39.83	36.08	10.53	6.83	55.60
NV	14.08	7.98	12.08	9.61	8.20	6.12	7.77	7.77	10.06	7.54	4.60	7.35	7.54	11.58	6.73	6.12	12.82	9.04	8.35	8.35	22.13	16.86	5.83	5.36	16.86
NH	26.15	11.92	19.85	24.52	14.29	7.33	13.00	13.00	10.32	5.63	9.25	21.17	21.90	7.88	10.06	10.99	45.41	8.04	15.28	15.28	64.41	23.21	11.64	6.10	64.41
NJ	14.79	8.84	14.79	15.65	9.31	4.38	8.18	8.18	8.97	12.55	10.73	13.44	11.40	16.59	10.73	5.79	37.97	6.78	9.33	9.33	20.26	16.09	7.96	6.20	21.39
NM	20.42	7.36	18.00	14.67	12.11	8.39	9.67	9.67	21.18	13.47	8.06	15.88	11.90	20.37	15.53	8.71	39.03	11.93	13.94	13.94	44.89	47.45	7.36	6.06	44.89
NY	7.96	3.39	7.17	8.25	6.46	3.17	4.63	4.63	5.71	3.73	6.49	8.74	5.50	8.04	4.21	3.98	14.20	5.50	5.44	5.44	13.31	7.90	3.26	2.83	5.51
NC	16.57	12.67	16.19	18.29	8.30	10.73	11.86	11.86	14.23	12.38	11.28	11.33	11.44	17.48	14.68	9.46	26.41	11.87	10.88	10.88	62.71	20.69	9.08	5.75	62.71
ND	10.87	10.87	10.87	6.50	6.50	3.55	5.58	5.58	10.87	10.87	7.25	17.45	6.87	19.11	7.25	5.19	21.35	5.19	19.14	19.14	19.14	19.14	10.87	21.35	10.90
OH	13.61	7.75	11.13	9.75	9.79	5.67	8.20	8.20	8.42	15.84	31.95	11.77	13.74	13.44	4.26	6.07	30.45	6.93	8.35	8.35	20.46	11.61	9.82	6.91	11.61
OK	13.42	9.17	11.76	11.76	6.34	5.73	10.83	10.83	16.07	19.47	8.45	10.02	8.47	20.25	12.19	7.09	21.13	9.05	13.76	13.76	39.72	23.97	6.50	5.71	39.72
OR	16.06	7.50	14.54	11.48	7.70	4.95	9.25	9.25	12.89	11.25	7.90	12.08	9.99	16.97	11.59	5.41	21.99	7.70	6.54	6.54	22.55	21.87	8.85	5.56	22.55
PA	14.68	14.68	12.10	15.30	11.22	6.32	9.13	9.13	12.06	12.10	12.06	13.10	14.00	16.59	12.06	7.77	29.88	8.29	15.11	15.11	22.64	15.11	8.45	13.10	22.64
RI	10.79	8.37	10.96	13.14	10.29	5.15	7.50	7.50	13.33	17.64	6.04	11.62	9.27	27.21	9.50	4.42	14.82	7.75	5.68	5.68	20.92	14.98	5.62	4.72	14.98
SC	21.43	18.13	22.49	16.89	9.59	11.45	11.78	11.78	16.99	18.58	11.06	13.93	15.50	20.61	15.75	12.36	50.08	13.06	13.91	13.91	41.22	32.53	10.50	6.33	41.22
SD	21.47	9.62	21.86	25.23	7.35	5.33	9.54	9.54	15.06	13.87	8.26	11.62	10.45	20.56	11.87	10.92	26.74	12.70	8.86	8.86	59.15	31.01	7.69	5.99	59.15
TN	14.91	11.20	13.03	12.97	6.75	5.79	13.34	13.34	12.18	9.21	8.32	12.35	11.84	28.35	13.71	6.79	24.19	9.81	8.15	8.15	25.81	25.53	8.36	4.08	25.81
TX	9.17	8.39	9.17	9.00	7.06	5.76	7.88	7.88	8.56	11.97	5.74	10.31	8.18	16.38	8.18	5.88	19.42	12.05	7.87	7.87	33.20	11.24	5.53	6.19	8.70
UT	11.93	6.73	9.82	7.43	6.42	3.47	8.78	8.78	11.14	9.47	4.81	8.51	6.34	9.45	5.50	4.68	18.93	6.11	5.43	5.43	19.45	11.11	5.16	3.24	13.92
VT	11.28	10.68	17.87	12.39	7.41	6.20	12.96	12.96	15.29	16.13	7.53	16.28	8.35	14.60	9.04	8.41	26.14	9.59	9.47	9.47	36.84	30.71	9.76	6.97	36.84
VA	10.22	7.61	9.05	11.32	5.11	4.92	7.70	7.70	10.13	7.82	11.06	8.34	9.02	9.49	9.78	5.63	22.11	6.32	6.90	6.90	32.08	20.56	5.08	2.91	32.08
WA	7.94	7.94	7.94	7.29	7.29	3.15	6.88	6.88	12.21	8.54	7.94	9.48	11.84	16.40	10.06	4.30	17.24	3.92	7.61	7.61	7.61	7.61	8.83	17.24	7.61
WV	14.45	9.91	13.88	11.18	5.99	6.38	8.45	8.45	10.70	9.85	8.29	10.83	10.72	16.15	10.77	6.65	24.50	7.80	7.57	7.57	32.81	7.91	7.67	3.60	17.91
WI	10.46	10.33	15.46	11.52	9.18	5.80	8.13	8.13	11.32	12.60	5.84	14.90	14.06	19.42	10.64	6.15	34.47	7.77	9.80	9.80	22.52	48.12	15.46	4.83	22.52
WY	4.77	4.77	4.77	4.77	4.77	4.77	4.77	4.77	4.77	4.77	4.77	4.77	4.77	4.77	4.77	4.77	4.77	4.77	4.77	4.77	4.77	4.77	4.77	4.77	4.77
AVG.	15.47	11.27	16.88	13.72	8.74	6.38	9.42	9.42	13.27	13.11	9.79	13.47	11.46	18.54	11.92	7.62	28.89	9.14	10.07	10.07	35.99	20.79	8.78	6.48	30.31

R013113-60 Workers' Compensation (cont.) (Canada in Canadian dollars)

Province		Alberta	British Columbia	Manitoba	Ontario	New Brunswick	Newfndld. & Labrador	Northwest Territories	Nova Scotia	Prince Edward Island	Quebec	Saskat-chewan	Yukon
Carpentry—3 stories or less	Rate	5.99	3.64	4.17	4.35	5.05	11.00	3.40	9.23	6.23	13.26	4.37	9.42
	Code	42143	721028	40102	723	238130	4226	4-41	4226	401	80110	1226	202
Carpentry—interior cab. work	Rate	1.86	2.62	4.17	4.35	4.59	4.37	3.40	4.33	3.82	13.26	2.57	9.42
	Code	42133	721021	40102	723	238350	4279	4-41	4274	402	80110	B1127	202
CARPENTRY—general	Rate	5.99	3.64	4.17	4.35	5.05	4.37	3.40	9.23	6.23	13.26	4.37	9.42
	Code	42143	721028	40102	723	238130	4299	4-41	4226	401	80110	12.02	202
CONCRETE WORK—NOC	Rate	4.01	3.87	7.01	16.50	5.05	11.00	3.40	5.19	6.23	15.41	5.44	5.23
	Code	42104	721010	40110	748	238110	4224	4-41	4224	401	80100	B1314	203
CONCRETE WORK—flat (flr. sidewalk)	Rate	4.01	3.87	7.01	16.50	5.05	11.00	3.40	5.19	6.23	15.41	5.44	5.23
	Code	42104	721010	40110	748	238110	4224	4-41	4224	401	80100	B1314	203
ELECTRICAL Wiring—inside	Rate	1.95	1.41	1.90	3.25	2.45	2.91	2.76	2.66	3.82	5.92	2.57	5.23
	Code	42124	721019	40203	704	238210	4261	4-46	4261	402	80170	B1105	206
EXCAVATION—earth NOC	Rate	2.34	3.25	4.02	4.68	5.05	3.84	3.57	2.88	3.13	7.55	2.76	5.23
	Code	40604	721031	40706	711	238190	4214	4-43	4214	404	80030	R1106	207
EXCAVATION—rock	Rate	2.34	3.25	4.02	4.68	5.05	3.84	3.57	2.88	3.13	7.55	2.76	5.23
	Code	40604	721031	40706	711	238190	4214	4-43	4214	404	80030	R1106	207
GLAZIERS	Rate	2.56	3.05	4.17	9.25	5.05	6.11	3.40	5.19	3.82	14.17	5.44	5.23
	Code	42121	715020	40109	751	238150	4233	4-41	4233	402	80150	B1304	212
INSULATION WORK	Rate	2.05	6.91	4.17	9.25	4.59	6.11	3.40	5.19	6.23	13.26	4.37	9.42
	Code	42184	721029	40102	751	238310	4234	4-41	4234	401	80110	B1207	202
LATHING	Rate	4.67	6.27	4.17	4.35	4.59	4.37	3.40	4.33	3.82	13.26	5.44	9.42
	Code	42135	721042	40102	723	238390	4279	4-41	4271	402	80110	B1316	202
MASONRY	Rate	4.01	3.50	4.17	11.15	5.05	6.11	3.40	5.19	6.23	15.41	5.44	9.42
	Code	42102	721037	40102	741	238140	4231	4-41	4231	401	80100	B1318	202
PAINTING & DECORATING	Rate	3.30	3.36	3.28	6.75	4.59	4.37	3.40	4.33	3.82	13.26	4.37	9.42
	Code	42111	721041	40105	719	238320	4275	4-41	4275	402	80110	B1201	202
PILE DRIVING	Rate	3.95	3.82	4.02	6.34	5.05	3.84	3.57	5.19	6.23	7.55	5.44	9.42
	Code	42159	722004	40706	732	238190	4129	4-43	4221	401	80030	B1310	202
PLASTERING	Rate	4.67	6.27	4.59	6.75	4.59	4.37	3.40	4.33	3.82	13.26	4.37	9.42
	Code	42135	721042	40108	719	238390	4271	4-41	4271	402	80110	B1221	202
PLUMBING	Rate	1.95	1.90	3.04	3.98	2.45	3.08	2.76	2.52	3.82	6.19	2.57	5.23
	Code	42122	721043	40204	707	238220	4241	4-46	4241	402	80160	B1101	214
ROOFING	Rate	6.41	6.18	7.37	13.30	5.05	11.00	3.40	9.23	6.23	19.71	5.44	9.42
	Code	42118	721036	40403	728	238160	4236	4-41	4236	401	80130	B1320	202
SHEET METAL WORK (HVAC)	Rate	1.95	1.90	7.37	3.98	2.45	3.08	2.76	2.52	3.82	6.19	2.57	5.23
	Code	42117	721043	40402	707	238220	4244	4-46	4244	402	80160	B1107	208
STEEL ERECTION—door & sash	Rate	2.05	12.66	10.05	16.50	5.05	11.00	3.40	5.19	6.23	21.98	5.44	9.42
	Code	42106	722005	40502	748	238120	4227	4-41	4227	401	80080	B1322	202
STEEL ERECTION—inter., ornam.	Rate	2.05	12.66	10.05	16.50	5.05	11.00	3.40	5.19	6.23	21.98	5.44	9.42
	Code	42106	722005	40502	748	238120	4227	4-41	4227	401	80080	B1322	202
STEEL ERECTION—structure	Rate	2.05	12.66	10.05	16.50	5.05	11.00	3.40	5.19	6.23	21.98	5.44	9.42
	Code	42106	722005	40502	748	238120	4227	4-41	4227	401	80080	B1322	202
STEEL ERECTION—NOC	Rate	2.05	12.66	10.05	16.50	5.05	11.00	3.40	5.19	6.23	21.98	5.44	9.42
	Code	42106	722005	40502	748	238120	4227	4-41	4227	401	80080	B1322	202
TILE WORK—inter. (ceramic)	Rate	2.89	3.62	1.95	6.75	4.59	4.37	3.40	4.33	3.82	13.26	5.44	9.42
	Code	42113	721054	40103	719	238340	4276	4-41	4276	402	80110	B1301	202
WATERPROOFING	Rate	3.30	3.35	4.17	4.35	5.05	4.37	3.40	5.19	3.82	19.71	4.37	9.42
	Code	42139	721016	40102	723	238190	4299	4-41	4239	402	80130	B1217	202
WRECKING	Rate	2.34	4.44	6.37	16.50	5.05	3.84	3.57	2.88	6.23	13.26	2.76	9.42
	Code	40604	721005	40106	748	238190	4211	4-43	4211	401	80110	R1125	202

General Requirements — R0131 Project Management & Coordination

R013113-80 Performance Bond

This table shows the cost of a Performance Bond for a construction job scheduled to be completed in 12 months. Add 1% of the premium cost per month for jobs requiring more than 12 months to complete. The rates are "standard" rates offered to contractors that the bonding company considers financially sound and capable of doing the work. Preferred rates are offered by some bonding companies based upon financial strength of the contractor. Actual rates vary from contractor to contractor and from bonding company to bonding company. Contractors should prequalify through a bonding agency before submitting a bid on a contract that requires a bond.

Contract Amount	Building Construction Class B Projects	Highways & Bridges: Class A New Construction	Highways & Bridges: Class A-1 Highway Resurfacing
First $ 100,000 bid	$25.00 per M	$15.00 per M	$9.40 per M
Next 400,000 bid	$ 2,500 plus $15.00 per M	$ 1,500 plus $10.00 per M	$ 940 plus $7.20 per M
Next 2,000,000 bid	8,500 plus 10.00 per M	5,500 plus 7.00 per M	3,820 plus 5.00 per M
Next 2,500,000 bid	28,500 plus 7.50 per M	19,500 plus 5.50 per M	15,820 plus 4.50 per M
Next 2,500,000 bid	47,250 plus 7.00 per M	33,250 plus 5.00 per M	28,320 plus 4.50 per M
Over 7,500,000 bid	64,750 plus 6.00 per M	45,750 plus 4.50 per M	39,570 plus 4.00 per M

General Requirements — R0154 Construction Aids

R015423-10 Steel Tubular Scaffolding

On new construction, tubular scaffolding is efficient up to 60′ high or five stories. Above this it is usually better to use a hung scaffolding if construction permits. Swing scaffolding operations may interfere with tenants. In this case, the tubular is more practical at all heights.

In repairing or cleaning the front of an existing building the cost of tubular scaffolding per S.F. of building front increases as the height increases above the first tier. The first tier cost is relatively high due to leveling and alignment.

The minimum efficient crew for erecting and dismantling is three workers. They can set up and remove 18 frame sections per day up to 5 stories high. For 6 to 12 stories high, a crew of four is most efficient. Use two or more on top and two on the bottom for handing up or hoisting. They can also set up and remove 18 frame sections per day. At 7' horizontal spacing, this will run about 800 S.F. per day of erecting and dismantling. Time for placing and removing planks must be added to the above. A crew of three can place and remove 72 planks per day up to 5 stories. For over 5 stories, a crew of four can place and remove 80 planks per day.

The table below shows the number of pieces required to erect tubular steel scaffolding for 1000 S.F. of building frontage. This area is made up of a scaffolding system that is 12 frames (11 bays) long by 2 frames high.

For jobs under twenty-five frames, add 50% to rental cost. Rental rates will be lower for jobs over three months duration. Large quantities for long periods can reduce rental rates by 20%.

Description of Component	Number of Pieces for 1000 S.F. of Building Front	Unit
5′ Wide Standard Frame, 6′-4″ High	24	Ea.
Leveling Jack & Plate	24	↓
Cross Brace	44	
Side Arm Bracket, 21″	12	
Guardrail Post	12	
Guardrail, 7′ section	22	
Stairway Section	2	
Stairway Starter Bar	1	
Stairway Inside Handrail	2	
Stairway Outside Handrail	2	
Walk-Thru Frame Guardrail	2	↓

Scaffolding is often used as falsework over 15′ high during construction of cast-in-place concrete beams and slabs. Two foot wide scaffolding is generally used for heavy beam construction. The span between frames depends upon the load to be carried with a maximum span of 5′.

Heavy duty shoring frames with a capacity of 10,000#/leg can be spaced up to 10′ O.C. depending upon form support design and loading.

Scaffolding used as horizontal shoring requires less than half the material required with conventional shoring.

On new construction, erection is done by carpenters.

Rolling towers supporting horizontal shores can reduce labor and speed the job. For maintenance work, catwalks with spans up to 70′ can be supported by the rolling towers.

R015433-10 Contractor Equipment

Rental Rates shown elsewhere in the book pertain to late model high quality machines in excellent working condition, rented from equipment dealers. Rental rates from contractors may be substantially lower than the rental rates from equipment dealers depending upon economic conditions; for older, less productive machines, reduce rates by a maximum of 15%. Any overtime must be added to the base rates. For shift work, rates are lower. Usual rule of thumb is 150% of one shift rate for two shifts; 200% for three shifts.

For periods of less than one week, operated equipment is usually more economical to rent than renting bare equipment and hiring an operator.

Costs to move equipment to a job site (mobilization) or from a job site (demobilization) are not included in rental rates, nor in any Equipment costs on any Unit Price line items or crew listings. These costs can be found elsewhere. If a piece of equipment is already at a job site, it is not appropriate to utilize mob/demob costs in an estimate again.

Rental rates vary throughout the country with larger cities generally having lower rates. Lease plans for new equipment are available for periods in excess of six months with a percentage of payments applying toward purchase.

Monthly rental rates vary from 2% to 5% of the cost of the equipment depending on the anticipated life of the equipment and its wearing parts. Weekly rates are about 1/3 the monthly rates and daily rental rates about 1/3 the weekly rate.

The hourly operating costs for each piece of equipment include costs to the user such as fuel, oil, lubrication, normal expendables for the equipment, and a percentage of mechanic's wages chargeable to maintenance. The hourly operating costs listed do not include the operator's wages.

The daily cost for equipment used in the standard crews is figured by dividing the weekly rate by five, then adding eight times the hourly operating cost to give the total daily equipment cost, not including the operator. This figure is in the right hand column of the Equipment listings under Equipment Cost/Day.

Pile Driving rates shown for pile hammer and extractor do not include leads, crane, boiler or compressor. Vibratory pile driving requires an added field specialist during set-up and pile driving operation for the electric model. The hydraulic model requires a field specialist for set-up only. Up to 125 reuses of sheet piling are possible using vibratory drivers. For normal conditions, crane capacity for hammer type and size are as follows.

Crane Capacity	Hammer Type and Size		
	Air or Steam	Diesel	Vibratory
25 ton	to 8,750 ft.-lb.		70 H.P.
40 ton	15,000 ft.-lb.	to 32,000 ft.-lb.	170 H.P.
60 ton	25,000 ft.-lb.		300 H.P.
100 ton		112,000 ft.-lb.	

Cranes should be specified for the job by size, building and site characteristics, availability, performance characteristics, and duration of time required.

Backhoes & Shovels rent for about the same as equivalent size cranes but maintenance and operating expense is higher. Crane operators rate must be adjusted for high boom heights. Average adjustments: for 150′ boom add 2% per hour; over 185′, add 4% per hour; over 210′, add 6% per hour; over 250′, add 8% per hour and over 295′, add 12% per hour.

Tower Cranes of the climbing or static type have jibs from 50′ to 200′ and capacities at maximum reach range from 4,000 to 14,000 pounds. Lifting capacities increase up to maximum load as the hook radius decreases.

Typical rental rates, based on purchase price are about 2% to 3% per month.

Erection and dismantling runs between 500 and 2000 labor hours. Climbing operation takes 10 labor hours per 20′ climb. Crane dead time is about 5 hours per 40′ climb. If crane is bolted to side of the building add cost of ties and extra mast sections. Climbing cranes have from 80′ to 180′ of mast while static cranes have 80′ to 800′ of mast.

Truck Cranes can be converted to tower cranes by using tower attachments. Mast heights over 400′ have been used.

A single 100′ high material **Hoist and Tower** can be erected and dismantled in about 400 labor hours; a double 100′ high hoist and tower in about 600 labor hours. Erection times for additional heights are 3 and 4 labor hours per vertical foot respectively up to 150′, and 4 to 5 labor hours per vertical foot over 150′ high. A 40′ high portable Buck hoist takes about 160 labor hours to erect and dismantle. Additional heights take 2 labor hours per vertical foot to 80′ and 3 labor hours per vertical foot for the next 100′. Most material hoists do not meet local code requirements for carrying personnel.

A 150′ high **Personnel Hoist** requires about 500 to 800 labor hours to erect and dismantle. Budget erection time at 5 labor hours per vertical foot for all trades. Local code requirements or labor scarcity requiring overtime can add up to 50% to any of the above erection costs.

Earthmoving Equipment: The selection of earthmoving equipment depends upon the type and quantity of material, moisture content, haul distance, haul road, time available, and equipment available. Short haul cut and fill operations may require dozers only, while another operation may require excavators, a fleet of trucks, and spreading and compaction equipment. Stockpiled material and granular material are easily excavated with front end loaders. Scrapers are most economically used with hauls between 300′ and 1-1/2 miles if adequate haul roads can be maintained. Shovels are often used for blasted rock and any material where a vertical face of 8′ or more can be excavated. Special conditions may dictate the use of draglines, clamshells, or backhoes. Spreading and compaction equipment must be matched to the soil characteristics, the compaction required and the rate the fill is being supplied.

Existing Conditions — R0241 Demolition

R024119-10 Demolition Defined

Whole Building Demolition - Demolition of the whole building with no concern for any particular building element, component, or material type being demolished. This type of demolition is accomplished with large pieces of construction equipment that break up the structure, load it into trucks and haul it to a disposal site, but disposal or dump fees are not included. Demolition of below-grade foundation elements, such as footings, foundation walls, grade beams, slabs on grade, etc., is not included. Certain mechanical equipment containing flammable liquids or ozone-depleting refrigerants, electric lighting elements, communication equipment components, and other building elements may contain hazardous waste, and must be removed, either selectively or carefully, as hazardous waste before the building can be demolished.

Foundation Demolition - Demolition of below-grade foundation footings, foundation walls, grade beams, and slabs on grade. This type of demolition is accomplished by hand or pneumatic hand tools, and does not include saw cutting, or handling, loading, hauling, or disposal of the debris.

Gutting - Removal of building interior finishes and electrical/mechanical systems down to the load-bearing and sub-floor elements of the rough building frame, with no concern for any particular building element, component, or material type being demolished. This type of demolition is accomplished by hand or pneumatic hand tools, and includes loading into trucks, but not hauling, disposal or dump fees, scaffolding, or shoring. Certain mechanical equipment containing flammable liquids or ozone-depleting refrigerants, electric lighting elements, communication equipment components, and other building elements may contain hazardous waste, and must be removed, either selectively or carefully, as hazardous waste, before the building is gutted.

Selective Demolition - Demolition of a selected building element, component, or finish, with some concern for surrounding or adjacent elements, components, or finishes (see the first Subdivision (s) at the beginning of appropriate Divisions). This type of demolition is accomplished by hand or pneumatic hand tools, and does not include handling, loading, storing, hauling, or disposal of the debris, scaffolding, or shoring. "Gutting" methods may be used in order to save time, but damage that is caused to surrounding or adjacent elements, components, or finishes may have to be repaired at a later time.

Careful Removal - Removal of a piece of service equipment, building element or component, or material type, with great concern for both the removed item and surrounding or adjacent elements, components or finishes. The purpose of careful removal may be to protect the removed item for later re-use, preserve a higher salvage value of the removed item, or replace an item while taking care to protect surrounding or adjacent elements, components, connections, or finishes from cosmetic and/or structural damage. An approximation of the time required to perform this type of removal is 1/3 to 1/2 the time it would take to install a new item of like kind (see Reference Number R220105-10). This type of removal is accomplished by hand or pneumatic hand tools, and does not include loading, hauling, or storing the removed item, scaffolding, shoring, or lifting equipment.

Cutout Demolition - Demolition of a small quantity of floor, wall, roof, or other assembly, with concern for the appearance and structural integrity of the surrounding materials. This type of demolition is accomplished by hand or pneumatic hand tools, and does not include saw cutting, handling, loading, hauling, or disposal of debris, scaffolding, or shoring.

Rubbish Handling - Work activities that involve handling, loading or hauling of debris. Generally, the cost of rubbish handling must be added to the cost of all types of demolition, with the exception of whole building demolition.

Minor Site Demolition - Demolition of site elements outside the footprint of a building. This type of demolition is accomplished by hand or pneumatic hand tools, or with larger pieces of construction equipment, and may include loading a removed item onto a truck (check the Crew for equipment used). It does not include saw cutting, hauling or disposal of debris, and, sometimes, handling or loading.

R024119-20 Dumpsters

Dumpster rental costs on construction sites are presented in two ways.

The cost per week rental includes the delivery of the dumpster; its pulling or emptying once per week, and its final removal. The assumption is made that the dumpster contractor could choose to empty a dumpster by simply bringing in an empty unit and removing the full one. These costs also include the disposal of the materials in the dumpster.

The Alternate Pricing can be used when actual planned conditions are not approximated by the weekly numbers. For example, these lines can be used when a dumpster is needed for 4 weeks and will need to be emptied 2 or 3 times per week. Conversely the Alternate Pricing lines can be used when a dumpster will be rented for several weeks or months but needs to be emptied only a few times over this period.

Existing Conditions — R0265 Underground Storage Tank Removal

R026510-20 Underground Storage Tank Removal

Underground Storage Tank Removal can be divided into two categories: Non-Leaking and Leaking. Prior to removing an underground storage tank, tests should be made, with the proper authorities present, to determine whether a tank has been leaking or the surrounding soil has been contaminated.

To safely remove Liquid Underground Storage Tanks:

1. Excavate to the top of the tank.
2. Disconnect all piping.
3. Open all tank vents and access ports.
4. Remove all liquids and/or sludge.
5. Purge the tank with an inert gas.
6. Provide access to the inside of the tank and clean out the interior using proper personal protective equipment (PPE).
7. Excavate soil surrounding the tank using proper PPE for on-site personnel.
8. Pull and properly dispose of the tank.
9. Clean up the site of all contaminated material.
10. Install new tanks or close the excavation.

Existing Conditions — R0282 Asbestos Remediation

R028213-20 Asbestos Removal Process

Asbestos removal is accomplished by a specialty contractor who understands the federal and state regulations regarding the handling and disposal of the material. The process of asbestos removal is divided into many individual steps. An accurate estimate can be calculated only after all the steps have been priced.

The steps are generally as follows:

1. Obtain an asbestos abatement plan from an industrial hygienist.
2. Monitor the air quality in and around the removal area and along the path of travel between the removal area and transport area. This establishes the background contamination.
3. Construct a two part decontamination chamber at entrance to removal area.
4. Install a HEPA filter to create a negative pressure in the removal area.
5. Install wall, floor and ceiling protection as required by the plan, usually 2 layers of fireproof 6 mil polyethylene.
6. Industrial hygienist visually inspects work area to verify compliance with plan.
7. Provide temporary supports for conduit and piping affected by the removal process.
8. Proceed with asbestos removal and bagging process. Monitor air quality as described in Step #2. Discontinue operations when contaminate levels exceed applicable standards.
9. Document the legal disposal of materials in accordance with EPA standards.
10. Thoroughly clean removal area including all ledges, crevices and surfaces.
11. Post abatement inspection by industrial hygienist to verify plan compliance.
12. Provide a certificate from a licensed industrial hygienist attesting that contaminate levels are within acceptable standards before returning area to regular use.

Thermal & Moist. Protec. — R0784 Firestopping

R078413-30 Firestopping

Firestopping is the sealing of structural, mechanical, electrical and other penetrations through fire-rated assemblies. The basic components of firestop systems are safing insulation and firestop sealant on both sides of wall penetrations and the top side of floor penetrations.

Pipe penetrations are assumed to be through concrete, grout, or joint compound and can be sleeved or unsleeved. Costs for the penetrations and sleeves are not included. An annular space of 1″ is assumed. Escutcheons are not included.

Metallic pipe is assumed to be copper, aluminum, cast iron or similar metallic material. Insulated metallic pipe is assumed to be covered with a thermal insulating jacket of varying thickness and materials.

Non-metallic pipe is assumed to be PVC, CPVC, FR Polypropylene or similar plastic piping material. Intumescent firestop sealant or wrap strips are included. Collars on both sides of wall penetrations and a sheet metal plate on the underside of floor penetrations are included.

Ductwork is assumed to be sheet metal, stainless steel or similar metallic material. Duct penetrations are assumed to be through concrete, grout or joint compound. Costs for penetrations and sleeves are not included. An annular space of 1/2″ is assumed.

Multi-trade openings include costs for sheet metal forms, firestop mortar, wrap strips, collars and sealants as necessary.

Structural penetrations joints are assumed to be 1/2″ or less. CMU walls are assumed to be within 1-1/2″ of metal deck. Drywall walls are assumed to be tight to the underside of metal decking.

Metal panel, glass or curtain wall systems include a spandrel area of 5′ filled with mineral wool foil-faced insulation. Fasteners and stiffeners are included.

Finishes — R0991 Painting

R099100-10 Painting Estimating Techniques

Proper estimating methodology is needed to obtain an accurate painting estimate. There is no known reliable shortcut or square foot method. The following steps should be followed:

- List all surfaces to be painted, with an accurate quantity (area) of each. Items having similar surface condition, finish, application method and accessibility may be grouped together.
- List all the tasks required for each surface to be painted, including surface preparation, masking, and protection of adjacent surfaces. Surface preparation may include minor repairs, washing, sanding and puttying.
- Select the proper Means line for each task. Review and consider all adjustments to labor and materials for type of paint and location of work. Apply the height adjustment carefully. For instance, when applying the adjustment for work over 8' high to a wall that is 12' high, apply the adjustment only to the area between 8' and 12' high, and not to the entire wall.

When applying more than one percent (%) adjustment, apply each to the base cost of the data, rather than applying one percentage adjustment on top of the other.

When estimating the cost of painting walls and ceilings remember to add the brushwork for all cut-ins at inside corners and around windows and doors as a LF measure. One linear foot of cut-in with brush equals one square foot of painting.

All items for spray painting include the labor for roll-back.

Deduct for openings greater than 100 SF, or openings that extend from floor to ceiling and are greater than 5' wide. Do not deduct small openings.

The cost of brushes, rollers, ladders and spray equipment are considered to be part of a painting contractor's overhead, and should not be added to the estimate. The cost of rented equipment such as scaffolding and swing staging should be added to the estimate.

Plumbing — R2201 Operation & Maintenance of Plumbing

R220105-10 Demolition (Selective vs. Removal for Replacement)

Demolition can be divided into two basic categories.

One type of demolition involves the removal of material with no concern for its replacement. The labor-hours to estimate this work are found under "Selective Demolition" in the Fire Protection, Plumbing and HVAC Divisions. It is selective in that individual items or all the material installed as a system or trade grouping such as plumbing or heating systems are removed. This may be accomplished by the easiest way possible, such as sawing, torch cutting, or sledge hammer as well as simple unbolting.

The second type of demolition is the removal of some item for repair or replacement. This removal may involve careful draining, opening of unions, disconnecting and tagging of electrical connections, capping of pipes/ducts to prevent entry of debris or leakage of the material contained as well as transport of the item away from its in-place location to a truck/dumpster. An approximation of the time required to accomplish this type of demolition is to use half of the time indicated as necessary to install a new unit. For example; installation of a new pump might be listed as requiring 6 labor-hours so if we had to estimate the removal of the old pump we would allow an additional 3 hours for a total of 9 hours. That is, the complete replacement of a defective pump with a new pump would be estimated to take 9 labor-hours.

Plumbing — R2205 Common Work Results for Plumbing

R220523-80 Valve Materials

VALVE MATERIALS

Bronze:
Bronze is one of the oldest materials used to make valves. It is most commonly used in hot and cold water systems and other non-corrosive services. It is often used as a seating surface in larger iron body valves to ensure tight closure.

Carbon Steel:
Carbon steel is a high strength material. Therefore, valves made from this metal are used in higher pressure services, such as steam lines up to 600 psi at 850° F. Many steel valves are available with butt-weld ends for economy and are generally used in high pressure steam service as well as other higher pressure non-corrosive services.

Forged Steel:
Valves from tough carbon steel are used in service up to 2000 psi and temperatures up to 1000° F in Gate, Globe and Check valves.

Iron:
Valves are normally used in medium to large pipe lines to control non-corrosive fluid and gases, where pressures do not exceed 250 psi at 450° F or 500 psi cold water, oil or gas.

Stainless Steel:
Developed steel alloys can be used in over 90% corrosive services.

Plastic PVC:
This is used in a great variety of valves generally in high corrosive service with lower temperatures and pressures.

VALVE SERVICE PRESSURES

Pressure ratings on valves provide an indication of the safe operating pressure for a valve at some elevated temperature. This temperature is dependent upon the materials used and the fabrication of the valve. When specific data is not available, a good "rule-of-thumb" to follow is the temperature of saturated steam on the primary rating indicated on the valve body. Example: The valve has the number 150S printed on the side indicating 150 psi and hence, a maximum operating temperature of 367° F (temperature of saturated steam and 150 psi).

DEFINITIONS

1. "WOG" – Water, oil, gas (cold working pressures).
2. "SWP" – Steam working pressure.
3. 100% area (full port) – means the area through the valve is equal to or greater than the area of standard pipe.
4. "Standard Opening" – means that the area through the valve is less than the area of standard pipe and therefore these valves should be used only where restriction of flow is unimportant.
5. "Round Port" – means the valve has a full round opening through the plug and body, of the same size and area as standard pipe.
6. "Rectangular Port" – valves have rectangular shaped ports through the plug body. The area of the port is either equal to 100% of the area of standard pipe, or restricted (standard opening). In either case it is clearly marked.
7. "ANSI" – American National Standards Institute.

R220523-90 Valve Selection Considerations

INTRODUCTION: In any piping application, valve performance is critical. Valves should be selected to give the best performance at the lowest cost.

The following is a list of performance characteristics generally expected of valves.

1. Stopping flow or starting it.
2. Throttling flow (Modulation).
3. Flow direction changing.
4. Checking backflow (Permitting flow in only one direction).
5. Relieving or regulating pressure.

In order to properly select the right valve, some facts must be determined.

A. What liquid or gas will flow through the valve?

B. Does the fluid contain suspended particles?

C. Does the fluid remain in liquid form at all times?

D. Which metals does fluid corrode?

E. What are the pressure and temperature limits? (As temperature and pressure rise, so will the price of the valve.)

F. Is there constant line pressure?

G. Is the valve merely an on-off valve?

H. Will checking of backflow be required?

I. Will the valve operate frequently or infrequently?

Valves are classified by design type into such classifications as Gate, Globe, Angle, Check, Ball, Butterfly and Plug. They are also classified by end connection, stem, pressure restrictions and material such as bronze, cast iron, etc. Each valve has a specific use. A quality valve used correctly will provide a lifetime of trouble- free service, but a high quality valve installed in the wrong service may require frequent attention.

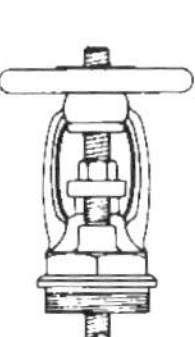

STEM TYPES
(OS & Y)—Rising Stem-Outside Screw and Yoke

Offers a visual indication of whether the valve is open or closed. Recommended where high temperatures, corrosives, and solids in the line might cause damage to inside-valve stem threads. The stem threads are engaged by the yoke bushing so the stem rises through the hand wheel as it is turned.

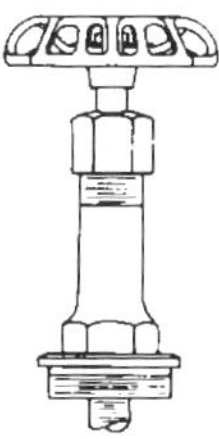

(R.S.)—Rising Stem-Inside Screw

Adequate clearance for operation must be provided because both the hand wheel and the stem rise.
The valve wedge position is indicated by the position of the stem and hand wheel.

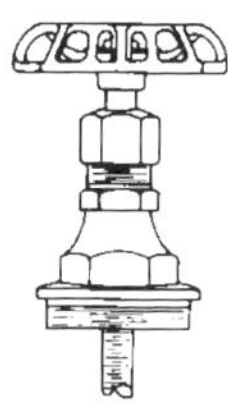

(N.R.S.)—Non-Rising Stem-Inside Screw

A minimum clearance is required for operating this type of valve. Excessive wear or damage to stem threads inside the valve may be caused by heat, corrosion, and solids. Because the hand wheel and stem do not rise, wedge position cannot be visually determined.

VALVE TYPES
Gate Valves

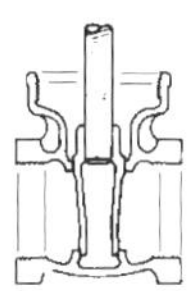

Provide full flow, minute pressure drop, minimum turbulence and minimum fluid trapped in the line.
They are normally used where operation is infrequent.

Globe Valves

Globe valves are designed for throttling and/or frequent operation with positive shut-off. Particular attention must be paid to the several types of seating materials available to avoid unnecessary wear. The seats must be compatible with the fluid in service and may be composition or metal. The configuration of the Globe valve opening causes turbulence which results in increased resistance. Most bronze Globe valves are rising stem-inside screw, but they are also available on O.S. & Y.

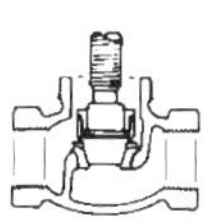

Angle Valves

The fundamental difference between the Angle valve and the Globe valve is the fluid flow through the Angle valve. It makes a 90° turn and offers less resistance to flow than the Globe valve while replacing an elbow. An Angle valve thus reduces the number of joints and installation time.

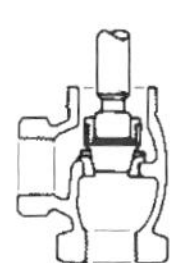

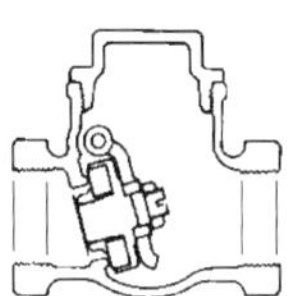

Check Valves

Check valves are designed to prevent backflow by automatically seating when the direction of fluid is reversed.
Swing Check valves are generally installed with Gate-valves, as they provide comparable full flow. Usually recommended for lines where flow velocities are low and should not be used on lines with pulsating flow. Recommended for horizontal installation, or in vertical lines only where flow is upward.

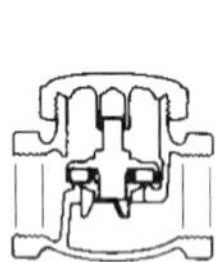

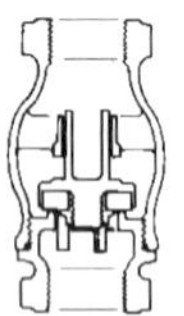

Lift Check Valves

These are commonly used with Globe and Angle valves since they have similar diaphragm seating arrangements and are recommended for preventing backflow of steam, air, gas and water, and on vapor lines with high flow velocities. For horizontal lines, horizontal lift checks should be used and vertical lift checks for vertical lines.

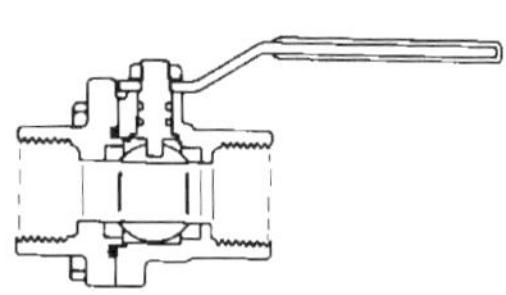

Ball Valves

Ball valves are light and easily installed, yet because of modern elastomeric seats, provide tight closure. Flow is controlled by rotating up to 90° a drilled ball which fits tightly against resilient seals. This ball seats with flow in either direction, and valve handle indicates the degree of opening. Recommended for frequent operation readily adaptable to automation, ideal for installation where space is limited.

Butterfly Valves

Butterfly valves provide bubble-tight closure with excellent throttling characteristics. They can be used for full-open, closed and for throttling applications.

The Butterfly valve consists of a disc within the valve body which is controlled by a shaft. In its closed position, the valve disc seals against a resilient seat. The disc position throughout the full 90° rotation is visually indicated by the position of the operator.

A Butterfly valve is only a fraction of the weight of a Gate valve and requires no gaskets between flanges in most cases. Recommended for frequent operation and adaptable to automation where space is limited.

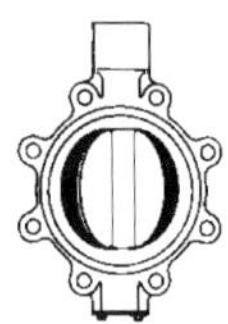

Wafer and Lug type bodies when installed between two pipe flanges, can be easily removed from the line. The pressure of the bolted flanges holds the valve in place.
Locating lugs makes installation easier.

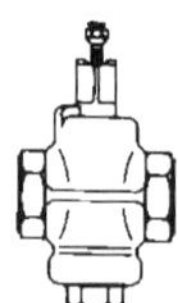

Plug Valves

Lubricated plug valves, because of the wide range of service to which they are adapted, may be classified as all purpose valves. They can be safely used at all pressure and vacuums, and at all temperatures up to the limits of available lubricants. They are the most satisfactory valves for the handling of gritty suspensions and many other destructive, erosive, corrosive and chemical solutions.

R221113-50 Pipe Material Considerations

1. Malleable fittings should be used for gas service.
2. Malleable fittings are used where there are stresses/strains due to expansion and vibration.
3. Cast fittings may be broken as an aid to disassembling of heating lines frozen by long use, temperature and minerals.
4. Cast iron pipe is extensively used for underground and submerged service.
5. Type M (light wall) copper tubing is available in hard temper only and is used for nonpressure and less severe applications than K and L.
6. Type L (medium wall) copper tubing, available hard or soft for interior service.
7. Type K (heavy wall) copper tubing, available in hard or soft temper for use where conditions are severe. For underground and interior service.
8. Hard drawn tubing requires fewer hangers or supports but should not be bent. Silver brazed fittings are recommended, however soft solder is normally used.
9. Type DMV (very light wall) copper tubing designed for drainage, waste and vent plus other non-critical pressure services.

Domestic/Imported Pipe and Fittings Cost

The prices shown in this publication for steel/cast iron pipe and steel, cast iron, malleable iron fittings are based on domestic production sold at the normal trade discounts. The above listed items of foreign manufacture may be available at prices of 1/3 to 1/2 those shown. Some imported items after minor machining or finishing operations are being sold as domestic to further complicate the system.

Caution: Most pipe prices in this book also include a coupling and pipe hangers which for the larger sizes can add significantly to the per foot cost and should be taken into account when comparing "book cost" with quoted supplier's cost.

R221113-70 Piping to 10′ High

When taking off pipe, it is important to identify the different material types and joining procedures, as well as distances between supports and components required for proper support.

During the takeoff, measure through all fittings. Do not subtract the lengths of the fittings, valves, or strainers, etc. This added length plus the final rounding of the totals will compensate for nipples and waste.

When rounding off totals always increase the actual amount to correspond with manufacturer's shipping lengths.

A. Both red brass and yellow brass pipe are normally furnished in 12′ lengths, plain end. The Unit Price section includes in the linear foot costs two field threads and one coupling per 10′ length. A carbon steel clevis type hanger assembly every 10′ is also prorated into the linear foot costs, including both material and labor.

B. Cast iron soil pipe is furnished in either 5′ or 10′ lengths. For pricing purposes, the Unit Price section features 10′ lengths with a joint and a carbon steel clevis hanger assembly every 5′ prorated into the per foot costs of both material and labor.

Three methods of joining are considered: lead and oakum poured joints or push-on gasket type joints for the bell and spigot pipe and a joint clamp for the no-hub soil pipe. The labor and material costs for each of these individual joining procedures are also prorated into the linear costs per foot.

C. Copper tubing covers types K, L, M, and DWV which are furnished in 20′ lengths. Means pricing data is based on a tubing cut each length and a coupling and two soft soldered joints every 10′. A carbon steel, clevis type hanger assembly every 10′ is also prorated into the per foot costs. The prices for refrigeration tubing are for materials only. Labor for full lengths may be based on the type L labor but short cut measures in tight areas can increase the installation labor-hours from 20 to 40%.

D. Corrosion-resistant piping does not lend itself to one particular standard of hanging or support assembly due to its diversity of application and placement. The several varieties of corrosion-resistant piping do not include any material or labor costs for hanger assemblies (See the Unit Price section for appropriate selection).

E. Glass pipe is furnished in standard lengths either 5′ or 10′ long, beaded on one end. Special orders for diverse lengths beaded on both ends are also available. For pricing purposes, R.S. Means features 10′ lengths with a coupling and a carbon steel band hanger assembly every 10′ prorated into the per foot linear costs.

Glass pipe is also available with conical ends and standard lengths ranging from 6″ through 3′ in 6″ increments, then up to 10′ in 12″ increments. Special lengths can be customized for particular installation requirements.

For pricing purposes, Means has based the labor and material pricing on 10′ lengths. Included in these costs per linear foot are the prorated costs for a flanged assembly every 10′ consisting of two flanges, a gasket, two insertable seals, and the required number of bolts and nuts. A carbon steel band hanger assembly based on 10′ center lines has also been prorated into the costs per foot for labor and materials.

F. Plastic pipe of several compositions and joining methods are considered. Fiberglass reinforced pipe (FRP) is priced, based on 10′ lengths (20′ lengths are also available), with coupling and epoxy joints every 10′. FRP is furnished in both "General Service" and "High Strength." A carbon steel clevis hanger assembly, 3 for every 10′, is built into the prorated labor and material costs on a per foot basis.

The PVC and CPVC pipe schedules 40, 80 and 120, plus SDR ratings are all based on 20′ lengths with a coupling installed every 10′, as well as a carbon steel clevis hanger assembly every 3′. The PVC and ABS type DWV piping is based on 10′ lengths with solvent weld couplings every 10′, and with carbon steel clevis hanger assemblies, 3 for every 10′. The rest of the plastic piping in this section is based on flexible 100′ coils and does not include any coupling or supports.

This section ends with PVC drain and sewer piping based on 10′ lengths with bell and spigot ends and 0-ring type, push-on joints.

G. Stainless steel piping includes both weld end and threaded piping, both in the type 304 and 316 specification and in the following schedules, 5, 10, 40, 80, and 160. Although this piping is usually furnished in 20′ lengths, this cost grouping has a joint (either heli-arc butt-welded or threads and coupling) every 10′. A carbon steel clevis type hanger assembly is also included at 10′ intervals and prorated into the linear foot costs.

H. Carbon steel pipe includes both black and galvanized. This section encompasses schedules 40 (standard) and 80 (extra heavy).

Several common methods of joining steel pipe — such as thread and coupled, butt welded, and flanged (150 lb. weld neck flanges) are also included.

For estimating purposes, it is assumed that the piping is purchased in 20′ lengths and that a compatible joint is made up every 10′. These joints are prorated into the labor and material costs per linear foot. The following hanger and support assemblies every 10′ are also included: carbon steel clevis for the T & C pipe, and single rod roll type for both the welded and flanged piping. All of these hangers are oversized to accommodate pipe insulation 3/4″ thick through 5″ pipe size and 1-1/2″ thick from 6″ through 12″ pipe size.

I. Grooved joint steel pipe is priced both black and galvanized, in schedules 10, 40, and 80, furnished in 20′ lengths. This section describes two joining methods: cut groove and roll groove. The schedule 10 piping is roll-grooved, while the heavier schedules are cut-grooved. The labor and material costs are prorated into per linear foot prices, including a coupled joint every 10′, as well as a carbon steel clevis hanger assembly.

Notes:

The pipe hanger assemblies mentioned in the preceding paragraphs include the described hanger; appropriately sized steel, box-type insert and nut; plus 18″ of threaded hanger rod.

C clamps are used when the pipe is to be supported from steel shapes rather than anchored in the slab. C clamps are slightly less costly than inserts. However, to save time in estimating, it is advisable to use the given line number cost, rather than substituting a C clamp for the insert.

Add to piping labor for elevated installation:

10′ to 14.5′ high	10%	30′ to 34.5′ high	40%
15′ to 19.5′ high	20%	35′ to 39.5′ high	50%
20′ to 24.5′ high	25%	Over 40′ and higher	55%
25′ to 29.5′ high	35%		

When using the percentage adds for elevated piping installations as shown above, bear in mind that the given heights are for the pipe supports, even though the insert, anchor, or clamp may be several feet higher than the pipe itself.

An allowance has been included in the piping installation time for testing and minor tightening of leaking joints, fittings, stuffing boxes, packing glands, etc. For extraordinary test requirements such as x-rays, prolonged pressure or demonstration tests, a percentage of the piping labor, based on the estimator's experience, must be added to the labor total. A testing service specializing in weld x-rays should be consulted for pricing if it is an estimate requirement. Equipment installation time includes start-up with associated adjustments.

R230500-10 Subcontractors

On the unit cost pages of the R.S. Means Cost Data books, the last column is entitled "Total Incl. O&P". This is normally the cost of the installing contractor. In the HVAC Division, this is the cost of the mechanical contractor. If the particular work being estimated is to be performed by a sub to the mechanical contractor, the mechanical's profit and handling charge (usually 10%) is added to the total of the last column.

R233100-10 Loudness Levels for Moving Air thru Fans, Diffusers, Register, Etc. (Measured in Sones)

Recommended Loudness Levels (Sones)											
Area	Very Quiet	Quiet	Noisy	Area	Very Quiet	Quiet	Noisy	Area	Very Quiet	Quiet	Noisy
Auditoriums				**Hotels**				**Offices**			
Auditorium lobbies	3	4	6	Banquet Rooms	1.5	3	6	Conference rooms	1	1.7	3
Concert and opera halls	0.8	1	1.5	Individual rooms, suites	1	2	4	Drafting	2	4	8
Courtrooms	2	3	4	Kitchens and laundries	4	7	10	General open offices	2	4	8
Lecture halls	1.5	2	3	Lobbies	2	4	8	Halls and corridors	2.5	5	10
Movie theaters	1.5	2	3	**Indoor Sports**				Professional offices	1.5	3	6
Churches and Schools				Bowling alleys	3	4	6	Reception room	1.5	3	6
Kitchens	4	6	8	Gymnasiums	2	4	6	Tabulation & computation	3	6	12
Laboratories	2	4	6	Swimming pools	4	7	10	**Public Buildings**			
Libraries	1.5	2	3	**Manufacturing Areas**				Banks	2	4	6
Recreation halls	2	4	8	Assembly lines	5	12	30	Court houses	2.5	4	6
Sanctuaries	1	1.7	3	Foreman's office	3	5	8	Museums	2	3	4
Schools and classrooms	1.5	2.5	4	Foundries	10	20	40	Planetariums	1.5	2	3
Hospital and Clinics				General storage	5	10	20	Post offices	2.5	4	6
Laboratories	2	4	6	Heavy machinery	10	25	60	Public libraries	1.5	2	4
Lobbies, waiting rooms	2	4	6	Light machinery	5	12	30	Waiting rooms	3	5	8
Halls and corridors	2	4	6	Tool maintenance	4	7	10	**Restaurants**			
Operating rooms	1.5	2.5	4	**Stores**				Cafeterias	3	6	10
Private rooms	1	1.7	3	Department stores	3	6	8	Night clubs	2.5	4	6
Wards	1.5	2.5	4	Supermarkets	4	7	10	Restaurants	2	4	8

R233100-20 Ductwork

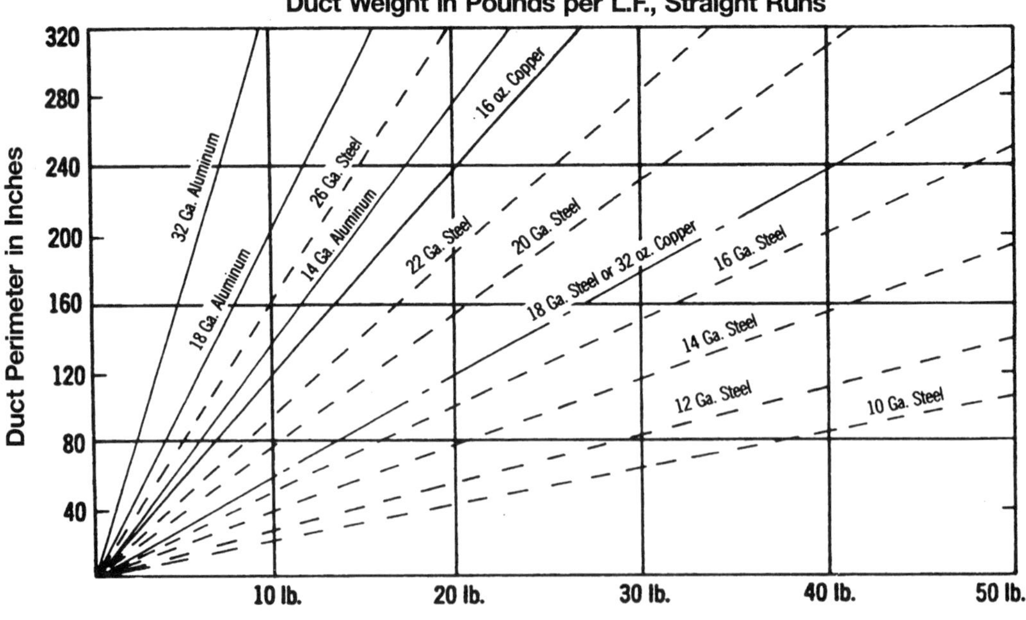

Add to the above for fittings; 90° elbow is 3 L.F.; 45° elbow is 2.5 L.F.; offset is 4 L.F.; transition offset is 6 L.F.; square-to-round transition is 4 L.F.; 90° reducing elbow is 5 L.F. For bracing and waste, add 20% to aluminum and copper, 15% to steel.

R233100-30 Duct Fabrication/Installation

The labor cost for fabricated sheet metal duct includes both the cost of fabrication and installation of the duct. The split is approximately 40% for fabrication, 60% for installation. It is for this reason that the percentage add for elevated installation of fabricated duct is less than the percentage add for prefabricated/preformed duct.

Example: assume a piece of duct cost $100 installed (labor only)

Sheet Metal Fabrication = 40% = $40

Installation = 60% = $60

The add for elevated installation is:

Based on total labor (fabrication & installation) = $100 x 6% = $6.00

Based on installation cost only (Material purchased prefabricated) = $60 x 10% = $6.00

The $6.00 markup (10′ to 15′ high) is the same.

R233100-40 Sheet Metal Calculator (Weight in Lb./Ft. of Length)

Gauge	26	24	22	20	18	16
Wt.-Lb./S.F.	.906	1.156	1.406	1.656	2.156	2.656
SMACNA Max. Dimension – Long Side		30″	54″	84″	85″ Up	
Sum-2 sides						
2	.3	.40	.50	.60	.80	.90
3	.5	.65	.80	.90	1.1	1.4
4	.7	.85	1.0	1.2	1.5	1.8
5	.8	1.1	1.3	1.5	1.9	2.3
6	1.0	1.3	1.5	1.7	2.3	2.7
7	1.2	1.5	1.8	2.0	2.7	3.2
8	1.3	1.7	2.0	2.3	3.0	3.6
9	1.5	1.9	2.3	2.6	3.4	4.1
10	1.7	2.2	2.5	2.9	3.8	4.5
11	1.8	2.4	2.8	3.2	4.2	5.0
12	2.0	2.6	3.0	3.5	4.6	5.4
13	2.2	2.8	3.3	3.8	4.9	5.9
14	2.3	3.0	3.5	4.1	5.3	6.3
15	2.5	3.2	3.8	4.4	5.7	6.8
16	2.7	3.4	4.0	4.6	6.1	7.2
17	2.8	3.7	4.3	4.9	6.5	7.7
18	3.0	3.9	4.5	5.2	6.8	8.1
19	3.2	4.1	4.8	5.5	7.2	8.6
20	3.3	4.3	5.0	5.8	7.6	9.0
21	3.5	4.5	5.3	6.1	8.0	9.5
22	3.7	4.7	5.5	6.4	8.4	9.9
23	3.8	5.0	5.8	6.7	8.7	10.4
24	4.0	5.2	6.0	7.0	9.1	10.8
25	4.2	5.4	6.3	7.3	9.5	11.3
26	4.3	5.6	6.5	7.5	9.9	11.7
27	4.5	5.8	6.8	7.8	10.3	12.2
28	4.7	6.0	7.0	8.1	10.6	12.6
29	4.8	6.2	7.3	8.4	11.0	13.1
30	5.0	6.5	7.5	8.7	11.4	13.5
31	5.2	6.7	7.8	9.0	11.8	14.0
32	5.3	6.9	8.0	9.3	12.2	14.4
33	5.5	7.1	8.3	9.6	12.5	14.9
34	5.7	7.3	8.5	9.9	12.9	15.3
35	5.8	7.5	8.8	10.2	13.3	15.8
36	6.0	7.8	9.0	10.4	13.7	16.2
37	6.2	8.0	9.3	10.7	14.1	16.7
38	6.3	8.2	9.5	11.0	14.4	17.1
39	6.5	8.4	9.8	11.3	14.8	17.6
40	6.7	8.6	10.0	11.6	15.2	18.0
41	6.8	8.8	10.3	11.9	15.6	18.5
42	7.0	9.0	10.5	12.2	16.0	18.9
43	7.2	9.2	10.8	12.5	16.3	19.4
44	7.3	9.5	11.0	12.8	16.7	19.8
45	7.5	9.7	11.3	13.1	17.1	20.3
46	7.7	9.9	11.5	13.3	17.5	20.7
47	7.8	10.1	11.8	13.6	17.9	21.2
48	8.0	10.3	12.0	13.9	18.2	21.6
49	8.2	10.5	12.3	14.2	18.6	22.1
50	8.3	10.7	12.5	14.5	19.0	22.5
51	8.5	11.0	12.8	14.8	19.4	23.0
52	8.7	11.2	13.0	15.1	19.8	23.4
53	8.8	11.4	13.3	15.4	20.1	23.9
54	9.0	11.6	13.5	15.7	20.5	24.3
55	9.2	11.8	13.8	16.0	20.9	24.8

Gauge	26	24	22	20	18	16
Wt.-Lb./S.F.	.906	1.156	1.406	1.656	2.156	2.656
SMACNA Max. Dimension – Long Side		30″	54″	84″	85″ Up	
Sum-2 Sides						
56	9.3	12.0	14.0	16.2	21.3	25.2
57	9.5	12.3	14.3	16.5	21.7	25.7
58	9.7	12.5	14.5	16.8	22.0	26.1
59	9.8	12.7	14.8	17.1	22.4	26.6
60	10.0	12.9	15.0	17.4	22.8	27.0
61	10.2	13.1	15.3	17.7	23.2	27.5
62	10.3	13.3	15.5	18.0	23.6	27.9
63	10.5	13.5	15.8	18.3	24.0	28.4
64	10.7	13.7	16.0	18.6	24.3	28.8
65	10.8	13.9	16.3	18.9	24.7	29.3
66	11.0	14.1	16.5	19.1	25.1	29.7
67	11.2	14.3	16.8	19.4	25.5	30.2
68	11.3	14.6	17.0	19.7	25.8	30.6
69	11.5	14.8	17.3	20.0	26.2	31.1
70	11.7	15.0	17.5	20.3	26.6	31.5
71	11.8	15.2	17.8	20.6	27.0	32.0
72	12.0	15.4	18.0	20.9	27.4	32.4
73	12.2	15.6	18.3	21.2	27.7	32.9
74	12.3	15.8	18.5	21.5	28.1	33.3
75	12.5	16.1	18.8	21.8	28.5	33.8
76	12.7	16.3	19.0	22.0	28.9	34.2
77	12.8	16.5	19.3	22.3	29.3	34.7
78	13.0	16.7	19.5	22.6	29.6	35.1
79	13.2	16.9	19.8	22.9	30.0	35.6
80	13.3	17.1	20.0	23.2	30.4	36.0
81	13.5	17.3	20.3	23.5	30.8	36.5
82	13.7	17.5	20.5	23.8	31.2	36.9
83	13.8	17.8	20.8	24.1	31.5	37.4
84	14.0	18.0	21.0	24.4	31.9	37.8
85	14.2	18.2	21.3	24.7	32.3	38.3
86	14.3	18.4	21.5	24.9	32.7	38.7
87	14.5	18.6	21.8	25.2	33.1	39.2
88	14.7	18.8	22.0	25.5	33.4	39.6
89	14.8	19.0	22.3	25.8	33.8	40.1
90	15.0	19.3	22.5	26.1	34.2	40.5
91	15.2	19.5	22.8	26.4	34.6	41.0
92	15.3	19.7	23.0	26.7	35.0	41.4
93	15.5	19.9	23.3	27.0	35.3	41.9
94	15.7	20.1	23.5	27.3	35.7	42.3
95	15.8	20.3	23.8	27.6	36.1	42.8
96	16.0	20.5	24.0	27.8	36.5	43.2
97	16.2	20.8	24.3	28.1	36.9	43.7
98	16.3	21.0	24.5	28.4	37.2	44.1
99	16.5	21.2	24.8	28.7	37.6	44.6
100	16.7	21.4	25.0	29.0	38.0	45.0
101	16.8	21.6	25.3	29.3	38.4	45.5
102	17.0	21.8	25.5	29.6	38.8	45.9
103	17.2	22.0	25.8	29.9	39.1	46.4
104	17.3	22.3	26.0	30.2	39.5	46.8
105	17.5	22.5	26.3	30.5	39.9	47.3
106	17.7	22.7	26.5	30.7	40.3	47.7
107	17.8	22.9	26.8	31.0	40.7	48.2
108	18.0	23.1	27.0	31.3	41.0	48.6
109	18.2	23.3	27.3	31.6	41.4	49.1
110	18.3	23.5	27.5	31.9	41.8	49.5

Example: If duct is 34″ x 20″ x 15′ long, 34″ is greater than 30″ maximum, for 24 ga. so must be 22 ga. 34″ + 20″ = 54″ going across from 54″ find 13.5 lb. per foot. 13.5 x 15′ = 202.5 lbs. For S.F. of surface area 202.5 ÷ 1.406 = 144 S.F.

Note: Figures include an allowance for scrap.

Heating, Ventilating & A.C. R2331 HVAC Ducts & Casings

R233100-50 Ductwork Packages (per Ton of Cooling)

System	Sheet Metal	Insulation	Diffusers	Return Register
Roof Top Unit Single Zone	120 Lbs.	52 S.F.	1	1
Roof Top Unit Multizone	240 Lbs.	104 S.F.	2	1
Self-contained Air or Water Cooled	108 Lbs.	—	2	—
Split System Air Cooled	102 Lbs.	—	2	—

Systems reflect most common usage.
Refer to system graphics for duct layout.

Heating, Ventilating & A.C. R2334 HVAC Fans

R233400-10 Recommended Ventilation Air Changes

Table below lists range of time in minutes per change for various types of facilities.

Facility	Minutes	Facility	Minutes	Facility	Minutes
Assembly Halls	2-10	Dance Halls	2-10	Laundries	1-3
Auditoriums	2-10	Dining Rooms	3-10	Markets	2-10
Bakeries	2-3	Dry Cleaners	1-5	Offices	2-10
Banks	3-10	Factories	2-5	Pool Rooms	2-5
Bars	2-5	Garages	2-10	Recreation Rooms	2-10
Beauty Parlors	2-5	Generator Rooms	2-5	Sales Rooms	2-10
Boiler Rooms	1-5	Gymnasiums	2-10	Theaters	2-8
Bowling Alleys	2-10	Kitchens-Hospitals	2-5	Toilets	2-5
Churches	5-10	Kitchens-Restaurant	1-3	Transformer Rooms	1-5

CFM air required for changes = Volume of room in cubic feet ÷ Minutes per change.

Heating, Ventilating & A.C. R2337 Air Outlets & Inlets

R233700-60 Diffuser Evaluation

CFM = V × An × K where V = Outlet velocity in feet per minute. An = Neck area in square feet and K = Diffuser delivery factor. An undersized diffuser for a desired CFM will produce a high velocity and noise level. When air moves past people at a velocity in excess of 25 FPM, an annoying draft is felt. An oversized diffuser will result in low velocity with poor mixing. Consideration must be given to avoid vertical stratification or horizontal areas of stagnation.

R235000-10 Heating Systems

Heating Systems

The basic function of a heating system is to bring an enclosed volume up to a desired temperature and then maintain that temperature within a reasonable range. To accomplish this, the selected system must have sufficient capacity to offset transmission losses resulting from the temperature difference on the interior and exterior of the enclosing walls in addition to losses due to cold air infiltration through cracks, crevices and around doors and windows. The amount of heat to be furnished is dependent upon the building size, construction, temperature difference, air leakage, use, shape, orientation and exposure. Air circulation is also an important consideration. Circulation will prevent stratification which could result in heat losses through uneven temperatures at various levels. For example, the most efficient use of unit heaters can usually be achieved by circulating the space volume through the total number of units once every 20 minutes or 3 times an hour. This general rule must, of course, be adapted for special cases such as large buildings with low ratios of heat transmitting surface to cubical volume. The type of occupancy of a building will have considerable bearing on the number of heat transmitting units and the location selected. It is axiomatic, however, that the basis of any successful heating system is to provide the maximum amount of heat at the points of maximum heat loss such as exposed walls, windows, and doors. Large roof areas, wind direction, and wide doorways create problems of excessive heat loss and require special consideration and treatment.

Heat Transmission

Heat transfer is an important parameter to be considered during selection of the exterior wall style, material and window area. A high rate of transfer will permit greater heat loss during the wintertime with the resultant increase in heating energy costs and a greater rate of heat gain in the summer with proportionally greater cooling cost. Several terms are used to describe various aspects of heat transfer. However, for general estimating purposes this book lists U values for systems of construction materials. U is the "overall heat transfer coefficient." It is defined as the heat flow per hour through one square foot when the temperature difference in the air on either side of the structure wall, roof, ceiling or floor is one degree Fahrenheit. The structural segment may be a single homogeneous material or a composite.

Total heat transfer is found using the following equation:

$\mathbf{Q = AU(T_2 - T_1)}$ where

- **Q** = Heat flow, BTU per hour
- **A** = Area, square feet
- **U** = Overall heat transfer coefficient
- $\mathbf{(T_2 - T_1)}$ = Difference in temperature of air on each side of the construction component. (Also abbreviated TD)

Note that heat can flow through all surfaces of any building and this flow is in addition to heat gain or loss due to ventilation, infiltration and generation (appliances, machinery, people).

R235000-20 Heating Approximations for Quick Estimating

Oil Piping & Boiler Room Piping

Small System	20 to 30% of Boiler
Complex System with Pumps, Headers, Etc.	80 to 110% of Boiler
Breeching With Insulation:	
Small	10 to 15% of Boiler
Large	15 to 25% of Boiler
Coils:	15 to 30% of Containing Unit
Balancing (Independent)	1/2% of H.V.A.C. Estimate

Quality/Complexity Adjustment: For all heating installations add these adjustments to the estimate to more closely allow for the equipment and conditions of the particular job under consideration.

Economy installation, add	0 to 5% of System
Good quality, medium complexity, add	5 to 15% of System
Above average quality and complexity, add	15 to 25% of System

R235000-30 The Basics of a Heating System

The function of a heating system is to achieve and maintain a desired temperature in a room or building by replacing the amount of heat being dissipated. There are four kinds of heating systems: hot-water, steam, warm-air and electric resistance. Each has certain essential and similar elements with the exception of electric resistance heating.

The basic elements of a heating system are:

A. A **combustion chamber** in which fuel is burned and heat transferred to a conveying medium.

B. The **"fluid"** used for conveying the heat (water, steam or air).

C. **Conductors** or pipes for transporting the fluid to specific desired locations.

D. A means of disseminating the heat, sometimes called **terminal units.**

A. The **combustion chamber** in a furnace heats air which is then distributed. This is called a warm-air system.

The combustion chamber in a boiler heats water which is either distributed as hot water or steam and this is termed a hydronic system.

The maximum allowable working pressures are limited by ASME "Code for Heating Boilers" to 15 PSI for steam and 160 PSI for hot water heating boilers, with a maximum temperature limitation of 250° F. Hot water boilers are generally rated for a working pressure of 30 PSI. High pressure boilers are governed by the ASME "Code for Power Boilers" which is used almost universally for boilers operating over 15 PSIG. High pressure boilers used for a combination of heating/process loads are usually designed for 150 PSIG.

Boiler ratings are usually indicated as either Gross or Net Output. The Gross Load is equal to the Net Load plus a piping and pickup allowance. When this allowance cannot be determined, divide the gross output rating by 1.25 for a value equal to or greater than the net heat loss requirement of the building.

B. Of the three **fluids** used, steam carries the greatest amount of heat per unit volume. This is due to the fact that it gives up its latent heat of vaporization at a temperature considerably above room temperature. Another advantage is that the pressure to produce a positive circulation is readily available. Piping conducts the steam to terminal units and returns condensate to the boiler.

The **steam system** is well adapted to large buildings because of its positive circulation, its comparatively economical installation and its ability to deliver large quantities of heat. Nearly all large office buildings, stores, hotels, and industrial buildings are so heated, in addition to many residences.

Hot water, when used as the heat carrying fluid, gives up a portion of its sensible heat and then returns to the boiler or heating apparatus for reheating. As the heat conveyed by each pound of water is about one-fiftieth of the heat conveyed by a pound of steam, it is necessary to circulate about fifty times as much water as steam by weight (although only one-thirtieth as much by volume). The hot water system is usually, although not necessarily, designed to operate at temperatures below that of the ordinary steam system and so the amount of heat transfer surface must be correspondingly greater. A temperature of 190° F to 200° F is normally the maximum. Circulation in small buildings may depend on the difference in density between hot water and the cool water returning to the boiler; circulating pumps are normally used to maintain a desired rate of flow. Pumps permit a greater degree of flexibility and better control.

In **warm-air** furnace systems, cool air is taken from one or more points in the building, passed over the combustion chamber and flue gas passages and then distributed through a duct system. A disadvantage of this system is that the ducts take up much more building volume than steam or hot water pipes. Advantages of this system are the relative ease with which humidification can be accomplished by the evaporation of water as the air circulates through the heater, and the lack of need for expensive disseminating units as the warm air simply becomes part of the interior atmosphere of the building.

C. Conductors (pipes and ducts) have been lightly treated in the discussion of conveying fluids. For more detailed information such as sizing and distribution methods, the reader is referred to technical publications such as the American Society of Heating, Refrigerating and Air-Conditioning Engineers "Handbook of Fundamentals."

D. Terminal units come in an almost infinite variety of sizes and styles, but the basic principles of operation are very limited. As previously mentioned, warm-air systems require only a simple register or diffuser to mix heated air with that present in the room. Special application items such as radiant coils and infrared heaters are available to meet particular conditions but are not usually considered for general heating needs. Most heating is accomplished by having air flow over coils or pipes containing the heat transporting medium (steam, hot-water, electricity). These units, while varied, may be separated into two general types, (1) radiator/convectors and (2) unit heaters.

Radiator/convectors may be cast, fin-tube or pipe assemblies. They may be direct, indirect, exposed, concealed or mounted within a cabinet enclosure, upright or baseboard style. These units are often collectively referred to as "radiatiors" or "radiation" although none gives off heat either entirely by radiation or by convection but rather a combination of both. The air flows over the units as a gravity "current." It is necessary to have one or more heat-emitting units in each room. The most efficient placement is low along an outside wall or under a window to counteract the cold coming into the room and achieve an even distribution.

In contrast to radiator/convectors which operate most effectively against the walls of smaller rooms, **unit heaters** utilize a fan to move air over heating coils and are very effective in locations of relatively large volume. Unit heaters, while usually suspended overhead, may be floor mounted. They also may take in fresh outside air for ventilation. The heat distributed by unit heaters may be from a remote source and conveyed by a fluid or it may be from the combustion of fuel in each individual heater. In the latter case the only piping required would be for fuel, however, a vent for the products of combustion would be necessary.

The following list gives may of the advantages of unit heaters for applications other than office or residential:

a. Large capacity so smaller number of units are required, **b.** Piping system simplified, **c.** Space saved where they are located overhead out of the way, **d.** Rapid heating directed where needed with effective wide distribution, **e.** Difference between floor and ceiling temperature reduced, **f.** Circulation of air obtained, and ventilation with introduction of fresh air possible, **g.** Heat output flexible and easily controlled.

R235000-50 Factor for Determining Heat Loss for Various Types of Buildings

General: While the most accurate estimates of heating requirements would naturally be based on detailed information about the building being considered, it is possible to arrive at a reasonable approximation using the following procedure:

1. Calculate the cubic volume of the room or building.
2. Select the appropriate factor from Table 1 below. Note that the factors apply only to inside temperatures listed in the first column and to 0° F outside temperature.
3. If the building has bad north and west exposures, multiply the heat loss factor by 1.1.
4. If the outside design temperature is other than 0° F, multiply the factor from Table 1 by the factor from Table 2.
5. Multiply the cubic volume by the factor selected from Table 1. This will give the estimated BTUH heat loss which must be made up to maintain inside temperature.

Table 1 — Building Type	Conditions	Qualifications	Loss Factor*
Factories & Industrial Plants General Office Areas at 70° F	One Story	Skylight in Roof	6.2
		No Skylight in Roof	5.7
	Multiple Story	Two Story	4.6
		Three Story	4.3
		Four Story	4.1
		Five Story	3.9
		Six Story	3.6
	All Walls Exposed	Flat Roof	6.9
		Heated Space Above	5.2
	One Long Warm Common Wall	Flat Roof	6.3
		Heated Space Above	4.7
	Warm Common Walls on Both Long Sides	Flat Roof	5.8
		Heated Space Above	4.1
Warehouses at 60° F	All Walls Exposed	Skylights in Roof	5.5
		No Skylight in Roof	5.1
		Heated Space Above	4.0
	One Long Warm Common Wall	Skylight in Roof	5.0
		No Skylight in Roof	4.9
		Heated Space Above	3.4
	Warm Common Walls on Both Long Sides	Skylight in Roof	4.7
		No Skylight in Roof	4.4
		Heated Space Above	3.0

*Note: This table tends to be conservative particularly for new buildings designed for minimum energy consumption.

Table 2 — Outside Design Temperature Correction Factor (for Degrees Fahrenheit)									
Outside Design Temperature	50	40	30	20	10	0	-10	-20	-30
Correction Factor	.29	.43	.57	.72	.86	1.00	1.14	1.28	1.43

R235000-70 Transmission of Heat

R235000-70 and R235000-80 provide a way to calculate heat transmission of various construction materials from their U values and the TD (Temperature Difference).

1. From the Exterior Enclosure Division or elsewhere, determine U values for the construction desired.
2. Determine the coldest design temperature. The difference between this temperature and the desired interior temperature is the TD (temperature difference).
3. Enter R235000-70 or R235000-80 at correct U Value. Cross horizontally to the intersection with appropriate TD. Read transmission per square foot from bottom of figure.
4. Multiply this value of BTU per hour transmission per square foot of area by the total surface area of that type of construction.

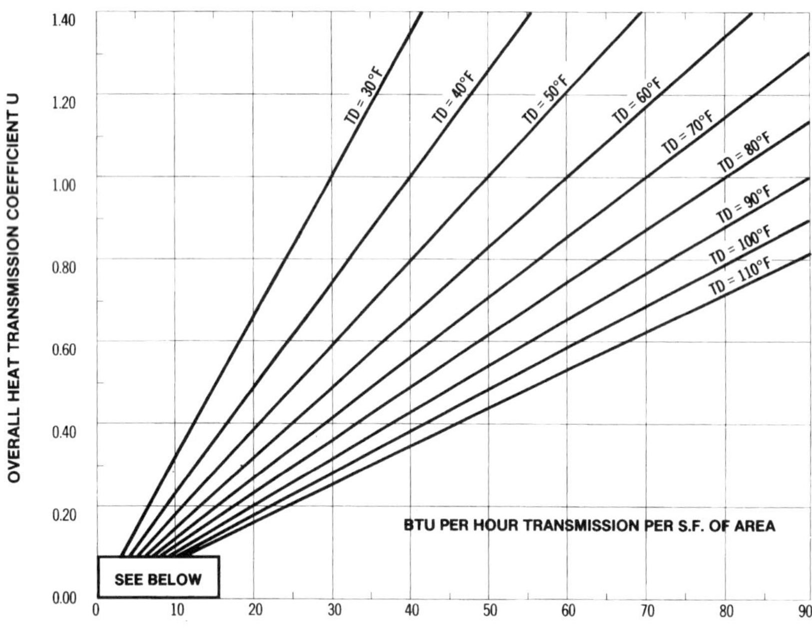

R235000-80 Transmission of Heat (Low Rate)

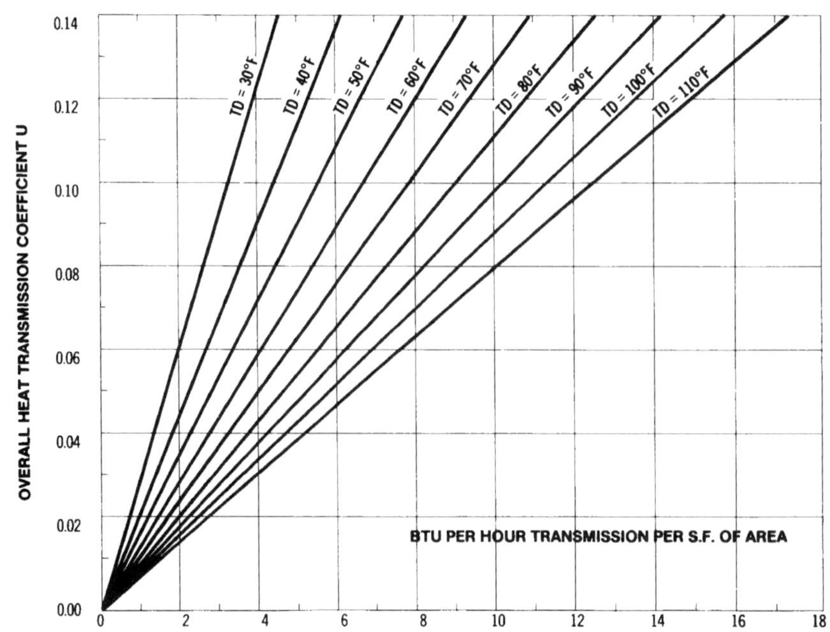

R235616-60 Solar Heating (Space and Hot Water)

Collectors should face as close to due South as possible, however, variations of up to 20 degrees on either side of true South are acceptable. Local climate and collector type may influence the choice between east or west deviations. Obviously they should be located so they are not shaded from the sun's rays. Incline collectors at a slope of latitude minus 5 degrees for domestic hot water and latitude plus 15 degrees for space heating.

Flat plate collectors consist of a number of components as follows: Insulation to reduce heat loss through the bottom and sides of the collector. The enclosure which contains all the components in this assembly is usually weatherproof and prevents dust, wind and water from coming in contact with the absorber plate. The cover plate usually consists of one or more layers of a variety of glass or plastic and reduces the reradiation by creating an air space which traps the heat between the cover and the absorber plates.

The absorber plate must have a good thermal bond with the fluid passages. The absorber plate is usually metallic and treated with a surface coating which improves absorptivity. Black or dark paints or selective coatings are used for this purpose, and the design of this passage and plate combination helps determine a solar system's effectiveness.

Heat transfer fluid passage tubes are attached above and below or integral with an absorber plate for the purpose of transferring thermal energy from the absorber plate to a heat transfer medium. The heat exchanger is a device for transferring thermal energy from one fluid to another.

Piping and storage tanks should be well insulated to minimize heat losses.

Size domestic water heating storage tanks to hold 20 gallons of water per user, minimum, plus 10 gallons per dishwasher or washing machine. For domestic water heating an optimum collector size is approximately 3/4 square foot of area per gallon of water storage. For space heating of residences and small commercial applications the collector is commonly sized between 30% and 50% of the internal floor area. For space heating of large commercial applications, collector areas less than 30% of the internal floor area can still provide significant heat reductions.

A supplementary heat source is recommended for Northern states for December through February.

The solar energy transmission per square foot of collector surface varies greatly with the material used. Initial cost, heat transmittance and useful life are obviously interrelated.

Heating, Ventilating & A.C. R2360 Central Cooling Equipment

R236000-10 Air Conditioning

General: The purpose of air conditioning is to control the environment of a space so that comfort is provided for the occupants and/or conditions are suitable for the processes or equipment contained therein. The several items which should be evaluated to define system objectives are:

- Temperature Control
- Humidity Control
- Cleanliness
- Odor, smoke and fumes
- Ventilation

Efforts to control the above parameters must also include consideration of the degree or tolerance of variation, the noise level introduced, the velocity of air motion and the energy requirements to accomplish the desired results.

The variation in **temperature** and **humidity** is a function of the sensor and the controller. The controller reacts to a signal from the sensor and produces the appropriate suitable response in either the terminal unit, the conductor of the transporting medium (air, steam, chilled water, etc.), or the source (boiler, evaporating coils, etc.).

The **noise level** is a by-product of the energy supplied to moving components of the system. Those items which usually contribute the most noise are pumps, blowers, fans, compressors and diffusers. The level of noise can be partially controlled through use of vibration pads, isolators, proper sizing, shields, baffles and sound absorbing liners.

Some **air motion** is necessary to prevent stagnation and stratification. The maximum acceptable velocity varies with the degree of heating or cooling which is taking place. Most people feel air moving past them at velocities in excess of 25 FPM as an annoying draft. However, velocities up to 45 FPM may be acceptable in certain cases. Ventilation, expressed as air changes per hour and percentage of fresh air, is usually an item regulated by local codes.

Selection of the system to be used for a particular application is usually a trade-off. In some cases the building size, style, or room available for mechanical use limits the range of possibilities. Prime factors influencing the decision are first cost and total life (operating, maintenance and replacement costs). The accuracy with which each parameter is determined will be an important measure of the reliability of the decision and subsequent satisfactory operation of the installed system.

Heat delivery may be desired from an air conditioning system. Heating capability usually is added as follows: A gas fired burner or hot water/steam/electric coils may be added to the air handling unit directly and heat all air equally. For limited or localized heat requirements the water/steam/electric coils may be inserted into the duct branch supplying the cold areas. Gas fired duct furnaces are also available.

Note: When water or steam coils are used the cost of the piping and boiler must also be added. For a rough estimate use the cost per square foot of the appropriate sized hydronic system with unit heaters. This will provide a cost for the boiler and piping, and the unit heaters of the system would equate to the approximate cost of the heating coils.

R236000-20 Air Conditioning Requirements

BTUs per hour per S.F. of floor area and S.F. per ton of air conditioning.

Type of Building	BTU/Hr per S.F.	S.F. per Ton	Type of Building	BTU/Hr per S.F.	S.F. per Ton	Type of Building	BTU/Hr per S.F.	S.F. per Ton
Apartments, Individual	26	450	Dormitory, Rooms	40	300	Libraries	50	240
Corridors	22	550	Corridors	30	400	Low Rise Office, Exterior	38	320
Auditoriums & Theaters	40	300/18*	Dress Shops	43	280	Interior	33	360
Banks	50	240	Drug Stores	80	150	Medical Centers	28	425
Barber Shops	48	250	Factories	40	300	Motels	28	425
Bars & Taverns	133	90	High Rise Office—Ext. Rms.	46	263	Office (small suite)	43	280
Beauty Parlors	66	180	Interior Rooms	37	325	Post Office, Individual Office	42	285
Bowling Alleys	68	175	Hospitals, Core	43	280	Central Area	46	260
Churches	36	330/20*	Perimeter	46	260	Residences	20	600
Cocktail Lounges	68	175	Hotel, Guest Rooms	44	275	Restaurants	60	200
Computer Rooms	141	85	Corridors	30	400	Schools & Colleges	46	260
Dental Offices	52	230	Public Spaces	55	220	Shoe Stores	55	220
Dept. Stores, Basement	34	350	Industrial Plants, Offices	38	320	Shop'g. Ctrs., Supermarkets	34	350
Main Floor	40	300	General Offices	34	350	Retail Stores	48	250
Upper Floor	30	400	Plant Areas	40	300	Specialty	60	200

*Persons per ton
12,000 BTU = 1 ton of air conditioning

R236000-30 Psychrometric Table

Dewpoint or Saturation Temperature (F)

Relative humidity (%)															
100	32	35	40	45	50	55	60	65	70	75	80	85	90	95	100
90	30	33	37	42	47	52	57	62	67	72	77	82	87	92	97
80	27	30	34	39	44	49	54	58	64	68	73	78	83	88	93
70	24	27	31	36	40	45	50	55	60	64	69	74	79	84	88
60	20	24	28	32	36	41	46	51	55	60	65	69	74	79	83
50	16	20	24	28	33	36	41	46	50	55	60	64	69	73	78
40	12	15	18	23	27	31	35	40	45	49	53	58	62	67	71
30	8	10	14	18	21	25	29	33	37	42	46	50	54	59	62
20	6	7	8	9	13	16	20	24	28	31	35	40	43	48	52
10	4	4	5	5	6	8	9	10	13	17	20	24	27	30	34
	32	**35**	**40**	**45**	**50**	**55**	**60**	**65**	**70**	**75**	**80**	**85**	**90**	**95**	**100**

Dry bulb temperature (F)

This table shows the relationship between RELATIVE HUMIDITY, DRY BULB TEMPERATURE AND DEWPOINT.

As an example, assume that the thermometer in a room reads 75° F, and we know that the relative humidity is 50%. The chart shows the dewpoint temperature to be 55° F. That is, any surface colder than 55° F will "sweat" or collect condensing moisture. This surface could be the outside of an uninsulated chilled water pipe in the summertime, or the inside surface of a wall or deck in the wintertime. After determining the extreme ambient parameters, the table at the left is useful in determining which surfaces need insulation or vapor barrier protection.

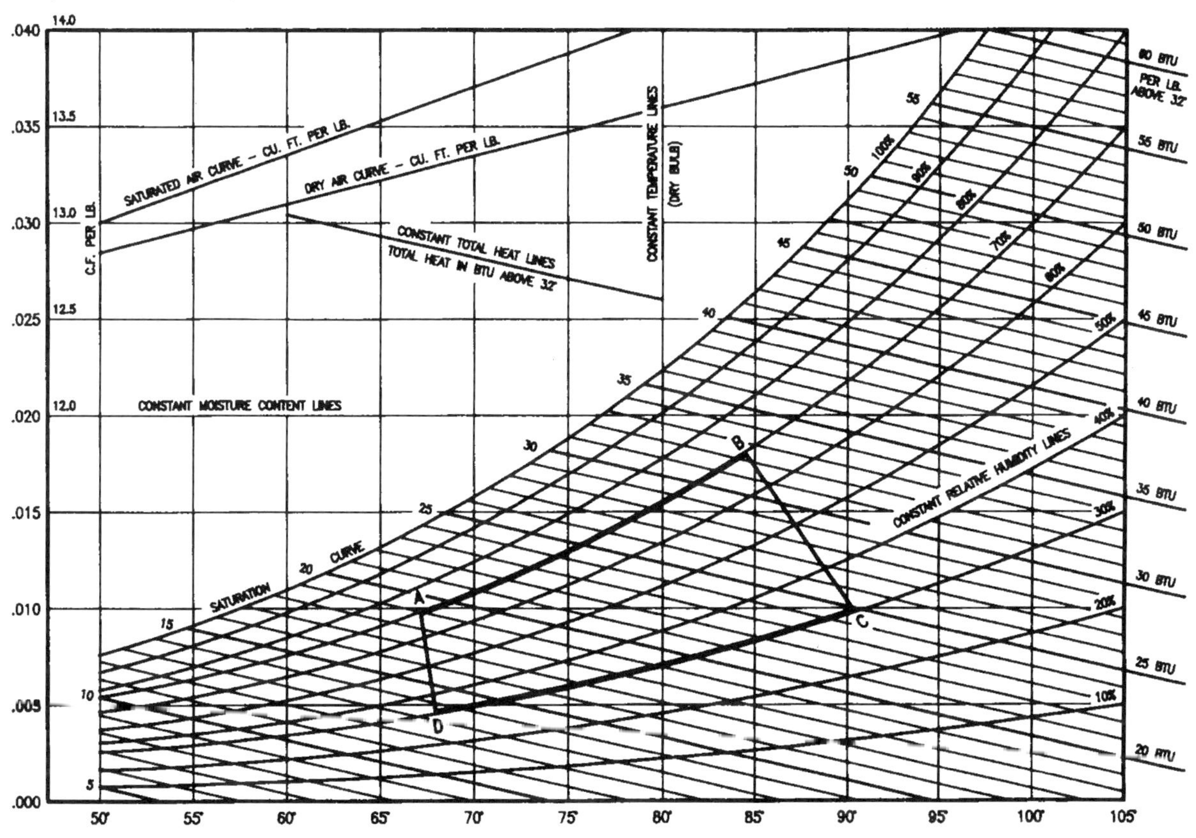

TEMPERATURE DEGREES FAHRENHEIT
TOTAL PRESSURE = 14.696 LB. PER SQ. IN. ABS.

Psychrometric chart showing different variables based on one pound of dry air. Space marked A B C D is temperature-humidity range which is most comfortable for majority of people.

R236000-90 Quality/Complexity Adjustment for Air Conditioning Systems

Economy installation, add	0 to 5%
Good quality, medium complexity, add	5 to 15%
Above average quality and complexity, add	15 to 25%

Add the above adjustments to the estimate to more closely allow for the equipment and conditions of the particular job under consideration.

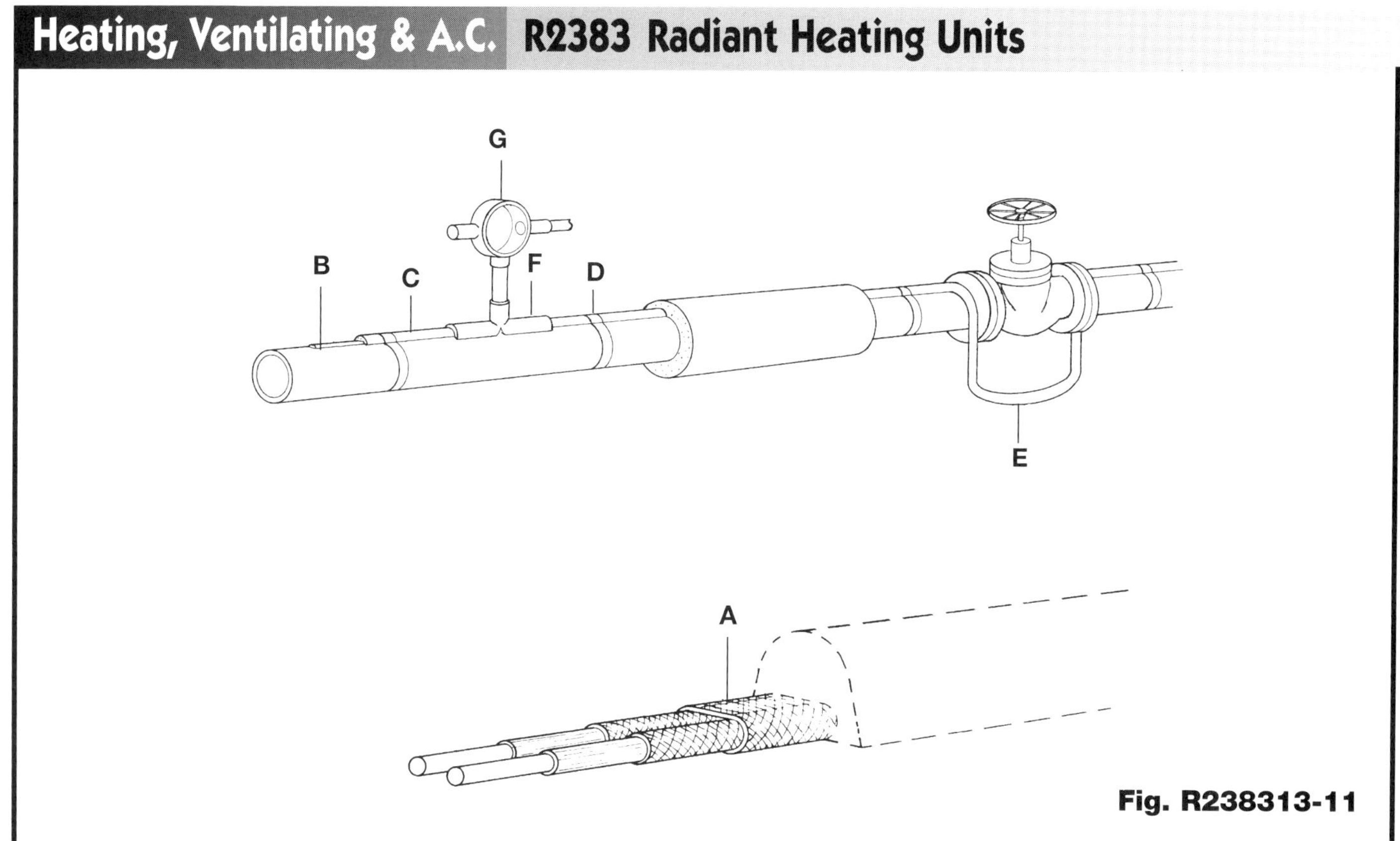

Fig. R238313-11

R238313-10 Heat Trace Systems

Before you can determine the cost of a HEAT TRACE installation, the method of attachment must be established. There are (4) common methods:

1. Cable is simply attached to the pipe with polyester tape every 12′.
2. Cable is attached with a continuous cover of 2″ wide aluminum tape.
3. Cable is attached with factory extruded heat transfer cement and covered with metallic raceway with clips every 10′.
4. Cable is attached between layers of pipe insulation using either clips or polyester tape.

Example: Components for method 3 must include:

A. Heat trace cable by voltage and watts per linear foot.
B. Heat transfer cement, 1 gallon per 60 linear feet of cover.
C. Metallic raceway by size and type.
D. Raceway clips by size of pipe.

When taking off linear foot lengths of cable add the following for each valve in the system. (E)

In all of the above methods each component of the system must be priced individually.

SCREWED OR WELDED VALVE:

Size		Length
1/2″	=	6″
3/4″	=	9″
1″	=	1′ -0″
1-1/2″	=	1′ -6″
2″	=	2′
2-1/2″	=	2′ -6″
3″	=	2′ -6″
4″	=	4′ -0″
6″	=	7′ -0″
8″	=	9′ -6″
10″	=	12′ -6″
12″	=	15′ -0″
14″	=	18′ -0″
16″	=	21′ -6″
18″	=	25′ -6″
20″	=	28′ -6″
24″	=	34′ -0″
30″	=	40′ -0″

FLANGED VALVE:

Size		Length
1/2″	=	1′ -0″
3/4″	=	1′ -6″
1″	=	2′ -0″
1-1/2″	=	2′ -6″
2″	=	2′ -6″
2-1/2″	=	3′ -0″
3″	=	3′ -6″
4″	=	4′ -0″
6″	=	8′ -0″
8″	=	11′ -0″
10″	=	14′ -0″
12″	=	16′ -6″
14″	=	19′ -6″
16″	=	23′ -0″
18″	=	27′ -0″
20″	=	30′ -0″
24″	=	36′ -0″
30″	=	42′ -0″

BUTTERFLY VALVES:

Size		Length
1/2″	-	0′
3/4″	=	0′
1″	=	1′ -0″
1-1/2″	=	1′ -6″
2″	=	2′ -0″
2-1/2″	=	2′ -6″
3″	=	2′ -6″
4″	=	3′ -0″
6″	=	3′ -6″
8″	=	4′ -0″
10″	=	4′ -0″
12″	=	5′ -0″
14″	=	5′ -6″
16″	=	6′ -0″
18″	=	6′ -6″
20″	=	7′ -0″
24″	=	8′ -0″
30″	=	10′ -0″

R238313-10 Heat Trace Systems (cont.)

Add the following quantities of heat transfer cement to linear foot totals for each valve:

Nominal Valve Size	Gallons of Cement per Valve
1/2″	0.14
3/4″	0.21
1″	0.29
1-1/2″	0.36
2″	0.43
2-1/2″	0.70
3″	0.71
4″	1.00
6″	1.43
8″	1.48
10″	1.50
12″	1.60
14″	1.75
16″	2.00
18″	2.25
20″	2.50
24″	3.00
30″	3.75

The following must be added to the list of components to accurately price HEAT TRACE systems:

1. Expediter fitting and clamp fasteners (F)
2. Junction box and nipple connected to expediter fitting (G)
3. Field installed terminal blocks within junction box
4. Ground lugs
5. Piping from power source to expediter fitting
6. Controls
7. Thermostats
8. Branch wiring
9. Cable splices
10. End of cable terminations
11. Branch piping fittings and boxes

Deduct the following percentages from labor if cable lengths in the same area exceed:

150′ to 250′	10%	351′ to 500′	20%
251′ to 350′	15%	Over 500′	25%

Add the following percentages to labor for elevated installations:

15′ to 20′ high	10%	31′ to 35′ high	40%
21′ to 25′ high	20%	36′ to 40′ high	50%
26′ to 30′ high	30%	Over 40′ high	60%

R238313-20 Spiral-Wrapped Heat Trace Cable (Pitch Table)

In order to increase the amount of heat, occasionally heat trace cable is wrapped in a spiral fashion around a pipe; increasing the number of feet of heater cable per linear foot of pipe.

Engineers first determine the heat loss per foot of pipe (based on the insulating material, its thickness, and the temperature differential across it). A ratio is then calculated by the formula:

$$\text{Feet of Heat Trace per Foot of Pipe} = \frac{\text{Watts/Foot of Heat Loss}}{\text{Watts/Foot of the Cable}}$$

The linear distance between wraps (pitch) is then taken from a chart or table. Generally, the pitch is listed on a drawing leaving the estimator to calculate the total length of heat tape required. An approximation may be taken from this table.

Feet of Heat Trace Per Foot of Pipe																
	Nominal Pipe Size in Inches															
Pitch In Inches	1	1¼	1½	2	2½	3	4	6	8	10	12	14	16	18	20	24
3.5	1.80															
4	1.65															
5	1.46	1.60	1.80													
6	1.34	1.45	1.55	1.75												
7	1.25	1.35	1.43	1.57	1.75											
8	1.20	1.28	1.34	1.45	1.60	1.80										
9	1.16	1.23	1.28	1.37	1.51	1.68										
10	1.13	1.19	1.24	1.32	1.44	1.57	1.82									
15	1.06	1.08	1.10	1.15	1.21	1.29	1.42	1.78								
20	1.04	1.05	1.06	1.08	1.13	1.17	1.25	1.49	1.73							
25		1.04	1.04	1.06	1.08	1.11	1.17	1.33	1.51	1.72						
30				1.04	1.05	1.07	1.12	1.24	1.37	1.54	1.70	1.80				
35					1.06	1.06	1.09	1.17	1.28	1.42	1.54	1.64	1.78			
40						1.05	1.07	1.14	1.22	1.33	1.44	1.52	1.64	1.75		
50							1.05	1.09	1.15	1.22	1.29	1.35	1.44	1.53	1.64	1.83
60								1.06	1.11	1.16	1.21	1.25	1.31	1.39	1.46	1.62
70								1.05	1.08	1.12	1.17	1.19	1.24	1.30	1.35	1.47
80									1.06	1.09	1.13	1.15	1.19	1.24	1.28	1.38
90									1.04	1.06	1.10	1.13	1.16	1.19	1.23	1.32
100										1.05	1.08	1.10	1.13	1.15	1.19	1.23

Note: Common practice would normally limit the lower end of the table to 5% of additional heat and above 80% an engineer would likely opt for two (2) parallel cables.

R312319-90 Wellpoints

A single stage wellpoint system is usually limited to dewatering an average 15′ depth below normal ground water level. Multi-stage systems are employed for greater depth with the pumping equipment installed only at the lowest header level. Ejectors with unlimited lift capacity can be economical when two or more stages of wellpoints can be replaced or when horizontal clearance is restricted, such as in deep trenches or tunneling projects, and where low water flows are expected. Wellpoints are usually spaced on 2-1/2′ to 10′ centers along a header pipe. Wellpoint spacing, header size, and pump size are all determined by the expected flow as dictated by soil conditions.

In almost all soils encountered in wellpoint dewatering, the wellpoints may be jetted into place. Cemented soils and stiff clays may require sand wicks about 12″ in diameter around each wellpoint to increase efficiency and eliminate weeping into the excavation. These sand wicks require 1/2 to 3 C.Y. of washed filter sand and are installed by using a 12″ diameter steel casing and hole puncher jetted into the ground 2′ deeper than the wellpoint. Rock may require predrilled holes.

Labor required for the complete installation and removal of a single stage wellpoint system is in the range of 3/4 to 2 labor-hours per linear foot of header, depending upon jetting conditions, wellpoint spacing, etc.

Continuous pumping is necessary except in some free draining soil where temporary flooding is permissible (as in trenches which are backfilled after each day's work). Good practice requires provision of a stand-by pump during the continuous pumping operation.

Systems for continuous trenching below the water table should be installed three to four times the length of expected daily progress to ensure uninterrupted digging, and header pipe size should not be changed during the job.

For pervious free draining soils, deep wells in place of wellpoints may be economical because of lower installation and maintenance costs. Daily production ranges between two to three wells per day, for 25′ to 40′ depths, to one well per day for depths over 50′.

Detailed analysis and estimating for any dewatering problem is available at no cost from wellpoint manufacturers. Major firms will quote "sufficient equipment" quotes or their affiliates offer lump sum proposals to cover complete dewatering responsibility.

Description for 200′ System with 8″ Header		Quantities
Equipment & Material	Wellpoints 25′ long, 2″ diameter @ 5′ O.C.	40 Each
	Header pipe, 8″ diameter	200 L.F.
	Discharge pipe, 8″ diameter	100 L.F.
	8″ valves	3 Each
	Combination jetting & wellpoint pump (standby)	1 Each
	Wellpoint pump, 8″ diameter	1 Each
	Transportation to and from site	1 Day
	Fuel for 30 days x 60 gal./day	1800 Gallons
	Lubricants for 30 days x 16 lbs./day	480 Lbs.
	Sand for points	40 C.Y.
Labor	Technician to supervise installation	1 Week
	Labor for installation and removal of system	300 Labor-hours
	4 Operators straight time 40 hrs./wk. for 4.33 wks.	693 Hrs.
	4 Operators overtime 2 hrs./wk. for 4.33 wks.	35 Hrs.

Earthwork — R3141 Shoring

R314116-40 Wood Sheet Piling

Wood sheet piling may be used for depths to 20′ where there is no ground water. If moderate ground water is encountered Tongue & Groove sheeting will help to keep it out. When considerable ground water is present, steel sheeting must be used.

For estimating purposes on trench excavation, sizes are as follows:

Depth	Sheeting	Wales	Braces	B.F. per S.F.
To 8′	3 x 12′s	6 x 8′s, 2 line	6 x 8′s, @ 10′	4.0 @ 8′
8′ x 12′	3 x 12′s	10 x 10′s, 2 line	10 x 10′s, @ 9′	5.0 average
12′ to 20′	3 x 12′s	12 x 12′s, 3 line	12 x 12′s, @ 8′	7.0 average

Sheeting to be toed in at least 2′ depending upon soil conditions. A five person crew with an air compressor and sheeting driver can drive and brace 440 SF/day at 8′ deep, 360 SF/day at 12′ deep, and 320 SF/day at 16′ deep. For normal soils, piling can be pulled in 1/3 the time to install. Pulling difficulty increases with the time in the ground. Production can be increased by high pressure jetting.

R314116-45 Steel Sheet Piling

Limiting weights are 22 to 38#/S.F. of wall surface with 27#/S.F. average for usual types and sizes. (Weights of piles themselves are from 30.7#/L.F. to 57#/L.F. but they are 15″ to 21″ wide.) Lightweight sections 12″ to 28″ wide from 3 ga. to 12 ga. thick are also available for shallow excavations. Piles may be driven two at a time with an impact or vibratory hammer (use vibratory to pull) hung from a crane without leads. A reasonable estimate of the life of steel sheet piling is 10 uses with up to 125 uses possible if a vibratory hammer is used. Used piling costs from 50% to 80% of new piling depending on location and market conditions. Sheet piling and H piles can be rented for about 30% of the delivered mill price for the first month and 5% per month thereafter. Allow 1 labor-hour per pile for cleaning and trimming after driving. These costs increase with depth and hydrostatic head. Vibratory drivers are faster in wet granular soils and are excellent for pile extraction. Pulling difficulty increases with the time in the ground and may cost more than driving. It is often economical to abandon the sheet piling, especially if it can be used as the outer wall form. Allow about 1/3 additional length or more for toeing into ground. Add bracing, waler and strut costs. Waler costs can equal the cost per ton of sheeting.

Utilities — R3311 Water Utility Distribution Piping

R331113-80 Piping Designations

There are several systems currently in use to describe pipe and fittings. The following paragraphs will help to identify and clarify classifications of piping systems used for water distribution.

Piping may be classified by schedule. Piping schedules include 5S, 10S, 10, 20, 30, Standard, 40, 60, Extra Strong, 80, 100, 120, 140, 160 and Double Extra Strong. These schedules are dependent upon the pipe wall thickness. The wall thickness of a particular schedule may vary with pipe size.

Ductile iron pipe for water distribution is classified by Pressure Classes such as Class 150, 200, 250, 300 and 350. These classes are actually the rated water working pressure of the pipe in pounds per square inch (psi). The pipe in these pressure classes is designed to withstand the rated water working pressure plus a surge allowance of 100 psi.

The American Water Works Association (AWWA) provides standards for various types of **plastic pipe.** C-900 is the specification for polyvinyl chloride (PVC) piping used for water distribution in sizes ranging from 4″ through 12″. C-901 is the specification for polyethylene (PE) pressure pipe, tubing and fittings used for water distribution in sizes ranging from 1/2″ through 3″. C-905 is the specification for PVC piping sizes 14″ and greater.

PVC pressure-rated pipe is identified using the standard dimensional ratio (SDR) method. This method is defined by the American Society for Testing and Materials (ASTM) Standard D 2241. This pipe is available in SDR numbers 64, 41, 32.5, 26, 21, 17, and 13.5. Pipe with an SDR of 64 will have the thinnest wall while pipe with an SDR of 13.5 will have the thickest wall. When the pressure rating (PR) of a pipe is given in psi, it is based on a line supplying water at 73 degrees F.

The National Sanitation Foundation (NSF) seal of approval is applied to products that can be used with potable water. These products have been tested to ANSI/NSF Standard 14.

Valves and strainers are classified by American National Standards Institute (ANSI) Classes. These Classes are 125, 150, 200, 250, 300, 400, 600, 900, 1500 and 2500. Within each class there is an operating pressure range dependent upon temperature. Design parameters should be compared to the appropriate material dependent, pressure-temperature rating chart for accurate valve selection.

Change Orders

Change Order Considerations

A Change Order is a written document, usually prepared by the design professional, and signed by the owner, the architect/engineer and the contractor. A change order states the agreement of the parties to: an addition, deletion, or revision in the work; an adjustment in the contract sum, if any; or an adjustment in the contract time, if any. Change orders, or "extras" in the construction process occur after execution of the construction contract and impact architects/engineers, contractors and owners.

Change orders that are properly recognized and managed can ensure orderly, professional and profitable progress for all who are involved in the project. There are many causes for change orders and change order requests. In all cases, change orders or change order requests should be addressed promptly and in a precise and prescribed manner. The following paragraphs include information regarding change order pricing and procedures.

The Causes of Change Orders

Reasons for issuing change orders include:

- Unforeseen field conditions that require a change in the work
- Correction of design discrepancies, errors or omissions in the contract documents
- Owner-requested changes, either by design criteria, scope of work, or project objectives
- Completion date changes for reasons unrelated to the construction process
- Changes in building code interpretations, or other public authority requirements that require a change in the work
- Changes in availability of existing or new materials and products

Procedures

Properly written contract documents must include the correct change order procedures for all parties—owners, design professionals and contractors—to follow in order to avoid costly delays and litigation.

Being "in the right" is not always a sufficient or acceptable defense. The contract provisions requiring notification and documentation must be adhered to within a defined or reasonable time frame.

The appropriate method of handling change orders is by a written proposal and acceptance by all parties involved. Prior to starting work on a project, all parties should identify their authorized agents who may sign and accept change orders, as well as any limits placed on their authority.

Time may be a critical factor when the need for a change arises. For such cases, the contractor might be directed to proceed on a "time and materials" basis, rather than wait for all paperwork to be processed—a delay that could impede progress. In this situation, the contractor must still follow the prescribed change order procedures, including but not limited to, notification and documentation.

All forms used for change orders should be dated and signed by the proper authority. Lack of documentation can be very costly, especially if legal judgments are to be made and if certain field personnel are no longer available. For time and material change orders, the contractor should keep accurate daily records of all labor and material allocated to the change. Forms that can be used to document change order work are available in *Means Forms for Building Construction Professionals.*

Owners or awarding authorities who do considerable and continual building construction (such as the federal government) realize the inevitability of change orders for numerous reasons, both predictable and unpredictable. As a result, the federal government, the American Institute of Architects (AIA), the Engineers Joint Contract Documents Committee (EJCDC) and other contractor, legal and technical organizations have developed standards and procedures to be followed by all parties to achieve contract continuance and timely completion, while being financially fair to all concerned.

In addition to the change order standards put forth by industry associations, there are also many books available on the subject.

Pricing Change Orders

When pricing change orders, regardless of their cause, the most significant factor is when the change occurs. The need for a change may be perceived in the field or requested by the architect/engineer *before* any of the actual installation has begun, or may evolve or appear *during* construction when the item of work in question is partially installed. In the latter cases, the original sequence of construction is disrupted, along with all contiguous and supporting systems. Change orders cause the greatest impact when they occur *after* the installation has been completed and must be uncovered, or even replaced. Post-completion changes may be caused by necessary design changes, product failure, or changes in the owner's requirements that are not discovered until the building or the systems begin to function.

Specified procedures of notification and record keeping must be adhered to and enforced regardless of the stage of construction: *before, during,* or *after* installation. Some bidding documents anticipate change orders by requiring that unit prices including overhead and profit percentages—for additional as well as deductible changes—be listed. Generally these unit prices do not fully take into account the ripple effect, or impact on other trades, and should be used for general guidance only.

When pricing change orders, it is important to classify the time frame in which the change occurs. There are two basic time frames for change orders: *pre-installation change orders,* which occur before the start of construction, and *post-installation change orders,* which involve reworking after the original installation. Change orders that occur between these stages may be priced according to the extent of work completed using a combination of techniques developed for pricing *pre-* and *post-installation* changes.

The following factors are the basis for a check list to use when preparing a change order estimate.

Factors To Consider When Pricing Change Orders

As an estimator begins to prepare a change order, the following questions should be reviewed to determine their impact on the final price.

General

- Is the change order work *pre-installation* or *post-installation?*

  Change order work costs vary according to how much of the installation has been completed. Once workers have the project scoped in their mind, even though they have not started, it can be difficult to refocus. Consequently they may spend more than the normal amount of time understanding the change. Also, modifications to work in place such as trimming or refitting usually take more time than was initially estimated. The greater the amount of work in place, the more reluctant workers are to change it. Psychologically they may resent the change and as a result the rework takes longer than normal. Post-installation change order estimates must include demolition of existing work as required to accomplish the change. If the work is performed at a later time, additional obstacles such as building finishes may be present which must be protected. Regardless of whether the change occurs pre-installation or post-installation, attempt to isolate the identifiable factors and price them separately. For example, add shipping costs that may be required pre-installation or any demolition required post-installation. Then analyze the potential impact on productivity of psychological and/or learning curve factors and adjust the output rates accordingly. One approach is to break down the typical workday into segments and quantify the impact on each segment. The following chart may be useful as a guide:

Activities (Productivity) Expressed as Percentages of a Workday			
Task	Means Mechanical Cost Data (for New Construction)	Pre-Installation Change Orders	Post-Installation Change Orders
1. Study plans	3%	6%	6%
2. Material procurement	3%	3%	3%
3. Receiving and storing	3%	3%	3%
4. Mobilization	5%	5%	5%
5. Site movement	5%	5%	8%
6. Layout and marking	8%	10%	12%
7. Actual installation	64%	59%	54%
8. Clean-up	3%	3%	3%
9. Breaks—non-productive	6%	6%	6%
Total	100%	100%	100%

Change Order Installation Efficiency

The labor-hours expressed (for new construction) are based on average installation time, using an efficiency level of approximately 60-65%. For change order situations, adjustments to this efficiency level should reflect the daily labor-hour allocation for that particular occurrence.

If any of the specific percentages expressed in the above chart do not apply to a particular project situation, then those percentage points should be reallocated to the appropriate task(s). Example: Using data for new construction, assume there is no new material being utilized. The percentages for Tasks 2 and 3 would therefore be reallocated to other tasks. If the time required for Tasks 2 and 3 can now be applied to installation, we can add the time allocated for *Material Procurement* and *Receiving and Storing* to the *Actual Installation* time for new construction, thereby increasing the Actual Installation percentage.

This chart shows that, due to reduced productivity, labor costs will be higher than those for new construction by 5% to 15% for pre-installation change orders and by 15% to 25% for post-installation change orders. Each job and change order is unique and must be examined individually. Many factors, covered elsewhere in this section, can each have a significant impact on productivity and change order costs. All such factors should be considered in every case.

- Will the change substantially delay the original completion date?

  A significant change in the project may cause the original completion date to be extended. The extended schedule may subject the contractor to new wage rates dictated by relevant labor contracts. Project supervision and other project overhead must also be extended beyond the original completion date. The schedule extension may also put installation into a new weather season. For example, underground piping scheduled for October installation was delayed until January. As a result, frost penetrated the trench area, thereby changing the degree of difficulty of the task. Changes and delays may have a ripple effect throughout the project. This effect must be analyzed and negotiated with the owner.

- What is the net effect of a deduct change order?

  In most cases, change orders resulting in a deduction or credit reflect only bare costs. The contractor may retain the overhead and profit based on the original bid.

Materials

- Will you have to pay more or less for the new material, required by the change order, than you paid for the original purchase?

  The same material prices or discounts will usually apply to materials purchased for change orders as new construction. In some instances, however, the contractor may forfeit the advantages of competitive pricing for change orders. Consider the following example:

  A contractor purchased over $20,000 worth of fan coil units for an installation, and obtained the maximum discount. Some time later it was determined the project required an additional matching unit. The contractor has to purchase this unit from the original supplier to ensure a match. The supplier at this time may not discount the unit because of the small quantity, and the fact that he is no longer in a competitive situation. The impact of quantity on purchase can add between 0% and 25% to material prices and/or subcontractor quotes.

- If materials have been ordered or delivered to the job site, will they be subject to a cancellation charge or restocking fee?

  Check with the supplier to determine if ordered materials are subject to a cancellation charge. Delivered materials not used as result of a change order may be subject to a restocking fee if returned to the supplier. Common restocking charges run between 20% and 40%. Also, delivery charges to return the goods to the supplier must be added.

Labor

- How efficient is the existing crew at the actual installation?

  Is the same crew that performed the initial work going to do the change order? Possibly the change consists of the installation of a unit identical to one already installed; therefore the change should take less time. Be sure to consider this potential productivity increase and modify the productivity rates accordingly.

- If the crew size is increased, what impact will that have on supervision requirements?

  Under most bargaining agreements or management practices, there is a point at which a working foreman is replaced by a nonworking foreman. This replacement increases project overhead by adding a nonproductive worker. If additional workers are added to accelerate the project or to perform changes while maintaining the schedule, be sure to add additional supervision time if warranted. Calculate the hours involved and the additional cost directly if possible.

- What are the other impacts of increased crew size?

  The larger the crew, the greater the potential for productivity to decrease. Some of the factors that cause this productivity loss are: overcrowding (producing restrictive conditions in the working space), and possibly a shortage of any special tools and equipment required. Such factors affect not only the crew working on the elements directly involved in the change order, but other crews whose movement may also be hampered.

  As the crew increases, check its basic composition for changes by the addition or deletion of apprentices or nonworking foreman and quantify the potential effects of equipment shortages or other logistical factors.

- As new crews, unfamiliar with the project, are brought onto the site, how long will it take them to become oriented to the project requirements?

  The orientation time for a new crew to become 100% effective varies with the site and type of project. Orientation is easiest at a new construction site, and most difficult at existing, very restrictive renovation sites. The type of work also affects orientation time. When all elements of the work are exposed, such as concrete or masonry work, orientation is decreased. When the work is concealed or less visible, such as existing electrical systems, orientation takes longer. Usually orientation can be accomplished in one day or less. Costs for added orientation should be itemized and added to the total estimated cost.

- How much actual production can be gained by working overtime?

  Short term overtime can be used effectively to accomplish more work in a day. However, as overtime is scheduled to run beyond several weeks, studies have shown marked decreases in output. The following chart shows the effect of long term overtime on worker efficiency. If the anticipated change requires extended overtime to keep the job on schedule, these factors can be used as a guide to predict the impact on time and cost. Add project overhead, particularly supervision, that may also be incurred.

Days per Week	Hours per Day	Production Efficiency					Payroll Cost Factors	
		1 Week	2 Weeks	3 Weeks	4 Weeks	Average 4 Weeks	@ 1-1/2 Times	@ 2 Times
5	8	100%	100%	100%	100%	100%	100%	100%
	9	100	100	95	90	96.25	105.6	111.1
	10	100	95	90	85	91.25	110.0	120.0
	11	95	90	75	65	81.25	113.6	127.3
	12	90	85	70	60	76.25	116.7	133.3
6	8	100	100	95	90	96.25	108.3	116.7
	9	100	95	90	85	92.50	113.0	125.9
	10	95	90	85	80	87.50	116.7	133.3
	11	95	85	70	65	78.75	119.7	139.4
	12	90	80	65	60	73.75	122.2	144.4
7	8	100	95	85	75	88.75	114.3	128.6
	9	95	90	80	70	83.75	118.3	136.5
	10	90	85	75	65	78.75	121.4	142.9
	11	85	80	65	60	72.50	124.0	148.1
	12	85	75	60	55	68.75	126.2	152.4

Effects of Overtime

Caution: Under many labor agreements, Sundays and holidays are paid at a higher premium than the normal overtime rate.

The use of long-term overtime is counterproductive on almost any construction job; that is, the longer the period of overtime, the lower the actual production rate. Numerous studies have been conducted, and while they have resulted in slightly different numbers, all reach the same conclusion. The figure above tabulates the effects of overtime work on efficiency.

As illustrated, there can be a difference between the *actual* payroll cost per hour and the *effective* cost per hour for overtime work. This is due to the reduced production efficiency with the increase in weekly hours beyond 40. This difference between actual and effective cost results from overtime work over a prolonged period. Short-term overtime work does not result in as great a reduction in efficiency, and in such cases, effective cost may not vary significantly from the actual payroll cost. As the total hours per week are increased on a regular basis, more time is lost because of fatigue, lowered morale, and an increased accident rate.

As an example, assume a project where workers are working 6 days a week, 10 hours per day. From the figure above (based on productivity studies), the average effective productive hours over a four-week period are:

$$0.875 \times 60 = 52.5$$

Depending upon the locale and day of week, overtime hours may be paid at time and a half or double time. For time and a half, the overall (average) *actual* payroll cost (including regular and overtime hours) is determined as follows:

$$\frac{40 \text{ reg. hrs.} + (20 \text{ overtime hrs.} \times 1.5)}{60 \text{ hrs.}} = 1.167$$

Based on 60 hours, the payroll cost per hour will be 116.7% of the normal rate at 40 hours per week. However, because the effective production (efficiency) for 60 hours is reduced to the equivalent of 52.5 hours, the effective cost of overtime is calculated as follows:

For time and a half:

$$\frac{40 \text{ reg. hrs.} + (20 \text{ overtime hrs.} \times 1.5)}{52.5 \text{ hrs.}} = 1.33$$

Installed cost will be 133% of the normal rate (for labor).

Thus, when figuring overtime, the actual cost per unit of work will be higher than the apparent overtime payroll dollar increase, due to the reduced productivity of the longer workweek. These efficiency calculations are true only for those cost factors determined by hours worked. Costs that are applied weekly or monthly, such as equipment rentals, will not be similarly affected.

Equipment

- What equipment is required to complete the change order?

  Change orders may require extending the rental period of equipment already on the job site, or the addition of special equipment brought in to accomplish the change work. In either case, the additional rental charges and operator labor charges must be added.

Summary

The preceding considerations and others you deem appropriate should be analyzed and applied to a change order estimate. The impact of each should be quantified and listed on the estimate to form an audit trail.

Change orders that are properly identified, documented, and managed help to ensure the orderly, professional and profitable progress of the work. They also minimize potential claims or disputes at the end of the project.

Estimating Tips

- The cost figures in this Square Foot Cost section were derived from approximately 11,200 projects contained in the RSMeans database of completed construction projects. They include the contractor's overhead and profit, but do not generally include architectural fees or land costs. The figures have been adjusted to January of the current year. New projects are added to our files each year, and outdated projects are discarded. For this reason, certain costs may not show a uniform annual progression. In no case are all subdivisions of a project listed.
- These projects were located throughout the U.S. and reflect a tremendous variation in square foot (S.F.) and cubic foot (C.F.) costs. This is due to differences, not only in labor and material costs, but also in individual owners' requirements. For instance, a bank in a large city would have different features than one in a rural area. This is true of all the different types of buildings analyzed. Therefore, caution should be exercised when using these Square Foot costs. For example, for court houses, costs in the database are local court house costs and will not apply to the larger, more elaborate federal court houses. As a general rule, the projects in the 1/4 column do not include any site work or equipment, while the projects in the 3/4 column may include both equipment and site work. The median figures do not generally include site work.
- None of the figures "go with" any others. All individual cost items were computed and tabulated separately. Thus, the sum of the median figures for Plumbing, HVAC and Electrical will not normally total up to the total Mechanical and Electrical costs arrived at by separate analysis and tabulation of the projects.
- Each building was analyzed as to total and component costs and percentages. The figures were arranged in ascending order with the results tabulated as shown. The 1/4 column shows that 25% of the projects had lower costs and 75% had higher. The 3/4 column shows that 75% of the projects had lower costs and 25% had higher. The median column shows that 50% of the projects had lower costs and 50% had higher.
- There are two times when square foot costs are useful. The first is in the conceptual stage when no details are available. Then square foot costs make a useful starting point. The second is after the bids are in and the costs can be worked back into their appropriate units for information purposes. As soon as details become available in the project design, the square foot approach should be discontinued and the project priced as to its particular components. When more precision is required, or for estimating the replacement cost of specific buildings, the current edition of *RSMeans Square Foot Costs* should be used.
- In using the figures in this section, it is recommended that the median column be used for preliminary figures if no additional information is available. The median figures, when multiplied by the total city construction cost index figures (see City Cost Indexes) and then multiplied by the project size modifier at the end of this section, should present a fairly accurate base figure, which would then have to be adjusted in view of the estimator's experience, local economic conditions, code requirements, and the owner's particular requirements. There is no need to factor the percentage figures, as these should remain constant from city to city. All tabulations mentioning air conditioning had at least partial air conditioning.
- The editors of this book would greatly appreciate receiving cost figures on one or more of your recent projects, which would then be included in the averages for next year. All cost figures received will be kept confidential, except that they will be averaged with other similar projects to arrive at Square Foot cost figures for next year's book. See the last page of the book for details and the discount available for submitting one or more of your projects.

		50 17 00 | S.F. Costs	UNIT	UNIT COSTS			% OF TOTAL			
				1/4	MEDIAN	3/4	1/4	MEDIAN	3/4	
01	0010	**APARTMENTS Low Rise (1 to 3 story)**	S.F.	66	83.50	111				01
	0020	Total project cost	C.F.	5.95	7.85	9.70				
	0100	Site work	S.F.	5.65	7.70	13.60	6.05%	10.55%	14.05%	
	0500	Masonry		1.30	3.03	5.25	1.54%	3.67%	6.35%	
	1500	Finishes		7	9.65	11.90	9.05%	10.75%	12.85%	
	1800	Equipment		2.16	3.28	4.87	2.73%	4.03%	5.95%	
	2720	Plumbing		5.15	6.60	8.40	6.65%	8.95%	10.05%	
	2770	Heating, ventilating, air conditioning		3.28	4.04	5.95	4.20%	5.60%	7.60%	
	2900	Electrical		3.82	5.10	6.85	5.20%	6.65%	8.40%	
	3100	Total: Mechanical & Electrical		13.25	16.90	21	15.90%	18.05%	23%	
	9000	Per apartment unit, total cost	Apt.	61,500	94,000	138,500				
	9500	Total: Mechanical & Electrical	"	11,600	18,300	23,900				
02	0010	**APARTMENTS Mid Rise (4 to 7 story)**	S.F.	87.50	106	131				02
	0020	Total project costs	C.F.	6.85	9.45	12.90				
	0100	Site work	S.F.	3.51	6.95	12.50	5.25%	6.70%	9.15%	
	0500	Masonry		5.85	8.05	11	5.10%	7.25%	10.50%	
	1500	Finishes		11.05	14.40	18.15	10.55%	13.45%	17.70%	
	1800	Equipment		2.55	3.82	5	2.54%	3.48%	4.31%	
	2500	Conveying equipment		1.89	2.41	2.92	1.94%	2.27%	2.69%	
	2720	Plumbing		5.15	8.25	8.70	5.70%	7.20%	8.95%	
	2900	Electrical		5.80	7.85	9.55	6.65%	7.20%	8.95%	
	3100	Total: Mechanical & Electrical		18.50	23	28	18.50%	21%	23%	
	9000	Per apartment unit, total cost	Apt.	99,000	117,000	193,500				
	9500	Total: Mechanical & Electrical	"	18,700	21,700	27,400				
03	0010	**APARTMENTS High Rise (8 to 24 story)**	S.F.	99.50	115	137				03
	0020	Total project costs	C.F.	9.65	11.25	14.35				
	0100	Site work	S.F.	3.61	5.85	8.15	2.58%	4.84%	6.15%	
	0500	Masonry		5.75	10.45	13	4.74%	9.65%	11.05%	
	1500	Finishes		11.05	13.80	16.30	9.75%	11.80%	13.70%	
	1800	Equipment		3.20	3.93	5.20	2.78%	3.49%	4.35%	
	2500	Conveying equipment		2.26	3.43	4.66	2.23%	2.78%	3.37%	
	2720	Plumbing		7.35	8.65	12.10	6.80%	7.20%	10.45%	
	2900	Electrical		6.85	8.65	11.65	6.45%	7.65%	8.80%	
	3100	Total: Mechanical & Electrical		20.50	26	31.50	17.95%	22.50%	24.50%	
	9000	Per apartment unit, total cost	Apt.	103,500	114,000	158,000				
	9500	Total: Mechanical & Electrical	"	22,400	25,500	27,000				
04	0010	**AUDITORIUMS**	S.F.	103	140	202				04
	0020	Total project costs	C.F.	6.45	9	12.90				
	2720	Plumbing	S.F.	6.25	9.05	11	5.85%	7.20%	8.70%	
	2900	Electrical		8	11.65	18.85	6.80%	9.05%	11.30%	
	3100	Total: Mechanical & Electrical		22.50	45.50	55	24.50%	29%	31.50%	
05	0010	**AUTOMOTIVE SALES**	S.F.	76.50	104	129				05
	0020	Total project costs	C.F.	5.05	6.05	7.85				
	2720	Plumbing	S.F.	3.49	6.05	6.60	2.89%	6.05%	6.50%	
	2770	Heating, ventilating, air conditioning		5.40	8.20	8.90	4.61%	10%	10.35%	
	2900	Electrical		6.15	9.65	13.10	7.25%	9.80%	12.15%	
	3100	Total: Mechanical & Electrical		19.30	27.50	33	19.15%	20.50%	22%	
06	0010	**BANKS**	S.F.	150	187	237				06
	0020	Total project costs	C.F.	10.70	14.55	19.20				
	0100	Site work	S.F.	17.20	26	38	7.85%	12.95%	16.95%	
	0500	Masonry		8.10	15.45	27	3.46%	7.65%	10.10%	
	1500	Finishes		13.90	20.50	25.50	5.85%	8.60%	11.55%	
	1800	Equipment		5.65	12.45	25.50	1%	5.55%	10.50%	
	2720	Plumbing		4.70	6.70	9.80	2.82%	3.90%	4.93%	
	2770	Heating, ventilating, air conditioning		8.95	11.95	15.90	4.86%	7.15%	8.50%	
	2900	Electrical		14.20	18.95	25	8.20%	10.20%	12.20%	
	3100	Total: Mechanical & Electrical		34	46	55	16%	19.40%	23%	
	3500	See also division 11020 & 11030 (MF2004 11 16 00 & 11 17 00)								

50 17 00 | S.F. Costs

			UNIT	UNIT COSTS 1/4	UNIT COSTS MEDIAN	UNIT COSTS 3/4	% OF TOTAL 1/4	% OF TOTAL MEDIAN	% OF TOTAL 3/4	
13	0010	**CHURCHES**	S.F.	101	128	166				13
	0020	Total project costs	C.F.	6.25	7.90	10.40				
	1800	Equipment	S.F.	1.21	2.89	6.15	.95%	2.11%	4.50%	
	2720	Plumbing		3.93	5.50	8.10	3.51%	4.96%	6.25%	
	2770	Heating, ventilating, air conditioning		9.20	11.95	16.95	7.50%	10%	12%	
	2900	Electrical		8.50	11.65	15.65	7.35%	8.75%	10.95%	
	3100	Total: Mechanical & Electrical	↓	26.50	34	45.50	18.30%	22%	24.50%	
	3500	See also division 11040 (MF2004 11 91 00)								
15	0010	**CLUBS, COUNTRY**	S.F.	108	131	164				15
	0020	Total project costs	C.F.	8.75	10.65	14.70				
	2720	Plumbing	S.F.	6.55	9.70	22	5.60%	7.90%	10%	
	2900	Electrical		8.50	11.65	15.20	7%	8.95%	11%	
	3100	Total: Mechanical & Electrical	↓	45.50	56.50	59.50	19%	26.50%	29.50%	
17	0010	**CLUBS, SOCIAL Fraternal**	S.F.	86.50	124	166				17
	0020	Total project costs	C.F.	5.40	8.20	9.75				
	2720	Plumbing	S.F.	5.45	6.75	10.25	5.60%	6.90%	8.55%	
	2770	Heating, ventilating, air conditioning		7.85	9.50	12.20	8.20%	9.25%	14.40%	
	2900	Electrical		6.50	10.70	12.25	6.50%	9.50%	10.55%	
	3100	Total: Mechanical & Electrical	↓	19.20	36.50	46	21%	23%	23.50%	
18	0010	**CLUBS, Y.M.C.A.**	S.F.	109	141	177				18
	0020	Total project costs	C.F.	5	8.35	12.45				
	2720	Plumbing	S.F.	6.85	13.65	15.30	5.65%	7.60%	10.85%	
	2900	Electrical		8.70	10.85	14.95	6.05%	7.60%	9.25%	
	3100	Total: Mechanical & Electrical	↓	33	37	50.50	18.40%	21.50%	28.50%	
19	0010	**COLLEGES Classrooms & Administration**	S.F.	110	150	199				19
	0020	Total project costs	C.F.	8	11.65	17.95				
	0500	Masonry	S.F.	8.05	15.15	18.60	5.65%	8.25%	10.50%	
	2720	Plumbing		5.60	11.40	20.50	5.10%	6.60%	8.95%	
	2900	Electrical		9.15	13.85	18.95	7.70%	9.85%	12%	
	3100	Total: Mechanical & Electrical	↓	36.50	51	60.50	24%	28%	31.50%	
21	0010	**COLLEGES Science, Engineering, Laboratories**	S.F.	204	238	276				21
	0020	Total project costs	C.F.	11.70	17.05	19.35				
	1800	Equipment	S.F.	11.35	25.50	28	2%	6.45%	12.65%	
	2900	Electrical		16.80	23	36.50	7.10%	9.40%	12.10%	
	3100	Total: Mechanical & Electrical	↓	62.50	74	115	28.50%	31.50%	41%	
	3500	See also division 11600 (MF2004 11 53 00)								
23	0010	**COLLEGES Student Unions**	S.F.	130	176	213				23
	0020	Total project costs	C.F.	7.25	9.50	11.75				
	3100	Total: Mechanical & Electrical	S.F.	49	53	62.50	23.50%	26%	29%	
25	0010	**COMMUNITY CENTERS**	S.F.	107	132	178				25
	0020	Total project costs	C.F.	7	10	12.95				
	1800	Equipment	S.F.	2.60	4.39	7	1.48%	3.01%	5.45%	
	2720	Plumbing		5.10	8.90	12.15	4.85%	7%	8.95%	
	2770	Heating, ventilating, air conditioning		8.15	11.90	17	6.80%	10.35%	12.90%	
	2900	Electrical		8.80	11.70	16.80	7.30%	9%	10.45%	
	3100	Total: Mechanical & Electrical	↓	30	37.50	53.50	20%	25%	31%	
28	0010	**COURT HOUSES**	S.F.	155	178	229				28
	0020	Total project costs	C.F.	11.85	14.20	17.90				
	2720	Plumbing	S.F.	7.35	10.30	14.80	5.95%	7.45%	8.20%	
	2900	Electrical		16.45	18.25	27	8.90%	10.45%	11.55%	
	3100	Total: Mechanical & Electrical	↓	41.50	57	63.50	22.50%	27.50%	30.50%	
30	0010	**DEPARTMENT STORES**	S.F.	57	77.50	98				30
	0020	Total project costs	C.F.	3.07	3.98	5.40				
	2720	Plumbing	S.F.	1.78	2.25	3.42	1.82%	4.21%	5.90%	
	2770	Heating, ventilating, air conditioning	↓	5.20	8.05	12.10	8.20%	9.10%	14.80%	

	50 17 00 | S.F. Costs		UNIT	UNIT COSTS 1/4	UNIT COSTS MEDIAN	UNIT COSTS 3/4	% OF TOTAL 1/4	% OF TOTAL MEDIAN	% OF TOTAL 3/4	
30	2900	Electrical	S.F.	6.55	9	10.65	9.05%	12.15%	14.95%	30
	3100	Total: Mechanical & Electrical		11.55	14.80	26	13.20%	21.50%	50%	
31	0010	**DORMITORIES Low Rise (1 to 3 story)**	S.F.	109	141	188				31
	0020	Total project costs	C.F.	6.15	9.90	14.85				
	2720	Plumbing	S.F.	6.55	8.75	11.05	8.05%	9%	9.65%	
	2770	Heating, ventilating, air conditioning		6.90	8.25	11	4.61%	8.05%	10%	
	2900	Electrical		7.20	10.95	14.95	6.40%	8.65%	9.50%	
	3100	Total: Mechanical & Electrical		37.50	40	62.50	22%	25%	27%	
	9000	Per bed, total cost	Bed	46,000	51,000	109,500				
32	0010	**DORMITORIES Mid Rise (4 to 8 story)**	S.F.	133	174	215				32
	0020	Total project costs	C.F.	14.70	16.15	19.30				
	2900	Electrical	S.F.	14.15	16.10	21.50	8.20%	10.20%	11.95%	
	3100	Total: Mechanical & Electrical	"	39.50	41	80.50	25.50%	34.50%	37.50%	
	9000	Per bed, total cost	Bed	19,000	43,300	109,500				
34	0010	**FACTORIES**	S.F.	50.50	75	116				34
	0020	Total project costs	C.F.	3.24	4.83	8				
	0100	Site work	S.F.	5.75	10.50	16.60	6.95%	11.45%	17.95%	
	2720	Plumbing		2.72	5.05	8.35	3.73%	6.05%	8.10%	
	2770	Heating, ventilating, air conditioning		5.30	7.60	10.25	5.25%	8.45%	11.35%	
	2900	Electrical		6.25	9.90	15.15	8.10%	10.50%	14.20%	
	3100	Total: Mechanical & Electrical		17.90	24	36.50	21%	28.50%	35.50%	
36	0010	**FIRE STATIONS**	S.F.	100	138	185				36
	0020	Total project costs	C.F.	5.85	8.05	10.70				
	0500	Masonry	S.F.	14.80	25	34	8.35%	11.55%	16.15%	
	1140	Roofing		3.26	8.85	10.05	1.90%	4.94%	5.05%	
	1580	Painting		2.53	3.78	3.87	1.37%	1.57%	2.07%	
	1800	Equipment		1.24	2.38	4.41	.62%	1.86%	3.54%	
	2720	Plumbing		5.60	9.05	13.05	5.85%	7.35%	9.45%	
	2770	Heating, ventilating, air conditioning		5.55	9	13.90	5.15%	7.40%	9.40%	
	2900	Electrical		7.15	12.55	17	6.80%	8.60%	10.60%	
	3100	Total: Mechanical & Electrical		36.50	47	53	18.40%	23%	26%	
37	0010	**FRATERNITY HOUSES & Sorority Houses**	S.F.	100	129	176				37
	0020	Total project costs	C.F.	9.95	10.40	12.50				
	2720	Plumbing	S.F.	7.55	8.65	15.85	6.80%	8%	10.85%	
	2900	Electrical		6.60	14.25	17.50	6.60%	9.90%	10.65%	
	3100	Total: Mechanical & Electrical		7.80	25	30		15.10%	15.90%	
38	0010	**FUNERAL HOMES**	S.F.	106	144	261				38
	0020	Total project costs	C.F.	10.75	12	23				
	2900	Electrical	S.F.	4.66	8.55	9.35	3.58%	4.44%	5.95%	
	3100	Total: Mechanical & Electrical		16.50	24	33.50	12.90%	12.90%	12.90%	
39	0010	**GARAGES, COMMERCIAL (Service)**	S.F.	60	92.50	128				39
	0020	Total project costs	C.F.	3.93	5.80	8.45				
	1800	Equipment	S.F.	3.37	7.60	11.80	2.21%	4.62%	6.80%	
	2720	Plumbing		4.14	6.40	11.65	5.45%	7.85%	10.65%	
	2730	Heating & ventilating		5.45	7.20	9.75	5.25%	6.85%	8.20%	
	2900	Electrical		5.70	8.65	12.50	7.15%	9.25%	10.85%	
	3100	Total: Mechanical & Electrical		12.55	24	35.50	12.35%	17.40%	26%	
40	0010	**GARAGES, MUNICIPAL (Repair)**	S.F.	88	117	165				40
	0020	Total project costs	C.F.	5.50	6.95	11.95				
	0500	Masonry	S.F.	8.25	16.10	25	5.60%	9.15%	12.50%	
	2720	Plumbing		3.94	7.55	14.25	3.59%	6.70%	7.95%	
	2730	Heating & ventilating		6.75	9.75	18.85	6.15%	7.45%	13.50%	
	2900	Electrical		6.50	10.20	14.70	6.65%	8.15%	11.15%	
	3100	Total: Mechanical & Electrical		30	41.50	60.50	21.50%	25.50%	28.50%	

50 17 00 | S.F. Costs

		50 17 00 | S.F. Costs	UNIT	UNIT COSTS 1/4	UNIT COSTS MEDIAN	UNIT COSTS 3/4	% OF TOTAL 1/4	% OF TOTAL MEDIAN	% OF TOTAL 3/4	
41	0010	**GARAGES, PARKING**	S.F.	34	50	85.50				41
	0020	Total project costs	C.F.	3.21	4.36	6.35				
	2720	Plumbing	S.F.	.97	1.50	2.32	1.72%	2.70%	3.85%	
	2900	Electrical		1.87	2.30	3.61	4.33%	5.20%	6.30%	
	3100	Total: Mechanical & Electrical	↓	3.84	5.35	6.65	7%	8.90%	11.05%	
	3200									
	9000	Per car, total cost	Car	14,400	18,100	23,100				
43	0010	**GYMNASIUMS**	S.F.	95.50	127	172				43
	0020	Total project costs	C.F.	4.74	6.45	7.90				
	1800	Equipment	S.F.	2.26	4.24	8.15	1.81%	3.30%	6.70%	
	2720	Plumbing		6	7.15	9.20	4.65%	6.40%	7.75%	
	2770	Heating, ventilating, air conditioning		6.45	9.85	19.80	5.15%	9.05%	11.10%	
	2900	Electrical		7.30	9.95	13.10	6.60%	8.30%	10.30%	
	3100	Total: Mechanical & Electrical	↓	26.50	36	43	19.75%	23.50%	29%	
	3500	See also division 11480 (MF2004 11 67 00)								
46	0010	**HOSPITALS**	S.F.	182	225	310				46
	0020	Total project costs	C.F.	13.85	17.20	24.50				
	1800	Equipment	S.F.	4.63	8.90	15.35	.96%	2.63%	5%	
	2720	Plumbing		15.70	22	28.50	7.60%	9.10%	10.85%	
	2770	Heating, ventilating, air conditioning		23	29	40	7.80%	12.95%	16.65%	
	2900	Electrical		20	27	38	9.90%	11.75%	14%	
	3100	Total: Mechanical & Electrical	↓	56	80.50	122	27%	33.50%	36.50%	
	9000	Per bed or person, total cost	Bed	211,500	291,000	335,500				
	9900	See also division 11700 (MF2004 11 71 00)								
48	0010	**HOUSING For the Elderly**	S.F.	89.50	114	140				48
	0020	Total project costs	C.F.	6.40	8.90	11.35				
	0100	Site work	S.F.	6.25	9.70	14.20	5.05%	7.90%	12.10%	
	0500	Masonry		2.74	10.20	14.95	1.30%	6.05%	11%	
	1800	Equipment		2.17	2.99	4.76	1.88%	3.23%	4.43%	
	2510	Conveying systems		2.19	2.93	3.98	1.78%	2.20%	2.81%	
	2720	Plumbing		6.65	8.50	10.70	8.15%	9.55%	10.50%	
	2730	Heating, ventilating, air conditioning		3.42	4.84	7.25	3.30%	5.60%	7.25%	
	2900	Electrical		6.70	9.10	11.65	7.30%	8.50%	10.25%	
	3100	Total: Mechanical & Electrical	↓	23	27.50	36.50	18.10%	22.50%	29%	
	9000	Per rental unit, total cost	Unit	83,500	97,500	108,500				
	9500	Total: Mechanical & Electrical	"	18,600	21,400	25,000				
50	0010	**HOUSING Public (Low Rise)**	S.F.	75.50	105	136				50
	0020	Total project costs	C.F.	6.70	8.40	10.40				
	0100	Site work	S.F.	9.60	13.85	22.50	8.35%	11.75%	16.50%	
	1800	Equipment		2.05	3.35	5.10	2.26%	3.03%	4.24%	
	2720	Plumbing		5.45	7.20	9.10	7.15%	9.05%	11.60%	
	2730	Heating, ventilating, air conditioning		2.73	5.30	5.80	4.26%	6.05%	6.45%	
	2900	Electrical		4.56	6.80	9.45	5.10%	6.55%	8.25%	
	3100	Total: Mechanical & Electrical	↓	21.50	28	31	14.50%	17.55%	26.50%	
	9000	Per apartment, total cost	Apt.	83,000	94,000	118,500				
	9500	Total: Mechanical & Electrical	"	17,700	21,800	24,100				
51	0010	**ICE SKATING RINKS**	S.F.	64.50	151	166				51
	0020	Total project costs	C.F.	4.74	4.85	5.60				
	2720	Plumbing	S.F.	2.41	4.52	4.62	3.12%	3.23%	5.65%	
	2900	Electrical		6.90	10.60	11.20	6.30%	10.15%	15.05%	
	3100	Total: Mechanical & Electrical	↓	11.45	16.20	20	18.95%	18.95%	18.95%	
52	0010	**JAILS**	S.F.	196	254	325				52
	0020	Total project costs	C.F.	17.70	24.50	29				
	1800	Equipment	S.F.	7.65	22.50	38.50	2.80%	5.55%	10.35%	
	2720	Plumbing		18.90	25.50	33.50	7%	8.90%	13.35%	
	2770	Heating, ventilating, air conditioning		17.70	23.50	45.50	7.50%	9.45%	17.75%	
	2900	Electrical	↓	20.50	27	34.50	8.20%	11.55%	14.95%	

	50 17 00 | S.F. Costs		UNIT	UNIT COSTS 1/4	UNIT COSTS MEDIAN	UNIT COSTS 3/4	% OF TOTAL 1/4	% OF TOTAL MEDIAN	% OF TOTAL 3/4	
52	3100	Total: Mechanical & Electrical	S.F.	53.50	98	116	28%	30%	34%	52
53	0010	**LIBRARIES**	S.F.	124	159	207				53
	0020	Total project costs	C.F.	8.50	10.65	13.60				
	0500	Masonry	S.F.	9.95	17.35	29	5.80%	7.80%	11.80%	
	1800	Equipment		1.70	4.58	6.90	.37%	1.50%	4.07%	
	2720	Plumbing		4.58	6.65	9.05	3.38%	4.60%	5.70%	
	2770	Heating, ventilating, air conditioning		10.15	17.25	22.50	7.80%	10.95%	12.80%	
	2900	Electrical		12.65	16.40	21	8.30%	10.25%	11.95%	
	3100	Total: Mechanical & Electrical		37.50	48	59	20.50%	23%	26.50%	
54	0010	**LIVING, ASSISTED**	S.F.	115	135	159				54
	0020	Total project costs	C.F.	9.65	11.30	12.80				
	0500	Masonry	S.F.	3.38	4.02	4.73	2.37%	3.16%	3.86%	
	1800	Equipment		2.61	3.03	3.89	2.12%	2.45%	2.66%	
	2720	Plumbing		9.60	12.90	13.35	6.05%	8.15%	10.60%	
	2770	Heating, ventilating, air conditioning		11.40	11.95	13.10	7.95%	9.35%	9.70%	
	2900	Electrical		11.25	12.40	14.35	9%	10%	10.70%	
	3100	Total: Mechanical & Electrical		31.50	37	42.50	26%	29%	31.50%	
55	0010	**MEDICAL CLINICS**	S.F.	117	144	183				55
	0020	Total project costs	C.F.	8.55	11.05	14.70				
	1800	Equipment	S.F.	3.16	6.65	10.35	1.05%	2.94%	6.35%	
	2720	Plumbing		7.75	10.95	14.60	6.15%	8.40%	10.10%	
	2770	Heating, ventilating, air conditioning		9.25	12.15	17.85	6.65%	8.85%	11.35%	
	2900	Electrical		10.05	14.25	18.65	8.10%	10%	12.25%	
	3100	Total: Mechanical & Electrical		32	43.50	59.50	22.50%	27%	33.50%	
	3500	See also division 11700 (MF2004 11 71 00)								
57	0010	**MEDICAL OFFICES**	S.F.	110	136	167				57
	0020	Total project costs	C.F.	8.20	11.10	15				
	1800	Equipment	S.F.	3.63	7.15	10.20	.70%	5.10%	7.05%	
	2720	Plumbing		6.05	9.35	12.60	5.60%	6.80%	8.50%	
	2770	Heating, ventilating, air conditioning		7.35	10.60	14	6.10%	8%	9.70%	
	2900	Electrical		8.95	12.80	17.85	7.60%	9.80%	11.70%	
	3100	Total: Mechanical & Electrical		24	34	49.50	19.30%	23%	28.50%	
59	0010	**MOTELS**	S.F.	69.50	100	131				59
	0020	Total project costs	C.F.	6.15	8.25	13.50				
	2720	Plumbing	S.F.	7.05	8.95	10.65	9.45%	10.60%	12.55%	
	2770	Heating, ventilating, air conditioning		4.28	6.40	11.45	5.60%	5.60%	10%	
	2900	Electrical		6.55	8.30	10.30	7.45%	9.05%	10.45%	
	3100	Total: Mechanical & Electrical		21	28	48	18.50%	24%	25.50%	
	5000									
	9000	Per rental unit, total cost	Unit	35,300	67,000	72,500				
	9500	Total: Mechanical & Electrical	"	6,875	10,400	12,100				
60	0010	**NURSING HOMES**	S.F.	109	140	172				60
	0020	Total project costs	C.F.	8.55	10.65	14.60				
	1800	Equipment	S.F.	3.42	4.54	7.55	2.02%	3.62%	4.99%	
	2720	Plumbing		9.30	14.20	17	8.75%	10.10%	12.70%	
	2770	Heating, ventilating, air conditioning		9.80	14.90	19.75	9.70%	11.45%	11.80%	
	2900	Electrical		10.75	13.45	18.30	9.40%	10.55%	12.45%	
	3100	Total: Mechanical & Electrical		25.50	36	60	26%	29.50%	30.50%	
	9000	Per bed or person, total cost	Bed	48,200	60,500	78,000				
61	0010	**OFFICES Low Rise (1 to 4 story)**	S.F.	91.50	119	154				61
	0020	Total project costs	C.F.	6.55	9	11.85				
	0100	Site work	S.F.	7.35	12.70	19.15	6.20%	9.70%	13.60%	
	0500	Masonry		3.54	7.25	13.05	2.62%	5.50%	8.60%	
	1800	Equipment		.97	1.91	5.20	.62%	1.50%	3.51%	
	2720	Plumbing		3.27	5.05	7.40	3.66%	4.50%	6.10%	
	2770	Heating, ventilating, air conditioning		7.25	10.10	14.75	7.20%	10.30%	11.70%	
	2900	Electrical		7.50	10.65	15.10	7.45%	9.60%	11.35%	

	50 17 00 | S.F. Costs		UNIT	UNIT COSTS 1/4	UNIT COSTS MEDIAN	UNIT COSTS 3/4	% OF TOTAL 1/4	% OF TOTAL MEDIAN	% OF TOTAL 3/4	
61	3100	Total: Mechanical & Electrical	S.F.	20.50	28	41	18.15%	22%	27%	61
62	0010	**OFFICES Mid Rise (5 to 10 story)**	S.F.	97	117	155				62
	0020	Total project costs	C.F.	6.85	8.75	12.45				
	2720	Plumbing	S.F.	2.93	4.54	6.55	2.83%	3.74%	4.50%	
	2770	Heating, ventilating, air conditioning		7.35	10.55	16.80	7.65%	9.40%	11%	
	2900	Electrical		7.20	9.35	12.80	6.35%	7.80%	10%	
	3100	Total: Mechanical & Electrical		18.65	24	45.50	19.15%	21.50%	27.50%	
63	0010	**OFFICES High Rise (11 to 20 story)**	S.F.	119	150	185				63
	0020	Total project costs	C.F.	8.30	10.40	14.95				
	2900	Electrical	S.F.	7.25	8.85	13.10	5.80%	7.85%	10.50%	
	3100	Total: Mechanical & Electrical		23.50	31.50	52.50	16.90%	23.50%	34%	
64	0010	**POLICE STATIONS**	S.F.	143	188	237				64
	0020	Total project costs	C.F.	11.35	13.95	19.10				
	0500	Masonry	S.F.	13.40	23.50	29.50	6.70%	9.10%	11.35%	
	1800	Equipment		2.27	9.95	15.80	.98%	3.35%	6.70%	
	2720	Plumbing		8	15.95	19.85	5.65%	6.90%	10.75%	
	2770	Heating, ventilating, air conditioning		12.50	16.60	22.50	5.85%	10.55%	11.70%	
	2900	Electrical		15.60	22.50	29.50	9.80%	11.85%	14.80%	
	3100	Total: Mechanical & Electrical		51	61.50	83.50	25%	31.50%	32.50%	
65	0010	**POST OFFICES**	S.F.	113	139	177				65
	0020	Total project costs	C.F.	6.80	8.60	9.75				
	2720	Plumbing	S.F.	5.10	6.30	7.95	4.24%	5.30%	5.60%	
	2770	Heating, ventilating, air conditioning		7.95	9.80	10.90	6.65%	7.15%	9.35%	
	2900	Electrical		9.30	13.10	15.55	7.25%	9%	11%	
	3100	Total: Mechanical & Electrical		27	35	40	16.25%	18.80%	22%	
66	0010	**POWER PLANTS**	S.F.	780	1,000	1,900				66
	0020	Total project costs	C.F.	21.50	47	100				
	2900	Electrical	S.F.	55.50	117	175	9.30%	12.75%	21.50%	
	8100	Total: Mechanical & Electrical		138	450	1,000	32.50%	32.50%	52.50%	
67	0010	**RELIGIOUS EDUCATION**	S.F.	91	119	146				67
	0020	Total project costs	C.F.	5.05	7.20	9.05				
	2720	Plumbing	S.F.	3.80	5.40	7.65	4.40%	5.30%	7.10%	
	2770	Heating, ventilating, air conditioning		9.60	10.85	15.35	10.05%	11.45%	12.35%	
	2900	Electrical		7.20	10.15	13.45	7.60%	9.05%	10.35%	
	3100	Total: Mechanical & Electrical		29.50	38.50	47	22%	23%	27%	
69	0010	**RESEARCH Laboratories & Facilities**	S.F.	136	195	284				69
	0020	Total project costs	C.F.	10.25	19.80	24				
	1800	Equipment	S.F.	6.05	11.90	29	.94%	4.58%	9.10%	
	2720	Plumbing		13.65	17.45	28	6.15%	8.30%	10.80%	
	2770	Heating, ventilating, air conditioning		12.25	41	48.50	7.25%	16.50%	17.50%	
	2900	Electrical		15.90	26	43	9.45%	11.15%	15.40%	
	3100	Total: Mechanical & Electrical		52	90.50	131	31.50%	37.50%	42%	
70	0010	**RESTAURANTS**	S.F.	132	170	221				70
	0020	Total project costs	C.F.	11.10	14.55	19.10				
	1800	Equipment	S.F.	8.50	21	31.50	6.10%	13%	15.65%	
	2720	Plumbing		10.45	12.65	16.65	6.10%	8.15%	9%	
	2770	Heating, ventilating, air conditioning		13.25	18.40	22.50	9.20%	12%	12.40%	
	2900	Electrical		14	17.20	22.50	8.35%	10.55%	11.55%	
	3100	Total: Mechanical & Electrical		43.50	46.50	60	21%	24.50%	29.50%	
	9000	Per seat unit, total cost	Seat	4,850	6,450	7,625				
	9500	Total: Mechanical & Electrical	"	1,225	1,600	1,900				
72	0010	**RETAIL STORES**	S.F.	61.50	82.50	110				72
	0020	Total project costs	C.F.	4.17	5.95	8.25				
	2720	Plumbing	S.F.	2.23	3.73	6.35	3.26%	4.60%	6.80%	
	2770	Heating, ventilating, air conditioning		4.82	6.60	9.90	6.75%	8.75%	10.15%	
	2900	Electrical		5.55	7.60	10.95	7.25%	9.90%	11.65%	
	3100	Total: Mechanical & Electrical		14.75	18.95	26	17.05%	21%	23.50%	

50 17 00 | S.F. Costs

		50 17 00 | S.F. Costs	UNIT	UNIT COSTS 1/4	UNIT COSTS MEDIAN	UNIT COSTS 3/4	% OF TOTAL 1/4	% OF TOTAL MEDIAN	% OF TOTAL 3/4	
74	0010	**SCHOOLS Elementary**	S.F.	99.50	123	152				74
	0020	Total project costs	C.F.	6.60	8.40	10.90				
	0500	Masonry	S.F.	9	15.50	23	5.80%	11%	14.95%	
	1800	Equipment		2.75	4.68	8.65	1.89%	3.32%	4.71%	
	2720	Plumbing		5.80	8.15	10.90	5.70%	7.15%	9.35%	
	2730	Heating, ventilating, air conditioning		8.65	13.80	19.25	8.15%	10.80%	14.90%	
	2900	Electrical		9.45	12.50	15.75	8.40%	10.05%	11.70%	
	3100	Total: Mechanical & Electrical	↓	33.50	42.50	52	25%	27.50%	30%	
	9000	Per pupil, total cost	Ea.	11,500	17,100	49,500				
	9500	Total: Mechanical & Electrical	"	3,250	4,125	14,300				
76	0010	**SCHOOLS Junior High & Middle**	S.F.	104	127	155				76
	0020	Total project costs	C.F.	6.60	8.50	9.55				
	0500	Masonry	S.F.	13.25	17.40	20.50	8.60%	11.60%	14.35%	
	1800	Equipment		3.31	5.35	8.05	1.79%	3.09%	4.86%	
	2720	Plumbing		6.05	7.45	9.25	5.30%	6.80%	7.25%	
	2770	Heating, ventilating, air conditioning		12.05	14.65	26	8.90%	11.55%	14.20%	
	2900	Electrical		10.15	12.25	15.80	7.90%	9.35%	10.60%	
	3100	Total: Mechanical & Electrical	↓	33	42.50	52.50	23.50%	27%	29.50%	
	9000	Per pupil, total cost	Ea.	13,100	17,200	23,200				
78	0010	**SCHOOLS Senior High**	S.F.	107	132	165				78
	0020	Total project costs	C.F.	6.50	9.60	15.40				
	1800	Equipment	S.F.	2.83	6.65	9.40	1.86%	2.91%	4.30%	
	2720	Plumbing		6.05	9.05	16.60	5.60%	6.90%	8.30%	
	2770	Heating, ventilating, air conditioning		12.35	14.15	27	8.95%	11.60%	15%	
	2900	Electrical		10.85	14.05	20.50	8.65%	10.15%	12.35%	
	3100	Total: Mechanical & Electrical	↓	36	42	70.50	23.50%	26.50%	29%	
	9000	Per pupil, total cost	Ea.	10,200	20,700	25,900				
80	0010	**SCHOOLS Vocational**	S.F.	87.50	127	157				80
	0020	Total project costs	C.F.	5.40	7.80	10.75				
	0500	Masonry	S.F.	5.10	12.65	19.35	3.53%	6.70%	10.95%	
	1800	Equipment		2.73	6.80	9.45	1.24%	3.10%	4.26%	
	2720	Plumbing		5.60	8.30	12.25	5.40%	6.90%	8.55%	
	2770	Heating, ventilating, air conditioning		7.80	14.55	24.50	8.60%	11.90%	14.65%	
	2900	Electrical		9.10	11.90	16.35	8.45%	10.95%	13.20%	
	3100	Total: Mechanical & Electrical	↓	31.50	35	60	23.50%	27.50%	31%	
	9000	Per pupil, total cost	Ea.	12,200	32,500	48,500				
83	0010	**SPORTS ARENAS**	S.F.	76.50	102	157				83
	0020	Total project costs	C.F.	4.14	7.40	9.55				
	2720	Plumbing	S.F.	4.42	6.70	14.15	4.35%	6.35%	9.40%	
	2770	Heating, ventilating, air conditioning		9.50	11.25	15.65	8.80%	10.20%	13.55%	
	2900	Electrical		7.95	10.80	13.95	8.60%	9.90%	12.25%	
	3100	Total: Mechanical & Electrical	↓	19.75	34.50	46.50	21.50%	25%	27.50%	
85	0010	**SUPERMARKETS**	S.F.	70.50	81.50	95.50				85
	0020	Total project costs	C.F.	3.92	4.74	7.20				
	2720	Plumbing	S.F.	3.93	4.96	5.75	5.40%	6%	7.45%	
	2770	Heating, ventilating, air conditioning		5.80	7.70	9.40	8.60%	8.65%	9.60%	
	2900	Electrical		8.80	10.15	12	10.40%	12.45%	13.60%	
	3100	Total: Mechanical & Electrical	↓	22.50	24.50	32	20.50%	26.50%	31%	
86	0010	**SWIMMING POOLS**	S.F.	114	191	405				86
	0020	Total project costs	C.F.	9.15	11.40	12.45				
	2720	Plumbing	S.F.	10.55	12.05	16.30	4.80%	9.70%	20.50%	
	2900	Electrical		8.55	13.90	20	5.75%	6.95%	7.60%	
	3100	Total: Mechanical & Electrical	↓	21	54	72.50	11.15%	14.10%	23.50%	
87	0010	**TELEPHONE EXCHANGES**	S.F.	151	222	281				87
	0020	Total project costs	C.F.	9.40	15.10	21				
	2720	Plumbing	S.F.	6.40	9.90	14.45	4.52%	5.80%	6.90%	
	2770	Heating, ventilating, air conditioning	↓	14.85	30	37	11.80%	16.05%	18.40%	

50 17 00 | S.F. Costs

			UNIT	UNIT COSTS 1/4	UNIT COSTS MEDIAN	UNIT COSTS 3/4	% OF TOTAL 1/4	% OF TOTAL MEDIAN	% OF TOTAL 3/4	
87	2900	Electrical	S.F.	15.40	24.50	43.50	10.90%	14%	17.85%	87
	3100	Total: Mechanical & Electrical	↓	45.50	86.50	122	29.50%	33.50%	44.50%	
91	0010	**THEATERS**	S.F.	95.50	118	180				91
	0020	Total project costs	C.F.	4.40	6.50	9.55				
	2720	Plumbing	S.F.	3.18	3.44	14.05	2.92%	4.70%	6.80%	
	2770	Heating, ventilating, air conditioning		9.25	11.20	13.85	8%	12.25%	13.40%	
	2900	Electrical		8.35	11.25	23	8.05%	9.35%	12.25%	
	3100	Total: Mechanical & Electrical	↓	21.50	29	34	21.50%	25.50%	27.50%	
94	0010	**TOWN HALLS City Halls & Municipal Buildings**	S.F.	106	135	176				94
	0020	Total project costs	C.F.	9.70	11.65	16.35				
	2720	Plumbing	S.F.	4.44	8.30	15.30	4.31%	5.95%	7.95%	
	2770	Heating, ventilating, air conditioning		8	15.95	23.50	7.05%	9.05%	13.45%	
	2900	Electrical		10.10	14.40	19.65	8.05%	9.45%	11.65%	
	3100	Total: Mechanical & Electrical	↓	35	44.50	68	22%	26.50%	31%	
97	0010	**WAREHOUSES & Storage Buildings**	S.F.	40	59	85				97
	0020	Total project costs	C.F.	2.08	3.25	5.40				
	0100	Site work	S.F.	4.10	8.15	12.30	6.05%	12.95%	19.85%	
	0500	Masonry		2.38	5.65	12.15	3.73%	7.40%	12.30%	
	1800	Equipment		.64	1.37	7.70	.91%	1.82%	5.55%	
	2720	Plumbing		1.32	2.37	4.44	2.90%	4.80%	6.55%	
	2730	Heating, ventilating, air conditioning		1.51	4.26	5.70	2.41%	5%	8.90%	
	2900	Electrical		2.35	4.42	7.30	5.15%	7.20%	10.10%	
	3100	Total: Mechanical & Electrical	↓	6.55	10.05	19.80	12.75%	18.90%	26%	
99	0010	**WAREHOUSE & OFFICES Combination**	S.F.	49	65	89				99
	0020	Total project costs	C.F.	2.50	3.62	5.35				
	1800	Equipment	S.F.	.85	1.64	2.43	.52%	1.20%	2.40%	
	2720	Plumbing		1.89	3.34	4.89	3.74%	4.76%	6.30%	
	2770	Heating, ventilating, air conditioning		2.98	4.65	6.50	5%	5.65%	10.05%	
	2900	Electrical		3.27	4.86	7.65	5.75%	8%	10%	
	3100	Total: Mechanical & Electrical	↓	9.10	14.05	22	14.40%	19.95%	24.50%	

Square Foot Project Size Modifier

One factor that affects the S.F. cost of a particular building is the size. In general, for buildings built to the same specifications in the same locality, the larger building will have the lower S.F. cost. This is due mainly to the decreasing contribution of the exterior walls plus the economy of scale usually achievable in larger buildings. The Area Conversion Scale shown below will give a factor to convert costs for the typical size building to an adjusted cost for the particular project.

The Square Foot Base Size lists the median costs, most typical project size in our accumulated data, and the range in size of the projects.

The Size Factor for your project is determined by dividing your project area in S.F. by the typical project size for the particular Building Type. With this factor, enter the Area Conversion Scale at the appropriate Size Factor and determine the appropriate cost multiplier for your building size.

Example: Determine the cost per S.F. for a 100,000 S.F. Mid-rise apartment building.

$$\frac{\text{Proposed building area} = 100{,}000 \text{ S.F.}}{\text{Typical size from below} = 50{,}000 \text{ S.F.}} = 2.00$$

Enter Area Conversion scale at 2.0, intersect curve, read horizontally the appropriate cost multiplier of .94. Size adjusted cost becomes .94 x $106.00 = $100.00 based on national average costs.

Note: For Size Factors less than .50, the Cost Multiplier is 1.1
For Size Factors greater than 3.5, the Cost Multiplier is .90

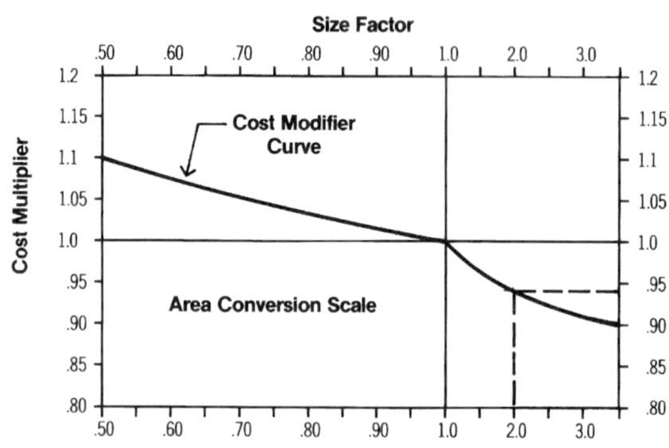

Square Foot Base Size

Building Type	Median Cost per S.F.	Typical Size Gross S.F.	Typical Range Gross S.F.	Building Type	Median Cost per S.F.	Typical Size Gross S.F.	Typical Range Gross S.F.
Apartments, Low Rise	$ 83.50	21,000	9,700 - 37,200	Jails	$ 254.00	40,000	5,500 - 145,000
Apartments, Mid Rise	106.00	50,000	32,000 - 100,000	Libraries	159.00	12,000	7,000 - 31,000
Apartments, High Rise	115.00	145,000	95,000 - 600,000	Living, Assisted	135.00	32,300	23,500 - 50,300
Auditoriums	140.00	25,000	7,600 - 39,000	Medical Clinics	144.00	7,200	4,200 - 15,700
Auto Sales	104.00	20,000	10,800 - 28,600	Medical Offices	136.00	6,000	4,000 - 15,000
Banks	187.00	4,200	2,500 - 7,500	Motels	100.00	40,000	15,800 - 120,000
Churches	128.00	17,000	2,000 - 42,000	Nursing Homes	140.00	23,000	15,000 - 37,000
Clubs, Country	131.00	6,500	4,500 - 15,000	Offices, Low Rise	119.00	20,000	5,000 - 80,000
Clubs, Social	124.00	10,000	6,000 - 13,500	Offices, Mid Rise	117.00	120,000	20,000 - 300,000
Clubs, YMCA	141.00	28,300	12,800 - 39,400	Offices, High Rise	150.00	260,000	120,000 - 800,000
Colleges (Class)	150.00	50,000	15,000 - 150,000	Police Stations	188.00	10,500	4,000 - 19,000
Colleges (Science Lab)	238.00	45,600	16,600 - 80,000	Post Offices	139.00	12,400	6,800 - 30,000
College (Student Union)	176.00	33,400	16,000 - 85,000	Power Plants	1000.00	7,500	1,000 - 20,000
Community Center	132.00	9,400	5,300 - 16,700	Religious Education	119.00	9,000	6,000 - 12,000
Court Houses	178.00	32,400	17,800 - 106,000	Research	195.00	19,000	6,300 - 45,000
Dept. Stores	77.50	90,000	44,000 - 122,000	Restaurants	170.00	4,400	2,800 - 6,000
Dormitories, Low Rise	141.00	25,000	10,000 - 95,000	Retail Stores	82.50	7,200	4,000 - 17,600
Dormitories, Mid Rise	174.00	85,000	20,000 - 200,000	Schools, Elementary	123.00	41,000	24,500 - 55,000
Factories	75.00	26,400	12,900 - 50,000	Schools, Jr. High	127.00	92,000	52,000 - 119,000
Fire Stations	138.00	5,800	4,000 - 8,700	Schools, Sr. High	132.00	101,000	50,500 - 175,000
Fraternity Houses	129.00	12,500	8,200 - 14,800	Schools, Vocational	127.00	37,000	20,500 - 82,000
Funeral Homes	144.00	10,000	4,000 - 20,000	Sports Arenas	102.00	15,000	5,000 - 40,000
Garages, Commercial	92.50	9,300	5,000 - 13,600	Supermarkets	81.50	44,000	12,000 - 60,000
Garages, Municipal	117.00	8,300	4,500 - 12,600	Swimming Pools	191.00	20,000	10,000 - 32,000
Garages, Parking	50.00	163,000	76,400 - 225,300	Telephone Exchange	222.00	4,500	1,200 - 10,600
Gymnasiums	127.00	19,200	11,600 - 41,000	Theaters	118.00	10,500	8,800 - 17,500
Hospitals	225.00	55,000	27,200 - 125,000	Town Halls	135.00	10,800	4,800 - 23,400
House (Elderly)	114.00	37,000	21,000 - 66,000	Warehouses	59.00	25,000	8,000 - 72,000
Housing (Public)	105.00	36,000	14,400 - 74,400	Warehouse & Office	65.00	25,000	8,000 - 72,000
Ice Rinks	151.00	29,000	27,200 - 33,600				

Abbreviations

A	Area Square Feet; Ampere
ABS	Acrylonitrile Butadiene Stryrene; Asbestos Bonded Steel
A.C.	Alternating Current; Air-Conditioning; Asbestos Cement; Plywood Grade A & C
A.C.I.	American Concrete Institute
AD	Plywood, Grade A & D
Addit.	Additional
Adj.	Adjustable
af	Audio-frequency
A.G.A.	American Gas Association
Agg.	Aggregate
A.H.	Ampere Hours
A hr.	Ampere-hour
A.H.U.	Air Handling Unit
A.I.A.	American Institute of Architects
AIC	Ampere Interrupting Capacity
Allow.	Allowance
alt.	Altitude
Alum.	Aluminum
a.m.	Ante Meridiem
Amp.	Ampere
Anod.	Anodized
Approx.	Approximate
Apt.	Apartment
Asb.	Asbestos
A.S.B.C.	American Standard Building Code
Asbe.	Asbestos Worker
ASCE.	American Society of Civil Engineers
A.S.H.R.A.E.	American Society of Heating, Refrig. & AC Engineers
A.S.M.E.	American Society of Mechanical Engineers
A.S.T.M.	American Society for Testing and Materials
Attchmt.	Attachment
Avg., Ave.	Average
A.W.G.	American Wire Gauge
AWWA	American Water Works Assoc.
Bbl.	Barrel
B&B	Grade B and Better; Balled & Burlapped
B.&S.	Bell and Spigot
B.&W.	Black and White
b.c.c.	Body-centered Cubic
B.C.Y.	Bank Cubic Yards
BE	Bevel End
B.F.	Board Feet
Bg. cem.	Bag of Cement
BHP	Boiler Horsepower; Brake Horsepower
B.I.	Black Iron
Bit., Bitum.	Bituminous
Bit., Conc.	Bituminous Concrete
Bk.	Backed
Bkrs.	Breakers
Bldg.	Building
Blk.	Block
Bm.	Beam
Boil.	Boilermaker
B.P.M.	Blows per Minute
BR	Bedroom
Brg.	Bearing
Brhe.	Bricklayer Helper
Bric.	Bricklayer
Brk.	Brick
Brng.	Bearing
Brs.	Brass
Brz.	Bronze
Bsn.	Basin
Btr.	Better
Btu	British Thermal Unit

BTUH	BTU per Hour
B.U.R.	Built-up Roofing
BX	Interlocked Armored Cable
°C	degree centegrade
c	Conductivity, Copper Sweat
C	Hundred; Centigrade
C/C	Center to Center, Cedar on Cedar
C-C	Center to Center
Cab.	Cabinet
Cair.	Air Tool Laborer
Calc	Calculated
Cap.	Capacity
Carp.	Carpenter
C.B.	Circuit Breaker
C.C.A.	Chromate Copper Arsenate
C.C.F.	Hundred Cubic Feet
cd	Candela
cd/sf	Candela per Square Foot
CD	Grade of Plywood Face & Back
CDX	Plywood, Grade C & D, exterior glue
Cefi.	Cement Finisher
Cem.	Cement
CF	Hundred Feet
C.F.	Cubic Feet
CFM	Cubic Feet per Minute
c.g.	Center of Gravity
CHW	Chilled Water; Commercial Hot Water
C.I.	Cast Iron
C.I.P.	Cast in Place
Circ.	Circuit
C.L.	Carload Lot
Clab.	Common Laborer
Clam	Common maintenance laborer
C.L.F.	Hundred Linear Feet
CLF	Current Limiting Fuse
CLP	Cross Linked Polyethylene
cm	Centimeter
CMP	Corr. Metal Pipe
C.M.U.	Concrete Masonry Unit
CN	Change Notice
Col.	Column
CO_2	Carbon Dioxide
Comb.	Combination
Compr.	Compressor
Conc.	Concrete
Cont.	Continuous; Continued, Container
Corr.	Corrugated
Cos	Cosine
Cot	Cotangent
Cov.	Cover
C/P	Cedar on Paneling
CPA	Control Point Adjustment
Cplg.	Coupling
C.P.M.	Critical Path Method
CPVC	Chlorinated Polyvinyl Chloride
C.Pr.	Hundred Pair
CRC	Cold Rolled Channel
Creos.	Creosote
Crpt.	Carpet & Linoleum Layer
CRT	Cathode-ray Tube
CS	Carbon Steel, Constant Shear Bar Joist
Csc	Cosecant
C.S.F.	Hundred Square Feet
CSI	Construction Specifications Institute
C.T.	Current Transformer
CTS	Copper Tube Size
Cu	Copper, Cubic
Cu. Ft.	Cubic Foot
cw	Continuous Wave
C.W.	Cool White; Cold Water

Cwt.	100 Pounds
C.W.X.	Cool White Deluxe
C.Y.	Cubic Yard (27 cubic feet)
C.Y./Hr.	Cubic Yard per Hour
Cyl.	Cylinder
d	Penny (nail size)
D	Deep; Depth; Discharge
Dis., Disch.	Discharge
Db.	Decibel
Dbl.	Double
DC	Direct Current
DDC	Direct Digital Control
Demob.	Demobilization
d.f.u.	Drainage Fixture Units
D.H.	Double Hung
DHW	Domestic Hot Water
DI	Ductile Iron
Diag.	Diagonal
Diam., Dia	Diameter
Distrib.	Distribution
Div.	Division
Dk.	Deck
D.L.	Dead Load; Diesel
DLH	Deep Long Span Bar Joist
Do.	Ditto
Dp.	Depth
D.P.S.T.	Double Pole, Single Throw
Dr.	Drive
Drink.	Drinking
D.S.	Double Strength
D.S.A.	Double Strength A Grade
D.S.B.	Double Strength B Grade
Dty.	Duty
DWV	Drain Waste Vent
DX	Deluxe White, Direct Expansion
dyn	Dyne
e	Eccentricity
E	Equipment Only; East
Ea.	Each
E.B.	Encased Burial
Econ.	Economy
E.C.Y	Embankment Cubic Yards
EDP	Electronic Data Processing
EIFS	Exterior Insulation Finish System
E.D.R.	Equiv. Direct Radiation
Eq.	Equation
EL	elevation
Elec.	Electrician; Electrical
Elev.	Elevator; Elevating
EMT	Electrical Metallic Conduit; Thin Wall Conduit
Eng.	Engine, Engineered
EPDM	Ethylene Propylene Diene Monomer
EPS	Expanded Polystyrene
Eqhv.	Equip. Oper., Heavy
Eqlt.	Equip. Oper., Light
Eqmd.	Equip. Oper., Medium
Eqmm.	Equip. Oper., Master Mechanic
Eqol.	Equip. Oper., Oilers
Equip.	Equipment
ERW	Electric Resistance Welded
E.S.	Energy Saver
Est.	Estimated
esu	Electrostatic Units
E.W.	Each Way
EWT	Entering Water Temperature
Excav.	Excavation
Exp.	Expansion, Exposure
Ext.	Exterior
Extru.	Extrusion
f.	Fiber stress
F	Fahrenheit; Female; Fill
Fab.	Fabricated

Abbreviations

FBGS	Fiberglass
F.C.	Footcandles
f.c.c.	Face-centered Cubic
f'c.	Compressive Stress in Concrete; Extreme Compressive Stress
F.E.	Front End
FEP	Fluorinated Ethylene Propylene (Teflon)
F.G.	Flat Grain
F.H.A.	Federal Housing Administration
Fig.	Figure
Fin.	Finished
Fixt.	Fixture
Fl. Oz.	Fluid Ounces
Flr.	Floor
F.M.	Frequency Modulation; Factory Mutual
Fmg.	Framing
Fdn.	Foundation
Fori.	Foreman, Inside
Foro.	Foreman, Outside
Fount.	Fountain
fpm	Feet per Minute
FPT	Female Pipe Thread
Fr.	Frame
F.R.	Fire Rating
FRK	Foil Reinforced Kraft
FRP	Fiberglass Reinforced Plastic
FS	Forged Steel
FSC	Cast Body; Cast Switch Box
Ft.	Foot; Feet
Ftng.	Fitting
Ftg.	Footing
Ft lb.	Foot Pound
Furn.	Furniture
FVNR	Full Voltage Non-Reversing
FXM	Female by Male
Fy.	Minimum Yield Stress of Steel
g	Gram
G	Gauss
Ga.	Gauge
Gal., gal.	Gallon
gpm, GPM	Gallon per Minute
Galv.	Galvanized
Gen.	General
G.F.I.	Ground Fault Interrupter
Glaz.	Glazier
GPD	Gallons per Day
GPH	Gallons per Hour
GPM	Gallons per Minute
GR	Grade
Gran.	Granular
Grnd.	Ground
H	High Henry
H.C.	High Capacity
H.D.	Heavy Duty; High Density
H.D.O.	High Density Overlaid
H.D.P.E.	high density polyethelene
Hdr.	Header
Hdwe.	Hardware
Help.	Helper Average
HEPA	High Efficiency Particulate Air Filter
Hg	Mercury
HIC	High Interrupting Capacity
HM	Hollow Metal
HMWPE	high molecular weight polyethylene
H.O.	High Output
Horiz.	Horizontal
H.P.	Horsepower; High Pressure
H.P.F.	High Power Factor
Hr.	Hour
Hrs./Day	Hours per Day
HSC	High Short Circuit
Ht.	Height
Htg.	Heating
Htrs.	Heaters
HVAC	Heating, Ventilation & Air-Conditioning
Hvy.	Heavy
HW	Hot Water
Hyd.;Hydr.	Hydraulic
Hz.	Hertz (cycles)
I.	Moment of Inertia
IBC	International Building Code
I.C.	Interrupting Capacity
ID	Inside Diameter
I.D.	Inside Dimension; Identification
I.F.	Inside Frosted
I.M.C.	Intermediate Metal Conduit
In.	Inch
Incan.	Incandescent
Incl.	Included; Including
Int.	Interior
Inst.	Installation
Insul.	Insulation/Insulated
I.P.	Iron Pipe
I.P.S.	Iron Pipe Size
I.P.T.	Iron Pipe Threaded
I.W.	Indirect Waste
J	Joule
J.I.C.	Joint Industrial Council
K	Thousand; Thousand Pounds; Heavy Wall Copper Tubing, Kelvin
K.A.H.	Thousand Amp. Hours
KCMIL	Thousand Circular Mils
KD	Knock Down
K.D.A.T.	Kiln Dried After Treatment
kg	Kilogram
kG	Kilogauss
kgf	Kilogram Force
kHz	Kilohertz
Kip.	1000 Pounds
KJ	Kiljoule
K.L.	Effective Length Factor
K.L.F.	Kips per Linear Foot
Km	Kilometer
K.S.F.	Kips per Square Foot
K.S.I.	Kips per Square Inch
kV	Kilovolt
kVA	Kilovolt Ampere
K.V.A.R.	Kilovar (Reactance)
KW	Kilowatt
KWh	Kilowatt-hour
L	Labor Only; Length; Long; Medium Wall Copper Tubing
Lab.	Labor
lat	Latitude
Lath.	Lather
Lav.	Lavatory
lb.; #	Pound
L.B.	Load Bearing; L Conduit Body
L. & E.	Labor & Equipment
lb./hr.	Pounds per Hour
lb./L.F.	Pounds per Linear Foot
lbf/sq.in.	Pound-force per Square Inch
L.C.L.	Less than Carload Lot
L.C.Y.	Loose Cubic Yard
Ld.	Load
LE	Lead Equivalent
LED	Light Emitting Diode
L.F.	Linear Foot
L.F. Nose	Linear Foot of Stair Nosing
L.F. Rsr	Linear Foot of Stair Riser
Lg.	Long; Length; Large
L & H	Light and Heat
LH	Long Span Bar Joist
L.H.	Labor Hours
L.L.	Live Load
L.L.D.	Lamp Lumen Depreciation
lm	Lumen
lm/sf	Lumen per Square Foot
lm/W	Lumen per Watt
L.O.A.	Length Over All
log	Logarithm
L-O-L	Lateralolet
long.	longitude
L.P.	Liquefied Petroleum; Low Pressure
L.P.F.	Low Power Factor
LR	Long Radius
L.S.	Lump Sum
Lt.	Light
Lt. Ga.	Light Gauge
L.T.L.	Less than Truckload Lot
Lt. Wt.	Lightweight
L.V.	Low Voltage
M	Thousand; Material; Male; Light Wall Copper Tubing
M^2CA	Meters Squared Contact Area
m/hr.; M.H.	Man-hour
mA	Milliampere
Mach.	Machine
Mag. Str.	Magnetic Starter
Maint.	Maintenance
Marb.	Marble Setter
Mat; Mat'l.	Material
Max.	Maximum
MBF	Thousand Board Feet
MBH	Thousand BTU's per hr.
MC	Metal Clad Cable
M.C.F.	Thousand Cubic Feet
M.C.F.M.	Thousand Cubic Feet per Minute
M.C.M.	Thousand Circular Mils
M.C.P.	Motor Circuit Protector
MD	Medium Duty
M.D.O.	Medium Density Overlaid
Med.	Medium
MF	Thousand Feet
M.F.B.M.	Thousand Feet Board Measure
Mfg.	Manufacturing
Mfrs.	Manufacturers
mg	Milligram
MGD	Million Gallons per Day
MGPH	Thousand Gallons per Hour
MH, M.H.	Manhole; Metal Halide; Man-Hour
MHz	Megahertz
Mi.	Mile
MI	Malleable Iron; Mineral Insulated
mm	Millimeter
Mill.	Millwright
Min., min.	Minimum, minute
Misc.	Miscellaneous
ml	Milliliter, Mainline
M.L.F.	Thousand Linear Feet
Mo.	Month
Mobil.	Mobilization
Mog.	Mogul Base
MPH	Miles per Hour
MPT	Male Pipe Thread

Abbreviations

MRT	Mile Round Trip
ms	Millisecond
M.S.F.	Thousand Square Feet
Mstz.	Mosaic & Terrazzo Worker
M.S.Y.	Thousand Square Yards
Mtd., mtd.	Mounted
Mthe.	Mosaic & Terrazzo Helper
Mtng.	Mounting
Mult.	Multi; Multiply
M.V.A.	Million Volt Amperes
M.V.A.R.	Million Volt Amperes Reactance
MV	Megavolt
MW	Megawatt
MXM	Male by Male
MYD	Thousand Yards
N	Natural; North
nA	Nanoampere
NA	Not Available; Not Applicable
N.B.C.	National Building Code
NC	Normally Closed
N.E.M.A.	National Electrical Manufacturers Assoc.
NEHB	Bolted Circuit Breaker to 600V.
N.L.B.	Non-Load-Bearing
NM	Non-Metallic Cable
nm	Nanometer
No.	Number
NO	Normally Open
N.O.C.	Not Otherwise Classified
Nose.	Nosing
N.P.T.	National Pipe Thread
NQOD	Combination Plug-on/Bolt on Circuit Breaker to 240V.
N.R.C.	Noise Reduction Coefficient/ Nuclear Regulator Commission
N.R.S.	Non Rising Stem
ns	Nanosecond
nW	Nanowatt
OB	Opposing Blade
OC	On Center
OD	Outside Diameter
O.D.	Outside Dimension
ODS	Overhead Distribution System
O.G.	Ogee
O.H.	Overhead
O&P	Overhead and Profit
Oper.	Operator
Opng.	Opening
Orna.	Ornamental
OSB	Oriented Strand Board
O.S.&Y.	Outside Screw and Yoke
Ovhd.	Overhead
OWG	Oil, Water or Gas
Oz.	Ounce
P.	Pole, Applied Load; Projection
p.	Page
Pape.	Paperhanger
P.A.P.R.	Powered Air Purifying Respirator
PAR	Parabolic Reflector
Pc., Pcs.	Piece, Pieces
P.C.	Portland Cement; Power Connector
P.C.F.	Pounds per Cubic Foot
P.C.M.	Phase Contrast Microscopy
P.E.	Professional Engineer; Porcelain Enamel; Polyethylene; Plain End
Perf.	Perforated
PEX	Cross linked polyethylene
Ph.	Phase
P.I.	Pressure Injected
Pile.	Pile Driver
Pkg.	Package
Pl.	Plate
Plah.	Plasterer Helper
Plas.	Plasterer
Pluh.	Plumbers Helper
Plum.	Plumber
Ply.	Plywood
p.m.	Post Meridiem
Pntd.	Painted
Pord.	Painter, Ordinary
pp	Pages
PP, PPL	Polypropylene
P.P.M.	Parts per Million
Pr.	Pair
P.E.S.B.	Pre-engineered Steel Building
Prefab.	Prefabricated
Prefin.	Prefinished
Prop.	Propelled
PSF, psf	Pounds per Square Foot
PSI, psi	Pounds per Square Inch
PSIG	Pounds per Square Inch Gauge
PSP	Plastic Sewer Pipe
Pspr.	Painter, Spray
Psst.	Painter, Structural Steel
P.T.	Potential Transformer
P. & T.	Pressure & Temperature
Ptd.	Painted
Ptns.	Partitions
Pu	Ultimate Load
PVC	Polyvinyl Chloride
Pvmt.	Pavement
Pwr.	Power
Q	Quantity Heat Flow
Qt.	Quart
Quan., Qty.	Quantity
Q.C.	Quick Coupling
r	Radius of Gyration
R	Resistance
R.C.P.	Reinforced Concrete Pipe
Rect.	Rectangle
Reg.	Regular
Reinf.	Reinforced
Req'd.	Required
Res.	Resistant
Resi.	Residential
Rgh.	Rough
RGS	Rigid Galvanized Steel
R.H.W.	Rubber, Heat & Water Resistant; Residential Hot Water
rms	Root Mean Square
Rnd.	Round
Rodm.	Rodman
Rofc.	Roofer, Composition
Rofp.	Roofer, Precast
Rohe.	Roofer Helpers (Composition)
Rots.	Roofer, Tile & Slate
R.O.W.	Right of Way
RPM	Revolutions per Minute
R.S.	Rapid Start
Rsr	Riser
RT	Round Trip
S.	Suction; Single Entrance; South
SC	Screw Cover
SCFM	Standard Cubic Feet per Minute
Scaf.	Scaffold
Sch., Sched.	Schedule
S.C.R.	Modular Brick
S.D.	Sound Deadening
S.D.R.	Standard Dimension Ratio
S.E.	Surfaced Edge
Sel.	Select
S.E.R., S.E.U.	Service Entrance Cable
S.F.	Square Foot
S.F.C.A.	Square Foot Contact Area
S.F. Flr.	Square Foot of Floor
S.F.G.	Square Foot of Ground
S.F. Hor.	Square Foot Horizontal
S.F.R.	Square Feet of Radiation
S.F. Shlf.	Square Foot of Shelf
S4S	Surface 4 Sides
Shee.	Sheet Metal Worker
Sin.	Sine
Skwk.	Skilled Worker
SL	Saran Lined
S.L.	Slimline
Sldr.	Solder
SLH	Super Long Span Bar Joist
S.N.	Solid Neutral
S-O-L	Socketolet
sp	Standpipe
S.P.	Static Pressure; Single Pole; Self-Propelled
Spri.	Sprinkler Installer
spwg	Static Pressure Water Gauge
S.P.D.T.	Single Pole, Double Throw
SPF	Spruce Pine Fir
S.P.S.T.	Single Pole, Single Throw
SPT	Standard Pipe Thread
Sq.	Square; 100 Square Feet
Sq. Hd.	Square Head
Sq. In.	Square Inch
S.S.	Single Strength; Stainless Steel
S.S.B.	Single Strength B Grade
sst, ss	Stainless Steel
Sswk.	Structural Steel Worker
Sswl.	Structural Steel Welder
St.;Stl.	Steel
S.T.C.	Sound Transmission Coefficient
Std.	Standard
Stg.	Staging
STK	Select Tight Knot
STP	Standard Temperature & Pressure
Stpi.	Steamfitter, Pipefitter
Str.	Strength; Starter; Straight
Strd.	Stranded
Struct.	Structural
Sty.	Story
Subj.	Subject
Subs.	Subcontractors
Surf.	Surface
Sw.	Switch
Swbd.	Switchboard
S.Y.	Square Yard
Syn.	Synthetic
S.Y.P.	Southern Yellow Pine
Sys.	System
t.	Thickness
T	Temperature; Ton
Tan	Tangent
T.C.	Terra Cotta
T & C	Threaded and Coupled
T.D.	Temperature Difference
Tdd	Telecommunications Device for the Deaf
T.E.M.	Transmission Electron Microscopy
TFE	Tetrafluoroethylene (Teflon)
T. & G.	Tongue & Groove; Tar & Gravel
Th., Thk.	Thick
Thn.	Thin
Thrded	Threaded
Tilf.	Tile Layer, Floor
Tilh.	Tile Layer, Helper
THHN	Nylon Jacketed Wire
THW.	Insulated Strand Wire

Abbreviations

THWN	Nylon Jacketed Wire
T.L.	Truckload
T.M.	Track Mounted
Tot.	Total
T-O-L	Threadolet
T.S.	Trigger Start
Tr.	Trade
Transf.	Transformer
Trhv.	Truck Driver, Heavy
Trlr	Trailer
Trlt.	Truck Driver, Light
TTY	Teletypewriter
TV	Television
T.W.	Thermoplastic Water Resistant Wire
UCI	Uniform Construction Index
UF	Underground Feeder
UGND	Underground Feeder
U.H.F.	Ultra High Frequency
U.I.	United Inch
U.L.	Underwriters Laboratory
Uld.	unloading
Unfin.	Unfinished
URD	Underground Residential Distribution
US	United States
USP	United States Primed
UTP	Unshielded Twisted Pair
V	Volt
V.A.	Volt Amperes
V.C.T.	Vinyl Composition Tile
VAV	Variable Air Volume
VC	Veneer Core
Vent.	Ventilation
Vert.	Vertical
V.F.	Vinyl Faced
V.G.	Vertical Grain
V.H.F.	Very High Frequency
VHO	Very High Output
Vib.	Vibrating
V.L.F.	Vertical Linear Foot
Vol.	Volume
VRP	Vinyl Reinforced Polyester
W	Wire; Watt; Wide; West
w/	With
W.C.	Water Column; Water Closet
W.F.	Wide Flange
W.G.	Water Gauge
Wldg.	Welding
W. Mile	Wire Mile
W-O-L	Weldolet
W.R.	Water Resistant
Wrck.	Wrecker
W.S.P.	Water, Steam, Petroleum
WT., Wt.	Weight
WWF	Welded Wire Fabric
XFER	Transfer
XFMR	Transformer
XHD	Extra Heavy Duty
XHHW, XLPE	Cross-Linked Polyethylene Wire Insulation
XLP	Cross-linked Polyethylene
Y	Wye
yd	Yard
yr	Year
Δ	Delta
%	Percent
~	Approximately
Ø	Phase; diameter
@	At
#	Pound; Number
<	Less Than
>	Greater Than
Z	zone

A

B

C

D

E

F

G

Q

R

T

U

X

Y

Z

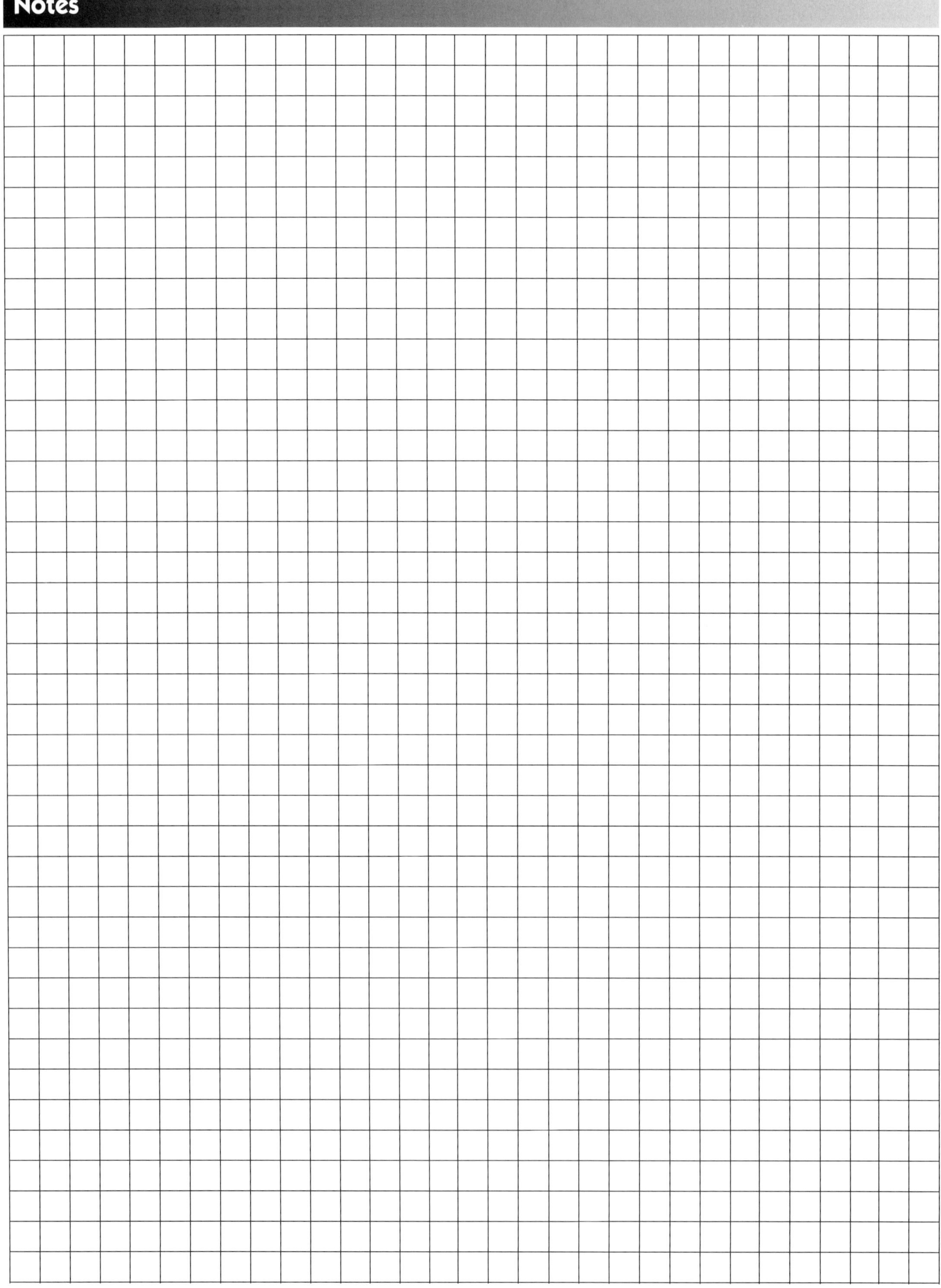

Notes

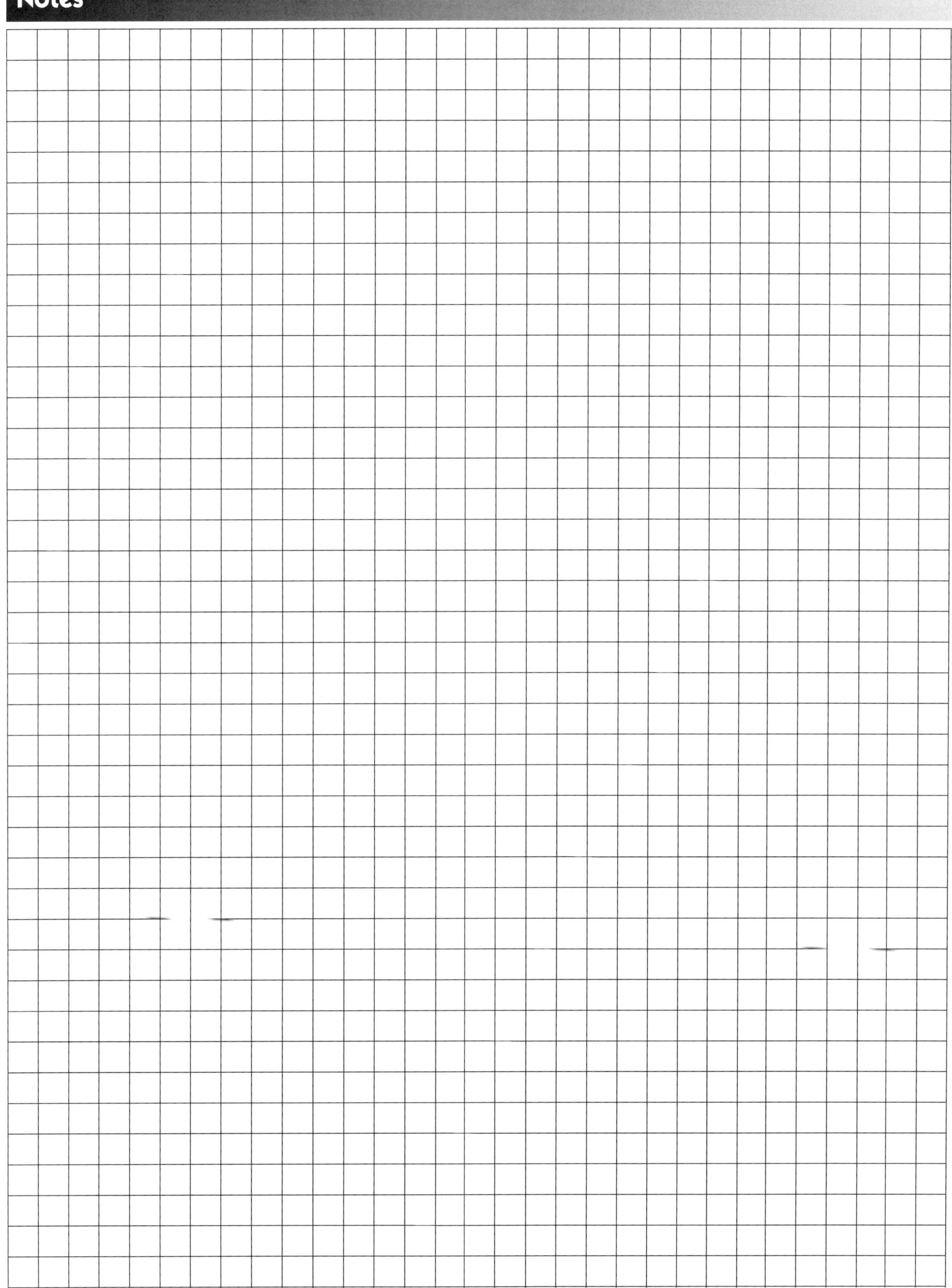

Notes

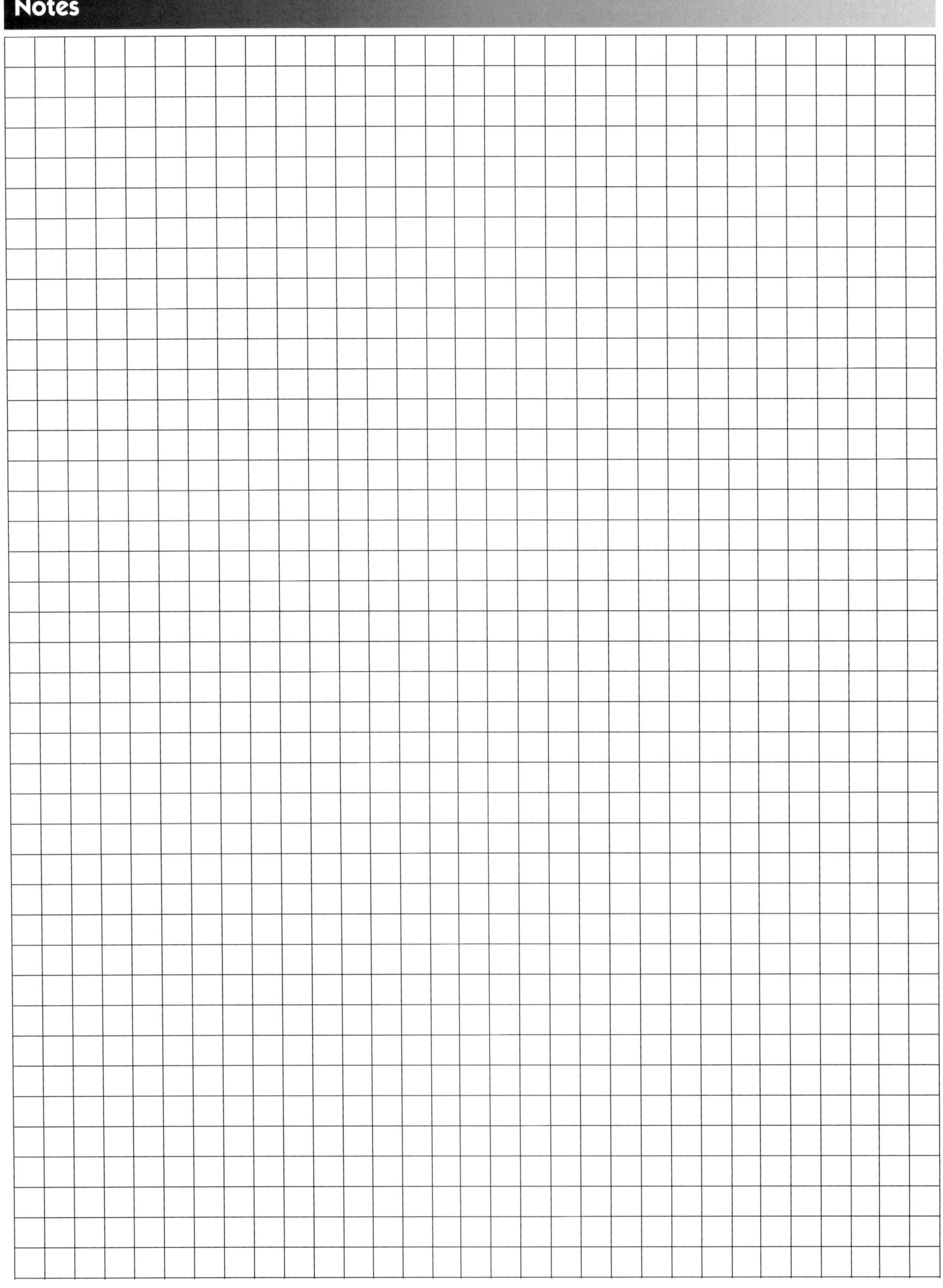

Notes

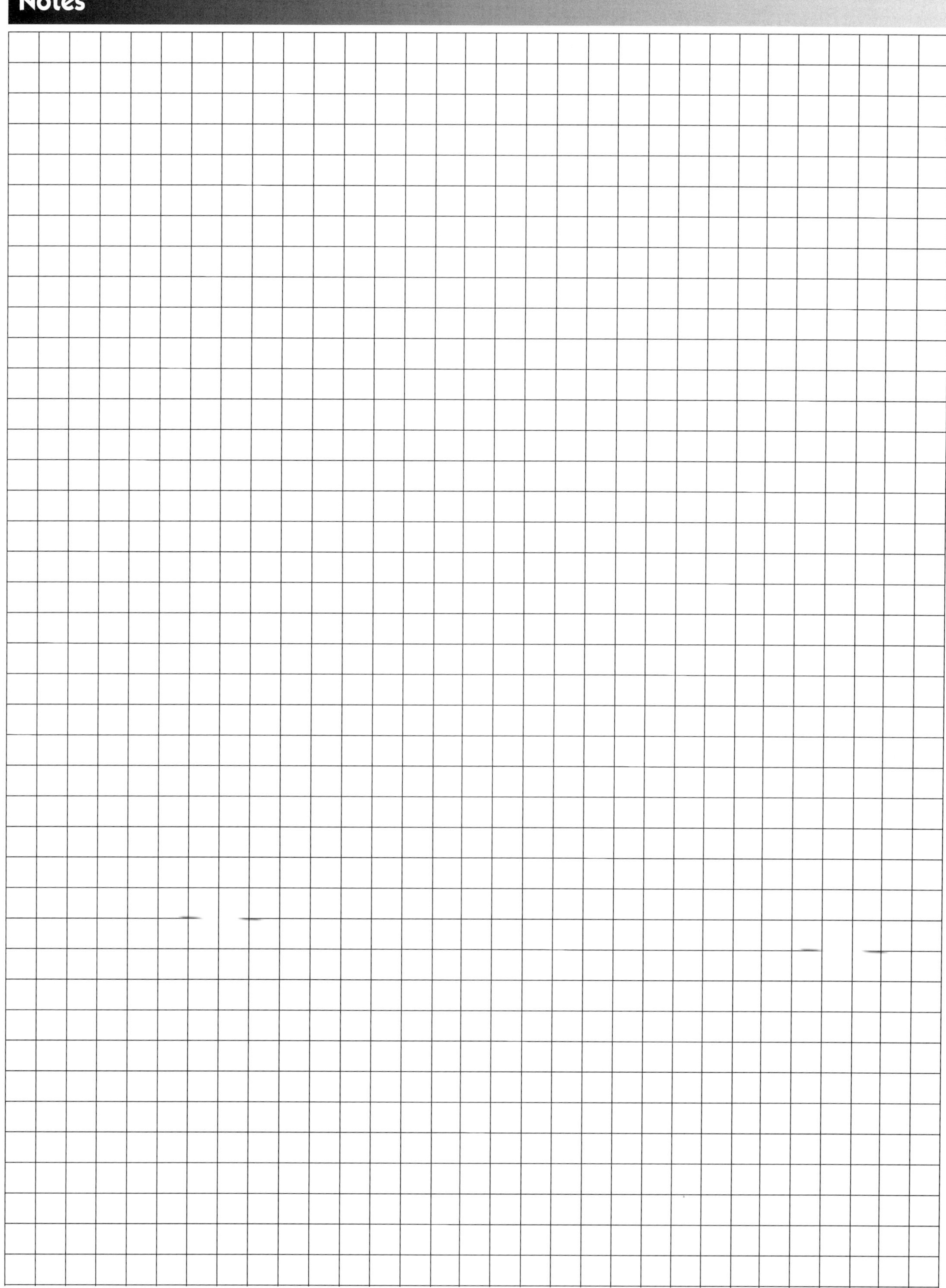

Notes

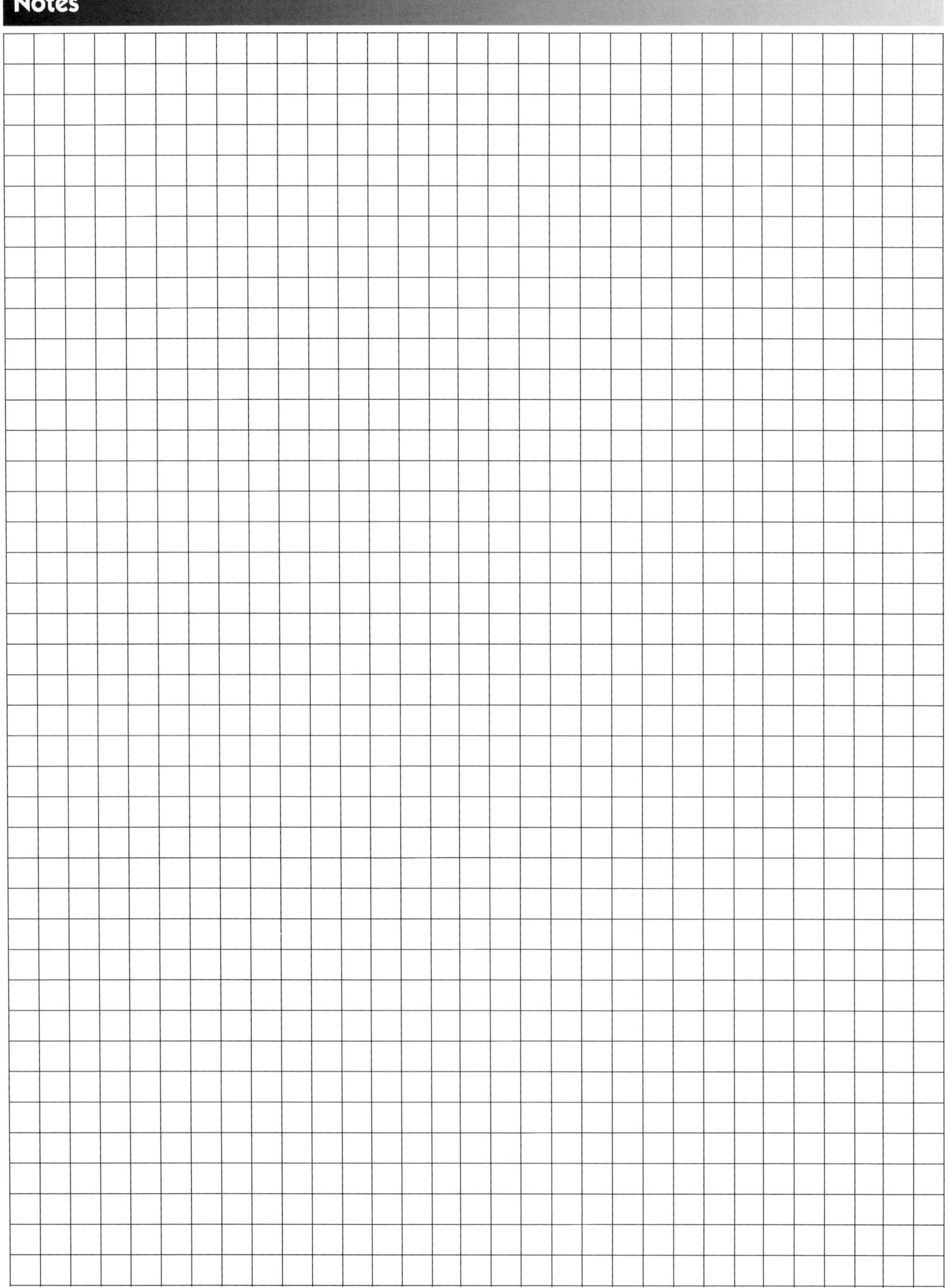

Notes

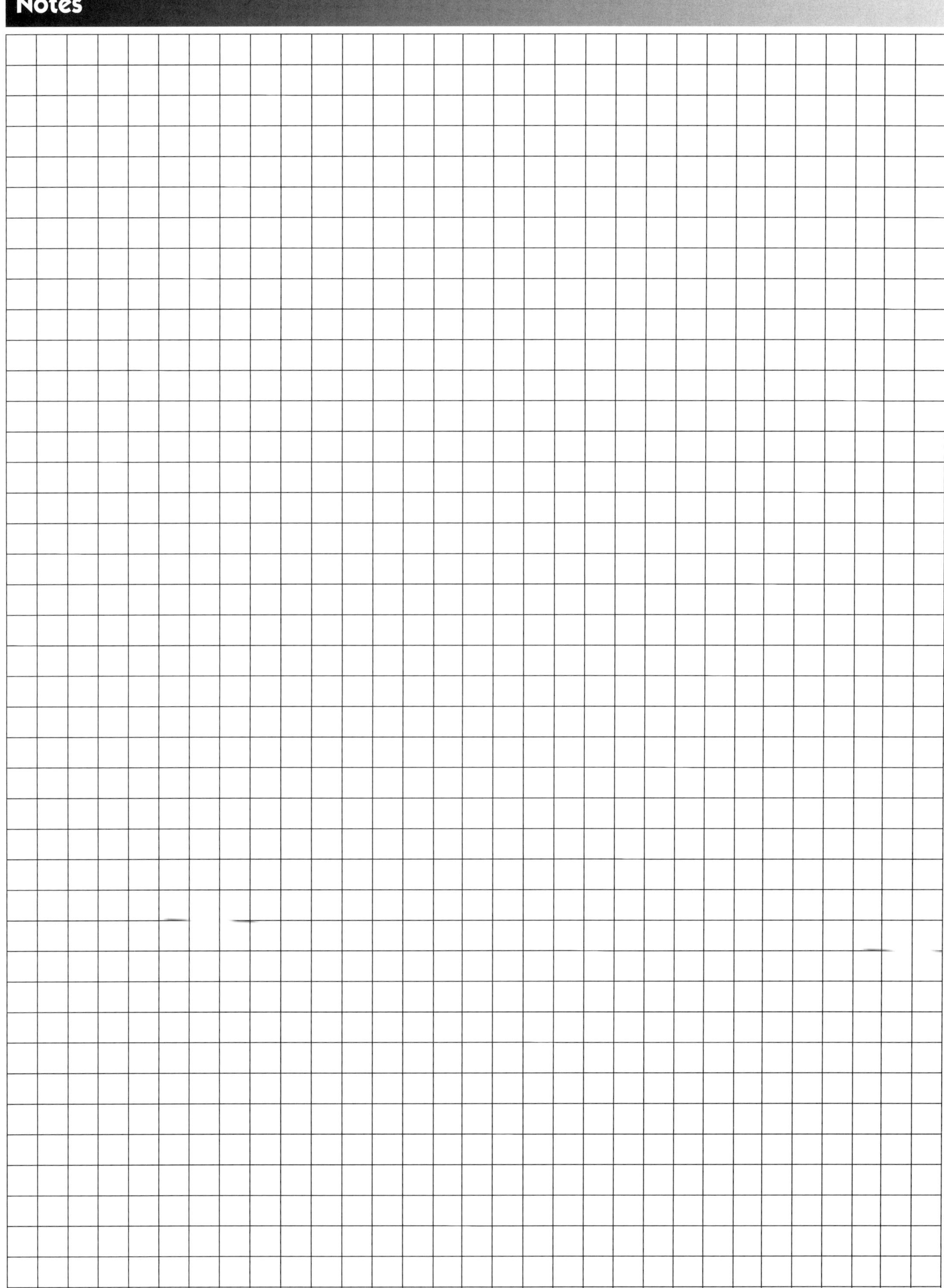

Reed Construction Data/RSMeans—a tradition of excellence in construction cost information and services since 1942

Table of Contents

For more information visit the RSMeans website at www.rsmeans.com

Book Selection Guide The following table provides definitive information on the content of each cost data publication. The number of lines of data provided in each unit price or assemblies division, as well as the number of reference tables and crews, is listed for each book. The presence of other elements such as equipment rental costs, historical cost indexes, city cost indexes, square foot models, or cross-referenced indexes is also indicated. You can use the table to help select the RSMeans book that has the quantity and type of information you most need in your work.

Unit Cost Divisions	Building Construction Costs	Mechanical	Electrical	Repair & Remodel	Square Foot	Site Work Landsc.	Assemblies	Interior	Concrete Masonry	Open Shop	Heavy Construction	Light Commercial	Facilities Construction	Plumbing	Residential
1	553	372	385	474		489		301	437	552	483	220	967	386	157
2	737	264	85	676		964		358	215	734	706	434	1154	288	242
3	1506	213	201	850		1357		277	1837	1506	1497	347	1422	182	359
4	897	24	0	697		699		572	1105	873	592	489	1145	0	405
5	1851	201	156	958		795		1021	758	1838	1050	826	1841	276	722
6	1841	67	69	1429		449		1160	299	1820	384	1530	1453	22	2043
7	1438	209	125	1462		524		546	466	1436	0	1143	1500	218	905
8	2048	61	10	2118		294		1737	694	1974	0	1527	2320	0	1397
9	1783	72	26	1598		268		1875	367	1737	15	1511	2027	54	1331
10	986	17	10	597		224		800	157	986	29	481	1072	231	214
11	1020	214	163	487		126		865	30	1005	0	231	1034	173	110
12	547	0	2	303		248		1691	12	536	19	349	1736	23	317
13	733	119	111	247		344		255	69	732	256	80	743	74	79
14	280	36	0	229		28		259	0	279	0	12	299	21	6
21	78	0	16	30		0		250	0	78	0	60	368	373	0
22	1148	7031	155	1106		1466		622	20	1103	1648	823	6940	8763	634
23	1185	7032	609	774		158		679	38	1105	110	660	5089	1869	418
26	1246	493	9875	1003		739		1053	55	1234	523	1080	9811	438	559
27	67	0	210	34		9		59	0	67	35	48	212	0	4
28	86	58	107	61		0		68	0	71	0	40	122	44	25
31	1477	735	608	798		2793		7	1208	1421	2865	600	1538	660	606
32	811	114	0	675		3967		359	274	784	1481	373	1443	226	422
33	519	1025	465	201		1967		0	225	518	1890	117	1553	1196	115
34	109	0	20	4		144		0	31	64	122	0	129	0	0
35	18	0	0	0		166		0	0	18	281	0	83	0	0
41	52	0	0	24		7		22	0	52	30	0	59	14	0
44	98	90	0	0		32		0	0	23	32	0	98	105	0
Totals	23114	18447	13408	16835		18257		14836	8297	22546	14048	12981	46158	15636	11070

Assembly Divisions	Building Construction Costs	Mechanical	Elec-trical	Repair & Remodel	Square Foot	Site Work Landscape	Assemblies	Interior	Concrete Masonry	Open Shop	Heavy Construction	Light Commercial	Facilities Construction	Plumbing	Asm Div	Residential
A		15	0	182	150	530	598	0	536		570	154	24	0	1	374
B		0	0	809	2471	0	5577	329	1914		368	2031	143	0	2	217
C		0	0	635	865	0	1235	1567	146		0	765	247	0	3	588
D		1066	811	694	1793	0	2422	767	0		0	1271	1025	1078	4	861
E		0	0	85	260	0	297	5	0		0	257	5	0	5	392
F		0	0	0	123	0	124	0	0		0	123	3	0	6	358
G		527	294	311	202	2366	668	0	535		1194	200	293	685	7	300
															8	760
															9	80
															10	0
															11	0
															12	0
Totals		1608	1105	2716	5864	2896	10921	2668	3131		2132	4801	1740	1763		3932

Reference Section	Building Construction Costs	Mechanical	Electrical	Repair & Remodel	Square Foot	Site Work Landscape	Assemblies	Interior	Concrete Masonry	Open Shop	Heavy Construction	Light Commercial	Facilities Construction	Plumbing	Residential
Reference Tables	yes	yes	yes	yes	no	yes	yes	yes	yes	yes	yes	yes	yes	yes	yes
Models					107							46			28
Crews	487	487	487	467		487		487	487	465	487	465	467	487	465
Equipment Rental Costs	yes	yes	yes	yes		yes		yes	yes	yes	yes	yes	yes	yes	yes
Historical Cost Indexes	yes	yes	yes	yes	yes	yes	yes	yes	yes	yes	yes	yes	yes	yes	no
City Cost Indexes	yes	yes	yes	yes	yes	yes	yes	yes	yes	yes	yes	yes	yes	yes	yes

Annual Cost Guides

For more information visit the RSMeans website at www.rsmeans.com

RSMeans Building Construction Cost Data 2010

Offers you unchallenged unit price reliability in an easy-to-use format. Whether used for verifying complete, finished estimates or for periodic checks, it supplies more cost facts better and faster than any comparable source. Over 21,000 unit prices have been updated for 2010. The City Cost Indexes and Location Factors cover over 930 areas, for indexing to any project location in North America. Order and get *RSMeans Quarterly Update Service* FREE. You'll have year-long access to the RSMeans Estimating **HOTLINE** FREE with your subscription. Expert assistance when using RSMeans data is just a phone call away.

$169.95 per copy
Available Sept. 2009
Catalog no. 60010

Unit prices organized according to the latest MasterFormat!

RSMeans Mechanical Cost Data 2010

- **HVAC**
- **Controls**

Total unit and systems price guidance for mechanical construction. . . materials, parts, fittings, and complete labor cost information. Includes prices for piping, heating, air conditioning, ventilation, and all related construction.

Plus new 2010 unit costs for:

- Thousands of installed HVAC/controls assemblies components
- "On-site" Location Factors for over 930 cities and towns in the U.S. and Canada
- Crews, labor, and equipment

$169.95 per copy
Available Oct. 2009
Catalog no. 60020

RSMeans Plumbing Cost Data 2010

Comprehensive unit prices and assemblies for plumbing, irrigation systems, commercial and residential fire protection, point-of-use water heaters, and the latest approved materials. This publication and its companion, *RSMeans Mechanical Cost Data*, provide full-range cost estimating coverage for all the mechanical trades.

Now contains updated costs for potable water and radiant heat systems and high efficiency fixtures.

$169.95 per copy
Available Oct. 2009
Catalog no. 60210

Unit prices organized according to the latest MasterFormat!

RSMeans Facilities Construction Cost Data 2010

For the maintenance and construction of commercial, industrial, municipal, and institutional properties. Costs are shown for new and remodeling construction and are broken down into materials, labor, equipment, and overhead and profit. Special emphasis is given to sections on mechanical, electrical, furnishings, site work, building maintenance, finish work, and demolition.

More than 43,000 unit costs, plus assemblies costs and a comprehensive Reference Section are included.

$419.95 per copy
Available Nov. 2009
Catalog no. 60200

Annual Cost Guides

RSMeans Square Foot Costs 2010

It's Accurate and Easy To Use!

- **Updated price information** based on nationwide figures from suppliers, estimators, labor experts, and contractors
- "How-to-Use" sections, with **clear examples** of commercial, residential, industrial, and institutional structures
- Realistic graphics, offering true-to-life illustrations of building projects
- Extensive information on using square foot cost data, including sample estimates and alternate pricing methods

$184.95 per copy
Available Oct. 2009
Catalog no. 60050

RSMeans Repair & Remodeling Cost Data 2010

Commercial/Residential

Use this valuable tool to estimate commercial and residential renovation and remodeling.

Includes: New costs for hundreds of unique methods, materials, and conditions that only come up in repair and remodeling, PLUS:

- Unit costs for over 15,000 construction components
- Installed costs for over 90 assemblies
- Over 930 "on-site" localization factors for the U.S. and Canada.

Unit prices organized according to the latest MasterFormat!

$144.95 per copy
Available Oct. 2009
Catalog no. 60040

RSMeans Electrical Cost Data 2010

Pricing information for every part of electrical cost planning. More than 13,000 unit and systems costs with design tables; clear specifications and drawings; engineering guides; illustrated estimating procedures; complete labor-hour and materials costs for better scheduling and procurement; and the latest electrical products and construction methods.

- A variety of special electrical systems including cathodic protection
- Costs for maintenance, demolition, HVAC/mechanical, specialties, equipment, and more

$169.95 per copy
Available Oct. 2009
Catalog no. 60030

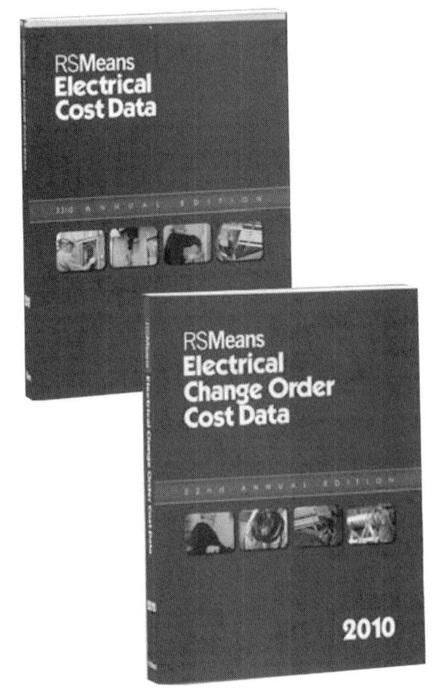

Unit prices organized according to the latest MasterFormat!

RSMeans Electrical Change Order Cost Data 2010

RSMeans Electrical Change Order Cost Data provides you with electrical unit prices exclusively for pricing change orders—based on the recent, direct experience of contractors and suppliers. Analyze and check your own change order estimates against the experience others have had doing the same work. It also covers productivity analysis and change order cost justifications. With useful information for calculating the effects of change orders and dealing with their administration.

$169.95 per copy
Available Dec. 2009
Catalog no. 60230

Annual Cost Guides

For more information visit the RSMeans website at www.rsmeans.com

RSMeans Assemblies Cost Data 2010

RSMeans Assemblies Cost Data takes the guesswork out of preliminary or conceptual estimates. Now you don't have to try to calculate the assembled cost by working up individual component costs. We've done all the work for you.

Presents detailed illustrations, descriptions, specifications, and costs for every conceivable building assembly—240 types in all—arranged in the easy-to-use UNIFORMAT II system. Each illustrated "assembled" cost includes a complete grouping of materials and associated installation costs, including the installing contractor's overhead and profit.

$279.95 per copy
Available Sept. 2009
Catalog no. 60060

RSMeans Open Shop Building Construction Cost Data 2010

The latest costs for accurate budgeting and estimating of new commercial and residential construction. . . renovation work. . . change orders. . . cost engineering.

RSMeans Open Shop "BCCD" will assist you to:

- Develop benchmark prices for change orders
- Plug gaps in preliminary estimates and budgets
- Estimate complex projects
- Substantiate invoices on contracts
- Price ADA-related renovations

Unit prices organized according to the latest MasterFormat!

$169.95 per copy
Available Dec. 2009
Catalog no. 60150

RSMeans Residential Cost Data 2010

Contains square foot costs for 30 basic home models with the look of today, plus hundreds of custom additions and modifications you can quote right off the page. With costs for the 100 residential systems you're most likely to use in the year ahead. Complete with blank estimating forms, sample estimates, and step-by-step instructions.

Now contains line items for cultured stone and brick, PVC trim lumber, and TPO roofing.

$144.95 per copy
Available Oct. 2009
Catalog no. 60170

Unit prices organized according to the latest MasterFormat!

RSMeans Site Work & Landscape Cost Data 2010

Includes unit and assemblies costs for earthwork, sewerage, piped utilities, site improvements, drainage, paving, trees & shrubs, street openings/repairs, underground tanks, and more. Contains 57 tables of assemblies costs for accurate conceptual estimates.

Includes:

- Estimating for infrastructure improvements
- Environmentally-oriented construction
- ADA-mandated handicapped access
- Hazardous waste line items

$169.95 per copy
Available Nov. 2009
Catalog no. 60280

RSMeans Facilities Maintenance & Repair Cost Data 2010

RSMeans Facilities Maintenance & Repair Cost Data gives you a complete system to manage and plan your facility repair and maintenance costs and budget efficiently. Guidelines for auditing a facility and developing an annual maintenance plan. Budgeting is included, along with reference tables on cost and management, and information on frequency and productivity of maintenance operations.

The only nationally recognized source of maintenance and repair costs. Developed in cooperation with the Civil Engineering Research Laboratory (CERL) of the Army Corps of Engineers.

$379.95 per copy
Available Nov. 2009
Catalog no. 60300

RSMeans Light Commercial Cost Data 2010

Specifically addresses the light commercial market, which is a specialized niche in the construction industry. Aids you, the owner/designer/contractor, in preparing all types of estimates—from budgets to detailed bids. Includes new advances in methods and materials.

Assemblies Section allows you to evaluate alternatives in the early stages of design/planning.

Over 11,000 unit costs ensure that you have the prices you need. . . when you need them.

Unit prices organized according to the latest MasterFormat!

$144.95 per copy
Available Nov. 2009
Catalog no. 60180

Annual Cost Guides

RSMeans Concrete & Masonry Cost Data 2010

Provides you with cost facts for virtually all concrete/masonry estimating needs, from complicated formwork to various sizes and face finishes of brick and block—all in great detail. The comprehensive Unit Price Section contains more than 7,500 selected entries. Also contains an Assemblies [Cost] Section, and a detailed Reference Section that supplements the cost data.

Unit prices organized according to the latest MasterFormat!

$154.95 per copy
Available Dec. 2009
Catalog no. 60110

RSMeans Labor Rates for the Construction Industry 2010

Complete information for estimating labor costs, making comparisons, and negotiating wage rates by trade for over 300 U.S. and Canadian cities. With 46 construction trades listed by local union number in each city, and historical wage rates included for comparison. Each city chart lists the county and is alphabetically arranged with handy visual flip tabs for quick reference.

$379.95 per copy
Available Dec. 2009
Catalog no. 60120

RSMeans Construction Cost Indexes 2010

What materials and labor costs will change unexpectedly this year? By how much?

- Breakdowns for 316 major cities
- National averages for 30 key cities
- Expanded five major city indexes
- Historical construction cost indexes

$362.00 per year (subscription)
Catalog no. 50140

$90.50 individual quarters
Catalog no. 60140 A,B,C,D

RSMeans Interior Cost Data 2010

Provides you with prices and guidance needed to make accurate interior work estimates. Contains costs on materials, equipment, hardware, custom installations, furnishings, and labor costs. . . for new and remodel commercial and industrial interior construction, including updated information on office furnishings, and reference information.

Unit prices organized according to the latest MasterFormat!

$169.95 per copy
Available Nov. 2009
Catalog no. 60090

RSMeans Heavy Construction Cost Data 2010

A comprehensive guide to heavy construction costs. Includes costs for highly specialized projects such as tunnels, dams, highways, airports, and waterways. Information on labor rates, equipment, and materials costs is included. Features unit price costs, systems costs, and numerous reference tables for costs and design.

$169.95 per copy
Available Nov. 2009
Catalog no. 60160

Unit prices organized according to the latest MasterFormat!

Value Engineering: Practical Applications

For Design, Construction, Maintenance & Operations

by Alphonse Dell'Isola, PE

A tool for immediate application—for engineers, architects, facility managers, owners, and contractors. Includes making the case for VE—the management briefing; integrating VE into planning, budgeting, and design; conducting life cycle costing; using VE methodology in design review and consultant selection; case studies; VE workbook; and a life cycle costing program on disk.

$79.95 per copy
Over 450 pages, illustrated, softcover
Catalog no. 67319A

Facilities Operations & Engineering Reference

by the Association for Facilities Engineering and RSMeans

An all-in-one technical reference for planning and managing facility projects and solving day-to-day operations problems. Selected as the official certified plant engineer reference, this handbook covers financial analysis, maintenance, HVAC and energy efficiency, and more.

$54.98 per copy
Over 700 pages, illustrated, hardcover
Catalog no. 67318

The Building Professional's Guide to Contract Documents

3rd Edition

by Waller S. Poage, AIA, CSI, CVS

A comprehensive reference for owners, design professionals, contractors, and students

- Structure your documents for maximum efficiency.
- Effectively communicate construction requirements.
- Understand the roles and responsibilities of construction professionals.
- Improve methods of project delivery.

$64.95 per copy, 400 pages
Diagrams and construction forms, hardcover
Catalog no. 67261A

Complete Book of Framing

by Scot Simpson

This straightforward, easy-to-learn method will help framers, carpenters, and handy homeowners build their skills in rough carpentry and framing. Shows how to frame all the parts of a house: floors, walls, roofs, door & window openings, and stairs—with hundreds of color photographs and drawings that show every detail.

The book gives beginners all the basics they need to go from zero framing knowledge to a journeyman level.

$29.95 per copy
352 pages, softcover
Catalog no. 67353

Cost Planning & Estimating for Facilities Maintenance

n this unique book, a team of facilities management uthorities shares their expertise on:

Evaluating and budgeting maintenance operations
Maintaining and repairing key building components
Applying *RSMeans Facilities Maintenance & Repair Cost Data* to your estimating

Covers special maintenance requirements of the ten major building types.

89.95 per copy
ver 475 pages, hardcover
atalog no. 67314

Life Cycle Costing for Facilities

by Alphonse Dell'Isola and Dr. Steven Kirk

Guidance for achieving higher quality design and construction projects at lower costs! Cost-cutting efforts often sacrifice quality to yield the cheapest product. Life cycle costing enables building designers and owners to achieve both. The authors of this book show how LCC can work for a variety of projects — from roads to HVAC upgrades to different types of buildings.

$99.95 per copy
450 pages, hardcover
Catalog no. 67341

Planning & Managing Interior Projects 2nd Edition

by Carol E. Farren, CFM

xpert guidance on managing renovation & relocation rojects

his book guides you through every step in relocating a new space or renovating an old one. From initial meeting through design and construction, to post-project administration, it helps you get the most for your company or client. Includes sample forms, spec lists, agreements, drawings, and much more!

34.98 per copy
00 pages, softcover
atalog no. 67245A

Construction Business Management

by Nick Ganaway

Only 43% of construction firms stay in business after four years. Make sure your company thrives with valuable guidance from a pro with 25 years of success as a commercial contractor. Find out what it takes to build all aspects of a business that is profitable, enjoyable, and enduring. With a bonus chapter on retail construction.

$49.95 per copy
200 pages, softcover
Catalog no. 67352

Interior Home Improvement Costs 9th Edition

Updated estimates for the most popular remodeling and repair projects—from small, do-it-yourself jobs to major renovations and new construction. Includes: kitchens & baths; new living space from your attic, basement, or garage; new floors, paint, and wallpaper; tearing out or building new walls; closets, stairs, and fireplaces; new energy-saving improvements, home theaters, and more!

$24.95 per copy
250 pages, illustrated, softcover
Catalog no. 67308E

Exterior Home Improvement Costs 9th Edition

Updated estimates for the most popular remodeling and repair projects—from small, do-it-yourself jobs, to major renovations and new construction. Includes: curb appeal projects—landscaping, patios, porches, driveways, and walkways; new windows and doors; decks, greenhouses, and sunrooms; room additions and garages; roofing, siding, and painting; "green" improvements to save energy & water.

$24.95 per copy
Over 275 pages, illustrated, softcover
Catalog no. 67309E

Builder's Essentials: Plan Reading & Material Takeoff

For Residential and Light Commercial Construction

by Wayne J. DelPico

A valuable tool for understanding plans and specs, and accurately calculating material quantities. Step-by-step instructions and takeoff procedures based on a full set of working drawings.

$35.95 per copy
Over 420 pages, softcover
Catalog no. 67307

Means Unit Price Estimating Methods

New 4th Edition

This new edition includes up-to-date cost data and estimating examples, updated to reflect changes to the CSI numbering system and new features of RSMeans cost data. It describes the most productive, universally accepted ways to estimate, and uses checklists and forms to illustrate shortcuts and timesavers. A model estimate demonstrates procedures. A new chapter explores computer estimating alternatives.

$59.95 per copy
Over 350 pages, illustrated, softcover
Catalog no. 67303B

Total Productive Facilities Management

by Richard W. Sievert, Jr.

Today, facilities are viewed as strategic resources. . . elevating the facility manager to the role of asset manager supporting the organization's overall business goals. Now, Richard Sievert Jr., in this well-articulated guidebook, sets forth a new operational standard for the facility manager's emerging role. . . a comprehensive program for managing facilities as a true profit center.

$29.98 per copy
275 pages, softcover
Catalog no. 67321

Green Building: Project Planning & Cost Estimating 2nd Edition

This new edition has been completely updated with the latest in green building technologies, design concepts, standards, and costs. Now includes a 2010 Green Building *CostWorks* CD with more than 300 green building assemblies and over 5,000 unit price line items for sustainable building. The new edition is also full-color with all new case studies—plus a new chapter on deconstruction, a key aspect of green building.

$129.95 per copy
350 pages, softcover
Catalog no. 67338A

Concrete Repair and Maintenance Illustrated

by Peter Emmons

Hundreds of illustrations show users how to analyze, repair, clean, and maintain concrete structures for optimal performance and cost effectiveness. From parking garages to roads and bridges to structural concrete, this comprehensive book describes the causes, effects, and remedies for concrete wear and failure. Invaluable for planning jobs, selecting materials, and training employees, this book is a must-have for concrete specialists, general contractors, facility managers, civil and structural engineers, and architects.

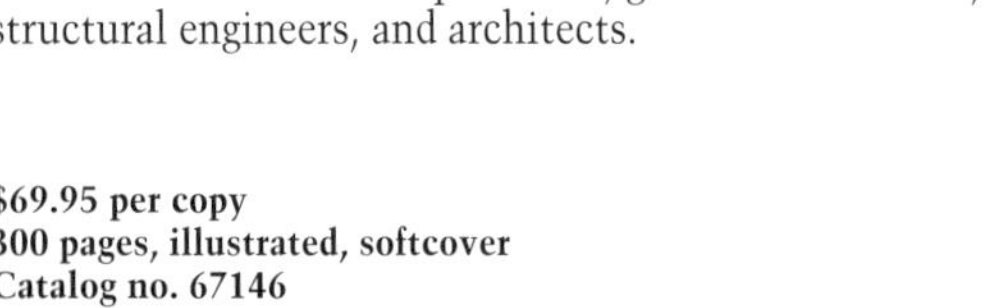

$69.95 per copy
300 pages, illustrated, softcover
Catalog no. 67146

Means Illustrated Construction Dictionary Condensed, 2nd Edition

The best portable dictionary for office or field use—an essential tool for contractors, architects, insurance and real estate personnel, facility managers, homeowners, and anyone who needs quick, clear definitions for construction terms. The second edition has been further enhanced with updates and hundreds of new terms and illustrations . . . in keeping with the most recent developments in the construction industry.

Now with a quick-reference Spanish section. Includes tools and equipment, materials, tasks, and more!

$59.95 per copy
Over 500 pages, softcover
Catalog no. 67282A

Means Repair & Remodeling Estimating 4th Edition

By Edward B. Wetherill and RSMeans

This important reference focuses on the unique problems of estimating renovations of existing structures, and helps you determine the true costs of remodeling through careful evaluation of architectural details and a site visit.

New section on disaster restoration costs.

$69.95 per copy
Over 450 pages, illustrated, hardcover
Catalog no. 67265B

Facilities Planning & Relocation

by David D. Owen

A complete system for planning space needs and managing relocations. Includes step-by-step manual, over 50 forms, and extensive reference section on materials and furnishings.

New lower price and user-friendly format.

$49.98 per copy
Over 450 pages, softcover
Catalog no. 67301

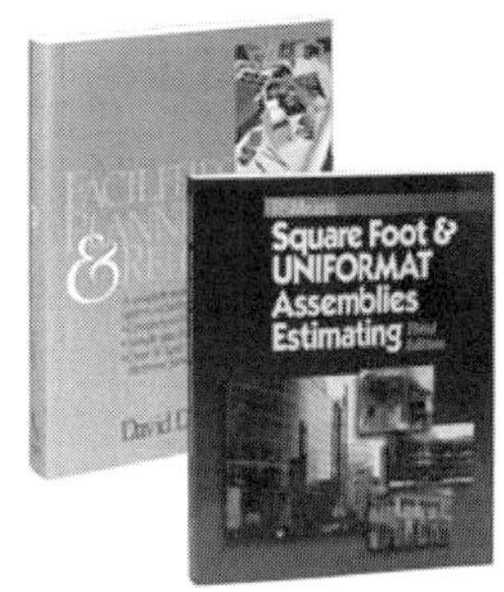

Means Square Foot & UNIFORMAT Assemblies Estimating Methods 3rd Edition

Develop realistic square foot and assemblies costs for budgeting and construction funding. The new edition features updated guidance on square foot and assemblies estimating using UNIFORMAT II. An essential reference for anyone who performs conceptual estimates.

$69.95 per copy
Over 300 pages, illustrated, softcover
Catalog no. 67145B

Means Electrical Estimating Methods 3rd Edition

Expanded edition includes sample estimates and cost information in keeping with the latest version of the CSI MasterFormat and UNIFORMAT II. Complete coverage of fiber optic and uninterruptible power supply electrical systems, broken down by components, and explained in detail. Includes a new chapter on computerized estimating methods. A practical companion to *RSMeans Electrical Cost Data.*

$64.95 per copy
Over 325 pages, hardcover
Catalog no. 67230B

Means Mechanical Estimating Methods 4th Edition

Completely updated, this guide assists you in making a review of plans, specs, and bid packages, with suggestions for takeoff procedures, listings, substitutions, and pre-bid scheduling for all components of HVAC. Includes suggestions for budgeting labor and equipment usage. Compares materials and construction methods to allow you to select the best option.

$64.95 per copy
Over 350 pages, illustrated, softcover
Catalog no. 67294B

Means ADA Compliance Pricing Guide 2nd Edition

by Adaptive Environments and RSMeans

Completely updated and revised to the new 2004 *Americans with Disabilities Act Accessibility Guidelines,* this book features more than 70 of the most commonly needed modifications for ADA compliance. Projects range from installing ramps and walkways, widening doorways and entryways, and installing and refitting elevators, to relocating light switches and signage.

$79.95 per copy
Over 350 pages, illustrated, softcover
Catalog no. 67310A

Project Scheduling & Management for Construction 3rd Edition

by David R. Pierce, Jr.

A comprehensive yet easy-to-follow guide to construction project scheduling and control—from vital project management principles through the latest scheduling, tracking, and controlling techniques. The author is a leading authority on scheduling, with years of field and teaching experience at leading academic institutions. Spend a few hours with this book and come away with a solid understanding of this essential management topic.

$64.95 per copy
Over 300 pages, illustrated, hardcover
Catalog no. 67247B

The Practice of Cost Segregation Analysis

by Bruce A. Desrosiers and Wayne J. DelPico

This expert guide walks you through the practice of cost segregation analysis, which enables property owners to defer taxes and benefit from "accelerated cost recovery" through depreciation deductions on assets that are properly identified and classified.

With a glossary of terms, sample cost segregation estimates for various building types, key information resources, and updates via a dedicated website, this book is a critical resource for anyone involved in cost segregation analysis.

$99.95 per copy
Over 225 pages
Catalog no. 67345

Preventive Maintenance for Multi-Family Housing

by John C. Maciha

Prepared by one of the nation's leading experts on multi-family housing.

This complete PM system for apartment and condominium communities features expert guidance, checklists for buildings and grounds maintenance tasks and their frequencies, a reusable wall chart to track maintenance, and a dedicated website featuring customizable electronic forms. A must-have for anyone involved with multi-family housing maintenance and upkeep.

$89.95 per copy
225 pages
Catalog no. 67346

How to Estimate with Means Data & CostWorks

New 3rd Edition

by RSMeans and Saleh A. Mubarak, Ph.D.

New 3rd Edition—fully updated with new chapters, plus new CD with updated *CostWorks* cost data and MasterFormat organization. Includes all major construction items—with more than 300 exercises and two sets of plans that show how to estimate for a broad range of construction items and systems—including general conditions and equipment costs.

$59.95 per copy
272 pages, softcover
Includes CostWorks CD
Catalog no. 67324B

Job Order Contracting

Expediting Construction Project Delivery

by Allen Henderson

Expert guidance to help you implement JOC—fast becoming the preferred project delivery method for repair and renovation, minor new construction, and maintenance projects in the public sector and in many states and municipalities. The author, a leading JOC expert and practitioner, shows how to:

- Establish a JOC program
- Evaluate proposals and award contracts
- Handle general requirements and estimating
- Partner for maximum benefits

$89.95 per copy
192 pages, illustrated, hardcover
Catalog no. 67348

Builder's Essentials: Estimating Building Costs

For the Residential & Light Commercial Contractor

by Wayne J. DelPico

Step-by-step estimating methods for residential and light commercial contractors. Includes a detailed look at every construction specialty—explaining all the components, takeoff units, and labor needed for well-organized, complete estimates. Covers correctly interpreting plans and specifications, and developing accurate and complete labor and material costs.

$29.95 per copy
Over 400 pages, illustrated, softcover
Catalog no. 67343

Building & Renovating Schools

This all-inclusive guide covers every step of the school construction process—from initial planning, needs assessment, and design, right through moving into the new facility. A must-have resource for anyone concerned with new school construction or renovation. With square foot cost models for elementary, middle, and high school facilities, and real-life case studies of recently completed school projects.

The contributors to this book—architects, construction project managers, contractors, and estimators who specialize in school construction—provide start-to-finish, expert guidance on the process.

$69.98 per copy
Over 425 pages, hardcover
Catalog no. 67342

The Homeowner's Guide to Mold By Michael Pugliese

Expert guidance to protect your health and your home.

Mold, whether caused by leaks, humidity or flooding, is a real health and financial issue—for homeowners and contractors. This full-color book explains:
- Construction and maintenance practices to prevent mold
- How to inspect for and remove mold
- Mold remediation procedures and costs
- What to do after a flood
- How to deal with insurance companies

$21.95 per copy
144 pages, softcover
Catalog no. 67344

Means Illustrated Construction Dictionary

Unabridged 3rd Edition, with CD-ROM

Long regarded as the industry's finest, *Means Illustrated Construction Dictionary* is now even better. With the addition of over 1,000 new terms and hundreds of new illustrations, it is the clear choice for the most comprehensive and current information. The companion CD-ROM that comes with this new edition adds many extra features: larger graphics, expanded definitions, and links to both CSI MasterFormat numbers and product information.

$99.95 per copy
Over 790 pages, illustrated, hardcover
Catalog no. 67292A

Means Landscape Estimating Methods

New 5th Edition

Answers questions about preparing competitive landscape construction estimates, with up-to-date cost estimates and the new MasterFormat classification system. Expanded and revised to address the latest materials and methods, including new coverage on approaches to green building. Includes:
- Step-by-step explanation of the estimating process
- Sample forms and worksheets that save time and prevent errors

$64.95 per copy
Over 350 pages, softcover
Catalog no. 67295C

Means Plumbing Estimating Methods 3rd Edition

by Joseph Galeno and Sheldon Greene

Updated and revised! This practical guide walks you through a plumbing estimate, from basic materials and installation methods through change order analysis. *Plumbing Estimating Methods* covers residential, commercial, industrial, and medical systems, and features sample takeoff and estimate forms and detailed illustrations of systems and components.

$29.98 per copy
330+ pages, softcover
Catalog no. 67283B

Understanding & Negotiating Construction Contracts

by Kit Werremeyer

Take advantage of the author's 30 years' experience in small-to-large (including international) construction projects. Learn how to identify, understand, and evaluate high risk terms and conditions typically found in all construction contracts—then negotiate to lower or eliminate the risk, improve terms of payment, and reduce exposure to claims and disputes. The author avoids "legalese" and gives real-life examples from actual projects.

$69.95 per copy
300 pages, softcover
Catalog no. 67350

The Gypsum Construction Handbook (Spanish)

The Gypsum Construction Handbook, Spanish edition, is an illustrated, comprehensive, and authoritative guide on all facets of gypsum construction. You'll find comprehensive guidance on available products, installation methods, fire- and sound-rated construction information, illustrated framing-to-finish application instructions, estimating and planning information, and more.

$29.95 per copy
Over 450 pages, softcover
Catalog No. 67357S

Means Spanish/English Construction Dictionary 2nd Edition

by RSMeans and the International Code Council

This expanded edition features thousands of the most common words and useful phrases in the construction industry with easy-to-follow pronunciations in both Spanish and English. Over 800 new terms, phrases, and illustrations have been added. It also features a new stand-alone "Safety & Emergencies" section, with colored pages for quick access.

Unique to this dictionary are the systems illustrations showing the relationship of components in the most common building systems for all major trades.

$23.95 per copy
Over 400 pages
Catalog no. 67327A

How Your House Works

by Charlie Wing

A must-have reference for every homeowner, handyman, and contractor—for repair, remodeling, and new construction. This book uncovers the mysteries behind just about every major appliance and building element in your house—from electrical, heating and AC, to plumbing, framing, foundations and appliances. Clear, "exploded" drawings show exactly how things should be put together and how they function—what to check if they don't work, and what you can do that might save you having to call in a professional.

$21.95 per copy
160 pages, softcover
Catalog no. 67351

Professional Development

Means CostWorks® CD Training

This one-day course helps users become more familiar with the functionality of *Means CostWorks* program. Each menu, icon, screen, and function found in the program is explained in depth. Time is devoted to hands-on estimating exercises.

Some of what you'll learn:
- Searching the database utilizing all navigation methods
- Exporting RSMeans Data to your preferred spreadsheet format
- Viewing crews, assembly components, and much more!
- Automatically regionalizing the database

You are required to bring your own laptop computer for this course.

When you register for this course you will receive an outline for your laptop requirements.

Also offering web training for CostWorks CD!

Unit Price Estimating

This interactive *two-day* seminar teaches attendees how to interpret project information and process it into final, detailed estimates with the greatest accuracy level.

The most important credential an estimator can take to the job is the ability to visualize construction, and estimate accurately.

Some of what you'll learn:
- Interpreting the design in terms of cost
- The most detailed, time-tested methodology for accurate pricing
- Key cost drivers—material, labor, equipment, staging, and subcontracts
- Understanding direct and indirect costs for accurate job cost accounting and change order management

Who should attend: Corporate and government estimators and purchasers, architects, engineers... and others needing to produce accurate project estimates.

Square Foot & Assemblies Estimating

This *two-day* course teaches attendees how to quickly deliver accurate square foot estimates using limited budget and design information.

Some of what you'll learn:
- How square foot costing gets the estimate done faster
- Taking advantage of a "systems" or "assemblies" format
- The RSMeans "building assemblies/square foot cost approach"
- How to create a reliable preliminary systems estimate using bare-bones design information

Who should attend: Facilities managers, facilities engineers, estimators, planners, developers, construction finance professionals... and others needing to make quick, accurate construction cost estimates at commercial, government, educational, and medical facilities.

Repair & Remodeling Estimating

This *two-day* seminar emphasizes all the underlying considerations unique to repair/remodeling estimating and presents the correct methods for generating accurate, reliable R&R project costs using the unit price and assemblies methods.

Some of what you'll learn:
- Estimating considerations—like labor-hours, building code compliance, working within existing structures, purchasing materials in smaller quantities, unforeseen deficiencies
- Identifying problems and providing solutions to estimating building alterations
- Rules for factoring in minimum labor costs, accurate productivity estimates, and allowances for project contingencies
- R&R estimating examples calculated using unit price and assemblies data

Who should attend: Facilities managers, plant engineers, architects, contractors, estimators, builders... and others who are concerned with the proper preparation and/or evaluation of repair and remodeling estimates.

Mechanical & Electrical Estimating

This *two-day* course teaches attendees how to prepare more accurate and complete mechanical/electrical estimates, avoiding the pitfalls of omission and double-counting, while understanding the composition and rationale within the RSMeans Mechanical/Electrical database.

Some of what you'll learn:
- The unique way mechanical and electrical systems are interrelated
- M&E estimates–conceptual, planning, budgeting, and bidding stages
- Order of magnitude, square foot, assemblies, and unit price estimating
- Comparative cost analysis of equipment and design alternatives

Who should attend: Architects, engineers, facilities managers, mechanical and electrical contractors... and others needing a highly reliable method for developing, understanding, and evaluating mechanical and electrical contracts.

Green Building: Planning & Construction

In this *two-day* course, learn about tailoring a building and its placement on the site to the local climate, site conditions, culture, and community. Includes information to reduce resource consumption and augment resource supply.

Some of what you'll learn:
- Green technologies, materials, systems, and standards
- Modeling tools and how to use them
- Cost vs. value of green products over their life cycle
- Low-cost green strategies, and economic incentives and funding
- Health, comfort, and productivity

Who should attend: Contractors, project managers, building owners, building officials, healthcare and insurance professionals.

Facilities Maintenance & Repair Estimating

This *two-day* course teaches attendees how to plan, budget, and estimate the cost of ongoing and preventive maintenance and repair for existing buildings and grounds.

Some of what you'll learn:
- The most financially favorable maintenance, repair, and replacement scheduling and estimating
- Auditing and value engineering facilities
- Preventive planning and facilities upgrading
- Determining both in-house and contract-out service costs
- Annual, asset-protecting M&R plan

Who should attend: Facility managers, maintenance supervisors, buildings and grounds superintendents, plant managers, planners, estimators... and others involved in facilities planning and budgeting.

Practical Project Management for Construction Professionals

In this *two-day* course, acquire the essential knowledge and develop the skills to effectively and efficiently execute the day-to-day responsibilities of the construction project manager.

Covers:
- General conditions of the construction contract
- Contract modifications: change orders and construction change directives
- Negotiations with subcontractors and vendors
- Effective writing: notification and communications
- Dispute resolution: claims and liens

Who should attend: Architects, engineers, owner's representatives, project managers.

Facilities Construction Estimating

In this *two-day* course, professionals working in facilities management can get help with their daily challenges to establish budgets for all phase of a project

Some of what you'll learn:
- Determine the full scope of a project
- Identify the scope of risks & opportunities
- Creative solutions to estimating issues
- Organizing estimates for presentation & discussion
- Special techniques for repair/remodel and maintenance projects
- Negotiating project change orders

Who should attend: Facility managers, engineers, contractors, facility trades-people, planners & project managers

Professional Development

For more information visit the RSMeans website at www.rsmeans.com

Scheduling and Project Management

This *two-day* course teaches attendees the most current and proven scheduling and management techniques needed to bring projects in on time and on budget.

Some of what you'll learn:
- Crucial phases of planning and scheduling
- How to establish project priorities and develop realistic schedules and management techniques
- Critical Path and Precedence Methods
- Special emphasis on cost control

Who should attend: Construction project managers, supervisors, engineers, estimators, contractors... and others who want to improve their project planning, scheduling, and management skills.

Assessing Scope of Work for Facility Construction Estimating

This *two-day* practical training program addresses the vital importance of understanding the SCOPE of projects in order to produce accurate cost estimates for facility repair and remodeling.

Some of what you'll learn:
- Discussions of site visits, plans/specs, record drawings of facilities, and site-specific lists
- Review of CSI divisions, including means, methods, materials, and the challenges of scoping each topic
- Exercises in SCOPE identification and SCOPE writing for accurate estimating of projects
- Hands-on exercises that require SCOPE, take-off, and pricing

Who should attend: Corporate and government estimators, planners, facility managers, and others who need to produce accurate project estimates.

Site Work & Heavy Construction Estimating

This *two-day* course teaches attendees how to estimate earthwork, site utilities, foundations, and site improvements, using the assemblies and the unit price methods.

Some of what you'll learn:
- Basic site work and heavy construction estimating skills
- Estimating foundations, utilities, earthwork, and site improvements
- Correct equipment usage, quality control, and site investigation for estimating purposes

Who should attend: Project managers, design engineers, estimators, and contractors doing site work and heavy construction.

2010 RSMeans Seminar Schedule

Note: call for exact dates and details.

Location	Dates
Las Vegas, NV	March
Washington, DC	April
Denver, CO	May
San Francisco, CA	June
Washington, DC	September
Dallas, TX	September
Las Vegas, NV	October
Atlantic City, NJ	October
Orlando, FL	November
San Diego, CA	December

2010 Regional Seminar Schedule

Location	Dates
Seattle, WA	January and August
Dallas/Ft. Worth, TX	January
Austin, TX	February
Anchorage, AK	March and September
Oklahoma City, OK	March
Phoenix, AZ	March
Aiea, HI	May
Kansas City, MO	May
Bethesda, MD	June
Columbus, GA	June
El Segundo, CA	August

Location	Dates
St. Louis, MO	August
Houston, TX	October
Colorado Springs, CO	October
Baltimore, MD	November
Sacramento, CA	November
San Antonio, TX	December
Raleigh, NC	December

1-800-334-3509, ext. 5115

Registration Information

Register early... Save up to $100! Register 30 days before the start date of a seminar and save $100 off your total fee. *Note: This discount can be applied only once per order. It cannot be applied to team discount registrations or any other special offer.*

How to register Register by phone today! The RSMeans toll-free number for making reservations is **1-800-334-3509, ext. 5115.**

Individual seminar registration fee - $935. *Means CostWorks*® training registration fee - $375. To register by mail, complete the registration form and return, with your full fee, to: RSMeans Seminars, 63 Smiths Lane, Kingston, MA 02364.

Government pricing All federal government employees save off the regular seminar price. Other promotional discounts cannot be combined with the government discount.

Team discount program for two to four seminar registrations. Call for pricing: 1-800-334-3509, ext. 5115

Multiple course discounts When signing up for two or more courses, call for pricing.

Refund policy Cancellations will be accepted up to ten business days prior to the seminar start. There are no refunds for cancellations received later than ten working days prior to the first day of the seminar. A $150 processing fee will be applied for all cancellations. Written notice of cancellation is required. Substitutions can be made at any time before the session starts. **No-shows are subject to the full seminar fee.**

AACE approved courses Many seminars described and offered here have been approved for 14 hours (1.4 recertification credits) of credit by the AACE International Certification Board toward meeting the continuing education requirements for recertification as a Certified Cost Engineer/Certified Cost Consultant.

AIA Continuing Education We are registered with the AIA Continuing Education System (AIA/CES) and are committed to developing quality learning activities in accordance with the CES criteria. Many seminars meet the AIA/CES criteria for Quality Level 2. AIA members may receive (14) learning units (LUs) for each two-day RSMeans course.

Daily course schedule The first day of each seminar session begins at 8:30 a.m. and ends at 4:30 p.m. The second day begins at 8:00 a.m. and ends at 4:00 p.m. Participants are urged to bring a hand-held calculator, since many actual problems will be worked out in each session.

Continental breakfast Your registration includes the cost of a continental breakfast, and a morning and afternoon refreshment break. These informal segments allow you to discuss topics of mutual interest with other seminar attendees. (You are free to make your own lunch and dinner arrangements.)

Hotel/transportation arrangements RSMeans arranges to hold a block of rooms at most host hotels. To take advantage of special group rates when making your reservation, be sure to mention that you are attending the RSMeans seminar. You are, of course, free to stay at the lodging place of your choice. **(Hotel reservations and transportation arrangements should be made directly by seminar attendees.)**

Important Class sizes are limited, so please register as soon as possible.

Note: *Pricing subject to change.*

Registration Form

ADDS-1000

Call 1-800-334-3509, ext. 5115 to register or FAX this form 1-800-632-6732. Visit our Web site: www.rsmeans.com

Please register the following people for the RSMeans construction seminars as shown here. We understand that we must make our own hotel reservations if overnight stays are necessary.

☐ Full payment of $________________ enclosed.

☐ Bill me

Please print name of registrant(s)

(To appear on certificate of completion)

____

____

____

____

P.O. #: ____________________

GOVERNMENT AGENCIES MUST SUPPLY PURCHASE ORDER NUMBER OR TRAINING FORM.

Firm name____________________

Address____________________

City/State/Zip____________________

Telephone no.____________ Fax no.____________

E-mail address____________________

Charge registration(s) to: ☐ MasterCard ☐ VISA ☐ American Express

Account no.____________ Exp. date____________

Cardholder's signature____________________

Seminar name____________________

Seminar City ____________________

Please mail check to: RSMeans Seminars, 63 Smiths Lane, P.O. Box 800, Kingston, MA 02364 USA

Harness the full power of RSMeans online with MeansCostWorks.com

MeansCostWorks.com provides 24/7 online access to the most used and quoted construction cost data—in tandem with comprehensive estimating and customization functions. Featuring the entire database of RSMeans unit, assembly, and square foot model costs in an online format, **MeansCostWorks.com** allows professionals to create reliable estimates quickly and easily. Estimates can be customized, shared with others, worked on collaboratively, exported to Excel, and saved for future use.

Construction Estimator
Get online access to the entire RSMeans database of unit costs, assemblies, repair and remodeling costs—covering every major category of construction. Create detailed construction cost estimates.

Square Foot Estimator
For easy-to-use conceptual estimating, this tool gives you online access to the extensive RSMeans library of commercial building models and systems. Get a preliminary square foot estimate in just minutes.

MeansCostWorks.com is an easy-to-use, one-stop estimating tool that provides online access to the most used and quoted construction cost data-in tandem with comprehensive estimating and customization functions.

Go to www.MeansCostWorks.com to view the online demo and sign up for your FREE trial subscription.
To speak with a representative call 1.800.334.3509.

MeansData™
Software Integration

CONSTRUCTION COSTS FOR SOFTWARE APPLICATIONS

Your construction estimating software is only as good as your cost data.

Profiler

Have you ever wondered why so many projects are over budget, over scope, and take longer to design and build than expected?

DProfiler with RSMeans is the solution that reduces your costs, the risks to the owner, and gives you a competitive advantage.

Use DProfiler with RSMeans cost data at the start of a project to establish the project scope and budget.

www.beck-technology.com

Reed Construction Data

The Gypsum Construction Handbook by USG

An invaluable reference of construction procedures for gypsum drywall, cement board, veneer plaster, and conventional plaster. This new edition includes the newest product developments, installation methods, fire- and sound-rated construction information, illustrated framing-to-finish application instructions, estimating and planning information, and more. Great for architects and engineers; contractors, builders, and dealers; apprentices and training programs; building inspectors and code officials; and anyone interested in gypsum construction. Features information on tools and safety practices, a glossary of construction terms, and a list of important agencies and associations. Also available in Spanish.

$29.95
Over 575 pages, illustrated, softcover
Catalog no. 67357A

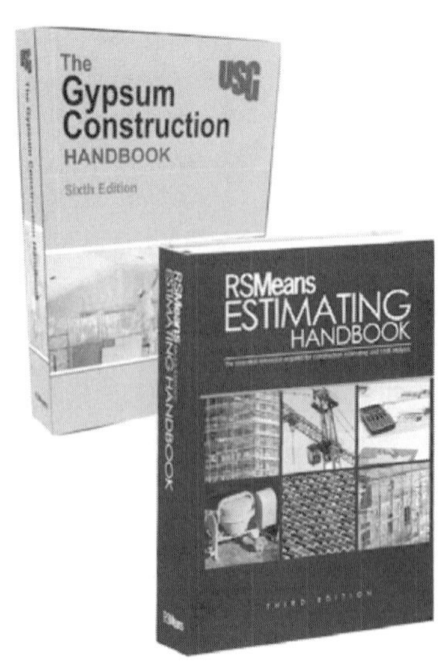

RSMeans Estimating Handbook, 3rd Edition

Widely used in the industry for tasks ranging from routine estimates to special cost analysis projects, this handbook has been completely updated and reorganized with new and expanded technical information.

RSMeans Estimating Handbook will help construction professionals:

- Evaluate architectural plans and specifications
- Prepare accurate quantity takeoffs
- Compare design alternatives and costs
- Perform value engineering
- Double-check estimates and quotes
- Estimate change orders

$99.95 per copy
Over 900 pages, hardcover
Catalog No. 67276B

BIM Handbook
A Guide to Building Information Modeling for Owners, Managers, Designers, Engineers and Contractors
by Chuck Eastman, Paul Teicholz, Rafael Sacks, Kathleen Liston

Building Information Modeling (BIM) is a new approach to design, construction, and facility management where a digital depiction of the building process is used to facilitate the exchange of information in digital format.

The BIM Handbook provides an in-depth understanding of BIM technologies, the business and organizational issues associated with its implementation, and the profound advantages that effective use of BIM can provide to all members of a project team. Also includes a rich set of informative BIM case studies.

$85.00
504 pages
Catalog no. 67356

Universal Design Ideas for Style, Comfort & Safety
by RSMeans and Lexicon Consulting, Inc.

Incorporating universal design when building or remodeling helps people of any age and physical ability more fully and safely enjoy their living spaces. This book shows how universal design can be artfully blended into the most attractive homes. It discusses specialized products like adjustable countertops and chair lifts, as well as simple ways to enhance a home's safety and comfort. With color photos and expert guidance, every area of the home is covered. Includes budget estimates that give an idea how much projects will cost.

$21.95
160 pages, illustrated, softcover
Catalog no. 67354

Green Home Improvement
by Daniel D. Chiras, PhD

With energy costs rising and environmental awareness increasing, many people are looking to make their homes greener. This book makes it easy, with 65 projects and actual costs and projected savings that will help homeowners prioritize their green improvements.

Projects range from simple water savers that cost only a few dollars, to bigger-ticket items such as efficient appliances and HVAC systems, green roofing, flooring, landscaping, and much more. With color photos and cost estimates, each project compares options and describes the work involved, the benefits, and the savings.

$24.95
320 pages, illustrated, softcover
Catalog no. 67355

Residential & Light Commercial Construction Standards, 3rd Edition, Updated
by RSMeans and contributing authors

This book provides authoritative requirements and recommendations compiled from leading professional associations, industry publications, and building code organizations. This all-in-one reference helps establish a standard for workmanship, quickly resolve disputes, and avoid defect claims. Updated third edition includes new coverage of green building, seismic, hurricane, and mold-resistant construction.

$59.95
Over 550 pages, illustrated, softcover
Catalog no. 67322B

ORDER TOLL FREE 1-800-334-3509
OR FAX 1-800-632-6732

2010 Order Form

Qty.	Book no.	COST ESTIMATING BOOKS	Unit Price	Total
	60060	Assemblies Cost Data 2010	$279.95	
	60010	Building Construction Cost Data 2010	169.95	
	60110	Concrete & Masonry Cost Data 2010	154.95	
	50140	Construction Cost Indexes 2010 (subscription)	362.00	
	60140A	Construction Cost Index–January 2010	90.50	
	60140B	Construction Cost Index–April 2010	90.50	
	60140C	Construction Cost Index–July 2010	90.50	
	60140D	Construction Cost Index–October 2010	90.50	
	60340	Contr. Pricing Guide: Resid. R & R Costs 2010	39.95	
	60330	Contr. Pricing Guide: Resid. Detailed 2010	39.95	
	60320	Contr. Pricing Guide: Resid. Sq. Ft. 2010	39.95	
	60230	Electrical Change Order Cost Data 2010	169.95	
	60030	Electrical Cost Data 2010	169.95	
	60200	Facilities Construction Cost Data 2010	419.95	
	60300	Facilities Maintenance & Repair Cost Data 2010	379.95	
	60160	Heavy Construction Cost Data 2010	169.95	
	60090	Interior Cost Data 2010	169.95	
	60120	Labor Rates for the Const. Industry 2010	379.95	
	60180	Light Commercial Cost Data 2010	144.95	
	60020	Mechanical Cost Data 2010	169.95	
	60150	Open Shop Building Const. Cost Data 2010	169.95	
	60210	Plumbing Cost Data 2010	169.95	
	60040	Repair & Remodeling Cost Data 2010	144.95	
	60170	Residential Cost Data 2010	144.95	
	60280	Site Work & Landscape Cost Data 2010	169.95	
	60050	Square Foot Costs 2010	184.95	
	62019	Yardsticks for Costing (2009)	154.95	
	62010	Yardsticks for Costing (2010)	169.95	
		REFERENCE BOOKS		
	67310A	ADA Compliance Pricing Guide, 2nd Ed.	79.95	
	67356	BIM Handbook	85.00	
	67330	Bldrs Essentials: Adv. Framing Methods	12.98	
	67329	Bldrs Essentials: Best Bus. Practices for Bldrs	29.95	
	67298A	Bldrs Essentials: Framing/Carpentry 2nd Ed.	15.98	
	67298AS	Bldrs Essentials: Framing/Carpentry Spanish	15.98	
	67307	Bldrs Essentials: Plan Reading & Takeoff	35.95	
	67342	Building & Renovating Schools	69.98	
	67339	Building Security: Strategies & Costs	44.98	
	67353	Complete Book of Framing	29.95	
	67146	Concrete Repair & Maintenance Illustrated	69.95	
	67352	Construction Business Management	49.95	
	67314	Cost Planning & Est. for Facil. Maint.	89.95	
	67328A	Designing & Building with the IBC, 2nd Ed.	99.95	
	67230B	Electrical Estimating Methods, 3rd Ed.	64.95	
	64777A	Environmental Remediation Est. Methods, 2nd Ed.	49.98	
	67343	Estimating Bldg. Costs for Resi. & Lt. Comm.	29.95	
	67276B	Estimating Handbook, 3rd Ed.	$99.95	
	67318	Facilities Operations & Engineering Reference	54.98	
	67301	Facilities Planning & Relocation	49.98	
	67338A	Green Building: Proj. Planning & Cost Est., 2nd Ed.	129.95	
	67355	Green Home Improvement	24.95	

Qty.	Book no.	REFERENCE BOOKS (Cont.)	Unit Price	Total
	67357A	The Gypsum Construction Handbook	29.95	
	67357S	The Gypsum Construction Handbook (Spanish)	29.95	
	67323	Historic Preservation: Proj. Planning & Est.	49.98	
	67308E	Home Improvement Costs–Int. Projects, 9th Ed.	24.95	
	67309E	Home Improvement Costs–Ext. Projects, 9th Ed.	24.95	
	67344	Homeowner's Guide to Mold	21.95	
	67324B	How to Est.w/Means Data & CostWorks, 3rd Ed.	59.95	
	67351	How Your House Works	21.95	
	67282A	Illustrated Const. Dictionary, Condensed, 2nd Ed.	59.95	
	67292A	Illustrated Const. Dictionary, w/CD-ROM, 3rd Ed.	99.95	
	67348	Job Order Contracting	89.95	
	67347	Kitchen & Bath Project Costs	19.98	
	67295C	Landscape Estimating Methods, 5th Ed.	64.95	
	67341	Life Cycle Costing for Facilities	99.95	
	67294B	Mechanical Estimating Methods, 4th Ed.	64.95	
	67245A	Planning & Managing Interior Projects, 2nd Ed.	34.98	
	67283B	Plumbing Estimating Methods, 3rd Ed.	29.98	
	67345	Practice of Cost Segregation Analysis	99.95	
	67337	Preventive Maint. for Higher Education Facilities	149.95	
	67346	Preventive Maint. for Multi-Family Housing	89.95	
	67326	Preventive Maint. Guidelines for School Facil.	149.95	
	67247B	Project Scheduling & Management for Constr. 3rd Ed.	64.95	
	67265B	Repair & Remodeling Estimating Methods, 4th Ed.	69.95	
	67322B	Resi. & Light Commercial Const. Stds., 3rd Ed.	59.95	
	67327A	Spanish/English Construction Dictionary, 2nd Ed.	23.95	
	67145B	Sq. Ft. & Assem. Estimating Methods, 3rd Ed.	69.95	
	67321	Total Productive Facilities Management	29.98	
	67350	Understanding and Negotiating Const. Contracts	69.95	
	67303B	Unit Price Estimating Methods, 4th Ed.	59.95	
	67354	Universal Design	21.95	
	67319A	Value Engineering: Practical Applications	79.95	
		MA residents add 6.25% state sales tax		
		Shipping & Handling**		
		Total (U.S. Funds)*		

Prices are subject to change and are for U.S. delivery only. *Canadian customers may call for current prices. **Shipping & handling charges: Add 7% of total order for check and credit card payments. Add 9% of total order for invoiced orders.

Send order to: **ADDV-1000**

Name (please print) ______________________________

Company ______________________________

☐ **Company**

☐ **Home** **Address** ______________________________

____

City/State/Zip ______________________________

Phone # ______________ P.O. # ______________

(Must accompany all orders being billed)

Mail to: RSMeans, P.O. Box 800, Kingston, MA 02364

RSMeans Project Cost Report

By filling out this report, your project data will contribute to the database that supports the RSMeans Project Cost Square Foot Data. When you fill out this form, RSMeans will provide a $30 discount off one of the RSMeans products advertised in the preceding pages. Please complete the form including all items where you have cost data, and all the items marked (✔).

$30.00 Discount per product for each report you submit.

Project Description (No remodeling projects, please.)

✔ Building Use (Office, School...) ____________

✔ Address (City, State) ____________

____

✔ Total Building Area (SF) ____________

✔ Ground Floor (SF) ____________

✔ Frame (Wood, Steel...) ____________

✔ Exterior Wall (Brick, Tilt-up...) ____________

✔ Basement: (check one) ☐ Full ☐ Partial ☐ None

✔ Number Stories ____________

✔ Floor-to-Floor Height ____________

✔ Volume (C.F.) ____________

% Air Conditioned ____________ Tons ______

Total Project Cost $ ____________

Owner ____________

Architect ____________

General Contractor ____________

✔ Bid Date ____________

✔ Typical Bay Size ____________

✔ Capacity ____________

✔ Labor Force: _____ % Union _____ % Non-Union

✔ Project Description (Circle one number in each line)

1. Economy 2. Average 3. Custom 4. Luxury
1. Square 2. Rectangular 3. Irregular 4. Very Irregular

Comments ____________

____

A	✔	General Conditions	$
B	✔	Site Work	$
C	✔	Concrete	$
D	✔	Masonry	$
E	✔	Metals	$
F	✔	Wood & Plastics	$
G	✔	Thermal & Moisture Protection	$
GR		Roofing & Flashing	$
H	✔	Doors and Windows	$
J	✔	Finishes	$
JP		Painting & Wall Covering	$

K	✔	Specialties	$
L	✔	Equipment	$
M	✔	Furnishings	$
N	✔	Special Construction	$
P	✔	Conveying Systems	$
Q	✔	Mechanical	$
QP		Plumbing	$
QB		HVAC	$
R	✔	Electrical	$
S	✔	Mech./Elec. Combined	$

Please specify the RSMeans product you wish to receive. Complete the address information.

Product Name ____________

Product Number ____________

Your Name ____________

Title ____________

Company ____________

☐ Company

☐ Home Street Address ____________

City, State, Zip ____________

Return by mail or Fax 888-492-6770.

Method of Payment:

Credit Card # ____________

Expiration Date ____________

Check ____________

Purchase Order ____________

RSMeans
a division of Reed Construction Data
Square Foot Costs Department
P.O. Box 800
Kingston, MA 02364-3008